Lecture Notes in Computer Science　　10625

Commenced Publication in 1973
Founding and Former Series Editors:
Gerhard Goos, Juris Hartmanis, and Jan van Leeuwen

More information about this series at http://www.springer.com/series/7410

Tsuyoshi Takagi · Thomas Peyrin (Eds.)

Advances in Cryptology – ASIACRYPT 2017

23rd International Conference on the Theory
and Applications of Cryptology and Information Security
Hong Kong, China, December 3–7, 2017
Proceedings, Part II

 Springer

Editors
Tsuyoshi Takagi
The University of Tokyo
Tokyo
Japan

Thomas Peyrin
Nanyang Technological University
Singapore
Singapore

ISSN 0302-9743   ISSN 1611-3349   (electronic)
Lecture Notes in Computer Science
ISBN 978-3-319-70696-2   ISBN 978-3-319-70697-9   (eBook)
https://doi.org/10.1007/978-3-319-70697-9

Library of Congress Control Number: 2017957984

LNCS Sublibrary: SL4 – Security and Cryptology

Printed on acid-free paper

This Springer imprint is published by Springer Nature
The registered company is Springer International Publishing AG
The registered company address is: Gewerbestrasse 11, 6330 Cham, Switzerland

Preface

ASIACRYPT 2017, the 23rd Annual International Conference on Theory and Application of Cryptology and Information Security, was held in Hong Kong, SAR China, during December 3–7, 2017.

The conference focused on all technical aspects of cryptology, and was sponsored by the International Association for Cryptologic Research (IACR).

ASIACRYPT 2017 received 243 submissions from all over the world. The Program Committee selected 67 papers (from which two were merged) for publication in the proceedings of this conference. The review process was made by the usual double-blind peer review by the Program Committee consisting of 48 leading experts of the field. Each submission was reviewed by at least three reviewers, and five reviewers were assigned to submissions co-authored by Program Committee members. This year, the conference operated a two-round review system with rebuttal phase. In the first-round review the Program Committee selected the 146 submissions that were considered of value for proceeding to the second round. In the second-round review the Program Committee further reviewed the submissions by taking into account their rebuttal letter from the authors. All the selection process was assisted by 334 external reviewers. These three-volume proceedings contain the revised versions of the papers that were selected. The revised versions were not reviewed again and the authors are responsible for their contents.

The program of ASIACRYPT 2017 featured three excellent invited talks. Dustin Moody gave a talk entitled "The Ship Has Sailed: The NIST Post-Quantum Cryptography 'Competition'," Wang Huaxiong spoke on "Combinatorics in Information-Theoretic Cryptography," and Pascal Paillier gave a third talk. The conference also featured a traditional rump session that contained short presentations on the latest research results of the field. The Program Committee selected the work "Identification Protocols and Signature Schemes Based on Supersingular Isogeny Problems" by Steven D. Galbraith, Christophe Petit, and Javier Silva for the Best Paper Award of ASIACRYPT 2017. Two more papers, "Kummer for Genus One over Prime Order Fields" by Sabyasachi Karati and Palash Sarkar, and "A Subversion-Resistant SNARK" by Behzad Abdolmaleki, Karim Baghery, Helger Lipmaa, and Michał Zając were solicited to submit the full versions to the *Journal of Cryptology*. The program chairs selected Takahiro Matsuda and Bart Mennink for the Best PC Member Award.

Many people have contributed to the success of ASIACRYPT 2017. We would like to thank the authors for submitting their research results to the conference. We are very grateful to all of the Program Committee members as well as the external reviewers for their fruitful comments and discussions on their areas of expertise. We are greatly indebted to Duncan Wong and Siu Ming Yiu, the general co-chairs, for their efforts and overall organization. We would also like to thank Allen Au, Catherine Chan, Sherman S.M. Chow, Lucas Hui, Zoe Jiang, Xuan Wang, and Jun Zhang, the local

Organizing Committee, for their continuous supports. We thank Duncan Wong and Siu Ming Yiu for expertly organizing and chairing the rump session.

Finally, we thank Shai Halevi for letting us use his nice software for supporting all the paper submission and review process. We also thank Alfred Hofmann, Anna Kramer, and their colleagues for handling the editorial process of the proceedings published at Springer LNCS.

December 2017   Tsuyoshi Takagi
                                                          Thomas Peyrin

ASIACRYPT 2017

The 23rd Annual International Conference on Theory and Application of Cryptology and Information Security

Sponsored by the International Association for Cryptologic Research (IACR)

December 3–7, 2017, Hong Kong, SAR China

General Co-chairs

Duncan Wong	CryptoBLK Limited
Siu Ming Yiu	The University of Hong Kong, SAR China

Program Co-chairs

Tsuyoshi Takagi	University of Tokyo, Japan
Thomas Peyrin	Nanyang Technological University, Singapore

Program Committee

Shweta Agrawal	IIT Madras, India
Céline Blondeau	Aalto University, Finland
Joppe W. Bos	NXP Semiconductors, Belgium
Chris Brzuska	TU Hamburg, Germany
Jie Chen	East China Normal University, China
Sherman S.M. Chow	The Chinese University of Hong Kong, SAR China
Kai-Min Chung	Academia Sinica, Taiwan
Nico Döttling	University of California, Berkeley, USA
Thomas Eisenbarth	Worcester Polytechnic Institute, USA
Dario Fiore	IMDEA Software Institute, Madrid, Spain
Georg Fuchsbauer	Inria and ENS, France
Steven Galbraith	Auckland University, New Zealand
Jian Guo	Nanyang Technological University, Singapore
Viet Tung Hoang	Florida State University, USA
Jérémy Jean	ANSSI, France
Jooyoung Lee	KAIST, South Korea
Dongdai Lin	Chinese Academy of Sciences, China
Feng-Hao Liu	Florida Atlantic University, USA
Stefan Mangard	Graz University of Technology, Austria
Takahiro Matsuda	AIST, Japan
Alexander May	Ruhr University Bochum, Germany
Bart Mennink	Radboud University, The Netherlands

Amir Moradi	Ruhr University Bochum, Germany
Pratyay Mukherjee	Visa Research, USA
Mridul Nandi	Indian Statistical Institute, India
Khoa Nguyen	Nanyang Technological University, Singapore
Miyako Ohkubo	NICT, Japan
Tatsuaki Okamoto	NTT Secure Platform Laboratories, Japan
Arpita Patra	Indian Institute of Science, India
Bart Preneel	KU Leuven, Belgium
Matthieu Rivain	CryptoExperts, France
Reihaneh Safavi-Naini	University of Calgary, Canada
Yu Sasaki	NTT Secure Platform Laboratories, Japan
Peter Schwabe	Radboud University, The Netherlands
Fang Song	Portland State University, USA
Francois-Xavier Standaert	UCL, Belgium
Damien Stehlé	ENS Lyon, France
Ron Steinfeld	Monash University, Australia
Rainer Steinwandt	Florida Atlantic University, USA
Mehdi Tibouchi	NTT Secure Platform Laboratories, Japan
Dominique Unruh	University of Tartu, Estonia
Gilles Van Assche	STMicroelectronics, Belgium
Serge Vaudenay	EPFL, Switzerland
Ingrid Verbauwhede	KU Leuven, Belgium
Ivan Visconti	University of Salerno, Italy
Lei Wang	Shanghai Jiaotong University, China
Meiqin Wang	Shandong University, China
Jiang Zhang	State Key Laboratory of Cryptology, China

Additional Reviewers

Masayuki Abe	Shi Bai	Begül Bilgin
Arash Afshar	Fatih Balli	Olivier Blazy
Divesh Aggarwal	Subhadeep Banik	Johannes Bloemer
Shashank Agrawal	Zhenzhen Bao	Sonia Mihaela Bogos
Ahmad Ahmadi	Hridam Basu	Sasha Boldyreva
Mamun Akand	Alberto Batistello	Charlotte Bonte
Gorjan Alagic	Balthazar Bauer	Raphael Bost
Joel Alwen	Carsten Baum	Leif Both
Abdelrahaman Aly	Georg T. Becker	Florian Bourse
Miguel Ambrona	Christof Beierle	Sébastien Canard
Elena Andreeva	Sonia Beläd	Brent Carmer
Diego Aranha	Fabrice Benhamouda	Wouter Castryck
Nuttapong Attrapadung	Francesco Berti	Dario Catalano
Sepideh Avizheh	Guido Bertoni	Gizem Çetin
Saikrishna	Sanjay Bhattacherjee	Avik Chakraborti
Badrinarayanan	Jean-Francois Biasse	Nishanth Chandran

Melissa Chase
Binyi Chen
Cong Chen
Long Chen
Yi-Hsiu Chen
Yu Chen
Yu-Chi Chen
Nai-Hui Chia
Gwangbae Choi
Wutichai Chongchitmate
Chi-Ning Chou
Ashish Choudhury
Chitchanok
Chuengsatiansup
Hao Chung
Michele Ciampi
Thomas De Cnudde
Katriel Cohn-Gordon
Henry Corrigan-Gibbs
Craig Costello
Geoffroy Couteau
Eric Crockett
Tingting Cui
Edouard Cuvelier
Joan Daemen
Wei Dai
Pratish Datta
Bernardo David
Marguerite Delcourt
Jeroen Delvaux
Yi Deng
David Derler
Julien Devigne
Claus Diem
Christoph Dobraunig
Yarkin Doroz
Léo Ducas
Dung H. Duong
Ratna Dutta
Stefan Dziembowski
Maria Eichlseder
Muhammed Esgin
Thomas Espitau
Xiong Fan
Antonio Faonio

Sebastian Faust
Björn Fay
Serge Fehr
Luca De Feo
Nils Fleischhacker
Jean-Pierre Flori
Tore Kasper Frederiksen
Thomas Fuhr
Marc Fyrbiak
Tommaso Gagliardoni
Chaya Ganesh
Flavio Garcia
Pierrick Gaudry
Rémi Géraud
Satrajit Ghosh
Irene Giacomelli
Benedikt Gierlichs
Junqing Gong
Louis Goubin
Alex Grilo
Hannes Gross
Vincent Grosso
Chun Guo
Hui Guo
Helene Haagh
Patrick Haddad
Harry Halpin
Shuai Han
Yoshikazu Hanatani
Jens Hermans
Gottfried Herold
Julia Hesse
Felix Heuer
Minki Hhan
Fumitaka Hoshino
Yin-Hsun Huang
Zhenyu Huang
Andreas Hülsing
Jung Yeon Hwang
Ilia Iliashenko
Mehmet Inci
Vincenzo Iovino
Ai Ishida
Takanori Isobe
Tetsu Iwata

Malika Izabachène
Michael Jacobson
Abhishek Jain
David Jao
Zhengfeng Ji
Dingding Jia
Shaoquan Jiang
Anthony Journault
Jean-Gabriel Kammerer
Sabyasachi Karati
Handan Kilinç
Dongwoo Kim
Jihye Kim
Jon-Lark Kim
Sam Kim
Taechan Kim
Elena Kirshanova
Ágnes Kiss
Fuyuki Kitagawa
Susumu Kiyoshima
Thorsten Kleinjung
Miroslav Knezevic
Alexander Koch
François Koeune
Konrad Kohbrok
Lisa Kohl
Ilan Komargodski
Yashvanth Kondi
Robert Kuebler
Frédéric Lafitte
Ching-Yi Lai
Russell W.F. Lai
Adeline Langlois
Gregor Leander
Changmin Lee
Hyung Tae Lee
Iraklis Leontiadis
Tancrède Lepoint
Debbie Leung
Yongqiang Li
Jyun-Jie Liao
Benoit Libert
Fuchun Lin
Wei-Kai Lin
Patrick Longa

Julian Loss
Steve Lu
Xianhui Lu
Atul Luykx
Chang Lv
Vadim Lyubashevsky
Monosij Maitra
Mary Maller
Giorgia Azzurra Marson
Marco Martinoli
Daniel Masny
Sarah Meiklejohn
Peihan Miao
Michele Minelli
Takaaki Mizuki
Ahmad Moghimi
Payman Mohassel
Maria Chiara Molteni
Seyyed Amir Mortazavi
Fabrice Mouhartem
Köksal Mus
Michael Naehrig
Ryo Nishimaki
Anca Nitulescu
Luca Nizzardo
Koji Nuida
Kaisa Nyberg
Adam O'Neill
Tobias Oder
Olya Ohrimenko
Emmanuela Orsini
Elisabeth Oswald
Elena Pagnin
Pascal Paillier
Jiaxin Pan
Alain Passelègue
Sikhar Patranabis
Roel Peeters
Chris Peikert
Alice Pellet-Mary
Ludovic Perret
Peter Pessl
Thomas Peters
Christophe Petit
Duong Hieu Phan
Antigoni Polychroniadou

Romain Poussier
Ali Poustindouz
Emmanuel Prouff
Kexin Qiao
Baodong Qin
Sebastian Ramacher
Somindu C. Ramanna
Shahram Rasoolzadeh
Divya Ravi
Francesco Regazzoni
Jean-René Reinhard
Ling Ren
Joost Renes
Oscar Reparaz
Joost Rijneveld
Damien Robert
Jérémie Roland
Arnab Roy
Sujoy Sinha Roy
Vladimir Rozic
Joeri de Ruiter
Yusuke Sakai
Amin Sakzad
Simona Samardjiska
Olivier Sanders
Pascal Sasdrich
Alessandra Scafuro
John Schanck
Tobias Schneider
Jacob Schuldt
Gil Segev
Okan Seker
Binanda Sengupta
Sourav Sengupta
Jae Hong Seo
Masoumeh Shafienejad
Setareh Sharifian
Sina Shiehian
Kazumasa Shinagawa
Dave Singelée
Shashank Singh
Javier Silva
Luisa Siniscalchi
Daniel Slamanig
Benjamin Smith
Ling Song

Pratik Soni
Koutarou Suzuki
Alan Szepieniec
Björn Tackmann
Mostafa Taha
Raymond K.H. Tai
Katsuyuki Takashima
Atsushi Takayasu
Benjamin Hong
    Meng Tan
Qiang Tang
Yan Bo Ti
Yosuke Todo
Ni Trieu
Roberto Trifiletti
Thomas Unterluggauer
John van de Wetering
Muthuramakrishnan
    Venkitasubramaniam
Daniele Venturi
Dhinakaran
    Vinayagamurthy
Vanessa Vitse
Damian Vizár
Satyanarayana Vusirikala
Sebastian Wallat
Alexandre Wallet
Haoyang Wang
Minqian Wang
Wenhao Wang
Xiuhua Wang
Yuyu Wang
Felix Wegener
Puwen Wei
Weiqiang Wen
Mario Werner
Benjamin Wesolowski
Baofeng Wu
David Wu
Keita Xagawa
Zejun Xiang
Chengbo Xu
Shota Yamada
Kan Yang
Kang Yang
Kan Yasuda

Donggeon Yhee
Kazuki Yoneyama
Kisoon Yoon
Yu Yu
Zuoxia Yu
Henry Yuen

Aaram Yun
Mahdi Zamani
Greg Zaverucha
Cong Zhang
Jie Zhang
Kai Zhang

Ren Zhang
Wentao Zhang
Yongjun Zhao
Yuqing Zhu

Local Organizing Committee

Co-chairs

Duncan Wong   CryptoBLK Limited
Siu Ming Yiu   The University of Hong Kong, SAR China

Members

Lucas Hui (Chair)   The University of Hong Kong, SAR China
Catherine Chan (Manager)   The University of Hong Kong, SAR China
Jun Zhang   The University of Hong Kong, SAR China
Xuan Wang   Harbin Institute of Technology, Shenzhen, China
Zoe Jiang   Harbin Institute of Technology, Shenzhen, China
Allen Au   The Hong Kong Polytechnic University, SAR China
Sherman S.M. Chow   The Chinese University of Hong Kong, SAR China

Invited Speakers

The Ship Has Sailed: the NIST Post-quantum Cryptography "Competition"

Dustin Moody

Computer Security Division, National Institute of Standards and Technology

Abstract. In recent years, there has been a substantial amount of research on quantum computers – machines that exploit quantum mechanical phenomena to solve mathematical problems that are difficult or intractable for conventional computers. If large-scale quantum computers are ever built, they will compromise the security of many commonly used cryptographic algorithms. In particular, quantum computers would completely break many public-key cryptosystems, including those standardized by NIST and other standards organizations.

Due to this concern, many researchers have begun to investigate post-quantum cryptography (also called quantum-resistant cryptography). The goal of this research is to develop cryptographic algorithms that would be secure against both quantum and classical computers, and can interoperate with existing communications protocols and networks. A significant effort will be required to develop, standardize, and deploy new post-quantum algorithms. In addition, this transition needs to take place well before any large-scale quantum computers are built, so that any information that is later compromised by quantum cryptanalysis is no longer sensitive when that compromise occurs.

NIST has taken several steps in response to this potential threat. In 2015, NIST held a public workshop and later published NISTIR 8105, Report on Post-Quantum Cryptography, which shares NIST's understanding of the status of quantum computing and post-quantum cryptography. NIST also decided to develop additional public-key cryptographic algorithms through a public standardization process, similar to the development processes for the hash function SHA-3 and the Advanced Encryption Standard (AES). To begin the process, NIST issued a detailed set of minimum acceptability requirements, submission requirements, and evaluation criteria for candidate algorithms, available at http://www.nist.gov/pqcrypto. The deadline for algorithms to be submitted was November 30, 2017.

In this talk, I will share the rationale on the major decisions NIST has made, such as excluding hybrid and (stateful) hash-based signature schemes. I will also talk about some open research questions and their potential impact on the standardization effort, in addition to some of the practical issues that arose while creating the API. Finally, I will give some preliminary information about the submitted algorithms, and discuss what we've learned during the first part of the standardization process.

Combinatorics in Information-Theoretic Cryptography

Huaxiong Wang

School of Physical and Mathematical Sciences,
Nanyang Technological University, Singapore
hxwang@ntu.edu.sg

Abstract. Information-theoretic cryptography is an area that studies crypto-graphic functionalities whose security does not rely on hardness assumptions from computational intractability of mathematical problems. It covers a wide range of cryptographic research topics such as one-time pad, authentication code, secret sharing schemes, secure multiparty computation, private information retrieval and post-quantum security etc., just to mention a few. Moreover, many areas in complexity-based cryptography are well known to benefit or stem from information-theoretic methods. On the other hand, combinatorics has been playing an active role in cryptography, for example, the hardness of Hamiltonian cycle existence in graph theory is used to design zero-knowledge proofs. In this talk, I will focus on the connections between combinatorics and information-theoretic cryptography. After a brief (incomplete) overview on their various connections, I will present a few concrete examples to illustrate how combinatorial objects and techniques are applied to the constructions and characterizations of information-theoretic schemes. Specifically, I will show

1. how perfect hash families and cover-free families lead to better performance in certain secret sharing schemes;
2. how graph colouring from planar graphs is used in constructing secure multiparty computation protocols over non-abelian groups;
3. how regular intersecting families are applied to the constructions of private information retrieval schemes.

Part of this research was funded by Singapore Ministry of Education under Research Grant MOE2016-T2-2-014(S).

Contents – Part II

Block Chains

Multi-party Protocols

Operating Modes Security Proofs

Asiacrypt 2017 Award Paper I

Kummer for Genus One over Prime Order Fields

Sabyasachi Karati[1][✉] and Palash Sarkar[2]

[1] iCIS Lab, Department of Computer Science,
University of Calgary, Calgary, Canada
sabyasachi.karati@ucalgary.ca
[2] Applied Statistics Unit, Indian Statistical Institute,
203, B.T. Road, Kolkata 700108, India
palash@isical.ac.in

Abstract. This work considers the problem of fast and secure scalar multiplication using curves of genus one defined over a field of prime order. Previous work by Gaudry and Lubicz in 2009 had suggested the use of the associated Kummer line to speed up scalar multiplication. In this work, we explore this idea in detail. The first task is to obtain an elliptic curve in Legendre form which satisfies necessary security conditions such that the associated Kummer line has small parameters and a base point with small coordinates. In turns out that the ladder step on the Kummer line supports parallelism and can be implemented very efficiently in constant time using the single-instruction multiple-data (SIMD) operations available in modern processors. For the 128-bit security level, this work presents three Kummer lines denoted as $K_1 :=$ KL2519(81, 20), $K_2 :=$ KL25519(82, 77) and $K_3 :=$ KL2663(260, 139) over the three primes $2^{251} - 9$, $2^{255} - 19$ and $2^{266} - 3$ respectively. Implementations of scalar multiplications for all the three Kummer lines using Intel intrinsics have been done and the code is publicly available. Timing results on the recent Skylake and the earlier Haswell processors of Intel indicate that both fixed base and variable base scalar multiplications for K_1 and K_2 are faster than those achieved by Sandy2x which is a highly optimised SIMD implementation in assembly of the well known Curve25519; for example, on Skylake, variable base scalar multiplication on K_1 is faster than Curve25519 by about 25%. On Skylake, both fixed base and variable base scalar multiplication for K_3 are faster than Sandy2x; whereas on Haswell, fixed base scalar multiplication for K_3 is faster than Sandy2x while variable base scalar multiplication for both K_3 and Sandy2x take roughly the same time. In fact, on Skylake, K_3 is both faster and also offers about 5 bits of higher security compared to Curve25519. In practical terms, the particular Kummer lines that are introduced in this work are serious candidates for deployment and standardisation.

Keywords: Elliptic curve cryptography · Kummer line · Montgomery curve · Scalar multiplication

S. Karati—Part of the work was done while the author was a post-doctoral fellow at the Turing Laboratory of the Indian Statistical Institute.
Part supported by Alberta Innovates in the Province of Alberta, Canada.

T. Takagi and T. Peyrin (Eds.): ASIACRYPT 2017, Part II, LNCS 10625, pp. 3–32, 2017.
https://doi.org/10.1007/978-3-319-70697-9_1

1   Introduction

Curve-based cryptography provides a platform for secure and efficient implementation of public key schemes whose security rely on the hardness of discrete logarithm problem. Starting from the pioneering work of Koblitz [29] and Miller [33] introducing elliptic curves and the work of Koblitz [30] introducing hyperelliptic curves for cryptographic use, the last three decades have seen an extensive amount of research in the area.

Appropriately chosen elliptic curves and genus two hyperelliptic curves are considered to be suitable for practical implementation. Table 1 summarises features for some of the concrete curves that have been proposed in the literature. Arguably, the two most well known curves proposed till date for the 128-bit security level are P-256 [37] and Curve25519 [2]. Also the secp256k1 curve [40] has become very popular due to its deployment in the Bitcoin protocol. All of these curves are in the setting of genus one over prime order fields. In particular, we note that Curve25519 has been extensively deployed for various applications. A listing of such applications can be found at [17]. So, from the point of view of deployment, practitioners are very familiar with genus one curves over prime order fields. Influential organisations, such as NIST, Brainpool, Microsoft (the NUMS curve) have concrete proposals in this setting. See [5] for a further listing of such primes and curves. It is quite likely that any future portfolio of proposals by standardisation bodies will include at least one curve in the setting of genus one over a prime field.

Our Contributions

The contribution of this paper is to propose new curves for the setting of genus one over a prime order field. Actual scalar multiplication is done over the Kummer line associated with such a curve. The idea of using Kummer line was proposed by Gaudry and Lubicz [22]. They, however, were not clear about whether competitive speeds can be obtained using this approach. Our main contribution is to show that this can indeed be done using the single-instruction multiple-data (SIMD) instructions available in modern processors. We note that the use of SIMD instructions to speed up computation has been earlier proposed for Kummer surface associated with genus two hyperelliptic curves [22]. The application of this idea, however, to Kummer line has not been considered in the literature. Our work fills this gap and shows that properly using SIMD instructions provides a competitive alternative to known curves in the setting of genus one and prime order fields.

As in the case of Montgomery curve [34], scalar multiplication on the Kummer line proceeds via a laddering algorithm. A ladder step corresponds to each bit of the scalar and each such step consists of a doubling and a differential addition irrespective of the value of the bit. As a consequence, it becomes easy to develop code which runs in constant time. We describe and implement a vectorised version of the laddering algorithm which is also constant time. Our target is the 128-bit security level.

Table 1. Features of some curves proposed in the last few years.

Reference	Genus	Form	Field order	Endomorphisms
NIST P-256 [37]	1	Weierstrass	Prime	No
Curve25519 [2]	1	Montgomery	Prime	No
secp256k1 [40]	1	Weierstrass	Prime	No
Brainpool [11]	1	Weierstrass	Prime	No
NUMS [41]	1	Twisted Edwards	Prime	No
Longa-Sica [32]	1	Twisted Edwards	p^2	Yes
Bos et al. [9]	2	Kummer	Prime	Yes
Bos et al. [10]	2	Kummer	p^2	yes
Hankerson et al. [26], Oliviera et al. [38]	1	Weierstrass/Koblitz	2^n	Yes
Longa-Sica [32], Faz-Hernández et al. [18]	1	Twisted Edwards	p^2	Yes
Costello et al. [15]	1	Montgomery	p^2	Yes
Gaudry-Schost [23], Bernstein et al. [4]	2	Kummer	Prime	No
Costello-Longa [14]	1	Twisted Edwards	p^2	Yes
Hankerson et al. [26], Oliviera et al. [39]	1	Weierstrass/Koblitz	2^n	Yes
This work	1	Kummer	Prime	No

Choice of the Underlying Field: Our target is the 128-bit security level. To this end, we consider three primes, namely, $2^{251} - 9$, $2^{255} - 19$ and $2^{266} - 3$. These primes are abbreviated as $p2519$, $p25519$ and $p2663$ respectively. The underlying field will be denoted as $\mathbb{F}_p$ where p is one of $p2519$, $p25519$ or $p2663$.

Choice of the Kummer Line: Following previous suggestions [3,9], we work in the square-only setting. In this case, the parameters of the Kummer line are given by two integers a^2 and b^2. We provide appropriate Kummer lines for all three of the primes $p2519$, $p25519$ and $p2663$. These are denoted as KL2519(81,20), KL25519(82,77) and KL2663(260,139) respectively. In each case, we identify a base point with small coordinates. The selection of the Kummer lines is done using a search for curves achieving certain desired security properties. Later we provide the details of these properties which indicate that the curves provide security at the 128-bit security level.

SIMD Implementation: On Intel processors, it is possible to pack 4 64-bit words into a single 256-bit quantity and then use SIMD instructions to simultaneously work on the 4 64-bit words. We apply this approach to carefully consider various aspects of field arithmetic over $\mathbb{F}_p$. SIMD instructions allow the simultaneous computation of 4 multiplications in $\mathbb{F}_p$ and also 4 squarings in $\mathbb{F}_p$.

The use of SIMD instructions dovetails very nicely with the scalar multiplication algorithm over the Kummer line as we explain below.

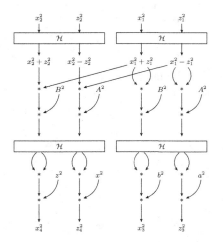

Fig. 1. One ladder step on the Kummer line.

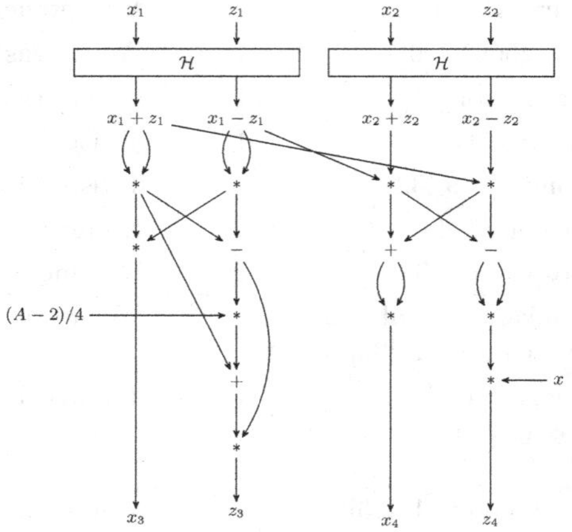

Fig. 2. One ladder step on the Montgomery curve.

Scalar Multiplication over the Kummer Line: A uniform, ladder style algorithm is used. In terms of operation count, each ladder step requires 2 field multiplications, 6 field squarings, 6 multiplications by parameters and 2 multiplications by base point coordinates [22]. In contrast, one ladder step on the Montgomery curve requires 4 field multiplications, 4 squarings, 1 multiplication by curve parameter and 1 multiplication by a base point coordinate. This had led to Gaudry and Lubicz [22] commenting that Kummer line can be advantageous provided that the advantage of trading off multiplications for squarings is not offset by the extra multiplications by the parameters and the base point coordinates.

Our choices of the Kummer lines ensure that the parameters and the base point coordinates are indeed very small. This is not to suggest that the Kummer line is only suitable for fixed based point scalar multiplication. The main advantage arises from the structure of the ladder step on the Kummer line versus that on the Montgomery curve.

An example of the ladder step on the Kummer line is shown in Fig. 1. In the figure, the Hadamard transform $\mathcal{H}(u, v)$ is defined to be $(u + v, u - v)$. Observe that there are 4 layers of 4 simultaneous multiplications. The first layer consists of 2 field multiplications and 2 squarings, while the third layer consists of 4 field squarings. Using 256-bit SIMD instructions, the 2 multiplications and the 2 squarings in the first layer can be computed simultaneously using an implementation of vectorised field multiplication while the third layer can be computed using an implementation of vectorised field squaring. The second layer consists only of multiplications by parameters and is computed using an implementation of vectorised multiplication by constants. The fourth layer consists of two multiplications by parameters and two multiplications by base point coordinates. For fixed base point, this layer can be computed using a single vectorised multiplication by constants while for variable base point, this layer requires a vectorised field multiplication. A major advantage of the ladder step on the Kummer line is that the packing and unpacking into 256-bit quantities is done once each. Packing is done at the start of the scalar multiplication and unpacking is done at the end. The entire scalar multiplication can be computed on the packed vectorised quantities.

In contrast, the ladder step on the Montgomery curve is shown in Fig. 2 which has been reproduced from [2]. The structure of this ladder is not as regular as the ladder step on the Kummer line. This makes it difficult to optimally group together the multiplications for SIMD implementation. Curve25519 is a Montgomery curve. SIMD implementations of Curve25519 have been reported in [7,12,16,19]. The work [16] forms four groups of independent multiplications/squarings with the first and the third group consisting of four multiplications/squarings each, the second group consisting of two multiplications and the fourth group consists of a single multiplication. Interspersed with these multiplications are two groups each consisting of four independent additions/subtractions. The main problem with this approach is that of repeated packing/unpacking of data within a ladder step. This drawback will outweigh the benefits of four simultaneous SIMD multiplications and this approach has not been followed in later works [7,12,19]. These later implementations grouped together only two independent multiplications. In particular, we note that the well known Sandy2x implementation of Curve25519 is an SIMD implementation which is based on [12] and groups together only two multiplications. AVX2 based implementation of Curve25519 in [19] also groups together only 2 multiplications/squarings.

At a forum[1] Tung Chou comments (perhaps oblivious of [16]) that it would better to find four independent multiplications/squarings and vectorise them. As discussed above, the previous works on SIMD implementation of Curve25519 do not seem to have been able to identify this. On the other hand, for the ladder step on the Kummer line shown in Fig. 1, performing vectorisation of 4 independent multiplications/squarings comes quite naturally. This indicates that the ladder step on the Kummer line is more SIMD friendly than the ladder step on the Montgomery curve.

Implementation: We report implementations of all the three Kummer lines KL2519(81,20), KL25519(82,77) and KL2663(260,139). The implementations are in Intel intrinsics and use AVX2 instructions. On the recent Skylake processor, both fixed and variable base scalar multiplications for all the three Kummer lines are faster than Sandy2x which is the presently the best known SIMD implementation in assembly of Curve25519. On the earlier Haswell processor, both fixed and variable base scalar multiplications for KL2519(81,20), KL25519(82,77) are

---

[1] https://moderncrypto.org/mail-archive/curves/2015/000637.html.

faster than that of Sandy2x; fixed base scalar multiplication for KL2663(260,139) is faster than that of Sandy2x while variable base scalar multiplication for both KL2663(260,139) and Sandy2x take roughly the same time. Detailed timing results are provided later.

At a broad level, the timing results reported in this work show that the availability of SIMD instructions leads to the following two practical consequences.

1. At the 128-bit security level, the choice of $\mathbb{F}_{2^{255}-19}$ as the base field is not the fastest. If one is willing to sacrifice about 2 bits of security, then using $\mathbb{F}_{2^{251}-9}$ as the base field leads to about 25% speed up on the Skylake processor.
2. More generally, the ladder step on the Kummer line is faster than the ladder step on the Montgomery curve. We have demonstrated this by implementing on the Intel processors. Future work can explore this issue on other platforms such as the ARM NEON architecture.

Due to page limit restrictions, we are unable to include all the details in this version. These are provided in the full version [28].

2   Background

In this section, we briefly describe theta functions over genus one, Kummer lines, Legendre form elliptic curves and their relations. In our description of the background material, the full version [28] provides certain details which are not readily available in the literature.

2.1   Theta Functions

In this and the next few sections, we provide a sketch of the mathematical background on theta functions over genus one and Kummer lines. Following previous works [22,27,36] we define theta functions over the complex field. For cryptographic purposes, our goal is to work over a prime field of large characteristic. All the derivations that are used have a good reduction [22] and so it is possible to use the Lefschetz principle [1,21] to carry over the identities proved over the complex to those over a large characteristic field.

Let $\tau \in \mathbb{C}$ having a positive imaginary part and $w \in \mathbb{C}$. Let $\xi_1, \xi_2 \in \mathbb{Q}$. Theta functions with characteristics $\vartheta[\xi_1, \xi_2](w, \tau)$ are defined to be the following:

$$\vartheta[\xi_1, \xi_2](w, \tau) = \sum_{n \in \mathbb{Z}} \exp\left[\pi i (n + \xi_1)^2 \tau + 2\pi i (n + \xi_1)(w + \xi_2)\right]. \tag{1}$$

For a fixed τ, the following theta functions are defined.

$$\vartheta_1(w) = \vartheta[0, 0](w, \tau) \quad \text{and} \quad \vartheta_2(w) = \vartheta[0, 1/2](w, \tau).$$
$$\Theta_1(w) = \vartheta[0, 0](w, 2\tau) \quad \text{and} \quad \Theta_2(w) = \vartheta[1/2, 0](w, 2\tau).$$

The following identities hold for the theta functions. Proofs are given in the appendix of the full version [28].

$$2\Theta_1(w_1 + w_2)\Theta_1(w_1 - w_2) = \vartheta_1(w_1)\vartheta_1(w_2) + \vartheta_2(w_1)\vartheta_2(w_2);$$
$$2\Theta_2(w_1 + w_2)\Theta_2(w_1 - w_2) = \vartheta_1(w_1)\vartheta_1(w_2) - \vartheta_2(w_1)\vartheta_2(w_2); \tag{2}$$

$$\vartheta_1(w_1 + w_2)\vartheta_1(w_1 - w_2) = \Theta_1(2w_1)\Theta_1(2w_2) + \Theta_2(2w_1)\Theta_2(2w_2);$$
$$\vartheta_2(w_1 + w_2)\vartheta_2(w_1 - w_2) = \Theta_1(2w_1)\Theta_1(2w_2) - \Theta_2(2w_1)\Theta_2(2w_2). \tag{3}$$

Putting $w_1 = w_2 = w$, we obtain

$$2\Theta_1(2w)\Theta_1(0) = \vartheta_1(w)^2 + \vartheta_2(w)^2; \quad 2\Theta_2(2w)\Theta_2(0) = \vartheta_1(w)^2 - \vartheta_2(w)^2; \tag{4}$$
$$\vartheta_1(2w)\vartheta_1(0) = \Theta_1(2w)^2 + \Theta_2(2w)^2; \quad \vartheta_2(2w)\vartheta_2(0) = \Theta_1(2w)^2 - \Theta_2(2w)^2. \tag{5}$$

Putting $w = 0$ in (4), we obtain

$$2\Theta_1(0)^2 = \vartheta_1(0)^2 + \vartheta_2(0)^2; \quad 2\Theta_2(0)^2 = \vartheta_1(0)^2 - \vartheta_2(0)^2. \tag{6}$$

2.2   Kummer Line

Let $\tau \in \mathbb{C}$ having a positive imaginary part and denote by $\mathbb{P}^1(\mathbb{C})$ the projective line over $\mathbb{C}$. The Kummer line ($\mathcal{K}$) associated with τ is the image of the map φ from $\mathbb{C}$ to $\mathbb{P}^1(\mathbb{C})$ defined by

$$\varphi : w \longmapsto (\vartheta_1(w), \vartheta_2(w)). \tag{7}$$

Suppose that $\varphi(w) = [\vartheta_1(w) : \vartheta_2(w)]$ is known for some $w \in \mathbb{F}_q$. Using (4) it is possible to compute $\Theta_1(2w)$ and $\Theta_2(2w)$ and then using (5) it is possible to compute $\vartheta_1(2w)$ and $\vartheta_2(2w)$. So, from $\varphi(w)$ it is possible to compute $\varphi(2w) = [\vartheta_1(2w) : \vartheta_2(2w)]$ without knowing the value of w.

Suppose that $\varphi(w_1) = [\vartheta_1(w_1) : \vartheta_2(w_1)]$ and $\varphi(w_2) = [\vartheta_1(w_2) : \vartheta_2(w_2)]$ are known for some $w_1, w_2 \in \mathbb{F}_q$. Using (4), it is possible to obtain $\Theta_1(2w_1)$, $\Theta_1(2w_2)$, $\Theta_2(2w_1)$ and $\Theta_2(2w_2)$. Then (3) allows the computation of $\vartheta_1(w_1 + w_2)\vartheta_1(w_1 - w_2)$ and $\vartheta_2(w_1 + w_2)\vartheta_2(w_1 - w_2)$. Further, if $\varphi(w_1 - w_2) = [\vartheta_1(w_1 - w_2) : \vartheta_2(w_1 - w_2)]$ is known, then it is possible to obtain $\varphi(w_1 + w_2) = [\vartheta_1(w_1 + w_2) : \vartheta_2(w_1 + w_2)]$ without knowing the values of w_1 and w_2.

The task of computing $\varphi(2w)$ from $\varphi(w)$ is called doubling and the task of computing $\varphi(w_1 + w_2)$ from $\varphi(w_1)$, $\varphi(w_2)$ and $\varphi(w_1 - w_2)$ is called differential (or pseudo) addition.

2.3   Square only Setting

Let $P = \varphi(w) = [x : z]$ be a point on the Kummer line. As described above, doubling computes the point $2P$ and suppose that $2P = [x_3 : z_3]$. Further, suppose that instead of $[x : z]$, we have the values x^2 and z^2 and after the doubling we are interested in the values x_3^2 and z_3^2. Then the doubling operation given by (8) and (9) only involves the squared quantities $\vartheta_1(0)^2, \vartheta_2(0)^2, \Theta_1(0)^2, \Theta_2(0)^2$

and x^2, z^2. As a consequence, the double of $[x : z]$ and $[x : -z]$ are same. We have

$$x_3^2 = b^2 \left(B^2(x^2 + z^2)^2 + A^2(x^2 - z^2)^2 \right)^2, \tag{8}$$

$$z_3^2 = a^2 \left(B^2(x^2 + z^2)^2 - A^2(x^2 - z^2)^2 \right)^2. \tag{9}$$

Similarly, consider that from $P_1 = \varphi(w_1) = [x_1 : z_1]$, $P_2 = \varphi(w_2) = [x_2 : z_2]$ and $P = P_1 - P2 = \varphi(w_1 - w_2) = [x : z]$ the requirement is to compute $P_1 + P_2 = \varphi(w_1 + w_2) = [x_3 : z_3]$. If we have the values $x_1^2, z_1^2, x_2^2, z_2^2$ and x^2, z^2 along with $\vartheta_1(0)^2, \vartheta_2(0)^2, \Theta_1(0)^2, \Theta_2(0)^2$ then we can compute the values x_3^2 and z_3^2 by Eqs. (10) and (11).

$$x_3^2 = z^2 \left(B^2(x_1^2 + z_1^2)(x_2^2 + z_2^2) + A^2(x_1^2 - z_1^2)(x_2^2 - z_2^2) \right)^2, \tag{10}$$

$$z_3^2 = x^2 \left(B^2(x_1^2 + z_1^2)(x_2^2 + z_2^2) - A^2(x_1^2 - z_1^2)(x_2^2 - z_2^2) \right)^2. \tag{11}$$

This approach requires only squared values, i.e., it starts with squared values and also returns squared values. Hence, this is called the square only setting. Note that in the square only setting, $[x^2 : z^2]$ represents two points $[x : \pm z]$ on the Kummer line. For the case of genus two, the square only setting was advocated in [3,9] (see also [13]). To the best of our knowledge, the details of the square only setting in genus one do not appear earlier in the literature.

Let

$$a^2 = \vartheta_1(0)^2, b^2 = \vartheta_2(0)^2, A^2 = a^2 + b^2 \text{ and } B^2 = a^2 - b^2.$$

Then from (6) we obtain $\Theta_1(0)^2 = A^2/2$ and $\Theta_2(0)^2 = B^2/2$. By $\mathcal{K}_{a^2,b^2}$ we denote the Kummer line having the parameters a^2 and b^2.

Table 2 shows the Algorithms dbl and diffAdd for doubling and differential addition. Details regarding correctness of the computation are provided in the full version [28].

Table 2. Double and differential addition in the square-only setting.

dbl(x^2, z^2) :	diffAdd$(x_1^2, z_1^2, x_2^2, z_2^2, x^2, z^2)$:
$s_0 = B^2(x^2 + z^2)^2$;	$s_0 = B^2(x_1^2 + z_1^2)(x_2^2 + z_2^2)$;
$t_0 = A^2(x^2 - z^2)^2$;	$t_0 = A^2(x_1^2 - z_1^2)(x_2^2 - z_2^2)$;
$x_3^2 = b^2(s_0 + t_0)^2$;	$x_3^2 = z^2(s_0 + t_0)^2$;
$z_3^2 = a^2(s_0 - t_0)^2$;	$z_3^2 = x^2(s_0 - t_0)^2$;
return (x_3^2, z_3^2).	return (x_3^2, z_3^2).

In $\mathcal{K}_{a^2,b^2}$, the point $[a^2 : b^2]$ (representing $[a : \pm b]$) in the square only setting acts as the identity element for the differential addition. The full version [28] provides further details.

In the rest of the paper, we will work in the square only setting over a Kummer line $\mathcal{K}_{a^2,b^2}$ for some values of the parameters a^2 and b^2.

Scalar Multiplication: Suppose $P = [x_1^2 : z_1^2]$ and n be a positive integer. We wish to compute $nP = [x_n^2 : z_n^2]$. The method for doing this is given by Algorithm scalarMult in Table 3. A conceptual description of a ladder step is given in Fig. 1.

Table 3. Scalar multiplication using a ladder.

scalarMult(P, n) :	ladder$(R, S, \mathfrak{b})$:
input: $P \in \mathcal{K}_{a,b}$;	if $(\mathfrak{b} = 0)$
$\quad$ ℓ-bit scalar $n = (1, n_{\ell-2}, \ldots, n_0)$;	$\quad$ $S = \mathsf{diffAdd}(R, S, P)$;
output: nP;	$\quad$ $R = \mathsf{dbl}(R)$;
$\quad$ set $R = P$ and $S = \mathsf{dbl}(P)$;	else
$\quad$ for $i = \ell - 2, \ell - 3, \ldots, 0$ do	$\quad$ $R = \mathsf{diffAdd}(R, S, P)$;
$\quad\quad$ $(R, S) = \mathsf{ladder}(R, S, n_i)$;	$\quad$ $S = \mathsf{dbl}(S)$;
$\quad$ return R.	return (R, S).

2.4 Legendre Form Elliptic Curve

Let E be an elliptic curve and $\sigma : E \to E$ be the automorphism which maps a point of E to its inverse, i.e., for $(a, b) \in E$, $\sigma(a, b) = (a, -b)$.

For $\mu \in \mathbb{F}_q$, let

$$E_\mu : Y^2 = X(X - 1)(X - \mu) \tag{12}$$

be an elliptic curve in the Legendre form. Let $\mathcal{K}_{a^2,b^2}$ be a Kummer line such that

$$\mu = \frac{a^4}{a^4 - b^4}. \tag{13}$$

An explicit map $\psi : \mathcal{K}_{a^2,b^2} \to E_\mu/\sigma$ has been given in [22]. In the square only setting, let $[x^2 : z^2]$ represent the points $[x : \pm z]$ of the Kummer line $\mathcal{K}_{a^2,b^2}$ such that $[x^2 : z^2] \neq [b^2 : a^2]$. Recall that $[a^2 : b^2]$ acts as the identity in $\mathcal{K}_{a^2,b^2}$. Then from [22],

$$\psi([x^2 : z^2]) = \begin{cases} \infty & \text{if } [x^2 : z^2] = [a^2 : b^2]; \\ \left(\frac{a^2 x^2}{a^2 x^2 - b^2 z^2}, \ldots\right) & \text{otherwise.} \end{cases} \tag{14}$$

Given $X = a^2 x^2/(a^2 x^2 - b^2 z^2)$, it is possible to find $\pm Y$ from the equation of E, though it is not possible to uniquely determine the sign of Y. The inverse ψ^{-1}, maps a point not of order two of E_μ/σ to the squared coordinates of points in $\mathcal{K}_{a^2,b^2}$. We have

$$\psi^{-1}(\mathbf{P}) = \begin{cases} [a^2 : b^2] & \text{if } \mathbf{P} = \infty; \\ \left[\frac{b^2 X}{a^2 (X-1)} : 1\right] & \text{if } \mathbf{P} = (X, \ldots). \end{cases} \tag{15}$$

Notation: We will use upper-case bold face letters to denote points of E_μ and upper case normal letters to denote points of $\mathcal{K}_{a^2,b^2}$.

Consistency: Let $\mathcal{K}_{a^2,b^2}$ and E_μ be such that (13) holds. Consider the point $\mathbf{T} = (\mu, 0)$ on E_μ. Note that $\mathbf{T}$ is a point of order two. Given any point $\mathbf{P} = (X, \dots)$ of E_μ, let $\mathbf{Q} = \mathbf{P} + \mathbf{T}$. Then it is easy to verify that

$$\mathbf{Q} = \left(\frac{\mu(X-1)}{X - \mu}, \dots \right).$$

Consider the map $\widehat{\psi} : \mathcal{K}_{a^2,b^2} \to E_\mu$ such that for points $[x : \pm z]$ represented by $[x^2 : z^2]$ in the square only setting

$$\widehat{\psi}([x^2 : z^2]) = \psi([x^2 : z^2]) + \mathbf{T}. \tag{16}$$

The inverse map $\widehat{\psi}^{-1}$ takes a point $\mathbf{P}$ of E_μ to squared coordinates in $\mathcal{K}_{a^2,b^2}$.

For any two points $\mathbf{P}_1, \mathbf{P}_2$ on E_μ which are not of order two and $\mathbf{P} = \mathbf{P}_1 - \mathbf{P}_2$ the following properties hold.

$$\left.\begin{aligned}
2 \cdot \widehat{\psi}([x^2 : z^2]) &= \widehat{\psi}(\mathsf{dbl}(x^2, z^2)); \\
\mathsf{dbl}\left(\widehat{\psi}^{-1}(\mathbf{P}_1)\right) &= \widehat{\psi}^{-1}(2\mathbf{P}_1); \\
\mathsf{diffAdd}\left(\widehat{\psi}^{-1}(\mathbf{P}_1), \widehat{\psi}^{-1}(\mathbf{P}_2), \widehat{\psi}^{-1}(\mathbf{P})\right) &= \widehat{\psi}^{-1}(\mathbf{P}_1 + \mathbf{P}_2).
\end{aligned}\right\} \tag{17}$$

The proofs for (17) can be derived from the formulas for $\widehat{\psi}, \widehat{\psi}^{-1}$; the formulas for addition and doubling on E_μ; and the formulas arising from dbl and $\mathsf{diffAdd}$. This involves simplifications of the intermediate expressions arising in these formulas. Such expressions become quite large. In the appendix of the full version [28] we provide a SAGE script which does the symbolic verification of the required calculations.

The relations given by (17) have the following important consequence to scalar multiplication. Suppose P is in $\mathcal{K}_{a^2,b^2}$ and $\mathbf{P} = \widehat{\psi}(P)$. Then $\widehat{\psi}(nP) = n\mathbf{P}$. Fig. 3 depicts this in pictorial form.

$$
\begin{array}{ccccccccc}
P & \xrightarrow{\psi} & \mathbf{P} & \xrightarrow{+\mathbf{T}} & \mathbf{Q} & \qquad & \mathbf{Q} & \xrightarrow{-\mathbf{T}} & \mathbf{P} & \xrightarrow{\psi^{-1}} & P \\
\downarrow{\scriptstyle *n} & & \downarrow{\scriptstyle *n} & & \downarrow{\scriptstyle *n} & & \downarrow{\scriptstyle *n} & & & & \downarrow{\scriptstyle *n} \\
P_n & \xrightarrow{\psi} & \mathbf{P}_n & \xrightarrow{+\mathbf{T}} & \mathbf{Q}_n & & \mathbf{Q}_n & \xrightarrow{-\mathbf{T}} & \mathbf{P}_n & \xrightarrow{\psi^{-1}} & P_n
\end{array}
$$

Fig. 3. Consistency of scalar multiplications on E_μ and $\mathcal{K}_{a^2,b^2}$.

Relation Between the Discrete Logarithm Problems: Suppose the Kummer line $\mathcal{K}_{a^2,b^2}$ is chosen such that the corresponding curve E_μ has a cyclic subgroup $\mathfrak{G} = \langle \mathbf{P} \rangle$ of large prime order. Given $\mathbf{Q} \in \mathfrak{G}$, the discrete logarithm problem in $\mathfrak{G}$ is to obtain an n such that $\mathbf{Q} = n\mathbf{P}$. This problem can be reduced to computing discrete logarithm problem in $\mathcal{K}_{a^2,b^2}$. Map the point $\mathbf{P}$ (resp. $\mathbf{Q}$) to $P \in \mathcal{K}_{a,b}$ (resp. $Q \in \mathcal{K}_{a,b}$) using $\widehat{\psi}^{-1}$ Find n such that $Q = nP$ and return n. Similarly, the discrete logarithm problem in $\mathcal{K}_{a,b}$ can be reduced to the discrete logarithm problem in E_μ.

The above shows the equivalence of the hardness of solving the discrete logarithm problem in either E_μ or in $\mathcal{K}_{a^2,b^2}$. So, if E_μ is a well chosen curve such that the discrete logarithm problem in E_μ is conjectured to be hard, then the discrete logarithm problem in the associated $\mathcal{K}_{a^2,b^2}$ will be equally hard. This fact forms the basis for using Kummer line for cryptographic applications.

2.5   Scalar Multiplication in E_μ

Let E_μ be a Legendre form curve and $\mathcal{K}_{a^2,b^2}$ be a Kummer line in the square only setting. Suppose $\mathfrak{G} = \langle \mathbf{P} = (X_P, Y_P) \rangle$ is a cryptographically relevant subgroup of E_μ. Further, suppose a point $P = [x^2 : z^2]$ in $\mathcal{K}_{a^2,b^2}$ is known such that $(X_P, \ldots) = \widehat{\psi}(P) = \psi(P) + \mathbf{T}$ where as before $\mathbf{T} = (\mu, 0)$. The point P is the base point on $\mathcal{K}_{a^2,b^2}$ which corresponds to the point $\mathbf{P}$ on E_μ.

Let n be a non-negative integer which is less than the order of $\mathfrak{G}$. The requirement is to compute the scalar multiplication $n\mathbf{P}$ via the laddering algorithm on the Kummer line $\mathcal{K}_{a^2,b^2}$. First, the ladder algorithm is applied to the inputs P and n. This results in a pair of points Q and R, where $Q = nP$ and $R = (n+1)P$ so that $Q - R = -P$. By the consistency of scalar multiplication, we have $\mathbf{Q} = n\mathbf{P}$. Let $\mathbf{Q} = (X_Q, Y_Q)$. From Q it is possible to directly recover X_Q and $\pm Y_Q$. Using Q, R and P, in the full version [28], we show that it is indeed possible to determine Y_Q so that a scalar multiplicationis possible in E_μ. The cost of recovering X_Q and Y_Q comes to a few finite field multiplications and one inversion.

3   Kummer Line over Prime Order Fields

Let p be a prime and $\mathbb{F}_p$ be the field of p elements. As mentioned earlier, using the Lefschetz principle, the theta identities also hold over $\mathbb{F}_p$. Consequently, it is possible to work over a Kummer line $\mathcal{K}_{a^2,b^2}$ and associated elliptic curve E_μ defined over the algebraic closure of $\mathbb{F}_p$. The only condition for this to be meaningful is that $a^4 - b^4 \neq 0 \bmod p$ so that $\mu = a^4/(a^4 - b^4)$ is defined over $\mathbb{F}_p$. We choose a^2 and b^2 to be small values while p is a large prime and so the condition $a^4 - b^4 \neq 0 \bmod p$ easily holds. Note that we will choose a^2 and b^2 to be in $\mathbb{F}_p$ without necessarily requiring a and b themselves to be in $\mathbb{F}_p$. Similarly, in the square only setting when we work with squared representation $[x^2 : z^2]$ of points $[x : \pm z]$, the values x^2, z^2 will be in $\mathbb{F}_p$ and it is not necessary for x and z themselves to be in $\mathbb{F}_p$.

Our target is the 128-bit security level. To this end, we consider the three primes $p2519$, $p25519$ and $p2663$. The choice of these three primes is motivated by the consideration that these are of the form $2^m - \delta$, where m is around 256 and δ is a small positive integer. For m in the range 250 to 270 and $\delta < 20$, the only three primes of the form $2^m - \delta$ are $p2519$, $p25519$ and $p2663$. We later discuss the comparative advantages and disadvantages of using Kummer lines based on these three primes.

3.1 Finding a Secure Kummer Line

For each prime p, the procedure for finding a suitable Kummer line is the following. The value of a^2 is increased from 2 onwards and for each value of a^2, the value of b^2 is varied from 1 to $a^2 - 1$; for each pair (a^2, b^2), the value of $\mu = a^4/(a^4 - b^4)$ is computed and the order of $E_\mu(\mathbb{F}_p)$ is computed. Let $t = p + 1 - \#E_\mu(\mathbb{F}_p)$. Let ℓ and ℓ_T be the largest prime factors of $p + 1 - t$ and $p + 1 + t$ respectively and let $h = (p+1-t)/\ell$ and $h_T = (p+1+t)/\ell_T$. Here h and h_T are the co-factors of the curve and its quadratic twists respectively. If both h and h_T are small, then (a^2, b^2) is considered. Among the possible (a^2, b^2) that were obtained, we have used the one with the minimum value of a^2. After fixing (a^2, b^2) the following parameters for E_μ have been computed.

1. Embedding degrees k and k_T of the curve and its twist. Here k (resp. k_T) is the smallest positive integer such that $\ell | p^k - 1$ (resp. $\ell_T | p^{k_T} - 1$). This is given by the order of p in $\mathbb{F}_\ell$ (resp. $\mathbb{F}_{\ell_T}$) and is found by checking the factors of $\ell - 1$ (resp. $\ell_T - 1$).
2. The complex multiplication field discriminant D. This is defined in the following manner (https://safecurves.cr.yp.to/disc.html): By Hasse's theorem, $|t| \leq 2\sqrt{p}$ and in the cases that we considered $|t| < 2\sqrt{p}$ so that $t^2 - 4p$ is a negative integer; let s^2 be the largest square dividing $t^2 - 4p$; define $D = (t^2 - 4p)/s^2$ if $t^2 - 4p \bmod 4 = 1$ and $D = 4(t^2 - 4p)/s^2$ otherwise. (Note that D is different from the discriminant of E_μ which is equal to $\mu^4 - 2\mu^3 + \mu^2$.)

Table 4 provides the three Kummer lines and (estimates of) the sizes of the various parameters of the associated Legendre form elliptic curves. As part of [20], we provide Magma code for computing these parameters and also their exact values. The Kummer line $\mathcal{K}_{a^2,b^2}$ over $p2519$ is compactly denoted as KL2519(a^2, b^2) and similarly for Kummer lines over $p25519$ and $p2663$. For each Kummer line reported in Table 4, the base point $[x^2 : z^2]$ is such that its order is ℓ. Table 4 also provides the corresponding details for Curve25519, P-256 and secp256k1 which have been collected from [5]. This will help in comparing the new proposals with some of the most important and widely used proposals over prime fields that are present in the literature.

The Four-$\mathbb{Q}$ proposal [14] is an elliptic curve over $\mathbb{F}_{p^2}$ where $p = 2^{127} - 1$. For this curve, the size ℓ of the cryptographic sub-group is 246 bits, the co-factor is 392 and the embedding degree is $(\ell - 1)/2$. The largest prime dividing the twist order is 158 bits and [14] does not consider twist security to be an issue. Note that the underlying field for Four-$\mathbb{Q}$ is composite and further endomorphisms are available to speed up scalar multiplication. So Four-$\mathbb{Q}$ is not directly comparable to the setting that we consider and hence we have not included it in Table 4.

Table 4. New Kummer lines and their parameters in comparison to Curve25519, P-256 and secp256k1.

	KL2519(81, 20)	KL25519(82, 77)	KL2663(260, 139)
$(\lg \ell, \lg \ell_T)$	$(248, 248)$	$(251.4, 252)$	$(262.4, 263)$
(h, h_T)	$(8, 8)$	$(12, 8)$	$(12, 8)$
(k, k_T)	$\left(\ell - 1, \frac{\ell_T - 1}{7}\right)$	$(\ell - 1, \ell_T - 1)$	$\left(\frac{\ell - 1}{2}, \ell_T - 1\right)$
$\lg(-D)$	246.3	255	266
base point	$[64 : 1]$	$[31 : 1]$	$[2 : 1]$

	Curve25519 [2]	P-256 [37]	secp256k1 [40]
$(\lg \ell, \lg \ell_T)$	$(252, 253)$	$(256, 240)$	$(256, 219.3)$
(h, h_T)	$(8, 4)$	$(1, 3 \cdot 5 \cdot 13 \cdot 179)$	$(1, 3^2 \cdot 13^2 \cdot 3319 \cdot 22639)$
(k, k_T)	$\left(\frac{\ell - 1}{6}, \ell_T - 1\right)$	$\left(\frac{\ell - 1}{3}, \frac{\ell_T - 1}{2}\right)$	$\left(\frac{\ell - 1}{6}, \frac{\ell_T - 1}{6}\right)$
$\lg(-D)$	254.7	258	1.58
base point	$(9, \ldots)$	large	large

For KL2519(81, 20), [15 : 1] is another choice of base point. Also, for $p2519$, KL2519(101, 61) is another good choice for which both h and h_T are 8, the other security parameters have large values and [4 : 1] is a base point. We have implementations of both KL2519(81, 20) and KL2519(101, 61) and the performance of both are almost the same. Hence, we report only the performance of KL2519(81, 20).

The points of order two on the Legendre form curve $Y^2 = X(X - 1)(X - \mu)$ are $(0, 0)$, $(1, 0)$ and $(\mu, 0)$. The sum of two distinct points of order two is also a point of order two and hence the sum is the third point of order two; as a result, the points of order two along with the identity form an order 4 subgroup of the group formed by the $\mathbb{F}_p$ rational points on the curve. Consequently, the group of $\mathbb{F}_p$ rational points has an order which is necessarily a multiple of 4, i.e., $p + 1 - t = 4a$ for some integer a.

1. If $p = 4m + 1$, then $p + 1 + t = 4a_T$ where $a_T = 2m - a + 1 \not\equiv a \bmod 2$. As a result, it is not possible to have both h and h_T to be equal to 4, or both of these to be equal to 8. So, the best possibilities for h and h_T are that one of them is 4 and the other is 8. The primes $p25519$ and $p2663$ are both $\equiv 1 \bmod 4$. For these two primes, searching for a^2 up to 512, we were unable to find any choice for which one of h and h_T is 4 and the other is 8. The next best possibilities for h and h_T are that one of them is 8 and the other is 12. We have indeed found such choices which are reported in Table 4.

2. If $p = 4m + 3$, then $p + 1 + t = 4a_T$ where $a_T = 2m - a + 2 \equiv a \bmod 2$. In this case, it is possible that both h and h_T are equal to 4. The prime $p2519$ is $\equiv 1 \bmod 3$. For this prime, searching for a^2 up to 512, we were unable to find any choice where $h = h_T = 4$. The next best possibility is $h = h_T = 8$ and we have indeed found such a choice which is reported in Table 4.

Gaudry and Lubicz [22] had remarked that for Legendre form curves, if $p \equiv 1 \bmod 4$, then the orders of the curve and its twist are divisible by 4 and 8 respectively; while if $p \equiv 3 \bmod 4$, then the orders of the curve and its twist are divisible by 8 and 16 respectively. The Legendre form curve corresponding to KL2519(81, 20) has $h = h_T = 8$ and hence shows that the second statement is incorrect. The discussion provided above clarifies the issue of divisibility by 4 of the order of the curve and its twist.

The effectiveness of small subgroup attacks [31] is determined by the size of the co-factor. Such attacks can be prevented by checking whether the order of a given point is equal to the co-factor before performing the actual scalar multiplication. This requires a scalar multiplication by h. In Table 4, the co-factors of the curve are either 8 or 12. A scalar multiplication by 8 requires 3 doublings whereas a scalar multiplication by 12 requires 3 doublings and one addition. Amortised over the cost of the actual scalar multiplication, this cost is negligible. Even without such protection, a small subgroup attack improves Pollard rho by a factor of $\sqrt{h}$ and hence degrades security by $\lg \sqrt{h}$ bits. So, as in the case of Curve25519, small subgroup attacks are not an issue for the proposed Kummer lines.

Let $\mathfrak{r}$ be a quadratic non-residue in $\mathbb{F}_p$ and consider the curve $\mathfrak{r}Y^2 = f(X) = X(X-1)(X-\mu)$. This is a quadratic twist of the original curve. For any $X \in \mathbb{F}_p$, either $f(X)$ is a quadratic residue or a quadratic non-residue. If $f(X)$ is a quadratic residue, then $(X, \pm\sqrt{f(X)})$ are points on the original curve; otherwise, $(X, \pm\sqrt{\mathfrak{r}^{-1}f(X)})$ are points on the quadratic twist. So, for each point X, there is a pair of points on the curve or on the quadratic twist. An x-coordinate only scalar multiplication algorithm does not distinguish between these two cases. One way to handle the problem is to check whether $f(X)$ is a quadratic residue before performing the scalar multiplication. This, however, has a significant cost. On the other hand, if this is not done, then an attacker may gain knowledge about the secret scalar modulo the co-factor of the twist. The twist co-factors of the new curves in Table 4 are all 8 which is only a little larger than the twist co-factor of 4 for Curve25519. Consequently, as in the case of Curve25519, attacks based on the co-factors of the twist are ineffective.

Note that the use of the square only setting for the Kummer line computation is not related to the twist security of the Legendre form elliptic curve. In particular, for the elliptic curve, computations are not in the square only setting.

To summarise, the three new curves listed in Table 4 provide security at approximately the 128-bit security level.

4   Field Arithmetic

As mentioned earlier, we consider three primes $p2519 = 2^{251} - 9$, $p25519 = 2^{255} - 19$ and $p2663 = 2^{266} - 3$. The general form of these primes is $p = 2^m - \delta$. Let η and ν be such that $m = \eta(\kappa - 1) + \nu$ with $0 \le \nu < \eta$. The values of m, δ, κ, η and ν for $p2519$, $p25519$ and $p2663$ are given in Table 5. The value of κ indicates the number of limbs used to represent elements of $\mathbb{F}_p$; the value of η

Table 5. The different values of κ, η and ν corresponding to the primes $p2519$, $p25519$ and $p2663$.

prime	m	δ	κ	η	ν	prime	m	δ	κ	η	ν	prime	m	δ	κ	η	ν
$p2519$	251	9	9	28	27	$p25519$	255	19	10	26	21	$p2663$	266	3	10	27	23
			5	51	47				5	51	51				5	54	50

represents the number of bits in the first $\kappa - 1$ limbs; and the value of ν is the number of bits in the last limb. For each prime, two sets of values of κ, η and ν are provided. This indicates that two different representations of each prime are used. The entire scalar multiplication is done using the longer representation (i.e., with $\kappa = 9$ or $\kappa = 10$); next the two components of the result are converted to the shorter representation (i.e., with $\kappa = 5$); and then the inversion and the single field multiplication are done using the representation with $\kappa = 5$. In the following sections, we describe methods to perform arithmetic over $\mathbb{F}_p$. Most of the description is in general terms of κ, η and ν. The specific values of κ, η and ν are required only to determine that no overflow occurs.

Representation of Field Elements: Let $\theta = 2^\eta$ and consider the polynomial $A(\theta)$ defined in the following manner: $A(\theta) = a_0 + a_1\theta + \cdots + a_{\kappa-1}\theta^{\kappa-1}$ where $0 \leq a_0, \ldots, a_{\kappa-1} < 2^\eta$ and $0 \leq a_{\kappa-1} < 2^\nu$. Such a polynomial will be called a *proper* polynomial. Note that proper polynomials are in 1-1 correspondence with the integers $0, \ldots, 2^m - 1$. This leads to non-unique representation of some elements of $\mathbb{F}_p$: specifically, the elements $0, \ldots, \delta - 1$ are also represented as $2^m - \delta, \ldots, 2^m - 1$. This, however, does not cause any of the computations to become incorrect. Conversion to unique representation using a simple constant time code is done once at the end of the computation. The issue of non-unique representation was already mentioned in [2] where the following was noted: 'Note that integers are not converted to a unique "smallest" representation until the end of the Curve25519 computation. Producing reduced representations is generally much faster than producing "smallest" representations.'

Representation of the Prime p: The representation of the prime p will be denoted by $\mathfrak{P}(\theta)$ where $\mathfrak{P}(\theta) = \sum_{i=0}^{\kappa-1} \mathfrak{p}_i\theta^i$ with $\mathfrak{p}_0 = 2^\eta - \delta$; $\mathfrak{p}_i = 2^\eta - 1$; $i = 1, \ldots, \kappa - 2$; and $\mathfrak{p}_{\kappa-1} = 2^\nu - 1$. This representation will only be required for the larger value of κ.

4.1 Reduction

This operation will be required for both values of κ.

Using $p = 2^m - \delta$, for $i \geq 0$, we have $2^{m+i} = 2^i \times 2^m = 2^i(2^m - \delta) + 2^i\delta \equiv 2^i\delta \bmod p$. So, multiplying by 2^{m+i} modulo p is the same as multiplying by $2^i\delta$ modulo p. Recall that we have set $\theta = 2^\eta$ and so $\theta^\kappa = 2^{\eta\kappa} = 2^{m+\eta-\nu}$ which implies that $\theta^\kappa \bmod p = 2^{\eta-\nu}\delta$. Suppose $C(\theta) = \sum_{i=0}^{\kappa-1} c_i\theta^i$ is a polynomial such that for some $\mathsf{m} \leq 64$, $c_i < 2^\mathsf{m}$ for all $i = 0, \ldots, 7$. If for some $i \in \{0, \ldots, \kappa - 2\}$, $c_i \geq 2^\eta$, or $c_{\kappa-1} \geq 2^\nu$, then $C(\theta)$ is not a proper polynomial. Following the

Table 6. The reduction algorithm.

---

reduce($C(\theta)$) :
input: $C(\theta) = c_0 + c_1\theta + \cdots + c_{\kappa-1}\theta^{\kappa-1}$, $c_i < 2^m$, $i = 0,\ldots,\kappa-1$;
output: polynomial $D(\theta)$ such that $D(\theta) \equiv C(\theta) \bmod p$;
1. $s_0 \leftarrow 0$;
2.   for $i = 0,\ldots,\kappa-2$ do
3. $d_i \leftarrow \mathsf{lsb}_\eta(c_i + s_i)$; $s_{i+1} \leftarrow (c_i + s_i)/2^\eta$;
4.   end for;
5. $d_{\kappa-1} \leftarrow \mathsf{lsb}_\nu(c_{\kappa-1} + s_{\kappa-1})$; $t_0 = (c_{\kappa-1} + s_{\kappa-1})/2^\nu$;
6. $e_0 \leftarrow \mathsf{lsb}_\eta(d_0 + 2^{\eta-\nu}\delta t_0)$; $t_1 \leftarrow (d_0 + 2^{\eta-\nu}\delta t_0)/2^\eta$; $[t_2 \leftarrow \lfloor(d_1 + t_1)/2^\eta\rfloor]$
7. $d_0 \leftarrow e_0$; $d_1 \leftarrow d_1 + t_1$;
8.   return $D(\theta)$.

---

idea in [2,7,12], Table 6 describes a method to obtain a polynomial $D(\theta) = \sum_{i=0}^{\kappa-1} d_i\theta^i$ such that $D(\theta) \equiv C(\theta) \bmod p$. For $i = 0,\ldots,\kappa-2$, Step 3 ensures $c_i + s_i = d_i + 2^\eta s_{i+1}$ and $d_i < 2^\eta$; Step 5 ensures $c_{\kappa-1} + s_{\kappa-1} = d_{\kappa-1} + 2^\nu t_0$ and $d_{\kappa-1} < 2^\nu$. In Step 6, t_2 is actually not computed, it is provided for the ease of analysis.

In the full version [28], we argue that there no overflows in the intermediate quantities arising in reduce. Also, we show that reduce($D(\theta)$) is indeed a proper polynomial. In other words, two successive invocations of reduce on $C(\theta)$ reduces it to a proper polynomial. In practice, however, this is not done at each step. Only one invocation is made. As observed above, reduce($C(\theta)$) returns $D(\theta)$ for which all coefficients $d_0, d_2, \ldots, d_{\kappa-1}$ satisfy the appropriate bounds and only d_1 can possibly require $\eta + 1$ bits to represent instead of the required η-bit representation. This does not cause any overflow in the intermediate computation and so we do not reduce $D(\theta)$ further. It is only at the end, that an additional invocation of reduce is made to ensure that a proper polynomial is obtained on which we apply the makeUnique procedure to ensure unique representation of elements of $\mathbb{F}_p$.

4.2   Field Negation

This operation will only be required for the representation using the longer value of κ and occurs only as part of the Hadamard operation.

Let $A(\theta) = \sum_{i=0}^{\kappa-1} a_i\theta^i$ be a polynomial. We wish to compute $-A(\theta) \bmod p$. Let $\mathfrak{n}$ be the least integer such that all the coefficients of $2^{\mathfrak{n}}\mathfrak{P}(\theta) - A(\theta)$ are non-negative. By negate($A(\theta)$) we denote $T(\theta) = 2^{\mathfrak{n}}\mathfrak{P}(\theta) - A(\theta)$. Reducing $T(\theta)$ modulo p gives the desired answer. Let $T(\theta) = \sum_{i=0}^{\kappa-1} t_i\theta^i$ so that $t_i = 2^{\mathfrak{n}}\mathfrak{p}_i - a_i \geq 0$. The condition of non-negativity on the coefficients of $T(\theta)$ eliminates the situation in two's complement subtraction where the result can be negative. Later we mention the appropriate values of $\mathfrak{n}$ that is to be used in different situations. Considering all values to be 64-bit quantities, the computation of t_i is done in the following manner: $t_i = ((2^{64} - 1) - a_i) + (1 + 2^{\mathfrak{n}}\mathfrak{p}_i) \bmod 2^{64}$.

The operation $(2^{64} - 1) - a_i$ is equivalent to taking the bitwise complement of a_i which is equivalent to $1^{64} \oplus a_i$.

4.3   Field Multiplication

This operation is required for both the larger and the smaller values of κ.

Suppose that $A(\theta) = \sum_{i=0}^{\kappa-1} a_i\theta^i$ and $B(\theta) = \sum_{i=0}^{\kappa-1} b_i\theta^i$ are to be multiplied. Two algorithms for multiplication called mult and multe are defined in Table 7.

Table 7. Field multiplication algorithms.

mult$(A(\theta), B(\theta))$:	multe$(A(\theta), B(\theta))$:
input: $A(\theta), B(\theta)$	input: $A(\theta), B(\theta)$
output: $C(\theta)$	output: $C(\theta)$
1. $C(\theta) \leftarrow$ polyMult$(A(\theta), B(\theta))$;	1. $C(\theta) \leftarrow$ polyMult$(A(\theta), B(\theta))$;
2. $C(\theta) \leftarrow$ fold$(C(\theta))$;	2. $C(\theta) \leftarrow$ expand$(C(\theta))$;
3. return reduce$(C(\theta))$.	3. $C(\theta) \leftarrow$ fold$(C(\theta))$;
	4. return reduce$(C(\theta))$.

Table 8. The expand procedure.

expand$(C(\theta))$:
input: $C(\theta) = c_0 + c_1\theta + \cdots + c_{2\kappa-2}\theta^{2\kappa-2}$
output: $D(\theta) = d_0 + d_1\theta + \cdots + d_{2\kappa-1}\theta^{2\kappa-1}$
1.   for $i = 0, \ldots, \kappa - 1$, $d_i \leftarrow c_i$;
2. $s_0 \leftarrow 0$;
3.   for $i = 0, \ldots, \kappa - 2$, $d_{\kappa+i} \leftarrow \mathsf{lsb}_\eta(c_{\kappa+i} + s_i)$; $s_{i+1} \leftarrow (c_{\kappa+i} + s_i)/2^\eta$;
4. $d_{2\kappa-1} \leftarrow s_{\kappa-1}$;
5.   return $D(\theta)$.

Let $C(\theta)$ be the result of polyMult$(A(\theta), B(\theta))$. Then $C(\theta)$ can be written as

$$C(\theta) = c_0 + c_1\theta + \cdots + c_{2\kappa-2}\theta^{2\kappa-2} \tag{18}$$

where $c_t = \sum_{s=0}^{t} a_s b_{t-s}$ with the convention that a_i, b_j is zero for $i, j > \kappa - 1$. For $s = 0, \ldots, \kappa - 1$, the coefficient $c_{\kappa-1\pm s}$ is the sum of $(\kappa - s)$ products of the form $a_i b_j$. Since $a_i, b_j < 2^\eta$, it follows that for $s = 0, \ldots, \kappa - 1$,

$$c_{\kappa-1\pm s} \leq (\kappa - s)(2^\eta - 1)^2. \tag{19}$$

Using the representation with the larger value of κ each c_t fits in a 64-bit word and using the representation with the smaller value of κ, each c_t fits in a 128-bit word.

The step polyMult multiplies $A(\theta)$ and $B(\theta)$ as polynomials in θ and returns the result polynomial of degree $2\kappa - 2$. In multe, the step expand is applied to this polynomial and returns a polynomial of degree $2\kappa - 1$. In mult, the step expand is not present and fold is applied to a polynomial of degree $2\kappa - 2$. For uniformity of description, we assume that the input to fold is a polynomial of degree $2\kappa - 1$ where for the case of mult the highest degree coefficient is 0.

The computation of $\mathsf{fold}(C(\theta))$ is the following.

$$
\begin{aligned}
C(\theta) &= c_0 + c_1\theta + \cdots + c_{\kappa-1}\theta^{\kappa-1} + \theta^\kappa \left(c_\kappa + c_{\kappa+1}\theta + \cdots + c_{2\kappa-1}\theta^{\kappa-1} \right) \\
&\equiv c_0 + c_1\theta + \cdots + c_{\kappa-1}\theta^{\kappa-1} + 2^{\eta-\nu}\delta \left(c_\kappa + c_{\kappa+1}\theta + \cdots + c_{2\kappa-1}\theta^{\kappa-1} \right) \bmod p \\
&= (c_0 + \mathfrak{h}c_\kappa) + (c_1 + \mathfrak{h}c_{\kappa+1})\theta + \cdots + (c_{\kappa-1} + \mathfrak{h}c_{2\kappa-1})\theta^{\kappa-1}
\end{aligned}
$$

where $\mathfrak{h} = 2^{\eta-\nu}\delta$. The polynomial in the last line is the output of $\mathsf{fold}(C(\theta))$.

The expand routine is shown in Table 8. For $D(\theta)$ that is returned by expand we have $d_\kappa, \ldots, d_{2\kappa-1} < 2^\eta$.

The situations where mult and multe are required are as follows.

1. For $\kappa = 5$, only mult is required.
2. For $p25519$ and $\kappa = 10$, mult will provide an incorrect result. This is because, in this case, some of the coefficients of $\mathsf{fold}(\mathsf{polyMult}(A(\theta), B(\theta)))$ do not fit into 64-bit words. This was already mentioned in [2] and it is for this reason that the "base 2^{26} representation" was discarded. So, for $p25519$ and $\kappa = 10$, only multe will be used.
3. For $p2519$ and $p2663$, both mult and multe will be used at separate places in the scalar multiplication algorithm. This may appear to be strange, since clearly mult is faster than multe. While this is indeed true, the speed improvement is not as much as seems to be apparent from the description of the two algorithms. We mention the following two points.
    - In both mult and multe, as part of fold, multiplication by $\mathfrak{h}$ is required. For the case of mult, the values to which $\mathfrak{h}$ is multiplied are all greater than 32 bits and so the multiplications have to be done using shifts and adds. On the other hand, in the case of multe, the values to which $\mathfrak{h}$ is multiplied are outputs of expand and are hence all less than 32 bits so that these multiplications can be done directly using unsigned integer multiplications. To a certain extent this mitigates the effect of having the expand operation in multe.
    - More importantly, multe is a better choice at one point of the scalar multiplication algorithm. There is a Hadamard operation which is followed by a multiplication. If we do not apply the reduce operation at the end of the Hadamard operation, then the polynomials which are input to the multiplication operation are no longer proper polynomials. Applying mult to these polynomials leads to an overflow after the fold step. Instead, multe is applied, where the expand ensures that there is no overflow at the fold step.

Due to the combination of the above two effects, the additional cost of the expand operation is more than offset by the savings in eliminating a prior reduce step.

Computation of polyMult: We discuss strategies for polynomial multiplication using the representation for the larger value of κ.

There are several strategies for multiplying two polynomials. For $p2519$, $\kappa = 9$, while for $p25519$ and $p2663$, $\kappa = 10$. Let $C(\theta) = \mathsf{polyMult}(A(\theta), B(\theta))$ where $A(\theta)$ and $B(\theta)$ are proper polynomials. Computing the coefficients of $C(\theta)$ involve 32-bit multiplications and 64-bit additions. The usual measure for assessing the efficacy of a polynomial multiplication algorithm is the number of 32-bit multiplications that would be required. Algorithms from [35] provide the smallest counts of 32-bit multiplication. This measure, however, does not necessarily provide the fastest implementation. Additions and dependencies do play a part and it turns out that an algorithm using a higher number of 32-bit multiplications turn out to be faster in practice. We discuss the cases of $\kappa = 9$ and $\kappa = 10$ separately. In the following, we abbreviate a 32-bit multiplication as [M].

Case $\kappa = 9$: Using 3-3 Karatsuba requires 36[M]. An algorithm given in [35] requires 34[M], but, this algorithm also requires multiplication by small constants which slows down the implementation. We have experimented with several variants and have found the following variant to provide the fastest speed (on the platform for implementation that we used). Consider the 9-limb multiplication to be 8-1 Karatsuba, i.e., the degree 8 polynomial is considered to be a degree 7 polynomial plus the term of degree 8. The two degree 7 (i.e., 8-limb) polynomials are multiplied by 3-level recursive Karatsuba: the 8-limb multiplication is done using 3 4-limb multiplications; each 4-limb multiplication is done using 3 2-limb multiplications; and finally the 2-limb multiplications are done using 4[M] using schoolbook. Using Karatsuba for the 2-limb multiplication is slower. The multiplication by the coefficients of the two degree 8 terms are done directly.

Case $\kappa = 10$: Using binary Karatsuba, this can be broken down into 3 5-limb multiplications. Two strategies for 5-limb multiplications in [35] require 13[M] and 14[M]. The strategy requiring 13[M] also requires multiplications by small constants and turns out to have a slower implementation than the strategy requiring 14[M].

Comparison to Previous Multiplication Algorithm for $p25519$: In the original paper [2] which introduced Curve25519, it was mentioned that for $p25519$, a 10-limb representation using base 2^{26} cannot be used as this leads to an overflow. Instead an approach called "base $2^{25.5}$" was advocated. This approach has been followed in later implementations [7,12] of Curve25519. In this representation, a 255-bit integer A is written as

$$A = a_0 + 2^{26}a_1 + 2^{51}a_2 + 2^{77}a_3 + 2^{102}a_4 + 2^{128}a_5 + 2^{153}a_6 + 2^{179}a_7 + 2^{204}a_8 + 2^{230}a_9$$

where $a_0, a_2, a_4, a_6, a_8 < 2^{26}$ and $a_1, a_3, a_5, a_7, a_9 < 2^{25}$. Note that this representation cannot be considered as a polynomial in some quantity and so the multiplication of two such representations cannot benefit from the various polynomial multiplication algorithms. Instead, multiplication of two integers A and B in this representation requires all the 100 pairwise multiplications of a_i and b_j

along with a few other multiplications by small constants. As mentioned in [12], a total of 109[M] are required to compute the product.

For $p25519$, we have described a 10-limb representation using base as $\theta = 2^{26}$ and have described a multiplication algorithm, namely multe, using this representation. Given the importance of Curve25519, this itself is of some interest. The advantage of multe is that it can benefit from the various polynomial multiplication strategies. On the other hand, the drawback is that the reduction requires a little more time, since the expand step has to be applied.

Following previous work [7], the Sandy2x implementation used SIMD instructions to simultaneously compute two field multiplications. The vpmuludq instruction is used to simultaneously carry out two 32-bit multiplications. As a result, the 109 multiplications can be implemented using 54.5 vpmuludq instructions per field multiplication.

The multiplication algorithm multe for $p25519$ can also be vectorised using vpmuludq to compute two simultaneous field multiplications. We have, however, not implemented this. Since our target is Kummer line computation, we used AVX2 instructions to simultaneously compute four field multiplications. It would be of independent interest to explore the 2-way vectorisation of the new multiplication algorithm for use in the Montgomery curve.

5-Limb Representation: For $\kappa = 5$, there is not much difference in the multiplication algorithm for $p2519$, $p25519$ and $p2663$. A previous work [6] showed how to perform field arithmetic for $p25519$ using the representation with $\kappa = 5$ and $\eta = \nu = 51$. The Sandy2x code provides an assembly implementation of the multiplication and squaring algorithm and a constant time implementation of the inversion algorithm for $p25519$. The Sandy2x software mentions that the code is basically from [6]. We have used this implementation to perform the inversion required after the Kummer line computation over KL25519(82, 77). We have modified the assembly code for multiplication and squaring over $p25519$ to obtain the respective routines for $p2519$ and $p2663$ which were then used to implement constant time inversion algorithms using fixed addition chains.

Multiplication by a Small Constant: This operation will only be required for the representation using the longer value of κ. Let $A(\theta) = \sum_{i=0}^{\kappa-1} a_i\theta^i$ be a polynomial and c be a small positive integer considered to be an element of $\mathbb{F}_p$. In our applications, c will be at most 9 bits. The operation constMult$(A(\theta), c)$ will denote the polynomial $C(\theta) = \sum_{i=0}^{\kappa-1}(ca_i)\theta^i$. We do not apply the algorithm reduce to $C(\theta)$. This is because in our application, multiplication by a constant will be followed by a Hadamard operation and the reduce algorithm is applied after the Hadamard operation. This improves efficiency.

Field Squaring: This operation is required for both the smaller and the larger values of κ. Let $A(\theta)$ be a proper polynomial. We define sqr$(A(\theta))$ (resp. sqre$(A(\theta))$) to be the proper polynomial $C(\theta)$ such that $C(\theta) \equiv A^2(\theta) \bmod p$. The computation of sqr (resp. sqre) is almost the same as that of mult (resp. sqre), except that polyMult$(A(\theta), B(\theta))$ is replaced by polySqr$(A(\theta))$ where polySqr$(A(\theta))$ returns $A^2(\theta)$ as the square of the polynomial $A(\theta)$. The algorithm sqre is required only for $p25519$ and $\kappa = 10$. In all other cases, the algorithm sqr is required. Unlike the situation for multiplication, there is no situation for either $p2519$ or $p2663$ where sqre is a better option compared to sqr.

4.4 Hadamard Transform

This operation is required only for the representation using the larger value of κ. Let $A_0(\theta)$ and $A_1(\theta)$ be two polynomials. By $\mathcal{H}(A_0(\theta), A_1(\theta))$ we denote the pair $(B_0(\theta), B_1(\theta))$ where $B_0(\theta) = \mathsf{reduce}(A_0(\theta) + A_1(\theta))$ and $B_1(\theta) = \mathsf{reduce}(A_0(\theta) - A_1(\theta)) = \mathsf{reduce}(A_0(\theta) + \mathsf{negate}(A_1(\theta)))$.

In our context, there is an application of the Hadamard transform to the output of multiplication by constant. Since the output of multiplication by constant is not reduced, the coefficients of the input polynomials to the Hadamard transform do not necessarily respect the bounds required for proper polynomials. As explained earlier, the procedure negate works correctly even with looser bounds on the coefficients of the input polynomial.

We define the operation $\mathsf{unreduced}\text{-}\mathcal{H}(A_0(\theta), A_1(\theta))$ which is the same as $\mathcal{H}(A_0(\theta), A_1(\theta))$ except that the reduce operations are dropped. If the inputs are proper polynomials, then it is not difficult to see that the first $\kappa - 1$ coefficients of the two output polynomials are at most $\eta + 1$ bits and the last coefficients are at most $\nu + 1$ bits. Leaving the output of the Hadamard operation unreduced saves time. In the scalar multiplication algorithm, in one case this can be done and is followed by the multe operation which ensures that there is no eventual overflow.

4.5 Field Inversion

This operation is required only for the representation using the smaller value of κ. Suppose the inversion of $A(\theta)$ is required. Inversion is computed in constant time using a fixed addition chain to compute $A(\theta)^{p-2} \bmod p$. This computation boils down to computing a fixed number of squarings and multiplications. In our context, field inversion is required only for conversion from projective to affine coordinates. The output of the scalar multiplication is in projective coordinates and if for some application the output is required in affine coordinates, then only a field inversion is required. The timing measurements that we report later includes the time required for inversion.

As mentioned earlier, the entire Kummer line scalar multiplication is done using the larger value of κ. Before performing the inversion, the operands are converted to the representation using the smaller value of κ. For $p25519$, the actual inversion is done using the constant time code for inversion used for Curve25519 in the Sandy2x implementation while for $p2519$ and $p2663$, appropriate modifications of this code are used.

5 Vector Operations

While considering vector operations, we consider the representation of field elements using the larger value of κ. To take advantage of SIMD instructions it is convenient to organise the data as vectors. The Intel instructions that we target apply to 256-bit registers which are considered to be 4 64-bit words (or, as 8 32-bit words). So, we consider vectors of length 4.

Let $\mathbf{A}(\theta) = (A_0(\theta), A_1(\theta), A_2(\theta), A_3(\theta))$ where $A_k(\theta) = \sum_{i=0}^{\kappa-1} a_{k,i}\theta^i$ are proper polynomials. We will say that such an $\mathbf{A}(\theta)$ is a proper vector. So, $\mathbf{A}(\theta)$ is a vector of 4 elements of $\mathbb{F}_p$. We describe a different way to consider $\mathbf{A}(\theta)$. Let $\mathbf{a}_i = (a_{0,i}, a_{1,i}, a_{2,i}, a_{3,i})$ and define $\mathbf{a}_i\theta^i = (a_{0,i}\theta^i, a_{1,i}\theta^i, a_{2,i}\theta^i, a_{3,i}\theta^i)$. Then we can write $\mathbf{A}(\theta)$ as $\mathbf{A}(\theta) = \sum_{i=0}^{\kappa-1} \mathbf{a}_i\theta^i$. Each $\mathbf{a}_i$ is stored as a 256-bit value. We define the following operations.

– pack(a_0, a_1, a_2, a_3): returns a 256-bit quantity $\mathbf{a}$. Here each a_i is a 64-bit quantity and $\mathbf{a}$ is obtained by concatenating a_0, a_1, a_2, a_3.
– pack($A_0(\theta), A_1(\theta), A_2(\theta), A_3(\theta)$): returns $\mathbf{A}(\theta) = \sum_{i=0}^{\kappa-1} \mathbf{a}_i\theta^i$, where $\mathbf{a}_i = $ pack($a_{i,0}, a_{i,1}, a_{i,2}, a_{i,3}$).

The corresponding operations unpack($\mathbf{a}$) and unpack($\mathbf{A}(\theta)$) are defined in the usual manner.

We define the following vector operations. The operands $\mathbf{A}(\theta)$ and $\mathbf{B}(\theta)$ represent $(A_0(\theta), A_1(\theta), A_2(\theta), A_3(\theta))$ and $(B_0(\theta), B_1(\theta), B_2(\theta), B_3(\theta),)$ respectively.

– reduce($\mathbf{A}(\theta)$): applies reduce to each component of $\mathbf{A}(\theta)$.
– $\mathcal{M}^4(\mathbf{A}(\theta), \mathbf{B}(\theta))$: uses mult to perform component-wise multiplication of the components of $\mathbf{A}(\theta)$ and $\mathbf{B}(\theta)$.
– $\mathcal{S}^4(\mathbf{A}(\theta))$: use sqr to square each component of $\mathbf{A}(\theta)$.
– $\mathcal{C}^4(\mathbf{A}(\theta), \mathbf{d})$: uses constMult to multiply each component of $\mathbf{A}(\theta)$ with the corresponding component of $\mathbf{d}$. Recall that the output of constMult is not reduced and so neither is the output of $\mathcal{C}^4$.

The operations $\mathcal{ME}^4$ and $\mathcal{SE}^4$ are defined in a manner similar to $\mathcal{M}^4$ and $\mathcal{S}^4$ with the only difference that mult and sqr are respectively replaced by multe and sqre.

The operation $\mathcal{H}^2$ is defined in Table 9 and computes two simultaneous Hadamard operations. The Hadamard operation involves a subtraction. As explained in Sect. 4.2 this is handled by first computing a negation followed by an addition. Negation of a polynomial is computed as subtracting the given polynomial from $2^n\mathfrak{P}(\theta)$ where n is chosen to ensure that all the coefficients of the result are positive. The operation $\mathcal{C}^4$ (which is the vector version of constMult) multiplies the input proper polynomials with constant and the result is not reduced (since the output of constMult is not reduced). The constant is one of the parameters A^2 and B^2 of the Kummer line. The output of $\mathcal{C}^4$ forms the input to $\mathcal{H}^2$. Choosing $n = \lceil\log_2\max(A^2, B^2)\rceil$ ensures the non-negativity condition for the subtraction operation.

We define unreduced-$\mathcal{H}^2$ to be a unreduced version of $\mathcal{H}^2$. This procedure is almost the same as $\mathcal{H}^2$ except that at the end instead of returning reduce($\mathbf{C}(\theta)$), $\mathbf{C}(\theta)$ is returned. Following the discussion in Sect. 4.2, to apply the procedure unreduced-$\mathcal{H}^2$ to a proper polynomial it is sufficient to choose $n = 1$.

Table 9. Vector Hadamard operation. For $\mathbf{a} = (a_0, a_1, a_2, a_3)$, the operations $\mathsf{dup}_1(\mathbf{a}) = (a_0, a_0, a_2, a_2)$ and $\mathsf{dup}_2(\mathbf{a}) = (a_1, a_1, a_3, a_3)$

$\mathcal{H}^2(\mathbf{A}(\theta))$:

input: $\mathbf{A}(\theta) = \sum_{i=0}^{\kappa-1} \mathbf{a}_i \theta^i$ representing $(A_0(\theta), A_1(\theta), A_2(\theta), A_3(\theta))$;

output: $\mathbf{C}(\theta) = \sum_{i=0}^{\kappa-1} \mathbf{c}_i \theta^i$ representing
$\qquad (A_0(\theta) + A_1(\theta), A_0(\theta) - A_1(\theta), A_2(\theta) + A_3(\theta), A_2(\theta) - A_3(\theta))$
$\qquad$ with each component reduced modulo p;

1.   for $i = 0, \ldots, \kappa - 1$ do
2. $\mathbf{s} = \mathsf{dup}_1(\mathbf{a}_i)$;
3. $\mathbf{t} = \mathsf{dup}_2(\mathbf{a}_i)$;
4. $\mathbf{t} = \mathbf{t} \oplus (0^{64}, 1^{64}, 0^{64}, 1^{64})$;
5. $\mathbf{t} = \mathbf{t} + (0^{64}, 2^{\mathsf{n}}\mathfrak{p}_i + 1, 0^{64}, 2^{\mathsf{n}}\mathfrak{p}_i + 1)$;
6. $\mathbf{c}_i = \mathbf{t} + \mathbf{s}$;
7.   end for;

return $\mathsf{reduce}(\mathbf{C}(\theta))$.

Let $\mathbf{a} = (a_0, a_1, a_2, a_3)$ and $\mathfrak{b}$ be a bit. We define an operation $\mathsf{copy}(\mathbf{a}, b)$ as follows: if $\mathfrak{b} = 0$, return (a_0, a_1, a_0, a_1); and if $\mathfrak{b} = 1$, return (a_2, a_3, a_2, a_3). The operation copy is implemented using the instruction _mm256_permutevar8x32_epi32. Let $\mathbf{A}(\theta) = \sum_{i=0}^{\kappa-1} \mathbf{a}_i \theta^i$ be a proper vector and $\mathfrak{b}$ be a bit. We define the operation $\mathcal{P}^4(\mathbf{A}, \mathfrak{b})$ to return $\sum_{i=0}^{\kappa-1} \mathsf{copy}(\mathbf{a}_i, \mathfrak{b})\theta^i$.

6   Vectorised Scalar Multiplication

Scalar multiplication on the Kummer line is computed from a base point represented as $[x^2 : z^2]$ in the square only setting and an ℓ-bit non-negative integer n. The quantities x^2 and z^2 are elements of $\mathbb{F}_p$ and we write their representations as $X(\theta)$ and $Z(\theta)$. If x^2 and z^2 are small as in the fixed base points of the Kummer lines, then $X(\theta)$ and $Z(\theta)$ have 1-limb representations. In general, the field elements $X(\theta)$ and $Z(\theta)$ will be arbitrary elements of $\mathbb{F}_p$ and will have a 9-limb (for $p2519$) or a 10-limb (for $p25519$ and $p2663$) representation.

The algorithm $\mathsf{scalarMult}(P, n)$ in Table 10 shows the scalar multiplication algorithm for $p2519$ and $p2663$ where the base point $[X(\theta) : Z(\theta)]$ is fixed and small. Modifications required for variable base scalar multiplications and $p25519$ are described later.

An inversion is required at Step 15. The representations of $U(\theta)$ and $V(\theta)$ are first converted to the one using the smaller value of κ. Let these be denoted as u and v. The computation of u/v is as follows: first $w = v^{-1}$ is computed and then $x = w \cdot u$ are computed. As mentioned in Sect. 4.5, the inversion is computed in constant time. The multiplications and squarings in this computation are performed using the representation with $\kappa = 5$ so that both w and x are also represented using $\kappa = 5$. A final reduce call is made on x followed by a $\mathsf{makeUnique}$ call whose output is returned.

Table 10. Vectorised scalar multiplication algorithm for $p2519$ and $p2663$ where the base point $[X(\theta) : Z(\theta)]$ is fixed and small. Recall that $A^2 = a^2 + b^2$ and $B^2 = a^2 - b^2$.

---

scalarMult(P, n) :
Input: base point $P = [X(\theta) : Z(\theta)]$;
        ℓ-bit scalar n given as $(1, n_{\ell-2}, \ldots, n_0)$;
Output: $U(\theta)/V(\theta)$ where $nP = [U(\theta) : V(\theta)]$;

1. $\mathfrak{a} = \mathsf{pack}(B^2, A^2, B^2, A^2)$;
2. $\mathfrak{c}_0 = \mathsf{pack}(b^2, a^2, Z, X)$;
3. $\mathfrak{c}_1 = \mathsf{pack}(Z, X, b^2, a^2)$;
4.   compute $2P = (X_2(\theta), Z_2(\theta))$;
5. $\mathbf{T}(\theta) = \mathsf{pack}(X(\theta), Z(\theta), X_2(\theta), Z_2(\theta))$;
6.   for $i = \ell - 2$ down to 0
7. $\mathbf{T}(\theta) = \mathcal{H}^2(\mathbf{T}(\theta))$;
8. $\mathbf{S}(\theta) = \mathcal{P}^4(\mathbf{T}(\theta), n_i)$;
9. $\mathbf{T}(\theta) = \mathcal{M}^4(\mathbf{T}(\theta), \mathbf{S}(\theta))$;
10. $\mathbf{T}(\theta) = \mathcal{C}^4(\mathbf{T}(\theta), \mathfrak{a})$;
11. $\mathbf{T}(\theta) = \mathcal{H}^2(\mathbf{T}(\theta))$;
12. $\mathbf{T}(\theta) = \mathcal{S}^4(\mathbf{T}(\theta))$;
13. $\mathbf{T}(\theta) = \mathcal{C}^4(\mathbf{T}(\theta), \mathfrak{c}_{n_i})$;
14.   end for;
15. $(U(\theta), V(\theta), \cdot, \cdot) = \mathsf{unpack}(\mathsf{reduce}(\mathbf{T}(\theta)))$;
16.   return $U(\theta)/V(\theta)$.

Modification for Variable Base Scalar Multiplication: The following modifications are made for variable base scalar multiplications.

1. In Step 13, the operation $\mathcal{M}^4$ is used instead of the operation $\mathcal{C}^4$.
2. In Step 7, $\mathcal{H}^2$ is replaced by unreduced-$\mathcal{H}^2$.
3. In Step 9, $\mathcal{M}^4$ is replaced by $\mathcal{M}\mathcal{E}^4$.

The first change is required since for variable base, $X(\theta)$ and $Z(\theta)$ are no longer small and a general multiplication is required in Step 13. On the other hand, the net effect of the last two changes is to reduce the number of operations.

Modifications for $p25519$:

1. For fixed base scalar multiplications, the operations $\mathcal{M}^4$ in Step 9 and $\mathcal{S}^4$ in Step 12 are replaced by $\mathcal{M}\mathcal{E}^4$ and $\mathcal{S}\mathcal{E}^4$ respectively.
2. For variable base scalar multiplication, the following are modifications are done:
   – The operations $\mathcal{M}^4$ in Step 9 and $\mathcal{S}^4$ in Step 12 are replaced by $\mathcal{M}\mathcal{E}^4$ and $\mathcal{S}\mathcal{E}^4$ respectively.
   – In Step 13, the operation $\mathcal{M}^4$ is used instead of the operation $\mathcal{C}^4$.
   – In Step 7, $\mathcal{H}^2$ is replaced by unreduced-$\mathcal{H}^2$.

Recall that for $p25519$, using mult leads to an overflow in the intermediate results and so multe has to be used for multiplication. This is reflected in the above modifications where $\mathcal{M}^4$ and $\mathcal{S}^4$ are replaced by $\mathcal{ME}^4$ and $\mathcal{SE}^4$ respectively. The last two changes for variable base scalar multiplication have the same rationale as in the case of $p2519$ and $p2663$.

7   Implementation and Timings

We have implemented the vectorised scalar multiplication algorithm in 64-bit AVX2 intrinsics instructions. The code implements the vectorised ladder algorithm which takes the same amount of time for all scalars. Consequently, our code also runs in constant time. The code is publicly available at [20].

Timing experiments were carried out on a single core of the following two platforms.

Haswell: Intel®Core™i7-4790 4-core CPU @ 3.60 GHz running
Skylake: Intel®Core™i7-6700 4-core CPU @ 3.40 GHz running

In both cases, the OS was 64-bit Ubuntu-16.04 LTS and the C code was complied using GCC version 5.4.0. During timing measurements, turbo boost and hyperthreading were turned off. An initial cache warming was done with 25000 iterations and then the median of 100000 iterations was recorded. The Time Stamp Counter (TSC) was read from the CPU to RAX and RDX registers by RDTSC instruction.

Table 11 compares the number of cycles required by our implementation with that of a few other concrete curve proposals. All the timings are for constant time code on the Haswell processor using variable base scalar multiplication. For Four-$\mathbb{Q}$, $\mathcal{K}_{11,-22,-19,-3}$ and the results from [25,39], the timings are obtained from the respective papers. For Curve25519, we downloaded the Sandy2x[2] library and measured the performance using the methodology from [24]. The cycle count of 140475 that we obtain for Curve25519 on Haswell is significantly faster than the 156076 cycles reported by Tung Chou at https://moderncrypto.org/mail-archive/curves/2015/000637.html and the count of about 156500 cycles reported in [19]. Further, EBACS (https://bench.cr.yp.to/results-dh.html) mentions about 156000 cycles on the machine titan0.

Timing results on Haswell and Skylake platforms for Curve25519 and the Kummer lines for both fixed base and variable base scalar multiplications are shown in Table 12.

Fixed base scalar multiplication can achieve efficiency improvements in two possible ways. One, by using a base point with small coordinates and two, by using pre-computation. We have used only the first method. Using pre-computed tables, [25] reports much faster timing for NIST P-256 and [12] reports much faster timing for Curve25519. We have not investigated the use of pre-computed tables to speed up fixed base scalar multiplication for Kummer lines.

---

[2] Downloaded from https://bench.cr.yp.to/supercop/supercop-20160910.tar.xz. We used crypto_scalarmult(q,n,p) to measure variable base scalar multiplication and crypto_scalarmult_base(q,n) to measure fixed base scalar multiplication.

Table 11. Timing comparison for variable base scalar multiplication on Haswell. The entries are cycle counts. The references point to the best known implementations. Curve25519 was proposed in [2]; NIST P-256 was proposed in [37]; the curve used in [39] was proposed in [32]; and $\mathcal{K}_{11,-22,-19,-3}$ was proposed in [23].

Curve	Genus	Security	Field	Endo	Cycles	Pre-comp tab
Curve25519 [12]	1	126	$\mathbb{F}_{2^{255}-19}$	No	140475	No
NIST P-256 [25]	1	128	$\mathbb{F}_{2^{256}-2^{224}+2^{192}+2^{96}-1}$	No	291000	No
Four-$\mathbb{Q}$ [14][a]	1	123	$\mathbb{F}_{(2^{127}-1)^2}$	Yes	59000	2048 bits
				No	109000	No
$\mathcal{K}_{11,-22,-19,-3}$ [4][b]	2	125	$\mathbb{F}_{2^{127}-1}$	No	60468	No
Koblitz [39]	1	128	$\mathbb{F}_{4^{149}}$	Yes	69656	4768 bits
KL2519(81, 20)	1	124	$\mathbb{F}_{2^{251}-9}$	No	98715	No
KL25519(82, 77)	1	125.7	$\mathbb{F}_{2^{255}-19}$	No	137916	No
KL2663(260, 139)	1	131.2	$\mathbb{F}_{2^{266}-3}$	No	143178	No

[a]Improved timing results of 54000 and 104000 respectively for implementation with and without endomorphism for Four-$\mathbb{Q}$ have been reported in the extended version http://eprint.iacr.org/2015/565.pdf

[b]The original speed reported in [4] was 54389. The Fig. 60468 is reported to be the median cycles per byte at https://bench.cr.yp.to/results-dh.html for the machine titan0. We refer to http://eprint.iacr.org/2015/565.pdf for a possible explanation of the discrepancy.

Based on entries in Table 12, we conclude the following. We use the shorthands K_1 := KL2519(81, 20), K_2 := KL25519(82, 77) and K_3 := KL2663(260, 139).

1. K_1 and K_2 are faster than Curve25519 on both the Haswell and the Skylake processors for both fixed base and variable base scalar multiplications. In particular, we note that even though Curve25519 and K_2 use the same underlying prime $p25519$, K_2 provides speed improvements over Curve25519. This points to the fact that the Kummer line is more SIMD friendly than the Montgomery curve.
2. On the recent Skylake processor, K_3 is faster than Curve25519 for both fixed base and variable base scalar multiplications. On the earlier Haswell processor, K_3 is faster than Curve25519 for fixed base scalar multiplication while both K_3 and Curve25519 take roughly the same time for variable base scalar multiplication. We note that speed improvements for fixed base scalar multiplication does not necessarily imply speed improvement for variable base scalar multiplication, since the code optimisations in the two cases are different.
3. In terms of security, K_3 offers the highest security followed by Curve25519, K_2 and K_1 in that order. The security gap between K_3 and Curve25519 is 5.2 bits; between Curve25519 and K_2 is 0.3 bits; and between Curve25519 and K_1 is 2 bits.

Multiplication and squaring using the 5-limb representation take roughly the same time for all the three primes $p2519$, $p25519$ and $p2663$. So, the comparative times for inversion modulo these three primes is determined by the comparative

Table 12. Timing comparison of Kummer lines with Curve25519 on Haswell and Skylake platforms. The entries are cycle counts.

Curve	Security	Haswell		Skylake	
		Fixed base	Var base	Fixed base	Var base
Curve25519 [12]	126	129825	140475	126518	136728
KL2519(81, 20)	124	80925	98715	74984	91392
KL25519(82, 77)	125.7	101358	137916	92694	120446
KL2663(260, 139)	131.2	98649	143178	91674	126770

sizes of the corresponding addition chains. As a result, the time for inversion is the maximum for $p2663$, followed by $p25519$ and $p2519$ in that order.

Curve25519 is based upon $p25519$ and so the inversion step for Curve25519 is faster than that for K_3. Further, the scalars for K_3 are about 10 bits longer than those for Curve25519. It is noticeable that despite these two facts, other than variable base scalar multiplication on Haswell, a scalar multiplication over K_3 is faster than that over Curve25519. This is due to the structure of the primes $p2663 = 2^{266} - 3$ and $p25519 = 2^{255} - 19$ where 3 being smaller than 19 allows significantly faster multiplication and squaring in the 10-limb representations of these two primes.

On the Skylake processor, K_3 provides both higher speed and higher security compared to Curve25519 If one is interested in obtaining the maximum security, then K_3 should be used. On the other hand, if one considers 124 bits of security to be adequate, then K_1 should be used. The only reason for considering the prime $p25519$ in comparison to either $p2519$ or $p2663$ is that 255 is closer to a multiple of 32 than either of 251 or 266. If public keys are transmitted as 32-bit words, then the wastage of bits would be minimum for $p25519$ compared to $p2519$ or $p2663$. Whether this is an overriding reason for discarding the higher security and higher speed offered by $p2663$ or the much higher speed and small loss in security offered by $p2519$ would probably depend on the application at hand. If for some reason, $p25519$ is preferred to be used, then K_2 offers higher speed than Curve25519 at a loss of only 0.3 bits of security.

We have comprehensively considered the different possibilities for algorithmic improvements to the basic idea leading to significant reductions in operations count. At this point of time, we do not see any way of further reducing the operation counts. On the other hand, we note that our implementations of the Kummer line scalar multiplications are based on Intel intrinsics. There is a possibility that a careful assembly implementation will further improve the speed.

8   Conclusion

This work has shown that compared to existing proposals, Kummer line based scalar multiplication for genus one curves over prime order fields offers competitive performance using SIMD operations. Previous works on implementation of Kummer arithmetic had focused completely on genus two. By showing competitive implementation also in genus one, our work fills a gap in the existing literature.

Acknowledgement. We would like to thank Pierrick Gaudry for helpful comments and clarifying certain confusion regarding conversion from Kummer line to elliptic curve. We would also like to thank Peter Schwabe for clarifying certain implementation issues regarding Curve25519 and Kummer surface computation on genus 2. Thanks to Alfred Menezes, René Struik, Patrick Longa and the reviewers of Asiacrypt 2017 for comments.

References

1. Barwise, J., Eklof, P.: Lefschetz's principle. J. Algebra **13**(4), 554–570 (1969)
2. Bernstein, D.J.: Curve25519: new Diffie-Hellman speed records. In: Yung, M., Dodis, Y., Kiayias, A., Malkin, T. (eds.) PKC 2006. LNCS, vol. 3958, pp. 207–228. Springer, Heidelberg (2006). https://doi.org/10.1007/11745853_14
3. Bernstein, D.J.: Elliptic vs. hyperelliptic, part I. Talk at ECC (2006)
4. Bernstein, D.J., Chuengsatiansup, C., Lange, T., Schwabe, P.: Kummer strikes back: new DH speed records. In: Sarkar, P., Iwata, T. (eds.) ASIACRYPT 2014. LNCS, vol. 8873, pp. 317–337. Springer, Heidelberg (2014). https://doi.org/10.1007/978-3-662-45611-8_17
5. Bernstein, D.J., Lange, T.: Safecurves: choosing safe curves for elliptic-curve cryptography. http://safecurves.cr.yp.to/index.html. Accessed 15 Sept 2016
6. Bernstein, D.J., Duif, N., Lange, T., Schwabe, P., Yang, B.-Y.: High-speed high-security signatures. In: Preneel, B., Takagi, T. (eds.) CHES 2011. LNCS, vol. 6917, pp. 124–142. Springer, Heidelberg (2011). https://doi.org/10.1007/978-3-642-23951-9_9
7. Bernstein, D.J., Schwabe, P.: NEON crypto. In: Prouff, E., Schaumont, P. (eds.) CHES 2012. LNCS, vol. 7428, pp. 320–339. Springer, Heidelberg (2012). https://doi.org/10.1007/978-3-642-33027-8_19
8. Bertoni, G., Coron, J.-S. (eds.): CHES 2013. LNCS, vol. 8086. Springer, Heidelberg (2013). https://doi.org/10.1007/978-3-642-40349-1
9. Bos, J.W., Costello, C., Hisil, H., Lauter, K.: Fast cryptography in genus 2. In: Johansson, T., Nguyen, P.Q. (eds.) EUROCRYPT 2013. LNCS, vol. 7881, pp. 194–210. Springer, Heidelberg (2013). https://doi.org/10.1007/978-3-642-38348-9_12
10. Bos, J.W., Costello, C., Hisil, H., Lauter, K.E.: High-performance scalar multiplication using 8-dimensional GLV/GLS decomposition. In: Bertoni, G., Coron, J.-S. (eds.) [8], pp. 331–348 (2013)
11. Brainpool: ECC standard. http://www.ecc-brainpool.org/ecc-standard.htm
12. Chou, T.: Sandy2x: new curve25519 speed records. In: Dunkelman, O., Keliher, L. (eds.) SAC 2015. LNCS, vol. 9566, pp. 145–160. Springer, Cham (2016). https://doi.org/10.1007/978-3-319-31301-6_8
13. Cosset, R.: Factorization with genus 2 curves. Math. Comput. **79**(270), 1191–1208 (2010)
14. Costello, C., Longa, P.: Fourℚ: four-dimensional decompositions on a ℚ-curve over the Mersenne Prime. In: Iwata, T., Cheon, J.H. (eds.) ASIACRYPT 2015. LNCS, vol. 9452, pp. 214–235. Springer, Heidelberg (2015). https://doi.org/10.1007/978-3-662-48797-6_10

15. Costello, C., Hisil, H., Smith, B.: Faster compact Diffie–Hellman: endomorphisms on the x-line. In: Nguyen, P.Q., Oswald, E. (eds.) EUROCRYPT 2014. LNCS, vol. 8441, pp. 183–200. Springer, Heidelberg (2014). https://doi.org/10.1007/978-3-642-55220-5_11

16. Costigan, N., Schwabe, P.: Fast elliptic-curve cryptography on the cell broadband engine. In: Preneel, B. (ed.) AFRICACRYPT 2009. LNCS, vol. 5580, pp. 368–385. Springer, Heidelberg (2009). https://doi.org/10.1007/978-3-642-02384-2_23

17. Curve25519: Wikipedia page on Curve25519. https://en.wikipedia.org/wiki/Curve25519. Accessed 15 Sept 2016

18. Faz-Hernández, A., Longa, P., Sánchez, A.H.: Efficient and secure algorithms for GLV-based scalar multiplication and their implementation on GLV-GLS curves. In: Benaloh, J. (ed.) CT-RSA 2014. LNCS, vol. 8366, pp. 1–27. Springer, Cham (2014). https://doi.org/10.1007/978-3-319-04852-9_1

19. Faz-Hernández, A., López, J.: Fast implementation of curve25519 using AVX2. In: Lauter, K., Rodríguez-Henríquez, F. (eds.) LATINCRYPT 2015. LNCS, vol. 9230, pp. 329–345. Springer, Cham (2015). https://doi.org/10.1007/978-3-319-22174-8_18

20. Code for Kummer Line Computations. Provided as part of the auxiliary supporting material corresponding to this submission. The code is also publicly available

21. Frey, G., Rück, H.-G.: The strong lefschetz principle in algebraic geometry. Manuscripta Math. **55**(3), 385–401 (1986)

22. Gaudry, P., Lubicz, D.: The arithmetic of characteristic 2 Kummer surfaces and of elliptic Kummer lines. Finite Fields Appl. **15**(2), 246–260 (2009)

23. Gaudry, P., Schost, É.: Genus 2 point counting over prime fields. J. Symb. Comput. **47**(4), 368–400 (2012)

24. Gueron, S.: Software optimizations for cryptographic primitives on general purpose x86_64 platforms. Tutorial at IndoCrypt (2011)

25. Gueron, S., Krasnov, V.: Fast prime field elliptic-curve cryptography with 256-bit primes. J. Cryptogr. Eng. **5**(2), 141–151 (2015)

26. Hankerson, D., Karabina, K., Menezes, A.: Analyzing the Galbraith-Lin-Scott point multiplication method for elliptic curves over binary fields. IEEE Trans. Comput. **58**(10), 1411–1420 (2009)

27. Igusa, J.: Theta Functions. Springer, Heidelberg (1972)

28. Karati, S., Sarkar, P.: Kummer for genus one over prime order fields. IACR Cryptology ePrint Archive 2016:938 (2016)

29. Koblitz, N.: Elliptic curve cryptosystems. Math. Comp. **48**(177), 203–209 (1987)

30. Koblitz, N.: Hyperelliptic cryptosystems. J. Cryptol. **1**(3), 139–150 (1989)

31. Lim, C.H., Lee, P.J.: A key recovery attack on discrete log-based schemes using a prime order subgroup. In: Kaliski, B.S. (ed.) CRYPTO 1997. LNCS, vol. 1294, pp. 249–263. Springer, Heidelberg (1997). https://doi.org/10.1007/BFb0052240

32. Longa, P., Sica, F.: Four-dimensional Gallant-Lambert-Vanstone scalar multiplication. In: Wang, X., Sako, K. (eds.) ASIACRYPT 2012. LNCS, vol. 7658, pp. 718–739. Springer, Heidelberg (2012). https://doi.org/10.1007/978-3-642-34961-4_43

33. Miller, V.S.: Use of elliptic curves in cryptography. In: Williams, H.C. (ed.) CRYPTO 1985. LNCS, vol. 218, pp. 417–426. Springer, Heidelberg (1986). https://doi.org/10.1007/3-540-39799-X_31

34. Montgomery, P.L.: Speeding the Pollard and elliptic curve methods of factorization. Math. Comput. **48**(177), 243–264 (1987)

35. Montgomery, P.L.: Five, six, and seven-term karatsuba-like formulae. IEEE Trans. Comput. **54**(3), 362–369 (2005)

36. Mumford, D.: Tata Lectures on Theta I. Progress in Mathematics 28. Birkh äuser, Basel (1983)
37. U.S. Department of Commerce/National Institute of Standards and Technology. Digital Signature Standard (DSS). FIPS-186-3 (2009). http://csrc.nist.gov/publications/fips/fips186-3/fips_186-3.pdf
38. Oliveira, T., López, J., Aranha, D.F., Rodríguez-Henríquez, F.: Lambda coordinates for binary elliptic curves. In: Bertoni, G., Coron, J.-S. (eds.) [8], pp. 311–330 (2013)
39. Oliveira, T., López, J., Rodríguez-Henríquez, F.: Software implementation of Koblitz curves over quadratic fields. In: Gierlichs, B., Poschmann, A.Y. (eds.) CHES 2016. LNCS, vol. 9813, pp. 259–279. Springer, Heidelberg (2016). https://doi.org/10.1007/978-3-662-53140-2_13
40. Certicom Research: SEC 2: Recommended elliptic curve domain parameters (2010). http://www.secg.org/sec2-v2.pdf
41. NUMS: Nothing up my sleeve. https://tools.ietf.org/html/draft-black-tls-numscurves-00

Pairing-based Protocols

ABE with Tag Made Easy
Concise Framework and New Instantiations in Prime-Order Groups

Jie Chen[1,2] and Junqing Gong[3(✉)]

[1] East China Normal University, Shanghai, China
S080001@e.ntu.edu.sg
[2] Jinan University, Guangzhou, China
[3] Laboratoire LIP (U. Lyon, CNRS, ENSL, INRIA, UCBL), ENS de Lyon, Lyon, France
junqing.gong@ens-lyon.fr

Abstract. Among all existing identity-based encryption (IBE) schemes in the bilinear group, Wat-IBE proposed by Waters [CRYPTO, 2009] and JR-IBE proposed by Jutla and Roy [AsiaCrypt, 2013] are quite special. A secret key and/or ciphertext in these two schemes consist of several group elements and an integer which is usually called *tag*. A series of prior work was devoted to extending them towards more advanced attribute-based encryption (ABE) including inner-product encryption (IPE), hierarchical IBE (HIBE). Recently, Kim *et al.* [SCN, 2016] introduced the notion of tag-based encoding and presented a generic framework for extending Wat-IBE. We may call these ABE schemes *ABE with tag* or *tag-based ABE*. Typically, a tag-based ABE construction is more efficient than its counterpart without tag. However the research on tag-based ABE severely lags—We do not know how to extend JR-IBE in a systematic way and there is no tag-based ABE for boolean span program even with Kim *et al.*'s generic framework.

In this work, we proposed a generic framework for tag-based ABE which is based on JR-IBE and compatible with Chen *et al.*'s (attribute-hiding) predicate encoding [EuroCrypt, 2015]. The adaptive security in the standard model relies on the k-linear assumption in the asymmetric prime-order bilinear group. This is the first framework showing how to extend JR-IBE systematically. In fact our framework and its simple extension are able to cover most concrete tag-based ABE constructions in previous literature. Furthermore, since Chen *et al.*'s predicate encoding supports a large number of predicates including boolean span program, we can now give the first (both key-policy and ciphertext-policy) tag-based ABE for boolean span program in the standard model. Technically our framework is based on a simplified version of JR-IBE. Both

---

J. Chen—School of Computer Science and Software Engineering. Supported by the National Natural Science Foundation of China (Nos. 61472142, 61632012) and the Science and Technology Commission of Shanghai Municipality (No. 14YF1404200). Homepage: http://www.jchen.top.

J. Gong—Partially supported by the French ANR ALAMBIC project (ANR-16-CE39-0006).

T. Takagi and T. Peyrin (Eds.): ASIACRYPT 2017, Part II, LNCS 10625, pp. 35–65, 2017.
https://doi.org/10.1007/978-3-319-70697-9_2

the description and its proof are quite similar to the prime-order IBE derived from Chen *et al.*'s framework. This not only allows us to work with Chen *et al.*'s predicate encoding but also provides us with a clear explanation of JR-IBE and its proof technique.

Keywords: Attribute-based encryption · Predicate encoding · Prime-order bilinear group · Attribute-hiding · Delegation

1   Introduction

An *attribute-based encryption* [BSW11] (ABE) is an advanced cryptographic primitive supporting fine-grained access control.[1] Such a system is established by an authority (a.k.a. key generation center, KGC for short). On a predicate $P : \mathcal{X} \times \mathcal{Y} \rightarrow \{0,1\}$ (and other system parameter), the authority publishes master public key mpk. Each user will receive a secret key sk_y associated with a policy $y \in \mathcal{Y}$ when he/she joins in the system. A sender can create a ciphertext ct_x associated with an attribute $x \in \mathcal{X}$. A user holding sk_y can decrypt the ciphertext ct_x if $P(x,y) = 1$ holds, otherwise he/she will infer nothing about the plaintext. This notion covers many concrete public-key encryptions such as identity-based encryption [Sha84] (IBE), fuzzy IBE [SW05], ABE for boolean span program [GPSW06] and inner-product encryption [KSW08] (IPE).

The basic security requirement is *collusion-resistance*. Intuitively, it is required that two (or more) users who are not authorized to decrypt a ciphertext individually can not do that by collusion either. The notion is formalized via the so-called *adaptive security model* [BF01] where the adversary holding mpk can get secret keys $sk_{y_1}, \ldots, sk_{y_q}$ for $y_1, \ldots, y_q \in \mathcal{Y}$ and a challenge ciphertext ct^* for target $x^* \in \mathcal{X}$ via *key extraction queries* and *challenge query* respectively. We emphasize that the adversary can make oracle queries in an adaptive way. Although several weaker security models [CHK03,CW14] were introduced and widely investigated, this paper will focus on the adaptive security model.

Dual System Methodology and Predicate Encoding. A recent breakthrough in this field is the *dual system methodology* invented by Waters [Wat09] in 2009. He obtained the first adaptively secure IBE construction with compact parameters under standard complexity assumptions in the standard model. Inspired by his novel proof technique, the community developed many ABE constructions in the next several years. More importantly, the dual system methodology finally led us to a clean and systematic understanding of ABE.

---

[1] The terminology introduced by Boneh *et al.* [BSW11] is *functional encryption*. However the notion we will consider here is restricted. Therefore we adopt a moderate terminology, i.e., *attribute-based encryption*. We note that the attribute-based encryption in our paper is more general than that introduced by Goyal *et al.* [GPSW06] which is named ABE for boolean formula/span program.

In 2014, Wee [Wee14] and Attrapadung [Att14] introduced the notion of *predicate encoding*[2] and proposed their respective frameworks in the composite-order bilinear group. For certain predicate, if we can construct a predicate encoding for it, their frameworks will immediately give us a full-fledged ABE scheme for the predicate. This significantly simplifies the process of designing an ABE scheme. In fact the powerful frameworks allow them to give many new concrete constructions.

As Wee pointed out in his landmark work [Wee14], the framework reflects the common structures and properties shared among a large group of dual-system ABE schemes while predicate encodings give predicate-dependent features. More concretely, investigating the IBE instance derived from a framework will show the basic construction and proof technique captured the framework, then developing and analysing various predicate encodings will tell us how to extend the IBE instance to more complex cases.

Based on pioneering work by Wee [Wee14] and Attrapadung [Att14], a series of progresses have been made to support more efficient prime-order bilinear groups, employ more advanced proof techniques and more complex predicate encodings [CGW15, AC16, Att16, AC17, AY15].

ABE with Tag. The proof technique behind all these frameworks can be traced back to the work from Lewko and Waters [LW10, LW12] and their variants. However there are two dual-system IBE schemes located beyond these frameworks: one is the first dual-system IBE by Waters [Wat09] and the other one is the IBE scheme based on quasi-adaptive non-interactive zero-knowledge (QA-NIZK) proof by Jutla and Roy [JR13]. In this paper, we call them Wat-IBE and JR-IBE, respectively, for convenience.

Apart from several group elements as usual, a ciphertext and/or a key of Wat-IBE and JR-IBE also include an integer which is called *tag*. This distinguishes them from other dual-system IBE schemes. We must emphasize that the tag plays an important role in the security proof and results in a different proof technique. Therefore we believe they deserve the terminology *IBE with tag* or *tag-based IBE*. Accordingly, an ABE with a tag in the ciphertext and/or key will be called *ABE with tag* or *tag-based ABE*.

There have been several concrete tag-based ABE schemes derived from Wat-IBE or JR-IBE such as [RCS12, RS14, Ram16, RS16]. Recently, Kim *et al.* [KSGA16] introduced the notion of *tag-based encoding* and developed a new framework based on Wat-IBE [Wat09], which is the first systematic study for tag-based ABE. All these work show that a tag-based ABE typically has shorter master public key and ciphertexts/keys, especially for complex predicates. This is a desirable advantage in most application settings of ABE.

---

[2] Attrapadung [Att14] actually introduced another terminology *pair encoding*. Although predicate encoding and pair encoding serve the same methodology, Attrapadung's framework is based on more advanced proof technique originated from [LW12].

1.1 Motivation

We review previous tag-based ABE in more detail. Ramanna *et al.* [RCS12] simplified Wat-IBE (with stronger assumptions) and extended it to build hierarchical IBE (HIBE) and broadcast encryption. Ramanna and Sarkar [RS14] described two HIBE constructions derived from JR-IBE. Then two IPE schemes were proposed by Ramanna [Ram16]. Although both of them were constructed from JR-IBE, the first one borrows some techniques from Wat-IBE. The recent work [RS16] by Ramanna and Sarkar provided us with two identity-based broadcast encryption schemes, both of which comes from JR-IBE. Kim *et al.*'s tag-based encoding and generic framework [KSGA16] allow them to present several new IPE, (doubly) spatial encryption with various features.

One may notice that there is no framework based on JR-IBE and all previous extensions were obtained in a somewhat ad-hoc manner. This immediately arises our first question.

Question 1: *Is it possible to propose a framework based on* JR-IBE*?*

We note that although both Wat-IBE and JR-IBE take the tag as an important component, their proof techniques are different. That is they use the tag in their own ways in the security proof. As a matter of fact, Wat-IBE requires distinct tags on ciphertext and secret key, respectively, while a secret key in JR-IBE has no tag. To our best knowledge, there is no explicit evidence demonstrating that a framework based on Wat-IBE (like that in [KSGA16]) implies a framework based on JR-IBE.

Furthermore, we surprisingly found that there is no tag-based ABE for boolean span program even with a generic framework! In fact, Kim *et al.* reported that their tag-based encoding is seemingly incompatible with linear secret sharing scheme which is a crucial ingredient of ABE for boolean span program. Hence it's natural to ask the following question.

Question 2: *Is there any limitation on the predicate for tag-based ABE?*

To some extent, we are asking *whether the tag-based proof techniques (used for* Wat-IBE *and* JR-IBE*) makes a trade-off between efficiency and expressiveness?* A very recent work [KSG+17] proposed an ABE with tag supporting boolean span program based on Wat-IBE. However the security analysis was given in the semi-adaptive model [CW14], which is much weaker than the standard adaptive security model (see [GKW16] for more discussions).

1.2 Our Contribution

In this paper, we propose a new framework for tag-based ABE. The framework is based on JR-IBE and can work with the predicate encoding defined in [CGW15]. The adaptive security in the standard model (without random oracles) relies on the k-linear assumption (k-Lin) in the prime-order bilinear group. Our framework is also compatible with *attribute-hiding predicate encoding* from [CGW15] and implies a family of tag-based ABE with *weak attribute-hiding* feature. Here *weak attribute-hiding* means that ciphertext ct_x reveals no information about x against an adversary who are not authorized to decrypt ct_x.

With this technical result, we are ready to answer the two questions:

Answer to Question 1: Our framework itself readily gives an affirmative answer to the question. Luckily, by defining new concrete predicate encodings motivated by [Ram16, RS16], our framework is able to cover these previous tag-based ABE. In order to capture the HIBE schemes proposed in [RS14], we need to extend both the framework and the predicate encoding to support *predicate with delegation*. (See Sect. 6.) However we note that Ramanna's first tag-based IPE [Ram16] and Ramanna and Sarkar's second identity-based broadcast encryption with tag [RS16] still fall out of our framework because they involve further developments of JR-IBE's proof technique which has not been captured by our framework.

Answer to Question 2: We highlight that both Chen *et al.*'s framework without tag [CGW15] and our tag-based framework are compatible with the predicate encoding described in [CGW15] (and its attribute-hiding variant). This answers **Question 2** with negation, that is there should be no restriction on predicates for tag-based ABE. Concretely, we can construct a series of new tag-based ABE schemes including ABE for boolean span program (for both key-policy and ciphertext-policy cases; see Sect. 5) thanks to concrete encodings listed in [CGW15].

We compare our framework with the CGW framework by Chen *et al.* [CGW15] and the KSGA framework by Kim *et al.* [KSGA16] in Table 1. Here we only focus on the space efficiency, i.e., the size of mpk, sk and ct. The comparison regarding the decryption time is analogous to that for the size of ct.

Table 1. Comparison among CGW [CGW15], KSGA [KSGA16], and our framework in terms of space efficiency. All of them work with asymmetric bilinear group (p, G_1, G_2, G_T, e). In the table, $|G_1|$, $|G_2|$ and $|G_T|$ represent the element sizes of three groups, respectively. Parameter n is the number of common parameter of the predicate encoding, $|\mathsf{sE}|$ and $|\mathsf{rE}|$ are the respective size of sender and receiver encodings.

	|mpk|		|sk|		|ct|		Sec.
	$\|G_1\|$	$\|G_T\|$	$\|G_2\|$	$\|\mathbb{Z}_p\|$	$\|G_1\|$	$\|\mathbb{Z}_p\|$	
CGW	$n(k^2+k)+k$	k	$(\|\mathsf{rE}\|+1)(k+1)$	0	$(\|\mathsf{sE}\|+1)(k+1)$	0	k-Lin
	$6n+2$	2	$3 \cdot \|\mathsf{rE}\|+3$	0	$3 \cdot \|\mathsf{sE}\|+3$	0	DLIN
	$2n+1$	1	$2 \cdot \|\mathsf{rE}\|+2$	0	$2 \cdot \|\mathsf{sE}\|+2$	0	SXDH
KSGA	$n+11$	1	$\|\mathsf{rE}\|+7$	$\|\mathsf{rE}\|$	$\|\mathsf{sE}\|+8$	$\|\mathsf{sE}\|$	DLIN
Ours	$(n+1)k^2+k$	k	$(\|\mathsf{rE}\|+1)(k+1)+k$	0	$\|\mathsf{sE}\| \cdot k + k + 1$	$\|\mathsf{sE}\|$	k-Lin
	$4n+6$	2	$3 \cdot \|\mathsf{rE}\|+5$	0	$2 \cdot \|\mathsf{sE}\|+3$	$\|\mathsf{sE}\|$	DLIN
	$n+2$	1	$2 \cdot \|\mathsf{rE}\|+3$	0	$\|\mathsf{sE}\|+2$	$\|\mathsf{sE}\|$	SXDH

Table 2. Comparison among CGW [CGW15], KSGA [KSGA16], and ours framework in terms of concrete instantiations. In the table, $|G_1|$, $|G_2|$ and $|G_T|$ have the same meanings as Table 1. The parameter ℓ is the dimension of vector space for IPE while it stands for the size of universe for ABE.

| | IPE | | | | | | KP-ABE-BSP | | | | | | Sec. |
| | |mpk| | | |sk| | | |ct| | | |mpk| | | |sk| | | |ct| | | |
	$|G_1|$	$|G_T|$	$|G_2|$	$|\mathbb{Z}_p|$	$|G_1|$	$|\mathbb{Z}_p|$	$|G_1|$	$|G_T|$	$|G_2|$	$|\mathbb{Z}_p|$	$|G_1|$	$|\mathbb{Z}_p|$	
CGW	$6\ell+2$	2	$3\ell+6$	0	6	0	$6\ell+2$	2	$3\ell+3$	0	$3\ell+3$	0	DLIN
	$2\ell+1$	1	$2\ell+4$	0	4	0	$2\ell+1$	1	$2\ell+2$	0	$2\ell+2$	0	SXDH
KSGA	$\ell+11$	1	$\ell+6$	$\ell-1$	9	1			—				DLIN
Ours	$4\ell+6$	2	$3\ell+8$	0	5	1	$4\ell+6$	2	$3\ell+5$	0	$2\ell+3$	ℓ	DLIN
	$\ell+2$	1	$2\ell+5$	0	3	1	$\ell+2$	1	$2\ell+3$	0	$\ell+2$	ℓ	SXDH

In general, our framework has shorter master public key and shorter ciphertexts than the CGW framework at the cost of slightly larger secret keys. We highlight that the cost we pay here is constant for specific assumption (i.e., consider k as a constant) while the improvement we gain will be proportional to n and $|\mathsf{sE}|$, respectively. In fact our work can be viewed as an improvement of CGW framework using the tag-based technique underlying JR-IBE and improves a family of concrete ABE constructions in a systematic way.

Superficially, the KSGA framework is much more efficient than ours (and CGW framework as well). However, as we have mentioned, this framework is less expressive. In particular, the KSGA framework works with *tag-based predicate encoding* [KSGA16] which fails to support many important predicates such as boolean/arithmetic span program. That is this framework does not imply more efficient ABE for these predicates. It's also worth noting that our framework has shorter ciphertext for concrete predicate encodings with $|\mathsf{sE}| < 5$, including predicate encoding for inner-product encryption with short ciphertext [CGW15]. Namely our framework also implies a more efficient IPE scheme. Actually the CGW framework also has a similar advantage but for predicate encodings with $|\mathsf{sE}| < 3$.

In Table 2, we compare concrete instantiations derived from CGW, KSGA, and our framework. Here we only take inner-product encryption (IPE) and key-policy ABE for boolean span program (KP-ABE-BSP) as examples. It is clear that our KP-ABE-BSP has the shortest master public key and ciphertexts (also fastest decryption algorithm). Under the DLIN assumption, our IPE has the shortest ciphertext but its master public key is larger than the IPE derived from KSGA framework. However if we are allowed to use stronger SXDH assumption, our IPE will have the shortest master public key.

1.3 Overview of Method: A Simplified JR-IBE

Our framework is based on JR-IBE. In Jutla and Roy's original paper [JR13], JR-IBE was derived from the QA-NIZK proof for a specific subspace language. Although it's important to describe/explain an IBE from the angle of NIZK proof, it's still a big challenge to work on it directly. Therefore the foundation of our framework is a simplified (and slightly generalized) version of JR-IBE. The simplified JR-IBE is similar to the prime-order IBE instantiated from the CGW framework [CGW15] and its proof analysis is cleaner and much easier to follow. With the benefits of these features, we are able to develop our new framework for tag-based ABE which is based on JR-IBE's proof technique [JR13] and is compatible with the predicate encoding from [CGW15]. The adaptive security relies on the k-Lin assumption, which is a generalized form of SXDH used in [JR13].

Simplified JR-IBE. We assume there is an asymmetric prime-order bilinear group (p, G_1, G_2, G_T, e). Let $g_1 \in G_1$, $g_2 \in G_2$ and $e(g_1, g_2) \in G_T$ be the respective generators, we will use the following notation: $[a]_s = g_s^a$ for all $a \in \mathbb{Z}_p$ and $s \in \{1, 2, T\}$. The notation can also be naturally applied to a matrix over $\mathbb{Z}_p$.

Let $\mathbb{Z}_p$ be the identity space. The JR-IBE can be re-written as

$$
\begin{aligned}
\mathsf{mpk} \;&:\; [\mathbf{A}]_1, \; [\mathbf{W}_0^\mathsf{T}\mathbf{A}]_1, \; [\mathbf{W}_1^\mathsf{T}\mathbf{A}]_1, \; \boxed{[\mathbf{W}^\mathsf{T}\mathbf{A}]_1,} \; [\mathbf{k}^\mathsf{T}\mathbf{A}]_T \\
\mathsf{ct}_{\mathsf{id}} \;&:\; [\mathbf{A}s]_1, \; [(\mathbf{W}_0 + \mathsf{id}\cdot\mathbf{W}_1 + \boxed{\tau\cdot\mathbf{W}})^\mathsf{T}\mathbf{A}s]_1, \; [\mathbf{k}^\mathsf{T}\mathbf{A}s]_T\cdot m, \; \tau \\
\mathsf{sk}_{\mathsf{id}} \;&:\; [\mathbf{r}]_2, \; [\mathbf{k} + (\mathbf{W}_0 + \mathsf{id}\cdot\mathbf{W}_1)\mathbf{r}]_2, \; \boxed{[\mathbf{Wr}]_2}
\end{aligned}
$$

Here $\mathbf{A} \leftarrow \mathbb{Z}_p^{(k+1)\times k}$ acts as the basis, matrices $\mathbf{W}_0, \mathbf{W}_1, \mathbf{W} \leftarrow \mathbb{Z}_p^{(k+1)\times k}$ are common parameters, the vector $\mathbf{k} \leftarrow \mathbb{Z}_p^{k+1}$ is the master secret value, vectors $\mathbf{s}, \mathbf{r} \leftarrow \mathbb{Z}_p^k$ are random coins for the ciphertext and the secret key, respectively, and the tag τ is a random element in $\mathbb{Z}_p$.

The boxed parts involving matrix $\mathbf{W}$ are relevant to the tag while the remaining structure is quite similar to the IBE implied by the CGW framework. The main difference is that we do not need another basis $\mathbf{B}$ in $\mathsf{sk}_{\mathsf{id}}$, which reduces the size of $\mathbf{W}, \mathbf{W}_0, \mathbf{W}_1$ and thus shortens mpk and $\mathsf{ct}_{\mathsf{id}}$. In fact this structure has been used in a recent tightly secure IBE [BKP14, GDCC16], and we indeed borrow some proof technique from the tight reduction method there.

Proof Blueprint. Our proof mainly follows the tag-based proof strategy given by Jutla and Roy [JR13] which basically employs the dual system methodology [Wat09]. The first step is to transform the normal challenge ciphertext (see the real system above) into the semi-functional (SF) one:

$$
\mathsf{ct}_{\mathsf{id}^*}^{\mathsf{SF}} \;:\; [\mathbf{A}s + \boxed{\mathbf{b}\hat{s}}\,]_1, \; [(\mathbf{W}_0 + \mathsf{id}^*\cdot\mathbf{W}_1 + \tau\cdot\mathbf{W})^\mathsf{T}(\mathbf{A}s + \boxed{\mathbf{b}\hat{s}}\,)]_1, \; [\mathbf{k}^\mathsf{T}(\mathbf{A}s + \boxed{\mathbf{b}\hat{s}}\,)]_T\cdot m, \; \tau
$$

where $\mathbf{b} \leftarrow \mathbb{Z}_p^{k+1}$ and $\hat{s} \leftarrow \mathbb{Z}_p$. One may prove such a transformation is not detectable by the adversary from the k-Lin assumption following [CGW15]. The next step is to convert secret keys revealed to adversary from normal form into semi-functional (SF) form defined as follows:

$$
\mathsf{sk}_{\mathsf{id}}^{\mathsf{SF}} \;:\; [\mathbf{r}]_2, \; [\mathbf{k} + \boxed{\alpha\mathbf{a}^\perp} + (\mathbf{W}_0 + \mathsf{id}\cdot\mathbf{W}_1)\mathbf{r}]_2, \; [\mathbf{Wr}]_2
$$

where $\mathbf{a}^{\perp} \leftarrow \mathbb{Z}_p^{k+1}$ with $\mathbf{A}^{\top}\mathbf{a}^{\perp} = \mathbf{0}$ and $\alpha \leftarrow \mathbb{Z}_p$. As usual, the conversion will be done in a one-by-one manner. That is we deal with single secret key each time which arises a security loss proportional to the number of secret keys sent to the adversary. When all secret keys and the challenge ciphertext become semi-functional, it would be quite direct to decouple the message m from the challenge ciphertext which implies the adaptive security.

Tag-Based Technique, Revisited. In fact what we have described still follows [CGW15] (and most dual-system proofs [Wat09]). However the proof for the indistinguishability between normal and SF secret key heavily relies on the tag as [JR13] and deviate from [CGW15].

Recall that a SF key is a normal key with additional entropy $\alpha\mathbf{a}^{\perp}$. To replace a normal key with a SF key, we use the following lemma in [BKP14, GDCC16] which states that

$$([\mathbf{r}]_2, [\mathbf{v}^{\top}\mathbf{r}]_2) \approx_c ([\mathbf{r}]_2, [\mathbf{v}^{\top}\mathbf{r} + \hat{r}]_2) \quad \text{given} \quad [\mathbf{Z}]_2, [\mathbf{v}^{\top}\mathbf{Z}]_2 \tag{1}$$

where $\mathbf{v}, \mathbf{r} \leftarrow \mathbb{Z}_p^k$, $\hat{r} \leftarrow \mathbb{Z}_p$ and $\mathbf{Z} \leftarrow \mathbb{Z}_p^{k \times k}$. Following [BKP14, GDCC16], we pick $\gamma_0, \gamma_1 \leftarrow \mathbb{Z}_p$ and embed the secret vector $\mathbf{v}$ into common parameter $\mathbf{W}_0$ and $\mathbf{W}_1$ as follows

$$\mathbf{W}_0 = \widetilde{\mathbf{W}}_0 + \gamma_0 \cdot \mathbf{a}^{\perp}\mathbf{v}^{\top} \quad \text{and} \quad \mathbf{W}_1 = \widetilde{\mathbf{W}}_1 + \gamma_1 \cdot \mathbf{a}^{\perp}\mathbf{v}^{\top}$$

where $\widetilde{\mathbf{W}}_0, \widetilde{\mathbf{W}}_1 \leftarrow \mathbb{Z}_p^{(k+1) \times k}$. The lemma shown in Eq. (1) will allow us to move from a normal key to the following transitional form

$$[\mathbf{r}]_2, \; [\mathbf{k} + (\mathbf{W}_0 + \mathsf{id} \cdot \mathbf{W}_1)\mathbf{r} + \boxed{(\gamma_0 + \mathsf{id} \cdot \gamma_1) \cdot \mathbf{a}^{\perp}\hat{r}}\,]_2, \; [\mathbf{Wr}]_2. \tag{2}$$

However we must ensure that $\mathbf{v}$ will not appear in mpk and $\mathsf{ct}_{\mathsf{id}}$, both of which consist of elements in G_1; otherwise, we can not apply the lemma at all since there is not element from G_1 with information on $\mathbf{v}$.

It is direct to see that $[\mathbf{W}_0^{\top}\mathbf{A}]_1$ and $[\mathbf{W}_1^{\top}\mathbf{A}]_1$ in mpk reveal nothing about $\mathbf{v}$ from the fact that $\mathbf{A}^{\top}\mathbf{a}^{\perp} = \mathbf{0}$, but the challenge ciphertext may include $\mathbf{v}$ since we have

$$\mathbf{W}_0 + \mathsf{id}^* \cdot \mathbf{W}_1 + \tau \cdot \mathbf{W} = \widetilde{\mathbf{W}}_0 + \mathsf{id}^* \cdot \widetilde{\mathbf{W}}_1 + \tau \cdot \mathbf{W} + \boxed{(\gamma_0 + \mathsf{id}^* \cdot \gamma_1)\mathbf{a}^{\perp}\mathbf{v}^{\top}}$$

and $\mathbf{As} + \mathbf{b}\hat{s} \in \mathbb{Z}_p^{k+1}$ with high probability. Fortunately, the tag can help us to circumvent the issue: set

$$\mathbf{W} = \widetilde{\mathbf{W}} - \mathbf{a}^{\perp}\mathbf{v}^{\top} \quad \text{and} \quad \tau = \gamma_0 + \mathsf{id}^* \cdot \gamma_1,$$

we can see that

$$\tau \cdot \mathbf{W} + (\gamma_0 + \mathsf{id}^* \cdot \gamma_1)\mathbf{a}^{\perp}\mathbf{v}^{\top} = \tau \cdot \widetilde{\mathbf{W}}$$

where there is no $\mathbf{v}$ anymore and the proof strategy now works well. We note that both $\mathbf{W}$ and τ are distributed correctly.

Then we can move from the transitional key (see Eq. 2) to the SF key using the statistical argument asserting that

$$\{\gamma_0 + \mathsf{id}^* \cdot \gamma_1, \ \gamma_0 + \mathsf{id} \cdot \gamma_1\}$$

are uniformly distributed over $\mathbb{Z}_p^2$. Chen *et al.*'s proof [CGW15] also involves this statistical argument. However, in Chen *et al.*'s proof [CGW15], both values appear "on the exponent" while $\gamma_0 + \mathsf{id}^* \cdot \gamma_1$ is given out "directly" as the tag in our case.

1.4 Related Work and Discussion

Since the work by Wee [Wee14] and Attrapadung [Att14], predicate/pair encodings and corresponding generic frameworks have been extended and improved via various methods [CGW15, AC16, Att16, AC17, ABS17]. Before that, there were many early-age ABE from pairing such as IBE [BF01, BB04b, BB04a, Wat05, Gen06], fuzzy IBE [SW05], inner-product encryption [KSW08, OT12], and ABE for boolean formula [GPSW06, OSW07, BSW07, OT10, LOS+10, Wat11, LW12]. Most of them has been covered by generic frameworks. Attrapadung *et al.* [AHY15] even gave a generic framework for *tightly secure* IBE based on broadcast encodings and reached a series of interesting constructions. However IBE schemes with exponent-inverse structure [Gen06, Wee16, CGW17] are out of the scope of current predicate encodings, and we still have no framework supporting *fully* attribute-hiding feature [OT12].

JR-IBE has been used to construct more advanced primitives [WS16, WES17]. We note that our framework is seemingly not powerful enough to cover them. However we believe our simplified JR-IBE (and our framework as well) can shed the light on more extensions in the future.

Organization. Our paper is organized as follows. The next section will give several basic notions. Our generic framework for ABE with tag will be given in Sect. 3. We also prove its adaptive security in the same section. We then show the compatibility of our framework with attribute-hiding encodings in Sect. 4. The next section, Sect. 5, illustrates the new ABE constructions derived from our framework. The last section, Sect. 6, shows how to extend our framework to support delegation.

2 Preliminaries

Notation. We use $s \leftarrow S$ to indicate that s is selected uniformly from finite set S. For a probability distribution $\mathcal{D}$, notation $x \leftarrow \mathcal{D}$ means that x is sampled according to $\mathcal{D}$. We consider λ as security parameter and a function $f(\lambda)$ is negligible in λ if, for each $c \in \mathbb{N}$, there exists λ_c such that $f(\lambda) < 1/\lambda^c$ for all $\lambda > \lambda_c$. "p.p.t." stands for "probabilistic polynomial time". For a matrix $\mathbf{A} \in \mathbb{Z}_p^{k \times k'}$ with $k > k'$, we let $\overline{\mathbf{A}}$ be the matrix consist of the first k' rows and $\underline{\mathbf{A}}$ be the matrix with all remaining rows.

2.1 Attribute-Based Encryptions

Syntax. An attribute-based encryption (ABE) scheme for predicate $P(\cdot, \cdot)$ consists of the following p.p.t. algorithms.

- $\mathsf{Setup}(1^\lambda, P) \to (\mathsf{mpk}, \mathsf{msk})$. The *setup* algorithm takes as input the security parameter λ and a description of predicate P and returns master public/secret key pair $(\mathsf{mpk}, \mathsf{msk})$. We assume that mpk contains descriptions of domains $\mathcal{X}$ and $\mathcal{Y}$ of P as well as message space $\mathcal{M}$.
- $\mathsf{Enc}(\mathsf{mpk}, x, m) \to \mathsf{ct}_x$. The *encryption* algorithm takes as input the master public key mpk, an index (attribute) $x \in \mathcal{X}$ and a message $m \in \mathcal{M}$ and outputs a ciphertext ct_x.
- $\mathsf{KeyGen}(\mathsf{mpk}, \mathsf{msk}, y) \to \mathsf{sk}_y$. The *key generation* algorithm takes as input the master public/secret key pair $(\mathsf{mpk}, \mathsf{msk})$ and an index (policy) $y \in \mathcal{Y}$ and generates a secret key sk_y.
- $\mathsf{Dec}(\mathsf{mpk}, \mathsf{sk}_y, \mathsf{ct}_x) \to m$. The *decryption* algorithm takes as input the master public key mpk, a secret key sk_y and a ciphertext ct_x with $P(x, y) = 1$ and outputs message m.

Correctness. For all $(\mathsf{mpk}, \mathsf{msk}) \leftarrow \mathsf{Setup}(1^\lambda, P)$, all $x \in \mathcal{X}$ and $y \in \mathcal{Y}$ satisfying $P(x, y) = 1$, and all $m \in \mathcal{M}$, it is required that

$$\Pr\left[\mathsf{Dec}(\mathsf{mpk}, \mathsf{sk}_y, \mathsf{ct}_x) = m \,\middle|\, \begin{array}{l} \mathsf{sk}_y \leftarrow \mathsf{KeyGen}(\mathsf{mpk}, \mathsf{msk}, y) \\ \mathsf{ct}_x \leftarrow \mathsf{Enc}(\mathsf{mpk}, x, m) \end{array}\right] = 1.$$

Security. For all adversary $\mathcal{A}$, define advantage function $\mathsf{Adv}_{\mathcal{A}}^{\mathrm{ABE}}(\lambda)$ as follows.

$$\mathsf{Adv}_{\mathcal{A}}^{\mathrm{ABE}}(\lambda) = \left|\Pr\left[\beta = \beta' \,\middle|\, \begin{array}{l} (\mathsf{mpk}, \mathsf{msk}) \leftarrow \mathsf{Setup}(1^\lambda, P), \ \beta \leftarrow \{0, 1\} \\ (x^*, m_0^*, m_1^*) \leftarrow \mathcal{A}^{\mathsf{KeyGen}(\mathsf{mpk}, \mathsf{msk}, \cdot)}(\mathsf{mpk}) \\ \mathsf{ct}^* \leftarrow \mathsf{Enc}(\mathsf{mpk}, x^*, m_\beta^*) \\ \beta' \leftarrow \mathcal{A}^{\mathsf{KeyGen}(\mathsf{mpk}, \mathsf{msk}, \cdot)}(\mathsf{mpk}, \mathsf{ct}^*) \end{array}\right] - \frac{1}{2}\right|.$$

An ABE scheme is said to be *adaptively secure* if $\mathsf{Adv}_{\mathcal{A}}^{\mathrm{ABE}}(\lambda)$ is negligible in λ and $P(x^*, y) = 0$ holds for each query y sent to oracle $\mathsf{KeyGen}(\mathsf{mpk}, \mathsf{msk}, \cdot)$ for all p.p.t. adversary $\mathcal{A}$. We may call x^* *the target attribute* and each y a *key extraction query*.

2.2 Prime-Order Bilinear Groups and Cryptographic Assumption

We assume a group generator GrpGen which takes as input security parameter 1^λ and outputs group description $\mathcal{G} = (p, G_1, G_2, G_T, e)$. Here G_1, G_2, G_T are cyclic groups of prime order p of $\Theta(\lambda)$ bits and $e : G_1 \times G_2 \to G_T$ is a non-degenerated bilinear map. We assume that descriptions of G_1 and G_2 contain respective generators g_1 and g_2.

Let $s \in \{1, 2, T\}$. For $\mathbf{A} = (a_{ij}) \in \mathbb{Z}_p^{k \times k'}$, we define the *implicit representation* [EHK+13] as

$$[\mathbf{A}]_s = \begin{pmatrix} g_s^{a_{11}} & \cdots & g_s^{a_{1k'}} \\ \vdots & & \vdots \\ g_s^{a_{k1}} & \cdots & g_s^{a_{kk'}} \end{pmatrix} \in G_s^{k \times k'}.$$

Given $[\mathbf{A}]_1 \in G_1^{k \times n}$ and $[\mathbf{B}]_2 \in G_2^{k \times n'}$, define $e([\mathbf{A}]_1, [\mathbf{B}]_2) = [\mathbf{A}^\top \mathbf{B}]_T \in G_T^{n \times n'}$. We also use the following notations: given $[\mathbf{a}]_s, [\mathbf{b}]_s \in G_s^k$ and $c \in \mathbb{Z}_p$, define

$$[\mathbf{a}]_s \cdot [\mathbf{b}]_s = [\mathbf{a} + \mathbf{b}]_s \in G_s^k \quad \text{and} \quad [\mathbf{a}]_s^c = [c\mathbf{a}]_s \in G_s^k; \tag{3}$$

for $([\mathbf{a}_1]_s, \ldots, [\mathbf{a}_n]_s), ([\mathbf{b}_1]_s, \ldots, [\mathbf{b}_n]_s) \in (G_s^k)^n$, we define

$$([\mathbf{a}_1]_s, \ldots, [\mathbf{a}_n]_s) \cdot ([\mathbf{b}_1]_s, \ldots, [\mathbf{b}_n]_s) = ([\mathbf{a}_1]_s \cdot [\mathbf{b}_1]_s, \ldots, [\mathbf{a}_n]_s \cdot [\mathbf{b}_n]_s) \in (G_s^k)^n;$$

for $\mathbf{a} = (a_1, \ldots, a_n) \in \mathbb{Z}_p^n$ and $[\mathbf{b}]_s \in G_s^k$, we define

$$[\mathbf{b}]_s^{\mathbf{a}} = ([\mathbf{b}]_s^{a_1}, \ldots, [\mathbf{b}]_s^{a_n}) \in (G_s^k)^n. \tag{4}$$

Let $\mathcal{D}_k$ be a matrix distribution sampling matrix $\mathbf{A} \in \mathbb{Z}_p^{(k+1) \times k}$ along with a non-zero vector $\mathbf{a}^\perp \in \mathbb{Z}_p^{k+1}$ satisfying $\mathbf{A}^\top \mathbf{a}^\perp = \mathbf{0}$. We need the following lemma with respect to $\mathcal{D}_k$.

Lemma 1 (Basic Lemma [GDCC16, GHKW16]**).** *With probability* $1 - 1/p$ *over* $(\mathbf{A}, \mathbf{a}^\perp) \leftarrow \mathcal{D}_k$ *and* $\mathbf{b} \leftarrow \mathbb{Z}_p^{k+1}$, *it holds that*

$$\mathbf{b} \notin \mathsf{span}(\mathbf{A}) \quad \text{and} \quad \mathbf{b}^\top \mathbf{a}^\perp \neq 0.$$

We review the matrix decisional Diffie-Hellman (MDDH) assumption in the prime-order bilinear groups as follows.

Assumption 1 ($\mathcal{D}_k$-MDDH [EHK+13]**).** *Let* $s \in \{1, 2\}$. *For any adversary* $\mathcal{A}$, *define the advantage function* $\mathsf{Adv}_{\mathcal{A}}^{\mathcal{D}_k}(\lambda)$ *as follows*

$$\mathsf{Adv}_{\mathcal{A}}^{\mathcal{D}_k}(\lambda) = |\Pr[\mathcal{A}(\mathcal{G}, [\mathbf{A}]_s, [\mathbf{A}\mathbf{s}]_s) = 1] - \Pr[\mathcal{A}(\mathcal{G}, [\mathbf{A}]_s, [\mathbf{u}]_s) = 1]|$$

where $\mathcal{G} \leftarrow \mathsf{GrpGen}(1^\lambda)$, $(\mathbf{A}, \mathbf{a}^\perp) \leftarrow \mathcal{D}_k$, $\mathbf{s} \leftarrow \mathbb{Z}_p^k$ *and* $\mathbf{u} \leftarrow \mathbb{Z}_p^{k+1}$. *The assumption says that* $\mathsf{Adv}_{\mathcal{A}}^{\mathcal{D}_k}(\lambda)$ *is negligible in* λ *for all p.p.t. adversary* $\mathcal{A}$.

2.3 Predicate Encodings

This subsection reviews the notion of predicate encoding [Wee14, CGW15] and shows some useful notations and facts.

Definition. A $\mathbb{Z}_p$-linear predicate encoding for $\mathsf{P} : \mathcal{X} \times \mathcal{Y} \to \{0, 1\}$ consists of five deterministic algorithms

$$\mathsf{sE} : \mathcal{X} \times \mathbb{Z}_p^n \to \mathbb{Z}_p^{n_s} \qquad\qquad \mathsf{sD} : \mathcal{X} \times \mathcal{Y} \times \mathbb{Z}_p^{n_s} \to \mathbb{Z}_p$$
$$\mathsf{rE} : \mathcal{Y} \times \mathbb{Z}_p^n \to \mathbb{Z}_p^{n_r} \qquad \mathsf{kE} : \mathcal{Y} \times \mathbb{Z}_p \to \mathbb{Z}_p^{n_r} \qquad \mathsf{rD} : \mathcal{X} \times \mathcal{Y} \times \mathbb{Z}_p^{n_r} \to \mathbb{Z}_p$$

for some $n, n_s, n_r \in \mathbb{N}$ with the following features:

(linearity). For all $(x, y) \in \mathcal{X} \times \mathcal{Y}$, $\mathsf{sE}(x, \cdot), \mathsf{rE}(y, \cdot), \mathsf{kE}(y, \cdot), \mathsf{sD}(x, y, \cdot), \mathsf{rD}(x, y, \cdot)$ are $\mathbb{Z}_p$-linear. A $\mathbb{Z}_p$-linear function $L : \mathbb{Z}_p^n \to \mathbb{Z}_p^{n'}$ can be encoded as a matrix $\mathbf{L} = (l_{i,j}) \in \mathbb{Z}_p^{n \times n'}$ such that

$$L : (w_1, \ldots, w_n) \mapsto (\textstyle\sum_{i=1}^n l_{i1} w_i, \ldots, \sum_{i=1}^n l_{in'} w_i). \tag{5}$$

(restricted α-reconstruction). For all $(x, y) \in \mathcal{X} \times \mathcal{Y}$ such that $\mathsf{P}(x, y) = 1$, all $\mathbf{w} \in \mathbb{Z}_p^n$ and all $\alpha \in \mathbb{Z}_p$, we have

$$\mathsf{sD}(x, y, \mathsf{sE}(x, \mathbf{w})) = \mathsf{rD}(x, y, \mathsf{rE}(y, \mathbf{w})) \quad \text{and} \quad \mathsf{rD}(x, y, \mathsf{kE}(y, \alpha)) = \alpha.$$

(α-privacy). For all $(x, y) \in \mathcal{X} \times \mathcal{Y}$ such that $\mathsf{P}(x, y) = 0$ and all $\alpha \in \mathbb{Z}_p$, the following distributions are identical.

$$\{x, y, \alpha, \mathsf{sE}(x, \mathbf{w}), \mathsf{kE}(y, \alpha) + \mathsf{rE}(y, \mathbf{w}) : \mathbf{w} \leftarrow \mathbb{Z}_p^n\} \quad \text{and}$$
$$\{x, y, \alpha, \mathsf{sE}(x, \mathbf{w}), \mathsf{rE}(y, \mathbf{w}) : \mathbf{w} \leftarrow \mathbb{Z}_p^n\}.$$

We call n the *parameter size* and use $|\mathsf{sE}|$ and $|\mathsf{rE}|$ to denote n_s and n_r, respectively, which indicate the *sizes of sender encodings and receiver encoding*. We note that $|\mathsf{sE}|$ and $|\mathsf{rE}|$ may depend on x and y respectively. For all $x \in \mathcal{X}$, we define the distribution

$$\mathsf{sE}(x) := \{\mathsf{sE}(x, \mathbf{w}) : \mathbf{w} \leftarrow \mathbb{Z}_p^n\}.$$

More Notations and Useful Facts. Assume $s \in \{1, 2, T\}$. We can naturally define a series of $\mathbb{Z}_p$-linear functions from $\mathbf{L}$ as follows:

$$L : \quad (G_s^k)^n \quad \to \quad (G_s^k)^{n'}$$
$$([\mathbf{w}_1]_s, \ldots, [\mathbf{w}_n]_s) \mapsto (\textstyle\prod_{i=1}^n [\mathbf{w}_i]_s^{l_{i1}}, \ldots, \prod_{i=1}^n [\mathbf{w}_i]_s^{l_{in'}}) \tag{6}$$

Because they essentially share the same structure (i.e., $\mathbf{L}$), we employ the same notation L. It should be clear from the context. Then we highlight three properties regarding functions L with respective to the same $\mathbf{L}$ as follows:

($L(\cdot)$ and pairing e are commutative). For any $\mathbf{a}, \mathbf{b}_1, \ldots, \mathbf{b}_n \in \mathbb{Z}_p^k$, we have

$$e([\mathbf{a}]_1, L([\mathbf{b}_1]_2, \ldots, [\mathbf{b}_n]_2)) = L(e([\mathbf{a}]_1, [\mathbf{b}_1]_2), \ldots, e([\mathbf{a}]_1, [\mathbf{b}_n]_2)) \tag{7}$$
$$e(L([\mathbf{b}_1]_1, \ldots, [\mathbf{b}_n]_1), [\mathbf{a}]_2) = L(e([\mathbf{b}_1]_1, [\mathbf{a}]_2), \ldots, e([\mathbf{b}_n]_1, [\mathbf{a}]_2)) \tag{8}$$

($L(\cdot)$ and $[\cdot]_s$ are commutative). For any $w_1, \ldots, w_n \in \mathbb{Z}_p$, we have

$$L([w_1]_s, \ldots, [w_n]_s) = [L(w_1, \ldots, w_n)]_s. \tag{9}$$

($L(\cdot)$ and "exponentiation" are commutative). For any $\mathbf{w} \in \mathbb{Z}_p^n$ and $[\mathbf{a}]_s \in G_s^k$, we have

$$[\mathbf{a}]_s^{L(\mathbf{w})} = L([\mathbf{a}]_s^{\mathbf{w}}). \tag{10}$$

Concrete Instantiations. As an example, we show the predicate encoding for equality predicate as below. This is the simplest encoding and extracted from classical Lewko-Waters IBE [LW10].

(encoding for equality [LW10]). Let $\mathcal{X} = \mathcal{Y} = \mathbb{Z}_p$ and $\mathsf{P}(x, y) = 1$ iff $x = y$. Let $n = 2$ and $w_1, w_2 \leftarrow \mathbb{Z}_p$. Define

$$
\begin{aligned}
\mathsf{sE}(x, (w_1, w_2)) &:= w_1 + x w_2 & & & \mathsf{sD}(x, y, c) := c \\
\mathsf{rE}(y, (w_1, w_2)) &:= w_1 + y w_2 & \mathsf{kE}(y, \alpha) &:= \alpha & \mathsf{rD}(x, y, k) := k
\end{aligned}
$$

Here we have $|\mathsf{sE}| = |\mathsf{rE}| = 1$.

3 ABE with Tags from Predicate Encodings

3.1 Construction

Our generic tag-based ABE from predicate encodings is described below.

- Setup(1^λ, P): Let n be parameter size of predicate encoding $(\mathsf{sE}, \mathsf{rE}, \mathsf{kE}, \mathsf{sD}, \mathsf{rD})$ for P. We sample

$$
\mathbf{A} \leftarrow \mathcal{D}_k, \quad \mathbf{W}_1, \ldots, \mathbf{W}_n, \mathbf{W} \leftarrow \mathbb{Z}_p^{(k+1) \times k}, \quad \mathbf{k} \leftarrow \mathbb{Z}_p^{k+1}
$$

and output the master public and secret key pair

$$
\begin{aligned}
\mathsf{mpk} &:= \{ [\mathbf{A}]_1, [\mathbf{W}_1^\top \mathbf{A}]_1, \ldots, [\mathbf{W}_n^\top \mathbf{A}]_1, [\mathbf{W}^\top \mathbf{A}]_1, [\mathbf{k}^\top \mathbf{A}]_T \} \\
\mathsf{msk} &:= \{ \mathbf{W}_1, \ldots, \mathbf{W}_n, \mathbf{W}; \mathbf{k} \}.
\end{aligned}
$$

- Enc(mpk, x, m): On input $x \in \mathcal{X}$ and $m \in G_T$, pick $\mathbf{s} \leftarrow \mathbb{Z}_p^k$ and $\boldsymbol{\tau} \leftarrow \mathsf{sE}(x)$. Output

$$
\mathsf{ct}_x := \left\{
\begin{array}{l}
C_0 := [\mathbf{As}]_1, \\
\mathbf{C}_1 := \mathsf{sE}(x, [\mathbf{W}_1^\top \mathbf{As}]_1, \ldots, [\mathbf{W}_n^\top \mathbf{As}]_1) \cdot [\mathbf{W}^\top \mathbf{As}]_1^\top, \\
C := [\mathbf{k}^\top \mathbf{As}]_T \cdot m, \ \boldsymbol{\tau}
\end{array}
\right\}
$$

- KeyGen(mpk, msk, y): On input $y \in \mathcal{Y}$, pick $\mathbf{r} \leftarrow_R \mathbb{Z}_p^k$ and output

$$
\mathsf{sk}_y := \left\{
\begin{array}{l}
K_0 := [\mathbf{r}]_2, \\
\mathbf{K}_1 := \mathsf{kE}(y, [\mathbf{k}]_2) \cdot \mathsf{rE}(y, [\mathbf{W}_1 \mathbf{r}]_2, \ldots, [\mathbf{W}_n \mathbf{r}]_2), \\
K_2 := [\mathbf{Wr}]_2
\end{array}
\right\}
$$

- Dec(mpk, sk_y, ct_x): Compute

$$
K \leftarrow e(C_0, \mathsf{rD}(x, y, \mathbf{K}_1)) \cdot e(C_0, K_2)^{\mathsf{sD}(x, y, \boldsymbol{\tau})} / e(\mathsf{sD}(x, y, \mathbf{C}_1), K_0)
$$

and recover the message as $m \leftarrow C/K \in G_T$.

Correctness. For all $(x, y) \in \mathcal{X} \times \mathcal{Y}$ with $\mathsf{P}(x, y) = 1$, we may use the following abbreviation

$$\mathsf{sE}(x, \cdot) = \mathsf{sE}(\cdot), \ \mathsf{rE}(y, \cdot) = \mathsf{rE}(\cdot), \ \mathsf{kE}(y, \cdot) = \mathsf{kE}(\cdot);$$
$$\mathsf{sD}(x, y, \cdot) = \mathsf{sD}(\cdot), \ \mathsf{rD}(x, y, \cdot) = \mathsf{rD}(\cdot)$$

and have

$$e(C_0, \mathsf{rD}(\mathbf{K}_1)) \cdot e(C_0, K_2)^{\mathsf{sD}(\tau)}$$
$$\overset{(a)}{=} e([\mathbf{As}]_1, \mathsf{rD}(\mathsf{kE}([\mathbf{k}]_2))) \cdot e([\mathbf{As}]_1, \mathsf{rD}(\mathsf{rE}([\mathbf{W}_1\mathbf{r}]_2, \ldots, [\mathbf{W}_n\mathbf{r}]_2))) \cdot [\mathbf{s}^\top \mathbf{A}^\top \mathbf{Wr}]_T^{\mathsf{sD}(\tau)}$$
$$\overset{(b)}{=} \mathsf{rD}(\mathsf{kE}([\mathbf{s}^\top \mathbf{A}^\top \mathbf{k}]_T)) \cdot \mathsf{rD}(\mathsf{rE}([\mathbf{s}^\top \mathbf{A}^\top \mathbf{W}_1\mathbf{r}]_T, \ldots, [\mathbf{s}^\top \mathbf{A}^\top \mathbf{W}_n\mathbf{r}]_T)) \cdot [\mathbf{s}^\top \mathbf{A}^\top \mathbf{Wr}]_T^{\mathsf{sD}(\tau)}$$
$$\overset{(c)}{=} [\mathsf{rD}(\mathsf{kE}(\mathbf{s}^\top \mathbf{A}^\top \mathbf{k}))]_T \cdot [\mathsf{rD}(\mathsf{rE}(\mathbf{s}^\top \mathbf{A}^\top \mathbf{W}_1\mathbf{r}, \ldots, \mathbf{s}^\top \mathbf{A}^\top \mathbf{W}_n\mathbf{r}))]_T \cdot [\mathbf{s}^\top \mathbf{A}^\top \mathbf{Wr}]_T^{\mathsf{sD}(\tau)}$$
$$\overset{(d)}{=} [\mathbf{s}^\top \mathbf{A}^\top \mathbf{k}]_T \cdot [\mathsf{sD}(\mathsf{sE}(\mathbf{s}^\top \mathbf{A}^\top \mathbf{W}_1\mathbf{r}, \ldots, \mathbf{s}^\top \mathbf{A}^\top \mathbf{W}_n\mathbf{r}))]_T \cdot [\mathbf{s}^\top \mathbf{A}^\top \mathbf{Wr}]_T^{\mathsf{sD}(\tau)}$$
$$\overset{(e)}{=} [\mathbf{s}^\top \mathbf{A}^\top \mathbf{k}]_T \cdot \mathsf{sD}(\mathsf{sE}([\mathbf{s}^\top \mathbf{A}^\top \mathbf{W}_1\mathbf{r}]_T, \ldots, [\mathbf{s}^\top \mathbf{A}^\top \mathbf{W}_n\mathbf{r}]_T)) \cdot [\mathbf{s}^\top \mathbf{A}^\top \mathbf{Wr}]_T^{\mathsf{sD}(\tau)}$$
$$\overset{(f)}{=} [\mathbf{s}^\top \mathbf{A}^\top \mathbf{k}]_T \cdot e(\mathsf{sD}(\mathsf{sE}([\mathbf{W}_1^\top \mathbf{As}]_1, \ldots, [\mathbf{W}_n^\top \mathbf{As}]_1)), [\mathbf{r}]_2) \cdot e([\mathbf{W}^\top \mathbf{As}]_1^{\mathsf{sD}(\tau)}, [\mathbf{r}]_2)$$
$$\overset{(g)}{=} [\mathbf{s}^\top \mathbf{A}^\top \mathbf{k}]_T \cdot e(\mathsf{sD}(\mathsf{sE}([\mathbf{W}_1^\top \mathbf{As}]_1, \ldots, [\mathbf{W}_n^\top \mathbf{As}]_1)), [\mathbf{r}]_2) \cdot e(\mathsf{sD}([\mathbf{W}^\top \mathbf{As}]_1^\tau), [\mathbf{r}]_2)$$
$$\overset{(h)}{=} [\mathbf{s}^\top \mathbf{A}^\top \mathbf{k}]_T \cdot e(\mathsf{sD}(\mathsf{sE}([\mathbf{W}_1^\top \mathbf{As}]_1, \ldots, [\mathbf{W}_n^\top \mathbf{As}]_1) \cdot [\mathbf{W}^\top \mathbf{As}]_1^\tau), [\mathbf{r}]_2)$$
$$= [\mathbf{s}^\top \mathbf{A}^\top \mathbf{k}]_T \cdot e(\mathsf{sD}(\mathbf{C}_1), K_0)$$

which is sufficient for the correctness. We list the properties and the facts justifying each (labelled) equality in Table 3.

Table 3. Properties and Facts for Correctness.

Eq.	Properties	Facts
(a)	Composition of two linear functions are linear	Both $e([\mathbf{As}]_1, \cdot)$ and $\mathsf{rD}(x, y, \cdot)$ are linear functions.
(b)	Linear functions and pairings are commutative. (cf. Eq. (7))	Both $\mathsf{rD}(x, y, \mathsf{kE}(y, \cdot))$ and $\mathsf{rD}(x, y, \mathsf{rE}(y, \cdot))$ are $\mathbb{Z}_p$-valued linear functions.
(c)	Linear functions and $[\cdot]_s$ are commutative. (cf. Eq. (9))	Both $\mathsf{rD}(x, y, \mathsf{kE}(y, \cdot))$ and $\mathsf{rD}(x, y, \mathsf{rE}(y, \cdot))$ are $\mathbb{Z}_p$-valued linear functions. $\mathbf{s}^\top \mathbf{A}^\top \mathbf{k}, \mathbf{s}^\top \mathbf{A}^\top \mathbf{W}_1\mathbf{r}, \ldots, \mathbf{s}^\top \mathbf{A}^\top \mathbf{W}_n\mathbf{r} \in \mathbb{Z}_p.$
(d)	Restricted α-reconstruction	$\mathbf{s}^\top \mathbf{A}^\top \mathbf{k}, \mathbf{s}^\top \mathbf{A}^\top \mathbf{W}_1\mathbf{r}, \ldots, \mathbf{s}^\top \mathbf{A}^\top \mathbf{W}_n\mathbf{r} \in \mathbb{Z}_p.$
(e)	Linear functions and $[\cdot]_s$ are commutative. (cf. Eq. (9))	$\mathsf{sD}(x, y, \mathsf{sE}(x, \cdot))$ is a $\mathbb{Z}_p$-valued linear function. $\mathbf{s}^\top \mathbf{A}^\top \mathbf{k}, \mathbf{s}^\top \mathbf{A}^\top \mathbf{W}_1\mathbf{r}, \ldots, \mathbf{s}^\top \mathbf{A}^\top \mathbf{W}_n\mathbf{r} \in \mathbb{Z}_p.$
(f)	Linear functions and pairings are commutative. (cf. Eq. (8))	$\mathsf{sD}(x, y, \mathsf{sE}(x, \cdot))$ is a $\mathbb{Z}_p$-valued linear function.
(g)	Linear functions and "exponentiations" are commutative. (cf. Eq. (10))	$\mathsf{sD}(x, y, \cdot)$ is a $\mathbb{Z}_p$-valued linear function.
(h)	Composition of two linear functions are linear	Both $e(\cdot, [\mathbf{r}]_2)$ and $\mathsf{sD}(x, y, \cdot)$ are linear functions

Security Result. We give the following main theorem stating that the above generic tag-based ABE scheme is adaptively secure under standard assumption in the standard model. The remaining of this section will be devoted to the proof of the main theorem.

Theorem 1 (Main Theorem). *For any p.p.t. adversary $\mathcal{A}$ making at most q key extraction queries, there exists algorithms $\mathcal{B}_1, \mathcal{B}_2, \mathcal{B}_3$ such that*

$$\mathsf{Adv}_{\mathcal{A}}^{\text{ABE}}(\lambda) \leq \mathsf{Adv}_{\mathcal{B}_1}^{\mathcal{D}_k}(\lambda) + q \cdot \mathsf{Adv}_{\mathcal{B}_2}^{\mathcal{D}_k}(\lambda) + q \cdot \mathsf{Adv}_{\mathcal{B}_3}^{\mathcal{D}_k}(\lambda) + 2^{-\Omega(\lambda)}$$

and $\max\{\mathsf{Time}(\mathcal{B}_1), \mathsf{Time}(\mathcal{B}_2), \mathsf{Time}(\mathcal{B}_3)\} \approx \mathsf{Time}(\mathcal{A}) + q \cdot k^2 \cdot \text{poly}(\lambda, n).$

3.2 Proving the Main Theorem: High-Level Roadmap

From a high level, the proof basically follows the common dual system methodology. We first define semi-functional distributions under the master public key

$$\mathsf{mpk} = \{[\mathbf{A}]_1, [\mathbf{W}_1^\top \mathbf{A}]_1, \ldots, [\mathbf{W}_n^\top \mathbf{A}]_1, [\mathbf{W}^\top \mathbf{A}]_1, [\mathbf{k}^\top \mathbf{A}]_T\}.$$

where $(\mathbf{A}, \mathbf{a}^\perp) \leftarrow \mathcal{D}_k, \mathbf{W}_1, \ldots, \mathbf{W}_n, \mathbf{W} \leftarrow \mathbb{Z}_p^{(k+1) \times k}, \mathbf{k} \leftarrow \mathbb{Z}_p^{k+1}$ as follows.

(semi-functional ciphertext). A semi-functional ciphertext for target attribute $x^* \in \mathcal{X}$ and challenge message pair $(m_0^*, m_1^*) \in \mathcal{M} \times \mathcal{M}$ is defined as follows:

$$\left\{ [\boxed{\mathbf{c}}]_1, \ \mathsf{sE}(x^*, [\mathbf{W}_1^\top \boxed{\mathbf{c}}]_1, \ldots, [\mathbf{W}_n^\top \boxed{\mathbf{c}}]_1) \cdot [\mathbf{W}^\top \boxed{\mathbf{c}}]_1^{\tau^*}, [\mathbf{k}^\top \boxed{\mathbf{c}}]_T \cdot m_\beta^*, \ \boldsymbol{\tau}^* \right\}$$

where $\mathbf{c} \leftarrow \mathbb{Z}_p^{k+1}$ and $\boldsymbol{\tau}^* \leftarrow \mathsf{sE}(x^*)$.

(semi-functional secret key). A semi-functional secret key for policy $y \in \mathcal{Y}$ is defined as follows:

$$\left\{ [\mathbf{r}]_2, \mathsf{kE}(y, [\mathbf{k} + \boxed{\alpha \mathbf{a}^\perp}]_2) \cdot \mathsf{rE}(y, [\mathbf{W}_1 \mathbf{r}]_2, \ldots, [\mathbf{W}_n \mathbf{r}]_2), [\mathbf{W} \mathbf{r}]_2 \right\}$$

where $\alpha \leftarrow \mathbb{Z}_p$. Note that all semi-functional secret keys in the system will share the same α.

Game Sequence. Our proof employs the following game sequence.

- G_0 is the real security game defined as Sect. 2.1.
- G_1 is identical to G_0 except that the challenge ciphertext is semi-functional.
- $\mathsf{G}_{2.i}$ (for $i \in [0, q]$) is identical to G_1 except that the first i key extraction queries are replied with semi-functional secret keys.
- G_3 is identical to $\mathsf{G}_{2.q}$ except that the challenge ciphertext is a semi-functional ciphertext for random message $m^* \in \mathcal{M}$.

Roughly speaking, we are going to prove that

$$G_0 \overset{\text{lem 4}}{\approx} G_1 = G_{2.0} \overset{\text{sec 3.3}}{\approx} G_{2.1} \overset{\text{sec 3.3}}{\approx} \cdots \overset{\text{sec 3.3}}{\approx} G_{2.q} \overset{\text{lem 5}}{=} G_3.$$

Here "$\approx$" indicates that two games are *computationally* indistinguishable while "$=$" means that they are *statistically* indistinguishable. Let $\mathsf{Adv}_{\mathcal{A}}^{i.j}(\lambda)$ be the advantage function of any p.p.t. adversary $\mathcal{A}$ making at most q key extraction queries in $G_{i.j}$ with security parameter λ.

We begin with two simple lemmas. First, it is not hard to see that G_1 and $G_{2.0}$ are actually the same and we have the following lemma.

Lemma 2 ($G_1 = G_{2.0}$). *For any adversary $\mathcal{A}$, we have $\mathsf{Adv}_{\mathcal{A}}^1(\lambda) = \mathsf{Adv}_{\mathcal{A}}^{2.0}(\lambda)$.*

Next, observe that the challenge ciphertext in the last game G_3 is created without secret bit $\beta \in \{0,1\}$. In other words, the challenge ciphertext reveals nothing about β. Therefore adversary has no advantage in guessing β and we have the lemma below.

Lemma 3. *For any adversary $\mathcal{A}$, we have $\mathsf{Adv}_{\mathcal{A}}^3(\lambda) = 0$.*

Following Chen *et al.*'s proof [CGW15], we can prove Lemma 4 showing $G_0 \approx G_1$ and Lemma 5 showing $G_{2.q} = G_3$. Due to the lack of space, we omit the proofs.

Lemma 4 ($G_0 \approx G_1$). *For any p.p.t. adversary $\mathcal{A}$ making at most q key extraction queries, there exists an algorithm $\mathcal{B}$ such that*

$$|\mathsf{Adv}_{\mathcal{A}}^0(\lambda) - \mathsf{Adv}_{\mathcal{A}}^1(\lambda)| \leq \mathsf{Adv}_{\mathcal{B}}^{\mathcal{D}_k}(\lambda) + 1/p$$

and $\mathsf{Time}(\mathcal{B}) \approx \mathsf{Time}(\mathcal{A}) + q \cdot k^2 \cdot \mathrm{poly}(\lambda, n)$.

Lemma 5 ($G_{2.q} = G_3$). *For any adversary $\mathcal{A}$, we have*

$$|\mathsf{Adv}_{\mathcal{A}}^{2.q}(\lambda) - \mathsf{Adv}_{\mathcal{A}}^3(\lambda)| = 1/p.$$

In order to complete the proof, we prove that $G_{2.i}$ is indistinguishable with $G_{2.i+1}$ for all $i \in [0, q-1]$. The details are deferred to the next subsection.

3.3   Proving the Theorem: Filling the Gap Between $G_{2.i}$ and $G_{2.i+1}$

This subsection proves the indistinguishability of $G_{2.i}$ and $G_{2.i+1}$. We introduce an auxiliary game sequence which is based on the proof idea from Jutla and Roy [JR13]. In particular, we need the following two auxiliary distributions.

(pesudo-normal secret key). A pseudo-normal secret key for policy $y \in \mathcal{Y}$ is defined as follows:

$$\left\{ \begin{array}{c} [\mathbf{r}]_2,\ \mathsf{kE}(y, [\mathbf{k}]_2) \cdot \mathsf{rE}(y, [\mathbf{W}_1\mathbf{r}]_2, \ldots, [\mathbf{W}_n\mathbf{r}]_2) \cdot \boxed{[\mathbf{a}^\perp \hat{r}]_2^{\mathsf{rE}(y, \gamma_1, \ldots, \gamma_n)}}, \\ [\mathbf{W}\mathbf{r}]_2 \cdot \boxed{[\mathbf{a}^\perp \hat{r}]_2^{-1}} \end{array} \right\}$$

where $\hat{r} \leftarrow \mathbb{Z}_p$ and $\gamma_1, \ldots, \gamma_n \leftarrow \mathbb{Z}_p$ are the random coins for tag $\boldsymbol{\tau}^*$ in the challenge ciphertext. Recall that we compute $\boldsymbol{\tau}^* = \mathsf{sE}(x^*, (\gamma_1, \ldots, \gamma_n))$ for target attribute x^*.

(**pesudo-semi-functional secret key**). A pseudo-semi-functional secret key for policy $y \in \mathcal{Y}$ is defined as follows:

$$\left\{ \begin{array}{c} [\mathbf{r}]_2, \ \mathsf{kE}(y, [\,\mathbf{k} + \boxed{\alpha \mathbf{a}^\perp}\,]_2) \cdot \mathsf{rE}(y, [\mathbf{W}_1\mathbf{r}]_2, \ldots, [\mathbf{W}_n\mathbf{r}]_2) \cdot [\mathbf{a}^\perp \hat{r}]_2^{\mathsf{rE}(y, \gamma_1, \ldots, \gamma_n)}, \\ [\mathbf{Wr}]_2 \cdot [\mathbf{a}^\perp \hat{r}]_2^{-1} \end{array} \right\}$$

where $\hat{r} \in \mathbb{Z}_p$ and $\gamma_1, \ldots, \gamma_n \in \mathbb{Z}_p$ are defined as before and $\alpha \in \mathbb{Z}_p$ is the one used in the semi-functional secret key.

We note that the random coins $\gamma_1, \ldots, \gamma_n$ for $\boldsymbol{\tau}^*$ are independent of x^* and thus we can pick them at the very beginning and use them to create secret keys of these two forms at *any* point.

Game Sub-sequence. For each $i \in [0, q-1]$, we define

- $\mathsf{G}_{2.i.1}$ is identical to $\mathsf{G}_{2.i}$ except that the $i+1$st key extraction query y is answered with a pseudo-normal secret key.
- $\mathsf{G}_{2.i.2}$ is identical to $\mathsf{G}_{2.i.1}$ except that the $i+1$st key extraction query y is answered with a pseudo-semi-functional secret key.

With this sub-sequence, we will prove that

$$\mathsf{G}_{2.i} \overset{\text{lem } 6}{\approx} \mathsf{G}_{2.i.1} \overset{\text{lem } 8}{=} \mathsf{G}_{2.i.2} \overset{\text{lem } 7}{\approx} \mathsf{G}_{2.i+1}.$$

We first prove Lemma 6 showing that $\mathsf{G}_{2.i} \approx \mathsf{G}_{2.i.1}$ and note that, following almost the same strategy, we can also prove that $\mathsf{G}_{2.i.2} \approx \mathsf{G}_{2.i+1}$. Hence we will show the corresponding result in Lemma 7 but omit the proof.

Lemma 6 ($\mathsf{G}_{2.i} \approx \mathsf{G}_{2.i.1}$). *For any p.p.t. adversary $\mathcal{A}$ making at most q key extraction queries, there exists an algorithm $\mathcal{B}$ such that*

$$|\mathsf{Adv}_{\mathcal{A}}^{2.i}(\lambda) - \mathsf{Adv}_{\mathcal{A}}^{2.i.1}(\lambda)| \ \leq \ \mathsf{Adv}_{\mathcal{B}}^{\mathcal{D}_k}(\lambda)$$

and $\mathsf{Time}(\mathcal{B}) \approx \mathsf{Time}(\mathcal{A}) + q \cdot k^2 \cdot \mathsf{poly}(\lambda, n)$.

Proof. Given $(\mathcal{G}, [\mathbf{M}]_2, [\mathbf{t}]_2 = [\mathbf{Mu} + e v]_2)$ where $\mathbf{u} \leftarrow \mathbb{Z}_p^k$, $\mathbf{e} = (0, \ldots, 0, 1)^\top \in \mathbb{Z}_p^{k+1}$ and either $v \leftarrow \mathbb{Z}_p$ or $v = 0$, algorithm $\mathcal{B}$ works as follows:

Initialize. Sample $(\mathbf{A}, \mathbf{a}^\perp) \leftarrow \mathcal{D}_k$, $\mathbf{k} \leftarrow \mathbb{Z}_p^{k+1}$ and $\beta \leftarrow \{0,1\}$. Pick

$$\widetilde{\mathbf{W}}_1, \cdots, \widetilde{\mathbf{W}}_n, \widetilde{\mathbf{W}} \leftarrow \mathbb{Z}_p^{(k+1) \times k} \quad \text{and} \quad \gamma_1, \ldots, \gamma_n \leftarrow \mathbb{Z}_p$$

and program "hidden parameter" $\gamma_1, \ldots, \gamma_n$ into $\mathbf{W}_1, \ldots, \mathbf{W}_n, \mathbf{W}$ as follows

$$\mathbf{W}_1 = \widetilde{\mathbf{W}}_1 + \gamma_1 \mathbf{V}, \ \ldots, \ \mathbf{W}_n = \widetilde{\mathbf{W}}_n + \gamma_n \mathbf{V}, \quad \mathbf{W} = \widetilde{\mathbf{W}} - \mathbf{V}$$

where $\mathbf{V} = \mathbf{a}^{\perp} \cdot (\underline{\mathbf{M}}\overline{\mathbf{M}}^{-1}) \in \mathbb{Z}_p^{(k+1)\times k}$. One may check that all $\mathbf{W}_i$ and $\mathbf{W}$ are uniformly distributed over $\mathbb{Z}_p^{(k+1)\times k}$ as required. Due to the fact that $\mathbf{A}^{\top}\mathbf{a}^{\perp} = \mathbf{0}$, we can return the master public key as follows

$$\mathsf{mpk} = \{[\mathbf{A}]_1, [\widetilde{\mathbf{W}_1^{\top}}\mathbf{A}]_1, \ldots, [\widetilde{\mathbf{W}_n^{\top}}\mathbf{A}]_1, [\widetilde{\mathbf{W}^{\top}}\mathbf{A}]_1, [\mathbf{k}^{\top}\mathbf{A}]_T\}.$$

We note that $\mathcal{B}$ can not compute $\mathbf{V}$ and thus does not know $\mathbf{W}_1, \ldots, \mathbf{W}_n, \mathbf{W}$.

Challenge ciphertext. For target attribute x^* and challenge message pair (m_0^*, m_1^*), we sample $\mathbf{c} \leftarrow \mathbb{Z}_p^{k+1}$, compute tag $\tau^* = \mathsf{sE}(x^*, (\gamma_1, \ldots, \gamma_n))$ using the "hidden parameter" and create the challenge ciphertext as follows

$$\{[\mathbf{c}]_1, \mathsf{sE}(x^*, [\widetilde{\mathbf{W}_1^{\top}}\mathbf{c}]_1, \ldots, [\widetilde{\mathbf{W}_n^{\top}}\mathbf{c}]_1) \cdot [\widetilde{\mathbf{W}^{\top}}\mathbf{c}]_1^{\tau^*}, [\mathbf{k}^{\top}\mathbf{c}]_T \cdot m_{\beta}^*, \tau^*\}$$

Let $\mathbf{v} = \underline{\mathbf{M}}\overline{\mathbf{M}}^{-1}$, we show that

$$\mathsf{sE}(x^*, [\mathbf{W}_1^{\top}\mathbf{c}]_1, \ldots, [\mathbf{W}_n^{\top}\mathbf{c}]_1) \cdot [\mathbf{W}^{\top}\mathbf{c}]_1^{\tau^*}$$
$$= \mathsf{sE}(x^*, [\widetilde{\mathbf{W}_1^{\top}}\mathbf{c}]_1, \ldots, [\widetilde{\mathbf{W}_n^{\top}}\mathbf{c}]_1) \cdot [\widetilde{\mathbf{W}^{\top}}\mathbf{c}]_1^{\tau^*} \cdot$$
$$\boxed{\mathsf{sE}(x^*, [\gamma_1\mathbf{v}^{\top}\mathbf{a}^{\perp\top}\mathbf{c}]_1, \ldots, [\gamma_n\mathbf{v}^{\top}\mathbf{a}^{\perp\top}\mathbf{c}]_1) \cdot [-\mathbf{v}^{\top}\mathbf{a}^{\perp\top}\mathbf{c}]_1^{\tau^*}}$$

from the linearity of $\mathsf{sE}(x^*, \cdot)$ and

$$\mathsf{sE}(x^*, [\gamma_1\mathbf{v}^{\top}\mathbf{a}^{\perp\top}\mathbf{c}]_1, \ldots, [\gamma_n\mathbf{v}^{\top}\mathbf{a}^{\perp\top}\mathbf{c}]_1) \cdot [-\mathbf{v}^{\top}\mathbf{a}^{\perp\top}\mathbf{c}]_1^{\tau^*} \quad \text{(the boxed part)}$$
$$= \mathsf{sE}(x^*, [\mathbf{v}^{\top}\mathbf{a}^{\perp\top}\mathbf{c}]_1^{(\gamma_1, \ldots, \gamma_n)}) \cdot [-\mathbf{v}^{\top}\mathbf{a}^{\perp\top}\mathbf{c}]_1^{\tau^*}$$
$$= [\mathbf{v}^{\top}\mathbf{a}^{\perp\top}\mathbf{c}]_1^{\mathsf{sE}(x^*, (\gamma_1, \ldots, \gamma_n))} \cdot [\mathbf{v}^{\top}\mathbf{a}^{\perp\top}\mathbf{c}]_1^{-\tau^*} = \mathbf{0},$$

where the first equality is mainly implied by Eqs. (3) and (4), and the second equality comes from the fact shown in Eq. (10). This is sufficient to see that our simulation is perfect.

Key extraction. We consider three cases: (1) For the first i queries y, we sample $\mathbf{r}' \leftarrow \mathbb{Z}_p^k$ and implicitly set

$$\mathbf{r} = \overline{\mathbf{M}}\mathbf{r}' \in \mathbb{Z}_p^k.$$

The vector $\mathbf{r}$ here is uniformly distributed as required and we can compute $[\mathbf{r}]_2$ and simulate

$$[\mathbf{V}\mathbf{r}]_2 = [\mathbf{a}^{\perp} \cdot (\underline{\mathbf{M}}\mathbf{r}')]_2.$$

These suffice for creating the secret key as

$$\{[\mathbf{r}]_2, \mathsf{kE}(y, [\mathbf{k}]_2) \cdot \mathsf{rE}(y, [\widetilde{\mathbf{W}_1}\mathbf{r}]_2 \cdot [\gamma_1\mathbf{V}\mathbf{r}]_2, \ldots, [\widetilde{\mathbf{W}_n}\mathbf{r}]_2 \cdot [\gamma_n\mathbf{V}\mathbf{r}]_2), [\widetilde{\mathbf{W}}\mathbf{r}]_2 \cdot [-\mathbf{V}\mathbf{r}]_2\}$$

because $\mathbf{k}, \widetilde{\mathbf{W}_1}, \ldots, \widetilde{\mathbf{W}_n}, \widetilde{\mathbf{W}}$ and $\gamma_1, \ldots, \gamma_n$ are all known to $\mathcal{B}$. (2) For the $i+1$st query y, we implicitly set

$$\mathbf{r} = \overline{\mathbf{M}}\mathbf{u} = \bar{\mathbf{t}} \in \mathbb{Z}_p^k.$$

The vector $\mathbf{r}$ is distributed properly and $[\mathbf{r}]_2 = [\bar{\mathbf{t}}]_2$ can be simulated. Then we may produce the secret key as follows

$$\left\{ \begin{array}{c} [\bar{\mathbf{t}}]_2, \ \mathsf{kE}(y, [\mathbf{k}]_2) \cdot \mathsf{rE}(y, [\widetilde{\mathbf{W}}_1\bar{\mathbf{t}}]_2 \cdot [\gamma_1\mathbf{a}^{\perp}\underline{\mathbf{t}}]_2, \ldots, [\widetilde{\mathbf{W}}_n\bar{\mathbf{t}}]_2 \cdot [\gamma_n\mathbf{a}^{\perp}\underline{\mathbf{t}}]_2), \\ [\widetilde{\mathbf{W}\mathbf{t}}]_2 \cdot [-\mathbf{a}^{\perp}\underline{\mathbf{t}}]_2 \end{array} \right\}.$$

(3) For the remaining $q - i - 1$ queries, we may work just as in the first case except that we employ $\mathbf{k} + \alpha\mathbf{a}^{\perp}$ in the place of $\mathbf{k}$.

Finalize. Output 1 when $\beta = \beta'$ and 0 otherwise.

Observe that, in the reply to the $i + 1$st key extraction query, we have

$$\mathbf{a}^{\perp}\underline{\mathbf{t}} = \mathbf{a}^{\perp}(\underline{\mathbf{M}}\mathbf{u} + v) = \mathbf{a}^{\perp}(\underline{\mathbf{M}}\overline{\mathbf{M}}^{-1})(\overline{\mathbf{M}}\mathbf{u}) + \mathbf{a}^{\perp}v = \mathbf{V}\bar{\mathbf{t}} + \mathbf{a}^{\perp}v.$$

Therefore we have

$$
\begin{aligned}
& \mathsf{kE}(y, [\mathbf{k}]_2) \cdot \mathsf{rE}(y, [\widetilde{\mathbf{W}}_1\bar{\mathbf{t}}]_2 \cdot [\gamma_1\mathbf{a}^{\perp}\underline{\mathbf{t}}]_2, \ldots, [\widetilde{\mathbf{W}}_n\bar{\mathbf{t}}]_2 \cdot [\gamma_n\mathbf{a}^{\perp}\underline{\mathbf{t}}]_2) \\
= & \mathsf{kE}(y, [\mathbf{k}]_2) \cdot \mathsf{rE}(y, [\mathbf{W}_1\mathbf{r}]_2 \cdot [\gamma_1\mathbf{a}^{\perp}v]_2, \ldots, [\mathbf{W}_n\mathbf{r}]_2 \cdot [\gamma_n\mathbf{a}^{\perp}v]_2) \\
= & \mathsf{kE}(y, [\mathbf{k}]_2) \cdot \mathsf{rE}(y, [\mathbf{W}_1\mathbf{r}]_2, \ldots, [\mathbf{W}_n\mathbf{r}]_2) \cdot \mathsf{rE}(y, [\gamma_1\mathbf{a}^{\perp}v]_2, \ldots, [\gamma_n\mathbf{a}^{\perp}v]_2) \\
= & \mathsf{kE}(y, [\mathbf{k}]_2) \cdot \mathsf{rE}(y, [\mathbf{W}_1\mathbf{r}]_2, \ldots, [\mathbf{W}_n\mathbf{r}]_2) \cdot \mathsf{rE}(y, [\mathbf{a}^{\perp}v]_2^{(\gamma_1, \ldots, \gamma_n)}) \\
= & \mathsf{kE}(y, [\mathbf{k}]_2) \cdot \mathsf{rE}(y, [\mathbf{W}_1\mathbf{r}]_2, \ldots, [\mathbf{W}_n\mathbf{r}]_2) \cdot [\mathbf{a}^{\perp}v]_2^{\mathsf{sE}(y, \gamma_1, \ldots, \gamma_n)} \\
& [\widetilde{\mathbf{W}\mathbf{t}}]_2 \cdot [-\mathbf{a}^{\perp}\underline{\mathbf{t}}]_2 \\
= & [\mathbf{W}\mathbf{r}]_2 \cdot [-\mathbf{a}^{\perp}v]_2 \\
= & [\mathbf{W}\mathbf{r}]_2 \cdot [\mathbf{a}^{\perp}v]_2^{-1}
\end{aligned}
$$

It is now clear that the simulation is identical to $\mathsf{G}_{2.i}$ when $v = 0$; and if v is a random element in $\mathbb{Z}_p$, the simulation is identical to $\mathsf{G}_{2.i.1}$ where $\hat{r} = v$. $\square$

Lemma 7 ($\mathsf{G}_{2.i.2} \approx \mathsf{G}_{2.i+1}$). *For any p.p.t. adversary $\mathcal{A}$ making at most q key extraction queries, there exists an algorithm $\mathcal{B}$ such that*

$$|\mathsf{Adv}_{\mathcal{A}}^{2.i.2}(\lambda) - \mathsf{Adv}_{\mathcal{A}}^{2.i+1}(\lambda)| \leq \mathsf{Adv}_{\mathcal{B}}^{\mathcal{D}_k}(\lambda)$$

and $\mathsf{Time}(\mathcal{B}) \approx \mathsf{Time}(\mathcal{A}) + q \cdot k^2 \cdot \mathsf{poly}(\lambda, n)$.

Proof. The proof is similar to that for Lemma 6. $\square$

We complete the proof by proving Lemma 8 which states that $\mathsf{G}_{2.i.1}$ and $\mathsf{G}_{2.i.2}$ are statistically indistinguishable. This is derived from the α-privacy of predicate encodings.

Lemma 8 ($\mathsf{G}_{2.i.1} = \mathsf{G}_{2.i.2}$). *For any adversary $\mathcal{A}$, we have*

$$\mathsf{Adv}_{\mathcal{A}}^{2.i.1}(\lambda) = \mathsf{Adv}_{\mathcal{A}}^{2.i.2}(\lambda).$$

Proof. We prove the lemma for any fixed

$- (\mathbf{A}, \mathbf{a}^{\perp}) \leftarrow \mathcal{D}_k, \mathbf{W}_1, \ldots, \mathbf{W}_n, \mathbf{W} \leftarrow \mathbb{Z}_p^{(k+1) \times k}, \mathbf{k} \leftarrow \mathbb{Z}_p^{k+1}, \beta \leftarrow \{0, 1\};$

- random coin $\mathbf{c} \leftarrow \mathbb{Z}_p^{k+1}$ for semi-functional challenge ciphertext ct^*;
- $\alpha \leftarrow \mathbb{Z}_p$, random coin $\mathbf{r} \leftarrow \mathbb{Z}_p^k$ for each key extraction query and the extra random coin $\hat{r} \leftarrow \mathbb{Z}_p$ for the $i + 1$st one.

Let $\mathsf{sk}_y^{(b)}$ be the reply to the $i + 1$st key extraction query y in $\mathsf{G}_{2.i.b}$ $(b = 1, 2)$ and ct^* is the challenge ciphertext for target attribute x^*. It is sufficient to show

$$\{(\mathsf{sk}_y^{(1)}, \mathsf{ct}^*)\} = \{(\mathsf{sk}_y^{(2)}, \mathsf{ct}^*)\}$$

where the probability space is defined by $\gamma_1, \ldots, \gamma_n \leftarrow \mathbb{Z}_p$. In fact, we can further reduce to the claim that

$$\{\mathsf{kE}(y, [\alpha \mathbf{a}^\perp]_2) \cdot [\mathbf{a}^\perp \hat{r}]_2^{\mathsf{rE}(y,\gamma_1,\ldots,\gamma_n)}, \quad \mathsf{sE}(x^*, \gamma_1, \ldots, \gamma_n)\} \text{ and}$$
$$\{[\mathbf{a}^\perp \hat{r}]_2^{\mathsf{rE}(y,\gamma_1,\ldots,\gamma_n)}, \quad \mathsf{sE}(x^*, \gamma_1, \ldots, \gamma_n)\}$$

are statistically close over the same probability space as before. One can rewrite the first distribution as

$$\{[\mathbf{a}^\perp \hat{r}]_2^{\mathsf{kE}(y,\alpha/\hat{r})+\mathsf{rE}(y,\gamma_1,\ldots,\gamma_n)}, \quad \mathsf{sE}(x^*, \gamma_1, \ldots, \gamma_n)\}.$$

which is statistically close to the second one by the α-privacy of predicate encoding. This readily proves the lemma. $\qquad \square$

4   Tag-Based ABE with Weak Attribute-Hiding Property

This section shows that our framework (presented in Sect. 3) is compatible with the *attribute-hiding* predicate encoding [CGW15]. This means that our framework can derive a series of attribute-hiding ABE with tag.

4.1   Preliminaries

Definition. We may call an ABE with (weak) attribute-hiding the *predicate encryption* (PE for short). A PE scheme is also defined by four p.p.t algorithms $\mathsf{Setup}, \mathsf{KeyGen}, \mathsf{Enc}, \mathsf{Dec}$ as in Sect. 2, but the security is defined in a slightly different way. For all adversary $\mathcal{A}$, define the advantage function $\mathsf{Adv}_{\mathcal{A}}^{\mathrm{PE}}(\lambda)$ as

$$\mathsf{Adv}_{\mathcal{A}}^{\mathrm{PE}}(\lambda) = \left| \Pr \left[\beta = \beta' \left| \begin{array}{c} (\mathsf{mpk}, \mathsf{msk}) \leftarrow \mathsf{Setup}(1^\lambda, \mathsf{P}), \; \beta \leftarrow \{0,1\} \\ (x_0^*, x_1^*, m_0^*, m_1^*) \leftarrow \mathcal{A}^{\mathsf{KeyGen}(\mathsf{mpk},\mathsf{msk},\cdot)}(\mathsf{mpk}) \\ \mathsf{ct}^* \leftarrow \mathsf{Enc}(\mathsf{mpk}, x_\beta^*, m_\beta^*) \\ \beta' \leftarrow \mathcal{A}^{\mathsf{KeyGen}(\mathsf{mpk},\mathsf{msk},\cdot)}(\mathsf{mpk}, \mathsf{ct}^*) \end{array} \right. \right] - \frac{1}{2} \right|.$$

A PE scheme is said to be *adaptively secure* and *weakly attribute-hiding* if $\mathsf{Adv}_{\mathcal{A}}^{\mathrm{PE}}(\lambda)$ is negligible in λ and $\mathsf{P}(x_0^*, y) = \mathsf{P}(x_1^*, y) = 0$ holds for each query y sent to oracle $\mathsf{KeyGen}(\mathsf{mpk}, \mathsf{msk}, \cdot)$ for all p.p.t. adversary $\mathcal{A}$.

Attribute-hiding Predicate Encoding. A $\mathbb{Z}_p$-linear predicate encoding $(\mathsf{sE}, \mathsf{rE}, \mathsf{kE}, \mathsf{sD}, \mathsf{rD})$ for $\mathsf{P} : \mathcal{X} \times \mathcal{Y} \to \{0, 1\}$ is *attribute-hiding* [CGW15] if it has the following two additional properties:

(*x*-oblivious α-reconstruction). $\mathsf{sD}(x, y, \cdot)$ and $\mathsf{rD}(x, y, \cdot)$ are independent of x.

(attribute-hiding). For all $(x, y) \in \mathcal{X} \times \mathcal{Y}$ such that $\mathsf{P}(x, y) = 0$, the following distributions are identical.

$$\{x, y, \mathsf{sE}(x, \mathbf{w}), \mathsf{rE}(y, \mathbf{w}) : \mathbf{w} \leftarrow \mathbb{Z}_p^n\} \quad \text{and} \quad \{x, y, \mathbf{r} : \mathbf{r} \leftarrow \mathbb{Z}_p^{|\mathsf{sE}|+|\mathsf{rE}|}\}.$$

4.2 Construction and Security Analysis

Assuming an attribute-hiding predicate encoding, we can construct a predicate encryption with tag as in Sect. 3.1. Technically we prove the following theorem stating that the generic tag-based PE scheme is adaptively secure and weakly attribute-hiding under standard assumption in the standard model.

Theorem 2 (Weak AH). *For any p.p.t. adversary $\mathcal{A}$ making at most q key extraction queries, there exists algorithms $\mathcal{B}_1, \mathcal{B}_2, \mathcal{B}_3$ such that*

$$\mathsf{Adv}_{\mathcal{A}}^{\mathrm{PE}}(\lambda) \leq \mathsf{Adv}_{\mathcal{B}_1}^{\mathcal{D}_k}(\lambda) + q \cdot \mathsf{Adv}_{\mathcal{B}_2}^{\mathcal{D}_k}(\lambda) + q \cdot \mathsf{Adv}_{\mathcal{B}_3}^{\mathcal{D}_k}(\lambda) + 2^{-\Omega(\lambda)}$$

and $\max\{\mathsf{Time}(\mathcal{B}_1), \mathsf{Time}(\mathcal{B}_2), \mathsf{Time}(\mathcal{B}_3)\} \approx \mathsf{Time}(\mathcal{A}) + q \cdot k^2 \cdot \mathrm{poly}(\lambda, n)$.

Proof Overview. We prove the theorem using almost the same game sequence as described in Sects. 3.2 and 3.3. We just describe the differences between them. Firstly, the challenge ciphertext will be generated for identity x_β^*. Secondly, we need to re-define *pseudo-semi-functional secret keys* and *semi-functional secret keys* as follows.

(pesudo-semi-functional secret key). A pseudo-semi-functional secret key for policy $y \in \mathcal{Y}$ is defined as follows:

$$\left\{ \begin{array}{l} [\mathbf{r}]_2, \\ \mathsf{kE}(y, [\mathbf{k} + \boxed{\alpha \mathbf{a}^\perp}]_2) \cdot \mathsf{rE}(y, [\mathbf{W}_1 \mathbf{r} + \boxed{\mathbf{a}^\perp \hat{u}_1}]_2, \ldots, [\mathbf{W}_n \mathbf{r} + \boxed{\mathbf{a}^\perp \hat{u}_n}]_2) \\ \hspace{6cm} \cdot [\mathbf{a}^\perp \hat{r}]_2^{\mathsf{rE}(y, \gamma_1, \ldots, \gamma_n)}, \\ [\mathbf{W} \mathbf{r}]_2 \cdot [\mathbf{a}^\perp \hat{r}]_2^{-1} \end{array} \right\}$$

where $\hat{r} \in \mathbb{Z}_p$, $\gamma_1, \ldots, \gamma_n \in \mathbb{Z}_p$, $\alpha \in \mathbb{Z}_p$ are defined as before and $\hat{u}_1, \ldots, \hat{u}_n \leftarrow \mathbb{Z}_p$ are fresh for each pseudo-semi-functional secret key.

(semi-functional secret key). A semi-functional secret key for policy $y \in \mathcal{Y}$ is defined as follows:

$$\{ [\mathbf{r}]_2, \mathsf{kE}(y, [\mathbf{k} + \boxed{\alpha \mathbf{a}^\perp}]_2) \cdot \mathsf{rE}(y, [\mathbf{W}_1 \mathbf{r} + \boxed{\mathbf{a}^\perp \hat{u}_1}]_2, \ldots, [\mathbf{W}_n \mathbf{r} + \boxed{\mathbf{a}^\perp \hat{u}_n}]_2), [\mathbf{W} \mathbf{r}]_2 \}$$

where $\alpha \in \mathbb{Z}_p$ are defined as before and $\hat{u}_1, \ldots, \hat{u}_n \leftarrow \mathbb{Z}_p$ are fresh for each semi-functional secret key.

Finally, we add an additional game G_4 shown below. Its preceding game G_3 (the final game in the previous game sequence) is restated so as to emphasize the difference between them.

- G_3 is identical to $\mathsf{G}_{2.q}$ except that the challenge ciphertext is a semi-functional ciphertext for attribute x^*_β and $\boxed{\text{random message } m^* \in \mathcal{M}}$.
- G_4 is identical to G_3 except that the challenge ciphertext is a semi-functional ciphertext for $\boxed{\text{random attribute } x^* \in \mathcal{X}}$ and random message $m^* \in \mathcal{M}$.

With the extended game sequence, we will prove that

$$\mathsf{G}_0 \approx \mathsf{G}_1 = \mathsf{G}_{2.0} \approx \mathsf{G}_{2.1} \approx \cdots \approx \mathsf{G}_{2.q} = \boxed{\mathsf{G}_3 = \mathsf{G}_4}.$$

where "$\mathsf{G}_{2.i} \approx \mathsf{G}_{2.i+1}$" for all $i \in [0, q]$ will be proved using the game sub-sequence

$$\mathsf{G}_{2.i} \approx \boxed{\mathsf{G}_{2.i.1} = \mathsf{G}_{2.i.2}} \approx \mathsf{G}_{2.i+1}.$$

Because of the similarity of game sequences, most lemmas we have presented in Sects. 3.2 and 3.3 still hold and can be proved in the same way. Due to the lack of space, we omit them. The proofs of "$\mathsf{G}_3 = \mathsf{G}_4$" and "$\mathsf{G}_{2.i.1} = \mathsf{G}_{2.i.2}$" (the boxed parts) mainly follow [CGW15] and our proof in previous section.

Finally, we point out the fact: The challenge ciphertext in G_3 still leaks information of $\beta \in \{0, 1\}$ via x^*_β, but such a dependence is removed in G_4. Therefore $\mathsf{Adv}^4_{\mathcal{A}}(\lambda) = 0$ for any $\mathcal{A}$.

5   New Tag-Based ABE

In this section we will exhibit two tag-based ABE for boolean span program derived from our generic tag-based ABE and two concrete encodings in [CGW15].

Boolean Span Program. Assume $n \in \mathbb{N}$. Let $[n]$ be the *attribute universe*. A span program over $[n]$ is defined by $(\mathbf{M}, \rho)$ where $\mathbf{M} \in \mathbb{Z}_p^{\ell \times \ell'}$ and $\rho : [\ell] \to [n]$. We use $\mathbf{M}_i$ to denote the ith row of $\mathbf{M}$. For an input $\mathbf{x} = (x_1, \ldots, x_n) \in \{0, 1\}^n$, we say

$$\mathbf{x} \text{ satisfies } (\mathbf{M}, \rho) \iff \mathbf{1} \in \mathsf{span}(\mathbf{M_x})$$

where $\mathbf{1} = (1, 0, \ldots, 0) \in \mathbb{Z}_p^{1 \times \ell'}$ and $\mathbf{M_x} = \{ \mathbf{M}_j : x_{\rho(j)} = 1 \}$. In this case one can efficiently find coefficients $\omega_1, \ldots, \omega_\ell \in \mathbb{Z}_p$ such that

$$\sum_{j \,:\, x_{\rho(j)}=1} \omega_j \mathbf{M}_j = \mathbf{1}.$$

Here we will assume $n = \ell$ and ρ is an identity map following [CGW15].

5.1   Key-Policy Construction

The corresponding predicate is defined as

$$P(\mathbf{x}, \mathbf{M}) = 1 \quad \Longleftrightarrow \quad \mathbf{x} \text{ satisfies } \mathbf{M}$$

where $\mathbf{x} \in \mathcal{X} = \{0,1\}^\ell$ and $\mathbf{M} \in \mathcal{Y} = \mathbb{Z}_p^{\ell \times \ell'}$. Our concrete KP-ABE scheme is described below:

– Setup$(1^\lambda, 1^\ell)$: Sample

$$\mathbf{A} \leftarrow \mathcal{D}_k, \quad \mathbf{W}_1, \ldots, \mathbf{W}_\ell, \mathbf{U}_2, \ldots, \mathbf{U}_{\ell'}, \mathbf{W} \leftarrow \mathbb{Z}_p^{(k+1) \times k}, \quad \mathbf{k} \leftarrow \mathbb{Z}_p^{k+1}$$

and output

$$\begin{aligned}
\mathsf{mpk} &:= \{[\mathbf{A}]_1, [\mathbf{W}_1^\top \mathbf{A}]_1, \ldots, [\mathbf{W}_\ell^\top \mathbf{A}]_1, [\mathbf{W}^\top \mathbf{A}]_1, [\mathbf{k}^\top \mathbf{A}]_T\} \\
\mathsf{msk} &:= \{\mathbf{W}_1, \ldots, \mathbf{W}_\ell, \mathbf{U}_2, \ldots, \mathbf{U}_{\ell'}, \mathbf{W}; \mathbf{k}\}
\end{aligned}$$

– Enc$(\mathsf{mpk}, \mathbf{x}, m)$: On input $\mathbf{x} \in \{0,1\}^\ell$ and $m \in G_T$, pick $\mathbf{s} \leftarrow \mathbb{Z}_p^k$ and $w_1, \ldots, w_\ell \leftarrow \mathbb{Z}_p$. Output

$$\mathsf{ct}_\mathbf{x} := \left\{ \begin{array}{l}
C_0 := [\mathbf{As}]_1, \\
C_1 := [x_1(\mathbf{W}_1 + w_1\mathbf{W})^\top \mathbf{As}]_1, \\
\quad \vdots \\
C_\ell := [x_\ell(\mathbf{W}_\ell + w_\ell\mathbf{W})^\top \mathbf{As}]_1, \\
C := [\mathbf{k}^\top \mathbf{As}]_T \cdot m, \\
\boldsymbol{\tau} := (\tau_1 := x_1 w_1, \ldots, \tau_\ell := x_\ell w_\ell)
\end{array} \right\} \in G_1^{k+1} \times (G_1^k)^\ell \times G_T \times \mathbb{Z}_p^\ell$$

– KeyGen$(\mathsf{mpk}, \mathsf{msk}, \mathbf{M})$: On input $\mathbf{M} \in \mathbb{Z}_p^{\ell \times \ell'}$, pick $\mathbf{r} \leftarrow_\mathrm{R} \mathbb{Z}_p^k$ and output

$$\mathsf{sk}_\mathbf{M} := \left\{ \begin{array}{l}
K_0 := [\mathbf{r}]_2, \\
K_1 := [(\mathbf{k}||\mathbf{U}_2\mathbf{r}||\cdots||\mathbf{U}_{\ell'}\mathbf{r})\mathbf{M}_1^\top + \mathbf{W}_1\mathbf{r}]_2, \\
\quad \vdots \\
K_\ell := [(\mathbf{k}||\mathbf{U}_2\mathbf{r}||\cdots||\mathbf{U}_{\ell'}\mathbf{r})\mathbf{M}_\ell^\top + \mathbf{W}_\ell\mathbf{r}]_2, \\
K_t := [\mathbf{Wr}]_2
\end{array} \right\} \in G_2^k \times (G_2^{k+1})^\ell \times G_2^{k+1}$$

– Dec$(\mathsf{mpk}, \mathsf{sk}_\mathbf{M}, \mathsf{ct}_\mathbf{x})$: Find out $\omega_1, \ldots, \omega_\ell \in \mathbb{Z}_p$ such that $\sum_{j:x_j=1} \omega_j \mathbf{M}_j = \mathbf{1}$ and compute

$$K \leftarrow e(C_0, \prod_{j:x_j=1} (K_j \cdot K_t^{\tau_j})^{\omega_j}) \cdot e(\prod_{j:x_j=1} C_j^{-\omega_j}, K_0)$$

Recover the message as $m \leftarrow C/K \in G_T$.

5.2   Ciphertext-Policy Construction

The corresponding predicate is defined as

$$P(\mathbf{M}, \mathbf{x}) = 1 \quad \Longleftrightarrow \quad \mathbf{x} \text{ satisfies } \mathbf{M}$$

where $\mathbf{M} \in \mathcal{X} = \mathbb{Z}_p^{\ell \times \ell'}$ and $\mathbf{x} \in \mathcal{Y} = \{0,1\}^\ell$. Our concrete CP-ABE scheme is described below:

– Setup($1^\lambda, 1^\ell$): Sample

$$\mathbf{A} \leftarrow \mathcal{D}_k, \quad \mathbf{W}_1, \ldots, \mathbf{W}_\ell, \mathbf{V}, \mathbf{W} \leftarrow \mathbb{Z}_p^{(k+1) \times k}, \quad \mathbf{k} \leftarrow \mathbb{Z}_p^{k+1}$$

and output

$$\begin{aligned} \mathsf{mpk} &:= \{[\mathbf{A}]_1, [\mathbf{W}_1^\top \mathbf{A}]_1, \ldots, [\mathbf{W}_\ell^\top \mathbf{A}]_1, [\mathbf{V}^\top \mathbf{A}]_1, [\mathbf{W}^\top \mathbf{A}]_1, [\mathbf{k}^\top \mathbf{A}]_T\} \\ \mathsf{msk} &:= \{\mathbf{W}_1, \ldots, \mathbf{W}_\ell, \mathbf{V}, \mathbf{W}; \mathbf{k}\} \end{aligned}$$

– Enc($\mathsf{mpk}, \mathbf{M}, m$): On input $\mathbf{M} \in \mathbb{Z}_p^{\ell \times \ell'}$ and $m \in G_T$, pick $\mathbf{s} \leftarrow \mathbb{Z}_p^k$ and

$$w_1, \ldots, w_\ell \leftarrow \mathbb{Z}_p, \ v \leftarrow \mathbb{Z}_p, \ \mathbf{u} \leftarrow \mathbb{Z}_p^{\ell'-1}, \ \mathbf{U}_2, \ldots, \mathbf{U}_{\ell'} \leftarrow \mathbb{Z}_p^{(k+1) \times k}.$$

Output

$$\mathsf{ct_M} :=$$

$$\left\{ \begin{aligned} C_0 &:= [\mathbf{As}]_1, \\ C_1 &:= [(\mathbf{W}_1 + w_1 \mathbf{W})^\top \mathbf{As} + ((\mathbf{V} + v\mathbf{W})^\top \mathbf{As} || \mathbf{U}_2^\top \mathbf{As} || \cdots || \mathbf{U}_{\ell'}^\top \mathbf{As}) \mathbf{M}_1^\top]_1, \\ &\vdots \\ C_\ell &:= [(\mathbf{W}_\ell + w_\ell \mathbf{W})^\top \mathbf{As} + ((\mathbf{V} + v\mathbf{W})^\top \mathbf{As} || \mathbf{U}_2^\top \mathbf{As} || \cdots || \mathbf{U}_{\ell'}^\top \mathbf{As}) \mathbf{M}_\ell^\top]_1, \\ C &:= [\mathbf{k}^\top \mathbf{As}]_T \cdot m, \\ \boldsymbol{\tau} &:= (\tau_1 := w_1 + \mathbf{M}_1 \left(\begin{smallmatrix} v \\ \mathbf{u} \end{smallmatrix}\right), \ \ldots, \ \tau_\ell := w_\ell + \mathbf{M}_\ell \left(\begin{smallmatrix} v \\ \mathbf{u} \end{smallmatrix}\right)) \end{aligned} \right\}$$

$$\in G_1^{k+1} \times (G_1^k)^\ell \times G_T \times \mathbb{Z}_p^\ell$$

– KeyGen($\mathsf{mpk}, \mathsf{msk}, \mathbf{x}$): On input $\mathbf{x} \in \{0,1\}^\ell$, pick $\mathbf{r} \leftarrow_{\mathrm{R}} \mathbb{Z}_p^k$ and output

$$\mathsf{sk_x} := \left\{ \begin{aligned} K_0 &:= [\mathbf{r}]_2, \\ K_1 &:= [x_1 \mathbf{W}_1 \mathbf{r}]_2 \\ &\vdots \\ K_\ell &:= [x_\ell \mathbf{W}_\ell \mathbf{r}]_2 \\ K_{\ell+1} &:= [\mathbf{k} + \mathbf{Vr}]_2 \\ K_t &:= [\mathbf{Wr}]_2 \end{aligned} \right\} \in G_2^k \times (G_2^{k+1})^{\ell+1} \times G_2^{k+1}$$

– Dec($\mathsf{mpk}, \mathsf{sk_x}, \mathsf{ct_M}$): Find out $\omega_1, \ldots, \omega_\ell \in \mathbb{Z}_p$ such that $\sum_{j:x_j=1} \omega_j \mathbf{M}_j = \mathbf{1}$ and compute

$$K \leftarrow e(C_0, K_{\ell+1} \cdot \prod_{j:x_j=1} K_j^{\omega_j} \cdot K_t^{\sum_{j:x_j=1} \omega_j \tau_j}) \cdot e(\prod_{j:x_j=1} C_j^{-\omega_j}, K_0)$$

Recover the message as $m \leftarrow C/K \in G_T$.

Using concrete encodings in [CGW15], we can derive more tag-based ABE instantiations. With our framework, we can also reproduce several previous concrete tag-based ABE schemes (such as those in [Ram16, RS16]) with simple proofs under the k-Lin assumption. In fact we just need to extract respective concrete encodings from them and apply our framework.

6   How to Support Predicate Encoding with Delegation

Our framework shown in Sect. 3 and extended in Sect. 4 cannot cover the HIBE schemes proposed by Ramanna and Sarkar [RS14]. This section further develops the framework and predicate encoding to accommodate the delegation mechanism in HIBE system.

We first recall a notion from [CGW15]. A predicate $\mathsf{P} : \mathcal{X} \times \mathcal{Y} \to \{0, 1\}$ is *delegatable* if there exists a partial ordering $\leq$ on $\mathcal{Y}$ such that

$$(y \leq y') \wedge \mathsf{P}(x, y) = 1 \implies \mathsf{P}(x, y') = 1 \quad \text{for all } x \in \mathcal{X}.$$

One of classical delegatable predicates is the predicate for HIBE: Let $\ell \in \mathbb{N}$ and $\mathcal{X} = \mathcal{Y} = \mathbb{Z}_p^{\leq \ell}$. The predicate is

$$\mathsf{P}(\mathbf{x}, \mathbf{y}) = 1 \quad \Longleftrightarrow \quad \mathbf{y} \text{ is a prefix of } \mathbf{x}.$$

The partial ordering is the prefix relation, that is $\mathbf{y} \leq \mathbf{y}'$ iff $\mathbf{y}'$ is a prefix of $\mathbf{y}$.

6.1   Syntax and Definition

An ABE scheme for a delegatable predicate P consists of algorithms Setup, KeyGen, Enc, Dec as defined in Sect. 2.1 and a delegation algorithm Del.

- Del$(\mathsf{mpk}, \mathsf{sk}_{y'}, y) \to \mathsf{sk}_y$. The *delegation* algorithm takes as input the master public key mpk, a secret key $\mathsf{sk}_{y'}$ for $y' \in \mathcal{Y}$ and index (policy) $y \in \mathcal{Y}$ with $y \leq y'$, and generates a secret key sk_y for y.

We require that the delegation algorithm is *path-oblivious* which means secret keys generated by KeyGen and Del have the same distribution, that is

$$\{(\mathsf{sk}_{y'}, \mathsf{sk}_y) : \mathsf{sk}_{y'} \leftarrow \mathsf{KeyGen}(\mathsf{mpk}, \mathsf{msk}, y'), \mathsf{sk}_y \leftarrow \mathsf{KeyGen}(\mathsf{mpk}, \mathsf{msk}, y)\}$$
$$= \{(\mathsf{sk}_{y'}, \mathsf{sk}_y) : \mathsf{sk}_{y'} \leftarrow \mathsf{KeyGen}(\mathsf{mpk}, \mathsf{msk}, y'), \mathsf{sk}_y \leftarrow \mathsf{Del}(\mathsf{mpk}, \mathsf{sk}_{y'}, y)\}$$

for all $y, y' \in \mathcal{Y}$ satisfying $y \leq y'$. The assumption is natural and allows us to continue working with the security model described in Sect. 2.1; otherwise one should turn to the model described in [SW08] where an adversary can decide how to create secret keys—using KeyGen or Del.

6.2 Predicate Encoding Supporting Delegation

A predicate encoding for delegatable predicate P is composed of five algorithms sE, sD, rE, kE, rD satisfying all requirements described in Sect. 2.3 and an extra algorithm

$$\mathsf{dE} : \mathcal{Y} \times \mathcal{Y} \times \mathbb{Z}_p^{n_r} \to \mathbb{Z}_p^{n_r'}$$

with the following features: (1) for all $\mathbf{w} \leftarrow \mathbb{Z}_p^n$, $\alpha \leftarrow \mathbb{Z}_p$ and $y, y' \in \mathcal{Y}$ with $y \le y'$, it holds that

$$\mathsf{dE}(y, y', \mathsf{kE}(y', \alpha) + \mathsf{rE}(y', \mathbf{w})) = \mathsf{kE}(y, \alpha) + \mathsf{rE}(y, \mathbf{w})$$

and (2) $\mathsf{dE}(y, y', \cdot)$ is $\mathbb{Z}_p$-linear. A predicate encoding for HIBE from Boneh-Boyen-Goh's HIBE with constant-size ciphertext [BBG05] is as follows:

(encoding for HIBE [BBG05]**).** Let $\mathbf{w} \leftarrow \mathbb{Z}_p^{1 \times (\ell+1)}$, $\mathbf{x} = (x_1, \ldots, x_{\ell_x})$ and $\mathbf{y} = (y_1, \ldots, y_{\ell_y})$ for $\ell_x, \ell_y \le \ell$. Define

$$\mathsf{sE}(\mathbf{x}, \mathbf{w}) := \mathbf{w}\,(1, \mathbf{x}, \mathbf{0})^\top \qquad \mathsf{sD}(\mathbf{x}, \mathbf{y}, c) := c$$

$$\mathsf{rE}(\mathbf{y}, \mathbf{w}) := \mathbf{w} \begin{pmatrix} 1 & \mathbf{y} \\ & \mathbf{I} \end{pmatrix}^\top \qquad \mathsf{rD}(\mathbf{x}, \mathbf{y}, \mathbf{k}) := \mathbf{k}\,(1, x_{\ell_y+1}, \ldots, x_{\ell_x}, \mathbf{0})^\top$$

$$\mathsf{kE}(\mathbf{y}, \alpha) := (\alpha, \mathbf{0})$$

As shown in [KSGA16], the encoding is linear and satisfies α-restriction and α-privacy. Besides that, for $\mathbf{y}' = (y_1, \ldots, y_{\ell_y'})$ and $\mathbf{y} = (y_1, \ldots, y_{\ell_y})$ with $\ell_y' \le \ell_y$, we also define

$$\mathsf{dE}(\mathbf{y}, \mathbf{y}', \mathbf{k}') = \mathbf{k}' \begin{pmatrix} 1 & y_{\ell_y'+1} & \cdots & y_{\ell_y} \\ & & \mathbf{I} \end{pmatrix}^\top .$$

It's straightforward to show that dE meets two requirements.

6.3 Generic Construction and Security Analysis

A direct way to support delegation in our framework in Sect. 3.1 is to apply dE to $\mathbf{K}_1$. However this delegation algorithm is not path-oblivious. Following [RS14], we publish $\mathbf{W}_1, \ldots, \mathbf{W}_n, \mathbf{W}$ in the master public key mpk in a proper form which makes it possible to publicly re-randomize any secret key.

Construction. Our tag-based ABE supporting delegation is as follows. We highlight all terms we add for delegation in the dashboxes.

– Setup(1^λ, P): Let n be the parameter size of predicate encoding supporting delegation (sE, rE, kE, rE, rD, $\lceil\mathsf{dE}\rceil$) for P. Sample

$$\mathbf{A} \leftarrow \mathcal{D}_k, \quad \boxed{\mathbf{Z} \leftarrow \mathbb{Z}_p^{k \times k},} \quad \mathbf{W}_1, \ldots, \mathbf{W}_n, \mathbf{W} \leftarrow \mathbb{Z}_p^{(k+1) \times k}, \quad \mathbf{k} \leftarrow \mathbb{Z}_p^{k+1}$$

and output the master public and secret key pair

$$\mathsf{mpk} := \left\{ \begin{array}{l} [\mathbf{A}]_1, [\mathbf{W}_1^\top\mathbf{A}]_1, \ldots, [\mathbf{W}_n^\top\mathbf{A}]_1, [\mathbf{W}^\top\mathbf{A}]_1 \\ \overline{[\mathbf{Z}]_2, [\mathbf{W}_1\mathbf{Z}]_2, \ldots, [\mathbf{W}_n\mathbf{Z}]_2, [\mathbf{W}\mathbf{Z}]_2} \end{array}, [\mathbf{k}^\top\mathbf{A}]_T \right\}$$

$$\mathsf{msk} := \{\mathbf{W}_1, \ldots, \mathbf{W}_n, \mathbf{W}; \mathbf{k}\}.$$

- $\mathsf{Del}(\mathsf{mpk}, \mathsf{sk}_{y'}, y)$: Let $\mathsf{sk}_{y'} = \{K_0', \mathbf{K}_1', K_2'\}$. Sample $\widetilde{\mathbf{r}} \leftarrow \mathbb{Z}_p^k$ and compute a re-randomizer

$$\{\widetilde{K}_0 := [\mathbf{Z}\widetilde{\mathbf{r}}], \ \widetilde{\mathbf{K}}_1 := \mathsf{rE}(y, [\mathbf{W}_1\mathbf{Z}\widetilde{\mathbf{r}}]_2, \ldots, [\mathbf{W}_n\mathbf{Z}\widetilde{\mathbf{r}}]_2), \ \widetilde{K}_2 := [\mathbf{W}\mathbf{Z}\widetilde{\mathbf{r}}]_2\}$$

and output

$$\mathsf{sk}_y := \{K_0 := K_0' \cdot \widetilde{K}_0, \ \mathbf{K}_1 := \mathsf{dE}(y, y', \mathbf{K}_1') \cdot \widetilde{\mathbf{K}}_1, \ K_2 := K_2' \cdot \widetilde{K}_2\}$$

The remaining algorithms KeyGen, Enc and Dec are defined as in Sect. 3.1. The algorithm Del is path-oblivious: if we let $\mathbf{r}'$ be the random coin for $\mathsf{sk}_{y'}$, the random coin in sk_y will be $\mathbf{r} = \mathbf{r}' + \mathbf{Z}\widetilde{\mathbf{r}}$ which is independent of $\mathbf{r}'$ thanks to $\widetilde{\mathbf{r}}$.

Security. Observe that the only difference in the security game here is that the mpk sent to the adversary also includes

$$[\mathbf{Z}]_2, [\mathbf{W}_1\mathbf{Z}]_2, \ldots, [\mathbf{W}_n\mathbf{Z}]_2, [\mathbf{W}\mathbf{Z}]_2,$$

since Del will not be involved. Therefore we can prove the adaptive security of our ABE scheme as in Sects. 3.2 and 3.3. In fact what we need to show here is how to simulate these extra entries in mpk in our previous proofs.

To prove Lemmas 4, 5 and 8, the simulator knows $\mathbf{W}_1, \ldots, \mathbf{W}_n$ and $\mathbf{W}$. It can sample matrix $\mathbf{Z} \leftarrow \mathbb{Z}_p^{k \times k}$ and simulate the extra entries directly. For Lemmas 6 and 7, we recall that the simulator received $(\mathcal{G}, [\mathbf{M}]_2, [\mathbf{t}]_2)$ and implicitly define

$$\mathbf{W}_1 = \widetilde{\mathbf{W}}_1 + \gamma_1\mathbf{V}, \ \ldots, \ \mathbf{W}_n = \widetilde{\mathbf{W}}_n + \gamma_n\mathbf{V}, \ \ \mathbf{W} = \widetilde{\mathbf{W}} - \mathbf{V}$$

where $\widetilde{\mathbf{W}}_1, \ldots, \widetilde{\mathbf{W}}_n, \widetilde{\mathbf{W}} \leftarrow \mathbb{Z}_p^{(k+1) \times k}$, $\gamma_1, \ldots, \gamma_n \leftarrow \mathbb{Z}_p$ and $\mathbf{V} = \mathbf{a}^\perp \cdot (\underline{\mathbf{M}}\overline{\mathbf{M}}^{-1})$. As we have mentioned, the simulator can not calculate $\mathbf{V}$ and does not know $\mathbf{W}_1, \ldots, \mathbf{W}_n, \mathbf{W}$. However it can still simulate the extra entries as follows: Sample $\widetilde{\mathbf{Z}} \leftarrow \mathbb{Z}_p^{k \times k}$ and define

$$\mathbf{Z} = \overline{\mathbf{M}}\widetilde{\mathbf{Z}}.$$

Since $\overline{\mathbf{M}}$ is full-rank with high probability, matrix $\mathbf{Z}$ is distributed correctly and we have

$$\mathbf{W}_i\mathbf{Z} = \widetilde{\mathbf{W}}_i\overline{\mathbf{M}}\widetilde{\mathbf{Z}} + \gamma_i\mathbf{a}^\perp\underline{\mathbf{M}}\widetilde{\mathbf{Z}} \text{ for all } i \in [n] \quad \text{and} \quad \mathbf{W}\mathbf{Z} = \widetilde{\mathbf{W}}\overline{\mathbf{M}}\widetilde{\mathbf{Z}} - \mathbf{a}^\perp\underline{\mathbf{M}}\widetilde{\mathbf{Z}}.$$

That means we can simulate all extra entries from $[\mathbf{M}]_2$. This is sufficient to finish our proof.

Acknowledgement. We want to thank Benoît Libert, Somindu C. Ramanna and Hoeteck Wee for their helpful discussion, and thank Jiangtao Li for his proof-reading at early stage. We also thank all anonymous reviewers of ASIACRYPT 2017 for their valuable comments.

References

[ABS17] Ambrona, M., Barthe, G., Schmidt, B.: Generic transformations of predicate encodings: constructions and applications. In: Katz, J., Shacham, H. (eds.) CRYPTO 2017. LNCS, vol. 10401, pp. 36–66. Springer, Cham (2017). https://doi.org/10.1007/978-3-319-63688-7_2

[AC16] Agrawal, S., Chase, M.: A study of pair encodings: predicate encryption in prime order groups. In: Kushilevitz, E., Malkin, T. (eds.) TCC 2016. LNCS, vol. 9563, pp. 259–288. Springer, Heidelberg (2016). https://doi.org/10.1007/978-3-662-49099-0_10

[AC17] Agrawal, S., Chase, M.: Simplifying design and analysis of complex predicate encryption schemes. In: Coron, J.-S., Nielsen, J.B. (eds.) EUROCRYPT 2017. LNCS, vol. 10210, pp. 627–656. Springer, Cham (2017). https://doi.org/10.1007/978-3-319-56620-7_22

[AHY15] Attrapadung, N., Hanaoka, G., Yamada, S.: A framework for identity-based encryption with almost tight security. In: Iwata, T., Cheon, J.H. (eds.) ASIACRYPT 2015. LNCS, vol. 9452, pp. 521–549. Springer, Heidelberg (2015). https://doi.org/10.1007/978-3-662-48797-6_22

[Att14] Attrapadung, N.: Dual system encryption via doubly selective security: framework, fully secure functional encryption for regular languages, and more. In: Nguyen, P.Q., Oswald, E. (eds.) EUROCRYPT 2014. LNCS, vol. 8441, pp. 557–577. Springer, Heidelberg (2014). https://doi.org/10.1007/978-3-642-55220-5_31

[Att16] Attrapadung, N.: Dual system encryption framework in prime-order groups via computational pair encodings. In: Cheon, J.H., Takagi, T. (eds.) ASIACRYPT 2016. LNCS, vol. 10032, pp. 591–623. Springer, Heidelberg (2016). https://doi.org/10.1007/978-3-662-53890-6_20

[AY15] Attrapadung, N., Yamada, S.: Duality in ABE: converting attribute based encryption for dual predicate and dual policy via computational encodings. In: Nyberg, K. (ed.) CT-RSA 2015. LNCS, vol. 9048, pp. 87–105. Springer, Cham (2015). https://doi.org/10.1007/978-3-319-16715-2_5

[BB04a] Boneh, D., Boyen, X.: Efficient selective-ID secure identity-based encryption without random oracles. In: Cachin, C., Camenisch, J.L. (eds.) EUROCRYPT 2004. LNCS, vol. 3027, pp. 223–238. Springer, Heidelberg (2004). https://doi.org/10.1007/978-3-540-24676-3_14

[BB04b] Boneh, D., Boyen, X.: Secure identity based encryption without random oracles. In: Franklin, M. (ed.) CRYPTO 2004. LNCS, vol. 3152, pp. 443–459. Springer, Heidelberg (2004). https://doi.org/10.1007/978-3-540-28628-8_27

[BBG05] Boneh, D., Boyen, X., Goh, E.-J.: Hierarchical identity based encryption with constant size ciphertext. In: Cramer, R. (ed.) EUROCRYPT 2005. LNCS, vol. 3494, pp. 440–456. Springer, Heidelberg (2005). https://doi.org/10.1007/11426639_26

[BF01] Boneh, D., Franklin, M.: Identity-based encryption from the weil pairing. In: Kilian, J. (ed.) CRYPTO 2001. LNCS, vol. 2139, pp. 213–229. Springer, Heidelberg (2001). https://doi.org/10.1007/3-540-44647-8_13

[BKP14] Blazy, O., Kiltz, E., Pan, J.: (Hierarchical) identity-based encryption from affine message authentication. In: Garay, J.A., Gennaro, R. (eds.) CRYPTO 2014. LNCS, vol. 8616, pp. 408–425. Springer, Heidelberg (2014). https://doi.org/10.1007/978-3-662-44371-2_23

[BSW07] Bethencourt, J., Sahai, A., Waters, B.: Ciphertext-policy attribute-based encryption. In: 2007 IEEE Symposium on Security and Privacy, pp. 321–334. IEEE Computer Society Press, May 2007

[BSW11] Boneh, D., Sahai, A., Waters, B.: Functional encryption: definitions and challenges. In: Ishai, Y. (ed.) TCC 2011. LNCS, vol. 6597, pp. 253–273. Springer, Heidelberg (2011). https://doi.org/10.1007/978-3-642-19571-6_16

[CGW15] Chen, J., Gay, R., Wee, H.: Improved dual system ABE in prime-order groups via predicate encodings. In: Oswald, E., Fischlin, M. (eds.) EUROCRYPT 2015. LNCS, vol. 9057, pp. 595–624. Springer, Heidelberg (2015). https://doi.org/10.1007/978-3-662-46803-6_20

[CGW17] Chen, J., Gong, J., Weng, J.: Tightly secure IBE under constant-size master public key. In: Fehr, S. (ed.) PKC 2017. LNCS, vol. 10174, pp. 207–231. Springer, Heidelberg (2017). https://doi.org/10.1007/978-3-662-54365-8_9

[CHK03] Canetti, R., Halevi, S., Katz, J.: A forward-secure public-key encryption scheme. In: Biham, E. (ed.) EUROCRYPT 2003. LNCS, vol. 2656, pp. 255–271. Springer, Heidelberg (2003). https://doi.org/10.1007/3-540-39200-9_16

[CW14] Chen, J., Wee, H.: Semi-adaptive attribute-based encryption and improved delegation for boolean formula. In: Abdalla, M., Prisco, R. (eds.) SCN 2014. LNCS, vol. 8642, pp. 277–297. Springer, Cham (2014). https://doi.org/10.1007/978-3-319-10879-7_16

[EHK+13] Escala, A., Herold, G., Kiltz, E., Ràfols, C., Villar, J.: An algebraic framework for Diffie-Hellman assumptions. In: Canetti, R., Garay, J.A. (eds.) CRYPTO 2013. LNCS, vol. 8043, pp. 129–147. Springer, Heidelberg (2013). https://doi.org/10.1007/978-3-642-40084-1_8

[GDCC16] Gong, J., Dong, X., Chen, J., Cao, Z.: Efficient IBE with tight reduction to standard assumption in the multi-challenge setting. In: Cheon, J.H., Takagi, T. (eds.) ASIACRYPT 2016. LNCS, vol. 10032, pp. 624–654. Springer, Heidelberg (2016). https://doi.org/10.1007/978-3-662-53890-6_21

[Gen06] Gentry, C.: Practical identity-based encryption without random oracles. In: Vaudenay, S. (ed.) EUROCRYPT 2006. LNCS, vol. 4004, pp. 445–464. Springer, Heidelberg (2006). https://doi.org/10.1007/11761679_27

[GHKW16] Gay, R., Hofheinz, D., Kiltz, E., Wee, H.: Tightly CCA-secure encryption without pairings. In: Fischlin, M., Coron, J.-S. (eds.) EUROCRYPT 2016. LNCS, vol. 9665, pp. 1–27. Springer, Heidelberg (2016). https://doi.org/10.1007/978-3-662-49890-3_1

[GKW16] Goyal, R., Koppula, V., Waters, B.: Semi-adaptive security and bundling functionalities made generic and easy. In: Hirt, M., Smith, A. (eds.) TCC 2016. LNCS, vol. 9986, pp. 361–388. Springer, Heidelberg (2016). https://doi.org/10.1007/978-3-662-53644-5_14

[GPSW06] Goyal, V., Pandey, O., Sahai, A., Waters, B.: Attribute-based encryption for fine-grained access control of encrypted data. In: ACM CCS 2006, pp. 89–98. ACM Press, October/November 2006. Available as Cryptology ePrint Archive Report 2006/309

[JR13] Jutla, C.S., Roy, A.: Shorter quasi-adaptive NIZK proofs for linear subspaces. In: Sako, K., Sarkar, P. (eds.) ASIACRYPT 2013. LNCS, vol. 8269, pp. 1–20. Springer, Heidelberg (2013). https://doi.org/10.1007/978-3-642-42033-7_1

[KSG+17] Kim, J., Susilo, W., Guo, F., Au, M.H., Nepal, S.: An efficient KP-ABE with short ciphertexts in prime ordergroups under standard assumption. In: Proceedings of the 2017 ACM on Asia Conference on Computer and Communications Security, AsiaCCS 2017, pp. 823–834 (2017)

[KSGA16] Kim, J., Susilo, W., Guo, F., Au, M.H.: A tag based encoding: an efficient encoding for predicate encryption in prime order groups. In: Zikas, V., Prisco, R. (eds.) SCN 2016. LNCS, vol. 9841, pp. 3–22. Springer, Cham (2016). https://doi.org/10.1007/978-3-319-44618-9_1

[KSW08] Katz, J., Sahai, A., Waters, B.: Predicate encryption supporting disjunctions, polynomial equations, and inner products. In: Smart, N. (ed.) EUROCRYPT 2008. LNCS, vol. 4965, pp. 146–162. Springer, Heidelberg (2008). https://doi.org/10.1007/978-3-540-78967-3_9

[LOS+10] Lewko, A., Okamoto, T., Sahai, A., Takashima, K., Waters, B.: Fully secure functional encryption: attribute-based encryption and (hierarchical) inner product encryption. In: Gilbert, H. (ed.) EUROCRYPT 2010. LNCS, vol. 6110, pp. 62–91. Springer, Heidelberg (2010). https://doi.org/10.1007/978-3-642-13190-5_4

[LW10] Lewko, A., Waters, B.: New techniques for dual system encryption and fully secure HIBE with short ciphertexts. In: Micciancio, D. (ed.) TCC 2010. LNCS, vol. 5978, pp. 455–479. Springer, Heidelberg (2010). https://doi.org/10.1007/978-3-642-11799-2_27

[LW12] Lewko, A., Waters, B.: New proof methods for attribute-based encryption: achieving full security through selective techniques. In: Safavi-Naini, R., Canetti, R. (eds.) CRYPTO 2012. LNCS, vol. 7417, pp. 180–198. Springer, Heidelberg (2012). https://doi.org/10.1007/978-3-642-32009-5_12

[OSW07] Ostrovsky, R., Sahai, A., Waters, B.: Attribute-based encryption with non-monotonic access structures. In: ACM CCS 2007, pp. 195–203. ACM Press, October 2007

[OT10] Okamoto, T., Takashima, K.: Fully secure functional encryption with general relations from the decisional linear assumption. In: Rabin, T. (ed.) CRYPTO 2010. LNCS, vol. 6223, pp. 191–208. Springer, Heidelberg (2010). https://doi.org/10.1007/978-3-642-14623-7_11

[OT12] Okamoto, T., Takashima, K.: Adaptively attribute-hiding (hierarchical) inner product encryption. In: Pointcheval, D., Johansson, T. (eds.) EUROCRYPT 2012. LNCS, vol. 7237, pp. 591–608. Springer, Heidelberg (2012). https://doi.org/10.1007/978-3-642-29011-4_35

[Ram16] Ramanna, S.C.: More efficient constructions for inner-product encryption. In: Manulis, M., Sadeghi, A.-R., Schneider, S. (eds.) ACNS 2016. LNCS, vol. 9696, pp. 231–248. Springer, Cham (2016). https://doi.org/10.1007/978-3-319-39555-5_13

[RCS12] Ramanna, S.C., Chatterjee, S., Sarkar, P.: Variants of waters' dual system primitives using asymmetric pairings. In: Fischlin, M., Buchmann, J., Manulis, M. (eds.) PKC 2012. LNCS, vol. 7293, pp. 298–315. Springer, Heidelberg (2012). https://doi.org/10.1007/978-3-642-30057-8_18

[RS14] Ramanna, S.C., Sarkar, P.: Efficient (anonymous) compact HIBE from standard assumptions. In: Chow, S.S.M., Liu, J.K., Hui, L.C.K., Yiu, S.M. (eds.) ProvSec 2014. LNCS, vol. 8782, pp. 243–258. Springer, Cham (2014). https://doi.org/10.1007/978-3-319-12475-9_17

[RS16] Ramanna, S.C., Sarkar, P.: Efficient adaptively secure IBBE from the SXDH assumption. IEEE Trans. Inf. Theory $62(10)$, 5709–5726 (2016)

[Sha84] Shamir, A.: Identity-based cryptosystems and signature schemes. In: Blakley, G.R., Chaum, D. (eds.) CRYPTO 1984. LNCS, vol. 196, pp. 47–53. Springer, Heidelberg (1985). https://doi.org/10.1007/3-540-39568-7_5

[SW05] Sahai, A., Waters, B.: Fuzzy identity-based encryption. In: Cramer, R. (ed.) EUROCRYPT 2005. LNCS, vol. 3494, pp. 457–473. Springer, Heidelberg (2005). https://doi.org/10.1007/11426639_27

[SW08] Shi, E., Waters, B.: Delegating capabilities in predicate encryption systems. In: Aceto, L., Damgård, I., Goldberg, L.A., Halldórsson, M.M., Ingólfsdóttir, A., Walukiewicz, I. (eds.) ICALP 2008. LNCS, vol. 5126, pp. 560–578. Springer, Heidelberg (2008). https://doi.org/10.1007/978-3-540-70583-3_46

[Wat05] Waters, B.: Efficient identity-based encryption without random oracles. In: Cramer, R. (ed.) EUROCRYPT 2005. LNCS, vol. 3494, pp. 114–127. Springer, Heidelberg (2005). https://doi.org/10.1007/11426639_7

[Wat09] Waters, B.: Dual system encryption: realizing fully secure IBE and HIBE under simple assumptions. In: Halevi, S. (ed.) CRYPTO 2009. LNCS, vol. 5677, pp. 619–636. Springer, Heidelberg (2009). https://doi.org/10.1007/978-3-642-03356-8_36

[Wat11] Waters, B.: Ciphertext-policy attribute-based encryption: an expressive, efficient, and provably secure realization. In: Catalano, D., Fazio, N., Gennaro, R., Nicolosi, A. (eds.) PKC 2011. LNCS, vol. 6571, pp. 53–70. Springer, Heidelberg (2011). https://doi.org/10.1007/978-3-642-19379-8_4

[Wee14] Wee, H.: Dual system encryption via predicate encodings. In: Lindell, Y. (ed.) TCC 2014. LNCS, vol. 8349, pp. 616–637. Springer, Heidelberg (2014). https://doi.org/10.1007/978-3-642-54242-8_26

[Wee16] Wee, H.: Déjà Q: encore! Un petit IBE. In: Kushilevitz, E., Malkin, T. (eds.) TCC 2016. LNCS, vol. 9563, pp. 237–258. Springer, Heidelberg (2016). https://doi.org/10.1007/978-3-662-49099-0_9

[WES17] Watanabe, Y., Emura, K., Seo, J.H.: New revocable IBE in prime-order groups: adaptively secure, decryption key exposure resistant, and with short public parameters. In: Handschuh, H. (ed.) CT-RSA 2017. LNCS, vol. 10159, pp. 432–449. Springer, Cham (2017). https://doi.org/10.1007/978-3-319-52153-4_25

[WS16] Watanabe, Y., Shikata, J.: Identity-based hierarchical key-insulated encryption without random oracles. In: Cheng, C.-M., Chung, K.-M., Persiano, G., Yang, B.-Y. (eds.) PKC 2016. LNCS, vol. 9614, pp. 255–279. Springer, Heidelberg (2016). https://doi.org/10.1007/978-3-662-49384-7_10

Towards a Classification of Non-interactive Computational Assumptions in Cyclic Groups

Essam Ghadafi[1(✉)] and Jens Groth[2]

[1] University of the West of England, Bristol, UK
essam.ghadafi@uwe.ac.uk
[2] University College London, London, UK
j.groth@ucl.ac.uk

Abstract. We study non-interactive computational intractability assumptions in prime-order cyclic groups. We focus on the broad class of computational assumptions which we call target assumptions where the adversary's goal is to compute concrete group elements.

Our analysis identifies two families of intractability assumptions, the q-Generalized Diffie-Hellman Exponent (q-GDHE) assumptions and the q-Simple Fractional (q-SFrac) assumptions (a natural generalization of the q-SDH assumption), that imply all other target assumptions. These two assumptions therefore serve as Uber assumptions that can underpin all the target assumptions where the adversary has to compute specific group elements. We also study the internal hierarchy among members of these two assumption families. We provide heuristic evidence that both families are necessary to cover the full class of target assumptions. We also prove that having (polynomially many times) access to an adversarial 1-GDHE oracle, which returns correct solutions with non-negligible probability, entails one to solve any instance of the Computational Diffie-Hellman (CDH) assumption. This proves equivalence between the CDH and 1-GDHE assumptions. The latter result is of independent interest. We generalize our results to the bilinear group setting. For the base groups, our results translate nicely and a similar structure of non-interactive computational assumptions emerges. We also identify Uber assumptions in the target group but this requires replacing the q-GDHE assumption with a more complicated assumption, which we call the bilinar gap assumption.

Our analysis can assist both cryptanalysts and cryptographers. For cryptanalysts, we propose the q-GDHE and the q-SDH assumptions are the most natural and important targets for cryptanalysis in prime-order groups. For cryptographers, we believe our classification can aid the choice of assumptions underpinning cryptographic schemes and be used as a guide to minimize the overall attack surface that different assumptions expose.

The research leading to these results has received funding from the European Research Council under the European Union's Seventh Framework Programme (FP/2007–2013)/ERC Grant Agreement n. 307937 and EPSRC grant EP/J009520/1.

E. Ghadafi—Part of the work was done while at University College London.

T. Takagi and T. Peyrin (Eds.): ASIACRYPT 2017, Part II, LNCS 10625, pp. 66–96, 2017.
https://doi.org/10.1007/978-3-319-70697-9_3

1   Introduction

Prime-order groups are widely used in cryptography because their clean mathematical structure enables the construction of many interesting schemes. However, cryptographers rely on an ever increasing number of intractability assumptions to prove their cryptographic schemes are secure. Especially after the rise of pairing-based cryptography, we have witnessed a proliferation of intractability assumptions. While some of those intractability assumptions, e.g. the discrete logarithm or the computational Diffie-Hellman assumptions, are well-studied, and considered by now "standard", the rest of the assumption wilderness has received less attention.

This is unfortunate both for cryptographers designing protocols and cryptanalysts studying the security of the underpinning assumptions. Cryptographers designing protocols are often faced with a trade-off between performance and security, and it would therefore be helpful for them to know how their chosen intractability assumptions compare to other assumptions. Moreover, when they are designing a suite of protocols, it would be useful to know whether the different assumptions they use increase the attack surface or whether the assumptions are related. Cryptanalysts facing the wilderness of assumptions are also faced with a problem: which assumptions should they focus their attention on? One option is to go for the most devastating attack and try to break the discrete logarithm assumption, but the disadvantage is that this is also the hardest assumption to attack and hence the one where the cryptanalyst is least likely to succeed. The other option is to try to attack an easier assumption but the question then is which assumption is the most promising target?

Our research vision is that a possible path out of the wilderness is to identify Uber assumptions that imply all the assumptions we use. An extreme Uber assumption would be that anything that cannot trivially be broken by generic group operations is secure, however, we already know that this is a too extreme position since there are schemes that are secure against generic attacks but insecure for any concrete instantiation of the groups [18]. Instead of trying to capture all of the generic group model, we therefore ask for a few concrete and plausible Uber assumptions that capture the most important part of the assumption landscape. Such a characterization of the assumption wilderness would help both cryptographers and crypanalysts. The cryptographic designer may choose assumptions that fall under the umbrella of a few of the Uber assumptions to minimize the attack surface on her schemes. The cryptanalyst can use the Uber assumptions as important yet potentially easy targets.

Related Work. The rapid development of cryptographic schemes has been accompanied by an increase in the number and complexity of intractability assumptions. Cryptographers have been in pursuit to study the relationship among existing assumptions either by means of providing templates which encompass assumptions in the same family, e.g. [13, 14, 27], or by studying direct implications or lack thereof among the different assumptions, e.g. [3, 7, 26, 31, 32].

A particular class of assumptions which has received little attention are fractional assumptions. Those include, for example, the q-SDH assumption [8] and many of its variants, e.g. the modified q-SDH assumption [12], and the hidden q-SDH (q-HSDH) assumption [12]. As posed by, e.g. [28], a subtle question that arises is how such class of assumptions, e.g. the q-SDH assumption, relate to other existing (discrete-logarithm related) computational and decisional intractability assumptions. For instance, while it is clear that the q-SDH assumption implies the computational Diffie-Hellman assumption, it is still unclear whether the q-SDH assumption is implied by the decisional Diffie-Hellman assumption. Another intriguing open question is if there is a hierarchy between fractional assumptions or the class of assumptions is inherently unstructured.

Sadeghi and Steiner [38] introduced a new parameter for discrete-logarithm related assumptions they termed *granularity* which deals with the choice of the underlying mathematical group and its respective generator. They argued such a parameter can influence the security of schemes based on such assumptions and showed that such a parameter influences the implications between assumptions.

Naor [35] classified assumptions based on the complexity of falsifying them. Informally speaking, an assumption is *falsifiable* if it is possible to efficiently decide whether an adversary against the assumption has successfully broken it. Very recently, Goldwasser and Kalai [24] provided another classification of intractability assumptions based on their complexity. They argued that classifications based merely on falsifiability of the assumptions might be too inclusive since they do not exclude assumptions which are too dependent on the underlying cryptographic construct they support.

Boneh et al. [9] defined a framework for proving that decisional and computational assumptions are secure in the generic group model [30,39] and formalized an Uber assumption saying that generic group security implies real security for these assumptions. Boyen [11] later highlighted extensions to the framework and informally suggested how some other families which were not encompassed by the original Uber assumption in [9] can be captured. In essence, the Uber assumption encompasses computational and decisional (discrete-logarithm related) assumptions with a fixed unique challenge. Unfortunately, the framework excludes some families of assumptions, in particular, those where the polynomial(s) used for the challenge are chosen adaptively by the adversary after seeing the problem instance. Examples of such assumptions include the q-SDH [8], the modified q-SDH [12], and the q-HSDH [12] assumptions. The statement of the assumption of the aforementioned yield exponentially many (mutually irreducible) valid solutions rather than a unique one. Another distinction, from the Uber assumption is that the exponent required for the solution involves a fraction of polynomials rather than a polynomial. Joux and Rojat [26] proved relationships between some instances of the Uber assumption [9]. In particular, they proved implications between some variants of the computational Diffie-Hellman assumption.

Cheon [17] observed that the computational complexity of the q-SDH assumption (and related assumptions) reduces by a factor of $O(\sqrt{q})$ from that of the discrete logarithm problem if some relation between the prime group order p and the parameter q, in particular, if either $q \mid p - 1$ or $q \mid p + 1$, holds.

Chase and Meiklejohn [16] showed that in composite-order groups some of the so-called q-type assumptions can be reduced to the standard subgroup hiding assumption. More recently, Chase et al. [15] extended their framework to cover more assumptions and get tighter reductions.

Barthe et al. [4] analyzed hardness of intractability assumptions in the generic group model by reducing them to solving problems related to polynomial algebra. They also provided an automated tool that verifies the hardness of a subclass of families of assumptions in the generic group model. More recently, Ambrona et al. [2] improved upon the results of [4] by allowing unlimited oracle queries.

Kiltz [27] introduced the poly-Diffie-Hellman assumption as a generalization of the computational Diffie-Hellman assumption. Bellare et al. [6] defined the general subgroup decision problem which is a generalization of many existing variants of the subgroup decision problem in composite-order groups. Escala et al. [19] proposed an algebraic framework as a generalization of Diffie-Hellman like decisional assumptions. Analogously to [19], Morillo et al. [34] extended the framework to computational assumptions.

Our Contribution. We focus on efficiently falsifiable computational assumptions in prime-order groups. More precisely, we define a target assumption as an assumption where the adversary has a specific target element that she is trying to compute. A well-known target assumption is the Computational Diffie-Hellman (CDH) assumption over a cyclic group $\mathbb{G}_p$ of prime order p, which states that given $(G, G^a, G^b) \in \mathbb{G}_p^3$, it is hard to compute the target $G^{ab} \in \mathbb{G}_p$. We define target assumptions quite broadly and also include assumptions where the adversary takes part in specifying the target to be computed. In the q-SDH assumption [8] for instance, the adversary is given $G, G^x, \ldots, G^{x^q}$ and has to output $(c, G^{\frac{1}{x+c}}) \in \mathbb{Z}_p \setminus \{-x\} \times \mathbb{G}_p$. Here c selected by the adversary is part of the specification of the target $G^{\frac{1}{x+c}}$. In other words, our work includes both assumptions in which the target element to be computed is either uniquely determined a priori by the instance, or a posteriori by the adversary. We note that the case of multiple target elements is also covered by our framework as long as all the target elements are uniquely determined. This is because a tuple of elements is hard to compute if any of its single elements is hard to compute.

Our main contribution is to identify two classes of assumptions that imply the security of all target assumptions. The first class of assumptions is the Generalized Diffie-Hellman Exponent (q-GDHE) assumption [9] that says given $(G, G^x, \ldots, G^{x^{q-1}}, G^{x^{q+1}}, \ldots, G^{x^{2q}}) \in \mathbb{G}_p^{2q}$, it is hard to compute $G^{x^q} \in \mathbb{G}_p$. The second class of assumptions, which is a straightforward generalization of the q-SDH assumption, we call the simple fractional (q-SFrac) assumption and it states that given $(G, G^x, \ldots, G^{x^q}) \in \mathbb{G}_p^{q+1}$, it is hard to output polynomials $r(X)$ and $s(X)$ together with the target $G^{\frac{r(x)}{s(x)}} \in \mathbb{G}_p$, where $0 \leq \deg(r(X)) < \deg(s(X)) \leq q$. We remark that the latter assumption is not totally new as variants thereof appeared in the literature. For instance, Fucshsbauer et al. [20] defined an identical variant over bilinear groups which they called the generalized q-co-SDH assumption. The assumption that the q-GDHE and q-SFrac assumptions both hold when q is polynomial in the security parameter can therefore be seen as an Uber assumption for the entire class of target assumptions.

Having identified the q-GDHE and q-SFrac assumptions as being central for the security of target assumptions in general, we investigate their internal structure. We first show that q-SFrac is unlikely to be able to serve as an Uber assumption on its own. More precisely, we show that for a generic group adversary the 2-GDHE assumption is not implied by q-SFrac assumptions. Second, we show that the q-GDHE assumptions appear to be strictly increasing as q grows, i.e., if the $(q+1)$-GDHE holds, then so does q-GDHE, but for a generic group adversary the $(q+1)$-GDHE may be false even though q-GDHE holds. We also analyze the particular case where $q = 1$ and prove that the 1-GDHE assumption is equivalent to the CDH assumption. We summarize the implications we prove in Fig. 1, where $\mathsf{A} \Rightarrow \mathsf{B}$ denotes that B is implied (in a black-box manner) by A, whereas $\mathsf{A} \not\Rightarrow \mathsf{B}$ denotes the absence of such an implication (cf. Sect. 2.1).

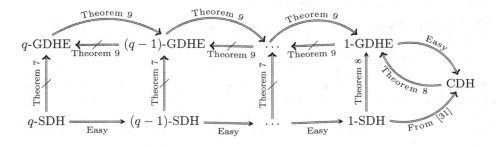

Fig. 1. Summary of Reductions

Based on these results we view the q-GDHE and q-SFrac assumptions as a bulwark. Whatever type of target assumptions a cryptographer bases her schemes on, they are secure as long as neither the q-GDHE nor the q-SFrac assumptions are broken.[1] Since the attacker has less leeway in the q-GDHE assumptions, the cryptographer may choose to rely exclusively on target assumpitons that are implied by the q-GDHE assumptions, and we therefore identify a large class of target assumptions that only need the q-GDHE assumptions to hold.

We also have advice for the cryptanalyst. We believe it is better to focus on canary in a coal mine assumptions than the discrete logarithm problem that has received the most attention so far. Based on our work, the easiest target assumptions to attack in single prime-order groups are the q-GDHE assumptions

---

[1] We note the caveat that our reductions are not tight. So it may be for concrete parameters a target assumption can be broken even if q-GDHE and q-SFrac hold for the same parameters. For all reductions, the concrete loss is stated explicitly in our proofs such that a cryptographer can work out a choice of parameters that yields security from the q-GDHE and q-SFrac assumptions.

and the q-SFrac assumptions. The class of q-SFrac assumptions allows more room for the adversary to maneuver in the choice of polynomials $r(X)$ and $s(X)$ and appears less structured than the q-GDHE assumptions. Pragmatically, we note that within the q-SFrac assumptions it is almost exclusively the q-SDH assumptions that are used. We therefore suggest the q-GDHE assumptions and the q-SDH assumptions to be the most suitable targets for cryptanalytic research.

Switching from single prime-order groups $\mathbb{G}_p$ to groups with a bilinear map $e : \mathbb{G}_1 \times \mathbb{G}_2 \to \mathbb{G}_T$, a similar structure emerges. For target assumptions in the base groups $\mathbb{G}_1$ and $\mathbb{G}_2$, we can again identify assumptions similar to the q-GDHE and q-SFrac assumptions that act as a joint Uber assumption. In the target group $\mathbb{G}_T$, a somewhat more complicated picture emerges under the influence of the pairing of source group elements. However, we can replace the q-GDHE assumption with an assumption we call the q-Bilinear Gap (q-BGap) assumption, and get that this assumption together with a natural generalization of the q-SFrac assumption to bilinear groups, which we name the q-BSFrac assumption, jointly act as an Uber assumption for all target assumptions in $\mathbb{G}_T$.

A natural question is whether our analysis extends to other assumptions as well, for instance "flexible" assumptions such as the q-HSDH assumption [12], where the adversary can choose secret exponents and therefore the target elements are no longer uniquely determined. Usually assumptions in the literature involve group elements that have discrete logarithms defined by polynomials in secret values in $\mathbb{Z}_p$ chosen by a challenger and/or public values in $\mathbb{Z}_p$. This gives rise to several classes of assumptions:

1. Non-interactive assumptions where the adversary's goal is to compute group elements defined by secret variables chosen by the challenger.
2. Non-interactive assumptions where the adversary's goal is to compute group elements defined by secret variables chosen by the challenger and public values chosen by the adversary.
3. Non-interactive assumptions where the adversary's goal is to compute group elements defined by secret variables chosen by the challenger, and public and secret values chosen by the adversary.
4. Interactive assumptions, where the challenger and adversary interact.

Target assumptions include all assumptions in classes 1 and 2. However, class 3 includes assumptions which are not falsifiable, e.g. knowledge-of-exponent assumptions [5]. Since q-GDHE is in class 1 and q-SFrac is in class 2, both of which only have falsifiable assumptions, we cannot expect them to capture non-falsifiable assumptions in class 3. We leave it as an interesting open problem which structure, if any, can be found in classes 3 and 4, and we hope our work will inspire research on this question.

We stress that our aim is to find concrete and precise reductions and/or prove separations among the different classes of assumptions which encompass a significant portion of existing assumptions. Thus, our approach is different from previous works such as [9,11,19,34] which aimed at defining algebraic frameworks or templates in generic groups to capture some families of assumptions.

The closest work to ours is Abdalla et al. [1], which provided an Uber assumption for decisional assumptions in cyclic groups (without a bilinear map, which invalidates many decisional assumptions). Other works, discussed above, have mostly dealt with specific relations among assumptions, e.g., the equivalence of CDH and square-CDH as opposed to the general approach we take.

Paper Organization. Our research contribution is organized into three parts. In Sect. 3, we define our framework for target assumptions in cyclic groups, and progressively seek reductions to simpler assumptions. In Sect. 4, we study the internal structure and the relationships among the families of assumptions we identify as Uber assumptions for our framework. In Sect. 5, we provide a generalization of our framework in the bilinear setting.

2   Preliminaries

Notation. We say a function f is *negligible* when $f(\kappa) = \kappa^{-\omega(1)}$ and it is *overwhelming* when $f(\kappa) = 1 - \kappa^{\omega(1)}$. We write $f(\kappa) \approx 0$, when $f(\kappa)$ is a negligible function. We will use κ to denote a security parameter, with the intuition that as κ grows we expect stronger security.

We write $y = A(x; r)$ when algorithm A on input x and randomness r, outputs y. We write $y \leftarrow A(x)$ for the process of picking randomness r at random and setting $y = A(x; r)$. We also write $y \leftarrow S$ for sampling y uniformly at random from the set S. We will assume it is possible to sample uniformly at random from sets such as $\mathbb{Z}_p$. We write PPT and DPT for probabilistic and deterministic polynomial time respectively.

Quadratic Residuosity. For an odd prime p and an integer $a \neq 0$, we say a is a *quadratic residue* modulo p if there exists a number x such that $x^2 \equiv a \pmod{p}$ and we say a is *quadratic non-residue* modulo p otherwise. We denote the set of quadratic residues modulo p by $\mathbb{QR}(p)$ and the set of quadratic non-residues modulo p by $\mathbb{QNR}(p)$. By Euler's theorem, we have $|\mathbb{QR}(p)| = |\mathbb{QNR}(p)| = \frac{p-1}{2}$.

2.1   Non-interactive Assumptions

In a non-interactive computational problem the adversary is given a problem instance and tries to find a solution. We say the adversary breaks the problem if it has non-negligible chance of finding a valid solution and we say the problem is hard if any PPT adversary has negligible chance of breaking it. We focus on non-interactive problems that are efficiently falsifiable, i.e., given the instance there is an efficient verification algorithm that decides whether the adversary won.

Definition 1 (Non-Interactive Computational Assumption). *A Non-interactive Computational Assumption consists of an instance generator $\mathcal{I}$ and a verifier $\mathcal{V}$ defined as follows:*

$(pub, priv) \leftarrow \mathcal{I}(1^\kappa)$: $\mathcal{I}$ *is a PPT algorithm that takes as input a security para-meter* 1^κ, *and outputs a pair of public/private information* $(pub, priv)$.

$b \leftarrow \mathcal{V}(pub, priv, sol)$: $\mathcal{V}$ *is a DPT algorithm that receives as input* $(pub, priv)$ *and a purported solution sol and returns* 1 *if it considers the answer correct and* 0 *otherwise.*

The assumption is that for all PPT adversaries $\mathcal{A}$, *the advantage* $\mathsf{Adv}_{\mathcal{A}}$ *is neg-ligible (in* κ), *where*

$$\mathsf{Adv}_{\mathcal{A}}(\kappa) := \Pr\left[(pub, priv) \leftarrow \mathcal{I}(1^\kappa); sol \leftarrow \mathcal{A}(pub) : \mathcal{V}(pub, priv, sol) = 1\right].$$

Relations Among Assumptions. For two non-interactive assumptions A and B, we will use the notation $\mathsf{A} \Rightarrow \mathsf{B}$ when assumption B is implied (in a black-box manner) by assumption A, i.e. given an efficient algorithm $\mathcal{B}$ for breaking assumption B, one can construct an efficient algorithm $\mathcal{A}$ that uses $\mathcal{B}$ as an oracle and breaks assumption A. The absence of implication will be denoted by $\mathsf{A} \not\Rightarrow \mathsf{B}$.

2.2 Non-interactive Assumptions over Cyclic Groups

We study non-interactive assumptions over prime-order cyclic groups. These assumptions are defined relative to a group generator $\mathcal{G}$.

Definition 2 (Group Generator). *A group generator is a PPT algorithm* $\mathcal{G}$, *which on input a security parameter* κ *(given in unary) outputs group parameters* $(\mathbb{G}_p, G)$, *where*

- $\mathbb{G}_p$ *is cyclic group of known prime order* p *with bitlength* $|p| = \Theta(\kappa)$.
- $\mathbb{G}_p$ *has a unique canonical representation of group elements, and polynomial time algorithms for carrying out group operations and deciding membership.*
- G *is a uniformly random generator of the group.*

In non-interactive assumptions over prime-order groups the instance genera-tor runs the group setup $(\mathbb{G}_p, G) \leftarrow \mathcal{G}(1^\kappa)$ and includes $\mathbb{G}_p$ in *pub*. Sadeghi and Steiner [38] distinguish between group setups with low, medium and high gran-ularity. In the low granularity setting the non-interactive assumption must hold with respect to random choices of $\mathbb{G}_p$ and G, in the medium granularity setting it must hold for all choices of $\mathbb{G}_p$ and a random G, and in the high granularity setting it must hold for all choices of $\mathbb{G}_p$ and G. Our definitions always assume G is chosen uniformly at random, so depending on $\mathcal{G}$ we are always working in the low or medium granularity setting.

We will use $[x]$ to denote the group element that has discrete logarithm x with respect to the group generator G. In this notation the group generator G is $[1]$ and the neutral element is $[0]$. We will find it convenient to use additive notation for the group operation, so we have $[x] + [y] = [x + y]$. Observe that given $\alpha \in \mathbb{Z}_p$ and $[x] \in \mathbb{G}_p$ it is easy to compute $[\alpha x]$ using the group operations. For a vector $\boldsymbol{x} = (x_1, \ldots, x_n) \in \mathbb{Z}_p^n$ we use $[\boldsymbol{x}]$ as a shorthand for the tuple

Computational Diffie-Hellman (CDH) Assumption

$\mathcal{I}_{\text{CDH}}(1^\kappa)$	$\mathcal{V}_{\text{CDH}}(pub, priv = (x,y), sol = [z])$
$(\mathbb{G}_p, [1]) \leftarrow \mathcal{G}(1^\kappa);\ x, y \leftarrow \mathbb{Z}_p$ Return $(pub = (\mathbb{G}_p, [1], [x], [y]), priv = (x,y))$	If $[z] = [xy]$ return 1 Else return 0

Square Computational Diffie-Hellman (SCDH) Assumption [32]

$\mathcal{I}_{\text{SCDH}}(1^\kappa)$	$\mathcal{V}_{\text{SCDH}}(pub, priv = x, sol = [z])$
$(\mathbb{G}_p, [1]) \leftarrow \mathcal{G}(1^\kappa);\ x \leftarrow \mathbb{Z}_p$ Return $(pub = (\mathbb{G}_p, [1], [x]), priv = x)$	If $[z] = [x^2]$ return 1 Else return 0

q-Generalized Diffie-Hellman Exponent (q-GDHE) Assumption [9, 10]

$\mathcal{I}_{q\text{-GDHE}}(1^\kappa)$	$\mathcal{V}_{q\text{-GDHE}}(pub, priv = x, sol = [z])$
$(\mathbb{G}_p, [1]) \leftarrow \mathcal{G}(1^\kappa);\ x \leftarrow \mathbb{Z}_p$ $pub = (\mathbb{G}_p, [1], [x], \ldots, [x^{q-1}], [x^{q+1}], \ldots, [x^{2q}])$ Return $(pub, priv = x)$	If $[z] = [x^q]$ return 1 Else return 0

q-Strong Diffie-Hellman (q-SDH) Assumption [8]

$\mathcal{I}_{q\text{-SDH}}(1^\kappa)$	$\mathcal{V}_{q\text{-SDH}}(pub, priv = x, sol = (c, [z]))$
$(\mathbb{G}_p, [1]) \leftarrow \mathcal{G}(1^\kappa);\ x \leftarrow \mathbb{Z}_p$ Return $(pub = (\mathbb{G}_p, [1], [x], \ldots, [x^q]), priv = x)$	If $[z] = \left[\frac{1}{x+c}\right]$ & $c \in \mathbb{Z}_p \setminus \{-x\}$ return 1 Else return 0

q-Diffie-Hellman Inversion (q-DHI) Assumption [33]

$\mathcal{I}_{q\text{-DHI}}(1^\kappa)$	$\mathcal{V}_{q\text{-DHI}}(pub, priv = x, sol = [z])$
$(\mathbb{G}_p, [1]) \leftarrow \mathcal{G}(1^\kappa);\ x \leftarrow \mathbb{Z}_p^\times$ Return $(pub = (\mathbb{G}_p, [1], [x], \ldots, [x^q]), priv = x)$	If $[z] = \left[\frac{1}{x}\right]$ return 1 Else return 0

Square Root Diffie-Hellman (SRDH) Assumption [29]

$\mathcal{I}_{\text{SRDH}}(1^\kappa)$	$\mathcal{V}_{\text{SRDH}}(pub, priv = x, sol = [z])$
$(\mathbb{G}_p, [1]) \leftarrow \mathcal{G}(1^\kappa);\ x \leftarrow \mathbb{Z}_p$ Return $\left(pub = (\mathbb{G}_p, [1], [x^2]), priv = x\right)$	If $[z] = [\pm x]$ return 1 Else return 0

Fig. 2. Some existing non-interactive intractability assumptions

$([x_1], \ldots, [x_n]) \in \mathbb{G}_p^n$. We will occasionally abuse the notation and let $[x_1, \ldots, x_n]$ denote the tuple $([x_1], \ldots, [x_n]) \in \mathbb{G}_p^n$.

There are many examples of non-interactive assumptions defined relative to a group generator $\mathcal{G}$. We list in Fig. 2 some of the existing non-interactive computational assumptions.

Generic Group Model. Obviously, if an assumption can be broken using generic-group operations, then it is false. The absence of a generic-group attack on an assumption does not necessarily mean the assumption holds [18,25] but is a necessary precondition for the assumption to be plausible.

We formalize the generic group model [36,39] where an adversary can only use generic group operations as follows. Given $\mathbb{G}_p$ we let $[\cdot]$ be a random bijection $\mathbb{Z}_p \rightarrow \mathbb{G}_p$. We give oracle access to the addition operation, i.e., $\mathcal{O}([x], [y])$ returns $[x + y]$. We say an assumption holds in the generic group model if an adversary with access to such an addition oracle has negligible chance of breaking the assumption. Note that the adversary gets $\mathbb{G}_p$ as input and hence is capable of deciding group membership. Also, given an arbitrary $[x]$, she can compute $[0] = [px]$ using the addition oracle. A generic adversary might be able to sample a random group element from $\mathbb{G}_p$, but since they are just random encodings we may without loss of generality assume she only generates elements as linear combinations of group elements she has already seen. All the assumptions listed in Fig. 2 are secure in the generic group model.

3 Target Assumptions

The assumptions that can be defined over cyclic groups are legion. We will focus on the broad class of non-interactive computational assumptions where the adversary's goal is to compute a particular group element. We refer to them as *target assumptions*.

The CDH assumption is an example of a target assumption where the adversary has to compute a specific group element. She is given $([1], [x], [y]) \in \mathbb{G}_p^3$ and is tasked with computing $[xy] \in \mathbb{G}_p$.

We aim for maximal generality of the class of assumptions and will therefore also capture assumptions where the adversary takes part in specifying the target element to be computed. In the q-SDH assumption for instance the adversary is given $([1], [x], \ldots, [x^q]) \in \mathbb{G}_p^{q+1}$ and is tasked with finding $c \in \mathbb{Z}_p$ and $[\frac{1}{x+c}] \in \mathbb{G}_p$. Here the problem instance in itself does not dictate which group element the adversary must compute but the output of the adversary includes c, which uniquely determines the target element to be computed.

We will now define target assumptions. The class will be defined very broadly in order to capture existing assumptions in the literature such as CDH and q-SDH as well as other assumptions that may appear in future works.

In a target assumption, the instance generator outputs *pub* that includes a prime-order group, a number of group elements, and possibly some additional information. Often, the group elements are of the form $[a(x)]$, where a is a polynomial and x is chosen uniformly at random from $\mathbb{Z}_p$. We generalize this by letting the instance generator output group elements of the form $[\frac{a(\boldsymbol{x})}{b(\boldsymbol{x})}]$, where $a(\boldsymbol{x})$ and $b(\boldsymbol{x})$ are multi-variate polynomials and $\boldsymbol{x}$ is chosen uniformly at random from $\mathbb{Z}_p^m$. We will assume all the polynomials are known to the adversary, i.e., they will be explicitly given in the additional information in *pub* in the form of their coefficients. The adversary will now specify a target group element. She does so by specifying polynomials $r(\boldsymbol{X})$ and $s(\boldsymbol{X})$ and making an attempt at computing the group element $[\frac{r(\boldsymbol{x})}{s(\boldsymbol{x})}]$.

If the target element can be computed using generic-group operations on the group elements in *pub*, then the problem is easy to solve and hence the assumption is trivially false. To exclude trivially false assumptions, the solution verifier will therefore check that for all fixed linear combinations $\alpha_1, \ldots, \alpha_n \in \mathbb{Z}_p$ there is a low probability over the choice of $\boldsymbol{x}$ that $\frac{r(\boldsymbol{x})}{s(\boldsymbol{x})} = \sum_i \alpha_i \frac{a_i(\boldsymbol{x})}{b_i(\boldsymbol{x})}$.

The solution tuple output by the adversary is $sol = (r(\boldsymbol{X}), s(\boldsymbol{X}), [y], sol')$, where sol' is some potential extra information the verifier may need to check, for instance about how the polynomials r and s were constructed.

Definition 3 (Target Assumption). *Given polynomials $d(\kappa), m(\kappa)$ and $n(\kappa)$ we say $(\mathcal{I}, \mathcal{V})$ is a (d, m, n)-target assumption for $\mathcal{G}$ if they can be defined by a PPT algorithm $\mathcal{I}^{\mathsf{core}}$ and a DPT algorithm $\mathcal{V}^{\mathsf{core}}$ such that*

$(pub, priv) \leftarrow \mathcal{I}(1^\kappa)$: *Algorithm $\mathcal{I}$ proceeds as follows:*

- $(\mathbb{G}_p, [1]) \leftarrow \mathcal{G}(1^\kappa)$
- $\left(\left\{ \frac{a_i(\boldsymbol{X})}{b_i(\boldsymbol{X})} \right\}_{i=1}^n, pub', priv' \right) \leftarrow \mathcal{I}^{\mathsf{core}}(\mathbb{G}_p)$
- $\boldsymbol{x} \leftarrow \mathbb{Z}_p^m$ *conditioned on $b_i(\boldsymbol{x}) \neq 0$*
- $pub := \left(\mathbb{G}_p, \left\{ \left[\frac{a_i(\boldsymbol{x})}{b_i(\boldsymbol{x})} \right] \right\}_{i=1}^n, \left\{ \frac{a_i(\boldsymbol{X})}{b_i(\boldsymbol{X})} \right\}_{i=1}^n, pub' \right)$
- *Return $(pub, priv = ([1], \boldsymbol{x}, priv'))$*

$0/1 \leftarrow \mathcal{V}\left(pub, priv, sol = \left(r(\boldsymbol{X}), s(\boldsymbol{X}), [y], sol' \right) \right)$: *Algorithm $\mathcal{V}$ returns 1 if all of the following checks pass and 0 otherwise:*

- $r(\boldsymbol{X}) \prod_{i=1}^n b_i(\boldsymbol{X}) \notin \mathrm{span}\left\{ \left\{ s(\boldsymbol{X}) a_j(\boldsymbol{X}) \prod_{i \neq j} b_i(\boldsymbol{X}) \right\}_{j=1}^n \right\}$
- $[y] = \frac{r(\boldsymbol{x})}{s(\boldsymbol{x})} [1]$
- $\mathcal{V}^{\mathsf{core}}(pub, priv, sol) = 1$

We require that the number of indeterminates in $\boldsymbol{X}$ is $m(\kappa)$ and each of the polynomials $a_1(\boldsymbol{X}), b_1(\boldsymbol{X}), \ldots, a_n(\boldsymbol{X}), b_n(\boldsymbol{X}), r(\boldsymbol{X}), s(\boldsymbol{X})$ has a total degree bounded by $d(\kappa)$, both of which can easily be checked by $\mathcal{V}^{\mathsf{core}}$.

It is easy to see that all assumptions in Fig. 2 are target assumptions. For CDH for instance, we have $d = 2, m = 2, n = 3$, $a_1(X_1, X_2) = 1, a_2(X_1, X_2) = X_1, a_3(X_1, X_2) = X_2$, and $b_1(X_1, X_2) = b_2(X_1, X_2) = b_3(X_1, X_2) = 1$. Algorithm $\mathcal{V}^{\mathsf{core}}$ then checks that the adversary's output is $r(X_1, X_2) = X_1 X_2$ and $s(X_1, X_2) = 1$, which means the adverary is trying to compute the target $[x_1 x_2] \in \mathbb{G}_p$.

In q-SDH we have $d = q, m = 1, n = q+1$, $a_i(X) = X^{i-1}$, and $b_i(X) = 1$ for $i = 1, \ldots, q+1$. Algorithm $\mathcal{V}^{\mathsf{core}}$ checks that the target polynomials are of the form $r(X) = 1$ and $s(X) = c + X$ for some $c \in \mathbb{Z}_p$, meaning the adversary it is trying to compute the target $[\frac{1}{x+c}] \in \mathbb{G}_p$.

3.1 Simple Target Assumptions

We now have a very general definition of target assumptions relating to the computation of group elements. In the following subsections, we go through progressively simpler classes of assumptions that imply the security of target assumptions. We start by defining simple target assumptions, where the divisor polynomials in the instance are trivial, i.e., $b_1(\boldsymbol{X}) = \cdots = b_n(\boldsymbol{X}) = 1$.

Algorithm $\mathcal{I}_\mathsf{B}(1^\kappa)$

- $(\mathbb{G}_p, [1]) \leftarrow \mathcal{G}(1^\kappa)$
- $\left(\left\{ \frac{a_i(\boldsymbol{X})}{b_i(\boldsymbol{X})} \right\}_{i=1}^n, pub'_\mathsf{A}, priv'_\mathsf{A} \right) \leftarrow \mathcal{I}_\mathsf{A}^\mathrm{core}(\mathbb{G}_p)$
- For $i = 1, \ldots, n$ set $c_i(\boldsymbol{X}) := a_i \prod_{j \neq i} b_j(\boldsymbol{X})$
- $\boldsymbol{x} \leftarrow \mathbb{Z}_p^m$
- $pub_\mathsf{B} = (\mathbb{G}_p, \{[c_i(\boldsymbol{x})]\}_{i=1}^n, \{c_i(\boldsymbol{X})\}_{i=1}^n, pub'_\mathsf{B} = (\{a_i(\boldsymbol{X}), b_i(\boldsymbol{X})\}_{i=1}^n, pub'_\mathsf{A}))$
- Return $(pub_\mathsf{B}, priv_\mathsf{B} = ([1], \boldsymbol{x}, priv'_\mathsf{A}))$

Algorithm $\mathcal{V}_\mathsf{B} \left(pub_\mathsf{B}, priv_\mathsf{B} = ([1], \boldsymbol{x}, priv'_\mathsf{A}), sol_\mathsf{B} = (t(\boldsymbol{X}), s(\boldsymbol{X}), [y], sol') \right)$

- Check $t(\boldsymbol{X}) \notin \mathrm{span}\{s(\boldsymbol{X})c_1(\boldsymbol{X}), \ldots, s(\boldsymbol{X})c_n(\boldsymbol{X})\}$
- Check $[y] = \frac{t(\boldsymbol{x})}{s(\boldsymbol{x})}[1]$
- Check $t(\boldsymbol{X}) = r(\boldsymbol{X}) \prod_{j=1}^n b_j(\boldsymbol{X})$ for a polynomial $r(\boldsymbol{X})$ of total degree $\leq d$
- Let $pub_\mathsf{A} = (\mathbb{G}_p, \{[c_i(\boldsymbol{x})]\}_{i=1}^n, \left\{ \frac{a_i(\boldsymbol{X})}{b_i(\boldsymbol{X})} \right\}_{i=1}^n, pub'_\mathsf{A})$
- Let $priv_\mathsf{A} = ([\prod_{j=1}^n b_j(\boldsymbol{x})], \boldsymbol{x}, priv'_\mathsf{A})$
- Check $\mathcal{V}_\mathsf{A}^\mathrm{core} \left(pub_\mathsf{A}, priv_\mathsf{A}, sol_\mathsf{A} = (r(\boldsymbol{X}), s(\boldsymbol{X}), [y], sol') \right) = 1$

Adversary $\mathcal{B} \left(\mathbb{G}_p, \{[c_i(\boldsymbol{x})]\}_{i=1}^n, \{c_i(\boldsymbol{X})\}_{i=1}^n, \{a_i(\boldsymbol{X}), b_i(\boldsymbol{X})\}_{i=1}^n, pub'_\mathsf{A} \right)$

- $(r(\boldsymbol{X}), s(\boldsymbol{X}), [y], sol') \leftarrow \mathcal{A} \left(\mathbb{G}_p, \{[c_i(\boldsymbol{x})]\}_{i=1}^n, \left\{ \frac{a_i(\boldsymbol{X})}{b_i(\boldsymbol{X})} \right\}_{i=1}^n, pub'_\mathsf{A} \right)$
- Return $sol_\mathsf{B} = \left(t(\boldsymbol{X}) = r(\boldsymbol{X}) \prod_{j=1}^n b_j(\boldsymbol{X}), s(\boldsymbol{X}), [y], sol' \right)$

Fig. 3. Simple target assumption $\mathsf{B} = (\mathcal{I}_\mathsf{B}, \mathcal{V}_\mathsf{B})$ and adversary $\mathcal{B}$ against it

Definition 4 (Simple Target Assumption). *We say a (d, m, n)-target assumption $(\mathcal{I}, \mathcal{V})$ for $\mathcal{G}$ is simple if the instance generator always picks polynomials $b_1(\boldsymbol{X}) = \ldots = b_n(\boldsymbol{X}) = 1$.*

Next, we will prove that the security of simple target assumptions implies the security of all target assumptions. The idea is to reinterpret the tuple the adversary gets using the random generator $[1]$ to having random generator $[\prod_{i=1}^n b_i(\boldsymbol{x})]$. Now all fractions of formal polynomials are scaled up by a factor $\prod_{i=1}^n b_i(\boldsymbol{X})$ and the divisor polynomials can be cancelled out.

Theorem 1. *For any (d, m, n)-target assumption $\mathsf{A} = (\mathcal{I}_\mathsf{A}, \mathcal{V}_\mathsf{A})$ there exists a $((n + 1)d, m, n)$-simple target assumption $\mathsf{B} = (\mathcal{I}_\mathsf{B}, \mathcal{V}_\mathsf{B})$ such that $\mathsf{B} \Rightarrow \mathsf{A}$.*

Proof. Given an assumption $\mathsf{A} = (\mathcal{I}_\mathsf{A}, \mathcal{V}_\mathsf{A})$ and an adversary $\mathcal{A}$ against it, we define a simple target assumption $\mathsf{B} = (\mathcal{I}_\mathsf{B}, \mathcal{V}_\mathsf{B})$ and an adversary $\mathcal{B}$ (that uses adversary $\mathcal{A}$ in a black-box manner) against it as illustrated in Fig. 3. The key observation is that as long as $\prod_{i=1}^n b_i(\boldsymbol{x}) \neq 0$, the two vectors of group elements $([c_1(\boldsymbol{x})], \ldots, [c_n(\boldsymbol{x})])$ and $\left([\frac{a_1(\boldsymbol{x})}{b_1(\boldsymbol{x})}], \ldots, [\frac{a_n(\boldsymbol{x})}{b_n(\boldsymbol{x})}] \right)$ are identically distributed. By the specification of assumption A, it follows that $b_i(\boldsymbol{X}) \not\equiv 0$ for all $i \in \{1, \ldots, n\}$.

By the Schwartz-Zippel lemma, the probability that $\prod_{i=1}^{n} b_i(\boldsymbol{x}) = 0$ is at most $\frac{dn}{p}$. Thus, if $\mathcal{A}$ has success probability $\epsilon_{\mathcal{A}}$, then $\mathcal{B}$ has success probability $\epsilon_{\mathcal{B}} \geq \epsilon_{\mathcal{A}} - \frac{dn}{p}$. □

3.2   Univariate Target Assumptions Imply Multivariate Target Assumptions

We will now show that security of target assumptions involving univariate polynomials imply security of target assumptions involving multivariate polynomials.

Theorem 2. *For any (d, m, n)-simple target assumption* $\mathsf{A} = (\mathcal{I}_\mathsf{A}, \mathcal{V}_\mathsf{A})$ *there exists a* $((n + 1)d, 1, n)$-*simple target assumption* $\mathsf{B} = (\mathcal{I}_\mathsf{B}, \mathcal{V}_\mathsf{B})$ *where* $\mathsf{B} \Rightarrow \mathsf{A}$.

Proof. Given $\mathsf{A} = (\mathcal{I}_\mathsf{A}, \mathcal{V}_\mathsf{A})$ and an adversary $\mathcal{A}$ with success probability $\epsilon_{\mathcal{A}}$ against it, we define a simple target assumption $(\mathcal{I}_\mathsf{B}, \mathcal{V}_\mathsf{B})$ with univariate polynomials and construct an adversary $\mathcal{B}$ (that uses $\mathcal{A}$ in a black-box manner) against it as illustrated in Fig. 4.

Without loss of generality we can assume $a_1(\boldsymbol{X}), \ldots, a_n(\boldsymbol{X})$ are linearly independent, and therefore the polynomials $r_\mathsf{A}(\boldsymbol{X})$ and $s_\mathsf{A}(\boldsymbol{X})a_1(\boldsymbol{X}), \ldots, s_\mathsf{A}(\boldsymbol{X})a_n(\boldsymbol{X})$ are all linearly independent. By Lemma 1 below this means

---

Algorithm $\mathcal{I}_\mathsf{B}(1^\kappa)$

- $(\mathbb{G}_p, [1]) \leftarrow \mathcal{G}(1^\kappa)$
- $\left(\{a_i(\boldsymbol{X})\}_{i=1}^{n}, pub'_\mathsf{A}, priv'_\mathsf{A} \right) \leftarrow \mathcal{I}_\mathsf{A}^{\mathsf{core}}(\mathbb{G}_p)$
- Randomly choose $\boldsymbol{c}(X) \leftarrow (\mathbb{Z}_p[X])^m$ where $\deg(c_i) = n + 1$
- For $i = 1, \ldots, n$ let $b_i(X) := a_i(\boldsymbol{c}(X))$
- $pub'_\mathsf{B} := \left(\{a_i(\boldsymbol{X})\}_{i=1}^{n}, \boldsymbol{c}(X), pub'_\mathsf{A} \right)$
- $x \leftarrow \mathbb{Z}_p$
- Return $\left(pub_\mathsf{B} = \left(\mathbb{G}_p, \{[b_i(x)]\}_{i=1}^{n}, \{b_i(X)\}_{i=1}^{n}, pub'_\mathsf{B} \right), priv_\mathsf{B} = ([1], x, priv'_\mathsf{A}) \right)$

Algorithm $\mathcal{V}_\mathsf{B}\left(pub_\mathsf{B}, priv_\mathsf{B}, sol_\mathsf{B} = \left(r(X), s(X), [y], (r_\mathsf{A}(\boldsymbol{X}), s_\mathsf{A}(\boldsymbol{X}), sol'_\mathsf{A}) \right) \right)$

- Check $r(X) \notin \mathrm{span}\{s(X)b_1(X), \ldots, s(X)b_n(X)\}$
- Check $[y] = \frac{r(x)}{s(x)}[1]$
- Let $pub_\mathsf{A} = \left(\mathbb{G}_p, \{[b_i(x)]\}_{i=1}^{n}, \{a_i(\boldsymbol{X})\}_{i=1}^{n}, pub'_\mathsf{A} \right)$
- Let $priv_\mathsf{A} = ([1], x, priv'_\mathsf{A})$
- Check $\mathcal{V}_\mathsf{A}^{\mathsf{core}}\left(pub_\mathsf{A}, priv_\mathsf{A}, sol_\mathsf{A} = (r_\mathsf{A}(\boldsymbol{X}), s_\mathsf{A}(\boldsymbol{X}), [y], sol'_\mathsf{A}) \right) = 1$

Adversary $\mathcal{B}\left(\mathbb{G}_p, \{[b_i(x)]\}_{i=1}^{n}, \{b_i(X)\}_{i=1}^{n}, pub'_\mathsf{B} \right)$

- $\left(r_\mathsf{A}(\boldsymbol{X}), s_\mathsf{A}(\boldsymbol{X}), [y], sol'_\mathsf{A} \right) \leftarrow \mathcal{A}\left(\mathbb{G}_p, \{[b_i(x)]\}_{i=1}^{n}, \{a_i(\boldsymbol{X})\}_{i=1}^{n}, pub'_\mathsf{A} \right)$
- Return $\left(r_\mathsf{A}(\boldsymbol{c}(X)), s_\mathsf{A}(\boldsymbol{c}(X)), [y], sol'_\mathsf{B} = (r_\mathsf{A}(\boldsymbol{X}), s_\mathsf{A}(\boldsymbol{X}), sol'_\mathsf{A}) \right)$

Fig. 4. Univariate simple target assumption B and adversary $\mathcal{B}$ against it

that with probability $1 - \frac{d(n+1)}{p}$ the univariate polynomials $r(c(X))$ and $s(c(X))a_1(c(X)), \ldots, s(c(X))a_n(c(X))$ output by $\mathcal{B}$ are also linearly independent since the only information about $c(X)$ that $\mathcal{B}$ passes on to $\mathcal{A}$ can be computed from $(x, c(x))$. This means $\mathcal{B}$ has advantage $\epsilon_{\mathcal{B}} \geq \epsilon_{\mathcal{A}} - \frac{d(n+1)}{p}$ against assumption B.   □

Lemma 1. *Let* $a_1(\boldsymbol{X}), \ldots, a_n(\boldsymbol{X}) \in \mathbb{Z}_p[\boldsymbol{X}]$ *be linearly independent m-variate polynomials of total degree bounded by d, and let $(x, \boldsymbol{x}) \in \mathbb{Z}_p \times \mathbb{Z}_p^m$. Pick a vector of m random univariate degree n polynomials $c(X) \in (\mathbb{Z}_p[X])^m$ that passes through $(x, \boldsymbol{x})$, i.e., $c(x) = \boldsymbol{x}$. The probability that $a_1(c(X)), \ldots, a_n(c(X))$ are linearly independent is at least $1 - \frac{dn}{p}$.*

Proof. Take n random points $\boldsymbol{x}_1, \ldots, \boldsymbol{x}_n \in \mathbb{Z}_p^m$ and consider the matrix

$$M = \begin{pmatrix} a_1(\boldsymbol{x}_1) & \cdots & a_1(\boldsymbol{x}_n) \\ \vdots & & \vdots \\ a_n(\boldsymbol{x}_1) & \cdots & a_n(\boldsymbol{x}_n) \end{pmatrix}.$$

We will argue by induction that with probability $1 - \frac{dn}{p}$ the matrix is invertible. For $n = 1$ it follows from the Schwartz-Zippel lemma that the probability $a_1(\boldsymbol{x}_1) = 0$ is at most $\frac{d}{p}$. Suppose now by induction hypothesis that we have probability $1 - \frac{d(n-1)}{p}$ that the top left $(n-1) \times (n-1)$ matrix is invertible. When it is invertible, the values $a_n(\boldsymbol{x}_1), \ldots, a_n(\boldsymbol{x}_{n-1})$ uniquely determine $\alpha_1, \ldots, \alpha_{n-1}$ such that for $j = 1, \ldots, n-1$ we have $a_n(\boldsymbol{x}_j) = \sum_{i=1}^{n-1} \alpha_i a_i(\boldsymbol{x}_j)$. Since the polynomials $a_1(\boldsymbol{X}), \ldots, a_n(\boldsymbol{X})$ are linearly independent, by the Schwartz-Zippel lemma, there is at most probability $\frac{d}{p}$ that we also have $a_n(\boldsymbol{x}_n) = \sum_{i=1}^{n-1} \alpha_i a_i(\boldsymbol{x}_n)$. So the row $(a_n(\boldsymbol{x}_1), \ldots, a_n(\boldsymbol{x}_n))$ is linearly independent of the other rows, and hence we have M is invertible with at least probability $1 - \frac{d(n-1)}{p} - \frac{d}{p} = 1 - \frac{dn}{p}$.

Finally, picking a vector of m random polynomials $c(X)$ of degree n such that $c(x) = \boldsymbol{x}$ and evaluating it in distinct points $x_1, \ldots, x_n \leftarrow \mathbb{Z}_p \setminus \{x\}$ gives us n random points $c(x_j) \in \mathbb{Z}_p^m$. So the matrix

$$\begin{pmatrix} a_1(c(x_1)) & \cdots & a_1(c(x_n)) \\ \vdots & & \vdots \\ a_n(c(x_1)) & \cdots & a_n(c(x_n)) \end{pmatrix}$$

has at least probability $1 - \frac{dn}{p}$ of being invertible. If $\sum_{i=1}^n \alpha_i a_i(c(X)) = 0$, then it must hold in the distinct points $x_1, \ldots, x_n$, and we can see there is only the trivial linear combination with $\alpha_1 = \cdots = \alpha_n = 0$.   □

Having reduced target assumptions to simple univariate target assumptions with $m = 1$, we will in the next two subsections consider two separate cases. First, the case where the polynomial $s(X)$ is fixed, i.e., it can be deterministically computed. Second, the case where the polynomial $s(X)$ may vary.

3.3 Polynomial Assumptions

We now consider simple target assumptions with univariate polynomials where $s(X)$ is fixed. We can without loss of generality assume this means $priv'$ output by the instance generator contains $s(X)$ and the solution verifier checks whether the adversary's solution matches $s(X)$. There are many assumptions where $s(X)$ is fixed, in the Diffie-Hellman inversion assumption we will for instance always have $s(X) = X$ and in the q-GDHE assumption we always have $s(X) = 1$. When the polynomial $s(X)$ is fixed, we can multiply it away as we did for the multivariate polynomials $b_1(\boldsymbol{X}), \ldots, b_n(\boldsymbol{X})$ when reducing target assumptions to simple target assumptions. This leads us to define the following class of assumptions:

Definition 5 (Polynomial Assumption). *We say a $(d, 1, n)$-simple target assumption $(\mathcal{I}, \mathcal{V})$ for $\mathcal{G}$ is a (d, n)-polynomial assumption if $\mathcal{V}$ only accepts a solution with $s(X) = 1$.*

We leave the proof of the following theorem to the reader.

Theorem 3. *For any $(d, 1, n)$-simple target assumption $\mathsf{A} = (\mathcal{I}_\mathsf{A}, \mathcal{V}_\mathsf{A})$ for $\mathcal{G}$ where the polynomial $s(X)$ is fixed, there is a $(2d, n)$-polynomial assumption $\mathsf{B} = (\mathcal{I}_\mathsf{B}, \mathcal{V}_\mathsf{B})$ where $\mathsf{B} \Rightarrow \mathsf{A}$.*

We will now show that all polynomial assumptions are implied by the generalized Diffie-Hellman exponent (q-GDHE) assumptions (cf. Fig. 2) along the lines of [22]. This means the q-GDHE assumptions are Uber assumptions that imply the security of a major class of target assumptions, which includes a majority of the non-interactive computational assumptions for prime-order groups found in the literature.

Theorem 4. *For any (d, n)-polynomial assumption $\mathsf{A} = (\mathcal{I}_\mathsf{A}, \mathcal{V}_\mathsf{A})$ for $\mathcal{G}$ we have that $(d + 1)$-GDHE $\Rightarrow \mathsf{A}$.*

Proof. Let $\mathcal{A}$ be an adversary against the (d, n)-polynomial assumption. We show how to build an adversary $\mathcal{B}$, which uses $\mathcal{A}$ in a black-box manner to break the $(d+1)$-GDHE assumption. Adversary $\mathcal{B}$ gets $[1], [x], \ldots, [x^d], [x^{d+2}], \ldots, [x^{2d+2}]$ from the $(d+1)$-GDHE instance generator and her aim is to output the element $[x^{d+1}] \in \mathbb{G}_p$. Adversary $\mathcal{B}$ uses algorithm $\mathcal{I}_\mathsf{A}^{\mathsf{core}}$ to generate a simulated polynomial problem instance as described below, which she then forwards to $\mathcal{A}$.

Adversary $\mathcal{B}\left(\mathbb{G}_p, [1], [x], \ldots, [x^d], [x^{d+2}], \ldots, [x^{2d+2}]\right)$
- $\left(\{a_i(X)\}_{i=1}^n, pub'_\mathsf{A}, priv'_\mathsf{A}\right) \leftarrow \mathcal{I}_\mathsf{A}^{\mathsf{core}}(\mathbb{G}_p)$
- Randomly choose $c(X) \leftarrow \mathbb{Z}_p[X]$ where $\deg(c) = d + 1$ and no $a_i(X)c(X)$ includes the term X^{d+1}
- For all i compute $[a_i(x)c(x)]$ using the $(d + 1)$-GDHE tuple
- Run $(r(X), [y], sol') \leftarrow \mathcal{A}\left(\mathbb{G}_p, \{[a_i(x)c(x)]\}_{i=1}^n, \{a_i(X)\}_{i=1}^n, pub'_\mathsf{A}\right)$
- Parse $r(X)c(X)$ as $\sum_{i=0}^{2d+1} t_i X^i$
- Return $\frac{1}{t_{d+1}}\left([y] - [\sum_{i \neq d+1} t_i x^i]\right)$

Assuming $c(x) \neq 0$, which happens with probability $1 - \frac{d+1}{p}$, the input to $\mathcal{A}$ looks identical to a normal problem instance for assumption A with generator $[c(x)]$. Furthermore, if $\mathcal{A}$ finds a satisfactory solution to this problem, we then have $[y] = r(x)[c(x)]$. By Lemma 2 below there is at most $\frac{1}{p}$ chance of returning $r(X)$ such that $r(X)c(X)$ has coefficient 0 for X^{d+1} and hence using the $(2d+2)$ elements from the $(d+1)$-GDHE tuple, we can recover $[x^{d+1}]$ from $[y]$. We now get that if $\mathcal{A}$ has advantage $\epsilon_{\mathcal{A}}$ against A, then $\mathcal{B}$ has advantage $\epsilon_{\mathcal{B}} \geq \epsilon_{\mathcal{A}} - \frac{d+2}{p}$ against the $(d+1)$-GDHE assumption. $\qquad\square$

Lemma 2 (Lemma 10 from [22]). *Let $\{a_i(X)\}_{i=1}^{n}$ be polynomials of degree at most d. Pick $x \leftarrow \mathbb{Z}_p$ and $c(X)$ as a random degree $d + 1$ polynomial such that all products $b_i(X) = a_i(X)c(X)$ have coefficient 0 for X^{d+1}. Given $(\{a_i(X)\}_{i=1}^{n}, x, c(x))$, the probability of guessing a non-trivial degree d polynomial $r(X)$ such that $r(X)c(X)$ has coefficient 0 for X^{d+1} is at most $\frac{1}{p}$.*

3.4 Fractional Assumptions

We now consider the alternative case of simple target assumptions with univariate polynomials, where $s(X)$ is not fixed. When $s(X)|r(X)$, we can without loss of generality divide out and get $s(X) = 1$. The remaining case is when $s(X) \nmid r(X)$, which we now treat.

Definition 6 ((d, n)-Fractional Assumption). *We say a $(d,1,n)$-simple target assumption $(\mathcal{I}, \mathcal{V})$ for $\mathcal{G}$ is an (d,n)-fractional assumption if $\mathcal{V}$ only accepts the solution if $s(X) \nmid r(X)$.*

Next we define a simple fractional assumption which we refer to for short as q-SFrac, which says given the tuple $([1], [x], [x^2], \ldots, [x^q])$ it is hard to compute $\left[\frac{r(x)}{s(x)}\right]$ when $\deg(r) < \deg(s)$. The simple fractional assumption is a straightforward generalization of the q-SDH assumption, where $\deg(r) = 0$ since $r(X) = 1$ and $\deg(s) = 1$. A proof for the intractability of the q-SFrac assumption in the generic group model can be found in the full version [23].

Definition 7 (q-SFrac Assumption). *The q-SFrac assumption is a simple target assumption where $n = q+1$, $a_i(X) = X^{i-1}$, and $0 \leq \deg(r) < \deg(s) \leq q$.*

We now prove the following theorem.

Theorem 5. *For any (d,n)-fractional assumption A $= (\mathcal{I}, \mathcal{V})$ for $\mathcal{G}$ we have d-SFrac $\Rightarrow$ A.*

Proof. Let $\mathcal{A}$ be an adversary against a (d, n)-fractional assumption A. We show how to use $\mathcal{A}$ to construct an adversary $\mathcal{B}$ against the d-SFracassumption. Adversary $\mathcal{B}$ gets $(\mathbb{G}_p, [1], [x], \ldots, [x^d])$ from her environment and her aim is to output a valid solution of the form $\left(r'(X), s'(X), \left[\frac{r'(x)}{s'(x)}\right]\right)$ where $\deg(r') < \deg(s') \leq d$.

Adversary $\mathcal{B}$ uses the instance generator algorithm $\mathcal{I}^{\mathsf{core}}$ of the fractional assumption as described below to generate a problem instance, which she then forwards to $\mathcal{A}$.

> Algorithm $\mathcal{B}(\mathbb{G}_p, [1], [x], \ldots, [x^d])$
> ─ $(\{a_i(X)\}_{i=1}^n, pub', priv') \leftarrow \mathcal{I}^{\mathsf{core}}(\mathbb{G}_p)$
> ─ For all i compute $[a_i(x)]$ using the d-SFrac tuple $([1][x], \ldots, [x^d])$
> ─ Run $(r(X), s(X), [y]) \leftarrow \mathcal{A}(\mathbb{G}_p, \{[a_i(x)]\}_{i=1}^n, \{a_i(X)\}_{i=1}^n, pub')$
> ─ Let $r'(X) = r(X) \bmod s(X)$ and write $r(X) = t(X)s(X) + r'(X)$
> ─ Use the tuple $([1], [x], \ldots, [x^d])$ to compute $t(x)$
> ─ Return $(r(X)', s(X), [y] - [t(x)])$

The advantage of adversary $\mathcal{B}$ against the d-SFracassumption is the same as that of adversary $\mathcal{A}$ against the fractional assumption A.   □

3.5 The q-SFracand q-GDHE Assumptions Together Imply All Target Assumptions in Cyclic Groups

We now prove that the q-SFracand q-GDHE assumptions together constitute an Uber assumption for all target assumptions in prime-order cyclic groups.

Theorem 6. *There is a polynomial $q(d, m, n)$ such that the joint q-SFrac and q-GDHE assumption implies all (d, m, n)-target assumptions.*

Proof. Let A be a (d, m, n)-target assumption. By Theorem 1, for any adversary $\mathcal{A}$ with advantage $\epsilon_{\mathcal{A}}$ against A, we can define a $(d(n+1), m, n)$-simple target assumption A_1 and an adversary $\mathcal{A}_1$ with advantage $\epsilon_{\mathcal{A}_1} \geq \epsilon_{\mathcal{A}} - \frac{dn}{p}$ against it. By Theorem 2, using $\mathcal{A}_1$ against A_1, we can define a $(d(n+1)^2, 1, n)$-simple target assumption A_2 and an adversary $\mathcal{A}_2$ against it with advantage $\epsilon_{\mathcal{A}_2} \geq \epsilon_{\mathcal{A}_1} - \frac{d(n+1)^2}{p} \geq \epsilon_{\mathcal{A}} - \frac{d(n+(n+1)^2)}{p}$. We now have two cases as follows:

- With non-negligible probability a successful solution has $s_{\mathsf{A}_2}(X) \nmid r_{\mathsf{A}_2}(X)$. By Theorem 5, we can use $\mathcal{A}_2$ to construct an adversary $\mathcal{A}_3$ against the $(d(n+1)^2)$-SFrac assumption where advantage $\epsilon_{\mathcal{A}_3} \geq \epsilon_{\mathcal{A}_2} \geq \epsilon_{\mathcal{A}} - \frac{d(n+(n+1)^2)}{p}$. Since by defintion $d, n \in \mathsf{Poly}(\kappa)$ and $\log p \in \theta(\kappa)$, it follows that $\frac{d(n+(n+1)^2)}{p}$ is negligible (in κ).

- With overwhelming probability a successful solution uses polynomials where $s_{\mathsf{A}_2}(X)|r_{\mathsf{A}_2}(X)$ which is equivalent to the case where $s_{\mathsf{A}_2}(X) = 1$. By Theorem 3, using $\mathcal{A}_2$ we can define a $(2d(n+1)^2, n)$-polynomial assumption A_3 and an adversary $\mathcal{A}_3$ with advantage $\epsilon_{\mathcal{A}_3} \geq \epsilon_{\mathcal{A}_2} - \frac{4d(n+1)^2}{p} \geq \epsilon_{\mathcal{A}} - \frac{d(n+5(n+1)^2)}{p}$. By Theorem 4, using adversary $\mathcal{A}_3$, we can construct an adversary $\mathcal{A}_4$ against the $(2d(n+1)^2+1)$-GDHE assumption with advantage $\epsilon_{\mathcal{A}_4} \geq \epsilon_{\mathcal{A}_3} - \frac{2d(n+1)^2+2}{p}$. From which it follows that $\epsilon_{\mathcal{A}_4} \geq \epsilon_{\mathcal{A}} - \frac{d(7(n+1)^2+n)+2}{p}$. Since by defintion $d, n \in \mathsf{Poly}(\kappa)$ and $\log p \in \theta(\kappa)$, it follows that $\frac{d(7(n+1)^2+n)+2}{p}$ is negligible (in κ).   □

4 The Relationship Between the GDHE and SFrac Assumptions

Having identified the q-GDHE and q-SFrac assumptions as Uber assumptions for all target assumptions, it is natural to investigate their internal structure and their relationship to each other. One obvious question is whether a further simplification is possible and one of the assumption classes imply the other. We first analyze the case where $q \geq 2$ and show that q-SFrac does not imply 2-GDHE for generic algorithms. This means that we need the q-GDHE assumptions to capture the polynomial target assumptions, the q-SFrac assumptions cannot act as an Uber assumption for all target assumptions on their own.

We also look at the lowest level of the q-SFrac and q-GDHE hierarchies. Observe that the 1-SFrac assumption is equivalent to the 1-SDH assumption. We prove that the 1-GDHE assumption is equivalent to the CDH assumption. This immediately also gives us that the 1-SFrac assumption implies the 1-GDHE assumption since the 1-SDH assumption implies the CDH assumption. A summary of the implications we prove can be found in Fig. 1.

4.1 The SFrac Assumptions Do Not Imply the 2-GDHE Assumption

We prove here that the i-GDHE assumption for $i \geq 2$ is not implied by any q-SFracassumption for generic adversaries, i.e. q-SFrac $\overset{\text{GGM}}{\not\Rightarrow}$ i-GDHE for all $i \geq 2$. More precisely, we show that providing an unbounded generic adversary $\mathcal{A}$ against a q-SFracassumption with a 2-GDHE oracle $\mathcal{O}_{\text{2-GDHE}}$, which on input $([a], [b], [c], [d])$ where $b = az$, $c = az^3$, $d = az^4$ returns the element $[az^2]$, and returns the symbol $\perp$ if the input is malformed, does not help the adversary.

Theorem 7. *The q-SFracassumption does not imply the 2-GDHE assumption in generic groups.*

Proof. Consider a generic adversary $\mathcal{A}$ which gets input $([1], [x], \ldots, [x^q])$ and is tasked with outputting $\left(\frac{r(X)}{s(X)}, \left[\frac{r(x)}{s(x)} \right] \right)$, where $0 \leq \deg(r) < \deg(s) \leq q$. We give $\mathcal{A}$ access to an oracle $\mathcal{O}_{\text{2-GDHE}}(\cdot, \cdot, \cdot, \cdot)$ as above, which can be queried polynomially many times.

Since $\mathcal{A}$ is generic, the tuple $([a], [b], [c], [d])$ she uses as input in her 1-st $\mathcal{O}_{\text{2-GDHE}}$ query can only be linear combinations of the elements $[1], [x], \ldots, [x^q]$. Thus, we have

$$a = \sum_{j=0}^{q} \alpha_j x^j \qquad b = \sum_{j=0}^{q} \beta_j x^j \qquad c = \sum_{j=0}^{q} \gamma_j x^j$$

for known $\alpha_j, \beta_j, \gamma_j$. Let the corresponding formal polynomials be $a(X)$, $b(X)$ and $c(X)$, respectively. We have that $\deg(a), \deg(b), \deg(c) \in \{0, \ldots, q\}$. By definition, for the input to the oracle $\mathcal{O}_{\text{2-GDHE}}$ to be well-formed, we must have $b = az$ and $c = az^3$ for some z. In the generic group model this has negligible probability of holding unless z corresponds to some (possibly rational) function $z(X)$ and we have $b(X) = a(X)z(X)$ and $c(X) = a(X)z(X)^3$ when viewed as formal polynomials.

If the adversary submits a query where $a(X) \equiv 0$ or $b(X) \equiv 0$, the oracle will just return $[0]$, which is useless to the adversary. So from now on let's assume that $a(X) \not\equiv 0$, $b(X) \not\equiv 0$ and $c(X) \not\equiv 0$.

We now have

$$c(X) = a(X)z(X)^3 = a(X)\left(\frac{b(X)}{a(X)}\right)^3 = \frac{b(X)^3}{a(X)^2}.$$

This means $a(X)^2 | b(X)^3$, which implies $a(X) | b(X)^2$. The answer returned by the oracle on a well-formed input corresponds to $a(X)z(X)^2 = a(X)\left(\frac{b(X)}{a(X)}\right)^2 = \frac{b(X)^2}{a(X)}$. Since $a(X) | b(X)^2$, the answer corresponds to a proper polynomial.

If $\deg(b) \leq \deg(a)$, we have $2\deg(b) - \deg(a) \leq \deg(b) \leq q$, and if $\deg(b) \geq \deg(a)$, we have $2\deg(b) - \deg(a) \leq 3\deg(b) - 2\deg(a) = \deg(c) \leq q$. Thus, the answer the oracle returns corresponds to a known polynomial of degree in $\{0, \ldots, q\}$ which could have been computed by the adversary herself using generic group operations on the tuple $[1], [x], \ldots, [x^q]$ without calling the oracle.   □

Since as we prove later the GDHE assumptions family is strictly increasingly stronger, we get the following corollary.

Corollary 1. *For all $i \geq 2$ it holds that $q\text{-SFrac} \overset{GGM}{\nRightarrow} i\text{-GDHE}$.*

4.2   CDH Implies the 1-GDHE Assumption

Since having access to a CDH oracle allows one to compute any polynomial in the exponent [31] (in fact, such an oracle provides more power as it allows computing even rational functions in the exponent for groups with known order [31]), it is clear that $1\text{-GDHE} \Rightarrow \text{CDH}$ and $q\text{-SDH} \Rightarrow \text{CDH}$. We prove in this section the implication $\text{CDH} \Rightarrow 1\text{-GDHE}$, which means that the assumptions CDH and 1-GDHE are equivalent. As a corollary $q\text{-SFrac} \Rightarrow 1\text{-GDHE}$ for all q.

We start by proving that the square computational Diffie-Hellman assumption (SCDH) (cf. Fig. 2), which is equivalent to the CDH assumption [3, 26], implies the square root Diffie-Hellman (SRDH) assumption [29] (cf. Fig. 2)[2]. Note that given a SCDH oracle, one can solve any CDH instance by making 2 calls to the SCDH oracle. Let $([1], [a], [b]) \in \mathbb{G}_p^3$ be a CDH instance, we have $[ab] = \frac{1}{4}([(a+b)^2] - ([(a-b)^2]))$.

We remark here that Roh and Hahn [37] also gave a reduction from the SCDH assumption to the SRDH assumption. However, their reduction relies on two assumptions: that the oracle will always (i.e. with probability 1) return a

---

[2] The SRDH assumption differs from the 1-GDHE assumption in that while the former accepts either $[-x]$ or $[x]$ as a valid answer, the latter only accepts $[x]$ as a valid answer.

correct answer when queried on quadratic-residue elements and uniformly random dom elements when queried on quadratic non-residue elements, and that the prime order of the group p has the special form $p = 2^t q + 1$ where $2^t = O(\kappa^{O(1)})$. Our reduction is more general since we do not place any restrictions on t or q and more efficient since it uses $4t + 2|q|$ oracle queries, whereas their reduction uses $O(t(2^t + |q|))$ oracle queries. Later on, we will also show how to boost an imperfect 1-GDHE oracle to get a SCDH oracle.

The Perfect Oracle Case. We prove that a perfect SRDH oracle $\mathcal{O}_{\mathrm{SRDH}}$ which on input a pair $([a], [az]) \in \mathbb{G}_p^2$ returns the symbol QNR (which for convenience we denote by [0]) if $z \notin \mathrm{QR}(p)$ and $[\pm a\sqrt{z}]$ otherwise, leads to a break of the SCDH assumption. The role of the exponent a is to allow queries on pairs w.r.t. a different group generator than the default one.

Let $p = 2^t q + 1$ for an odd positive integer q be the prime order of the group $\mathbb{G}_p$. Note that when $p \equiv 3 \pmod 4$ (this is the case when $-1 \in \mathrm{QNR}(p)$), we have $t = 1$, and in the special case where p is a safe prime, q is also a prime. On the other hand, when $p \equiv 1 \pmod 4$ (this is the case when $-1 \in \mathrm{QR}(p)$), we have $t > 1$.

In the following let $\omega \in \mathrm{QNR}(p)$ be an arbitrary 2^t-th root of unity of $\mathbb{Z}_p^\times$, i.e., $\omega^{2^{t-1}} \equiv -1 \bmod (p-1)$ and $\omega^{2^t} \equiv 1 \bmod (p-1)$. Note that there are $\phi(2^t) = 2^{t-1}$ roots of unity and finding one is easy since for any generator g of $\mathbb{Z}_p^\times$, g^q is a 2^t-th root of unity. Observe that all elements in $\mathbb{Z}_p^\times$ can be written in the form $\omega^i \beta$, where β has odd order $k|q$. The quadratic residues are those where i is even, and the quadratic non-residues are the ones where i is odd.

Theorem 8. *Given a perfect $\mathcal{O}_{SRDH}$ oracle, we can solve any SCDH instance using at most $4t + 2|q|$ oracle calls when the group order is $p = 2^t q + 1$ for odd q.*

Proof. Given a SCDH instance $([1], [x]) \in \mathbb{G}_p^2$, our task is to compute $[x^2] \in \mathbb{G}_p$. The task is trivial when $x = 0$, so let's from now on assume $x \neq 0$. Since any $x \in \mathbb{Z}_p^\times$ can be written as $x = \omega^i \beta$ where $\omega \in \mathbb{Z}_p^\times$ is a 2^t-th root of unity and $\beta \in \mathbb{Z}_p^\times$ has an odd order k where $k|q$, our task is to compute $[x^2] = [\omega^{2i} \beta^2]$. In the following, we will first describe an algorithm FindExpi that uses the square-root oracle to determine i. Next, we describe an algorithm Square that computes $[y] = [\omega^j \beta^2]$ for some j, and then use FindExpi to clean it up to get $[x^2] = [\omega^{2i-j} y] = [\omega^{2i} \beta^2]$. Both of these algorithms are given in Fig. 5.

Recall the perfect SRDH oracle responds with QNR, i.e. [0], whenever it gets a quadratic non-residue as input, i.e., whenever it gets input $([1], [\omega^i \beta])$ for an odd i and β has an odd order. When it gets a quadratic residue as input, i.e., when i is even, it returns $[\pm \omega^{\frac{i}{2}} \beta^{\frac{1}{2}}]$. Since $\omega^{2^{t-1}} = -1$ this means it returns either $[\omega^{\frac{i}{2}} \beta^{\frac{1}{2}}]$ or $[\omega^{2^{t-1}+\frac{i}{2}} \beta^{\frac{1}{2}}]$.

Let the binary expansion of the exponent be $i = i_{t-1} i_{t-2} \ldots i_0$. In the FindExpi algorithm we use the oracle to learn the least significant bit and also to right-shift the bits. Consider running the oracle on $([1], [x])$ and on $([1], [\omega^{-1} x])$. If $i_0 = 0$, then i is even and on the first input the oracle returns a new element

```
1  Algorithm FindExpi([1], [x], ω)
2    [y] ← [x]
3    i ← 0
4    For j = 0 to t − 1:
5      [z] ← O_SRDH([1], [y])
6      If [z] = [0]   /* QNR */
7        i ← i + 2^j
8        [y] ← O_SRDH([1], [ω^{-1}y])
9      Else
10       [y] ← [z]
11   Return i
```

```
1  Algorithm Square([1], [x], ω)
2    [y] ← O_SRDH([1], [x])
3    If [y] = [0]   /* QNR */
4      [y] ← O_SRDH([1], [ωy])
5    h ← (q+1)/2
6    While h > 2:
7      If h is even
8        [z] ← O_SRDH([1], [y])
9        If [z] = [0]   /* QNR */
10         [y] ← O_SRDH([1], [ωy])
11       Else
12         [y] ← [z]
13       h ← h/2
14     Else   /* h is odd */
15       [z] ← O_SRDH([x], [y])
16       If [z] = [0]   /* QNR */
17         [y] ← O_SRDH([x], [ωy])
18       Else
19         [y] ← [z]
20       h ← (h+1)/2
21   i ← FindExpi([1], [x], ω)
22   j ← FindExpi([1], [y], ω)
23   Return [ω^{2i−j}y]
```

Fig. 5. Algorithms FindExpi and Square used in the proof of Theorem 8

with exponent $i_t i_{t-1} \ldots i_1$. If $i_1 = 1$ then i is odd, and the oracle returns a new element with exponent $i_t i_{t-1} \ldots i_1$ on the second input. Which call returns a non-trivial group element tells us what i_0 is and in both cases we get a new element where the bits have been shifted right and a new most significant bit i_t has been added. Repeating t times allows us to learn all of $i = i_{t-1} \ldots i_0$.

Next we describe the Square algorithm. The idea behind this algorithm is that given $[\omega^i \beta]$ we want to compute $[\omega^j \beta^2]$ for some j, but we do not care much about the root of unity part, i.e., what j is, since we can always determine that by calling FindExpi and clean it up later. As a first step one of the inputs $([1], [x])$ or $([1], [\omega x])$ will correspond to a quadratic residue and the square root oracle will return some $[y] = [\omega^j \beta^{\frac{1}{2}}] = [\omega^j \beta^{\frac{q+1}{2}}]$. Let's define $h = \frac{q+1}{2}$, which is a positive integer, so we have $[y] = [\omega^j \beta^h]$.

The idea now is that we will use repeated applications of the SRDH oracle to halve h until we get down to $h = 2$. If h is even, this works fine as one of the pairs $([1], [\omega^j \beta^h])$ or $([1], [\omega^{j+1} \beta^h])$ will correspond to a quadratic residue and we get a new element of the form $[\omega^{j'} \beta^{h'}]$, where $h' = \frac{h}{2}$.

If h is odd, this strategy does not work directly. However, in this case we can use $[x]$ as generator instead of $[1]$ and have that one of the pairs $([\omega^i \beta], [\omega^j \beta^h])$ or $([\omega^i \beta], [\omega^{j+1} \beta^h])$ is a quadratic residue. By applying the square-root oracle, we get a new element of the form $[\omega^{j'} \beta^{h'}]$, where $h' = \frac{h+1}{2}$.

Repeated application of these two types of calls, depending on whether h is even or odd, eventually gives us an element of the form $[\omega^j \beta^2]$. At this stage, we can use the FindExpi algorithm to determine i and j, which makes it easy to compute $[x^2] = \omega^{2i-j}[\omega^j \beta^2]$.

We now analyse the time complexity of the algorithms. Algorithm FindExpi makes at most $2t$ oracle calls, whereas Square makes at most $2|q|$ oracle calls plus two invocations of FindExpi, i.e., at most $4t + 2|q|$ oracle calls in total.   □

Using an Adversarial $\mathcal{O}^*_{1-GDHE}$ Oracle. In Theorem 8, we assumed the reduction had a perfect $\mathcal{O}_{\text{SRDH}}$ oracle. Here we weaken the assumption used in the reduction and consider an $\mathcal{O}^*_{1\text{-GDHE}}$ oracle that returns a correct answer with a non-negligible probability ϵ when queried on a quadratic residue element. More precisely, let $([a], [b]) \in \mathbb{G}_p^2$ be the input we are about to query the $\mathcal{O}^*_{1\text{-GDHE}}$ oracle on. Since we can easily detect if $b = 0$, we can assume that we never need to query to oracle on any input where $b = 0$. When queried on $([a], [b]) \in \mathbb{G}_p^\times \times \mathbb{G}_p^\times$, the oracle will return either the symbol QNR, i.e. $[0] \in \mathbb{G}_p$ or $[c] \in \mathbb{G}_p$ for some $c \in \mathbb{Z}_p^\times$. Our assumption about correctness is

$$\Pr\left[(\mathbb{G}_p, [1]) \leftarrow \mathcal{G}(1^\kappa); a, z \leftarrow \mathbb{Z}_p^\times : \mathcal{O}^*_{1\text{-GDHE}}([a], [az^2]) = \pm[az]\right] \geq \epsilon(\kappa).$$

The oracle can behave arbitrarily when it does not return a correct answer or when the input is not a quadratic residue.

We will now show that we can rectify the adversarial behaviour of the oracle so that it cannot adapt its answer based on the instance input. The idea is to randomize the inputs to be queried to the oracle so that they are uniformly distributed over the input space and we get ϵ chance of getting a correct square-root when the input is a quadratic residue. To check the solution, we then randomize an element related to the answer, which we can use to detect when the oracle is misbehaving. The result is a Monte Carlo algorithm described in Fig. 6, which with probability $\epsilon' \geq \epsilon^2$ returns a correct square-root when queried on a quadratic residue and $\perp$ in all other cases.

```
1  Algorithm MOracle([a], [b]):
2      Sample  α, β, γ, δ, r, s ← Z_p^×
3      [y] ← O*_1-GDHE(α[a], αβ²[b])
4      [z] ← O*_1-GDHE(γ[a], γδ²(r²[a] + 2rs 1/αβ [y] + s²[b]))
5      If  1/γδ [z] = ±(r[a] + s/αβ [y])
6          Return  1/αβ [y]
7      Else
8          Return  ⊥
```

Fig. 6. Monte Carlo Algorithm using $\mathcal{O}^*_{1\text{-GDHE}}$

Lemma 3. *Using $\mathcal{O}^*_{1\text{-}GDHE}$ which returns a correct answer with probability ϵ, algorithm* MOracle *returns a correct answer with probability $\epsilon' \geq \epsilon^2$ when queried on a well-formed pair $([a], [b]) \in \mathbb{G}_p^\times \times \mathbb{G}_p^\times$ and $\perp$ otherwise.*

Proof. If $\frac{b}{a} \in \mathbb{QR}(p)$, we also have $\frac{\beta^2 b}{a} \in \mathbb{QR}(p)$, and when $\frac{b}{a} \notin \mathbb{QR}(p)$, we also have $\frac{\beta^2 b}{a} \notin \mathbb{QR}(p)$. We have probability ϵ that the answer $[y]$ is a correct answer when $\frac{b}{a} \in \mathbb{QR}(p)$ in which case $y = \pm \alpha\beta a \sqrt{\frac{b}{a}}$, we can thus recover $[\pm a\sqrt{\frac{b}{a}} = \pm\sqrt{ab}]$ by computing $\frac{1}{\alpha\beta}[y]$. Let $y' = \frac{y}{\alpha\beta} = \pm\sqrt{ab}$. We have probability ϵ that $[z]$ is a correct answer. Now let $z' = \frac{z}{\gamma\delta}$.

Note that $r^2 a + 2rsy' + s^2 b = (ra + sy')^2 + s^2(b - y'^2)$ and if $[z]$ is a correct answer then we have $z' = \pm(ra + sy')$. Thus, we have probability at least ϵ^2 that algorithm MOracle will return a correct square root when the input is well-formed.

We now argue that if $y' \neq \pm\sqrt{ab}$, with overwhelming probability the algorithm will return $\perp$. Let $\tau = a(r^2 a + 2rsy' + s^2 b) = r^2 a^2 + 2arsy' + s^2 ab$.

Since $a(r^2 a + 2rsy' + s^2 b) = (ra + sy')^2 + s^2(ab - y'^2)$, the query to the oracle is determined by a and τ, and there are roughly p^2 pairs (r, s) mapping into a maximum of p choices of τ. Therefore, for the same oracle query there are many possible values s could have. Now if $y' \neq \pm\sqrt{ab}$, i.e. y is an incorrect answer, then the oracle has negligible chance of passing the test $z' = \pm(ra + sy')$ in line 5. If the test passes, then $z'^2 = (ra + sy')^2 = r^2 a^2 + 2arsy' + s^2 ab = \tau - s^2(ab - y'^2)$. Since s is information-theoretically undetermined from a and τ, there is negligible chance over the choice of s that this equality holds unless $ab - y'^2 = 0$.   □

Since ϵ is non-negligible, there must be a constant $c > 0$ such that for infinitely many κ we have $\epsilon' \geq \kappa^{-c}$. We can use repetitions to boost the oracle to give the correct answer with overwhelming probability on these κ values, i.e., on quadratic residues it returns square-roots and on quadratic non-residues it returns $\perp$ or equivalently $[0]$. Chernoff-bounds ensure we only need $\frac{\kappa}{\epsilon^2}$ polynomially iterations of the Monte Carlo algorithm to build a good SRDH oracle.

4.3   The q-GDHE Family Structure

We say a family of assumptions $\{q\text{-A}\}$ is a *strictly increasingly stronger* family if for all polynomials $q \leq q'$ it holds that $q'\text{-A} \Rightarrow q\text{-A}$ but $q\text{-A} \not\Rightarrow (q'+1)\text{-A}$. A proof for the following theorem can be found in the full version [23].

Theorem 9. *The q-GDHE family is a strictly increasingly stronger family.*

5   Target Assumptions over Bilinear Groups

We now turn our attention to prime-order bilinear groups. Our reductions from the cyclic group setting translate into a bilinear framework that captures existing computational bilinear assumptions where the adversary's task is to compute a specific group element in the base groups or the target group. For instance, the bilinear variants of the matrix computational Diffie-Hellman assumption in [34] (which implies the (computational bilinear) k-linear assumptions), the (bilinear) q-SDH assumptions, and the bilinear assumptions studied in [26] are all examples of target assumptions in bilinear groups.

Definition 8 (Bilinear Group Generator). *A bilinear group generator is a PPT algorithm* $\mathcal{BG}$, *which on input a security parameter* κ *(given in unary) outputs bilinear group parameters* $(\mathbb{G}_1, \mathbb{G}_2, \mathbb{G}_T, G_1, G_2)$, *where*

- $\mathbb{G}_1, \mathbb{G}_2, \mathbb{G}_T$ *are cyclic groups of prime order* p *with bitlength* $|p| = \Theta(\kappa)$.
- $\mathbb{G}_1, \mathbb{G}_2, \mathbb{G}_T$ *have polynomial-time algorithms for carrying out group operations and unique representations for group elements.*
- *There is an efficiently computable bilinear map (pairing)* $e : \mathbb{G}_1 \times \mathbb{G}_2 \to \mathbb{G}_T$.
- G_1 *and* G_2 *are independently chosen uniformly random generators of* $\mathbb{G}_1$ *and* $\mathbb{G}_2$, *respectively, and* $e(G_1, G_2)$ *generates* $\mathbb{G}_T$.

Again, we will be working in the low/medium granularity setting so we always assume uniformly random generators of the base groups.

According to [21], bilinear groups of prime order can be classified into 3 main types depending on the existence of efficiently computable isomorphisms between the groups. In Type-1, $\mathbb{G}_1 = \mathbb{G}_2$. In Type-2 $\mathbb{G}_1 \neq \mathbb{G}_2$ and there is an isomorphism $\psi : \mathbb{G}_2 \to \mathbb{G}_1$ that is efficiently computable in one direction, whereas in Type-3 no efficient isomorphism between the groups in either direction exists. Type-3 bilinear groups are the most efficient and hence practically relevant and we therefore restrict our focus to this type, although much of this section also applies to Type-1 and Type-2 bilinear groups.

We will use $[x]_1, [y]_2, [z]_T$ to denote the group elements in the respective groups and as in the previous sections use additive notation for all group operations. This means the generators are $[1]_1, [1]_2$ and $e([1]_1, [1]_2) = [1]_T$. We will often denote the pairing with multiplicative notation, i.e., $[x]_1 \cdot [y]_2 = [xy]_T$.

A bilinear group generator can be seen as a particular example of a cyclic group generator generating $\mathbb{G}_1, \mathbb{G}_2$ or $\mathbb{G}_T$. All our results regarding non-interactive computational assumptions therefore still apply in the respective groups but in this section we will also cover the case where exponents are shared between the groups. The presence of the pairing e makes it possible for elements in the base groups $\mathbb{G}_1, \mathbb{G}_2$ to combine in the target group $\mathbb{G}_T$, so we can formulate assumptions that involve several groups, e.g., that given $[1]_1, [1]_2, [x]_T$ it is hard to compute $[x]_1$. In the following sections, we define and analyze non-interactive target assumptions in the bilinear group setting.

5.1 Target Assumptions in Bilinear Groups

We now define and analyze target assumptions, where the adversary's goal is to compute a group element in $\mathbb{G}_j$, where $j \in \{1, 2, T\}$. When we defined target assumptions over a single cyclic group, we gave the adversary group elements of the form $\left[\frac{a(x)}{b(x)}\right]$. In the bilinear group setting, the adversary may get a mix of group elements in all three groups. We note that if the adversary has $\left[\frac{a^{(1)}(x)}{b^{(1)}(x)}\right]_1$ and $\left[\frac{a^{(2)}(x)}{b^{(2)}(x)}\right]_2$ she can obtain $\left[\frac{a^{(1)}(x)a^{(2)}(x)}{b^{(1)}(x)b^{(2)}(x)}\right]_T$ via the pairing operation. When we define target assumptions in bilinear groups, we will therefore without loss of generality assume the fractional polynomials the instance generator outputs for the target group $\mathbb{G}_T$ include all products $\frac{a_i^{(1)}(X)a_j^{(2)}(X)}{b_i^{(1)}(X)b_j^{(2)}(X)}$ of fractional polynomials for elements in the base groups.

Definition 9 (Bilinear Target Assumption in $\mathbb{G}_j$). *Given polynomials $d(\kappa)$, $m(\kappa)$, $n_1(\kappa)$, $n_2(\kappa)$, and $n_T(\kappa)$ we say $(\mathcal{I}, \mathcal{V})$ is a (d, m, n_1, n_2, n_T)-bilinear target assumption in $\mathbb{G}_j$ for $\mathcal{BG}$ if it works as follows:*

$(pub, priv) \leftarrow \mathcal{I}(1^\kappa)$: *There is a PPT algorithm $\mathcal{I}^{\mathsf{core}}$ defining $\mathcal{I}$ as follows:*

1. $\left(\mathbb{G}_1, \mathbb{G}_2, \mathbb{G}_T, [1]_1, [1]_2\right) \leftarrow \mathcal{BG}(1^\kappa); \mathsf{bgp} := (\mathbb{G}_1, \mathbb{G}_2, \mathbb{G}_T)$
2. $\left(\left\{\frac{a_i^{(1)}(X)}{b_i^{(1)}(X)}\right\}_{i=1}^{n_1}, \left\{\frac{a_i^{(2)}(X)}{b_i^{(2)}(X)}\right\}_{i=1}^{n_2}, \left\{\frac{a_i^{(T)}(X)}{b_i^{(T)}(X)}\right\}_{i=1}^{n_T}, pub', priv'\right) \leftarrow \mathcal{I}^{\mathsf{core}}(\mathsf{bgp})$
3. $x \leftarrow \mathbb{Z}_p^m$ *conditioned on* $b_i^{(j)}(x) \neq 0$ *for all choices of i and j*
4. $pub := \left(\mathbb{G}_1, \mathbb{G}_2, \mathbb{G}_T, \left\{\left\{\left[\frac{a_i^{(j)}(x)}{b_i^{(j)}(x)}\right]_j\right\}_{i=1}^{n_j}, \left\{\frac{a_i^{(j)}(X)}{b_i^{(j)}(X)}\right\}_{i=1}^{n_j}\right\}_{j=1,2,T}, pub'\right)$
5. *Return* $(pub, priv := ([1]_1, [1]_2, x, priv'))$

$b \leftarrow \mathcal{V}\left(pub, priv, sol = \left(r(X), s(X), [y]_j, sol'\right)\right)$: *There is a DPT algorithm $\mathcal{V}^{\mathsf{core}}$ such that $\mathcal{V}$ returns 1 if all of the following checks pass and 0 otherwise:*

1. $r(X) \prod_{i=1}^{n_j} b_i^{(j)}(X) \notin \mathrm{span}\left(\left\{s(X)a_i^{(j)}(X) \prod_{\ell \neq i} b_\ell^{(j)}(X)\right\}_{i=1}^{n_j}\right)$
2. $[y]_j = \frac{r(x)}{s(x)}[1]_j$
3. $\mathcal{V}^{\mathsf{core}}(pub, priv, sol) = 1$

We require that the number of variables in X is $m(\kappa)$, the total degrees of the polynomials are bounded by $d(\kappa)$, and that all products of polynomial fractions in $\mathbb{G}_1$ and $\mathbb{G}_2$ are included in the polynomial fractions in $\mathbb{G}_T$.

Also, since the pairing function allows one to obtain the product of any two polynomials from the opposite source groups in the target group, for assumptions where the required target element is in $\mathbb{G}_T$, the degree of the polynomials $r(X)$ and $s(X)$ the adversary specifies is upper bounded by $2d$ instead of d.

Similarly to the single group setting, we can reduce any bilinear target assumption to a simple bilinear target assumption, where all $b_i^{(1)}(X) = b_i^{(2)}(X) = b_i^{(T)}(X) = 1$. Also, we can reduce bilinear target assumptions with multivariate polynomials to bilinear target assumptions with univariate polynomials.

We can then consider two cases depending on whether or not the polynomial $s(X)$ divides $r(X)$. Just as in the single cyclic group setting, we get that all bilinear target assumptions can be reduced to the following two assumptions.

Definition 10 (Bilinear Polynomial Assumption in $\mathbb{G}_j$). *We say a $(d, 1, n_1, n_2, n_T)$-simple bilinear target assumption $(\mathcal{I}, \mathcal{V})$ in $\mathbb{G}_j$ for $\mathcal{BG}$ is a (d, n_1, n_2, n_T)-bilinear polynomial assumption in $\mathbb{G}_j$ if $\mathcal{V}$ only accepts solutions where $s(X) = 1$.*

Definition 11 (Bilinear Fractional Assumption in $\mathbb{G}_j$ (BFrac$_j$)). *We say a $(d, 1, n_1, n_2, n_T)$-simple bilinear target assumption $(\mathcal{I}, \mathcal{V})$ in $\mathbb{G}_j$ for $\mathcal{BG}$ is a (d, n_1, n_2, n_T)-bilinear fractional assumption in $\mathbb{G}_j$ if $\mathcal{V}$ only accepts solutions where $s(X) \nmid r(X)$.*

It is straightforward to prove the following theorem, which states that the (d, n_1, n_2, n_T)-BFrac$_i$ assumption can be further simplified to only consider the target group case.

Theorem 10. *If the (d, n_1, n_2, n_T)-BFrac$_T$ assumption over $\mathcal{BG}$ holds, then the assumptions (d, n_1, n_2, n_T)-BFrac$_1$ and (d, n_1, n_2, n_T)-BFrac$_2$ over $\mathcal{BG}$ also hold.*

5.2   Bilinear Target Assumptions in the Base Groups

All the results in this section are assuming a Type-3 bilinear group. Intuitively, bilinear target assumptions in $\mathbb{G}_1$ and $\mathbb{G}_2$ are very similar to the cyclic group case because the generic computations one can do in a base group are not affected by group elements in the other groups. We will now formalize this intuition by generalizing the q-GDHE and the q-SFrac assumptions to the bilinear setting.

Definition 12 (q-Bilinear GDHE Assumption in $\mathbb{G}_1$ (q-BGDHE$_1$)). *The q-BGDHE assumption in $\mathbb{G}_1$ over $\mathcal{BG}$ is that for all PPT adversaries $\mathcal{A}$*

$$\Pr\left[\begin{array}{l}(\mathbb{G}_1, \mathbb{G}_2, \mathbb{G}_T, [1]_1, [1]_2) \leftarrow \mathcal{BG}(1^\kappa); x \leftarrow \mathbb{Z}_p : \\ [x^q]_1 \leftarrow \mathcal{A}\left(\mathbb{G}_1, \mathbb{G}_2, \mathbb{G}_T, [1, x, \ldots, x^{q-1}, x^{q+1}, \ldots, x^{2q}]_1, [1, x, \ldots, x^{2q}]_2\right)\end{array}\right] \approx 0.$$

The q-BGDHE assumption in $\mathbb{G}_2$ (q-BGDHE$_2$) over $\mathcal{BG}$ is defined similarly.

Definition 13 (q-Bilinear SFrac in $\mathbb{G}_i$ Assumption[3] (q–BSFrac$_i$)). *The q-BSFrac$_i$ for $i \in \{1, 2, T\}$ over $\mathcal{BG}$ is that for all PPT adversaries $\mathcal{A}$*

$$\Pr\left[\begin{array}{l}(\mathbb{G}_1, \mathbb{G}_2, \mathbb{G}_T, [1]_1, [1]_2) \leftarrow \mathcal{BG}(1^\kappa); x \leftarrow \mathbb{Z}_p; \\ (r(X), s(X), [y]_i) \leftarrow \mathcal{A}\left(\mathbb{G}_1, \mathbb{G}_2, \mathbb{G}, [1, x, \ldots, x^q]_1, [1, x, \ldots, x^q]_2\right) : \\ q \geq \deg(s) > \deg(r) \geq 0 \text{ and } [y]_i = \left[\frac{r(x)}{s(x)}\right]_i\end{array}\right] \approx 0.$$

---

[3] The q-BSFrac$_1$ assumption is the same as the q-co-SDH assumption in [20].

Similarly to Theorem 4, we can prove that all bilinear polynomial assumptions in the base group $\mathbb{G}_j$ for $j \in \{1, 2\}$ are implied by the q-BGDHE$_j$ assumption. Also, similarly to Theorem 5, we can show that all bilinear fractional assumptions in the base group $\mathbb{G}_j$ for $j \in \{1, 2\}$ are implied by the q-BSFrac$_j$ assumption. For bilinear target assumptions in the base groups, we therefore get a situation very similar to the single cyclic group case, which we express in the following theorem:

Theorem 11. *For $j \in \{1, 2\}$ there is a polynomial $q(d, m, n_1, n_2, n_T)$ such that the joint q-BSFrac$_j$ and q-BGDHE$_j$ assumption imply all (d, m, n_1, n_2, n_T) bilinear target assumptions in $\mathbb{G}_j$.*

By the above theorem, the joint q-BGDHE$_j$ and q-BSFrac$_j$ assumption serves as an Uber assumption for all bilinear target assumptions in the base group $\mathbb{G}_j$.

5.3   Bilinear Target Assumptions in the Target Group

We now consider bilinear target assumptions in the target group $\mathbb{G}_T$. This is different from the case of bilinear target assumptions in the base groups since the bilinear map allow one to map elements from the base groups into the target group resulting new elements in that group.

Consider a polynomial assumption. We can use matrices $\mathbf{A}$ and $\mathbf{B}$ to represent the polynomials in $\mathbb{G}_1$ and $\mathbb{G}_2$ as $[1 \ X \ \ldots \ X^d]_1 \mathbf{A}$ and $[X^d \ \ldots \ X \ 1]_2 \mathbf{B}$. Given group elements with evaluations of these polynomials one may use the pairing to compute products of the polynomial evaluations in the target group $\mathbb{G}_T$. Observe that $\mathbf{A}^T\mathbf{B}$ represents the coefficients of the X^d terms in all the possible pairings of two elements in $\mathbb{G}_1$ and $\mathbb{G}_2$. We will use this to formulate a target group analogue of the q-BGDHE assumptions.

Definition 14 (q-Bilinear Gap Assumption (q-BGap)). *We say the q-BGapassumption holds when for all PPT $\mathcal{A}$*

$$\Pr \left[\begin{array}{l} (\mathbb{G}_1, \mathbb{G}_2, \mathbb{G}_T, [1]_1, [1]_2) \leftarrow \mathcal{BG}(1^\kappa); (\mathbf{A}, \mathbf{B}) \leftarrow \mathcal{A}(\mathbb{G}_1, \mathbb{G}_2, \mathbb{G}_T); \ x \leftarrow \mathbb{Z}_p \\ : \quad [x^q]_T \leftarrow \mathcal{A} \left(\begin{array}{l} [1 \ x \ \ldots \ x^q]_1 \mathbf{A}, [x^q \ldots \ x \ 1]_2 \mathbf{B}, \\ [1, x, \ldots, x^{q-1}]_T, [x^{q+1}, \ldots, x^{2q}]_{\{1,2,T\}} \end{array} \right) \\ \text{and } \mathbf{A}^T\mathbf{B} = \mathbf{0} \text{ and } rank(\mathbf{A}) + rank(\mathbf{B}) = q + 1 \end{array} \right] \approx 0 \cdot$$

Let $\mathbf{A} \in \mathbb{Z}_p^{(q+1) \times n_1}$ and $\mathbf{B} \in \mathbb{Z}_p^{(q+1) \times n_2}$ where without loss of generality we can assume their ranks are n_1 and n_2, respectively, and $n_1 + n_2 = q + 1$. Note that for smaller dimensions we can always add a column in the null space of $\mathbf{A}^T$ to $\mathbf{B}$ and still have $\mathbf{A}^T\mathbf{B} = \mathbf{0}$.

Theorem 12. *For any (d, n_1, n_2, n_T)-polynomial assumption $\mathsf{A} = (\mathcal{I}_\mathsf{A}, \mathcal{V}_\mathsf{A})$ in $\mathbb{G}_T$ over $\mathcal{BG}$, we have that $(3d + 1)$-BGap $\Rightarrow \mathsf{A}$.*

Proof. Let $\mathcal{A}$ be an adversary against the (d, n_1, n_2, n_T)-polynomial assumption. We build an adversary $\mathcal{B}$ against the $(3d + 1)$-BGap assumption which uses $\mathcal{A}$ in a black-box manner. Adversary $\mathcal{B}$ runs $\mathcal{I}_{\mathcal{A}}^{\mathsf{core}}(\mathbb{G}_1, \mathbb{G}_2, \mathbb{G}_T)$ to get the tuple $\left(\left\{ \left\{ a_i^{(j)}(X) \right\}_{i=1}^{n_j} \right\}_{j=1,2,T}, pub_{\mathsf{A}}', priv_{\mathsf{A}}' \right)$. Now, $\mathcal{B}$ randomly chooses a polynomial $c(X) \leftarrow \mathbb{Z}_p[X]$ where $\deg(c) \leq 2d + 1$ conditioned on all $a_i^{(T)}(X)c(X)X^d$ having coefficient 0 in the term for X^{3d+1}. Define the following polynomials

$$a_i'^{(1)}(X) = a_i^{(1)}(X)c(X) = \sum_{j=0}^{3d+1} \alpha_{i,j}X^j, \text{ for } i = 1, \ldots, n_1$$

$$a_i'^{(2)}(X) = a_i^{(2)}(X)X^d = \sum_{j=0}^{2d} \beta_{i,j}X^j, \text{ for } i = 1, \ldots, n_2$$

$$a_i'^{(T)}(X) = a_i^{(T)}(X)c(X)X^d = \sum_{j=0}^{5d+1} \gamma_{i,j}X^j, \text{ for } i = 1, \ldots, n_T$$

and define the matrices $\mathbf{A} \in \mathbb{Z}_p^{(3d+2) \times n_1}$ and $\mathbf{B} \in \mathbb{Z}_p^{(3d+2) \times (3d+2-n_1)}$ as follows:

$$\mathbf{A} = \begin{pmatrix} \alpha_{1,0} & \cdots & \alpha_{n_1,0} \\ \vdots & & \vdots \\ \alpha_{1,3d+1} & \cdots & \alpha_{n_1,3d+1} \end{pmatrix} \text{ and } \mathbf{B} = \begin{pmatrix} \beta_{1,3d+1} & \cdots & \beta_{n_2,3d+1} & \cdots \\ \vdots & & \vdots & \vdots & \beta_{i,j}' & \vdots \\ \beta_{1,0} & \cdots & \beta_{n_2,0} & \cdots \end{pmatrix},$$

where the values $\beta_{i,j}' \in \mathbb{Z}_p$ are chosen as an arbitrary extension to give us $\text{rank}(\mathbf{A}) + \text{rank}(\mathbf{B}) = 3d + 2$ in a way such that we still have $\mathbf{A}^T\mathbf{B} = 0$. Adversary $\mathcal{B}$ now forwards $\mathbf{A}$ and $\mathbf{B}$ to her environment and gets back the tuple

$$\left(\begin{matrix} \left[1\ x\ \cdots\ x^{3d+1} \right]_1 \mathbf{A}, \left[x^{3d+1}\ \cdots\ x\ 1 \right]_2 \mathbf{B}, \\ \left[1, x, \ldots, x^{3d} \right]_T, \left[x^{3d+2}, \ldots, x^{6d+2} \right]_{\{1,2,T\}} \end{matrix} \right),$$

using which she can compute all of the group elements $[a_i'^{(1)}(x)]_1$, $[a_i'^{(2)}(x)]_2$, and $[a_i'^{(T)}(x)]_T$. Adversary $\mathcal{B}$ now starts $\mathcal{A}$ on the tuple

$$pub := \left(\left\{ \left\{ \left[a_i'^{(j)}(x) \right]_j \right\}_{i=1}^{n_j}, \left\{ a_i^{(j)}(X) \right\}_{i=1}^{n_j}, \right\}_{j=1,2,T}, pub_{\mathsf{A}}' \right)$$

and gets $(r(X), [y]_T, sol')$ as an answer, where $r(X) \notin \text{span}\left(\left\{ a_i^{(T)}(X) \right\}_{i=1}^{n_T} \right)$. Parse $r(X)c(X)X^d$ as $\sum_{i=0}^{5d+1} \gamma_i X^i$. Using the tuple available to her, $\mathcal{B}$ can recover $[x^{3d+1}]_T$ by computing $\frac{1}{\gamma_{3d+1}}([y]_T - [\sum_{i \neq 3d+1} \gamma_i x^i]_T)$.

Assume $c(x) \neq 0$, which happens with probability $1 - \frac{2d+1}{p}$, and $x^d \neq 0$, which happens with probability $1 - \frac{d}{p}$. The instance $\mathcal{A}$ sees is indistinguishable from an instance she gets directly from the instance generator with generators $[c(x)]_1$, $[x^d]_2$ and $[c(x)x^d]_T$. By Lemma 2, the probability that $r(X)c(X)X^d$ has a 0 coefficient of X^{3d+1} is $\frac{1}{p}$. Thus, if $\mathcal{A}$ has probability $\epsilon_{\mathcal{A}}$ against assumption A, we have $\epsilon_{\mathcal{B}} = \epsilon_{\mathcal{A}} - \frac{3d+2}{p}$ is the probability of $\mathcal{B}$ against the $(3d+1)$-BGap assumption. □

We conclude with the following theorem.

Theorem 13. *There is a polynomial $q(d, m, n_1, n_2, n_T)$ such that the joint q-BSFrac$_T$ and q-BGap assumption imply all (d, m, n_1, n_2, n_T) bilinear target assumptions in $\mathbb{G}_T$.*

References

1. Abdalla, M., Benhamouda, F., Passelègue, A.: An algebraic framework for pseudorandom functions and applications to related-key security. In: Gennaro, R., Robshaw, M. (eds.) CRYPTO 2015. LNCS, vol. 9215, pp. 388–409. Springer, Heidelberg (2015). https://doi.org/10.1007/978-3-662-47989-6_19
2. Ambrona, M., Barthe, G., Schmidt, B.: Automated unbounded analysis of cryptographic constructions in the generic group model. In: Fischlin, M., Coron, J.-S. (eds.) EUROCRYPT 2016. LNCS, vol. 9666, pp. 822–851. Springer, Heidelberg (2016). https://doi.org/10.1007/978-3-662-49896-5_29
3. Bao, F., Deng, R.H., Zhu, H.F.: Variations of Diffie-Hellman problem. In: Qing, S., Gollmann, D., Zhou, J. (eds.) ICICS 2003. LNCS, vol. 2836, pp. 301–312. Springer, Heidelberg (2003). https://doi.org/10.1007/978-3-540-39927-8_28
4. Barthe, G., Fagerholm, E., Fiore, D., Mitchell, J., Scedrov, A., Schmidt, B.: Automated analysis of cryptographic assumptions in generic group models. In: Garay, J.A., Gennaro, R. (eds.) CRYPTO 2014. LNCS, vol. 8616, pp. 95–112. Springer, Heidelberg (2014). https://doi.org/10.1007/978-3-662-44371-2_6
5. Bellare, M., Palacio, A.: The knowledge-of-exponent assumptions and 3-round zero-knowledge protocols. In: Franklin, M. (ed.) CRYPTO 2004. LNCS, vol. 3152, pp. 273–289. Springer, Heidelberg (2004). https://doi.org/10.1007/978-3-540-28628-8_17
6. Bellare, M., Waters, B., Yilek, S.: Identity-based encryption secure against selective opening attack. In: Ishai, Y. (ed.) TCC 2011. LNCS, vol. 6597, pp. 235–252. Springer, Heidelberg (2011). https://doi.org/10.1007/978-3-642-19571-6_15
7. Boer, B.: Diffie-Hellman is as strong as discrete log for certain primes. In: Goldwasser, S. (ed.) CRYPTO 1988. LNCS, vol. 403, pp. 530–539. Springer, New York (1990). https://doi.org/10.1007/0-387-34799-2_38
8. Boneh, D., Boyen, X.: Short signatures without random oracles and the SDH assumption in bilinear groups. J. Crypt. **21**(2), 149–177 (2008)
9. Boneh, D., Boyen, X., Goh, E.-J.: Hierarchical identity based encryption with constant size ciphertext. In: Cramer, R. (ed.) EUROCRYPT 2005. LNCS, vol. 3494, pp. 440–456. Springer, Heidelberg (2005). https://doi.org/10.1007/11426639_26
10. Boneh, D., Gentry, C., Waters, B.: Collusion resistant broadcast encryption with short ciphertexts and private keys. In: Shoup, V. (ed.) CRYPTO 2005. LNCS, vol. 3621, pp. 258–275. Springer, Heidelberg (2005). https://doi.org/10.1007/11535218_16
11. Boyen, X.: The Uber-assumption family. In: Galbraith, S.D., Paterson, K.G. (eds.) Pairing 2008. LNCS, vol. 5209, pp. 39–56. Springer, Heidelberg (2008). https://doi.org/10.1007/978-3-540-85538-5_3

12. Boyen, X., Waters, B.: Full-domain subgroup hiding and constant-size group signatures. In: Okamoto, T., Wang, X. (eds.) PKC 2007. LNCS, vol. 4450, pp. 1–15. Springer, Heidelberg (2007). https://doi.org/10.1007/978-3-540-71677-8_1
13. Bresson, E., Chevassut, O., Pointcheval, D.: The group Diffie-Hellman problems. In: Nyberg, K., Heys, H. (eds.) SAC 2002. LNCS, vol. 2595, pp. 325–338. Springer, Heidelberg (2003). https://doi.org/10.1007/3-540-36492-7_21
14. Bresson, E., Lakhnech, Y., Mazaré, L., Warinschi, B.: A generalization of DDH with applications to protocol analysis and computational soundness. In: Menezes, A. (ed.) CRYPTO 2007. LNCS, vol. 4622, pp. 482–499. Springer, Heidelberg (2007). https://doi.org/10.1007/978-3-540-74143-5_27
15. Chase, M., Maller, M., Meiklejohn, S.: Déjà Q all over again: tighter and broader reductions of q-type assumptions. In: Cheon, J.H., Takagi, T. (eds.) ASIACRYPT 2016. LNCS, vol. 10032, pp. 655–681. Springer, Heidelberg (2016). https://doi.org/10.1007/978-3-662-53890-6_22
16. Chase, M., Meiklejohn, S.: Déjà Q: using dual systems to revisit q-type assumptions. In: Nguyen, P.Q., Oswald, E. (eds.) EUROCRYPT 2014. LNCS, vol. 8441, pp. 622–639. Springer, Heidelberg (2014). https://doi.org/10.1007/978-3-642-55220-5_34
17. Cheon, J.H.: Security analysis of the strong Diffie-Hellman problem. In: Vaudenay, S. (ed.) EUROCRYPT 2006. LNCS, vol. 4004, pp. 1–11. Springer, Heidelberg (2006). https://doi.org/10.1007/11761679_1
18. Dent, A.W.: Adapting the weaknesses of the random oracle model to the generic group model. In: Zheng, Y. (ed.) ASIACRYPT 2002. LNCS, vol. 2501, pp. 100–109. Springer, Heidelberg (2002). https://doi.org/10.1007/3-540-36178-2_6
19. Escala, A., Herold, G., Kiltz, E., Ràfols, C., Villar, J.: An algebraic framework for Diffie-Hellman assumptions. In: Canetti, R., Garay, J.A. (eds.) CRYPTO 2013. LNCS, vol. 8043, pp. 129–147. Springer, Heidelberg (2013). https://doi.org/10.1007/978-3-642-40084-1_8
20. Fuchsbauer, G., Hanser, C., Slamanig, D.: Structure-preserving signatures on equivalence classes and constant-size anonymous credentials. Cryptology ePrint Archive, Report 2014/944 (2014)
21. Galbraith, S.D., Paterson, K.G., Smart, N.P.: Pairings for cryptographers. Discrete Appl. Math. 156(16), 3113–3121 (2008)
22. Gennaro, R., Gentry, C., Parno, B., Raykova, M.: Quadratic span programs and succinct NIZKs without PCPs. In: Johansson, T., Nguyen, P.Q. (eds.) EURO-CRYPT 2013. LNCS, vol. 7881, pp. 626–645. Springer, Heidelberg (2013). https://doi.org/10.1007/978-3-642-38348-9_37
23. Ghadafi, E., Groth, J.: Towards a classification of non-interactive computational assumptions in cyclic groups. Cryptology ePrint Archive, Report 2017/343 (2017). http://eprint.iacr.org/2017/343
24. Goldwasser, S., Tauman Kalai, Y.: Cryptographic assumptions: a position paper. In: Kushilevitz, E., Malkin, T. (eds.) TCC 2016. LNCS, vol. 9562, pp. 505–522. Springer, Heidelberg (2016). https://doi.org/10.1007/978-3-662-49096-9_21
25. Jager, T., Schwenk, J.: On the analysis of cryptographic assumptions in the generic ring model. J. Crypt. 26(2), 225–245 (2012)
26. Joux, A., Rojat, A.: Security ranking among assumptions within the *Uber Assumption* framework. In: Desmedt, Y. (ed.) ISC 2013. LNCS, vol. 7807, pp. 391–406. Springer, Cham (2015). https://doi.org/10.1007/978-3-319-27659-5_28
27. Kiltz, E.: A tool box of cryptographic functions related to the Diffie-Hellman function. In: Rangan, C.P., Ding, C. (eds.) INDOCRYPT 2001. LNCS, vol. 2247, pp. 339–349. Springer, Heidelberg (2001). https://doi.org/10.1007/3-540-45311-3_32

28. Koblitz, N., Menezes, A.: Another look at generic groups. Adv. Math. Commun. **1**(1), 13–28 (2007)
29. Konoma, C., Mambo, M., Shizuya, H.: Complexity analysis of the cryptographic primitive problems through square-root exponent. IEICE Trans. **E87–A**(5), 1083–1091 (2004)
30. Maurer, U.: Abstract models of computation in cryptography. In: Smart, N.P. (ed.) Cryptography and Coding 2005. LNCS, vol. 3796, pp. 1–12. Springer, Heidelberg (2005). https://doi.org/10.1007/11586821_1
31. Maurer, U.M.: Towards the equivalence of breaking the Diffie-Hellman protocol and computing discrete logarithms. In: Desmedt, Y.G. (ed.) CRYPTO 1994. LNCS, vol. 839, pp. 271–281. Springer, Heidelberg (1994). https://doi.org/10.1007/3-540-48658-5_26
32. Maurer, U.M., Wolf, S.: Diffie-Hellman oracles. In: Koblitz, N. (ed.) CRYPTO 1996. LNCS, vol. 1109, pp. 268–282. Springer, Heidelberg (1996). https://doi.org/10.1007/3-540-68697-5_21
33. Mitsunari, S., Sakai, R., Kasahara, M.: A new traitor tracing. IEICE Trans. **E85–A**(2), 481–484 (2002)
34. Morillo, P., Ràfols, C., Villar, J.L.: The kernel matrix Diffie-Hellman assumption. In: Cheon, J.H., Takagi, T. (eds.) ASIACRYPT 2016. LNCS, vol. 10031, pp. 729–758. Springer, Heidelberg (2016). https://doi.org/10.1007/978-3-662-53887-6_27
35. Naor, M.: On cryptographic assumptions and challenges. In: Boneh, D. (ed.) CRYPTO 2003. LNCS, vol. 2729, pp. 96–109. Springer, Heidelberg (2003). https://doi.org/10.1007/978-3-540-45146-4_6
36. Nechaev, V.I.: Complexity of a determinate algorithm for the discrete logarithm. Math. Zametki **55**(2), 91–101 (1994)
37. Roh, D., Hahn, S.G.: The square root Diffie-Hellman problem. Des. Codes Crypt. **62**(2), 179–187 (2012)
38. Sadeghi, A.-R., Steiner, M.: Assumptions related to discrete logarithms: why subtleties make a real difference. In: Pfitzmann, B. (ed.) EUROCRYPT 2001. LNCS, vol. 2045, pp. 244–261. Springer, Heidelberg (2001). https://doi.org/10.1007/3-540-44987-6_16
39. Shoup, V.: Lower bounds for discrete logarithms and related problems. In: Fumy, W. (ed.) EUROCRYPT 1997. LNCS, vol. 1233, pp. 256–266. Springer, Heidelberg (1997). https://doi.org/10.1007/3-540-69053-0_18

An Efficient Pairing-Based Shuffle Argument

Prastudy Fauzi[1]([✉]), Helger Lipmaa[2], Janno Siim[2,3], and Michał Zając[2]

[1] Aarhus University, Aarhus, Denmark
`prastudy.fauzi@gmail.com`
[2] Institute of Computer Science, University of Tartu, Tartu, Estonia
[3] STACC, Tartu, Estonia

Abstract. We construct the most efficient known pairing-based NIZK shuffle argument. It consists of three subarguments that were carefully chosen to obtain optimal efficiency of the shuffle argument:
1. A same-message argument based on the linear subspace QANIZK argument of Kiltz and Wee,
2. A (simplified) permutation matrix argument of Fauzi, Lipmaa, and Zając,
3. A (simplified) consistency argument of Groth and Lu.

We prove the knowledge-soundness of the first two subarguments in the generic bilinear group model, and the culpable soundness of the third subargument under a KerMDH assumption. This proves the soundness of the shuffle argument. We also discuss our partially optimized implementation that allows one to prove a shuffle of $100\,000$ ciphertexts in less than a minute and verify it in less than 1.5 min.

Keywords: Common Reference String · Generic group model · Mix-net · Shuffle argument · Zero knowledge

1 Introduction

Consider the case of using mix-networks [9] in e-voting, where n voters individually encrypt their vote using a blindable public-key cryptosystem and send the encrypted votes to a bulletin board. After the vote casting period ends, the first mix-server gets all encrypted votes from the bulletin board. The mix-servers are ordered sequentially, creating a mix-network, and it is assumed that some of them are honest. The kth mix-server obtains input ciphertexts $(\mathfrak{M}_i)_{i=1}^n$, shuffles them, and sends the resulting ciphertext tuple $(\mathfrak{M}_i')_{i=1}^n$ to the next mix-server. Shuffling means that the mix-server generates a random permutation $\sigma \leftarrow_r S_n$ and a vector t of randomizers, and sets $\mathfrak{M}_i' = \mathfrak{M}_{\sigma(i)} + \mathsf{Enc}_{\mathsf{pk}}(\mathsf{o}; t_i)$.[1]

If at least one of the mix-servers behaves honestly, the link between a voter and his votes is completely hidden. However, in the malicious model, a corrupt

---

[1] Throughout this paper, we use additive notation combined with the bracket notation of [13]. We also denote group elements by using the Fraktur script as in $\mathfrak{M}_i$ or o. Thus, adding $\mathsf{Enc}_{\mathsf{pk}}(\mathsf{o}; t_i)$ results in a blinded version of $\mathfrak{M}_{\sigma(i)}$.

© International Association for Cryptologic Research 2017
T. Takagi and T. Peyrin (Eds.): ASIACRYPT 2017, Part II, LNCS 10625, pp. 97–127, 2017.
https://doi.org/10.1007/978-3-319-70697-9_4

mix-server can do an incorrect shuffle, resulting in a set of decrypted votes that do not reflect the original voters' votes. Hence there needs to be some additional steps to achieve security against corruption.

The cryptographically prudent way to proceed is to get each mix-server to prove in zero-knowledge [18] that her shuffle was done correctly. The resulting proof is known as a *(zero-knowledge) shuffle argument*. Based on earlier work [25,34], in CT-RSA 2016, Fauzi and Lipmaa (FL, [14]) proposed the then most efficient shuffle argument in the common reference string (CRS, [8]) model in terms of prover's computation.[2] Importantly, the FL shuffle argument is based on the standard Elgamal cryptosystem. The culpable soundness [25,26] of the FL shuffle argument is proven under a knowledge assumption [10] and three computational assumptions. Intuitively, culpable soundness means that if a cheating adversary produces an invalid shuffle (yet accepted by the verifier) together with the secret key, then one can break one of the underlying knowledge or computational assumptions.

While the FL shuffle argument is quite efficient for the prover, it is quite inefficient for the verifier. More precisely, while the prover's online cost is only dominated by $4n$ exponentiations in $\mathbb{G}_1$, the verifier's online cost is dominated by $8n$ pairings. (See Table 1.) Depending on the concrete implementation, a pairing can be up to 8 times slower than a $\mathbb{G}_1$ exponentiation. Such a large gap is non-satisfactory since verification time is more important in practice.

In ACNS 2016, González and Ràfols [20] proposed a new shuffle argument that importantly relies only on standard (falsifiable) assumptions. However, they achieve this (and efficient computation) by allowing the CRS to be quadratically long in n. Since in e-voting applications one could presumably have $n > 2^{20}$, quadratic CRS length will not be acceptable in such applications.

In Asiacrypt 2016, Fauzi, Lipmaa and Zając (FLZ, [15]) improved on the efficiency of the FL shuffle argument by proving knowledge-soundness (not culpable soundness) in the generic bilinear group model (GBGM, [35,39]). By additionally using batching techniques, they sped up the verification time of the FL shuffle argument approximately 3.5 times and the online verification time approximately twice. Here, the precise constants depend on how efficient operations such pairings and exponentiations in different groups are relative to each other; they do not provide an implementation of their shuffle.

However, due to the construction of the FLZ argument, they need each message to be encrypted twice, in $\mathbb{G}_1$ and $\mathbb{G}_2$. Since Elgamal cannot guarantee security in such a case, they use the non-standard ILin cryptosystem of Escala *et al.* [13]. This means that each ciphertext in this case will consist of 6 group elements, which makes storing and transmitting them more burdensome.

This results in several concrete drawbacks. First, since the prover's online complexity includes shuffling the ciphertexts and uses a non-standard cryptosystem (amongst other things), the prover's online complexity in the FLZ shuffle

---

[2] Many random-oracle model shuffle arguments are known, such as [3,16,22]. We will not provide comparisons with such arguments or discussions about the benefits of the CRS vs the random oracle model. We remark that the CRS can be created by using multi-party computation, see, e.g., [5].

argument is more than 4 times worse (using the comparison given in [15]) than in the FL shuffle argument. Second, since plaintexts in this shuffle argument are elements of $\mathbb{Z}_q$, decryption requires computing a discrete logarithm, which means that plaintexts must be small elements of $\mathbb{Z}_q$ (e.g. only 40 bits long). This rules out voting mechanisms with sufficiently complex ballots, such as the single transferable vote. Third, the GBGM knowledge-soundness proof of this shuffle argument means that there must exist a generic adversary that knows the discrete logarithm of each ballot submitted by each individual voter. Such a version of soundness (named *white-box soundness* in [14]) seems to be more dubious than culpable soundness achieved by the Groth-Lu [25] and FL shuffle arguments. (See [14] for more thorough comparison of these two soundness definitions.) Fourth, the CRS of this shuffle argument has public keys in both $\mathbb{G}_1$ and $\mathbb{G}_2$, which makes it more difficult to design an efficient shuffle or to prove its soundness in the GBGM. Indeed, [15] used a computer algebra system to derive knowledge-soundness of their shuffle argument.

This brings us to the main question of this paper:

Is it possible to construct a NIZK shuffle argument that shares the best features of the FL and the FLZ shuffle arguments? That is, it would use standard Elgamal (and in only one group), be (non-whitebox) sound, have linear-length CRS, have prover as efficient or better than in the FL shuffle argument, and have verifier as efficient or better than in the FLZ shuffle argument. Moreover, can one simplify (significantly?) the soundness proof of the FLZ shuffle argument while not losing in efficiency?

Our Constructions. We answer the main question positively, constructing a new pairing-based NIZK shuffle argument that is more efficient than prior work in essentially all parameters. As in [14], we use the Elgamal cryptosystem (with plaintexts in $\mathbb{G}_2$), which means that unlike [15], compatible voting mechanisms are not restricted by the size of the plaintext space. We construct more efficient subarguments, which sometimes leads to a significant efficiency gain. Since the CRS has very few elements from $\mathbb{G}_2$, the new shuffle has much simpler soundness proofs than in the case of the FLZ shuffle argument. Moreover, as in [14,25] (but not in [15,34]), we do not give the generic adversary in our soundness proof access to the discrete logarithms of encrypted messages. Our high-level approach in the shuffle argument is as follows; it is similar to the approach in the FL shuffle argument except that we use (significantly) more efficient subarguments.

We first let the prover choose a permutation matrix and commit separately to its every row. The prover then proves that the committed matrix is a permutation matrix, by proving that each row is a unit vector, including the last row which is computed explicitly, see Sect. 4.2. We construct a new unit vector argument based on the square span programs of Danezis *et al.* [11]; it is similar to but somewhat simpler than the 1-sparsity argument of [15]. Basically, to show that a vector a is unit vector, we choose polynomials $(P_i(X))_{i \in [0 \ldots n]}$ that interpolate a certain matrix (and a certain vector) connected to the definition of "unit vectorness", and then commit to a by using a version of the extended

Pedersen commitment scheme, $\mathfrak{c} = \sum_{i=1}^{n} a_i [P_i(\chi)]_1 + r[\varrho]_1$ for trapdoor values (χ, ϱ) and randomizer r. (This commitment scheme, though for different polynomials $P_i(X)$, was implicitly used first by Groth [24] in EUROCRYPT 2016, and then used in the FLZ shuffle argument; similar commitment schemes have been used before [19,23,32].) The new unit vector argument differs from the corresponding (1-sparsity) argument in [15] by a small optimization that makes it possible to decrease the number of trapdoor elements by one. If the unit vector argument for each row is accepting, it follows that the committed matrix is a permutation matrix [14]. The knowledge-soundness proof of the new unit vector argument is almost trivial, in contrast to the very complex machine-assisted knowledge-soundness proof in [15].

We then use the same high-level idea as previous NIZK shuffle arguments [14, 15,25,34] to obtain a shuffle argument from a permutation matrix argument. Namely, we construct a verification equation that holds tautologically under a corresponding KerMDH [36] assumption. That is, if the permutation matrix argument is knowledge-sound, the mentioned verification equation holds, and the KerMDH assumption holds, then the prover has used his committed permutation matrix to also shuffle the ciphertexts.

However, as in [14,15,25,34], the resulting KerMDH assumption itself will not be secure if we use here the same commitment scheme as before. Intuitively, this is since the polynomials $P_i(X)$ were carefully chosen to make the permutation matrix argument as efficient as possible. Therefore, we define an alternative version of the extended Pedersen commitment scheme with the commitment computed as $\hat{\mathfrak{c}} = \sum a_i [\hat{P}_i(\chi)]_1 + r[\hat{\varrho}]_1$ for trapdoor values $(\chi, \hat{\varrho})$ and randomizer r. Here, $\hat{P}_i(X)$ are well-chosen polynomials that satisfy a small number of requirements, including that $\{P_i(X)\hat{P}_j(X)\}_{1 \le i,j \le n}$ is linearly independent.

Before going on, we obviously need an efficient argument (that we call, following [14], a *same-message argument*) to show that $\mathfrak{c}$ and $\hat{\mathfrak{c}}$ commit to the same vector $\boldsymbol{a}$ (and use the same randomness r) while using different shrinking commitment schemes. We first write down the objective of this argument as the requirement that $\binom{a}{r}$ belongs to a subspace generated by a certain matrix $\boldsymbol{M}$. After doing that, we use the quasi-adaptive NIZK (QANIZK, [28,29]) argument of Kiltz and Wee (EUROCRYPT 2015, [30]) for linear subspaces to construct an efficient same-message argument. Since we additionally need it to be knowledge-sound, we give a proof in GBGM.

The new consistency argument is similar to but again more efficient than the consistency arguments of previous pairing-based shuffles. Here, we crucially use the fact that neither the permutation matrix argument nor the same-message argument add "too many" $\mathbb{G}_2$ elements to the CRS. Hence, while the Groth-Lu and FL shuffle arguments require two consistency verification equations, for us it suffices to only have one. (The Lipmaa-Zhang [34] and FLZ shuffle arguments have only one consistency verification equation, but this was compensated by using a non-standard cryptosystem with ciphertexts of length 6.)

In fact, we generalize the consistency argument to prove that given a committed matrix $\boldsymbol{E}$ and two tuples of ciphertexts $\mathfrak{M}'$ and $\mathfrak{M}$, it holds that

$\mathsf{Dec_{sk}}(\mathfrak{M}') = \boldsymbol{E} \cdot \mathsf{Dec_{sk}}(\mathfrak{M})$. Moreover, we prove that the consistency argument is culpably sound [25,26] under a suitable KerMDH [36] assumption, and we prove that the concrete KerMDH assumption holds in the GBGM.

Finally, we will give a standard (i.e., non-culpable) soundness proof for the full shuffle argument, assuming that the used commitment scheme is computationally binding, the same-message argument and the permutation matrix argument are knowledge-sound, and the consistency argument is culpably sound. Additionally, as in the FLZ shuffle argument, we use batching techniques [4] to speed up verification time. However, we use batching in a more aggressive manner than in the FLZ shuffle argument.

Efficiency Comparison. We provide the first implementation of a pairing-based shuffle argument. Our implementation is built on top of the freely available libsnark library, [6]. In fact, we implement two versions of the new shuffle argument, where in the second version we switch the roles of the groups $\mathbb{G}_1$ and $\mathbb{G}_2$. In the first case we get better overall prover's computation, while in the second case we get the most efficient online computation for both prover and verifier, and overall the most efficient verifier.

Table 1 shows a comparison between both versions of the new shuffle argument and prior state of the art CRS-based shuffle arguments with either the best prover's computational complexity or best verifier's computational complexity. Hence, we for instance do not include in this comparison table prior work by Groth and Lu [25] or Lipmaa and Zhang [34], since their shuffle arguments are slower than [14,15] in both prover's and verifier's computation. We also do not include the shuffle argument of González and Ràfols [20] since it has quadratic CRS length. In each row, the argument with best efficiency or best security property is highlighted.

Units (the main parameter) are defined in Table 3 in Sect. 7. One should compare the number of units, which is a weighted sum of different exponentiations and pairings, and hence takes into account the fact that (say) computations in $\mathbb{G}_1$ and $\mathbb{G}_2$ take different time. Moreover, this table counts separately the number of general exponentiations, multi-exponentiations, and fixed-base exponentiations, since the last two can be much more efficient than general exponentiations. We take this into account by using different unit values for these three types of exponentiations, see Table 3 for the number of units required for each type of operation. Note that we use our implementation of each operation in libsnark to compute the number of units.

Table 2 gives the running time of the new shuffle argument (without and with switching the groups) on our test machine. As seen from this table, our preliminary implementation enables one to prove a shuffle argument in less than 1 min and verify it in less than 1.5 min for $n = 100\,000$. After switching the groups, the prover's online computation takes less than 15 s and online verification takes less than 3 min for $n = 300\,000$. This means that the new shuffle argument is actually ready to be used in practice. Section 7 provides more information about implementation, including the definition of units and data about the test machine.

Table 1. A comparison of the new NIZK shuffle argument and prior work by Fauzi and Lipmaa (FL, [14]), and Fauzi, Lipmaa and Zając (FLZ, [15]). We include shuffling itself to the efficiency analysis of communication and prover's computation.

	FL	FLZ	Current Work	Current Work ($\mathbb{G}_1/\mathbb{G}_2$ switched)
$|\text{CRS}|$ in $(\mathbb{G}_1, \mathbb{G}_2, \mathbb{G}_T)$	$(6n+8, 2n+8, 1)$	$(2n+6, n+7, 1)$	$(4n+7, n+7, 1)$	$(n+7, 4n+7, 1)$
Communication	$(7n+2, 2n, 0)$	$(5n+1, 4n+2, 0)$	$(4n-1, 3n+1, 0)$	$(3n+1, 4n-1, 0)$
		Prover's computation		
Exp. in $(\mathbb{G}_1, \mathbb{G}_2)$	$(2n-1, 0)$	$(2n, 0)$	$(n, 0)$	$(0, n)$
Fb-exp. in $(\mathbb{G}_1, \mathbb{G}_2)$	$(8n-2, 2n-2)$	$(4n-1, 4n-1)$	$(3n-1, 3n-1)$	$(3n-1, 3n-1)$
M. exp. in $(\mathbb{G}_1, \mathbb{G}_2)$	$(6n+6, 2n+2)$	$(3n+3, 5n+5)$	$(n+1, 2n+2)$	$(2n+2, n+1)$
Units	4.84	5.34	2.87	4.25
		Prover's online computation		
M. exp. in $(\mathbb{G}_1, \mathbb{G}_2)$	$(2n+2, 0)$	$(3n+3, 3n+3)$	$(0, 2n+2)$	$(2n+2, 0)$
Units	0.26	1.2	0.54	0.26
		Verifier's computation		
Exp. in $(\mathbb{G}_1, \mathbb{G}_2, \mathbb{G}_T)$	$(0, 0, 0)$	$(7n+6, 7, 1)$	$(n, 2n+3, 1)$	$(2n+3, n, 1)$
M. exp. in $(\mathbb{G}_1, \mathbb{G}_2)$	$(0, 0)$	$(4n, 3n)$	$(4n-4, 0)$	$(0, 4n-4)$
Pairing product	$18n+6$	$3n+6$	$3n+6$	$3n+6$
Units	38.52	14.75	12.98	12.02
		Verifier's online computation		
Exp. in $(\mathbb{G}_1, \mathbb{G}_2)$	$(0, 0)$	$(6n+3, 3)$	$(0, 2n+1)$	$(2n+1, 0)$
M. exp. in $(\mathbb{G}_1, \mathbb{G}_2)$	$(0, 0)$	$(3n, 3n)$	$(0, 0)$	$(0, 0)$
Pairing product	$8n+4$	$2n+3$	$2n+1$	$2n+1$
Units	17.12	11.48	9.32	6.28
Lifted encryption	No	Yes	No	No
Soundness	Culpable	White-box	Full	Full

2 Preliminaries

Let S_n be the symmetric group on n elements, i.e., all elements of the group are permutations on n elements. All vectors will be by default column vectors. By $\boldsymbol{a} = (a_i)_{i=1}^n$ we denote column vectors and by $\boldsymbol{a} = (a_1, \ldots, a_n)$ we denote row vectors. For a matrix $\boldsymbol{A}$, $\boldsymbol{A}_i$ is its ith row vector and $\boldsymbol{A}^{(i)}$ is its ith column vector. A Boolean $n \times n$ matrix $\boldsymbol{A}$ is a permutation matrix representing $\sigma \in S_n$, when $A_{ij} = 1$ iff $\sigma(i) = j$. Clearly, in this case $\boldsymbol{Ax} = (x_{\sigma(i)})_{i=1}^n$ for any vector $\boldsymbol{x}$.

For a field $\mathbb{F}$, let $\mathbb{F}[\boldsymbol{X}]$ be the ring of multivariate polynomials over $\mathbb{F}$, and let $\mathbb{F}[\boldsymbol{X}^{\pm 1}]$ be the ring of multivariate Laurent polynomials over $\mathbb{F}$. For any (Laurent) polynomials $f_i(X)$, $i \in [1 .. n]$, we denote $\boldsymbol{f}(X) = (f_i(X))_{i=1}^n$.

Let $(\omega_i)_{i=1}^{n+1}$ be $n+1$ different values in $\mathbb{Z}_q$. For example, one can define $\omega_i = i$. Define the following polynomials:

- $Z(X) = \prod_{i=1}^{n+1}(X - \omega_i)$: the unique degree $n+1$ monic polynomial such that $Z(\omega_i) = 0$ for all $i \in [1 .. n]$.
- $\ell_i(X) = \prod_{j=1, j \neq i}^{n+1} \frac{X - \omega_j}{\omega_i - \omega_j}$: the ith Lagrange basis polynomial, i.e., the unique degree n polynomial such that $\ell_i(\omega_i) = 1$ and $\ell_i(\omega_j) = 0$ for $j \neq i$.

Table 2. Efficiency of the shuffle implementation (in minutes and seconds) using the libsnark library: the original argument (left) and the one with switched groups (right), for various values of n.

	10,000	100,000	300,000		10,000	100,000	300,000
CRS generation	1.7s	13.4s	37.0s	CRS generation	2.6s	20.5s	55.0s
Prover	6.4s	56.7s	2m38.3s	Prover	9.1s	1m24.0s	4m2.4s
Prover (online)	1.1s	10.2s	32.3s	Prover (online)	0.5s	4.1s	13.5s
Verifier	8.5s	1m27.8s	5m18.8s	Verifier	8.3s	1m22.0s	4m49.2s
Verifier (online)	5.7s	1m0.5s	3m29s	Verifier (online)	5.0s	49.9s	2m55.5s

Cryptography. Let κ be the security parameter; intuitively it should be difficult to break a hardness assumption or a protocol in time $O(2^\kappa)$. If $f(\kappa)$ is a negligible function then we write $f(\kappa) \approx_\kappa 0$. We use type-III, asymmetric, pairings [17]. Assume we use a secure bilinear group generator BG that on input (1^κ) returns $(q, \mathbb{G}_1, \mathbb{G}_2, \mathbb{G}_T, \mathfrak{g}_1, \mathfrak{g}_2, \bullet)$, where $\mathbb{G}_1$, $\mathbb{G}_2$, and $\mathbb{G}_T$ are three groups of prime order q, $\bullet : \mathbb{G}_1 \times \mathbb{G}_2 \to \mathbb{G}_T$, $\mathfrak{g}_1$ generates $\mathbb{G}_1$, $\mathfrak{g}_2$ generates $\mathbb{G}_2$, and $\mathfrak{g}_T = \mathfrak{g}_1 \bullet \mathfrak{g}_2$ generates $\mathbb{G}_T$. It is required that $\bullet$ is efficiently computable, bilinear, and non-degenerate.

To implement pairing-based cryptography, we use the libsnark library [6] which currently provides (asymmetric) Ate pairings [27] over Barreto-Naehrig curves [2,37] with 256 bit primes. Due to recent advances in computing discrete logarithms [31] this curve does not guarantee 128 bits of security, but still roughly achieves 100 bits of security [1].

Within this paper, we use additive notation combined with the bracket notation [13] and denote the elements of $\mathbb{G}_z$, $z \in \{1, 2, T\}$, as in $[a]_z$ (even if a is unknown). Alternatively, we denote group elements by using the Fraktur font as in $\mathfrak{b}$. We assume that $\cdot$ has higher precedence than $\bullet$; for example, $b a \bullet c = (ba) \bullet c$; while this is not important mathematically, it makes a difference in implementation since exponentiation in $\mathbb{G}_1$ is cheaper than in $\mathbb{G}_T$. In this notation, we write the generator of $\mathbb{G}_z$ as $\mathfrak{g}_z = [1]_z$ for $z \in \{1, 2, T\}$. Hence, $[a]_z = a\mathfrak{g}_z$, so the bilinear property can be written as $[a]_1 \bullet [b]_2 = (ab)[1]_T = [ab]_T$.

We freely combine additive notation with vector notation, by defining say $[a_1, \ldots, a_s]_z = ([a_1]_z, \ldots, [a_s]_z)$, and $[\boldsymbol{A}]_1 \bullet [\boldsymbol{B}]_2 = [\boldsymbol{AB}]_T$ for matrices $\boldsymbol{A}$ and $\boldsymbol{B}$ of compatible size. We sometimes misuse the notation and, say, write $[\boldsymbol{A}]_2 \boldsymbol{B}$ instead of the more cumbersome $(\boldsymbol{B}^\top [\boldsymbol{A}]_2^\top)^\top$. Hence, if $\boldsymbol{A}$, $\boldsymbol{B}$ and $\boldsymbol{C}$ have compatible dimensions, then $(\boldsymbol{B}^\top [\boldsymbol{A}]_1^\top)^\top \bullet [\boldsymbol{C}]_2 = [\boldsymbol{A}]_1 \boldsymbol{B} \bullet [\boldsymbol{C}]_2 = [\boldsymbol{A}]_1 \bullet \boldsymbol{B}[\boldsymbol{C}]_2$.

Recall that a distribution $\mathcal{D}_{par}$ that outputs matrices of group elements $[\boldsymbol{M}]_1 \in \mathbb{G}_1^{n \times t}$ is *witness-sampleable* [28], if there exists a distribution $\mathcal{D}'_{par}$ that outputs matrices of integers $\boldsymbol{M}' \in \mathbb{Z}_q^{n \times t}$, such that $[\boldsymbol{M}]_1$ and $[\boldsymbol{M}']_1$ have the same distribution.

We use the Elgamal cryptosystem [12] (Gen, Enc, Dec) in group $\mathbb{G}_2$. In Elgamal, the key generator $\mathsf{Gen}(1^\kappa)$ chooses a secret key $\mathsf{sk} \leftarrow_r \mathbb{Z}_q$ and a public key pk $\leftarrow [1, \mathsf{sk}]_2$ for a generator $[1]_2$ of $\mathbb{G}_2$ fixed by BG. The encryption algorithm sets $\mathsf{Enc}_{\mathsf{pk}}(\mathsf{m}; r) = ([0]_2, \mathsf{m}) + r \cdot \mathsf{pk}$ for $\mathsf{m} \in \mathbb{G}_2$ and $r \leftarrow_r \mathbb{Z}_q$. The decryption algorithm sets $\mathsf{Dec}_{\mathsf{sk}}(\mathfrak{M}_1, \mathfrak{M}_2) = \mathfrak{M}_2 - \mathsf{sk} \cdot \mathfrak{M}_1$. Note that $\mathsf{Enc}_{\mathsf{pk}}(\mathsf{o}; r) = r \cdot \mathsf{pk}$. The Elgamal cryptosystem is blindable, with $\mathsf{Enc}_{\mathsf{pk}}(\mathsf{m}; r_1) + \mathsf{Enc}_{\mathsf{pk}}(\mathsf{o}; r_2) = \mathsf{Enc}_{\mathsf{pk}}(\mathsf{m}; r_1 + r_2)$. Clearly, if r_2 is uniformly random, then $r_1 + r_2$ is also uniformly random.

Finally, for $[a]_1 \in \mathbb{G}_1$, and $[b]_2 \in \mathbb{G}_2^{1 \times 2}$, let $[a]_1 \circ [b]_2 := [a \cdot b]_T = [a \cdot b_1, a \cdot b_2]_T \in \mathbb{G}_T^{1 \times 2}$. Analogously, for $[a]_1 \in \mathbb{G}_1^n$, and $[B]_2 \in \mathbb{G}_2^{n \times 2}$, let $[a]_1^\top \circ [B]_2 := \sum_{i=1}^n [a_i]_1 \circ [B_i]_2 \in \mathbb{G}_T^2$. Intuitively, here $[b]_2$ is an Elgamal ciphertext and $[B]_2$ is a vector of Elgamal ciphertexts.

Kernel Matrix Assumptions [36]. Let $k \in \mathbb{N}$. We call $\mathcal{D}_{k+d,k}$ a *matrix distribution* [13] if it outputs matrices in $\mathbb{Z}_q^{(k+d) \times k}$ of full rank k in polynomial time. W.l.o.g., we assume that the first k rows of $A \leftarrow \mathcal{D}_{k+d,k}$ form an invertible matrix. We denote $\mathcal{D}_{k+1,k}$ as $\mathcal{D}_k$.

Let $\mathcal{D}_{k+d,k}$ be a matrix distribution and $z \in \{1, 2\}$. The $\mathcal{D}_{k+d,k}$-*KerMDH* *assumption* [36] holds in $\mathbb{G}_z$ relative to algorithm BG, if for all probabilistic polynomial-time $\mathcal{A}$,

$$\Pr\left[\begin{array}{l} \mathsf{gk} \leftarrow \mathsf{BG}(1^\kappa), M \leftarrow_r \mathcal{D}_{k+d,k}, [c]_{3-z} \leftarrow \mathcal{A}(\mathsf{gk}, [M]_z) : \\ M^\top c = 0 \wedge c \neq 0 \end{array}\right] \approx_\kappa 0.$$

By varying the distribution $\mathcal{D}_{k+d,k}$, one can obtain various assumptions (such as the SP assumption of Groth and Lu [25] or the PSP assumption of Fauzi and Lipmaa [14]) needed by some previous shuffle arguments.

Since we use KerMDH to prove soundness of subarguments of the shuffle argument, we define a version of this assumption with an auxiliary input that corresponds to the CRS of the shuffle argument. We formalize it by defining a *valid CRS distribution* $\mathcal{D}_{n+d,d}$ to be a joint distribution of an $(n+d) \times d$ matrix distribution and a distribution of auxiliary inputs aux, such that

1. aux contains only elements of the groups $\mathbb{G}_1$, $\mathbb{G}_2$, and $\mathbb{G}_T$,
2. $\mathcal{D}_{n+d,d}$ outputs as a trapdoor td all the used random coins from $\mathbb{Z}_q$.

We denote this as $(M, \mathsf{aux}; \mathsf{td}) \leftarrow_r \mathcal{D}_{n+d,d}$. We prove the culpable soundness of the consistency argument under a variant of the KerMDH assumption that allows for an auxiliary input.

Definition 1 (KerMDH with an auxiliary input). *Let $\mathcal{D}_{n+d,d}$ be a valid CRS distribution. The $\mathcal{D}_{n+d,d}$-KerMDH assumption with an auxiliary input holds in $\mathbb{G}_z$, $z \in \{1, 2\}$, relative to algorithm BG, if for all probabilistic polynomial-time $\mathcal{A}$,*

$$\Pr\left[\begin{array}{l} \mathsf{gk} \leftarrow \mathsf{BG}(1^\kappa), ([M]_z, \mathsf{aux}) \leftarrow_r \mathcal{D}_{n+d,d}, [c]_{3-z} \leftarrow \mathcal{A}(\mathsf{gk}, [M]_z, \mathsf{aux}) : \\ M^\top c = 0 \wedge c \neq 0 \end{array}\right] \approx_\kappa 0.$$

Commitment Schemes. A (pairing-based) trapdoor commitment scheme is a pair of efficient algorithms $(\mathsf{K}, \mathsf{com})$, where $\mathsf{K}(\mathsf{gk})$ (for $\mathsf{gk} \leftarrow \mathsf{BG}(1^\kappa)$) outputs a commitment key ck and a trapdoor td, and the commitment algorithm outputs $\mathfrak{a} \leftarrow \mathsf{com}(\mathsf{gk}, \mathsf{ck}, \boldsymbol{a}; r)$. A commitment scheme is *computationally binding* if for any ck output by K, it is computationally infeasible to find $(\boldsymbol{a}, r) \neq (\boldsymbol{a}', r)$, such that $\mathsf{com}(\mathsf{gk}, \mathsf{ck}, \boldsymbol{a}; r) = \mathsf{com}(\mathsf{gk}, \mathsf{ck}, \boldsymbol{a}'; r')$. A commitment scheme is *perfectly hiding* if for any ck output by K, the distribution of the output of $\mathsf{com}(\mathsf{gk}, \mathsf{ck}, \boldsymbol{a}; r)$ does not depend on $\boldsymbol{a}$, assuming that r is uniformly random. A commitment scheme is *trapdoor*, if given access to td it is trivial to open the commitment to any value.

Throughout this paper, we use the following trapdoor commitment scheme, first implicitly used by Groth [24]. Fix some linearly independent polynomials $P_i(X)$. Let $\mathsf{gk} \leftarrow \mathsf{BG}(1^\kappa)$, and $\mathsf{td} = (\chi, \varrho) \leftarrow_r \mathbb{Z}_q^2$. Denote $\boldsymbol{P} \triangleq (P_i(\chi))_{i=1}^n$. Let $\mathsf{ck} \leftarrow [\begin{smallmatrix} \boldsymbol{P} \\ \varrho \end{smallmatrix}]_1$, and $\mathsf{com}(\mathsf{gk}, \mathsf{ck}, \boldsymbol{a}; r) := (\begin{smallmatrix} \boldsymbol{a} \\ r \end{smallmatrix})^\top \cdot \mathsf{ck} = [\sum_{i=1}^n a_i P_i(\chi) + r\varrho]_1$. We will call it the $((P_i(X))_{i=1}^n, X_\varrho)$ *-commitment scheme*.

Several variants of this commitment scheme are known to be perfectly hiding and computational binding under a suitable computational assumption [19, 23, 32]. Since this commitment scheme is a variant of the extended Pedersen commitment scheme, it can be proven to be computationally binding under a suitable KerMDH assumption [36].

Theorem 1. *Let* $\mathcal{D}_n^{((P_i(X))_{i=1}^n, X_\varrho)} = \{[\begin{smallmatrix} \boldsymbol{P} \\ \varrho \end{smallmatrix}]_1 : (\chi, \varrho) \leftarrow_r \mathbb{Z}_q \times \mathbb{Z}_q^*\}$ *be the distribution of* ck *in this commitment scheme. The* $((P_i(X))_{i=1}^n, X_\varrho)$*-commitment scheme is perfectly hiding, and computationally binding under the* $\mathcal{D}_n^{((P_i(X))_{i=1}^n, X_\varrho)}$*-KerMDH assumption. It is trapdoor with* $\mathsf{td} = (\chi, \varrho)$*.*

Proof (Sketch). Given two different openings $(\boldsymbol{a}, r)$ and $(\boldsymbol{a}', r')$ to a commitment, $(\boldsymbol{a} - \boldsymbol{a}', r - r')$ is a solution to the KerMDH problem. Perfect hiding is obvious, since ϱ is not 0, and hence $r[\varrho]_1$ is uniformly random. Given $(\boldsymbol{a}, r)$, one can open $\mathsf{com}(\mathsf{gk}, \mathsf{ck}, \boldsymbol{a}; r)$ to $(\boldsymbol{a}', r')$ by taking $r' = r + (\sum_{i=1}^n (a_i - a_i') P_i(\chi))/\varrho$. $\qquad\square$

Generic Bilinear Group Model. A *generic* algorithm uses only generic group operations to create and manipulate group elements. Shoup [39] formalized it by giving algorithms access to random injective encodings $\langle\langle\mathfrak{a}\rangle\rangle$ instead of real group elements $\mathfrak{a}$. Maurer [35] considered a different formalization, where the group elements are given in memory cells, and the adversary is given access to the address (but not the contents) of the cells. For simplicity, let's consider Maurer's formalization. The memory cells initially have some input (in our case, the random values generated by the CRS generator together with other group elements in the CRS). The generic group operations are handled through an oracle that on input $(op, i_1, \ldots, i_n)$, where op is a generic group operation and i_j are addresses, first checks that memory cells in i_j have compatible type $z \in \{1, 2, T\}$ (e.g., if $op = +$, then i_1 and i_2 must belong to the same group $\mathbb{G}_z$), performs op on inputs $(i_1, \ldots, i_n)$, stores the output in a new memory cell, and then returns the address of this cell. In the generic bilinear group model (GBGM), the oracle can also compute the pairing $\bullet$. In addition, the oracle can answer equality queries.

We will use the following, slightly less formal way of thinking about the GBGM. Each memory cell has an implicit polynomial attached to it. A random value generated by the CRS generator is assigned a new indeterminate. If an $(+, i_1, i_2)$ (resp., $(\bullet, i_1, i_2)$) input is given to the oracle that successfully returns address i_3, then the implicit polynomial attached to i_3 will be the sum (resp., product) of implicit polynomials attached to i_1 and i_2. Importantly, since one can only add together elements from the same group (or, from the memory cells with the same type), this means that any attached polynomials with type 1 or 2 can only depend on the elements of the CRS that belong to the same group.

A generic algorithm does not know the values of the indeterminates (for her they really are indeterminates), but she will know all attached polynomials. At any moment, one can ask the oracle for an equality test between two memory cells i_1 and i_2. A generic algorithm is considered to be successful if she makes — in polynomial time — an equality test query $(=, i_1, i_2)$ that returns success (the entries are equal) but i_1 and i_2 have different polynomials $F_1(\boldsymbol{X})$ and $F_2(\boldsymbol{X})$ attached to them. This is motivated by the Schwartz-Zippel lemma [38,40] that states that if $F_1(\boldsymbol{X}) \neq F_2(\boldsymbol{X})$ as a polynomial, then there is negligible probability that $F_1(\boldsymbol{\chi}) = F_2(\boldsymbol{\chi})$ for uniformly random $\boldsymbol{\chi}$.

Zero Knowledge. Let $\mathcal{R} = \{(u, w)\}$ be an efficiently computable binary relation with $|w| = \mathsf{poly}(|u|)$. Here, u is a statement, and w is a witness. Let $\mathcal{L} = \{u : \exists w, (u, w) \in \mathcal{R}\}$ be an **NP**-language. Let $n = |u|$ be the input length. For fixed n, we have a relation $\mathcal{R}_n$ and a language $\mathcal{L}_n$. Since we argue about group elements, both $\mathcal{L}_n$ and $\mathcal{R}_n$ are group-dependent and thus we add gk (output by BG) as an input to $\mathcal{L}_n$ and $\mathcal{R}_n$. Let $\mathcal{R}_n(\mathsf{gk}) := \{(u, w) : (\mathsf{gk}, u, w) \in \mathcal{R}_n\}$.

A *non-interactive argument* for a group-dependent relation family $\mathcal{R}$ consists of four probabilistic polynomial-time algorithms: a setup algorithm BG, a common reference string (CRS) generator K, a prover P, and a verifier V. Within this paper, BG is always the bilinear group generator that outputs $\mathsf{gk} \leftarrow (q, \mathbb{G}_1, \mathbb{G}_2, \mathbb{G}_T, \mathfrak{g}_1, \mathfrak{g}_2, \bullet)$. For $\mathsf{gk} \leftarrow \mathsf{BG}(1^\kappa)$ and $(\mathsf{crs}, \mathsf{td}) \leftarrow \mathsf{K}(\mathsf{gk}, n)$ (where n is input length), $\mathsf{P}(\mathsf{gk}, \mathsf{crs}, u, w)$ produces an argument π, and $\mathsf{V}(\mathsf{gk}, \mathsf{crs}, u, \pi)$ outputs either 1 (accept) or 0 (reject). The verifier may be probabilistic, to speed up verification time by the use of batching techniques [4].

A non-interactive argument Ψ is *perfectly complete*, if for all $n = \mathsf{poly}(\kappa)$,

$$\Pr \left[\begin{array}{l} \mathsf{gk} \leftarrow \mathsf{BG}(1^\kappa), (\mathsf{crs}, \mathsf{td}) \leftarrow \mathsf{K}(\mathsf{gk}, n), (u, w) \leftarrow \mathcal{R}_n(\mathsf{gk}) : \\ \mathsf{V}(\mathsf{gk}, \mathsf{crs}, u, \mathsf{P}(\mathsf{gk}, \mathsf{crs}, u, w)) = 1 \end{array} \right] = 1 .$$

Ψ is adaptively *computationally sound* for $\mathcal{L}$, if for all $n = \mathsf{poly}(\kappa)$ and all non-uniform probabilistic polynomial-time adversaries $\mathcal{A}$,

$$\Pr \left[\begin{array}{l} \mathsf{gk} \leftarrow \mathsf{BG}(1^\kappa), (\mathsf{crs}, \mathsf{td}) \leftarrow \mathsf{K}(\mathsf{gk}, n), \\ (u, \pi) \leftarrow \mathcal{A}(\mathsf{gk}, \mathsf{crs}) : (\mathsf{gk}, u) \notin \mathcal{L}_n \wedge \mathsf{V}(\mathsf{gk}, \mathsf{crs}, u, \pi) = 1 \end{array} \right] \approx_\kappa 0.$$

We recall that in situations where the inputs have been committed to using a computationally binding trapdoor commitment scheme, the notion of computational soundness does not make sense (since the commitments could be to any input messages). Instead, one should either prove culpable soundness or knowledge-soundness.

Ψ is adaptively *computationally culpably sound* [25,26] for $\mathcal{L}$ using a polynomial-time decidable binary relation $\mathcal{R}^{\mathsf{glt}} = \{\mathcal{R}_n^{\mathsf{glt}}\}$ consisting of elements from $\bar{\mathcal{L}}$ and witnesses w^{glt}, if for all $n = \mathsf{poly}(\kappa)$ and all non-uniform probabilistic polynomial-time adversaries $\mathcal{A}$,

$$\Pr\left[\begin{array}{l} \mathsf{gk} \leftarrow \mathsf{BG}(1^\kappa), (\mathsf{crs},\mathsf{td}) \leftarrow \mathsf{K}(\mathsf{gk},n), (u,\pi,w^{\mathsf{glt}}) \leftarrow \mathcal{A}(\mathsf{gk},\mathsf{crs}): \\ (\mathsf{gk},u,w^{\mathsf{glt}}) \in \mathcal{R}_n^{\mathsf{glt}} \wedge \mathsf{V}(\mathsf{gk},\mathsf{crs},u,\pi) = 1 \end{array}\right] \approx_\kappa 0 \ .$$

For algorithms $\mathcal{A}$ and $\mathsf{Ext}_{\mathcal{A}}$, we write $(y; y') \leftarrow (\mathcal{A}\|\mathsf{Ext}_{\mathcal{A}})(\chi)$ if $\mathcal{A}$ on input χ outputs y, and $\mathsf{Ext}_{\mathcal{A}}$ on the same input (including the random tape of $\mathcal{A}$) outputs y'. Ψ is *knowledge-sound*, if for all $n = \mathsf{poly}(\kappa)$ and all non-uniform probabilistic polynomial-time adversaries $\mathcal{A}$, there exists a non-uniform probabilistic polynomial-time extractor $\mathsf{Ext}_{\mathcal{A}}$, such that for every auxiliary input $\mathsf{aux} \in \{0,1\}^{\mathsf{poly}(\kappa)}$,

$$\Pr\left[\begin{array}{l} \mathsf{gk} \leftarrow \mathsf{BG}(1^\kappa), (\mathsf{crs},\mathsf{td}) \leftarrow \mathsf{K}(\mathsf{gk},n), ((u,\pi); w) \leftarrow (\mathcal{A}\|\mathsf{Ext}_{\mathcal{A}})(\mathsf{crs},\mathsf{aux}): \\ (\mathsf{gk},u,w) \notin \mathcal{R}_n \wedge \mathsf{V}(\mathsf{gk},\mathsf{crs},u,\pi) = 1 \end{array}\right] \approx_\kappa 0 \ .$$

Here, aux can be seen as the common auxiliary input to $\mathcal{A}$ and $\mathsf{Ext}_{\mathcal{A}}$ that is generated by using benign auxiliary input generation [7].

Ψ is *perfectly (composable) zero-knowledge* [21], if there exists a probabilistic polynomial-time simulator S, such that for all stateful non-uniform probabilistic adversaries $\mathcal{A}$ and $n = \mathsf{poly}(\kappa)$, $\varepsilon_0 = \varepsilon_1$, where

$$\varepsilon_b := \Pr\left[\begin{array}{l} \mathsf{gk} \leftarrow \mathsf{BG}(1^\kappa), (\mathsf{crs},\mathsf{td}) \leftarrow \mathsf{K}(\mathsf{gk},n), (u,w) \leftarrow \mathcal{A}(\mathsf{gk},\mathsf{crs}), \\ \text{if } b = 0 \text{ then } \pi \leftarrow \mathsf{P}(\mathsf{gk},\mathsf{crs},u,w) \text{ else } \pi \leftarrow \mathsf{S}(\mathsf{gk},\mathsf{crs},u,\mathsf{td}) \text{ endif}: \\ (\mathsf{gk},u,w) \in \mathcal{R}_n \wedge \mathcal{A}(\mathsf{gk},\mathsf{crs},\pi) = 1 \end{array}\right] \ .$$

Shuffle Argument. In a (pairing-based) shuffle argument [25], the prover aims to convince the verifier that, given system parameters gk output by BG, a public key pk, and two tuples of ciphertexts $\mathfrak{M}$ and $\mathfrak{M}'$, the second tuple is a permutation of rerandomized versions of the first. More precisely, we will construct a shuffle argument that is sound with respect to the following relation:

$$\mathcal{R}_{sh} = \left\{ \begin{array}{l} ((\mathsf{gk}, \mathsf{pk}, \mathfrak{M}, \mathfrak{M}'), (\sigma, \boldsymbol{r})) : \sigma \in S_n \wedge \\ \boldsymbol{r} \in \mathbb{Z}_q^n \wedge (\forall i : \mathfrak{M}'_i = \mathfrak{M}_{\sigma(i)} + \mathsf{Enc}_{\mathsf{pk}}(o; r_i)) \end{array} \right\} \ .$$

A number of pairing-based shuffle arguments have been proposed in the literature, [14,15,20,25,34].

3   New Shuffle Argument

Intuitively, in the new shuffle argument the prover first commits to the permutation σ (or more precisely, to the corresponding permutation matrix), then executes three subarguments (the same-message, the permutation matrix, and the consistency arguments). Each of the subarguments corresponds to one check performed by the verifier (see Protocol 2). However, since all subarguments use the same CRS, they are not independent. For example, the permutation matrix argument uses the $((P_i(X))_{i=1}^n, X_\varrho)$-commitment scheme and the consistency argument uses the $((\hat{P}_i(X))_{i=1}^n, X_{\hat{\varrho}})$-commitment scheme for different polynomials $(\hat{P}_i(X))_{i=1}^n$. (See Eqs. (3) and (5) for the actual definition of $P_i(X)$ and $\hat{P}_i(X)$). Both commitment schemes share a part of their trapdoor (χ), while the second part of the trapdoor is different (either ϱ or $\hat{\varrho}$). Moreover, the knowledge-soundness of the same-message argument is a prerequisite for the knowledge-soundness of the permutation matrix argument. The verifier recovers explicitly the commitment to the last row of the permutation matrix (this guarantees that the committed matrix is left stochastic), then verifies the three subarguments.

The full description of the new shuffle argument is given in Protocol 1 (the CRS generation and the prover) and in Protocol 2 (the verifier). The CRS has entries that allow to efficiently evaluate all subarguments, and hence also both commitment schemes. The CRS in Protocol 1 includes three substrings, crs_{sm}, crs_{pm}, and crs_{con}, that are used in the three subarguments. To prove and verify (say) the first subargument (the same-message argument), one needs access to crs_{sm}. However, the adversary of the same-message argument will get access to the full CRS. For the sake of exposition, the verifier's description in Protocol 2 does not include batching. In Sect. 6, we will explain how to speed up the verifier considerably by using batching techniques.

We will next briefly describe the subarguments. In Sect. 4, we will give more detailed descriptions of each subargument, and in Sect. 5, we will prove the security of the shuffle argument.

Same-Message Argument. Consider the subargument of the new shuffle argument where the verifier only computes $\mathfrak{a}_n$ and then performs the check on Step 4 of Protocol 2 for one concrete i. We will call it the *same-message argument* [14]. In Sect. 4.1 we motivate this name, by showing that if the same-message argument accepts, then the prover knows a message $\boldsymbol{a}$ and a randomizer r, such that $\mathfrak{a}_i = [\sum a_i P_i(\chi) + r\varrho]_1$ and $\hat{\mathfrak{a}}_i = [\sum a_i \hat{P}_i(\chi) + r\hat{\varrho}]_1$ both commit to $\boldsymbol{a}$ with randomizer r, by using respectively the $((P_i(X))_{i=1}^n, X_\varrho)$-commitment scheme and the $((\hat{P}_i(X))_{i=1}^n, X_{\hat{\varrho}})$-commitment scheme.

For the same-message argument to be knowledge-sound, we will require that $\{P_i(X)\}_{i=1}^n$ and $\{\hat{P}_i(X)\}_{i=1}^n$ are both linearly independent sets.

Permutation Matrix Argument. Consider the subargument of Protocols 1 and 2, where (i) the prover computes $\mathfrak{a}$ and π_{pm}, and (ii) the verifier computes $\mathfrak{a}_n$ and then checks the verification equation on Step 5 of Protocol 2. We will call it the *permutation matrix argument*. In Sect. 4.2 we motivate this name, by proving in the GBGM that if the verifier accepts the permutation matrix argument, then either the prover knows how to open $(\mathfrak{a}_1, \ldots, \mathfrak{a}_n)$ as a $((P_i(X))_{i=1}^n, X_\varrho)$-

$\mathsf{K}(\mathsf{gk}, n)$: Generate random $(\chi, \beta, \hat{\beta}, \varrho, \hat{\varrho}, \mathsf{sk}) \leftarrow_r \mathbb{Z}_q^3 \times (\mathbb{Z}_q^*)^2 \times \mathbb{Z}_q$. Denote $\boldsymbol{P} = (P_i(\chi))_{i=1}^n$, $P_0 = P_0(\chi)$, and $\hat{\boldsymbol{P}} = (\hat{P}_i(\chi))_{i=1}^n$. Let $\mathsf{crs}_{sm} \leftarrow$ $\left([(\beta P_i + \hat{\beta}\hat{P}_i)_{i=1}^n, \beta\varrho + \hat{\beta}\hat{\varrho}]_1, [\beta, \hat{\beta}]_2^\top\right)$,

$$\mathsf{crs}_{pm} \leftarrow \begin{pmatrix} [1, P_0, (((P_i + P_0)^2 - 1)/\varrho)_{i=1}^n, \sum_{i=1}^n P_i, \sum_{i=1}^n \hat{P}_i]_1, \\ [P_0, \sum_{i=1}^n P_i]_2, [1]_T \end{pmatrix},$$

$\mathsf{crs}_{con} \leftarrow [\frac{\hat{\boldsymbol{P}}}{\hat{\varrho}}]_1$. Set $\mathsf{crs} \leftarrow \left(\mathsf{pk} = [1, \mathsf{sk}]_2, [\frac{\boldsymbol{P}}{\varrho}]_1, [\frac{\boldsymbol{P}}{\varrho}]_2, \mathsf{crs}_{sm}, \mathsf{crs}_{pm}, \mathsf{crs}_{con}\right)$. Set $\mathsf{td} \leftarrow$ $(\chi, \hat{\varrho})$. Return $(\mathsf{crs}, \mathsf{td})$.

$\mathsf{P}(\mathsf{gk}, \mathsf{crs}, \mathfrak{M} \in \mathbb{G}_2^{n \times 2}; \sigma \in S_n, \boldsymbol{t} \in \mathbb{Z}_q^n)$:
 1. For $i = 1$ to $n - 1$: // commits to the permutation σ
    (a) $r_i \leftarrow_r \mathbb{Z}_q$; $\mathfrak{r}_i \leftarrow r_i[\varrho]_1$;
    (b) $\mathfrak{a}_i \leftarrow [P_{\sigma^{-1}(i)}]_1 + \mathfrak{r}_i$; $\mathfrak{b}_i \leftarrow [P_{\sigma^{-1}(i)}]_2 + r_i[\varrho]_2$; $\hat{\mathfrak{a}}_i \leftarrow [\hat{P}_{\sigma^{-1}(i)}]_1 + r_i[\hat{\varrho}]_1$;
 2. $\mathfrak{a}_n \leftarrow [\sum_{i=1}^n P_i]_1 - \sum_{j=1}^{n-1} \mathfrak{a}_j$; $\mathfrak{b}_n \leftarrow [\sum_{i=1}^n P_i]_2 - \sum_{j=1}^{n-1} \mathfrak{b}_j$;
 3. $\hat{\mathfrak{a}}_n \leftarrow [\sum_{i=1}^n \hat{P}_i]_1 - \sum_{j=1}^{n-1} \hat{\mathfrak{a}}_j$;
 4. $r_n \leftarrow -\sum_{i=1}^{n-1} r_i$; $\mathfrak{r}_n \leftarrow r_n[\varrho]_1$;
 5. For $i = 1$ to n:
    (a) $\mathfrak{d}_i \leftarrow [\beta P_{\sigma^{-1}(i)} + \hat{\beta}\hat{P}_{\sigma^{-1}(i)}]_1 + r_i[\beta\varrho + \hat{\beta}\hat{\varrho}]_1$;
    (b) $\mathfrak{c}_i \leftarrow r_i \cdot (2(\mathfrak{a}_i + [P_0]_1) - \mathfrak{r}_i) + [((P_{\sigma^{-1}(i)} + P_0)^2 - 1)/\varrho]_1$;
 6. $r_t \leftarrow_r \mathbb{Z}_q$; $\mathfrak{t} \leftarrow \boldsymbol{t}^\top[\hat{\boldsymbol{P}}]_1 + r_t[\hat{\varrho}]_1$;
 7. For $i = 1$ to n: $\mathfrak{t}_i' \leftarrow t_i \cdot \mathsf{pk}$;
 8. $\mathfrak{M}' \leftarrow (\mathfrak{M}_{\sigma(i)} + \mathfrak{t}_i')_{i=1}^n$; // Shuffling, online
 9. $\mathfrak{N} \leftarrow \boldsymbol{r}^\top\mathfrak{M} + r_t \cdot \mathsf{pk}$; // Online
 10. $\pi_{sm} \leftarrow \mathfrak{d}$; // Same-message argument
 11. $\pi_{pm} \leftarrow ((\mathfrak{b}_i)_{i=1}^{n-1}, \mathfrak{c})$; // Permutation matrix argument
 12. $\pi_{con} \leftarrow ((\hat{\mathfrak{a}}_i)_{i=1}^{n-1}, \mathfrak{t}, \mathfrak{N})$; // Consistency argument
 13. Return $\pi_{sh} \leftarrow (\mathfrak{M}', (\mathfrak{a}_j)_{j=1}^{n-1}, \pi_{sm}, \pi_{pm}, \pi_{con})$.

Protocol 1: The CRS generation and the prover of the new shuffle argument.

$\mathsf{V}(\mathsf{gk}, \mathsf{crs}, \mathfrak{M}; \mathfrak{M}', (\mathfrak{a}_j)_{j=1}^{n-1}, \pi_{sm}, \pi_{pm}, \pi_{con})$:
 1. Parse $(\pi_{sm}, \pi_{pm}, \pi_{con})$ as in the prover's Steps 11–12, abort if unsuccessful;
 2. Compute $\mathfrak{a}_n$, $\mathfrak{b}_n$, and $\hat{\mathfrak{a}}_n$ as in the prover's Steps 2 and 3;
 3. $\alpha \leftarrow_r \mathbb{Z}_q$;
 4. For $i = 1$ to n: check that $\mathfrak{d}_i \bullet [1]_2 \stackrel{?}{=} (\mathfrak{a}_i, \hat{\mathfrak{a}}_i) \bullet \begin{bmatrix} \beta \\ \hat{\beta} \end{bmatrix}_2$; // Same-message argument
 5. For $i = 1$ to n: check that // Permutation matrix argument
    $(\mathfrak{a}_i + \alpha[1]_1 + [P_0]_1) \bullet (\mathfrak{b}_i - \alpha[1]_2 + [P_0]_2) \stackrel{?}{=} \mathfrak{c}_i \bullet [\varrho]_2 + (1 - \alpha^2)[1]_T$;
 6. Check that // Consistency argument
    $[\hat{\boldsymbol{P}}]_1^\top \circ \mathfrak{M}' - \hat{\mathfrak{a}}^\top \circ \mathfrak{M} \stackrel{?}{=} \mathfrak{t} \circ \mathsf{pk} - [\hat{\varrho}]_1 \circ \mathfrak{N}$;

Protocol 2: The non-batched verifier of the new shuffle argument.

commitment to a permutation matrix or we can break the same-message argument. For this, we first prove the security of a subargument of the permutation matrix argument — the unit vector argument [14] — where the verifier performs the verification Step 5 for exactly one i.

For the unit vector argument to be efficient, we need to make a specific choice of the polynomials $P_i(X)$ (see Eq. (3)). For the knowledge-soundness of the unit vector argument, we additionally need that $\{P_i(X)\}_{i=0}^n$ and $\{P_i(X)\}_{i=1}^n \cup \{1\}$ are linearly independent.

In [15], the verifier adds $[\alpha]_1 + [P_0]_1$ from the CRS to $\mathfrak{a}_i$ (and adds $[\alpha]_2 - [P_0]_2$ from the CRS to $\mathfrak{b}_i$), while in our case, the verifier samples α itself during verification. Due to this small change, we can make the CRS independent of α. (In fact, it suffices if the verifier chooses α once and then uses it at each verification.) This makes the CRS shorter, and also simplifies the latter soundness proof. For this optimization to be possible, one has to rely on the same-message argument (see Sect. 5).

Consistency Argument. Consider the subargument of the new shuffle argument where the prover only computes π_{con} and the verifier performs the check on Step 6 of Protocol 2. We will call it the *consistency argument*. In Sect. 4.3 we motivate this name, by showing that if $\hat{\mathfrak{a}}$ ($\{\hat{P}_i(X)\}, X_{\hat{\varrho}}$)-commits to a permutation, then $\mathsf{Dec}(\mathfrak{M}_i') = \mathsf{Dec}(\mathfrak{M}_{\sigma(i)})$ for the same permutation σ that the prover committed to earlier. We show that the new consistency argument is culpably sound under a (novel) variant of the KerMDH computational assumption [36] that we describe in Sect. 4.3. In particular, the KerMDH assumption has to hold even when the adversary is given access to the full CRS of the shuffle argument.

For the consistency argument to be sound (and in particular, for the KerMDH variant to be secure in the GBGM), we will require that $\{\hat{P}_i(X)\}_{i=1}^n$ and $\{P_i(X)\hat{P}_j(X)\}_{1 \le i,j \le n}$ are both linearly independent sets.

4   Subarguments

Several of the following knowledge-soundness proofs use the GBGM and therefore we will first give common background for those proofs. In the GBGM, the generic adversary in each proof has only access to generic group operations, pairings, and equality tests. However, she will have access to the full CRS of the new shuffle argument.

Let $\boldsymbol{\chi} = (\chi, \alpha, \beta, \hat{\beta}, \varrho, \hat{\varrho}, \mathsf{sk})$ be the tuple of all random values generated by either K or V. Note that since α is sampled by the verifier each time, for an adversary it is essentially an indeterminate. Thus, for the generic adversary each element of $\boldsymbol{\chi}$ will be an indeterminate. Let us denote the tuple of corresponding indeterminates by $\boldsymbol{X} = (X, X_\alpha, X_\beta, X_{\hat{\beta}}, X_\varrho, X_{\hat{\varrho}}, X_S)$.

The adversary is given oracle access to the full CRS. This means that for each element of the CRS, she knows the attached (Laurent) polynomial in $\boldsymbol{X}$. E.g., for the CRS element $[\beta\varrho + \hat{\beta}\hat{\varrho}]_1$, she knows that the attached polynomial is $X_\beta X_\varrho + X_{\hat{\beta}} X_{\hat{\varrho}}$. Each element output by the adversary can hence be seen as a polynomial in $\boldsymbol{X}$. Moreover, if this element belongs to $\mathbb{G}_z$ for $z \in \{1, 2\}$, then this polynomial has to belong to the span of attached polynomials corresponding to the elements of CRS from the same group. Observing the definition of crs in Protocol 1, we see that this means that each $\mathbb{G}_z$ element output by the adversary must have an attached polynomial of the form $\mathsf{crs}_z(\boldsymbol{X}, T, t)$ for symbolic values T and t:

$$crs_1(\boldsymbol{X}, T, t) = t(X) + T_0 P_0(X) + T_\varrho X_\varrho + T^\dagger(X) Z(X)/X_\varrho + T^*(X)$$
$$+ T_{\hat{\varrho}} X_{\hat{\varrho}} + \sum_{i=1}^{n} T_{\beta,i}(X_\beta P_i(X) + X_{\hat{\beta}} \hat{P}_i(X)) + T_{\beta\varrho}(X_\beta X_\varrho + X_{\hat{\beta}} X_{\hat{\varrho}}),$$
$$crs_2(\boldsymbol{X}, T, t) = t(X) + T_0 P_0(X) + T_\varrho X_\varrho + T_S X_S + T_\beta X_\beta + T_{\hat{\beta}} X_{\hat{\beta}},$$

where $T^\dagger(X)$ is in the span of $\{((P_i(X) + P_0(X))^2 - 1)/Z(X)\}_{i=1}^{n}$ (it will be a polynomial due to the definition of $P_i(X)$ and $P_0(X)$), $T^*(X)$ is in the span of $\{\hat{P}_i(X)\}_{i=1}^{n}$, and $t(X)$ is in the span of $\{P_i(X)\}_{i=1}^{n}$. (Here we use Lemma 1, given below, that states that $\{P_i(X)\}_{i=1}^{n} \cup \{1\}$ and $\{P_i(X)\}_{i=0}^{n}$ are two bases of degree-$\leq n$ polynomials.) We will follow the same notation in the rest of the paper. E.g., polynomials with a star (like $b^*(X)$) are in the span of $\{\hat{P}_i(X)\}_{i=1}^{n}$.

4.1 Same-Message Argument

For the sake of this argument, let $\boldsymbol{P}(X) = (P_i(X))_{i=1}^{n}$ and $\hat{\boldsymbol{P}}(X) = (\hat{P}_i(X))_{i=1}^{n}$ be two families of linearly independent polynomials. We do not specify the parameters X, X_ϱ and $X_{\hat{\varrho}}$ in the case when they take their canonical values χ, ϱ, and $\hat{\varrho}$.

In the *same-message argument*, the prover aims to prove that given $\mathfrak{a}, \hat{\mathfrak{a}} \in \mathbb{G}_1$, she knows $\boldsymbol{a}$ and r, such that $\binom{\mathfrak{a}}{\hat{\mathfrak{a}}} = [M]_1 \binom{\boldsymbol{a}}{r}$ for

$$[M]_1^\top := (\mathsf{ck}, \widehat{\mathsf{ck}}) = \begin{bmatrix} \boldsymbol{P} & \hat{\boldsymbol{P}} \\ \varrho & \hat{\varrho} \end{bmatrix}_1 \in \mathbb{G}_1^{(n+1)\times 2} \tag{1}$$

and $\mathsf{ck} = [\begin{smallmatrix} \boldsymbol{P} \\ \varrho \end{smallmatrix}]_1$ and $\widehat{\mathsf{ck}} = [\begin{smallmatrix} \hat{\boldsymbol{P}} \\ \hat{\varrho} \end{smallmatrix}]_1$. That is, $\mathfrak{a}$ and $\hat{\mathfrak{a}}$ are commitments to the same vector $\boldsymbol{a}$ with the same randomness r, but using the $((P_i(X))_{i=1}^{n}, X_\varrho)$-commitment scheme and the $((\hat{P}_i(X))_{i=1}^{n}, X_{\hat{\varrho}})$-commitment scheme correspondingly.

We construct the same-message argument by essentially using the (second) QANIZK argument of Kiltz and Wee [30] for the *linear subspace*

$$\mathcal{L}_M = \{(\begin{smallmatrix} \mathfrak{a} \\ \hat{\mathfrak{a}} \end{smallmatrix}) : \exists (\begin{smallmatrix} \boldsymbol{a} \\ r \end{smallmatrix}) \in \mathbb{Z}_q^{n+1} : (\begin{smallmatrix} \mathfrak{a} \\ \hat{\mathfrak{a}} \end{smallmatrix}) = [M]_1 (\begin{smallmatrix} \boldsymbol{a} \\ r \end{smallmatrix})\} .$$

However, as we will see in the proof of the permutation matrix argument, we need knowledge-soundness of the same-message argument. Therefore, while we use exactly the QANIZK argument of Kiltz and Wee, we prove its knowledge-soundness in the GBGM: we show that if the verifier accepts then the prover knows a witness $\boldsymbol{w}$ such that $(\mathfrak{a}, \hat{\mathfrak{a}}) = \boldsymbol{w}^\top \cdot [M]_1$. Moreover, we need it to stay knowledge-sound even when the adversary has access to an auxiliary input.

More precisely, denote by $\mathcal{D}_{n,2}^{sm}$ the distribution of matrices M in Eq. (1) given that $(\chi, \varrho, \hat{\varrho}) \leftarrow_r \mathbb{Z}_q \times (\mathbb{Z}_q^*)^2$. For $k \geq 1$, let $\mathcal{D}_k$ be a distribution such that the $\mathcal{D}_k$-KerMDH assumption holds. Clearly, $\mathcal{D}_{n,2}^{sm}$ is witness-sampleable.

For a matrix $A \in \mathbb{Z}_q^{(k+1) \times k}$, let $\bar{A} \in \mathbb{Z}_q^{k \times k}$ denote the upper square matrix of A. The same-message argument (i.e., the Kiltz-Wee QANIZK argument that $\left(\begin{smallmatrix} \mathfrak{a} \\ \hat{\mathfrak{a}} \end{smallmatrix}\right) = [M]_1 \left(\begin{smallmatrix} a \\ r \end{smallmatrix}\right)$) for witness-sampleable distributions is depicted as follows:

$\mathsf{K}_{sm}(\mathsf{gk}, [M]_1 \in \mathbb{G}_1^{2 \times (n+1)})$: $A \leftarrow_r \mathcal{D}_k$; $K \leftarrow_r \mathbb{Z}_q^{2 \times k}$; $[Q]_1 \leftarrow [M]_1^\top K \in$
$\quad \mathbb{G}_1^{(n+1) \times k}$; $C \leftarrow K\bar{A} \in \mathbb{Z}_q^{2 \times k}$; $\mathsf{crs}_{sm} \leftarrow ([Q]_1, [C]_2, [\bar{A}]_2)$; $\mathsf{td}_{sm} \leftarrow K$;
$\quad$ Return $(\mathsf{crs}_{sm}, \mathsf{td}_{sm})$;
$\mathsf{P}_{sm}(\mathsf{gk}, \mathsf{crs}_{sm}, \left(\begin{smallmatrix} \mathfrak{a} \\ \hat{\mathfrak{a}} \end{smallmatrix}\right), \left(\begin{smallmatrix} a \\ r \end{smallmatrix}\right))$: Return $\pi_{sm} \leftarrow \left(\begin{smallmatrix} a \\ r \end{smallmatrix}\right)^\top [Q]_1 \in \mathbb{G}_1^{1 \times k}$;
$\mathsf{S}_{sm}(\mathsf{gk}, \mathsf{crs}_{sm}, \mathsf{td}_{sm}, \left(\begin{smallmatrix} \mathfrak{a} \\ \hat{\mathfrak{a}} \end{smallmatrix}\right))$: Return $\pi_{sm} \leftarrow \left(\begin{smallmatrix} \mathfrak{a} \\ \hat{\mathfrak{a}} \end{smallmatrix}\right)^\top K \in \mathbb{G}_1^{1 \times k}$;
$\mathsf{V}_{sm}(\mathsf{gk}, \mathsf{crs}_{sm}, \left(\begin{smallmatrix} \mathfrak{a} \\ \hat{\mathfrak{a}} \end{smallmatrix}\right), \pi_{sm})$: Check that $\pi_{sm} \bullet [\bar{A}]_2 \stackrel{?}{=} \left(\begin{smallmatrix} \mathfrak{a} \\ \hat{\mathfrak{a}} \end{smallmatrix}\right)^\top \bullet [C]_2$;

Clearly, the verification accepts since $\pi_{sm} \bullet [\bar{A}]_2 = \left(\begin{smallmatrix} a \\ r \end{smallmatrix}\right)^\top [Q]_1 \bullet [\bar{A}]_2 = \left(\begin{smallmatrix} a \\ r \end{smallmatrix}\right)^\top [M]_1^\top K \bullet [\bar{A}]_2 = \left(\begin{smallmatrix} \mathfrak{a} \\ \hat{\mathfrak{a}} \end{smallmatrix}\right)^\top \bullet [C]_2$.

For the sake of efficiency, we will assume $k = 1$ and $\mathcal{D}_1 = \mathcal{L}_1 = \{\left(\begin{smallmatrix} 1 \\ a \end{smallmatrix}\right) : a \leftarrow_r \mathbb{Z}_q\}$. Then, $\bar{A} = 1$, $K = (\beta, \hat{\beta})^\top$, $Q = (\beta P_1 + \hat{\beta}\hat{P}_1, \ldots, \beta P_n + \hat{\beta}\hat{P}_n, \beta \varrho + \hat{\beta}\hat{\varrho})^\top$, and $C = K$. Thus, crs_{sm} and td_{sm} are as in Protocol 1. In the case of a shuffle, $\left(\begin{smallmatrix} a \\ r \end{smallmatrix}\right) = \left(\begin{smallmatrix} e_{\sigma^{-1}(i)} \\ r \end{smallmatrix}\right)$, and thus $\pi_{sm} \leftarrow [\beta P_{\sigma^{-1}(i)} + \hat{\beta}\hat{P}_{\sigma^{-1}(i)}]_1 + r[\beta\varrho + \hat{\beta}\hat{\varrho}]_1$ as in Protocol 1. The verifier has to check that $\pi_{sm} \bullet [1]_2 = \mathfrak{a} \bullet [\beta]_2 + \hat{\mathfrak{a}} \bullet [\hat{\beta}]_2$, as in Protocol 2. The simulator, given the trapdoor $\mathsf{td}_{sm} = (\beta, \hat{\beta})^\top$, sets $\pi_{sm} \leftarrow \beta\mathfrak{a} + \hat{\beta}\hat{\mathfrak{a}}$.

Theorem 2. *Assume* $\mathsf{crs} = (\mathsf{crs}_{sm}, \mathsf{aux})$ *where* aux *does not depend on* β *or* $\hat{\beta}$. *The same-message argument has perfect zero knowledge for* $\mathcal{L}_M$. *It has adaptive knowledge-soundness in the GBGM.*

Proof. ZERO KNOWLEDGE: follows from $\pi_{sm} = [Q^\top \chi]_1 = [K^\top M^\top \chi]_1 = [K^\top y]_1$.

KNOWLEDGE-SOUNDNESS: In the generic group model, the adversary knows polynomials $A(X) = crs_1(X, A, a)$, $\hat{A}(X) = crs_1(X, \hat{A}, \hat{a})$, and $\pi(X) = crs_1(X, \Pi, \pi)$, such that $\mathfrak{a} = [A(\chi)]_1$, $\hat{\mathfrak{a}} = [\hat{A}(\chi)]_1$, $\pi_{sm} = [\pi(\chi)]_1$.

Because the verification accepts, by the Schwartz–Zippel lemma, from this it follows (with all but negligible probability) that $\pi(X) = X_\beta A(X) + X_{\hat{\beta}}\hat{A}(X)$ as a polynomial. Now, the only elements in crs in group $\mathbb{G}_1$ that depend on X_β and $X_{\hat{\beta}}$ are the elements from crs_{sm}: (a) $X_\beta P_i(X) + X_{\hat{\beta}}\hat{P}_i(X)$ for each i, and (b) $X_\beta X_\varrho + X_{\hat{\beta}}X_{\hat{\varrho}}$. Thus, we must have that for some a and r,

$$\pi(X) = \sum_{i=1}^{n} a_i(X_\beta P_i(X) + X_{\hat{\beta}}\hat{P}_i(X)) + r(X_\beta X_\varrho + X_{\hat{\beta}}X_{\hat{\varrho}})$$

$$= X_\beta \left(\sum_{i=1}^{n} a_i P_i(X) + rX_\varrho\right) + X_{\hat{\beta}}\left(\sum_{i=1}^{n} a_i \hat{P}_i(X) + rX_{\hat{\varrho}}\right)$$

Hence, $A(X) = \sum_{i=1}^{n} a_i P_i(X) + rX_\varrho$ and $\hat{A}(X) = \sum_{i=1}^{n} a_i \hat{P}_i(X) + rX_{\hat{\varrho}}$. Thus, $\mathfrak{a}$ and $\hat{\mathfrak{a}}$ commit to the same vector a using the same randomness r. $\qquad\square$

Remark 1. Here, the only thing we require from the distribution $\mathcal{D}_{n,2}^{sm}$ is witness-sampleability and therefore exactly the same zero-knowledge argument can be used with any two shrinking commitment schemes. □

4.2   Permutation Matrix Argument

In this section, we show that a subargument of the new shuffle argument (see Protocols 1 and 2), where the verifier only computes $\mathfrak{a}_n$ as in prover Step 2 and then checks verification Step 5, gives us a permutation matrix argument. However, we first define the unit vector argument, prove its knowledge-soundness, and then use this to prove knowledge-soundness of the permutation matrix argument. The resulting permutation matrix argument is significantly simpler than in the FLZ shuffle argument [15].

Unit Vector Argument. In a unit vector argument [14], the prover aims to convince the verifier that he knows how to open a commitment $\mathfrak{a}$ to $(\boldsymbol{a}, r)$, such that exactly one coefficient a_I, $I \in [1 .. n]$, is equal to 1, while other coefficients of $\boldsymbol{a}$ are equal to 0. Recall [14,15] that if we define $\boldsymbol{V} := \begin{pmatrix} 2 \cdot I_{n \times n} \\ 1_n^\top \end{pmatrix} \in \mathbb{Z}_q^{(n+1) \times n}$ and $\boldsymbol{b} := \begin{pmatrix} 0_n \\ 1 \end{pmatrix} \in \mathbb{Z}_q^{n+1}$, then $\boldsymbol{a}$ is a unit vector iff

$$(\boldsymbol{V}\boldsymbol{a} + \boldsymbol{b} - 1_{n+1}) \circ (\boldsymbol{V}\boldsymbol{a} + \boldsymbol{b} - 1_{n+1}) = 1_{n+1}, \tag{2}$$

where $\circ$ denotes the Hadamard (entry-wise) product of two vectors. Really, this equation states that $a_i \in \{0, 1\}$ for each $i \in [1 .. n]$, and that $\sum a_i = 1$.

Similar to the 1-sparsity argument in [15], we construct the unit vector argument by using a variant of square span programs (SSP-s, [11]). To proceed, we need to define the following polynomials. For $i \in [1 .. n]$, set $P_i(X)$ to be the degree n polynomial that interpolates the ith column of the matrix $\boldsymbol{V}$, i.e.,

$$P_i(X) := 2\ell_i(X) + \ell_{n+1}(X). \tag{3}$$

Set

$$P_0(X) := \ell_{n+1}(X) - 1, \tag{4}$$

i.e., $P_0(X)$ is the polynomial that interpolates $\boldsymbol{b} - 1_{n+1}$. Define $Q(X) = (\sum_{i=1}^{n} a_i P_i(X) + P_0(X))^2 - 1$. Due to the choice of the polynomials, Eq. (2) holds iff $Q(\omega_i) = 0$ for all $i \in [1 .. n + 1]$, which holds iff $Z(X) \mid Q(X)$, [15].

Lemma 1 ([14,15]). *The sets* $\{P_i(X)\}_{i=1}^{n} \cup \{P_0(X)\}$ *and* $\{P_i(X)\}_{i=1}^{n} \cup \{1\}$ *are both linearly independent.*

Clearly, both $\{P_i(X)\}_{i=1}^{n} \cup \{P_0(X)\}$ and $\{P_i(X)\}_{i=1}^{n} \cup \{1\}$ are a basis of all polynomials of degree $\leq n$.

Let

$$\hat{P}_i(X) := X^{(i+1)(n+1)}. \tag{5}$$

As explained before, $\{\hat{P}_i(X)\}_{i=1}^n$ are needed in the same-message argument and in the consistency argument. Due to that, the CRS has entries allowing to efficiently compute $[\hat{P}_i(\chi)]_1$, which means that a generic adversary of the unit vector argument of the current section has access to those polynomials.

Let $\mathcal{U}_n$ be the set of all unit vectors of length n. The unit vector argument is the following subargument of the new shuffle argument:

$\mathsf{K}_{uv}(\mathsf{gk}, n)$: the same as in Protocol 1.
$\mathsf{P}_{uv}(\mathsf{gk}, \mathsf{crs}, \mathfrak{a}_j, (\boldsymbol{a} \in \mathcal{U}_n, r))$: Compute $(\mathfrak{b}_j, \mathfrak{c}_j)$ as in Protocol 1.
$\mathsf{V}_{uv}(\mathsf{gk}, \mathsf{crs}, \mathfrak{a}_j, (\mathfrak{b}_j, \mathfrak{c}_j))$: $\alpha \leftarrow_r \mathbb{Z}_q$; Check that $(\mathfrak{a}_j + [\alpha]_1 + [P_0]_1) \bullet (\mathfrak{b}_j + [-\alpha]_2 + [P_0]_2) \stackrel{?}{=} \mathfrak{c}_i \bullet [\varrho]_2 + [1 - \alpha^2]_T$;

This argument is similar to the 1-sparsity argument presented in [15], but with a different CRS, meaning we cannot directly use their knowledge-soundness proof. Moreover, here the verifier generates α randomly, while in the argument of [15], α is generated by K. Fortunately, the CRS of the new shuffle argument is in a form that facilitates writing down a human readable and verifiable knowledge-soundness proof while [15] used a computer algebra system to solve a complicated system of polynomial equations.

The only problem of this argument is that it guarantees that the committed vector is a unit vector only under the condition that $A_0 = 0$ (see the statement of the following theorem). However, this will be fine since in the soundness proof of the shuffle argument, we can use the same-message argument to guarantee that $A_0 = 0$.

Theorem 3. *The described unit vector argument is perfectly complete and perfectly witness-indistinguishable. Assume that $\{P_i(X)\}_{i=1}^n \cup \{1\}$, and $\{P_i(X)\}_{i=1}^n \cup \{P_0(X)\}$ are two linearly independent sets. Assume that the same-message argument accepts. The unit vector argument is knowledge-sound in the GBGM in the following sense: there exists an extractor Ext such that if the verifier accepts Eq. 6 for $j = i$, then Ext returns $(r_j, I_j \in [1 .. n])$, such that*

$$\mathfrak{a}_j = [P_{I_j}(\chi) + r_j X_\varrho]_1. \tag{6}$$

We note that the two requirements for linear independence follow from Lemma 1.

Proof. COMPLETENESS: For an honest prover, and $I = \sigma^{-1}(j)$, $\mathfrak{a}_j = [P_I(\chi) + r_j\varrho]_1$, $\mathfrak{b}_j = [P_I(\chi) + r_j\varrho]_2$, and $\mathfrak{c}_j = [r_j(2(P_I(\chi) + r_j\varrho + P_0(\chi)) - r_j\varrho) + h(\chi)Z(\chi)/\varrho]_1$ for $r_j \in \mathbb{Z}_q$. Hence, the verification equation assesses that $(P_I(\chi) + r_j\varrho + \alpha + P_0(\chi)) \cdot (P_I(\chi) + r_j\varrho - \alpha + P_0(\chi)) - (r_j(2(P_I(\chi) + r_j\varrho + P_0(\chi)) - r_j\varrho) + h(\chi)Z(\chi)/\varrho) \cdot \varrho - (1 - \alpha^2) = 0$. This simplifies to the claim that $(P_I(\chi) + P_0(\chi))^2 - 1 - h(\chi)Z(\chi) = 0$, or $h(\chi) = ((P_I(\chi) + P_0(\chi))^2 - 1)/Z(\chi)$ which holds since the prover is honest.

KNOWLEDGE-SOUNDNESS: Assume a generic adversary has returned $\mathfrak{a} = [A(\chi)]_1, \mathfrak{b} = [B(\chi)]_2, \mathfrak{c} = [C(\chi)]_2$, for attached polynomials $A(\boldsymbol{X}) = crs_1(\boldsymbol{X}, A, a)$, $B(\boldsymbol{X}) = crs_2(\boldsymbol{X}, B, b)$, and $C(\boldsymbol{X}) = crs_1(\boldsymbol{X}, C, c)$, such that verification in Step 5 of Protocol 2 accepts. Observing this verification equation, it is easy to see that for the polynomial $V_{uv}(\boldsymbol{X}) := (A(\boldsymbol{X}) + X_\alpha + P_0(X)) \cdot (B(\boldsymbol{X}) - X_\alpha + P_0(X)) - C(\boldsymbol{X}) \cdot X_\varrho - (1 - X_\alpha^2)$, the verification equation assesses that $V_{uv}(\boldsymbol{\chi}) = 0$. Since we are in the GBGM, the adversary knows all coefficients of $A(\boldsymbol{X})$, $B(\boldsymbol{X})$, $C(\boldsymbol{X})$, and $V_{uv}(\boldsymbol{X})$.

Moreover, due to the knowledge-soundness of the same-message argument, we know that $A(\boldsymbol{X}) = a(X) + A_\varrho X_\varrho$ for $a(X) \in \mathsf{span}\{P_i(X)\}_{i=1}^n$. This simplifies correspondingly the polynomial $V_{uv}(\boldsymbol{X})$.

Now, let $V_{uv}(\boldsymbol{X} \setminus X)$ be equal to $V_{uv}(\boldsymbol{X})$ but without X being considered as an indeterminate; in particular, this means that the coefficients of $V_{uv}(\boldsymbol{X} \setminus X)$ can depend on X. Since the verifier accepts, $V_{uv}(\boldsymbol{\chi}) = 0$, so by the Schwartz–Zippel lemma, with all but negligible probability $V_{uv}(\boldsymbol{X}) \cdot X_\varrho = 0$ and hence also $V_{uv}(\boldsymbol{X} \setminus X) \cdot X_\varrho = 0$ as a polynomial. The latter holds iff all coefficients of $V_{uv}(\boldsymbol{X} \setminus X) \cdot X_\varrho$ are equal to 0. We will now consider the corollaries from the fact that coefficients C_M of the following monomials M of $V_{uv}(\boldsymbol{X} \setminus X) \cdot X_\varrho$ are equal to 0, and use them to prove the theorem:

$\underline{M = X_\alpha X_\varrho}$: $C_M = b(X) - a(X) + B_0 P_0(X) = 0$. Since $\{P_i(X)\}_{i=0}^n$ is linearly independent, we get that $B_0 = 0$, $b(X) = a(X)$.

$\underline{M = X_\varrho}$: $C_M = -Z(X)c^\dagger(X) - 1 + (a(X) + P_0(X))(b(X) + (B_0 + 1)P_0(X)) = 0$: since $B_0 = 0$ and $b(X) = a(X)$, we get that $C_M = -Z(X)c^\dagger(X) - 1 + (a(X) + P_0(X))^2 = 0$, or alternatively

$$C^\dagger(X) = \frac{(a(X) + P_0(X))^2 - 1}{Z(X)}$$

and hence, $Z(X) \mid ((a(X) + P_0(X))^2 - 1)$.

Therefore, due to the definition of $P_i(X)$ and the properties of square span programs, $a(X) = P_{I_j}(X)$ for some $I_j \in [1 \mathinner{.\,.} n]$, and Eq. (6) holds. Denote $r_j := A_\varrho$. The theorem follows from the fact that we have a generic adversary who knows all the coefficients, and thus we can build an extractor that just outputs two of them, (r_j, I_j).

WITNESS-INDISTINGUISHABILITY: Consider a witness $(\boldsymbol{a}, r_j)$. If $\boldsymbol{a}$ is fixed and the prover picks $r_j \leftarrow_r \mathbb{Z}_q$ (as in the honest case), then an accepting argument $(\mathfrak{a}_j, \mathfrak{b}_j)$ has distribution $\{([a']_1, [a']_2) : a' \leftarrow_r \mathbb{Z}_q\}$ and $\mathfrak{c}_j$ is uniquely fixed by $(\mathfrak{a}_j, \mathfrak{b}_j)$ and the verification equation. Hence an accepting argument $(\mathfrak{a}_j, \mathfrak{b}_j, \mathfrak{c}_j)$ has equal probability of being constructed from any valid $\boldsymbol{a} \in \mathcal{U}_n$. □

Remark 2. While a slight variant of this subargument was proposed in [15], they did not consider its privacy separately. It is easy to see that neither the new argument nor the 1-sparsity argument [15] is zero-knowledge in the case of type-III pairings. Knowing the witness, the prover can produce $\mathfrak{b}_j$ such that $\mathfrak{a}_j \bullet [1]_2 = [1]_1 \bullet \mathfrak{b}_j$. On the other hand, given an arbitrary input $\mathfrak{a} \in \mathbb{G}_1$ as input, the simulator cannot construct such $\mathfrak{b}$ since there is no efficient isomorphism from $\mathbb{G}_1$ to $\mathbb{G}_2$. Witness-indistinguishability suffices in our application. □

Permutation Matrix Argument. A left stochastic matrix (i.e., its row vectors add up to $\mathbf{1}^{\top}$) where every row is 1-sparse is a permutation matrix, [34]. To guarantee that the matrix is left stochastic, it suffices to compute $\mathfrak{a}_n$ explicitly, i.e., $\mathfrak{a}_n = [\sum_{j=1}^{n} P_j(\chi)]_1 - \sum_{j=1}^{n-1} \mathfrak{a}_j$. After that, we need to perform the unit vector argument on every $\mathfrak{a}_j$; this guarantees that Eq. (6) holds for each row. Since a unit vector is also a 1-sparse vector, we get the following result.

Theorem 4. *The permutation matrix argument of this section is perfectly complete and perfectly witness-indistinguishable. Assume that $\{P_i(X)\}_{i=1}^{n} \cup \{1\}$ is a linearly independent set. The permutation matrix argument is knowledge-sound in the GBGM in the following sense: there exists an extractor* Ext *such that if the verifier accepts the verification equation on Step 5 of Protocol 2 for all $j \in [1 .. n]$, and $\mathfrak{a}_n$ is explicitly computed as in Protocol 2, then* Ext *outputs $(\sigma \in S_n, \mathbf{r})$, such that for all $j \in [1 .. n]$, $\mathfrak{a}_j = [P_{\sigma^{-1}(j)}(\chi) + r_j \varrho]_1$.*

Proof. KNOWLEDGE-SOUNDNESS follows from the explicit construction of $\mathfrak{a}_n$, and from the fact that we have a generic adversary that knows all the coefficients, and thus also knows $(\sigma, \mathbf{r})$. More precisely, from the knowledge-soundness of the unit vector argument, for each $j \in [1 .. n]$ there exists an extractor that outputs $(\mathbf{a}_j, r_j)$ such that $\mathbf{a}_j$ is a unit vector with a 1 at some position $I_j \in [1 .. n]$ and $\mathfrak{a}_j = [P_{I_j}(\chi) + r_j \varrho]_1$. Since $[\sum_{j=1}^{n} P_{I_j}(\chi)]_1 = \sum_{j=1}^{n} \mathfrak{a}_j = [\sum_{j=1}^{n} P_j(\chi)]_1$, by the Schwartz-Zippel lemma we have that with overwhelming probability $\sum_{j=1}^{n} P_{I_j}(X) = \sum_{j=1}^{n} P_j(X)$ as a polynomial. Since due to Lemma 1 $\{P_i(X)\}_{i=1}^{n}$ is linearly independent, this means that $(I_1, \ldots, I_n)$ is a permutation of $[1 .. n]$, so $(\mathbf{a}_j)_{j=1}^{n}$ is a permutation matrix.

WITNESS-INDISTINGUISHABILITY   follows   from   the   witness-indistinguishability of the unit vector argument.   □

4.3 Consistency Argument

The last subargument of the shuffle argument is the consistency argument. In this argument the prover aims to show that, given $\hat{\mathbf{a}}$ that commits to a matrix $\mathbf{E} \in \mathbb{Z}_q^{n \times n}$ and two tuples ($\mathfrak{M}$ and $\mathfrak{M}'$) of Elgamal ciphertexts, it holds that $\mathsf{Dec}_{\mathsf{sk}}(\mathfrak{M}') = \mathbf{E} \cdot \mathsf{Dec}_{\mathsf{sk}}(\mathfrak{M})$. Since we use a shrinking commitment scheme, each $\hat{\mathbf{a}}$ can commit to any matrix $\mathbf{E}$. Because of that, we use the fact that the permutation matrix argument is knowledge-sound in the GBGM, and thus there exists an extractor (this is formally provern in Sect. 3) that, after the same-message and the permutation matrix argument, extracts $\mathbf{E}$ and randomizer vectors $\mathbf{r}$ and $\mathbf{t}$, such that $\hat{\mathbf{a}} = \left(\frac{\mathbf{E}}{\mathbf{r}^{\top}} \right)^{\top} [\frac{\hat{P}}{\hat{\varrho}}]_1$. Hence, the guilt relation is

$$\mathcal{R}_{con,n}^{\mathsf{glt}} = \left\{ \begin{array}{l} (\mathsf{gk}, \mathsf{pk}, (\mathfrak{M}, \mathfrak{M}', \hat{\mathbf{a}}), (\mathbf{E}, \mathbf{r})) : \\ \hat{\mathbf{a}} = \left(\frac{\mathbf{E}}{\mathbf{r}^{\top}} \right)^{\top} [\frac{\hat{P}}{\hat{\varrho}}]_1 \wedge \mathsf{Dec}_{\mathsf{sk}}(\mathfrak{M}') \neq \mathbf{E} \cdot \mathsf{Dec}_{\mathsf{sk}}(\mathfrak{M}) \end{array} \right\}.$$

We note that in the case of a shuffle argument, $\mathbf{E}$ is a permutation matrix. However, we will prove the soundness of the consistency argument for the general case of arbitrary matrices.

The new (general) consistency argument for a valid CRS distribution $\mathcal{D}_{n+1,1}$, such that $\mathsf{td}_{con} = (\chi, \hat{\varrho})$, works as follows. Note that the prover should be able to work without knowing the Elgamal secret key, and that $[M]_1 = \mathsf{ck}$.

$\mathsf{K}_{con}(\mathsf{gk}, n)$: Return $(\mathsf{crs}_{con}; \mathsf{td}) = ([M]_1 = [\begin{smallmatrix}\hat{P}\\\hat{\varrho}\end{smallmatrix}]_1, \mathsf{aux}; (\chi, \hat{\varrho})) \leftarrow \mathcal{D}_{n+1,1}$.

$\mathsf{P}_{con}(\mathsf{gk}, \mathsf{crs}, (\mathfrak{M}, \mathfrak{M}'), (E, r))$: // $\mathfrak{M}' = E\mathfrak{M} + (t_i \cdot \mathsf{pk})_{i=1}^n$

$\quad \hat{\mathbf{a}} \leftarrow \left(\begin{smallmatrix}E\\r^\top\end{smallmatrix}\right)^\top [\begin{smallmatrix}\hat{P}\\\hat{\varrho}\end{smallmatrix}]_1; \quad r_t \leftarrow_r \mathbb{Z}_q; \quad t \leftarrow t^\top[\hat{P}]_1 + r_t[\hat{\varrho}]_1; \quad \mathfrak{N} \leftarrow r^\top \mathfrak{M} + r_t \cdot \mathsf{pk};$

$\quad$ Return $(\pi_{con} \leftarrow (\hat{\mathbf{a}}, \mathsf{t}, \mathfrak{N}))$.

$\mathsf{V}_{con}(\mathsf{gk}, \mathsf{crs}, (\mathfrak{M}, \mathfrak{M}'), \pi_{con})$: Check that $[\hat{P}]_1^\top \circ \mathfrak{M}' - \hat{\mathbf{a}}^\top \circ \mathfrak{M} \overset{?}{=} \mathsf{t} \circ \mathsf{pk} - [\hat{\varrho}]_1 \circ \mathfrak{N}$.

Next, we prove that the consistency argument is culpably sound under a suitable KerMDH assumption (with an auxiliary input). After that, we prove that this variant of the KerMDH assumption holds in the GBGM, given that the auxiliary input satisfies some easily verifiable conditions.

Theorem 5. *Assume that $\mathcal{D}_n^{con}$ is a valid CRS distribution, where the matrix distribution outputs $[M]_1 = [\begin{smallmatrix}\hat{P}(\chi)\\\hat{\varrho}\end{smallmatrix}]_1 \in \mathbb{Z}_q^{n+1}$ for $(\chi, \hat{\varrho}) \leftarrow_r \mathbb{Z}_q \times \mathbb{Z}_q^*$. The consistency argument is perfectly complete and perfectly zero knowledge. Assume that the $\mathcal{D}_n^{con}$-KerMDH assumption with an auxiliary input holds in $\mathbb{G}_1$. Then the consistency argument is culpably sound using $\mathcal{R}_{con}^{glt}$ with the CRS $\mathsf{crs} = ([M]_1, \mathsf{aux})$.*

Proof. PERFECT COMPLETENESS: In the case of the honest prover, $[\hat{P}]_1^\top \circ \mathfrak{M}' - \hat{\mathbf{a}}^\top \circ \mathfrak{M} = [\hat{P}]_1^\top \circ (E\mathfrak{M} + (t_i \cdot \mathsf{pk})_{i=1}^n) - \left(\begin{smallmatrix}E\\r^\top\end{smallmatrix}\right)^\top [\begin{smallmatrix}\hat{P}\\\hat{\varrho}\end{smallmatrix}]_1^\top \circ \mathfrak{M} = [\hat{P}]_1^\top \circ ((t_i \cdot \mathsf{pk})_{i=1}^n) + r_t[\hat{\varrho}]_1 \circ \mathsf{pk} - [\hat{\varrho}]_1 \circ r_t\mathsf{pk} - [\hat{\varrho}]_1 \circ r^\top \mathfrak{M} = \mathsf{t} \circ \mathsf{pk} - [\hat{\varrho}]_1 \circ \mathfrak{N}$.

CULPABLE SOUNDNESS: Assume that $\mathcal{A}_{con}$ is an adversary that, given input $(\mathsf{gk}, \mathsf{crs}_{con})$ for $\mathsf{crs}_{con} = (M, \mathsf{aux}) \leftarrow_r \mathcal{D}_n^{con}$, returns $u = (\mathfrak{M}, \mathfrak{M}')$, $\pi_{con} = (\hat{\mathbf{a}}, \mathsf{t}, \mathfrak{N})$ and $w^{glt} = (E, r)$. $\mathcal{A}_{con}$ succeeds iff (i) the verifier of the consistency argument accepts, (ii) $\hat{\mathbf{a}} = \left(\begin{smallmatrix}E\\r^\top\end{smallmatrix}\right)^\top [\begin{smallmatrix}\hat{P}\\\hat{\varrho}\end{smallmatrix}]_1$, and (iii) $\mathsf{Dec}_{\mathsf{sk}}(\mathfrak{M}') = E \cdot \mathsf{Dec}_{\mathsf{sk}}(\mathfrak{M})$. Assume $\mathcal{A}_{con}$ succeeds with probability ε.

We construct the following adversary $\mathcal{A}_{ker}$ that breaks the $\mathcal{D}_n^{con}$-KerMDH assumption with auxiliary input. $\mathcal{A}_{ker}$ gets an input $(\mathsf{gk}, [M]_1, \mathsf{aux})$ where $\mathsf{gk} \leftarrow \mathsf{BG}(1^\kappa)$ and $\mathsf{crs}_{ker} = ([M]_1, \mathsf{aux}) \leftarrow_r \mathcal{D}_n^{con}$, and is supposed to output $[c]_2$, such that $M^\top c = 0$ but $c \neq 0$.

On such an input, $\mathcal{A}_{ker}$ first parses the auxiliary input as $\mathsf{aux} = (\mathsf{aux}', \mathsf{pk})$, then picks $\mathsf{sk}' \leftarrow_r \mathbb{Z}_q$ and creates $\mathsf{crs}_{con} = ([M]_1, (\mathsf{aux}', \mathsf{pk}'))$, where $\mathsf{pk}' = ([1]_2, [\mathsf{sk}']_2)$. Note that crs_{ker} and crs_{con} have the same distribution. $\mathcal{A}_{ker}$ makes a query to $\mathcal{A}_{con}$ with input $(\mathsf{gk}, \mathsf{crs}_{con})$.

After obtaining the answer $u = (\mathfrak{M}, \mathfrak{M}')$, $\pi_{con} = (\hat{\mathbf{a}}, \mathsf{t}, \mathfrak{N})$ and $w^{glt} = (E, r)$ from $\mathcal{A}_{con}$, he does the following:

1. If $[\hat{P}]_1^\top \circ \mathfrak{M}' - \hat{\mathbf{a}}^\top \circ \mathfrak{M} \neq \mathsf{t} \circ \mathsf{pk}' - [\hat{\varrho}]_1 \circ \mathfrak{N}$ then abort.
2. Use sk' to decrypt: $\mathfrak{m} \leftarrow \mathsf{Dec}_{\mathsf{sk}'}(\mathfrak{M})$, $\mathfrak{m}' \leftarrow \mathsf{Dec}_{\mathsf{sk}'}(\mathfrak{M}')$, $\mathfrak{n} \leftarrow \mathsf{Dec}_{\mathsf{sk}'}(\mathfrak{N})$.
3. Return $[c]_2 \leftarrow \left(\begin{smallmatrix}\mathfrak{m}' - E\mathfrak{m}\\\mathfrak{n} - r^\top \mathfrak{m}\end{smallmatrix}\right)$.

Let us now analyze $\mathcal{A}_{ker}$'s success probability. With probability $1 - \varepsilon$, $\mathcal{A}_{ker}$ fails, in which case $\mathcal{A}_{ker}$ will abort. Otherwise, the verification equation holds. Decrypting the right-hand sides of each $\circ$ in the verification equation in Step 6, we get that $[M]_1^\top \bullet \left(\begin{smallmatrix} \mathfrak{m}' \\ \mathfrak{n} \end{smallmatrix} \right) - \hat{\mathbf{a}}^\top \bullet \mathbf{m} = [0]_T$.

Since $\mathcal{A}_{con}$ is successful, then $\mathbf{a} = \left(\begin{smallmatrix} E \\ r^\top \end{smallmatrix} \right)^\top [M]_1$, hence we get

$$
\begin{aligned}
[0]_T &= [M]_1^\top \bullet \left(\begin{smallmatrix} \mathfrak{m}' \\ \mathfrak{n} \end{smallmatrix} \right) - [M]_1^\top \left(\begin{smallmatrix} E \\ r^\top \end{smallmatrix} \right) \bullet \mathbf{m} \\
&= [M]_1^\top \bullet \left(\begin{smallmatrix} \mathfrak{m}' \\ \mathfrak{n} \end{smallmatrix} \right) - [M]_1^\top \left(\begin{smallmatrix} E \\ r^\top \end{smallmatrix} \right) \bullet \mathbf{m} \\
&= [M]_1^\top \bullet \left(\begin{smallmatrix} \mathfrak{m}' \\ \mathfrak{n} \end{smallmatrix} \right) - [M]_1^\top \bullet \left(\begin{smallmatrix} E \\ r^\top \end{smallmatrix} \right) \mathbf{m} \\
&= [M]_1^\top \bullet \left(\begin{matrix} \mathfrak{m}' - E\mathbf{m} \\ \mathfrak{n} - r^\top \mathbf{m} \end{matrix} \right).
\end{aligned}
$$

Since $\mathcal{A}_{con}$ is successful, then $\mathfrak{m}' \neq E\mathfrak{m}$, which means that $c \neq 0$ but $M^\top c = 0$. Thus, $\mathcal{A}_{ker}$ solves the $\mathcal{D}_n^{con}$-KerMDH problem with probability ε.

PERFECT ZERO-KNOWLEDGE: The simulator $\mathsf{S}_{con}(\mathsf{gk}, \mathsf{crs}, (\mathfrak{M}, \mathfrak{M}'), \mathsf{td}_{con} = (\chi, \hat{\varrho}))$ proceeds like the prover in the case $E = I$ (the identity matrix) and $t = 0$, and then computes an $\mathfrak{N}$ that makes the verifier accept. More precisely, the simulator sets $r \leftarrow_r \mathbb{Z}_q$, $\hat{\mathbf{a}} \leftarrow [\hat{P}]_1 + r[\hat{\varrho}]_1$, $r_t \leftarrow_r \mathbb{Z}_q$, $t \leftarrow r_t[\hat{\varrho}]_1$, and $\mathfrak{N} \leftarrow (\hat{P}/\hat{\varrho} + r)^\top \mathfrak{M} - (\hat{P}/\hat{\varrho})^\top \mathfrak{M}' + r_t \cdot \mathsf{pk}$. S_{con} then outputs $\pi_{con} \leftarrow (\hat{\mathbf{a}}, t, \mathfrak{N})$.

Due to the perfect hiding property of the commitment scheme, $(\hat{\mathbf{a}}, t)$ has the same distribution as in the real protocol. Moreover,

$$
\begin{aligned}
[\hat{P}]_1^\top \circ \mathfrak{M}' - \hat{\mathbf{a}}^\top \circ \mathfrak{M} &= [\hat{P}]_1^\top \circ \mathfrak{M}' - ([\hat{P}]_1 + r[\hat{\varrho}]_1)^\top \circ \mathfrak{M} \\
&= [\hat{\varrho}]_1 \circ (\hat{P}/\hat{\varrho})^\top \mathfrak{M}' - [\hat{\varrho}]_1 \circ (\hat{P}/\hat{\varrho} + r)^\top \mathfrak{M} \\
&= [\hat{\varrho}]_1 \circ (r_t \cdot \mathsf{pk} - r_t \cdot \mathsf{pk} + (\hat{P}/\hat{\varrho})^\top \mathfrak{M}' - (\hat{P}/\hat{\varrho} + r)^\top \mathfrak{M}) \\
&= t \circ \mathsf{pk} - [\hat{\varrho}]_1 \circ \mathfrak{N}
\end{aligned}
$$

and thus this choice of $\mathfrak{N}$ makes the verifier accept. Since t is uniformly random and $\mathfrak{N}$ is uniquely defined by $(\hat{\mathbf{a}}, t)$ and the verification equation, we have a perfect simulation. □

Example 1. In the case of the shuffle argument, E is a permutation matrix with $E_{ij} = 1$ iff $j = \sigma(i)$. Thus, $[\hat{\mathbf{a}}]_1 = \left(\begin{smallmatrix} E \\ r^\top \end{smallmatrix} \right)^\top [\begin{smallmatrix} \hat{P} \\ \hat{\varrho} \end{smallmatrix}]_1 = [(\hat{P}_{\sigma^{-1}(i)})_{i=1}^n + r\hat{\varrho}]_1$, and $E\mathbf{m} = (\mathfrak{m}_{\sigma(i)})_{i=1}^n$. Hence in the shuffle argument, assuming that it has been already established that E is a permutation matrix, then after the verification in Step 6 one is assured that $\mathfrak{m}' = (\mathfrak{m}_{\sigma(i)})_{i=1}^n$. □

We will now prove that the used variant of KerMDH with auxiliary input is secure in the GBGM. Before going on, we will establish the following linear independence result.

Lemma 2. *The set $\Psi_\times := \{P_i(X)\hat{P}_j(X)\}_{1 \leq i,j \leq n} \cup \{\hat{P}_i(X)\}_{i=1}^n$ is linearly independent.*

Proof. By Lemma 1, for each $j \in [1 .. n]$ we have that $\{P_i(X)\hat{P}_j(X)\}_{1 \leq i \leq n} \cup \{\hat{P}_j(X)\}$ is linearly independent. Hence to show that $\Psi_\times$ is linearly independent, it suffices to show that for all $1 \leq j < k \leq n$, the span of sets $\{P_i(X)\hat{P}_j(X)\}_{1 \leq i \leq n} \cup \{\hat{P}_j(X)\}$ and $\{P_i(X)\hat{P}_k(X)\}_{1 \leq i \leq n} \cup \{\hat{P}_k(X)\}$ only intersect at 0. This holds since for non-zero vectors $(\boldsymbol{a}, \boldsymbol{b})$ and integers (A, B), $\sum_{i=1}^{n} a_i P_i(X)\hat{P}_j(X) + A\hat{P}_j(X) = (\sum_{i=1}^{n} a_i P_i(X) + A)X^{(j+1)(n+1)}$ has degree at most $n + (j+1)(n+1) < (k+1)(n+1)$, while $\sum_{i=1}^{n} b_i P_i(X)\hat{P}_k(X) + B\hat{P}_k(X) = (\sum_{i=1}^{n} b_i P_i(X) + B)X^{(k+1)(n+1)}$ has degree at least $(k+1)(n+1)$. $\qquad\square$

Theorem 6. *Assume that $(\hat{P}_i(X))_{i=1}^{n}$ satisfy the following properties:*

- *$\{\hat{P}_i(X)\}_{i=1}^{n}$ is linearly independent,*
- *$\{P_i(X)\hat{P}_j(X)\}_{1 \leq i,j \leq n}$ is linearly independent,*

Assume also that the second input aux *output by* $\mathcal{D}_n^{con}$ *satisfies the following property. The subset* aux$_2$ *of* $\mathbb{G}_2$*-elements in* aux *must satisfy that*

- aux$_2$ *does not depend on* $\hat{\varrho}$,
- *the only element in* aux$_2$ *that depends on* ϱ *is* $[\varrho]_2$,
- *the only elements in* aux$_2$ *that depend on* χ *are* $[P_i(\chi)]_2$ *for* $i \in [0 .. n]$.

Then in the GBGM, the $\mathcal{D}_n^{con}$*-KerMDH assumption with an auxiliary input holds in* $\mathbb{G}_1$.

Clearly, $(P_i(X))_{i=1}^{n}$, $(\hat{P}_i(X))_{i=1}^{n}$, and crs in Protocol 1 (if used as aux) satisfy the required properties.

Proof. Consider a generic group adversary $\mathcal{A}_{ker}$ who, given $([\begin{smallmatrix}\hat{P}(\chi)\\\hat{\varrho}\end{smallmatrix}]_1, \text{aux})$ as an input, outputs a non-zero solution $(\begin{smallmatrix}\mathfrak{m}\\\mathfrak{n}\end{smallmatrix}) \in \mathbb{G}_2^{n+1}$ to the KerMDH problem, i.e., $[\boldsymbol{M}]_1^\top \bullet (\begin{smallmatrix}\mathfrak{m}\\\mathfrak{n}\end{smallmatrix}) = [0]_T$. In the generic model, each element in aux$_2$ has some polynomial attached to it. Since $\mathcal{A}_{ker}$ is a generic adversary, he knows polynomials $M_i(\boldsymbol{X})$ (for each i) and $N(\boldsymbol{X})$, such that $\mathfrak{m}_i = [M_i(\chi)]_2$ and $\mathfrak{n} = [N(\chi)]_2$. Those polynomials are linear combinations of the polynomials involved in aux$_2$.

Hence, if $(\begin{smallmatrix}\mathfrak{m}\\\mathfrak{n}\end{smallmatrix}) \in \mathbb{G}_2^{n+1}$ is a solution to the KerMDH problem, then $[\boldsymbol{M}]_1^\top \bullet (\begin{smallmatrix}\mathfrak{m}\\\mathfrak{n}\end{smallmatrix}) = [0]_T$ or equivalently, $\hat{V}_{ker}(\chi) = 0$, where

$$\hat{V}_{ker}(\boldsymbol{X}) := \sum_{i=1}^{n} \hat{P}_i(X)M_i(\boldsymbol{X}) + X_{\hat{\varrho}} \cdot N(\boldsymbol{X}), \tag{7}$$

for coefficients known to the generic adversary. By the Schwartz–Zippel lemma, from this it follows (with all but negligible probability) that $\hat{V}_{ker}(\boldsymbol{X}) = 0$ as a polynomial.

Due to the assumptions on aux$_2$, and since $\{P_i(X)\}_{i=0}^{n} \cup \{1\}$ and $\{P_i(X)\}_{i=0}^{n} \cup \{P_0(X)\}$ are interchangeable bases of degree-$(\leq n)$ polynomials, we can write

$$M_i(\boldsymbol{X}) = \sum_{j=1}^{n} M_{ij}P_j(X) + M_i'X_\varrho + m_i(\boldsymbol{X}),$$

where $m_i(\boldsymbol{X})$ does not depend on X, X_ϱ or $X_{\hat{\varrho}}$.

Since $\sum \hat{P}_i(X) M_i(\boldsymbol{X})$ does not depend on $X_{\hat{\varrho}}$, we get from $\hat{V}_{ker}(\boldsymbol{X}) = 0$ that $\sum_{i=1}^{n} \hat{P}_i(X) M_i(\boldsymbol{X}) = N(\boldsymbol{X}) = 0$.

Next, we can rewrite $\sum_{i=1}^{n} \hat{P}_i(X) M_i(\boldsymbol{X}) = 0^{\bullet}$ as

$$\sum_{i=1}^{n} \sum_{j=1}^{n} M_{ij} \hat{P}_i(X) P_j(X) + X_{\varrho} \sum_{i=1}^{n} M_i' \hat{P}_i(X) + \sum_{i=1}^{n} \hat{P}_i(X) m_i(\boldsymbol{X}) = 0.$$

Due to assumptions on aux_2, this means that

- $\sum_{i=1}^{n} \sum_{j=1}^{n} M_{ij} \hat{P}_i(X) P_j(X) = 0$. Since $\{\hat{P}_i(X) P_j(X)\}_{i,j \in [1..n]}$ is linearly independent, $M_{ij} = 0$ for all $i, j \in [1..n]$,
- $\sum_{i=1}^{n} M_i' \hat{P}_i(X) = 0$. Since $\{\hat{P}_i(X)\}_{i \in [1..n]}$ is linearly independent, $M_i' = 0$ for all $i \in [1..n]$.
- $\sum_{i=1}^{n} m_i(\boldsymbol{X}) \hat{P}_i(X) = 0$. Since $m_i(\boldsymbol{X})$ does not depend on X and $\{\hat{P}_i(X)\}_{i \in [1..n]}$ is linearly independent, we get $m_i(\boldsymbol{X}) = 0$ for all $i \in [1..n]$.

Hence, $M_i(\boldsymbol{X}) = 0$ for each $i \in [1..n]$. Thus, with all but negligible probability, we have that $N(\boldsymbol{X}) = M_i(\boldsymbol{X}) = 0$, which means $\binom{\mathsf{m}}{\mathsf{n}} = [\boldsymbol{0}]_2$. $\square$

5   Security Proof of Shuffle

Theorem 7. *The new shuffle argument is perfectly complete.*

Proof. We showed the completeness of the same-message argument (i.e., that the verification equation on Step 4 holds) in Sect. 4.1, the completeness of the unit vector argument (i.e., that the verification equation on Step 5 holds) in Theorem 3, and the completeness of the consistency argument (i.e., that the verification equation on Step 6 holds) in Sect. 4.3. $\square$

Theorem 8. (Soundness of Shuffle Argument). *Assume that the following sets are linearly independent:*

- $\{P_i(X)\}_{i=0}^{n}$,
- $\{P_i(X)\}_{i=1}^{n} \cup \{1\}$,
- $\{P_i(X) \hat{P}_j(X)\}_{1 \le i,j \le n}$.

Let $\mathcal{D}_n^{con}$ be as before with $([\boldsymbol{M}]_1, \mathsf{aux})$, such that aux is equal to the CRS in Protocol 1 minus the elements already in $[\boldsymbol{M}]_1$. If the $\mathcal{D}_1$-KerMDH assumption holds in $\mathbb{G}_2$, the $\mathcal{D}_n^{((P_i(X))_{i=1}^{n}, X_{\varrho})}$-KerMDH and $\mathcal{D}_n^{((\hat{P}_i(X))_{i=1}^{n}, X_{\hat{\varrho}})}$-KerMDH assumptions hold and the $\mathcal{D}_n^{con}$-KerMDH assumption with auxiliary input holds in $\mathbb{G}_1$, then the new shuffle argument is sound in the GBGM.

Proof. First, from the linear independence of $\{P_i(X) \hat{P}_j(X)\}_{1 \le i,j \le n}$, we get straightforwardly that $\{\hat{P}_j(X)\}_{j=1}^{n}$ is also linearly independent. We construct the following simple sequence of games.

---

1. Generate a new random tape r' for $\mathcal{A}_{sh}$
2. Run $\mathcal{A}_{sh}(\mathsf{gk}, \mathsf{crs}_{sh}; r')$ to obtain a shuffle argument $\pi_{sh} = (\mathfrak{M}', (\mathfrak{a}_i)_{i=1}^{n-1}, \pi_{sp}, \pi_{sm}, (\hat{\mathfrak{a}}_i)_{i=1}^{n-1}, \mathsf{t}, \mathfrak{N})$. Compute $\mathfrak{a}_n$ and $\hat{\mathfrak{a}}_n$ as in Protocol 1.
3. Abort when π_{sh} is not accepting.
4. For $i = 1$ to n:
   (a) Extract $(\mathbf{E}'_i, r'_i) \leftarrow \mathsf{Ext}^{\mathcal{A}}_{sm:i}(\mathsf{gk}, \mathsf{crs}_{sh}; r')$.
   (b) Abort when $(\mathbf{E}'_i, r'_i)$ is not a valid opening of $\mathfrak{a}_i$ or $\hat{\mathfrak{a}}_i$.
5. Extract $(\mathbf{E}, r) \leftarrow \mathsf{Ext}^{\mathcal{A}}_{pm}(\mathsf{gk}, \mathsf{crs}_{sh}; r')$.
6. Abort when $\mathbf{E}' \neq \mathbf{E}$ or $r' \neq r$.
7. // At this point $\mathcal{A}_{con}$ knows a permutation matrix $\mathbf{E}$ and vector r, such that $\hat{\mathfrak{a}} = [\mathbf{E}^\top \hat{\mathbf{P}} + r\hat{\varrho}]_1$ and $\mathsf{Dec}_{\mathsf{sk}}(\mathfrak{M}') \neq \mathbf{E} \cdot \mathsf{Dec}_{\mathsf{sk}}(\mathfrak{M})$.
8. Return $u = (\mathfrak{M}, \mathfrak{M}')$, $\pi_{con} = (\hat{\mathfrak{a}}, \mathsf{t}, \mathfrak{N})$, and $w^{\mathsf{glt}} = (\mathbf{E}, r)$.

Protocol 3: Adversary $\mathcal{A}_{con}$ on input $(\mathsf{gk}, \mathsf{crs}_{con})$

$\underline{\mathsf{GAME}_0}$: This is the original soundness game. Let $\mathcal{A}_{sh}$ be an adversary that breaks the soundness of the new shuffle argument. That is, given $(\mathsf{gk}, \mathsf{crs})$, $\mathcal{A}_{sh}$ outputs $u = (\mathfrak{M}, \mathfrak{M}')$ and an accepting proof π, such that for all permutations σ, there exists $i \in [1..n]$, such that $\mathsf{Dec}_{\mathsf{sk}}(\mathfrak{M}'_i) \neq \mathsf{Dec}_{\mathsf{sk}}(\mathfrak{M}_{\sigma(i)})$.

$\underline{\mathsf{GAME}_1}$: Here, we let $\mathcal{A}_{con}$ herself generate a new secret key $\mathsf{sk} \leftarrow_r \mathbb{Z}_q$ and set $\mathsf{pk} \leftarrow [1, \mathsf{sk}]_2$. Since the distribution of pk is witness-sampleable, then for any (even omnipotent) adversary $\mathcal{B}$, $|\Pr[\mathsf{GAME}_1(\mathcal{B}) = 1] - \Pr[\mathsf{GAME}_0(\mathcal{B}) = 1]| = 0$ (i.e., GAME_1 and GAME_0 are indistinguishable).

Now, consider GAME_1 in more detail. Clearly, under given assumptions and due to the construction of crs in Protocol 1 the same-message and permutation matrix arguments are knowledge-sound. Hence there exist extractors $\mathsf{Ext}^{\mathcal{A}}_{sm:i}$ and $\mathsf{Ext}^{\mathcal{A}}_{pm}$ such that the following holds:

- If the same-message argument verifier accepts $(\mathfrak{a}_i, \hat{\mathfrak{a}}_i, \pi_{sm:i}) \leftarrow \mathcal{A}_{sm}(\mathsf{gk}, \mathsf{crs}_{sh}; r')$, then $\mathsf{Ext}^{\mathcal{A}}_{sm:i}(\mathsf{gk}, \mathsf{crs}_{sh}; r')$ outputs the common opening $(\mathbf{E}'_i, r_i)$ of $\mathfrak{a}_i$ and $\hat{\mathfrak{a}}_i$.
- If the permutation matrix argument verifier accepts $((\mathfrak{a}_i)_{i=1}^{n-1}, \pi_{pm}) \leftarrow \mathcal{A}_{pm}(\mathsf{gk}, \mathsf{crs}_{sh}; r')$, then $\mathsf{Ext}^{\mathcal{A}}_{pm}(\mathsf{gk}, \mathsf{crs}_{sh}; r')$ outputs a permutation matrix $\mathbf{E}$ and vector r such that $\mathfrak{a} = [\mathbf{E}^\top \mathbf{P} + r\varrho]_1$ (here, $\mathfrak{a}_n$ is computed as in Protocol 1).

We construct the following adversary $\mathcal{A}_{con}$, see Protocol 3, that breaks the culpable soundness of the consistency argument using $\mathcal{R}^{\mathsf{glt}}_{con}$ with auxiliary input. $\mathcal{A}_{con}$ gets an input $(\mathsf{gk}, \mathsf{crs}_{con})$ where $\mathsf{gk} \leftarrow \mathsf{BG}(1^\kappa)$, and $\mathsf{crs}_{con} \leftarrow ([\mathbf{M}]_1, \mathsf{aux}) \leftarrow_r \mathcal{D}^{con}_n$. $\mathcal{A}_{con}$ is supposed to output $(u, \pi_{con}, w^{\mathsf{glt}})$ where $u = (\mathfrak{M}, \mathfrak{M}')$, $\pi_{con} = (\hat{\mathfrak{a}}, \mathsf{t}, \mathfrak{N})$, and $w^{\mathsf{glt}} = (\mathbf{E}, r)$ such that $\hat{\mathfrak{a}} = [\mathbf{E}^\top \hat{\mathbf{P}} + r\hat{\varrho}]_1$, π_{con} is accepting, and $\mathsf{Dec}_{\mathsf{sk}}(\mathfrak{M}') \neq \mathbf{E} \cdot \mathsf{Dec}_{\mathsf{sk}}(\mathfrak{M})$.

It is clear that if $\mathcal{A}_{con}$ does not abort, then he breaks the culpable soundness of the consistency argument. Let us now analyze the probability that $\mathcal{A}_{con}$ does abort in any step.

1. $\mathcal{A}_{con}$ aborts in Step 3 if $\mathcal{A}_{sh}$ fails, i.e., with probability $1 - \varepsilon_{sh}$.
2. $\mathcal{A}_{con}$ will never abort in Step 4b since by our definition of knowledge-soundness, $\mathsf{Ext}^{\mathcal{A}}_{sm:i}$ always outputs a valid opening as part of the witness.
3. $\mathcal{A}_{con}$ aborts in Step 6 only when he has found two different openings of $\mathfrak{a}_i$ for some $i \in [1..n]$. Hence, the probability of abort is upper bounded by $n\varepsilon_{binding}$, where $\varepsilon_{binding}$ is the probability of breaking the binding property of the $((P_i(X))_{i=1}^{n}, \varrho)$-commitment scheme.

Thus, $\mathcal{A}_{con}$ aborts with probability $\leq 1 - \varepsilon_{sh} + n\varepsilon_{binding}$, or succeeds with probability $\varepsilon_{con} \geq \varepsilon_{sh} - n\varepsilon_{binding}$. Hence $\varepsilon_{sh} \leq \varepsilon_{con} + n\varepsilon_{binding}$. Since under the assumptions of this theorem, both ε_{con} and $\varepsilon_{binding}$ are negligible, we have that ε_{sh} is also negligible. $\qquad\square$

Theorem 9. *The new shuffle argument is perfectly zero-knowledge.*

Proof. We proved that the permutation matrix argument is witness-indistinguishable (Theorem 4), and that the same-message argument is zero-knowledge (Theorem 2) and hence also witness-indistinguishable.

We will construct a simulator S_{sh}, that given a CRS crs and trapdoor $\mathsf{td} = (\chi, \hat{\varrho})$ simulates the prover in Protocol 1. S_{sh} takes any pair $(\mathfrak{M}, \mathfrak{M}')$ as input and performs the steps in Protocol 4. Clearly, the simulator computes $((\mathfrak{a}_i)_{i=1}^{n-1}, (\hat{\mathfrak{a}}_i)_{i=1}^{n-1}, \pi_{uv}, \pi_{sm})$ exactly as an honest prover with permutation $\sigma = \mathrm{Id}$ would, which ensures correct distribution of these values. Moreover, since the commitment scheme is perfectly hiding and the permutation matrix and same-message arguments are witness-indistinguishable, the distribution of these values is identical to that of an honest prover that uses a (possibly different) permutation σ' and randomness values to compute $\mathfrak{M}'$ from $\mathfrak{M}$.

On the other hand, on the last step, S_{sh} uses the simulator S_{con} of the consistency argument from Theorem 5. Since π_{con} output by S_{sh} has been computed exactly as by S_{con} in Theorem 5, the verification equation in Step 6 of Protocol 2 accepts. Moreover, as in Theorem 5, $(\hat{\mathfrak{a}}, \mathsf{t})$ has the same distribution as in the real protocol due to the perfect hiding of the commitment scheme, and $\mathfrak{N}$ is uniquely fixed by $(\hat{\mathfrak{a}}, \mathsf{t})$ and the verification equation. Hence the simulation is perfect. $\qquad\square$

6   Batching

The following lemma, slightly modified from [15,33] (and relying on the batching concept introduced in [4]), shows that one can often use a batched version of the verification equations.

Lemma 3. *Assume $1 < t < q$. Assume $\boldsymbol{y}$ is a vector chosen uniformly random from $[1..t]^{k-1} \times \{1\}$, χ is a vector of integers in $\mathbb{Z}_q$, and $(f_i)_{i \in [1..k]}$ are some polynomials of arbitrary degree. If there exists $i \in [1..k]$ such that $f_i(\chi)([1]_1 \bullet [1]_2) \neq [0]_T$ then $\sum_{i=1}^{k} f_i(\chi)y_i \cdot ([1]_1 \bullet [1]_2) = [0]_T$ with probability $\leq \frac{1}{t}$.*

---

1. For $i = 1$ to $n - 1$: // commits to the permutation $\sigma = \mathrm{Id}$
   (a) $r_i \leftarrow_r \mathbb{Z}_q$; $\mathfrak{r}_i \leftarrow r_i[\varrho]_1$;
   (b) $\mathfrak{a}_i \leftarrow [P_i]_1 + \mathfrak{r}_i$; $\mathfrak{b}_i \leftarrow [P_i]_2 + r_i[\varrho]_2$; $\hat{\mathfrak{a}}_i \leftarrow [\hat{P}_i]_1 + r_i[\hat{\varrho}]_1$;
2. $\mathfrak{a}_n \leftarrow [\sum_{i=1}^{n} P_i]_1 - \sum_{j=1}^{n-1} \mathfrak{a}_j$; $\mathfrak{b}_n \leftarrow [\sum_{i=1}^{n} P_i]_2 - \sum_{j=1}^{n-1} \mathfrak{b}_j$;
3. $\hat{\mathfrak{a}}_n \leftarrow [\sum_{i=1}^{n} \hat{P}_i]_1 - \sum_{j=1}^{n-1} \hat{\mathfrak{a}}_j$;
4. $r_n \leftarrow -\sum_{i=1}^{n-1} r_i$; $\mathfrak{r}_n \leftarrow r_n[\varrho]_1$;
5. For $i = 1$ to n:
   (a) $\mathfrak{c}_i \leftarrow r_i \cdot (2(\mathfrak{a}_i + [P_0]_1) - \mathfrak{r}_i) + [((P_i + P_0)^2 - 1)/\varrho]_1$;
   (b) $\mathfrak{d}_i \leftarrow [\beta P_i + \hat{\beta}\hat{P}_i]_1 + r_i[\beta\varrho + \hat{\beta}\hat{\varrho}]_1$;
6. Set $r_t \leftarrow_r \mathbb{Z}_q$;
7. $\mathfrak{t} \leftarrow r_t[\hat{\varrho}]_1$; // Commits to $\mathbf{0}$
8. $\mathfrak{N} \leftarrow (\hat{\boldsymbol{P}}/\hat{\varrho} + \boldsymbol{r})^\top \mathfrak{M} - (\hat{\boldsymbol{P}}/\hat{\varrho})^\top \mathfrak{M}' + r_t \cdot \mathsf{pk}$;
9. $\pi_{sm} \leftarrow \mathfrak{d}$; // Same-message argument
10. $\pi_{uv} \leftarrow ((\mathfrak{b}_i)_{i=1}^{n-1}, \mathfrak{c})$; // Unit vector argument
11. $\pi_{con} \leftarrow ((\hat{\mathfrak{a}}_i)_{i=1}^{n-1}, \mathfrak{t}, \mathfrak{N})$; // Consistency argument
12. Return $\pi_{sh} \leftarrow ((\mathfrak{M}'_i)_{i=1}^{n}, (\mathfrak{a}_i)_{i=1}^{n-1}, \pi_{uv}, \pi_{sm}, \pi_{con})$.

Protocol 4: The simulator of the shuffle argument

$\mathsf{V}(\mathsf{gk}, \mathsf{crs}, \mathfrak{M}; (\mathfrak{M}'_i)_{i=1}^{n}, (\mathfrak{a}_j)_{j=1}^{n-1}, \pi_{uv}, \pi_{sm}, \pi_{con})$:

1. Parse $(\pi_{uv}, \pi_{sm}, \pi_{con})$ as in the prover's Steps 11–12
2. Compute $\mathfrak{a}_n$, $\mathfrak{b}_n$, and $\hat{\mathfrak{a}}_n$ as in the prover's Step 2,
3. Set $(y_{1j})_{j \in [1 .. n-1]} \leftarrow_r [1 .. t]^{n-1}$, Set $y_{1n} \leftarrow 1$,
4. Set $y_{21} \leftarrow_r [1 .. t]$, $y_{22} \leftarrow 1$,
5. $\alpha \leftarrow_r \mathbb{Z}_q$;
6. Check that // Permutation matrix argument
   $\sum_{j=1}^{n} ((y_{1j}(\mathfrak{a}_j + \alpha[1]_1 + [P_0(\chi)]_1)) \bullet (\mathfrak{b}_j - \alpha[1]_2 + [P_0(\chi)]_2)) = (\boldsymbol{y}_1^\top \mathfrak{c}) \bullet [\varrho]_2 + (\sum_{j=1}^{n} y_{1j})(1 - \alpha^2)[1]_T$;
7. Check that $(\boldsymbol{y}_1^\top \mathfrak{d}) \bullet [1]_2 = (\boldsymbol{y}_1^\top (\mathfrak{a}, \hat{\mathfrak{a}})) \bullet \begin{bmatrix} \beta \\ \hat{\beta} \end{bmatrix}_2$ // Same-message argument
8. Set $\mathfrak{q} \leftarrow \mathfrak{t} \bullet (\mathsf{pk} \cdot \boldsymbol{y}_2)$.
9. Check that // Consistency argument, online
   $[\hat{\boldsymbol{P}}]_1^\top \bullet (\mathfrak{M}' \boldsymbol{y}_2) - \hat{\mathfrak{a}}^\top \bullet (\mathfrak{M} \boldsymbol{y}_2) = \mathfrak{q} - [\hat{\varrho}]_1 \bullet (\mathfrak{N} \boldsymbol{y}_2)$.

Protocol 5: The batched verifier of the new shuffle argument.

Proof. Let us define a multivariate polynomial $V(Y_1, \ldots, Y_{k-1}) = \sum_{i=1}^{k-1} f_i(\boldsymbol{\chi})Y_i + f_k(\boldsymbol{\chi}) \neq 0$. By the Schwartz–Zippel lemma $V(y_1, \ldots, y_{k-1}) = 0$ with probability $\leq \frac{1}{t}$. Since $[1]_1 \bullet [1]_2 \neq [0]_T$, we get that $\sum_{i=1}^{k} f_i(\boldsymbol{\chi})y_i \cdot ([1]_1 \bullet [1]_2) = [0]_T$ with probability $\leq \frac{1}{t}$. $\square$

For the sake of concreteness, the full batched version of the verifier of the new shuffle argument is described in Protocol 5. The following corollary follows immediately from Lemma 3. It implies that if the non-batched verifier in Protocol 2 does not accept then the batched verifier in Protocol 5 only accepts with probability $\leq \frac{3}{t}$. This means that with all but negligible probability, the non-batched verifier accepts iff the batched verifier accepts. In practice a verifier could even set (say) $t = 3 \cdot 2^{40}$.

Corollary 1. *Let* $1 < t < q$. *Assume* $\chi = (\chi, \alpha, \beta, \hat{\beta}, \varrho, \hat{\varrho}, \mathsf{sk}) \in \mathbb{Z}_q^4 \times (\mathbb{Z}_q^*)^2 \times \mathbb{Z}_q$. *Assume* $(y_{1j})_{j \in [1 \,..\, n-1]}$, *and* y_{21} *are values chosen uniformly random from* $[1 \,..\, t]$, *and set* $y_{1n} = y_{22} = 1$.

- *If there exists* $i \in [1 \,..\, n]$ *such that the* ith *verification equation on Step 5 of Protocol 2 does not hold, then the verification equation on Step 6 of Protocol 5 holds with probability* $\leq \frac{1}{t}$.
- *If there exists* $i \in [1 \,..\, n]$ *such that the* ith *verification equation on Step 4 of Protocol 2 does not hold, then the verification equation on Step 7 of Protocol 5 holds with probability* $\leq \frac{1}{t}$.
- *If the verification equation on Step 6 of Protocol 2 does not hold, then the verification equation on Step 9 of Protocol 5 holds with probability* $\leq \frac{1}{t}$.

7   Implementation

We implement the new shuffle argument in C++ using the libsnark library [6]. This library provides an efficient implementation of pairings over several different elliptic curves. We run our measurements on a machine with the following specifications: (i) Intel Core i5-4200U CPU with 1.60 GHz and 4 threads; (ii) 8 GB RAM; (iii) 64bit Linux Ubuntu 16.10; (iv)Compiler GCC version 6.2.0.

In addition, libsnark provides algorithms for multi-exponentiation and fixed-base exponentiation. Here, an n-wide multi-exponentiation means evaluating an expression of the form $\sum_{i=1}^n L_i \mathfrak{a}_i$, and n fixed-base exponentiations means evaluating an expression of the form $(L_1 \mathfrak{b}, \ldots, L_n \mathfrak{b})$, where $L_i \in \mathbb{Z}_q$ and $\mathfrak{b}, \mathfrak{a}_i \in \mathbb{G}_k$. Both can be computed significantly faster than n individual scalar multiplications. (See Table 3.) Importantly, most of the scalar multiplications in the new shuffle argument can be computed by using either a multi-exponentiation or a fixed-base exponentiation.

We also optimize a sum of n pairings by noting that the final exponentiation of a pairing can be done only once instead of n times. All but a constant number of pairings in our protocol can use this optimization. We use parallelization to further optimize computation of pairings and exponentiations.

To generate the CRS, we have to evaluate Lagrange basis polynomials $(\ell_i(X))_{i \in [1 \,..\, n+1]}$ at $X = \chi$. In our implementation, we pick $\omega_j = j$ and make two optimizations. First, we precompute $Z(\chi) = \prod_{j=1}^{n+1}(\chi - j)$. This allows us to write $\ell_i(\chi) = Z(\chi)/((\chi - i) \prod_{j=1, j \neq i}^{n+1}(i - j))$. Second, denote $F_i = \prod_{\substack{j=1 \\ j \neq i}}^{n+1}(i - j)$. Then for $i \in [1 \,..\, n]$, we have $F_{i+1} = (iF_i)/(i - n - 1)$. This allows us to compute all F_i with $2n$ multiplications and n divisions. Computing all $\ell_i(\chi) = \frac{Z(\chi)}{(\chi - i)F_i}$ takes $4n + 2$ multiplications and $2n + 1$ divisions.

Table 3. Efficiency comparison of various operations based on libsnark. Units in the last column show efficiency relative to n exponentiations in $\mathbb{G}_1$ for $n = 100,000$.

	10,000	100,000	1000,000	units
Multi-exp. $\mathbb{G}_1$	0.26 s	2.54 s	24.2 s	0.13
Fixed-base exp. $\mathbb{G}_1$	0.28 s	2.40 s	18.44 s	0.12
Multi-exp. $\mathbb{G}_2$	0.65 s	5.54 s	48.04 s	0.27
Fixed-base exp. $\mathbb{G}_2$	0.75 s	5.62 s	44.34 s	0.28
Exp. $\mathbb{G}_1$	2.15 s	20.29 s	207.11 s	1
Exp. $\mathbb{G}_2$	5.54 s	51.10 s	506.26 s	2.52
Pairing product	4.38 s	43.37 s	471.72 s	2.14
Pairings	10.24 s	97.07 s	915.21 s	4.78
Exp. $\mathbb{G}_T$	10.65 s	100.20 s	1110.53 s	4.94

Acknowledgment. The majority of this work was done while the first author was working at the University of Tartu, Estonia. This work was supported by the European Union's Horizon 2020 research and innovation programme under grant agreement No 653497 (project PANORAMIX) and grant agreement No 731583 (project SODA), by institutional research funding IUT2-1 of the Estonian Ministry of Education and Research, and by the Danish Independent Research Council, Grant-ID DFF-6108-00169.

References

1. Barbulescu, R., Duquesne, S.: Updating Key Size Estimations for Pairings. Technical Report 2017/334, IACR (2017). http://eprint.iacr.org/2017/334. Revision from 26 April 2017
2. Barreto, P.S.L.M., Naehrig, M.: Pairing-friendly elliptic curves of prime order. In: Preneel, B., Tavares, S. (eds.) SAC 2005. LNCS, vol. 3897, pp. 319–331. Springer, Heidelberg (2006). https://doi.org/10.1007/11693383_22
3. Bayer, S., Groth, J.: Efficient zero-knowledge argument for correctness of a shuffle. In: Pointcheval, D., Johansson, T. (eds.) EUROCRYPT 2012. LNCS, vol. 7237, pp. 263–280. Springer, Heidelberg (2012). https://doi.org/10.1007/978-3-642-29011-4_17
4. Bellare, M., Garay, J.A., Rabin, T.: Batch verification with applications to cryptography and checking. In: Lucchesi, C.L., Moura, A.V. (eds.) LATIN 1998. LNCS, vol. 1380, pp. 170–191. Springer, Heidelberg (1998). https://doi.org/10.1007/BFb0054320
5. Ben-Sasson, E., Chiesa, A., Green, M., Tromer, E., Virza, M.: Secure sampling of public parameters for succinct zero knowledge proofs. In: IEEE SP 2015, pp. 287–304 (2015)
6. Ben-Sasson, E., Chiesa, A., Tromer, E., Virza, M.: Succinct non-interactive zero knowledge for a von Neumann architecture. In: USENIX 2014, pp. 781–796 (2014)
7. Bitansky, N., Canetti, R., Paneth, O., Rosen, A.: On the existence of extractable one-way functions. In: STOC 2014, pp. 505–514 (2014)

8. Blum, M., Feldman, P., Micali, S.: Non-interactive zero-knowledge and its applications. In: STOC 1988, pp. 103–112 (1988)
9. Chaum, D.: Untraceable electronic mail, return addresses, and digital pseudonyms. Commun. ACM **24**(2), 84–88 (1981)
10. Damgård, I.: Towards practical public key systems secure against chosen ciphertext attacks. In: Feigenbaum, J. (ed.) CRYPTO 1991. LNCS, vol. 576, pp. 445–456. Springer, Heidelberg (1992). https://doi.org/10.1007/3-540-46766-1_36
11. Danezis, G., Fournet, C., Groth, J., Kohlweiss, M.: Square span programs with applications to succinct NIZK arguments. In: Sarkar, P., Iwata, T. (eds.) ASIACRYPT 2014. LNCS, vol. 8873, pp. 532–550. Springer, Heidelberg (2014). https://doi.org/10.1007/978-3-662-45611-8_28
12. Elgamal, T.: A public key cryptosystem and a signature scheme based on discrete logarithms. IEEE Trans. Inf. Theor. **31**(4), 469–472 (1985)
13. Escala, A., Herold, G., Kiltz, E., Ràfols, C., Villar, J.L.: An algebraic framework for Diffie-Hellman assumptions. In: Canetti, R., Garay, J.A. (eds.) CRYPTO 2013. LNCS, vol. 8043, pp. 129–147. Springer, Heidelberg (2013). https://doi.org/10.1007/978-3-642-40084-1_8
14. Fauzi, P., Lipmaa, H.: Efficient culpably sound NIZK shuffle argument without random oracles. In: Sako, K. (ed.) CT-RSA 2016. LNCS, vol. 9610, pp. 200–216. Springer, Cham (2016). https://doi.org/10.1007/978-3-319-29485-8_12
15. Fauzi, P., Lipmaa, H., Zając, M.: A shuffle argument secure in the generic model. In: Cheon, J.H., Takagi, T. (eds.) ASIACRYPT 2016. LNCS, vol. 10032, pp. 841–872. Springer, Heidelberg (2016). https://doi.org/10.1007/978-3-662-53890-6_28
16. Furukawa, J., Sako, K.: An efficient scheme for proving a shuffle. In: Kilian, J. (ed.) CRYPTO 2001. LNCS, vol. 2139, pp. 368–387. Springer, Heidelberg (2001). https://doi.org/10.1007/3-540-44647-8_22
17. Galbraith, S.D., Paterson, K.G., Smart, N.P.: Pairings for cryptographers. Discrete Appl. Math. **156**(16), 3113–3121 (2008)
18. Goldwasser, S., Micali, S., Rackoff, C.: The knowledge complexity of interactive proof-systems. In: STOC 1985, pp. 291–304 (1985)
19. Golle, P., Jarecki, S., Mironov, I.: Cryptographic primitives enforcing communication and storage complexity. In: Blaze, M. (ed.) FC 2002. LNCS, vol. 2357, pp. 120–135. Springer, Heidelberg (2003). https://doi.org/10.1007/3-540-36504-4_9
20. González, A., Ràfols, C.: New techniques for non-interactive shuffle and range arguments. In: Manulis, M., Sadeghi, A.-R., Schneider, S. (eds.) ACNS 2016. LNCS, vol. 9696, pp. 427–444. Springer, Cham (2016). https://doi.org/10.1007/978-3-319-39555-5_23
21. Groth, J.: Simulation-sound NIZK proofs for a practical language and constant size group signatures. In: Lai, X., Chen, K. (eds.) ASIACRYPT 2006. LNCS, vol. 4284, pp. 444–459. Springer, Heidelberg (2006). https://doi.org/10.1007/11935230_29
22. Groth, J.: A verifiable secret shuffle of homomorphic encryptions. J. Cryptol. **23**(4), 546–579 (2010)
23. Groth, J.: Short pairing-based non-interactive zero-knowledge arguments. In: Abe, M. (ed.) ASIACRYPT 2010. LNCS, vol. 6477, pp. 321–340. Springer, Heidelberg (2010). https://doi.org/10.1007/978-3-642-17373-8_19
24. Groth, J.: On the size of pairing-based non-interactive arguments. In: Fischlin, M., Coron, J.-S. (eds.) EUROCRYPT 2016. LNCS, vol. 9666, pp. 305–326. Springer, Heidelberg (2016). https://doi.org/10.1007/978-3-662-49896-5_11
25. Groth, J., Lu, S.: A non-interactive shuffle with pairing based verifiability. In: Kurosawa, K. (ed.) ASIACRYPT 2007. LNCS, vol. 4833, pp. 51–67. Springer, Heidelberg (2007). https://doi.org/10.1007/978-3-540-76900-2_4

26. Groth, J., Ostrovsky, R., Sahai, A.: New techniques for noninteractive zero-knowledge. J. ACM **59**(3), 11 (2012)
27. Hess, F., Smart, N.P., Vercauteren, F.: The Eta pairing revisited. IEEE Trans. Inf. Theor. **52**(10), 4595–4602 (2006)
28. Jutla, C.S., Roy, A.: Shorter quasi-adaptive NIZK proofs for linear subspaces. In: Sako, K., Sarkar, P. (eds.) ASIACRYPT 2013. LNCS, vol. 8269, pp. 1–20. Springer, Heidelberg (2013). https://doi.org/10.1007/978-3-642-42033-7_1
29. Jutla, C.S., Roy, A.: Switching lemma for bilinear tests and constant-size NIZK proofs for linear subspaces. In: Garay, J.A., Gennaro, R. (eds.) CRYPTO 2014. LNCS, vol. 8617, pp. 295–312. Springer, Heidelberg (2014). https://doi.org/10.1007/978-3-662-44381-1_17
30. Kiltz, E., Wee, H.: Quasi-adaptive NIZK for linear subspaces revisited. In: Oswald, E., Fischlin, M. (eds.) EUROCRYPT 2015. LNCS, vol. 9057, pp. 101–128. Springer, Heidelberg (2015). https://doi.org/10.1007/978-3-662-46803-6_4
31. Kim, T., Barbulescu, R.: Extended tower number field sieve: a new complexity for the medium prime case. In: Robshaw, M., Katz, J. (eds.) CRYPTO 2016. LNCS, vol. 9814, pp. 543–571. Springer, Heidelberg (2016). https://doi.org/10.1007/978-3-662-53018-4_20
32. Lipmaa, H.: Progression-free sets and sublinear pairing-based non-interactive zero-knowledge arguments. In: Cramer, R. (ed.) TCC 2012. LNCS, vol. 7194, pp. 169–189. Springer, Heidelberg (2012). https://doi.org/10.1007/978-3-642-28914-9_10
33. Lipmaa, H.: Prover-efficient commit-and-prove zero-knowledge SNARKs. In: Pointcheval, D., Nitaj, A., Rachidi, T. (eds.) AFRICACRYPT 2016. LNCS, vol. 9646, pp. 185–206. Springer, Cham (2016). https://doi.org/10.1007/978-3-319-31517-1_10
34. Lipmaa, H., Zhang, B.: A more efficient computationally sound non-interactive zero-knowledge shuffle argument. In: Visconti, I., De Prisco, R. (eds.) SCN 2012. LNCS, vol. 7485, pp. 477–502. Springer, Heidelberg (2012). https://doi.org/10.1007/978-3-642-32928-9_27
35. Maurer, U.M.: Abstract models of computation in cryptography. In: Smart, N.P. (ed.) Cryptography and Coding 2005. LNCS, vol. 3796, pp. 1–12. Springer, Heidelberg (2005). https://doi.org/10.1007/11586821_1
36. Morillo, P., Ràfols, C., Villar, J.L.: The kernel matrix Diffie-Hellman assumption. In: Cheon, J.H., Takagi, T. (eds.) ASIACRYPT 2016. LNCS, vol. 10031, pp. 729–758. Springer, Heidelberg (2016). https://doi.org/10.1007/978-3-662-53887-6_27
37. Pereira, G.C.C.F., Simplício Jr., M.A., Naehrig, M., Barreto, P.S.L.M.: A family of implementation-friendly BN elliptic curves. J. Syst. Softw. **84**(8), 1319–1326 (2011)
38. Schwartz, J.T.: Fast probabilistic algorithms for verification of polynomial identities. J. ACM **27**(4), 701–717 (1980)
39. Shoup, V.: Lower bounds for discrete logarithms and related problems. In: Fumy, W. (ed.) EUROCRYPT 1997. LNCS, vol. 1233, pp. 256–266. Springer, Heidelberg (1997). https://doi.org/10.1007/3-540-69053-0_18
40. Zippel, R.: Probabilistic algorithms for sparse polynomials. In: Ng, E.W. (ed.) Symbolic and Algebraic Computation. LNCS, vol. 72, pp. 216–226. Springer, Heidelberg (1979). https://doi.org/10.1007/3-540-09519-5_73

Efficient Ring Signatures in the Standard Model

Giulio Malavolta$^{(\boxtimes)}$ and Dominique Schröder

Friedrich-Alexander-University Erlangen-Nürnberg, Erlangen, Germany
`malavolta@cs.fau.de`

Abstract. A ring signature scheme allows one party to sign messages on behalf of an arbitrary set of users, called the ring. The anonymity of the scheme guarantees that the signature does not reveal which member of the ring signed the message. The ring of users can be selected "on-the-fly" by the signer and no central coordination is required. Ring signatures have made their way into practice in the area of privacy-enhancing technologies and they build the core of several cryptocurrencies. Despite their popularity, almost all ring signature schemes are either secure in the random oracle model or in the common reference string model. The only candidate instantiations in the plain model are either impractical or not fully functional.

In this work, we close this gap by proposing a new construction paradigm for ring signatures without random oracles: We show how to efficiently instantiate full-fledged ring signatures from signature schemes with re-randomizable keys and non-interactive zero-knowledge. We obtain the following results:
- The first almost practical ring signature in the plain model from standard assumptions in bilinear groups.
- The first efficient ring signature in the plain model secure under a generalization of the knowledge of exponent assumption.

1 Introduction

Ring signatures were envisioned by Rivest et al. [36] as a tool to leak a secret information in an authenticated manner while being anonymous in within a crowd of users. The idea behind this primitive is that a signer can choose a set of users via their public-keys and sign a message on behalf of this set, also called a ring. Signing in the name of the users means that it is infeasible to tell which of the users signed the message. Ring signatures guarantee great flexibility: Rings can be formed arbitrarily and "on-the-fly" by the signer and no trusted authority is required. In fact, users do not even have to be aware of each other. The widely accepted security notions of anonymity against full key exposure and unforgeability with respect to insider corruption were formalized by Bender et al. [4].

In the past years many applications of ring signatures were suggested, such as the ability to leak secrets while staying anonymous within a certain set [36]. Recently, certain types of ring signatures made it into practice being a building block in the cryptocurrency Monero. In Monero, to spend a certain amount of coins, the user searches for other public-keys sharing the same amount and it

T. Takagi and T. Peyrin (Eds.): ASIACRYPT 2017, Part II, LNCS 10625, pp. 128–157, 2017.
https://doi.org/10.1007/978-3-319-70697-9_5

issues a ring signature for this set of users. Since the ring signature is anonymous, nobody can tell which of the users in the ring spent their coin.

Despite being one of the classical problems of cryptography and being deployed in practice, almost all ring signatures are either secure in the random oracle model or the common reference string model. Even worse, among the construction without random oracles, the asymptotically most efficient instance is the scheme of Bose et al. [8] with a signature of 95 group elements for a composite order bilinear group. Notable exceptions to what discussed above are the scheme of Bender et al. [4] and the one of Chow et al. [19]. However, the former is only a feasibility result that relies on generic ZAPs [23], whereas the latter supports only rings of *constant size* and it is secure against a tailored assumption.

In this work, we close this gap presenting a new generic framework to construct efficient ring signatures without random oracles: Our abstraction gives us the first efficient scheme secure in the plain model. As a corollary of our transformation, we also obtain the most efficient scheme with constant size signatures in the common reference string model.

1.1   Our Contribution

At the core of our contributions is a novel generic construction of ring signatures from signatures with re-randomizable keys, a property that was recently leveraged by Fleischhacker et al. [26] to build efficient sanitizable signature scheme [12,13,35]. In addition to that, our scheme requires a non-interactive zero-knowledge proof. Our generic transformation is secure in the common reference string model, without random oracles. In the process of instantiating our construction we propose a modification to the signature scheme of Hofheinz and Kiltz [34] that significantly simplifies the statement to be proven in zero-knowledge, thereby boosting the efficiency of our ring signature[1]. A nice feature of our transformation is that the resulting signature size does not depend on the size of the ring, except for the statement to be proven in zero-knowledge. Therefore, instantiating the zero-knowledge proof with SNARKs automatically yields a ring signature of constant size. As an example, we can implement the proof of knowledge with the scheme recently proposed by Groth [31], which adds only three group elements to our signatures.

RING SIGNATURES IN THE PLAIN MODEL. Our next observation is that, if the zero-knowledge scheme uses a common reference string with some homomorphic properties, then we can include a different string in each user's verification key. The proof for a ring of users is then computed against a combination of all the corresponding strings, that results in a correctly distributed reference string. This effectively lifts the resulting scheme to the plain model, given a suitable zero-knowledge protocol. We show that the scheme of Groth [29] partially satisfies our

---

[1] We choose Hofheinz-Kiltz signatures for efficiency reasons, although our transformation can also be instantiated from Waters' scheme [41], whose hardness relies on the Computational Diffie-Hellman (CDH) assumption.

constraints and we demonstrate how to integrate it in our system. The resulting ring signature scheme relies on standard assumptions for bilinear groups and it has somewhat efficient signature size, although still large for practical usage. The caveat here is that the scheme achieves only a weak form of anonymity.

ACHIEVING EFFICIENCY AND FULL SECURITY. The last step towards our main result is a novel instantiation of a zero-knowledge proof system for proving the knowledge of discrete logarithms. The scheme relies on asymmetric bilinear groups and its efficiency is comparable to schemes derived from the Fiat-Shamir heuristic, although it does not use random oracles. We prove the security of our scheme under a generalization of the knowledge of exponent assumption and confirm its hardness in the generic group model [40]. When combined with our variant of the scheme of Hofheinz and Kiltz, the resulting ring signature is fully secure in the standard model (by using a similar trick as described above) and the signatures are composed by roughly 4 group elements per user in the ring.

A comparison of our results with existing schemes is summarized in Table 1. Our construction instantiated with [31] improves the signature size by and order of magnitude with respect to the most efficient scheme without random oracles. The scheme of [4] relies on generic ZAPs and therefore statements need to go through an NP-reduction before being proven, making it hard for us to estimate the real cost of the resulting signature. Although the ring signature of [19] has a very small signature size, it only supports rings of constant size, hindering its the practical deployment. Our two instantiations in the standard model offer a tradeoff between efficiency and assumptions, broadening the landscape of instances without any setup.

ON THE KNOWLEDGE OF EXPONENT ASSUMPTION. Although the knowledge of exponent assumption is clearly non-standard, we believe that it is slightly better than assuming the existence of random oracles - at least from a theoretical point of view. The reason is that it is well known that the random oracle is not sound [15], whereas it might be possible that certain assumptions hold in practice. As an example, the SNARKs used in the real-world cryptocurrency Zerocash [3] rely on a variant of the knowledge-of-exponent assumptions.

Table 1. Comparison amongst ring signature schemes without random oracles

Ring signature	Model	Anon.	Unforg.	Assumptions	Ring size	Signature size
[39]	crs	✓	✓	CDH + SubD	$\text{poly}(\lambda)$	$(2n + 2)\mathbb{G}$
[9]	crs	✓	✓	$(q,\ell,1)$-Pluri-SDH	$\text{poly}(\lambda)$	$(n + 1)\mathbb{G} + (n + 1)\mathbb{F}_p$
[37]	crs	Basic	Sub	CDH	$\text{poly}(\lambda)$	$(n + 1)\mathbb{G}$
[16]	crs	✓	✓	SDH + SubD	$\text{poly}(\lambda)$	$O(\sqrt{n})$
[8]	crs	✓	✓	q-SDH + SXDH + SQROOT	$\text{poly}(\lambda)$	$92\mathbb{G} + 3\mathbb{Z}_p$
This work + [31]	crs	✓	✓	q-SDH + GGM	$\text{poly}(\lambda)$	$6\mathbb{G} + \mathbb{Z}_p$
[19]	std	✓	✓	(q,n)-DsjSDH	$O(1)$	$n\mathbb{G} + n\mathbb{Z}_p$
[4]	std	✓	✓	Enc + ZAP	$\text{poly}(\lambda)$	$O(n)$
This work + [29]	std	Basic	✓	q-SDH + DLIN	$\text{poly}(\lambda)$	$\sim 10^3 n\mathbb{G}$
This work	std	✓	✓	q-SDH + L-KEA	$\text{poly}(\lambda)$	$(4n + 3)\mathbb{G} + \mathbb{Z}_p$

1.2   Related Work

The notion of ring signatures has been introduced in the visionary work of Rivest et al. [36], as a way to leak secrets while staying anonymous within the crowd. The authors proposed a construction based on trapdoor permutations and several other schemes have followed based on different assumptions such as discrete logarithms [7], bilinear maps [33], factoring [22], and hybrids [2]. Remarkably, the size of the signatures in the scheme of [22] is independent from the size of the ring. Such a surprising result is achieved with a clever usage of RSA accumulators [14] and the Fiat-Shamir transformation. A practical scheme constructed from a combination of Σ protocols and the Fiat-Shamir heuristic was recently proposed by Groth and Kohlweiss in [32]. The security of all of the aforementioned constructions is based on the existence of random oracles.

There has been a considerable effort in the community in building ring signature schemes from more classical assumptions. In particular, we know how to instantiate ring signatures efficiently admitting the existence of a common reference string model: Shacham and Waters [39] proposed the first efficient scheme in composite order groups with a pairing, whose performance were later on improved in the work of Boyen on mesh signatures [9]. Recently, Schäge and Schwenk [37] have shown a very efficient instantiation based on CDH-hard groups with extremely appealing signatures size, but at the cost of a weaker notion of unforgeability (chosen subring attacks in the terminology of [4]). Derler and Slamanig [21] suggested an efficient linear-size scheme from key-homomorphic signatures and zero-knowledge proofs. The first scheme with sublinear size signatures has been proposed by Chandran et al. [16], which exhibits signatures that grow linearly with the square root of the size of the ring. To the best of our knowledge, the most (asymptotically) efficient construction without random oracles is due to Bose et al. [8], where a signature accounts for 95 group elements for a composite order bilinear group. We shall mention that in the work on signatures of knowledge [17] the authors claim that one can use their primitive combined with the techniques of Dodis et al. [22] to construct ring signatures of constant size, but this is not supported by any formal analysis. For fairness, we must say that the security model of ring signatures was not yet well established, as the seminal work of Bender et al. [4] has been published concurrently in the same year.

In contrast to the common reference string settings, ring signature schemes in the plain model (without any setup assumption) have been surprisingly understudied. The first work that considered this problem was [4], where the authors proposed a construction from non-interctive ZAPs [23]. Such a scheme represents the first proof of feasibility of ring signature schemes in the standard model. Concurrently, Chow et al. [19] published a scheme for constant-size rings based on a custom assumption. At today, these two instantiations were the only known candidates for a ring signature scheme in the plain model.

A notion related to ring signatures is that of group signatures, originally envisioned by Chaum and Van Heyst [18]: The main difference here is that a group manager controls the enrolment within the group of users and can revoke anonymity. Efficient realizations are known in the random oracle model [5] and in the standard model [10]. The absence of a trusted authority in ring signatures, makes the two primitives incomparable.

2   Preliminaries

We denote by $\lambda \in \mathbb{N}$ the security parameter and by $\mathsf{poly}(\lambda)$ any function that is bounded by a polynomial in λ. We denote any function that is *negligible* in the security parameter by $\mathsf{negl}(\lambda)$. We say that an algorithm is PPT if it is modelled as a probabilistic Turing machine whose running time is bounded by some function $\mathsf{poly}(\lambda)$. Given a set S, we denote by $x \leftarrow S$ the sampling of and element uniformly at random in S.

2.1   Bilinear Maps

Let $\mathbb{G}_1$ and $\mathbb{G}_2$ be two cyclic groups of large prime order p. Let $g_1 \in \mathbb{G}_1$ and $g_2 \in \mathbb{G}_2$ be respective generators of $\mathbb{G}_1$ and $\mathbb{G}_2$. Let $e : \mathbb{G}_1 \times \mathbb{G}_2$ be a function that maps pairs of elements $\in (\mathbb{G}_1, \mathbb{G}_2)$ to elements of some cyclic group $\mathbb{G}_T$ of order p. Throughout the following sections we write all of the group operations mutiplicatively, with identity elements denoted by 1. We further require that:

- The map e and all the group operations in $\mathbb{G}_2$, $\mathbb{G}_2$, and $\mathbb{G}_T$ are efficiently computable.
- The map e is non degenerate, i.e., $e(g_1, g_2) \neq 1$.
- The map e is bilinear, i.e., $\forall u \in \mathbb{G}_1, \forall v \in \mathbb{G}_2, \forall (a, b) \in \mathbb{Z}^2, e(u^a, v^b) = e(u, v)^{ab}$.
- There exists an efficiently computable function $\phi : \mathbb{G}_1 \rightarrow \mathbb{G}_2$ such that $\forall (u, v) \in \mathbb{G}_1$ it holds that $\phi(u \cdot v) = \phi(u) \cdot \phi(v)$.

2.2   Ring Signatures

In the following we recall the notion of ring signatures. Our definitions follow the strongest security model proposed by Bender et al. [4].

Definition 1 (Ring Signature). *A ring signature scheme* RSig $=$ (RGen, RSig, RVer) *is a triple of the following* PPT *algorithms:*

$(\mathsf{sk}, \mathsf{vk}) \leftarrow \mathsf{RGen}(1^\lambda) :$ *On input the security parameter* 1^λ *outputs a key pair* $(\mathsf{sk}, \mathsf{vk})$.

$\sigma \leftarrow \mathsf{RSig}(\mathsf{sk}, m, R) :$ *On input a secret key* sk, *a message* m, *and a ring* $R = (\mathsf{vk}_1, \ldots, \mathsf{vk}_n)$, *outputs a signature* σ.

$b \leftarrow \mathsf{RVer}(R, \sigma, m) :$ *On input a ring* $R = (\mathsf{vk}_1, \ldots, \mathsf{vk}_n)$, *a signature* σ *and a message* m *outputs a bit* b.

For completeness we require that for all $\lambda \in \mathbb{N}$, for all $\{(\mathsf{sk}_i, \mathsf{vk}_i)\}_{i \in n}$ output by $\mathsf{RGen}(1^\lambda)$, any $i \in \{1, \ldots, n\}$, and any m, we have that $\mathsf{RVer}(R, \mathsf{RSig}(\mathsf{sk}_i, m, R), m) = 1$, where $R = (\mathsf{vk}_1, \ldots, \mathsf{vk}_n)$.

SECURITY OF RING SIGNATURES. The security of a ring signature scheme is captured by the notions of anonymity against full key exposure and unforgeability with respect to insider corruption. We refer to [4] for a comprehensive discussion on the matter. In the following definitions, we assume without loss of generality that the adversary never submits a query where the ring consists only of maliciously generated keys.

Definition 2 (Anonymity). *Let ℓ be a polynomial function, $\mathcal{A}$ a PPT adversary, and $\mathsf{RSig} = (\mathsf{RGen}, \mathsf{RSig}, \mathsf{RVer})$ a ring signature scheme. Consider the following game:*

1. *For all $i \in \{1, \ldots \ell(\lambda)\}$ the challenger runs $(\mathsf{sk}_i, \mathsf{vk}_i) \leftarrow \mathsf{RGen}(1^\lambda; \omega_i)$, recording each randomness ω_i. The adversary $\mathcal{A}$ is provided with the verification keys $(\mathsf{vk}_1, \ldots, \mathsf{vk}_{\ell(\lambda)})$.*
2. *$\mathcal{A}$ is allowed to make queries of the form (j, R, m), where m is the message to be signed, R is a set of verification keys and j is and index such that $\mathsf{vk}_j \in R$. The challenger responds with $\mathsf{RSig}(\mathsf{sk}_j, m, R)$.*
3. *$\mathcal{A}$ requests a challenge by sending the tuple (i_0, i_1, R, m), where i_0 and i_1 are indices such that $(\mathsf{vk}_{i_0}, \mathsf{vk}_{i_1}) \in R$. The challenger samples a random bit $b \leftarrow \{0, 1\}$ and sends $\mathsf{RSig}(\mathsf{sk}_{i_b}, m, R)$ to $\mathcal{A}$. Additionally, the adversary is provided with the randomnesses $(\omega_1, \ldots, \omega_{\ell(\lambda)})$.*
4. *$\mathcal{A}$ outputs a guess b' and succeeds if $b' = b$.*

A ring signature scheme RSig achieves anonymity if, for all PPT $\mathcal{A}$ and for all polynomial functions ℓ, the success probability of $\mathcal{A}$ in the above experiment is negligibly close to $1/2$.

Definition 3 (Unforgeability). *Given a ring signature scheme $\mathsf{RSig} = (\mathsf{RGen}, \mathsf{RSig}, \mathsf{RVer})$, a polynomial function ℓ, and a PPT adversary $\mathcal{A}$, consider the following game:*

1. *For all $i \in \{1, \ldots, \ell(\lambda)\}$ the challenger runs $(\mathsf{sk}_i, \mathsf{vk}_i) \leftarrow \mathsf{RGen}(1^\lambda; \omega_i)$. The adversary $\mathcal{A}$ is provided with the verification keys $\mathbf{vk} := (\mathsf{vk}_1, \ldots, \mathsf{vk}_{\ell(\lambda)})$. Additionally, the challenger initializes an empty list of corrupted users $\mathcal{C}$.*
2. *$\mathcal{A}$ is allowed to make signatures and corruption queries. A signature query is of the form (j, R, m), where m is the message to be signed, R is a set of verification keys and j is and index such that $\mathsf{vk}_j \in R$. The challenger responds with $\mathsf{RSig}(\mathsf{sk}_j, m, R)$. A corruption query is of the form $j \in \{1, \ldots, \ell(\lambda)\}$. The challenger sends sk_j to $\mathcal{A}$ and appends vk_j to $\mathcal{C}$.*
3. *$\mathcal{A}$ outputs a tuple (R^*, m^*, σ^*) and wins if it never made any signing query of the form $(\cdot, R^*, m^*)$ and $R^* \subseteq \mathbf{vk} \setminus \mathcal{C}$ and $\mathsf{RVer}(R^*, \sigma^*, m^*) = 1$.*

A ring signature scheme RSig achieves unforgeability if, for all PPT $\mathcal{A}$ and for all polynomial functions ℓ, the success probability of $\mathcal{A}$ in the above experiment is negligible.

2.3   Non-Interactive Zero-Knowledge

We recall the definitions and the security properties of non-interactive zero-knowledge proof systems as defined in [29].

Definition 4 (Non-Interactive Zero-Knowledge Proof System). *Let $\mathcal{L}$ be an NP language and let $\mathcal{R}$ be the corresponding relation. A non-interactive zero-knowledge proof system* NIZK *consists of the following* PPT *algorithms:*

crs $\leftarrow \mathcal{G}(1^\lambda)$: *The setup algorithm takes as input the security parameter 1^λ and generates a random common reference string* crs *and a trapdoor α.*

$\pi \leftarrow \mathcal{P}(\text{crs}, w, x)$: *The prover algorithm takes as input the common reference string* crs, *a statement x, and a witness w and outputs a zero-knowledge proof π.*

$b \leftarrow \mathcal{V}(\text{crs}, x, \pi)$: *The verifier algorithm takes as input the common reference string* crs, *a statement x, and a proof π and returns either 0 or 1.*

Definition 5 (Completeness). *A* NIZK *system is* complete *if for all $\lambda \in \mathbb{N}$ and all $x \in \mathcal{L}$ it holds that*

$$\Pr\big[(\text{crs}, \alpha) \leftarrow \mathcal{G}(1^\lambda), \pi \leftarrow \mathcal{P}(\text{crs}, w, x) : 1 \leftarrow \mathcal{V}(\text{crs}, x, \pi)\big] = 1.$$

Soundness, zero-knowledge and proof of knowledge are defined in the following.

Definition 6 (Soundness). *A* NIZK *system is* computationally sound *if for all $\lambda \in \mathbb{N}$ and all* PPT *adversaries $\mathcal{A}$ there exists a negligible function* negl *such that*

$$\Pr\big[(\text{crs}, \alpha) \leftarrow \mathcal{G}(1^\lambda); (x, \pi) \leftarrow \mathcal{A}(\text{crs}); 1 \leftarrow \mathcal{V}(\text{crs}, x, \pi) \mid x \notin \mathcal{L}\big] \leq \text{negl}(n).$$

Definition 7 (Zero-Knowledge). *An* NIZK *system is* statistically zero-knowledge *if there exists a* PPT *simulator $\mathcal{S}$ and a negligible function such that for all $\lambda \in \mathbb{N}$, for all $x \in \mathcal{L}$ with witness w, and for $(\text{crs}, \alpha) \leftarrow \mathcal{G}(1^\lambda)$ it holds that*

$$\mathcal{P}(\text{crs}, w, x) \approx \mathcal{S}(\text{crs}, \alpha, x).$$

where $\approx$ denotes statistical indistinguishability.

Additionally, we say that a NIZK system is *unconditionally* zero-knowledge if the condition above holds for any choice of (crs, α).

Definition 8 (Argument of Knowledge). *A* NIZK *system is an* argument of knowledge *if for all $\lambda \in \mathbb{N}$ and for all* PPT *adversaries $\mathcal{A}$ there exists a* PPT *extractor $\mathcal{E}$ running on the same random tape of $\mathcal{A}$ and a negligible function* negl *such that*

$$\Pr\left[\begin{array}{l}(\text{crs}, \alpha) \leftarrow \mathcal{G}(1^\lambda), (x, \pi) \leftarrow \mathcal{A}(\text{crs}), \\ (x, \pi, w) \leftarrow \mathcal{E}(\text{crs}, \alpha) : (x, w) \in \mathcal{R}\end{array}\middle| \mathcal{V}(\text{crs}, x, \pi) = 1\right] \geq (1 - \text{negl}(\lambda)).$$

Additionally, we call a proof *succinct* if the size of the proof is constant, in particular it must be independent from statement to be proven and corresponding witness. In literature this primitive is known as succinct non-interactive argument of knowledge (SNARK) and several efficient realizations are known to exist without random oracles, among the others see [30,31].

2.4 Signatures with Re-Randomizable Keys

We recall the notion of signatures with re-randomizable keys, as defined in [26]. This primitive allows one to consistently re-randomize private and public keys of a signature scheme.

Definition 9 (Signatures with Re-Randomizable Keys). *A digital signature scheme* $\mathsf{Sig} = (\mathsf{SGen}, \mathsf{Sig}, \mathsf{Ver}, \mathsf{RndSK}, \mathsf{RndVK})$ *with perfectly re-randomizable keys is composed by the following* PPT *algorithms:*

$(\mathsf{sk}, \mathsf{vk}) \leftarrow \mathsf{SGen}(1^\lambda)$: *The key generation algorithm takes as input the security parameter* 1^λ *and generates a key pair* $(\mathsf{sk}, \mathsf{vk})$.

$\sigma \leftarrow \mathsf{Sig}(\mathsf{sk}, m)$: *The signing algorithm takes as input a signing key* sk *and a message* m *and outputs a signature* σ.

$b \leftarrow \mathsf{Ver}(\mathsf{vk}, m, \sigma)$: *The verification algorithm takes as input a verification key* vk, *a message* m, *and a candidate signature* σ *and outputs a bit* b.

$\mathsf{sk}' \leftarrow \mathsf{RndSK}(\mathsf{sk}, \rho)$: *The secret key re-randomization algorithm takes as input a signing key* sk *and a randomness* ρ *and outputs a new signing key* sk'.

$\mathsf{vk}' \leftarrow \mathsf{RndVK}(\mathsf{vk}, \rho)$: *The public key re-randomization algorithm takes as input a verification key* vk *and a randomness* ρ *and outputs a new verification key* vk'.

The scheme is *complete* if for all $\lambda \in \mathbb{N}$, all key-pairs $(\mathsf{sk}, \mathsf{vk}) \leftarrow \mathsf{SGen}(1^\lambda)$, all messages m we have that $\mathsf{Ver}(\mathsf{vk}, m, \mathsf{Sig}(\mathsf{sk}, m)) = 1$. Additionally we require that for all ρ it holds that $\mathsf{Ver}(\mathsf{RndVK}(\mathsf{vk}, \rho), m, \mathsf{Sig}(\mathsf{RndSK}(\mathsf{sk}, \rho), m)) = 1$. The formal definition of re-randomizable keys follows.

Definition 10 (Re-Randomizable Keys). *A signature scheme* Sig *has perfectly re-randomizable keys if for all* $\lambda \in \mathbb{N}$, *all key-pairs* $(\mathsf{sk}, \mathsf{vk}) \leftarrow \mathsf{SGen}(1^\lambda)$, *and a* ρ *chosen uniformly at random we have that the following distributions are identical:*

$$\{(\mathsf{sk}, \mathsf{vk}, \mathsf{sk}', \mathsf{vk}')\} = \{(\mathsf{sk}, \mathsf{vk}, \mathsf{sk}'', \mathsf{vk}'')\}$$

where $(\mathsf{sk}', \mathsf{vk}') \leftarrow \mathsf{SGen}(1^\lambda)$, $\mathsf{sk}'' \leftarrow \mathsf{RndSK}(\mathsf{sk}, \rho)$, *and* $\mathsf{vk}'' \leftarrow \mathsf{RndVK}(\mathsf{vk}, \rho)$.

The notion of existential unforgeability for signatures with re-randomizable keys is an extension of the standard definition where the attacker is provided with an additional oracle that signs messages under re-randomized keys. The formal definition follows.

Definition 11 (Unforgeability Under Re-Randomizable Keys). *Given a signature scheme* Sig *and a* PPT *adversary* $\mathcal{A}$, *consider the following game:*

1. *The challenger runs* $(\mathsf{sk}, \mathsf{vk}) \leftarrow \mathsf{SGen}(1^\lambda)$ *and provides* $\mathcal{A}$ *with* vk.
2. $\mathcal{A}$ *is allowed to make signature and randomized signature queries. A signature query is for the form* $(m, \perp)$, *where* m *is the message to be signed, and the challenger responds with* $\mathsf{Sig}(\mathsf{sk}, m)$. *A randomized signature query is of the form* (m, ρ), *where* m *is the message to be signed and* ρ *is a randomness. The challenger replies with* $\mathsf{Sig}(\mathsf{RndSK}(\mathsf{sk}, \rho), m)$.
3. $\mathcal{A}$ *outputs a tuple* (m^*, σ^*, ρ^*) *and wins the game if it never made a query of the form* $(m^*, \cdot)$ *and either* $\mathsf{Ver}(\mathsf{vk}, m^*, \sigma^*) = 1$ *or* $\mathsf{Ver}(\mathsf{RndVK}(\mathsf{vk}, \rho^*), m^*, \sigma^*) = 1$.

A signature scheme Sig *achieves* unforgeability under re-randomizable keys *if, for all* PPT $\mathcal{A}$, *the success probability of* $\mathcal{A}$ *in the above experiment is negligible.*

3 Ring Signatures from Signatures with Re-Randomizable Keys

In this section we describe our framework for designing efficient ring signature schemes without random oracles. For the ease of the exposition and for a more modular presentation we propose a generic construction in the common reference string model. Jumping ahead, we will show later how to upgrade it to the standard model (i.e., without any setup assumptions) for some specific instantiations. The idea behind our construction based on signatures with re-randomizable keys is the following: The signer generates a randomized version of its own public key and signs the message under the secret key re-randomized with the same factor. The signature is then composed by the output of the signing algorithm and a disjunctive argument of knowledge of the randomization factor of the new verification key with respect to a ring of public keys. More formally, our building blocks are a signature scheme with re-randomizable keys Sig and a non-interactive zero-knowledge argument NIZK for the following language $\mathcal{L}$:

$$\mathcal{L} = \{(\mathsf{vk}_1, \ldots, \mathsf{vk}_n, \mathsf{vk}^*) : \exists (\rho, i) : \mathsf{vk}^* = \mathsf{RndVK}(\mathsf{vk}_i, \rho)\} .$$

Our ring signature scheme $\mathsf{RSig} = (\mathsf{RGen}, \mathsf{RSig}, \mathsf{RVer})$ is shown in Fig. 1. The completeness of the scheme follows directly from the completeness of the zero knowledge argument and the signature scheme. The security analysis is elaborated below.

Theorem 1. *Let* NIZK *be a statistically zero-knowledge argument and let* (SGen, Sig, Ver, RndSK, RndVK) *be a signature with perfectly re-randomizable keys, then the construction in Fig. 1 is an anonymous ring signature scheme in the common reference string model.*

RGen(1^λ)	RSig(sk, m, R)	RVer(R, σ, m)
(sk, vk) $\leftarrow$ SGen(1^λ)	**parse** $R = (\text{vk}_1, \ldots, \text{vk}_n)$	**parse** $R = (\text{vk}_1, \ldots, \text{vk}_n)$
return (sk, vk)	**if** $\nexists i : \text{vk} = \text{vk}_i$	**parse** $\sigma = (\sigma', \pi, \text{vk}')$
	return $\bot$	$x := R\|\text{vk}'$
	$\rho \leftarrow \{0,1\}^\lambda$	$b \leftarrow \mathcal{V}(\text{crs}, x, \pi)$
	$\text{vk}' \leftarrow \text{RndVK}(\text{vk}, \rho)$	$b' \leftarrow \text{Ver}(\text{vk}', (m\|R), \sigma')$
	$\text{sk}' \leftarrow \text{RndSK}(\text{sk}, \rho)$	**return** $(b = b' = 1)$
	$x := R\|\text{vk}'$	
	$\pi \leftarrow \mathcal{P}(\text{crs}, (\rho, i), x)$	
	$\sigma \leftarrow \text{Sig}(\text{sk}', m\|R)$	
	return $(\sigma, \pi, \text{vk}')$	

Fig. 1. A generic construction of a ring signature

Proof. Let us consider the following sequence of hybrids:

H_0 : Is the original experiment as defined in Definition 2.

H_1 : Is defined as H_0 except that the proof π in the challenge signature is computed as $\pi \leftarrow \mathcal{S}(\text{crs}, \alpha, x)$, where $x = R^*\|\text{vk}_{i_b}$.

H_2 : Is defined as H_1 except that the challenge signature is substituted with the tuple $(\text{Sig}(\text{sk}', m^*\|R^*), \pi, \text{vk}')$, where $(\text{sk}', \text{vk}') \leftarrow \text{SGen}(1^\lambda)$.

We now argue about the indistinguishability of adjacent experiments.

$H_0 \approx H_1$. The two hybrids are identical except for the proof π that is honestly generated in H_0, while in H_1 it is computed by the simulator $\mathcal{S}$ of NIZK. By the perfect zero knowledge of NIZK, the two simulations are statistically close.

$H_1 \approx H_2$. The two experiments differ only in the sampling procedure of the key pair (sk', vk') used to compute the challenge signature. In the former case (sk', vk') is the re-randomization of the pair $(\text{sk}_{i_b}, \text{vk}_{i_b})$, while in the latter the pair is freshly sampled. Since the signature scheme has perfectly re-randomizable keys, we can conclude that the two simulations are identical.

We observe that in the hybrid H_2 the computation of the challenge signature is completely independent from the bit b. Therefore the success probability of $\mathcal{A}$ in H_2 is exactly $1/2$. By the argument above we have that $H_0 \approx H_2$ and therefore we can conclude that any unbounded $\mathcal{A}$ cannot win the experiment with probability negligibly greater than guessing. $\square$

We now need to show that our ring signature scheme achieves unforgeability against full key exposure. For our formal argument we need to assume the existence of an extractor for the NIZK scheme that is successful even in presence of a signing oracle. For the case of black-box extraction, such a property is trivially guaranteed by any construction. However, as shown in [25], for the case of non-blackbox extraction the definition itself does not necessarily cover this additional requirement. For a more comprehensive discussion on the matter we refer the reader to [25]. In order to be as generic as possible, we are going to explicitly assume the existence of a NIZK that is extractable also in presence of a signing oracle. As discussed in [25], such an assumption as been (implicitly) adopted in several seminal works such as [6,11,27,28].

Theorem 2. *Let* NIZK *be a computationally sound argument of knowledge that is extractable in presence of a signing oracle and let* Sig *be a signature with perfectly re-randomizable keys, then the construction in Fig. 1 is an unforgeable ring signature scheme in the common reference string model.*

Proof. Assume towards contradiction that there exists a PPT adversary $\mathcal{A}$ that succeeds in the experiment of Definition 3 with probability $\epsilon(\lambda)$, for some non negligible function ϵ. Then we can construct the following reduction $\mathcal{R}$ against the unforgeability under re-randomizable keys of the signature scheme (SGen, Sig, Ver, RndSK, RndVK).

$\mathcal{R}(1^\lambda, vk)$. On input the security parameter and the verification key vk the reduction samples an $i \leftarrow \{1, \ldots, \ell(\lambda)\}$ and sets $vk_i = vk$. For all $j \in \{1, \ldots, \ell(\lambda)\} \backslash i$, the reduction sets $(sk_j, vk_j) \leftarrow SGen(1^\lambda)$. Upon a corruption query from $\mathcal{A}$ on index $k \neq i$, the reduction sends sk_k to $\mathcal{A}$, if $k = i$ then $\mathcal{R}$ aborts. Signing queries of the form (k, R, m), for $k \neq i$ are handled as specified in the experiment. On the other hand, for queries of the form (i, R, m), the reduction samples a random ρ, and computes $vk' \leftarrow RndVK(vk, \rho)$ and $\pi \leftarrow \mathcal{P}(crs, (\rho, i), R \| vk')$. Then it queries the signing oracle on input $(m \| R, \rho)$. Let σ be the response of the challenger, $\mathcal{R}$ returns (σ, π, vk') to $\mathcal{A}$. At some point of the execution $\mathcal{A}$ outputs a tuple (R^*, m^*, σ^*). $\mathcal{R}$ parses $\sigma^* = (\theta^*, \pi^*, vk^*)$ and runs $(R^* \| vk^*, w^*, \pi^*) \leftarrow \mathcal{E}_\mathcal{A}$ on the same inputs and random tape of $\mathcal{A}$. $\mathcal{R}$ parses $w^* = (\rho^*, i^*)$ and aborts if $i^* \neq i$. Otherwise $\mathcal{R}$ returns the tuple $(m^* \| R^*, \theta^*, \rho^*)$ and interrupts the simulation.

Since $\mathcal{A}$ and $\mathcal{E}_\mathcal{A}$ are PPT machines it follows that $\mathcal{R}$ runs in polynomial time. Let us assume for the moment that $i = i^*$. In this case the successful $\mathcal{A}$ never queries the corruption oracle on i, since $vk_i \in R^*$ (by the soundness of the NIZK) and $R^* \subseteq vk \backslash \mathcal{C}$. Therefore the reduction does not aborts. It is also easy to see that all the signing queries on i are answered correctly by the reduction since the signing oracle of the challenger returns some signature $\sigma \leftarrow Sig(RndSK(sk_i, \rho), m \| R)$, which is exactly what $\mathcal{A}$ is expecting. It follows that, for the case $i = i^*$, $\mathcal{R}$ correctly simulates the inputs for $\mathcal{A}$. We now show that a successful forgery by $\mathcal{A}$ implies a successful forgery by $\mathcal{R}$. Let (R^*, m^*, σ^*) be the outputs of $\mathcal{A}$, by the winning condition of the unforgeability experiment we have that $RVer(R^*, m^*, \sigma^*) = 1$. Let $\sigma^* = (\theta^*, \pi^*, vk^*)$, this implies that $\mathcal{V}(crs, R^* \| vk^*, \pi^*) = 1$ and that $Ver(vk^*, m^* \| R^*, \theta^*) = 1$. Since the extractor is successful with overwhelming probability we can rewrite $vk^* = RndVK(vk_{i^*}, \rho^*) = RndVK(vk_i, \rho^*)$, for the case $i^* = i$. It follows that

$\mathsf{Ver}(\mathsf{RndVK}(\mathsf{vk}_i, \rho^*), m^* \| R^*, \theta^*) = 1$. Recall that $\mathsf{vk}_i = \mathsf{vk}$, which means that $(m^* \| R^*, \theta^*, \rho^*)$ is a valid signature under vk. Since $\mathcal{A}$ is required to never query the signing oracle on $(\cdot, R^*, m^*)$, then $\mathcal{R}$ never queried the challenger on some tuple $(m^* \| R^*, \cdot)$. We can conclude that, if $i^* = i$, $\mathcal{R}$ returns a valid forgery with the same probability of $\mathcal{A}$. That is

$$\Pr[\mathcal{R} \text{ wins}] = \Pr[\mathcal{R} \text{ wins}|i^* = i]\Pr[i^* = i] + \Pr\left[\mathcal{R} \text{ wins}|\overline{i^* = i}\right]\Pr\left[\overline{i^* = i}\right]$$
$$\geq \Pr[\mathcal{R} \text{ wins}|i^* = i]\Pr[i^* = i]$$
$$\gtrsim \frac{\epsilon(\lambda)}{\ell(\lambda)},$$

which is a non-negligible probability. This represents a contradiction to the existential unforgeability of signatures with re-randomizable keys and concludes our proof. $\qquad\qquad\qquad\qquad\qquad\qquad\qquad\qquad\qquad\qquad\qquad\qquad\qquad\qquad\Box$

4   Bilinear Groups Instantiation

With the goal of an efficient standard model scheme in mind, in this section we show how to efficiently instantiate the construction presented in Sect. 3 in prime order groups with an efficiently computable pairing. As it was shown in [26], signatures with perfectly re-randomizable keys are known to exist in the standard model due to a construction by Hofheinz and Kiltz [34]. We proceed by describing a slightly modified version of the scheme and then we show how to deploy it in our generic framework.

4.1   A Variation of [34]

We recall the digital signature scheme from Hofheinz and Kiltz [34] as described in [26]. The scheme assumes the existence of a group with an efficiently computable bilinear map and a programmable hash function $\mathsf{H} = (\mathsf{HGen}, \mathsf{HEval})$ with domain $\{0, 1\}^*$ and range $\mathbb{G}_1$. The common reference string contains the key k of the programmable hash function H, that is assumed to be honestly generated. Signing keys are random elements $\mathsf{sk} \in \mathbb{Z}_p$ and verification keys are of the form $\mathsf{vk} = g_2^{\mathsf{sk}}$. To compute a signature on a message m, the signer returns $\left(s, y = \mathsf{HEval}(k, m)^{\frac{1}{\mathsf{sk}+s}}\right)$, where s is a randomly chosen element of $\mathbb{Z}_p$. The verification of a signature (s, y) consists of checking whether $e\left(y, \mathsf{vk} \cdot g_2^s\right) = e\left(\mathsf{HEval}(k, m), g_2\right)$. Keys can be efficiently re-randomized by computing $\mathsf{sk}' = \mathsf{sk} + \rho \bmod p$ and $\mathsf{vk}' = \mathsf{vk} \cdot g_2^\rho$, respectively. The scheme is shown to be existentially unforgeable under re-randomizable key under the q-strong Diffie-Hellman assumption in [26], in the common reference string model.

TOWARDS MORE EFFICIENT RING SIGNATURES. We now propose a slight modification of the scheme that allows us for a more efficient instantiation of our generic ring signature scheme. The changes are minimal: We introduce an additional randomness in the signing algorithm that is used in the computation of

$\mathsf{SSetup}(1^\lambda)$	$\mathsf{SGen}(1^\lambda)$	$\mathsf{Sig}(\mathsf{sk}, m)$
$k \leftarrow \mathsf{HGen}(1^\lambda)$	$x \leftarrow \mathbb{Z}_p$	$s \leftarrow \mathbb{Z}_p$
return k	**return** (x, g_2^x)	$\delta \leftarrow \mathbb{Z}_p$
		$c \leftarrow \mathsf{HEval}(k, m)^\delta$
		$y \leftarrow c^{\frac{1}{\mathsf{sk}+s}}$
		return (s, y, c, δ)

$\mathsf{Ver}(\mathsf{vk}, \sigma, m)$	$\mathsf{RndSK}(\mathsf{sk}, \rho)$	$\mathsf{RndVK}(\mathsf{vk}, \rho)$
parse $\sigma = (s, y, c, \delta)$	**return** $\mathsf{sk} + \rho$	**return** $(\mathsf{vk} \cdot g_2^\rho, k)$
if $\mathsf{HEval}(k, m)^\delta = c$ **return**		
$\quad e(y, \mathsf{vk} \cdot g_2^s) = e(c, g_2)$		

Fig. 2. Our variation of the [34] signature scheme

$\mathsf{H} = (\mathsf{HGen}, \mathsf{HEval})$. Such a randomness is included in the plain signature and it is used in the verification algorithm to check the validity of the output of the algorithm HEval. Our variation of the signature scheme can be found in Fig. 2. The correctness of the scheme is trivial to show. Note that the keys of our construction are identical to the ones of the signature scheme in [34], therefore the scheme has also perfectly re-randomizable keys. What is left to be shown is that signatures are still unforgeable.

Theorem 3. *Let* $\mathsf{H} = (\mathsf{HGen}, \mathsf{HEval})$ *be a programmable hash function, then the construction in Fig. 2 is unforgeable under re-randomizable keys under the q-strong Diffie-Hellman assumption.*

Proof. Our proof strategy consists in showing that a forgery in our signature scheme implies a forgery in the scheme of Hofheinz and Kiltz [34]. Since the scheme is proven to be secure against the q-strong Diffie-Hellman assumption, the theorem follows. More formally, assume that there exists an attacker that succeeds in the experiment as described in Definition 11 with probability $\epsilon(\lambda)$, for some non-negligible function ϵ. Then we can construct the following reduction against the unforgeability of the scheme [34].

$\mathcal{R}(1^\lambda, \mathsf{vk})$. On input vk the reduction samples $k \leftarrow \mathsf{HGen}(1^\lambda)$ and forwards (vk, k) to $\mathcal{A}$. The adversary is allowed to issue randomized signature queries of the form (m, ρ). $\mathcal{R}$ forwards (m, ρ) to its own oracle and receives (s, y) as a response. Then it samples a random $\delta \leftarrow \mathbb{Z}_p$ and hands over $(s, y^\delta, \mathsf{HEval}(k, m)^\delta, \delta)$ to $\mathcal{A}$. Standard signature queries are handled analogously. At some point of the execution the adversary outputs a challenge tuple $(m^*, \sigma^* = (s^*, y^*, c^*, \delta^*))$ and $\mathcal{R}$ returns $\left(m^*, \left(s^*, (y^*)^{\frac{1}{\delta^*}}\right)\right)$ to the challenger and terminates the execution.

The reduction is clearly efficient. To see that the queries of $\mathcal{A}$ are correctly answered it is enough to observe that the tuple

$$\left(s, y^\delta, \mathsf{HEval}(k, m)^\delta, \delta\right) = \left(s, \left(\mathsf{HEval}(k, m)^{\frac{1}{s+(\mathsf{sk}+\rho)}}\right)^\delta, \mathsf{HEval}(k, m)^\delta, \delta\right)$$

$$= \left(s, \mathsf{HEval}(k, m)^{\frac{\delta}{s+(\mathsf{sk}+\rho)}}, \mathsf{HEval}(k, m)^\delta, \delta\right)$$

$$= \left(s, c^{\frac{1}{s+(\mathsf{sk}+\rho)}}, c, \delta\right)$$

is a correctly distributed key if and only if (s, y) is correctly distributed. To conclude we need to show that $\left(m^*, \left(s^*, (y^*)^{\frac{1}{\delta^*}}\right)\right)$ is a valid forgery if and only if $(m^*, \sigma^* = (s^*, y^*, c^*, \delta^*))$ is a valid forgery. It is easy to see that since m^* was not queried by the adversary then it holds that m^* was not queried by the reduction either. Furthermore, by the winning condition of the experiment we have that

$$e\left(y^*, \mathsf{vk} \cdot g_2^{s^*}\right) = e(c^*, g_2),$$

which we can rewrite as

$$e\left((y^*)^{\frac{1}{\delta^*}}, \mathsf{vk} \cdot g_2^{s^*}\right) = e\left((c^*)^{\frac{1}{\delta^*}}, g_2\right).$$

Since it must be the case that $c^* = \mathsf{HEval}(k, m^*)^{\delta^*}$, then we have that

$$e\left((y^*)^{\frac{1}{\delta^*}}, \mathsf{vk} \cdot g_2^{s^*}\right) = e(\mathsf{HEval}(k, m^*), g_2),$$

which implies that $\left(m^*, \left(s^*, (y^*)^{\frac{1}{\delta^*}}\right)\right)$ is a valid message-signature tuple for the scheme in [34]. By initial assumption this happens with non-negligible probability $\epsilon(\lambda)$, which is a contradiction to the existential unforgeability of the scheme in [34] under re-randomizable keys and it concludes our proof. $\square$

4.2   A Ring Signature Scheme

Now we show how to deploy our scheme as described above in our framework for the construction of an efficient ring signature scheme. As our ultimate goal is to remove the common reference string, the first challenge in using the scheme of Fig. 2 in our generic construction of Sect. 3 is that it assumes a trusted setup of a key k for a programable hash function. The natural strategy to solve this issue is to include the key k in the verification key and show a re-randomized version of k in the ring signature. However, this comes at the price of proving the knowledge of a randomization factor of k, which is typically very expensive. In fact, the most efficient instance of a programmable hash function is due to Waters [41] and the size of the key is linear in the security parameter. Instead, we leverage some non-blackbox properties of the scheme presented in Sect. 4.1 to re-randomize the output of the programmable hash function. Loosely speaking,

$\mathsf{RGen}(1^\lambda)$	$\mathsf{RSig}(\mathsf{sk}, m, R)$	$\mathsf{RVer}(R, \sigma, m)$
$x \leftarrow \mathbb{Z}_p$	**parse** $R = (\mathsf{vk}_1, \ldots, \mathsf{vk}_n)$	**parse** $R = (\mathsf{vk}_1, \ldots, \mathsf{vk}_n)$
$k \leftarrow \mathsf{HGen}(1^\lambda)$	**if** $\not\exists i : \mathsf{vk} = \mathsf{vk}_i$	**parse** $\mathsf{vk}_i = (z_i, k_i)$
return $(x, (g_2^x, k))$	**return** $\bot$	**parse** $\sigma = (\sigma', \pi, z')$
	parse $\mathsf{vk} = (z, k)$	**parse** $\sigma' = (s, y, c)$
	$(s, \rho, \delta) \leftarrow \mathbb{Z}_p^3$	$x := R\|z'\|c\|(m, R)$
	$z' \leftarrow z \cdot g_2^\rho$	$b \leftarrow \mathcal{V}(\mathsf{crs}, x, \pi)$
	$x' \leftarrow \mathsf{sk} + \rho$	$b' = 1$ **if**
	$c \leftarrow \mathsf{HEval}(k, m\|R)^\delta$	$e(y, \mathsf{vk}' \cdot g_2^s) = e(c, g_2)$
	$x := R\|z'\|c\|(m, R)$	**return** $(b = b' = 1)$
	$\pi \leftarrow \mathcal{P}(\mathsf{crs}, (\rho, \delta, i), x)$	
	$y \leftarrow c^{\frac{1}{x'}}$	
	$\sigma = (s, y, c)$	
	return (σ, π, z')	

Fig. 3. A ring signature scheme in the common reference string model

this allows us to remove the evaluation of H from the statement to be proven in zero knowledge thus greatly improving the efficiency of the resulting scheme. A complete description of the resulting ring signature can be found in Fig. 3. Here the language $\mathcal{L}$ of our proof system is defined as

$$\mathcal{L} = \left\{ \begin{array}{l} ((\mathsf{vk}_1, \ldots, \mathsf{vk}_n, \mathsf{vk}^*), (k_1, \ldots, k_n), c, m) \in \mathbb{G}_2^{n+1} \times \mathbb{G}_1^{\lambda \cdot n + 1} \times \{0,1\}^* : \\ \exists (\delta, \rho, i) : \frac{\mathsf{vk}^*}{\mathsf{vk}_i} = g_2^\rho \wedge c = \mathsf{HEval}(k_i, m)^\delta \end{array} \right\},$$

which is essentially a disjunctive proof of two discrete logarithms over two vectors of group elements.

Since the scheme in Fig. 3 is not a direct instantiation of the construction discussed in Sect. 3, we shall prove that the construction satisfies the requirements for a ring signature scheme. First we show that our scheme is statistically anonymous.

Theorem 4. *Let* NIZK *be a statistically zero-knowledge argument, and let* $\mathsf{H} = (\mathsf{HGen}, \mathsf{HEval})$ *be a programmable hash function, then the construction in Fig. 3 is an anonymous ring signature scheme in the common reference string model.*

Proof. The first steps of the proof are identical to the ones of the proof of Theorem 1. We only need to introduce an extra hybrid H_3 where we substitute c^* with a random element in $\mathbb{G}_1$. For the indistinguishability $H_2 \approx H_3$ we argue that for all $m \in \{0,1\}^*$, for a random key $k \leftarrow \mathsf{HGen}(1^\lambda)$, and for a random $\delta \in \mathbb{Z}_p$, the element $\mathsf{HEval}(k, m)^\delta$ is uniformly distributed in $\mathbb{G}_1$, except with negligible probability. This is clearly the case as long as $\mathsf{HEval}(k, m) \neq 1$, which, in the construction of [41], happens only with negligible probability over the random choice of k. With this in mind, the argument follows along the same lines.   □

Finally, we show that our scheme is unforgeable against full key exposure.

Theorem 5. *Let* NIZK *be a computationally sound argument of knowledge that is extractable in presence of a signing oracle and let* H $=$ (HGen, HEval) *be a programmable hash function, then the construction in Fig. 3 is an unforgeable ring signature scheme under the* q-*strong Diffie-Hellman assumption in the common reference string model.*

Proof. Our proof strategy is to reduce against the unforgeability of the scheme described in Fig. 2. The proof outline is identical to the proof of Theorem 2, but there are some subtleties to address due to the non-blackbox usage of the signature scheme. More formally, assume towards contradiction that there exists a PPT adversary $\mathcal{A}$ that succeeds in the experiment of Definition 3 with probability $\epsilon(\lambda)$, for some non negligible function ϵ. Then we can construct the following reduction $\mathcal{R}$ against the unforgeability under re-randomizable keys of the signature scheme in Fig. 2.

$\mathcal{R}(1^\lambda, (\mathsf{vk}, k))$. On input the security parameter and the key (vk, k) the reduction samples an $i \leftarrow \{1, \dots, \ell(\lambda)\}$ and sets $\mathsf{vk}_i = (\mathsf{vk}, k)$. For all $j \in \{1, \dots, \ell(\lambda)\}\backslash i$, the reduction executes $(\mathsf{sk}'_j, \mathsf{vk}'_j) \leftarrow \mathsf{SGen}(1^\lambda)$ (as defined in Fig. 2) and $k_i \leftarrow \mathsf{HGen}(1^\lambda)$. $\mathcal{R}$ sets $\mathsf{vk}_j = (\mathsf{vk}'_j, k_j)$. Upon a corruption query from $\mathcal{A}$ on index $j \neq i$, the reduction sends sk_j to $\mathcal{A}$, if $j = i$ then $\mathcal{R}$ aborts. Signing queries of the form (j, R, m), for $j \neq i$ are handled as specified in the experiment. Upon queries of the form (i, R, m), the reduction sends $(m\|R, \rho)$ to the signing oracle, for a random $\rho \in \mathbb{Z}_p$. $\mathcal{R}$ parses the response as (s, y, c, δ), sets $x := (R, \mathsf{vk} \cdot g_2^\rho, c, m\|R)$ and computes $\pi \leftarrow \mathcal{P}(\mathsf{crs}, (\rho, \delta, i), x)$. $\mathcal{A}$ is provided with the tuple $((s, y, c), \pi, \mathsf{vk} \cdot g_2^\rho)$. At some point of the execution $\mathcal{A}$ outputs a tuple (R^*, m^*, σ^*). $\mathcal{R}$ parses $\sigma^* = (\theta^*, \pi^*, z^*)$ and $\theta^* = (s^*, y^*, c^*)$ and runs $(x^*, w^*, \pi^*) \leftarrow \mathcal{E}_\mathcal{A}$ on the same inputs and random tape of $\mathcal{A}$. $\mathcal{R}$ parses $w^* = (\rho^*, \delta^*, i^*)$ and aborts if $i^* \neq i$. Otherwise $\mathcal{R}$ returns the tuple $(m^*\|R^*, (s^*, y^*, c^*, \delta^*), \rho^*)$ and interrupts the simulation.

Since it runs only PPT machines, it follows that $\mathcal{R}$ terminates in polynomial time. For the case $i = i^*$ it is easy to see that the queries are answered correctly: π is clearly a correct argument of knowledge and for a tuple $((s, y, c), \pi, \mathsf{vk} \cdot g_2^\rho)$ it holds that $e(y, \mathsf{vk} \cdot g_2^\rho \cdot g_2^s) = e(c, g_2)$, since $y = c^{\frac{1}{\mathsf{sk}+\rho+s}}$, as expected. We now have to argue that a successful forgery by $\mathcal{A}$ implies a successful forgery of $\mathcal{R}$. First we note that, in order to win the experiment, $\mathcal{A}$ must not have queried the signing oracle on input $(\cdot, R^*, m^*)$, it follows that a valid signature on $m^*\|R^*$ by $\mathcal{R}$ must be a forgery. Let $\sigma^* = (\theta^*, \pi^*, z^*)$ be the challenge signature of $\mathcal{A}$ and let $\theta^* = (s^*, y^*, c^*)$. Since the extractor is successful with overwhelming probability we have that $c = \mathsf{HEval}(k_i, m^*\|R^*)^{\delta^*}$ and that $z^* = \mathsf{vk} \cdot g_2^{\rho^*}$, and by the winning condition of the game we have that $e\left(y^*, z^* \cdot g_2^{s^*}\right) = e\left(y^*, \mathsf{vk} \cdot g_2^{\rho^*} \cdot g_2^{s^*}\right) = e(c^*, g_2)$.

It follows that $(m^* \| R^*, (s^*, y^*, c^*, \delta^*), \rho^*)$ is a valid message-signature pair for the scheme of Fig. 2. Since $i^* = i$ with probability at least $\frac{1}{\ell(\lambda)}$, this represents a contradiction to the q-strong Diffie-Hellman assumption and it concludes our proof.   □

CONSTANT SIZE SIGNATURES. An interesting feature of our construction is that the computation of a signature under a re-randomized key is completely independent from the size of the ring. The only element that potentially grows with the size of the ring is therefore the proof π. However, if we implement the NIZK as a SNARK (such as in [31]) then we obtain a ring signature scheme with constant size signatures without random oracles. For the particular instantiation of [31], the proof adds only three group elements to the signature, independently of the size of the ring.

5   Efficient Non-Interactive Zero-Knowledge Without Random Oracles

In this section we put forward a novel NIZK system for languages of the class of the discrete logarithms in groups that admit an efficient bilinear map and a homomorphism $\phi : \mathbb{G}_1 \rightarrow \mathbb{G}_2$. Our constructions enjoy very simple algorithms and extremely short proofs (of size comparable to proofs derived from the Fiat-Shamir heuristic [24]) and do not rely on random oracles.

5.1   Complexity Assumptions

To prove the existence of an extractor for our main protocol, we need to assume that the knowledge of the exponent assumption holds in bilinear groups of prime order. On a high level, the knowledge of exponent ensures that, for a random $h \in \mathbb{G}_1$ and for all $x \in \mathbb{Z}_p$ it is hard to compute (h^x, g^x) without knowing x. This assumption was introduced by Damgård in [20] and proven to be secure in the generic group model by Abe and Fehr in [1].

Assumption 1 (Knowledge of Exponent (KEA)). *For all* PPT *adversaries* $\mathcal{A}$ *there exists a non-uniform* PPT *algorithm* $\mathcal{E}_{\mathcal{A}}$ *running on the same random tape of* $\mathcal{A}$ *and a negligible function* negl *such that*

$$\Pr\left[\begin{array}{l} (Y, Z) \leftarrow \mathcal{A}(p, e, g_1, g_2, g_2^x), \\ (y, Y, Z) \leftarrow \mathcal{E}_{\mathcal{A}}(p, e, g_1, g_2, g_2^x) \end{array} \middle| Z = Y^x \wedge g^y \neq Y \right] \leq \mathsf{negl}(\lambda).$$

We define a new variant of the knowledge of exponent assumption to guarantee the existence of an extractor for disjunctive statements. This modified version of the assumption allows the adversary to choose multiple bases $h_1, \dots, h_n$ as long as they satisfy the constraint $\prod_{i \in n} h_i = h$. Then $\mathcal{A}$ has to output $\{h_i^{x_i}, g^{x_i}\}_{i \in n}$ without knowing any x_i. The intuition why we believe this assumption to be as hard as the standard knowledge of exponent is that there must be at least one h_i that is not sampled independently from h, and therefore we conjecture that a tuple of the form (h^x, g^x) can be recovered from the code of the adversary. To back up this intuition, we show that such an assumption holds against generic algorithms in Sect. 5.3.

Assumption 2 (Linear Knowledge of Exponent (L-KEA)). *For all* $n \in$ poly(λ) *and for all* PPT *adversaries* $\mathcal{A}$ *there exists a non-uniform* PPT *algorithm* $\mathcal{E}_{\mathcal{A}}$ *running on the same random tape of* $\mathcal{A}$ *and a negligible function* negl *such that*

$$
\Pr\left[
\begin{array}{l}
(\{Y_i, V_i, Z_i\}_{i \in n}) \leftarrow \mathcal{A}(p, e, g_1, g_2, g_2^x), \\
(y, \{Y_i, V_i, Z_i\}_{i \in n}) \leftarrow \mathcal{E}_{\mathcal{A}}(p, e, g_1, g_2, g_2^x)
\end{array}
\;\middle|\;
\begin{array}{l}
\forall i \in \{1, \ldots, n\}: \\
\quad \mathsf{Dlog}_{g_1}(Y_i) \cdot \mathsf{Dlog}_{g_1}(V_i) \\
\quad = \mathsf{Dlog}_{g_1}(Z_i) \\
\wedge \prod_{i \in n} V_i = g_2^x \\
\wedge \forall i : g_1^y \neq Y_i
\end{array}
\right] \leq \mathsf{negl}(\lambda).
$$

5.2 Our Construction

Here we introduce our construction for efficient NIZK (without random oracles) for proving the knowledge of discrete logarithms. Although we sample statements from $\mathbb{G}_1$ it is easy to extend our techniques to handle the same languages in $\mathbb{G}_2$.

BASE PROTOCOL. The starting point for our system is the proof suggested by Abe and Fehr [1] for the knowledge of discrete logarithms: To prove the knowledge of an x such that $g_1^x = h$ the prover computes $\pi = \phi(T^x)$, where T is a randomly sampled group element and it is part of the common reference string. Such a proof is publicly verifiable by computing $e(h, T) = e(\pi, g_2)$. Note that the proving algorithm is deterministic and the proofs are uniquely determined by the statements and the common reference string. In some application, such as anonymous credentials, we would like to be able to re-randomize the proof. To address this issue, we present our first protocol. More formally, we describe a NIZK scheme for the language $\mathcal{L}_B$ defined as follows:

$$
\mathcal{L}_B = \{A \in \mathbb{G}_1 : \exists(a) : g_1^a = A\}.
$$

Clearly every element of a cyclic group have a well defined discrete logarithm when the base is the generator of the group, therefore the validity of such a statement can be simply verified by checking that A is a valid encoding of an element of $\mathbb{G}_1$. For groups of prime order this is a trivial task. However, it is unclear whether the party that outputs A *knows* the value of its discrete logarithm with respect to g_1. The scheme is depicted in Fig. 4.

In the following we prove that our scheme is a secure NIZK protocol.

Theorem 6. *The protocol in Fig. 4 is a non-interactive statistically sound unconditionally zero-knowledge argument of knowledge for the language* $\mathcal{L}_B$ *under the knowledge of exponent assumption.*

$\mathcal{G}(1^\lambda)$	$\mathcal{P}(\mathsf{crs}, x, w)$	$\mathcal{V}(\mathsf{crs}, x, \pi)$
$\alpha \leftarrow \mathbb{Z}_p$	**parse** $\mathsf{crs} = T \in \mathbb{G}_2$	**parse** $\mathsf{crs} = T \in \mathbb{G}_2$
$\mathsf{crs} \leftarrow g_2^\alpha$	$r \leftarrow \mathbb{Z}_p$	**parse** $x = A \in \mathbb{G}_1$
return (crs, α)	$R \leftarrow \phi(g_2^r)$	**parse** $\pi = (P_A, R, P_R) \in \mathbb{G}_1^2 \times \mathbb{G}_2$
	$P_R \leftarrow T^r$	**return** 1 **iff**
	$P_A \leftarrow \phi(T^{r \cdot w})$	$e(R, T) = e(\phi(P_R), g_2) \wedge$
	return (P_A, R, P_R)	$e(A, P_R) = e(P_A, g_2)$

Fig. 4. The base NIZK protocol.

Proof. To check whether and element A belongs to an group of order p (for some prime p) one can simply test whether A is relatively prime to p. Therefore soundness is trivial to prove. For the zero-knowledge property consider the following simulator:

$\mathcal{S}(\mathsf{crs}, x, \alpha)$. The simulator parses crs as $T \in \mathbb{G}_2$ and the statement as $A \in \mathbb{G}_1$, then it samples a random $r \leftarrow \mathbb{Z}_p$ and computes $R \leftarrow \phi(g_2^r)$, $P_R \leftarrow T^r$, and $P_A \leftarrow A^{(\alpha \cdot r)}$, The output of the simulator is (P_A, R, P_R).

The simulator is clearly efficient. It is easy to show that the proof $\pi = (P_A, R, P_R)$ is a correctly distributed proof for $x = A$, therefore the protocol is perfectly zero-knowledge. Note that this holds for any choice of the common reference string crs, as long as it is an element of $\mathbb{G}_2$ (which can be efficiently checked). It follows that the proof is unconditionally zero-knowledge. Argument of knowledge is proven by constructing an extractor for any valid proof output by any (possibly malicious) algorithm $\mathcal{A}$. Consider the following algorithm $\mathcal{B}$:

$\mathcal{B}(\mathsf{crs})$. The algorithm runs the adversary $\mathcal{A}$ on crs, which returns a tuple of the form $(x = A, \pi = (P_A, R, P_R))$ and it runs the corresponding extractor the retrieve $(x, \pi, r) \leftarrow \mathcal{E}_\mathcal{A}(\mathsf{crs})$. $\mathcal{B}$ returns the tuple $\left(A, \pi = (P_A)^{r^{-1}}\right)$.

Let extract be the event such that $R = g_2^r$. Since we consider only valid arguments, it must be the case that $\mathcal{V}(\mathsf{crs}, x, \pi) = 1$, and therefore we have that $P_R = R^\alpha$. Thus, by the knowledge of exponent assumption, we have that

$$\Pr[\text{extract}] \geq (1 - \mathsf{negl}(\lambda)).$$

Let us now consider the following algorithm extractor:

$\mathcal{E}(\mathsf{crs})$. The extractor exexutes $\mathcal{B}$ (defined as describe above) on input crs to receive the tuple $(x = A, \pi = P_A)$. Concurrently it runs the corresponding extractor on the same input $(x, \pi, w) \leftarrow \mathcal{E}_\mathcal{B}(\mathsf{crs})$ and it returns w.

The algorithm $\mathcal{E}$ is obviously efficient. We now have to show that the extraction is successful with overwhelming probability, i.e., $g_1^w = A$. We split the probability as follows:

$$\Pr[g_1^w = A] = \Pr[g_1^w = A|\mathsf{extract}]\,\Pr[\mathsf{extract}] + \Pr[g_1^w = A|\overline{\mathsf{extract}}]\,\Pr[\overline{\mathsf{extract}}].$$

By the argument above we have that

$$\Pr[g_1^w = A] \geq \Pr[g_1^w = A|\mathsf{extract}]\,(1 - \mathsf{negl}(\lambda)) + \Pr[g_1^w = A|\overline{\mathsf{extract}}]\,\mathsf{negl}(\lambda).$$

It follows that
$$\Pr[g_1^w = A] \gtrsim \Pr[g_1^w = A|\mathsf{extract}].$$

Note that when $\mathsf{extract}$ happens it holds that $P_A = A^\alpha$. Therefore, by the knowledge of exponent assumption, the extraction is successful with probability negligibly close to 1. This concludes our proof. $\qquad\square$

LOGICAL CONJUNCTIONS. We now describe a protocol to prove conjunction of discrete-logarithm proofs. Specifically we define the following language:

$$\mathcal{L}_C = \left\{ \begin{array}{l} (A_0, \ldots, A_n, h_1, \ldots, h_n) \in \mathbb{G}_1^{n+1} \times \mathbb{G}_2^n : \\ \exists(a) : g_1^a = A_0 \wedge \forall i \in \{1, \ldots, n\} : h_i^a = A_i \end{array} \right\}.$$

We show how to modify the previous scheme to handle statements in this language in Fig. 5. We omit the description of the setup algorithm, since it is unchanged.

Theorem 7. *The protocol in Fig. 5 is a non-interactive statistically sound unconditionally zero-knowledge argument of knowledge for the language $\mathcal{L}_C$ under the knowledge of exponent assumption.*

$\mathcal{P}(\mathsf{crs}, x, w)$	$\mathcal{V}(\mathsf{crs}, x, \pi)$
parse $\mathsf{crs} = T \in \mathbb{G}_2$	**parse** $\mathsf{crs} = T \in \mathbb{G}_2$
$r \leftarrow \mathbb{Z}_p$	**parse** $x = (A_0, \ldots, A_n) \in \mathbb{G}_1^{n+1}$,
$R \leftarrow \phi(g_2^r)$	$(h_1, \ldots, h_n) \in \mathbb{G}_2^n$
$P_R \leftarrow T^r$	**parse** $\pi = (P_A, R, P_R) \in \mathbb{G}_1^2 \times \mathbb{G}_2$
$P_A \leftarrow \phi(T^{r \cdot w})$	**return** 1 **iff**
return (P_A, R, P_R)	$e(R, T) = e(\phi(P_R), g_2) \wedge$
	$e(A_0, P_R) = e(P_A, g_2) \wedge$
	$\forall i \in \{1, \ldots, n\} :$
	$e(A_i, P_R) = e(P_A, h_i)$

Fig. 5. NIZK for conjunctive statements.

Proof. The proof is the simple observation that the algorithm $\mathcal{P}$ is the same algorithm as the one for the language $\mathcal{L}_B$. The formal argument follows along the same lines of the one for Theorem 6. □

LOGICAL DISJUNCTIONS. We now show a protocol that handles disjunctive statements over the family of discrete logarithms. We formally define the corresponding language $\mathcal{L}_D$ as follows:

$$\mathcal{L}_D = \{(A_1, \ldots, A_n) \in \mathbb{G}_1^n : \exists (a, i) : g_1^a = A_i\}.$$

The protocol can be found in Fig. 6. The basic idea here is to let the adversary cheat on $n-1$ statements by letting it choose the corresponding value of T_i (which can be seen as a temporary reference string). However, since we require that the relation $\prod_{i \in n} T_i = T$ is satisfied, at least one T_{i*} must depend on the common T. By the linear knowledge of exponent assumption, computing a proof over such a T_{i*} is as hard as computing it over T. In the following we show that our protocol is a secure NIZK.

Theorem 8. *The protocol in Fig. 6 is a non-interactive statistically sound unconditionally zero-knowledge argument of knowledge for the language $\mathcal{L}_D$ under the linear knowledge of exponent assumption.*

Proof. As argued in the proof of Theorem 6, soundness is trivial to show for this class of statements. In order to prove zero-knowledge we construct the following simulator:

$\mathcal{S}(\mathsf{crs}, x, \alpha)$. The simulator parses crs as $T \in \mathbb{G}_2$ and x as $(A_1, \ldots, A_n) \in \mathbb{G}_1^n$. Then it samples some $i \in \{1, \ldots, n\}$ and for all $j \in \{1, \ldots, n\} \backslash i$ it samples a random $t_j \leftarrow \mathbb{Z}_p$ and sets $P_j \leftarrow (A_j)^{t_j}$ and $T_j \leftarrow (g_2)^{t_j}$. Then it computes

$\mathcal{P}(\mathsf{crs}, x, w)$	$\mathcal{V}(\mathsf{crs}, x, \pi)$
parse $\mathsf{crs} = T \in \mathbb{G}_2$	**parse** $\mathsf{crs} = T \in \mathbb{G}_2$
parse $w = (a, i)$	**parse** $x = (A_1, \ldots, A_n) \in \mathbb{G}_1^n$
$\forall j \in \{1, \ldots, n\} \backslash i :$	**parse** $\pi = (T_1, \ldots, T_n) \in \mathbb{G}_2^n$
$\quad t_j \leftarrow \mathbb{Z}_p$	$\quad\quad\quad (P_1, \ldots, P_n) \in \mathbb{G}_1^n$
$\quad T_j \leftarrow g_2^{t_j}$	**return** 1 **iff**
$\quad P_j \leftarrow (A_j)^{t_j}$	$\prod_{i \in n} T_i = T \wedge$
$T_i \leftarrow T \cdot (\prod_{j \in n \backslash i} g_2^{t_j})^{-1}$	$\forall i \in \{1, \ldots, n\} :$
$P_i \leftarrow \phi(T_i^a)$	$\quad e(A_i, T_i) = e(P_i, g_2)$
return $(T_1, \ldots, T_n, P_1, \ldots, P_n)$	

Fig. 6. NIZK for disjunctive statements.

$T_i = T \cdot (\prod_{j \in n \setminus i} g_2^{t_j})^{-1}$ and $P_i \leftarrow A_i^{\frac{\alpha}{\sum_{j \in n \setminus i} t_j}}$. The algorithm returns the tuple $(T_1, \ldots, T_n, P_1, \ldots, P_n)$.

The simulation is clearly efficient. To show that the proof is correctly distributed it is enough to observe that the tuple $(T_1, \ldots, T_n)$ is sampled identically to $\mathcal{P}$ and that for all $i \in \{1, \ldots, n\} : P_i = A_i^{\mathsf{Dlog}_{g_1}(T_i)}$. Hence the scheme is a perfect zero-knowledge proof. As before, one can easily show that the proof is correctly distributed for all crs as long as crs is an element of $\mathbb{G}_2$. Therefore the scheme achieves unconditional zero-knowledge. The formal argument to show that our protocol is an argument of knowledge consists of the following extractor for any PPT $\mathcal{A}$:

$\mathcal{E}(\mathsf{crs})$. The extractor runs $\mathcal{A}$ on the common reference string and it receives the tuple $(x = (A_1, \ldots, A_n), \pi = (T_1, \ldots, T_n, P_1, \ldots, P_n))$. Then it executes the extractor $\mathcal{E}_{\mathcal{A}}$ on the same random tape to obtain (x, π, w). The extractor checks for all $i \in \{1, \ldots, n\}$ whether $A_i = g_1^w$, if this is the case it returns (w, i).

The algorithm runs in polynomial time. To show that the extraction is successful whenever the proof correctly verifies, we observe that it must hold that $\prod_{i \in n} T_i = T = g_1^\alpha$ and that for all $i \in \{1, \ldots, n\} : \mathsf{Dlog}_{g_1}(P_i) = \mathsf{Dlog}_{g_1}(T_i) \cdot \mathsf{Dlog}_{g_1}(A_i)$. Let fail be the event such that there exists no i such that $A_i = g_1^w$, where w is defined as above. Assume towards contradiction that

$$\Pr[\mathsf{fail}] \geq \epsilon(\lambda)$$

for some non-negligible function ϵ. Then it is easy to see that $\mathcal{E}$ contradicts the linear knowledge of exponent assumption with the same probability. It follows that

$$\Pr[\mathsf{fail}] \leq \mathsf{negl}(\lambda)$$

which implies that the extraction is successful with overwhelming probability and it concludes our proof. $\square$

5.3   The Hardness of L-KEA Against Generic Attacks

In the following we show that L-KEA holds against generic algorithms. We model the notion of generic group algorithms as introduced by Shoup [40]. In this abstraction, algorithms can solve hard problems only by using operations and the structure of the group. In particular, generic algorithms cannot exploit the encoding of the elements. Although a hardness proof in the generic group model shall not be interpreted as a comprehensive proof of security, is also states that an algorithm that solves that specific problem will have to necessarily use the encoding of the group in some non-trivial way.

Our generic model for groups with a bilinear map is taken almost in verbatim from [5]: In such a model elements of $\mathbb{G}_1$, $\mathbb{G}_2$, and $\mathbb{G}_T$ are encoded as unique random strings. Given such an encoding one can only test whether two strings correspond to the same element. The group operations between elements are substituted by oracle queries to five oracles: Three oracles to compute the group operation in each of the three groups $(\mathbb{G}_2, \mathbb{G}_2, \mathbb{G}_T)$, one for the homomorphism $\phi : \mathbb{G}_2 \to \mathbb{G}_1$, and one for the bilinear map $e : \mathbb{G}_1 \times \mathbb{G}_2 \to \mathbb{G}_T$. We denote such set of oracles by $\mathcal{O}$. The encoding of elements of $\mathbb{G}_1$ is defined as an injective function $\delta_1 : \mathbb{Z}_p \to S_1$, where $S_1 \subset \{0,1\}^*$. Mappings $\delta_2 : \mathbb{Z}_p \to S_2$ for $\mathbb{G}_2$ and $\delta_T : \mathbb{Z}_p \to S_T$ for $\mathbb{G}_T$ are defined analogously. The main result of this section is the proof of the following theorem.

Theorem 9. *Let $(\delta_1, \delta_2, \delta_T)$ be random encoding functions for $(\mathbb{G}_1, \mathbb{G}_2, \mathbb{G}_T)$, respectively. For all $n \in \mathsf{poly}(\lambda)$ and for all generic algorithms $\mathcal{A}$ with oracle access to $\mathcal{O}$ there exists a non-uniform PPT algorithm $\mathcal{E}_{\mathcal{A}}$ running on the same random tape of $\mathcal{A}$ and a negligible function negl such that*

$$\Pr\left[\begin{array}{l} (\{\delta_1(y_i), \delta_2(v_i), \delta_1(z_i)\}_{i\in n}) \leftarrow \mathcal{A}(p, \delta_1(1), \delta_2(1), \delta_2(x)), \\ (y, \{\delta_1(y_i), \delta_2(v_i), \delta_1(z_i)\}_{i\in n}) \leftarrow \mathcal{E}_{\mathcal{A}}(p, \delta_1(1), \delta_2(1), \delta_2(x)) \end{array} \middle| \begin{array}{l} \forall i \in \{1, \ldots, n\} : \\ y_i \cdot v_i = z_i \\ \wedge \sum_{i\in n} v_i = x \\ \wedge \forall i : y \neq y_i \end{array}\right]$$
$$\leq \mathsf{negl}(\lambda).$$

Proof. In the following we describe an extractor $\mathcal{E}$ that simulates the oracle set $\mathcal{O}$ to extract a well-formed y. At the beginning of the simulation, $\mathcal{E}$ initializes three empty lists $(\mathcal{Q}_1, \mathcal{Q}_2, \mathcal{Q}_T)$, then it samples three random strings $(r_1, r_2, r_x) \in \{0,1\}^*$ and it appends $(1, r_1)$ and (x, r_x) to $\mathcal{Q}_1$ and $(1, r_2)$ to $\mathcal{Q}_2$. Note that we can express all of the entries in all of the lists as (F, r), where r is a random string and $F \in \mathbb{Z}_p[x]$ is a polynomial in the indeterminate x with coefficients in $\mathbb{Z}_p$. We assume without loss of generality that $\mathcal{A}$ makes oracle queries only on encodings previously received by $\mathcal{E}$, since we can set the range $\{0,1\}^*$ to be arbitrarily large. $\mathcal{E}$ provides $\mathcal{A}$ with the tuple (p, r_1, r_2, r_x) and simulates the queries of $\mathcal{A}$ to the different oracles as follows.

- Group Operation: On input two strings (r_i, r_j), $\mathcal{E}$ parses $\mathcal{Q}_1$ to retrieve the corresponding polynomials F_i and F_j, then it computes $F_k = F_i \pm F_j$ (depending on whether a multiplication or a division is requested). If an entry (F_k, r_k) is present in $\mathcal{Q}_1$ then $\mathcal{E}$ returns r_k, else samples a random $r'_k \in \{0,1\}^*$, adds (F_k, r'_k) to $\mathcal{Q}_1$ and returns r'_k. Group operation queries for $\mathbb{G}_2$ and $\mathbb{G}_T$ are treated analogously.
- Homomorphism: On input a string r_i, $\mathcal{E}$ fetches the corresponding F_i from $\mathcal{Q}_2$. If there exists an entry $(F_i, r_j) \in \mathcal{Q}_1$, then it returns r_j. Otherwise it samples an $r'_j \in \{0,1\}^*$, appends (F_i, r'_j) to $\mathcal{Q}_1$, and returns r'_j.
- Pairing: On input two strings (r_i, r_j), $\mathcal{E}$ retrieves the corresponding F_i and F_j from $\mathcal{Q}_1$ and $\mathcal{Q}_2$, respectively. Then it computes $F_k = F_i \cdot F_j$. If $(F_k, r_k) \in \mathcal{Q}_T$ then $\mathcal{E}$ returns r_k. Otherwise it samples a random $r'_k \in \{0,1\}^*$, adds (F_k, r'_k) to $\mathcal{Q}_T$ and returns r'_k.

At some point of the execution, $\mathcal{A}$ outputs a list of encodings of the form $((y_1, v_1, z_1), \ldots, (y_n, v_n, z_n))$. As argued above, we can assume that such a list is composed only by valid encodings, i.e., returned by $\mathcal{E}$ as a response to some oracle query. For all $i \in \{1, \ldots, n\}$ $\mathcal{E}$ parses $\mathcal{Q}_1$ to retrieve the F_{y_i} corresponding to y_i. If there exists an F_{y_i} that is a constant (a polynomial of degree 0 in x), then $\mathcal{E}$ returns such an F_{y_i}. Otherwise it aborts. The simulation is clearly efficient. Note that whenever $\mathcal{E}$ does not abort, its output is an element o such that $\delta_1(o) = y_i$, for some $i \in \{1, \ldots, n\}$. What is left to be shown is that $\mathcal{E}$ does not abort with all but negligible probability. First we introduce the following technical lemma from Schwarz [38].

Lemma 1. *Let $F(x_1, \ldots, x_m)$ be a polynomial of total degree $d \geq 1$. Then the probability that $F(x_1, \ldots, x_m) = 0 \bmod n$ for randomly chosen values $(x_1, \ldots, x_m) \in \mathbb{Z}_n^m$ is bounded above by d/q, where q is the largest prime dividing n.*

Observe that at any point of the execution, for all elements $(F_i, r_i) \in \mathcal{Q}_1$, F_i is a polynomial of degree at most 1 and the same holds for the elements of $\mathcal{Q}_2$. Consequently, for each element $(F_j, r_j) \in \mathcal{Q}_T$, F_j is a polynomial of degree at most 2 in x. We can now show the following:

Lemma 2. *For all $i \in \{1, \ldots, n\}$:*

$$\Pr[degree \ of \ F_{y_i} \ in \ x \ is \ \geq 1 \wedge degree \ of \ F_{v_i} \ in \ x \ is \ \geq 1] \leq \mathsf{negl}(\lambda)$$

where F_{y_i} and F_{v_i} are the polynomials corresponding to y_i in $\mathcal{Q}_1$ and v_i in $\mathcal{Q}_2$, respectively.

Proof (Lemma 2). Let F_{z_i} be the polynomial corresponding to z_i in $\mathcal{Q}_1$. As we argued above F_{z_i} is a polynomial of degree at most 1 in x. Therefore if $F_{z_i} = F_{y_i} \cdot F_{v_i}$, then it is clear that either F_{y_i} or F_{v_i} must be a constant. Note that we require that $F_{z_i}(x) = F_{y_i}(x) \cdot F_{v_i}(x)$, for a randomly chosen x. By Lemma 1 we can bound the probability of the case $(F_{z_i} \neq F_{y_i} \cdot F_{v_i}) \wedge (F_{z_i}(x) = F_{y_i}(x) \cdot F_{v_i}(x))$ to happen to $1/p$, which is a negligible function. It follows that if $F_{z_i}(x) = F_{y_i}(x) \cdot F_{v_i}(x)$ then $F_{z_i} = F_{y_i} \cdot F_{v_i}$, with overwhelming probability and thus either F_{y_i} or F_{v_i} is a polynomial of degree 0 with the same probability. $\square$

We proved that for all i at least one between F_{y_i} and F_{v_i} is a polynomial of degree x in 0. We now want to show that in at least one case F_{y_i} is a constant. More formally:

Lemma 3. *Let F_{v_i} be the polynomial corresponding to v_i in $\mathcal{Q}_2$. Then*

$$\Pr[\forall i : degree \ of \ F_{v_i} \ in \ x \ is \ 0] \leq \mathsf{negl}(\lambda).$$

Proof (Lemma 3). Assume towards contradiction that for all i it holds that F_{v_i} is a polynomial of degree 0 in x with probability $\epsilon(\lambda)$, for some non-negligible function ϵ. Note that we require that $\sum_i F_{v_i}(x) = x$, or equivalently $a - x = 0$ for some constant $a = F_{v_i}(x)$. By Lemma 1 this happens only with negligible probability over the random choice of x. Therefore we have that ϵ is a negligible function, which is a contradiction. $\square$

Combining Lemmas 2 and 3 we have that with all but negligible probability there exists an i such that F_{v_i} is a polynomial of degree 1 in x and that the corresponding F_{y_i} is a constant. It follows that the extractor $\mathcal{E}$ does not abort with the same probability. This concludes our proof. $\square$

6   A Ring Signature Scheme in the Standard Model

We now have all the tools to upgrade to the standard model our construction presented in Sect. 4.2. Our first observation is that our ring signature scheme uses the common reference string only for the computation of the zero-knowledge argument. In particular, the signature scheme is completely independent from the crs. Therefore we can potentially use different strings for different rings without compromising the correctness of the scheme. Consider our instantiation of a NIZK scheme as presented in Sect. 5: If we include a different $crs_i = g_2^{\alpha_i}$ in each verification key vk_i, the linear combination of the crs_i for a given ring $R = (vk_1, \ldots, vk_n)$ is still a correctly distributed common reference string $crs = \prod_{i \in n} g_2^{\alpha_i} = g_2^{\sum_{i \in n} \alpha_i}$. The crux of this transformation is that the common reference string has no underlying hidden structure, other than being composed by a uniformly chosen group element, and that the corresponding group is closed under composition. We can modify our construction to obtain a scheme without setup as follows: Each verification key has extra element $C_i = g_2^{\alpha_i}$. On input a ring $R = (vk_1, \ldots, vk_n)$, the signing algorithm sets $crs = \prod_{i \in n} C_i$. The verification algorithm is modified analogously and the rest of the scheme is unchanged. A complete description of the scheme can be found in Fig. 7. Note that the unconditional zero-knowledge of the proof system is a fundamental component for the security of our scheme, as it guarantees that the construction stays anonymous even in presence of adversarially chosen keys. The formal analysis is elaborated below.

Theorem 10. *Let* NIZK *be an unconditional zero-knowledge argument, and let* H = (HGen, HEval) *be a programmable hash function, then the construction in Fig. 7 is an anonymous ring signature scheme.*

Proof. The proof is essentially equivalent to the one for Theorem 4. The only subtlety that we need to address is that the simulator does not necessarily know the trapdoor for the common reference string corresponding to the challenge ring, since it may contain adversarially generated keys. However, note that any crs has a well defined discrete logarithm that the simulator can compute and use as a trapdoor to output the simulated proof. We stress that, since we are proving statistical anonymity, we do not require the simulator to run in polynomial time. By the unconditional zero-knowledge of the NIZK, the indistinguishability argument follows.   □

Theorem 11. *Let* NIZK *be a computationally sound argument of knowledge that is extractable in presence of a signing oracle and let* H = (HGen, HEval) *be a programmable hash function, then the construction in Fig. 7 is an unforgeable ring signature scheme under the q-strong Diffie-Hellman assumption.*

Proof. The proof of unforgeability follows along the same lines of the one for Theorem 5: For the extraction it is enough to observe that the challenge ring R^* must be composed exclusively by honest verification keys. It follows that the corresponding crs is a random element of $\mathbb{G}_2$ of the form $g_2^{\sum_{i \in n} \alpha_i}$. We can

$\mathsf{RGen}(1^\lambda)$	$\mathsf{RSig}(\mathsf{sk}, m, R)$	$\mathsf{RVer}(R, \sigma, m)$
$(x, \alpha) \leftarrow \mathbb{Z}_p^2$	**parse** $R = (\mathsf{vk}_1, \ldots, \mathsf{vk}_n)$	**parse** $R = (\mathsf{vk}_1, \ldots, \mathsf{vk}_n)$
$k \leftarrow \mathsf{HGen}(1^\lambda)$	**if** $\nexists i : \mathsf{vk} = \mathsf{vk}_i$	**parse** $\mathsf{vk}_i = (z_i, k_i, C_i)$
$C \leftarrow g_2^\alpha$	**return** $\perp$	**parse** $\sigma = (\sigma', \pi, z')$
return $(x, (g_2^x, k, C))$	**parse** $\mathsf{vk} = (z, k, C)$	**parse** $\sigma' = (s, y, c)$
	parse $\mathsf{vk}_i = (z_i, k_i, C_i)$	$x := R\|z'\|c\|(m, R)$
	$(s, \rho, \delta) \leftarrow \mathbb{Z}_p^3$	$b \leftarrow \mathcal{V}\left(\prod_i C_i, x, \pi\right)$
	$z' \leftarrow z \cdot g_2^\rho$	
	$x' \leftarrow \mathsf{sk} + \rho$	$b' = 1$ **if**
	$c \leftarrow \mathsf{HEval}(k, m\|R)^\delta$	$\quad e(y, \mathsf{vk}' \cdot g_2^s) = e(c, g_2)$
	$x := R\|z'\|c\|(m, R)$	**return** $(b = b' = 1)$
	$\pi \leftarrow \mathcal{P}\left(\prod_i C_i, (\rho, \delta, i), x\right)$	
	$y \leftarrow c^{\frac{1}{x'}}$	
	$\sigma = (s, y, c)$	
	return (σ, π, z')	

Fig. 7. A ring signature in the standard model

therefore execute the extractor $\mathcal{E}$ on input $\sum_{i \in n} \alpha_i$ and learn a correct witness with overwhelming probability. $\qquad\square$

EFFICIENCY. For our standard model ring signature scheme as defined in Fig. 7 a signing key sk is composed by a single integer in $\mathbb{Z}_p$ and a verification key is a collection of λ elements of $\mathbb{G}_1$ and two elements of $\mathbb{G}_2$. For a ring of size n, signatures are composed by $(2 \cdot n + 2)$ elements of $\mathbb{G}_1$, $(2 \cdot n + 1)$ elements of $\mathbb{G}_2$ and an integer in $\mathbb{Z}_p$. Signing requires $(4 \cdot n + 3)$ modular exponentiations and n computations of a programmable hash. The verification algorithm is roughly as efficient as $(4 \cdot n + 2)$ pairings, a modular exponentiation, and n computations of a programmable hash function.

6.1 Alternative Instantiations

We observe that our techniques are generically applicable to all NIZK systems whose common reference string has suitable homomorphic properties. Let us consider the NIZK of Groth [29], here the common reference string comes in two forms: The honestly generated string $(g, h = g^x, f = g^y, f^r, h^s, g^t)$ gives perfectly sound proofs, whereas the simulated string $(g, h = g^x, f = g^y, f^r, h^s, g^{r+s})$ gives perfectly zero-knowledge proofs and allows one to simulate proofs with the knowledge of (x, y, r, s). Assume that an independently sampled simulated string $(g, h_i, f_i, u_i, v_i, w_i)$ is included in each key vk_i and that the each argument of knowledge for a ring $(\mathsf{vk}_1, \ldots, \mathsf{vk}_n)$ is proven against the reference string $\mathsf{crs} := \left(g, \prod_{i \in n} h_i, \prod_{i \in n} f_i, \prod_{i \in n} u_i, \prod_{i \in n} v_i, \prod_{i \in n} w_i\right)$, similarly as what has been done before. Our observation is that, if all of the strings are distributed according to the simulated variant, then one can simulate and extract proofs with the knowledge of $\left(\sum_{i \in n} x_i, \sum_{i \in n} y_i, \sum_{i \in n} r_i, \sum_{i \in n} s_i\right)$ and, since crs is a simulated string, the resulting proof is correctly distributed. Thus, the resulting ring signature scheme is anonymous as long as all the keys in the challenge ring are honestly generated. This weaker variant of the property is called *basic anonymity* in [4]. For unforgeability we leverage the fact that the challenge ring has to be composed exclusively of honestly generated keys. It follows that the relation $\mathsf{Dlog}_f \left(\prod_{i \in n} u_i\right) + \mathsf{Dlog}_h \left(\prod_{i \in n} v_i\right) = \mathsf{Dlog}_g \left(\prod_{i \in n} w_i\right)$ holds. Since the simulator knows the trapdoor, we can conclude that the extraction succeeds with overwhelming probability. The rest of the argument stays unchanged.

It follows that one could also implement our construction of Sect. 4.2 with the zero-knowledge argument in [29] to obtain a standard model instantiation provably secure against the Decisional-Linear assumption (DLIN). This yields a ring signature scheme without setup from more "classical" assumptions, although at the cost of a slower running time of the signing and verification algorithms, an increased size of the signatures, and weaker anonymity guarantees.

Acknowledgement. This research is based upon work supported by the German research foundation (DFG) through the collaborative research center 1223, by the German Federal Ministry of Education and Research (BMBF) through the project PROMISE (16KIS0763), and by the state of Bavaria at the Nuremberg Campus of Technology (NCT). NCT is a research cooperation between the Friedrich-Alexander-Universität Erlangen-Nürnberg (FAU) and the Technische Hochschule Nürnberg Georg Simon Ohm (THN). We thank Sherman S.M. Chow and the reviewers for for valuable comments that helped to improve our paper.

References

1. Abe, M., Fehr, S.: Perfect NIZK with adaptive soundness. In: Vadhan, S.P. (ed.) TCC 2007. LNCS, vol. 4392, pp. 118–136. Springer, Heidelberg (2007). https://doi.org/10.1007/978-3-540-70936-7_7

2. Abe, M., Ohkubo, M., Suzuki, K.: 1-out-of-n signatures from a variety of keys. In: Zheng, Y. (ed.) ASIACRYPT 2002. LNCS, vol. 2501, pp. 415–432. Springer, Heidelberg (2002). https://doi.org/10.1007/3-540-36178-2_26

3. Ben-Sasson, E., Chiesa, A., Garman, C., Green, M., Miers, I., Tromer, E., Virza, M.: Zerocash: decentralized anonymous payments from bitcoin. In: 2014 IEEE Symposium on Security and Privacy, Berkeley, CA, USA, 18–21 May 2014, pp. 459–474. IEEE Computer Society Press (2014)

4. Bender, A., Katz, J., Morselli, R.: Ring signatures: stronger definitions, and constructions without random oracles. In: Halevi, S., Rabin, T. (eds.) TCC 2006. LNCS, vol. 3876, pp. 60–79. Springer, Heidelberg (2006). https://doi.org/10.1007/11681878_4

5. Boneh, D., Boyen, X., Shacham, H.: Short group signatures. In: Franklin, M. (ed.) CRYPTO 2004. LNCS, vol. 3152, pp. 41–55. Springer, Heidelberg (2004). https://doi.org/10.1007/978-3-540-28628-8_3

6. Boneh, D., Freeman, D.M.: Homomorphic signatures for polynomial functions. In: Paterson, K.G. (ed.) EUROCRYPT 2011. LNCS, vol. 6632, pp. 149–168. Springer, Heidelberg (2011). https://doi.org/10.1007/978-3-642-20465-4_10

7. Boneh, D., Gentry, C., Lynn, B., Shacham, H.: Aggregate and verifiably encrypted signatures from bilinear maps. In: Biham, E. (ed.) EUROCRYPT 2003. LNCS, vol. 2656, pp. 416–432. Springer, Heidelberg (2003). https://doi.org/10.1007/3-540-39200-9_26

8. Bose, P., Das, D., Rangan, C.P.: Constant size ring signature without random oracle. In: Foo, E., Stebila, D. (eds.) ACISP 2015. LNCS, vol. 9144, pp. 230–247. Springer, Cham (2015). https://doi.org/10.1007/978-3-319-19962-7_14

9. Boyen, X.: Mesh signatures. In: Naor, M. (ed.) EUROCRYPT 2007. LNCS, vol. 4515, pp. 210–227. Springer, Heidelberg (2007). https://doi.org/10.1007/978-3-540-72540-4_12

10. Boyen, X., Waters, B.: Full-domain subgroup hiding and constant-size group signatures. In: Okamoto, T., Wang, X. (eds.) PKC 2007. LNCS, vol. 4450, pp. 1–15. Springer, Heidelberg (2007). https://doi.org/10.1007/978-3-540-71677-8_1

11. Boyle, E., Goldwasser, S., Ivan, I.: Functional signatures and pseudorandom functions. In: Krawczyk, H. (ed.) PKC 2014. LNCS, vol. 8383, pp. 501–519. Springer, Heidelberg (2014). https://doi.org/10.1007/978-3-642-54631-0_29

12. Brzuska, C., Fischlin, M., Freudenreich, T., Lehmann, A., Page, M., Schelbert, J., Schröder, D., Volk, F.: Security of sanitizable signatures revisited. In: Jarecki, S., Tsudik, G. (eds.) PKC 2009. LNCS, vol. 5443, pp. 317–336. Springer, Heidelberg (2009). https://doi.org/10.1007/978-3-642-00468-1_18

13. Brzuska, C., Fischlin, M., Lehmann, A., Schröder, D.: Unlinkability of sanitizable signatures. In: Nguyen, P.Q., Pointcheval, D. (eds.) PKC 2010. LNCS, vol. 6056, pp. 444–461. Springer, Heidelberg (2010). https://doi.org/10.1007/978-3-642-13013-7_26

14. Camenisch, J., Lysyanskaya, A.: Dynamic accumulators and application to efficient revocation of anonymous credentials. In: Yung, M. (ed.) CRYPTO 2002. LNCS, vol. 2442, pp. 61–76. Springer, Heidelberg (2002). https://doi.org/10.1007/3-540-45708-9_5

15. Canetti, R., Goldreich, O., Halevi, S.: The random oracle methodology, revisited (preliminary version). In: 30th Annual ACM Symposium on Theory of Computing, Dallas, TX, USA, 23–26 May 1998, pp. 209–218. ACM Press (1998)

16. Chandran, N., Groth, J., Sahai, A.: Ring signatures of sub-linear size without random oracles. In: Arge, L., Cachin, C., Jurdziński, T., Tarlecki, A. (eds.) ICALP 2007. LNCS, vol. 4596, pp. 423–434. Springer, Heidelberg (2007). https://doi.org/10.1007/978-3-540-73420-8_38

17. Chase, M., Lysyanskaya, A.: On signatures of knowledge. In: Dwork, C. (ed.) CRYPTO 2006. LNCS, vol. 4117, pp. 78–96. Springer, Heidelberg (2006). https://doi.org/10.1007/11818175_5

18. Chaum, D., van Heyst, E.: Group signatures. In: Davies, D.W. (ed.) EUROCRYPT 1991. LNCS, vol. 547, pp. 257–265. Springer, Heidelberg (1991). https://doi.org/10.1007/3-540-46416-6_22

19. Chow, S.S.M., Wei, V.K., Liu, J.K., Yuen, T.H.: Ring signatures without random oracles. In Proceedings of the 2006 ACM Symposium on Information, Computer and Communications Security, ASIACCS 2006, pp. 297–302. ACM (2006)

20. Damgård, I.: Towards practical public key systems secure against chosen ciphertext attacks. In: Feigenbaum, J. (ed.) CRYPTO 1991. LNCS, vol. 576, pp. 445–456. Springer, Heidelberg (1992). https://doi.org/10.1007/3-540-46766-1_36

21. Derler, D., Slamanig, D.: Key-homomorphic signatures and applications to multiparty signatures. Cryptology ePrint Archive, Report 2016/792 (2016). http://eprint.iacr.org/2016/792

22. Dodis, Y., Kiayias, A., Nicolosi, A., Shoup, V.: Anonymous identification in Ad Hoc groups. In: Cachin, C., Camenisch, J.L. (eds.) EUROCRYPT 2004. LNCS, vol. 3027, pp. 609–626. Springer, Heidelberg (2004). https://doi.org/10.1007/978-3-540-24676-3_36

23. Dwork, C., Naor, M.: Zaps and their applications. In: 41st Annual Symposium on Foundations of Computer Science, Redondo Beach, CA, USA, 12–14 November 2000, pp. 283–293. IEEE Computer Society Press (2000)

24. Fiat, A., Shamir, A.: How to prove yourself: practical solutions to identification and signature problems. In: Odlyzko, A.M. (ed.) CRYPTO 1986. LNCS, vol. 263, pp. 186–194. Springer, Heidelberg (1987). https://doi.org/10.1007/3-540-47721-7_12

25. Fiore, D., Nitulescu, A.: On the (In)Security of SNARKs in the presence of oracles. In: Hirt, M., Smith, A. (eds.) TCC 2016. LNCS, vol. 9985, pp. 108–138. Springer, Heidelberg (2016). https://doi.org/10.1007/978-3-662-53641-4_5

26. Fleischhacker, N., Krupp, J., Malavolta, G., Schneider, J., Schröder, D., Simkin, M.: Efficient unlinkable sanitizable signatures from signatures with rerandomizable keys. In: Cheng, C.-M., Chung, K.-M., Persiano, G., Yang, B.-Y. (eds.) PKC 2016. LNCS, vol. 9614, pp. 301–330. Springer, Heidelberg (2016). https://doi.org/10.1007/978-3-662-49384-7_12

27. Gennaro, R., Wichs, D.: Fullyhomomorphic message authenticators. In: Sako, K., Sarkar, P. (eds.) ASIACRYPT 2013. LNCS, vol. 8270, pp. 301–320. Springer, Heidelberg (2013). https://doi.org/10.1007/978-3-642-42045-0_16

28. Gorbunov, S., Vaikuntanathan, V., Wichs, D.: Leveled fully homomorphic signatures from standard lattices. In: Servedio, R.A., Rubinfeld, R. (eds) 47th Annual ACM Symposium on Theory of Computing, Portland, OR, USA, 14–17 June 2015, pp. 469–477. ACM Press (2015)

29. Groth, J.: Simulation-sound NIZK proofs for a practical language and constant size group signatures. In: Lai, X., Chen, K. (eds.) ASIACRYPT 2006. LNCS, vol. 4284, pp. 444–459. Springer, Heidelberg (2006). https://doi.org/10.1007/11935230_29

30. Groth, J.: Short non-interactive zero-knowledge proofs. In: Abe, M. (ed.) ASIACRYPT 2010. LNCS, vol. 6477, pp. 341–358. Springer, Heidelberg (2010). https://doi.org/10.1007/978-3-642-17373-8_20

31. Groth, J.: On the size of pairing-based non-interactive arguments. In: Fischlin, M., Coron, J.-S. (eds.) EUROCRYPT 2016. LNCS, vol. 9666, pp. 305–326. Springer, Heidelberg (2016). https://doi.org/10.1007/978-3-662-49896-5_11

32. Groth, J., Kohlweiss, M.: One-out-of-many proofs: or how to leak a secret and spend a coin. In: Oswald, E., Fischlin, M. (eds.) EUROCRYPT 2015. LNCS, vol. 9057, pp. 253–280. Springer, Heidelberg (2015). https://doi.org/10.1007/978-3-662-46803-6_9

33. Herranz, J., Sáez, G.: Forking lemmas for ring signature schemes. In: Johansson, T., Maitra, S. (eds.) INDOCRYPT 2003. LNCS, vol. 2904, pp. 266–279. Springer, Heidelberg (2003). https://doi.org/10.1007/978-3-540-24582-7_20

34. Hofheinz, D., Kiltz, E.: Programmable hash functions and their applications. In: Wagner, D. (ed.) CRYPTO 2008. LNCS, vol. 5157, pp. 21–38. Springer, Heidelberg (2008). https://doi.org/10.1007/978-3-540-85174-5_2

35. Lai, R.W.F., Zhang, T., Chow, S.S.M., Schröder, D.: Efficient sanitizable signatures without random oracles. In: Askoxylakis, I., Ioannidis, S., Katsikas, S., Meadows, C. (eds.) ESORICS 2016. LNCS, vol. 9878, pp. 363–380. Springer, Cham (2016). https://doi.org/10.1007/978-3-319-45744-4_18
36. Rivest, R.L., Shamir, A., Tauman, Y.: How to leak a secret. In: Boyd, C. (ed.) ASIACRYPT 2001. LNCS, vol. 2248, pp. 552–565. Springer, Heidelberg (2001). https://doi.org/10.1007/3-540-45682-1_32
37. Schäge, S., Schwenk, J.: A CDH-based ring signature scheme with short signatures and public keys. In: Sion, R. (ed.) FC 2010. LNCS, vol. 6052, pp. 129–142. Springer, Heidelberg (2010). https://doi.org/10.1007/978-3-642-14577-3_12
38. Schwartz, J.T.: Fast probabilistic algorithms for verification of polynomial identities. J. ACM **27**(4), 701–717 (1980)
39. Shacham, H., Waters, B.: Efficient ring signatures without random oracles. In: Okamoto, T., Wang, X. (eds.) PKC 2007. LNCS, vol. 4450, pp. 166–180. Springer, Heidelberg (2007). https://doi.org/10.1007/978-3-540-71677-8_12
40. Shoup, V.: Lower bounds for discrete logarithms and related problems. In: Fumy, W. (ed.) EUROCRYPT 1997. LNCS, vol. 1233, pp. 256–266. Springer, Heidelberg (1997). https://doi.org/10.1007/3-540-69053-0_18
41. Waters, B.: Efficient identity-based encryption without random oracles. In: Cramer, R. (ed.) EUROCRYPT 2005. LNCS, vol. 3494, pp. 114–127. Springer, Heidelberg (2005). https://doi.org/10.1007/11426639_7

Quantum Algorithms

Grover Meets Simon – Quantumly Attacking the FX-construction

Gregor Leander[✉] and Alexander May

Horst Görtz Institute for IT-Security, Faculty of Mathematics,
Ruhr-University Bochum, Bochum, Germany
{gregor.leander,alex.may}@rub.de

Abstract. Using whitening keys is a well understood mean of increasing the key-length of any given cipher. Especially as it is known ever since Grover's seminal work that the effective key-length is reduced by a factor of two when considering quantum adversaries, it seems tempting to use this simple and elegant way of extending the key-length of a given cipher to increase the resistance against quantum adversaries. However, as we show in this work, using whitening keys does not increase the security in the quantum-CPA setting significantly. For this we present a quantum algorithm that breaks the construction with whitening keys in essentially the same time complexity as Grover's original algorithm breaks the underlying block cipher. Technically this result is based on the combination of the quantum algorithms of Grover and Simon for the first time in the cryptographic setting.

Keywords: Symmetric cryptography · Quantum attacks · Grover's algorithm · Simon's algorithm · FX-construction

1 Introduction

The existence of sufficiently large quantum computers has a major impact on the security of many cryptographic schemes we are using today. In particular the seminal work of Shor [24] showed that such computers would allow to factor numbers and compute discrete logs in abelian groups in polynomial time. As almost all public key schemes currently in use are build upon the assumption that those problems are intractable, the existence of quantum computers would seriously compromise the security of most of our digital communication.

This situation has triggered a whole new line of research, namely post-quantum cryptography (or quantum-resistant cryptography), that aims at developing new cryptographic primitives that would (hopefully) withstand even attackers that are equipped with quantum computers. Recently, NIST has announced a competition to eventually standardize one or several quantum-resistant public-key cryptographic algorithms [22], underlining the importance of the problem. Indeed, as NIST points out in their call for candidates, the roll out of new cryptographic schemes is a long time process and it is therefore

© International Association for Cryptologic Research 2017
T. Takagi and T. Peyrin (Eds.): ASIACRYPT 2017, Part II, LNCS 10625, pp. 161–178, 2017.
https://doi.org/10.1007/978-3-319-70697-9_6

important to start this switch to quantum resistant cryptography well before quantum computers are actually available.[1]

In the case of symmetric cryptography, the situation seems less critical – but is also significantly less studied. For almost 20 years time, it was believed, that the only advantage an attacker would have by using a quantum computer when attacking symmetric cryptography is due to Grover's algorithm [11] for speeding up brute force search. Indeed, Grover's algorithm reduces the effective key-length of any cryptographic scheme, and thus in particular of any block-cipher, by a factor of two.

Given an m bit key, Grover's algorithm allows to recover the key using $\mathcal{O}(2^{m/2})$ quantum steps.

To counter that attack, it seems to be sufficient to just double the key-length of the block cipher to achieve the same security against quantum attackers. For doing so, the two main generic options are using either whitening keys or using multiple encryptions.

More recently, starting with the initial works by Kuwakado and Morii [17,18] and followed by the work by Kaplan et al. [13], it was stressed that Grover's algorithm might not be the only threat for symmetric cryptography. In particular, Kuwakado and Morii showed that the so-called Even-Mansour construction can be broken in polynomial time in the quantum CPA-setting. In this setting, the attacker is allowed to make quantum queries, that is queries to the encryption function in quantum superposition. The quantum CPA setting was first defined by Boneh, Zhandry in [5], and further intensively discussed in Kaplan et al. [13] and Anand et al. [4].

The Even-Mansour construction [10] consists of a public permutation P on n bits and two secret keys k_1 and k_2 that are used as pre- (resp. post-) whitening keys for the encryption $\text{Enc}_{EM}(m)$ of some message m.

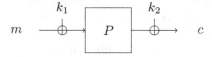

In a nutshell, Even and Mansour proved that, in an ideal world, where P is randomly chosen amongst all possible permutations, an attacker's advantage of distinguishing between the encryption and a random permutation is bounded by $q^2/2^n$ where q is the number of queries to P or to the encryption/decryption oracle. However, in the quantum CPA-model the scheme is completely insecure. The main idea of [18] was to consider the function

$$f(x) := \text{Enc}_{EM}(x) + P(x) = P(x + k_1) + k_2 + P(x),$$

---

[1] Note that NIST states that it is "primarily concerned with attacks that use classical (rather than quantum) queries to the decryption oracle or other private-key functionality."

where $+$ is the bitwise XOR.

As this function fulfills $f(x) = f(x + k_1)$ for all x, one can use Simon's quantum algorithm [7, 25], that allows to compute the unknown period k_1 of function f in linear time. Once k_1 is computed, computing k_2 is trivial even on a classical computer. It should be pointed out that Kaplan et al. [13] and Santoli, Schaffner [23] solved the technical issue of dealing with a function that does not fulfill Simon's promise, namely that $f(x) = f(y)$ iff $y \in \{x, x + k_1\}$, see Sect. 2 for more details.

The same idea was then used by Kaplan et al. [13] (and independently in [23]) to construct polynomial time quantum-CPA attacks on many modes of operations. Kaplan et al. further showed how slide attacks can profit from using a quantum computer.

The natural question that arises from the attacks on a generic cipher using Grover's algorithm and the attack on the Even-Mansour scheme using Simon's algorithm is the following: How secure is the FX construction against quantum adversaries?

This construction, proposed by Killian and Rogaway in [15, 16], is an elegant way of extending the key-length of a given block cipher and is the natural combination of the Even-Mansour construction and a generic cipher. For this, we assume we are given a (secure) block cipher E, encrypting n bit messages under an m bit key k_0, and we introduce two more n bit keys k_1 and k_2 as pre- and post-whitening keys. The new block cipher is given as

$$\text{Enc}(x) = E_{k_0}(x + k_1) + k_2.$$

From an efficiency point of view, the overhead of this modification is negligible. Moreover, in an idealized model, one can prove that (using classical computers) in order to attack the FX construction scheme, the success probability of an attacker is bounded by $\frac{q^2}{2^{n+m}}$, where q is the number of queries to the encryption scheme and to the underlying block cipher.

Initially, when considering Grover's algorithm only, this scheme seems to provide significantly more resistance against quantum computers, since now $(k_0, k_1, k_2) \in \mathbb{F}_2^{m+2n}$ define the key space. Moreover, Simon's algorithm does not apply either, as the function $\text{Enc}(x) + E_k(x)$ is periodic only for the correct guess of $k = k_0$.

Our Contribution

As the main result of our paper, we show that the FX construction as described above can be broken in the quantum-CPA model in basically the same time as the scheme without whitening keys, namely in $\mathcal{O}(m + n) \cdot 2^{m/2}$ quantum steps. Thus, using whitening keys does not help at all against quantum-CPA attackers.

Technically we have to combine the quantum algorithms of Simon and Grover for this attack. Thus, in contrast to most of the other works mentioned above, we actually have to define a new quantum algorithm, rather than applying known ones to new problems. While merging Simon and Grover might seem straightforward on a high level, the main technical obstacle is that in its original form, Simon's algorithm extracts information on the secret period bit by bit while Grover's algorithm, or more generally quantum amplitude amplification [7] inherently requires all the information to be available at once. We solve this issue by running several instances of Simon's quantum circuit in parallel, which in turn comes at the price of a linear growth of the size of the quantum computer. Furthermore, we postpone the measurements in Simon's algorithm to the very end of our entire quantum algorithm using the general *deferred measurement principle* of quantum computation.

DESX, PRINCE and PRIDE. To illustrate our results on actual ciphers, we like to mention that our work implies, among others, heuristic key recovery attacks on DESX, proposed by Rivest in 1984 and formally treated in [15] as well as on PRINCE [6] and PRIDE [2]. DESX, using a 56 bit internal key and two 64 bit whitening keys, can thus be attacked in the quantum-CPA model with complexity roughly 2^{28}, while PRINCE and PRIDE, both using a 64 bit internal key and two 64 bit whitening keys, can be attacked in the quantum-CPA model with complexity roughly 2^{32}.

We like to point out that, in the analysis of the success probabilities for our attack, we actually assume that the underlying functions are random functions, which is clearly not the case for the mentioned ciphers. However, it would be very surprising if this heuristic would fail for any reasonable block cipher.

Related Work

Besides the works on Simon's algorithm already mentioned above, we like to highlight in particular the work of Kaplan [12] who shows that multiple encryption is significantly weaker in the quantum setting than in the classical setting. This work is based on quantum random walks (cf e.g. [3]).

Together with our result presented here this implies that the two most common methods for extending the key-length are far from being optimal in a quantum-CPA setting. Finally, a very recent work [1] at EUROCRYPT 2017 already explores other means of mixing the key into the state that are (potentially) more resistant against quantum attacks. It remains to be seen if our idea of combining Grover with Simon, or related algorithms, allow to attack those recent proposals as well.

Another interesting approach is to use quantum computers to improve classical attacks on block ciphers, such as linear or differential cryptanalysis. This approach has been treated for the first time in [14].

Organization of the Paper

Before we formulate and prove our main theorem (Theorem 2) in Sect. 3 in all technical details, we outline the high level idea in Sect. 2. The latter section also contains some technical lemmata that are needed in the proof of our main result. We conclude by discussing some topics for future work in Sect. 4.

2 Main Ideas of Our Quantum Algorithm

Throughout the paper, we assume that the reader is familiar with the basics of quantum algorithms, although this section is supposed to be comprehensible without deeper quantum knowledge. For a comprehensive introduction into quantum algorithms we recommend the textbooks of Mermin [20] and Lipton, Regan [19].

Recall that we are attacking the FX-construction $\text{Enc}(x) = E_{k_0}(x + k_1) + k_2$. Let us look at the function

$$f(k, x) = \text{Enc}(x) + E_k(x) = E_{k_0}(x + k_1) + k_2 + E_k(x).$$

For the correct key $k = k_0$, we have $f(k, x) = f(k, x + k_1)$ for all x, and thus $f(k_0, \cdot)$ has period k_1 in its second argument. However, for $k \neq k_0$ the function $f(k, \cdot)$ is not periodic with high probability.

Main Idea. We define a Grover search over $k \in \mathbb{F}_2^m$, where we test for every $f(k, \cdot)$ periodicity via Simon's algorithm. Thus, we have Grover as an outer loop with running time roughly $2^{\frac{m}{2}}$, and Simon as an inner loop with polynomial complexity.

Classically, we could define an outer loop that guesses $k = k_0$ correctly with probability $p = 2^{-m}$. This would require an expected $\frac{1}{p} = 2^m$ number of iterations until we hit the correct key. Hence, each iteration roughly increases the success probability linearly by an amount of $\frac{1}{p}$. By contrast, in a quantum setting each iteration roughly increases the *amplitude* of success by a constant. Since the probabilities are the square of their respective amplitudes, we only have to repeat approximately $\sqrt{\frac{1}{p}} = 2^{\frac{m}{2}}$ times.

This process is called *amplitude amplification* and can be seen as a natural generalization of the original Grover search [11]. The results of amplitude amplification are more accurately captured in the following theorem by Brassard et al. [8, Theorem 2].

Theorem 1 (Brassard, Hoyer, Mosca and Tapp). *Let $\mathcal{A}$ be any quantum algorithm on q qubits that uses no measurement. Let $\mathcal{B} : \mathbb{F}_2^q \to \{0, 1\}$ be a function that classifies outcomes of $\mathcal{A}$ as good or bad. Let $p > 0$ be the initial success probability that a measurement of $\mathcal{A}|0\rangle$ is good. Set $k = \lceil \frac{\pi}{4\theta} \rceil$, where θ is* *defined via* $\sin^2(\theta) = p$. *Moreover, define the unitary operator* $Q = -AS_0A^{-1}S_B$, *where the operator* S_B *changes the sign of the good state*

$$|x\rangle \mapsto \begin{cases} -|x\rangle & \text{if } B(x) = 1 \\ |x\rangle & \text{if } B(x) = 0 \end{cases},$$

while S_0 *changes the sign of the amplitude only for the zero state* $|0\rangle$.

Then after the computation of $Q^k A|0\rangle$, *a measurement yields good with probability at least* $\max\{1 - p, p\}$.

Let us describe in a high-level fashion the process of amplitude amplification behind Theorem 1. The classifier B partitions the Hilbert space $\mathcal{H}$ of our quantum system in a direct sum of two orthogonal subspaces, the good subspace and the bad subspace. The good one is the subspace defined by all basis states $|x\rangle$ with $B(x) = 1$, the bad one is its orthogonal complement in $\mathcal{H}$.

Let $|\psi\rangle = A|0\rangle$ be the initial vector, and denote by $|\psi_1\rangle$, $|\psi_0\rangle$ its projection on the good and the bad subspace, respectively. Now look at the two-dimensional plane $\mathcal{H}_\psi$ spanned by $|\psi_1\rangle$, $|\psi_0\rangle$. In $\mathcal{H}_\psi$, the state $|\psi\rangle = A|0\rangle$ has angle θ (defined by $\sin^2(\theta) = p$) with the bad subspace. Each iteration via Q increases this angle by 2θ via the two reflections S_0 and S_B. Thus, after k iterations we have angle $(2k + 1)\theta$. If this angle roughly equals $\frac{\pi}{2}$, then the resulting state is almost co-linear with the good subspace, and thus a measurement yields a good vector with high probability. This explains the choice of the number of iterations $k \approx \frac{\pi}{4\theta}$ in Theorem 1.

Let us now assume that $p = 2^{-m}$ is the probability of guessing the correct key in the FX-construction. Then $\theta = \arcsin(2^{-\frac{m}{2}}) \approx 2^{-\frac{m}{2}}$, since $\arcsin(x) \approx x$ for small x. This implies $k = \Theta(2^{-\frac{m}{2}})$, as desired. Moreover, by Theorem 1 we obtain overwhelming success probability $1 - 2^{-m}$.

Ideally, we would choose A as Simon's algorithm and directly apply Theorem 1 for our setting. Although Theorem 1 excludes the use of measurements, while Simon's algorithm uses measurements to extract information about the period, this slight technical problem can be easily resolved by the quantum principle of deferred measurement that postpones all measurements until the very end of the computation.

However, we still have to resolve the following problems for an application of Theorem 1.

1. **Classifier.** We need to define some classifier B that identifies states as good iff they correspond to the correct key $k = k_0$. However, we do not see any way of efficiently checking correctness of k_0 without the knowledge of k_1. Simon's algorithm *iteratively* computes information about the period k_1 in a bit-wise fashion, where we need a *complete* candidate k_1' for the period in order to classify states as good or bad.
2. **Simon's promise.** Simon's algorithm is originally defined for periodic functions with the promise $f(x) = f(x + k_1)$ for all x, i.e., f is a $(2 : 1)$-mapping. However, our function $f(k_0, \cdot)$ does not fulfill the promise, since some function values might have more than two preimages.

3. Success probability. Assume that we are able to define a suitable classifier $\mathcal{B}$, then we might still only be capable of lower bounding the initial success probability p (which is the case for our algorithm), instead of exactly determining it. This causes problems in properly setting the number of iterations k.

Let us briefly give an outlook how we address these problems.

Classifier. In Simon's algorithm one computes a period $k_1 \in \mathbb{F}_2^n$ bit by bit. Namely, each iteration gives a random vector u_i from the subspace $U = \{u \in \mathbb{F}_2^n \mid \langle u, k_1 \rangle = 0\}$ of all vectors orthogonal to k_1. Thus, in each iteration we obtain a linear relation $\langle u_i, k_1 \rangle = 0$. After $\mathcal{O}(n)$ iterations, we can compute the unique solution k_1.

However, there is no need to compute the u_i sequentially. In our algorithm, we compute $u_1, \ldots, u_\ell$ for some sufficiently large ℓ in parallel. We choose ℓ such that for the periodic function $f(k_0, \cdot)$ the linear span $\langle u_1, \ldots, u_\ell \rangle$ is identical to U with high probability.

Where Simon's algorithm requires $\mathcal{O}(n)$ input bits, our parallel version of Simon's algorithm $\mathcal{A}$ requires $\mathcal{O}(n^2)$ many qubits. We leave it as an open problem whether this quadratic blowup can be avoided.

Our classifier $\mathcal{B}(x)$ should now identify states $|x\rangle$ with $k = k_0$ as good. We know that $f(k_0, \cdot)$ is periodic. Thus we compute for any $f(k, \cdot)$ sufficiently many u_i's with our parallel version $\mathcal{A}$ of Simon's algorithm.

Then $\mathcal{B}$ does the classical post-processing for Simon's algorithm. Namely, we compute from the u_i's some candidate period k_1'. If $\mathcal{B}(x)$ fails in any step, we classify state $|x\rangle$ as bad.

Otherwise $\mathcal{B}$ succeeds in computing some candidate values (k, k_1') for (k_0, k_1). This allows us to check via sufficiently many plaintext/ciphertext pairs (m_i, c_i), (m_i', c_i'), whether for all i the following identity holds

$$c_i + c_i' = \text{Enc}(m) + \text{Enc}(m') = E_{k_0}(m_i + k_1) + E_{k_0}(m_i' + k_1)$$
$$\overset{?}{=} E_k(m_i + k_1') + E_k(m_i' + k_1').$$

Checking this identity for sufficiently many plaintext/ciphertext pairs allows us to classify all incorrect states with $(k, k_1') \neq (k_0, k_1)$ as bad.

Simon's promise. It was shown recently by Kaplan et al. [13] and Santoli, Schaffner [23] that Simon's promise can be weakened at the cost of computing more u_i. We will use yet another technique for dealing with general functions. Namely, under the mild assumption that $f(k_0, x)$ behaves as a random periodic function with period k_1, we show that any function value $f(k_0, x)$ has only two preimages with probability at least $\frac{1}{2}$. We then only argue about these proper function values $f(k_0, x)$.

This provides a simple and clean way to use only a limited number ℓ of u_i. For comparison, $\ell = 2(n + \sqrt{n})$ is sufficient for our purpose, whereas a direct application of the techniques of Kaplan et al. [13] requires $\ell > 3n$.

Success probability. We define some $\mathcal{B}$ that classifies states with $k \neq k_0$ as bad with overwhelming probability. However, for states $|x\rangle$ with $k = k_0$ our $\mathcal{B}$ classifies $|x\rangle$ as good only with a probability that is lower bounded by some constant.

Still, we choose the number k of iterations analogous to Theorem 1, basically assuming that we would classify all states with $k = k_0$ as good. This in turn implies that our choice of k might be too small to fully rotate towards the subspace of good states. Nevertheless, by adapting the analysis of Brassard et al. [8] we are still able to show that we succeed in our case with probability at least $\frac{2}{5}$.

The following two basic lemmata will be frequently used in our further analysis. The first one shows, that any $n-1$ vectors obtained from Simon's algorithm form a basis of the $n-1$-dimensional vector space U with probability at least $\frac{1}{4}$. The second one shows that this basis allows us to compute its unique orthogonal complement, and therefore the period in Simon's algorithm, in polynomial time.

Lemma 1. *Let $U \subset \mathbb{F}_2^n$ be an $(n-1)$-dimensional subspace. Suppose we obtain $\mathbf{u}_1, \ldots, \mathbf{u}_{n-1} \in U$ drawn independently at uniform from U. Then $\mathbf{u}_1, \ldots, \mathbf{u}_{n-1}$ are linearly independent and form a basis of U with probability greater than $\frac{1}{4}$.*

Proof. Let E_i, $0 \leq i < n$ be the event that the first i vectors $\mathbf{u}_1, \ldots, \mathbf{u}_i$ form an i-dimensional space. Define $\Pr[E_0] := 1$.

Let $p_1 = \Pr[E_1]$ and $p_i = \Pr[E_i \mid E_{i-1}]$ for $2 \leq i < n$. Then $p_1 = 1 - \frac{1}{2^{n-1}}$, since we only have to exclude $\mathbf{u}_1 = 0^n \in U$. Moreover for $1 < i < n$, we have

$$p_i = 1 - \frac{2^{i-1}}{2^{n-1}},$$

since $\mathbf{u}_i$ should not lie in the $(i-1)$-dimensional span $\langle \mathbf{u}_1, \ldots, \mathbf{u}_{i-1}\rangle$. We obtain

$$\Pr[E_{n-1}] = \Pr[E_{n-1} \mid E_{n-2}] \cdot \Pr[E_{n-2}] = \ldots = \prod_{i=1}^{n-1} \Pr[E_i \mid E_{i-1}]$$

$$= \prod_{i=1}^{n-1} p_i = \prod_{i=1}^{n-1} 1 - 2^{i-n} = \prod_{i=1}^{n-1} 1 - 2^{-i}.$$

Since $\Pr[E_{n-1}] \geq \lim_{n \to \infty} \prod_{i=1}^{n-1} 1 - 2^{-i} \geq 0.288$, the claim follows. $\square$

Lemma 2. *Let $\mathbf{u}_1, \ldots, \mathbf{u}_{n-1} \in \mathbb{F}_2^n$ be linearly independent. Then one can compute in time $\mathcal{O}(n^3)$ the unique vector $\mathbf{v} \in \mathbb{F}_2^n \setminus \{\mathbf{0}\}$ such that $\langle \mathbf{v}, \mathbf{u}_i\rangle = 0$ for all $i = 1, \ldots, n-1$.*

Proof. Define the matrix $U \in \mathbb{F}_2^{(n-1)\times n}$, whose rows consist of the vectors $\mathbf{u}_i$. Transform U via Gaussian elimination into

$$U' = (I_{n-1}|\bar{\mathbf{v}}) \text{ for some column vector } \bar{\mathbf{v}} \in \mathbb{F}_2^{n-1}.$$

This costs time $\mathcal{O}(n^3)$. Notice that we have $\text{span}(U) = \text{span}(U')$, since we only applied elementary row operations.

Let $\mathbf{v} = (\bar{\mathbf{v}}^t | 1) \in \mathbb{F}_2^n \setminus \{\mathbf{0}\}$. Let $\mathbf{e}_i$ denote the i-th $(n-1)$-dimensional unit vector. Then the i-th basis vector $\mathbf{u}'_i = (\mathbf{e}_i | \bar{v}_i)$ of U' satisfies

$$\langle \mathbf{v}, \mathbf{u}'_i \rangle = \langle (\bar{\mathbf{v}}^t | 1), (\mathbf{e}_i, \bar{v}_i) \rangle = \bar{v}_i + \bar{v}_i = 0 \quad \text{for } i = 1, \ldots, n-1.$$

Since $\mathbf{v}$ is orthogonal to all $\mathbf{u}'_i$, it is also orthogonal to all $\mathbf{u}_i$. $\square$

3   Combining the Algorithms of Grover and Simon

Let us now proof our main theorem, whose statement is formulated in a slightly more general fashion than in Sect. 2 to make it useful also outside the FX construction context. In the FX construction, $f_{k_0,k_1,k_2}(x)$ is $\text{Enc}(x)$, and $g(k,x)$ is $E_k(x)$.

Theorem 2. *Let* $f : \mathbb{F}_2^m \times \mathbb{F}_2^{3n} \to \mathbb{F}_2^n$ *with*

$$(k_0, k_1, k_2, x) \mapsto g(k_0, x + k_1) + k_2,$$

where $g : \mathbb{F}_2^m \times \mathbb{F}_2^n \to \mathbb{F}_2^n$ *and* $g(k, \cdot)$ *is a random function* $\mathbb{F}_2^n \to \mathbb{F}_2^n$ *for any fixed* $k \in \mathbb{F}_2^m$. *Given quantum oracle access to* $f_{k_0,k_1,k_2}(\cdot)$ *and* $g(\cdot, \cdot)$, *the tuple* (k_0, k_1, k_2) *can be computed with success probability at least* $\frac{2}{5}$ *using* $m + 4n(n + \sqrt{n})$ *qubits and*

$$2^{\frac{m}{2}} \cdot \mathcal{O}(m + n) \text{ oracle queries.}$$

Proof. Let us define the function $f' : \mathbb{F}_2^m \times \mathbb{F}_2^n \to \mathbb{F}_2^n$ with

$$(k, x) \mapsto f_{k_0,k_1,k_2}(x) + g(k, x).$$

Notice that

$$f'(k_0, x) = g(k_0, x + k_1) + k_2 + g(k_0, x).$$

Hence $f'(k_0, x) = f'(k_0, x + k_1)$, and therefore $f'(k_0, \cdot)$ is periodic with period k_1 in its second component. We use amplitude amplification to search for k_0. A generalized version of Simon's algorithm then tells us which $f'(k, \cdot)$ is periodic.

However notice that we have a non-trivial period only if $k_1 \neq 0^n$. For $k_1 = 0^n$ we obtain a constant function with $f'(k_0, \cdot) = k_2$ for all inputs x. In the case of $k \neq k_0$, by the randomness of $g(k, \cdot)$ the function $f'(k, \cdot)$ is almost balanced in each output bit. This implies that we could use a generalized version of Deutsch-Joszas's algorithm [9] to decide which $f'(k, \cdot)$ is constant.

For simplicity, we will instead describe a basic Grover search for k_0. Notice that once k_0 is found, we can compute $k_2 = f_{k_0,k_1,k_2}(x) + g(k_0, x)$ for some arbitrary value of x.

Lemma 3. *Let* $k_1 = 0^n$. *Then one can determine* k_0, k_2 *with probability at least* $1 - 2^{-m}$ *using* $2^{\frac{m}{2}} \cdot \mathcal{O}(m)$ *oracle queries and* $m + 1$ *qubits.*

Proof. In the case $k_1 = 0^n$, we have

$$f'(k_0, x) = f_{k_0, 0^n, k_2}(x) + g(k_0, x) = g(k_0, x) + k_2 + g(k_0, x) = k_2 \text{ for any } x.$$

Thus, $f'(k_0, \cdot)$ is a constant function. Let us define a test, which checks whether $f_{k_0, 0^n, k_2}(x) + g(k, x)$ is constant to decide if $k = k_0$.

By evaluating $f_{k_0, 0^n, k_2}(\cdot)$, we compute for $1 + 2\lceil \frac{m}{n} \rceil$ random $x_i \in_R \mathbb{F}_2^n$ the function values

$$y_i = f_{k_0, 0^n, k_2}(x_i) = g(k_0, x_i) + k_2.$$

Moreover, we define the classical test $h : \mathbb{F}_2^m \to \mathbb{F}_2$ that takes as input a value $k \in \mathbb{F}_2^m$. We map k to 1 iff

$$y_i + g(k, x_i) = y_{i+1} + g(k, x_{i+1}) \text{ for all } 1 \le i \le 2\lceil \frac{m}{n} \rceil.$$

For $k = k_0$ we have $y_i + g(k, x_i) = k_2$, and therefore all identities are satisfied with probability 1. In the case $k \ne k_0$ by the randomness of $g(k, \cdot)$ any identity is fulfilled with probability 2^{-n}. Hence, all $2\lceil \frac{m}{n} \rceil$ identities are simultaneously fulfilled with probability at most 2^{-2m}.

Therefore, the probability that there is an incorrect $k \ne k_0$ that passes the test by h is at most $(2^m - 1)2^{-2m} < 2^{-m}$.

Altogether, we can use the quantum embedding of h on $m + 1$ qubits as a Grover oracle $\mathcal{B}$ from Theorem 1 that takes value 1 iff $k = k_0$ (with overwhelming probability) using $\mathcal{O}(m)$ oracle queries to $f_{k_0, k_1, k_2}(\cdot)$ and $g(\cdot, \cdot)$. Define $\mathcal{A}$ from Theorem 1 as the m-fold Hadamard transform

$$H^{\otimes m} : \mathcal{H} \to \mathcal{H} \text{ that maps } |x\rangle \mapsto \frac{1}{2^{m/2}} \sum_{k \in \mathbb{F}_2^m} (-1)^{xk} |k\rangle. \tag{1}$$

By Theorem 1, we start with $\mathcal{A}|0\rangle = H^{\otimes m}|0^m\rangle = 2^{-m/2} \sum_{k \in \mathbb{F}_2^m} |k\rangle$, the uniform superposition of all keys $k \in \mathbb{F}_2^m$. Then we have initial success probability $p = 2^{-m}$ to measure the good key $|k_0\rangle$. After $k = \lceil \frac{\pi}{4 \arcsin(\sqrt{p})} \rceil \le 2^{\frac{m}{2}}$ Grover iterations we measure $|k_0\rangle$ with probability at least $1 - 2^{-m}$.

Finally, we compute $k_2 = f_{k_0, 0^n, k_2}(x) + g(k_0, x)$ for some arbitrary value of x. □

For the remainder of the proof of Theorem 2, let us now assume $k_1 \ne 0^n$.

Let $\ell = 2(n + \sqrt{n})$. The following function h evaluates f' in parallel on ℓ arguments in the second component. Let $h : \mathbb{F}_2^m \times (\mathbb{F}_2^n)^\ell \mapsto (\mathbb{F}_2^n)^\ell$ with

$$(k, x_1, \ldots x_\ell) \mapsto f'(k, x_1) || f'(k, x_2) || \ldots || f'(k, x_\ell).$$

Let U_h be the universal bijective quantum embedding of h on $m + 2n\ell$ qubits, i.e. we map

$$|k, x_1, \ldots x_\ell, 0, \ldots, 0\rangle \mapsto |k, x_1, \ldots x_\ell, h(k, x_1, \ldots, x_\ell)\rangle.$$

Since we have quantum access to $f_{k_0, k_1, k_2}(\cdot)$ and $g(\cdot, \cdot)$, we can realize f' with two function queries and therefore also U_h with 2ℓ queries.

We now describe a quantum process $\mathcal{A}$, which is a parallel version of Simon's algorithm, that succeeds with initial probability $p \geq \frac{1}{5} \cdot 2^{-m}$ using $2\ell = \mathcal{O}(n)$ queries and $m + 2n\ell = m + 4n(n + \sqrt{n})$ qubits. Afterwards, we amplify $\mathcal{A}$'s success probability to at least $\frac{2}{5}$ using roughly $2^{\frac{m}{2}}$ iterations, where each iteration consumes $\mathcal{O}(m + n)$ oracle queries.

Quantum algorithm $\mathcal{A}$ on input $|0\rangle$

1. Prepare the initial $m + 2n\ell$-qubit state $|0\rangle$.
2. Apply Hadamard $H^{\otimes m+n\ell}$, as defined in Eq. (1), on the first $m + n\ell$ qubits resulting in

$$\sum_{k\in\mathbb{F}_2^m,\ x_1,\ldots,x_\ell\in\mathbb{F}_2^n} |k\rangle|x_1\rangle..|x_\ell\rangle|0\rangle,$$

where we omit the amplitudes $2^{-(m+n\ell)/2}$ for ease of exposition.
3. An application of U_h yields

$$\sum_{k\in\mathbb{F}_2^m,\ x_1,\ldots,x_\ell\in\mathbb{F}_2^n} |k\rangle|x_1\rangle..|x_\ell\rangle|h(k, x_1, \ldots, x_\ell)\rangle$$

4. We now apply Hadamard on the qubits in position $m + 1 \ldots m + n\ell$ (i.e. for $|x_1\rangle..|x_\ell\rangle$), which results in

$$|\psi\rangle = \sum_{\substack{k\in\mathbb{F}_2^m,\ u_1,\ldots,u_\ell\in\mathbb{F}_2^n, \\ x_1,\ldots,x_\ell\in\mathbb{F}_2^n}} |k\rangle(-1)^{\langle u_1,x_1\rangle}|u_1\rangle \ldots (-1)^{\langle u_\ell,x_\ell\rangle}|u_{n-1}\rangle|h(k, x_1, \ldots, x_\ell)\rangle. \quad (2)$$

Assume that we would measure the last $n\ell$ qubits of state $|\psi\rangle$ from step 4. Then these qubits would collapse into

$$|h(k, x_1, \ldots, x_\ell)\rangle = |f'(k, x_1)||f'(k, x_2)|| \ldots ||f'(k, x_\ell)\rangle,$$

for some *fixed values* of $k, x_1, \ldots, x_\ell \in \mathbb{F}_2^n$. Assume further that $k = k_0$.

Let us look at an arbitrary n-qubit state $|z_i\rangle = (-1)^{\langle u_i,x_i\rangle}|u_i\rangle$ from $|\psi\rangle$ that is entangled with $|f'(k_0, x_i)\rangle$. Hence, $|z_i\rangle$ collapses into a superposition that is consistent with the measured $f'(k_0, x_i)$.

We call the state $|z_i\rangle$ *proper* if x_i and $x_i + k_1$ are the only preimages of $f'(k_0, x_i)$. Notice that a proper $|z_i\rangle$ collapses into the superposition

$$\left((-1)^{\langle u_i,x_i\rangle} + (-1)^{\langle u_i,x_i+k_1\rangle}\right)|u_i\rangle = (-1)^{\langle u_i,x_i\rangle}\left(1 + (-1)^{\langle u_i,k_1\rangle}\right)|u_i\rangle. \quad (3)$$

As one can see from the right-hand side of Eq. (3), the qubits $|u_i\rangle$ have a non-vanishing amplitude iff $\langle u_i, k_1\rangle = 0$. Therefore, a measurement of a proper state yields some uniformly random $u_i \in U$, where $U = \{u \in \mathbb{F}_2 \mid \langle u, k_1\rangle = 0\}$.

Notice, that in general one can have more than two preimages of $f'(k_0, x_i)$, which results in $u_i \in \mathbb{F}_2^n$ that are with a certain probability chosen from some subset of U. Although such u_i usually still provide useful information about k_1 – whenever the subset is not too small – their probability distribution is somewhat cumbersome to deal with.

For ease of exposition, we want to work with proper states $|z_i\rangle$ only. In the following lemma, we show that any $|z_i\rangle$ is proper with probability at least $\frac{1}{2}$. This in turn means that on expectation a set of vectors $S = \{u_1, \ldots, u_{2(n-1)}\}$ derived from measuring $2(n-1)$ states contains at least $n-1$ vectors $u_{i_1}, \ldots, u_{i_{n-1}}$ that are chosen independently uniformly at random from U. Notice that a priori we are not able to identify these $n-1$, since we are not able to tell which state is proper. Nevertheless, we can easily compute from S a maximal set of independent vectors. Since $u_{i_1}, \ldots, u_{i_{n-1}} \in S$, by Lemma 1 these vectors form a basis of U with probability greater than $\frac{1}{4}$.

Moreover, if we increase the cardinality of S from $2(n-1)$ to $\ell = 2(n + \sqrt{n})$, as it is done in algorithm $\mathcal{A}$, then the above does not only hold on expectation, but with constant probability.

Lemma 4. *Any state $|z_i\rangle = (-1)^{\langle u_i, x_i \rangle} |u_i\rangle$ is proper with probability at least $\frac{1}{2}$. Any set of $\ell = 2(n + \sqrt{n})$ states contains at least $n-1$ proper states with probability greater than $\frac{4}{5}$.*

Proof. Recall that $|z_i\rangle$ is proper if $f'(k_0, x_i)$ has only two preimages. By definition of f', we have

$$f'(k_0, x_i) = g(k_0, x_i + k_1) + k_2 + g(k_0, x_i).$$

Let us denote by $S_i = i \times \{0, 1\}^{n-1}$ for $i = 0, 1$ the set of all n-dimensional vectors that start with bit i. Since $k_1 \neq 0^n$, assume wlog that the first bit of k_1 is 1, i.e., $k_1 \in S_1$.

Since $f'(k_0, \cdot)$ is periodic with k_1 in its second argument, the values of $f'(k_0, x)$ with $x \in S_0$ already determine the values of $f'(k_0, x + k_1)$ with $x + k_1 \in S_1$.

Let us fix some $x \in S_0$ and thus also $f'(k_0, x)$. The state $|z\rangle = (-1)^{\langle u, x \rangle} |u\rangle$ is proper if there is no other $x' \in S_0 \setminus \{x\}$ that collides with x under $f'(k_0, \cdot)$. By the randomness of $g(k_0, \cdot)$ this happens with probability

$$\Pr[|z\rangle \text{ is proper }] = 1 - \Pr[\exists x' \in S_0 \setminus \{x\} \text{ with } f'(k_0, x') = f'(k_0, x)]$$
$$\geq 1 - \frac{2^{n-1} - 1}{2^n} \geq \frac{1}{2},$$

where the first inequality follows from a union bound.

It remains to count the number of proper states within a set of $\ell = 2(n + \sqrt{n})$ states. Let X_i be an indicator variable that takes value 1 iff $|z_i\rangle$ is proper. Let $X = X_1 + \ldots X_{2(n+\sqrt{n})}$. Furthermore, define $\mu := n + \sqrt{n}$, which implies $\mathbb{E}[X] \geq \mu$. Now we apply the following Chernoff bound (see [21], Corollary 4.9 and Exercise 4.7)

$$\Pr[X \leq \mu - a] \leq e^{\frac{-2a^2}{n}} \quad \text{for all } a < \mu.$$

This implies in our case

$$\Pr[X \geq n - 1] = 1 - \Pr[X \leq n] = 1 - \Pr[X \leq \mu - \sqrt{n}] \geq 1 - e^{-2} \geq \frac{4}{5}. \qquad \square$$

Let us now go back to the quantum superposition $|\psi\rangle$ of qubits in positions $1, \ldots, m + n\ell$ (i.e. for the qubits $|k\rangle|u_1\rangle \ldots |u_{n-1}\rangle$) after applying algorithm $\mathcal{A}$, i.e. *without* measurement. Assume we had a classifier $\mathcal{B} : \mathbb{F}_2^{m+n\ell} \to \{0, 1\}$ that partitions $|\psi\rangle$ in a good subspace and a bad subspace, where the good subspace is spanned by the set of basis states $|x\rangle$ for which $\mathcal{B}(x) = 1$. We split

$$|\psi\rangle = |\psi_1\rangle + |\psi_0\rangle,$$

where $|\psi_1\rangle$ and $|\psi_0\rangle$ denotes the projection onto the good and onto the bad subspace, respectively.

Ideally we would like to define $|\psi_1\rangle$ as the sum of those basis states for which $|k\rangle = |k_0\rangle$. Unfortunately, we cannot check correctness of k_0 directly. Instead, with our classifier $\mathcal{B}$ we compute from $k, u_1, \ldots, u_\ell$ a candidate k_1' for the period k_1. This allows for an easy test $(k, k_1') \overset{?}{=} (k_0, k_1)$.

Classifier $\mathcal{B}$ (polynomial time computable Boolean function). Let us define the following classical Boolean function

$$\mathcal{B} : \mathbb{F}_2^{m+n\ell} \to \{0, 1\} \text{ that maps } (k, u_1, \ldots, u_\ell) \mapsto \{0, 1\}.$$

In $\mathcal{B}$, we hardwire for $\lceil \frac{3m+n\ell}{n} \rceil$ random pairs $m_i, m_i' \in_R \mathbb{F}_2^n$ with $m_i \neq m_i'$ the values

$$y_i = f_{k_0,k_1,k_2}(m_i) + f_{k_0,k_1,k_2}(m_i') = g(k_0, m_i + k_1) + g(k_0, m_i' + k_1),$$

which can be computed via $2\lceil \frac{3m+n\ell}{n} \rceil$ function evaluations of $f_{k_0,k_1,k_2}(\cdot)$. Now we check the following two steps.

1. Let $\bar{U} = \langle u_1, \ldots, u_\ell \rangle$ be the linear span of all u_i. If $\dim(\bar{U}) \neq n - 1$, output 0. Otherwise compute a basis of $\bar{U}$, and use Lemma 2 to compute the unique vector $k_1' \in \mathbb{F}_2^n \setminus \{\mathbf{0}\}$ orthogonal to $\bar{U}$.
2. Check via $2\lceil \frac{3m+n\ell}{n} \rceil$ function evaluations of $g(\cdot, \cdot)$ whether

$$y_i \overset{?}{=} g(k, m_i + k_1') + g(k, m_i' + k_1') \text{ for all } i = 1, \ldots, \left\lceil \frac{3m + n\ell}{n} \right\rceil.$$

If all identities hold, output GOOD. Else output BAD.

The following lemma shows that our test $\mathcal{B}$ classifies basis states with the correct $k = k_0$ as GOOD, respectively 1, with probability at least $\frac{1}{5}$. We could easily boost this into a probability close to 1 by increasing the number of qubits of $\mathcal{A}$. However, for the ease of description and for minimizing the number of qubits, we keep it this way, which eventually only slightly lowers the overall success probability of our algorithm.

More important is that $\mathcal{B}$ almost never gives false positives. Namely, whenever $\mathcal{B}$ declares a state as GOOD then indeed $k = k_0$ with overwhelming probability. This in turn implies that by our choice of parameters we safely sort out *all of the exponentially many* incorrect keys $k \neq k_0$.

Lemma 5. *If $k = k_0$ then test $\mathcal{B}$ outputs 1 with probability at least $\frac{1}{5}$.*
Vice versa, if $\mathcal{B}$ outputs 1 then $k_0 = k$ with probability at least $1 - \frac{1}{2^{2m+n\ell-4}}$.

Proof. Let us denote by GOOD the event that $\mathcal{B}$ outputs 1. We first compute

$$p_0 = \Pr[\text{GOOD} \mid k = k_0].$$

If $k = k_0$, then $f'(k, \cdot)$ is periodic with k_1 in its second argument. Moreover, we know by Lemma 4 that $u_1, \ldots, u_\ell$ contain with probability at least $\frac{4}{5}$ at least $n-1$ vectors that are uniformly at random from the subspace $U \subset \mathbb{F}_2^n$ orthogonal to k_1. From Lemma 1, these vectors form a basis of U with probability at least $\frac{1}{4}$.

In total, we pass step (1) of $\mathcal{B}$ with probability at least $\frac{4}{5} \cdot \frac{1}{4} = \frac{1}{5}$. Moreover, in the case $k = k_0$ we also have $k_1' = k_1$, i.e., $\mathcal{B}$ computes the correct k_1. Therefore, we pass all tests in step (2) of $\mathcal{B}$ with probability 1. Altogether, we obtain $p_0 \geq \frac{1}{5}$, which proves the first claim.

In order to prove the second claim, let us compute a lower bound for the probability

$$p_1 = \Pr[k = k_0 \mid \text{GOOD}] = \frac{p_0 \cdot \Pr[k = k_0]}{\Pr[\text{GOOD}]}.$$

Since $\Pr[k = k_0] = 2^{-m}$, it remains to compute

$$
\begin{aligned}
\Pr[\text{GOOD}] &= \Pr[k = k_0] \cdot \Pr[\text{GOOD} \mid k = k_0] + \Pr[k \neq k_0] \cdot \Pr[\text{GOOD} \mid k \neq k_0] \\
&= 2^{-m} \cdot p_0 + (1 - 2^{-m}) \cdot \Pr[\text{GOOD} \mid k \neq k_0] \\
&\leq 2^{-m} \cdot p_0 + \Pr[\text{GOOD} \mid k \neq k_0].
\end{aligned}
$$

Let us further bound the probability $\Pr[\text{GOOD} \mid k \neq k_0]$. This event means that we pass steps (1) and (2) of $\mathcal{B}$, even though we have the incorrect k. Since we need an upper bound only, we can assume that we always pass step (1) and compute some k_1'. In step (2), we check the identities

$$y_i \overset{?}{=} g(k, m_i + k_1') + g(k, m_i' + k_1') \text{ for all } i = 1, \ldots, \left\lceil \frac{3m + n\ell}{n} \right\rceil.$$

By our assumption $g(k, \cdot) : \mathbb{F}_2^n \to \mathbb{F}_2^n$ is random for any fixed $k \in \mathbb{F}_2^m$. Thus, each of these identities holds independently with probability 2^{-n}. Notice that this probability is even independent of the value of k_1' computed in step (2).

Thus, *all* the $\left\lceil \frac{3m+n\ell}{n} \right\rceil$ identities are simultaneously fulfilled with probability at most $2^{-3m-n\ell}$, which is an upper bound for $\Pr[\text{GOOD} \mid k \neq k_0]$.

This in turn implies

$$
\begin{aligned}
p_1 = \Pr[k = k_0 \mid \text{GOOD}] &= \frac{p_0 \cdot \Pr[k = k_0]}{\Pr[\text{GOOD}]} \geq \frac{p_0 \cdot 2^{-m}}{p_0 \cdot 2^{-m}(1 + \frac{2^{-2m-n\ell}}{p_0})} \\
&> \frac{1}{1 + 2^{3-2m-n\ell}} = \frac{1 - 2^{3-2m-n\ell}}{1 - 2^{2(3-2m-n\ell)}} \\
&= \frac{1 - 2^{2(3-2m-n\ell)} - (2^{3-2m-n\ell} - 2^{2(3-2m-n\ell)})}{1 - 2^{2(3-2m-n\ell)}} \geq 1 - \frac{1}{2^{2m+n\ell-4}},
\end{aligned}
$$

where the last inequality holds for $2m + n\ell > 3$. This concludes the proof. $\quad\square$

By Lemma 5 our test $\mathcal{B}$ classifies a bad state $|k\rangle|u_1\rangle\ldots|u_\ell\rangle$ with $k \neq k_0$ as good with probability at most $2^{-2m-n\ell+4}$. Notice that there are at most $2^{m+n\ell}$ states in any superposition. Therefore, $\mathcal{B}$ classifies any bad state in a superposition as good with probability at most $2^{m+n\ell}\cdot 2^{-2m-n\ell+4} = 2^{-m+4}$. We will have at most $2^{\frac{m}{2}}$ iterations of $\mathcal{A}$. By a union bound, the probability that $\mathcal{B}$ classifies any bad state in a superposition in any of these iterations as good is at most $2^{\frac{m}{2}}\cdot 2^{-m+4} = 2^{-\frac{m}{2}+4}$, which is exponentially small.

This implies that $\mathcal{B}$ (almost) never yields false positives. Hence, we classify a state $|k\rangle|u_1\rangle\ldots|u_\ell\rangle$ as good iff and only if $\mathcal{B}(k, u_1, \ldots, u_\ell) = 1$. The initial success probability p of $\mathcal{A}$ in producing a good state is by Lemma 5

$$p = \Pr[|k\rangle|u_1\rangle\ldots|u_\ell\rangle \text{ is good}]$$

$$= \Pr[k = k_0] \cdot \Pr[\mathcal{B}(k, u_1, \ldots, u_\ell) = 1 \mid k = k_0] \geq \frac{1}{5}\cdot 2^{-m}. \tag{4}$$

Our Boolean function $\mathcal{B}$ defines a unitary operator $S_\mathcal{B}$ that conditionally changes the sign of the amplitudes of the good states

$$|k\rangle|u_1\rangle\ldots|u_\ell\rangle \mapsto \begin{cases} -|k\rangle|u_1\rangle\ldots|u_\ell\rangle & \text{if } \mathcal{B}(k, u_1, \ldots, u_\ell) = 1 \\ |k\rangle|u_1\rangle\ldots|u_\ell\rangle & \text{if } \mathcal{B}(k, u_1, \ldots, u_\ell) = 0 \end{cases}.$$

The complete amplification process is realized by repeatedly applying the unitary operator $Q = -AS_0 A^{-1} S_\mathcal{B}$ to the initial state $|\psi\rangle = A|0\rangle$, i.e., we compute $Q^k A|0\rangle$ and measure the system for some suitable number of iterations k.

Let $|\psi_1\rangle$, $|\psi_0\rangle$ be the projection of $|\psi\rangle$ onto the good and the bad subspace, respectively. Denote by $\mathcal{H}_\psi$ the 2-dimensional subspace spanned by $|\psi_1\rangle$, $|\psi_0\rangle$. Initially, in $\mathcal{H}_\psi$ the angle between $A|0\rangle$ and the bad subspace is θ, where

$$\sin^2(\theta) = p = \Big\langle \psi_1 \mid \psi_1 \Big\rangle.$$

Thus, $\theta = \arcsin(\sqrt{p}) \geq \arcsin(\sqrt{\frac{1}{5}\cdot 2^{-\frac{m}{2}}})$, where the lower bound follows from (4).

Now every Grover iteration by Q increases the angle by 2θ, i.e., to $(2k+1)\theta$ after k iterations. If this angle is roughly $\frac{\pi}{2}$, then we are almost parallel to $|\psi_1\rangle$ in $\mathcal{H}_\psi$ and measure a good state with high probabilty. Therefore, let us choose

$$k = \Big\lceil \frac{\pi}{4\arcsin(2^{-\frac{m}{2}})} \Big\rceil.$$

After k iterations a final measurement produces a good state with probability $p_{\text{good}} = \sin^2((2k+1)\theta)$. Thus, we obtain

$$p_{\text{good}} \geq \sin^2\left(\frac{\pi}{2}\cdot \frac{\arcsin\left(\sqrt{\frac{1}{5}}\cdot 2^{-\frac{m}{2}}\right)}{\arcsin(2^{-\frac{m}{2}})} \right). \tag{5}$$

Notice that $\arcsin(x) \approx x$ for small x. So for m sufficiently large, the right hand side of (5) quickly converges to $\sin^2(\frac{\pi}{2\sqrt{5}}) \approx 0.42$. Namely, for $m \geq 12$ the right hand side is already larger than $\frac{2}{5}$, which shows that we measure a good state after amplification with the claimed success probability.

This measurement reveals k_0, and an application of $\mathcal{B}$ on a good state also reveals the correct value for k_1. We can then easily compute for an arbitrary x the value

$$k_2 = f_{k_0,k_1,k_2}(x) + g(k_0, x + k_1).$$

This completes the proof of our main theorem. $\square$ Theorem 2

3.1   Potential Improvements

As already pointed out, we did not try to optimize all constants in Theorem 2. Many parameters are chosen in a way that allows for smooth and simple proofs. Let us comment which parameters might be subject to (small) improvements.

Memory. We use $m + 4n(n + \sqrt{n})$ many qubits. However, we use only proper states in our proof, which gives a factor 2 loss for the u_i. But states that are not proper should usually still provide enough information. Hence, roughly $m + 2n^2$ qubits should be sufficient. An open question is whether this can be lowered to $m + o(n^2)$.

Notice that one can solely consider the projection of $f'(k_0, \cdot)$ to one output bit, which is also periodic. This however still requires $m + n^2$ input qubits.

Success probability. If we have $n + \mathcal{O}(1)$ values u_i, then in the periodic case $k = k_0$ one would expect to obtain a basis of U with probability close to 1, whereas in the non-periodic case $k \neq k_0$ one would expect to obtain an n-dimensional basis with probability close to 1. This means that our classifier $\mathcal{B}$ works almost perfect, which in turn implies that our success probability is in fact close to 1.

4   Two Open Problems

Our new quantum algorithm shows that the use of whitening keys in the FX-construction does not increase security in the quantum-CPA model. This result raises at least two more natural questions to be investigated in the future.

The first and maybe most important question is the security of key-alternating ciphers against quantum-CPA attacks. Key-alternating ciphers can be seen as a multiple round generalization of the Even-Mansour construction and many popular ciphers, most importantly the AES, follow this general design principle. It would thus be of great interest to design quantum algorithms that break those ciphers, or show that this is not possible in general.

The second question is the investigation of sound techniques for extending the key length of a given cipher in a quantum setting. Of course, it is always possible to design new key-schedulings (and potentially increase the number of rounds slightly) for larger key-sizes. The most well-known example is again the AES with its different variants of 128, 192 or 256 key bits. However, this requires an exact understanding of the internal behaviour of the cipher and it is thus of interest to investigate generic ways of increasing the key length. That is, given a cipher with an m bit key, how can we extend its key size to obtain a cipher that achieves m bit security against quantum adversaries, while tolerating only a minimal performance penalty. Initial ideas along these lines have recently been presented in [1], where the key-addition in Even-Mansour has been replaced by a different group operation.

References

1. Alagic, G., Russell, A.: Quantum-secure symmetric-key cryptography based on hidden shifts. In: Coron, J.-S., Nielsen, J.B. (eds.) EUROCRYPT 2017. LNCS, vol. 10212, pp. 65–93. Springer, Cham (2017). https://doi.org/10.1007/978-3-319-56617-7_3
2. Albrecht, M.R., Driessen, B., Kavun, E.B., Leander, G., Paar, C., Yalçın, T.: Block ciphers – focus on the linear layer (feat. PRIDE). In: Garay, J.A., Gennaro, R. (eds.) CRYPTO 2014. LNCS, vol. 8616, pp. 57–76. Springer, Heidelberg (2014). https://doi.org/10.1007/978-3-662-44371-2_4
3. Ambainis, A., Bach, E., Nayak, A., Vishwanath, A., Watrous, J.: One-dimensional quantum walks. In: Vitter, J.S., Spirakis, P.G., Yannakakis, M. (eds.) Proceedings on 33rd Annual ACM Symposium on Theory of Computing, Heraklion, Crete, Greece, 6–8 July 2001, pp. 37–49. ACM (2001)
4. Anand, M.V., Targhi, E.E., Tabia, G.N., Unruh, D.: Post-quantum security of the CBC, CFB, OFB, CTR, and XTS modes of operation. In: Takagi, T. (ed.) PQCrypto 2016. LNCS, vol. 9606, pp. 44–63. Springer, Cham (2016). https://doi.org/10.1007/978-3-319-29360-8_4
5. Boneh, D., Zhandry, M.: Secure signatures and chosen ciphertext security in a quantum computing world. In: Canetti, R., Garay, J.A. (eds.) CRYPTO 2013. LNCS, vol. 8043, pp. 361–379. Springer, Heidelberg (2013). https://doi.org/10.1007/978-3-642-40084-1_21
6. Borghoff, J., et al.: PRINCE – a low-latency block cipher for pervasive computing applications. In: Wang, X., Sako, K. (eds.) ASIACRYPT 2012. LNCS, vol. 7658, pp. 208–225. Springer, Heidelberg (2012). https://doi.org/10.1007/978-3-642-34961-4_14
7. Brassard, G., Høyer, P.: An exact quantum polynomial-time algorithm for simon's problem. In: Proceedings of the Fifth Israel Symposium on Theory of Computing and Systems, ISTCS 1997, Ramat-Gan, Israel, 17–19 June 1997, pp. 12–23. IEEE Computer Society (1997)
8. Brassard, G., Hoyer, P., Mosca, M., Tapp, A.: Quantum amplitude amplification and estimation. Contemp. Math. **305**, 53–74 (2002)
9. Deutsch, D., Jozsa, R.: Rapid solution of problems by quantum computation. In: Proceedings of the Royal Society of London A: Mathematical, Physical and Engineering Sciences, vol. 439, pp. 553–558. The Royal Society (1992)
10. Even, S., Mansour, Y.: A construction of a cipher from a single pseudorandom permutation. J. Cryptol. **10**(3), 151–162 (1997)

11. Grover, L.K.: A fast quantum mechanical algorithm for database search. In: Miller, G.L. (ed.) Proceedings of the Twenty-Eighth Annual ACM Symposium on the Theory of Computing, Philadelphia, Pennsylvania, USA, May 22–24, pp. 212–219. ACM (1996)

12. Kaplan, M.: Quantum attacks against iterated block ciphers. CoRR (2014). http://arxiv.org/abs/1410.1434

13. Kaplan, M., Leurent, G., Leverrier, A., Naya-Plasencia, M.: Breaking symmetric cryptosystems using quantum period finding. In: Robshaw, M., Katz, J. (eds.) CRYPTO 2016. LNCS, vol. 9815, pp. 207–237. Springer, Heidelberg (2016). https://doi.org/10.1007/978-3-662-53008-5_8

14. Kaplan, M., Leurent, G., Leverrier, A., Naya-Plasencia, M.: Quantum differential and linear cryptanalysis. IACR Trans. Symmetric Cryptol. **2016**(1), 71–94 (2016)

15. Kilian, J., Rogaway, P.: How to protect DES against exhaustive key search. In: Koblitz, N. (ed.) CRYPTO 1996. LNCS, vol. 1109, pp. 252–267. Springer, Heidelberg (1996). https://doi.org/10.1007/3-540-68697-5_20

16. Kilian, J., Rogaway, P.: How to protect DES against exhaustive key search (an analysis of DESX). J. Cryptology **14**(1), 17–35 (2001)

17. Kuwakado, H., Morii, M.: Quantum distinguisher between the 3-round feistel cipher and the random permutation. In: Proceedings of the IEEE International Symposium on Information Theory, ISIT 2010, 13–18 June 2010, Austin, Texas, USA, pp. 2682–2685. IEEE (2010)

18. Kuwakado, H., Morii, M.: Security on the quantum-type even-mansour cipher. In: Proceedings of the International Symposium on Information Theory and its Applications, ISITA 2012, Honolulu, HI, USA, 28–31 October 2012, pp. 312–316. IEEE (2012)

19. Lipton, R.J., Regan, K.W.: Quantum Algorithms via Linear Algebra: A Primer. The MIT Press, Boston (2014)

20. Mermin, N.: Quantum Computer Science: An Introduction. Cambridge University Press, New York (2007)

21. Mitzenmacher, M., Upfal, E.: Probability and Computing - Randomized Algorithms and Probabilistic Analysis. Cambridge University Press, New York (2005)

22. NIST: Post quantum project. http://csrc.nist.gov/groups/ST/post-quantum-crypto/ Accessed 6 Feb 2017

23. Santoli, T., Schaffner, C.: Using simon's algorithm to attack symmetric-key cryptographic primitives. Quantum Inf. Comput. **17**(1&2), 65–78 (2017)

24. Shor, P.W.: Algorithms for quantum computation: discrete logarithms and factoring. In: 35th Annual Symposium on Foundations of Computer Science, Santa Fe, New Mexico, USA, 20–22 November 1994, pp. 124–134. IEEE Computer Society (1994)

25. Simon, D.R.: On the power of quantum computation. SIAM J. Comput. **26**(5), 1474–1483 (1997)

Quantum Multicollision-Finding Algorithm

Akinori Hosoyamada$^{(\boxtimes)}$, Yu Sasaki$^{(\boxtimes)}$, and Keita Xagawa$^{(\boxtimes)}$

NTT Secure Platform Laboratories, 3-9-11, Midori-cho, Musashino-shi,
Tokyo 180-8585, Japan
{hosoyamada.akinori,sasaki.yu,xagawa.keita}@lab.ntt.co.jp

Abstract. The current paper presents a new quantum algorithm for
finding multicollisions, often denoted by l-collisions, where an l-collision
for a function is a set of l distinct inputs having the same output value.
Although it is fundamental in cryptography, the problem of finding multicollisions has not received much attention *in a quantum setting*. The
tight bound of quantum query complexity for finding 2-collisions of random functions has been revealed to be $\Theta(N^{1/3})$, where N is the size of
a codomain. However, neither the lower nor upper bound is known for
l-collisions. The paper first integrates the results from existing research
to derive several new observations, e.g. l-collisions can be generated only
with $O(N^{1/2})$ quantum queries for a small constant l. Then a new quantum algorithm is proposed, which finds an l-collision of any function that
has a domain size l times larger than the codomain size. A rigorous proof
is given to guarantee that the expected number of quantum queries is
$O\left(N^{(3^{l-1}-1)/(2\cdot 3^{l-1})}\right)$ for a small constant l, which matches the tight
bound of $\Theta(N^{1/3})$ for $l = 2$ and improves the known bounds, say, the
above simple bound of $O(N^{1/2})$.

Keywords: Post-quantum cryptography · Multicollision · Quantum
algorithm · Grover · BHT · Rigorous complexity evaluation · State-of-art

1 Introduction

Finding collisions or multicollisions is a fundamental problem in theoretical computer sciences and one of the most critical problems especially in cryptography.
For given finite sets X and Y with $|Y| = N$, and a function $H\colon X \to Y$, an
l-collision finding problem is to find a set of l distinct inputs $x_1, \ldots, x_l$ such
that $H(x_1) = \cdots = H(x_l)$. Both upper and lower bounds of query and time
complexity of the l-collision finding problem are fundamental and have several
applications in cryptography.

Applications of Multicollisions. We often use the lower bound of query complexity (or the upper bound of the success probability) to prove the security of
cryptographic schemes. Let us consider a cryptographic scheme based on Pseudo-
Random Functions (PRFs). In the security proof, we replace the PRFs with truly

© International Association for Cryptologic Research 2017
T. Takagi and T. Peyrin (Eds.): ASIACRYPT 2017, Part II, LNCS 10625, pp. 179–210, 2017.
https://doi.org/10.1007/978-3-319-70697-9_7

random functions (or random oracles) and show the security of the scheme with the random oracles by information-theoretic arguments. In the latter security arguments, we often use *the lower bound of queries* for finding multicollisions of random functions. For example, Chang and Nandi [CN08] proved the indifferentiability of the chopMD hash function construction; Jaulmes et al. [JJV02] proved the indistinguishability of RMAC; Hirose et al [HIK+10] proved the indifferentiability of the ISO standard lightweight hash function Lesamnta-LW; Naito and Ohta [NO14] improved the indifferentiability of PHOTON and Parazoa hash functions; and Javanovic et al. [JLM14] greatly improved the security lower bounds of authenticated-encryption mode of KeyedSponge. The upper bound of the probability to obtain multicollisions after q queries plays an important role in their proofs.

In addition, studying and improving the upper bound for the l-collision finding problem also help our understanding, which often leads to the complexity of generic attacks. For example, l-collisions are exploited in the collision attack on the MDC-2 hash function construction by Knudsen et al [KMRT09], the preimage attack on the JH hash function by Mendel and Thomsen [MT08], the internal state recovery attack on HMAC by Naito et al [NSWY13], the key recovery attack on iterated Even-Mansour by Dinur et al [DDKS14], and the key recovery attack on LED block cipher by Nikolić et al. [NWW13].

Furthermore, multicollisions also have applications in protocols. An interesting example is a micro-payment scheme, MicroMint [RS96]. Here, a coin is a bit-string the validity of which can be easily checked but hard to produce. In MicroMint, coins are 4-collisions of a function. If 4-collisions can be produced quickly, a malicious user can counterfeit coins.

Existing Results for Multicollisions in Classical Setting. The problem of finding (multi-)collisions has been extensively discussed in the classical setting. Suppose that we can access the function H in the *classical* query; that is, we can send $x \in X$ to the oracle H and obtain $y \in Y$ as $H(x)$. For a random function H, making q queries to H can find the collision of H with probability at most q^2/N. The birthday bound shows when $q \approx N^{1/2}$, we obtain a collision with probability $1/2$. This can be extended to the l-collision case. Suzuki et al. [STKT08] showed that with $N^{(l-1)/l}$ queries the probability of finding an l-collision is upper bounded by $1/l!$ and lower bounded by $1/l! - 1/2(l!)^2$, which shows that the query complexity can be approximated to $N^{(l-1)/l}$ for a small constant l. To be more precise, it is shown that $O\big((l!)^{1/l}N^{(l-1)/l}\big)$ evaluations of the function H finds an l-collision with probability about $1/2$ if $H \colon X \to Y$ is a random function.

The above argument only focuses on the number of queries. To implement the l-collision finding algorithm, the computational cost, T, and the memory amount, S, or their tradeoff should be considered. The simple method needs to store all the results of the queries. Hence, it requires $T = S = N^{1/2}$ for collisions and $T = S = O(N^{(l-1)/l})$ for l-collisions. The collision finding algorithm can be made memoryless by using Floyd's cycle detecting algorithm [Flo67]. However, no such memoryless algorithm is known for l-collisions, thus the researcher's goal is to achieve better complexity with respect to $T \times S$ or to trade T and S for a given $T \times S$.

An l-collision can be found with $T = l \cdot N$ and $S = O(1)$ by running a brute-force preimage attack l times for a fixed target. Although this method achieves better $T \times S$ than the simple method, it cannot trade T for S Joux and Lucks [JL09] discovered the 3-collision finding algorithm with $T = N^{1-\alpha}$ and $S = N^{\alpha}$ for $\alpha < 1/3$ by using the parallel collision search technique. Nikolić and Sasaki [NS16] achieved the same complexity as Joux and Lucks by using an unbalanced meet-in-the-middle attack.

1.1 Collisions and Multicollisions in Quantum Setting

Algorithmic speedup using quantum computers has been actively discussed recently. For example, Grover's seminal result [Gro96] attracted cryptographers' attention because of the quantum speedup of database search. Given a function $F \colon X \to \{0,1\}$ such that there exists a unique $x_0 \in X$ that satisfies $F(x_0) = 1$, Grover's algorithm finds x_0 in $O(|X|^{1/2})$ queries.

This paper discusses the complexity of quantum algorithms in *the quantum query model*. In this model, a function H is given as a black box, and the complexity of quantum algorithms is measured as the number of quantum queries to H. A quantum query model is widely adopted, and previous studies on finding collisions in the quantum setting follow this model [BHT97, Amb07, Bel12, Yue14, Zha15].

Previous research on finding collisions and multicollisions can be classified with respect to two types of dichotomies.

Domain size and codomain size. The domain size and codomain size of the function $H \colon X \to Y$ is a sensitive problem for quantum algorithms. Some quantum algorithms aim to find collisions and multicollisions of H with $|X| \geq |Y|$, while others target H with $|X| < |Y|$. The former algorithms can be directly applied to find collisions and multicollisions of real *hash functions* such as SHA-3. The latter ones mainly target *database search* rather than hash functions. The (multi-)collision search on database can still be converted for hash functions, but it generally requires a huge complexity increase. (On the other hand, the (multi-)collision search for hash functions with $|X| \geq |Y|$ cannot be converted for a database with $|X| < |Y|$.)
Hereafter, we use "H" and "D" to denote the cases with $|X| \geq |Y|$ and $|X| < |Y|$, respectively. We note that our goal is finding a new multicollision algorithm that can be applied to real hash functions, namely the H setting.

Random function and any function. Both in classical and quantum settings, existing algorithms often assume randomness: they can find collisions only on average when H is chosen uniformly at random from $\mathrm{Map}(X, Y) := \{f \mid f \colon X \to Y\}$. If an algorithm finds collisions of *any* function $H \in \mathrm{Map}(X, Y)$, it also finds collisions of randomly chosen functions. Hence algorithms applied to any function are stronger than ones only applied to a random function. Hereafter, we use "Rnd" and "Arb" to denote the cases in which H is chosen uniformly at random and H is chosen arbitrarily, respectively. We note that the Rnd setting is sufficient for our goal and will show that our new algorithm can be applied to the Arb setting.

In the following, we revisit the existing results of collision and multicollision-finding algorithms in the quantum setting.

- Brassard et al. [BHT97] proposed a quantum algorithm **BHT**, which can be classified as H-Arb for 2-collisions. To be more precise, **BHT** finds a 2-collision of any l-to-one function with $O(N^{1/3})$ quantum queries and a memory amount of $O(N^{1/3})$.
- Ambainis [Amb07] studied an element distinctness problem rather than the collision finding problem, but his algorithm can be directly applied to find (multi)collisions of functions. The algorithm is for D-Arb for l-collisions with $O(M^{l/(l+1)})$ quantum queries, where M is the domain size.
- Belovs [Bel12] improved the complexity of Ambainis' algorithm [Amb07].
- Zhandry [Zha15] observed that Ambainis' algorithm [Amb07] can be modified to H-Rnd for 2-collisions with $O(N^{1/3})$ quantum queries, when $|X| = \Omega(N^{1/2})$ and $N = |Y|$.
- Yuen [Yue14] discussed the application of **BHT** when $|X| = |Y|$ and the target function H is weakened to Rnd. The complexity is $O(N^{1/3})$ quantum queries. We do not discuss its details because the discussed case in Yuen's work [Yue14] is a subset of Zhandry's extension of Ambainis' algorithm.
- Regarding the lower bound, $O((N/l)^{1/3})$ of **BHT** to find 2-collisions against l-to-one function was proved to be tight by several researchers [AS04, Amb05, Kut05]. Zhandry proved that $O(N^{1/3})$ for 2-collisions against random function is tight. That is, any quantum algorithm that finds a 2-collision against a random function requires $\Omega(N^{1/3})$ quantum queries [Zha15].[1] Obviously, $\Omega(N^{1/3})$ can also be a lower bound for $l > 2$, but no advanced lower bound is known for $l > 2$. Hülsing et al. [HRS16] studied *quantum generic security* of hash functions by considering quantum query complexity in the quantum random-oracle model. They successfully showed the upper and lower bound of *quantum query complexity* to solve the *one-wayness, second-preimage resistance, extended target-collision resistance*, and their variants.[2] Unfortunately, they did not treat collision and multicollision resistances.

The classifications of the existing algorithms are shown in Table 1. As mentioned earlier, Ambainis' algorithm [Amb07] and its improvement by Belovs [Bel12] originally focused on the database search, but they can be converted into the hash function setting with extra complexity. However, all the other approaches for the hash function setting only analyze 2-collisions. Hence, we can conclude that no quantum algorithm exists that is optimized to find l-collisions for hash functions.

---

[1] Zhandry showed that any quantum algorithm with q quantum queries finds a 2-collision with probability at most $O((q+2)^3/N)$ [Zha15].

[2] For example, they showed that any quantum algorithm with q quantum queries finds a preimage with probability at most $O((q+1)^2/N)$ [HRS16].

Table 1. Summary of existing quantum algorithms to find (multi-)collisions.

	Random function "Rnd"	Arbitrary function "Arb"
Database "D"	Zhandry + Ambainis (2-col)	Ambainis (l-col)
		Belovs (l-col)
Hash "H"	Zhandry + Ambainis (2-col)	BHT (2-col)
	Yuen (2-col)	Converted Ambainis (l-col)
		Converted Belovs (l-col)
		Ours (l-col)

1.2   Our Contributions

In this paper, we study quantum algorithms to find l-collisions against a function $H\colon X \to Y$.

First, the problem of finding l-collisions against hash functions has not received much attention in the literature. Even if the previous work can be directly applied to l-collisions against hash functions, nobody has considered this problem and no generic attack is known. This motivates us to provide a systematization of knowledge about existing quantum algorithms. Namely, we, for the first time in this field, provide the state of the art of the complexity of finding l-collisions against hash functions with a direct application, trivial extension, and simple combination of existing results.

This state of the art sheds light on the problems that require further investigation. For the second but main contribution of this paper, we present a new quantum algorithm to find l-collisions against hash functions.

Our contributions in each part are detailed below.

Systematization of Knowledge (combination of Existing Results)

- Our first observation is that, when H is a random function and $|X| = l|Y|$ for a small constant l, the query complexity of the l-collision finding problem is lowered to $O(N^{1/2})$ by simply applying Grover's algorithm. Hence, any meaningful generic attack in the quantum setting must achieve the query complexity below $O(N^{1/2})$. Intuitively, a preimage of the hash value can be generated with $O(N^{1/2})$ queries in the quantum setting and l-collisions are generated by generating l preimages. This corresponds to the upper bound of $O(N)$ complexity in the classical setting. (Note that this upper bound is for the Rnd setting and does not hold for the Arb setting.)
- The above observation is quite straightforward but useful to measure the effect of other attacks. For example, Ambainis' l-collision search for database [Amb07] can be converted for hash functions with $O(M^{l/(l+1)})$ complexity where M is the domain size. However, this cannot be below $O(N^{1/2})$ for any l. The same applies to the improvement by Belovs [Bel12]. Those converted algorithms can be meaningful only in the Arb setting.

– Zhandry [Zha15] discussed the application of Ambainis' l-collision search in H-Rnd and D-Rnd only for $l = 2$, although it can trivially be extended to $l > 2$. If it is extended, the complexity for $l = 3$ reaches $O(N^{1/2})$. Thus, Zhandry's idea only works for $l = 2$.

– Zhandry [Zha15] considered Ambainis' l-collision search rather than Belovs' improvement [Bel12]. If we consider Zhandry + Belovs, the complexity in H-Rnd for $l = 3$ becomes $O(N^{10/21})$, which is faster than the simple application of Grover's algorithm. Thus, it is a meaningful generic attack. For $l \geq 4$, the complexity of Zhandry + Belovs reaches $O(N^{1/2})$.

– In summary, for the Rnd setting, the tight algorithm with $O(N^{1/2})$ complexity exists for $l = 2$. There is a better generic attack than the simple application of Grover's algorithm for $l = 3$, although the lower bound is unknown. For $l \geq 4$, there is no known algorithm better than the application of Grover's algorithm, and the lower bound is also unknown. For the Arb setting, direct application of Belovs' algorithm is the existing best attack.

New Quantum Multicollision-Finding Algorithm

– Given the above state of the art, our main contribution is a new l-collision finding algorithm with $O\left(N^{(3^{l-1}-1)/2 \cdot 3^{l-1}}\right)$ quantum queries against an arbitrary function $H: X \to Y$ with $|X| = l|Y|$. By applying this algorithm in the Rnd setting, we achieve a speedup compared with the simple upper bound of $O(N^{1/2})$ for any l. The complexity of our algorithm matches the tight bound of $O(N^{1/3})$ for $l = 2$ and is faster than $O(10/21)$ of Zhandry + Belovs for $l = 3$. The complexity of our algorithm for a small constant l is shown in Table 2. The complexities are compared in Fig. 1.

– Unlike other algorithms for Arb, our algorithm asymptotically approaches to $O(N^{1/2})$ as l increases. The previous results by Ambainis [Amb07] asymptotically approaches to $O(M)$, and Belovs [Bel12] asymptotically approaches to $O(M^{3/4})$, respectively, where $M = |X|$. Our algorithm improves these results for $M \geq l \cdot N$. The complexities are compared in Fig. 2 for $M = l \cdot N$.

– The core idea of our algorithm is a sophisticated combination of the 3-collision algorithm in the classical setting by Joux and Lucks [JL09] and the generalized Grover algorithm for the quantum setting [BBHT98].

In short, we recursively call a collision finding algorithm and Grover's algorithm. For example, to generate 3-collisions, we first iterate the 2-collision finding algorithm of $O(N^{1/3})$ complexity $O(N^{1/9})$ times. Then, we search for the preimage of one of $O(N^{1/9})$ 2-collisions by using Grover's algorithm, which runs with $O(N^{4/9})$ complexity. To generate 4-collisions, we iterate the 3-collision finding algorithm of $O(N^{4/9})$ complexity $O(N^{1/27})$ times, then search for the preimage of one of $O(N^{1/27})$ 3-collisions with $O(N^{13/27})$ complexity.

In classical setting, the recursive application of the algorithm of [JL09] has never been discussed in literature. This is because the resulting complexity easily exceeds the information theoretically upper bound of $O(N^{(l-1)n/l})$.

Table 2. Quantum query complexity of our l-collision finding algorithm. Query denotes $\log_N(\text{query})$, which asymptotically approaches $1/2$ as l increases.

l	2	3	4	5	6	7	8
Query	$\frac{1}{3}$	$\frac{4}{9}$	$\frac{13}{27}$	$\frac{40}{81}$	$\frac{121}{243}$	$\frac{364}{729}$	$\frac{1093}{2187}$

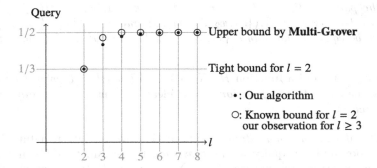

Fig. 1. Quantum query complexity needed to find l-collision in H-Rnd setting. Query denotes $\log_N(\text{query})$.

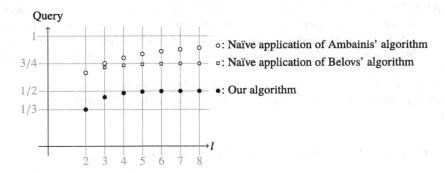

Fig. 2. Quantum query complexity for finding an l-collision in H-Arb setting. Query denotes $\log_N(\text{query})$.

In contrast, no such upper bounds are known in quantum complexity, thus we can obtain advantages with the recursive application.

– Finally, we provide a rigorous complexity evaluation of our algorithm, which is another main focus of this paper. The point of our proof is that lower and upper bounding the number of collisions of H is necessary for lower bound success probability. Our evaluation suggests that our algorithm finds a 2-collision of SHA3-512 with 2^{179} quantum queries and finds a 3-collision with 2^{238} quantum queries.

2   Preliminaries

Notation. We define *l-collision* as follows.

Definition 2.1 (*l*-collision). *Let l be a positive integer. Let X and Y be finite spaces. Let H : X → Y be a function from X to Y. Let $\{x_1, x_2, \ldots, x_l\}$ be a subset of X and let y be an element of Y. We define $(\{x_1, x_2, \ldots, x_l\}, y)$ as an l-collision of H if the pair satisfies $x_i \neq x_j$ for $i \neq j$ and $y = H(x_1) = H(x_2) = \cdots = H(x_l)$. Two l-collisions $c = (\{x_1, x_2, \ldots, x_l\}, y)$ and $c' = (\{x'_1, x'_2, \ldots, x'_l\}, y')$ are said to be equal if and only if $\{x_1, x_2, \ldots, x_l\} = \{x'_1, x'_2, \ldots, x'_l\}$ as sets and $y = y'$.*

If $|X| = l \cdot |Y|$ and a function $H : X \to Y$ satisfies $|H^{-1}(H(x))| = l$ for any $x \in X$, we call H *l-to-one function*. If l is clear by context, we simply call H *regular function*.

Complexity of quantum algorithm. Suppose that we are given a function H as a black box and can query *a quantum state* to the function H; that is, we can send a quantum superposition, say, $\sum_{x \in X} \alpha_x |x\rangle |b\rangle$ to the oracle H and obtain $\sum_{x \in X} \alpha_x |x\rangle |b \oplus H(x)\rangle$. In the *quantum query model*, the complexity of a quantum algorithm is measured by the number of quantum queries to H that the algorithm makes. Many existing studies on collision problems in quantum setting follow this model [BHT97, Amb07, Bel12, Zha15], and the quantum query complexity of collision problems must be understood when we make security proofs in the *quantum random oracle model* [BDF+11], which corresponds to the *random oracle model* in a classical setting. As for time complexity, we will discuss it in Sect. 6.

In the rest of the paper, we assume that readers already have sufficient basic knowledge about the quantum circuit model and omit a detailed explanation of it.

2.1   Grover's Algorithm and Its Generalization

Grover's algorithm [Gro96] was proposed for fast database search in a quantum setting. The problem of database search is modeled as follows:

Problem 2.1 (Quantum Database Search). Suppose that there is a function $F : X \to \{0, 1\}$ such that there is only one element $x_0 \in X$ that satisfies $F(x_0) = 1$. The problem is to find x_0 under the condition that we are allowed to access a quantum oracle *of F.*

Grover's algorithm can solve this problem with high probability, making quantum queries to F for roughly $\sqrt{|X|}$ times. This means that the complexity needed for an exhaustive search in a quantum setting is the square root of one in the classical setting. For example, an exhaustive key search against AES-128 will succeed with approximately 2^{64} quantum queries.

The database search problem described above is naturally extended so that F has more than one preimage of 1. A formal description is given below.

Problem 2.2 (Generalized Quantum Database Search). Suppose that there is a function $F\colon X \to \{0,1\}$ and we are allowed to make quantum queries to F. Then, find x_0 that satisfies $F(x_0) = 1$.

Boyer et al. proposed a quantum algorithm solving this problem [BBHT98]. The advantage of their algorithm is that it can be applied without knowing the number of $x \in X$ that satisfies $F(x) = 1$ in advance.

Theorem 2.1 ([BBHT98] Theorem 3). *Let X be a finite set and $F\colon X \to \{0,1\}$ be a function. Let $t = |\{x \in X \mid F(x) = 1\}|$. If $t \le \frac{3}{4}|X|$, there exists a quantum algorithm **BBHT** that finds $x \in X$ that satisfies $F(x) = 1$ with an expected number of quantum queries to F at most $9/2 \cdot \sqrt{|X|/t}$, without knowing t in advance. When $t = 0$, this algorithm will never abort.*

The above algorithm **BBHT** is applicable only to the case $t \le \frac{3}{4}|X|$, but we want an algorithm that is also applicable to the case $t > \frac{3}{4}|X|$. Now we consider the following algorithm $\mathcal{A}$. $\mathcal{A}$ runs **BBHT**, and simultaneously choose random elements from X independently and uniformly at random, and make queries to F. $\mathcal{A}$ makes exactly one query when **BBHT** makes one query, and $\mathcal{A}$ stops at once if it finds $x \in X$ such that $F(x) = 1$. This algorithm $\mathcal{A}$ is also applicable to the case $t > \frac{3}{4}|X|$, and it finds $x \in X$ such that $F(x) = 1$ with an expected number of quantum queries to F at most

$$\max\left\{2 \cdot \frac{9}{2}\sqrt{\frac{|X|}{t}}, 2 \cdot \frac{4}{3}\right\} = 9\sqrt{\frac{|X|}{t}}.$$

We also call this algorithm **BBHT**. Now we have the following corollary.

Corollary 2.1. *Let X be a finite set and $F\colon X \to \{0,1\}$ be a function. Let $t = |\{x \in X \mid F(x) = 1\}|$. There exists a quantum algorithm **BBHT** that finds $x \in X$ that satisfies $F(x) = 1$ with an expected number of quantum queries to F at most $9 \cdot \sqrt{|X|/t}$, without knowing t in advance. When $t = 0$, this algorithm will never abort.*

3 Systematization of Knowledge on Quantum Multicollision Algorithms

In the classical setting, l-collision on hash functions can be found with $O(N^{(l-1)/l})$ queries for a small constant l.

However, the problem has not received much attention in the quantum setting. This section surveys previous work and integrates the findings of different researchers to make several new observations on this topic.

3.1 Survey of Previous Work

We review the algorithm **BHT** [BHT97] because our new algorithm explained in Sect. 4 is an extension of it. We also survey previous studies, classifying them in two types: element l-distinctness problem (D-Arb), and collision finding problem on random functions (D-Rnd and H-Rnd).

BHT: Collision Finding Problem on l-to-one functions. For simplicity, we describe **BHT** only for the case $l = 2$. Let X, Y be sets that satisfy $|X| = 2 \cdot |Y|$, $|Y| = N$, and $H \colon X \to Y$ be a 2-to-one function.

The basic idea of **BHT** is as follows. First, we choose a parameter k ($k = N^{1/3}$ will turn out to be optimal) and a subset $X' \subset X$ of cardinality k. We then make a list $L = \{(x, H(x))\}_{x \in X'}$. Second, we use the **BBHT** algorithm to find an element $x \in X$ such that there exists $x_0 \in X'$ that satisfies $(x_0, H(x)) \in L$ and $x \neq x_0$, i.e., we try to find a pair $(x_0, H(x_0)) \in L$ that can be extended to a collision $(\{x, x_0\}, H(x_0))$. The precise description of **BHT** is as follows.

Definition 3.1 (BHT(H, k))

1. *Choose an arbitrary subset $X' \subset X$ of cardinality k.*
2. *Make a list $L = \big\{(x, H(x))\big\}_{x \in X'}$ by querying $x \in X'$ to H.*
3. *Sort L in accordance with $H(x)$.*
4. *Check whether L contains a 2-collision, i.e., there exist $(x, H(x)), (y, H(y)) \in L$ such that $x \neq y$ and $H(x) = H(y)$. If so, output the 2-collision $(\{x, y\}, H(x))$. Otherwise proceed to the next step.*
5. *Construct the oracle $F \colon X \to \{0, 1\}$ by defining $F(x) = 1$ if and only if there exists $x_0 \in X'$ such that $(x, H(x_0)) \in L$ and $x \neq x_0$.*
6. *Run **BBHT**(F) to find $\tilde{x} \in X'$ such that $F(\tilde{x}) = 1$.*
7. *Find $x_0 \in X'$ that satisfies $H(\tilde{x}) = H(x_0)$ from the list L. Output the 2-collision $(\{\tilde{x}, x_0\}, H(x_0))$.*

This algorithm makes k quantum queries in Step 2 and $O(\sqrt{N/k})$ quantum queries in Step 6 (in fact, in constructing the list L, we need no advantage of quantum calculation, so queries in Step 2 can also be made *classically* if we are allowed to access a classical oracle of H). Thus, the total number of quantum queries is $O(k + \sqrt{N/k})$, which is minimized when $k = N^{1/3}$. Brassard et al gave the following theorem [BHT97].

Theorem 3.1 ([BHT97, **Theorem 1**]). *Suppose that X and Y are finite sets that satisfy $|X| = 2 \cdot |Y|$, and $H \colon X \to Y$ is a 2-to-one function. Let $N = |Y|$ and k be an integer such that $1 \leq k \leq N$. **BHT** finds a 2-collision of H with an expected quantum query complexity $O(k + \sqrt{N/k})$ and memory complexity $O(k)$. In particular, when $k = N^{1/3}$, **BHT** finds a 2-collision of H with expected quantum query complexity $O(N^{1/3})$ and memory complexity $O(N^{1/3})$.*

Element l-distinctness problem (l-collisions in D-Arb). Consider *the element l-distinctness problem*, in which we are given access to the oracle $H \colon X' \to Y$ to find whether there exist distinct $x_1, \ldots, x_l$ such that $H(x_1) = \cdots = H(x_l)$, i.e., there exits an l-collision of H. Note that H obviously has an l-collision if $|X'| \geq (l-1)|Y|$, and the element l-distinctness problem considers the collision detecting problem on the *database* rather than the hash function.

Ambainis [Amb07] proposed a quantum algorithm based on quantum walks that solves the element l-distinctness problem. His algorithm finds not only whether there exists an l-collision but also an actual l-collision value

$(\{x_1, \ldots, x_l\}, y)$ and can be applied even for finding collisions in $|X'| \geq (l-1)|Y|$. His algorithm requires $O(|X'|^{l/(l+1)})$ quantum queries to H. This algorithm was later improved by Belovs [Bel12], who developed an algorithm that requires $O(|X'|^{1-2^{l-2}/(2^l-1)}) = o(|X'|^{3/4})$ quantum queries.[3]

Although the algorithms by Ambainis and Belovs can be applied to find an l-collision for $|X'| \geq (l-1)|Y|$, the complexity increases as the domain size $|X'|$ increases. These algorithms are inefficient to find collisions of hash functions, since the domain size of cryptographic hash functions is exponentially larger than the codomain size, and we often regard the problem size as dependent on the codomain size $|Y|$ not the domain size $|X'|$. Hence we need another dedicated quantum algorithm to efficiently find collisions of hash functions. The black circles and rectangles in Fig. 2 correspond to the query complexity for naïve applications of Ambainis' algorithm and Belovs' algorithm for hash functions, respectively.

Collision Finding Problem on Random Functions (l-collisions in D-Rnd and H-Rnd). Among variants of the collision problem, the *collision finding problem on random functions* is the most significant problem in the context of cryptography. We introduce algorithms for $l = 2$ in the following.

A modification of **BHT**. Let us consider a modification of **BHT**, denoted **BHT′**, in which we choose a subset X' uniformly at random. This small modification yields two important improvements of **BHT**:

- Brassard et al [BHT97] mentioned that if $|X| \geq l|Y|$, then **BHT′**$(H, N^{1/3})$ finds a collision with quantum query complexity $O(N^{1/3})$ with constant probability.
- Yuen [Yue14] showed that if $|X| = |Y|$ and H is random, then **BHT′**$(H, N^{1/3})$ finds a collision with quantum query complexity $O(N^{1/3})$ with constant probability.

Zhandry's algorithm. Zhandry [Zha15] proposed a quantum algorithm finding a collision with $O(N^{1/3})$-quantum queries even if $|X| = \Omega(N^{1/2})$ and H is random. This improves the restrictions of **BHT** and **BHT′**, $|X| \geq 2|Y|$ [BHT97], or $|X| = |Y|$ and H is random [Yue14].

His algorithm is summarized as follows:

1. Choose a random subset $X' \subset X$ of size $N^{1/2}$.
2. Invoke Ambainis' algorithm for $H|_{X'}: X' \to Y$ and obtain a collision.

The collision exists if H is random because of the birthday bound and the query complexity is $O(|X'|^{2/3}) = O((N^{1/2})^{2/3}) = O(N^{1/3})$.

---

[3] For $l = 3$, there exists a further improvement of time complexity by Belovs et al. [BCJ+13] and Jeffery [Jef14]. However, the quantum query complexity is still $\tilde{O}(|X'|^{1-2^{3-2}/(2^3-1)}) = \tilde{O}(|X'|^{5/7})$.

3.2   New Observations

This section gives our new observations, which are summarized as:

1. In quantum setting, the trivial upper bound for finding an l-collision of a random function is $O(N^{1/2})$.
2. We can find a 3-collision of a random function with quantum query complexity $O(N^{10/21})$.

Observation 1 is obtained by applying a generalized Grover algorithm, and Observation 2 is obtained by combining the idea of Zhandry [Zha15] and the result of Belovs [Bel12].

Trivial Upper-Bound for Finding l-collisions in quantum setting. In the classical setting, the trivial upper bound for finding an l-collision is $O(N)$ because of the following algorithm:

1. Choose an element $x_1 \in X$ uniformly at random.
2. Operate exhaustive search to find x_i for $i = 2, \ldots, l$ that satisfies $H(x_i) = H(x)$.
3. Output $(\{x_1, \ldots, x_l\}, H(x_1))$ as an l-collision.

In the quantum setting, we can replace the exhaustive search with **BBHT**. We call this algorithm **Multi-Grover**, described as follows:

Definition 3.2 (Multi-Grover(H))

1. *Choose an element $x_1 \in X$ uniformly at random and set $L = \{x_1\}$.*
2. *While $|L| < l$, do:*
   (a) *Invoke **BBHT**(F) to find $x \in X$ such that $H(x) = H(x_1)$, where we implement $F \colon X \to \{0, 1\}$ as $F(x) = 1$ if and only if $H(x) = H(x_1)$.*
   (b) *If $x \notin L$, then $L \leftarrow L \cup \{x\}$.*
3. *Output $(L, H(x_1))$ as an l-collision.*

Roughly speaking, each step in the loop requires $O(N^{1/2})$ queries to find x_i. Thus, the total query complexity is $O(N^{1/2})$ for a small constant l. Therefore, to achieve a meaningful improvement, we need to find an l-collision with fewer than $O(N^{1/2})$ quantum queries.

We note that the lower bound of 2-collisions in [Zha15] also applies to multi-collisions. Hence, complexity of any multicollision-finding algorithm is between $O(N^{n/3})$ by 2-collisions and $O(N^{n/2})$ by the trivial upper bound. This corresponds to between birthday bound and preimage bound in the classical setting.

Extension of Element l-distinctness to l-collision. We observe that algorithms for l-distinctness problem can be used to find l-collisions of a random function $H \colon X \to Y$ by extending Zhandry's idea. Let X, Y be finite sets with $|Y| = N$ and $|X| \geq (l!)^{1/l} N^{(l-1)/l}$. Let $H \colon X \to Y$ be a random function.

1. Choose a random subset $X' \subset X$ of size $(l!)^{1/l} N^{(l-1)/l}$
2. Invoke Belovs' algorithm for $H|_{X'}: X' \to Y$ and obtain an l-collision

According to Suzuki et al. [STKT08], $H|_{X'}$ has an l-collision with probability approximately $1/2$. Thus, we observe that Belovs' algorithm can find an l-collision of $H|_{X'}$ with quantum query complexity $O\left((N^{1-2^{l-2}/(2^l-1)})^{(l-1)/l}\right)$.[4] This matches the tight bound $\Theta(N^{1/3})$ for $l = 2$ [Zha15] and gives a new upper bound $O(N^{10/21})$ for $l = 3$, which is crucially lower than the trivial bound $O(N^{1/2})$ (see Sect. 3.2). The white rectangles for $l = 2, 3$ in Fig. 1 correspond to this algorithm.

Note that for the case of H-Rnd, if $l \geq 4$, $(N^{1-2^{l-2}/(2^l-1)})^{(l-1)/l}$ becomes greater than or equal to $N^{1/2}$, which matches the trivial bound for finding l-collisions. Therefore, we have to make another quantum algorithm if we want to find l-collisions for $l \geq 4$ with fewer than $N^{1/2}$ quantum queries.

Our algorithm given in the next section finds an l-collision with the same query complexity as existing work for $l = 2$, and less query complexity than observations above for $l \geq 3$.

4   New Quantum Algorithm for Finding Multicollisions

Now we describe our algorithm for finding multicollisions. We begin with intuitive arguments about how to come up with an algorithm for finding multicollisions by extending the **BHT** algorithm and then give a formal description of our algorithm.

4.1   Intuitive Discussion from 2-Collisions to l-Collisions

First, we intuitively assume that **BHT**(H, k) can find a collision for a function $H: X \to Y$ if $|X| = 2N$ without any modification, because the expected number of preimages $|H^{-1}(y)|$ for each $y \in Y$ is 2 when H is chosen uniformly at random from $\mathrm{Map}(X, Y)$. (See Sect. 3.1 and the original paper [BHT97] for this justification.) Recall that the principle of **BHT**(H, k) is to make a list L of 1-collisions the size of which is k and to extend 1-collisions in L to 2-collisions with the **BBHT** algorithm. Constructing the list L requires k quantum queries, and **BBHT** makes $O(\sqrt{N/k})$ quantum queries, so the total number of quantum queries is $O(k + \sqrt{N/k})$. The optimal k that minimizes $k + \sqrt{N/k}$ satisfies $k = \sqrt{N/k}$, which is $k = N^{1/3}$ and then $O(k + \sqrt{N/k}) = O(N^{1/3})$.

Next we consider to find a 3-collision of a function $H: X \to Y$ under the condition $|X| = 3N$. We take a similar strategy to that of **BHT**, i.e., we make

---

[4] The approach is not improved by picking a smaller random subset. For example, consider finding 3-collision of random function H. If we pick a smaller random subset X' of size N^b with $b < 2/3$,, then the probability that X' contains a 3-collision is roughly N^{3b}/N^2. Thus, we need to iterate Belovs' algorithm N^2/N^{3b} times, where each iteration makes $N^{5b/7}$ queries. Therefore, the total number of queries is $N^{(14-16b)/7} > N^{10/21}$ for $b < 2/3$.

a list L of 2-collisions the size of which is k, and extend 2-collisions in L to 3-collisions with the **BBHT** algorithm. We can find a 2-collision of H with **BHT**$(H, N^{1/3})$, which makes $O(N^{1/3})$ quantum queries. Constructing the list L requires $k \cdot N^{1/3}$ queries, and **BBHT** makes $O(\sqrt{N/k})$ quantum queries, so the total number of quantum queries is $O(k \cdot N^{1/3} + \sqrt{N/k})$. The optimal k that minimizes $k \cdot N^{1/3} + \sqrt{N/k}$ satisfies $k \cdot N^{1/3} = \sqrt{N/k}$. This is $k = N^{1/9}$ and then $O(k \cdot N^{1/3} + \sqrt{N/k}) = O(N^{4/9})$. Hence, our new algorithm improves the bound $O\left(N^{10/21}\right)$ for $l = 3$, which we observed in the previous section.

Similarly to above, we can find l-collisions of a function $H \colon X \to Y$ under the condition $|X| = lN$, i.e., we construct a list L of $(l-1)$-collisions of the size k, and extend $(l-1)$-collisions in L to l-collisions using **BBHT**. By inductive argument, we can find that constructing the list L requires $k \cdot N^{(3^{l-2}-1)/(2 \cdot 3^{l-2})}$ queries, and **BBHT** makes $O(\sqrt{N/k})$ quantum queries, so the total number of quantum queries is $O(k \cdot N^{(3^{l-2}-1)/(2 \cdot 3^{l-2})} + \sqrt{N/k})$. The optimal k that minimizes $k \cdot N^{(3^{l-2}-1)/(2 \cdot 3^{l-2})} + \sqrt{N/k}$ satisfies $k \cdot N^{(3^{l-2}-1)/(2 \cdot 3^{l-2})} = \sqrt{N/k}$, which is $k = N^{1/3^{l-1}}$, and then

$$O\left(k \cdot N^{(3^{l-2}-1)/(2 \cdot 3^{l-2})} + \sqrt{N/k}\right) = O\left(N^{(3^{l-1}-1)/(2 \cdot 3^{l-1})}\right)$$

holds. Again, our new algorithm improves the trivial bound $N^{1/2}$ for l-collisions, $l \geq 4$.

If there exists an algorithm finding l-collisions in the case $|X| = lN$, then we can use it to find l-collisions in the case $|X| > lN$ with the same number of queries and the same memory size, by choosing a subset $X' \subset X$ of size lN and by operating the algorithm on $H|_{X'}$.

4.2   Formal Description of Our Algorithm

Formalizing the above arguments, we obtain a quantum algorithm that finds l-collisions of any function $H \colon X \to Y$ with $|Y| = N$ and $|X| \geq lN$. As briefly introduced in Sect. 4.1, our main idea is to construct a recursive algorithm **MColl**. The algorithm below focuses on the procedure. Complexity analysis of **MColl** will be given in the next section. Although our algorithm is an extension of **BHT**, the definition of the function F is slightly modified to simplify the complexity analysis.

MColl(H, l)

1. If $|X| > lN$, then choose a subset $X' \subset X$ such that $|X'| = lN$ uniformly at random and operate **MColl**$(H|_{X'}, l)$. Otherwise proceed to the next step.
2. If $l = 1$, then choose x from X uniformly at random and output $(\{x\}, H(x))$. Otherwise proceed to the next step.
3. Operate **MColl**$(H, l-1)$ repeatedly for $N^{1/3^{l-1}}$ times and obtain $(l-1)$-collisions $c^{(i)} = (\{x_1^{(i)}, x_2^{(i)}, \ldots, x_{l-1}^{(i)}\}, y^{(i)})$. Store these $(l-1)$-collisions in a list L.

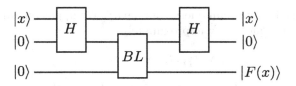

Fig. 3. Quantum circuit of F. H is the function we want to find collisions, and $BL\colon Y \to \{0,1\}$ is the binary function that is defined by $BL(y) = 1$ if and only if there exists $c^{(i)} = \left(\{x_1^{(i)}, \ldots, x_{l-1}^{(i)}\}, y^{(i)}\right) \in L$ such that $y = y^{(i)}$. Here BL corresponds to a quantum circuit $|x\rangle\,|y\rangle \mapsto |x\rangle\,|y \oplus BL(x)\rangle$.

4. Sort L in accordance with $y^{(i)}$.
5. Check whether L contains duplication, i.e., there exist indices $i \neq j$ such that $c^{(i)} = c^{(j)}$. If it does, then stop and restart from Step 3. Otherwise proceed to the next step.
6. Check whether L contains an l-collision. If there is an l-collision, then output it. Otherwise proceed to the next step.
7. Define $F\colon X \to \{0,1\}$ by $F(x) = 1$ if and only if $H(x) = y^{(i)}$ holds for $1 \leq \exists i \leq N^{1/3^{l-1}}$ (F can be implemented in a quantum circuit by calling H twice as shown in Fig. 3).
8. Operate **BBHT**(F). Let $\tilde{x} \in X$ be the obtained answer, which satisfies $F(\tilde{x}) = 1$.
9. Find i_0 that satisfies $H(\tilde{x}) = y^{(i_0)}$ from the list L. If $\tilde{x} \in \{x_1^{(i_0)}, x_2^{(i_0)}, \ldots, x_{l-1}^{(i_0)}\}$, then stop and restart from Step 3. Otherwise output an l-collision $(\{x_1^{(i_0)}, x_2^{(i_0)}, \ldots, x_{l-1}^{(i_0)}, \tilde{x}\}, y^{(i_0)})$.

5 Complexity Analysis of MColl

In this section we analyze the complexity of **MColl**. First, we discuss complexity intuitively in Sect. 5.1 and then give formal arguments and proofs in Sect. 5.2.

5.1 Intuitive Analysis

We intuitively discuss the complexity of our algorithm. In the following, we show that **MColl**(H, l) finds that an l-collision with memory complexity is approximately $N^{1/3}$ and the expected quantum query complexity is at most approximately $l! \cdot N^{(3^{l-1}-1)/(2\cdot 3^{l-1})}$.

First, we consider memory complexity. The claim obviously holds for $l = 1$. In the case $l \geq 2$, the algorithm uses memory only for storing the list L. The memory size needed for L is $N^{1/3^{l-1}}$, which is less than or equal to $N^{1/3}$. Thus, the memory complexity is at most $N^{1/3}$.

Next, we consider quantum query complexity. We upper bound the expected number of quantum queries by approximately

$$Q_l := (N/P_l) \cdot l! \cdot N^{(3^{l-1}-1)/(2 \cdot 3^{l-1})},$$

where P_l is the number of the points in Y that have at least l preimages for a fixed H. Regarding (N/P_l) as constant, we obtain the desired bound.

The claim obviously holds for $l = 1$. Assume that the claim holds for $(l-1)$. Since Step 3 makes $N^{1/3^{l-1}}$-times calls of $\mathbf{MColl}(H, l-1)$, the number of queries made in operating Step 3 once is approximately

$$N^{1/3^{l-1}} \cdot Q_{l-1} = N^{1/3^{l-1}} \cdot \left((N/P_{l-1}) \cdot (l-1)! \cdot N^{(3^{l-2}-1)/(2 \cdot 3^{l-2})}\right)$$
$$= (N/P_{l-1}) \cdot (l-1)! \cdot N^{(3^{l-1}-1)/(2 \cdot 3^{l-1})}. \tag{1}$$

Note that $\mathbf{BBHT}(F)$ finds an element x that satisfies $F(x) = 1$ with approximately $\sqrt{1/p}$ queries to F, where $p = \mathrm{Pr}_{x \leftarrow X}[F(x) = 1]$. Since L contains $N^{1/3^{l-1}}$ elements, here we approximately argue that $p \approx N^{1/3^{l-1}}/|X| \approx N^{(1-3^{l-1})/3^{l-1}}$ and thus the number of queries to F is approximately $\sqrt{1/p}$, which is further approximated to $N^{(3^{l-1}-1)/(2 \cdot 3^{l-1})}$. From the construction of F, the number of queries to H in Step 8 is twice the number of queries to F (see Fig. 3), so the number of queries to H is

$$2 \cdot N^{(3^{l-1}-1)/(2 \cdot 3^{l-1})}. \tag{2}$$

Summing up the numbers of queries in Steps 3 and 8 in Eqs. (1) and (2), we obtain the number of queries to H in the case that $\mathbf{MColl}(H, l)$ does not stop in Steps 5 or 9 as

$$((N/P_{l-1}) \cdot (l-1)! + 2) \cdot N^{(3^{l-1}-1)/(2 \cdot 3^{l-1})} \approx (N/P_{l-1}) \cdot (l-1)! \cdot N^{(3^{l-1}-1)/(2 \cdot 3^{l-1})}. \tag{3}$$

Now let q denote the probability that $\mathbf{MColl}(H, l)$ outputs without being terminated at Steps 5 or 9. Then the overall quantum query complexity is approximately $(1/q) \cdot \left((N/P_{l-1}) \cdot (l-1)! \cdot N^{(3^{l-1}-1)/2 \cdot 3^{l-1}}\right)$. We assume that q equals the probability that an l-collision is outputted in Step 9, since the probability that Step 5 finds a duplication in L is very small when l is a small constant, and ignoring Step 6 only decreases q and increases the overall complexity. Intuitively, we can assume that q equals the product of two probabilities in Step 9:

1. The probability that the $(l-1)$-collision $\{x_1^{(i_0)}, x_2^{(i_0)}, \ldots, x_{l-1}^{(i_0)}\}$ can be extended to an l-collision.
2. The probability that $\tilde{x} \notin \{x_1^{(i_0)}, x_2^{(i_0)}, \ldots, x_{l-1}^{(i_0)}\}$ holds, under the condition in which $\{x_1^{(i_0)}, x_2^{(i_0)}, \ldots, x_{l-1}^{(i_0)}\}$ can be extended to an l-collision.

The probability of 1. is approximately P_l/P_{l-1}, and the probability of 2. is lower bounded by $1/l$. Thus, we have $q \geq (P_l/P_{l-1}) \cdot (1/l)$. Consequently, we have overall approximated complexity

$$(1/q) \cdot \left((N/P_{l-1}) \cdot (l-1)! \cdot N^{(3^{l-1}-1)/(2 \cdot 3^{l-1})}\right),$$

which is at most

$$Q_l = (N/P_l) \cdot l! \cdot N^{(3^{l-1}-1)/2 \cdot 3^{l-1}}.$$

This validates the claim.

5.2   Precise Analysis

The discussion in the previous section is very informal with many approximations. This section gives the precise bound and proof. The main theorem in this section is as follows:

Theorem 5.1. *Let X and Y be finite sets with $|Y| = N$ and $|X| \geq l \cdot |Y|$. Let $H: X \rightarrow Y$ be an arbitrary function. For $l \geq 1$, $\mathbf{MColl}(H, l)$ finds an l-collision with expected quantum query complexity at most*

$$\left(1 + 18\sqrt{2}e\right) \cdot \left(\frac{2lN^{1/3}}{2lN^{1/3} - 1}\right)^{l-1} \cdot l \cdot l! \cdot N^{(3^{l-1}-1)/(2 \cdot 3^{l-1})}$$

and memory complexity $N^{1/3}$.

Remark 5.1. Expected time complexity of $\mathbf{MColl}(H, l)$ is upper bounded by the product of expected quantum query complexity and $O(T_H + \lg N)$, where T_H is the time needed to make a quantum query to H once, using $O(N^{1/3})$ qubits. See Sect. 6 for details.

Proof. It suffices to show that the claim holds in the case $|X| = l \cdot |Y|$. The proof for memory complexity is the same as that we described in the previous section. In the following, we consider quantum query complexity.

For $l \geq 1$, define A_l as

$A_l :=$ the total number of quantum queries to H that $\mathbf{MColl}(H, l)$ makes,

and for $l \geq 2$, define B_l, C_l as

$B_l :=$ the number of quantum queries to H made in Step 3,

$C_l :=$ the number of quantum queries to H made in Step 8.

For $l \geq 2$, we consider a modification of $\mathbf{MColl}(H, l)$, denoted by $\mathbf{MColl}'(H, l)$, which never restarts from Step 3 once it stops in Steps 5 or 9. Let D_l be the total number of quantum queries to H that $\mathbf{MColl}'(H, l)$ makes. Let

success denote the event such that $\mathbf{MColl}'(H, l)$ outputs an l-collision. Then we have

$$E[D_l] = E[B_l] + E[C_l]$$

and

$$E[A_l] = \frac{E[D_l]}{\Pr[\text{success}]} = \frac{E[B_l] + E[C_l]}{\Pr[\text{success}]} \tag{4}$$

for $l \geq 2$. In addition, since $E[B_l] = N^{1/3^{l-1}} \cdot E[A_{l-1}]$ for $l \geq 2$, we have

$$E[A_l] = \frac{N^{1/3^{l-1}} \cdot E[A_{l-1}] + E[C_l]}{\Pr[\text{success}]}. \tag{5}$$

We will show two lemmas on bounds for $E[C_l]$ and $\Pr[\text{success}]$ in Sects. 5.3 and 5.4, respectively:

Lemma 5.1. *For $l \geq 2$, $E[C_l] \leq 18 \cdot \sqrt{\frac{l}{l-1}} \cdot N^{\frac{3^{l-1}-1}{2 \cdot 3^{l-1}}}$ holds.*

Lemma 5.2. *For $l \geq 2$, $\Pr[\text{success}] \geq \frac{l-1}{l} \cdot \frac{1}{l} \cdot \left(1 - \frac{1}{2l} \cdot N^{\frac{2}{3^{l-1}}-1}\right)$ holds.*

Putting them in the inequality (5), we obtain

$$E[A_l] \leq \left(N^{1/3^{l-1}} E[A_{l-1}] + 18\sqrt{\frac{l}{l-1}} N^{\frac{3^{l-1}-1}{2 \cdot 3^{l-1}}}\right) \cdot \frac{l^2 f_l}{l-1}, \tag{6}$$

where

$$f_l = \frac{1}{1 - \frac{1}{2l} \cdot N^{\frac{2}{3^{l-1}}-1}}.$$

Let $\{g_l\}_{1 \leq l}$ be a sequence of numbers defined by $g_1 = 1$ and

$$g_l = \left(g_{l-1} + \frac{18\sqrt{l/(l-1)}}{(l-1) \cdot (l-1)!}\right) f_l$$

for $l \geq 2$.

We show the following claims:

Claim. For $l \geq 1$, $E[A_l] \leq g_l \cdot l \cdot l! \cdot N^{\frac{3^{l-1}-1}{2 \cdot 3^{l-1}}}$ holds.

Claim. For $l \geq 1$, $g_l \leq \left(1 + 18\sqrt{2}e\right) \cdot \left(\frac{2lN^{1/3}}{2lN^{1/3}-1}\right)^{l-1}$ holds.

Combining them, we obtain for $l \geq 1$,

$$E[A_l] \leq \left(1 + 18\sqrt{2}e\right) \cdot \left(\frac{2lN^{1/3}}{2lN^{1/3}-1}\right)^{l-1} \cdot l \cdot l! \cdot N^{\frac{3^{l-1}-1}{2 \cdot 3^{l-1}}}$$

as we wanted. $\qquad\qquad\qquad\qquad\qquad\qquad\qquad\qquad\qquad\qquad\qquad\qquad\qquad\qquad\square$

Proof (Proof of Claim). We give a proof of this claim by induction on l. Since $E[A_1] = 1$, the claim holds for $l = 1$. Now we assume that the claim holds for $(l-1)$. By the induction, we have

$$E[A_l] \leq \left(N^{1/3^{l-1}} E[A_{l-1}] + 18\sqrt{\frac{l}{l-1}} N^{\frac{3^{l-1}-1}{2\cdot3^{l-1}}} \right) \cdot \frac{l^2 f_l}{l-1}$$

$$\leq \left(N^{1/3^{l-1}} \left(g_{l-1} \cdot (l-1) \cdot (l-1)! \cdot N^{\frac{3^{l-2}-1}{2\cdot3^{l-2}}} \right) + 18\sqrt{\frac{l}{l-1}} N^{\frac{3^{l-1}-1}{2\cdot3^{l-1}}} \right) \cdot \frac{l^2 f_l}{l-1}$$

$$= \left(g_{l-1} + \frac{18\sqrt{l/(l-1)}}{(l-1)\cdot(l-1)!} \right) \cdot f_l \cdot l \cdot l! \cdot N^{\frac{3^{l-1}-1}{2\cdot3^{l-1}}}$$

$$= g_l \cdot l \cdot l! \cdot N^{\frac{3^{l-1}-1}{2\cdot3^{l-1}}}$$

and the claim also holds for any $l \geq 1$. $\square$

Proof (Proof of Claim). Finally, we upper bound g_l. Letting $h_l = \frac{18\sqrt{l/(l-1)}}{(l-1)\cdot(l-1)!}$, we have $g_l = (g_{l-1} + h_l)f_l$. Since $f_l \geq 1$ holds for $l \geq 2$, we have

$$g_l = (g_{l-1} + h_l) f_l = ((g_{l-2} + h_{l-1}) f_{l-1} + h_l) f_l \leq (g_{l-2} + h_{l-1} + h_l) f_{l-1} f_l.$$

Continuing calculations, we obtain $g_l \leq \left(1 + \sum_{i=2}^{l} h_i\right) \cdot \prod_{i=2}^{l} f_i$. Thus, we have

$$g_l \leq \left(1 + \sum_{i=2}^{l} h_i\right) \cdot \prod_{i=2}^{l} f_i$$

$$= \left(1 + \sum_{i=2}^{l} \frac{18\sqrt{i/(i-1)}}{(i-1)\cdot(i-1)!}\right) \prod_{i=2}^{l} \frac{1}{1 - \frac{1}{2l} \cdot N^{\frac{2}{3^i-1}-1}}$$

$$\leq \left(1 + \sum_{i=2}^{l} \frac{18\sqrt{2}}{(i-1)!}\right) \prod_{i=2}^{l} \frac{1}{1 - \frac{1}{2l} \cdot N^{-1/3}}$$

$$\leq \left(1 + 18\sqrt{2} \left(\sum_{i=2}^{l} \frac{1}{(i-1)!}\right)\right) \prod_{i=2}^{l} \frac{2lN^{1/3}}{2lN^{1/3} - 1}$$

$$\leq \left(1 + 18\sqrt{2} \left(\sum_{i=0}^{\infty} \frac{1}{i!}\right)\right) \left(\frac{2lN^{1/3}}{2lN^{1/3} - 1}\right)^{l-1} = \left(1 + 18\sqrt{2}e\right) \left(\frac{2lN^{1/3}}{2lN^{1/3} - 1}\right)^{l-1},$$

as we wanted. $\square$

5.3   Proof of Lemma 5.1

Note that

$$E[C_l] \leq E[C_l \mid \text{Step 8 is operated}]$$

holds, and we upper bound the conditional expectation $E[C_l \mid \text{Step 8 is operated}]$. When the algorithm operates in Step 8, it has already passed Steps 5 and 6. Thus, L has neither duplication nor l-collision. In particular, we can assume that L is a list of completely distinct $(l-1)$ collisions of H, i.e., $y^{(i_1)} = y^{(i_2)}$ holds if and only if $i_1 = i_2$. Thus, we have

$$|F^{-1}(1)| = \left| \bigcup_{i=1}^{N^{1/3^{l-1}}} H^{-1}(y^{(i)}) \right| = \sum_{i=1}^{N^{1/3^{l-1}}} |H^{-1}(y^{(i)})| \geq (l-1) \cdot N^{1/3^{l-1}}$$

and

$$\frac{|F^{-1}(1)|}{|X|} \geq \frac{(l-1) \cdot N^{1/3^{l-1}}}{l \cdot N}.$$

Since **MColl** makes two quantum queries to H while making one query to F (See Fig. 3), we have

$$E[C_l \mid \text{Step 8 is operated}] \leq 2 \cdot 9 \cdot \sqrt{\frac{l \cdot N}{(l-1) \cdot N^{1/3^{l-1}}}} = 18 \cdot \sqrt{\frac{l}{l-1}} \cdot N^{\frac{3^{l-1}-1}{2 \cdot 3^{l-1}}}$$

by Corollary 2.1 as we wanted.

5.4   Proof of Lemma 5.2

Next, we lower bound $\Pr[\mathsf{success}]$. Note that

$$\Pr[\mathsf{success}] = \Pr\left[c^{(i)} \neq c^{(j)} \text{ for } i \neq j \right] \cdot \Pr\left[\mathsf{success} \middle| c^{(i)} \neq c^{(j)} \text{ for } i \neq j \right]$$

holds.

We need two lemmas. For the proof of Lemma 5.3, we refer readers to Shoup's textbook [Sho08]. The proof of Lemma 5.4 is given in Appendix A.

Lemma 5.3 ([Sho08, Theorem 8.26]). *Let $[d]$ be the set of integers $\{1, 2, \ldots, d\}$, and $[d]^{\times n}$ be the n-array Cartesian power set of $[d]$ for positive integers d, n. If $s = (s_1, s_2, \ldots, s_n)$ is chosen uniformly at random from $[d]^{\times n}$, then the probability that $s_i \neq s_j$ holds for all $i \neq j$ is lower bounded by $1 - n^2/(2d)$.*

Lemma 5.4. *Let X and Y be finite sets with $|Y| = N$ and $|X| = lN$. Let H be a function from X to Y. Then the number of l-collisions and $(l-1)$-collisions of H are greater than or equal to N and lN, respectively.*

First, we lower bound $\Pr\left[c^{(i)} \neq c^{(j)} \text{ for } i \neq j \right]$. From the construction of **MColl**, we can assume that $\mathbf{MColl}(H, l-1)$ outputs an $(l-1)$-collision of H uniformly at random. Thus, we can assume that elements $c^{(i)} \in L$ are chosen independently and uniformly at random from the set of $(l-1)$-collisions of H. By Lemma 5.4, the number of $(l-1)$-collisions of H is at least $l \cdot N$. Moreover, if n is fixed,

$1 - n^2/2d$ is a monotonically increasing function on d. Therefore, by Lemma 5.3, we have

$$\Pr\left[c^{(i)} \neq c^{(j)} \text{ for } i \neq j\right] \geq 1 - \frac{(N^{1/3^{l-1}})^2}{2lN} = 1 - \frac{1}{2l} \cdot N^{\frac{2}{3^{l-1}}-1}. \tag{7}$$

Second, we lower bound $\Pr\left[\mathsf{success}|c^{(i)} \neq c^{(j)} \text{ for } i \neq j\right]$. Note that the event $\mathsf{success}$ occurs if and only if

$$y^{(i)} = y^{(j)} \text{ for some } i \neq j, \tag{8}$$

or

$$c^{(i_0)} \text{ can be extended to an } l\text{-collision, and } \tilde{x} \notin \left\{x_1^{(i_0)}, x_2^{(i_0)}, \ldots, x_{l-1}^{(i_0)}\right\}. \tag{9}$$

occurs. Recall that $\tilde{x}$ is the output of Step 8 and i_0 is an index satisfying $H(\tilde{x}) = y^{(i_0)}$. The event (8) corresponds to the event in which **MColl** finds an l-collision in Step 6, and the event (9) corresponds to the event in which **MColl** finds an l-collision in Step 9.

Now, let $\mathcal{L}$ be all the possible lists L that satisfy $c^{(i)} \neq c^{(j)}$ for $i \neq j$. Let $\mathcal{L}_1 \subset \mathcal{L}$ denote the set of lists in which there exists l-collisions, i.e., there are two indices $i \neq j$ such that $y^{(i)} = y^{(j)}$, and $\mathcal{L}_2 \subset \mathcal{L}$ denotes the set of lists in which there is no l-collision, i.e., $y^{(i)} \neq y^{(j)}$ holds for $i \neq j$. Then we have $\mathcal{L} = \mathcal{L}_1 \coprod \mathcal{L}_2$. **MColl** finds an l-collision in Step 6 if and only if $L \in \mathcal{L}_1$. In the following, we ignore Step 4 and consider that L is not sorted for simplicity.

For a fixed $L \in \mathcal{L}$, let A^L and B^L denote the sets of elements in L that can and cannot be extended to l-collisions, respectively. We have that $L = A^L \coprod B^L$, $|A^L|$ equals the number of $y^{(i)}$ such that $|H^{-1}(y^{(i)})| \geq l$, and $|B^L|$ equals the number of $y^{(i)}$ such that $|H^{-1}(y^{(i)})| = l - 1$. Define $\langle A^L \rangle$, $\langle B^L \rangle$ by

$$\langle A^L \rangle := \left| \bigcup_{c^{(i)}=(\ldots,y^{(i)}) \in A^L} H^{-1}(y^{(i)}) \right| \text{ and } \langle B^L \rangle := \left| \bigcup_{c^{(i)}=(\ldots,y^{(i)}) \in B^L} H^{-1}(y^{(i)}) \right|,$$

which are the numbers of preimages of $y^{(i)}$'s in A^L and B^L, respectively. Note that

$$\Pr[\mathsf{success} \mid c^{(i)} \neq c^{(j)} \text{ for } i \neq j] = \sum_{L \in \mathcal{L}} \Pr[\mathsf{success} \mid L] \Pr[L].$$

holds.

If $L \in \mathcal{L}_2$, then $\mathsf{success}$ occurs if and only if the event (9) occurs, that is, $\tilde{x}$ can be used to construct an l-collision with an $(l-1)$-collision in L. Note that $\tilde{x}$ is chosen uniformly at random from the set

$$\left(\bigcup_{c^{(i)}=(\ldots,y^{(i)}) \in A^L} H^{-1}(y^{(i)}) \right) \bigcup \left(\bigcup_{c^{(i)}=(\ldots,y^{(i)}) \in B^L} H^{-1}(y^{(i)}) \right),$$

and the event (9) occurs if and only if

$$\tilde{x} \in \bigcup_{c^{(i)}=(\ldots,y^{(i)})\in A^L} H^{-1}(y^{(i)}) \wedge \tilde{x} \neq x_j^{(i)} \text{ for all } i \text{ and } j$$

holds. Now we have

$$\Pr\left[\tilde{x} \in \bigcup_{c^{(i)}\in A^L} H^{-1}(y^{(i)}) \,\middle|\, L\right] = \frac{\langle A^L \rangle}{\langle A^L \rangle + \langle B^L \rangle},$$

and

$$\Pr\left[\tilde{x} \neq x_j^{(i)} \text{ for all } i \text{ and } j \,\middle|\, L, \tilde{x} \in \bigcup_{c^{(i)}\in A^L} H^{-1}(y^{(i)})\right] \geq \frac{1}{l},$$

which suggests that

$$\Pr[\text{success} \mid L] = \Pr\left[\tilde{x} \in \bigcup_{c^{(i)}\in A^L} H^{-1}(y^{(i)}) \wedge \tilde{x} \neq x_j^{(i)} \text{ for all } i \text{ and } j \,\middle|\, L\right]$$

$$= \Pr\left[\tilde{x} \in \bigcup_{c^{(i)}\in A^L} H^{-1}(y^{(i)}) \,\middle|\, L\right]$$

$$\cdot \Pr\left[\tilde{x} \neq x_j^{(i)} \text{ for all } i \text{ and } j \,\middle|\, L, \tilde{x} \in \bigcup_{c^{(i)}\in A^L} H^{-1}(y^{(i)})\right]$$

$$\geq \frac{\langle A^L \rangle}{\langle A^L \rangle + \langle B^L \rangle} \cdot \frac{1}{l}.$$

In addition, we have $\langle A^L \rangle \geq l \cdot |A^L|$ since $y^{(i)} \neq y^{(j)}$ holds for $i \neq j$ if $L \in \mathcal{L}_2$. Thus, we have $\langle A^L \rangle \geq l \cdot |A^L| \geq (l-1) \cdot |A^L|$ and $\langle B^L \rangle = (l-1) \cdot |B^L|$. This yields that

$$\frac{\langle A^L \rangle}{\langle A^L \rangle + \langle B^L \rangle} = \frac{1}{1 + \frac{\langle B^L \rangle}{\langle A^L \rangle}} \geq \frac{1}{1 + \frac{(l-1)|B^L|}{(l-1)|A^L|}} = \frac{|A^L|}{|A^L| + |B^L|}.$$

Thus, we have

$$\Pr[\text{success} \mid L] \geq \frac{|A^L|}{|A^L| + |B^L|} \cdot \frac{1}{l} \tag{10}$$

for $L \in \mathcal{L}_2$. Moreover, since $\Pr[\text{success} \mid L] = 1$ for $L \in \mathcal{L}_1$, the inequality (10) also holds for $L \in \mathcal{L}_1$. Therefore, we have

$$\Pr\left[\text{success} \,\middle|\, c^{(i)} \neq c^{(j)} \text{ for } i \neq j\right] = \sum_{L\in\mathcal{L}} \Pr[\text{success} \mid L]\Pr[L]$$

$$\geq \frac{1}{l} \sum_{L\in\mathcal{L}} \frac{|A^L|}{|A^L| + |B^L|} \cdot \Pr[L].$$

Now we use the following lemmas the proofs of which are given in Appendices B and C, respectively.

Lemma 5.5. *Let X, Y be finite sets such that $|X| = l \cdot |Y|$, and H be a function from X to Y. Let A, B denote the sets of $(l-1)$-collisions of H that can and cannot be extended to l-collisions, respectively. Then we have*

$$\frac{|A|}{|A| + |B|} \geq \frac{l-1}{l}.$$

Lemma 5.6. *Let X, Y be finite sets such that $|X| = l \cdot |Y|$, and H be a function from X to Y. Let A, B denote the sets of $(l-1)$-collisions of H that can and cannot be extended to l-collisions, respectively. Then we have*

$$\sum_{L \in \mathcal{L}} \frac{|A^L|}{|A^L| + |B^L|} \cdot \Pr[L] = \frac{|A|}{|A| + |B|}.$$

By the above lemmas, we have

$$\Pr\left[\text{success} \middle| c^{(i)} \neq c^{(j)} \text{ for } i \neq j\right] \geq \frac{l-1}{l} \cdot \frac{1}{l}.$$

Consequently, $\Pr[\text{success}]$ is lower bounded as $\Pr[\text{success}] \geq \frac{l-1}{l} \cdot \frac{1}{l} \cdot \left(1 - \frac{1}{2l} \cdot N^{\frac{2}{3^{l-1}} - 1}\right)$, that completes the proof.

6   Discussions on Time Complexity

The previous section only focused on quantum query complexity. This section discusses time complexity of **MColl**. We measure the unit of time complexity by the number of executions of quantum gates, which operate primary binary calculations on $\lg N$-bit strings such as NOT, AND, OR, and XOR. For a function $F\colon X \to \{0,1\}$, **BBHT** finds an x_0 such that $F(x_0) = 1$ in time $O(\sqrt{|X|/t} \cdot T')$, where $t = |\{x \in X \mid F(x) = 1\}|$ and T' is the time for evaluating F once.

To begin with, we show the following theorem.

Theorem 6.1. *Let X, Y be finite sets with $|Y| = N$ and $|X| \geq l \cdot |Y|$. For any function $H\colon X \to Y$, $\mathbf{MColl}(H, l)$ runs in expected time*

$$C \cdot \left(\frac{2lN^{1/3}}{2lN^{1/3} - 1}\right)^{l-1} \cdot l \cdot l! \cdot N^{(3^{l-1} - 1)/2 \cdot 3^{l-1}} \cdot (T_H + \lg N)$$

for some constant C, using $O(N^{1/3})$ qubits, where T_H denotes the time needed to make a quantum query to H.

Proof. Let A'_l be the running time of $\mathbf{MColl}(H, l)$ for $l \geq 1$. For $l \geq 2$, let B'_l, G'_l, C'_l, K'_l be the running time of Steps 3, 4, 8, and 9, respectively. Similarly to the inequality 4, we have

$$\mathrm{E}[A'_l] = \frac{\mathrm{E}[B'_l] + \mathrm{E}[G'_l] + \mathrm{E}[C'_l] + \mathrm{E}[K'_l]}{\Pr[\text{success}]} \tag{11}$$

for $l \geq 2$. We have

$$E[B_l'] = N^{1/3^{l-1}} \cdot E[A_{l-1}'],$$

$$E[G_l'] = O\left(N^{1/3^{l-1}} \lg N^{1/3^{l-1}}\right) = O\left(N^{1/3^{l-1}} \lg N\right),$$

$$E[K_l'] = O\left(\lg N^{1/3^{l-1}}\right) = O\left(\lg N\right),$$

since Steps 4 and 9 can be done classically. In addition, we have that

$$E[C_l'] = E[C_l] \cdot \left(T_H + O\left(\lg N^{1/3^{l-1}}\right)\right) = O\left(E[C_l] \cdot \left(T_H + \lg N^{1/3^{l-1}}\right)\right)$$

$$= O\left(\sqrt{\frac{l}{l-1}} \cdot N^{\frac{3^{l-1}-1}{2 \cdot 3^{l-1}}} \cdot (T_H + \lg N)\right),$$

which follows from the construction of the quantum circuit of F (See Fig. 3) and the claim below. See Appendix D for details of this claim.

Claim. The quantum circuit BL can be constructed so that it runs in time $O(\lg N^{1/3^{l-1}})$ using $O(N^{1/3^{l-1}})$ qubits.

Eventually, we have

$$E[A_l'] = O\left(\frac{N^{1/3^{l-1}} \cdot E[A_{l-1}'] + N^{1/3^{l-1}} \lg N + \sqrt{\frac{l}{l-1}} \cdot N^{\frac{3^{l-1}-1}{2 \cdot 3^{l-1}}} \cdot (T_H + \lg N) + \lg N}{\Pr[\text{success}]}\right)$$

$$\leq O\left(\frac{N^{1/3^{l-1}} \cdot E[A_{l-1}'] + \sqrt{\frac{l}{l-1}} \cdot N^{\frac{3^{l-1}-1}{2 \cdot 3^{l-1}}}}{\Pr[\text{success}]} \cdot (T_H + \lg N)\right)$$

$$= O\left(\left(N^{1/3^{l-1}} E[A_{l-1}'] + \sqrt{\frac{l}{l-1}} N^{\frac{3^{l-1}-1}{2 \cdot 3^{l-1}}}\right) \cdot \frac{l^2 f_l}{l-1} \cdot (T_H + \lg N)\right).$$

The above equation yields the claim of Theorem 6.1 due to the same argument as that in the proof of Theorem 5.1.

Remark 6.1. From the viewpoint of time complexity, there are a few criticisms of existing quantum 2-collision finding algorithms [GR03, Ber09]. They are based on the observation that *memory size* is essentially the same as *machine size* for quantum machines, since we have to embed data that we use in a quantum algorithm into the quantum circuit of the algorithm.

Note that these criticisms only focus on collisions of *random* functions and thus are invalid when we consider finding collisions of *any* function. Furthermore, the target of these criticisms is *time complexity*, and our main result (Theorem 5.1), which focuses on *quantum query complexity*, is out of the scope of these criticisms.

7   Conclusion

Finding multicollisions is one of the most important problems in cryptology, both for attack and provable security. In the post-quantum era, this problem needs to be studied in a quantum setting to realize quantum-secure cryptographic schemes. This paper systematized knowledge on the multicollision-finding problem in a quantum setting and proposed a new quantum multicollision-finding algorithm. Our algorithm finds an l-collision of *any* function $H\colon X \to Y$, where $|X| \geq l \cdot |Y|$, with expected quantum query complexity $O(N^{(3^{l-1}-1)/2 \cdot 3^{l-1}}) = o(N^{1/2})$ and memory complexity $O(N^{1/3})$ for a small constant l. If our algorithm is applied to the *random* function, the complexity matches the known tight bound for $l = 2$, improves the simple combination of Zhandry and Belovs' results for $l = 3$, and for the first time improves the simple bound of $O(N^{1/2})$ for $l \geq 4$. Getting rid of the condition $|X| \geq l \cdot |Y|$ and proving a lower bound to find an l-collision are left for future work.

The quantum stuff in this paper is encapsulated in Grover's algorithm, and the results can equally well be understood as query complexity given a "Grover black-box" without assuming any knowledge of quantum theory on the reader. We hope this paper encourages researchers in classical setting to actively discuss quantum algorithms.

A   Proof of Lemma 5.4

Define sequence of functions $\{H_i\colon X \to Y\}_{i \geq 0}$ as follows. First, define H_0 by $H_0 = H$. For each $i \geq 0$, if $\left|H_i^{-1}(y)\right| = l$ holds for all $y \in Y$ (i.e. H_i is a regular function), then define H_{i+1} by $H_{i+1} = H_i$. Otherwise, choose $x_1 \in X$ that satisfies $\left|H_i^{-1}(H_i(x_1))\right| > l$ and $y_2 \in Y$ that satisfies $\left|H_i^{-1}(y_2)\right| < l$ and define H_{i+1} by

$$H_{i+1}(x) = \begin{cases} H_i(x) & (x \neq x_1), \\ y_2 & (x = x_1). \end{cases}$$

Note that there exists an index i_0 such that $H_{i_0} = H_{i_0+j}$ holds for all $j \geq 0$ since $|X|, |Y| < \infty$, and $|X| = l \cdot |Y|$.

Let a_i be the number of l-collisions of H_i and k_i be $\left|H_i^{-1}(H_i(x_1))\right|$. When i is incremented, then the number of the preimages of y_2 is incremented, and the number of l-collisions is increased at most 1 accordingly. On the other hand, the number of the preimages of $H_i(x_1)$ is decremented, and the number of l-collisions is decreased by $\binom{k_i}{l} - \binom{k_i-1}{l}$ accordingly. Therefore, we have

$$a_{i+1} \leq a_i - \left(\binom{k_i}{l} - \binom{k_i-1}{l}\right) + 1.$$

Since $k_i = \left|H_i^{-1}(H_i(x_1))\right| > l$,

$$\binom{k_i}{l} - \binom{k_i-1}{l} = \binom{k_i-1}{l} + \binom{k_i-1}{l-1} - \binom{k_i-1}{l-1} = \binom{k_i-1}{l-1} \geq 1$$

holds, and thus we have $a_i \geq a_{i+1}$ for $i \geq 0$. By constructing the sequence $\{H_i\}_{i\geq 0}$, there exists an integer i_0 such that $H_{i_0} = H_{i_0+j}$ holds for all $j \geq 0$. Since H_{i_0} satisfies $\left|H_{i_0}^{-1}(y)\right| = l$ for all $y \in Y$, we have $a_{i_0} = N$. Therefore, $a_0 \geq a_1 \geq \cdots \geq a_{i_0} = N$ holds, which completes the proof for l-collisions.

Next, let b_i denote the number of $(l-1)$-collisions of H_i. Similarly, for the proof for l-collisions, we have

$$b_{i+1} \leq b_i - \left(\binom{k_i}{l-1} - \binom{k_i-1}{l-1} \right) + (l-1).$$

Since $k_i = \left|H_i^{-1}(H_i(x_1))\right| > l$,

$$\binom{k_i}{l-1} - \binom{k_i-1}{l-1} = \binom{k_i-1}{l-1} + \binom{k_i-1}{l-2} - \binom{k_i-1}{l-2} = \binom{k_i-1}{l-2} \geq \binom{l-1}{l-2} = l-1$$

holds, and thus we have $b_i \geq b_{i+1}$ for $i \geq 0$. Since H_{i_0} satisfies $\left|H_{i_0}^{-1}(y)\right| = l$ for all $y \in Y$, we have $b_{i_0} = l \cdot N$. Therefore, $b_0 \geq b_1 \geq \cdots \geq b_{i_0} = l \cdot N$ holds, which completes the proof.

B   Proof of Lemma 5.5

First, we have $|B| \leq N - 1$, since it contradicts the condition $|X| = l \cdot |Y|$ if we assume $|B| \geq N$. Since $|A| + |B| \geq l \cdot N$ holds by Lemma 5.4, we have

$$\frac{|A|}{|A|+|B|} = 1 - \frac{|B|}{|A|+|B|} \geq 1 - \frac{N-1}{l \cdot N} = \frac{l-1+\frac{1}{N}}{l} \geq \frac{l-1}{l},$$

which completes the proof.

C   Proof of Lemma 5.6

Let S be the direct union set of A and B, that is, $S := A \coprod B$. Consider a trial $\mathsf{T}(a, b; k)$ in which we choose k elements independently and uniformly at random from S, and make an list of chosen elements (here we consider k-permutations of $|S|$, rather than a combination), where $a = |A|$ and $b = |B|$.

Since L is an element of $\mathcal{L}$, which is the set of all the possible lists L that satisfy $c^{(i)} \neq c^{(j)}$ for $i \neq j$, L can be regarded as the list made by operating this trial with $k = N^{1/3^{l-1}}$. The sets A^L and B^L correspond to the sets of elements in the list L chosen from A and B, respectively.

We consider trial $\mathsf{T}(a, b; k)$ for non-negative integers a, b, k such that $a \geq 1$ or $b \geq 1$, and $1 \leq k \leq a + b$. In considering the trial $\mathsf{T}(a, b; k)$, we focus on cardinality of sets $a = |A|$ and $b = |B|$, rather than sets A, B themselves. We show the following claim:

Claim. For non-negative integers a, b, k such that $a \geq 1$ or $b \geq 1$, and $1 \leq k \leq a + b$,

$$\mathop{\mathrm{E}}_{a,b,k}\left[|A^L|\right] = \frac{ka}{a+b}$$

holds, where $\mathrm{E}_{a,b,k}$ denotes the expected value corresponding to the trial $\mathsf{T}(a, b; k)$.

If the above claim holds, then we can finish the proof of our lemma, because the statement of our lemma corresponds to the trial $\mathsf{T}(|A|, |B|; N^{1/3^{l-1}})$ and we have

$$\sum_{L \in \mathcal{L}} \frac{|A^L|}{|A^L| + |B^L|} \cdot \Pr[L] = \mathop{\mathrm{E}}_{|A|,|B|,N^{1/3^{l-1}}}\left[\frac{|A^L|}{|A^L| + |B^L|}\right] = \mathop{\mathrm{E}}_{|A|,|B|,N^{1/3^{l-1}}}\left[\frac{|A^L|}{N^{1/3^{l-1}}}\right]$$

$$= \frac{1}{N^{1/3^{l-1}}} \cdot \frac{N^{1/3^{l-1}}|A|}{|A| + |B|} = \frac{|A|}{|A| + |B|}.$$

Now, we prove the claim by induction on $|S| = |A| + |B| = a + b$. If $a + b = 1$, then it is obvious. Assume that the claim holds for $a + b - 1$. For each element $s \in S$, let $\mathcal{L}_s \subset \mathcal{L}$ denote the set of lists the first element of which is s. Then, since $\mathcal{L} = \coprod_{s \in S} \mathcal{L}_s$, we have

$$\mathop{\mathrm{E}}_{k,a,b}\left[|A^L|\right] = \sum_{s \in S} \mathop{\mathrm{E}}_{k,a,b}\left[|A^L| \big| L \in \mathcal{L}_s\right] \cdot \Pr[L \in \mathcal{L}_s]$$

$$= \sum_{s \in A} \mathop{\mathrm{E}}_{k,a,b}\left[|A^L| \big| L \in \mathcal{L}_s\right] \cdot \Pr[L \in \mathcal{L}_s]$$

$$+ \sum_{s \in B} \mathop{\mathrm{E}}_{k,a,b}\left[|A^L| \big| L \in \mathcal{L}_s\right] \cdot \Pr[L \in \mathcal{L}_s].$$

Note that

$$\mathop{\mathrm{E}}_{k,a,b}\left[|A^L| - 1 \big| L \in \mathcal{L}_s\right] = \mathop{\mathrm{E}}_{k-1,a-1,b}\left[|A^L|\right] \text{ for } s \in A,$$

$$\mathop{\mathrm{E}}_{k,a,b}\left[|A^L| \big| L \in \mathcal{L}_s\right] = \mathop{\mathrm{E}}_{k-1,a,b-1}\left[|A^L|\right] \text{ for } s \in B,$$

holds, and by assumption we have

$$\mathop{\mathrm{E}}_{k-1,a-1,b}\left[|A^L|\right] = \frac{(k-1)(a-1)}{a+b-1} \text{ and } \mathop{\mathrm{E}}_{k-1,a,b-1}\left[|A^L|\right] = \frac{(k-1)a}{a+b-1}.$$

Therefore, we have

$$
\mathop{\mathrm{E}}_{k,a,b}\left[|A^L|\right] = \sum_{s \in A} \left(\frac{(k-1)(a-1)}{a+b-1}+1\right) \cdot \Pr[L \in \mathcal{L}_s] + \sum_{s \in B}\left(\frac{(k-1)a}{a+b-1}\right) \cdot \Pr[L \in \mathcal{L}_s]
$$

$$
= a \cdot \left(\frac{(k-1)(a-1)}{a+b-1}+1\right) \cdot \frac{1}{a+b} + b \cdot \frac{(k-1)a}{a+b-1} \cdot \frac{1}{a+b}
$$

$$
= \frac{a(k-1)(a-1)+a(a+b-1)}{(a+b-1)(a+b)} + \frac{ab(k-1)}{(a+b-1)(a+b)}
$$

$$
= \frac{a(k-1)\left((a-1)+b\right)+a(a+b-1)}{(a+b-1)(a+b)}
$$

$$
= \frac{(a(k-1)+a)\,(a+b-1)}{(a+b-1)(a+b)} = \frac{ka}{a+b},
$$

which completes the proof.

D   Constructing Quantum Circuit of *BL*

This section shows that the quantum circuit BL can be constructed so that it runs in time $O(\lg N^{1/3^{l-1}})$, using $O(N^{1/3^{l-1}})$ qubits. We regard that primary operations on $\lg_2 N$-bit strings such as NOT, AND, OR, and XOR take unit time. In the following, we regard that a function f corresponds to a quantum circuit that calculates $|x\rangle\,|y\rangle \mapsto |x\rangle\,|y \oplus f(x)\rangle$.

The quantum circuit BL is constructed as illustrated in Fig. 4, and consists of three kinds of gates: Expand, J, and OR$_{\text{all}}$ (See Fig. 5). Expand: $Y \rightarrow Y^{\times N^{1/3^{l-1}}}$ is the iteration function Expand: $y \mapsto (y, \ldots, y)$, and $J: Y^{\times N^{1/3^{l-1}}} \rightarrow \{0,1\}^{\times N^{1/3^{l-1}}}$ is defined by

$$
J \colon \left(y_1, \cdots, y_{N^{1/3^{l-1}}}\right) \mapsto \left(J_1(y_1), \ldots, J_{N^{1/3^{l-1}}}(y_{N^{1/3^{l-1}}})\right),
$$

where $J_i \colon Y \rightarrow \{0,1\}$ is the function defined by $J_i(y) = 1$ if and only if $y = y^{(i)}$. The binary function OR$_{\text{all}} \colon \{0,1\}^{\times N^{1/3^{l-1}}} \rightarrow \{0,1\}$ calculates the OR of all $N^{1/3^{l-1}}$ input bits.

The circuit of BL illustrated in Fig. 5 runs as follows. The input string y is first sent to the gate Expand, which expands y to $(y, \ldots, y)$ and the output is sent to J. Recall that $y^{(i)}$ is in the list of collisions $c^{(i)} = (\{x_1, \ldots, x_{l-1}\}, y^{(i)})$ that is made by **MColl**. The gate J runs gates J_i in parallel, each of which can access the data $y^{(i)}$ and checks whether y equals $y^{(i)}$, and sends outputs to OR$_{\text{all}}$. The output of J is sent to OR$_{\text{all}}$, which calculates $b_1 \vee \cdots \vee b_{N^{1/3^{l-1}}}$, here each b_i corresponds to the output of J_i. Consequently, OR$_{\text{all}}$ outputs 1 if and only if there exists an index i such that $y = y^{(i)}$, which is the value $BL(y)$. The last two gates, J and Expand, are used to reset ancilla qubits.

Next, we prove that Expand, J, and OR$_{\text{all}}$ run in $O(\lg N^{1/3^{l-1}})$ time, using $O(N^{1/3^{l-1}})$ qubits. Note that if a function can be *classically* calculated by a boolean circuit that has a binary tree structure of depth D and width W, then

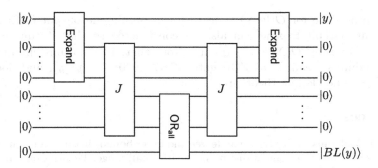

Fig. 4. Quantum circuit of BL.

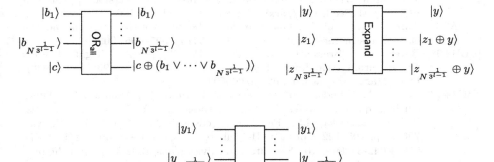

Fig. 5. Quantum circuits of OR_{all}, Expand, and J.

it can be implemented as a quantum circuit that runs in time $O(D)$, using $O(W)$ qubits. This quantum circuit runs primary gates such as NOT, AND, OR, and XOR up to $O(W)$ in parallel and operates $O(D)$ steps.

The function OR_{all} can be classically calculated by a boolean circuit that has a binary tree structure, which has leaves corresponding to inputs and a root corresponding to output, with depth $O(\lg N^{1/3^{l-1}})$ and width $O(N^{1/3^{l-1}})$. Thus, OR_{all} can be constructed as a *quantum* circuit that runs in time $O(\lg N^{1/3^{l-1}})$, using $O(N^{1/3^{l-1}})$ qubits. As for J, it calculates all J_i at the same time in parallel. Since each J_i can be classically calculated in time $O(1)$ using $O(1)$ primary operations, the quantum circuit of J_i can be constructed so that it runs in time $O(1)$, using $O(1)$ qubits. Therefore, J runs in time $O(1)$, using $O(N^{1/3^{l-1}})$ qubits. The function Expand can also be classically calculated using a binary tree structure, which has a root corresponding to the input and leaves corresponding to the output, with depth $O(\lg N^{1/3^{l-1}})$ and width $O(N^{1/3^{l-1}})$. We first copy the input y to obtain (y, y), and then copy (y, y) to obtain (y, y, y, y). Repeating the copy procedure for $O(\lg N)$ times, we can obtain the output of Expand. , the quantum circuit Expand can also be constructed so that it runs in time $O(\lg N^{1/3^{l-1}})$, using $O(N^{1/3^{l-1}})$ qubits. Note that, in quantum circuits, this copy procedure is realized as XOR gate XOR: $|y\rangle\,|0\rangle \mapsto |y\rangle\,|y\rangle$, and does not contradict the no-cloning theorem.

References

[Amb05] Ambainis, A.: Polynomial degree and lower bounds in quantum complexity: collision and element distinctness with small range. Theory Comput. **1**, 37–46 (2005). https://arxiv.org/abs/quant-ph/0305179v3

[Amb07] Ambainis, A.: Quantum walk algorithm for element distinctness. SIAM J. Comput. **37**(1), 210–239 (2007). The preliminary version appeared in FOCS 2004. See https://arxiv.org/abs/quant-ph/0311001

[AS04] Aaronson, S., Shi, Y.: Quantum lower bounds for the collision and the element distinctness problems. J. ACM **51**(4), 595–605 (2004)

[BBHT98] Boyer, M., Brassard, G., Høyer, P., Tapp, A.: Tight bounds on quantum searching. Fortsch. Phys. **46**(4–5), 493–505 (1998). https://arxiv.org/abs/quant-ph/9605034

[BCJ+13] Belovs, A., Childs, A.M., Jeffery, S., Kothari, R., Magniez, F.: Time-efficient quantum walks for 3-distinctness. In: Fomin, F.V., Freivalds, R., Kwiatkowska, M., Peleg, D. (eds.) ICALP 2013. LNCS, vol. 7965, pp. 105–122. Springer, Heidelberg (2013). https://doi.org/10.1007/978-3-642-39206-1_10. See http://arxiv.org/abs/1302.3143 and http://arxiv.org/abs/1302.7316

[BDF+11] Boneh, D., Dagdelen, Ö., Fischlin, M., Lehmann, A., Schaffner, C., Zhandry, M.: Random oracles in a quantum world. In: Lee, D.H., Wang, X. (eds.) ASIACRYPT 2011. LNCS, vol. 7073, pp. 41–69. Springer, Heidelberg (2011). https://doi.org/10.1007/978-3-642-25385-0_3. https://eprint.iacr.org/2010/428

[Bel12] Belovs, A.: Learning-graph-based quantum algorithm for k-distinctness. In: FOCS 2012, pp. 207–216 (2012). https://arxiv.org/abs/1205.1534v2

[Ber09] Bernstein, D.J.: Cost analysis of hash collisions: will quantum computers make SHARCS obsolete? In: SHARCS 2009 (2009)

[BHT97] Brassard, G., Høyer, P., Tapp, A.: Quantum algorithm for the collision problem. CoRR, quant-ph/9705002 (1997). See also Quantum cryptanalysis of hash and claw-free functions. In: LATIN 1998, pp. 163–169 (1998). See https://arxiv.org/abs/quant-ph/9705002

[CN08] Chang, D., Nandi, M.: Improved indifferentiability security analysis of chopMD hash function. In: Nyberg, K. (ed.) FSE 2008. LNCS, vol. 5086, pp. 429–443. Springer, Heidelberg (2008). https://doi.org/10.1007/978-3-540-71039-4_27

[DDKS14] Dinur, I., Dunkelman, O., Keller, N., Shamir, A.: Cryptanalysis of iterated even-mansour schemes with two keys. In: Sarkar, P., Iwata, T. (eds.) ASIACRYPT 2014. LNCS, vol. 8873, pp. 439–457. Springer, Heidelberg (2014). https://doi.org/10.1007/978-3-662-45611-8_23

[Flo67] Floyd, R.W.: Nondeterministic algorithms. J. ACM **14**(4), 636–644 (1967)

[GR03] Grover, L., Rudolph, T.: How significant are the known collision and element distinctness quantum algorithms? CoRR, quant-ph/0309123 (2003). See GR04

[Gro96] Grover, L.K.: A fast quantum mechanical algorithm for database search. In: STOC 1996, pp. 212–219 (1996). https://arxiv.org/abs/quant-ph/9605043

[HIK+10] Hirose, S., Ideguchi, K., Kuwakado, H., Owada, T., Preneel, B., Yoshida, H.: A lightweight 256-bit hash function for hardware and low-end devices: Lesamnta-LW. In: Rhee, K.-H., Nyang, D.H. (eds.) ICISC 2010. LNCS, vol. 6829, pp. 151–168. Springer, Heidelberg (2011). https://doi.org/10.1007/978-3-642-24209-0_10

[HRS16] Hülsing, A., Rijneveld, J., Song, F.: Mitigating multi-target attacks in hash-based signatures. In: Cheng, C.-M., Chung, K.-M., Persiano, G., Yang, B.-Y. (eds.) PKC 2016. LNCS, vol. 9614, pp. 387–416. Springer, Heidelberg (2016). https://doi.org/10.1007/978-3-662-49384-7_15

[Jef14] Jeffery, S.: Frameworks for Quantum Algorithms. Ph.D. thesis, University of Waterloo (2014)

[JJV02] Jaulmes, É., Joux, A., Valette, F.: On the security of randomized CBC-MAC beyond the birthday paradox limit a new construction. In: Daemen, J., Rijmen, V. (eds.) FSE 2002. LNCS, vol. 2365, pp. 237–251. Springer, Heidelberg (2002). https://doi.org/10.1007/3-540-45661-9_19

[JL09] Joux, A., Lucks, S.: Improved generic algorithms for 3-collisions. In: Matsui, M. (ed.) ASIACRYPT 2009. LNCS, vol. 5912, pp. 347–363. Springer, Heidelberg (2009). https://doi.org/10.1007/978-3-642-10366-7_21. https://eprint.iacr.org/2009/305

[JLM14] Jovanovic, P., Luykx, A., Mennink, B.: Beyond $2^{c/2}$ security in sponge-based authenticated encryption modes. In: Sarkar, P., Iwata, T. (eds.) ASIACRYPT 2014. LNCS, vol. 8873, pp. 85–104. Springer, Heidelberg (2014). https://doi.org/10.1007/978-3-662-45611-8_5. https://eprint.iacr.org/2014/373

[KMRT09] Knudsen, L.R., Mendel, F., Rechberger, C., Thomsen, S.S.: Cryptanalysis of MDC-2. In: Joux, A. (ed.) EUROCRYPT 2009. LNCS, vol. 5479, pp. 106–120. Springer, Heidelberg (2009). https://doi.org/10.1007/978-3-642-01001-9_6

[Kut05] Kutin, S.: Quantum lower bound for the collision problem with small range. Theory Comput. 1, 29–36 (2005). https://arxiv.org/abs/quant-ph/0304162

[MT08] Mendel, F., Thomsen, S.S.: An observation on JH-512 (2008). http://ehash.iaik.tugraz.at/uploads/d/da/Jh_preimage.pdf

[NO14] Naito, Y., Ohta, K.: Improved indifferentiable security analysis of PHOTON. In: Abdalla, M., Prisco, R. (eds.) SCN 2014. LNCS, vol. 8642, pp. 340–357. Springer, Cham (2014). https://doi.org/10.1007/978-3-319-10879-7_20

[NS16] Nikolić, I., Sasaki, Y.: A new algorithm for the unbalanced meet-in-the-middle problem. In: Cheon, J.H., Takagi, T. (eds.) ASIACRYPT 2016. LNCS, vol. 10031, pp. 627–647. Springer, Heidelberg (2016). https://doi.org/10.1007/978-3-662-53887-6_23

[NSWY13] Naito, Y., Sasaki, Y., Wang, L., Yasuda, K.: Generic state-recovery and forgery attacks on ChopMD-MAC and on NMAC/HMAC. In: Sakiyama, K., Terada, M. (eds.) IWSEC 2013. LNCS, vol. 8231, pp. 83–98. Springer, Heidelberg (2013). https://doi.org/10.1007/978-3-642-41383-4_6

[NWW13] Nikolić, I., Wang, L., Wu, S.: Cryptanalysis of round-reduced LED. In: Moriai, S. (ed.) FSE 2013. LNCS, vol. 8424, pp. 112–129. Springer, Heidelberg (2014). https://doi.org/10.1007/978-3-662-43933-3_7

[RS96]   Rivest, R.L., Shamir, A.: PayWord and MicroMint: two simple micro-payment schemes. In: Lomas, M. (ed.) Security Protocols 1996. LNCS, vol. 1189, pp. 69–87. Springer, Heidelberg (1997). https://doi.org/10.1007/3-540-62494-5_6

[Sho08]   Shoup, V.: A Computational Introduction to Number Theory and Algebra, 2nd edn. Cambridge University Press, Cambridge (2008)

[STKT08] Suzuki, K., Tonien, D., Kurosawa, K., Toyota, K.: Birthday paradox for multi-collisions. IEICE Trans. **91-A**(1), 39–45 (2008). The preliminary version is in ICISC 2006

[Yue14]   Yuen, H.: A quantum lower bound for distinguishing random functions from random permutations. Quant. Inf. Comput. **14**(13–14), 1089–1097 (2014). https://arxiv.org/abs/1310.2885

[Zha15]   Zhandry, M.: A note on the quantum collision and set equality problems. Quantum Inf. Comput. **15**(7–8), 557–567 (2015)

An Efficient Quantum Collision Search Algorithm and Implications on Symmetric Cryptography

André Chailloux[(✉)], María Naya-Plasencia, and André Schrottenloher

Inria, Paris, France
{Andre.Chailloux,Maria.Naya-Plasencia,Andre.Schrottenloher}@inria.fr

Abstract. The cryptographic community has widely acknowledged that the emergence of large quantum computers will pose a threat to most current public-key cryptography. Primitives that rely on order-finding problems, such as factoring and computing Discrete Logarithms, can be broken by Shor's algorithm ([49]).

Symmetric primitives, at first sight, seem less impacted by the arrival of quantum computers: Grover's algorithm [31] for searching in an unstructured database finds a marked element among 2^n in time $\widetilde{O}(2^{n/2})$, providing a quadratic speedup compared to the classical exhaustive search, essentially optimal. Cryptographers then commonly consider that doubling the length of the keys used will be enough to maintain the same level of security.

From similar techniques, quantum collision search is known to attain $\widetilde{O}(2^{n/3})$ *query complexity* [20], compared to the classical $O(2^{n/2})$. However this quantum speedup is illusory: the actual quantum computation performed is actually more expensive than in the classical algorithm.

In this paper, we investigate quantum collision and multi-target preimage search and present a new algorithm, that uses the amplitude amplification technique. As such, it relies on the same principle as Grover's search. Our algorithm is the first to propose a time complexity that improves upon $O(2^{n/2})$, in a simple setting with a single processor. This time complexity is $\widetilde{O}(2^{2n/5})$ (equal to its query complexity), with a polynomial quantum memory needed ($O(n)$), and a small classical memory complexity of $\widetilde{O}(2^{n/5})$. For multi-target preimage attacks, these complexities become $\widetilde{O}(2^{3n/7})$, $O(n)$ and $\widetilde{O}(2^{n/7})$ respectively. To the best of our knowledge, this is the first proof of an actual quantum time speedup for collision search. We also propose a parallelization of these algorithms. This result has an impact on several symmetric cryptography scenarios: we detail how to improve upon previous attacks for hash function collisions and multi-target preimages, how to perform an improved key recovery in the multi-user setting, how to improve the collision attacks on operation modes, and point out that these improved algorithms can serve as basic tools for some families of cryptanalytic techniques.

In the end, we discuss the implications of these new attacks on post-quantum security.

© International Association for Cryptologic Research 2017
T. Takagi and T. Peyrin (Eds.): ASIACRYPT 2017, Part II, LNCS 10625, pp. 211–240, 2017.
https://doi.org/10.1007/978-3-319-70697-9_8

212 A. Chailloux et al.

Keywords: Post-quantum cryptography · Symmetric cryptography · Collision search · Amplitude amplification

1 Introduction

The emergence of large-scale quantum computing devices would have enormous consequences in physics, mathematics and computer science.

While *quantum hegemony* has yet to be achieved by these machines, the field of post-quantum cryptography has become very active in the last twenty years, as it is of foremost importance to achieve *today* security against possible adversaries from tomorrow. As a consequence, post-quantum asymmetric primitives are being developed and standardized, to protect public-key cryptography against the ravages of Shor's period-finding algorithm ([49]), that provides an exponential advantage to a quantum adversary compared to all known classical factorization algorithms.

Symmetric Cryptography in the Quantum World. In the symmetric setting, Grover's algorithm can speed up quadratically the classical exhaustive key search. As a result, ideal ciphers with k-bit keys would provide only $k/2$-bit security in a post-quantum world. The confidence we have on real symmetric primitives is based on cryptanalysis, i.e. the more we analyze a primitive without finding any weakness, the more trust we can put in it. Until recently, little was known on how quantum adversaries could try to attack symmetric primitives. Therefore, as little was known about their security and confidence they should inspire in a quantum world.

This is why turning classical attacks into quantum attacks was studied in [34,36]. By transposing the weaknesses of an encryption function to the post-quantum world, it is indeed possible to improve on the naive, all-purpose Grover search. How classical attacks can be "quantized" requires, however, a careful analysis.

Besides, if the adversary has stronger capacities than the mere access to a quantum computing device (e.g., if she can ask superposition chosen-plaintext queries), an exponential speedup has been shown to occur for some constructions. This was first noted by Kuwanado and Morii against the Even-Mansour construction ([42]) and the three-round Feistel ([41]), later extended to slide attacks and modes of operation for MACs ([35]). All these attacks use Simon's algorithm [50].

Quantum Collision and Multi-target Preimage Search. In a classical setting, it is well known that finding a collision for a random function H on n bits, i.e. a pair x, y with $x \neq y$ such that $H(x) = H(y)$, costs $O\left(2^{n/2}\right)$ in time and queries [48]. A parallelization of this algorithm was proposed in [47], that has a product of time and memory complexities of also $O\left(2^{n/2}\right)$.

In a quantum setting, an algorithm was presented by Brassard, Høyer and Tapp in [20] that, given superposition query access to a (2-to-1) function H

on n bits, outputs a collision using $\widetilde{O}\left(2^{n/3}\right)$ superposition queries to H. This algorithm is optimal, but only in terms of *query complexity*, while its product of time and memory complexities is as high as $\widetilde{O}\left(2^{2n/3}\right)$ and makes it non-competitive when compared to the classical attack.

Regarding the multi-target preimage search, i.e. given t values of $H(x_i)$ for i from 1 to t, find one out of the t values of the x_i, the best classical algorithm finds a preimage in time $O\left(2^{n-t}\right)$. In the quantum setting, the best algorithm takes time $\widetilde{O}(2^{n/2})$, it consists in fact of finding the preimage of a single *chosen* target with Grover's algorithm.

1.1 Contributions

The contributions we present in this paper are two folded:

Improved Algorithms for Collision and Multi-target Preimage Search. First, we propose a new quantum algorithm for collision search, based on amplitude amplification, which runs in real time $\widetilde{O}\left(2^{2n/5}\right)$ with a single quantum processor, uses $O(n)$ qubits of memory, and $\widetilde{O}\left(2^{n/5}\right)$ bits of classical storage, accessed via a classical processor. The algorithm can be adapted to solve the multi-target preimage problem, with a running time $\widetilde{O}\left(2^{3n/7}\right)$, the same quantum requirements and $\widetilde{O}\left(2^{n/7}\right)$ bits of classical storage.

We also extend these results if quantum parallelization is allowed. These quantum algorithms are the first ones to significantly improve on the best classical algorithms for those two problems. These results also solve an open problem and contradict a conjecture on the complexity of quantum collision and multi-target preimage search, as we will detail in Sect. 7.2.

Implications of these Algorithms. We have studied the impact of these new algorithms on several cryptographic settings, and obtained the following conclusions:

- Hash functions: We are able to improve the best known collision and multi-target preimage attacks when the attacker has only access to a quantum computer.
- Multi-user setting: We are able to improve the best known attacks in a multi-user setting, i.e. recover one key out of many, thanks to the multi-target preimage search improved algorithm. The model for the attacker here is also not very strong, and we only suppose she has access to a quantum computer.
- Operation Modes: Considering collision attacks on operation modes, we are able to improve them with our new algorithms. In this case, the attacker is placed in a more powerful model: she can make superposition queries to a quantum cryptographic oracle. The question of a new data limit enforcement is raised.
- Bricks for cryptanalysis techniques: we show how these algorithms can be used as building blocks in more complex cryptanalytic techniques.

We also discuss the implications of these attacks with respect to security bounds for symmetric cryptographic primitives.

Organisation. In the next section, we detail some security notions that will be considered in our applications and some basic notions of quantum computing. In Sect. 3, we present the considered problems: collision and multi-target preimage search, and we report the state-of-the-art of the previous best known quantum and classical algorithms for solving them. We also present some cryptographic scenarios where these problems are shown to be useful. In Sect. 4, we develop a prerequisite building block of our algorithms, while Sect. 5 is dedicated to detail the new algorithms and possible trade-offs. In Sect. 6 we analyze the impact of these algorithms with respect to the cryptographic scenarios previously presented. A discussion on our results and a conclusion are provided in Sect. 7. In the auxiliary supporting material appended to this submission, we deal with the algorithmic imperfections of the amplitude amplification algorithm.

2   Preliminaries

This section describes some concepts that will be needed for presenting our results: we first provide some classical security notions. Next we describe the two models most commonly considered for quantum adversaries, as both will be considered in applications (Sect. 6). Finally, we will briefly describe the basic quantum computing notions that we will need in order to explain our new algorithms in Sect. 5.

2.1   Some Classical Security Notions

In this section we briefly describe some notions from symmetric key cryptography that will be used through the paper.

Key Recovery Attack. Consider a cipher E_K, that is a pseudo-random permutation parameterized by a secret key K of size k. This cipher takes as input a plaintext of size n and generates a ciphertext of the same size. In the common known-plaintext setting (KPA), the attacker gets access to some pairs of plaintexts and ciphertexts (P_i, C_i). Sometimes the attacker is also allowed to choose the plaintexts: this is called chosen-plaintext attacks (CPA).

It is always possible to perform an exhaustive search on the key and to find the correct one as the one that verifies $C_i = E_K(P_i)$ for all i. The cost of this is 2^k encryptions, and this is the security an ideal cipher provides: the best attack is the generic attack. Therefore, a cipher is *broken* if we are able to recover its key with a complexity smaller than 2^k. The data complexity will be the number of calls to the cryptographic oracle E_K, i.e. the number of pairs (P_i, C_i) needed to successfully perform the attack; the time complexity is the overall time needed to recover the key, and the memory complexity is the amount of memory needed to perform the attack.

Distinguisher and Plaintext Recovery. Key recovery attacks are the strongest, but being able to distinguish the generated ciphertexts from random values is also considered as a weakness. Moreover, when the attacker only captures ciphertexts, she shouldn't able to recover information of any kind on the corresponding plaintexts.

Modes of Operations. In order to be able to treat messages of different lengths and to provide specific security properties, as confidentiality and integrity, block ciphers are typically used in *operation modes*. One of the security properties that these modes should offer is, for instance, not to allow an attacker to identify when the same two blocks have been encrypted under the same key, without having to change the key for each block (which wouldn't be very efficient). Some popular modes are Cipher Block Chaining, CBC [26], or Counter Mode, CTR [25]. It is also possible to build authenticated encryption primitives by using authentication modes, as the Offset Codebook Mode, OCB [39] proposed by Krovetz and Rogaway. Their securities have been widely studied in the classical setting ([8]), as well as recently in a post-quantum setting ([5]).

A plaintext m is split in blocks $m_0 \ldots m_{l-1}$, that will be encrypted with the help of the cipher E_K and combined; the ciphertext is $c = c_0 \ldots c_{l-1}$.

CBC. The Code Block Chaining (CBC) mode of operation defines the ciphertext blocks as follows: $c_0 = E_K(m_0 \oplus IV)$ and for all $i \leq l-1$:

$$c_i = E_K(m_i \oplus c_{i-1})$$

where IV is a public initial value.

The block size being n, some restrictions on the maximal number of blocks encrypted under the same key must be enforced. Indeed, the birthday paradox implies that after recovering $2^{n/2}$ encrypted blocks, there is a high (and constant) probability that two of them are equal, leading to:

$$E_K(m_i \oplus c_{i-1}) = E_K(m_j \oplus c_{j-1}) \ .$$

And since E_K is a permutation, we get $m_i \oplus c_{i-1} = m_j \oplus c_{j-1}$ hence $m_i \oplus m_j$, the XOR of two plaintext blocks, from the knowledge of the ciphertexts.

CTR. In the counter mode (CTR), blocks m_i are encrypted as $c_i = E_K(IV \oplus i) \oplus m_i$ where IV is an initial public value, and i is a counter initialized to zero. As all the inputs of the encryption function are different, we won't have collisions due to the birthday paradox as in the CBC case, but this lack of collisions can exploited to distinguish the construction if more than the $2^{n/2}$ recommended bound of data was generated with the same key.

2.2 Quantum Adversary Models

In this section we describe and justify the two models most commonly considered for quantum adversaries. The application scenarios described in Sect. 6 will use both of them.

Model Q_1. The adversary has access to a quantum computer: this is the case, for instance, in [15,18,51,54]. The adversary can query a quantum random oracle with arbitrary superpositions of the inputs, but is only able to make classical queries to a classical encryption oracle (and therefore no quantum superposition queries to the cryptographic oracle).

Model Q_2. In this case, the adversary is allowed to perform quantum superposition queries to a remote quantum cryptographic oracle (qCPA): she obtains a superposition of the outputs. This model has been considered for instance in [15,16,23,29,35,52]. This is a strong model, but it has the advantages of being simple, inclusive of any possible intermediate and more realistic model, and achievable. In particular, in several of these publications, secure constructions were provided for this scenario.

2.3 Quantum Computing

In this section we provide some basic notions from quantum computing that will be used through the paper. The interested reader can see [46] for a detailed introduction to quantum computing.

Quantum Oracles for Functions. Any function $f : \{0,1\}^n \to \{0,1\}^m$ with a known circuit description can be efficiently implemented as a quantum unitary O_f on $n + m$ qubits, with:

$$O_f : |x\rangle |y\rangle \mapsto |x\rangle |y \oplus f(x)\rangle \ .$$

The quantum running time of O_f is twice[1] the running time of f.

Projection Oracle. Let P a projector acting on n qubits. We define O_P as the following unitary acting on $n + 1$ qubits

$$O_P(|\psi\rangle |b\rangle) := \begin{cases} |\psi\rangle |b \oplus 1\rangle & \text{if } |\psi\rangle \in Im(P) \\ |\psi\rangle |b\rangle & \text{if } |\psi\rangle \in Ker(P) \end{cases} .$$

The above expression defines O_P on a basis of the $n + 1$ qubit pure states and O_P is therefore defined for all states by linearity.

Amplitude Amplification. One of the main tools we will use in our algorithms is the quantum amplitude amplification routine.

---

[1] Computing f makes use of ancillary (additional) qubits. Properly initialized to $|0\rangle$, those end up in a state $|g(x)\rangle$ that cannot be simply dismissed: instead, by *uncomputing*, we can restore these qubits to their initial state $|0\rangle$ and make sure that the oracle has no side-effects.

Theorem 1 ([19], **Quantum amplitude amplification**). Let P a projector acting on n qubits and O_P a projection oracle for P. Let $\mathcal{A}$ be a quantum unitary that produces a state $|\phi\rangle = \alpha |\phi_P\rangle + \beta |\phi_P^\perp\rangle$ where $|\phi_P\rangle \in Im(P)$ and $|\phi_P^\perp\rangle \in Ker(P)$. Notice that $tr(P |\phi\rangle\langle\phi|) = |\alpha|^2$. We note $|\alpha| = \sin(\theta)$ for some $\theta \in [0, \pi/2]$. There exists a quantum algorithm that:

- Consists of exclusively $N = \lfloor \frac{\pi}{4\theta} - \frac{1}{2} \rfloor$ calls to $O_P, O_P^\dagger, \mathcal{A}, \mathcal{A}^\dagger$ and a final measurement.
- Produces a quantum state close to $|\phi_P\rangle$.

The algorithm $\mathcal{A}$ is called the setup and the projection P the projector of the quantum amplification algorithm. This whole procedure will be denoted

$$\texttt{QAA}(\texttt{setup}, \texttt{proj}) = \texttt{QAA}(\mathcal{A}, P)$$

and its running time is

$$N \left(|\mathcal{A}|_{RT} + |O_P|_{RT} \right) \ .$$

where the notation $|\cdot|_{RT}$ represents the running time of the respective algorithms. If both $\mathcal{A}$ and O_P can be done in time polynomial in n, the above is $\widetilde{O}(N)$.

This projection P can be sometimes characterized by a test function f, such that $|x\rangle \in Im(P)$ when $f(x) = 1$ and $|x\rangle \in Ker(P)$ when $f(x) = 0$. Amplitude amplification can be seen as a generalization of Grover's algorithm. Let us briefly show how to retrieve it.

Grover's Algorithm. We are given an efficiently computable function $f : \{0,1\}^n \mapsto \{0,1\}$ and we want to find an element x such that $f(x) = 1$. We take P such that $|x\rangle \in Im(P)$ when $f(x) = 1$ and $|x\rangle \in Ker(P)$ when $f(x) = 0$. O_P can be constructed with a single call to O_f. We use as setup the algorithm $\mathcal{A}$ that produces the state $|\phi\rangle = \frac{1}{2^{n/2}} \sum_{x \in \{0,1\}^n} |x\rangle$. In order to produce $|\phi\rangle$, we perform a Hadamard operation on each qubit, which is very efficient.

We write $|\phi\rangle = \frac{1}{2^{n/2}} \sum_{x:f(x)=1} |x\rangle + \frac{1}{2^{n/2}} \sum_{x:f(x)=0} |x\rangle$. We have $tr(P |\phi\rangle\langle\phi|) = \frac{|\{x:f(x)=1\}|}{2^n}$. Using the above quantum amplitude amplification procedure $\texttt{QAA}(\mathcal{A}, P)$, and by measuring the obtained state, we can find with high probability an element x such that $f(x) = 1$ in time $\widetilde{O}\left(\sqrt{\frac{2^n}{|\{x:f(x)=1\}|}} \right)$.

For most applications, e.g. quantum exhaustive key search, there is only one "marked" element x such that $f(x) = 1$ (e.g., the key). Then Grover search attains a complexity $\widetilde{O}(\sqrt{2^n})$.

Quantum Query, Memory and Time Complexity. Most of the complexity lower bounds on quantum algorithms in the literature, as well as the algorithms that meet these bounds, are based on *query complexity*. As such, they count the number of oracle queries O_f used, where O_f is a quantum oracle for a function f or more generally the data that is being accessed.

Notice that classical queries are a particular case of superposition queries, so we consider them alike in what follows.

However, *query complexity* can be completely different from *time complexity*: the latter represents the number of elementary quantum gates (unitaries) successively applied to a qubit or a qubit register. It has the same flavor as classical time complexity, since it counts elementary operations applied sequentially.

We emphasize that *memory complexity* has a different meaning in the quantum and the classical setting. While classical memory is thought of as a database with fast access, *quantum memory* denotes the number of qubits in the circuit. Having more qubits means that more operations can be applied in parallel, hence decreases the time complexity: it rather corresponds to classical *parallelization*.

3   State-of-the-Art on Collision and Multi-target Preimage Search

The two problems that we consider in a quantum setting, collision search and multi-target preimage search, are described in this section. We also briefly describe the best known classical algorithms for solving them and their complexities, as well as the previously best known quantum algorithms, that we will improve in Sect. 5. We will provide a discussion on the comparison of both previous algorithms. In the end of this section we additionally provide some examples of common applications of this problems on cryptanalysis.

3.1   Studied Problems

In this work we consider the two following problems:

Problem 1 (Collision finding). Given access to a random function $H : \{0,1\}^n \to \{0,1\}^n$, find $x, y \in \{0,1\}^n$ with $x \neq y$ such that $H(x) = H(y)$.

We consider here a random function which models best the cryptographic functions (encryption functions or hash functions) that we want to study.

Problem 2 (Multi-target preimage search). Given access to a random permutation $H : \{0,1\}^n \to \{0,1\}^n$ and a set $T = \{y_1, \ldots, y_{2^t}\}$, find the preimage of one of the y_i by E *i.e.* find $i \in \{1, \ldots, \ell\}$ and $x \in \{0,1\}^n$ such that $E(x) = y_i$.

The above problem can also be considered when replacing H with a random function.

Previous quantum algorithms in the literature ([19]) for Problems 1 and 2 consider sometimes the case of r-to-1 functions. Although we restrict ourselves to the case of random functions and permutations, which is relevant in cryptographic applications, we remark that the algorithms presented below could be rewritten for r-to-1 functions.

3.2   Classical Algorithms to Solve Them

Collision search. The birthday paradox states that if we draw at random $2^{n/2}$ elements $x_i \in \{0,1\}^n$ we will find a collision between two of their images, i.e. $H(x_i) = H(x_j)$, with good probability (i.e. 0.5), and a collision can be found with $O(2^{n/2})$ time and memory. Pollard's rho algorithm [48] allows to reduce the memory complexity to a negligible amount while keeping the same time complexity. No classical algorithm with a single processor exists for finding collisions on a set of 2^n elements with a lower time complexity than $O(2^{n/2})$.

Parallelizing collision search. In [47] a method for reducing the time complexity efficiently through parallelization is proposed. The total amount of computations is slightly increased, and the time-space product is not smaller than $O(2^{n/2})$, but the speed up will be linear. The method is based in considering a common list were all found distinguished points will be stored, until a collision on them is found. The time complexity becomes $O(2^{n/2}/m + 2.5\theta)$, for a case where all collisions are useful and must be located when considering m processors, and θ is the proportion of distinguished points, that will have a direct impact in the memory needs.

Multi-target preimage attacks. With respect to this second problem, the best classical algorithm finds one out of $\ell = 2^t$ targets with an exhaustive search in $\Omega(2^{(n-t)})$ (see for instance [6]). This complexity can be trivially derived by the fact that the probability of finding one out of the 2^t preimages is $\frac{2^t}{2^n}$.

3.3   Previous Quantum Algorithms

Quantum Algorithms for the Collision Problem. The quantum collision search problem was first studied by Brassard, Høyer and Tapp ([20]). Using Grover's algorithm as a subroutine, they showed that the collision problem for a 2-to-1 function f could be solved using $\widetilde{O}(2^{n/3})$ queries to O_f and $\widetilde{O}(2^{n/3})$ *quantum memory*. After, there has been many results on query lower bounds for the collision problem, ([1,2,40]), until a bound $\Omega(2^{n/3})$ was reached. Zhandry also extended the collision problem to random functions, which is relevant in a cryptographic setting ([53]), and proved that this bound still held.

Another related and well studied problem is *element distinctness*, where the question is to decide whether the outputs of the function f are all distinct (or, equivalently, to find a collision if there is at most one). In particular, Ambainis ([3]) presented a quantum walk algorithm for this problem and showed a *time complexity* of $\widetilde{O}(2^{2n/3})$, using $\widetilde{O}(2^{2n/3})$ quantum bits of memory. In [1] $\Omega(2^{2n/3})$ was shown to be a query lower bound for this problem, so those results are essentially optimal. It is known that element distinctness can be reduced to collision by gaining a root in the time complexity, which gives an essentially optimal quantum time and memory of $\widetilde{O}(2^{n/3})$.

Here, we show the original algorithm for collision search from ([20]), that uses Grover. This algorithm has query complexity $\widetilde{O}(2^{n/3})$ but running time

$\widetilde{O}(2^{2n/3})$. It is also possible to reduce the running time of the algorithm below to $\widetilde{O}(2^{n/3})$ by using $\widetilde{O}(2^{n/3})$ quantum processors in parallel.

Algorithm 1. Collision search in a 2-to-1 function using Grover ([20])

Input. Quantum query access to the 2-to-1 function $H : \{0,1\}^n \to \{0,1\}^n$ (oracle O_H).

Membership oracle. We query H on an arbitrary set of $2^{n/3}$ values T, obtaining a set $H(T)$. The algorithm queries a quantum oracle O for testing if $H(x) \in H(T)$. Either each query to O needs quantum time $\widetilde{O}(|T|) = \widetilde{O}(2^{n/3})$, either each is performed in $O(1)$ but the implementation requires quantum memory $\widetilde{O}(2^{n/3})$.

Grover instance. We find x such that $x \notin T \wedge H(x) \in H(T)$ using Grover's algorithm. The search space is the set $\{0,1\}^n$ of size 2^n, while the test function is $g(x) = 1 \iff x \notin T \wedge H(x) \in T$. Computing g requires a query to O_H and a query to O.

The number of good states is $|\{x, H(x) \in H(T) \wedge x \notin T\}|$, expected to be $|T| = 2^{n/3}$, since H is 2-to-1. Hence, this Grover instance needs $\widetilde{O}(\sqrt{2^{n-n/3}}) = \widetilde{O}(2^{n/3})$ iterations.

Limits of existing work. The practical downside of the currently available algorithms for collision is that, although they might require as little as $\widetilde{O}(2^{n/3})$ time, they would need $\widetilde{O}(2^{n/3})$ quantum memory, as in ([3]) or even sometimes $\widetilde{O}(2^{n/3})$ quantum processors as in ([20]), see also ([33]). Contrarily to the classical memory, which is cheap, *quantum memory* is a very costly part in quantum computers. It was argued by Grover and Rudolph ([33]) that a large amount of quantum memory is almost equivalent to a large amount of quantum processors. Even if one disagrees with this statement, it is widely believed that if any implementations of such algorithms will ever exist, they cannot use a large amount of quantum memory. A general discussion on the impracticality of known quantum algorithms for collision was also made by Bernstein in [9].

In summary, even if the collision problem can be solved in quantum time $\widetilde{O}(2^{n/3})$, the current algorithms require the same amount of quantum memory: the quantum time-memory product of such algorithms is $O(2^{2n/3})$, and are arguably considered impractical, even with a functioning quantum computer. The goal of our work is to present a quantum algorithm for those problems with a small number of qubits required, which will clarify the real advantage of a quantum adversary.

Quantum Algorithms for Multi-target Preimage Search. The multi-target preimage search has been much less studied than quantum collisions. As said before, in the classical setting, the best known algorithm requires time $\Omega(2^{n-t})$. In the quantum setting, we present here a basic algorithm, that uses Grover search, inspired by [20]. Independently of our work, Banegas and Bernstein presented at SAC 2017 a method to perform quantum parallel multi-target preimage search ([7]). It has however little to do with the techniques studied in this paper.

Algorithm 2. Multi-target preimage search using Grover ([20])

Input. The set $T = \{y_1, \ldots y_{2^t}\}$ of targets, quantum query access to the permutation $H : \{0,1\}^n \to \{0,1\}^n$ (oracle O_H).

Membership oracle. The algorithm needs a quantum oracle O_T for membership in T. A query to O_T costs quantum time $\widetilde{O}(|T|) = \widetilde{O}(2^t)$ (it can be turned to a cost in quantum memory).

Grover instance. We find x such that $H(x) \in T$ using Grover's algorithm. The search space is the set of possible plaintexts $\{0,1\}^n$ of size 2^n, while the test function is $g(x) = 1 \iff H(x) \in T$. Computing g requires a query to O_H and a query to O_T.

The number of good states is $|\{x, H(x) \in T\}|$, expected to be $|T|$, since H is a permutation. Hence, this Grover instance needs $O(\sqrt{2^{n-t}})$ iterations.

Algorithm 2 has a query complexity of $\widetilde{O}(2^{\frac{n-t}{2}})$. However, the actual time complexity can be much larger. Given a classical description of the set T, the membership oracle O_T costs either $\widetilde{O}(2^t)$ quantum memory, either $\widetilde{O}(2^t)$ quantum time. In any case, the time-memory product of this algorithm is at least $\widetilde{O}(2^t 2^{\frac{n-t}{2}}) = \widetilde{O}(2^{\frac{n+t}{2}})$. Surprisingly (and quite annoyingly), the best tradeoff would be obtained for $t = 0$, i.e. one preimage only.

Comparison of Our Work and Existing Work Using Different Benchmarks. The comparisons between quantum and classical time-memory products are summarized in Tables 1 and 2. Let us now consider different benchmarking scenarios and compare our work with existing work for the collision problem. When considering multiple processors in parallel, we will use the variable s, such that we have access to 2^s processors in parallel.

- If quantum memory is expensive: our quantum algorithms are the only ones that beat classical algorithms with only $O(n)$ quantum bits, with a single quantum processor. Our algorithms also beat existing quantum algorithms if we compare in terms of quantum time-space product.
- If quantum memory becomes as cheap as classical memory, but parallelization is hard then Ambainis' algorithm will have better performances than ours.
- When comparing to classical algorithms, how should we treat classical vs. quantum memory? If we consider just a time-space product (including classical space) then our single processor algorithm has a time-space product of $\widetilde{O}(2^{3n/5})$. However, if this is the quantity of interest then we can take $s = n/5$ in our quantum parallel algorithm and we will obtain a time-space product of $\widetilde{O}(2^{12n/25}) < \widetilde{O}(2^{n/2})$ which again beats the best classical algorithms with

Table 1. Algorithms for collision search. The last line is valid for $s \le n/4$.

	Time	Q-memory	C-storage	# Processors
Improved Grover search ([20])	$2^{n/3}$	$2^{n/3}$	-	$2^{n/3}$
Ambainis' algorithm ([3])	$2^{n/3}$	$2^{n/3}$	-	no
Classical parallelization ([47])	$2^{n/2-s}$	-	2^s	2^s
Our work - single processor	$2^{2n/5}$	$O(n)$	$2^{n/5}$	no
Our work - parallelization	$2^{2n/5-3s/5}$	$O(2^s n)$	$2^{n/5+s/5}$	2^s

Table 2. Algorithms for multi-target preimage search. We consider 2^s processors for the two parallelized algorithms and a single one for the rest.

	Time	Q-memory	C-storage
Classical algorithm	2^{n-t}	-	2^t
Classical algorithm - parallel	2^{n-t-s}	-	$2^t + 2^s$
Naive quantum algorithm	$2^{n/2}$	$O(n)$	-
Our work - single processor	$2^{n/2-t/6} + \min\{2^t, 2^{3n/7}\}$	$O(n)$	$\min\{2^{t/3}, 2^{n/7}\}$
Our work - parallelization	$2^{n/2-t/6-s/2} + \min\{2^t, 2^{\frac{3n-4s}{7}}\}$	$O(2^s n)$	$\min\{2^{t/3}, 2^{n/7+s/7}\}$

this benchmarking. If we consider that classical memory is very cheap then our algorithms compare even better to the classical ones (if we still reasonably consider the parallelization cost).

3.4   Cryptographic Applications of the Problems

Searching for collisions and (multi-target) preimages are recurrent generic problems in symmetric cryptanalysis. We describe here several scenarios whose security would be considerately affected by an improvement in the resolution of these problems by quantum adversaries. The improvements permitted by our algorithms will be detailed in Sect. 6.

Hash Functions. A hash function is a function H that, given a message M of an arbitrary length, returns a value $H(M) = h$ of a fixed length n. They have many applications in computer security, as in message authentication codes, digital signatures and user authentication. Hash functions must be easy to compute. An "ideal" hash function verifies the following properties:

- Collision resistance: Finding two messages M and $M' \ne M$ such that $H(M) = H(M')$ should cost $\Omega(2^{n/2})$ by the birthday paradox [52].
- Second preimage resistance: Given a message M and its hash $H(M)$, finding another message M' such that $H(M) = H(M')$ should cost $\Theta(2^n)$ by exhaustive search.[2]

---

[2] For single pipe constructions this is reduced by the blocks of length of the message M.

- Preimage resistance: From a hash h, finding a message M so that $H(M) = h$ should cost $\Theta(2^n)$ by exhaustive search.

We can see how, if the algorithms for solving collision search or preimages are improved, the offered security of hash functions would be reduced.

Multi-user Setting. In what follows, E_K will always denote a symmetric cipher under key K of length k, of block size n. We consider E_K as a random permutation of bit-strings $E_K : \{0,1\}^n \rightarrow \{0,1\}^n$. We consider the setting where an adversary tries to recover the keys of many users of E_K in parallel. One of the considered scenarios [13,14,22,45] tries to recover one key out of the 2^t more efficiently than in the single key setting. It is easy to see how this can be associated to the multi-target preimage problem: we can for instance consider that all the 2^t users are encrypting the same message, each with a different key, and we recover the corresponding encrypted blocks. This setting seems realistic: it could be the case of users using the CTR operation mode [25], which is one of the two most popular and recommended modes (see for instance [43]), in protocols like for instance TLS [24]. The users consider $IV = 0$ and different secret keys. Recovering one key out of the 2^t would cost in a generic and classical way 2^{k-t} encryptions, for a key of length k. Similar scenarios have been studied in [28] with respect to the Even-Mansour cipher [27] and the Prince block cipher [17].

Collision Attacks on Operation Modes. Using operation modes such as CBC or CTR, block ciphers are secure up to $2^{n/2}$ encryptions with the same key [44], where collisions start to occur and reveal information about the plaintexts (see Sect. 2.1). The recommendation from the community is to limit the number of blocks encrypted with the same key to $\ell \ll 2^{n/2}$, but this is not always respected by standards or actual applications. Such an attack scenario is not merely theoretical, as the authors of [11] pointed out.

They proved that when the birthday bound was only weakly enforced, collision attacks were practical against 64-bit block ciphers when using CBC. In their scenario, the attacker lures the user into sending a great number of HTTP requests to a target website, then captures the server's replies: blocks of sensitive data encrypted under the same key. This attack has time and data complexity $O(2^{n/2})$ (practical when $n = 64$).

Bricks for Cryptanalysis Techniques. Both collision search and multi-target preimage search are often bricks used in some evolved cryptanalysis techniques, as for instance in truncated differential attacks [38] or in impossible differential attacks [12,37] where the adversary needs to find partial output collisions to perform the attacks. Consequently, any acceleration of the algorithms solving these problems would be directly translated in an acceleration of one of the terms of the complexity, and potentially, on an improvement of the complexity of the cryptanalysis technique.

4   The Membership Oracle

In the algorithm of Brassard *et al.*, as well as in the algorithm that will be detailed in Sect. 5, a quantum oracle is needed for computing membership in a large, unstructured set. We formalize here this essential building block.

Definition 1. *Given a set T of 2^t n-bit strings, a classical membership oracle is a function f_T that computes: $f_T(x) = 1$ if $x \in T$ and 0 otherwise.*
   A quantum membership oracle for T is an operator O_T that computes f_T:

$$O_T(|x\rangle |b\rangle) = |x\rangle |b \oplus f_T(x)\rangle \ .$$

The model of computation and memory. The set $T = \{x_1, \dots, x_{2^t}\}$ for which we want to construct a quantum membership oracle is stored in some classical memory, and we require only classical access to it, meaning that for any $i \in [1, \dots, 2^t]$, we can efficiently obtain element x_i. Notice that all x_i are distinct; this is ensured e.g. by the data structure itself or by a preliminary in-place sort in $\widetilde{O}(2^t)$. We use the following quantum operations:

- A quantum creation algorithm that takes a classical input x of n bits, and n qubits initialized at $|0\rangle$ and outputs $|x\rangle$ in this register. This can be done in time n by constructing each qubit of $|x\rangle$ separately.
- A quantum unitary COMP defined as follows:

$$\forall x, y \in \{0, 1\}^n, \forall b \in \{0, 1\}, \ COMP(|x\rangle |y\rangle |b\rangle) := |x\rangle |y\rangle |b \oplus (\delta_{xy})\rangle \ .$$

- A quantum deletion algorithm that takes a classical input x and $|x\rangle$ and outputs $|0\rangle$. This is just done by inverting the creation algorithm.

Using those operations, we describe now how to construct O_T. We start from an input $|x\rangle |b\rangle$ and want to construct $|x\rangle |b \oplus f_T(x)\rangle$. Our construction will be clearly linear and will correspond to a quantum unitary. The idea is simple: for each $x_i \in T$, we will check whether $x = x_i$ and update the second register accordingly.

Algorithm 3. Quantum algorithm for set membership

Start from the input $|\phi_1\rangle := |x\rangle |b\rangle$ on $n + 1$ qbits. For $i = 1$ to 2^t :

- Get element x_i from T and construct a quantum register $|x_i\rangle$ using the creation operator to which we concatenate the current state $|\phi_i\rangle$.
- Apply COMP on the state $|x_i\rangle |\phi_i\rangle$.
- Discard the first register using the deletion operator. Let $|\phi_{i+1}\rangle$ be the remaining state.

The final state $|\phi_{2^t+1}\rangle$ is exactly equal to $|x\rangle |b \oplus f_T(x)\rangle$.

Proposition 1. *The above procedure implements O_T perfectly, in time $n2^t$ using $2n + 1$ bits of quantum memory.*

Proof. The proof is by a straightforward induction. It is easy to see that $|\phi_{i+1}\rangle$ is the state:

$$|x\rangle\, |b \oplus (\delta_{xx_1}) \oplus (\delta_{xx_2}) \ldots \oplus (\delta_{xx_i})\rangle \ .$$

By definition:

$$f_T(x) = 1 \iff x \in T \iff (x = x_1 \vee \ldots \vee x = x_{2^t})$$

which implies (all x_i are distinct):

$$\delta_{xx_1} \oplus \delta_{xx_2} \ldots \oplus \delta_{xx_i} = (x = x_1 \vee \ldots \vee x = x_i) \ .$$

The result follows:

$$|\phi_{2^t+1}\rangle = |x\rangle\, |b \oplus \delta_{xx_1} \oplus \delta_{xx_2} \ldots \oplus \delta_{xx_{2^t}} \rangle = |x\rangle\, |b \oplus f_T(x)\rangle \ .$$

5 Description of Our Quantum Algorithms

In this section we describe our new algorithms for collision and multi-target preimage search. They use three (resp. two) instances of the amplitude amplification procedure (see Theorem 1 in Sect. 2).

5.1 Quantum Algorithm for Collision Finding

Our algorithm, described in Algorithm 4, relies on a balance between the cost of queries to the function and queries to the membership oracle. This balance principle was in fact already considered in [32] to improve the running time of Grover's algorithm. In the algorithm of Brassard *et al.*, when using only logarithmic quantum memory, each query costs $O(2^{n/3})$ time, so there is much room for improvement.

The way we construct the input space S_r^H and the membership oracle f_L^H allow us to decrease the projecting time while increasing the setup time. Independently from the choice of t and r, the quantum memory complexity remains $O(n)$.

Analysis of the Algorithm. In this section, we make some simplifying assumptions. These assumptions are the following:

- The QAA used in our setting outputs exactly the desired state.
- $|S_r^H| \approx 2^{n-r}$.
- Let us define $Sol_f := \{x : f_L^H(x) = 1\}$. We have $|Sol_f| \approx 2 \times 2^{t-r}$. Indeed, each element of L can be mapped with its first coordinate to an element of Sol_f which corresponds to 2^{t-r} elements. Each x such that $(x, H(x)) \notin L$ is in Sol_f with probability $2^{-n+(t-r)}$. Since there are $2^n - 2^{t-r}$ such elements, this corresponds to approximately $2^{t-r} - 2^{2(t-r)-n} \approx 2^{t-r}$ elements.
- We omit all factors polynomial in n and consider that the running time of O_H is 1.

With those assumptions, we get a running time of $2^{\frac{2n}{5}}$ as we show below. If we remove all the above assumptions, we will still obtain a running time of $2^{\frac{2n}{5}}(|O_H|_{RT} + O(n))$.

Algorithm 4. Quantum algorithm for collision finding

The input is a random function $H : \{0,1\}^n \to \{0,1\}^n$ to which we have quantum oracle access. The output is a collision (x, x') such that $x \neq x'$ and $H(x) = H(x')$. The parameters r and t are fixed and will be optimized later. For $r \in [1, \ldots, n]$, let $S_r^H := \{(x, H(x)) : \exists z \in \{0,1\}^{n-r}, H(x) = \underbrace{0 \ldots 0}_{r\ times} \| z\}$.

S_r^H consists of input/output pairs $(x, H(x))$ such that $H(x)$ starts with r zeros. The algorithm works as follows:

1. Construct a list L consisting of 2^{t-r} elements from S_r^H. Let $f_L^H(x) := 1$ if $\exists (x', H(x')) \in L, \ H(x) = H(x')$ and $f_L^H(x) := 0$ otherwise.
2. Apply a quantum amplification algorithm where
   - The setup is the construction of $|\phi_r\rangle := \frac{1}{\sqrt{|S_r^H|}} \sum_{x \in S_r^H} |x, H(x)\rangle$.
   - The projector is a quantum oracle query to $O_{f_L^H}$ meaning that

$$O_{f_L^H}(|x, H(x)\rangle |b\rangle) = |x, H(x)\rangle \left| b \oplus f_L^H(x) \right\rangle.$$

The above quantum amplification algorithm is essentially a Grover search algorithm for f_L^H but on input space S_r^H. The algorithm will output an element $(x, H(x))$ such that $f_L^H(x) = 1$, which means that $\exists (x', H(x')) \in L, \ H(x) = H(x')$. We will finally argue that with constant probability, $(x, H(x)) \notin L$ which will imply $x \neq x'$ and a collision for H.

Probability of success. We constructed a list L of 2^{t-r} elements of the form $(x, H(x))$. The algorithm outputs a random $x \in Sol_f$. Our protocol succeeds if that element is not in L. Since $|L| = 2^{t-r}$ and $|Sol_f| \approx 2 \times 2^{t-r}$, we get a good outcome with probability $\frac{1}{2}$.

Time analysis. Recall that an amplification procedure QAA uses two algorithms: a projection oracle O_P as well as a setup setup that produces a state $|\phi\rangle = \alpha |\phi_P\rangle + \beta |\phi_P^\perp\rangle$.

We decompose our algorithm into four subroutines.

1. Constructing the list L: an element of L can be constructed in time $2^{r/2}$ by applying Grover's search algorithm on the function $f(x) := 1$ if $x \in S_r^H$ and $f(x) := 0$ otherwise. Since the whole list L contains 2^{t-r} elements, it can be constructed in time $2^{t-\frac{r}{2}}$.
2. Constructing $|\phi_r\rangle$: we use an algorithm $\mathcal{A} = \text{QAA}(\text{setup}_\mathcal{A}, \text{proj}_\mathcal{A})$, where $\text{setup}_\mathcal{A}$ builds the superposition $|\phi_0\rangle = \frac{1}{2^{n/2}} \sum_{x \in \{0,1\}^n} |x\rangle$ using a query to O_H and $\text{proj}_\mathcal{A} = \sum_{x \in S_r^H} |x\rangle\langle x|$.
   $tr(P |\phi_0\rangle\langle\phi_0|) = 2^{-r}$ so we have to perform $2^{r/2}$ iterations, *i.e.* make $2^{r/2}$ calls to $\text{setup}_\mathcal{A}$ and $\text{proj}_\mathcal{A}$. Algorithm $\mathcal{A}$ takes therefore time $2^{r/2}$.
3. Constructing $O_{f_L^H}$. The details of this construction appear in Sect. 4. In particular, we saw that $O_{f_L^H}$ runs in time 2^{t-r} by testing sequentially against the elements of L (recall we dismissed the factor n for simplicity).

4. Performing the main amplitude amplification: Algorithm $\mathcal{B} = \text{QAA}(\text{setup}_{\mathcal{B}} = \mathcal{A}, \text{proj}_{\mathcal{B}})$, where $\mathcal{A}$ is the setup routine that constructs state $|\phi_r\rangle$, and $\text{proj}_{\mathcal{B}} = \sum_{x:f_L^H(x)=1} |x\rangle\langle x|$. $O_{\text{proj}_{\mathcal{B}}}$ can be done with 2 calls to $O_{f_L^H}$.

The probability that a random $x \in S_r^H$ satisfies $f_L^H(x) = 1$ is $\frac{|Sol_f|}{|S_r|} = \frac{2*2^{t} - r}{2^{n} - r}$ so $tr(\text{proj}_{\mathcal{B}} |\phi_r\rangle\langle\phi_r|) = \frac{2*2^{t} - r}{2^{n} - r} = 2^{-n+t+1}$ and Algorithm $\mathcal{B}$ makes $2^{\frac{n-t-1}{2}}$ calls to $\mathcal{A}$ and $O_{f_L^H}$. As a result, algorithm $\mathcal{B}$ runs in time:

$$2^{\frac{n-t-1}{2}} (\text{setup}_{\mathcal{B}} + \text{proj}_{\mathcal{B}}) = 2^{\frac{n-t-1}{2}} \left(2^{r/2} + 2^{t-r}\right) .$$

The running time of the procedure is the time to create the list L plus the time to run algorithm $\mathcal{B}$, which is

$$2^{\frac{n-t-1}{2}} \left(2^{r/2} + 2^{t-r}\right) + 2^{t - \frac{r}{2}} .$$

A quick optimization of the above expression imposes $t = \frac{3n}{5}$ and $r = \frac{2t}{3} = \frac{2n}{5}$. This realizes a balance in $\mathcal{B}$ between the cost of the setup and the cost of a projection, and between the cost of $\mathcal{B}$ and the cost of computing L.

This gives a total running time of $2^{\frac{2n}{5}}$, up to a small multiplicative factor in n.

Memory analysis. The quantum amplitude amplification algorithms and the circuit $O_{f_L^H}$ only require quantum circuits of size $O(n)$: the quantum memory (number of qubits) needed is low. As for the classical memory required, the only data we need to store is the list L that contains $2^{t-r} = 2^{\frac{n}{5}}$ elements.

Theorem 2. *Let $H : \{0,1\}^n \to \{0,1\}^n$ be a random function computable efficiently. There exists a quantum algorithm running in time $\tilde{O}\left(2^{\frac{2n}{5}}\right)$, using $\tilde{O}\left(2^{\frac{n}{5}}\right)$ classical memory and $O(n)$ quantum space, that outputs a collision of H.*

5.2   Quantum Algorithm for Multi-target Preimage Search

Here, we are given a function H and a list $L' = \{y_1, \ldots, y_{2^t}\}$ of elements of size 2^t. The goal is to find x such that $\exists y_i, H(x) = y_i$, the preimage of one of them. The algorithm used is very similar to Algorithm 4.

The only difference with respect to the previous algorithm is that the list L' of targets has to be read, even in an online manner, to create the sublist L. This operation will take time 2^t.

Because the rest of the algorithm remains unchanged, the total running time is:

$$2^{\frac{n-t}{2}} \left(2^{\frac{r}{2}} + 2^{t-r}\right) + 2^t$$

which is minimized for $r = \frac{2t}{3}$ and $t = \frac{3n}{7}$. We distinguish 2 cases:

- if $t \le \frac{3n}{7}$, we take $r = \frac{2t}{3}$ and the above running time becomes

$$2^{n/2-t/6} + 2^t \le 2^{n/2-t/6+1}.$$

Algorithm 5. Quantum algorithm for multi-target preimage search

The input is a random permutation $H : \{0,1\}^n \to \{0,1\}^n$ to which we have quantum oracle access, and a list $L' = \{y_1, \ldots, y_{2^t}\}$. The output is the preimage of one of the y_i. For $r \in [1, \ldots, n]$, let
$S_r^H := \{x : \exists z \in \{0,1\}^{n-r}, H(x) = \underbrace{0 \ldots 0}_{r \text{ times}} ||z\}$. S_r^H consists of elements x such
that $H(x)$ starts with r zeros. The algorithm works as follows:

1. Construct a list L consisting of all elements of L' that start with r zeros. L contains 2^{t-r} elements on average (and the deviation is actually small). Let $f_L^H(x) := 1$ if $H(x) \in L$ and $f_L^H(x) := 0$ otherwise.
2. Apply a quantum amplification algorithm where
   - The setup is the construction of $|\phi_r\rangle := \frac{1}{\sqrt{|S_r^H|}} \sum_{x \in S_r^H} |x\rangle$.
   - The projector is a quantum oracle query to $O_{f_L^H}$ meaning that

$$O_{f_l^H}(|x\rangle |b\rangle) = |x\rangle \left|b \oplus f_L^H(x)\right\rangle.$$

The above quantum amplification algorithm is essentially a Grover search algorithm for f_L^H but on input space S_r^H. The algorithm will output an element x such that $f_L^H(x) = 1$, which means that $H(x) \in L$. Notice that we slightly changed the definitions here. In particular, we didn't need to keep the couples $(x, H(x))$ and we could just work on x.

- if $t \geq \frac{3n}{7}$, we truncate the list L' beforehand so that it has $2^{3n/7}$ elements and we apply our algorithm on this list. The running time will therefore be $2^{3n/7}$.

Memory analysis. The only data we need to store is the list L that contains $2^{t-r} = 2^{\frac{t}{3}}$ elements. The reason why we do not have to store all elements of L' is that we can discard all elements of L' that are not in S_r^H as soon as we receive them and locally (L' is analyzed in an online way). The quantum algorithm is still a circuit of size $O(n)$, without external quantum memory.

Theorem 3. *Let* $H : \{0,1\}^n \to \{0,1\}^n$ *be a random permutation. Given a list of* 2^t *elements, with* $t \leq \frac{3n}{7}$, *there exists a quantum algorithm running in time* $\widetilde{O}\left(2^{n/2 - t/6}\right)$, *using* $\widetilde{O}\left(2^{\frac{t}{3}}\right)$ *classical memory and* $O(n)$ *quantum space, that finds the preimage of one of them.*

Theorem 4. *Let* $H : \{0,1\}^n \to \{0,1\}^n$ *be a random permutation. Given a list of* $2^{\frac{3n}{7}}$ *elements, there exists a quantum algorithm running in time* $\widetilde{O}\left(2^{\frac{3n}{7}}\right)$, *using* $\widetilde{O}\left(2^{\frac{n}{7}}\right)$ *classical memory and* $O(n)$ *quantum space, that finds the preimage of one of them.*

A similar analysis can be done with only marginal differences if we replace the random permutation by a random function.

5.3   Parallelization and Time-Space Tradeoff

Assume that the adversary has now 2^s registers of n qubits available. A simple way to trade space (more qubits) for time is to run in parallel multiple instances of the algorithm. We call this process *outer parallelization*, and emphasize that *quantum memory* corresponds to the number of quantum processors working in parallel.

List computation. In the case of collision search, computing the list L now costs only $2^{t-r/2-s}$ time. Notice, however, that the number of queries to O_H remains $2^{t-r/2}$.

Outer parallelization. Our algorithm consists of iterations of an operator that amplifies the amplitude of the good states (recall that $2^{\frac{n-t}{2}}$ such iterations are performed). So, instead of running only one instance and getting a good result with probability close to 1, we can run multiple instances in parallel with less iterations for each. The number of queries made to O_H will be the same.

By running $O(2^s)$ instances, to ensure success probability $O(1)$, we need each of them to have success probability $O(2^{-s})$. So instead of running $2^{\frac{n-t}{2}}$ iterations of the outer amplification procedure, only $2^{\frac{n-t-s}{2}}$ suffice. The running time for collision becomes

$$2^{\frac{n-t-s}{2}}\left(2^{r/2}+2^{t-r}\right)+2^{t-r/2-s} .$$

In collision search, this is $t=\frac{3n}{5}+\frac{3s}{5}$ which gives $r=\frac{2n}{5}+\frac{2s}{5}$, a classical memory $t-r=\frac{n}{5}+\frac{s}{5}$ and a time complexity exponent $t-\frac{r}{2}-s=\frac{2n}{5}-\frac{3s}{5}$.

In order for those parameters to be valid for collision, we need $n-t-s\geq 0$ with $t=\frac{3n}{5}+\frac{3s}{5}$ which gives $s\leq\frac{n}{4}$.

For multi-target preimage, the running time becomes

$$2^{\frac{n-t-s}{2}}\left(2^{r/2}+2^{t-r}\right)+2^{t-s} .$$

The optimal value of r is still $r=\frac{2}{3}t$. In multi-target preimage search, the optimal value of t is achieved for $\frac{n}{2}-\frac{t}{6}-\frac{s}{2}=t-s$ or equivalently $t=\frac{3n}{7}+\frac{3s}{7}$. The running time becomes $2^{3n/7-4s/7}$ and the used classical memory is $2^{\frac{n+s}{7}}$.

Theorem 5 (Outer parallelization). *Let* $H\;:\{0,1\}^n\rightarrow\{0,1\}^n$ *be a random permutation. Given a list of* 2^t *elements, with* $t\leq\frac{3n+3s}{7}$, *there exists a quantum algorithm with* 2^s *quantum processors running in time* $\widetilde{O}\left(2^{n/2-t/6-s/2}\right)$, *using* $\widetilde{O}\left(2^{\frac{t}{3}}\right)$ *classical memory, that finds the preimage of one of them.*

Theorem 6 (Outer parallelization). *Let* $H\;:\{0,1\}^n\rightarrow\{0,1\}^n$ *be a random permutation. Given a list of* 2^t *elements, with* $t=\frac{3n+3s}{7}$, *there exists a quantum algorithm with* 2^s *quantum processors running in time* $\widetilde{O}\left(2^{3n/7-4s/7}\right)$, *using* $\widetilde{O}\left(2^{\frac{n+s}{7}}\right)$ *classical memory, that finds the preimage of one of them.*

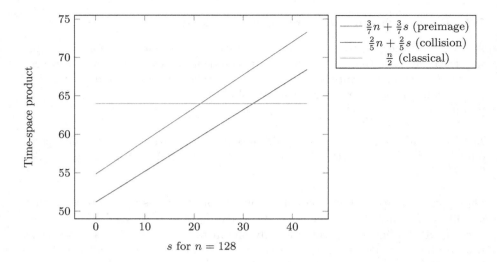

Fig. 1. Quantum time-space product for outer parallelization

As shown on Fig. 1 , there is a range of values of s where the time-space product is effectively smaller than in the classical setting (where all current algorithms obtain an exponent $\frac{n}{2}$). The limit value is $s = \frac{n}{6}$ for preimage search and $s = \frac{n}{4}$ for collisions.

Inner parallelization. It is also possible to parallelize computations in the algorithm itself, especially its most costly building block: the membership oracle $O_{f_L^H}$. We studied this and concluded that this way of parallelizing is not as efficient as outer parallelization.

5.4 Accurate Computations and Parameters

In what precedes, we didn't take into account four sources of possible differences between theory and practice. First, hidden constants: we dismiss the $\pi/4$ factor that stems from amplitude amplification. Second, the logarithmic factor n that appears in the membership oracle. Third, the errors that propagate in the amplitude amplification procedure. Fourth, the cost of a query to the oracle O_H. This last one is actually the most relevant parameter.

Let 2^c be the time complexity of a query. We adapt the parameters r and t as follows:

- In any case, $r = \frac{2}{3}(t - c + \ln_2(n))$ balances the costs;
- In multi-target preimage search, $t = \frac{3n}{7} + \frac{4c}{7} + \frac{2\ln_2(n)}{7}$ is the new complexity exponent. Notice that our method also amortizes the cost of O_H w.r.t a simple Grover search.
- In collision search, $t = \frac{3n}{5} - \frac{4c}{5} + \frac{4\ln_2(n)}{5}$ and time complexity exponent is $\frac{2n}{5} + \frac{4c}{5} + \frac{\ln_2(n)}{5}$.

These computations mean that there is no surprise: the n factor missing above does no more than multiplying the time complexity by 4 ($n = 128$) or 16 ($n = 256$), and by taking into account the cost of a query 2^c, the time complexity does not exceed the previous one multiplied by 2^c. It even behaves better.

In Tables 4 and 3, we give some complexity results *without* taking into account the n and 2^c factors. We do *not* take into account *ancilla qubits*, i.e. additional qubits used during the computation. Detailed studies on the quantum cost of implementing Grover's algorithm have been made, e.g. in [30] for an AES exhaustive key search and [4] for preimage search on SHA-2 and SHA-3 using Grover. Due to space constraints, we cannot go into the technicalities of quantum implementations and restrict ourselves to high-level comparisons; notice, however, that the two aforementioned articles could help in deriving precise hardware costs for our algorithms.

Errors that Propagate in the Amplitude Amplification Procedure. We perform many instances of QAA in our algorithm so it is important to understand how the errors propagate and see if it doesn't create a large cost in the algorithm. We want to briefly describe here the behavior of those errors in our algorithm; more detailed computations are available in the long version of this paper [21]. Let us consider our first QAA algorithm to construct $|\phi_r\rangle$. There are 2 factors that can induce errors here: (1) the fact that we do not know exactly $|S_r^H|$ and (2) the fact that even with perfect knowledge of the angle used in QAA, the algorithm will construct a state close to $|\phi_r\rangle$ but can't hit it exactly. The second problem is solved by using a construction from [19]. In order to solve the first problem, we will use the fact that H is random so that the uncertainty will remain very small. To give a rough idea, the first QAA will give a state $|\phi_{output}\rangle$ such that

$$|\langle \phi_{output}|\phi_r\rangle| \geq \cos(2^{r/2 - n/2} + o(2^{r/2 - n/2})) \ .$$

This error will then propagate to the second QAA. We have two possible scenarios:

- For the collision problem, we have $r = 2n/5$ and we repeat the second QAA $2^{n/5}$ times. The error in the angle will increase by this factor so the final error will be $\approx 2^{n/5}2^{r/2 - n/2} \ll 1$.

Table 3. Quantum collision attack – rounded exponents

n	Space (registers)	Classical memory	Quantum time comp.	Quantum time-space prod.	Classical time-space prod.
128	$O(1)(s = 0)$	26	51	51	64
128	$s = n/6 = 21$	30	39	60	64
256	$O(1)(s = 0)$	51	102	102	128
256	$s = n/6 = 43$	60	77	119	128

Table 4. Quantum multi-target preimage attack – rounded exponents

n	Space (registers)	Targets	Classical memory	Quantum time comp.	Quantum time-space prod.	Classical time comp.
128	$O(1)(s = 0)$	55	18	55	55	73
128	$s = n/8 = 16$	62	21	47	63	66
256	$O(1)(s = 0)$	110	37	110	110	146
256	$s = n/8 = 32$	124	41	92	124	132

- For the preimage problem, we have $r = 2n/7$ and we repeat the second QAA $2^{2n/7}$ times. The error in the angle will increase by this factor so the final error will be $\approx 2^{2n/7}2^{r/2 - n/2} \ll 1$.

This means that the final probability of success will be reduced only marginally.

5.5   Many Collisions

For some purposes, it happens that we do not want to retrieve only one collision pair, but many of them. Suppose 2^c are needed. We modify the parameters in our algorithm to take this into account: now $t = 3n/5 + 6c/5$ and $r = 2n/5 + 4c/5$. Each call returns a collision involving one element of the arbitrary list L of size 2^{t-r}. Hence, we expect 2^{t-r} such collisions to be found by repeating our algorithm and sharing the list L: this forces $t - r > c \Rightarrow c < \frac{n}{3}$. Outside this range, c constraints the size of L: we must have $t - r = c$, $t = 3c$, computing L now costs $2^{t-(t-c)/2} = 2^{2c}$ and the list has 2^c elements. The time complexity exponent becomes $\frac{n}{2} + \frac{c}{2}$; it still presents an advantage over classical collision search.

Theorem 7 (Searching many collisions). *Given a random function H : $\{0,1\}^n \to \{0,1\}^n$ on n bits, there exists a quantum algorithm using $O(n)$ qubits and outputting 2^c collisions:*

- *If $c < \frac{n}{3}$, in time $\widetilde{O}\left(2^{2n/5 + 4c/5}\right)$, using $2^{n/5+2c/5}$ classical memory;*
- *If $c > \frac{n}{3}$, in time $\widetilde{O}\left(2^{n/2 + c/2}\right)$, using 2^c classical memory.*

To ensure that the collisions found are all distinct, one should also multiply this requirements by a small (logarithmic) factor.

6   Impact on Symmetric Cryptography

We discuss below the applications of our new algorithms on the cryptographic scenarios detailed in Sect. 3.4.

We ask the reader to keep in mind that these results seemed particularly surprising as it was conjectured that quantum algorithms for solving these problems wouldn't be more efficient than classical ones.

6.1   On Hash Functions

We consider the setting presented in Sect. 3.4: finding collisions and multi-target preimages on hash functions in a post-quantum world can be considerably accelerated by using the new algorithms. It is important to point out that this can be done considering the Q1 setting for the attacker described in Sect. 2.2: that is, just having access to a quantum computer will allow her to accelerate the attack, and she has no need of access to quantum encryption oracles with superposition queries.

To correspond precisely to the description of the problem, we can consider messages with the same length as the hash value.[3] Indeed, to find a collision, the attacker just has to provide the hash function itself as input for Algorithm 4 (Sect. 5.1). Algorithm 4 will output a collision with a complexity of $\widetilde{O}(2^{2n/5})$ in time and queries, using $\widetilde{O}(2^{n/5})$ classical memory and $O(n)$ qubits. This is to compare with the previous best time complexity of $O(2^{n/2})$.

Finding a preimage out of a set of 2^t generated hash values can be done with Algorithm 5 from Sect. 5.2. It is optimal for $t = 3n/7$ with a cost of $\widetilde{O}(2^{3n/7})$ in time and queries, using $\widetilde{O}(2^{n/7})$ classical memory and $O(n)$ qubits. This should be compared to a classical time complexity of $2^{n-t} = 2^{4n/7}$, or to the previous best quantum attack in $2^{n/2}$, ours being the most performant one. Tables 3 and 4 give concrete values when the time-space tradeoff is used.

6.2   On the Multi-user Setting

The scenario that we presented in Sect. 3.4 can also be accelerated by Algorithm 5 of Sect. 5.2. In this case, the attacker recovers a list of ciphertexts generated from the same plaintext, each encrypted under a different key on size k (one key per user).

The goal is to recover one key out of the total 2^t. In this case, we can consider the attacker scenario Q1: we do not need access to a quantum *encryption* oracle, but instead implement the function that encrypts a fixed plaintext under the key in argument (as we would do for an exhaustive search with a Grover attack). In this case though, the target ciphertexts must be recovered classically. When the key has the same size as the ciphertext, we can directly apply the multi-target preimage search algorithm, that will be optimal for a value of $2^t = 2^{3k/7}$ users. The best time complexity we can achieve here is $\widetilde{O}(2^{3k/7})$, compared to the previous best classical $O(2^{4k/7})$ and the previous best quantum $\widetilde{O}(2^{k/2})$.

Bigger Key than the State. If the key is bigger than the ciphertext, i.e. $k = mn$, we re-construct the problem solved by Algorithm 5 by considering that each user encrypts not one, but m fixed plaintexts.

---

[3] Some blocks fixed to a random value can be considered previously for randomization.

Less multi-users than optimal. If the number of multiusers is smaller than $2^{3k/7}$, we will obtain less gain in the complexities, but still considerable with respect to previous attacks. We can consider, to illustrate this, the attack in [28] presented at Asiacrypt 2014 on the Prince block cipher [17]: In this attack, the authors proposed a technique that provided improved complexity regarding already the best known previously classical multi-target preimage attacks, and they were able to recover one key of size 128 bits out of 2^{32} in time 2^{65} (already improved with respect to the naive $2^{128-32} = 2^{96}$ given by the best generic algorithm). If we apply in this case our algorithm we recover a time complexity of

$$2^{\frac{k}{2}-\frac{t}{6}} + 2^t = 2^{\frac{128}{2}-\frac{32}{6}} + 2^3 2 = 2^{64-5.33} = 2^{58.67},$$

which improves upon previous results. Our results only need a classical memory of $2^{18.3}$ and $O(n)$ qubits (compared to a memory need of 2^{32}). Parallelization can also reduce the time complexity in this scenario, but for the sake of simplicity, we won't go into the details and remit to Sect. 5.3.

6.3   On Operation Modes

As a quantum adversary, we can improve the classical collision attack on CBC introduced in Sect. 3.4 with the help of our algorithm from Sect. 5.2. In this scenario, the attacker has to be placed in the Q2 model from Sect. 2.2: she has access to 2^t classically encrypted blocks, to a quantum computing device and also quantum encryption oracle using the same secret key K.[4] After recovering the 2^t ciphertexts that form the list $L' = C_1, \ldots, C_{2^t}$, we try to find a preimage x of one of them, i.e., find x such that $E_K(x) = C_i$ for $i \in \{1, \ldots, 2^t\}$. This can be done by directly applying Algorithm 5.

Once we find such an x, we can recover P_i, i.e. the plaintext that generates C_i through encryption. Due to the CBC construction, we know that $E_K(P_i \oplus C_{i-1}) = C_i$. Therefore, and as C_{i-1} is known for the attacker, if we recover $x = P_i \oplus C_{i-1}$, we also recover P_i. This can be done by a quantum adversary with a cost in time of $\widetilde{O}(2^{3n/7})$, compared to the classical $O(2^{n/2})$.

In Sect. 7 we discuss the impact this attack should have on maximal data length enforcement.

Frequent rekeying. If we consider a scenario where the user could be forced to change his key after a few encryptions, this previous attack could be translated in a key-recovery one, in the Q1 model, with a similar procedure as in the multi-user case. We first recover classically 2^t ciphertexts, generated by the encryption of one plaintext with 2^t different keys, and next search for a preimage of one of these multitargets.

---

[4] See Sect. 2.2 for a justification of the Q2 model.

6.4   On Bricks for Cryptanalysis Techniques

The last scenario proposed in Sect. 3.4 is less concrete but of great importance. Being very general, the algorithms that we presented here may be used as building blocks by cryptanalysts. With powerful black-box tools and available trade-offs, quantized classical cryptanalysis might become indeed more efficient.

Let us consider as an example the analysis of quantum truncated differential distinguishers from [36]. The aim of the attack is to find a pair of plaintexts with a difference belonging to a certain set Δ_{in}, that generate a pair of ciphertexts belonging to another particular set of differences Δ_{out}, which is equivalent to colliding in a part of the output state. The attack succeeds if such a pair is found quicker than for a random permutation. The probability of this happening for the attack cipher is denoted by 2^{-h_T}.

We consider the case where a single structure[5] is enough for finding the good pair statistically, i.e. if $2^{h_T} \leq 2^{2|\Delta_{in}|-1}$. The authors remark that finding the good pair will cost $O(2^{h_T/3})$ queries for a quantum adversary. But this would also cost the same amount in space. We could, instead, apply our new algorithm, allowing the quantum space needed to remain polynomial in n with a time complexity still improved over the classic one.

7   Conclusion

7.1   Efficient Algorithms for Collision and Multi-target Preimage Search

We have presented a quantum algorithm for collision and preimage search, based on the amplitude amplification procedure, that is sub-optimal in terms of query complexity but beats the current best algorithm in terms of time complexity with small quantum memory.

To the best of our knowledge, this is the first generic procedure that solves this problem effectively faster than classically, when only linear quantum space is available. Our algorithm can also be parallelized, providing better performance than classical algorithms for a wide range or parameters.

7.2   Impact on Symmetric Primitives

From the applications presented in Sect. 6, we can obtain the following conclusions:

Open Problem on Best Quantum Multi-target Preimages. In Eurocrypt 2015 [10], Sect. 3.2, the authors notice that the best known post-quantum cost for multi-target preimage attacks is also of $2^{n/2}$, and they provide the following example: for $n = 256$ and $2^t = 2^{56}$, they claim that the best quantum algorithm has a cost of 2^{128}, though it only needs 2^{100} queries. With our algorithms, this

---

[5] A structure is a set of plaintexts of size $2^{|\Delta_{in}|}$ belonging to the same truncated difference Δ_{in}, which means that they allow to build $2^{2|\Delta_{in}|-1}$ pairs.

implication does not hold anymore: it is possible to attack their example with a time complexity of

$$2^{100}(2^{56/3} + 2^{56/3}) + 2^{56} \approx 2^{119.6}$$

by applying Algorithm 5, which is clearly better than the classical attack, and using a polynomial amount of qubits.

On Maximal Data Length to Enforce. While attacking operation modes via collisions, 2^t data is recovered classically. This 2^t can be significantly smaller than $2^{n/2}$, and the attack would still have an advantage over the birthday paradox. In fact, when more data is available, the time complexity of the quantum computations decreases up to the limit $\widetilde{O}\left(2^{5n/12}\right)$ (when $t = n/2$).

We can forget about the term 2^t, as we are considering $t < n/2$, and the quantum procedure has complexity $\widetilde{O}\left(2^{\frac{n}{2} - \frac{t}{6}}\right)$, which offers a factor $2^{-t/6}$ compared to classical collision search, independent of the block size n. The security requirements will determine the maximal amount of data that can be generated with a given key.

Is doubling the key length enough? The multi-user scenario, as well as the re-keying one make us wonder about the actual security offered by symmetric ciphers in a post-quantum world. By accelerating collision search, we showed that Grover's exhaustive key search is not the only general threat posed to them: the block size is also a relevant parameter in quantum attacks.

These results increase our impression that many scenarios and settings should be carefully studied before being able to promise certain security levels in a post-quantum world.

7.3   Open Problems and Future Work

Our result fills a gap that existed between the theoretical query complexity of collision search and the actual requirements of an attack. It follows recent non-intuitive results in quantum symmetric cryptanalysis (see e.g. [35]), that have shown the necessity of a better understanding of how symmetric primitives could be affected by quantum adversaries. To our opinion, many such counter-intuitive results are yet to appear.

This work reopens the direction of designing improved quantum algorithms for collisions and preimage finding when the quantum computer does not have access to an exponential amount of quantum memory. The algorithm we propose will not be dismissed as implausible if we want to prove security against quantum attackers: the quantum memory needed is reasonably small. We have been able to propose several significant complexity improvements thanks to this result. Although $2n/5$ is the optimal exponent of our collision algorithm, it introduces additional structure (a prefix of the image is chosen) that is not relevant in many applications: is it possible to get rid of these specificities and bring the exponent down to $n/3$?

Acknowledgements. This project has received funding from the European Research Council (ERC) under the European Union's Horizon 2020 research and innovation programme (grant agreement no. 714294 - acronym QUASYModo).

References

1. Aaronson, S., Shi, Y.: Quantum lower bounds for the collision and the element distinctness problems. J. ACM **51**(4), 595–605 (2004)
2. Ambainis, A.: Polynomial degree and lower bounds in quantum complexity: collision and element distinctness with small range. Theor. Comput. **1**(1), 37–46 (2005). http://dx.doi.org/10.4086/toc.2005.v001a003
3. Ambainis, A.: Quantum walk algorithm for element distinctness. SIAM J. Comput. **37**(1), 210–239 (2007). http://dx.doi.org/10.1137/S0097539705447311
4. Amy, M., Di Matteo, O., Gheorghiu, V., Mosca, M., Parent, A., Schanck, J.M.: Estimating the cost of generic quantum pre-image attacks on SHA-2 and SHA-3. In: IACR Cryptology ePrint Archive, p. 992 (2016)
5. Anand, M.V., Targhi, E.E., Tabia, G.N., Unruh, D.: Post-quantum security of the CBC, CFB, OFB, CTR, and XTS modes of operation. In: Takagi, T. (ed.) PQCrypto 2016. LNCS, vol. 9606, pp. 44–63. Springer, Cham (2016). https://doi.org/10.1007/978-3-319-29360-8_4
6. Andreeva, E., Bouillaguet, C., Fouque, P.-A., Hoch, J.J., Kelsey, J., Shamir, A., Zimmer, S.: Second preimage attacks on dithered hash functions. In: Smart, N. (ed.) EUROCRYPT 2008. LNCS, vol. 4965, pp. 270–288. Springer, Heidelberg (2008). https://doi.org/10.1007/978-3-540-78967-3_16
7. Banegas, G., Bernstein, D.J.: Low-communication parallel quantum multi-target preimage search. In: SAC 2017 (2017)
8. Bellare, M., Desai, A., Jokipii, E., Rogaway, P.: A concrete security treatment of symmetric encryption. In: FOCS, pp. 394–403. IEEE Computer Society (1997)
9. Bernstein, D.J.: Cost analysis of hash collisions: will quantum computers make SHARCS obsolete? In: SHARCS 2009 Special-Purpose Hardware for Attacking Cryptographic Systems, p. 105 (2009)
10. Bernstein, D.J., Hopwood, D., Hülsing, A., Lange, T., Niederhagen, R., Papachristodoulou, L., Schneider, M., Schwabe, P., Wilcox-O'Hearn, Z.: SPHINCS: practical stateless hash-based signatures. In: Oswald, E., Fischlin, M. (eds.) EUROCRYPT 2015. LNCS, vol. 9056, pp. 368–397. Springer, Heidelberg (2015). https://doi.org/10.1007/978-3-662-46800-5_15
11. Bhargavan, K., Leurent, G.: On the Practical (In-)Security of 64-bit Block Ciphers: Collision Attacks on HTTP over TLS and OpenVPN. IACR Cryptology ePrint Archive 2016, 798 (2016). http://eprint.iacr.org/2016/798
12. Biham, E., Biryukov, A., Shamir, A.: Cryptanalysis of skipjack reduced to 31 rounds using impossible differentials. In: Stern, J. (ed.) EUROCRYPT 1999. LNCS, vol. 1592, pp. 12–23. Springer, Heidelberg (1999). https://doi.org/10.1007/3-540-48910-X_2
13. Biham, E.: How to decrypt or even substitute des-encrypted messages in 2^{28} steps. Inf. Process. Lett. **84**(3), 117–124 (2002). http://dx.doi.org/10.1016/S0020-0190(02)00269-7
14. Biryukov, A., Mukhopadhyay, S., Sarkar, P.: Improved time-memory trade-offs with multiple data. In: Preneel, B., Tavares, S. (eds.) SAC 2005. LNCS, vol. 3897, pp. 110–127. Springer, Heidelberg (2006). https://doi.org/10.1007/11693383_8

15. Boneh, D., Dagdelen, Ö., Fischlin, M., Lehmann, A., Schaffner, C., Zhandry, M.: Random oracles in a quantum world. In: Lee, D.H., Wang, X. (eds.) ASIACRYPT 2011. LNCS, vol. 7073, pp. 41–69. Springer, Heidelberg (2011). https://doi.org/10.1007/978-3-642-25385-0_3

16. Boneh, D., Zhandry, M.: Secure signatures and chosen ciphertext security in a quantum computing world. In: Canetti, R., Garay, J.A. (eds.) CRYPTO 2013. LNCS, vol. 8043, pp. 361–379. Springer, Heidelberg (2013). https://doi.org/10.1007/978-3-642-40084-1_21

17. Borghoff, J., et al.: PRINCE – a low-latency block cipher for pervasive computing applications. In: Wang, X., Sako, K. (eds.) ASIACRYPT 2012. LNCS, vol. 7658, pp. 208–225. Springer, Heidelberg (2012). https://doi.org/10.1007/978-3-642-34961-4_14

18. Brassard, G., Høyer, P., Kalach, K., Kaplan, M., Laplante, S., Salvail, L.: Merkle puzzles in a quantum world. In: Rogaway, P. (ed.) CRYPTO 2011. LNCS, vol. 6841, pp. 391–410. Springer, Heidelberg (2011). https://doi.org/10.1007/978-3-642-22792-9_22

19. Brassard, G., Hoyer, P., Mosca, M., Tapp, A.: Quantum amplitude amplification and estimation. Contemp. Math. **305**, 53–74 (2002)

20. Brassard, G., Høyer, P., Tapp, A.: Quantum cryptanalysis of hash and claw-free functions. In: Lucchesi, C.L., Moura, A.V. (eds.) LATIN 1998. LNCS, vol. 1380, pp. 163–169. Springer, Heidelberg (1998). https://doi.org/10.1007/BFb0054319

21. Chailloux, A., Naya-Plasencia, M., Schrottenloher, A.: An Efficient Quantum Collision Search Algorithm and Implications on Symmetric Cryptography. IACR Cryptology ePrint Archive 2017, 847 (2017). http://eprint.iacr.org/2017/847

22. Chatterjee, S., Menezes, A., Sarkar, P.: Another look at tightness. In: Miri, A., Vaudenay, S. (eds.) SAC 2011. LNCS, vol. 7118, pp. 293–319. Springer, Heidelberg (2012). https://doi.org/10.1007/978-3-642-28496-0_18

23. Damgård, I., Funder, J., Nielsen, J.B., Salvail, L.: Superposition attacks on cryptographic protocols. In: Padró, C. (ed.) ICITS 2013. LNCS, vol. 8317, pp. 142–161. Springer, Cham (2014). https://doi.org/10.1007/978-3-319-04268-8_9

24. Dierks, T., Rescorla, E.: The transport layer security (TLS) protocol version 1.2. In: IETF RFC 5246 (2008)

25. Diffie, W., Hellman, M.: Privacy and authentication: an introduction to cryptography. In: Proceedings of the IEEE, vol. 67, pp. 397–427 (1979)

26. Ehrsam, W.R., Meyer, C.H., Smith, J.L., Tuchman, W.L.: Message verification and transmission error detection by block chaining. US Patent 4074066 (1976)

27. Even, S., Mansour, Y.: A construction of a cipher from a single pseudorandom permutation. J. Cryptol. **10**(3), 151–162 (1997). http://dx.doi.org/10.1007/s001459900025

28. Fouque, P.-A., Joux, A., Mavromati, C.: Multi-user collisions: applications to discrete logarithm, even-mansour and PRINCE. In: Sarkar, P., Iwata, T. (eds.) ASIACRYPT 2014. LNCS, vol. 8873, pp. 420–438. Springer, Heidelberg (2014). https://doi.org/10.1007/978-3-662-45611-8_22

29. Gagliardoni, T., Hülsing, A., Schaffner, C.: Semantic security and indistinguishability in the quantum world. In: Robshaw, M., Katz, J. (eds.) CRYPTO 2016. LNCS, vol. 9816, pp. 60–89. Springer, Heidelberg (2016). https://doi.org/10.1007/978-3-662-53015-3_3

30. Grassl, M., Langenberg, B., Roetteler, M., Steinwandt, R.: Applying Grover's algorithm to AES: quantum resource estimates. In: Takagi, T. (ed.) PQCrypto 2016. LNCS, vol. 9606, pp. 29–43. Springer, Cham (2016). https://doi.org/10.1007/978-3-319-29360-8_3

31. Grover, L.K.: A fast quantum mechanical algorithm for database search. In: Miller, G.L. (ed.) Proceedings of the Twenty-Eighth Annual ACM Symposium on the Theory of Computing, Philadelphia, Pennsylvania, USA, 22–24 May 1996, pp. 212–219. ACM (1996). http://doi.acm.org/10.1145/237814.237866

32. Grover, L.K.: Trade-offs in the quantum search algorithm. In: Physical Review A, vol. 66 (2002)

33. Grover, L.K., Rudolph, T.: How significant are the known collision and element distinctness quantum algorithms? Quantum Inf. Comput. **4**(3), 201–206 (2004). http://portal.acm.org/citation.cfm?id=2011622

34. Kaplan, M.: Quantum attacks against iterated block ciphers. CoRR abs/1410.1434 (2014). http://arxiv.org/abs/1410.1434

35. Kaplan, M., Leurent, G., Leverrier, A., Naya-Plasencia, M.: Breaking symmetric cryptosystems using quantum period finding. In: Robshaw, M., Katz, J. (eds.) CRYPTO 2016. LNCS, vol. 9815, pp. 207–237. Springer, Heidelberg (2016). https://doi.org/10.1007/978-3-662-53008-5_8

36. Kaplan, M., Leurent, G., Leverrier, A., Naya-Plasencia, M.: Quantum differential and linear cryptanalysis. IACR Trans. Symmetric Cryptol. **2016**(1), 71–94 (2016). http://tosc.iacr.org/index.php/ToSC/article/view/536

37. Knudsen, L.R.: DEAL - A 128-bit cipher. Technical Report, Department of Informatics, University of Bergen, Norway (1998)

38. Knudsen, L.R.: Truncated and higher order differentials. In: Preneel, B. (ed.) FSE 1994. LNCS, vol. 1008, pp. 196–211. Springer, Heidelberg (1995). https://doi.org/10.1007/3-540-60590-8_16

39. Krovetz, T., Rogaway, P.: The software performance of authenticated-encryption modes. In: Joux, A. (ed.) FSE 2011. LNCS, vol. 6733, pp. 306–327. Springer, Heidelberg (2011). https://doi.org/10.1007/978-3-642-21702-9_18

40. Kutin, S.: Quantum lower bound for the collision problem with small range. Theor. Comput. **1**(2), 29–36 (2005). http://www.theoryofcomputing.org/articles/v001a002

41. Kuwakado, H., Morii, M.: Quantum distinguisher between the 3-round feistel cipher and the random permutation. In: IEEE International Symposium on Information Theory, ISIT 2010, Austin, Texas, USA, Proceedings, pp. 2682–2685. IEEE, 13–18 June 2010. http://dx.doi.org/10.1109/ISIT.2010.5513654

42. Kuwakado, H., Morii, M.: Security on the quantum-type even-mansour cipher. In: Proceedings of the International Symposium on Information Theory and its Applications, ISITA 2012, Honolulu, HI, USA, pp. 312–316. IEEE, 28–31 October 2012. http://ieeexplore.ieee.org/xpl/freeabs_all.jsp?arnumber=6400943

43. Lipmaa, H., Rogaway, P., Wagner, D.: Comments to nist concerning aes modes of operations: Ctr-mode encryption (2000). http://csrc.nist.gov/groups/ST/toolkit/BCM/documents/proposedmodes/ctr/ctr-spec.pdf

44. McGrew, D.A.: Impossible plaintext cryptanalysis and probable-plaintext collision attacks of 64-bit block cipher modes. IACR Cryptology ePrint Archive 2012, 623 (2012). http://eprint.iacr.org/2012/623

45. Menezes, A.: Another look at provable security. In: Pointcheval, D., Johansson, T. (eds.) EUROCRYPT 2012. LNCS, vol. 7237, p. 8. Springer, Heidelberg (2012). https://doi.org/10.1007/978-3-642-29011-4_2

46. Nielsen, M.A., Chuang, I.: Quantum Computation and Quantum Information. Cambridge University Press, New York (2002)

47. van Oorschot, P.C., Wiener, M.J.: Parallel collision search with application to hash functions and discrete logarithms. In: CCS 1994, Proceedings of the 2nd ACM Conference on Computer and Communications Security, Fairfax, Virginia, USA, pp. 210–218. ACM, 2–4 November 1994

48. Pollard, J.M.: A Monte Carlo method for factorization. BIT Numer. Math. **15**(3), 331–334 (1975). http://dx.doi.org/10.1007/BF01933667

49. Shor, P.W.: Algorithms for quantum computation: discrete logarithms and factoring. In: 35th Annual Symposium on Foundations of Computer Science, Santa Fe, New Mexico, USA, pp. 124–134. IEEE Computer Society, 20–22 November 1994. http://dx.doi.org/10.1109/SFCS.1994.365700

50. Simon, D.R.: On the power of quantum cryptography. In: 35th Annual Symposium on Foundations of Computer Science, Santa Fe, New Mexico, USA, pp. 116–123. IEEE Computer Society, 20–22 November 1994. http://dx.doi.org/10.1109/SFCS.1994.365701

51. Unruh, D.: Non-interactive zero-knowledge proofs in the quantum random oracle model. In: Oswald, E., Fischlin, M. (eds.) EUROCRYPT 2015. LNCS, vol. 9057, pp. 755–784. Springer, Heidelberg (2015). https://doi.org/10.1007/978-3-662-46803-6_25. Preprint on IACR ePrint 2014/587

52. Yuval, G.: How to swindle rabin. Cryptologia **3**(3), 187–191 (1979). http://dx.doi.org/10.1080/0161-117991854025

53. Zhandry, M.: A note on the quantum collision and set equality problems. Quantum Info. Comput. **15**(7–8), 557–567 (2015). http://dl.acm.org/citation.cfm?id=2871411.2871413

54. Zhandry, M.: Secure identity-based encryption in the quantum random oracle model. Int. J. Quantum Inf. **13**(04), 1550014 (2015)

Quantum Resource Estimates for Computing Elliptic Curve Discrete Logarithms

Martin Roetteler[(✉)], Michael Naehrig, Krysta M. Svore, and Kristin Lauter

Microsoft Research, Redmond, USA
martinro@microsoft.com

Abstract. We give precise quantum resource estimates for Shor's algorithm to compute discrete logarithms on elliptic curves over prime fields. The estimates are derived from a simulation of a Toffoli gate network for controlled elliptic curve point addition, implemented within the framework of the quantum computing software tool suite LIQ$Ui|\rangle$. We determine circuit implementations for reversible modular arithmetic, including modular addition, multiplication and inversion, as well as reversible elliptic curve point addition. We conclude that elliptic curve discrete logarithms on an elliptic curve defined over an n-bit prime field can be computed on a quantum computer with at most $9n + 2\lceil \log_2(n) \rceil + 10$ qubits using a quantum circuit of at most $448n^3 \log_2(n) + 4090n^3$ Toffoli gates. We are able to classically simulate the Toffoli networks corresponding to the controlled elliptic curve point addition as the core piece of Shor's algorithm for the NIST standard curves P-192, P-224, P-256, P-384 and P-521. Our approach allows gate-level comparisons to recent resource estimates for Shor's factoring algorithm. The results also support estimates given earlier by Proos and Zalka and indicate that, for current parameters at comparable classical security levels, the number of qubits required to tackle elliptic curves is less than for attacking RSA, suggesting that indeed ECC is an easier target than RSA.

Keywords: Quantum cryptanalysis · Elliptic curve cryptography · Elliptic curve discrete logarithm problem

1 Introduction

Elliptic Curve Cryptography (ECC). Elliptic curves are a fundamental building block of today's cryptographic landscape. Thirty years after their introduction to cryptography [27,32], they are used to instantiate public key mechanisms such as key exchange [11] and digital signatures [17,23] that are widely deployed in various cryptographic systems. Elliptic curves are used in applications such as transport layer security [5,10], secure shell [47], the Bitcoin digital currency system [34], in national ID cards [22], the Tor anonymity network [12], and the WhatsApp messaging app [53], just to name a few. Hence, they play a significant role in securing our data and communications.

© International Association for Cryptologic Research 2017
T. Takagi and T. Peyrin (Eds.): ASIACRYPT 2017, Part II, LNCS 10625, pp. 241–270, 2017.
https://doi.org/10.1007/978-3-319-70697-9_9

Different standards (e.g., [8,50]) and standardization efforts (e.g., [13,36]) have identified elliptic curves of different sizes targeting different levels of security. Notable curves with widespread use are the NIST curves P-256, P-384, P-521, which are curves in Weierstrass form over special primes of size 256, 384, and 521 bits respectively, the Bitcoin curve secp256k1 from the SEC2 [8] standard and the Brainpool curves [13]. More recently, Bernstein's Curve25519 [54], a Montgomery curve over a 255-bit prime field, has seen more and more deployment, and it has been recommended to be used in the next version of the TLS protocol [29] along with another even more recent curve proposed by Hamburg called Goldilocks [20].

The security of elliptic curve cryptography relies on the hardness of computing discrete logarithms in elliptic curve groups, i.e. the difficulty of the Elliptic Curve Discrete Logarithm Problem (ECDLP). Elliptic curves have the advantage of relatively small parameter and key sizes in comparison to other cryptographic schemes, such as those based on RSA [41] or finite field discrete logarithms [11], when compared at the same security level. For example, according to NIST recommendations from 2016, a 256-bit elliptic curve provides a similar resistance against classical attackers as an RSA modulus of size 3072 bits[1]. This advantage arises from the fact that the currently known best algorithms to compute elliptic curve discrete logarithms are exponential in the size of the input parameters[2], whereas there exist subexponential algorithms for factoring [9,30] and finite field discrete logarithms [18,24].

The Quantum Computer Threat. In his famous paper [44], Peter Shor presented two polynomial-time quantum algorithms, one for integer factorization and another one for computing discrete logarithms in a finite field of prime order. Shor notes that the latter algorithm can be generalized to other fields. It also generalizes to the case of elliptic curves. Hence, given the prerequisite that a large enough general purpose quantum computer can be built, the algorithms in Shor's paper completely break all current crypto systems based on the difficulty of factoring or computing discrete logarithms. Scaling up the parameters for such schemes to sizes for which Shor's algorithm becomes practically infeasible will most likely lead to highly impractical instantiations.

Recent years have witnessed significant advances in the state of quantum computing hardware. Companies have invested in the development of qubits, and the field has seen an emergence of startups, with some focusing on quantum hardware, others on software for controlling quantum computers, and still others offering consulting services to ready for the quantum future. The predominant approach to quantum computer hardware focuses on physical implementations that are scalable, digital, programmable, and universal. With the amount of

---

[1] Opinions about such statements of equivalent security levels differ, for an overview see https://www.keylength.com. There is consensus about the fact that elliptic curve parameters can be an order of magnitude smaller than parameters for RSA or finite field discrete logarithm systems to provide similar security.

[2] For a recent survey, see [16].

investment in quantum computing hardware, the pace of scaling is increasing and underscoring the need to understand the scaling of the difficulty of ECDLP.

Language-Integrated Quantum Operations: LIQ$Ui|\rangle$. As quantum hardware advances towards larger-scale systems of upwards of tens to hundreds of qubits, there is a critical need for a software architecture to program and control the device. We use the LIQ$Ui|\rangle$ software architecture [52] to determine the resource costs of solving the ECDLP. LIQ$Ui|\rangle$ is a high-level programming language for quantum algorithms embedded in F#, a compilation stack to translate and compile quantum algorithms into quantum circuits, and a simulator to test and run quantum circuits[3]. LIQ$Ui|\rangle$ can simulate roughly 32 qubits in 32 GB RAM, however, we make use of the fact that reversible circuits can be simulated efficiently on classical input states for thousands of qubits.

Gate Sets and Toffoli Gate Networks. The basic underlying fault-tolerant architecture and coding scheme of a quantum computer determine the universal gate set, and hence by extension also the synthesis problems that have to be solved in order to compile high-level, large-scale algorithms into a sequence of operations that an actual physical quantum computer can then execute. A gate set that arises frequently and that has been studied often in the literature, but by no means the only conceivable gate set, is the so-called Clifford+T gate set [35]. This gate set consists of the Hadamard gate $H = \frac{1}{\sqrt{2}}\left[\begin{smallmatrix} 1 & 1 \\ 1 & -1 \end{smallmatrix}\right]$, the phase gate $P = \text{diag}(1, i)$, and the controlled NOT (CNOT) gate which maps $(x, y) \mapsto (x, x \oplus y)$ as generators of the Clifford group, along with the T gate given by $T = \text{diag}(1, \exp(\pi i/4))$. The Clifford+$T$ gate set is known to be universal [35]. This means that it can be used to approximate any given target unitary single qubit operation to within precision ε using sequences of length $4 \log_2(1/\varepsilon)$ [26,43], and using an entangling gate such as the CNOT gate, the Clifford+T gate set can approximate any unitary operation. When assessing the complexity of a quantum circuit built from Clifford+T gates, often only T-gates are counted as many fault-tolerant implementations of the Clifford+T gate set at the logical gate level require much more resources for T-gates than for Clifford gates [15].

In this paper, we base reversible computations entirely on the Toffoli gate. The Toffoli gate $|x, y, z\rangle \mapsto |x, y, z \oplus xy\rangle$ is known to be universal for reversible computing [35] and can be implemented exactly over the Clifford+T gate set, see [42] for a T-depth 1 implementation using a total of 7 qubits and [1] for a T-depth 3 realization using a total of 3 qubits. As discussed in [21, Sect. 5], there are two main reasons for focusing on Toffoli gate networks as our preferred realization of quantum circuits. The first is that because the Toffoli gate can be implemented exactly over the Clifford+T gate set, Toffoli networks do not have gate synthesis overhead. The second is testability and debugging. Toffoli gate networks can be simulated using classical reversible simulators. While a fully functional simulation of a quantum circuit could be deemed feasible for

---

[3] See http://stationq.github.io/Liquid/ and https://github.com/StationQ/Liquid.

circuits on up to 50 qubits, classical simulation of Toffoli gate-based circuits can deal with a lot more qubits. Also, for implementations on actual quantum hardware, Toffoli gate circuits can be debugged efficiently and faults can be localized through binary search [21].

Estimating Quantum Resources for Shor's ECDLP Algorithm. Understanding the concrete requirements for a quantum computer that is able to run Shor's algorithm helps to put experimental progress in quantum computing into perspective. Although it is clear that the polynomial runtime asymptotically breaks ECC, constant factors can make an important difference when actually implementing the algorithm.

In [39], Proos and Zalka describe how Shor's algorithm can be implemented for the case of elliptic curve groups. They conclude with a table of resource estimates for the number of logical qubits and time (measured in "1-qubit additions") depending on the bitsize of the elliptic curve. Furthermore, they compare these estimates to those for Shor's factoring algorithm and argue that computing elliptic curve discrete logarithms is significantly easier than factoring RSA moduli at comparable classical security levels. However, some questions remained unanswered by [39], the most poignant of which being whether it is actually possible to construct and simulate the circuits to perform elliptic curve point addition in order to get confidence in their correctness. Another question that remained open is whether it is possible to determine constants that were left in terms of asymptotic scaling and whether some of the proposed circuit constructions to compress registers and to synchronize computations can actually be implemented in code that can then be automatically generated for arbitrary input curves.

Here we build on their work and fully program and simulate the underlying arithmetic. We verify the correctness of our algorithms and obtain concrete resource costs measured by the overall number of logical qubits, the number of Toffoli gates and the depth of a quantum circuit for implementing Shor's algorithm.

Contributions. In this paper, we present precise resource estimates for quantum circuits that implement Shor's algorithm to solve the ECDLP. In particular, our contributions are as follows:

- We describe reversible algorithms for modular quantum arithmetic. This includes modular addition, subtraction, negation and doubling of integers held in quantum registers, modular multiplication, squaring and inversion.
- For modular multiplication, we consider two different approaches, besides an algorithm based on modular doublings and modular additions, we also give a circuit for Montgomery multiplication.
- Based on our implementations it transpired that using Montgomery arithmetic is beneficial as the cost for the multiplication can be seen to be lower than that of the double-and-add method. The latter requires less ancillas,

however, in the given algorithm there are always enough ancillas available as overall a relatively large number of ancillas must be provided.

- Our modular inversion algorithm is a reversible implementation of the Montgomery inverse via the binary extended Euclidean (binary GCD) algorithm. To realize this algorithm as a circuit, we introduce tools that might be of independent interest for other reversible algorithms.
- We describe a quantum circuit for elliptic curve point addition in affine coordinates and describe how it can be used to implement scalar multiplication within Shor's algorithm.
- We have implemented all of the above algorithms in F# within the framework of the quantum computing software tool suite LIQU$i|\rangle$ [52] and have simulated and tested all of these algorithms for real-world parameters of up to 521 bits[4].
- Derived from our implementation, we present concrete resource estimates for the total number of qubits, the number of Toffoli gates and the depth of the Toffoli gate networks to realize Shor's algorithm and its subroutines. We compare the quantum resources for solving the ECDLP to those required in Shor's factoring algorithm that were obtained in the recent work [21].

Results. Our implementation realizes a reversible circuit for controlled elliptic curve point addition on an elliptic curve defined over a field of prime order with n bits and needs at most $9n + 2\lceil \log_2(n) \rceil + 10$ qubits. An interpolation of the data points for the number of Toffoli gates shows that the quantum circuit can be implemented with at most roughly $224n^2 \log_2(n) + 2045n^2$ Toffoli gates. For Shor's full algorithm, the point addition needs to be run $2n$ times sequentially and does not need additional qubits. The overall number of Toffoli gates is thus about $448n^3 \log_2(n) + 4090n^3$. For example, our simulation of the point addition quantum circuit for the NIST standardized curve P-256 needs 2330 logical qubits and the full Shor algorithm would need about $1.26 \cdot 10^{11}$ Toffoli gates. In comparison, Shor's factoring algorithm for a 3072-bit modulus needs 6146 qubits and $1.86 \cdot 10^{13}$ Toffoli gates[5], which aligns with results by Proos and Zalka showing that it is easier to break ECC than RSA at comparable classical security.

Our estimates provide a data point that allows a better understanding of the requirements to run Shor's quantum ECDLP algorithm and we hope that they will serve as a basis to make better predictions about the time horizon until which elliptic curve cryptography can still be considered secure. Besides helping to gain a better understanding of the post-quantum (in-) security of elliptic curve cryptosystems, we hope that our reversible algorithms (and their LIQU$i|\rangle$ implementations) for modular arithmetic and the elliptic curve group law are of independent interest to some, and might serve as building blocks for other quantum algorithms.

---

[4] Our code will be made publicly available at http://microsoft.com/quantum.

[5] These estimates are interpolated from the results in [21].

2   Elliptic Curves and Shor's Algorithm

This section provides some background on elliptic curves over finite fields, the elliptic curve discrete logarithm problem (ECDLP) and Shor's quantum algorithm to solve the ECDLP. Throughout, we restrict to the case of curves defined over prime fields of large characteristic.

2.1   Elliptic Curves and the ECDLP

Let $p > 3$ be a prime. Denote by $\mathbb{F}_p$ the finite field with p elements. An elliptic curve over $\mathbb{F}_p$ is a projective, non-singular curve of genus 1 with a specified base point. It can be given by an affine Weierstrass model, i.e. it can be viewed as the set of all solutions (x, y) to the equation $E : y^2 = x^3 + ax + b$ with two curve constants $a, b \in \mathbb{F}_p$, together with a point at infinity $\mathcal{O}$. The set of $\mathbb{F}_p$-rational points consists of $\mathcal{O}$ and all solutions $(x, y) \in \mathbb{F}_p \times \mathbb{F}_p$ and is denoted by $E(\mathbb{F}_p) = \{(x, y) \in \mathbb{F}_p \times \mathbb{F}_p \mid y^2 = x^3 + ax + b\} \cup \{\mathcal{O}\}$. The set $E(\mathbb{F}_p)$ is an abelian group with respect to a group operation "+" that is defined via rational functions in the point coordinates with $\mathcal{O}$ as the neutral element. Similarly, for a field extension $\mathbb{F} \supseteq \mathbb{F}_p$, one similarly defines the group of $\mathbb{F}$-rational points $E(\mathbb{F})$ and if $\mathbb{F}$ is an algebraic closure of $\mathbb{F}_p$, we simply denote $E = E(\mathbb{F})$. For an extensive treatment of elliptic curves, we refer the reader to [46].

The elliptic curve group law on an affine Weierstrass curve can be computed as follows. Let $P_1, P_2 \in E$ and let $P_3 = P_1 + P_2$. If $P_1 = \mathcal{O}$ then $P_3 = P_2$ and if $P_2 = \mathcal{O}$, then $P_3 = P_1$. Now let $P_1 \neq \mathcal{O} \neq P_2$ and write $P_1 = (x_1, y_1)$ and $P_2 = (x_2, y_2)$ for $x_1, y_1, x_2, y_2 \in \mathbb{F}$. If $P_2 = -P_1$, then $x_1 = x_2$, $y_2 = -y_1$ and $P_3 = \mathcal{O}$. If neither of the previous cases occurs, then $P_3 = (x_3, y_3)$ is an affine point and can be computed as

$$x_3 = \lambda^2 - x_1 - x_2, \quad y_3 = (x_1 - x_3)\lambda - y_1,$$

where $\lambda = \frac{y_2 - y_1}{x_2 - x_1}$ if $P_1 \neq P_2$, i.e. $x_1 \neq x_2$, and $\lambda = \frac{3x_1^2 + a}{2y_1}$ if $P_1 = P_2$. For a positive integer m, denote by $[m]P$ the m-fold sum of P, i.e. $[m]P = P + \cdots + P$, where P occurs m times. Extended to all $m \in \mathbb{Z}$ by $[0]P = \mathcal{O}$ and $[-m]P = [m](-P)$, the map $[m] : E \to E, P \mapsto [m]P$ is called the multiplication-by-m map or simply scalar multiplication by m. Scalar multiplication (or group exponentiation in the multiplicative setting) is one of the main ingredients for discrete-logarithm-based cryptographic protocols. It is also an essential operation in Shor's ECDLP algorithm. The order $\mathrm{ord}(P)$ of a point P is the smallest positive integer r such that $[r]P = \mathcal{O}$.

Curves that are most widely used in cryptography are defined over large prime fields. One works in a cyclic subgroup of $E(\mathbb{F}_p)$ of large prime order r, where $\#E(\mathbb{F}_p) = h \cdot r$. The group order can be written as $\#E(\mathbb{F}_p) = p + 1 - t$, where t is called the trace of Frobenius and the Hasse bound ensures that $|t| \leq 2\sqrt{p}$. Thus $\#E(\mathbb{F}_p)$ and p are of roughly the same size. The most efficient instantiations of ECC are achieved for small cofactors h. For example, the above mentioned NIST curves have prime order, i.e. $h = 1$, and Curve25519 has cofactor $h = 8$. Let $P \in E(\mathbb{F}_p)$ be an $\mathbb{F}_p$-rational point on E of order r and let $Q \in \langle P \rangle$ be an element of the cyclic subgroup generated by P. The Elliptic Curve Discrete Logarithm Problem (ECDLP) is the problem to find the integer $m \in \mathbb{Z}/r\mathbb{Z}$ such that $Q = [m]P$. The bit security of an elliptic curve is estimated by extrapolating the runtime of the most efficient algorithms for the ECDLP.

The currently best known classical algorithms to solve the ECDLP are based on parallelized versions of Pollard's rho algorithm [37,38,51]. When working in a group of order n, the expected running time for solving a single ECDLP is $(\sqrt{\pi/2} + o(1))\sqrt{n}$ group operations based on the birthday paradox. This is exponential in the input size $\log(n)$. See [16] for further details and [6] for a concrete, implementation-based security assessment.

2.2 Shor's Quantum Algorithm for Solving the ECDLP

In [44], Shor presented two polynomial time quantum algorithms, one for factoring integers, the other for computing discrete logarithms in finite fields. The second one can naturally be applied for computing discrete logarithms in the group of points on an elliptic curve defined over a finite field.

We are given an instance of the ECDLP as described above. Let $P \in E(\mathbb{F}_p)$ be a fixed generator of a cyclic subgroup of $E(\mathbb{F}_p)$ of known order $\mathrm{ord}(P) = r$, let $Q \in \langle P \rangle$ be a fixed element in the subgroup generated by P; our goal is to find the unique integer $m \in \{1, \ldots, r\}$ such that $Q = [m]P$. Shor's algorithm proceeds as follows. First, two registers of length $n + 1$ qubits[6] are created and each qubit is initialized in the $|0\rangle$ state. Then a Hadamard transform H is applied to each qubit, resulting in the state $\frac{1}{2^{n+1}} \sum_{k,\ell=0}^{2^{n+1}-1} |k, \ell\rangle$. Next, conditioned on the content of the register holding the label k or ℓ, we add the corresponding multiple of P and Q, respectively, i.e., we implement the map

$$\frac{1}{2^{n+1}} \sum_{k,\ell=0}^{2^{n+1}-1} |k, \ell\rangle \mapsto \frac{1}{2^{n+1}} \sum_{k,\ell=0}^{2^{n+1}-1} |k, \ell\rangle |[k]P + [\ell]Q\rangle.$$

Hereafter, the third register is discarded and a quantum Fourier transform $\mathrm{QFT}_{2^{n+1}}$ on $n + 1$ qubits is computed on each of the two registers. Finally, the state of the first two registers—which hold a total of $2(n + 1)$ qubits—is measured. As shown in [45], the discrete logarithm m can be computed from this measurement data via classical post-processing. The corresponding quantum circuit is shown in Fig. 1.

Using Kitaev's phase estimation framework [35], Beauregard [2] obtained a quantum algorithm for factoring an integer N from a circuit that performs a conditional multiplication of the form $x \mapsto ax \bmod N$, where $a \in \mathbb{Z}_N$ is a random constant integer modulo N. The circuit uses only $2n + 3$ qubits, where n is the bitlength of the integer to be factored. An implementation of this algorithm on $2n + 2$ qubits, using Toffoli-gate-based modular multiplication is described

---

[6] Hasse's bound guarantees that the order of P can be represented with $n + 1$ bits.

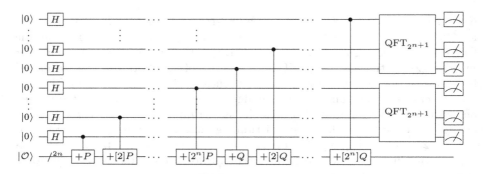

Fig. 1. Shor's algorithm to compute the discrete logarithm in the subgroup of an elliptic curve generated by a point P.

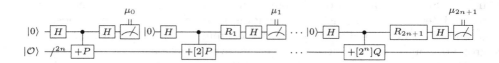

Fig. 2. Shor's algorithm to compute the discrete logarithm in the subgroup of an elliptic curve generated by a point P, analogous to the algorithms from [2,19,21]. The gates R_k are phase shift gates given by $\mathrm{diag}(1, e^{i\theta_k})$, where $\theta_k = -\pi \sum_{j=0}^{k-1} 2^{k-j} \mu_j$ and the sum runs over all previous measurements j with outcome $\mu_j \in \{0, 1\}$.

in [21]. Following the semiclassical Fourier transform method [19], one can modify Shor's ECDLP algorithm, resulting in the circuit shown in Fig. 2. The phase shift matrices $R_i = \begin{pmatrix} 1 & 0 \\ 0 & e^{i\theta_k} \end{pmatrix}$, $\theta_k = -\pi \sum_{j=0}^{k-1} 2^{k-j} \mu_j$, depend on all previous measurement outcomes $\mu_j \in \{0, 1\}$, $j \in \{0, \ldots, k-1\}$.

3   Reversible Modular Arithmetic

Shor's algorithm for factoring actually only requires modular multiplication of a quantum integer with classically known constants. In contrast, the elliptic curve discrete logarithm algorithm requires elliptic curve scalar multiplications to compute $[k]P + [\ell]Q$ for a superposition of values for the scalars k and ℓ. These scalar multiplications are comprised of elliptic curve point additions, which in turn consist of a sequence of modular operations on the coordinates of the elliptic curve points. This requires the implementation of full modular arithmetic, which means that one needs to add and multiply two integers held in quantum registers modulo the constant integer modulus p.

This section presents quantum circuits for reversible modular arithmetic on n-bit integers that are held in quantum registers. We provide circuit diagrams for the modular operations, in which black triangles on the right side of gate symbols indicate qubit registers that are modified and hold the result of the computation. Essential tools for implementing modular arithmetic are integer addition and bit shift operations on integers, which we describe first.

3.1   Integer Addition and Binary Shifts

The algorithms for elliptic curve point addition as described below need integer addition and subtraction in different variants: standard integer addition and subtraction of two n-bit integers, addition and subtraction of a classical constant integer, as well as controlled versions of those.

For adding two integers, we take the quantum circuit described by Takahashi et al. [49]. The circuit works on two registers holding the input integers, the first of size n qubits and the second of size $n + 1$ qubits. It operates in place, i.e. the contents of the second register are replaced to hold the sum of the inputs storing a possible carry bit in the additionally available qubit. To obtain a subtraction circuit, we implement an inverse version of this circuit. The carry bit in this case indicates whether the result of the subtraction is negative. Controlled versions of these circuits can be obtained by using partial reflection symmetry to save controls, which compares favorably to a generic version where simply all gates are controlled. For the constant addition circuits, we take the algorithms described in [21]. Binary doubling and halving circuits are needed for the Montgomery multiplication and inversion algorithms. They are implemented essentially as cyclic bit shifts realized by sequences of symmetric bit swap operations built from CNOT gates.

3.2   Modular Addition and Doubling

We now turn to modular arithmetic. The circuit shown in Fig. 3 computes a modular addition of two integers x and y held in n-qubit quantum registers $|x\rangle$ and $|y\rangle$, modulo the constant integer modulus p. It performs the operation in place $|x\rangle|y\rangle \mapsto |x\rangle|(x+y) \mod p\rangle$ and replaces the second input with the result. It uses quantum circuits for plain integer addition and constant addition and subtraction of the modulus. It uses two auxiliary qubits, one of which is used as an ancilla qubit in the constant addition and subtraction and can be in an unknown state to which it will be returned at the end of the circuit. The other qubit stores the bit that determines whether a modular reduction in form of a modulus subtraction actually needs to be performed or not. It is uncomputed at the end by a strict comparison circuit between the result and the first input. Modular subtraction is implemented by reversing the circuit.

The modular doubling circuit for a constant odd integer modulus p in Fig. 4 follows the same principle. There are two changes that make it more efficient than the addition circuit. First of all it works in place on only one n-qubit input integer $|x\rangle$, it computes $|x\rangle \mapsto |2x \mod p\rangle$. Therefore it uses only $n + 2$ qubits. The first integer addition in the modular adder circuit is replaced by a more efficient multiplication by 2 implemented via a cyclic bit shift as described in the previous subsection. Since we assume that the modulus p is odd in this circuit, the auxiliary reduction qubit can be uncomputed by checking whether

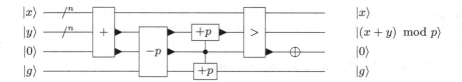

Fig. 3. `add_modp`: Quantum circuit for in-place modular addition $|x\rangle|y\rangle \mapsto |x\rangle|(x+y) \bmod p\rangle$. The circuit uses integer addition $+$, addition $+p$ and subtraction $-p$ of the constant modulus p, and strict comparison of two n-bit integers in the registers $|x\rangle$ and $|y\rangle$, where the output bit flips the carry qubit in the last register. The constant adders use an ancilla qubit in an unknown state $|g\rangle$, which is returned to the same state at the end of the circuit. To implement controlled modular addition `ctrl_add_modp`, one simply controls all operations in this circuit.

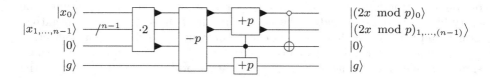

Fig. 4. `dbl_modp`: Quantum circuit for in-place modular doubling $|x\rangle \mapsto |2x \bmod p\rangle$ for an odd constant modulus p. The circuit uses a binary doubling operation $\cdot 2$ and addition $+p$ and subtraction $-p$ of the constant modulus p. The constant adders use an ancilla qubit in an unknown state $|g\rangle$, which is returned to the same state at the end of the circuit.

the least significant bit of the result is 0 or 1. A subtraction of the modulus has taken place if, and only if, this bit is 1.

For adding a classical constant to a quantum register modulo a classical constant modulus, we use the in-place modular addition circuit described in [21, Sect. 2]. The circuit operates on the n-bit input and requires only 1 ancilla qubit initialized in the state $|0\rangle$ and $n-1$ dirty ancillas that are given in an unknown state and will be returned in the same state at the end of the computation.

3.3   Modular Multiplication

Multiplication by Modular Doubling and Addition. Modular multiplication can be computed by repeated modular doublings and conditional modular additions. Figure 5 shows a circuit that computes the product $z = x \cdot y \bmod p$ for constant modulus p as described by Proos and Zalka [39, Sect. 4.3.2] by using a simple expansion of the product along a binary decomposition of the first multiplicand, i.e.

$$x \cdot y = \sum_{i=0}^{n-1} x_i 2^i \cdot y = x_0 y + 2(x_1 y + 2(x_2 y + \cdots + 2(x_{n-2} y + 2(x_{n-1} y)) \ldots)).$$

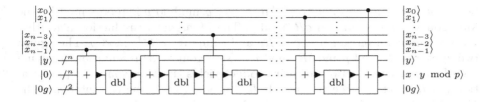

Fig. 5. dbl_modp: Quantum circuit for modular multiplication $|x\rangle|y\rangle|0\rangle \mapsto |x\rangle|y\rangle|x \cdot y \bmod p\rangle$ built from modular doublings dbl $\leftarrow$ dbl_modp and controlled modular additions $+ \leftarrow$ ctrl_add_modp.

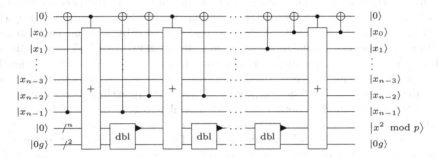

Fig. 6. squ_modp: Quantum circuit for modular squaring $|x\rangle|0\rangle \mapsto |x\rangle|x^2 \bmod p\rangle$ built from modular doublings dbl $\leftarrow$ dbl_modp and controlled modular additions $+ \leftarrow$ ctrl_add_modp.

The circuit runs on $3n + 2$ qubits, $2n$ of which are used to store the inputs, n to accumulate the result and 2 ancilla qubits are needed for the modular addition and doubling operations, one of which can be dirty. The latter could be taken to be one of the x_i, for example x_0 except in the last step, when the modular addition gate is conditioned on x_0. For simplicity, we assume it to be a separate qubit.

Figure 6 shows the corresponding specialization to compute a square $z = x^2 \bmod p$. It uses $2n + 3$ qubits by removing the n qubits for the second input multiplicand, and adding one ancilla qubit, which is used in round i to copy out the current bit x_i of the input in order to add x to the accumulator conditioned on the value of x_i.

Montgomery Multiplication. In classical applications, Montgomery multiplication [33] is often the most efficient choice for modular multiplication if the modulus does not have a special shape such as being close to a power of 2. Here, we explore it as an alternative to the algorithm using modular doubling and addition as described above.

In [33], Montgomery introduced a representation for an integer modulo p he called a p-residue that is now called the Montgomery representation. Let $R > p$

be an integer radix coprime to p. An integer a modulo p is represented by the Montgomery representation $aR \mod p$. The Montgomery reduction algorithm takes as input an integer $0 \leq c < Rp$ and computes $cR^{-1} \mod p$. Thus given two integers $aR \mod p$ and $bR \mod p$ in Montgomery representation, applying the Montgomery reduction to their product yields the Montgomery representation $(ab)R \mod p$ of the product. If R is a power of 2, one can interleave the Montgomery reduction with school-book multiplication, obtaining a combined Montgomery multiplication algorithm. The division operations usually needed for computing remainders are replaced by binary shifts in each round of the multiplication algorithm.

The multiplication circuit using modular doubling and addition operations described in the previous subsection contains two modular reductions in each round of the algorithm. Each of those is realized here by at least two integer additions. In contrast, the Montgomery algorithm shown in Fig. 7 avoids these and uses only one integer addition per round. This reduces the circuit depth in comparison to the double-and-add approach. However, it comes at the cost of requiring more qubits. The main issue is that the algorithm stores the information for each round, whether the odd modulus p had to be added to the intermediate result to make it even or not. This is done to allow divisions by 2 through a simple bit shift of an even number. These bits are still set at the end of the circuit shown in Fig. 7. To uncompute these values, we copy the result to another n-qubit register, and run the algorithm backwards, which essentially doubles the depth of the algorithm. But this still leads to a lower overall depth than the one of the double-and-add algorithm. Hence, switching to Montgomery

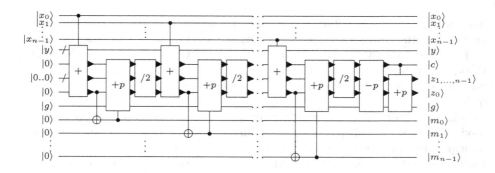

Fig. 7. mul_modp: Quantum circuit for the forward Montgomery modular multiplication $|x\rangle|y\rangle|0\rangle \mapsto |x\rangle|y\rangle|z = x \cdot y \mod p\rangle$. The qubit $|g\rangle$ is a dirty ancilla qubit in an unknown state. The qubit labeled $|m_i\rangle$ holds the information whether the intermediate result in round i was odd and thus whether p was added to it. The circuit uses integer addition $+$, integer addition $+p$ and subtraction $-p$ of the constant modulus p and a halving circuit $/2$ that performs a cyclic qubit shift. The last two gates reflect the standard conditional subtraction of p. To uncompute the qubits $|m_i\rangle$, one copies out the result z and runs the circuit backwards.

multiplication presents a trade-off between the required number of qubits and the multiplication circuit depth.

The same optimization as shown in the previous section allows to save $n - 1$ qubits when implementing a Montgomery squaring circuit that computes $z = x^2$ mod p.

3.4   Modular Inversion

Performing the modular inversion on a quantum computer is by far the most costly operation required in order to implement the affine group law of an elliptic curve. We use a reversible circuit for the extended binary greatest common divisor algorithm [48] that implements Kaliski's algorithm [25] for inverting a number xR mod p given in Montgomery representation for $R = 2^n$; i.e. an algorithm (i) which only uses elementary reversible operations such as Toffoli gates, (ii) whose sequence of instructions does not depend on the given input $x2^n$ mod p, and (iii) whose output is again in Montgomery form $x^{-1}2^n$ mod p.

We use the extended binary GCD algorithm to compute the representation of the greatest common divisor between x and p, with the added requirement to ensure property (ii), namely to make sure that the sequence of operations that carries out the Euclidean algorithm is the same, independent of x. In particular, an issue is that for different inputs $x \neq x'$ the usual, irreversible Euclidean algorithm can terminate in a different number of steps. To fix this, we include a counter register which is incremented upon termination of the loop to ensure the algorithm is run for $2n$ rounds, which is the worst-case runtime.

In the following algorithm to compute the Montgomery inverse the inputs are a prime p and a value x where $0 \leq x < p$. The output is $x^{-1}2^n$ mod p. In functional programming style (here, using F# syntax), Kaliski's algorithm is described as follows:

```
let MGinverse p x =
  let rec xmg u v r s k =
    match u, v, r, s with
      | _,0,r,_                      -> r
      | u,_,_,_ when u\%2=0 -> xmg (u >>> 1) v r (s <<< 1) (k+1)
      | _,v,_,_ when v\%2=0 -> xmg u (v >>> 1) (r <<< 1) s (k+1)
      | u,v,_,d when u > v  -> xmg ((u-v) >>> 1) v (r+s) (s <<< 1) (k+1)
      | _,_,_,_                      -> xmg u ((v-u) >>> 1) (r <<< 1) (r+s) (k+1)
    xmg p x 0 1 0
```

The algorithm actually computes only the so-called "almost inverse" which is of the form $x^{-1}2^k$, i.e., there is a secondary step necessary to convert to the correct form (not shown here). Note that here k depends on x, i.e., it is necessary to uncompute k. Two example executions are shown in Fig. 8.

As shown in Fig. 8, the actual number of steps that need to be executed until the gcd is obtained, depends on the actual input x: in the first example the usual Kaliski algorithm terminates after $k = 7$ steps, whereas in the second example the usual algorithm would terminate after $k = 5$ steps. To make the algorithm reversible, we must find an implementation that carries out the same operations, irrespective of the input. The two main ingredients to obtain such

u	11	11	11	11	5	2	1	1	1
v	8	4	2	1	1	1	1	0	0
r	0	0	0	0	1	3	3	6	6
s	1	1	1	1	2	4	8	11	11
k	0	1	2	3	4	5	6	7	7
ℓ	0	0	0	0	0	0	0	0	1

(a)

u	11	2	1	1	1	1	1	1	1
v	7	7	7	3	1	0	0	0	0
r	0	1	1	2	4	8	8	8	8
s	1	2	4	5	7	11	11	11	11
k	0	1	2	3	4	5	5	5	5
ℓ	0	0	0	0	0	0	1	2	3

(b)

Fig. 8. Two example runs of the reversible extended binary Euclidean algorithm to compute the Montgomery inverse modulo $p = 11$. Shown in (a) is the execution for input $x = 8$ which leads to termination of the usual irreversible algorithm after $k = 7$ steps. The algorithm is always executed for $2n$ rounds, where n is the bit-size of p which is an upper bound on the maximum number of steps required for general input x. Once the final step $v = 0$ has been reached, a counter register ℓ is incremented. Shown in (b) is the execution for input $x = 7$ which leads to termination after 5 steps after which the counter is incremented three times.

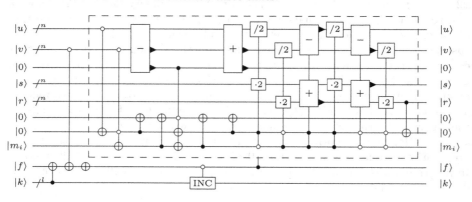

Fig. 9. Quantum circuit for the Montgomery-Kaliski round function. The function is repeated $2n$ times to obtain a reversible modular inversion algorithm. The n-qubit registers $|u\rangle$, $|v\rangle$, $|r\rangle$, $|s\rangle$ represent the usual coefficients in the binary Euclidean algorithm. The circuit uses integer subtraction $-$ and addition $+$, as well as multiplication and division by 2 functions $\cdot 2$ and $/2$ and an incrementer circuit INC. The circuits $\cdot 2$ and $/2$ are implemented as cyclic qubit shifts.

an implementation are (a) an upper bound of $2n$ steps that Kaliski's algorithm can take in the worst case [25] and (b) the introduction of a counter that ensures that either the computation is propagated forward or, in case the usual Kaliski algorithm has terminated, the counter is incremented. Such a counter can be implemented using $O(\log(n))$ qubits.

The circuit shown in Fig. 9 implements the Kaliski algorithm in a reversible fashion. We next describe the various registers used in this circuit and explain why this algorithm actually computes the same output as the Kaliski algorithm. The algorithm uses n-bit registers for inputs u and v, where u is initially set to the underlying prime p. As p is constant, the register can be prepared using bit flips corresponding to the binary representation of p. Moreover, v is initially set to the input x of which we would like to compute the inverse. Moving downward from the top, the next line represents a single ancilla qubit which is used to store an intermediate value which is the result of a comparison. Next is an $n + 1$-bit register for r and likewise an $n + 1$-bit register for s, so that the loop invariant $p = rv + su$ holds at each stage of the algorithm. Eventually, when $v = 0$ is reached, register r will hold the almost inverse and register s will be equal to p. The next 2 lines represent ancilla qubits which are used as scratch space to store an intermediate computation. The technically most interesting part is the next register which consists of a single qubit labeled m_i. This indicates that in round i, where $1 \leq i \leq 2n$, a fresh qubit is introduced, then acted upon by the circuit and then kept around.

After the maximum number of $2n$ rounds is executed, hence $2n$ qubits have been introduced and entangled in this way. The purpose of the qubit m_i is to remember which of the 4 branches in Kaliski's algorithm was taken in step i. As there are 4 branches, this choice could be naively encoded into 2 qubits, which however would lead to a space overhead of $4n$ instead of $2n$. The fact that one of these two qubits is actually redundant is shown below. The next qubit, labeled f in the figure, is part of a mechanism to unroll the entire algorithm which drives precisely one of two processes forward: either the Kaliski algorithm itself, or a counter, here represented as the "INC" operation. The flag f starts out in state 1 which indicates that the algorithm is in Kaliski-mode. Once the terminating condition $v = 0$ is reached, the flag switches to 0, indicating that the algorithm is in counter-mode. Finally, the register k holds the state of the counter. As the counter can take values between n and $2n$ only [25], it can be implemented using $\lceil \log_2(n) + 1 \rceil$ qubits only.

Having covered all registers that are part of the circuit, we next explain how the circuit is actually unraveled to compute the almost inverse. Shown in Fig. 9 is only one round. The circuit is applied over and over to the same set of qubit registers, with the sole exception of qubit m_i which depends on round i and which is initialized, acted upon, and then stored. In each round there are 4 possible branches. These are dispatched using the gates inside the dashed box. The first gate is a controlled NOT that acts only on the least significant bit of u, checking whether u is even. The next gate does the same for v, which flips the target bit in case u was odd and v was even. If both u and v are odd, the difference $u - v$ respectively $v - u$ has to be computed, depending on whether $u > v$ or $u \leq v$. To figure out which case actually holds, we use a subtractor and store the most significant qubit in the mentioned ancilla. The sequences of 5 gates underneath the two subtractors/adders serve as an encoder that prepares the following correspondence: '10' for the case u even, '01' for the case u odd, v even, '11' for the case both odd and $u > v$, and '00' for the case both odd and $u \leq v$. Denote the two bits involved in this encoding as 'ab' we see that b is the round qubit m_i. The fact that a can be immediately uncomputed is a consequence of the following observation.

In each step of Kaliski's algorithm, precisely one of r and s is even and the other is odd. If the updated value of r is even, then the branch must be the result of either the case v even or the case both u and v odd and $u \leq v$. Correspondingly, if the updated value of s is even, then the branch must have been the result of either the case u even or the case both u and v odd and $u > v$. Indeed, an even value of r arises only from the mentioned two branches v even or u and v both odd and $u \leq v$. Similarly, the other statement is obtained for s. The invariant $p = rv + su$ implies inductively that precisely one of r and s is even and the other henceforth must be odd.

Coming back to the dashed circuit, the next block of 6 gates is to dispatch the appropriate case, depending on the 2 bits a and b, which corresponds to the 4 branches in the match statement. Finally, the last CNOT gate between the least significant bit of r (indicating whether r is even) is used to uncompute 'a'.

The shown circuit is then applied precisely $2n$ times. At this juncture, the computation of the almost inverse will have stopped after k steps where $n \leq k \leq 2n$ and the counter INC will have been advanced precisely $2n - k$ times. The counter INC could be implemented using a simple increment $x \mapsto x+1$, however in our implementation we chose a finite state machine that has a transition function requiring less Toffoli gates.

Next, the register r which is known to hold $-x^{-1}2^k$ is converted to $x^{-1}2^n$. This is done by performing precisely $n - k$ controlled modular doublings and a sign flip. Finally, the result is copied out into another register and the entire circuit is run backwards.

4   Reversible Elliptic Curve Operations

Based on the reversible algorithms for modular arithmetic from the previous section, we now turn to implementing a reversible algorithm for adding two points on an elliptic curve. Next, we describe a reversible point addition in the generic case in which none of the exceptional cases of the simple affine Weierstrass group law occurs. After that, we describe a reversible algorithm for computing a scalar multiplication $[m]P$.

4.1   Point Addition

The reversible point addition we implement is very similar to the one described in Sect. 4.3 of [39]. It uses affine coordinates. As was also mentioned in [39] (and as we argue below), it is enough to consider the generic case of an addition. This means that we assume the following situation. Let $P_1, P_2 \in E(\mathbb{F}_p)$, $P_1, P_2 \neq \mathcal{O}$ such that $P_1 = (x_1, y_1)$ and $P_2 = (x_2, y_2)$. Furthermore let, $x_1 \neq x_2$ which means that $P_1 \neq \pm P_2$. Recall that then $P_3 = P_1 + P_2 \neq \mathcal{O}$ and it is given by $P_3 = (x_3, y_3)$, where $x_3 = \lambda^2 - x_1 - x_2$ and $y_3 = \lambda(x_1 - x_3) - y_1$ for $\lambda = (y_1 - y_2)/(x_1 - x_2)$.

As explained in [39], for computing the sum P_3 reversibly and in place (replacing the input point P_1 by the sum), the algorithm makes essential use of the fact that the slope λ can be re-computed from the result P_3 via the point addition $P_3 + (-P_2)$ independent of P_1 using the equation

$$\frac{y_1 - y_2}{x_1 - x_2} = -\frac{y_3 + y_2}{x_3 - x_2}.$$

Algorithm 1 depicts our algorithm for computing a controlled point addition. As input it takes the four point coordinates for P_1 and P_2, a control bit ctrl, and replaces the coordinates holding P_1 with the result $P_3 = (x_3, y_3)$. Note that we assume P_2 to be a constant point that has been classically precomputed, because we compute scalar multiples of the input points P and Q to Shor's algorithm by conditionally adding together precomputed 2-power multiples of these points as shown in Fig. 1 above. The point P_2 will thus always be one of these values. Therefore, operations involving the coordinates x_2 and y_2 are implemented as constant operations. Algorithm 1 uses two additional temporary variables λ and t_0. All the point coordinates and the temporary variables fit in n-bit registers and thus the algorithm can be implemented with a circuit on a quantum register $|x_1\ y_1\ \text{ctrl}\ \lambda\ t_0\ \text{tmp}\rangle$, where the register tmp holds auxiliary registers that are needed by the modular arithmetic operations used in Algorithm 1 as described in Sect. 3.

The algorithm is given as a straight line program of (controlled) arithmetic operations on the point coefficients and auxiliary variables. The comments at the end of the line after each operation show the current values held in the variable that is possibly changed. The notation $[\cdot]_1$ shows the value of the variable in case the control bit is ctrl $= 1$, if it is ctrl $= 0$ instead, the value is shown with $[\cdot]_0$. In the latter case, it is easy to check that the algorithm indeed returns the original state of the register.

The functions in the algorithm all use the fact that the modulus p is known as a classical constant. They relate to the algorithms described in Sect. 3 as follows:

- add_const_modp is a modular addition of a constant to a quantum state, sub_const_modp is its reverse, a modular subtraction of a constant.
- ctrl_add_const_modp is single qubit controlled modular addition of a constant to a qubit register, i.e. the controlled version of the above. Its reverse ctrl_sub_const_modp performs the controlled modular subtraction.
- ctrl_sub_modp is a single qubit controlled modular subtraction on two qubit registers, implemented as the reverse of the corresponding modular addition.
- ctrl_neg_modp is a single qubit controlled modular negation on a register.
- mul_modp, squ_modp, inv_modp are the out-of-place modular multiplication, squaring and inversion algorithms on two input qubit registers, respectively.

Figure 10 shows a quantum circuit that implements Algorithm 1. The quantum registers $|x_1\rangle, |y_1\rangle, |t_0\rangle, |\lambda\rangle$ all consist of n logical qubits, whereas $|\text{ctrl}\rangle$ is a single logical qubit. For simplicity in the circuit diagram, we do not show the register $|\text{tmp}\rangle$ with the auxiliary qubits. These qubits are used as needed by the modular arithmetic operations and are returned to their original state after each operation. The largest amount of ancilla qubits is needed by the modular inversion algorithm, which determines that we require $5n$ qubits in the register $|\text{tmp}\rangle$.

Algorithm 1. Reversible, controlled elliptic curve point addition. This algorithm operates on a quantum register holding the point $P_1 = (x_1, y_1)$, a control bit ctrl, and two auxiliary values λ and t_0. In addition it needs auxiliary registers for the functions that are called as described for those functions. The second point $P_2 = (x_2, y_2)$ is assumed to be a precomputed classical constant. For $P_1, P_2 \neq \mathcal{O}$, $P_1 \neq \pm P_2$, if ctrl $= 1$, the algorithm correctly computes $P_1 \leftarrow P_1 + P_2$ in the register holding P_1; if ctrl $= 0$ it returns the register in the received state.

1: `sub_const_modp` x_1 x_2;   // $x_1 \leftarrow x_1 - x_2$
2: `ctrl_sub_const_modp` y_1 y_2 ctrl;   // $y_1 \leftarrow [y_1 - y_2]_1, [y_1]_0$
3: `inv_modp` x_1 t_0;   // $t_0 \leftarrow 1/(x_1 - x_2)$t
4: `mul_modp` y_1 t_0 λ;   // $\lambda \leftarrow [\frac{y_1 - y_2}{x_1 - x_2}]_1, [\frac{y_1}{x_1 - x_2}]_0$
5: `mul_modp` λ x_1 y_1;   // $y_1 \leftarrow 0$
6: `inv_modp` x_1 t_0;   // $t_0 \leftarrow 0$
7: `squ_modp` λ t_0;   // $t_0 \leftarrow \lambda^2$
8: `ctrl_sub_modp` x_1 t_0 ctrl;   // $x_1 \leftarrow [x_1 - x_2 - \lambda^2]_1, [x_1 - x_2]_0$
9: `ctrl_add_const_modp` x_1 $3x_2$ ctrl;   // $x_1 \leftarrow [x_2 - x_3]_1, [x_1 - x_2]_0$
10: `squ_modp` λ t_0;   // $t_0 \leftarrow 0$
11: `mul_modp` λ x_1 y_1;   // $y_1 \leftarrow [y_3 + y_2]_1, [y_1]_0$
12: `inv_modp` x_1 t_0;   // $t_0 \leftarrow [\frac{1}{x_2 - x_3}]_1, [\frac{1}{x_1 - x_2}]_0$
13: `mul_modp` t_0 y_1 λ;   // $\lambda \leftarrow 0$
14: `inv_modp` x_1 t_0;   // $t_0 \leftarrow 0$
15: `ctrl_neg_modp` x_1 ctrl;   // $x_1 \leftarrow [x_3 - x_2]_1, [x_1 - x_2]_0$
16: `ctrl_sub_const_modp` y_1 y_2 ctrl;   // $y_1 \leftarrow [y_3]_1, [y_1]_0$
17: `add_const_modp` x_1 x_2;   // $x_1 \leftarrow [x_3]_1, [x_1]_0$

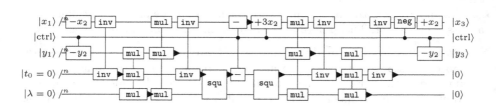

Fig. 10. Quantum circuit for controlled elliptic curve point addition. All operations are modulo p and we use the abbreviations $+ \leftarrow$ add_modp, $- \leftarrow$ sub_modp, mul $\leftarrow$ mul_modp, squ $\leftarrow$ squ_modp, inv $\leftarrow$ inv_modp.

To avoid permuting the wires between gates, we have used a split gate notation for some modular operations. For all gates, the black triangles mark the output wire that contains the result. As described in Sect. 3, addition and subtraction gates carry out their operations in place, meaning that one of the input registers is overwritten with the result. Modular multiplication, squaring and inversion operate out of place and store the result in a separate output register.

Remark 1 (Projective coordinates). As can be seen from Sect. 3, modular inversion is by far the most complex and resource consuming part of the elliptic curve point addition. The need for computing and uncomputing the slope λ leads to four calls to the inversion in Algorithm 1. In accordance with the observations provided in [39], it accounts for the main cost of the algorithm.

Unsurprisingly, this situation resembles the one for classical modular arithmetic. For example, in elliptic curve cryptography, a modular inversion can be two orders of magnitudes more costly than a modular multiplication, depending on the specific prime field. A significant speed-up can be achieved by using some form of projective coordinates[7], which allow to avoid almost all modular inversions in cryptographic protocols by essentially multiplying through with all denominators. This comes at the relatively small cost of storing more coefficients and a moderate increase in addition and multiplication operations and has proved highly effective. It is thus a natural question to ask whether the use of projective coordinates can also make Shor's algorithm more efficient.

There are several obstacles that make it non-trivial to use projective coordinates for quantum algorithms, such as the fact that each point is represented by an equivalence class of coordinate vectors and the increased number of temporary variables, which need to be uncomputed [31]. In this work, we thus refrained from investigating projective coordinate systems any further and leave it as an open problem to explore their benefits in the context of Shor's algorithm.

4.2   Scalar Multiplication

In order to compute a scalar multiplication $[m]P$ of a known base point P, we also follow the approach outlined in [39]. Namely, by classically precomputing all n 2-power multiples of P, the scalar multiple can be computed by a sequence of n controlled additions of those constant points to an accumulator in a quantum register along the binary representation of the scalar. Namely, let $m = \sum_{i=0}^{n-1} m_i 2^i$, $m_i \in \{0, 1\}$, be the binary representation of the n-bit scalar m. Then,

$$[m]P = [\sum_{i=0}^{n-1} m_i 2^i]P = \sum_{i=0}^{n-1} m_i([2^i]P).$$

This has the advantage that all doubling operations can be carried out on a classical computer and the quantum circuit only requires the generic point addition, which simplifies the overall implementation.

The latter has been argued by Proos and Zalka in [39, Sect. 4.2]. They say that on average, for any addition step, the probability of an exceptional case is negligibly low, and hence this will only have a negligible influence on the fidelity of the algorithm. To prevent the addition with the point at infinity in the first step, they suggest to initialize the register with a non-zero multiple of the point P. For the purpose of estimating resources for Shor's algorithm, we follow the

---

[7] A collection of possible coordinate systems and the corresponding formulas to carry out the group law is provided at https://www.hyperelliptic.org/EFD/.

approach by Proos and Zalka and only consider the generic group law. We will have a closer look at the details next.

Counting Scalars with Exceptional Cases. As explained in Sect. 2, Shor's algorithm involves generating a superposition over all possible pairs of $(n+1)$-bit strings k and ℓ, i.e. the state $\frac{1}{2^{n+1}} \sum_{k,\ell=0}^{2^{n+1}-1} |k,\ell\rangle$. Then over this superposition, involving two additional n-qubit registers to hold an elliptic curve point, one computes a double scalar multiplication $\frac{1}{2^{n+1}} \sum_{k,\ell=0}^{2^{n+1}-1} |k,\ell\rangle |[k]P + [\ell]Q\rangle$ of the input points given by the ECDLP instance.

Figure 1 depicts the additional elliptic curve point register to be initialized with a representation of the neutral element $\mathcal{O}$. But if we only consider the generic case of the group law, the first addition of P would already involve an exceptional case due to one of the inputs being $\mathcal{O}$. Proos and Zalka [39] propose to solve this issue by instead initializing the register with a uniform random non-zero multiple of P, say $[a]P$ for a random $a \in \{1, 2, \ldots, r-1\}$. Recall that r is the order of P which we assume to be a large prime. Now, if $a \notin \{1, r-1\}$, the first point addition with P works as a generic point addition. With high probability, this solves the issue of an exception in the first addition, but still exceptions occur along the way for many of the possibilities for bit strings k and ℓ. Whenever a bit string leads to an exceptional case in the group law, it produces a wrong result for the double scalar multiplication and pollutes the quantum register. We call such a scalar invalid. For Shor's algorithm to work, the overall number of such invalid scalars must be small enough. In the following we count these scalars, similar to the reasoning in [39].

Exceptional Additions of a Point to Itself. Let $a \in \{1, 2, \ldots, r-1\}$ be fixed and write $k = \sum_{i=0}^{n} k_i 2^i$, $k_i \in \{0,1\}$. We first consider the exceptional case in which both input points are the same, which we call an exceptional doubling. If $a = 1$, this occurs in the first iteration for $k_0 = 1$, because we attempt to add P to itself. This means that for $a = 1$, all scalars k with $k_0 = 1$ lead to a wrong result and therefore half of the scalars are invalid, i.e. in total 2^n.

For $a = 2$, the case $k_0 = 1$ is not a problem since the addition $[2]P + P$ is a generic addition, but $(k_0, k_1) = (0, 1)$ leads to an exceptional doubling operation in the second controlled addition. This means that all scalars $(0, 1, k_2, \ldots, k_n)$ are invalid. These are one quarter of all scalars, i.e. 2^{n-1}.

For general a, assume that k is a scalar such that the first $i - 1$ additions, $i \in \{1, \ldots, n\}$, controlled on the bits $k_0, \ldots, k_{i-1}$ do not encounter any exceptional doubling cases. The i-th addition means the addition of $[2^i]P$ for $0 \le i \le n$. Then the i-th addition is an exceptional doubling if, and only if

$$a + (k_0 + k_1 \cdot 2 + \cdots + k_{i-1} \cdot 2^{i-1}) = 2^i \pmod{r}.$$

If i is such that $2^i < r$. Then, the above condition is equivalent to the condition $a = 2^i - \sum_{j=0}^{i-1} k_j \cdot 2^j$ over the integers. This means that an a can only lead to an exceptional doubling in the i-th addition if $a \in \{1, \ldots, 2^i\}$. Furthermore, if i

is the smallest integer, such that there exist $k_0, \ldots, k_{i-1}$ such that this equation holds, we can conclude that $a \in \{2^{i-1} + 1, \ldots, 2^i\}$ and $k_{i-1} = 0$. In that case, any scalar of the form $(k_0, \ldots, k_{i-2}, 0, 1, *, \ldots, *)$ is invalid. The number of such scalars is 2^{n-i}. If i is instead such that $2^i \geq r$ and if $a \leq 2^i - \mu r$ for some positive integer $\mu \leq \lfloor 2^i/r \rfloor$, then in addition to the solutions given by the equation over the integers as above, there exist additional solutions given by the condition $a = (2^i - \mu r) - \sum_{j=0}^{i-1} k_j \cdot 2^j$, namely $(k_0, \ldots, k_{i-1}, 1, *, \ldots, *)$. The maximal number of such scalars is $\lfloor (2^i - a)/r \rfloor 2^{n-i}$, but we might have counted some of these already.

For a given $a \in \{1, 2, \ldots, r - 1\}$, denote by S_a the set of scalars that contain an exceptional doubling, i.e. the set of all $k = (k_0, k_1, \ldots, k_n) \in \{0, 1, \}^{n+1}$ such that there occurs an exceptional doubling when executing the addition $[a + \sum_{j=0}^{i-1} k_j \cdot 2^j]P + [2^i]P$ for any $i \in \{0, 1, \ldots, n\}$. Let $i_a = \lceil \log(a) \rceil$. Then, an upper bound for the number of invalid scalars is given by

$$\#S_a \leq 2^{n-i_a} + \sum_{i=\lceil \log(r) \rceil}^{n} \lfloor (2^i - a)/r \rfloor 2^{n-i}.$$

Hasse's bound gives $\lceil \log(r) \rceil \geq n - 1$, which means that $\#S_a \leq 2^{n-i_a} + 2\lfloor (2^{n-1} - a)/r \rfloor + \lfloor (2^n - a)/r \rfloor \leq 2^{n-i_a} + 8$. Hence on average, the number of invalid scalars over a uniform choice of $k \in \{1, \ldots, r - 1\}$ can be bounded as

$$\sum_{a=1}^{r-1} \Pr(a) \cdot \#S_a \leq \frac{1}{r-1} \sum_{a=1}^{r-1} 2^{n-\lceil \log(a) \rceil} + 8.$$

Grouping values of a with the same $\lceil \log(a) \rceil$ and possibly adding terms at the end of the sum, the first term can be simplified and further bounded by $\frac{1}{r-1}(2^n + \lceil \log(r-1) \rceil 2^{n-1}) = (2 + \lceil \log(r-1) \rceil)\frac{2^{n-1}}{r-1}$. For large enough bitsizes, we use that $r - 1 \geq 2^{n-1}$ and obtain the upper bound on the expected number of invalid scalars of roughly $\lceil \log(r) \rceil + 10 \approx n + 10$. This corresponds to a negligible fraction of about $n/2^{n+1}$ of all scalars.

Exceptional Additions of a Point to Its Negative. To determine the number of invalid scalars arising from the second possibility of exceptions, namely the addition of a point to its negative, we carry out the same arguments. An invalid scalar is a scalar that leads to an addition $[-2^i]P + [2^i]P$. The condition on the scalar a is slightly changed with 2^i replaced by $r - 2^i$, i.e.

$$a + (k_0 + k_1 \cdot 2 + \cdots + k_{i-1} \cdot 2^{i-1}) = r - 2^i \pmod{r}.$$

Whenever this equation holds over the integers, i.e. $r - a = 2^i + (k_0 + k_1 \cdot 2 + \cdots + k_{i-1} \cdot 2^{i-1})$ holds, we argue analogously as above. If $2^i < r$ and $r - a \in \{2^i, \ldots, 2^{i+1} - 1\}$, there are 2^{n-i} invalid scalars. Similar arguments as above for the steps such that $2^i > r$ lead to similar counts. Overall, we conclude that in this case the fraction of invalid scalars can also be approximated by $n/2^{n+1}$.

Exceptional Additions of the Point at Infinity. Since the quantum register holding the elliptic curve point is initialized with a non-zero point and the multiples of P added during the scalar multiplication are also non-zero, the point at infinity can only occur as the result of an exceptional addition of a point to its negative. Therefore, all scalars for which this occurs have been excluded previously and we do not further consider this case.

Overall, an approximate upper bound for the fraction of invalid scalars among the superposition of all scalars due to exceptional cases in the addition law is $2n/2^{n+1} = n/2^n$.

Double Scalar Multiplication. In Shor's algorithm with the above modification, one needs to compute a double scalar multiplication $[a + k]P + [\ell]Q$ where P and Q are the points given by the ECDLP instance we are trying to solve and a is a fixed uniformly random non-zero integer modulo r. We are trying to find the integer m modulo r such that $Q = [m]P$. Since r is a large prime, we can assume that $m \in \{1, \ldots, r - 1\}$ and we can write $P = [m^{-1}]Q$. Multiplication by m^{-1} on the elements modulo r is a bijection, simply permuting these scalars. Hence, after having dealt with the scalar multiplication to compute $[a + k]P$ above, we can now apply the same treatment to the second part, the addition of $[\ell]Q$ to this result.

Let a be chosen uniformly at random. For any k, we write $[a + k]P = [m^{-1}(a + k)]Q$. Assume that k is a valid scalar for this fixed choice of a. Then, the computation of $[a + k]P$ did not involve any exceptional cases and thus $[a + k]P \neq \mathcal{O}$, which means that $a + k \neq 0 \pmod{r}$. If we assume that the unknown discrete logarithm m has been chosen from $\{1, \ldots, r - 1\}$ uniformly at random, then the value $b = m^{-1}(a+k) \mod r$ is uniform random in $\{1, \ldots, r-1\}$ as well, and we have the same situation as above when we were looking at the choice of a and the computation of $[a + k]P$.

Using the rough upper bound for the fraction of invalid scalars from above, for a fixed random choice of a, the probability that a random scalar k is valid, is at least $1 - n/2^n$. Further, the probability that (k, ℓ) is a pair of valid scalars for computing $[a + k]P + [\ell]Q$, conditioned on k being valid for computing $[a + k]P$ is also at least $1 - n/2^n$. Hence, for a fixed uniform random a, the probability for (k, ℓ) being valid is at least $(1 - n/2^n)^2 = 1 - n/2^{n-1} + n^2/2^{2n} \approx 1 - n/2^{n-1}$. This result confirms the rough estimate by Proos and Zalka [39, Sect. 4.2] of a fidelity loss of $4n/p \geq 4n/2^{n+1}$.

Remark 2 (Complete addition formulas). There exist complete formulas for the group law on an elliptic curve in Weierstrass form [7]. This means that there is a single formula that can evaluate the group law on any pair of $\mathbb{F}_p$-rational points on the curve and thus avoids the occurrence of exceptional cases altogether. For classical computations, this comes at the cost of a relatively small slowdown [40]. Using such formulas would increase the algorithm's fidelity in comparison to the above method. Furthermore, there exist alternative curve models for elliptic curves which allow coordinate systems that offer even more efficient complete formulas. One such example is the twisted Edwards form of an elliptic curve [4].

However, not all elliptic curves allow a curve model in twisted Edwards form, like, for example, the prime order NIST curves. We leave it as an open problem to investigate the use of a complete group law, or more generally the use of different curve models and coordinate systems in Shor's ECDLP algorithm.

5 Cost and Resource Estimates for Shor's Algorithm

We implemented the reversible algorithm for elliptic curve point addition on elliptic curves E in short Weierstrass form defined over a prime field $\mathbb{F}_p$, where p has n bits, as shown in Algorithm 1 and Fig. 10 in Sect. 4 in F# within the quantum computing software tool suite LIQ$Ui|\rangle$ [52]. This allows us to test and simulate the circuit and all its components and obtain precise counts of the number of qubits, the number of Toffoli gates and the Toffoli gate depth for a working simulation. We thus do not have to rely on mere estimates obtained by pen-and-paper calculations and thus gain a higher confidence in the results. When implementing the algorithms, our overall emphasis was to minimize first the number of required logical qubits and second the Toffoli gate count.

We have simulated and tested our implementation for cryptographically relevant parameter sizes and were able to simulate the elliptic curve point addition circuit for curves over prime fields of size up to 521 bits. For each case, we computed the number of qubits required to implement the circuit, and its size and depth in terms of Toffoli gates.

Number of Logical Qubits. The number of logical qubits of the modular arithmetic circuits in our simulation that are needed in the elliptic curve point addition are given in Table 1. We list each function with its total required number of qubits and the number of ancilla qubits included in that number. All ancilla qubits are expected to be input in the state $|0\rangle$ and are returned in that state, except for the circuits in the first two rows, which only require one or two such ancilla qubits and $n - 1$ or $n - 2$ ancillas in an unknown state to which they will be returned. The addition, subtraction and negation circuits all work in place, such that one n-qubit input register is replaced with the result. The multiplication, squaring and inversion circuits require an n-qubit register with which the result of the computation is XOR-ed.

Although the modular multiplication circuit based on modular doubling and additions uses less qubits than Montgomery multiplication, we have used the Montgomery approach to report the results of our experiments. Since the lower bound on the overall required number of qubits is dictated by the modular inversion circuit, neither multiplication approach adds qubit registers to the elliptic curve addition circuit since they can use ancilla qubits provided by the inversion algorithm. We therefore find that the Montgomery circuit is the better choice then because it reduces the number of Toffoli gates substantially.

Table 1. Total number of qubits and number of Toffoli gates needed for the modular arithmetic circuits used in the elliptic curve point addition on $E/\mathbb{F}_p$ with n-bit prime p. The column labeled "ancilla" denotes the number of required ancilla qubits included in the total count. Except for the first two rows (un-/controlled constant addition/subtraction), they are expected to be input in state $|0\ldots0\rangle$ and are returned in that state. The constant addition/subtraction circuits in the first row only need one clean ancilla qubit and can take $n-1$ dirty ancilla qubits in an unknown state, in which they are returned. The controlled constant addition/subtraction circuits in the second row use two dirty ancillas.

Modular arith. circuit	# of qubits		# Toffoli gates
	Total	Ancillas	
add_const_modp, sub_const_modp	$2n$	n	$16n\log_2(n) - 26.9n$
ctrl_add_const_modp, ctrl_sub_const_modp	$2n+1$	n	$16n\log_2(n) - 26.9n$
ctrl_sub_modp	$2n+4$	3	$16n\log_2(n) - 23.8n$
ctrl_neg_modp	$n+3$	2	$8n\log_2(n) - 14.5n$
mul_modp (dbl/add)	$3n+2$	2	$32n^2\log_2(n) - 59.4n^2$
mul_modp (Montgomery)	$5n+4$	$2n+4$	$16n^2\log_2(n) - 26.3n^2$
squ_modp (dbl/add)	$2n+3$	3	$32n^2\log_2(n) - 59.4n^2$
squ_modp (Montgomery)	$4n+5$	$2n+5$	$16n^2\log_2(n) - 26.3n^2$
inv_modp	$7n+2\lceil\log_2(n)\rceil+9$	$5n+2\lceil\log_2(n)\rceil+9$	$32n^2\log_2(n)$

Because the maximum amount of qubits is used during an inversion operation, the overall number of logical qubits for the controlled elliptic curve point addition in our simulation is

$$9n + 2\lceil\log_2(n)\rceil + 10.$$

In addition to the $7n + 2\lceil\log_2(n)\rceil + 9$ required by the inversion, an additional qubit is needed for the control qubit $|\text{ctrl}\rangle$ of the overall operation and $2n$ more qubits are needed since two n-qubit registers need to hold intermediate results during each inversion.

Number of Toffoli Gates and Depth. Perhaps surprisingly, the precise resource count of the number of Toffoli gates in the constructed circuits is not a trivial matter. There are two main reasons for this: first, as constants are folded, the actual value of the constants matter as different bit-patterns (e.g., of the underlying prime p) give rise to different circuits. This effect, however, is not large and does not impact the leading order coefficients for the functions in the table. Second, the asymptotically dominating cost arises from the incrementer construction based on [21]. For the basic functions reported in Table 1 one can determine the number of incrementers used, which is either of the form $an\log_2(n)$ or $an^2\log_2(n)$ with a constant a. We determined the leading order term by inspection of the circuit and determining how many constant incrementers occur. Then we computed a regression of the next order term by solving a standard polynomial interpolation problem. The results are summarized in the last column of Table 1.

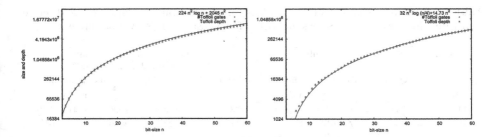

Fig. 11. Shown on the left are resource estimates for the number of Toffoli gates and the Toffoli gate depth for the implementation of elliptic curve point addition $|P\rangle \mapsto |P + Q\rangle$, where Q is a constant point. Shown on the right are resource estimates for the same metrics for modular multiplication $|x\rangle \mapsto |ax \bmod N\rangle$, where a and N are constants. Fitting the data for the elliptic curve case we obtain a scaling as $224n^2 \log_2(n) + 2045n^2$ up to lower order terms. The cost for the entire Shor algorithm over the elliptic curve scales as $2n$ the cost for a single point addition, i.e. $448n^3 \log_2(n)$ up to lower order terms. As shown in [21], the cost for modular multiplication scales as $32n^2(\log_2(n) - 2) + 14.73n^2$ and the cost of the entire Shor factoring algorithm scales as $64n^2 \log_2(n)$.

Putting everything together, we now obtain an estimate for the entire group law as computed by Algorithm 1: As there are a total of 4 inverters, 2 squarers, and 4 multipliers, we obtain that the leading order coefficient of a single point addition is $224 = 4 \cdot 32 + 2 \cdot 16 + 4 \cdot 16$. We then again perform a regression to determine the next coefficient. As a result, we estimate that the number of Toffoli gates in the point addition circuit scales as $224n^2 \log_2(n) + 2045n^2$. Figure 11 shows the scaling of the estimates for the Toffoli gate count and the Toffoli gate depth of the circuit for a range of relatively small bit sizes n. To estimate the overall resource requirements for Shor's algorithm, one simply multiplies by $2n$, since the controlled point addition is iterated $2n$ times. This leads to the overall estimate for the scaling of the number of Toffoli gates in Shor's ECDLP algorithm as

$$(448 \log_2(n) + 4090)n^3.$$

With respect to a given circuit, the Toffoli depth is computed as follows: we sweep all gates in the circuits and keep a running counter for each qubit on which time step it was acted upon last by a Toffoli gate. The depth is then the maximum of these quantities over all qubits. As the number of qubits is comparatively small in the circuits considered here, we can perform these updates efficiently, leading to an algorithm to compute the depth in time linear in the number of gates in the circuit. Note that whenever we encounter a CNOT or NOT gate, we do not increase the counter as by our assumption these gates do not contribute to the overall depth as they are Clifford gates. Overall, we find that the circuit Toffoli depth is a little bit smaller than the total number of Toffoli gates which shows that there is some parallelism in the circuit that can be exploited when implementing it on a quantum computer that facilitates parallel application of quantum gates.

Table 2. Resource estimates of Shor's algorithm for computing elliptic curve discrete logarithms in $E(\mathbb{F}_p)$ versus Shor's algorithm for factoring an RSA modulus N.

ECDLP in $E(\mathbb{F}_p)$ simulation results					Factoring of RSA modulus N interpolation from [21]		
$\lceil \log_2(p) \rceil$ bits	#Qubits	#Toffoli gates	Toffoli depth	Sim time (s)	$\lceil \log_2(N) \rceil$ (bits)	#Qubits	#Toffoli gates
110	1014	$9.44 \cdot 10^9$	$8.66 \cdot 10^9$	273	512	1026	$6.41 \cdot 10^{10}$
160	1466	$2.97 \cdot 10^{10}$	$2.73 \cdot 10^{10}$	711	1024	2050	$5.81 \cdot 10^{11}$
192	1754	$5.30 \cdot 10^{10}$	$4.86 \cdot 10^{10}$	1 149	–	–	–
224	2042	$8.43 \cdot 10^{10}$	$7.73 \cdot 10^{10}$	1 881	2048	4098	$5.20 \cdot 10^{12}$
256	2330	$1.26 \cdot 10^{11}$	$1.16 \cdot 10^{11}$	3 848	3072	6146	$1.86 \cdot 10^{13}$
384	3484	$4.52 \cdot 10^{11}$	$4.15 \cdot 10^{11}$	17 003	7680	15362	$3.30 \cdot 10^{14}$
521	4719	$1.14 \cdot 10^{12}$	$1.05 \cdot 10^{12}$	42 888	15360	30722	$2.87 \cdot 10^{15}$

We compare our results to the corresponding simulation results for Shor's factoring algorithm presented in [21], where the corresponding function is modular constant multiplication. In this case, the number of Toffoli gates scales as $32n^2(\log_2(n) - 2) + 14.73n^2$, where n is the bitsize of the modulus to be factored. As above, to estimate the overall resource requirements, one again multiplies by $2n$, which gives $(64(\log_2(n) - 2) + 29.46)n^3$.

Table 2 contains the resources required in our simulated circuits for parameters of cryptographic magnitude that are used in practice. The simulation time only refers to our implementation of the elliptic curve group law. The simulation timings were measured when running our LIQ$Ui|\rangle$ implementation on an HP ProLiant DL580 Gen8 machine consisting of 4 Intel Xeon processors @ 2.20 GHz and 3 TB of memory. The rows for ECC and RSA are aligned so that the corresponding parameters provide a similar classical security level according to NIST recommendations from 2016.

Remark 3. Recently, a variation of Shor's quantum algorithms for computing discrete logarithms and factoring was developed in [14]. The basic observation of this paper is that the quantum circuit size can be reduced at the cost of a more expensive classical post-processing. Inasmuch as the argument given in [14] applies only to the case of discrete logarithms that are guaranteed to be small, it does not apply to the ECC column in Table 2, however, the argument given about factoring *does* apply. This means that if one is willing to perform a classical post-processing based on lattice enumeration algorithms, one can reduce the rounds of phase estimation from $2n$ to $n/2$. While this does not save qubits, it does lead to a division by 4 of all reported resource counts on the number of Toffoli gates in the RSA column of Table 2.

Remark 4. In [3] it was shown that even in the presence of a quantum computer with limited size, the heuristic time complexity of the Number Field Sieve can be reduced from $L^{1.901+o(1)}$ to $L^{1.386+o(1)}$. Specifically, it is shown that a sub-linear number of $n^{2/3+o(1)}$ qubits is enough for a hybrid quantum-classical algorithm to work. Taking this result into account, shifts the alignment of security parameters in Table 2, however, it does so in a way that significantly complicates the analysis as constants for the algorithm given in [3] would have to be worked out.

6   Discussion

Comparing to the theoretical estimates by Proos and Zalka in [39], our results provide additional evidence that for cryptographically relevant sizes, elliptic curve discrete logarithms can be computed more easily on a quantum computer than factoring an RSA modulus of similar classical security level. However, neither the Toffoli gate counts for factoring that were provided in [21], nor the ones for elliptic curves that were provided here are as low as the theoretically predicted "time" estimates in [39]. Also, the number of qubits in our simulation-based estimates is higher than the ones conjectured in [39].

The reasons for the larger number of qubits lie in the implementation of the modular inversion algorithm. Proos and Zalka describe a version of the standard Euclidean algorithm which requires divisions with remainder. We chose to implement the binary GCD algorithm, which only requires additions, subtractions and binary bit shifts. One optimization that applies to both algorithms is register sharing as proposed in [39, Sect. 5.3.5]. The standard Euclidean algorithm as well as the binary GCD work on four intermediate variables, requiring $4n$ bits in total. In our description in Sect. 3.4, these are the variables u, v, r, s. However, Proos and Zalka use a heuristic argument to show that they actually only need about $2n + 8\sqrt{n}$ bits at any time during the algorithm. A complication for implementing this optimization is that the boundaries between variables change during the course of the algorithm. We leave it for future work to implement and simulate a reversible modular inversion algorithm that makes use of register sharing to reduce the number of qubits.

Since the basis for register sharing in [39] is an experimental analysis, Proos and Zalka provide a space analysis that does not take into account the register sharing optimization. With this space analysis, we still need about $2n$ qubits more than their Euclidean algorithm. These qubits come from the fact that our extended binary GCD algorithm generates one bit of garbage in each of the $2n$ rounds. In contrast, [39] only needs n carry qubits. Furthermore, we need an additional n-qubit register to copy out the result and run the algorithm in reverse to clean-up all garbage and ancilla qubits. We could not see how to avoid this and how to achieve step-wise reversibility for the extended binary Euclidean algorithm. We leave it as a future challenge to match or even lower the number of qubits for reversible modular inversion from [39].

To summarize, we presented quantum circuits to implement Shor's algorithm to solve the ECDLP. We analyzed the resources required to implement these circuits and simulated large parts of them on a classical machine. Indeed, the overwhelming majority of gates in our circuits are Toffoli, CNOT, and NOT gates, which implement the controlled addition of an elliptic curve point that is known at circuit generation time. We were able to classically simulate the point addition circuit and hence test it for implementation bugs. Our findings imply that attacking elliptic curve cryptography is indeed easier than attacking RSA, even for relatively small key sizes.

Acknowledgments. We thank Christof Zalka for feedback and discussions and the anonymous reviewers for their valuable comments.

References

1. Amy, M., Maslov, D., Mosca, M., Roetteler, M.: A meet-in-the-middle algorithm for fast synthesis of depth-optimal quantum circuits. IEEE Trans. Comput. Aided Des. Integr. Circ. Syst. **32**(6), 818–830 (2013)
2. Beauregard, S.: Circuit for Shor's algorithm using 2n+3 qubits. Quantum Inf. Comput. **3**(2), 175–185 (2003)
3. Bernstein, D.J., Biasse, J.-F., Mosca, M.: A low-resource quantum factoring algorithm. In: Lange and Takagi [28], pp. 330–346
4. Bernstein, D.J., Birkner, P., Joye, M., Lange, T., Peters, C.: Twisted edwards curves. In: Vaudenay, S. (ed.) AFRICACRYPT 2008. LNCS, vol. 5023, pp. 389–405. Springer, Heidelberg (2008). https://doi.org/10.1007/978-3-540-68164-9_26
5. Blake-Wilson, S., Bolyard, N., Gupta, V., Hawk, C., Moeller, B.: Elliptic curve cryptography (ECC) cipher suites for Transport Layer Security (TLS). RFC 4492, RFC Editor (2006)
6. Bos, J.W., Costello, C., Miele, A.: Elliptic and hyperelliptic curves: a practical security analysis. In: Krawczyk, H. (ed.) PKC 2014. LNCS, vol. 8383, pp. 203–220. Springer, Heidelberg (2014). https://doi.org/10.1007/978-3-642-54631-0_12
7. Bosma, W., Lenstra, H.W.: Complete system of two addition laws for elliptic curves. J. Number Theor. **53**(2), 229–240 (1995)
8. Certicom Research: Standards for efficient cryptography 2: recommended elliptic curve domain parameters. Standard SEC2, Certicom (2000)
9. Crandall, R., Pomerance, C. (eds.): Prime Numbers - A Computational Perspective. Springer, New York (2005). https://doi.org/10.1007/0-387-28979-8
10. Dierks, T., Rescorla, E.: The Transport Layer Security (TLS) protocol version 1.2. RFC 5246, RFC Editor (2008)
11. Diffie, W., Hellman, M.E.: New directions in cryptography. IEEE Trans. Inf. Theor. **22**(6), 644–654 (1976)
12. Dingledine, R., Mathewson, N., Syverson, P.F.: Tor: the second-generation onion router. In: Blaze, M. (ed.) USENIX Security 2004, pp. 303–320. USENIX (2004)
13. ECC Brainpool: ECC brainpool standard curves and curve generation (2005). http://www.ecc-brainpool.org/download/Domain-parameters.pdf
14. Ekerå, M., Håstad, J.: Quantum algorithms for computing short discrete logarithms and factoring RSA integers. In: Lange and Takagi [28], pp. 347–363
15. Fowler, A.G., Mariantoni, M., Martinis, J.M., Cleland, A.N.: Surface codes: towards practical large-scale quantum computation. Phys. Rev. A **86**, 032324 (2012). arXiv:1208.0928
16. Galbraith, S.D., Gaudry, P.: Recent progress on the elliptic curve discrete logarithm problem. Des. Codes Crypt. **78**(1), 51–72 (2016)
17. ElGamal, T.: A public key cryptosystem and a signature scheme based on discrete logarithms. In: Blakley, G.R., Chaum, D. (eds.) CRYPTO 1984. LNCS, vol. 196, pp. 10–18. Springer, Heidelberg (1985). https://doi.org/10.1007/3-540-39568-7_2

18. Gordon, D.M.: Discrete logarithms in GF(P) using the number field sieve. SIAM J. Discrete Math. **6**(1), 124–138 (1993)

19. Griffiths, R., Niu, C.: Semiclassical Fourier transform for quantum computation. Phys. Rev. Lett. **76**(17), 3228–3231 (1996)

20. Hamburg, M.: Ed448-Goldilocks, a new elliptic curve. IACR Cryptology ePrint Archive 2015:625 (2015)

21. Haner, T., Roetteler, M., Svore, K.M.: Factoring using $2n + 2$ qubits with Toffoli based modular multiplication. Quantum Inf. Comput. **18**(7&8), 673–684 (2017)

22. Hollosi, A., Karlinger, G., Rossler, T., Centner, M., et al.: Die osterreichische Burgerkarte (2008). http://www.buergerkarte.at/konzept/securitylayer/spezifikation/20080220/

23. Johnson, D., Menezes, A., Vanstone, S.A.: The elliptic curve digital signature algorithm (ECDSA). Int. J. Inf. Secur. **1**(1), 36–63 (2001)

24. Joux, A., Lercier, R.: Improvements to the general number field sieve for discrete logarithms in prime fields. A comparison with the Gaussian integer method. Math. Comput. **72**(242), 953–967 (2003)

25. Kaliski Jr., B.S.: The Montgomery inverse and its applications. IEEE Trans. Comput. **44**(8), 1064–1065 (1995)

26. Kliuchnikov, V., Maslov, D., Mosca, M.: Practical approximation of single-qubit unitaries by single-qubit quantum Clifford and T circuits. IEEE Trans. Comput. **65**(1), 161–172 (2016)

27. Koblitz, N.: Elliptic curve cryptosystems. Math. Comput. **48**(177), 203–209 (1987)

28. Lange, T., Takagi, T. (eds.): PQCrypto 2017. LNCS, vol. 10346. Springer, Cham (2017). https://doi.org/10.1007/978-3-319-59879-6

29. Langley, A., Hamburg, M., Turner, S.: Elliptic curves for security. RFC 7748, RFC Editor (2016)

30. Lenstra, A.K., Lenstra, H.W. (eds.): The Development of the Number Field Sieve. LNM, vol. 1554. Springer, Heidelberg (1993). https://doi.org/10.1007/BFb0091534

31. Maslov, D., Mathew, J., Cheung, D., Pradhan, D.K.: An $O(m^2)$-depth quantum algorithm for the elliptic curve discrete logarithm problem over $GF(2^m)^a$. Quantum Inf. Comput. **9**(7), 610–621 (2009)

32. Miller, V.S.: Use of elliptic curves in cryptography. In: Williams, H.C. (ed.) CRYPTO 1985. LNCS, vol. 218, pp. 417–426. Springer, Heidelberg (1986). https://doi.org/10.1007/3-540-39799-X_31

33. Montgomery, P.L.: Modular multiplication without trial division. Math. Comput. **44**(170), 519–521 (1985)

34. Nakamoto, S.: Bitcoin: a peer-to-peer electronic cash system (2009). http://bitcoin.org/bitcoin.pdf

35. Nielsen, M.A., Chuang, I.L.: Quantum Computation and Quantum Information. Cambridge University Press, Cambridge (2000)

36. Paterson, K.G.: Formal request from TLS WG to CFRG for new elliptic curves. CFRG mailing list, 14 July 2014. http://www.ietf.org/mail-archive/web/cfrg/current/msg04655.html

37. Pollard, J.M.: Monte Carlo methods for index computation mod p. Math. Comput. **32**(143), 918–924 (1978)

38. Pollard, J.M.: Kangaroos, monopoly and discrete logarithms. J. Cryptol. **13**(4), 437–447 (2000)

39. Proos, J., Zalka, C.: Shor's discrete logarithm quantum algorithm for elliptic curves. Quantum Inf. Comput. **3**(4), 317–344 (2003)

40. Renes, J., Costello, C., Batina, L.: Complete addition formulas for prime order elliptic curves. In: Fischlin, M., Coron, J.-S. (eds.) EUROCRYPT 2016. LNCS, vol. 9665, pp. 403–428. Springer, Heidelberg (2016). https://doi.org/10.1007/978-3-662-49890-3_16
41. Rivest, R.L., Shamir, A., Adleman, L.M.: A method for obtaining digital signatures and public-key cryptosystems. Commun. ACM **21**(2), 120–126 (1978)
42. Selinger, P.: Quantum circuits of T-depth one. Phys. Rev. A **87**, 042302 (2013)
43. Selinger, P.: Efficient Clifford+T approximation of single-qubit operators. Quantum Inf. Comput. **15**(1–2), 159–180 (2015)
44. Shor, P.W.: Algorithms for quantum computation: discrete logarithms and factoring. In: FOCS 1994, pp. 124–134. IEEE Computer Society (1994)
45. Shor, P.W.: Polynomial-time algorithms for prime factorization and discrete logarithms on a quantum computer. SIAM J. Comput. **26**(5), 1484–1509 (1997)
46. Silverman, J.H.: The Arithmetic of Elliptic Curves. Graduate Texts in Mathematics, vol. 106, 2nd edn. Springer, New York (2009). https://doi.org/10.1007/978-0-387-09494-6
47. Stebila, D., Green, J.: Elliptic curve algorithm integration in the Secure Shell Transport Layer. RFC 5656, RFC Editor (2009)
48. Stein, J.: Computational problems associated with Racah algebra. J. Comput. Phys. **1**(3), 397–405 (1967)
49. Takahashi, Y., Tani, S., Kunihiro, N.: Quantum addition circuits and unbounded fan-out. Quantum Inf. Comput. **10**(9&10), 872–890 (2010)
50. U.S. Department of Commerce/National Institute of Standards and Technology: Digital Signature Standard (DSS). FIPS-186-4 (2013). http://nvlpubs.nist.gov/nistpubs/FIPS/NIST.FIPS.186-4.pdf
51. van Oorschot, P.C., Wiener, M.J.: Parallel collision search with cryptanalytic applications. J. Cryptol. **12**(1), 1–28 (1999)
52. Wecker, D., Svore, K.M.: LIQU$i|\rangle$: a software design architecture and domain-specific language for quantum computing (2014). https://arxiv.org/abs/1402.4467
53. WhatsApp Inc.: Whatsapp encryption overview. Technical White Paper (2016)
54. Yung, M., Dodis, Y., Kiayias, A., Malkin, T. (eds.): PKC 2006. LNCS, vol. 3958. Springer, Heidelberg (2006). https://doi.org/10.1007/11745853

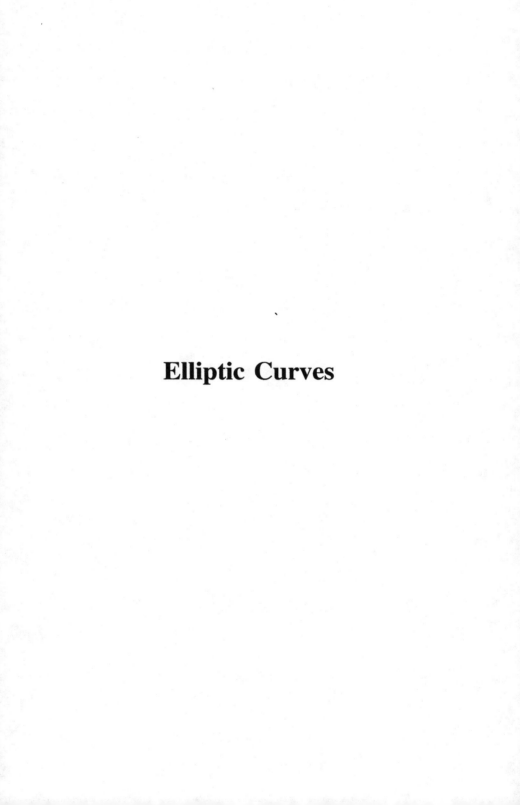

Elliptic Curves

qDSA: Small and Secure Digital Signatures with Curve-Based Diffie–Hellman Key Pairs

Joost Renes[1(✉)] and Benjamin Smith[2]

[1] Digital Security Group, Radboud University, Nijmegen, The Netherlands
j.renes@cs.ru.nl
[2] INRIA and Laboratoire d'Informatique de l'École polytechnique (LIX),
Université Paris–Saclay, Palaiseau, France
smith@lix.polytechnique.fr

Abstract. qDSA is a high-speed, high-security signature scheme that facilitates implementations with a very small memory footprint, a crucial requirement for embedded systems and IoT devices, and that uses the same public keys as modern Diffie–Hellman schemes based on Montgomery curves (such as Curve25519) or Kummer surfaces. qDSA resembles an adaptation of EdDSA to the world of Kummer varieties, which are quotients of algebraic groups by ± 1. Interestingly, qDSA does not require any full group operations or point recovery: all computations, including signature verification, occur on the quotient where there is no group law. We include details on four implementations of qDSA, using Montgomery and fast Kummer surface arithmetic on the 8-bit AVR ATmega and 32-bit ARM Cortex M0 platforms. We find that qDSA significantly outperforms state-of-the-art signature implementations in terms of stack usage and code size. We also include an efficient compression algorithm for points on fast Kummer surfaces, reducing them to the same size as compressed elliptic curve points for the same security level.

Keywords: Signatures · Kummer · Curve25519 · Diffie–Hellman · Elliptic curve · Hyperelliptic curve

1 Introduction

Modern asymmetric cryptography based on elliptic and hyperelliptic curves [29, 31] achieves two important goals. The first is efficient key exchange using the Diffie–Hellman protocol [16], using the fact that the (Jacobian of the) curve carries the structure of an abelian group. But in fact, as Miller observed [31], we do not need the full group structure for Diffie–Hellman: the associated *Kummer variety* (the quotient by ± 1) suffices, which permits more efficiently-computable arithmetic [21,32]. Perhaps the most well-known example is Curve25519 [5], which offers fast scalar multiplications based on x-only arithmetic.

J. Renes—This work has been supported by the Technology Foundation STW (project 13499 - TYPHOON & ASPASIA), from the Dutch government.

© International Association for Cryptologic Research 2017
T. Takagi and T. Peyrin (Eds.): ASIACRYPT 2017, Part II, LNCS 10625, pp. 273–302, 2017.
https://doi.org/10.1007/978-3-319-70697-9_10

The second objective is efficient digital signatures, which are critical for authentication. There are several group-based signature schemes, the most important of which are ECDSA [1], Schnorr [40], and now EdDSA [8] signatures. In contrast to the Diffie–Hellman protocol, all of these signature schemes explicitly require the group structure of the (Jacobian of the) curve. An unfortunate side-effect of this is that users essentially need two public keys to support both curve-based protocols. Further, basic cryptographic libraries need to provide implementations for arithmetic on both the Jacobian and the Kummer variety, thus complicating and increasing the size of the trusted code base. For example, the NaCl library [9] uses Ed25519 [8] for signatures, and Curve25519 [5] for key exchange. This problem is worse for genus-2 hyperelliptic curves, where the Jacobian is significantly harder to use safely than its Kummer surface.

There have been several partial solutions to this problem. By observing that elements of the Kummer variety are elements of the Jacobian *up to sign*, one can build scalar multiplication on the Jacobian based on the fast Kummer arithmetic [14,35]. This avoids the need for a separate scalar multiplication on the Jacobian, but does not avoid the need for its group law; it also introduces the need for projecting to and recovering from the Kummer. In any case, it does not solve the problem of having different public key types.

Another proposal is XEdDSA [36], which uses the public key on the Kummer variety to construct EdDSA signatures. In essence, it creates a key pair on the Jacobian by appending a sign bit to the public key on the Kummer variety, which can then be used for signatures. In [23] Hamburg shows that one can actually verify signatures using only the x-coordinates of points on an elliptic curve, which is applied in the recent STROBE framework [24]. We generalize this approach to allow Kummer varieties of curves of higher genera, and naturally adapt the scheme by only allowing challenges up to sign. This allows us to provide a proof of security, which has thus far not been attempted (in [23] Hamburg remarks that verifying up to sign does "probably not impact security at all"). Similar techniques have been applied for batch verification of ECDSA signatures [28], using the theory of summation polynomials [41].

In this paper we show that there is no intrinsic reason why Kummer varieties cannot be used for signatures. We present qDSA, a signature scheme relying only on Kummer arithmetic, and prove it secure in the random oracle model. It should not be surprising that the reduction in our proof is slightly weaker than the standard proof of security of Schnorr signatures [37], but not by more than we should expect. There is no difference between public keys for qDSA and Diffie–Hellman. After an abstract presentation in Sect. 2, we give a detailed description of elliptic-curve qDSA instances in Sect. 3. We then move on to genus-2 instances based on fast Kummer surfaces, which give better performance. The necessary arithmetic appears in Sect. 4, before Sect. 5 describes the new verification algorithm.

We also provide an efficient compression method for points on fast Kummer surfaces in Sect. 6, solving a long-standing open problem [6]. Our technique means that qDSA public keys for $g = 2$ can be efficiently compressed to 32 bytes, and that qDSA signatures fit into 64 bytes; it also finally reduces the size of Kummer-based Diffie–Hellman public keys from 48 to 32 bytes.

Finally, we provide constant-time software implementations of genus-1 and genus-2 qDSA instances for the AVR ATmega and ARM Cortex M0 platforms. The performance of all four qDSA implementations, reported in Sect. 7, comfortably beats earlier implementations in terms of stack usage and code size.

Source code. We place all of the software described here into the public domain, to maximize the reusability of our results. The software is available at http://www.cs.ru.nl/jrenes/.

2   The qDSA Signature Scheme

In this section we define qDSA, the *quotient Digital Signature Algorithm*. We start by recalling the basics of Kummer varieties in Sect. 2.1 and defining key operations in Sect. 2.2. The rest of the section is dedicated to the definition of the qDSA signature scheme, which is presented in full in Algorithm 1, and its proof of security, which follows Pointcheval and Stern [37,38]. qDSA closely resembles the Schnorr signature scheme [40], as it results from applying the Fiat–Shamir heuristic [19] to an altered Schnorr identification protocol, together with a few standard changes as in EdDSA [8]. We comment on some special properties of qDSA in Sect. 2.5.

Throughout, we work over finite fields $\mathbb{F}_p$ with $p > 3$. Our low-level algorithms include costs in terms of basic $\mathbb{F}_p$-operations: **M**, **S**, **C**, **a**, **s**, **I**, and **E** denote the unit costs of computing a single multiplication, squaring, multiplication by a small constant, addition, subtraction, inverse, and square root, respectively.

2.1   The Kummer Variety Setting

Let $\mathcal{C}$ be a (hyper)elliptic curve and $\mathcal{J}$ its Jacobian[1]. The Jacobian is a commutative algebraic group with group operation $+$, inverse $-$, and identity 0. We assume $\mathcal{J}$ has a subgroup of large prime order N. The associated *Kummer variety* $\mathcal{K}$ is the quotient $\mathcal{K} = \mathcal{J}/\pm$. By definition, working with $\mathcal{K}$ corresponds to working on $\mathcal{J}$ *up to sign*. If P is an element of $\mathcal{J}$, we denote its image in $\mathcal{K}$ by $\pm P$. In this paper we take $\log_2 N \approx 256$, and consider two important cases.

Genus 1. Here $\mathcal{J} = \mathcal{C}/\mathbb{F}_p$ is an elliptic curve with $\log_2 p \approx 256$, while $\mathcal{K} = \mathbb{P}^1$ is the x-line. We choose $\mathcal{C}$ to be Curve25519 [5], which is the topic of Sect. 3.

Genus 2. Here $\mathcal{J}$ is the Jacobian of a genus-2 curve $\mathcal{C}/\mathbb{F}_p$, where $\log_2 p \approx 128$, and $\mathcal{K}$ is a *Kummer surface*. We use the Gaudry–Schost parameters [22] for our implementations. Kummer arithmetic, including some new constructions we need for signature verification and compression, is described in Sects. 4, 5 and 6.

---

[1] In what follows, we could replace $\mathcal{J}$ by an arbitrary abelian group and all the proofs would be completely analogous. For simplicity we restrict to the cryptographically most interesting case of a Jacobian.

A point $\pm P$ in $\mathcal{K}(\mathbb{F}_p)$ is the image of a pair of points $\{P, -P\}$ on $\mathcal{J}$. It is important to note that P and $-P$ are not necessarily in $\mathcal{J}(\mathbb{F}_p)$; if not, then they are conjugate points in $\mathcal{J}(\mathbb{F}_{p^2})$, and correspond to points in $\mathcal{J}'(\mathbb{F}_p)$, where $\mathcal{J}'$ is the *quadratic twist* of $\mathcal{J}$. Both $\mathcal{J}$ and $\mathcal{J}'$ always have the same Kummer variety; we return to this fact, and its implications for our scheme, in Sect. 2.5 below.

2.2   Basic Operations

While a Kummer variety $\mathcal{K}$ has no group law, the operation

$$\{\pm P, \pm Q\} \mapsto \{\pm(P + Q), \pm(P - Q)\} \tag{1}$$

is well-defined. We can therefore define a *pseudo-addition* operation by xADD : $(\pm P, \pm Q, \pm(P - Q)) \mapsto \pm(P + Q)$. The special case where $\pm(P - Q) = \pm 0$ is the *pseudo-doubling* xDBL : $\pm P \mapsto \pm[2]P$. In our applications we can often improve efficiency by combining two of these operations in a single function

$$\text{xDBLADD} : (\pm P, \pm Q, \pm(P - Q)) \longmapsto (\pm[2]P, \pm(P + Q)) .$$

For any integer m, the scalar multiplication $[m]$ on $\mathcal{J}$ induces the key cryptographic operation of *pseudomultiplication* on $\mathcal{K}$, defined by

$$\text{Ladder} : (m, \pm P) \longmapsto \pm[m]P .$$

As its name suggests, we compute Ladder using Montgomery's famous ladder algorithm [32], which is a uniform sequence of xDBLADDs and constant-time conditional swaps.[2] This constant-time nature will be important for signing.

Our signature verification requires a function Check on $\mathcal{K}^3$ defined by

$$\text{Check} : (\pm P, \pm Q, \pm R) \longmapsto \begin{cases} \textbf{True} & \text{if } \pm R \in \{\pm(P + Q), \pm(P - Q)\} \\ \textbf{False} & \text{otherwise} \end{cases}$$

Since we are working with projective points, we need a way to uniquely represent them. Moreover, we want this representation to be as small as possible, to minimize communication overhead. For this purpose we define the functions

$$\text{Compress} : \mathcal{K}(\mathbb{F}_p) \longrightarrow \{0, 1\}^{256} ,$$

writing $\overline{\pm P} := \text{Compress}(\pm P)$, and

$$\text{Decompress} : \{0, 1\}^{256} \longrightarrow \mathcal{K}(\mathbb{F}_p) \cup \{\perp\}$$

such that $\text{Decompress}(\overline{\pm P}) = \pm P$ for $\pm P$ in $\mathcal{K}(\mathbb{F}_p)$ and $\text{Decompress}(X) = \perp$ for $X \in \{0, 1\}^{256} \setminus \text{Im}(\text{Compress})$.

For the remainder of this section we assume that Ladder, Check, Compress, and Decompress are defined. Their implementation depends on whether we are in the genus 1 or 2 setting; we return to this in later sections.

---

[2] In contemporary implementations such as NaCl, the Ladder function is sometimes named crypto_scalarmult.

2.3   The qID Identification Protocol

Let P be a generator of a prime order subgroup of $\mathcal{J}$, of order N, and let $\pm P$ be its image in $\mathcal{K}$. Let $\mathbb{Z}_N^+$ denote the subset of $\mathbb{Z}_N$ with zero least significant bit (where we identify elements of $\mathbb{Z}_N$ with their representatives in $[0, N-1]$). Note that since N is odd, $\mathtt{LSB}(-x) = 1 - \mathtt{LSB}(x)$ for all $x \in \mathbb{Z}_N^*$. The private key is an element $d \in \mathbb{Z}_N$. Let $Q = [d]P$ and let the public key be $\pm Q$. Now consider the following Schnorr-style identification protocol, which we call qID:

(1) The **prover** sets $r \leftarrow_R \mathbb{Z}_N^*$, $\pm R \leftarrow \pm[r]P$ and sends $\pm R$ to the **verifier**;
(2) The **verifier** sets $c \leftarrow_R \mathbb{Z}_N^+$ and sends c to the **prover**;
(3) The **prover** sets $s \leftarrow (r - cd) \mod N$ and sends s to the **verifier**;
(4) The **verifier** accepts if and only if $\pm R \in \{\pm([s]P + [c]Q), \pm([s]P - [c]Q)\}$.

There are some important differences between qID and the basic Schnorr identification protocol in [40].

Scalar multiplications on $\mathcal{K}$. It is well-known that one can use $\mathcal{K}$ to perform the scalar multiplication [14,35] within a Schnorr identification or signature scheme, but with this approach one must always lift back to an element of a group. In contrast, in our scheme this recovery step is not necessary.

Verification on $\mathcal{K}$. The original verification [40] requires checking that $R = [s]P + [c]Q$ for some $R, [s]P, [c]Q \in \mathcal{J}$. Working on $\mathcal{K}$, we only have these values up to sign (i. e. $\pm R$, $\pm[s]P$ and $\pm[c]Q$), which is not enough to check that $R = [s]P + [c]Q$. Instead, we only verify that $\pm R = \pm([s]P \pm [c]Q)$.

Challenge from $\mathbb{Z}_N^+$. A Schnorr protocol using the weaker verification above would not satisfy the special soundness property: the transcripts $(\pm R, c, s)$ and $(\pm R, -c, s)$ are both valid, and do not allow us to extract a witness. Choosing c from $\mathbb{Z}_N^+$ instead of $\mathbb{Z}$ eliminates this possibility, and allows a security proof. (This is the main difference with Hamburg's STROBE [24].)

Proposition 1. *The qID identification protocol is a sigma protocol.*

Proof. We prove the required properties (see [25, Sect. 6]).

Completeness: If the protocol is followed, then $r = s + cd$, and therefore $[r]P = [s]P + [c]Q$ on $\mathcal{J}$. Mapping to $\mathcal{K}$, it follows that $\pm R = \pm([s]P + [c]Q)$.

Special soundness: Let $(\pm R, c_0, s_0)$ and $(\pm R, c_1, s_1)$ be two valid transcripts such that $c_0 \neq c_1$. By verification, each $s_i \equiv \pm r \pm c_i d \pmod{N}$, so $s_0 \pm s_1 \equiv (c_0 \pm c_1)d \pmod{N}$, where the signs are chosen to cancel r. Now $c_0 \pm c_1 \not\equiv 0 \pmod{N}$ because c_0 and c_1 are both in $\mathbb{Z}_N^+$, so we can extract a witness $d \equiv (s_0 \pm s_1)(c_0 \pm c_1)^{-1} \pmod{N}$.

Honest-verifier zero-knowledge: A simulator $\mathcal{S}$ generates $c \leftarrow_R \mathbb{Z}_N^+$ and sets $s \leftarrow_R \mathbb{Z}_N$ and $R \leftarrow [s]P + [c]Q$.[3] If $R = \mathcal{O}$, it restarts. It outputs $(\pm R, c, s)$. As in [38, Lemma 5], we let

$$\delta = \left\{ (\pm R, c, s) : c \in_R \mathbb{Z}_N^+, r \in_R \mathbb{Z}_N^*, \pm R = \pm[r]P, s = r - cd \right\},$$

$$\delta' = \left\{ (\pm R, c, s) : c \in_R \mathbb{Z}_N^+, s \in_R \mathbb{Z}_N, R = [s]P + [c]Q, R \neq \mathcal{O} \right\}$$

---

[3] As we only know Q up to sign, we may need two attempts to construct $\mathcal{S}$.

be the distributions of honest and simulated signatures, respectively. The elements of δ and δ' are the same. First, consider δ. There are exactly $N - 1$ choices for r, and exactly $(N + 1)/2$ for c; all of them lead to distinct tuples. There are thus $(N^2 - 1)/2$ possible tuples, all of which have probability $2/(N^2 - 1)$ of occurring. Now consider δ'. Again, there are $(N + 1)/2$ choices for c. We have N choices for s, exactly one of which leads to $R = \mathcal{O}$. Thus, given c, there are $N - 1$ choices for s. We conclude that δ' also contains $(N^2 - 1)/2$ possible tuples, which all have probability $2/(N^2 - 1)$ of occurring.   □

2.4 Applying Fiat–Shamir

Applying the Fiat–Shamir transform [19] to qID yields a signature scheme qSIG. We will need a hash function $\overline{H} : \{0,1\}^* \to \mathbb{Z}_N^+$, which we define by taking a hash function $H : \{0,1\}^* \to \mathbb{Z}_N$ and then setting $\overline{H}$ by

$$\overline{H}(M) \longmapsto \begin{cases} H(M) & \text{if } \mathsf{LSB}(H(M)) = 0 \\ -H(M) & \text{if } \mathsf{LSB}(H(M)) = 1 \end{cases}.$$

The qSIG signature scheme is defined as follows:

(1) To sign a message $M \in \{0,1\}^*$ with private key $d \in \mathbb{Z}_N$ and public key $\pm Q \in \mathcal{K}$, the prover sets $r \leftarrow_R \mathbb{Z}_N^*$, $\pm R \leftarrow \pm[r]R$, $h \leftarrow \overline{H}(\pm R \parallel M)$, and $s \leftarrow (r - hd) \bmod N$, and sends $(\pm R \parallel s)$ to the verifier.
(2) To verify a signature $(\pm R \parallel s) \in \mathcal{K} \times \mathbb{Z}_N$ on a message $M \in \{0,1\}^*$ with public key $\pm Q \in \mathcal{K}$, the verifier sets $h \leftarrow \overline{H}(\pm R \parallel M)$, $\pm T_0 \leftarrow \pm[s]P$, and $\pm T_1 \leftarrow \pm[h]Q$, and accepts if and only if $\pm R \in \{\pm(T_0 + T_1), \pm(T_0 - T_1)\}$.

Proposition 2 asserts that the security properties of qID carry over to qSIG.

Proposition 2. *In the random oracle model, if an existential forgery of the qSIG signature scheme under an adaptive chosen message attack has non-negligible probability of success, then the DLP in $\mathcal{J}$ can be solved in polynomial time.*

Proof. This is the standard proof of applying the Fiat–Shamir transform to a sigma protocol: see [37, Theorem 13] or [38, Sect. 3.2].   □

2.5 The qDSA Signature Scheme

Moving towards the real world, we slightly alter the qSIG protocol with some pragmatic choices, following Bernstein et al. [8]:

(1) We replace the randomness r by the output of a pseudo-random function, which makes the signatures deterministic.
(2) We include the public key $\pm Q$ in the generation of the challenge, to prevent attackers from attacking multiple public keys at the same time.
(3) We compress and decompress points on $\mathcal{K}$ where necessary.

The resulting signature scheme, qDSA, is summarized in Algorithm 1.

Algorithm 1. The qDSA signature scheme

---

1 **function** keypair

    Input: ()

    Output: $(\overline{\pm Q} \mathbin{\|} (d' \mathbin{\|} d''))$: a compressed public key $\overline{\pm Q} \in \{0,1\}^{256}$
                where $\pm Q \in \mathcal{K}$, and a private key $(d' \mathbin{\|} d'') \in \left(\{0,1\}^{256}\right)^2$

2    $d \leftarrow \mathtt{Random}(\{0,1\}^{256})$

3    $(d' \mathbin{\|} d'') \leftarrow H(d)$

4    $\pm Q \leftarrow \mathtt{Ladder}(d', \pm P)$ // $\pm Q = \pm[d']P$

5    $\overline{\pm Q} \leftarrow \mathtt{Compress}(\pm Q)$

6   **return** $(\overline{\pm Q} \mathbin{\|} (d' \mathbin{\|} d''))$

7 **function** sign

    Input: $d', d'' \in \{0,1\}^{256}, \overline{\pm Q} \in \{0,1\}^{256}, M \in \{0,1\}^*$

    Output: $(\overline{\pm R} \mathbin{\|} s) \in \left(\{0,1\}^{256}\right)^2$

8    $r \leftarrow H(d'' \mathbin{\|} M)$

9    $\pm R \leftarrow \mathtt{Ladder}(r, \pm P)$ // $\pm R = \pm[r]P$

10    $\overline{\pm R} \leftarrow \mathtt{Compress}(\pm R)$

11    $h \leftarrow \overline{H}(\overline{\pm R} \mathbin{\|} \overline{\pm Q} \mathbin{\|} M)$

12    $s \leftarrow (r - hd') \bmod N$

13   **return** $(\overline{\pm R} \mathbin{\|} s)$

14 **function** verify

    Input: $M \in \{0,1\}^*$, the compressed public key $\overline{\pm Q} \in \{0,1\}^{256}$, and a
                putative signature $(\overline{\pm R} \mathbin{\|} s) \in \left(\{0,1\}^{256}\right)^2$

    Output: True if $(\overline{\pm R} \mathbin{\|} s)$ is a valid signature on M under $\overline{\pm Q}$,
                False otherwise

15    $\pm Q \leftarrow \mathtt{Decompress}(\overline{\pm Q})$

16   **if** $\pm Q = \perp$ **then**

17   **return** False

18    $h \leftarrow \overline{H}(\overline{\pm R} \mathbin{\|} \overline{\pm Q} \mathbin{\|} M)$

19    $\pm \mathcal{T}_0 \leftarrow \mathtt{Ladder}(s, \pm P)$ // $\pm \mathcal{T}_0 = \pm[s]P$

20    $\pm \mathcal{T}_1 \leftarrow \mathtt{Ladder}(h, \pm Q)$ // $\pm \mathcal{T}_1 = \pm[h]Q$

21    $\pm R \leftarrow \mathtt{Decompress}(\overline{\pm R})$

22   **if** $\pm R = \perp$ **then**

23   **return** False

24    $v \leftarrow \mathtt{Check}(\pm \mathcal{T}_0, \pm \mathcal{T}_1, \pm R)$ // is $\pm R = \pm(\mathcal{T}_0 \pm \mathcal{T}_1)$?

25   **return** v

Unified keys. Signatures are entirely computed and verified on $\mathcal{K}$, which is also the natural setting for Diffie–Hellman key exchange. We can therefore use identical key pairs for Diffie–Hellman and for qDSA signatures. This significantly simplifies the implementation of cryptographic libraries, as we no longer need arithmetic for the two distinct objects $\mathcal{J}$ and $\mathcal{K}$. Technically, there is no reason not to use a single key pair for both key exchange and signing; but one should be very careful in doing so, as using one key across multiple protocols could potentially lead to attacks. The primary interest of this aspect of qDSA is not necessarily in reducing the number of keys, but in unifying key formats and reducing the size of the trusted code base.

Security level. The security reduction to the discrete logarithm problem is almost identical to the case of Schnorr signatures [37]. Notably, the challenge space has about half the size ($\mathbb{Z}_N^+$ versus $\mathbb{Z}_N$) while the proof of soundness computes either $s_0 + s_1$ or $s_0 - s_1$. This results in a slightly weaker reduction, as should be expected by moving from $\mathcal{J}$ to $\mathcal{K}$ and by weakening verification. By choosing $\log_2 N \approx 256$ we obtain a scheme with about the same security level as state-of-the-art schemes (eg. EdDSA combined with Ed25519). This could be made more precise (cf. [38]), but we do not provide this analysis here.

Key and signature sizes. Public keys fit into 32 bytes in both the genus 1 and genus 2 settings. This is standard for Montgomery curves; for Kummer surfaces it requires a new compression technique, which we present in Sect. 6. In both cases $\log_2 N < 256$, which means that signatures ($\pm R \parallel s$) fit in 64 bytes.

Twist security. Rational points on $\mathcal{K}$ correspond to pairs of points on either $\mathcal{J}$ or its quadratic twist. As opposed to Diffie–Hellman, in qDSA scalar multiplications with secret scalars are *only* performed on the public parameter $\pm P$, which is chosen as the image of large prime order element of $\mathcal{J}$. Therefore $\mathcal{J}$ is not technically required to have a secure twist, unlike in the modern Diffie–Hellman setting. But if $\mathcal{K}$ is also used for key exchange (which is the whole point!), then twist security is crucial. We therefore strongly recommend twist-secure parameters for qDSA implementations.

Hash function. The function H can be any hash function with at least a $\log_2 \sqrt{N}$-bit security level and at least $2 \log_2 N$-bit output. Throughout this paper we take H to be the extendable output function SHAKE128 [18] with fixed 512-bit output. This enables us to implicitly use H as a function mapping into either $\mathbb{Z}_N \times \{0,1\}^{256}$ (eg. Line 3 of Algorithm 1), $\mathbb{Z}_N$ (eg. Line 8 of Algorithm 1), or $\mathbb{Z}_N^+$ (eg. Line 11 of Algorithm 1, by combining it with a conditional negation) by appropriately reducing (part of) the output modulo N.

Signature compression. Schnorr mentions in [40] that signatures ($R \parallel s$) may be compressed to ($H(R \parallel Q \parallel M) \parallel s$), taking only the first 128 bits of the hash, thus reducing signature size from 64 to 48 bytes. This is possible because we can recompute R from P, Q, s, and $H(R \parallel Q \parallel M)$. However, on $\mathcal{K}$ we cannot recover $\pm R$ from $\pm P$, $\pm Q$, s, and $H(\pm R \parallel \pm Q \parallel M)$, so Schnorr's compression technique is not an option for us.

Batching. Proposals for batch signature verification typically rely on the group structure, verifying random linear combinations of points [8,33]. Since $\mathcal{K}$ has no group structure, these batching algorithms are not possible.

Scalar multiplication for verification. Instead of computing the full point $[s]P + [c]Q$ with a two-dimensional multiscalar multiplication operation, we have to compute $\pm[s]P$ and $\pm[c]Q$ separately. As a result we are unable to use standard tricks for speeding up two-dimensional scalar multiplications (eg. [20]), resulting in increased run-time. On the other hand, it has the benefit of relying on the already implemented Ladder function, mitigating the need for a separate algorithm, and is more memory-friendly. Our implementations show a significant decrease in stack usage, at the cost of a small loss of speed (see Sect. 7).

3   Implementing qDSA with Elliptic Curves

Our first concrete instantiation of qDSA uses the Kummer variety of an elliptic curve, which is just the x-line $\mathbb{P}^1$.

3.1   Montgomery Curves

Consider the elliptic curve in Montgomery form

$$E_{AB}/\mathbb{F}_p : By^2 = x(x^2 + Ax + 1),$$

where $A^2 \neq 4$ and $B \neq 0$. The map $E_{AB} \to \mathcal{K} = \mathbb{P}^1$ defined by

$$P = (X : Y : Z) \longmapsto \pm P = \begin{cases} (X : Z) & \text{if } Z \neq 0 \\ (1 : 0) & \text{if } Z = 0 \end{cases}$$

gives rise to efficient x-only arithmetic on $\mathbb{P}^1$ (see [32]). We use the Ladder specified in [17, Algorithm 1]. Compression uses Bernstein's map

$$\texttt{Compress} : (X : Z) \in \mathbb{P}^1(\mathbb{F}_p) \longmapsto XZ^{p-2} \in \mathbb{F}_p,$$

while decompression is the near-trivial

$$\texttt{Decompress} : x \in \mathbb{F}_p \longmapsto (x : 1) \in \mathbb{P}^1(\mathbb{F}_p).$$

Note that Decompress never returns $\perp$, and that $\texttt{Decompress}(\texttt{Compress}((X : Z))) = (X : Z)$ whenever $Z \neq 0$ (however, the points $(0 : 1)$ and $(1 : 0)$ should never appear as public keys or signatures).

3.2   Signature Verification

It remains to define the `Check` operation for Montgomery curves. In the final step of verification we are given $\pm R$, $\pm P$, and $\pm Q$ in $\mathbb{P}^1$, and we need to check whether $\pm R \in \{\pm(P+Q), \pm(P-Q)\}$. Proposition 3 reduces this to checking a quadratic relation in the coordinates of $\pm R$, $\pm P$, and $\pm Q$.

Proposition 3. *Writing $(X^P : Z^P) = \pm P$ for P in E_{AB}, etc.: If P, Q, and R are points on E_{AB}, then $\pm R \in \{\pm(P+Q), \pm(P-Q)\}$ if and only if*

$$B_{ZZ}(X^R)^2 - 2B_{XZ}X^R Z^R + B_{XX}(Z^R)^2 = 0 \qquad (2)$$

where

$$B_{XX} = \left(X^P X^Q - Z^P Z^Q\right)^2, \qquad (3)$$
$$B_{XZ} = \left(X^P X^Q + Z^P Z^Q\right)\left(X^P Z^Q + Z^P X^Q\right) + 2A X^P Z^P X^Q Z^Q, \qquad (4)$$
$$B_{ZZ} = \left(X^P Z^Q - Z^P X^Q\right)^2. \qquad (5)$$

Proof. Let $S = (X^S : Z^S) = \pm(P+Q)$ and $D = (X^D : Z^D) = \pm(P-Q)$. If we temporarily assume $\pm 0 \neq \pm P \neq \pm Q \neq \pm 0$ and put $x_P = X^P / Z^P$, etc., then the group law on E_{AB} gives us $x_S x_D = (x_P x_Q - 1)^2 / (x_P - x_Q)^2$ and $x_S + x_D = 2((x_P x_Q + 1)(x_P + x_Q) + 2A x_P x_Q)$. Homogenizing, we obtain

$$\left(X^S X^D : X^S Z^D + Z^S X^D : Z^S Z^D\right) = (\lambda B_{XX} : \lambda 2 B_{XZ} : \lambda B_{ZZ}). \qquad (6)$$

One readily verifies that Eq. (6) still holds even when the temporary assumption does not (that is, when $\pm P = \pm Q$ or $\pm P = \pm 0$ or $\pm Q = \pm 0$). Having degree 2, the homogeneous polynomial $B_{ZZ}X^2 - B_{XZ}XZ + B_{XX}Z^2$ cuts out two points in $\mathbb{P}^1$ (which may coincide); by Eq. (6), they are $\pm(P+Q)$ and $\pm(P-Q)$, so if $(X^R : Z^R)$ satisfies Eq. (2) then it must be one of them. $\qquad \square$

3.3   Using Cryptographic Parameters

We use the elliptic curve $E/\mathbb{F}_p : y^2 = x^3 + 486662x^2 + x$ where $p = 2^{255} - 19$, which is commonly referred to as `Curve25519` [5]. Let $P \in E(\mathbb{F}_p)$ be such that $\pm P = (9 : 1)$. Then P has order $8N$, where

$$N = 2^{252} + 27742317777372353535851937790883648493$$

is prime. The `xDBLADD` operation requires us to store $(A+2)/4 = 121666$, and we implement optimized multiplication by this constant. In [5, Sect. 3] Bernstein sets and clears some bits of the private key, also referred to as "clamping". This is not necessary in qDSA, but we do it anyway in `keypair` for compatibility.

Algorithm 2. Checking the verification relation for $\mathbb{P}^1$

---

1 **function** Check
  Input: $\pm P, \pm Q, \pm R = (x : 1)$ in $\mathbb{P}^1$ images of points of $E_{AB}(\mathbb{F}_p)$
  Output: **True** if $\pm R \in \{\pm(P + Q), \pm(P - Q)\}$, **False** otherwise
  Cost: $8\mathbf{M} + 3\mathbf{S} + 1\mathbf{C} + 8\mathbf{a} + 4\mathbf{s}$
2   $(B_{XX}, B_{XZ}, B_{ZZ}) \leftarrow \mathtt{BValues}(\pm P, \pm Q)$
3   **if** $B_{XX}x^2 - B_{XZ}x + B_{ZZ} = 0$ **then** **return** **True**
4   **else return** **False**

5 **function** BValues
  Input: $\pm P = (X^P : Z^P), \pm Q = (X^Q : Z^Q)$ in $\mathcal{K}(\mathbb{F}_p)$
  Output: $(B_{XX}(\pm P, \pm Q), B_{XZ}(\pm P, \pm Q), B_{ZZ}(\pm P, \pm Q))$ in $\mathbb{F}_p^3$
  Cost: $6\mathbf{M} + 2\mathbf{S} + 1\mathbf{C} + 7\mathbf{a} + 3\mathbf{s}$
  // Use Eq. (3), (4) and (5) in Proposition 3

---

4   Implementing qDSA with Kummer Surfaces

A number of cryptographic protocols that have been successfully implemented with Montgomery curves have seen substantial practical improvements when the curves are replaced with *Kummer surfaces*. From a general point of view, a Kummer surface is the quotient of some genus-2 Jacobian $\mathcal{J}$ by ± 1; geometrically it is a surface in $\mathbb{P}^3$ with sixteen point singularities, called *nodes*, which are the images in $\mathcal{K}$ of the 2-torsion points of $\mathcal{J}$ (since these are precisely the points fixed by -1). From a cryptographic point of view, a Kummer surface is just a 2-dimensional analogue of the x-coordinate used in Montgomery curve arithmetic.

The algorithmic and software aspects of efficient Kummer surface arithmetic have already been covered in great detail elsewhere (see eg. [7,21,39]). Indeed, the Kummer scalar multiplication algorithms and software that we use in our signature implementation are identical to those described in [39], and use the cryptographic parameters proposed by Gaudry and Schost [22].

This work includes two entirely new Kummer algorithms that are essential for our signature scheme: verification relation testing (Check, Algorithm 3) and compression/decompression (Compress and Decompress, Algorithms 4 and 5). Both of these new techniques require a fair amount of technical development, which we begin in this section by recalling the basic Kummer equation and constants, and deconstructing the pseudo-doubling operation into a sequence of surfaces and maps that will play important roles later. Once the scene has been set, we will describe our signature verification algorithm in Sect. 5 and our point compression scheme in Sect. 6. The reader primarily interested in the resulting performance improvements may wish to skip directly to Sect. 7 on first reading.

The Check, Compress, and Decompress algorithms defined below require the following subroutines:

- Had implements a Hadamard transform. Given a vector (x_1, x_2, x_3, x_4) in $\mathbb{F}_p^4$, it returns $(x_1+x_2+x_3+x_4, x_1+x_2-x_3-x_4, x_1-x_2+x_3-x_4, x_1-x_2-x_3+x_4)$.

– `Dot` computes the sum of a 4-way multiplication. Given a pair of vectors (x_1, x_2, x_3, x_4) and (y_1, y_2, y_3, y_4) in $\mathbb{F}_p^4$, it returns $x_1 y_1 + x_2 y_2 + x_3 y_3 + x_4 y_4$.

4.1   Constants

Our Kummer surfaces are defined by four fundamental constants $\alpha_1, \alpha_2, \alpha_3, \alpha_4$ and four dual constants $\widehat{\alpha}_1, \widehat{\alpha}_2, \widehat{\alpha}_3,$ and $\widehat{\alpha}_4$, which are related by

$$2\widehat{\alpha}_1^2 = \alpha_1^2 + \alpha_2^2 + \alpha_3^2 + \alpha_4^2\,,$$
$$2\widehat{\alpha}_2^2 = \alpha_1^2 + \alpha_2^2 - \alpha_3^2 - \alpha_4^2\,,$$
$$2\widehat{\alpha}_3^2 = \alpha_1^2 - \alpha_2^2 + \alpha_3^2 - \alpha_4^2\,,$$
$$2\widehat{\alpha}_4^2 = \alpha_1^2 - \alpha_2^2 - \alpha_3^2 + \alpha_4^2\,.$$

We require all of the α_i and $\widehat{\alpha}_i$ to be nonzero. The fundamental constants determine the dual constants up to sign, and vice versa. These relations remain true when we exchange the α_i with the $\widehat{\alpha}_i$; we call this "swapping x with $\widehat{x}$" operation "dualizing". To make the symmetry in what follows clear, we define

$$
\begin{aligned}
&\mu_1 := \alpha_1^2\,, &\quad &\epsilon_1 := \mu_2 \mu_3 \mu_4\,, &\quad &\kappa_1 := \epsilon_1 + \epsilon_2 + \epsilon_3 + \epsilon_4\,, \\
&\mu_2 := \alpha_2^2\,, &\quad &\epsilon_2 := \mu_1 \mu_3 \mu_4\,, &\quad &\kappa_2 := \epsilon_1 + \epsilon_2 - \epsilon_3 - \epsilon_4\,, \\
&\mu_3 := \alpha_3^2\,, &\quad &\epsilon_3 := \mu_1 \mu_2 \mu_4\,, &\quad &\kappa_3 := \epsilon_1 - \epsilon_2 + \epsilon_3 - \epsilon_4\,, \\
&\mu_4 := \alpha_4^2\,, &\quad &\epsilon_4 := \mu_1 \mu_2 \mu_3\,, &\quad &\kappa_4 := \epsilon_1 - \epsilon_2 - \epsilon_3 + \epsilon_4\,,
\end{aligned}
$$

along with their respective duals $\widehat{\mu}_i, \widehat{\epsilon}_i,$ and $\widehat{\kappa}_i$. Note that

$$(\epsilon_1 : \epsilon_2 : \epsilon_3 : \epsilon_4) = (1/\mu_1 : 1/\mu_2 : 1/\mu_3 : 1/\mu_4)$$

and $\mu_i \mu_j - \mu_k \mu_l = \widehat{\mu}_i \widehat{\mu}_j - \widehat{\mu}_k \widehat{\mu}_l$ for $\{i, j, k, l\} = \{1, 2, 3, 4\}$. There are many clashing notational conventions for theta constants in the cryptographic Kummer literature; Table 1 provides a dictionary for converting between them.

Our applications use only the squared constants μ_i and $\widehat{\mu}_i$, so only they need be in $\mathbb{F}_p$. In practice we want them to be as "small" as possible, both to reduce the cost of multiplying by them and to reduce the cost of storing them. In fact, it follows from their definition that it is much easier to find simultaneously small μ_i and $\widehat{\mu}_i$ than it is to find simultaneously small α_i and $\widehat{\alpha}_i$ (or a mixture of the two); this is ultimately why we prefer the squared surface for scalar multiplication. We note that if the μ_i are very small, then the ϵ_i and κ_i are also small, and the same goes for their duals. While we will never actually compute with the unsquared constants, we need them to explain what is happening in the background below.

Finally, the Kummer surface equations involve some derived constants

$$E := \frac{16 \alpha_1 \alpha_2 \alpha_3 \alpha_4 \widehat{\mu}_1 \widehat{\mu}_2 \widehat{\mu}_3 \widehat{\mu}_4}{(\mu_1 \mu_4 - \mu_2 \mu_3)(\mu_1 \mu_3 - \mu_2 \mu_4)(\mu_1 \mu_2 - \mu_3 \mu_4)}\,,$$

$$F := 2\frac{\mu_1 \mu_4 + \mu_2 \mu_3}{\mu_1 \mu_4 - \mu_2 \mu_3}\,, \qquad G := 2\frac{\mu_1 \mu_3 + \mu_2 \mu_4}{\mu_1 \mu_3 - \mu_2 \mu_4}\,, \qquad H := 2\frac{\mu_1 \mu_2 + \mu_3 \mu_4}{\mu_1 \mu_2 - \mu_3 \mu_4}\,,$$

and their duals $\widehat{E}, \widehat{F}, \widehat{G}, \widehat{H}$. We observe that $E^2 = F^2 + G^2 + H^2 + FGH - 4$ and $\widehat{E}^2 = \widehat{F}^2 + \widehat{G}^2 + \widehat{H}^2 + \widehat{F}\widehat{G}\widehat{H} - 4$.

Table 1. Relations between our theta constants and others in selected related work

Source	Fundamental constants	Dual constants
[21] and [7]	$(a : b : c : d) = (\alpha_1 : \alpha_2 : \alpha_3 : \alpha_4)$	$(A : B : C : D) = (\widehat{\alpha}_1 : \widehat{\alpha}_2 : \widehat{\alpha}_3 : \widehat{\alpha}_4)$
[11]	$(a : b : c : d) = (\alpha_1 : \alpha_2 : \alpha_3 : \alpha_4)$	$(A : B : C : D) = (\widehat{\mu}_1 : \widehat{\mu}_2 : \widehat{\mu}_3 : \widehat{\mu}_4)$
[39]	$(a : b : c : d) = (\mu_1 : \mu_2 : \mu_3 : \mu_4)$	$(A : B : C : D) = (\widehat{\mu}_1 : \widehat{\mu}_2 : \widehat{\mu}_3 : \widehat{\mu}_4)$
[15]	$(\alpha : \beta : \gamma : \delta) = (\mu_1 : \mu_2 : \mu_3 : \mu_4)$	$(A : B : C : D) = (\widehat{\mu}_1 : \widehat{\mu}_2 : \widehat{\mu}_3 : \widehat{\mu}_4)$

4.2 Fast Kummer Surfaces

We compute all of the pseudoscalar multiplications in qDSA on the so-called **squared Kummer surface**.

$$\mathcal{K}^{\mathrm{Sqr}} : 4E^2 \cdot X_1 X_2 X_3 X_4 = \left(\begin{array}{c} X_1^2 + X_2^2 + X_3^2 + X_4^2 - F(X_1 X_4 + X_2 X_3) \\ - G(X_1 X_3 + X_2 X_4) - H(X_1 X_2 + X_3 X_4) \end{array} \right)^2 ,$$

which was proposed for factorization algorithms by the Chudnovskys [13], then later for Diffie–Hellman by Bernstein [6]. Since E only appears as a square, $\mathcal{K}^{\mathrm{Sqr}}$ is defined over $\mathbb{F}_p$. The zero point on $\mathcal{K}^{\mathrm{Sqr}}$ is $\pm 0 = (\mu_1 : \mu_2 : \mu_3 : \mu_4)$. In our implementations we used the xDBLADD and Montgomery ladder exactly as they were presented in [39, Algorithms 6–7] The pseudo-doubling xDBL on $\mathcal{K}^{\mathrm{Sqr}}$ is

$$\pm P = \left(X_1^P : X_2^P : X_3^P : X_4^P \right) \longmapsto \left(X_1^{[2]P} : X_2^{[2]P} : X_3^{[2]P} : X_4^{[2]P} \right) = \pm[2]P$$

where

$$X_1^{[2]P} = \epsilon_1 (U_1 + U_2 + U_3 + U_4)^2, \qquad U_1 = \widehat{\epsilon}_1 (X_1^P + X_2^P + X_3^P + X_4^P)^2, \qquad (7)$$

$$X_2^{[2]P} = \epsilon_2 (U_1 + U_2 - U_3 - U_4)^2, \qquad U_2 = \widehat{\epsilon}_2 (X_1^P + X_2^P - X_3^P - X_4^P)^2, \qquad (8)$$

$$X_3^{[2]P} = \epsilon_3 (U_1 - U_2 + U_3 - U_4)^2, \qquad U_3 = \widehat{\epsilon}_3 (X_1^P - X_2^P + X_3^P - X_4^P)^2, \qquad (9)$$

$$X_4^{[2]P} = \epsilon_4 (U_1 - U_2 - U_3 + U_4)^2, \qquad U_4 = \widehat{\epsilon}_4 (X_1^P - X_2^P - X_3^P + X_4^P)^2 \qquad (10)$$

for $\pm P$ with all $X_i^P \neq 0$; more complicated formulæ exist for other $\pm P$ (cf. Sect. 5.1).

4.3 Deconstructing Pseudo-doubling

Figure 1 deconstructs the pseudo-doubling on $\mathcal{K}^{\mathrm{Sqr}}$ from Sect. 4.2 into a cycle of atomic maps between different Kummer surfaces, which form a sort of hexagon. Starting at any one of the Kummers and doing a complete cycle of these maps carries out pseudo-doubling on that Kummer. Doing a half-cycle from a given Kummer around to its dual computes a $(2, 2)$-isogeny splitting pseudo-doubling.

Six different Kummer surfaces may seem like a lot to keep track of—even if there are really only three, together with their duals. However, the new surfaces are important, because they are crucial in deriving our Check routine (of course,

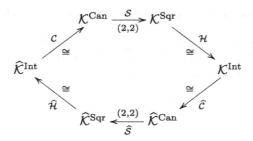

Fig. 1. Decomposition of pseudo-doubling on fast Kummer surfaces into a cycle of morphisms. Here, $\mathcal{K}^{\mathrm{Sqr}}$ is the "squared" surface we mostly compute with; $\mathcal{K}^{\mathrm{Can}}$ is the related "canonical" surface; and $\mathcal{K}^{\mathrm{Int}}$ is a new "intermediate" surface which we use in signature verification. (The surfaces $\widehat{\mathcal{K}}^{\mathrm{Sqr}}$, $\widehat{\mathcal{K}}^{\mathrm{Can}}$, and $\widehat{\mathcal{K}}^{\mathrm{Int}}$ are their duals.)

once the algorithm has been written down, the reader is free to forget about the existence of these other surfaces).

The cycle actually begins one step before $\mathcal{K}^{\mathrm{Sqr}}$, with the **canonical surface**

$$\mathcal{K}^{\mathrm{Can}} : 2E \cdot T_1 T_2 T_3 T_4 = \begin{aligned} & T_1^4 + T_2^4 + T_3^4 + T_4^4 - F(T_1^2 T_4^2 + T_2^2 T_3^2) \\ & - G(T_1^2 T_3^2 + T_2^2 T_4^2) - H(T_1^2 T_2^2 + T_3^2 T_4^2). \end{aligned}$$

This was the model proposed for cryptographic applications by Gaudry in [21]; we call it "canonical" because it is the model arising from a canonical basis of theta functions of level $(2,2)$.

Now we can begin our tour around the hexagon, moving from $\mathcal{K}^{\mathrm{Can}}$ to $\mathcal{K}^{\mathrm{Sqr}}$ via the **squaring** map

$$\mathcal{S} : (T_1 : T_2 : T_3 : T_4) \longmapsto (X_1 : X_2 : X_3 : X_4) = (T_1^2 : T_2^2 : T_3^2 : T_4^3),$$

which corresponds to a $(2,2)$-isogeny of Jacobians. Moving on from $\mathcal{K}^{\mathrm{Sqr}}$, the **Hadamard transform** isomorphism

$$\mathcal{H} : (X_1 : X_2 : X_3 : X_4) \longmapsto (Y_1 : Y_2 : Y_3 : Y_4) = \begin{pmatrix} X_1 + X_2 + X_3 + X_4 \\ : X_1 + X_2 - X_3 - X_4 \\ : X_1 - X_2 + X_3 - X_4 \\ : X_1 - X_2 - X_3 + X_4 \end{pmatrix}$$

takes us into a third kind of Kummer, which we call the **intermediate surface**:

$$\mathcal{K}^{\mathrm{Int}} : \frac{2\widehat{E}}{\alpha_1 \alpha_2 \alpha_3 \alpha_4} \cdot Y_1 Y_2 Y_3 Y_4 = \begin{aligned} & \frac{Y_1^4}{\mu_1^4} + \frac{Y_2^4}{\mu_2^4} + \frac{Y_3^4}{\mu_3^4} + \frac{Y_4^4}{\mu_4^4} - \widehat{F}\left(\frac{Y_1^2 Y_4^2}{\mu_1 \mu_4} + \frac{Y_2^2 Y_3^2}{\mu_2 \mu_3}\right) \\ & - \widehat{G}\left(\frac{Y_1^2 Y_3^2}{\mu_1 \mu_3} + \frac{Y_2^2 Y_4^2}{\mu_2 \mu_4}\right) - \widehat{H}\left(\frac{Y_1^2 Y_2^2}{\mu_1 \mu_2} + \frac{Y_3^2 Y_4^2}{\mu_3 \mu_4}\right). \end{aligned}$$

We will use $\mathcal{K}^{\mathrm{Int}}$ for signature verification. Now the **dual scaling** isomorphism

$$\widehat{\mathcal{C}} : (Y_1 : Y_2 : Y_3 : Y_4) \longmapsto (\widehat{T}_1 : \widehat{T}_2 : \widehat{T}_3 : \widehat{T}_4) = (Y_1/\widehat{\alpha}_1 : Y_2/\widehat{\alpha}_2 : Y_3/\widehat{\alpha}_3 : Y_4/\widehat{\alpha}_4)$$

takes us into the **dual canonical surface**

$$\widehat{\mathcal{K}}^{\mathrm{Can}} : 2\widehat{E} \cdot \widehat{T}_1\widehat{T}_2\widehat{T}_3\widehat{T}_4 = \begin{array}{l} \widehat{T}_1^4 + \widehat{T}_2^4 + \widehat{T}_3^4 + \widehat{T}_4^4 - \widehat{F}(\widehat{T}_1^2\widehat{T}_4^2 + \widehat{T}_2^2\widehat{T}_3^2) \\ - \widehat{G}(\widehat{T}_1^2\widehat{T}_3^2 + \widehat{T}_2^2\widehat{T}_4^2) - \widehat{H}(\widehat{T}_1^2\widehat{T}_2^2 + \widehat{T}_3^2\widehat{T}_4^2). \end{array}$$

We are now halfway around the hexagon; the return journey is simply the dual of the outbound trip. The **dual squaring** map

$$\widehat{\mathcal{S}} : (\widehat{T}_1 : \widehat{T}_2 : \widehat{T}_3 : \widehat{T}_4) \longmapsto (\widehat{X}_1 : \widehat{X}_2 : \widehat{X}_3 : \widehat{X}_4) = (\widehat{T}_1^2 : \widehat{T}_2^2 : \widehat{T}_3^2 : \widehat{T}_4^3),$$

another $(2,2)$-isogeny, carries us into the **dual squared surface**

$$\widehat{\mathcal{K}}^{\mathrm{Sqr}} : 4\widehat{E}^2 \cdot \widehat{X}_1\widehat{X}_2\widehat{X}_3\widehat{X}_4 = \left(\begin{array}{l} \widehat{X}_1^2 + \widehat{X}_2^2 + \widehat{X}_3^2 + \widehat{X}_4^2 - \widehat{F}(\widehat{X}_1\widehat{X}_4 + \widehat{X}_2\widehat{X}_3) \\ - \widehat{G}(\widehat{X}_1\widehat{X}_3 + \widehat{X}_2\widehat{X}_4) - \widehat{H}(\widehat{X}_1\widehat{X}_2 + \widehat{X}_3\widehat{X}_4) \end{array} \right)^2,$$

before the **dual Hadamard transform**

$$\widehat{\mathcal{H}} : (\widehat{X}_1 : \widehat{X}_2 : \widehat{X}_3 : \widehat{X}_4) \longmapsto (\widehat{Y}_1 : \widehat{Y}_2 : \widehat{Y}_3 : \widehat{Y}_4) = \left(\begin{array}{l} \widehat{X}_1 + \widehat{X}_2 + \widehat{X}_3 + \widehat{X}_4 \\ : \widehat{X}_1 + \widehat{X}_2 - \widehat{X}_3 - \widehat{X}_4 \\ : \widehat{X}_1 - \widehat{X}_2 + \widehat{X}_3 - \widehat{X}_4 \\ : \widehat{X}_1 - \widehat{X}_2 - \widehat{X}_3 + \widehat{X}_4 \end{array} \right)$$

takes us into the **dual intermediate surface**

$$\widehat{\mathcal{K}}^{\mathrm{Int}} : \frac{2E}{\alpha_1\alpha_2\alpha_3\alpha_4} \cdot \widehat{Y}_1\widehat{Y}_2\widehat{Y}_3\widehat{Y}_4 = \begin{array}{l} \frac{\widehat{Y}_1^4}{\mu_1^2} + \frac{\widehat{Y}_2^4}{\mu_2^2} + \frac{\widehat{Y}_3^4}{\mu_3^2} + \frac{\widehat{Y}_4^4}{\mu_4^2} - \widehat{F}\left(\frac{\widehat{Y}_1^2}{\mu_1} \frac{\widehat{Y}_4^2}{\mu_4} - \frac{\widehat{Y}_2^2}{\mu_2} \frac{\widehat{Y}_3^2}{\mu_3} \right) \\ - \widehat{G}\left(\frac{\widehat{Y}_1^2}{\mu_1} \frac{\widehat{Y}_3^2}{\mu_3} - \frac{\widehat{Y}_2^2}{\mu_2} \frac{\widehat{Y}_4^2}{\mu_4} \right) - \widehat{H}\left(\frac{\widehat{Y}_1^2}{\mu_1} \frac{\widehat{Y}_2^2}{\mu_2} - \frac{\widehat{Y}_3^2}{\mu_3} \frac{\widehat{Y}_4^2}{\mu_4} \right). \end{array}$$

A final **scaling** isomorphism

$$\mathcal{C} : (\widehat{Y}_1 : \widehat{Y}_2 : \widehat{Y}_3 : \widehat{Y}_4) \longmapsto (T_1 : T_2 : T_3 : T_4) = (\widehat{Y}_1/\alpha_1 : \widehat{Y}_2/\alpha_2 : \widehat{Y}_3/\alpha_3 : \widehat{Y}_4/\alpha_4)$$

takes us from $\widehat{\mathcal{K}}^{\mathrm{Int}}$ back to $\mathcal{K}^{\mathrm{Can}}$, where we started.

The canonical surfaces $\mathcal{K}^{\mathrm{Can}}$ resp. $\widehat{\mathcal{K}}^{\mathrm{Can}}$ are only defined over $\mathbb{F}_p(\alpha_1\alpha_2\alpha_3\alpha_4)$ resp. $\mathbb{F}_p(\widehat{\alpha}_1\widehat{\alpha}_2\widehat{\alpha}_3\widehat{\alpha}_4)$, while the scaling isomorphisms $\widehat{\mathcal{C}}$ resp. $\mathcal{C}$ are defined over $\mathbb{F}_p(\widehat{\alpha}_1, \widehat{\alpha}_2, \widehat{\alpha}_3, \widehat{\alpha}_4)$ resp. $\mathbb{F}_p(\alpha_1, \alpha_2, \alpha_3, \alpha_4)$. Everything else is defined over $\mathbb{F}_p$.

We confirm that one cycle around the hexagon, starting and ending on $\mathcal{K}^{\mathrm{Sqr}}$, computes the pseudo-doubling of Eqs. (7), (8), (9), and (10). Similarly, one cycle around the hexagon starting and ending on $\mathcal{K}^{\mathrm{Can}}$ computes Gaudry's pseudo-doubling from [21, Sect. 3.2].

5   Signature Verification on Kummer Surfaces

To verify signatures in the Kummer surface implementation, we need to supply a **Check** algorithm which, given $\pm P$, $\pm Q$, and $\pm R$ on $\mathcal{K}^{\mathrm{Sqr}}$, decides whether $\pm R \in \{\pm(P+Q), \pm(P-Q)\}$. For the elliptic version of qDSA described in Sect. 3, we saw that this came down to checking that $\pm R$ satisfied one quadratic relation whose three coefficients were biquadratic forms in $\pm P$ and $\pm Q$. The same principle extends to Kummer surfaces, where the pseudo-group law is similarly defined by biquadratic forms; but since Kummer surfaces are defined in terms of four coordinates (as opposed to the two coordinates of the x-line), this time there are six simple quadratic relations to verify, with a total of ten coefficient forms.

5.1   Biquadratic Forms and Pseudo-addition

Let $\mathcal{K}$ be a Kummer surface. If $\pm P$ is a point on $\mathcal{K}$, then we write $(Z_1^P : Z_2^P : Z_3^P : Z_4^P)$ for its projective coordinates. The classical theory of abelian varieties tells us that there exist biquadratic forms B_{ij} for $1 \le i, j \le 4$ such that for all $\pm P$ and $\pm Q$, if $\pm S = \pm(P + Q)$ and $\pm D = \pm(P - Q)$ then

$$\left(Z_i^S Z_j^D + Z_j^S Z_i^D\right)^4_{i,j=1} = \lambda \left(B_{ij}(Z_1^P, Z_2^P, Z_3^P, Z_4^P, Z_1^Q, Z_2^Q, Z_3^Q, Z_4^Q)\right)^4_{i,j=1} \tag{11}$$

where $\lambda \in \Bbbk^\times$ is some common projective factor depending only on the affine representatives chosen for $\pm P$, $\pm Q$, $\pm(P+Q)$ and $\pm(P-Q)$. These biquadratic forms are the foundation of pseudo-addition and doubling laws on $\mathcal{K}$: if the "difference" $\pm D$ is known, then we can use the B_{ij} to compute $\pm S$.

Proposition 4. *Let $\{B_{ij} : 1 \le i, j \le 4\}$ be a set of biquadratic forms on $\mathcal{K} \times \mathcal{K}$ satisfying Eq. (11) for all $\pm P$, $\pm Q$, $\pm(P + Q)$, and $\pm(P - Q)$. Then*

$$\pm R = (Z_1^R : Z_2^R : Z_3^R : Z_4^R) \in \{\pm(P + Q), \pm(P - Q)\}$$

if and only if (writing B_{ij} for $B_{ij}(Z_1^P, \ldots, Z_4^Q)$) we have

$$B_{jj} \cdot (Z_i^R)^2 - 2B_{ij} \cdot Z_i^R Z_j^R + B_{ii} \cdot (Z_j^R)^2 = 0 \quad \text{for all } 1 \le i < j \le 4. \tag{12}$$

Proof. Looking at Eq. (11), we see that the system of six quadratics from Eq. (12) cuts out a zero-dimensional degree-2 subscheme of $\mathcal{K}$: that is, the pair of points $\{\pm(P + Q), \pm(P - Q)\}$ (which may coincide). Hence, if $(Z_1^R : Z_2^R : Z_3^R : Z_4^R) = \pm R$ satisfies all of the equations, then it must be one of them. $\qquad\square$

5.2   Deriving Efficiently Computable Forms

Proposition 4 is the exact analogue of Proposition 3 for Kummer surfaces. All that we need to turn it into a `Check` algorithm for qDSA is an explicit and efficiently computable representation of the B_{ij}. These forms depend on the projective model of the Kummer surface; so we write B_{ij}^{Can}, B_{ij}^{Sqr}, and B_{ij}^{Int} for the forms on the canonical, squared, and intermediate surfaces.

On the canonical surface, the forms B_{ij}^{Can} are classical (see e.g. [3, Sect. 2.2]). The on-diagonal forms B_{ii}^{Can} are

$$B_{11}^{\text{Can}} = \frac{1}{4}\left(\frac{V_1}{\widehat{\mu}_1} + \frac{V_2}{\widehat{\mu}_2} + \frac{V_3}{\widehat{\mu}_3} + \frac{V_4}{\widehat{\mu}_4}\right), \quad B_{22}^{\text{Can}} = \frac{1}{4}\left(\frac{V_1}{\widehat{\mu}_1} + \frac{V_2}{\widehat{\mu}_2} - \frac{V_3}{\widehat{\mu}_3} - \frac{V_4}{\widehat{\mu}_4}\right), \tag{13}$$

$$B_{33}^{\text{Can}} = \frac{1}{4}\left(\frac{V_1}{\widehat{\mu}_1} - \frac{V_2}{\widehat{\mu}_2} + \frac{V_3}{\widehat{\mu}_3} - \frac{V_4}{\widehat{\mu}_4}\right), \quad B_{44}^{\text{Can}} = \frac{1}{4}\left(\frac{V_1}{\widehat{\mu}_1} - \frac{V_2}{\widehat{\mu}_2} - \frac{V_3}{\widehat{\mu}_3} + \frac{V_4}{\widehat{\mu}_4}\right), \tag{14}$$

where

$$V_1 = \big((T_1^P)^2 + (T_2^P)^2 + (T_3^P)^2 + (T_4^P)^2\big)\big((T_1^Q)^2 + (T_2^Q)^2 + (T_3^Q)^2 + (T_4^Q)^2\big),$$
$$V_2 = \big((T_1^P)^2 + (T_2^P)^2 - (T_3^P)^2 - (T_4^P)^2\big)\big((T_1^Q)^2 + (T_2^Q)^2 - (T_3^Q)^2 - (T_4^Q)^2\big),$$
$$V_3 = \big((T_1^P)^2 - (T_2^P)^2 + (T_3^P)^2 - (T_4^P)^2\big)\big((T_1^Q)^2 - (T_2^Q)^2 + (T_3^Q)^2 - (T_4^Q)^2\big),$$
$$V_4 = \big((T_1^P)^2 - (T_2^P)^2 - (T_3^P)^2 + (T_4^P)^2\big)\big((T_1^Q)^2 - (T_2^Q)^2 - (T_3^Q)^2 + (T_4^Q)^2\big),$$

while the off-diagonal forms B_{ij} with $i \neq j$ are

$$B_{ij}^{\mathrm{Can}} = \frac{2}{\widehat{\mu}_i\widehat{\mu}_j - \widehat{\mu}_k\widehat{\mu}_l}\left(\begin{array}{c}\alpha_i\alpha_j\big(T_i^P T_j^P T_i^Q T_j^Q + T_k^P T_l^P T_k^Q T_l^Q\big)\\ -\alpha_k\alpha_l\big(T_i^P T_j^P T_k^Q T_l^Q + T_k^P T_l^P T_i^Q T_j^Q\big)\end{array}\right) \tag{15}$$

where $\{i, j, k, l\} = \{1, 2, 3, 4\}$.

All of these forms can be efficiently evaluated. The off-diagonal B_{ij}^{Can} have a particularly compact shape, while the symmetry of the on-diagonal B_{ii}^{Can} makes them particularly easy to compute simultaneously: indeed, that is exactly what we do in Gaudry's fast pseudo-addition algorithm for $\mathcal{K}^{\mathrm{Can}}$ [21, Sect. 3.2].

Ideally, we would like to evaluate the B_{ij}^{Sqr} on $\mathcal{K}^{\mathrm{Sqr}}$, since that is where our inputs $\pm P$, $\pm Q$, and $\pm R$ live. We can compute the B_{ij}^{Sqr} by dualizing the B_{ij}^{Can}, then pulling the $\widehat{B}_{ij}^{\mathrm{Can}}$ on $\widehat{\mathcal{K}}^{\mathrm{Can}}$ back to $\mathcal{K}^{\mathrm{Sqr}}$ via $\widehat{\mathcal{C}} \circ \mathcal{H}$. But while the resulting on-diagonal B_{ii}^{Sqr} maintain the symmetry and efficiency of the B_{ii}^{Can},[4] the off-diagonal B_{ij}^{Sqr} turn out to be much less pleasant, with less apparent exploitable symmetry. For our applications, this means that evaluating B_{ij}^{Sqr} for $i \neq j$ implies taking a significant hit in terms of stack and code size, not to mention time.

We could avoid this difficulty by mapping the inputs of Check from $\mathcal{K}^{\mathrm{Sqr}}$ into $\widehat{\mathcal{K}}^{\mathrm{Can}}$, and then evaluating the $\widehat{B}_{ij}^{\mathrm{Can}}$. But this would involve using—and, therefore, storing— the four large unsquared $\widehat{\alpha}_i$, which is an important drawback.

Why do the nice $\widehat{B}_{ij}^{\mathrm{Can}}$ become so ugly when pulled back to $\mathcal{K}^{\mathrm{Sqr}}$? The map $\widehat{\mathcal{C}} : \mathcal{K}^{\mathrm{Int}} \to \widehat{\mathcal{K}}^{\mathrm{Can}}$ has no impact on the shape or number of monomials, so most of the ugliness is due to the Hadamard transform $\mathcal{H} : \mathcal{K}^{\mathrm{Sqr}} \to \mathcal{K}^{\mathrm{Int}}$. In particular, if we only pull back the $\widehat{B}_{ij}^{\mathrm{Can}}$ as far as $\mathcal{K}^{\mathrm{Int}}$, then the resulting B_{ij}^{Int} retain the nice form of the B_{ij}^{Can} but do not involve the $\widehat{\alpha}_i$. This fact prompts our solution: we map $\pm P$, $\pm Q$, and $\pm R$ through $\mathcal{H}$ onto $\mathcal{K}^{\mathrm{Int}}$, and verify using the forms B_{ij}^{Int}.

Theorem 1. *Up to a common projective factor, the on-diagonal biquadratic forms on the intermediate surface $\mathcal{K}^{\mathrm{Int}}$ are*

$$B_{11}^{\mathrm{Int}} = \widehat{\mu}_1\left(\kappa_1 F_1 + \kappa_2 F_2 + \kappa_3 F_3 + \kappa_4 F_4\right), \tag{16}$$
$$B_{22}^{\mathrm{Int}} = \widehat{\mu}_2\left(\kappa_2 F_1 + \kappa_1 F_2 + \kappa_4 F_3 + \kappa_3 F_4\right), \tag{17}$$
$$B_{33}^{\mathrm{Int}} = \widehat{\mu}_3\left(\kappa_3 F_1 + \kappa_4 F_2 + \kappa_1 F_3 + \kappa_2 F_4\right), \tag{18}$$
$$B_{44}^{\mathrm{Int}} = \widehat{\mu}_4\left(\kappa_4 F_1 + \kappa_3 F_2 + \kappa_2 F_3 + \kappa_1 F_4\right), \tag{19}$$

---

[4] As they should, since they are the basis of the efficient pseudo-addition on $\mathcal{K}^{\mathrm{Sqr}}$!

where

$$F_1 = P_1Q_1 + P_2Q_2 + P_3Q_3 + P_4Q_4, \quad F_2 = P_1Q_2 + P_2Q_1 + P_3Q_4 + P_4Q_3,$$
$$F_3 = P_1Q_3 + P_3Q_1 + P_2Q_4 + P_4Q_2, \quad F_4 = P_1Q_4 + P_4Q_1 + P_2Q_3 + P_3Q_2,$$

where $P_i = \widehat{\epsilon}_i(Y_i^P)^2$ and $Q_i = \widehat{\epsilon}_i(Y_i^Q)^2$ for $1 \le i \le 4$. Up to the same common projective factor, the off-diagonal forms are

$$B_{ij}^{\mathrm{Int}} = C \cdot C_{ij} \cdot \left(\widehat{\mu}_k\widehat{\mu}_l (Y_{ij}^P - Y_{kl}^P)(Y_{ij}^Q - Y_{kl}^Q) + (\widehat{\mu}_i\widehat{\mu}_j - \widehat{\mu}_k\widehat{\mu}_l)Y_{kl}^P Y_{kl}^Q \right) \quad (20)$$

for $\{i,j,k,l\} = \{1,2,3,4\}$ where $C_{ij} := \widehat{\mu}_i\widehat{\mu}_j(\widehat{\mu}_i\widehat{\mu}_k - \widehat{\mu}_j\widehat{\mu}_l)(\widehat{\mu}_i\widehat{\mu}_l - \widehat{\mu}_j\widehat{\mu}_k)$, $Y_{ij}^P := Y_i^P Y_j^P$, $Y_{ij}^Q := Y_i^Q Y_j^Q$, and

$$C := \frac{8(\mu_1\mu_2\mu_3\mu_4)(\widehat{\mu}_1\widehat{\mu}_2\widehat{\mu}_3\widehat{\mu}_4)}{(\widehat{\mu}_1\widehat{\mu}_2 - \widehat{\mu}_3\widehat{\mu}_4)(\widehat{\mu}_1\widehat{\mu}_3 - \widehat{\mu}_2\widehat{\mu}_4)(\widehat{\mu}_1\widehat{\mu}_4 - \widehat{\mu}_2\widehat{\mu}_3)} .$$

Proof. By definition, $\widehat{T}_i^S\widehat{T}_i^D + \widehat{T}_j^S\widehat{T}_i^D = \widehat{B}_{ij}^{\mathrm{Can}}(\widehat{T}_1^P, \ldots, \widehat{T}_4^Q)$. Pulling back via $\widehat{C}$ using $\widehat{T}_i = Y_i/\widehat{\alpha}_i$ yields

$$B_{ij}^{\mathrm{Int}}(Y_1^P, \ldots, Y_4^Q) = Y_i^S Y_j^D + Y_j^S Y_i^D = \widehat{\alpha}_i\widehat{\alpha}_j \left(\widehat{T}_i^S\widehat{T}_j^D + \widehat{T}_j^S\widehat{T}_i^D \right)$$
$$= \widehat{\alpha}_i\widehat{\alpha}_j \cdot \widehat{B}_{ij}^{\mathrm{Can}}(\widehat{T}_1^P, \ldots, \widehat{T}_4^Q)$$
$$= \widehat{\alpha}_i\widehat{\alpha}_j \cdot \widehat{B}_{ij}^{\mathrm{Can}}(Y_1^P/\widehat{\alpha}_1, \ldots, Y_4^Q/\widehat{\alpha}_4) .$$

Dualizing the B_{ij}^{Can} from Eqs. (13), (14), and (15), we find

$$B_{11}^{\mathrm{Int}} = \widehat{\mu}_1 / \left(4\mu_1\mu_2\mu_3\mu_4(\widehat{\mu}_1\widehat{\mu}_2\widehat{\mu}_3\widehat{\mu}_4)^2 \right) \cdot (\kappa_1 F_1 + \kappa_2 F_2 + \kappa_3 F_3 + \kappa_4 F_4),$$
$$B_{22}^{\mathrm{Int}} = \widehat{\mu}_2 / \left(4\mu_1\mu_2\mu_3\mu_4(\widehat{\mu}_1\widehat{\mu}_2\widehat{\mu}_3\widehat{\mu}_4)^2 \right) \cdot (\kappa_2 F_1 + \kappa_1 F_2 + \kappa_4 F_3 + \kappa_3 F_4),$$
$$B_{33}^{\mathrm{Int}} = \widehat{\mu}_3 / \left(4\mu_1\mu_2\mu_3\mu_4(\widehat{\mu}_1\widehat{\mu}_2\widehat{\mu}_3\widehat{\mu}_4)^2 \right) \cdot (\kappa_3 F_1 + \kappa_4 F_2 + \kappa_1 F_3 + \kappa_2 F_4),$$
$$B_{44}^{\mathrm{Int}} = \widehat{\mu}_4 / \left(4\mu_1\mu_2\mu_3\mu_4(\widehat{\mu}_1\widehat{\mu}_2\widehat{\mu}_3\widehat{\mu}_4)^2 \right) \cdot (\kappa_4 F_1 + \kappa_3 F_2 + \kappa_2 F_3 + \kappa_1 F_4),$$

while the off-diagonal forms B_{ij} with $i \ne j$ are

$$B_{ij}^{\mathrm{Int}} = \frac{2}{\widehat{\mu}_k\widehat{\mu}_l(\widehat{\mu}_i\widehat{\mu}_j - \widehat{\mu}_k\widehat{\mu}_l)} \left(\begin{array}{c} \widehat{\mu}_k\widehat{\mu}_l(Y_{ij}^P - Y_{kl}^P)(Y_{ij}^Q - Y_{kl}^Q) \\ + (\widehat{\mu}_i\widehat{\mu}_j - \widehat{\mu}_k\widehat{\mu}_l)Y_{kl}^P Y_{kl}^Q \end{array} \right)$$

for $\{i,j,k,l\} = \{1,2,3,4\}$. Multiplying all of these forms by a common projective factor of $4(\mu_1\mu_2\mu_3\mu_4)(\widehat{\mu}_1\widehat{\mu}_2\widehat{\mu}_3\widehat{\mu}_4)^2$ eliminates the denominators in the coefficients, and yields the forms of the theorem. □

5.3   Signature Verification

We are now finally ready to implement the `Check` algorithm for $\mathcal{K}^{\mathrm{Sqr}}$. Algorithm 3 does this by applying $\mathcal{H}$ to its inputs, then using the biquadratic forms of Theorem 1. Its correctness is implied by Proposition 4.

Algorithm 3. Checking the verification relation for points on $\mathcal{K}^{\mathrm{Sqr}}$

---

1 **function** Check
   Input: $\pm P, \pm Q, \pm R$ in $\mathcal{K}^{\mathrm{Sqr}}(\mathbb{F}_p)$
   Output: True if $\pm R \in \{\pm(P+Q), \pm(P-Q)\}$, **False** otherwise
   Cost: $76\mathbf{M} + 8\mathbf{S} + 88\mathbf{C} + 42\mathbf{a} + 42\mathbf{s}$
2    $(\mathsf{Y}^P, \mathsf{Y}^Q) \leftarrow (\mathtt{Had}(\pm P), \mathtt{Had}(\pm Q))$
3    $(\mathsf{B}_{11}, \mathsf{B}_{22}, \mathsf{B}_{33}, \mathsf{B}_{44}) \leftarrow \mathtt{BiiValues}(\mathsf{Y}^P, \mathsf{Y}^Q)$
4    $\mathsf{Y}^R \leftarrow \mathtt{Had}(\pm R)$
5   **for** (i, j) **in** $\{(1,2), (1,3), (1,4), (2,3), (2,4), (3,4)\}$ **do**
6      $\mathsf{LHS} \leftarrow \mathsf{B}_{ii} \cdot (\mathsf{Y}_j^R)^2 + \mathsf{B}_{jj} \cdot (\mathsf{Y}_i^R)^2$
7      $\mathsf{B}_{ij} \leftarrow \mathtt{BijValue}(\mathsf{Y}^P, \mathsf{Y}^Q, (i, j))$
8      $\mathsf{RHS} \leftarrow 2\mathsf{B}_{ij} \cdot \mathsf{Y}_i^R \cdot \mathsf{Y}_j^R$
9   **if** $\mathsf{LHS} \neq \mathsf{RHS}$ **then return False**
10   **return True**

11 **function** BiiValues
   Input: $\pm P, \pm Q$ in $\mathcal{K}^{\mathrm{Int}}(\mathbb{F}_p)$
   Output: $(B_{ii}^{\mathrm{Int}}(\pm P, \pm Q))_{i=1}^4$ in $\mathbb{F}_p^4$
   Cost: $16\mathbf{M} + 8\mathbf{S} + 28\mathbf{C} + 24\mathbf{a}$
   `// Use Eqs. (16), (17), (8) and (19) in Theorem 1`

12 **function** BijValue
   Input: $\pm P, \pm Q$ in $\mathcal{K}^{\mathrm{Int}}(\mathbb{F}_p)$ and (i, j) with $1 \leq i, j \leq 4$ and $i \neq j$
   Output: $B_{ij}^{\mathrm{Int}}(\pm P, \pm Q)$ in $\mathbb{F}_p$
   Cost: $10\mathbf{M} + 10\mathbf{C} + 1\mathbf{a} + 5\mathbf{s}$
   `// Use Eq. (20) in Theorem 1`

5.4 Using Cryptographic Parameters

Gaudry and Schost take $p = 2^{127} - 1$ and $(\mu_1 : \mu_2 : \mu_3 : \mu_4) = (-11 : 22 : 19 : 3)$ in [22]. We also need the constants $(\widehat{\mu}_1 : \widehat{\mu}_2 : \widehat{\mu}_3 : \widehat{\mu}_4) = (-33 : 11 : 17 : 49)$, $(\kappa_1 : \kappa_2 : \kappa_3 : \kappa_4) = (-4697 : 5951 : 5753 : -1991)$, and $(\widehat{\epsilon}_1 : \widehat{\epsilon}_2 : \widehat{\epsilon}_3 : \widehat{\epsilon}_4) = (-833 : 2499 : 1617 : 561)$.[5] In practice, where these constants are "negative", we reverse their sign and amend the formulæ above accordingly. All of these constants are small, and fit into one or two bytes each (and the $\widehat{\epsilon}_i$ are already stored for use in Ladder). We store one large constant

$$C = \mathtt{0x40F50EEFA320A2DD46F7E3D8CDDDA843},$$

and recompute the C_{ij} on the fly.

---

[5] Following the definitions of Sect. 4.1, the $\widehat{\mu}_i$ are scaled by -2, the $\widehat{\epsilon}_i$ by $1/11$, and C by $2/11^2$. These changes influence the B_{ij}^{Int}, but only up to the same projective factor.

6   Kummer Point Compression

Our public keys are points on $\mathcal{K}^{\mathrm{Sqr}}$, and each signature includes one point on $\mathcal{K}^{\mathrm{Sqr}}$. Minimizing the space required by Kummer points is therefore essential.

A projective Kummer point is composed of four field elements; normalizing by dividing through by a nonzero coordinate reduces us to three field elements (this can also be achieved using Bernstein's "wrapping" technique [6], as in [7, 39]). But we are talking about Kummer *surfaces*—two-dimensional objects—so we might hope to compress to two field elements, plus a few bits to enable us to correctly recover the whole Kummer point. This is analogous to elliptic curve point compression, where we compress projective points $(X : Y : Z)$ by normalizing to $(x, y) = (X/Z, Y/Z)$, then storing (x, σ), where σ is a bit indicating the "sign" of y. Decompressing the datum (x, σ) to $(X : Y : Z) = (x : y : 1)$ then requires solving a simple quadratic to recover the correct y-coordinate.

For some reason, no such Kummer point compression method has explicitly appeared in the literature. Bernstein remarked in 2006 that if we compress a Kummer point to two coordinates, then decompression appears to require solving a complicated quartic equation [6]. This would be much more expensive than computing the single square root required for elliptic decompression; this has perhaps discouraged implementers from attempting to compress Kummer points.

But while it may not always be obvious from their defining equations, the classical theory tells us that every Kummer is in fact a double cover of $\mathbb{P}^2$, just as elliptic curves are double covers of $\mathbb{P}^1$. We use this principle below to show that we can always compress any Kummer point to two field elements plus two auxiliary bits, and then decompress by solving a quadratic. In our applications, this gives us a convenient packaging of Kummer points in exactly 256 bits.

6.1   The General Principle

First, we sketch a general method for Kummer point compression that works for any Kummer presented as a singular quartic surface in $\mathbb{P}^3$.

Recall that if N is any point in $\mathbb{P}^3$, then projection away from N defines a map $\pi_N : \mathbb{P}^3 \to \mathbb{P}^2$ sending points in $\mathbb{P}^3$ on the same line through N to the same point in $\mathbb{P}^2$. (The map π_N is only a rational map, and not a morphism; the image of N itself is not well-defined.) Now, let N be a node of a Kummer surface $\mathcal{K}$: that is, N is one of the 16 singular points of $\mathcal{K}$. The restriction of π_N to $\mathcal{K}$ forms a double cover of $\mathbb{P}^2$. By definition, π_N maps the points on $\mathcal{K}$ that lie on the same line through N to the same point of $\mathbb{P}^2$. Now $\mathcal{K}$ has degree 4, so each line in $\mathbb{P}^3$ intersects $\mathcal{K}$ in four points; but since N is a double point of $\mathcal{K}$, every line through N intersects $\mathcal{K}$ at N *twice*, and then in two other points. These two remaining points may be "compressed" to their common image in $\mathbb{P}^2$ under π_N, plus a single bit to distinguish the appropriate preimage.

To make this more concrete, let L_1, L_2, and L_3 be linearly independent linear forms on $\mathbb{P}^3$ vanishing on N; then N is the intersection of the three planes in $\mathbb{P}^3$ cut out by the L_i. We can now realise the projection $\pi_N : \mathcal{K} \to \mathbb{P}^2$ as

$$\pi_N : (P_1 : \cdots : P_4) \longmapsto \left(L_1(P_1, \ldots, P_4) : L_2(P_1, \ldots, P_4) : L_3(P_1, \ldots, P_4) \right).$$

Replacing (L_1, L_2, L_3) with another basis of $\langle L_1, L_2, L_3 \rangle$ yields another projection, which corresponds to composing π_N with a linear automorphism of $\mathbb{P}^2$.

If L_1, L_2, and L_3 are chosen as above to vanish on N, and L_4 is any linear form not in $\langle L_1, L_2, L_3 \rangle$, then the fact that π_N is a double cover of the (L_1, L_2, L_3)-plane implies that the defining equation of $\mathcal{K}$ can be rewritten in the form

$$\mathcal{K} : K_2(L_1, L_2, L_3)L_4^2 - 2K_3(L_1, L_2, L_3)L_4 + K_4(L_1, L_2, L_3) = 0$$

where each K_i is a homogeneous polynomial of degree i in L_1, L_2, and L_3. This form, quadratic in L_4, allows us to replace the L_4-coordinate with a single bit indicating the "sign" in the corresponding root of this quadratic; the remaining three coordinates can be normalized to an affine plane point. The net result is a compression to two field elements, plus one bit indicating the normalization, plus another bit to indicate the correct value of L_4.

Remark 1. Stahlke gives a compression algorithm in [42] for points on genus-2 Jacobians in the usual Mumford representation. The first step can be seen as a projection to the most general model of the Kummer (as in [12, Chap. 3]), and then the second is an implicit implementation of the principle above.

6.2   From Squared Kummers to Tetragonal Kummers

We want to define an efficient point compression scheme for $\mathcal{K}^{\mathrm{Sqr}}$. The general principle above makes this possible, but it leaves open the choice of node N and the choice of forms L_i. These choices determine the complexity of the resulting K_i, and hence the cost of evaluating them; this in turn has a non-negligible impact on the time and space required to compress and decompress points, as well as the number of new auxiliary constants that must be stored.

In this section we define a choice of L_i reflecting the special symmetry of $\mathcal{K}^{\mathrm{Sqr}}$. A similar procedure for $\mathcal{K}^{\mathrm{Can}}$ appears in more classical language[6] in [26, Sect. 54]. The trick is to distinguish not one node of $\mathcal{K}^{\mathrm{Sqr}}$, but rather the four nodes forming the kernel of the $(2,2)$-isogeny $\widehat{\mathcal{S}} \circ \widehat{\mathcal{C}} \circ \mathcal{H} : \mathcal{K}^{\mathrm{Sqr}} \rightarrow \widehat{\mathcal{K}}^{\mathrm{Sqr}}$, namely

$$\pm 0 = N_0 = (\mu_1 : \mu_2 : \mu_3 : \mu_4), \qquad N_1 = (\mu_2 : \mu_1 : \mu_4 : \mu_3),$$
$$N_2 = (\mu_3 : \mu_4 : \mu_1 : \mu_2), \qquad N_3 = (\mu_4 : \mu_3 : \mu_2 : \mu_1).$$

We are going to define a coordinate system where these four nodes become the vertices of a coordinate tetrahedron; then, projection onto any three of the four coordinates will represent a projection away from one of these four nodes. The result will be an isomorphic Kummer $\mathcal{K}^{\mathrm{Tet}}$ whose defining equation is quadratic in *all four* of its variables. This might seem like overkill for point compression— quadratic in just one variable would suffice—but it has the agreeable effect of

---

[6] The analogous model of $\mathcal{K}^{\mathrm{Can}}$ in [26, Sect. 54] is called "the equation referred to a Rosenhain tetrad", whose defining equation "...may be deduced from the fact that Kummer's surface is the focal surface of the congruence of rays common to a tetrahedral complex and a linear complex." Modern cryptographers will understand why we have chosen to give a little more algebraic detail here.

dramatically reducing the overall complexity of the defining equation, saving time and memory in our compression and decompression algorithms.

The key is the matrix identity

$$
\begin{pmatrix}
\widehat{\kappa}_4 & \widehat{\kappa}_3 & \widehat{\kappa}_2 & \widehat{\kappa}_1 \\
\widehat{\kappa}_3 & \widehat{\kappa}_4 & \widehat{\kappa}_1 & \widehat{\kappa}_2 \\
\widehat{\kappa}_2 & \widehat{\kappa}_1 & \widehat{\kappa}_4 & \widehat{\kappa}_3 \\
\widehat{\kappa}_1 & \widehat{\kappa}_2 & \widehat{\kappa}_3 & \widehat{\kappa}_4
\end{pmatrix}
\begin{pmatrix}
\mu_1 & \mu_2 & \mu_3 & \mu_4 \\
\mu_2 & \mu_1 & \mu_4 & \mu_3 \\
\mu_3 & \mu_4 & \mu_1 & \mu_2 \\
\mu_4 & \mu_3 & \mu_2 & \mu_1
\end{pmatrix}
= 8\widehat{\mu}_1\widehat{\mu}_2\widehat{\mu}_3\widehat{\mu}_4
\begin{pmatrix}
0 & 0 & 0 & 1 \\
0 & 0 & 1 & 0 \\
0 & 1 & 0 & 0 \\
1 & 0 & 0 & 0
\end{pmatrix},
\tag{21}
$$

which tells us that the projective isomorphism $\mathcal{T}: \mathbb{P}^3 \to \mathbb{P}^3$ defined by

$$
\mathcal{T}:
\begin{pmatrix}
X_1 \\
: X_2 \\
: X_3 \\
: X_4
\end{pmatrix}
\mapsto
\begin{pmatrix}
L_1 \\
: L_2 \\
: L_3 \\
: L_4
\end{pmatrix}
=
\begin{pmatrix}
\widehat{\kappa}_4 X_1 + \widehat{\kappa}_3 X_2 + \widehat{\kappa}_2 X_3 + \widehat{\kappa}_1 X_4 \\
: \widehat{\kappa}_3 X_1 + \widehat{\kappa}_4 X_2 + \widehat{\kappa}_1 X_3 + \widehat{\kappa}_2 X_4 \\
: \widehat{\kappa}_2 X_1 + \widehat{\kappa}_1 X_2 + \widehat{\kappa}_4 X_3 + \widehat{\kappa}_3 X_4 \\
: \widehat{\kappa}_1 X_1 + \widehat{\kappa}_2 X_2 + \widehat{\kappa}_3 X_3 + \widehat{\kappa}_4 X_4
\end{pmatrix}
$$

maps the four "kernel" nodes to the corners of a coordinate tetrahedron:

$$
\begin{aligned}
\mathcal{T}(N_0) &= (0:0:0:1), & \mathcal{T}(N_2) &= (0:1:0:0), \\
\mathcal{T}(N_1) &= (0:0:1:0), & \mathcal{T}(N_3) &= (1:0:0:0).
\end{aligned}
$$

The image of $\mathcal{K}^{\mathrm{Sqr}}$ under $\mathcal{T}$ is the **tetragonal surface**

$$
\mathcal{K}^{\mathrm{Tet}} : 4tL_1L_2L_3L_4 =
\begin{aligned}
& r_1^2(L_1L_2 + L_3L_4)^2 + r_2^2(L_1L_3 + L_2L_4)^2 + r_3^2(L_1L_4 + L_2L_3)^2 \\
& - 2r_1s_1((L_1^2 + L_2^2)L_3L_4 + L_1L_2(L_3^2 + L_4^2)) \\
& - 2r_2s_2((L_1^2 + L_3^2)L_2L_4 + L_1L_3(L_2^2 + L_4^2)) \\
& - 2r_3s_3((L_1^2 + L_4^2)L_2L_3 + L_1L_4(L_2^2 + L_3^2))
\end{aligned}
$$

where $t = 16\mu_1\mu_2\mu_3\mu_4\widehat{\mu}_1\widehat{\mu}_2\widehat{\mu}_3\widehat{\mu}_4$ and

$$
\begin{aligned}
r_1 &= (\mu_1\mu_3 - \mu_2\mu_4)(\mu_1\mu_4 - \mu_2\mu_3), & s_1 &= (\mu_1\mu_2 - \mu_3\mu_4)(\mu_1\mu_2 + \mu_3\mu_4), \\
r_2 &= (\mu_1\mu_2 - \mu_3\mu_4)(\mu_1\mu_4 - \mu_2\mu_3), & s_2 &= (\mu_1\mu_3 - \mu_2\mu_4)(\mu_1\mu_3 + \mu_2\mu_4), \\
r_3 &= (\mu_1\mu_2 - \mu_3\mu_4)(\mu_1\mu_3 - \mu_2\mu_4), & s_3 &= (\mu_1\mu_4 - \mu_2\mu_3)(\mu_1\mu_4 + \mu_2\mu_3).
\end{aligned}
$$

As promised, the defining equation of $\mathcal{K}^{\mathrm{Tet}}$ is quadratic in all four of its variables.

For compression we project away from $\mathcal{T}(\pm 0) = (0:0:0:1)$ onto the $(L_1 : L_2 : L_3)$-plane. Rewriting the defining equation as a quadratic in L_4 gives

$$
\mathcal{K}^{\mathrm{Tet}} : K_4(L_1, L_2, L_3) - 2K_3(L_1, L_2, L_3)L_4 + K_2(L_1, L_2, L_3)L_4^2 = 0
$$

where

$$
\begin{aligned}
K_2 &:= r_3^2 L_1^2 + r_2^2 L_2^2 + r_1^2 L_3^2 - 2\left(r_3s_3L_2L_3 + r_2s_2L_1L_3 + r_1s_1L_1L_2\right), \\
K_3 &:= r_1s_1(L_1^2 + L_2^2)L_3 + r_2s_2(L_1^2 + L_3^2)L_2 + r_3s_3(L_2^2 + L_3^2)L_1 \\
& \quad + (2t - (r_1^2 + r_2^2 + r_3^2))L_1L_2L_3, \\
K_4 &:= r_3^2 L_2^2 L_3^2 + r_2^2 L_1^2 L_3^2 + r_1^2 L_1^2 L_2^2 - 2\left(r_3s_3L_1 + r_2s_2L_2 + r_1s_1L_3\right)L_1L_2L_3.
\end{aligned}
$$

Lemma 1. *If $(l_1 : l_2 : l_3 : l_4)$ is a point on $\mathcal{K}^{\mathrm{Tet}}$, then*

$$K_2(l_1, l_2, l_3) = K_3(l_1, l_2, l_3) = K_4(l_1, l_2, l_3) = 0 \iff l_1 = l_2 = l_3 = 0.$$

Proof. Write k_i for $K_i(l_1, l_2, l_3)$. If $(l_1, l_2, l_3) = 0$ then $(k_2, k_3, k_4) = 0$, because each K_i is nonconstant and homogeneous. Conversely, if $(k_2, k_3, k_4) = 0$ and $(l_1, l_2, l_3) \neq 0$ then we could embed a line in $\mathcal{K}^{\mathrm{Tet}}$ via $\lambda \mapsto (l_1 : l_2 : l_3 : \lambda)$; but this is a contradiction, because $\mathcal{K}^{\mathrm{Tet}}$ contains no lines. □

6.3   Compression and Decompression for $\mathcal{K}^{\mathrm{Sqr}}$

In practice, we compress points on $\mathcal{K}^{\mathrm{Sqr}}$ to tuples (l_1, l_2, τ, σ), where l_1 and l_2 are field elements and τ and σ are bits. The recipe is

(1) Map $(X_1 : X_2 : X_3 : X_4)$ through $\mathcal{T}$ to a point $(L_1 : L_2 : L_3 : L_4)$ on $\mathcal{K}^{\mathrm{Tet}}$.
(2) Compute the unique (l_1, l_2, l_3, l_4) in one of the forms $(*, *, 1, *)$, $(*, 1, 0, *)$, $(1, 0, 0, *)$, or $(0, 0, 0, 1)$ such that $(l_1 : l_2 : l_3 : l_4) = (L_1 : L_2 : L_3 : L_4)$.
(3) Compute $k_2 = K_2(l_1, l_2, l_3)$, $k_3 = K_3(l_1, l_2, l_3)$, and $k_4 = K_4(l_1, l_2, l_3)$.
(4) Define the bit $\sigma = \mathtt{Sign}(k_2 l_4 - k_3)$; then (l_1, l_2, l_3, σ) determines l_4. Indeed, $q(l_4) = 0$, where $q(X) = k_2 X^2 - 2k_3 X + k_4$; and Lemma 1 tells us that $q(X)$ is either quadratic, linear, or identically zero.
   - If q is a nonsingular quadratic, then l_4 is determined by (l_1, l_2, l_3) and σ, because $\sigma = \mathtt{Sign}(R)$ where R is the correct square root in the quadratic formula $l_4 = (k_3 \pm \sqrt{k_3^2 - k_2 k_4})/k_2$.
   - If q is singular or linear, then (l_1, l_2, l_3) determines l_4, and σ is redundant.
   - If $q = 0$ then $(l_1, l_2, l_3) = (0, 0, 0)$, so $l_4 = 1$; again, σ is redundant.
   Setting $\sigma = \mathtt{Sign}(k_2 l_4 - k_3)$ in every case, regardless of whether or not we need it to determine l_4, avoids ambiguity and simplifies code.
(5) The normalization in Step 2 forces $l_3 \in \{0, 1\}$; so encode l_3 as a single bit τ.

The datum (l_1, l_2, τ, σ) completely determines (l_1, l_2, l_3, l_4), and thus determines $(X_1 : X_2 : X_3 : X_4) = \mathcal{T}^{-1}((l_1 : l_2 : l_2 : l_4))$. Conversely, the normalization in Step 2 ensures that (l_1, l_2, τ, σ) is uniquely determined by $(X_1 : X_2 : X_3 : X_4)$, and is independent of the representative values of the X_i.

   Algorithm 4 carries out the compression process above; the most expensive step is the computation of an inverse in $\mathbb{F}_p$. Algorithm 5 is the corresponding decompression algorithm; its cost is dominated by computing a square root in $\mathbb{F}_p$.

Proposition 5. *Algorithms 4 and 5 (*Compress *and* Decompress*) satisfy the following properties: given (l_1, l_2, τ, σ) in $\mathbb{F}_p^2 \times \{0, 1\}^2$,* Decompress *always returns either a valid point in $\mathcal{K}^{\mathrm{Sqr}}(\mathbb{F}_p)$ or $\perp$; and for every $\pm P$ in $\mathcal{K}^{\mathrm{Sqr}}(\mathbb{F}_p)$, we have*

$$\mathtt{Decompress}(\mathtt{Compress}(\pm P)) = \pm P.$$

Proof. In Algorithm 5 we are given (l_1, l_2, τ, σ). We can immediately set $l_3 = \tau$, viewed as an element of $\mathbb{F}_p$. We want to compute an l_4 in $\mathbb{F}_p$, if it exists, such that $k_2 l_4^2 - 2k_3 l_4 + k_4 = 0$ and $\mathtt{Sign}(k_2 l_4 - l_3) = \sigma$ where $k_i = K_i(l_1, l_2, l_3)$.

Algorithm 4. Kummer point compression for $\mathcal{K}^{\mathrm{Sqr}}$

---

1 **function** Compress

  Input: $\pm P$ in $\mathcal{K}^{\mathrm{Sqr}}(\mathbb{F}_p)$

  Output: (l_1, l_2, τ, σ) with $l_1, l_2 \in \mathbb{F}_p$ and $\sigma, \tau \in \{0, 1\}$

  Cost: $8\mathbf{M} + 5\mathbf{S} + 12\mathbf{C} + 8\mathbf{a} + 5\mathbf{s} + 1\mathbf{I}$

2   $\begin{pmatrix} \mathsf{L}_1, \mathsf{L}_2, \\ \mathsf{L}_3, \mathsf{L}_4 \end{pmatrix} \leftarrow \begin{pmatrix} \mathrm{Dot}(\pm P, (\widehat{\kappa}_4, \widehat{\kappa}_3, \widehat{\kappa}_2, \widehat{\kappa}_1)), \mathrm{Dot}(\pm P, (\widehat{\kappa}_3, \widehat{\kappa}_4, \widehat{\kappa}_1, \widehat{\kappa}_2)), \\ \mathrm{Dot}(\pm P, (\widehat{\kappa}_2, \widehat{\kappa}_1, \widehat{\kappa}_4, \widehat{\kappa}_3)), \mathrm{Dot}(\pm P, (\widehat{\kappa}_1, \widehat{\kappa}_2, \widehat{\kappa}_3, \widehat{\kappa}_4)) \end{pmatrix}$

3   **if** $\mathsf{L}_3 \neq 0$ **then**

4 | $(\tau, \lambda) \leftarrow (1, \mathsf{L}_3^{-1})$ // Normalize to $(* : * : 1 : *)$

5   **else if** $\mathsf{L}_2 \neq 0$ **then**

6 | $(\tau, \lambda) \leftarrow (0, \mathsf{L}_2^{-1})$ // Normalize to $(* : 1 : 0 : *)$

7   **else if** $\mathsf{L}_1 \neq 0$ **then**

8 | $(\tau, \lambda) \leftarrow (0, \mathsf{L}_1^{-1})$ // Normalize to $(1 : 0 : 0 : *)$

9   **else**

10      $(\tau, \lambda) \leftarrow (0, \mathsf{L}_4^{-1})$ // Normalize to $(0 : 0 : 0 : 1)$

11 $(\mathsf{l}_1, \mathsf{l}_2, \mathsf{l}_4) \leftarrow (\mathsf{L}_1 \cdot \lambda, \mathsf{L}_2 \cdot \lambda, \mathsf{L}_4 \cdot \lambda)$ // $(\mathsf{l}_1 : \mathsf{l}_2 : \tau : \mathsf{l}_4) = (\mathsf{L}_1 : \mathsf{L}_2 : \mathsf{L}_3 : \mathsf{L}_4)$

12 $(\mathsf{k}_2, \mathsf{k}_3) \leftarrow (K_2(\mathsf{l}_1, \mathsf{l}_2, \tau), K_3(\mathsf{l}_1, \mathsf{l}_2, \tau))$

13 $\mathsf{R} \leftarrow \mathsf{k}_2 \cdot \mathsf{l}_4 - \mathsf{k}_3$

14 $\sigma \leftarrow \mathrm{Sign}(\mathsf{R})$

15 **return** $(\mathsf{l}_1, \mathsf{l}_2, \tau, \sigma)$

If such an l_4 exists, then we will have a preimage $(l_1 : l_2 : l_3 : l_4)$ in $\mathcal{K}^{\mathrm{Tet}}(\mathbb{F}_p)$, and we can return the decompressed $\mathcal{T}^{-1}((l_1 : l_2 : l_3 : l_4))$ in $\mathcal{K}^{\mathrm{Sqr}}$.

If $(k_2, k_3) = (0, 0)$ then $k_4 = 2k_3l_4 - k_2l_4^2 = 0$, so $l_1 = l_2 = \tau = 0$ by Lemma 1. The only legitimate datum in this form is is $(l_1 : l_2 : \tau : \sigma) = (0 : 0 : 0 : \mathrm{Sign}(0))$. If this was the input, then the preimage is $(0 : 0 : 0 : 1)$; otherwise we return $\perp$.

If $k_2 = 0$ but $k_3 \neq 0$, then $k_4 = 2k_3l_4$, so $(l_1 : l_2 : \tau : l_4) = (2k_3l_1 : 2k_3l_2 : 2k_3\tau : k_4)$. The datum is a valid compression unless $\sigma \neq \mathrm{Sign}(-k_3)$, in which case we return $\perp$; otherwise, the preimage is $(2k_3l_1 : 2k_3l_2 : 2k_3\tau : k_4)$.

If $k_2 \neq 0$, then the quadratic formula tells us that any preimage satisfies $k_2l_4 = k_3 \pm \sqrt{k_3^2 - k_2k_4}$, with the sign determined by $\mathrm{Sign}(k_2l_4 - k_3)$. If $k_3^2 - k_2k_4$ is not a square in $\mathbb{F}_p$ then there is no such l_4 in $\mathbb{F}_p$; the input is illegitimate, so we return $\perp$. Otherwise, we have a preimage $(k_2l_1 : k_2l_2 : k_2l_3 : l_3 \pm \sqrt{k_3^2 - k_2k_4})$.

Line 17 maps the preimage $(l_1 : l_2 : l_3 : l_4)$ in $\mathcal{K}^{\mathrm{Tet}}(\mathbb{F}_p)$ back to $\mathcal{K}^{\mathrm{Sqr}}(\mathbb{F}_p)$ via $\mathcal{T}^{-1}$, yielding the decompressed point $(X_1 : X_2 : X_3 : X_4)$. $\qquad\square$

6.4   Using Cryptographic Parameters

Our compression scheme works out particularly nicely for the Gaudry–Schost Kummer over $\mathbb{F}_{2^{127}-1}$. First, since every field element fits into 127 bits, every compressed point fits into exactly 256 bits. Second, the auxiliary constants are small: we have $(\widehat{\kappa}_1 : \widehat{\kappa}_2 : \widehat{\kappa}_3 : \widehat{\kappa}_4) = (-961 : 128 : 569 : 1097)$, each of which fits

---

Algorithm 5. Kummer point decompression to $\mathcal{K}^{\mathrm{Sqr}}$

---

1 **function** Decompress
> **Input:** (l_1, l_2, τ, σ) with $l_1, l_2 \in \mathbb{F}_p$ and $\tau, \sigma \in \{0, 1\}$
> **Output:** The point $\pm P$ in $\mathcal{K}^{\mathrm{Sqr}}(\mathbb{F}_p)$ such that
> $\quad\quad$ Compress$(\pm P) = (l_1, l_2, \tau, \sigma)$, or $\perp$ if no such $\pm P$ exists
> **Cost:** $10\mathbf{M} + 9\mathbf{S} + 18\mathbf{C} + 13\mathbf{a} + 8\mathbf{s} + 1\mathbf{E}$

2 $\quad$ $(\mathsf{k}_2, \mathsf{k}_3, \mathsf{k}_4) \leftarrow (K_2(l_1, l_2, \tau), K_3(l_1, l_2, \tau), K_4(l_1, l_2, \tau))$
3 $\quad$ **if** $\mathsf{k}_2 = 0$ **and** $\mathsf{k}_3 = 0$ **then**
4 $\quad\quad$ **if** $(l_1, l_2, \tau, \sigma) \neq (0, 0, 0, \mathrm{Sign}(0))$ **then**
5 $\quad\quad\quad$ **return** $\perp$ $\quad\quad\quad\quad\quad\quad\quad$ // Invalid compression
6 $\quad\quad$ $\mathsf{L} \leftarrow (0, 0, 0, 1)$
7 $\quad$ **else if** $\mathsf{k}_2 = 0$ **and** $\mathsf{k}_3 \neq 0$ **then**
8 $\quad\quad$ **if** $\sigma \neq \mathrm{Sign}(-\mathsf{k}_3)$ **then**
9 $\quad\quad\quad$ **return** $\perp$ $\quad\quad\quad\quad\quad\quad\quad$ // Invalid compression
10 $\quad\quad$ $\mathsf{L} \leftarrow (2 \cdot l_1 \cdot \mathsf{k}_3, 2 \cdot l_2 \cdot \mathsf{k}_3, 2 \cdot \tau \cdot \mathsf{k}_3, \mathsf{k}_4)$ $\quad\quad$ // $\mathsf{k}_4 = 2\mathsf{k}_3 l_4$
11 $\quad$ **else**
12 $\quad\quad$ $\Delta \leftarrow \mathsf{k}_3^2 - \mathsf{k}_2 \mathsf{k}_4$
13 $\quad\quad$ $\mathsf{R} \leftarrow$ HasSquareRoot(Δ, σ) $\quad$ // $\mathsf{R} = \perp$ or $\mathsf{R}^2 = \Delta$, $\mathrm{Sign}(\mathsf{R}) = \sigma$
14 $\quad\quad$ **if** $\mathsf{R} = \perp$ **then**
15 $\quad\quad\quad$ **return** $\perp$ $\quad\quad\quad\quad\quad\quad\quad$ // No preimage in $\mathcal{K}^{\mathrm{Tet}}(\mathbb{F}_p)$
16 $\quad\quad$ $\mathsf{L} \leftarrow (\mathsf{k}_2 \cdot l_1, \mathsf{k}_2 \cdot l_2, \mathsf{k}_2 \cdot \tau, \mathsf{k}_3 + \mathsf{R})$ $\quad\quad$ // $\mathsf{k}_3 + \mathsf{R} = \mathsf{k}_2 l_4$
17 $\quad$ $\begin{pmatrix} X_1, X_2, \\ X_3, X_4 \end{pmatrix} \leftarrow \begin{pmatrix} \mathrm{Dot}(\mathsf{L}, (\mu_4, \mu_3, \mu_2, \mu_1)), \mathrm{Dot}(\mathsf{L}, (\mu_3, \mu_4, \mu_1, \mu_2)), \\ \mathrm{Dot}(\mathsf{L}, (\mu_2, \mu_1, \mu_4, \mu_3)), \mathrm{Dot}(\mathsf{L}, (\mu_1, \mu_2, \mu_3, \mu_4)) \end{pmatrix}$
18 $\quad$ **return** $(X_1 : X_2 : X_3 : X_4)$

---

into well under 16 bits. Computing the polynomials K_2, K_3, K_4 and dividing them all through by 11^2 (which does not change the roots of the quadratic) gives

$$K_2(l_1, l_2, \tau) = (q_5 l_1)^2 + (q_3 l_2)^2 + (q_4 \tau)^2 - 2q_3 \big(q_2 l_1 l_2 + \tau(q_0 l_1 - q_1 l_2)\big), \quad (22)$$

$$K_3(l_1, l_2, \tau) = q_3 \big(q_0 (l_1^2 + \tau) l_2 - q_1 l_1 (l_2^2 + \tau) + q_2 (l_1^2 + l_2^2)\tau\big) - q_6 q_7 l_1 l_2 \tau, \quad (23)$$

$$K_4(l_1, l_2, \tau) = ((q_3 l_1)^2 + (q_5 l_2)^2 - 2q_3 l_1 l_2 (q_0 l_2 - q_1 l_1 + q_2))\tau + (q_4 l_1 l_2)^2, \quad (24)$$

where $(q_0, \ldots, q_7) = (3575, 9625, 4625, 12259, 11275, 7475, 6009, 43991)$; each of the q_i fits into 16 bits. In total, the twelve new constants we need for Compress and Decompress together fit into less than two field elements' worth of space.

7 Implementation

In this section we present the results of the implementation of the scheme on the AVR ATmega and ARM Cortex M0 platforms. We have a total of four implementations: on both platforms we implemented both the Curve25519-based scheme and the scheme based on a fast Kummer surface in genus 2. The benchmarks for the AVR software are obtained from the Arduino MEGA development board containing an ATmega2560 MCU, compiled with GCC v4.8.1. For the Cortex M0, they are measured on the STM32F051R8 MCU on the STMF0Discovery board, compiled with Clang v3.5.0. We refer to the (publicly available) code for more detailed compiler settings. For both Diffie–Hellman and signatures we follow the eBACS [4] API.

7.1   Core Functionality

The arithmetic of the underlying finite fields is well-studied and optimized, and we do not reinvent the wheel. For field arithmetic in $\mathbb{F}_{2^{255}-19}$ we use the highly optimized functions presented by Hutter and Schwabe [27] for the AVR ATmega, and the code from Düll et al. [17] for the Cortex M0. For arithmetic in $\mathbb{F}_{2^{127}-1}$ we use the functions from Renes et al. [39], which in turn rely on [27] for the AVR ATmega, and on [17] for the Cortex M0.

The SHAKE128 functions for the ATmega are taken from [10], while on the Cortex M0 we use a modified version from [2]. Cycle counts for the main functions defined in the rest of this paper are presented in Table 2. Notably, the Ladder routine is by far the most expensive function. In genus 1 the Compress function is relatively costly (it is essentially an inversion), while in genus 2 Check, Compress and Decompress have only minor impact on the total run-time. More interestingly, as seen in Tables 3 and 4, the simplicity of operating only on the Kummer variety allows smaller code and less stack usage.

Table 2. Cycle counts for the four key functions of qDSA at the 128-bit security level.

Genus	Function	Ref.	AVR ATmega	ARM Cortex M0
1	Ladder	–	12 539 098	3 338 554
	Check	Algorithm 2	46 546	17 044
	Compress	Sect. 3.1	1 067 004	270 867
	Decompress	Sect. 3.1	694	102
2	Ladder	–	9 624 637	2 683 371
	Check[a]	Algorithm 3	84 424	24 249
	Compress	Algorithm 4	212 374	62 165
	Decompress	Algorithm 5	211 428	62 471

[a] The implementation decompresses $\pm R$ within Check, while Algorithm 3 assumes $\pm R$ to be decompressed. We have subtracted the cost of the Decompress function once.

7.2 Comparison to Previous Work

There are not many implementations of complete signature and key exchange schemes on microcontrollers. On the other hand, there are implementations of scalar multiplication on elliptic curves. The current fastest on our platforms are presented by Düll et al. [17], and since we are relying on exactly the same arithmetic, we have essentially the same results. Similarly, the current records for scalar multiplication on Kummer surfaces are presented by Renes et al. [39]. Since we use the same underlying functions, we have similar results.

More interestingly, we compare the speed and memory usage of signing and verification to best known results of implementations of complete signature schemes. To the best of our knowledge, the only other works are the Ed25519-based scheme by Nascimento et al. [34], the FourQ-based scheme (obtaining fast scalar multiplication by relying on easily computable endomorphisms) by Liu et al. [30], and the genus-2 implementation from [39].

AVR ATmega. As we see in Table 3, our implementation of the scheme based on Curve25519 outperforms the Ed25519-based scheme from [34] in every way. It reduces the number of clock cycles needed for `sign` resp. `verify` by more than 26% resp. 17%, while reducing stack usage by more than 65% resp. 47%. Code size is not reported in [34]. Comparing against the FourQ implementation of [30], we see a clear trade-off between speed and size: FourQ has a clear speed advantage, but qDSA on Curve25519 requires only a fraction of the stack space.

The implementation based on the Kummer surface of the genus-2 Gaudry–Schost Jacobian does better than the Curve25519-based implementation across the board. Compared to [39], the stack usage of `sign` resp. `verify` decreases by more than 54% resp. 38%, while decreasing code size by about 11%. On the other

Table 3. Performance comparison of the qDSA signature scheme against the current best implementations, on the AVR ATmega platform.

Ref.	Object	Function	Clock cycles	Stack	Code size[a]
[34]	Ed25519	sign	19 047 706	1 473 bytes	–
		verify	30 776 942	1 226 bytes	
[30]	FourQ	sign	5 174 800	1 572 bytes	25 354 bytes
		verify	11 003 800	4 957 bytes	33 372 bytes
This work	Curve25519	sign	14 067 995	512 bytes	21 347 bytes
		verify	25 355 140	644 bytes	
[39]	Gaudry– Schost $\mathcal{J}$	sign	10 404 033	926 bytes	20 242 bytes
		verify	16 240 510	992 bytes	
This work	Gaudry– Schost $\mathcal{K}$	sign	10 477 347	417 bytes	17 880 bytes
		verify	20 423 937	609 bytes	

[a] All reported code sizes except those from [30, Table 6] include support for both signatures and key exchange

Table 4. Performance comparison of the qDSA signature scheme against the current best implementations, on the ARM Cortex M0 platform.

Ref.	Object	Function	Clock cycles	Stack	Code size[a]
This work	Curve25519	sign	3 889 116	660 bytes	18 443 bytes
		verify	6 793 695	788 bytes	
[39]	Gaudry–Schost $\mathcal{J}$	sign	2 865 351	1 360 bytes	19 606 bytes
		verify	4 453 978	1 432 bytes	
This work	Gaudry–Schost $\mathcal{K}$	sign	2 908 215	580 bytes	18 064 bytes
		verify	5 694 414	808 bytes	

[a] In this work 8 448 bytes come from the SHAKE128 implementation, while [39] uses 6 938 bytes. One could probably reduce this significantly by optimizing the implementation, or by using a more memory-friendly hash function

hand, verification is about 26% slower. This is explained by the fact that in [39] the signature is compressed to 48 bytes (following Schnorr's suggestion), which means that one of the scalar multiplications in verification is only half length. Comparing to the FourQ implementation of [30], again we see a clear trade-off between speed and size, but this time the loss of speed is less pronounced than in the comparison with Curve25519-based qDSA.

ARM Cortex M0. In this case there is no elliptic-curve-based signature scheme to compare to, so we present the first. As we see in Table 4, it is significantly slower than its genus-2 counterpart in this paper (as should be expected), while using a similar amount of stack and code.

The genus-2 signature scheme has similar trade-offs on this platform when compared to the implementation by Renes et al. [39]. The stack usage for sign resp. verify is reduced by about 57% resp. 43%, while code size is reduced by about 8%. For the same reasons as above, verification is about 28% slower.

Acknowledgements. We thank Peter Schwabe for his helpful contributions to discussions during the creation of this paper.

References

1. Accredited Standards Committee X9: American National Standard X9.62-1999, Public key cryptography for the financial services industry: the elliptic curve digital signature algorithm (ECDSA). Technical report. ANSI (1999)
2. Alkim, E., Jakubeit, P., Schwabe, P.: NEWHOPE on ARM cortex-M. In: Carlet, C., Hasan, M.A., Saraswat, V. (eds.) SPACE 2016. LNCS, vol. 10076, pp. 332–349. Springer, Cham (2016). https://doi.org/10.1007/978-3-319-49445-6_19
3. Baily Jr., W.L.: On the theory of θ-functions, the moduli of abelian varieties, and the moduli of curves. Ann. Math. **2**(75), 342–381 (1962)
4. Bernstein, D.J., Lange, T.: eBACS: ECRYPT Benchmarking of Cryptographic Systems. https://bench.cr.yp.to/index.html

5. Bernstein, D.J.: Curve25519: new Diffie-Hellman speed records. In: Yung, M., Dodis, Y., Kiayias, A., Malkin, T. (eds.) PKC 2006. LNCS, vol. 3958, pp. 207–228. Springer, Heidelberg (2006). https://doi.org/10.1007/11745853_14
6. Bernstein, D.J.: Elliptic vs. hyperelliptic, part 1 (2006)
7. Bernstein, D.J., Chuengsatiansup, C., Lange, T., Schwabe, P.: Kummer strikes back: new DH speed records. In: Sarkar, P., Iwata, T. (eds.) ASIACRYPT 2014. LNCS, vol. 8873, pp. 317–337. Springer, Heidelberg (2014). https://doi.org/10.1007/978-3-662-45611-8_17
8. Bernstein, D.J., Duif, N., Lange, T., Schwabe, P., Yang, B.-Y.: High-speed high-security signatures. J. Cryptogr. Eng. **2**(2), 77–89 (2012)
9. Bernstein, D.J., Lange, T., Schwabe, P.: The security impact of a new cryptographic library. In: Hevia, A., Neven, G. (eds.) LATINCRYPT 2012. LNCS, vol. 7533, pp. 159–176. Springer, Heidelberg (2012). https://doi.org/10.1007/978-3-642-33481-8_9
10. Bertoni, G., Daemen, J., Peeters, M., Assche, G.V.: The KECCAK sponge function family (2016)
11. Bos, J.W., Costello, C., Hisil, H., Lauter, K.E.: Fast cryptography in genus 2. In: Johansson, T., Nguyen, P.Q. (eds.) EUROCRYPT 2013. LNCS, vol. 7881, pp. 194–210. Springer, Heidelberg (2013). https://doi.org/10.1007/978-3-642-38348-9_12
12. Cassels, J.W.S., Flynn, E.V.: Prolegomena to a Middlebrow Arithmetic of Curves of Genus 2, vol. 230. Cambridge University Press, Cambridge (1996)
13. Chudnovsky, D.V., Chudnovsky, G.V.: Sequences of numbers generated by addition in formal groups and new primality and factorization tests. Adv. Appl. Math. **7**, 385–434 (1986)
14. Chung, P.-N., Costello, C., Smith, B.: Fast, uniform, and compact scalar multiplication for elliptic curves and genus 2 jacobians with applications to signature schemes. Cryptology ePrint Archive, Report 2015/983 (2015)
15. Cosset, R.: Applications des fonctions theta à la cryptographie sur les courbes hyperelliptiques. Ph.D. thesis, Université Henri Poincaré - Nancy I (2011)
16. Diffie, W., Hellman, M.E.: New directions in cryptography. IEEE Trans. Inf. Theor. **22**(6), 644–654 (1976)
17. Düll, M., Haase, B., Hinterwälder, G., Hutter, M., Paar, C., Sánchez, A.H., Schwabe, P.: High-speed curve25519 on 8-bit, 16-bit and 32-bit microcontrollers. Des. Codes Cryptogr. **77**(2), 493–514 (2015)
18. Dworkin, M.J.: SHA-3 standard: Permutation-based hash and extendable-output functions. Technical report. National Institute of Standards and Technology (NIST) (2015)
19. Fiat, A., Shamir, A.: How to prove yourself: practical solutions to identification and signature problems. In: Odlyzko, A.M. (ed.) CRYPTO 1986. LNCS, vol. 263, pp. 186–194. Springer, Heidelberg (1987). https://doi.org/10.1007/3-540-47721-7_12
20. Gamal, T.E.: A public key cryptosystem and a signature scheme based on discrete logarithms. In: Blakley, G.R., Chaum, D. (eds.) CRYPTO 1984. LNCS, vol. 196, pp. 10–18. Springer, Heidelberg (1985). https://doi.org/10.1007/3-540-39568-7_2
21. Gaudry, P.: Fast genus 2 arithmetic based on theta functions. J. Math. Cryptol. **1**(3), 243–265 (2007)
22. Gaudry, P., Schost, E.: Genus 2 point counting over prime fields. J. Symb. Comput. **47**(4), 368–400 (2012)
23. Hamburg, M.: Fast and compact elliptic-curve cryptography. Cryptology ePrint Archive, Report 2012/309 (2012)
24. Hamburg, M.: The STROBE protocol framework. Cryptology ePrint Archive, Report 2017/003 (2017)

25. Hazay, C., Lindell, Y.: Efficient Secure Two-Party Protocols. Springer, Heidelberg (2010)
26. Hudson, R.W.H.T.: Kummer's Quartic Surface. Cambridge University Press, Cambridge (1905)
27. Hutter, M., Schwabe, P.: NaCl on 8-Bit AVR microcontrollers. In: Youssef, A., Nitaj, A., Hassanien, A.E. (eds.) AFRICACRYPT 2013. LNCS, vol. 7918, pp. 156–172. Springer, Heidelberg (2013). https://doi.org/10.1007/978-3-642-38553-7_9
28. Karati, S., Das, A.: Faster batch verification of standard ECDSA signatures using summation polynomials. In: Boureanu, I., Owesarski, P., Vaudenay, S. (eds.) ACNS 2014. LNCS, vol. 8479, pp. 438–456. Springer, Cham (2014). https://doi.org/10.1007/978-3-319-07536-5_26
29. Koblitz, N.: Elliptic curve cryptosystems. Math. Comput. **48**, 203–209 (1987)
30. Liu, Z., Longa, P., Pereira, G., Reparaz, O., Seo, H.: FourQ on embedded devices with strong countermeasures against side-channel attacks. Cryptology ePrint Archive, Report 2017/434 (2017)
31. Miller, V.S.: Use of elliptic curves in cryptography. In: Williams, H.C. (ed.) CRYPTO 1985. LNCS, vol. 218, pp. 417–426. Springer, Heidelberg (1986). https://doi.org/10.1007/3-540-39799-X_31
32. Montgomery, P.L.: Speeding the Pollard and elliptic curve methods of factorization. Math. Comput. **48**, 243–264 (1987)
33. Naccache, D., M'Raïhi, D., Vaudenay, S., Raphaeli, D.: Can D.S.A. be improved? — complexity trade-offs with the digital signature standard —. In: De Santis, A. (ed.) EUROCRYPT 1994. LNCS, vol. 950, pp. 77–85. Springer, Heidelberg (1995). https://doi.org/10.1007/BFb0053426
34. Nascimento, E., López, J., Dahab, R.: Efficient and secure elliptic curve cryptography for 8-bit AVR microcontrollers. In: Chakraborty, R.S., Schwabe, P., Solworth, J. (eds.) SPACE 2015. LNCS, vol. 9354, pp. 289–309. Springer, Cham (2015). https://doi.org/10.1007/978-3-319-24126-5_17
35. Okeya, K., Sakurai, K.: Efficient elliptic curve cryptosystems from a scalar multiplication algorithm with recovery of the y-coordinate on a montgomery-form elliptic curve. In: Koç, Ç.K., Naccache, D., Paar, C. (eds.) CHES 2001. LNCS, vol. 2162, pp. 126–141. Springer, Heidelberg (2001). https://doi.org/10.1007/3-540-44709-1_12
36. Perrin, T.: The XEdDSA and VXEdDSA Signature Schemes. https://whispersystems.org/docs/specifications/xeddsa/
37. Pointcheval, D., Stern, J.: Security proofs for signature schemes. In: Maurer, U. (ed.) EUROCRYPT 1996. LNCS, vol. 1070, pp. 387–398. Springer, Heidelberg (1996). https://doi.org/10.1007/3-540-68339-9_33
38. Pointcheval, D., Stern, J.: Security arguments for digital signatures and blind signatures. J. Cryptol. **13**(3), 361–396 (2000)
39. Renes, J., Schwabe, P., Smith, B., Batina, L.: µKummer: efficient hyperelliptic signatures and key exchange on microcontrollers. In: Gierlichs, B., Poschmann, A.Y. (eds.) CHES 2016. LNCS, vol. 9813, pp. 301–320. Springer, Heidelberg (2016). https://doi.org/10.1007/978-3-662-53140-2_15
40. Schnorr, C.P.: Efficient identification and signatures for smart cards. In: Brassard, G. (ed.) CRYPTO 1989. LNCS, vol. 435, pp. 239–252. Springer, New York (1990). https://doi.org/10.1007/0-387-34805-0_22
41. Semaev, I.A.: Summation polynomials and the discrete logarithm problem on elliptic curves. IACR Cryptology ePrint Archive **2004**, 31 (2004)
42. Stahlke, C.: Point compression on jacobians of hyperelliptic curves over $\mathbb{F}_q$. Cryptology ePrint Archive, Report 2004/030 (2004)

A Simple and Compact Algorithm for SIDH with Arbitrary Degree Isogenies

Craig Costello[1](✉) and Huseyin Hisil[2]

[1] Microsoft Research, Redmond, USA
craigco@microsoft.com
[2] Yasar University, Izmir, Turkey
huseyin.hisil@yasar.edu.tr

Abstract. We derive a new formula for computing arbitrary odd-degree isogenies between elliptic curves in Montgomery form. The formula lends itself to a simple and compact algorithm that can efficiently compute any low odd-degree isogenies inside the supersingular isogeny Diffie-Hellman (SIDH) key exchange protocol. Our implementation of this algorithm shows that, beyond the commonly used 3-isogenies, there is a moderate degradation in relative performance of $(2d + 1)$-isogenies as d grows, but that larger values of d can now be used in practical SIDH implementations.

We further show that the proposed algorithm can be used to both compute isogenies of curves and evaluate isogenies at points, unifying the two main types of functions needed for isogeny-based public-key cryptography. Together, these results open the door for practical SIDH on a much wider class of curves, and allow for simplified SIDH implementations that only need to call one general-purpose function inside the fundamental computation of the large degree secret isogenies.

As an additional contribution, we also give new explicit formulas for 3- and 4-isogenies, and show that these give immediate speedups when substituted into pre-existing SIDH libraries.

Keywords: Post-quantum cryptography · Isogeny-based cryptography · SIDH · Montgomery curves

1 Introduction

Post-quantum Key Establishment. The existence of a quantum computer that is capable of implementing Shor's algorithm [36] at a large enough scale would have devastating consequences on the current public-key cryptographic standards and thus on the current state of cybersecurity [32]. Subsequently, the field of *post-quantum cryptography* (PQC) [4] is rapidly growing as cryptographers look for public-key solutions that can resist large-scale quantum adversaries. Recently, the USA's National Institute of Standards and Technology (NIST) began a process to develop new cryptographic standards and announced a call for PQC proposals with a deadline of November 30, 2017 [40].

© International Association for Cryptologic Research 2017
T. Takagi and T. Peyrin (Eds.): ASIACRYPT 2017, Part II, LNCS 10625, pp. 303–329, 2017.
https://doi.org/10.1007/978-3-319-70697-9_11

Although the PQC community is currently examining alternatives to replace both traditional key establishment and traditional digital signature algorithms, there is an argument for scrutinising proposals in the former category with more haste than those in the latter. While digital signatures only need to be quantum-secure at the moment a powerful enough quantum adversary is realised, the realistic threat of long-term archival of sensitive data and a retroactive quantum break means that, ideally, key establishment protocols will offer quantum resistance long before such a quantum adversary exists [38].

Post-quantum key establishment proposals typically fall under one of three umbrellas:

(i) *Code-based.* Based on the McEliece cryptosystem [28] and its variants [28], modern proposals include Bernstein, Chou and Schwabe's McBits [5] and Misoczki et al.'s specialised MDPC-McEliece [10,29].

(ii) *Lattice-based.* Proposals here began with Hoffstein, Pipher and Silverman's standardised NTRUEncrypt [20], and in more recent times have been based on either Regev's learning with errors (LWE) problem [34] or Lyubashevsky, Peikert and Regev's ring variant (R-LWE) [27]. Peikert brought these problems to life in [33], and his protocols served as a basis for a number of recent implementations, including Bos et al.'s R-LWE key establishment software [7], Alkim et al.'s R-LWE successor NewHope [1], and Bos et al.'s LWE key establishment software Frodo [6].

(iii) *Isogeny-based.* Starting with the work of Couveignes [13] and with later work by Rostovsev and Stolbunov [35,39], Jao and De Feo proposed and implemented supersingular isogeny Diffie-Hellman (SIDH) key exchange [21]. In recent times a number of improvements and optimisations of their SIDH protocol have been proposed and implemented [2,11,12,15,26].

To date there is no clear frontrunner among the post-quantum key establishment proposals. In terms of functionality, all of the public implementations resulting from (i), (ii) and (iii) suffer the same drawback of requiring modifications (e.g., the Fujisaka-Okamoto transformation [17]) to achieve active security[1]. However, there are bandwidth versus performance trade-offs to consider when examining the above proposals; while SIDH affords significantly smaller public keys than its code- and lattice-based counterparts, the performance of the state-of-the-art SIDH software is currently orders of magnitude slower than the state-of-the-art implementations mentioned in (i) and (ii) above. The reason for this wide performance gap is that well-chosen code- and lattice-based instantiations typically involve simple matrix/vector operations over special, and comparatively tiny, implementation-friendly moduli that are either powers of 2 or very close to a power of 2. On the other hand, in addition to SIDH inheriting several of the more complex operations from traditional curve-based cryptography like scalar multiplications and pairings, it also involves a new style of isogeny arithmetic and requires a new breed of significantly larger underlying

---

[1] For (i), see [5, Sect. 6] and [23]; for (ii), see [33, Sect 5.3], [16] and [22]; for (iii), see [19,22].

finite fields. Whereas classical elliptic curve cryptography affords implementers the flexibility to cherry-pick the fastest underlying finite fields of sizes as small as 256 bits, most of the SIDH implementations to date have required extension fields of over one thousand bits whose underlying characteristic are of the form $p = 2^i 3^j - 1$. Imposing this special form of prime restricts both the number of SIDH-friendly fields available at a given security level and the number of field arithmetic optimisations possible for implementers.

Our Contributions. This paper presents a new algorithm for computing the fundamental operation in isogeny-based public-key cryptography, and in particular, within the SIDH protocol.

- *Odd-degree Montgomery isogenies.* We derive a new formula for odd-degree isogenies between Montgomery curves – see Theorem 1. Compared to Vélu's formulas for isogenies between Weierstrass curves, this formula is elegant and simple, both to write down and to implement. This formula immediately lends itself to a compact algorithm that computes arbitrary odd-degree isogenies.
- *Unifying the two isogeny operations.* SIDH operations require isogeny computations to be applied to elliptic curves within the isogeny class and to the points that lie on those curves. These two operations are typically different and require independent functions. For odd-degree isogenies, we show that both of these operations can be performed using the same core function by exploiting the simple connection between 2-torsion points and the Montgomery curve coefficient. This streamlines SIDH code, and for isogenies of degree 5 and above, has the added benefit of being significantly faster than performing the computations independently.
- *Simplified algorithm.* Together, the above two improvements culminate in a general-purpose algorithm that can efficiently compute isogenies of any odd degree. Coupled with specialised code for 2- and/or 4-isogenies, this allows arbitrary SIDH computations and gives rise to new possibilities within the SIDH framework. Our implementation benchmarks show that practitioners can lift the restriction of primes of the form $p = 2^i 3^j - 1$ without paying a huge performance penalty.
- *Faster 3- and 4-isogenies.* While the contributions mentioned above broaden the scope of curves that can be considered SIDH-friendly, they do not give an immediate speedup to existing SIDH implementations because the pre-existing formulas for 3-isogenies are a special case of Theorem 1. Nevertheless, as an auxiliary result, we give new dedicated 3- and 4-isogeny algorithms that do give immediate speedups. When plugging these new algorithms into Microsoft's recent v2.0 release of their SIDH library[2], Alice and Bob's key generations are both sped up by a factor 1.18x, while their shared secret computations are both sped up by a factor 1.11x.

Although this paper is largely geared towards SIDH key exchange, we note that almost all of the discussion applies analogously to other supersingular

---

[2] See https://github.com/Microsoft/PQCrypto-SIDH.

isogeny-based cryptographic schemes, e.g., to the other schemes proposed by De Feo *et al.* [15], and to the recent isogeny-based signature scheme from Yoo *et al.* [43].

Organisation. We give the preliminaries in Sect. 2. We provide the new formula for odd-degree Montgomery isogenies in Sect. 3 and discuss its connection to related works. We show how the point and curve isogeny computations can be performed using the same function in Sect. 4, before presenting the general-purpose odd-degree isogeny algorithm in Sect. 5. We provide implementation benchmarks and conclude with some potential implications in Sect. 6. The faster explicit formulas for 3- and 4-isogenies are presented in Appendix A.

Remark 1 (Even degree isogenies). Since any separable isogeny can be written as a *chain* of prime degree isogenies [18, Theorem 25.1.2], our claim of treating arbitrary degree isogenies on Montgomery curves follows from the coupling of Theorem 1 (which covers isogenies of any odd degree) with the prior treatment of 2-isogenies on Montgomery curves by De Feo *et al.* [15]. It is worth noting that a technicality arises in the treatment of 2-isogenies on Montgomery curves: there is currently no known way of computing a 2-isogeny directly from a generic 2-torsion point without extracting a square root to transform the image curve into Montgomery form. De Feo, Jao and Plût overcome this obstruction by making use of a special 8-torsion point *lying above* the 2-torsion point in the kernel, which is already available for use in the SIDH framework. In broader contexts, however, the preservation of the Montgomery form under general 2-isogenies might become problematic; in these cases even powers of 2 can be treated by the application of 4-isogenies which do not need to compute square roots in order to preserve the Montgomery form [15]. In Remark 2 we discuss the related work of Moody and Shumow [31] on the (twisted) Edwards model [31]; in their case the 2-isogeny formula also requires a square root computation to preserve the Edwards form. Although Vélu's formulas for 2-isogenies between short Weierstrass curves do not require square root computations, we believe it worthwhile to pose the open question of finding efficient 2-isogenies that preserve either of the faster Montgomery and/or twisted Edwards models (on input of a generic 2-torsion point).

2 Preliminaries

Montgomery Curves. Unless stated otherwise, all elliptic curves E/K in this paper are assumed to be written in Montgomery form [30]

$$E/K : by^2 = x^3 + ax^2 + x.$$

We will be dealing with the group of K-rational points on E, denoted $E(K)$, which is the set of solutions $(x, y) \in K \times K$ to the above equation, furnished with a point at infinity, $\mathcal{O}_E$. This point looks different under different projective embeddings of E. The usual embedding into $\mathbb{P}^2$ via $x = X/Z$ and $y = Y/Z$ gives $\mathcal{O}_E = (0 : 1 : 0)$, but the proof of Theorem 1 makes use of the alternative embedding into $\mathbb{P}(1, 2, 1)$ via $x = X/Z$ and $y = Y/Z^2$, under which $\mathcal{O}_E = (1 : 0 : 0)$. The inverse of a point (x, y) is $(x, -y)$, and the number of points in $E(K)$ is always divisible by 4.

Let $P = (x_P, y_P)$ and $Q = (x_Q, y_Q)$ be such that $x_P \neq x_Q$. Then the coordinates of these points and the x-coordinates of their sum $P + Q$ and difference $P - Q$ are related by Montgomery's group law identities [30, p. 261]

$$x_{P+Q}(x_P - x_Q)^2 x_P x_Q = b(x_P y_Q - x_Q y_P)^2, \quad \text{and}$$
$$x_{P-Q}(x_P - x_Q)^2 x_P x_Q = b(x_P y_Q + x_Q y_P)^2. \tag{1}$$

Montgomery multiplies these equations to produce his celebrated differential arithmetic formulas [30, p. 262]

$$x_{P+Q} x_{P-Q} = (x_P x_Q - 1)^2 / (x_P - x_Q)^2, \quad \text{and}$$
$$x_{[2]P} = (x_P^2 - 1)^2 / (4x_P(x_P^2 + ax_P + 1)). \tag{2}$$

Assuming the usual embedding of E into $\mathbb{P}^2$, then following [30], we use $\mathbf{x}$ throughout to denote the subsequent projection of points into $\mathbb{P}^1$ that *drops* the Y-coordinate, i.e.,

$$\mathbf{x}: \quad E \setminus \{\mathcal{O}_E\} \to \mathbb{P}^1, \quad (X : Y : Z) \mapsto (X : Z).$$

Applying this to (2) gives Montgomery's two algorithms for differential arithmetic in $\mathbb{P}^1$, i.e.,

$$\text{xDBL}: \quad (\mathbf{x}(P), a) \mapsto \mathbf{x}([2]P), \quad \text{and}$$
$$\text{xADD}: \quad (\mathbf{x}(P), \mathbf{x}(Q), \mathbf{x}(P - Q)) \mapsto \mathbf{x}(P + Q). \tag{3}$$

If ℓ is odd, then the ℓ-th division polynomial of an elliptic curve E/K is written as $\psi_\ell(x) \in K[x]$, and this vanishes precisely at the nontrivial ℓ-torsion points, i.e., the points P such that $[\ell]P = \mathcal{O}_E$. The first two nontrivial odd division polynomials on the Montgomery curve $E/K : by^2 = x^3 + ax^2 + x$ are

$$\psi_3(x) = 3x^4 + 4ax^3 + 6x^2 - 1,$$
$$\psi_5(x) = 5x^{12} + 20ax^{11} + (16a^2 + 62)x^{10} + 80ax^9 - 105x^8 - 360ax^7$$
$$- 60(5 + 4a^2)x^6 - 16a(23 + 4a^2)x^5 - 5(25 + 32a^2)x^4 - 140ax^3 - 50x^2 + 1. \tag{4}$$

SIDH. Let $p = f \cdot n_A n_B \pm 1$ be a large prime where $\gcd(n_A, n_B) = 1$ and f is a small cofactor. SIDH [21] works in the isogeny class of supersingular elliptic curves over $\mathbb{F}_{p^2}$, all of which have cardinality $(p \mp 1)^2 = (f \cdot n_A n_B)^2$. Let E be a public starting curve in this isogeny class. To generate her public key, Alice chooses a secret subgroup G_A of order n_A on E and computes her public key as E/G_A. Likewise, Bob chooses a secret subgroup G_B of order n_B and computes his public key as E/G_B. The shared secret is then $E/\langle G_A, G_B \rangle$, and so long as computing this from E, E/G_A and E/G_B is hard, this offers an alternative instantiation of the Diffie-Hellman protocol [14]. The key to the SIDH construction is ensuring that both parties exchange enough information to allow the mutual computation of $E/\langle G_A, G_B \rangle$, while still hiding their secret keys.

To achieve this, Jao and De Feo [21] propose that the public keys also contain the images of certain public points under the isogenies defined by their secret subgroups. If $\phi_A : E \rightarrow E/G_A$ is the secret isogeny corresponding to the subgroup G_A, then Alice not only sends Bob the curve E/G_A, but also the image of ϕ_A on two points P_B and Q_B whose linear combinations generate the set of subgroups chosen by Bob, i.e., Alice's public key is $\mathrm{PK}_A = (E/G_A\,,\,\phi_A(P_B)\,,\,\phi_A(Q_B))$. Similarly, if linear combinations of P_A and Q_A generate the set of subgroups chosen by Alice, then Bob's public key is $\mathrm{PK}_B = (E/G_B\,,\,\phi_B(P_A)\,,\,\phi_B(Q_A))$. In this way Alice's key generation amounts to randomly choosing two secret integers $u_A, v_A \in \mathbb{Z}_{n_A}$, computing $G_A = \langle [u_A]P_A + [v_A]Q_A \rangle$, and upon receipt of Bob's public key, she can then compute $E/\langle G_A, G_B \rangle = (E/G_B)/\langle [u_A]\phi_B(P_A) + [v_A]\phi_B(Q_A)\rangle$. Bob proceeds analogously, and both parties compute the shared secret as the j-invariant of $E/\langle G_A, G_B \rangle$.

In order for SIDH to be secure, n_A and n_B must be exponentially large so that Alice and Bob have an exponentially large keyspace. On the other hand, in order for SIDH to be practical, the computation of the n_A- and n_B-isogenies must be manageable. To achieve this, Jao and De Feo propose that $n_A = \ell_A^{e_A}$ and $n_B = \ell_B^{e_B}$ for ℓ_A and ℓ_B small; in this way there are $\ell_A^{e_A-1}(\ell_A + 1)$ secret cyclic subgroups of order n_A for Alice to choose from, and her secret isogeny computations can be performed as the composition of e_A low-degree ℓ_A-isogenies (the analogous statement applies to Bob). In all of the SIDH implementations to date [2,11,12,15,26], $\ell_A = 2$ and $\ell_B = 3$, and Alice computes her 2^{e_A}-isogeny as a composition of 2- and/or 4-isogenies (see [12,15]), while Bob computes his 3^{e_B}-isogeny as a composition of 3-isogenies. One consequence of this paper is to facilitate practical ℓ^e-isogenies where $\ell \geq 5$.

Following [21, Fig. 2], one way to compute an ℓ^e cyclic isogeny ϕ on the curve E_0 is to start with a point P_0 of order ℓ^e, compute the point $[\ell^{e-1}]P_0$ of order ℓ, then use Vélu's formulas [41] to compute the ℓ-isogeny $\phi_0 : E_0 \rightarrow E_1$ with $\ker(\phi_0) = \langle [\ell^{e-1}]P_0 \rangle$, and evaluate it at P_0 to give $\phi_0(P_0) = P_1 \in E_1$. Note that pushing P_0 through the ℓ-isogeny reduces the order of its image point, P_1, by a factor of ℓ on E_1. This process is then repeated at each new iteration by first computing the order ℓ point $[\ell^{e-1-i}]P_i$, then the ℓ-isogeny $\phi_i : E_i \rightarrow E_{i+1}$, and finally the computation of the new point $P_{i+1} = \phi_i(P_i)$; this is done until $i = e - 1$ and we have the final curve $E_e = \phi_{e-1} \circ \cdots \circ \phi_0(E_0) = \phi(E_0)$.

In their extended article, De Feo et al. [15] detailed a much faster approach towards the computation described above. Roughly speaking, they achieve large speedups by storing intermediate multiples of the P_i at each step and evaluating ϕ_i at these multiples in such a way that the length of the scalar multiplication to find an order-ℓ point on E_{i+1} is reduced. They aim to minimise the overall cost of the ℓ^e-isogeny computation by comparing the costs of point multiplications and isogeny computations and studying the optimisation problem in a combinatorial context – see [15, Sect. 4.2.2].

Following [12], in order to thwart simple timing attacks [24], the fastest way to compute SIDH operations in a constant-time fashion is to (i) perform point operations on the projective line $\mathbb{P}^1$ associated to Montgomery's x-coordinate, i.e., using the map **x** above, and (ii) to also perform isogeny operations projectively in $\mathbb{P}^1$ by ignoring the b coefficient (in the same way the Y-coordinate is ignored in the point arithmetic). The reasoning here is that the j-invariant (i.e., the isomorphism class) of the Montgomery curve $E/K : by^2 = x^3 + ax^2 + x$ is $j(E) = 256(a^2 - 3)^3/(a^2 - 4)$, which is independent of b, as is the differential arithmetic arising from (3). All of the formulas and algorithms we describe in the remainder of this paper fit into this same framework.

3   Coordinate Maps for Odd-Degree Montgomery Isogenies

At the heart of this paper is the coordinate maps in Eq. (6) of Theorem 1 below. Although we are mostly concerned with the SIDH-specific applications to come in the following sections, we believe that the simplicity and usability of the formula may be of interest outside the realm of SIDH, so we leave the underlying field unspecified and state the isogeny formula in full. We follow the theorem with a discussion of the related work of Moody and Shumow [31].

Theorem 1. *For a field K with $\mathrm{char}(K) \neq 2$, let $P \in E(\bar{K})$ be a point of order $\ell = 2d + 1$ on the Montgomery curve $E/K : by^2 = x^3 + ax^2 + x$ and write $\sigma = \sum_{i=1}^{d} x_{[i]P}$, $\tilde{\sigma} = \sum_{i=1}^{d} 1/x_{[i]P}$ and $\pi = \prod_{i=1}^{d} x_{[i]P}$. The Montgomery curve*

$$E'/K : b'y^2 = x^3 + a'x^2 + x \tag{5}$$

with

$$a' = (6\tilde{\sigma} - 6\sigma + a) \cdot \pi^2 \qquad \text{and} \qquad b' = b \cdot \pi^2$$

is the codomain of the ℓ-isogeny $\phi : E \to E'$ with $\ker(\phi) = \langle P \rangle$, which is defined by the coordinate maps

$$\phi : (x, y) \mapsto (f(x), y \cdot f'(x)), \tag{6}$$

where

$$f(x) = x \cdot \prod_{i=1}^{d} \left(\frac{x \cdot x_{[i]P} - 1}{x - x_{[i]P}} \right)^2,$$

and $f'(x)$ is its derivative.

Proof. The proof follows along the lines of Washington's proof of Vélu's formulas on general Weiestrass curves [42, Theorem 12.16]. Write

$$\phi : (x,y) \mapsto (X,Y),$$

where $X = f(x) = x \cdot u(x)^2/w(x)^2$ and $Y = y \cdot f'(x)$, with $u(x) = \prod_{i=1}^{d}(x \cdot x_i - 1)$ and $w(x) = \prod_{i=1}^{d}(x - x_i)$, and write $G = \langle P \rangle$. Since X and Y are rational functions of x and y, they are functions on E, and it is clear that the only poles of X and Y are at the points in G. Our main goal is to show that the function

$$F(X,Y) = b'Y^2 - (X^3 + a'X^2 + X) \tag{7}$$

is 0. The idea is to introduce a uniformising parameter t at $\mathcal{O}_E$ and ultimately show that $F(X,Y) \in O(t)$, i.e., that $F(X,Y)$ vanishes at $\mathcal{O}_E$. Now, since $x_{[i]P} = x_{[\ell-i]P}$ for $i = 1 \ldots d$, it follows from (2) that for any $Q = (x_Q, y_Q) \notin G$,

$$f(x_Q) = \prod_{T \in G} x_{Q+T},$$

and therefore that the functions X and Y are invariant under translation by elements of G. Thus, if we can show that $F(X,Y)$ vanishes at $\mathcal{O}_E$, we will have also shown that it vanishes at all of the points in G. Since the function $F(X,Y)$ can only have poles at the points in G, the only possibility is that $F(X,Y)$ has no poles, which means that it is constant [42, Proposition 11.1(3)]. Furthermore, since it is 0 at infinity, it must be that $F(X,Y)$ is identically zero, and therefore that X and Y satisfy the Montgomery curve equation in (5).

To show that $F(X,Y)$ vanishes at $\mathcal{O}_E$, let $t = x/y$ be a uniformising parameter at $\mathcal{O}_E$ and let $s = 1/y$. Dividing $by^2 = x^3 + ax^2 + x$ by y^3 and rearranging yields

$$s = (t^3 + at^2 s + ts^2)/b.$$

Continually substituting this value for s into the above equation eventually yields

$$s = t_3 \cdot t^3 + t_5 \cdot t^5 + t_7 \cdot t^7 + t_9 \cdot t^9 + \ldots,$$

where

$$t_3 = 1/b, \qquad t_5 = a/b^2, \qquad t_7 = (1+a^2)/b^3, \qquad t_9 = a(3+a^2)/b^4,$$
$$t_{11} = (6a^2 + a^4 + 2)/b^5, \qquad t_{13} = a(a^4 + 10a^2 + 10)/b^6.$$

Since $y = 1/s$ and $x = ty$, we invert the above equation to give

$$y = bt^{-3} \cdot (y_0 + y_2 \cdot t^2 + y_4 \cdot t^4 + y_6 \cdot t^6 + \ldots), \quad \text{and}$$
$$x = bt^{-2} \cdot (y_0 + y_2 \cdot t^2 + y_4 \cdot t^4 + y_6 \cdot t^6 + \ldots) \tag{8}$$

where

$$y_0 = 1, \qquad y_2 = -a/b, \qquad y_4 = -1/b^2, \qquad y_6 = -a/b^3,$$
$$y_8 = -(a^2 + 1)/b^4, \qquad y_{10} = -(a^3 + 3a)/b^5. \qquad y_{12} = (a^6 + 14a^4 + 24a^2 + 3)/b^6.$$

From (6), we write

$$X = x \cdot f_1(x)^2 \cdot f_2(x)^2 \cdot \cdots \cdot f_d(x)^2, \tag{9}$$

where

$$f_i(x) = \frac{x_{[i]P} \cdot x - 1}{x - x_{[i]P}},$$

for $1 \le i \le d$. Substitution of (8) gives

$$f_i(t) = \frac{x_{[i]P} \cdot \left(bt^{-2} \cdot (y_0 + y_2 \cdot t^2 + y_4 \cdot t^4 + y_6 \cdot t^6 + \ldots)\right) - 1}{(bt^{-2} \cdot (y_0 + y_2 \cdot t^2 + y_4 \cdot t^4 + y_6 \cdot t^6 + \ldots)) - x_{[i]P}},$$

$$= x_{[i]P} + \left[\frac{x_{[i]P}^2 - 1}{b}\right] \cdot t^2 + \left[\frac{(x_{[i]P}^2 - 1)(a + x_{[i]P})}{b^2}\right] \cdot t^4$$

$$+ \left[\frac{(x_{[i]P}^2 - 1)((a + x_{[i]P})^2 + 1)}{b^3}\right] \cdot t^6 + O(t^8).$$

Squaring the above equation yields

$$f_i(t)^2 = x_{[i]P}^2 + \left[\frac{2x_{[i]P}(x_{[i]P}^2 - 1)}{b}\right] \cdot t^2 + \left[\frac{(x_{[i]P}^2 - 1)(3x_{[i]P}^2 + 2ax_{[i]P} - 1)}{b^2}\right] \cdot t^4$$

$$+ \left[\frac{2(x_{[i]P}^2 - 1)(a^2 x_{[i]P} + 3ax_{[i]P}^2 + 2x_{[i]P}^3 - a)}{b^3}\right] \cdot t^6 + O(t^8). \tag{10}$$

Substitution of (8) and (10) into (9) gives

$$X(t) = X_{-2} \cdot t^{-2} + X_0 + X_2 \cdot t^2 + X_4 \cdot t^4 + O(t^6). \tag{11}$$

where

$$X_{-2} = b\pi^2$$
$$X_0 = -\pi^2(2(\tilde{\sigma} - \sigma) + a)$$
$$X_2 = -\frac{4\pi^4((\sigma - \tilde{\sigma})(a + 3(\tilde{\sigma} - \sigma)) + 1) + 1}{5b\pi^2}$$
$$X_4 = -\frac{12\pi^4(\sigma - \tilde{\sigma})a^2 + (3 - 16\pi^4((\sigma - \tilde{\sigma})^2 - 2))a - 10(\sigma - \tilde{\sigma})(8\pi^4(\sigma - \tilde{\sigma})^2 - 1)}{35b^2\pi^2}.$$

Now, the product rule gives $X'(x) = X'(t) \cdot (x'(t))^{-1}$, so from (11) we have

$$X'(t) = t^{-3} \cdot \left(-2X_{-2} + 2X_2 t^4 + 4X_4 t^6 + O(t^8)\right), \tag{12}$$

and from (8) we have

$$x'(t) = bt^{-3} \cdot (-2y_0 + 2y_4 \cdot t^4 + 4y_6 \cdot t^6 + 6y_8 \cdot t^8 \ldots).$$

Inverting the above equation yields

$$x'(t)^{-1} = -\frac{t^3}{2by_0^2} \cdot \left(y_0 + y_4 \cdot t^4 + 2y_6 \cdot t^6 + \left[\frac{3y_0y_8 + y_4^2}{y_0}\right] \cdot t^8 \right.$$
$$\left. + \left[\frac{4y_0y_{10} + 4y_4y_6}{y_0}\right] \cdot t^{10} + O(t^{12})\right). \qquad (13)$$

We now have all the ingredients to write $F(X, Y)$ entirely in terms of t. We write

$$F(t) = b' \left(y(t) \cdot X'(t) \cdot x'(t)^{-1}\right)^2 - \left(X(t)^3 + a'X(t)^2 + X(t)\right),$$

Substituting (8), (11), (12) and (13) into the above equation, and collecting coefficients, yields

$$F(X, Y) = F_{-6} \cdot t^{-6} + F_{-4} \cdot t^{-4} + F_{-2} \cdot t^{-2} + F_0 + O(t), \qquad (14)$$

where

$$F_{-6} = X_{-2}^2 \cdot (b' - X_{-2}),$$
$$F_{-4} = -X_{-2}^2 \cdot \frac{2ab' + b(a' + 3X_0)}{b},$$
$$F_{-2} = X_{-2} \cdot \frac{((a^2 - 4)X_{-2} - 2b^2 X_2) \cdot b' - b^2(2a'X_0 + 3X_0^2 + 3X_2X_{-2} + 1)}{b^2},$$
$$F_0 = \frac{4X_{-2}(aX_2 - bX_4)b' - b \cdot (a'X_0^2 + 2a'X_2X_{-2} + X_0^3 + 6X_{-2}X_0X_2 + 3X_4X_{-2}^2 + X_0)}{b}.$$

With X_{-2}, X_0, X_2 and X_4 as in (11), and with a' and b' from (5), we get $F_{-6} = F_{-4} = F_{-2} = F_0 = 0$. Thus, we have $F(X, Y) \in O(t)$, which means that $F(X, Y)$ vanishes at $\mathcal{O}_E$, and thus (as detailed above) that $F(X, Y)$ is identically zero. It follows that X and Y satisfy the equation for E' in (5), and thus that ϕ is a rational map from E to E'. Since E is a smooth curve, we have that ϕ is a morphism [37, Proposition II.2.1]. To show that ϕ is an isogeny, we project into $\mathbb{P}(1, 2, 1)$, where $\mathcal{O}_E = (1:0:0)$. Substitution of $x = X/Z$ and $y = Y/Z^2$ into (6) reveals that $\phi(\mathcal{O}_E) = (1:0:0) = \mathcal{O}_{E'}$, and we have established that ϕ is an isogeny [37, III.4]. This completes the proof.   □

In comparison to Vélu's formulas [41] on general Weierstrass curves, the simplicity of Eq. (6) lies in the fact that it factors neatly across the different multiples of P. This lends itself to the simple algorithm we describe in Sect. 5.

Remark 2. To our knowledge, the work of Moody and Shumow [31] is the only prior work to investigate arbitrary degree isogenies on non-Weierstrass models. They managed to successfully derive general isogenies on both (twisted) Edwards curves [31, Theorem 3] and on Huff curves [31, Theorem 5] without passing back and forth to Vélu's formulas on Weierstrass models. Given the 'uniform-variable' formulas in [31, Sect. 4.4], we could have presumably arrived at (6) by exploiting the birational equivalence between twisted Edwards and Montgomery curves [3, Theorem 3.2]. In particular, there is a simple relationship between the twisted Edwards y-coordinate and the Montgomery x-coordinate, and subsequently, there are Edwards y-only analogues of Montgomery's x-only differential arithmetic that offer favourable trade-offs in certain ECC scenarios[3]. However, our experiments seemed to suggest that these trade-offs evaporate in SIDH when the curve constants are treated projectively. Nevertheless, given the similarities between the y-only isogeny formula in [31, Theorem 4] and our Theorem 1, it could be that there are savings to be gained in a twisted Edwards version of SIDH, or perhaps in some sort of hybrid that passes back and forth between the two models – see [9]. We leave this investigation open, pointing out that the sorts of trade-offs discussed in [9] can become especially favourable in SIDH, due to the large field sizes and the nature of arithmetic in quadratic extension fields.

We conclude this section by pointing out that (in our case) it is a simple exercise to transform Eq. (6) into an analogue that, rather than writing the isogeny map in terms of the coordinates of the torsion points à la Vélu, instead writes it in terms of the coefficients of the polynomial defining the kernel subgroup à la Kohel [25, Sect. 2.4]. While a Kohel-style formulation of our formula is arguably more natural from a mathematical perspective, the way it is factored and written in (6) is more natural from an algorithmic perspective.

4   Computing the Isogenous Curve Using the 2-Torsion

Let $\phi : E \mapsto E'$ be the isogeny of Montgomery curves in Theorem 1 and let $Q \in E$ be any point where $Q \notin \ker(\phi)$. All supersingular isogeny-based cryptosystems, and in particular all known implementations of SIDH [2,11,12,15,26], require separate functions for computing isogenous curves, i.e., iso_curve : $E \mapsto \phi(E)$, and for evaluating the isogeny at points, i.e., iso_point : $Q \mapsto \phi(Q)$. In this brief section we show that these two functions can be unified in the computation of odd-degree isogenies. The idea is to exploit the correspondence between the 2-torsion points and the curve-twist isomorphism class, and to replace calls to the iso_curve function with calls to iso_point on the input of 2-torsion points. Pushing 2-torsion points through an odd-degree isogeny preserves their order on the image curve, and so the correspondence between 2-torsion points and the isogenous curves they lie on remains an invariant throughout the SIDH algorithm.

On the Montgomery curve $E/K : y^2 = x^3 + ax^2 + x$, the three affine points of order 2 in $E(\bar{K})$ are

$$P_0 = (0,0), \qquad P_\alpha = (\alpha, 0), \qquad \text{and} \qquad P_{1/\alpha} = (1/\alpha, 0),$$

where $a = -(\alpha^2 + 1)/\alpha$. Note that the full 2-torsion is K-rational if $x^2 + ax + 1$ is reducible in $K[x]$, i.e., if $\alpha \in K$; this is typically the case in SIDH and is therefore assumed in this section.

---

[3] See http://hyperelliptic.org/EFD/g1p/auto-edwards-yz.html.

Under the **x** map from Sect. 2, the 2-torsion points are then

$$\mathbf{x}(P_0) = (0 : 1), \qquad \mathbf{x}(P_\alpha) = (X_\alpha : Z_\alpha), \qquad \text{and} \qquad \mathbf{x}(P_{1/\alpha}) = (Z_\alpha : X_\alpha),$$

and in $\mathbb{P}^1$ we now have

$$(a : 1) = (X_\alpha^2 + Z_\alpha^2 : -X_\alpha Z_\alpha). \tag{15}$$

Observe that for the isogeny ϕ described in Theorem 1, the point $P_0 = (0,0) \in E$ is mapped to the point $\phi(P_0) = (0,0) \in E'$. Since E' is a Montgomery curve and 2-torsion points preserve their order under odd isogenies, it must be that

$$\mathbf{x}(\phi(P_0)) = (0 : 1), \qquad \mathbf{x}(\phi(P_\alpha)) = (X_\alpha' : Z_\alpha'), \qquad \text{and} \qquad \mathbf{x}(\phi(P_{1/\alpha})) = (Z_\alpha' : X_\alpha'),$$

so that the relation in (15) between 2-torsion coordinates and the curve coefficient holds on the new curve.

Rather than thinking of the Montgomery curve as being represented by the coefficient $(a : 1) = (A : C)$, we can (without loss of generality) think of it as being represented by the 2-torsion point $(X_\alpha : Z_\alpha)$. A close inspection of Theorem 1 reveals that, for values of d greater than 3, computing the isogenous curve via (5) becomes increasingly more expensive than passing a 2-torsion point through (6). In these cases a function for computing (5) is no longer needed. If, during the current iteration, the curve constant $(A : C)$ is needed for point operations (e.g., the multiplication-by-ℓ map), then we can recover A using (15) at a cost[4] of $2\mathbf{S} + 5\mathbf{a}$. In fact, the general multiplication-by-ℓ routine is the Montgomery ladder [30] that calls xDBL as a subroutine, and (in SIDH) xDBL makes use of the constant $(a - 2 : 4) = ((A - 2C)/4 : C)$. Parsing directly to this format is slightly faster than parsing to $(A : C)$, since from (15) we have $((A - 2C)/4 : C) = ((X_\alpha + Z_\alpha)^2 : (X_\alpha - Z_\alpha)^2 - (X_\alpha + Z_\alpha)^2)$, which can be computed in $2\mathbf{S} + 3\mathbf{a}$.

Although parsing from α to $(a : 1) = (1 + \alpha^2 : -\alpha)$ is trivial, parsing in the other direction requires a square root computation in general. We never have to do this, however, since (i) during key generation, the starting curve is fixed and so the corresponding 2-torsion point(s) can be thought of as system parameters[5], and (ii) for the subsequent shared secret computations, we can happily replace a with α as the description of the supersingular curve in the (compressed or uncompressed) public key[6]. In the next section we use the notation a_from_alpha to represent the function that performs this cheap parsing.

---

[4] As usual, we use $\mathbf{M}$, $\mathbf{S}$, $\mathbf{a}$ and $\mathbf{I}$ to denote the costs of multiplications, squarings, additions/subtractions and inversions in the field $\mathbb{F}_{p^2}$.

[5] All public implementations of SIDH currently take the starting curve to be $E/\mathbb{F}_{p^2} : y^2 = x^3 + x$, where $\mathbb{F}_{p^2} = \mathbb{F}_p(i)$ with $i^2 + 1$. In these scenarios the starting 2-torsion point can be defined by setting $\alpha = i$.

[6] We note that the uncompressed keys in the SIDH library associated with [12] do not send the curve constant a explicitly, so would require modest modifications to take advantage of this 2-torsion technique. The compressed key format from the subsequent work in [11] does send a in the public key so has immediately compatibility.

Remark 3. If P and $Q \neq \pm P$ are two points on the Montgomery curve $E/K : by^2 = x^3 + ax^2 + x$, then the curve constant a relates to their x-coordinates and the x-coordinate of their difference via [12, Remark 4]

$$a = \frac{(1 - x_P x_Q - x_P x_{Q-P} - x_Q x_{Q-P})^2}{4 x_P x_Q x_{Q-P}} - x_P - x_Q - x_{Q-P}. \qquad (16)$$

Thus, if we ever have three points on an isogenous curve whose coefficient has not been computed, we can use the projective version of the above equation to recover $(a : 1) = (A : C)$ from $\mathbf{x}(P)$, $\mathbf{x}(Q)$ and $\mathbf{x}(Q - P)$. Since the cost of computing the isogenous curve using the 2-torsion technique grows as the degree of the isogeny grows, while the cost of computing the isogenous curve via (16) is fixed, there will obviously be a crossover point when taking advantage of three available points becomes faster. Based on the cost of computing one ℓ-isogeny presented in the next section, and on the projective version of (16) costing $8\mathbf{M} + 5\mathbf{S} + 11\mathbf{a}$, it will be faster to use the above after $d \geq 2$. Following [12], we note that there is always three such x-coordinates that can be exploited during key generation, namely the three x-coordinates whose image under the isogeny forms (part of) the public key. During the shared secret computation, however, there will not always be three points available at each stage. Thus, we recommend that unless d is very large (so that the performance benefits of using (16) over an additional isogeny evaluation will be visible), it is most simple to stick to the 2-torsion approach in this section through the SIDH algorithm.

5   A General-Purpose Algorithm for Arbitrary Odd-Degree Isogenies

We now turn to deriving an optimised algorithm for arbitrary odd-degree isogeny evaluation based on Theorem 1. Since we are working exclusively within the $\mathbb{P}^1$ framework under the map $\mathbf{x}$, the only equation we need to recall is (6), which we rewrite as

$$f(x) = x \cdot \left(\prod_{i=1}^{d} \left(\frac{x \cdot x_{[i]P} - 1}{x - x_{[i]P}} \right) \right)^2.$$

We begin working this into an algorithm by first projectivising into $\mathbb{P}^1$, writing $(X_i : Z_i) = (x_{[i]P} : 1)$ for $i = 1 \ldots d$, $(X : Z) = (x : 1)$ for the indeterminate coordinate where the isogeny is evaluated, and $(X' : Z') = \mathbf{x}(\phi(x, y))$ for the result. Then

$$X' = X \cdot \left(\prod_{i=1}^{d} (X \cdot X_i - Z_i \cdot Z) \right)^2, \quad \text{and}$$

$$Z' = Z \cdot \left(\prod_{i=1}^{d} (X \cdot Z_i - X_i \cdot Z) \right)^2$$

At first glance it appears that computing the pairs $(X \cdot X_i - Z_i \cdot Z)$ and $(X \cdot Z_i - X_i \cdot Z)$ will cost $4\mathbf{M} + 2\mathbf{a}$ each, but following Montgomery [30], we can achieve this in $2\mathbf{M} + 6\mathbf{a}$ by rewriting the above as

$$X' = X \cdot \left(\prod_{i=1}^{d} \left[(X - Z)(X_i + Z_i) + (X + Z)(X_i - Z_i) \right] \right)^2, \quad \text{and}$$

$$Z' = Z \cdot \left(\prod_{i=1}^{d} \left[(X - Z)(X_i + Z_i) - (X + Z)(X_i - Z_i) \right] \right)^2. \tag{17}$$

Observe that when $d > 1$ the values of $X - Z$ and $X + Z$ can be reused across the d expressions in both of the products above. Furthermore, the isogeny ϕ is usually going to be evaluated at multiple *points* of the form $(X : Z)$, and this will always be the case if the 2-torsion technique from the previous section is employed. Thus, suppose the isogeny is to be evaluated at the n elements $(\mathbf{X}_1 : \mathbf{Z}_1), \ldots, (\mathbf{X}_n : \mathbf{Z}_n)$, where we use boldface to distinguish these points and the coordinates of the i-th multiples of the kernel generator P. We note at once that the values of $(X_i + Z_i)$ and $(X_i - Z_i)$ can now also be reused across the n elements evaluated by the isogeny. This mutual recycling across both sets of points suggests a simple subroutine that merely computes the sum and difference of these pairwise products as in (17): we dub this routine CrissCross and present it in Algorithm 1 for completeness.

Algorithm 1. CrissCross: $K^4 \rightarrow K^2$.

Input: $(\alpha, \beta, \gamma, \delta) \in K^4$
Output: $(\alpha\delta + \beta\gamma, \alpha\delta - \beta\gamma) \in K^2$
Cost: $2\mathbf{M} + 2\mathbf{a}$.
1 $(t_1, t_2) \leftarrow (\alpha \cdot \delta, \beta \cdot \gamma)$ // 2M
2 **return** $(t_1 + t_2, t_1 - t_2)$ // 2a

Now, on input of the kernel generator $\mathbf{x}(P) = (X_1 : Z_1)$, the first step of the main algorithm will be to generate the $d - 1$ additional elements $\mathbf{x}([i]P) = (X_i : Z_i)$. This subroutine is called KernelPoints and we present it in Algorithm 2. Since it must start with a call to xDBL[7], we also need to input the modified curve constant $(\hat{A} : \hat{C}) = (a - 2 : 4)$.

Looking back at (17), we can see that once the $(X_i : Z_i)$ have been computed for $i = 1, \ldots, d$, they can immediately be overwritten by their sum and difference pairs through assigning $(\hat{X}_i, \hat{Z}_i) \leftarrow (X_i + Z_i, X_i - Z_i)$ in preparation for CrissCross. Based on (17), we now present an algorithm for evaluating a

---

[7] For many values of d, all of the kernel elements can be generated by repeated calls to xDBL, which is slightly cheaper than xADD in our context. However, for the sake of general applicability, we make repeated calls to xADD after the initial xDBL in KernelPoints.

Algorithm 2. KernelPoints: $\mathbb{P}^1 \times \mathbb{P}^1 \to (\mathbb{P}^1)^d$.

Input: $(X_1 : Z_1) = \mathbf{x}(P) \in \mathbb{P}^1$ and $(\hat{A} : \hat{C}) = (a - 2 : 4) \in \mathbb{P}^1$
Output: $((X_1 : Z_1), \ldots, (X_d : Z_d)) = (\mathbf{x}(P), \mathbf{x}([2]P), \ldots, \mathbf{x}([d]P)) \in (\mathbb{P}^1)^d$
Cost: $4(d - 1)\mathbf{M} + 2(d - 1)\mathbf{S} + 2(3d - 4)\mathbf{a}$.
1 **if** $d \geq 2$ **then** $(X_2 : Z_2) \leftarrow$ xDBL$((X_1 : Z_1), (\hat{A} : \hat{C}))$
  // 4M+2S+4a
2 **for** $i = 3$ **to** d **do**
3 $\quad$ | $(X_i : Z_i) \leftarrow$ xADD$((X_{i-1} : Z_{i-1}), (X_1 : Z_1), (X_{i-2} : Z_{i-2}))$ // 4M+2S+6a
4 **end**
5 **return** $((X_1 : Z_1), \ldots, (X_d : Z_d))$

Algorithm 3. OddIsogeny: $(K^2)^d \times \mathbb{P}^1 \to \mathbb{P}^1$.

Input: $((\hat{X}_1, \hat{Z}_1), \ldots, (\hat{X}_d, \hat{Z}_d))$ and $(X : Z) \in \mathbb{P}^1$
Output: $(X' : Z') \in \mathbb{P}^1$ corresponding to $\mathbf{x}(\phi(Q))$ where $\mathbf{x}(Q) = (X : Z)$
Cost: $4d\mathbf{M} + 2\mathbf{S} + 2(d + 1)\mathbf{a}$.
1 $(\hat{X}, \hat{Z}) \leftarrow (X + Z, X - Z)$ // 2a
2 $(X', Z') \leftarrow$ CrissCross$(\hat{X}_1, \hat{Z}_1, \hat{X}, \hat{Z})$ // Algorithm 1
3 **for** $i = 2$ **to** d **do**
4 $\quad$ | $(t_0, t_1) \leftarrow$ CrissCross$(\hat{X}_i, \hat{Z}_i, \hat{X}, \hat{Z})$ // Algorithm 1
5 $\quad$ | $(X', Z') \leftarrow (t_0 \cdot X', t_1 \cdot Z')$ // 2M
6 **end**
7 $(X', Z') \leftarrow (X \cdot (X')^2, Z \cdot (Z')^2)$ // 2M+2S
8 **return** $(X' : Z')$

---

single isogeny that takes as input the modified set of kernel point coordinates: OddIsogeny is given in Algorithm 3.

We are now in a position to present SimultaneousOddIsogeny, which is the main algorithm – see Algorithm 4. It takes as input $\mathbf{x}(P) = (X_1 : Z_1) \in \mathbb{P}^1$ and $(\hat{A} : \hat{C}) = (a - 2 : 4)$, which correspond to a point P of order ℓ on $E/K : by^2 = x^3 + ax^2 + x$, as well as an n-tuple $(\mathbf{x}(Q_1), \ldots \mathbf{x}(Q_n)) = ((\mathbf{X}_1 : \mathbf{Z}_1), \ldots, (\mathbf{X}_n : \mathbf{Z}_n)) \in (\mathbb{P}^1)^n$ where the $Q_i \in E$ are such that $Q \notin \langle P \rangle$. The output is an n-tuple corresponding to $(\mathbf{x}(\phi(Q_1)), \ldots \mathbf{x}(\phi(Q_n))) \in (\mathbb{P}^1)^n$, where $\ker(\phi) = \langle P \rangle$.

Simplified Odd-Degree Isogenies in SIDH. Together with an algorithm for computing the multiplication-by-ℓ map, Algorithm 4 is essentially all that is needed to compute an odd ℓ^e-degree isogeny in the context of SIDH. Regardless of which high-level strategy is used to compute the ℓ^e-isogeny (i.e., whether it be the multiplication-based approach [21, Fig. 2] or the *optimal* strategy [15, Sect. 4.2.2]), Algorithm 4 will be called e times to compute e isogenies of degree ℓ. In Algorithm 5 we show how SimultaneousOddIsogeny is to be used in conjunction with the simple conversion function a_from_alpha from Sect. 4 and the Montgomery ladder for computing the multiplication-by-ℓ map. We assume the use

---

Algorithm 4. SimultaneousOddIsogeny: $\mathbb{P}^1 \times \mathbb{P}^1 \times (\mathbb{P}^1)^n \to (\mathbb{P}^1)^n$.

Input: $(X_1 : Z_1) \in \mathbb{P}^1$, $(\hat{A} : \hat{C}) \in \mathbb{P}^1$, and $((\mathbf{X}_1 : \mathbf{Z}_1), \ldots, (\mathbf{X}_n : \mathbf{Z}_n)) \in (\mathbb{P}^1)^n$
Output: $((\mathbf{X}_1' : \mathbf{Z}_1'), \ldots, (\mathbf{X}_n' : \mathbf{Z}_n')) \in (\mathbb{P}^1)^n$
Cost: $4(d(n+1) - 1)\mathbf{M} + 2(n + d - 1)\mathbf{S} + 2((d+1)n + (3d-4))\mathbf{a}$.
1 $((X_1 : Z_1), \ldots, (X_d : Z_d)) \leftarrow \text{KernelPoints}((X_1 : Z_1), (\hat{A}, \hat{C}))$ // Algorithm 2
2 $((\hat{X}_1, \hat{Z}_1), \ldots, (\hat{X}_d, \hat{Z}_d)) \leftarrow ((X_1 + Z_1, X_1 - Z_1), \ldots, (X_d + Z_d, X_d - Z_d))$ // 2da
3 **for** $j = 1$ **to** n **do**
4    $(\mathbf{X}_j' : \mathbf{Z}_j') \leftarrow \text{OddIsogeny}((\hat{X}_1, \hat{Z}_1), \ldots, (\hat{X}_d, \hat{Z}_d)), (\mathbf{X}_j : \mathbf{Z}_j))$
      // Algorithm 3
5 **end**
6 **return** $((\mathbf{X}_1' : \mathbf{Z}_1'), \ldots, (\mathbf{X}_n' : \mathbf{Z}_n'))$

---

of the function LADDER as discussed in Sect. 4, where the Montgomery coefficient a is passed in projectively as $(\hat{A} : \hat{C}) = (a - 2 : 4)$; i.e.,

$$\text{LADDER} : \mathbb{P}^1 \times \mathbb{P}^1 \times \mathbb{Z} \to \mathbb{P}^1, \quad (\mathbf{x}(P), (\hat{A} : \hat{C}), \ell^z) \mapsto \mathbf{x}([\ell^z]P). \qquad (18)$$

For ease of exposition, we adopt the multiplication-based approach [21, Fig. 2] for computing the degree ℓ^e-isogeny, but note that the way in which the proposed algorithms are called in Lines 4–6 of Algorithm 5 is analogous if the optimal strategy mentioned above is used; the only difference worth mentioning is that the length of the list of the $(\mathbf{X}_i' : \mathbf{Z}_i')$ passed in and out of SimultaneousOddIsogeny on Line 6 can change when it is called within the code executing the optimal strategy.

In the notation of Sect. 2, let $E/\mathbb{F}_{p^2} : y^2 = x(x - \alpha)(x - 1/\alpha)$ be the public starting curve in the SIDH protocol. For public key generation, Alice would compute her secret kernel generator as $R_A = [u_A]P_A + [v_A]Q_A$ of order $\ell_A^{e_A}$ (see [15, Algorithm 1]), and with Bob's public basis P_B and Q_B, she can then compute her public key by calling Algorithm 5 as

$$\big((X_{\alpha,A} : Z_{\alpha,A}), ((\mathbf{x}(\phi_A(P_B)), \mathbf{x}(\phi_A(Q_B)), \mathbf{x}(\phi_A(Q_B - P_B))))\big)$$
$$= \text{SIDH\_Isogeny}\big(\mathbf{x}(R_A), (\ell_A, e_A), (\alpha : 1), (\mathbf{x}(P_B), \mathbf{x}(Q_B), \mathbf{x}(Q_B - P_B))\big),$$

where $\ker(\phi_A) = \langle R_A \rangle$, and where $\mathbf{x}(Q_B - P_B)$ is included as an input to avoid sign ambiguities in the subsequent shared secret computations – see [12]. Alice would normalise each of these elements, i.e., convert them all from $\mathbb{P}^1(\mathbb{F}_{p^2})$ into $\mathbb{F}_{p^2}$ via a simultaneous inversion [30], then send them to Bob. Writing $\alpha_A = X_{\alpha,A}/Z_{\alpha,A}$, Bob can then compute $\mathbf{x}(S_B) = \mathbf{x}([u_B]\phi_A(P_B) + [v_B]\phi_A(Q_B))$, and compute the shared secret by calling Algorithm 5 as

$$(X_{\alpha,AB} : Z_{\alpha,AB}) = \text{SIDH\_Isogeny}\big(\mathbf{x}(S_B), (\ell_B, e_B), (\alpha_A : 1)\big),$$

before computing the j-invariant of the Montgomery curve whose coefficient is the output of the function a_from_alpha$((X_{\alpha,AB} : Z_{\alpha,AB}))$. Note that, during the shared secret computation, the $(\mathbb{P}^1)^k$ input to SIDH_Isogeny is empty, i.e., has $k = 0$.

Algorithm 5. SIDH_Isogeny: $\mathbb{P}^1 \times \mathbb{Z}^2 \times \mathbb{P}^1 \times (\mathbb{P}^1)^k \rightarrow \mathbb{P}^1 \times (\mathbb{P}^1)^k$.

Input: $\mathbf{x}(P) = (X_1 : Z_1) \in \mathbb{P}^1$, and $(\ell, e) \in \mathbb{Z}^2$, where $|\langle P \rangle| = \ell^e$ on E.
$(\alpha : 1) \in \mathbb{P}^1$, where $\text{ord}((\alpha, 0)) = 2$ on E and $\alpha \neq 0$.
$(\mathbf{x}(Q_1), \ldots \mathbf{x}(Q_k)) = ((\mathbf{X}_1 : \mathbf{Z}_1), \ldots, (\mathbf{X}_k : \mathbf{Z}_k)) \in (\mathbb{P}^1)^k$, where $Q_i \in$
E and $Q \notin \langle P \rangle$.

Output: $(X_{\alpha'} : Z_{\alpha'}) \in \mathbb{P}^1$, where $\text{ord}((X_{\alpha'}/Z_{\alpha'}, 0)) = 2$ on $\phi(E')$ for $\ker(\phi) =$
$\langle P \rangle$ and $X_{\alpha'} \neq 0$.
$(\mathbf{x}(\phi(Q_1)), \ldots \mathbf{x}(\phi(Q_k))) = ((\mathbf{X}'_1 : \mathbf{Z}'_1), \ldots, (\mathbf{X}'_k : \mathbf{Z}'_k)) \in (\mathbb{P}^1)^k$

---

1   $((X_{\alpha'} : Z_{\alpha'}), (\mathbf{X}'_1 : \mathbf{Z}'_1), \ldots, (\mathbf{X}'_k : \mathbf{Z}'_k)) \leftarrow ((\alpha : 1), (\mathbf{X}_1 : \mathbf{Z}_1), \ldots, (\mathbf{X}_k : \mathbf{Z}_k))$
   // Initialise
2   $(X_R : Z_R) \leftarrow (X_1 : Z_1)$ // Initialise
3   **for** $z = e - 1$ **downto** 0 **do**
4      $(\hat{A} : \hat{C}) \leftarrow$ a_from_alpha$((X_{\alpha'} : Z_{\alpha'}))$ // See Sect. 4
5      $(X_S : Z_S) \leftarrow$ LADDER$((X_R : Z_R), (\hat{A} : \hat{C}), \ell^z)$ // See Eq. (18)
6      $((X_R : Z_R), (X_{\alpha'} : Z_{\alpha'}), (\mathbf{X}'_1 : \mathbf{Z}'_1), \ldots, (\mathbf{X}'_k : \mathbf{Z}'_k)) \leftarrow$
      SimultaneousOddIsogeny   // Algorithm 4
      $((X_S : Z_S), (\hat{A} : \hat{C}), ((X_R : Z_R), (X_{\alpha'} : Z_{\alpha'}), (\mathbf{X}'_1 : \mathbf{Z}'_1), \ldots, (\mathbf{X}'_k : \mathbf{Z}'_k)))$
7   **end**
8   **return** $(X_{\alpha'} : Z_{\alpha'}), ((\mathbf{X}'_1 : \mathbf{Z}'_1), \ldots, (\mathbf{X}'_n : \mathbf{Z}'_n))$

We note that the operation counts presented in Algorithms 2–4 do not apply to the special case of $d = 1$. Although Algorithm 4 still performs the 3-isogeny computations in the same number of operations as the dedicated formulas in Appendix A, the claimed operation counts only hold if KernelPoints is called, which is not the case for 3-isogenies (where no additional kernel elements are required).

We conclude this section with a remark on a more compact version of Algorithm 4.

Remark 4 (A low-storage version). The description of the general odd-degree isogeny function in Algorithm 4 aims to minimise the total number of field operations needed for an ℓ-isogeny computation and its evaluation at an arbitrary number of points. However, the recycling of the additions computed in (17) requires us to generate the entire list of d kernel elements before entering the loop that repeatedly calls Algorithm 3. If d is large, the space required to store d elements in $\mathbb{F}_{p^2}$ might become infeasible, especially given the size of the fields used in real-world SIDH implementations. Moreover, this recycling only saves $\mathbb{F}_{p^2}$ additions, and our benchmarking of the SIDH v2.0 software accompanying [12] in the following section revealed that their software has $1\mathbf{M} \approx 20\mathbf{a}$, which means the above recycling will only have a minor benefit on the overall performance. Thus, a more streamlined version of Algorithm 4 would simply compute one of the elements $\mathbf{x}([i]P) = (X_i : Z_i)$ at a time and absorb its contribution to (17) immediately before calling xADD to replace it by $\mathbf{x}([i + 1]P) = (X_{i+1} : Z_{i+1})$, and so on, with no need for Algorithm 2. Since the required storage would then remain fixed as d increases, this would give a much more compact algorithm for larger d, both in its description and in terms of the storage required to implement it.

6   Implementation Results and Implications

In this section we provide benchmarks for $\texttt{SimultaneousOddIsogeny}$, i.e., the general odd-degree isogeny function in Algorithm 4. We stress that we are not aiming to outperform the relative performance of the 3- and 4-isogenies, by pointing out that the relative performance of odd ℓ-isogenies decreases as ℓ grows larger. The point of this paper is to broaden the class of curves for which SIDH is practical in all of the relevant aspects, i.e., memory requirements, code size, simplicity of the implementation, as well as efficiency. Nevertheless, there are scenarios where larger odd-degree isogenies could be preferred over the low degree ones, as we will discuss later in this section.

Table 1 presents benchmarks for the evaluation of isogenies of degree $\ell \in \{3, 5, \ldots, 15\}$ at $n \in \{1, 2, 5, 8\}$ input points. These timings were obtained by wrapping Algorithm 4 around the SIDH v2.0 software[8] accompanying [11, 12]; this software uses the supersingular isogeny class containing the curve $E/\mathbb{F}_{p^2}: y^2 = x^3 + x$ where $p = 2^{372} \cdot 3^{239} - 1$, where all curves in the class have cardinality $(2^{372} \cdot 3^{239})^2$. We note that this curve does not have $\mathbb{F}_{p^2}$-rational points of order ℓ for odd $\ell > 3$, but this is immaterial; the timings for Algorithm 4 would be exactly the same when working with a curve with rational ℓ-torsion over the same field. We benchmarked in this way in order to get a fair comparison of the cost of different values of ℓ and n when the field arithmetic stays fixed at a size that is relevant to real-world SIDH implementations. We discuss the influence of needing rational ℓ-torsion on the field arithmetic later in this section. The reason we chose to benchmark $n \in \{1, 2, 5, 8\}$ is based on the average number of isogeny evaluations for both Alice and Bob at each step of the SIDH v2.0 software that uses the optimal *tree traversal* (see Sect. 2 or [15, Sect. 4.2.2]) in the main loop: Alice and Bob use roughly 7.15 and 7.70 evaluations of every 4- and 3-isogeny (respectively) during key generation, and this would include one more evaluation if our 2-torsion technique from Sect. 4 was employed (hence $n = 8$), and they use 4.15 and 4.70 respective isogeny evaluations per step during the shared secret phase (hence $n = 5$). We also include $n = 1$ to benchmark the cost of a single isogeny evaluation and $n = 2$ assuming a single isogeny evaluation is included alongside the 2-torsion technique from Sect. 4; this is to view the relative performance of the simple SIDH loop in Algorithm 5 that evaluates each isogeny at one point during the main loop.

Table 1 shows a natural increase in latency as ℓ grows. A single 5-isogeny evaluation costs around 2.71x that of a single 3-isogeny, and the cost of a 15-isogeny evaluation is around 11.40x that of a 3-isogeny. Due to the multiple isogeny evaluations sharing computations performed on the kernel elements (see Sect. 5), naturally these ratios become slightly more favourable for larger ℓ as n

---

[8] See https://github.com/Microsoft/PQCrypto-SIDH.

Table 1. Cycle counts for `SimultaneousOddIsogeny` for different values of ℓ and n. Timing benchmarks were taken on an Intel Core i7-6500U Skylake processor running Ubuntu 14.04.5 LTS with TurboBoost disabled and all cores but one are switched-off. To obtain the executables, we used GNU-`gcc` version 4.8.4 with the `-O2` flag set and GNU assembler version 2.24.

d	ℓ	$n = 1$	$n = 2$	$n = 5$	$n = 8$
1	3	9,780	19,420	48,270	76,930
2	5	26,450	43,420	93,400	143,670
3	7	43,310	67,490	139,280	219,270
4	9	60,170	91,390	184,480	277,700
5	11	77,000	115,490	230,070	344,220
6	13	93,710	139,370	275,170	411,800
7	15	110,510	163,480	320,460	477,980

increases: the evaluation of a 5-isogeny (resp. 15-isogeny) at $n = 8$ points costs around 2.03x (resp. 7.18x) the same number of 3-isogeny evaluations. These numbers are depicted graphically on the left of Fig. 1, and the approximate relative slowdown of using ℓ-isogenies within the SIDH framework is depicted on the right. An analogous version of Fig. 1 for ℓ up to $\ell = 301$ is given in Fig. 2. In both figures the cycle counts have been divided by n in order to give a cost per isogeny evaluation.

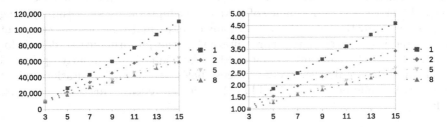

Fig. 1. Average cycle counts for `SimultaneousOddIsogeny` for different values of ℓ and n. Timing benchmarks were taken on an Intel Core i7-6500U Skylake processor running Ubuntu 14.04.5 LTS with TurboBoost disabled and all cores but one are switched-off. To obtain the executables, we used GNU-`gcc` version 4.8.4 with the `-O2` flag set and GNU assembler version 2.24. Raw cycle counts per isogeny evaluation are given on the left, while on the right they are scaled by the factor $\log(3)/(\log(\ell) \cdot \mathbf{C}_3)$, where $\mathbf{C}_3$ is the cost of a 3-isogeny, in order to approximate the relative factor slowdown within the SIDH framework.

The right graphs in Figs. 1 and 2 aim to depict the relative factor slowdowns of computing an ℓ^e isogeny versus a 3^{e_3} isogeny assuming that $\ell^e \approx 3^{e_3}$. However, we must note that a more accurate depiction of the relative slowdown in the

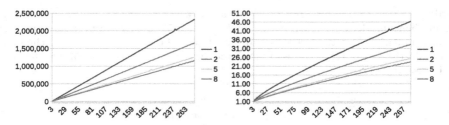

Fig. 2. Average cycle counts for SimultaneousOddIsogeny for different values of ℓ and n. Timing benchmarks were taken on an Intel Core i7-6500U Skylake processor running Ubuntu 14.04.5 LTS with TurboBoost disabled and all cores but one are switched-off. To obtain the executables, we used GNU-gcc version 4.8.4 with the -O2 flag set and GNU assembler version 2.24. Raw cycle counts per isogeny evaluation are given on the left, while on the right they are scaled by the factor $\log(3)/(\log(\ell) \cdot \mathbf{C}_3)$, where $\mathbf{C}_3$ is the cost of a 3-isogeny, in order to approximate the relative factor slowdown within the SIDH framework.

SIDH framework would incorporate the relative costs of the multiplication-by-ℓ functions, since these are called almost as frequently as the ℓ-isogeny functions in an optimised implementation (and significantly more times than the ℓ-isogeny functions in the simple SIDH loop – see [12, Sect. 6]). To that end, we point out that the relative slowdown of using ℓ-isogenies would be much less than these graphs depict (as ℓ increases), under the assumption that the Montgomery ladder is called to compute $\mathbf{x}(P) \mapsto \mathbf{x}([\ell]P)$. Table 2 and Fig. 3 exhibit the obvious trend in LADDER's performance as ℓ increases: unlike the linear increase in ℓ-isogeny latencies, the performance of ladder is asymptotically logarithmic, being (roughly) fixed by the value $\lceil \log_2(\ell) \rceil$. In any case, we make the obvious comment that a practically meaningful representation of the performance trade-offs for different values of ℓ can only be obtained by benchmarking similarly optimised implementations in all cases. As we discuss below, such implementations might call for vastly different styles of field arithmetic, so we leave this open for future work.

Implications. At a first glance, Table 1 and Figs. 1 and 2 seem to suggest that, unless faster isogenies of degree $\ell \geq 5$ are found, such higher degree isogenies will not find any meaningful real-world application. However, the ability to compute arbitrary degree isogenies in SIDH already opens up some interesting possibilities as we now discuss.

Firstly, recent work by Bos and Friedberger [8] studied SIDH-friendly primes of the form $p = 2^i r^j - 1$, where r can be any small prime. They investigated a number of different arithmetic techniques, and interestingly, when implementing arithmetic over the field with $p = 2^{372}3^{329} - 1$ above, found that arithmetic over a comparably sized field $p = 2^{391}19^{88} - 1$ was actually significantly faster [8, Table 3]. The more severe slowdown of 19-isogenies versus 3-isogenies means that, overall, the performance of 3-isogenies will still be preferred. However,

Table 2. Cycle counts for $[\ell](X : Z)$ on Montgomery Ladder with projective inputs: $(X : Z)$ and $(A_{24} : C_{24})$. Timing benchmarks were taken on an Intel Core i7-6500U Skylake processor running Ubuntu 14.04.5 LTS with TurboBoost disabled and all cores but one are switched-off. To obtain the executables, we used GNU-**gcc** version 4.8.4 with the **-O2** flag set and GNU assembler version 2.24. For the fixed odd low degrees of $\ell \in \{3, 5, 7\}$, we also present the cycle counts of our own optimised, dedicated algorithms for computing the multiplication-by-ℓ maps, since this might be of interest for future implementers; see Appendix A for justification.

Operation	ladder	Optimized
$[2](X : Z)$	-	9,608
$[3](X : Z)$	28,954	18,622
$[5](X : Z)$	48,603	27,346
$[7](X : Z)$	49,086	36,110
$[9](X : Z)$	67,610	-
$[11](X : Z)$	68,429	-
$[13](X : Z)$	68,125	-
$[15](X : Z)$	68,848	-
$[17](X : Z)$	86,717	-

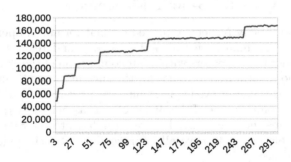

Fig. 3. Cycle counts chart for $[\ell](X : Z)$ using Montgomery ladder with projective inputs: $(X : Z)$ and $(A_{24} : C_{24})$. Timing benchmarks were taken on an Intel Core i7-6500U Skylake processor running Ubuntu 14.04.5 LTS with TurboBoost disabled and all cores but one are switched-off. To obtain the executables, we used GNU-**gcc** version 4.8.4 with the **-O2** flag set and GNU assembler version 2.24.

in real-world applications like the transport layer security (TLS) protocol, it is typically one side of the protocol (i.e., the server, who is processing many SIDH instances) where performance is the bottleneck, while the performance of a single SIDH instance on the client side is ultimately a non-issue. In such a situation, we could envision affording the server the luxury of the faster prime $p = 2^{391}19^{88} - 1$ *and* the faster 4-isogenies in order to get the best of both worlds, while the client could put up with the 19-isogenies and not be noticibly hampered by the increased latency on their side.

Another possibility opened up by Algorithm 4 is the abandonment of even-degree isogenies on either side of the protocol, in the name of implementation simplicity. For example, SIDH implementations using primes of the form[9] $p = 4 \cdot 3^i 5^j - 1$ could be implemented using Algorithm 4 for isogenous curve and point operations on both sides. This would make for a much simpler and more compact code-base, and could be an attractive option if the relatively modest slowdown from 4- to 5-isogenies (and the possible slowdown of the new shaped primes) is justifiable.

Finally, we leave it as an open question to see whether primes *not* of the form $p = f \cdot 2^i 3^j - 1$ can be found where arithmetic is fast enough to justify isogenies of $\ell \geq 5$. It could even be possible to find fast primes where $p \pm 1$ is smooth but contains many small, unique prime factors, and where isogeny walks on either or both sides of the protocol involve isogenies of different degrees. Of course, the security implications of such a choice are also left as open.

Acknowledgements. We are especially grateful to Steven Galbraith for his help in correcting the proof of Theorem 1. We thank Dustin Moody for his detailed comments on an earlier version of this paper, as well as Joppe Bos, Patrick Longa, Michael Naehrig and the anonymous reviewers for their useful comments.

A   Improvements for 3- and 4-isogenies

In this section, we briefly present improved explicit formulas, operation counts, and performance benchmarks for 3- and 4-isogeny and related operations in SIDH; this provides implementers with a fair comparison of the general ℓ-isogeny algorithm and more optimised formulas for the currently used 3- and 4-isogenies. A complete list of the improved operations are presented in Table 3 and the cycle counts are compared in Table 4; all of the associated explicit formulas are presented thereafter. We plugged these formulas into the SIDH v2.0 library from [12] and Table 5 gives the overall improvements of each stage of the SIDH key exchange protocol.

Below we use $(X_3 : Z_3)$ and $(X_4 : Z_4)$ to denote the coordinates of points P_3 and P_4 of orders 3 and 4 respectively, under the **x** map. We use $(A_{24} : C_{24})$ to denote the projective version of the Montgomery curve constant $(a - 2)/4$, i.e., $(A_{24} : C_{24}) = (a - 2 : 4)$ in $\mathbb{P}^1$.

The 3_iso_curve operation

$$(A'_{24} : C'_{24}) = ((X_3 + Z_3)(Z_3 - 3X_3)^3 : 16X_3 Z_3^3)$$

takes $2\mathbf{M} + 3\mathbf{S} + 14\mathbf{a}$ and produces the common subexpressions $K_1 = X_3 - Z_3$ and $K_2 = X_3 + Z_3$. The justification of the claimed operation count is as follows:

$K_1 = X_3 - Z_3,\ R_1 = K_1^2,\ K_2 = X_3 + Z_3,\ R_2 = K_2^2,\ R_3 = R_2 + R_1,$
$R_4 = K_1 + K_2,\ R_4 = R_4^2,\ R_4 = R_4 - R_3,\ R_3 = R_4 + R_2,\ R_4 = R_4 + R_1,$
$R_5 = R_1 + R_4,\ R_5 = 2R_5,\ R_5 = R_5 + R_2,\ A'_{24} = R_5 \cdot R_3,\ R_5 = R_2 + R_3,$
$R_5 = 2R_5,\ R_5 = R_5 + R_1,\ R_5 = R_5 \cdot R_4,\ C'_{24} = R_5 - A'_{24}.$

---

[9] We still need the cofactor of 4 in the group order to be able to exploit the Montgomery form – see Sect. 2.

Table 3. Operation counts comparison for common elliptic curve and isogeny functions within existing SIDH implementations.

	xTPL	3_iso_point	3_iso_curve	4_iso_point	4_iso_curve
CLN2016	8M+4S+8a	6M+2S+2a	3M+3S+8a	9M+1S+6a	5S+7a
This work	9M+2S+17a	4M+2S+4a	2M+3S+14a	6M+2S+6a	4S+4a
	7M+5S+11a				

Table 4. Cycle counts comparison for common elliptic curve and isogeny functions within existing SIDH implementations. Timing benchmarks were taken on an Intel Core i7-6500U Skylake processor running Ubuntu 14.04.5 LTS with TurboBoost disabled and all cores but one are switched-off. To obtain the executables, we used GNU-**gcc** version 4.8.4 with the -O2 flag set and GNU assembler version 2.24.

Operation	CLN2016	This work	Speed-up
xTPL	19,226	18,622	1.032x
3_iso_curve	9,264	8,202	1.129x
3_iso_point	12,901	9,581	1.346x
4_iso_curve	6,276	5,095	1.232x
4_iso_point	17,432	13,545	1.287x

Table 5. Cycle counts for Ephemeral isogeny-based key exchange. Timing benchmarks were taken on an Intel Core i7-6500U Skylake processor running Ubuntu 14.04.5 LTS with TurboBoost disabled and all cores but one are switched-off. To obtain the executables, we used GNU-**gcc** version 4.8.4 with the -O2 flag set and GNU assembler version 2.24.

Operation	CLN2016	This work	Speed-up
Alice's key generation	39,043,000	33,266,000	1.174x
Bob's key generation	44,289,000	37,430,000	1.183x
Alice's shared key computation	36,716,000	33,240,000	1.105x
Bob's shared key computation	42,576,000	38,046,000	1.120x

The 3_iso_point operation

$$(X' : Z') = \left(X(X_3 X - Z_3 Z)^2 : Z(Z_3 X - X_3 Z)^2\right)$$

takes $4\mathbf{M} + 2\mathbf{S} + 4\mathbf{a}$ and uses the common subexpressions $K_1 = X_3 - Z_3$ and $K_2 = X_3 + Z_3$. The justification of the claimed operation count is as follows:

$R_2 := X + Z, \ R_3 := R_2 \cdot K_1, \ R_2 := X - Z, \ R_1 := R_2 \cdot K_2, \ R_2 := R_1 + R_3,$
$R_2 := R_2^2, \ X' := R_2 \cdot X, \ R_1 := R_3 - R_1, \ R_1 := R_1^2, \ Z' := R_1 \cdot Z.$

The 4_iso_curve operation

$$(A'_{24} : C'_{24}) = \left(X_4^4 - Z_4^4 : Z_4^4\right)$$

takes $4\mathbf{S} + 4\mathbf{a}$ and produces the common subexpressions $K_1 = 4Z_4^2$, $K_2 = X_4 - Z_4$, and $K_3 = X_4 + Z_4$. The justification of the claimed operation count is as follows:

$$K_1 = Z_4^2, \; R_1 = X_4^2, \; R_1 = R_1^2, \; C'_{24} = K_1^2, \; A'_{24} = R_1 - C'_{24}, \; K_1 = 4K_1,$$
$$K_2 = X_4 - Z_4, \; K_3 = X_4 + Z_4.$$

The `4_iso_point` operation

$$(X' : Z') = \left(X \left(2X_4 Z_4 Z - (X_4^2 + Z_4^2)X \right) (X_4 X - Z_4 Z)^2 : \right.$$
$$\left. Z \left(2X_4 Z_4 X - (X_4^2 + Z_4^2)Z \right) (Z_4 X - X_4 Z)^2 \right)$$

takes $6\mathbf{M} + 2\mathbf{S} + 6\mathbf{a}$ and benefits from the common subexpressions $K_1 = 4Z_4^2$, $K_2 = X_4 - Z_4$, and $K_3 = X_4 + Z_4$ generated by `4_iso_curve`. The justification of the claimed operation count is as follows:

$$R_2 = X + Z, \; R_3 = R_2 \cdot K_2, \; R_4 = X - Z, \; R_1 = R_4 \cdot K_3, \; R_2 = R_4 \cdot R_2,$$
$$R_4 = R_1 + R_3, \; R_4 = R_4^2, \; R_3 = R_1 - R_3, \; R_3 = R_3^2, \; R_2 = K_1 \cdot R_2, \; R_1 = R_4 + R_2,$$
$$R_2 = R_3 - R_2, \; X' = R_4 \cdot R_1, \; Z' = R_3 \cdot R_2.$$

The `xTPL` operation

$$[3](X : Z) = \left(X(16A_{24}XZ^3 - C_{24}(X - 3Z)(X + Z)^3)^2 : \right.$$
$$\left. Z(16A_{24}X^3Z + C_{24}(3X - Z)(X + Z)^3)^2 \right)$$

takes $9\mathbf{M} + 2\mathbf{S} + 16\mathbf{a}$ assuming that $K_1 = A_{24} + C_{24}$ is cached. The justification of the claimed operation count is as follows:

$$R_1 = X - Z, \; R_2 = R_1^2, \; R_3 = X + Z, \; R_4 = R_3^2, \; R_5 = R_4 + R_2, \; R_6 = R_2 - R_4,$$
$$R_7 = R_4 \cdot K_1, \; R_8 = R_2 \cdot A_{24}, \; R_4 = R_8 + R_7, \; R_2 = R_7 - R_8, \; R_4 = R_4 \cdot R_6,$$
$$R_5 = R_2 \cdot R_5, \; R_2 = R_2 \cdot R_6, \; R_2 = 2R_2, \; R_6 = R_4 + R_5, \; R_5 = R_4 - R_5,$$
$$R_4 = R_6 + R_2, \; R_6 = R_6 - R_2, \; R_4 = R_4 \cdot R_6, \; R_6 = R_2 \cdot R_5, \; R_6 = 2R_6,$$
$$R_5 = R_4 - R_6, \; R_4 = R_4 + R_6, \; R_2 = R_4 \cdot R_3, \; R_1 = R_1 \cdot R_5, \; X_{out} = R_2 + R_1,$$
$$Z_{out} = R_2 - R_1.$$

Alternatively, the `xTPL` operation takes $7\mathbf{M} + 5\mathbf{S} + 10\mathbf{a}$ assuming that $K_1 = A_{24} + C_{24}$ is cached. The justification of the claimed operation count is as follows:

$$R_1 = X - Z, \; R_3 = R_1^2, \; R_2 = X + Z, \; R_4 = R_2^2, \; R_5 = R_2 + R_1, \; R_1 = R_2 - R_1,$$
$$R_2 = R_5^2, \; R_2 = R_2 - R_4, \; R_2 = R_2 - R_3, \; R_6 = R_4 \cdot K_1, \; R_4 = R_6 \cdot R_4,$$
$$R_7 = R_3 \cdot A_{24}, \; R_3 = R_3 \cdot R_7, \; R_4 = R_3 - R_4, \; R_3 = R_6 - R_7, \; R_2 = R_3 \cdot R_2,$$
$$R_3 = R_4 + R_2, \; R_3 = R_3^2, \; X_{out} = R_3 \cdot R_5, \; R_2 = R_4 - R_2, \; R_2 = R_2^2,$$
$$Z_{out} = R_2 \cdot R_1.$$

The dedicated explicit formulas for the multiplication-by-5 map referred to in Table 2 take $(X_{out} : Z_{out}) = \mathbf{x}([5]P)$, where $\mathbf{x}(P) = (X : Z)$, and are as below. The cost is $11\mathbf{M} + 6\mathbf{S} + 14\mathbf{a}$, assuming that $K_1 = A_{24} + C_{24}$ is cached.

$$R_1 = X - Z, \; R_2 = R_1^2, \; R_3 = X + Z, \; R_4 = R_3^2, \; R_5 = R_4 + R_2, \; R_1 = R_1 + R_3,$$
$$R_1 = R_1^2, \; R_3 = R_1 - R_5, \; R_2 = R_2 \cdot A_{24}, \; R_4 = K_1 \cdot R_4, \; R_4 = R_4 - R_2,$$
$$R_5 = R_4 \cdot R_5, \; R_2 = R_3^2, \; R_1 = R_2 \cdot C_{24}, \; R_2 = R_1/4, \; R_2 = R_5 - R_2,$$
$$R_5 = R_4 \cdot R_3, \; R_3 = R_2 + R_5, \; R_4 = R_2 - R_5, \; R_3 = R_3 \cdot R_4, \; R_4 = R_2 \cdot R_3,$$
$$R_1 = R_1 \cdot R_2, \; R_1 = R_1 + R_3, \; R_3 = R_1 \cdot R_5, \; R_2 = R_4 + R_3, \; R_2 = R_2^2,$$
$$X_{out} = R_2 \cdot X_1, \; R_1 = R_4 - R_3, \; R_1 = R_1^2, \; Z_{out} = R_1 \cdot Z_1.$$

The dedicated explicit formulas for the multiplication-by-7 map referred to in Table 2 take $(X_{out} : Z_{out}) = \mathbf{x}([7]P)$, where $\mathbf{x}(P) = (X : Z)$, and are as below. The cost is $14\mathbf{M} + 9\mathbf{S} + 18\mathbf{a}$, assuming that $K_1 = A_{24} + C_{24}$ is cached.

$$R_1 = X - Z, \; R_3 = R_1^2, \; R_2 = X + Z, \; R_4 = R_2^2, \; R_1 = R_1 + R_2, \; R_1 = R_1^2,$$
$$R_1 = R_1 - R_4, \; R_2 = R_1 - R_3, \; R_1 = K_1 \cdot R_4, \; R_4 = R_1 \cdot R_4, \; R_5 = R_3 \cdot A_{24},$$
$$R_1 = R_1 - R_5, \; R_3 = R_3 \cdot R_5, \; R_3 = R_4 - R_3, \; R_4 = R_2 \cdot R_1, \; R_1 = R_3^2, \; R_5 = R_4^2,$$
$$R_4 = R_3 + R_4, \; R_4 = R_4^2, \; R_2 = R_2^2, \; R_2 = R_2 \cdot R_3, \; R_3 = R_1 - R_5, \; R_1 = R_1 + R_5,$$
$$R_4 = R_1 - R_4, \; R_1 = R_1 \cdot R_3, \; R_3 = 2R_3, \; R_2 = R_2 \cdot C_{24}, \; R_5 = R_2 \cdot R_5,$$
$$R_5 = R_5 + R_1, \; R_1 = R_3 + R_2, \; R_3 = R_3 \cdot R_5, \; R_2 = R_1 + R_2, \; R_2 = R_2 \cdot R_1,$$
$$R_1 = 2R_3, \; R_2 = R_4 \cdot R_2, \; R_3 = R_1 + R_2, \; R_3 = R_3^2, \; X_{out} = X_1 \cdot R_3,$$
$$R_3 = R_1 - R_2, \; R_3 = R_3^2, \; Z_{out} = Z_1 \cdot R_3.$$

References

1. Alkim, E., Ducas, L., Pöppelmann, T., Schwabe, P.: Post-quantum key exchange - A new hope. In: Holz, T., Savage, S. (eds.) 25th USENIX Security Symposium, USENIX Security 16, Austin, TX, USA, 10–12 August 2016, pp. 327–343. USENIX Association (2016)
2. Azarderakhsh, R., Jao, D., Kalach, K., Koziel, B., Leonardi, C.: Key compression for isogeny-based cryptosystems. In: Emura, K., Hanaoka, G., Zhang, R. (eds.) Proceedings of the 3rd ACM International Workshop on ASIA Public-Key Cryptography, AsiaPKC@AsiaCCS, Xi'an, China, 30 May–03 June 2016, pp. 1–10. ACM (2016)
3. Bernstein, D.J., Birkner, P., Joye, M., Lange, T., Peters, C.: Twisted edwards curves. In: Vaudenay, S. (ed.) AFRICACRYPT 2008. LNCS, vol. 5023, pp. 389–405. Springer, Heidelberg (2008). https://doi.org/10.1007/978-3-540-68164-9_26
4. Bernstein, D.J., Buchmann, J., Dahmen, E.: Post-Quantum Cryptography. Springer Science & Business Media, Heidelberg (2009). https://doi.org/10.1007/978-3-540-88702-7
5. Bernstein, D.J., Chou, T., Schwabe, P.: McBits: Fast constant-time code-based cryptography. In: Bertoni, G., Coron, J.-S. (eds.) CHES 2013. LNCS, vol. 8086, pp. 250–272. Springer, Heidelberg (2013). https://doi.org/10.1007/978-3-642-40349-1_15
6. Bos, J.W., Costello, C., Ducas, L., Mironov, I., Naehrig, M., Nikolaenko, V., Raghunathan, A., Stebila, D.: Frodo: Take off the ring! Practical, quantum-secure key exchange from LWE. In: Weippl, E.R., Katzenbeisser, S., Kruegel, C., Myers, A.C., Halevi, S. (eds.) Proceedings of the 2016 ACM SIGSAC Conference on Computer and Communications Security, Vienna, Austria, 24–28 October 2016, pp. 1006–1018. ACM (2016)
7. Bos, J.W., Costello, C., Naehrig, M., Stebila, D.: Post-quantum key exchange for the TLS protocol from the ring learning with errors problem. In: 2015 IEEE Symposium on Security and Privacy, SP 2015, San Jose, CA, USA, 17–21 May 2015, pp. 553–570. IEEE Computer Society (2015)
8. Bos, J.W., Friedberger, S.: Fast arithmetic modulo $2^x p^y \pm 1$. In: Burgess, N., Bruguera, J.D., de Dinechin, F. (eds.) 24th IEEE Symposium on Computer Arithmetic, ARITH 2017, London, United Kingdom, 24–26 July 2017, pp. 148–155. IEEE Computer Society (2017)

9. Castryck, W., Galbraith, S., Farashahi, R.R.: Efficient arithmetic on elliptic curves using a mixed Edwards-Montgomery representation. Cryptology ePrint Archive, Report 2008/218 (2008). http://eprint.iacr.org/2008/218

10. Chou, T.: QcBits: Constant-time small-key code-based cryptography. In: Gierlichs, B., Poschmann, A.Y. (eds.) CHES 2016. LNCS, vol. 9813, pp. 280–300. Springer, Heidelberg (2016). https://doi.org/10.1007/978-3-662-53140-2_14

11. Costello, C., Jao, D., Longa, P., Naehrig, M., Renes, J., Urbanik, D.: Efficient compression of SIDH public keys. In: Coron, J.-S., Nielsen, J.B. (eds.) EUROCRYPT 2017. LNCS, vol. 10210, pp. 679–706. Springer, Cham (2017). https://doi.org/10.1007/978-3-319-56620-7_24

12. Costello, C., Longa, P., Naehrig, M.: Efficient algorithms for supersingular isogeny diffie-hellman. In: Robshaw, M., Katz, J. (eds.) CRYPTO 2016. LNCS, vol. 9814, pp. 572–601. Springer, Heidelberg (2016). https://doi.org/10.1007/978-3-662-53018-4_21

13. Couveignes, J.: Hard homogeneous spaces. Cryptology ePrint Archive, Report 2006/291 (2006). http://eprint.iacr.org/2006/291

14. Diffie, W., Hellman, M.E.: New directions in cryptography. IEEE Trans. Inf. Theor. 22(6), 644–654 (1976)

15. De Feo, L., Jao, D., Plût, J.: Towards quantum-resistant cryptosystems from supersingular elliptic curve isogenies. J. Math. Cryptol. 8(3), 209–247 (2014)

16. Fluhrer, S.: Cryptanalysis of ring-LWE based key exchange with key share reuse. Cryptology ePrint Archive, Report 2016/085 (2016). http://eprint.iacr.org/2016/085

17. Fujisaki, E., Okamoto, T.: How to enhance the security of public-key encryption at minimum cost. In: Imai, H., Zheng, Y. (eds.) PKC 1999. LNCS, vol. 1560, pp. 53–68. Springer, Heidelberg (1999). https://doi.org/10.1007/3-540-49162-7_5

18. Galbraith, S.D.: Mathematics of Public Key Cryptography. Cambridge University Press, New York (2012)

19. Galbraith, S.D., Petit, C., Shani, B., Ti, Y.B.: On the security of supersingular isogeny cryptosystems. In: Cheon, J.H., Takagi, T. (eds.) ASIACRYPT 2016. LNCS, vol. 10031, pp. 63–91. Springer, Heidelberg (2016). https://doi.org/10.1007/978-3-662-53887-6_3

20. Hoffstein, J., Pipher, J., Silverman, J.H.: NTRU: A ring-based public key cryptosystem. In: Buhler, J.P. (ed.) ANTS 1998. LNCS, vol. 1423, pp. 267–288. Springer, Heidelberg (1998). https://doi.org/10.1007/BFb0054868

21. Jao, D., Feo, L.: Towards quantum-resistant cryptosystems from supersingular elliptic curve isogenies. In: Yang, B.-Y. (ed.) PQCrypto 2011. LNCS, vol. 7071, pp. 19–34. Springer, Heidelberg (2011). https://doi.org/10.1007/978-3-642-25405-5_2

22. Kirkwood, D., Lackey, B.C., McVey, J., Motley, M., Solinas, J.A., Tuller, D.: Failure is not an option: Standardization issues for post-quantum key agreement. Talk at NIST workshop on Cybersecurity in a Post-Quantum World, April 2015. http://www.nist.gov/itl/csd/ct/post-quantum-crypto-workshop-2015.cfm

23. Kobara, K., Imai, H.: Semantically secure McEliece public-key cryptosystems - conversions for McEliece PKC -. In: Kim, K. (ed.) PKC 2001. LNCS, vol. 1992, pp. 19–35. Springer, Heidelberg (2001). https://doi.org/10.1007/3-540-44586-2_2

24. Kocher, P.C.: Timing attacks on implementations of diffie-hellman, RSA, DSS, and other systems. In: Koblitz, N. (ed.) CRYPTO 1996. LNCS, vol. 1109, pp. 104–113. Springer, Heidelberg (1996). https://doi.org/10.1007/3-540-68697-5_9

25. Kohel, D.R.: Endomorphism rings of elliptic curves over finite fields. PhD thesis, University of California, Berkeley (1996)

26. Koziel, B., Azarderakhsh, R., Kermani, M.M., Jao, D.: Post-quantum cryptography on FPGA based on isogenies on elliptic curves. IEEE Trans. Circuits Syst. **64**(1), 86–99 (2017)

27. Lyubashevsky, V., Peikert, C., Regev, O.: On ideal lattices and learning with errors over rings. J. ACM **60**(6), 43:1–43:35 (2013)

28. McEliece, R.J.: A public-key cryptosystem based on algebraic coding theory. Coding Thv **4244**, 114–116 (1978)

29. Misoczki, R., Tillich, J., Sendrier, N., Barreto, P.S.L.M.: MDPC-McEliece: New McEliece variants from moderate density parity-check codes. In: Proceedings of the 2013 IEEE International Symposium on Information Theory, Istanbul, Turkey, 7–12 July 2013, pp. 2069–2073. IEEE (2013)

30. Montgomery, P.L.: Speeding the Pollard and elliptic curve methods of factorization. Math. Comput. **48**(177), 243–264 (1987)

31. Moody, D., Shumow, D.: Analogues of Vélu's formulas for isogenies on alternate models of elliptic curves. Math. Comput. **85**(300), 1929–1951 (2016)

32. Mosca, M.: Cybersecurity in an ERA with quantum computers: will we be ready? Cryptology ePrint Archive, Report 2015/1075 (2015). http://eprint.iacr.org/2015/1075

33. Peikert, C.: Lattice cryptography for the internet. In: Mosca, M. (ed.) PQCrypto 2014. LNCS, vol. 8772, pp. 197–219. Springer, Cham (2014). https://doi.org/10.1007/978-3-319-11659-4_12

34. Regev, O.: On lattices, learning with errors, random linear codes, and cryptography. In: Gabow, H.N., Fagin, R. (ed.) Proceedings of the 37th Annual ACM Symposium on Theory of Computing, Baltimore, MD, USA, 22–24 May 2005, pp. 84–93. ACM (2005)

35. Rostovtsev, A., Stolbunov, A.: Public-key cryptosystem based on isogenies. Cryptology ePrint Archive, Report 2006/145 (2006). http://eprint.iacr.org/

36. Shor, P.W.: Algorithms for quantum computation: Discrete logarithms and factoring. In: 1994 Proceedings and 35th Annual Symposium on Foundations of Computer Science, pp. 124–134. IEEE (1994)

37. Silverman, J.H.: The Arithmetic of Elliptic Curves. Graduate Texts in Mathematics, 2nd edn. Springer, New York (2009). https://doi.org/10.1007/978-1-4757-1920-8

38. Stebila, D., Mosca, M.: Post-quantum key exchange for the Internet and the open quantum safe project. Cryptology ePrint Archive, Report 2016/1017 (2016). http://eprint.iacr.org/2016/1017

39. Stolbunov, A.: Cryptographic Schemes Based on Isogenies. PhD thesis, Norwegian University of Science and Technology (2012)

40. The National Institute of Standards and Technology (NIST). Submission requirements and evaluation criteria for the post-quantum cryptography standardization process, December 2016

41. Vélu, J.: Isogénies entre courbes elliptiques. CR Acad. Sci. Paris Sér. AB **273**, A238–A241 (1971)

42. Washington, L.C.: Elliptic Curves: Number Theory and Cryptography. CRC Press, Boca Raton (2008)

43. Yoo, Y., Azarderakhsh, R., Jalali, A., Jao, D., Soukharev, V.: A post-quantum digital signature scheme based on supersingular isogenies (2017). http://eprint.iacr.org/2017/186. To appear in Financial Cryptography and Data Security

Faster Algorithms for Isogeny Problems Using Torsion Point Images

Christophe Petit[✉]

School of Computer Science, University of Birmingham, Birmingham, UK
christophe.f.petit@gmail.com

Abstract. There is a recent trend in cryptography to construct protocols based on the hardness of computing isogenies between supersingular elliptic curves. Two prominent examples are Jao-De Feo's key exchange protocol and the resulting encryption scheme by De Feo-Jao-Plût. One particularity of the isogeny problems underlying these protocols is that some additional information is given as input, namely the image of some torsion points with order coprime to the isogeny. This additional information was used in several active attacks against the protocols but the current best passive attacks make no use of it at all.

In this paper, we provide new algorithms that exploit the additional information provided in isogeny protocols to speed up the resolution of the underlying problems. Our techniques lead to heuristic polynomial-time key recovery on two non-standard variants of De Feo-Jao-Plût's protocols in plausible attack models. This shows that at least some isogeny problems are easier to solve when additional information is leaked.

1 Introduction

Following calls from major national security and standardization agencies, the next cryptographic standards will have to be "post-quantum secure", namely they will have to rely on computational problems that will (at least to the best of our knowledge) remain hard for quantum computers. Several directions are currently explored for post-quantum cryptography, including lattice-based cryptography, code-based cryptography, multivariate cryptography, hash-based cryptography and most recently cryptography based on isogeny problems. The latter are appealing for their mathematical elegance but also for the relatively small key sizes compared to other post-quantum candidates.

The interest in isogeny problems as potential cryptographic building blocks is relatively new, and there has therefore not been much cryptanalytic work on them. The most established isogeny problem is the endomorphism ring computation problem, which was already considered by Kohel in his PhD thesis [12]. In the supersingular case this problem is (heuristically at least) equivalent to the problem of computing an isogeny between two randomly chosen curves [15], and it remains exponential time even for quantum algorithms today.

The supersingular key exchange protocol of Jao-De Feo [11] and the encryption scheme and signature schemes that are derived from it [7,9,24] rely on

© International Association for Cryptologic Research 2017
T. Takagi and T. Peyrin (Eds.): ASIACRYPT 2017, Part II, LNCS 10625, pp. 330–353, 2017.
https://doi.org/10.1007/978-3-319-70697-9_12

variants of these problems, where special primes and relatively small degree isogenies are used. More importantly for this paper, the attacker is also provided with the image by the isogeny of a large torsion group, in addition to the origin and image curves. Although it was observed that this additional information could a priori make the problems easier, all security evaluations against passive attacks were based on a meet-in-the-middle strategy that makes no use at all of it.

1.1   Contributions

In this paper, we study the impact of revealing the images of torsion points on the hardness of isogeny problems. We provide new techniques to successively exploit this additional information and improve on the best previous attacks, namely meet-in-the-middle attacks (see Sect. 2). Among other results, these techniques lead to polynomial-time algorithms to compute isogenies between two curves E_0 and E_1 assuming

1. Some non scalar endomorphisms of E_0 are known and/or are of small degree.
2. The images of N_2 torsion points are revealed, where N_2 is significantly larger than the degree of the isogeny N_1.

So far our techniques do not invalidate the parameters proposed in the original protocol (where $N_1 \approx N_2$). However, we describe two natural variants, which we call unbalanced variant and optimal degree variant, which can be attacked by our methods in plausible attack scenarios. We believe these generalizations are of independent interest, as they have some advantages over the original protocol when appropriate parameters are chosen.

Our main contribution in this paper is our new attack techniques. We illustrate their potential with the following results:

1. (Sect. 3.) A nearly square root speedup on the problem of computing an endomorphism of a supersingular elliptic curve of a certain degree, when provided with some torsion point images through this endomorphism.
2. (Sect. 4.4.) A polynomial time key recovery attack on our optimal degree variant, provided $N_2 > N_1^4$ and E_0 is "special" (such special curves were suggested in previous implementations [4,7] for efficiency reasons).
3. (Sect. 4.5.) A polynomial time key recovery attack on both variants, provided $\log N_2 = O(\log^2 N_1)$ and E_0 has a small degree non scalar endomorphism.

These attacks show that (at least some) isogeny problems are easier to solve when the images of torsion points through the isogeny are revealed. Some of these attacks require further assumptions on N_2; we refer to the next sections for details. We provide a heuristic analysis for all these attacks. The heuristics used involve factorization patterns and other properties of integers of particular forms appearing in our algorithms, which we treat as random numbers of the same size. For the first two attacks these heuristics are very plausible, and we believe that they can either be proved or made unnecessary (though any of those options would require significant work). For the third attack they are still a priori plausible, but they may be very hard to prove or remove. Indeed the attack involves a recursive step, and a rigorous result would have to take into account correlations between successive steps. For this reason we additionally provide some experimental support for our third attack.

We believe the three attacks we develop here are only some examples of what our new techniques can achieve, and we leave further developments to further work.

1.2 Background Reading

We refer to the books of Silverman [18] and Vignéras [21] for background results on elliptic curves and quaternion algebras. Recent cryptographic constructions based on isogeny problems include [2,7,9,11,17,23]. Computational aspects related to isogenies are covered in David Kohel's PhD thesis [12] and more recently in [8,9].

1.3 Complexity Model

Unless otherwise stated all complexity estimations in this paper use elementary bitwise operations as units. We use the standard "big O notation" to describe asymptotic complexities of algorithms. Recall that for any two functions $f, g : \mathbb{N} \to \mathbb{Z}_+$ we have $f = O(g)$ if and only if there exists $N \in \mathbb{N}$ and $c \in \mathbb{Z}_+$ such that $g(n) \geq cf(n)$ for any $n \geq N$. We also use the "big O tilde notation" to hide any polylogarithmic factors in our complexity statements: namely for any two functions $f, g : \mathbb{N} \to \mathbb{Z}_+$ we have $f = \tilde{O}(g)$ when there exists $d \in \mathbb{Z}_+$ such that $f = O(g \log^d g)$. The security levels of the protocols studied in this paper are functions of one or several security parameters. When we refer to "polynomial time" complexity we mean complexity $O(f)$, where f is a polynomial function of these security parameters.

1.4 Outline

In Sect. 2 we first describe the supersingular key exchange protocol of Jao-De Feo [11] and our two variants of this protocol, then we recall the most relevant cryptanalysis results on it. In Sect. 3 we describe faster algorithms to compute an endomorphism of a given supersingular elliptic curve, given the image of torsion points by this endomorphism. In Sect. 4 we turn to the problem of computing an isogeny between two supersingular elliptic curves given the images of torsion points by this isogeny, and we describe two attacks faster than the state-of-the-art meet-in-the-middle algorithm in this context. Finally, we summarize the impact of our techniques and results in Sect. 5, and we give perspectives for further work.

2   Supersingular Isogeny Key Exchange

2.1   Jao-De Feo's Key Exchange

We recall the supersingular key exchange protocol of Jao-De Feo [11].

Setup. Let ℓ_1, ℓ_2 be two small primes. Given a security parameter λ, let e_1, e_2 be the smallest integers such that $\ell_1^{e_1}, \ell_2^{e_2} \geq 2^{2\lambda}$ (or $2^{3\lambda}$ for post-quantum security). Let f be the smallest integer such that $p = \ell_1^{e_1} \ell_2^{e_2} f - 1$ is prime. Let E_0 be a supersingular elliptic curve over $\mathbb{F}_{p^2}$. Let P_1, Q_1 and P_2, Q_2 be respectively bases of the $\ell_1^{e_1}$ and $\ell_2^{e_2}$ torsions on E_0.

First round. Alice chooses a random cyclic subgroup of order $\ell_1^{e_1}$, say $G_1 = \langle \alpha_1 P_1 + \beta_1 Q_1 \rangle$ with at least one of α_1, β_1 coprime to ℓ_1. She computes the corresponding isogeny ϕ_1 and image curve E_1, as well as $\phi_1(P_2)$ and $\phi_1(Q_2)$. She sends E_1, $\phi_1(P_2)$ and $\phi_1(Q_2)$ to Bob. Bob proceeds similarly, permuting the roles of ℓ_1 and ℓ_2.

Second round. Upon receiving E_2, $\phi_2(P_1)$ and $\phi_2(Q_1)$, Alice computes $G_1' = \langle \alpha_1 \phi_2(P_1) + \beta_1 \phi_2(Q_1) \rangle$, the corresponding isogeny ϕ_1', the image curve $E_{12} = E/\langle G_1, G_2 \rangle$ and its j-invariant j_{12}. Bob computes $j_{21} = j_{12}$ similarly with the information sent by Alice. The shared secret is the value $j_{12} = j_{21}$, or the result of applying some key derivation function to this value.

The protocol is summarized in the following commutative diagram:

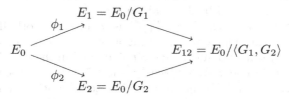

This protocol can be broken if one can compute isogenies between two given curves. However we stress that the curves appearing in this protocol are closer to each other in the isogeny graphs than random curves would be: indeed for any fixed E_0 there are only $(\ell_i + 1)\ell_i^{e_i - 1} \approx \sqrt{p}$ possible curves for E_1, while there are roughly $p/12$ supersingular j-invariants over $\mathbb{F}_{p^2}$. This allows more efficient meet-in-the-middle attacks in complexity $\tilde{O}(\sqrt[4]{p})$ instead of $\tilde{O}(\sqrt{p})$ for a generic curve pair. More importantly for this paper, some information on the isogenies is leaked by the protocol, as the image of a full torsion coprime with the isogeny degree is revealed. Finally, special primes are used to ensure that the $\ell_i^{e_i}$ torsions are defined over $\mathbb{F}_{p^2}$. For arbitrary p these torsions subgroups would be defined over large field extensions, resulting in an inefficient protocol.

Remark. Let $N_1 = \ell_1^{e_1}$. If the image of the N_1 torsion by a degree N_1 isogeny was revealed it would be straightforward to recompute the isogeny, as this image would be the kernel of the dual isogeny. More generally if N_1 is not coprime with the degree then part of the isogeny can be recovered efficiently.

2.2   Unbalanced and Optimal Degree Variants

We now present two variants of the protocol, which we call unbalanced and optimal degree variants.

Unbalanced variant. In their paper Jao and De Feo suggested parameters such that $\ell_1^{e_1} \approx \ell_2^{e_2}$. We suggest to generalize the setup to allow for unbalanced parameters $\ell_2^{e_2} \gg \ell_1^{e_1}$ in some contexts. The size of $\ell_i^{e_i}$ determines the security of the corresponding secret key G_i with respect to all previous attacks (see next subsection), while the size of p would influence efficiency. Jao and De Feo therefore chose $\ell_1^{e_1} \approx \ell_2^{e_2}$ to provide the same security level on both Alice and Bob's ephemeral keys. However in some contexts as in the public key encryption scheme [7] one secret key is static and it may therefore make sense to protect it more strongly. This is achieved by our unbalanced variant.

In the unbalanced variant, the setup procedure takes two security parameters λ_1 and λ_2 as input. For $i = 1, 2$ it computes the smallest integer e_i such that $\ell_i^{e_i}, \geq 2^{2\lambda_i}$ (or $2^{3\lambda_i}$ for post-quantum security), and then the smallest integer f such that $p = \ell_1^{e_1} \ell_2^{e_2} f - 1$ is prime. The rest of the protocol is as in Jao and De Feo.

Optimal degree variant. We now generalize the parameters such that the isogeny degrees are large enough to ensure uniform distribution of E_i among all curves on the isogeny graphs: we call the resultant protocol "optimal degree variant" for this reason. In addition, this variant allows for arbitrary primes p rather than the very special primes used by Jao and De Feo.

We recall that a number $N = \prod p_i^{e_i}$ is B-powersmooth if for all i we have $p_i^{e_i} < B$. In this paper we say that a number is powersmooth if it is B-powersmooth for some bound B that is polynomial in the security parameter.

For an arbitrary prime p, we replace $\ell_1^{e_1}$ and $\ell_2^{e_2}$ in the protocol by any powersmooth numbers N_1 and N_2 that are coprime to each other and of size about p^2. Note that the N_1 and N_2 torsions are a priori not defined over $\mathbb{F}_{p^2}$; however the powersmooth requirement ensures that they can be efficiently represented in a Chinese remainder manner (see [9]). On the other hand, the coprimality requirement ensures that the isogeny diagram commutes as in the original protocol. Finally, the condition $N_i \approx p^2$ on the isogeny degrees guarantees that E_1 and E_2 are close to uniformly distributed [9], while for the original parameters and the unbalanced variant above we have $N_1 N_2 \approx p$.

In the optimal degree variant, the setup procedure takes a security parameter λ. It chooses a random prime p with 2λ bits (or 3λ bits for post-quantum security). Then N_1 and N_2 are chosen coprime to each other, such that both of them are powersmooth and have at least $2 \log p$ bits. Then for each maximal prime power $\ell_j^{e_j}$ dividing either N_1 or N_2 we fix a basis for the $\ell_j^{e_j}$ torsion. Note that this is defined over an extension field of degree at most $2\ell_j^{e_j}$, which has a size polynomial in λ.

If $N_1 = \prod p_j^{e_j}$ then in the first round Alice chooses for each j one cyclic subgroup $G_{1j} = \langle \alpha_j P_j + \beta_j Q_j \rangle$ with at least one of α_j, β_j coprime to p_j. This implicitly defines a cyclic subgroup G_1 of order N_1 such that $G_1 = G_{1j} \bmod E_0[\ell_j^{e_j}]$. She computes the corresponding isogeny ϕ_1 as a composition of isogenies of prime degrees, the image curve $E_1 = E_0/G_1$, and the image by ϕ_1 of the $\ell_j^{e_j}$ torsion basis points, for each $\ell_j^{e_j}$ dividing N_2. Alice sends E_1 and all torsion point images to Bob. Note that although the torsion points and their images are defined over some field extensions, all isogenies computed are defined over $\mathbb{F}_{p^2}$. Moreover the degree of any extension field involved is bounded by $2\ell_j^{e_j}$ which is polynomial in the security parameter, so all elements can be efficiently represented and the computation runs in polynomial time. Bob proceeds similarly.

In the second round, Alice computes $\phi_2(G_{1j})$ using the information sent by the other party (as in the original protocol), then she computes $E_2/\phi_2(G_1)$ as above, and finally the j-invariant of this curve. Bob proceeds similarly.

A first implementation of this variant is given in [3]. Because it allows both for arbitrary primes and for "large enough" degree isogenies, the optimal degree variant can a priori be more secure than the original protocol. On the other hand, working over field extensions, even of moderate degrees, has a significant efficiency cost in practice. We leave a precise complexity estimation and a thorough comparison of this variant with the original protocol to further work.

Remark. Of course, one could also allow intermediate parameters where $\gcd(N_1 N_2, (p+1)^2)$ is a medium size factor of $(p+1)^2$ to ensure that the primes are not too special and at the same limit the size of the extension fields needed.

2.3 State-of-the Art on Cryptanalysis

We refer to [9] for a thorough discussion of existing cryptanalysis results, and only describe the most relevant work for this paper. With the exception of active attacks in [8,10,20], previous cryptanalysis results have ignored the additional information revealed in De Feo-Jao-Plût's protocols. They therefore considered the following problem:

Problem 1. *Let N be a positive integer, let p be a prime and let E_1, E_2 be two supersingular elliptic curves defined over $\mathbb{F}_{p^2}$, such that there exists an isogeny ϕ of degree N such that $E_2 = E_1/\ker\phi$. Compute ϕ.*

Remark. The most natural representation of ϕ is some canonical representation as two elements of the function field $E_1(x, y)$. In cryptographic contexts the degree of ϕ is of exponential size so this representation is not efficient. However in these contexts the degree is often a smooth number so that the isogeny can be efficiently returned as a composition of rational maps.

When N is large enough any pair of elliptic curves are connected by an isogeny of degree N, and this problem is heuristically equivalent to the endomorphism ring computation problem [15]. In De Feo-Jao-Plût's protocols, however, $N = O(\sqrt{p})$ is too small to ensure this, and as N is moreover smooth one can do a meet-in-the-middle attack with complexity $\tilde{O}(\sqrt[4]{p})$ (respectively $\tilde{O}(\sqrt[6]{p})$ with a quantum computer) even if the endomorphism ring computation problem remains of complexity $\tilde{O}(\sqrt{p})$ (respectively $\tilde{O}(\sqrt[3]{p})$ with a quantum computer). We stress that the optimal degree variant we introduced above does not suffer from this problem, as the isogeny degrees are chosen large enough to ensure a uniform distribution of E_2.

The following lemma generalizes the meet-in-the-middle strategy when the smoothness bound on N is not polynomial in $\log p$.

Lemma 1. *Assume* $N = N_1 \cdot N_2$ *where both* N_1 *and* N_2 *are* B-*smooth. Then the meet-in-the-middle strategy has a time and memory complexity* $\tilde{O}(B \max(N_1, N_2))$, *neglecting log factors.*

Proof: The factorization of N can be obtained in subexponential time, which is negligible with respect to $\max(N_1, N_2)$. Isogenies of prime degree can be computed in quasilinear time in the degree. The meet-in-the-middle strategy computes $O(N_1)$ isogenies of degree N_1 and $O(N_2)$ isogenies of degree N_2, each of them as a composition of isogenies of degrees at most B.   □

The active attack presented in [8] runs $O(\log p)$ executions of the key exchange protocol with the same party. Assuming this party uses a static secret key G_1, the attacker provides them with incorrect values for $\phi_2(P_1), \phi_2(Q_1)$, observes variations in the resulting shared key $j(E_{12})$, and progressively deduces the key G_1. The loop-abort fault attack developed in [10] is similar to this attack. A fault attack is also used in [20] to replace $\phi_2(P_1)$ and $\phi_2(Q_1)$ by points whose order is not coprime with the isogeny degree. Our goal in this paper is to show how to exploit the "torsion image" information revealed in De Feo-Jao-Plût's protocols but using only passive attacks, namely for normal executions of the protocols.

3   Computing an Endomorphism from Additional Information

From a computational number theory point of view, computing endomorphisms of a curve is a somewhat more natural task than computing isogenies between two curves. At the same time, there are strong relations between the two problems (see [9,15]). In this section we define an "endomorphism computation" counterpart to De Feo-Jao-Plût's isogeny problem, and we show how leaking the image of torsion points helps in solving this problem.

3.1   Endomorphism Computation Problem with Additional Information

We consider the following problem:

Problem 2. *Let* p *be a prime and let* E *be a supersingular elliptic curve defined over* $\mathbb{F}_{p^2}$. *Let* ϕ *be a non scalar endomorphism of* E *with smooth degree* N_1. *Let* N_2 *be a smooth integer with* $\gcd(N_1, N_2) = 1$, *and let* P, Q *be a basis of* $E[N_2]$. *Let* R *be a subring of* $\mathrm{End}(E)$ *that is either easy to compute, or given. Given* E, P, Q, $\phi(P)$, $\phi(Q)$, $\deg \phi$, R, *compute* ϕ.

Remark. This problem is similar to the problem appearing in De Feo-Jao-Plût protocols, with the additional requirement $E_1 = E_2$.

Remark. When no endomorphism subring is explicitly given one can take for R the subring of scalar multiplications, which we will denote $R = \mathbb{Z}$.

Remark. If we do not use the additional information the best algorithm for this problem will be a meet-in-the-middle approach: compute all isogenies of degree approximately $\sqrt{N_1}$ from E and search for a collision. As N_1 is smooth the cost for each isogeny is polynomial, resulting in an algorithm with roughly $\tilde{O}(\sqrt{N_1})$ complexity.

3.2   General Strategy

Our general strategy is summarized in Algorithm 1.

Algorithm 1. Computing an Endomorphism from Additional Information

---

Require: As in Problem 2, plus parameter B.
Ensure: A description of ϕ as a composition of low degree maps.
 1: Find $N_1' \in \mathbb{N}$ and $\theta_1, \theta_2 \in R$ such that $\deg(\theta_1\phi+\theta_2) = N_1'N_2$ and $\gcd(\deg\theta_1, N_1) = 1$, and such that N_1' is B-smooth and as small as possible.
 2: Compute $\ker \psi_{N_2}$ using the additional information, where $\theta_1\phi + \theta_2 = \psi_{N_1'}\psi_{N_2}$ and $\psi_{N_1'}, \psi_{N_2}$ are respectively of degrees N_1' and N_2.
 3: Compute $\psi_{N_1'}$ using a meet-in-the-middle approach.
 4: Compute $\ker \phi = \ker(\theta_1^{-1}(\psi_{N_1'}\psi_{N_2} - \theta_2))$ by evaluating all maps on the N_1 torsion.
 5: Compute ϕ from $\ker \phi$.

From what is given in the problem we can compute the image of ϕ on any point in $E[N_2]$. Let $\theta_1, \theta_2 \in R$ be known endomorphisms of E, to which we associate another endomorphism

$$\psi := \theta_1\phi + \theta_2.$$

Of course we do not know ϕ so far, but since we know θ_1, θ_2, and the action of ϕ on $E[N_2]$ we can nevertheless evaluate ψ on any point of $E[N_2]$.

Let us now assume that the maps θ_1, θ_2 are chosen such that $\deg \psi = N_1'N_2$ for some $N_1' \in \mathbb{Z}$. An algorithm to achieve that together with an additional smoothness condition on N_1' will be described in the next subsection for the case $R = \mathbb{Z}$. The endomorphism ψ can then be written as a composition of two isogenies

$$\psi = \psi_{N_1'}\psi_{N_2}$$

with $\psi_{N_1'}$ and ψ_{N_2} respectively of degrees N_1' and N_2.

By computing ψ on a basis of $E[N_2]$ and solving some discrete logarithm problems in $E[N_2]$ we deduce the kernel of ψ_{N_2} and then deduce ψ_{N_2} itself. This is efficient since N_2 is smooth by assumption.

At this point, the map $\psi_{N_1'}$ is an isogeny of degree N_1' between two known j-invariants, namely the curve image of ψ_{N_2} and the original curve E. We recover this isogeny using the meet-in-the-middle approach analyzed in Lemma 1. The efficiency of this step depends on the factorization of N_1'.

At this point, we have computed the map ψ as a composition $\psi = \psi_{N_1'}\psi_{N_2}$. We deduce an expression for ϕ, namely $\theta_1^{-1}(\psi_{N_1'}\psi_{N_2} - \theta_2)$, and assuming $\gcd(\deg\theta_1, N_1) = 1$ we evaluate this map on the N_1 torsion to identify $\ker\phi$, from which we recompute a more canonical description of ϕ. This is efficient as N_1 is smooth.

Remark. We do not use the additional information to compute $\psi_{N_1'}$. Note that part of the N_2 torsion is annihilated by ψ_{N_2}, so we only know $\psi_{N_1'}$ and its dual on one cyclic subgroup of the respective N_2 torsions.

Remark. There is no gain in generality in considering maps of the form $\psi := \theta_1\phi\theta_3 + \theta_2$ for $\theta_1, \theta_2, \theta_3 \in R$. Indeed we have $\theta_1\phi\theta_3 + \theta_2 = \hat{\phi}\hat{\theta}_1\theta_3 + \theta_2 + \mathrm{Tr}(\theta_1\phi)\theta_3$. Taking conjugates we obtain an element $\hat{\psi} = \hat{\theta}_3\theta_1\phi + \hat{\theta}_2 + \mathrm{Tr}(\theta_1\phi)\hat{\theta}_3 \in R\phi + R$ with the same norm. Similarly, there is no gain in generality in using powers of ϕ since $\phi^2 = -(\mathrm{Tr}\,\phi)\phi - \deg\phi$.

3.3   Attack When $R = \mathbb{Z}$

We first consider the most generic case where the only known endomorphisms of E are scalar multiplications. We define

$$\psi = \psi_{a,b} = a\phi + b$$

for $a, b \in \mathbb{Z}$, which has degree

$$\deg\psi_{a,b} = a^2 \deg\phi + b^2 + ab\,\mathrm{Tr}\,\phi = \left(b + a\frac{\mathrm{Tr}\,\phi}{2}\right)^2 + a^2\left(\deg\phi - \left(\frac{\mathrm{Tr}\,\phi}{2}\right)^2\right).$$

Our goal is to find a, b such that $\deg\psi_{a,b} = N_1'N_2$, where N_1' is as small and as smooth as possible.

Parameter Restriction. The attack we describe below requires two assumptions on the parameters.

1. We require $N_2 > 2\sqrt{N_1}$.
2. We also require that $-D$ is a square modulo N_2, where $D = \deg\phi - \left(\frac{\mathrm{Tr}(\phi)}{2}\right)^2$.

Note that in the Jao-De Feo key exchange protocol we have $N_2 \approx N_1$ so the first assumption does not look too strong. By Hensel's lifting lemma the second condition is equivalent to $-D$ being a square modulo every odd prime factor of N_2 and congruent to 1 modulo 8 when 2 divides N_2. If we consider the endomorphism ϕ as fixed, this condition restricts N_2 values as follows:

- If $N_2 = \ell^e$ is a prime power (as for Jao-De Feo's parameters and the unbalanced variant), the second condition is satisfied if and only if $-D$ is a quadratic residue modulo ℓ, and heuristically we expect this to occur for half of the primes ℓ.
- In our optimal degree variant, N_2 is a powersmooth number whose prime factors will be as small as possible. We can heuristically expect that about one half of these prime factors ℓ_i will be such that $-D$ is a quadratic residue modulo ℓ_i. There will therefore exist a factor N_2' of N_2 such that $N_2' \approx \sqrt{N_2}$ and $-D$ is a quadratic residue modulo N_2'. Moreover if $N_2 \approx N_1$ or bigger, we can use N_2' in the attack instead of N_2, and still satisfy the first condition.

This suggests that the conditions above are relatively mild, in the sense that they are satisfied for a large set of parameters with the expected forms. In the remaining of this section we assume that both conditions above are satisfied.

Algorithm. Remember that from the additional information given in the problem we can compute the image of ϕ on any point in $E[N_2]$. Note that since N_1 and N_2 are coprime, ϕ is a one-to-one map on $E[N_2]$. From the relation $\phi\hat{\phi} = [\deg \phi]$ we can also compute the image of any point in $E[N_2]$ by the dual map $\hat{\phi}$. We can therefore also evaluate $\mathrm{Tr}(\phi)$ on $E[N_2]$. By solving a discrete logarithm problem in $E[N_2]$ we deduce $\mathrm{Tr}(\phi) \bmod N_2$. By the Cauchy-Schwarz inequality we also have $\mathrm{Tr}(\phi) \leq 2\sqrt{\deg \phi}$ so under our first parameter restriction that $N_2 > 2\sqrt{N_1}$ we actually recover $\mathrm{Tr}(\phi)$ exactly.

Let $D = \deg \phi - \frac{1}{4}(\mathrm{Tr}(\phi))^2$ and let τ such that $\tau^2 = -D \bmod N_2$. Such a τ exists under our second parameter restriction, and can be efficiently computed using Tonelli-Shanks algorithm and Hensel's lifting lemma. Points (x, y) in the lattice generated by the two vectors $(N_2, 0)$ and $(\tau, 1)$ correspond to solutions of the equation $x^2 + Dy^2 = 0 \bmod N_2$. We compute a reduced basis for the lattice, with respect to a weighted inner product norm where the second component is weighted by $\sqrt{D}$. This can be done in polynomial time. Finally we let $a := y_0$, $b = x_0 - \frac{\mathrm{Tr}(\phi)}{2} y_0$ and $N_1' = \frac{x_0^2 + Dy_0^2}{N_2}$, where (x_0, y_0) is a well-chosen short vector in the lattice.

To choose (x_0, y_0) we proceed as follows. Using the short basis computed above we generate short vectors and compute the corresponding N_1' values, until we obtain N_1' such that the meet-in-the-middle strategy is efficient enough (see Lemma 1).

Complexity analysis. We first analyze the expected norms of minimal lattice vectors.

Lemma 2. *Under plausible heuristic assumptions, the shortest vectors in the lattice have norm $N_1' N_2$ where $N_1' \approx \sqrt{N_1}$.*

Proof: Heuristically, a proportion about $1/N_2$ pairs (x, y) will satisfy the congruence $x^2 + Dy^2 = 0 \bmod N_2$ so we expect $xy \approx N_2$. We can also expect that minimal vectors (x, y) in the lattice have their coefficients balanced such that $x^2 \approx Dy^2 \approx N_1' N_2$. (If $N_2^2 < D$ then the smallest element will of course be

$(N_2, 0)$, however by our parameter restriction we have $D \approx N_1 < N_2^2$.) Combining all these approximations gives $(N_1' N_2)^2 \approx x^2 D y^2 = D(xy)^2 \approx D N_2^2$, hence $N_1' \approx \sqrt{D} \approx \sqrt{N_1}$. □

By construction any lattice vector will have a norm divisible by N_2. In our algorithm we generate random short vectors until the cofactor N_1' is smooth enough. To estimate the number of random trials needed, we (heuristically) approximate the smoothness probability of N_1' by the smoothness probability of a random number of the same size.

For any positive integers X, Y, let $\pi(X, Y)$ be the proportion of integers smaller than X that are Y-smooth. For any positive integer X and any $0 \leq \alpha \leq 1$, let $L_X(\alpha) = \exp^{(\log X)^\alpha (\log \log X)^{1-\alpha}}$ be the subexponential function. We recall the following well-known fact [16]:

Lemma 3. *For any $0 \leq \alpha \leq 1$ and any large enough X, we have $\pi(X, L_X(\alpha)) \approx (L_X(1 - \alpha))^{-1}$.*

We deduce the following result:

Proposition 1. *Subject to the above parameter restrictions and under plausible heuristic assumptions, Problem 2 can be solved in time $O(N_1^{1/4+\epsilon})$ for any $\epsilon > 0$.*

Proof: The algorithm and heuristic assumptions required have been described earlier in this section. The cost of the algorithm depends mostly on the smoothness bound required on lattice vectors, which decides both the cost of finding a suitable vector and the meet-in-the-middle cost needed to compute $\psi_{N_1'}$.

Using a smoothness bound $L_{N_1}(\alpha)$ for any $0 < \alpha < 1$, the cost of finding a suitable vector in the lattice is bounded by $(L_{N_1}(1 - \alpha))^{-1} << O(N_1^{1/4+\epsilon})$, and the meet-in-the-middle computation has a cost $\tilde{O}(N_1^{1/4} L_{N_1}(\alpha)) = O(N_1^{1/4+\epsilon})$ by Lemma 1. □

By exploiting the images of torsion points, our algorithm provides a near-square root speedup for Problem 2 over the previous state-of-the-art algorithm.

Improvement when $\gcd(D, N_2) \neq 1$. For any $r | \gcd(D, N_2)$ there exist $a, b \in \mathbb{Z}$ with $(a\phi + b)/r \in \mathrm{End}(E)$ and $r \nmid \gcd(a, b)$. Moreover we can normalize pairs of this form such that $a = 1$. We can identify the corresponding correct b by trying every possibility until $\phi + b$ annihilates the r torsion. This has a cost $O(r)$. Alternatively we can solve some discrete logarithm problems to find b in at most $O(\sqrt{r})$ operations. In any case since N_2 is smooth we can process small factors one at the time, and efficiently deduce $\phi' \in \frac{1}{r}\mathbb{Z}[1, \phi] \cap \mathrm{End}(E)$, with a new D value $D' = D/\gcd(D, N_2)$. Moreover we can evaluate ϕ' on the $N_2/\gcd(D, N_2)$ torsion. Following the analysis above, we expect that this will reduce the complexity by a factor $\sqrt{\gcd(D, N_2)}$.

3.4   Variants and Extensions

We could consider variants of Problem 2 where information is given on several endomorphisms of a single curve, and develop similar attacks.

A particular case of interest is the case of subfield curves, namely curves defined over $\mathbb{F}_p$, when we can use $R = \mathbb{Z} + \pi_p \mathbb{Z}$ where $\pi_p : (x, y) \rightarrow (x^p, y^p)$.

Remark: When E is defined over $\mathbb{F}_p$ one can compute the full endomorphism ring in expected time $\tilde{O}(p^{1/4})$ using the techniques of Delfs and Galbraith [6].

Under the (reasonable) parameter restriction that $N_2 > 2\sqrt{N_1}$ we can compute $\mathrm{Tr}\, \phi$ as above, and substitute ϕ by $\phi' = \phi - \frac{\mathrm{Tr}\, \phi}{2}$ in the problem so that $\mathrm{Tr}\, \phi' = 0$. Let $\Delta := \deg \phi' = N_1 - \frac{1}{4}(\mathrm{Tr}\, \phi)^2$. We can consider an endomorphism of the form

$$\psi = (a\phi' + b)\pi_p + c\phi' + d,$$

with degree

$$\deg \psi = (a^2\Delta + b^2)p + (c^2\Delta + d^2) + \mathrm{Tr}\left((a\phi' + b)\pi_p(-c\phi' + d)\right)$$
$$= (a^2\Delta + b^2)p + (c^2\Delta + d^2) + (ad - bc)\,\mathrm{Tr}(\phi'\pi_p).$$

If $N_2 > 2\sqrt{N_1 p}$ we can evaluate $\mathrm{Tr}(\pi_p\phi')$. We are then left with finding $a, b, c, d, N_1' \in \mathbb{Z}$ such that $\deg \psi = N_1' N_2$ and moreover N_1' is both small and smooth such that the meet-in-the-middle strategy (Lemma 1) is efficient.

Note that for the minimal solution we expect $a^2 p N_1 \approx b^2 N_1 \approx c^2 p \approx d^2 \approx N_1' N_2$ and $abcd \approx N_2$, hence $d^4 \approx N_2 N_1 p$ and $N_1' \approx N_1^{1/2} p^{1/2} N_2^{-1/2}$. This means that if $N_2 \approx N_1 p$ we can expect a solution with $N_1' = O(1)$.

Remark: The discussion in this section provides a reduction from an isogeny problem to a Diophantine equation problem, arguably a step forward in the cryptanalysis. We leave the construction of an efficient (classical or quantum) algorithm to solve this Diophantine equation to further work.

Remark: Efficient solutions for quaternary quadratic form equations exist over the rationals [5,19]; however we are not aware of any efficient algorithm that would return integer solutions.

4   Attacks on (Variants of) the Key Exchange Protocol

We now turn to isogeny problems with additional information, as in De Feo-Jao-Plût's protocols.

4.1   Problem Statement

In this section we consider the following problem.

Problem 3. *Let p be a prime. Let $N_1, N_2 \in \mathbb{Z}$ be coprime. Let E_0 be a super-singular elliptic curve over $\mathbb{F}_{p^2}$. Let $\phi_1 : E_0 \to E_1$ be an isogeny of degree N_1. Let R_0, R_1 be subrings of $\mathrm{End}(E_0)$, $\mathrm{End}(E_1)$ respectively. Given N_1, E_1, R_0, R_1 and the image of ϕ_1 on the whole N_2 torsion, compute ϕ_1.*

Remark. The most generic case for this problem is $R_0 = R_1 = \mathbb{Z}$, namely only the scalar multiplications are known (and do not need to be explicitly given). If E_0 is defined over $\mathbb{F}_p$ we can take $R_0 = \mathbb{Z}[\pi_p]$ where π_p is the Frobenius. In some previous implementation works [4,7] it was suggested for efficiency reasons to use special curves in the key exchange protocol, such as a curve with j-invariant $j = 1728$. In this case we have $R_0 = \mathrm{End}(E_0)$, and moreover R_0 contains some non scalar elements of small degrees.

4.2   Attack Model and General Strategy

We provide algorithms that use the additional information provided by the image of torsion points to solve Problem 3 with dramatic speedups compared to the basic meet-in-the-middle strategy.

All our attacks assume that the subring of endomorphisms R_0 contains more than the scalar multiplications. They are particularly efficient when special curves E_0 are used, such as in [4,7].

Another current limitation of our attacks is that they require N_2 significantly larger than N_1. This condition could plausibly be met in practice (should this paper not have warned against them!) in the following scenarios:

- In the unbalanced variant of the original protocol. We recall that this variant could a priori have been used when one party uses a static key and the other party uses an ephemeral key, as is the case for example in the public key encryption scheme.
- In the optimal degree variant of the protocol, a server might have used a static key and published the images of a very large torsion group $E_0[N_2]$, for example to allow connections with a wide range of clients using different sets of parameters.

Our basic strategy is as follows. For any known endomorphism $\theta \in \mathrm{End}(E_0)$ we can consider the endomorphism $\phi = \phi_1 \theta \hat{\phi}_1 \in \mathrm{End}(E_1)$. Moreover if θ is non scalar then ϕ is also non scalar. Using our knowledge of how ϕ_1 acts on the N_2 torsion we can also evaluate ϕ on the N_2 torsion, and hence apply the techniques from the previous section. Once we have an expression for ϕ we can use it to evaluate $\phi_1 \theta \hat{\phi}_1$ on the N_1 torsion. Since N_1 is smooth an easy discrete logarithm computation gives generators for $\ker(\phi_1 \theta \hat{\phi}_1) \cap E_1[N_1]$. The latest group contains $\ker \hat{\phi}_1$ as a cyclic subgroup of order N_1. When it is cyclic we directly recover $\ker \hat{\phi}_1$ and deduce ϕ_1; in Sect. 4.3 below we show how to do it in the general case.

Remark. Our resolution strategy requires that R_0 contains more than the scalar multiplications, as otherwise ϕ is just a scalar multiplication.

In Sects. 4.4 and 4.5 below we give two examples of attacks that can be developed using our techniques.

- The first attack assumes that E_0 is defined over $\mathbb{F}_p$, and moreover that E_0 has a small degree endomorphism ι such that $\mathrm{Tr}(\iota) = \mathrm{Tr}(\iota\pi_p) = 0$. This is the case for example if $j(E_0) = 1728$. Currently the attack applies only to our optimal degree variant. For well-chosen values of N_2 larger than N_1^4 the attack recovers the secret key G_1 in polynomial time.
- The second attack only requires that E_0 has a small degree endomorphism, but on the other hand it needs $\log N_2 = O(\log^2 N_1)$ to recover the secret key G_1 in polynomial time. This attack deviates from the basic strategy explained above and it instead uses some recursive step. We provide a heuristic analysis and some experimental support for this attack for both the unbalanced and the optimal degree variants.

Both attacks are heuristic, as their analysis makes unproven assumptions on factorization properties of certain numbers. We leave a better analysis, further variants and improvements to further work.

4.3 Recovering ϕ_1 from $\mathrm{ker}(\phi_1\theta\hat{\phi}_1)$

In the strategy outlined above we need to recover ϕ_1 from $\mathrm{ker}(\phi_1\theta\hat{\phi}_1)$. Here we give a method to do this and we show that the method is efficient. For simplicity we assume without loss of generality that $\deg\theta$ is coprime with N_1.

Let $G := \mathrm{ker}(\phi_1\theta\hat{\phi}_1) \cap E_1[N_1]$. Clearly $\mathrm{ker}\,\hat{\phi}_1$ is a cyclic subgroup of order N_1 in G. When G is cyclic this immediately gives $\mathrm{ker}\,\hat{\phi}_1$. When G is not cyclic, let $M|N_1$ be the largest integer such that $E_1[M] \subset G$. The isogeny $\phi_1 : E_0 \to E_1$ can be decomposed as an isogeny ϕ_M of degree M from E_0 to a curve E_M, and a second isogeny of degree N_1/M from E_M to E_1. We denote by $\phi_{N_1/M}$ the *dual* of this second isogeny, namely $\phi_{N_1/M} : E_1 \to E_M$ and $\phi_1 = \hat{\phi}_{N_1/M}\phi_M$. This is represented in the picture below:

$$E_0 \xrightarrow[\phi_1]{\quad\phi_M\quad} E_M \xleftarrow{\quad\phi_{N_1/M}\quad} E_1$$

Clearly, recovering ϕ_M and $\phi_{N_1/M}$, or equivalently their kernels, is sufficient to recover ϕ_1. The second isogeny $\phi_{N_1/M}$ is the easiest one to recover:

Lemma 4. *We have* $\mathrm{ker}\,\phi_{N_1/M} = M\left(\mathrm{ker}(\phi_1\theta\hat{\phi}_1) \cap E_1[N_1]\right)$.

Proof: Clearly $\mathrm{ker}\,\phi_{N_1/M} = M\,\mathrm{ker}\,\hat{\phi}_1$. The later is a cyclic subgroup of $M\left(\mathrm{ker}(\phi_1\theta\hat{\phi}_1) \cap E_1[N_1]\right)$ of order N_1/M. By our definition of M, the group $M\left(\mathrm{ker}(\phi_1\theta\hat{\phi}_1) \cap E_1[N_1]\right)$ is cyclic, hence equal to $M\,\mathrm{ker}\,\hat{\phi}_1$ as well. □

We now focus on ϕ_M, and we first identify a property that its kernel must satisfy:

Lemma 5. *We have $\theta(\ker \phi_M) = \ker \phi_M$.*

Proof: Equivalently, we want to prove $\theta^{-1}(\ker \phi_M) = \ker \phi_M$. We have $\ker \phi_M = \ker \phi_1 \cap E_0[M] = \hat{\phi}_1(E_1[M])$ and similarly $\theta^{-1}(\ker \phi_M) = \theta^{-1}(\ker \phi_1) \cap E_0[M] = \ker(\phi_1 \theta) \cap E_0[M]$, so we can rephrase the lemma as $\hat{\phi}_1(E_1[M]) = \ker(\phi_1 \theta) \cap E_0[M]$.

Since $\hat{\phi}_1(E_1[N_1])$ is cyclic, so is $\hat{\phi}_1(E_1[M])$. Therefore $E_1[M] \subset \ker(\phi_1 \theta \hat{\phi}_1) \cap E_1[M]$ if and only if $\hat{\phi}_1(E_1[M]) \subset \ker \phi_1 \theta$.

By the definition of M we have $E_1[M] \subset \ker(\phi_1 \theta \hat{\phi}_1) \cap E_1[M]$ so $\hat{\phi}_1(E_1[M]) \subset \ker \phi_1 \theta$. Moreover M is the largest such integer and $\hat{\phi}_1(E_1[M])$ is cyclic, so the equality holds. $\qquad\square$

As we know the endomorphism θ, we can evaluate its action on the M torsion and identify potential candidates for $\ker \phi_M$.

Lemma 6. *Let k be the number of distinct prime factors of M. Then there are at most 2^k cyclic subgroups H of order M in $E_0[M]$ such that $\theta(H) = H$.*

Proof: Let $\{P, Q\}$ be a basis for $E_0[M]$, and let α, β be integers such that $\ker \phi_M = \langle \alpha P + \beta Q \rangle$. We have $\gcd(\alpha, \beta, M) = 1$.

The action of θ on $E_0[M]$ can be described by a matrix $m = \left(\begin{smallmatrix} a & b \\ c & d \end{smallmatrix} \right) \in GL(2, \mathbb{Z}_M)$ such that $\theta(P) = aP + bQ$ and $\theta(Q) = cP + dQ$. Moreover we have $\det(m) = ad - bc = \deg \theta \bmod M$ and $\mathrm{Tr}(m) = a + d = \mathrm{Tr}(\theta) \bmod M$.

The condition $\theta(\ker \phi_M) = \ker \phi_M$ from Lemma 5 now becomes

$$\langle \alpha P + \beta Q \rangle = \langle (a\alpha + c\beta)P + (b\alpha + d\beta)Q \rangle$$

or equivalently

$$(a\alpha + c\beta)\beta = (b\alpha + d\beta)\alpha \bmod M,$$

or

$$c\beta^2 + (a - d)\alpha\beta - b\alpha^2 = 0 \bmod M.$$

The latest has solutions if and only the discriminant

$$(a - d)^2 - 4bc = (\mathrm{Tr}(\theta))^2 - 4 \deg \theta \bmod M$$

is a quadratic residue, and this is the case by assumption. Clearly there are at most two solutions modulo any prime $\ell | M$, and by Hensel's lifting lemma a solution modulo a prime $\ell | M$ determines a unique solution modulo any power of ℓ dividing M. $\qquad\square$

We remark that when N_1 is smooth, our proof implicitly provides an efficient algorithm to identify all the candidate kernels. When N_1 is a prime power then k is at most one, and we are done. Our last lemma shows that for powersmooth numbers, the expected value of k is small enough to allow a polynomial time exhaustive search of all candidate kernels.

Lemma 7. *Let N_1 be a powersmooth number. Assume ϕ_1 be chosen uniformly at random among all isogenies of degree N_1 from E_0. Then the expected value of k is bounded by $2\log\log N_1$.*

Proof: Clearly the number of distinct prime factors of N_1 is smaller than $\log_2 N_1$. In the proof of the previous lemma we showed that for every prime ℓ dividing $M|N_1$, there are at most two candidate cyclic subgroups H_ℓ such that $\theta(H_\ell) = H_\ell$. We can therefore bound the expected value of k by

$$E[k] \leq \sum_{\substack{\ell|N_1,\ell\text{ prime}\\ \ell\leq\log N_1}} \frac{2}{\ell+1} < \sum_{\ell\leq\log N_1} \frac{2}{\ell} \approx 2\int_1^{\log N_1} \frac{1}{\ell} \approx 2\log\log N_1.$$

$\square$

4.4   Attack When E_0 Is Special

In this section we focus on the optimal degree variant of the protocol. We assume E_0 is defined over $\mathbb{F}_p$, so that $\text{End}(E_0)$ contains the Frobenius endomorphism $\pi_p : (x,y) \to (x^p, y^p)$. Moreover we assume $\text{End}(E_0)$ contains some non scalar element ι with small norm q such that $\text{Tr}(\iota) = \text{Tr}(\iota\pi_p) = 0$. (Maximal orders with minimal such ι were called special in [13].) Then clearly the attacker knows π_p and they can efficiently compute ι by testing all isogenies of small degree. We consider an endomorphism of E_1 defined by

$$\phi = \phi_1(a\iota\pi_p + b\pi_p + c\iota)\hat{\phi}_1 + d,$$

with degree

$$\deg\phi = N_1^2 pqa^2 + N_1^2 pb^2 + N_1^2 qc^2 + d^2.$$

Remark: There is no gain of generality in allowing scalar components in R_0: indeed $\phi_1\mathbb{Z}\hat{\phi}_1 = N_1\mathbb{Z} \subset R_1$.

Similarly as before, our goal is now to find tuples of integers (a,b,c,d) such that $\deg\phi = N_1'N_2$ and N_1' is small. We first discuss some elementary properties of the solutions.

Lemma 8. *Let (a,b,c,d) defining ϕ as above, with $\deg\phi = N_1'N_2$ for some N_1'. Then*

- *$N_1'N_2$ is a square modulo N_1^2;*
- *except for "exceptional" parameters, $N_1'N_2$ is not much smaller than N_1^4;*
- *except for "exceptional" parameters, N_1 is not much smaller than p.*

Proof: We have $d^2 = N_1'N_2 \bmod N_1^2$. For any N_1' this defines d modulo N_1^2 up to sign, hence except for exceptional parameters d^2 will not be much smaller than N_1^4. We then have

$$pqa^2 + pb^2 + qc^2 = \frac{N_1'N_2 - d^2}{N_1^2} \approx N_1^2,$$

and the value of c is defined modulo p up to sign (assuming such a value exists). Except for exceptional parameters c^2 will not be much smaller than p^2, hence N_1 will not be much smaller than p.   □

Parameter restriction. Recall that in this section we focus on the optimal degree variant, hence N_1 and N_2 are powersmooth numbers. From now on, we assume that $N_1 > p$, that $N_2 \approx N_1^4$ and that N_2 is a square modulo N_1^2. This ensures that all the conditions identified in Lemma 8 are satisfied provided N_1' is a square modulo N_1^2.

Note that we can always ensure that a powersmooth number N_2 is also a square modulo N_1^2 by dividing and/or multiplying it by a well-chosen small prime. In the first case we will have to work with a slightly smaller N_2 value in our attack, and in the second case we will have to perform some small guess on the images of the full N_2 torsion.

Algorithm. We now describe an algorithm that computes a tuple (a, b, c, d) that can be used in our attack. We first attempt to find a solution with $N_1' = 1$, and when this fails we successively increase N_1' to the next square. For a given N_1', the value of d is determined modulo N_1^2 up to the sign. We try possible values of d until we find one such that $q \cdot \frac{N_1' N_2 - d^2}{N_1^2}$ is a square modulo p. At this point we try random values for c satisfying the congruence condition, until the equation

$$a^2 q + b^2 = \frac{N_1' N_2 - d^2 - pc^2}{pN_1^2}$$

has a solution, which we compute with Cornacchia's algorithm. This algorithm is detailed below in Algorithm 2.

The complexity of this algorithm is analyzed in the following lemma.

Lemma 9. *Let all parameters be restricted as above. Under plausible heuristic assumptions Algorithm 2 terminates in polynomial time.*

Proof: Computing quadratic residues (Step 5), modular square roots (Steps 3 and 12) and Cornacchia's algorithm (Step 17) all run in polynomial time.

Heuristically, the quadratic residuosity condition in Step 5 will be satisfied for every other value of d, so the algorithm will reach Step 12 with a very small value of N_1'. Consequently in Step 4 we expect $m = \frac{N_1' N_2 - d^2}{N_1^2} \approx N_1^2$ and in Step 13 we expect $m/p \approx N_1^2/p > p$. In Step 15 we expect $n = \frac{N_1' N_2 - d^2 - c^2 N_1^2 q}{N_1^2 p} \approx \frac{N_1' N_2}{N_1^2 p} \approx \frac{N_1' N_1^2}{p} > N_1' p \approx p$.

As long as $\log N_1 = O(\log p)$, the expected number of random trials on c until n is prime is therefore $\log n \approx O(\log p)$. Moreover by Dirichlet's density theorem the density of primes represented by the norm form $a^2 q + b^2$ is $1/2H(q) > \sqrt{q}/2$ where $H(q) < \sqrt{q}$ is the class number of $\mathbb{Q}[\sqrt{-q}]$. Finally under the Generalized Riemann Hypothesis we have $q = O((\log p)^2)$ [1]. This shows that a polynomial number (in the security parameter λ) of values r must be tested in Step 13 until

---

Algorithm 2. Finding attack parameters when E_0 is special

Require: N_1, N_2, q as above.
Ensure: Parameters (a, b, c, d) and N_1' for an attack.
1: $i \leftarrow 1$.
2: $N_1' \leftarrow i^2$.
3: Let d such that $0 \leq d \leq N_1^2$ and $d^2 = N_1' N_2 \bmod N_1^2$.
4: $m \leftarrow \frac{N_1' N_2 - d^2}{N_1^2}$.
5: **if** mq is not a square modulo p **then**
6:     **if** $d < N_1' N_2 - N_1^2$ **then**
7:         $d \leftarrow d + N_1^2$.
8:         **go to** Step 4.
9:     **else**
10:         $i \leftarrow i + 1$.
11:         **go to** Step 2.
12: Let $\hat{c}$ such that $0 \leq \hat{c} < p$ and $q\hat{c}^2 = m \bmod p$.
13: Let r be a random integer in $[0, m/p]$.
14: $c \leftarrow \hat{c} + rp$.
15: $n \leftarrow \frac{N_1' N_2 - d^2 - c^2 N_1^2 q}{N_1^2 p}$.
16: **if** n has an easy factorization (for example if n is prime) **then**
17:     Solve equation $a^2 q + b^2 = n$ with Cornacchia's algorithm
18:     **if** there is no solution **then**
19:         **go to** Step 13.
20: **return** (a, b, c, d, N_1').

---

a suitable one is found. Since the expected size of m/p is bigger than p, the algorithm is expected to terminate in polynomial time with a solution.   □

We deduce the following:

Proposition 2. *Let N_1 and N_2 be powersmooth numbers as in our optimal degree variant of Jao-De Feo's protocol. Assume moreover that $N_1 > p$, that $N_2 \approx N_1^4$ and that N_2 is a square modulo N_1^2. Then under plausible heuristic assumptions Problem 3 can be solved in polynomial time when the initial curve E_0 has j invariant $j = 1728$, or more generally when the curve is "special" in the sense of [13].*

Remark. In the original and unbalanced variants of the protocol we have $N_1 N_2 < p$ so $N_1' > N_1^2 p / N_2 > N_1$, unless $a = b = 0$. In the next section we provide an attack that works in this setting.

4.5   Attack When $R_0 = \mathbb{Z} + \theta\mathbb{Z}$ (with deg θ Small) and $R_1 = \mathbb{Z}$

An algorithm to recover ϕ using only the scalar multiplications of E_1 and the image of ϕ on the N_2 torsion was described in Sect. 3.3. However this in combination with our basic strategy above does not a priori provide any speedup on the straighforward meet-in-the-middle approach. Indeed we have

$\deg \phi = N_1^2 \deg \theta \approx N_1^2$ in the most favorable case (when $\deg \theta = 1$) so by the analysis of Sect. 3.3 we expect to have at best $N_1' \approx \sqrt{D} \approx \sqrt{\deg \phi} \approx N_1$. We therefore modify the basic strategy.

Modified Strategy. We adapt the techniques of Sect. 3.3 to reduce Problem 3 to another instance of itself with smaller parameters $N_1' < N_1/2$ and N_2' some factor of N_2. After repeating this reduction step $O(\log N_1)$ times we end up with an instance of Problem 3 where N_1 is sufficiently small that it can be solved in polynomial time with a meet-in-the-middle approach.

Parameter Restriction. We will require that $\mathrm{End}(E_0)$ has some non scalar element θ of small degree (which does not need to be explicitly given, as it can then be computed efficiently by trying all isogenies of this degree). This is for example the case in Costello et al.'s implementation [4] where $j = 1728$. In our reduction we will also require $N_2/N_2' > 2N_1\Delta_\theta$ where $\Delta_\theta = \deg \theta - \frac{1}{4}\mathrm{Tr}^2 \theta$. This implies that we will need to start with parameters such that $\log N_2$ is at least $O(\log^2 N_1)$. Note that in the original De Feo-Jao-Plût protocols we had $N_1 \approx N_2$.

Reduction Step. We fix some $\theta \in \mathrm{End}(E_0)$ with small norm q, and let $D_\theta := \deg \theta - \frac{1}{4}\mathrm{Tr}^2 \theta$. Then we choose some factor $\tilde{N}_2$ of N_2 such that $\tilde{N}_2 > KN_1q$ for some $K > 1$, and $-D_\theta$ is a square modulo $\tilde{N}_2$. We proceed as in Sect. 3.3 to compute a, b and N_1' such that $\deg(a\phi_1\theta\hat{\phi}_1 + b) = N_1'\tilde{N}_2$ and N_1' is as small as possible. Namely, we choose τ such that $\tau^2 = -D_\theta \mod \tilde{N}_2$, then we compute a short vector in a two-dimensional lattice generated by two vectors $(\tilde{N}_2, 0)$ and $(\tau, 1)$ with a weighted norm $||(x, y)|| = (x^2 + D_\theta y^2)^{1/2}$, and we deduce a, b and N_1'. If $N_1' > N_1/2$ we start again with a new square root of $-D_\theta$ modulo $\tilde{N}_2$, or with a new $\tilde{N}_2$ value.

If $N_1' < N_1/2$ we define $\phi_{N_1'}, \phi_{\tilde{N}_2}$ two (still unknown) isogenies of degrees N_1' and $\tilde{N}_2$ such that $a\phi_1\theta\hat{\phi}_1 + b = \phi_{N_1'}\phi_{\tilde{N}_2}$. We evaluate $a\phi_1\theta\hat{\phi}_1 + b$ on the $\tilde{N}_2$ torsion to identify the $\tilde{N}_2$ part of the kernel of $a\phi_1\theta\hat{\phi}_1 + b$, then the corresponding isogeny. We evaluate this isogeny on the $N_2' = N_2/\tilde{N}_2$ torsion, and deduce the action of $\phi_{N_1'}$ on the N_2' torsion. We then apply the reduction step recursively to compute some representation of $\phi_{N_1'}$. Finally, we evaluate $(\phi_{N_1'}\phi_{\tilde{N}_2} - b)/a$ on the N_1 torsion to identify $\ker \hat{\phi}_1$, and from there we compute a more canonical expression for ϕ.

Complexity analysis. Our reduction procedure implicitly relies on the following informal assumption:

Assumption 1. *Let $K > 1$ be a "small" constant, and suppose that D_θ is "small". The probability that a "random" powersmooth value $\tilde{N}_2 > KN_1q$ leads to $N_1' < N_1/2$ is "large".*

Note that following the analysis of Lemma 2 we expect to find N_1' of size at most $N_1\sqrt{D_\theta}$. Assumption 1 tells that with some probability on the choice of $\tilde{N}_2$, we can find a value N_1' smaller than this bound by at least a (small) factor $2\sqrt{D_\theta}$. This assumption seems very plausible. Using lattice terminology, the expectation on N_1' comes from the well-known Gaussian heuristic, and Assumption 1 tells that the proportion of lattices with small deviations from this heuristic is significant. In continued fraction terminology, Assumption 1 considers the proportion of values $\tilde{N}_2$ such that some rational fraction approximation of τ/N_1 is a little bit better than what is guaranteed by the bounds, and tells that this proportion is significant. Finally, Assumption 1 receives further support from our experiments described below.

We deduce the following result:

Proposition 3. *Let N_1 and N_2 be coprime smooth numbers, with $\log N_2 = O(\log^2 N_1)$. Then under plausible heuristic assumptions Problem 3 can be solved in polynomial time when the initial curve E_0 has a small degree endomorphism.*

Proof: All subroutines in our reduction procedure require at most polynomial time, and under Assumption 1 these steps will only be executed a polynomial number of times.

Remark: Suppose $p = 3 \bmod 4$ and suppose $j_0 = 1728$ as in Costello et al.'s implementation [4]. In this case there exists a non scalar endomorphism $\iota \in \mathrm{End}(E_0)$ with norm 1 and trace 0. Any $\theta \in \mathrm{End}(E_0)$ must either have a large norm or be of the form $\theta = a\iota + b$ for two small $a, b \in \mathbb{Z}$. In the last case we then have $\Delta_\theta = a^2$, so $-\Delta_\theta$ is a square modulo some prime r if and only if -1 is a square modulo r. This implies that no prime factor r of N_2 with $r = 3 \bmod 4$ can be used in our attack. On the other hand, any prime factors with $r = 1 \bmod 4$ can be used in the attack.

Experiments for the optimal degree variant. We wrote a small Magma program [22] to compute the successive pairs of parameters (a, b) to use in our attack, and test the heuristic assumptions involved in our analysis (the code is available in the eprint version of this paper [14]). In our experiments we generate random p values, choose N_1 powersmooth and then search for a coprime $\tilde{N}_2 > 2qN_1$ leading to $N_1' < N_1/2q$. We repeat this recursively until N_1' is small enough (smaller than some polylog bound in p). We used $K = 2$ in these experiments. For 80-bit security parameters our program gives the parameters of an attack in a few seconds. The full attack requires isogenies of degree at most about 36000.

Experiments for the unbalanced variant. We also ran attack experiments for the unbalanced protocol variant. In all experiments we took $\ell_1 = 2$ and $\ell_2 = 5$. We considered values of e_1 between 20 and 100, and we searched for the minimal value of e_2 such that the attack could reduce N_1 to a value smaller than 100. Table 1 provides some successful attack parameters. In addition to e_1 and e_2 it shows the value $\left\lceil \frac{e_2 \log_2 5}{e_1^2} \right\rceil$ (which seems close to a constant $1/2$, as expected), the value K used for these parameters, and the number of reduction steps used. Our Magma code is provided in the eprint version of this paper [14].

Table 1. Some successful attack parameters against the unbalanced variant ($\ell_1 = 2$ and $\ell_2 = 5$)

e_1	e_2	$\left\lceil \frac{e_2 \log_2 5}{e_1^2} \right\rceil$	K	# steps
20	102	0.59	50	11
30	194	0.50	50	17
40	330	0.48	50	22
50	405	0.38	10	30
60	610	0.39	10	38
70	1047	0.50	2	61
80	1473	0.53	2	72
90	1775	0.51	2	80
100	2180	0.51	2	90

Remark: The parameter K must be larger than 1 in our attack as $N_1' > N_1^2 q / \tilde{N}_2$ for any $a \neq 0$. We experimentally observed that $K = 2$ was sufficient in the optimal degree variant to make N_1 decrease by a factor 2 at each reduction step. The unbalanced variant leaves less flexibility in the parameter choice, so we did not impose a factor 2 decrease on N_1 (and in fact we even allowed it to increase in some reduction steps). We observed that lower values of K were then sufficient. We have also observed experimentally that the value of K has some moderate impact on the overall performances of the attack (required size for N_2, number of reduction steps). We leave a thorough investigation of optimal parameter choices for our attack to further work.

Remark: When N_2 is too small to execute $O(\log N_1)$ reduction steps, then we may replace the missing last reduction steps by a final meet-in-the-middle strategy. Depending on the final size of N_1' and on its largest prime factor this may still provide some exponential speedup over the basic meet-in-the-middle strategy. We note, however, that for the original parameters proposed by Jao and De Feo, at most one recursive step can be performed. In this case it might be possible to find some (exceptional) set of parameters that would improve the best attack by a few bits, but for most parameters we do not expect any savings.

Possible Extensions. One can vary R_0 and R_1 depending on the attack model, or consider variants of Problem 3 involving several isogenies, and derive similar attacks. We leave details to the reader and further work.

5 Impact and Perspectives

The techniques developed in this paper solve some isogeny problems using the images of certain torsion points by the isogenies. Such images are revealed in De Feo-Jao supersingular key exchange protocol as well as the public key encryption and signature scheme that derive from it (see [7,24] and the first signature scheme of [9]). Until now all existing attacks against these protocols made no use at all of this auxiliary information.

At the moment our techniques do not apply to the parameters originally proposed in these protocols. However they apply on some natural variants of them, and they issue a warning that the auxiliary information might weaken isogeny problems. One could also fear that further developments of our techniques and particular attack models will be able to threaten the original protocol itself.

In anticipation of potential future improvements of our attacks, we recommend to avoid the use of special E_0 in the protocols, as any (partial) knowledge of the endomorphism ring of E_0 may a priori be useful to the attacker with our techniques. We stress, however, that the only known algorithm to avoid special curves for E_0 consists in generating a special curve and then performing a random walk from there to obtain a truly random curve; depending on the context this procedure might still allow some form of backdoor attack. An algorithm that could generate a random supersingular j-invariant without performing a random walk from a curve with known endomorphism ring would be a handy tool for designing cryptosystems based on supersingular isogeny problems. Of course, the algorithm may come with additional insight on the underlying Mathematics, which might also help further cryptanalysis. We would like to encourage research in this direction.

We note that the hash function proposed by Charles-Goren-Lauter [2] can also be attacked when starting from a curve with known endomorphism ring. There is also a corresponding "backdoor collision attack"; however the attack is less powerful than above as it can be detected and any use of the backdoor will leak it. We refer to [15] for details of this attack.

The second signature scheme of [9] relies on the endomorphism ring computation problem for random curves, with no extra information leaked, and is not affected by our techniques. In contrast to the isogeny problem variants considered in this paper, we are not aware of any cryptanalysis result that affects the endomorphism ring computation problem, and we believe that cryptosystems based on this problem offer the strongest security guarantees in the area of isogeny-based cryptography. Of course, cryptanalysis research in this direction is also fairly scarce despite some early work by Kohel [12], and more cryptanalysis will be needed to gain confidence on their security.

Acknowledgments. We thank Bryan Birch, Jonathan Bootle, Luca De Feo, Steven Galbraith, Chloe Martindale, Lorenz Panny and Yan Bo Ti, as well as the anonymous reviewers of the Asiacrypt 2017 conference for their useful comments on preliminary versions of this paper. This work was developed while the author was at the Mathematical Institute of the University of Oxford, funded by a research grant from the UK government.

References

1. Ankeny, N.C.: The least quadratic non residue. Ann. Math. **55**(1), 65–72 (1952)
2. Charles, D.X., Lauter, K.E., Goren, E.Z.: Cryptographic hash functions from expander graphs. J. Cryptol. **22**(1), 93–113 (2009)
3. Coggia, D.: Implémentation d'une variante du protocole de key-exchange SIDH (2017). https://github.com/dnlcog/sidh_variant
4. Costello, C., Longa, P., Naehrig, M.: Efficient algorithms for supersingular isogeny Diffie-Hellman. In: Robshaw, M., Katz, J. (eds.) CRYPTO 2016. LNCS, vol. 9814, pp. 572–601. Springer, Heidelberg (2016). https://doi.org/10.1007/978-3-662-53018-4_21
5. Cremona, J.E., Rusin, D.: Efficient solution of rational conics. Math. Comput. **72**(243), 1417–1441 (2003)
6. Delfs, C., Galbraith, S.D.: Computing isogenies between supersingular elliptic curves over $\mathbb{F}_p$. Des. Codes Crypt. **78**(2), 425–440 (2016)
7. Feo, L.D., Jao, D., Plût, J.: Towards quantum-resistant cryptosystems from supersingular elliptic curve isogenies. J. Math. Cryptol. **8**(3), 209–247 (2014)
8. Galbraith, S.D., Petit, C., Shani, B., Ti, Y.B.: On the security of supersingular isogeny cryptosystems. In: Cheon, J.H., Takagi, T. (eds.) ASIACRYPT 2016. LNCS, vol. 10031, pp. 63–91. Springer, Heidelberg (2016). https://doi.org/10.1007/978-3-662-53887-6_3
9. Galbraith, S.D., Petit, C., Silva, J.: Identification protocols and signature schemes based on supersingular isogeny problems. In: Takagi, T., Peyrin, T. (eds.) ASIACRYPT 2017, Part I. LNCS, vol. 10624, pp. 3–33. Springer, Cham (2017). https://doi.org/10.1007/978-3-319-70694-8_1
10. Gélin, A., Wesolowski, B.: Loop-abort faults on supersingular isogeny cryptosystems. In: Lange, T., Takagi, T. (eds.) PQCrypto 2017. LNCS, vol. 10346, pp. 93–106. Springer, Cham (2017). https://doi.org/10.1007/978-3-319-59879-6_6
11. Jao, D., De Feo, L.: Towards quantum-resistant cryptosystems from supersingular elliptic curve isogenies. In: Yang, B.-Y. (ed.) PQCrypto 2011. LNCS, vol. 7071, pp. 19–34. Springer, Heidelberg (2011). https://doi.org/10.1007/978-3-642-25405-5_2
12. Kohel, D.: Endomorphism rings of elliptic curves over finite fields. PhD thesis, University of California, Berkeley (1996)
13. Kohel, D., Lauter, K., Petit, C., Tignol, J.-P.: On the quaternion ℓ-isogeny path problem. LMS J. Comput. Math. **17A**, 418–432 (2014)
14. Petit, C.: Faster algorithms for isogeny problems using torsion point images. IACR Cryptology ePrint Archive, 2017:571 (2017)
15. Petit, C., Lauter, K.: Hard and easy problems in supersingular isogeny graphs (2017)
16. Canfield, R., Erdös, P., Pomerance, C.: On a problem of Oppenheim concerning "factorisatio numerorum". J. Number Theory **17**, 1–28 (1983)
17. Rostovtsev, A., Stolbunov, A.: Public-key cryptosystem based on isogenies. Cryptology ePrint Archive, Report 2006/145 (2006). http://eprint.iacr.org/
18. Silverman, J.: The Arithmetic of Elliptic Curves. Springer Verlag, New York (1986)
19. Simon, D.: Quadratic equations in dimensions 4, 5 and more. Preprint (2005). http://www.math.unicaen.fr/~simon/
20. Ti, Y.B.: Fault attack on supersingular isogeny cryptosystems. In: Lange, T., Takagi, T. (eds.) PQCrypto 2017. LNCS, vol. 10346, pp. 107–122. Springer, Cham (2017). https://doi.org/10.1007/978-3-319-59879-6_7

21. Vignéras, M.-F.: Arithmétique des Algèbres de Quaternions. Springer, Heidelberg (2006). https://doi.org/10.1007/BFb0091027
22. Fieker, C., Steel, A., Bosma, W., Cannon, J.J. (eds.): Handbook of Magma functions, edition 2.20 (2013). http://magma.maths.usyd.edu.au/magma/
23. Xi, S., Tian, H., Wang, Y.: Toward quantum-resistant strong designated verifier signature from isogenies. Int. J. Grid Util. Comput. 5(2), 292–296 (2012)
24. Yoo, Y., Azarderakhsh, R., Jalali, A., Jao, D., Soukharev, V.: A post-quantum digital signature scheme based on supersingular isogenies. Financial Crypto (2017)

Block Chains

Beyond Hellman's Time-Memory Trade-Offs with Applications to Proofs of Space

Hamza Abusalah[1]($\boxtimes$), Joël Alwen[1], Bram Cohen[2], Danylo Khilko[3], Krzysztof Pietrzak[1], and Leonid Reyzin[4]

[1] Institute of Science and Technology Austria, Klosterneuburg, Austria
{habusalah,jalwen,pietrzak}@ist.ac.at
[2] Chia Network, San Francisco, CA, USA
bram@chia.network
[3] ENS Paris, Paris, France
dkhilko@ukr.net
[4] Boston University, Boston, USA
reyzin@cs.bu.edu

Abstract. Proofs of space (PoS) were suggested as more ecological and economical alternative to proofs of work, which are currently used in blockchain designs like Bitcoin. The existing PoS are based on rather sophisticated graph pebbling lower bounds. Much simpler and in several aspects more efficient schemes based on inverting random functions have been suggested, but they don't give meaningful security guarantees due to existing time-memory trade-offs.

In particular, Hellman showed that any *permutation* over a domain of size N can be inverted in time T by an algorithm that is given S bits of auxiliary information whenever $S \cdot T \approx N$ (e.g. $S = T \approx N^{1/2}$). For *functions* Hellman gives a weaker attack with $S^2 \cdot T \approx N^2$ (e.g., $S = T \approx N^{2/3}$). To prove lower bounds, one considers an adversary who has access to an oracle $f : [N] \to [N]$ and can make T oracle queries. The best known lower bound is $S \cdot T \in \Omega(N)$ and holds for random functions and permutations.

We construct functions that provably require more time and/or space to invert. Specifically, for any constant k we construct a function $[N] \to [N]$ that cannot be inverted unless $S^k \cdot T \in \Omega(N^k)$ (in particular, $S = T \approx N^{k/(k+1)}$). Our construction does not contradict Hellman's time-memory trade-off, because it cannot be efficiently evaluated in forward direction. However, its entire function table can be computed in time quasilinear in N, which is sufficient for the PoS application.

Our simplest construction is built from a random function oracle $g : [N] \times [N] \to [N]$ and a random permutation oracle $f : [N] \to [N]$ and is defined as $h(x) = g(x, x')$ where $f(x) = \pi(f(x'))$ with π being any involution without a fixed point, e.g. flipping all the bits. For this function we prove that any adversary who gets S bits of auxiliary information, makes at most T oracle queries, and inverts h on an ϵ fraction of outputs must satisfy $S^2 \cdot T \in \Omega(\epsilon^2 N^2)$.

T. Takagi and T. Peyrin (Eds.): ASIACRYPT 2017, Part II, LNCS 10625, pp. 357–379, 2017.
https://doi.org/10.1007/978-3-319-70697-9_13

1   Introduction

A proof of work (PoW), introduced by Dwork and Naor [DN93], is a proof system in which a prover $\mathcal{P}$ convinces a verifier $\mathcal{V}$ that he spent some computation with respect to some statement x. A simple PoW can be constructed from a function $H(\cdot)$, where a proof with respect to a statement x is simply a salt s such that $H(s, x)$ starts with t 0's. If H is modelled as a random function, $\mathcal{P}$ must evaluate H on 2^t values (in expectation) before he finds such an s.

The original motivation for PoWs was prevention of email spam and denial of service attacks, but today the by far most important application for PoWs is securing blockchains, most prominently the Bitcoin blockchain, whose security is based on the assumption that the majority of computing power dedicated towards the blockchain comes from honest users. This results in a massive waste of energy and other resources, as this mining is mostly done on dedicated hardware (ASICs) which has no other use than Bitcoin mining. In [DFKP15] proofs of space (PoS) have been suggested as an alternative to PoW. The idea is to use disk space rather than computation as the main resource for mining. As millions of users have a significant amount of unused disk space available (on laptops etc.), dedicating this space towards securing a blockchain would result in almost no waste of resources.

Let $[N]$ denote some domain of size N. For convenience we'll often assume that $N = 2^n$ is a power of 2 and identify $[N]$ with $\{0,1\}^n$, but this is never crucial and $[N]$ can be any other efficiently samplable domain. A simple idea for constructing a PoS is to have the verifier specify a random function $f : [N] \rightarrow [N]$ during the initialization phase, and have the prover compute the function table of f and sort it by the output values.[1] Then, during the proof phase, to convince the verifier that he really stores this table, the prover must invert f on a random challenge. We will call this approach "simple PoS"; we will discuss it in more detail in Sect. 1.3 below, and explain why it miserably fails to provide any meaningful security guarantees.

Instead, existing PoS [DFKP15, RD16] are based on pebbling lower bounds for graphs. These PoS provide basically the best security guarantees one could hope for: a cheating prover needs $\Theta(N)$ space or time after the challenge is known to make a verifier accept. Unfortunately, compared to the (insecure) simple PoS, they have two drawbacks which make them more difficult to use as replacement for PoW in blockchains. First, the proof size is quite large (several MB instead of a few bytes as in the simple PoS or Bitcoin's PoW). Second, the initialization phase requires two messages: the first message, like in the simple PoS, is sent from the verifier to the prover specifying a random function f, and second message, unlike in the simple PoS, is a "commitment" from the prover to the verifier.[2]

---

[1] f must have a short description, so it cannot be actually random. In practice the prover would specify f by, for example, a short random salt s for a cryptographic hash function H, and set $f(x) = H(s, x)$.

[2] Specifically, the prover computes a "graph labelling" of the vertices of a graph (which is specified by the PoS) using f, and then a Merkle tree commitment to this entire labelling, which must be sent back to the verifier.

If such a pebbling-based PoS is used as a replacement for PoW in a blockchain design, the first message can be chosen non-interactively by the miner (who plays the role of the prover), but the commitment sent in the second message is more tricky. In Spacemint (a PoS-based decentralized cryptocurrency [PPK+15]), this is solved by having a miner put this commitment into the blockchain itself before he can start mining. As a consequence, Spacemint lacks the nice property of the Bitcoin blockchain where miners can join the effort by just listening to the network, and only need to speak up once they find a proof and want to add it to the chain.

1.1 Our Results

In this work we "resurrect" the simple approach towards constructing PoS based on inverting random functions. This seems impossible, as Hellman's time-memory trade-offs — which are the reason for this approach to fail — can be generalized to apply to all functions (see Sect. 1.4). For Hellman's attacks to apply, one needs to be able to evaluate the function efficiently in forward direction. At first glance, this may not seem like a real restriction at all, as inverting functions which cannot be efficiently computed in forward direction is undecidable in general.[3] However, we observe that for functions to be used in the simple PoS outlined above, the requirement of efficient computability can be relaxed in a meaningful way: we only need to be able to compute the entire function table in time linear (or quasilinear) in the size of the input domain. We construct functions satisfying this relaxed condition for which we prove lower bounds on time-memory trade-offs beyond the upper bounds given by Hellman's attacks.

Our most basic construction of such a function $g_f : [N] \to [N]$ is based on a function $g : [N] \times [N] \to [N]$ and a permutation $f : [N] \to [N]$. For the lower bound proof g and f are modelled as truly random, and all parties access them as oracles. The function is now defined as $g_f(x) = g(x, x')$ where $f(x) = \pi(f(x'))$ for any involution π without fixed points. For concreteness we let π simply flip all bits, denoted $f(x) = \overline{f(x')}$. Let us stress that f does not need to be a permutation – it can also be a random function[4] – but we'll state and prove our main result for a permutation as it makes the analysis cleaner. In practice — where one has to instantiate f and g with something efficient — one would rather use a function, because it can be instantiated with a cryptographic hash function like

---

[3] Consider the function $f : \mathbb{N} \times \{0,1\}^* \to \{0,1\} \times \{0,1\}^*$ where $f(s, T) = (b, T)$ with $b = 1$ iff the Turing machine T stops in s steps. Here deciding if $(1, T)$ has a pre-image at all requires to solve the halting problem.

[4] If f is a function, we don't need π; the condition $f(x) = \pi(f(x'))$ can be replaced with simply $f(x) = f(x'), x \neq x'$. Note than now for some x there's no output $g_f(x)$ at all (i.e., if $\forall x' \neq x \; f(x) \neq f(x')$), and for some x there's more than one possible value for $g_f(x)$. This is a bit unnatural, but such a g_f can be used for a PoS in the same way as if f were a permutation.

SHA-3 or (truncated) AES,[5] whereas we don't have good candidates for suitable permutations (at the very least f needs to be one-way; and, unfortunately, all candidates we have for one-way permutations are number-theoretic and thus much less efficient).

In Theorem 2 we state that for g_f as above, any algorithm which has a state of size S (that can arbitrarily depend on g and f), and inverts g_f on an ϵ fraction of outputs, must satisfy $S^2T \in \Omega(\epsilon^2 N^2)$. This must be compared with the best lower bound known for inverting random functions (or permutations) which is $ST = \Omega(\epsilon N)$. We can further push the lower bound to $S^kT \in \Omega(\epsilon^k N^k)$ by "nesting" the construction; in the first iteration of this nesting one replaces the inner function f with g_f.[6] These lower bounds are illustrated in Fig. 1.

In this paper we won't give a proof for the general construction, as the proof for the general construction doesn't require any new ideas, but just gets more technical. We also expect the basic construction to be already sufficient for constructing a secure PoS. Although for g_f there exists a time-memory trade-off $S^4T \in O(N^4)$ (say, $S = T \approx N^{4/5}$), which is achieved by "nesting" Hellman's attack,[7] we expect this attack to only be of concern for extremely large N.[8]

A caveat of our lower bound is that it only applies if $T \le N^{2/3}$. We don't see how to break our lower bound if $T > N^{2/3}$, and the restriction $T \le N^{2/3}$ seems to be mostly related to the proof technique. One can improve the bound to $T \le N^{t/(t+1)}$ for any t by generalizing our construction to t-wise collisions. One way to do this – if f is a permutation and t divides N – is as follows: let $g : [N]^t \to [N]$ and define $g_f(x) = g(x, x_1, \ldots, x_{t-1})$ where for some partition $S_1, \ldots, S_{N/t}, |S_i| = t$ of $[N]$ the values $f(x), f(x_1), \ldots, f(x_{t-1})$ contain all elements of a partition S_i and $x_1 < x_2 < \ldots < x_{t-1}$.

---

[5] As a concrete proposal, let $\mathsf{AES}_n : \{0,1\}^{128} \times \{0,1\}^{128} \to \{0,1\}^n$ denote AES with the output truncated to n bits. We can now define f, g by a random key $k \leftarrow \{0,1\}^{128}$ as $f(x) = \mathsf{AES}_n(k, 0\|x\|0^{128-n-1})$ and $g(x) = \mathsf{AES}_n(k, 1\|x\|0^{128-2n-1})$. As in practice n will be something like $30 - 50$, which corresponds to space (which is $\approx n \cdot 2^n$ bits) in the Gigabyte to Petabyte range. Using AES with the smallest 128 bit blocksize is sufficient as $2n \ll 128$.

[6] The dream version would be a result showing that one needs either $S = \Omega(N)$ or $T = \Omega(N)$ to invert. Our results approach this as k grows showing that $S = T = \Omega(N^{k/(k+1)})$ is required.

[7] Informally, nesting Hellman's attack works as follows. Note that if we could efficiently evaluate $g_f(.)$, we could use Hellman's attack. Now to evaluate g_f we need to invert f. For this make a Hellman table to invert f, and use this to "semi-efficiently" evaluate $g_f(.)$. More generally, for our construction with nesting parameter k (when the lower bound is $S^kT \in \Omega(N^k)$) the nested Hellman attack applies if $S^{2k}T \in O(N^{2k})$.

[8] The reason is that for this nested attack to work, we need tables which allow to invert with very high probability, and in this case the tables will store many redundant values. So the hidden constant in the $S^4T \in O(N^4)$ bound of the nested attack will be substantial.

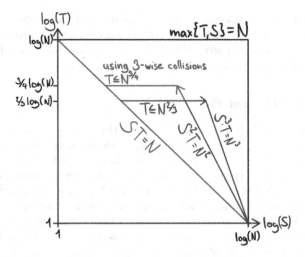

Fig. 1. Illustration of lower bounds. Orange : the $ST = \Omega(N)$ lower bound for inverting random permutations or functions. Dark green : the ideal bound where either T or S is $\Omega(N)$ as achieved by the pebbling-based PoS [DFKP15, RD16] (more precisely, the bound approaches the dark green line for large N). Light green : the lower bound $S^2T = \Omega(N^2)$ for $T \leq N^{2/3}$ for our most basic construction as stated in Theorem 2. Pink : the restriction $T \leq N^{2/3}$ on T we need for our proof to go through can be relaxed to $T \leq N^{t/(t+1)}$ by using t-wise collisions instead of pairwise collisions in our construction. The pink arrow shows how the bound improves by going from $t = 2$ to $t = 3$. Purple : we can push the $S^2T = \Omega(N^2)$ lower bound of the basic construction to $S^kT = \Omega(N^k)$ by using $k - 1$ levels of nesting. The purple arrow shows how the bound improves by going from $k = 2$ to $k = 3$. (Color figure online)

1.2 Proofs of Space

A proof of space as defined in [DFKP15] is a two-phase protocol between a prover $\mathcal{P}$ and a verifier $\mathcal{V}$, where after an initial phase $\mathcal{P}$ holds a file F of size N,[9] whereas $\mathcal{V}$ only needs to store some small value. The running time of $\mathcal{P}$ during this phase must be at least N as $\mathcal{P}$ has to write down F which is of size N, and we require that it's not much more, quasilinear in N at most. $\mathcal{V}$ on the other hand must be very efficient, in particular, its running time can be polynomial in a security parameter, but must be basically independent of N.

Then there's a proof execution phase — which typically will be executed many times over a period of time — in which $\mathcal{V}$ challenges $\mathcal{P}$ to prove it stored F. The security requirement states that a cheating prover $\tilde{\mathcal{P}}$ who only stores a file F' of size significantly smaller than N either fails to make $\mathcal{V}$ accept, or

---

[9] We use the same letter N for the space committed by an honest prover in a PoS as we used for the domain size of the functions discussed in the previous section to emphasize that in our construction of a PoS from a function these will be roughly the same. We'll come back to this in Remark 2 in the next section.

must invest a significant amount of computation, ideally close to $\mathcal{P}$'s cost during initialization. Note that we cannot hope to make it more expensive than that as a cheating $\tilde{\mathcal{P}}$ can always just store the short communication during initialization, and then reconstruct all of F before the execution phase.

1.3 A Simple PoS that Fails

Probably the first candidate for a PoS scheme that comes to mind is to have — during the initalization phase — $\mathcal{V}$ send the (short) description of a "random behaving" function $f : [N] \rightarrow [N]$ to $\mathcal{P}$, who then computes the entire function table of f and stores it sorted by the outputs. During proof execution $\mathcal{V}$ will pick a random $x \in [N]$, and then challenge $\mathcal{P}$ to invert f on $y = f(x)$.[10]

An honest prover can answer any challenge y by looking up an entry (x', y) in the table, which is efficient as the table is sorted by the y's. At first one might hope this provides good security against any cheating prover; intuitively, a prover who only stores $\ll N \log N$ bits (i.e., uses space sufficient to only store $\ll N$ output labels of length $\log N$) will not have stored a value $x \in f^{-1}(y)$ for most y's, and thus must invert by brute force which will require $\Theta(N)$ invocations to f. Unfortunately, even if f is modelled as a truly random function, this intuition is totally wrong due to Hellman's time-memory trade-offs, which we'll discuss in the next section.

The goal of this work is to save this elegant and simple approach towards constructing PoS. As discussed before, for our function $g_f : [N] \rightarrow [N]$ (defined as $g_f(x) = g(x, x')$ where $f(x) = \overline{f(x')}$) we can prove better lower bounds than for random functions. Instantiating the simple PoS with g_f needs some minor adaptions. $\mathcal{V}$ will send the description of a function $g : [N] \times [N] \rightarrow [N]$ and a permutation $f : [N] \rightarrow [N]$ to $\mathcal{P}$. Now $\mathcal{P}$ first computes the entire function table of f and sorts it by the output values. Note that with this table $\mathcal{P}$ can efficiently invert f. Then $\mathcal{P}$ computes (and sorts) the function table of g_f (using that $g_f(x) = g(x, f^{-1}(\overline{f(x)}))$). Another issue is that in the execution phase $\mathcal{V}$ can no longer compute a challenge as before – i.e. $y = g_f(x)$ for a random x – as it cannot evaluate g_f. Instead, we let $\mathcal{V}$ just pick a random $y \in [N]$. The prover $\mathcal{P}$ must answer this challenge with a tuple (x, x') s.t. $f(x) = \overline{f(x')}$ and $g(x, x') = y$ (i.e., $g_f(x) = y$). Just sending the preimage x of g_f for y is no longer sufficient, as $\mathcal{V}$ is not able to verify if $g_f(x) = y$ without x'.

Remark 1 (Completeness and Soundness Error). This protocol has a significant soundness and completeness error. On the one hand, a cheating prover $\tilde{\mathcal{P}}$ who only stores, say 10%, of the function table, will still be able to make $\mathcal{V}$ accept in 10% of the cases. On the other hand, even if g_f behaves like a random function,

---

[10] Instead of storing all N tuples $(x, f(x))$ (sorted by the 2nd entry), which takes $2N \log N$ bits, one can compress this list by almost a factor 2 using the fact that the 2nd entry is sorted, and another factor $\approx 1 - 1/e \approx 0.632$ by keeping only one entry whenever there are multiple tuples with the same 2nd entry, thus requiring $\approx 0.632N \log N$ bits.

an honest prover $\mathcal{P}$ will only be able to answer a $1 - 1/e$ fraction ($\approx 63\%$) of the challenges $y \in [N]$, as some will simply not have a preimage under g_f.[11]

When used as a replacement for PoW in cryptocurrencies, neither the soundness nor the completeness error are an issue. If this PoS is to be used in a context where one needs negligible soundness and/or completeness, one can use standard repetition tricks to amplify the soundness and completeness, and make the corresponding errors negligible.[12]

Remark 2 (Domain vs. Space). When constructing a PoS from a function with a domain of size N, the space the honest prover requires is around $N \log N$ bits for the simple PoS outlined above (where we store the sorted function table of a function $f : [N] \to [N]$), and roughly twice that for our basic construction (where we store the function tables of $g_f : [N] \to [N]$ and $f : [N] \to [N]$). Thus, for a given amount N' of space the prover wants to commit to, it must use a function with domain $N \approx N'/\log(N')$. In particular, the time-memory trade-offs we can prove on the hardness of inverting the underlying function translate directly to the security of the PoS.

1.4   Hellman's Time-Memory Trade Offs

Hellman [Hel80] showed that any permutation $p : [N] \to [N]$ can be inverted using an algorithm that is given S bits of auxiliary information on p and makes at most T oracle queries to $p(\cdot)$, where ($\tilde{O}$ below hides $\log(N)^{O(1)}$ factors)

$$S \cdot T \in \tilde{O}(N) \quad \text{e.g. when} \quad S = T \approx N^{1/2} . \tag{1}$$

Hellman also presents attacks against *random* functions, but with worse parameters. A rigorous bound was only later proven by Fiat and Naor [FN91] where they show that Hellman's attack on *random* functions satisfies

$$S^2 \cdot T \in \tilde{O}(N^2) \quad \text{e.g. when} \quad S = T \approx N^{2/3} . \tag{2}$$

Fiat and Naor [FN91] also present an attack with worse parameters which works for *any* (not necessarily random) function, where

$$S^3 \cdot T \in \tilde{O}(N^3) \quad \text{e.g. when} \quad S = T \approx N^{3/4} . \tag{3}$$

The attack on a permutation $p : [N] \to [N]$ for a given T is easy to explain: Pick any $x \in [N]$ and define $x_0, x_1, \ldots$ as $x_0 = x$, $x_{i+1} = p(x_i)$, let $\ell \leq N - 1$

---

[11] Throwing N balls in N bins at random will leave around N/e bins empty, so g_f's outputs will miss $N/e \approx 0.37 \cdot N$ values in $[N]$.

[12] To decrease the soundness error from 0.37 to negligible, the verifier can ask the prover to invert g_f on $t \in \mathbb{N}$ independent random challenges in $[N]$. In expectation g_f will have a preimage on $0.63 \cdot t$ challenges. The probability that – say at least $0.5 \cdot t$ – of the challenges have a preimage is then exponentially (in t) close to 1 by the Chernoff bound. So if we only require the prover to invert half the challenges, the soundness error becomes negligible.

be minimal such that $x_0 = x_\ell$. Now store the values $x_T, x_{2T}, \ldots, x_{(\ell \mod T)T}$ in a sorted list. Let us assume for simplicity that $\ell - 1 = N$, so $x_0, \ldots, x_{\ell-1}$ cover the entire domain (if this is not the case, one picks some x' not yet covered and makes a new table for the values $x_0 = x'$, $x_1 = p(x_0), \ldots$). This requires storing $S = N/T$ values. If we have this table, given a challenge y to invert, we just apply p to y until we hit some stored value x_{iT}, then continue applying p to $x_{(i-1)T}$ until we hit y, at which point we found the inverse $p^{-1}(y)$. By construction this attack requires T invocations to p. The attack on general functions is more complicated and gives worse bounds as we don't have such a nice cycle structure. In a nutshell, one computes several different chains, where for the jth chain we pick some random $h_j : [N] \to [N]$ and compute $x_0, x_1, \ldots, x_n$ as $x_i = f(h_j(x_{i-1}))$. Then, every T'th value of the chain is stored. To invert a challenge y we apply $f(h_1(\cdot))$ sequentially on input y up to T times. If we hit a value x_{iT} we stored in the first chain, we try to invert by applying $f(h_1(\cdot))$ starting with $x_{(i-1)T}$.[13] If we don't succeed, continue with the chains generated by $f(h_2(\cdot))$, $f(h_3(\cdot)), \ldots$ until the inverse is found or all chains are used up. This attack will be successful with high probability if the chains cover a large fraction of f's output domain.

1.5   Samplability is Sufficient for Hellman's Attack

One reason the lower bound for our function $g_f : [N] \to [N]$ (defined as $g_f(x) = g(x, x')$ where $f(x) = \overline{f(x')}$) does not contradict Hellman's attacks is the fact that g_f cannot be efficiently evaluated in forward direction. One can think of simpler constructions such as $g'_f(x) = g(x, f^{-1}(x))$ which also have this property, but observe that Hellman's attack is easily adapted to break g'_f. More generally, Hellman's attack doesn't require that the function can be efficiently computed in forward direction, it is sufficient to have an algorithm that efficiently samples random input/output tuples of the function. This is possible for g'_f as for a random z the tuple $f(z), g(f(z), z)$ is a valid input/output: $g'_f(f(z)) = g(f(z), f^{-1}(f(z))) = g(f(z), z)$. To adapt Hellman's attack to this setting – where we just have an efficient input/output sampler σ_f for f – replace the $f(h_i(\cdot))$'s in the attack described in the previous section with $\sigma_f(h_i(\cdot))$.

1.6   Lower Bounds

De, Trevisan and Tulsiani [DTT10] (building on work by Yao [Yao90], Gennaro-Trevisan [GT00] and Wee [Wee05]) prove a lower bound for inverting random permutations, and in particular show that Hellman's attack as stated in Eq. (1) is optimal: For any oracle-aided algorithm $\mathcal{A}$, it holds that for most permutations $p : [N] \to [N]$, if $\mathcal{A}$ is given advice (that can arbitrarily depend on p) of size S, makes at most T oracle queries and inverts p on ϵN values, we have $S \cdot T \in \Omega(\epsilon N)$. Their lower bound proof can easily be adapted to random functions

---

[13] Unlike for permutations, there's no guarantee we'll be successful, as the challenge might lie on a branch of the function graph different from the one that includes $x_{(i-1)T}$.

$f : [N] \rightarrow [N]$, but note that in this case it is no longer tight, i.e., matching Eq. (2). Barkan, Biham, and Shamir [BBS06] show a matching $S^2 \cdot T \in \tilde{\Omega}(N^2)$ lower bound for a restricted class of algorithms.

1.7   Proof Outline

The starting point of our proof is the $S \cdot T \in \Omega(\epsilon N)$ lower bound for inverting random permutations by De, Trevisan and Tulsiani [DTT10] mentioned in the previous section. We sketch their simple and elegant proof, with a minor adaption to work for functions rather than permutations in Appendix A.

The high level idea of their lower bound proof is as follows: Assume an adversary $\mathcal{A}$ exists, which is given an auxiliary string aux, makes at most T oracle queries and can invert a random permutation $p : [N] \rightarrow [N]$ on an ϵ fraction of $[N]$ with high probability (aux can depend arbitrarily on p). One then shows that given (black box access to) $\mathcal{A}_{\text{aux}}(\cdot) \stackrel{\text{def}}{=} \mathcal{A}(\text{aux}, \cdot)$ it's possible to "compress" the description of p from $\log(N!)$ to $\log(N!) - \Delta$ bits for some $\Delta > 0$. As a random permutation is incompressible (formally stated as Fact 1 in Sect. 2 below), the Δ bits we saved must come from the auxiliary string given, so $S = |\text{aux}| \gtrsim \Delta$.

To compress p, one now finds a subset $G \subset [N]$ where (1) $\mathcal{A}$ inverts successfully, i.e., for all $y \in p(G) = \{p(x) \ : \ x \in G\}$ we have $\mathcal{A}_{\text{aux}}^p(y) = p^{-1}(y)$ and (2) $\mathcal{A}$ never makes a query in G, i.e., for all $y \in G$ all oracle queries made by $\mathcal{A}_{\text{aux}}^p(y)$ are in $[N] - G$ (except for the last query which we always assume is $p^{-1}(y)$).

The compression now exploits the fact that one can learn the mapping $G \rightarrow p(G)$ given aux, an encoding of the set $p(G)$, and the remaining mapping $[N] - G \rightarrow p([N] - G)$. While decoding, one recovers $G \rightarrow p(G)$ by invoking $\mathcal{A}_{\text{aux}}^p(\cdot)$ on all values $p(G)$ (answering all oracle queries using $[N] - G \rightarrow p([N] - G)$, the first query outside $[N] - G$ will be the right value by construction).

Thus, we compressed by not encoding the mapping $G \rightarrow p(G)$, which will save us $|G| \log(N)$ bits, however we have to pay an extra $|G| \log(eN/|G|)$ bits to encode the set $p(G)$, so overall we compressed by $|G| \log(|G|/e)$ bits, and therefore $S \geq |G|$ assuming $|G| \geq 2e$. Thus the question is how large a set G can we choose. A simple probabilistic argument, basically picking values at random until it's no longer possible to extend G, shows that we can always pick a G of size at least $|G| \geq \epsilon N/T$, and we conclude $S \geq \epsilon N/T$ assuming $T \leq \epsilon N/2e$.

In the De et al. proof, the size of the good set G will always be close to $\epsilon N/T$, no matter how $\mathcal{A}_{\text{aux}}$ actually behaves. In this paper we give a more fine grained analysis introducing a new parameter T_g as discussed next.

The T_g parameter. Informally, our compression algorithm for a function $g : [N] \rightarrow [N]$ goes as follows: Define the set $I = \{x \ : \ \mathcal{A}_{\text{aux}}^g(g(x)) = x\}$ of values where $\mathcal{A}_{\text{aux}}^g$ inverts $g(I)$, by assumption $|I| = \epsilon N$. Now we can add values from I to G as long as possible, every time we add a value x, we "spoil" up to T values in I, where we say x' gets spoiled if $\mathcal{A}_{\text{aux}}^g(g(x))$ makes oracle query x', and thus we will not be able to add x' to G in the future. As we start with $|I| = \epsilon N$, and spoil at most T values for every value added to G, we can add at least $\epsilon N/T$ values to G.

This is a worst case analysis assuming $\mathcal{A}_{\mathsf{aux}}^g$ really spoils close to T values every time we add a value to G, but potentially $\mathcal{A}_{\mathsf{aux}}^g$ behaves nicer and on average spoils less. In the proof of Lemma 1 we take advantage of this and extend G as long as possible, ending up with a good set G of size at least $\epsilon N/2T_g$ for some $1 \leq T_g \leq T$. Here T_g is the average number of elements we spoiled for every element added to G.

This doesn't help to improve the De et al. lower bound, as in general T_g can be as large as T in which case our lower bound $S \cdot T_g \in \Omega(\epsilon N)$ coincides with the De et al. $S \cdot T \in \Omega(\epsilon N)$ lower bound.[14] But this more fine grained bound will be a crucial tool to prove the lower bound for g_f.

Lower Bound for g_f. We now outline the proof idea for our lower bound $S^2 \cdot T \in \Omega(\epsilon^2 N^2)$ for inverting $g_f(x) = g(x, x')$, $f(x) = \overline{f(x')}$ assuming $g : [N] \times [N] \rightarrow [N]$ is a random function and $f : [N] \rightarrow [N]$ is a random permutation. We assume an adversary $\mathcal{A}_{\mathsf{aux}}^{g,f}$ exists which has oracle access to f, g and inverts $g_f : [N] \rightarrow [N]$ on a set $J = \{y : g_f(\mathcal{A}_{\mathsf{aux}}^{f,g}(y)) = y\}$ of size $J = \lfloor \epsilon N \rfloor$.

If the function table of f is given, $g_f : [N] \rightarrow [N]$ is a random function that can be efficiently evaluated, and we can prove a lower bound $S \cdot T_g \in \Omega(\epsilon N)$ as outlined above.

At this point, we make a case distinction, depending on whether T_g is below or above $\sqrt{T}$.

If $T_g < \sqrt{T}$ our $S \cdot T_g \in \Omega(\epsilon N)$ bound becomes $S^2 \cdot T \in \Omega(\epsilon^2 N^2)$ and we are done.

The more complicated case is when $T_g \geq \sqrt{T}$ where we show how to use the existence of $\mathcal{A}_{\mathsf{aux}}^{f,g}$ to compress f instead of g. Recall that T_g is the average number of values that got "spoiled" while running the compression algorithm for g_f, that means, for every value added to the good set G, $\mathcal{A}_{\mathsf{aux}}^{f,g}$ made on average T_g "fresh" queries to g_f. Now making fresh g_f queries isn't that easy, as it requires finding x, x' where $f(x) = \overline{f(x')}$. We can use $\mathcal{A}_{\mathsf{aux}}^{f,g}$ which makes many such fresh g_f queries to "compress" f: when $\mathcal{A}_{\mathsf{aux}}^{f,g}$ makes two f queries x, x' where $f(x) = \overline{f(x')}$, we just need to store the first output $f(x)$, but won't need the second $f(x')$ as we know it is $\overline{f(x)}$. For decoding we also must store when exactly $\mathcal{A}_{\mathsf{aux}}^{f,g}$ makes the f queries x and x', more on this below.

Every time we invoke $\mathcal{A}_{\mathsf{aux}}^{f,g}$ for compression as just outlined, up to T outputs of f may get "spoiled" in the sense that $\mathcal{A}_{\mathsf{aux}}^{f,g}$ makes an f query that we need to answer at this point, and thus it's no longer available to be compressed later.

As $\mathcal{A}_{\mathsf{aux}}^{f,g}$ can spoil up to T queries on every invocation, we can hope to invoke it at least $\epsilon N/T$ times before all the f queries are spoiled. Moreover, on average $\mathcal{A}_{\mathsf{aux}}^{f,g}$ makes T_g fresh g_f queries, so we can hope to compress around T_g outputs of f with every invocation of $\mathcal{A}_{\mathsf{aux}}^{f,g}$, which would give us around $T_g \cdot \epsilon N/T$ compressed

---

[14] Note that for the adversary as specified by Hellman's attack against permutations as outlined in Sect. 1.4 we do have $T_g = T$, which is not surprising given that for permutations the De et al. lower bound matches Hellman's attack.

values. This assumes that a large fraction of the fresh g_f queries uses values of f that were not spoiled in previous invocations. The technical core of our proof is a combinatorial lemma which we state and prove in Sect. 5, which implies that it's always possible to find a sequence of inputs to $\mathcal{A}_{\text{aux}}^{f,g}$ such that this is the case. Concretely, we can always find a sequence of inputs such that at least $T_g \cdot \epsilon N / 32T$ values can be compressed.[15]

2   Notation and Basic Facts

We use brackets like $(x_1, x_2, \ldots)$ and $\{x_1, x_2, \ldots\}$ to denote ordered and unordered sets, respectively. We'll usually refer to unordered sets simply as sets, and to ordered sets as lists. $[N]$ denotes some domain of size N, for notational convenience we assume $N = 2^n$ is a power of two and identify $[N]$ with $\{0,1\}^n$. For a function $f : [N] \to [M]$ and a set $S \subseteq [N]$ we denote with $f(S)$ the set $\{f(S[1]), \ldots, f(S[|S|])\}$, similarly for a list $L \subseteq [N]$ we denote with $f(L)$ the list $(f(L[1]), \ldots, f(L[|L|]))$. For a set $\mathcal{X}$, we denote with $x \leftarrow \mathcal{X}$ that x is assigned a value chosen uniformly at random from $\mathcal{X}$.

Fact 1 (from [DTT10]). *For any randomized encoding procedure* $\mathsf{Enc} : \{0,1\}^r \times \{0,1\}^n \to \{0,1\}^m$ *and decoding procedure* $\mathsf{Dec} : \{0,1\}^r \times \{0,1\}^m \to \{0,1\}^n$ *where*

$$\Pr_{x \leftarrow \{0,1\}^n, \rho \leftarrow \{0,1\}^r} [\mathsf{Dec}(\rho, \mathsf{Enc}(\rho, x)) = x] \geq \delta$$

we have $m \geq n - \log(1/\delta)$.

Fact 2. *If a set X is at least ϵ dense in Y, i.e., $X \subset Y$, $|X| \geq \epsilon|Y|$, and Y is known, then X can be encoded using $|X| \cdot \log(e/\epsilon)$ bits. To show this we use the inequality $\binom{n}{\epsilon n} \leq (en/\epsilon n)^{\epsilon n}$, which implies $\log\binom{n}{\epsilon n} \leq \epsilon n \log(e/\epsilon)$.*

3   A Lower Bound for Functions

The following theorem is basically from [DTT10], but stated for functions not permutations.

Theorem 1. *Fix some $\epsilon \geq 0$ and an oracle algorithm $\mathcal{A}$ which on any input makes at most T oracle queries. If for every function $f : [N] \to [N]$ there exists a string aux of length $|\mathsf{aux}| = S$ such that*

$$\Pr_{y \leftarrow [N]} [f(\mathcal{A}_{\mathsf{aux}}^f(y)) = y] \geq \epsilon$$

then

$$T \cdot S \in \Omega(\epsilon N) . \tag{4}$$

---

[15] The constant 32 here can be decreased with a more fine-grained analysis, we opted for a simpler proof rather than optimising this constant.

The theorem follows from Lemma 1 below using Fact 1 as follows: in Fact 1 let $\delta = 0.9$ and $n = N \log N$, think of x as the function table of a function $f : [N] \rightarrow [N]$. Then $|\mathsf{Enc}(\rho, \mathsf{aux}, f)| \geq N \log N - \log(1/0.9)$, together with the upper bound on the encoding from Eq. (6) this implies Eq. (4). Note that the extra assumption that $T \leq \epsilon N/40$ in the lemma below doesn't matter, as if it's not satisfied the theorem is trivially true. For now the value T_g in the lemma below is not important and the reader can just assume $T_g = T$.

Lemma 1. *Let $\mathcal{A}$, T, S, ϵ and f be as in Theorem 1, and assume $T \leq \epsilon N/40$. There are randomized encoding and decoding procedures Enc, Dec such that if $f : [N] \rightarrow [N]$ is a function and for some aux of length $|\mathsf{aux}| = S$*

$$\Pr_{y \leftarrow [N]}[f(\mathcal{A}_{\mathsf{aux}}^f(y)) = y] \geq \epsilon$$

then

$$\Pr_{\rho \leftarrow \{0,1\}^r}[\mathsf{Dec}(\rho, \mathsf{Enc}(\rho, \mathsf{aux}, f)) = f] \geq 0.9 \tag{5}$$

and the length of the encoding is at most

$$|\mathsf{Enc}(\rho, \mathsf{aux}, f)| \leq \underbrace{N \log N}_{=|f|} - \frac{\epsilon N}{2T_g} + S + \log(N) \tag{6}$$

for some T_g, $1 \leq T_g \leq T$.

3.1   Proof of Lemma 1

The Encoding and Decoding Algorithms. In Algorithms 1 and 2, we always assume that if $\mathcal{A}_{\mathsf{aux}}^f(y)$ outputs some value x, it makes the query $f(x)$ at some point. This is basically w.l.o.g. as we can turn any adversary into one satisfying this by making at most one extra query. If at some point $\mathcal{A}_{\mathsf{aux}}^f(y)$ makes an oracle query x where $f(x) = y$, then we also w.l.o.g. assume that right after this query $\mathcal{A}$ outputs x and stops. Note that if $\mathcal{A}$ is probabilistic, it uses random coins which are given as input to Enc, Dec, so we can make sure the same coins are used during encoding and decoding.

The Size of the Encoding. We will now upper bound the size of the encoding of G, $f(Q')$, $(|q_1|, \ldots, |q_{|G|}|)$, $f([N] - \{G^{-1} \cup Q'\})$ as output in line (15) of the Enc algorithm.

Let $T_g := |B|/|G|$ be the average number of elements we added to the bad set B for every element added to the good set G, then

$$|G| \geq \epsilon N/2T_g . \tag{7}$$

To see this we note that when we leave the while loop (see line (8) of the algorithm Enc) it holds that $|B| \geq |J|/2 = \epsilon N/2$, so $|G| = |B|/T_g \geq |J|/2T_g = \epsilon N/2T_g$.

Algorithm 1. Enc

1: **Input:** $\mathcal{A}$, aux, randomness ρ and a function $f : [N] \to [N]$ to compress.
2: Initialize: $B, G := \emptyset, c := -1$
3: Throughout we identify $[N]$ with $\{0, \ldots, N-1\}$.
4: Pick a random permutation $\pi : [N] \to [N]$ (using random coins from ρ)
5: Let $J := \{y \; : \; f(\mathcal{A}_{\mathsf{aux}}^f(y)) = y\}, |J| = \epsilon N$ ▷ The set J where $\mathcal{A}$ inverts. If $\mathcal{A}$ is probabilistic, use random coins from ρ.
6: For $i = 0, \ldots, N-1$ define $y_i := \pi(i)$.   ▷ Randomize the order
7: For $y \in J$ let the list $q(y)$ contain all queries made by $A^f(y)$ except the last query (which is x s.t. $f(x) = y$).
8: **while** $|B| < |J|/2$ **do**   ▷ While the bad set contains less than half of J
9:      $c := \min\{c' > c \; : \; y_{c'} \in \{J \setminus B\}\}$ ▷ Increase c to the next y_c in $J \setminus B$
10:      $G := G \cup y_c$ ▷ Add this y_c to good set
11:      $B := B \cup (f(q(y_c)) \cap J)$ ▷ Add spoiled queries to bad set
12: **end while**
13: Let $G = \{g_1, \ldots, g_{|G|}\}, Q = (q(g_1), \ldots, q(g_{|G|}))$, and define $Q' = (q'_1, \ldots, q'_{|G|}), q'_i \subseteq q(g_i)$ to contain only the "fresh" queries in Q by deleting all but the first occurrence of every element. E.g. if $(q(g_1), q(g_2)) = ((1, 2, 3, 1), (2, 4, 5, 4))$ then $(q'_1, q'_2) = ((1, 2, 3), (4, 5))$.
14: Let $G^{-1} = \{\mathcal{A}_{\mathsf{aux}}^f(y) \; : \; y \in G\}$
15: Output an encoding of (the set) G, (the lists) $f(Q'), (|q'_1|, \ldots, |q'_{|G|}|), f([N] - \{G^{-1} \cup Q'\})$ and (the string) aux.

G: Instead of G we will actually encode the set $\pi^{-1}(G) = \{c_1, \ldots, c_{|G|}\}$. From this the decoding Dec (who gets ρ, and thus knows π) can then reconstruct $G = \pi(\pi^{-1}(G))$. We claim that the elements in $c_1 < c_2 < \ldots < c_{|G|}$ are whp. at least $\epsilon/2$ dense in $[c_{|G|}]$ (equivalently, $c_{|G|} \leq 2|G|/\epsilon$). By Fact 2 we can thus encode $\pi^{-1}(G)$ using $|G| \log(2e/\epsilon) + \log N$ bits (the extra $\log N$ bits are used to encode the size of G which is required so decoding later knows how to parse the encoding). To see that the c_i's are $\epsilon/2$ dense whp. consider line (9) in Enc which states $c := \min\{c' > c \; : \; y_{c'} \in \{J \setminus B\}\}$. If we replace $J \setminus B$ with J, then the c_i's would be whp. close to ϵ dense as J is ϵ dense in $[N]$ and the y_i are uniformly random. As $|B| < |J|/2$, using $J \setminus B$ instead of J will decrease the density by at most a factor 2. If we don't have this density, i.e., $c_{|G|} > 2|G|/\epsilon$, we consider encoding to have failed.

$f(Q')$: This is a list of Q' elements in $[N]$ and can be encoded using $|Q'| \log N$ bits.

$(|q'_1|, \ldots, |q'_{|G|}|)$: Require $|G| \log T$ bits as $|q'_i| \leq |q_i| \leq T$. A more careful argument (using that the q'_i are on average at most T_g) requires $|G| \log(eT_g)$ bits.

$f([N] - \{G^{-1} \cup Q'\})$: Requires $(N - |G| - |Q'|) \log N$ bits (using that $G^{-1} \cap Q' = \emptyset$ and $|G^{-1}| = |G|$).

aux: Is S bits long.

Summing up we get

$$|\mathsf{Enc}(\rho, \mathsf{aux}, f)| = |G| \log(2e^2 T_g/\epsilon) + (N - |G|) \log N + S + \log N$$

Algorithm 2. Dec

1: **Input:** $\mathcal{A}, \rho$ and the encoding $(G, f(Q'), (|q_1'|, \ldots, |q_{|G|}'|), f([N] - \{G^{-1} \cup Q'\}), \mathsf{aux})$.
2: Let π be as in Enc.
3: Let $(g_1, \ldots, g_{|G|})$ be the elements of G ordered as they were added by Enc (i.e., $\pi^{-1}(g_i) < \pi^{-1}(g_{i+1})$ for all i).
4: Invoke $\mathcal{A}_{\mathsf{aux}}^{(\cdot)}(\cdot)$ sequentially on inputs $g_1, \ldots, g_{|G|}$ using $f(Q')$ to answer $\mathcal{A}_{\mathsf{aux}}$'s oracle queries.   ▷ If $\mathcal{A}$ is probabilistic, use the same random coins from ρ as in Enc.
5: Combine the mapping $G^{-1} \cup Q' \to f(G^{-1} \cup Q')$ (which we learned in the previous step) with $[N] - \{G^{-1} \cup Q'\} \to f([N] - \{G^{-1} \cup Q'\})$ to learn the entire $[N] \to f([N])$
6: Output $f([N])$

---

as by assumption $T_g \le T \le \epsilon N/40$, we get $\log N - \log(2e^2 T_g/\epsilon) \ge 1$, and further using (7) we get

$$|\mathsf{Enc}(\rho, \mathsf{aux}, f)| \le N \log N - \frac{\epsilon N}{2T_g} + S + \log N$$

as claimed.

4   A Lower Bound for $g(x, f^{-1}(\overline{f(x)}))$

For a permutation $f : [N] \to [N]$ and a function $g : [N] \times [N] \to [N]$ we define $g_f : [N] \to [N]$ as

$$g_f(x) = g(x, x') \text{ where } f(x) = \overline{f(x')} \text{ or equivalently } g_f(x) = g(x, f^{-1}(\overline{f(x)}))$$

Theorem 2. *Fix some $\epsilon > 0$ and an oracle algorithm $\mathcal{A}$ which makes at most*

$$T \le (N/4e)^{2/3} \tag{8}$$

oracle queries and takes an advice string aux of length $|\mathsf{aux}| = S$. If for all functions $f : [N] \to [N]$, $g : [N] \times [N] \to [N]$ and some aux of length $|\mathsf{aux}| = S$ we have

$$\Pr_{y \leftarrow [N]}[g_f(A_{\mathsf{aux}}^{f,g}(y)) = y] \ge \epsilon \tag{9}$$

then

$$TS^2 \in \Omega(\epsilon^2 N^2) . \tag{10}$$

The theorem follows from Lemma 2 below as we'll prove thereafter.

Lemma 2. *Fix some $\epsilon \ge 0$ and an oracle algorithm $\mathcal{A}$ which makes at most $T \le (N/4e)^{2/3}$ oracle queries. There are randomized encoding and decoding procedures $\mathsf{Enc}_g, \mathsf{Dec}_g$ and $\mathsf{Enc}_f, \mathsf{Dec}_f$ such that if $f : [N] \to [N]$ is a permutation, $g : [N] \times [N] \to [N]$ is a function and for some advice string aux of length $|\mathsf{aux}| = S$ we have*

$$\Pr_{y \leftarrow [N]}[g_f(A_{\mathsf{aux}}^{f,g}(y)) = y] \ge \epsilon$$

then

$$\Pr_{\rho \leftarrow \{0,1\}^r}[\mathsf{Dec}_g(\rho, f, \mathsf{Enc}_g(\rho, \mathsf{aux}, f, g)) = g] \geq 0.9 \tag{11}$$

$$\Pr_{\rho \leftarrow \{0,1\}^r}[\mathsf{Dec}_f(\rho, g, \mathsf{Enc}_f(\rho, \mathsf{aux}, f, g)) = f] \geq 0.9 \ . \tag{12}$$

Moreover for every ρ, aux, f, g there is a $T_g, 1 \leq T_g \leq T$, such that

$$|\mathsf{Enc}_g(\rho, \mathsf{aux}, f, g)| \leq \underbrace{N^2 \log N}_{=|g|} - \frac{\epsilon N}{2T_g} + S + \log N \tag{13}$$

and if $T_g \geq \sqrt{T}$

$$|\mathsf{Enc}_f(\rho, \mathsf{aux}, f, g)| \leq \underbrace{\log N!}_{=|f|} - \frac{\epsilon N T_g}{64T} + S + \log N \ . \tag{14}$$

We first explain how Theorem 2 follows from Lemma 2 using Fact 1.

Proof (of Theorem 2). The basic idea is to make a case analysis; if $T_g < \sqrt{T}$ we compress g, otherwise we compress f. Intuitively, our encoding for g achieving Eq. (13) makes both f and g queries, but only g queries "spoil" g values. As the compression runs until all g values are spoiled, it compresses better the smaller T_g is. On the other hand, the encoding for f achieving Eq. (12) is derived from our encoding for g, and it manages to compresses in the order of T_g values of f for every invocation (while "spoiling" at most T of the f values), so the larger T_g the better it compresses f.

Concretely, pick f, g uniformly at random (and assume Eq. (9) holds). By a union bound for at least a 0.8 fraction of the ρ Eqs. (11) and (12) hold simultaneously. Consider any such good ρ, which together with f, g fixes some $T_g, 1 \leq T_g \leq T$ as in the statement of Lemma 2. Now consider an encoding $\mathsf{Enc}_{f,g}$ where $\mathsf{Enc}_{f,g}(\rho, \mathsf{aux}, f, g)$ outputs $(f, \mathsf{Enc}_g(\rho, \mathsf{aux}, f, g))$ if $T_g < \sqrt{T}$, and $(g, \mathsf{Enc}_f(\rho, \mathsf{aux}, f, g))$ otherwise.

- If $T_g < \sqrt{T}$ we use (13) to get

$$|\mathsf{Enc}_{f,g}(\rho, \mathsf{aux}, f, g)| = |f| + |\mathsf{Enc}_g(\rho, \mathsf{aux}, f, g)| \leq |f| + |g| - \epsilon N/2T_g + S + \log N$$

and now using Fact 1 (with $\delta = 0.8$) we get

$$S \geq \epsilon N/2T_g - \log N - \log(1/0.8) > \epsilon N/2\sqrt{T} - \log N - \log(1/0.8)$$

and thus $TS^2 \in \Omega(\epsilon^2 N^2)$ as claimed in Eq. (10).
- If $T_g \geq \sqrt{T}$ then we use Eq. (14) and Fact 1 and again get $S \geq \epsilon N T_g/64T - \log N - \log(1/0.8)$ which implies Eq. (10) as $T_g \geq \sqrt{T}$. $\qquad \square$

Algorithm 3. Enc_g

1: **Input:** $\mathcal{A}, \rho, \mathsf{aux}, f, g$
2: Compute the function table of $g_f : [N] \to [N]$, $g_f(x) = g(x, x')$ where $f(x) = \overline{f(x')}$.
3: Invoke $E_{g_f} \leftarrow \mathsf{Enc}(\mathcal{A}, g_f, aux, \rho)$
4: Let g' be the function table of $g([N]^2) = g(1,1)\| \ldots \|g(N,N)$, but with the N entries (x, x') where $f(x) = \overline{f(x')}$ deleted.
5: Output $E_{g_f}, g', \mathsf{aux}$.

Algorithm 4. Dec_g

1: **Input:** $\mathcal{A}, \rho, f$ and the encoding $(E_{g_f}, g', \mathsf{aux})$ of g.
2: Invoke $g_f \leftarrow \mathsf{Dec}(\mathcal{A}, \rho, \mathsf{aux}, E_{g_f})$.
3: Reconstruct g from g' and g_f (this is possible as f is given).
4: Output $g([N]^2)$

Algorithm 5. Enc_f

1: **Input:** $\mathcal{A}, \rho, \mathsf{aux}, f, g$
2: Invoke $E_{g_f} \leftarrow \mathsf{Enc}(\mathcal{A}, g_f, \mathsf{aux}, \rho)$ ▷ Compute the same encoding of g_f as Enc_g did.
3: For $G \in E_{g_f}$, let $G_f \subset G, G_f = \{z_1, \ldots, z_{|G_f|}\}$ be as defined in proof of Lemma 2.
4: Initialize empty lists $L_f, T_f, C_f := \emptyset$.
5: **for** $i = 1$ to $|G_f|$ **do**
6:     Invoke $\mathcal{A}_{\mathsf{aux}}^{f,g}(z_i)$.   ▷ Using random coins from ρ if $\mathcal{A}$ is probabilistic.
7:     For each pair of f queries x, x' (made in this order during invocation) where $f(x) = \overline{f(x')}$ and neither $f(x')$ nor $f(x)$ is in $L_f \cup C_f$, let (t, t') be the indices $(1 \le t < t' \le T)$ specifying when during invocation these queries were made. Append (t, t') to T_f and append $f(x')$ to C_f.
8:     Append all images of oracle queries to f made during invocation of $\mathcal{A}_{\mathsf{aux}}^{f,g}(z_i)$ to L_f, except if the value is in $L_f \cup C_f$. ▷ Append the images of all fresh f queries which were not compressed.
9: **end for**
10: Let L_f^{-1} (similarly C_f^{-1}) contain the inputs corresponding to L_f, i.e., add x to L_f^{-1} when adding $f(x)$ to L_f.
11: Output an encoding of G_f, the list of values of f queries L_f, the list of tuples T_f and the remaining outputs $f([N - L_f^{-1} - C_f^{-1}])$ which were neither in the list L_f nor compressed.

Algorithm 6. Dec$_f$

1: **Input:** $\mathcal{A}, \rho, \mathsf{aux}, g$ and encoding $(G_f, L_f, T_f, f([N - L_f^{-1} - C_f^{-1}]))$ of f.
2: Let $G_f = \{z_1, \dots, z_{|G_f|}\}$.
3: **for** $i = 1$ to $|G_f|$ **do**
4:     Invoke $\mathcal{A}_{\mathsf{aux}}^{(\cdot),g}(z_i)$ reconstructing the answers to the first oracle (which should be f) using the lists L_f and T_f.
5: **end for**
6: For L_f^{-1}, C_f^{-1} as in Enc$_f$, we have learned the mapping $(L_f^{-1} \cup C_f^{-1}) \to f((L_f^{-1} \cup C_f^{-1}))$. Reconstruct all of $f([N])$ by combining this with $f([N] - (L_f^{-1} \cup C_f^{-1}))$.
7: **Output** $f([N])$

Proof (of Lemma 2).

The Encoding and Decoding Algorithms. The encoding and decoding of g are depicted in Algorithms 3 and 4, and those of f in Algorithms 5 and 6. $\mathcal{A}_{\mathsf{aux}}^{f,g}(\cdot)$ can make up to T queries in total to its oracles $f(.)$ and $g(.)$. We will assume that whenever a query $g(x, x')$ is made, the adversary made queries $f(x)$ and $f(x')$ before. This is basically without loss of generality as we can turn any adversary into one adhering to this by at most tripling the number of queries. It will also be convenient to assume that $\mathcal{A}_{\mathsf{aux}}^{f,g}$ only queries g on its restriction to g_f, that is, for all $g(x, x')$ queries it holds that $f(x) = \overline{f(x')}$, but the proof is easily extended to allow all queries to g as our encoding will store the function table of g on all "uninteresting" inputs (x, x'), $f(x) \neq \overline{f(x')}$ and thus can directly answer any such query.

As in the proof of Lemma 1, we don't explicitly show the randomness in case $\mathcal{A}$ is probabilistic.

The Size of the Encodings. We will now upper bound the size of the encodings output by Enc$_g$ and Enc$_f$ in Algorithms 3 and 5 and hence prove Eqs.(13) and (14).

Equation (13) now follows almost directly from Theorem 1 as our compression algorithm Enc$_g$ for $g : [N] \times [N] \to [N]$ simply uses Enc to compress g restricted to $g_f : [N] \to [N]$, and thus compresses by exactly the same amount as Enc.

It remains to prove an upper bound on the length of the encoding of f by our algorithm Enc$_f$ as claimed in Eq. (14). Recall that Enc (as used inside Enc$_g$) defines a set G such that for every $y \in G$ we have (1) $A_{\mathsf{aux}}^{f,g}(y)$ inverts, i.e., $g_f(A_{\mathsf{aux}}^{f,g}(y)) = y$ and (2) never makes a g_f query x where $g_f(x) \in G$. Recall that T_g in Eq. (13) satisfies $T_g = \epsilon N/2|G|$, and corresponds to the average number of "fresh" g_f queries made by $A_{\mathsf{aux}}^{f,g}(\cdot)$ when invoked on the values in G.

Enc$_f$ invokes $A_{\mathsf{aux}}^{f,g}(\cdot)$ on a carefully chosen subset $G_f = (z_1, \dots, z_{|G_f|})$ of G (to be defined later). It keeps lists L_f, C_f and T_f such that after invoking $A_{\mathsf{aux}}^{f,g}(\cdot)$ on G_f, $L_f \cup C_f$ holds the outputs to all f queries made. Looking ahead, the decoding Dec$_f$ will also invoke $A_{\mathsf{aux}}^{f,g}(\cdot)$ on G_f, but will only need L_f and T_f (but not C_f) to answer all f queries.

The lists L_f, T_f, C_f are generated as follows. On the first invocation $A_{\mathsf{aux}}^{f,g}(z_1)$ we observe up to T oracle queries made to g and f. Every g query (x, x') must be preceded by f queries x and x' where $f(x) = \overline{f(x')}$. Assume x and x' are the queries number t, t' ($1 \leq t < t' \leq T$). A key observation is that by just storing (t, t') and $f(x)$, Dec_f will later be able to reconstruct $f(x')$ by invoking $A_{\mathsf{aux}}^{f,g}(z_1)$, and when query t' is made, looking up the query $f(x)$ in L_f (its position in L_f is given by t), and set $f(x') = \overline{f(x)}$. Thus, every time a fresh query $f(x')$ is made we append it to L_f, unless earlier in this invocation we made a fresh query $f(x)$ where $f(x') = \overline{f(x)}$. In this case we append the indices (t, t') to the list T_f. We also add $f(x')$ to a list C_f just to keep track of what we already compressed. Enc_f now continues this process by invoking $A_{\mathsf{aux}}^{f,g}(\cdot)$ on inputs $z_2, z_3, \ldots, z_{|G_f|} \in G_f$ and finally outputs and encoding of G_f, an encoding of the list of images of fresh queries L_f, an encoding of the list of colliding indices T_f, aux, and all values of f that were neither compressed nor queried.

In the sequel we show how to choose $G_f \in G$ such that $|G_f| \geq \epsilon N/8T$ and hence it can be encoded using $|G_f| \log N + \log N$ where the extra $\log N$ is used to encode $|G_f|$. We also show that $|T_f| \geq |G_f| \cdot T_g/4$ and furthermore that we can compress at least one bit per element of T_f. Putting things together we get

$$|\mathsf{Enc}_f(\rho, \mathsf{aux}, f, g)| \leq \log N! - |G_f|(T_g/4 - \log N) + S + \log N \ .$$

And if $\log N \leq T_g/8$, we get Eq. (14)

$$|\mathsf{Enc}_f(\rho, \mathsf{aux}, f, g)| \leq \log N! - \epsilon N T_g/64T + S + \log N \ .$$

Given G such that $|G| \geq \epsilon N/2T_g$, the subset G_f can be constructed by carefully applying Lemma 3 which we prove in Sect. 5. Let $(X_1, \ldots, X_{|G|})$, $(Y_1, \ldots, Y_{|G|})$ be two sequences of sets such that $Y_i \subseteq X_i \subseteq [N]$ and $|X_i| \leq T$ such that Y_i and X_i respectively correspond to g and f queries in $|G|$ consecutive executions of $A_{\mathsf{aux}}^{f,g}(\cdot)$ on G.[16] Given such sequences Lemma 3 constructs a subsequence of executions $G_f \subseteq G$ whose corresponding g queries $(Y_{i_1}, \ldots, Y_{i_{|G_f|}})$ are fresh. As a g query is preceded by two f queries, such a subsequence induces a sequence $(Z_{i_1}, \ldots, Z_{i_{|G_f|}})$ of queries that are not only fresh for g but also fresh for f. Furthermore, such a sequence covers $y \cdot |I|/16T$ where $y = |I|/|G|$ is the average coverage of Y_i's and $I \subseteq [N]$ is their total coverage.

However, Lemma 3 considers a g query $(x, x') \in Y_i$ to be fresh if either $x \notin \cup_{j=1}^{i-1} X_j$ or $x' \notin \cup_{j=1}^{i-1} X_j$, i.e., if at least one of x, x' is fresh in the ith execution, then the pair is considered fresh. For compressing f both x, x' need to be fresh. To enforce that and apply Lemma 3 directly, we apply Lemma 3 on augmented sets $X_1, \ldots, X_{|G|}$ such that whenever X_i, Y_i are selected, the corresponding Z_i

---

[16] Here is how these sets are compiled. Note that if q is an f query then $q \in [N]$, and if q is a g query then $q \in [N]^2$. In the ith execution, both X_i, Y_i are initially empty and later will contain only elements in $[N]$. Therefore for each query q, if $q = (x, x')$ is a g query we add two elements x and x' to Y_i, and if $q = x$ is an f query we add the single element x to X_i. Furthermore as a g query (x, x') is preceded by two f queries x, x', then $Y_i \subseteq X_i$, and as the max number of queries is T we have $|X_i| \leq T$.

contains exactly $|Z_i|/2$ pairs of queries that are fresh for both g and f. We augment X_i as follows. For every X_i and every f query x made in the ith step, add $f^{-1}(\overline{f(x)})$ to X_i. This augmentation results in X_i such that $|X_i| \leq 2T$ as originally we have $|X_i| \leq T$.

Applying Lemma 3 on $Y_1, \ldots, Y_{|G|}$ and such augmented sets $X_1, \ldots, X_{|G|}$ yields G_f such that the total number of fresh colliding *queries* is of size at least

$$y \cdot \frac{|I|}{16 \cdot 2T} = \frac{\epsilon N}{|G|} \cdot \frac{\epsilon N}{32T} = \frac{\epsilon N T_g}{16T} \ .$$

Therefore the total number of fresh colliding *pairs*, or equivalently $|T_f|$, is $\epsilon N T_g / 32T$ as claimed. Furthermore, Lemma 3 guarantees that $|G_f| \geq \epsilon N / 8T$.[17]

What remains to show is that for each colliding pair in T_f we compress by at least one bit. Recall that the list T_f has exactly as many entries as C_f. However entries in T_f are colliding pairs of indices (t, t') and entries in C_f are images of size $\log N$. Per each entry (t, t') in T_f we compress if the encoding size of (t, t') is strictly less than $\log N$. Here is an encoding of T_f that achieves this. Instead of encoding each entry (t, t') as two indices which costs $2 \log T$ and therefore we save one bit per element in T_f assuming $T \leq \sqrt{N/2}$, we encode the set of colliding pairs among all possible query pairs. Concretely, for each $z \in G_f$ we obtain a set of colliding indices of size at least $T_g / 4$. Then we encode this set of colliding pairs $T_g / 4$ among all possible pairs[18], which is upper bounded by T^2, using

$$\log \binom{T^2}{T_g / 4} \leq \frac{T_g}{4} \log \frac{4eT^2}{T_g}$$

bits, and therefore, given that $T_g \geq \sqrt{T}$ and $T \leq (N/4e)^{2/3}$, we have that $\log N - \log 4eT^2 / T_g \geq 1$ and therefore we compress by at least one bit for each pair, i.e., for each element in T_f, and that concludes the proof. □

5   A Combinatorial Lemma

In this section we state and prove a lemma which can be cast in terms of the following game between Alice and Bob. For some integers n, N, M, Alice can choose a partition $(Y_1, \ldots, Y_n)$ of $I \subseteq [N]$, and for every Y_i also a superset $X_i \supseteq Y_i$ of size $|X_i| \leq M$. The goal of Bob is to find a subsequence $1 \leq b_1 < b_2 < \ldots < b_\ell$ such that $Y_{b_1}, Y_{b_2}, \ldots, Y_{b_\ell}$ contains as many "fresh" elements as possible, where after picking Y_{b_i} the elements $\bigcup_{k=1}^{i} X_{b_k}$ are not fresh, i.e., picking Y_{b_i} "spoils" all of X_{b_i}. How many fresh elements can Bob expect to hit in the worst case? Intuitively, as every Y_{b_i} added spoils up to M elements, he can hope

---

[17] $|G_f|$ corresponds to ℓ in the proof of Lemma 3.

[18] Note that T^2 is an upper bound on all possible pairs of queries, however as we have that $t < t'$ for each pair (t, t'), we can cut T^2 by at least a factor of 2. Other optimizations are possible. This extra saving one can use to add extra dummy pairs of indices to separate executions for decoding. The details are tedious and do not affect the bound as we were generous to consider T^2 to be the size of possible pairs.

to pick up to $\ell \approx |I|/M$ of the Y_i's before most of the elements are spoiled. As the Y_i are on average of size $y := |I|/n$, this is also an upper bound on the number of fresh elements he can hope to get with every step. This gives something in the order of $y \cdot (|I|/M)$ fresh elements in total. By the lemma below a subsequence that contains about that many fresh elements always exists.

Lemma 3. *For $M, N \in \mathbb{N}, M \leq N$ and any disjoint sets $Y_1, \ldots, Y_n \subset [N]$*

$$\bigcup_{i=1}^{n} Y_i = I, \quad \forall i \neq j : Y_i \cap Y_j = \emptyset$$

and supersets $(X_1, \ldots, X_n)$ where

$$\forall i \in [n] \; : \; Y_i \subseteq X_i \subseteq [N], \quad |X_i| \leq M$$

there exists a subsequence $1 \leq b_1 < b_2 < \ldots < b_\ell \leq n$ such that the sets

$$Z_{b_j} = Y_{b_j} \setminus \cup_{k<j} X_{b_k} \tag{15}$$

have total size

$$\sum_{j=1}^{\ell} |Z_{b_j}| = |\bigcup_{j=1}^{\ell} Z_{b_j}| \geq y \cdot \frac{|I|}{16M}$$

where $y = |I|/n$ denotes the average size of the Y_i's.

Proof. Let $(Y_{a_1}, \ldots, Y_{a_m})$ be a subquence of $(Y_1, \ldots, Y_n)$ that contains all the sets of size at least $y/2$. By a Markov bound, these large Y_{a_i}'s cover at least half of the domain I, i.e.

$$|\cup_{i\in[m]} Y_{a_i}| > |I|/2 . \tag{16}$$

We now choose the subsequence $(Y_{b_1}, \ldots, Y_{b_\ell})$ from the statement of the lemma as a subsequence of $(Y_{a_1}, \ldots, Y_{a_m})$ in a greedy way: for $i = 1, \ldots, m$ we add Y_{a_i} to the sequence if it adds a lot of "fresh" elements, concretely, assume we are in step i and so far have added $Y_{b_1}, \ldots, Y_{b_{j-1}}$, then we'll pick the next element, i.e., $Y_{b_j} := Y_{a_i}$, if the fresh elements $Z_{b_j} = Y_{b_j} \setminus \cup_{k<j} X_{b_k}$ contributed by Y_{b_j} are of size at least $|Z_{b_j}| > |Y_{b_j}|/2$.

We claim that we can always add at least one more Y_{b_j} as long as we haven't yet added at least $|I|/4M$ sets, i.e., $j < |I|/4M$. Note that this then proves the lemma as

$$\sum_{j=1}^{\ell} |Z_{b_j}| \geq \sum_{j=1}^{\ell} |Y_{b_j}|/2 \geq \ell y/4 \geq |I|/4M \cdot y/4 = y|I|/16M .$$

It remains to prove the claim. For contradiction assume our greedy algorithm picked $(Y_{b_1}, \ldots, Y_{b_\ell})$ with $\ell < |I|/4M$. We'll show that there is a Y_{a_t} (with $a_t > b_\ell$) with

$$|Y_{a_t} \setminus \cup_{i=1}^{j} X_{b_i}| \geq |Y_{a_t}|/2$$

which is a contradiction as this means the sequence could be extended by $Y_{b_{\ell+1}} = Y_{a_t}$. We have

$$| \cup_{i=1}^{\ell} X_{b_i} | \leq |I|/4M \cdot M = |I|/4 \; .$$

This together with (16) implies

$$| \cup_{i \in [m]} Y_{a_i} \setminus \cup_{i=1}^{\ell} X_{a_i} | > | \cup_{i \in [m]} Y_{a_i} |/2 \; .$$

By Markov there must exist some Y_{a_t} with

$$|Y_{a_t} \setminus \cup_{i=1}^{\ell} X_{a_i}| \geq |Y_{a_t}|/2$$

as claimed. $\square$

6   Conclusions

In this work we showed that existing time-memory trade-offs for inverting functions can be overcome, if one relaxes the requirement that the function is efficiently computable, and just asks for the function table to be computed in (quasi)linear time. We showed that such functions have interesting applications towards constructing proofs of space. The ideas we introduced can potentially also be used for related problems, like memory-bound or memory-hard functions.

Acknowledgements. Hamza Abusalah, Joël Alwen, and Krzysztof Pietrzak were supported by the European Research Council, ERC consolidator grant (682815 - TOC-NeT).

Leonid Reyzin gratefully acknowledges the hospitality and support of IST Austria, where much of this work was performed. He was also supported, in part, by US NSF grants 1012910, 1012798, and 1422965.

A   Proof of Lemma 1 Following [DTT10]

In this section we sketch the proof of Theorem 1 following the proof from [DTT10], just marginally adapting it so it applied to functions not just permutations. Let

$$I = \{x \; : \; A_{\mathsf{aux}}^f(f(x)) = x\}, \quad J = f(I) = \{y \; : \; f(A_{\mathsf{aux}}^f(y)) = y\}$$

For $x \in I$ let $q(f(x))$ denote the queries made by $A^f f_{\mathsf{aux}}(f(x))$ but without the final query x. By assumption $|I| = \epsilon N$. Pick a random subset $R \subset [N]$ of size $|R| = N/T$ (for this use the randomness ρ given to the encoding) (Fig. 2). Let G denote the set of x's in R where A_{aux}^f on input $f(x)$ finds x and makes no queries in R, i.e.,

$$G = \{x \in I \cap R \text{ and } q(f(x)) \cap R = \emptyset\}$$

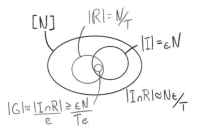

Fig. 2. The different subsets of the input domain of $f : [N] \to [N]$. R is a random subset of size N/T, I are the values x where $A_{\mathsf{aux}}^f(f(x)) = x$. G is the subset of $I \cap R$ of x where $A_{\mathsf{aux}}^f(f(x))$ makes no oracle queries to values in R (except for the last query x).

The expected size of $I \cap R$ is $\epsilon N/T$ (as for any $x \in I$, $\mathrm{Pr}_R[x \in R] = 1/T$). For any $x \in I$, the probability that $q(f(x)) \cap R = \emptyset$ is at least $(1 - 1/T)^T \approx 1/e > 1/e$, so the expected size of G is $> \epsilon N/eT$. In this proof sketch we assume the sets have exactly their expected size, i.e.,

$$|R| = N/T, \quad |I| = \epsilon N, \quad |I \cap R| = \epsilon N/T, \quad |G| = \epsilon N/eT$$

The encoding of $\mathsf{Enc}(\rho, f)$ now contains (the random coins r are used to sample the set R)

aux: The auxiliary input of size S.

$E_{f([N]-R)}$: The list of values $f([N] - R)$ using $(N - |R|) \log N$ bits.

$E_{\{f(G)\}}$: An encoding of the set $\{f(G)\} = \{f(x) : x \in G\}$. As $\{f(G)\}$ is ϵ/Te dense in N, so by Fact 2 this requires only $|G|(\log(e^2 T/\epsilon))$ bits.

$E_{f(R-G)}$: The list of values $f(R - G)$ using $(|R| - |G|) \log N$ bits.

Overall, the encoding size is $S + |G|(\log(e^2/T\epsilon)) + (N - |G|) \log N$

$$N \log N + S - \frac{\epsilon N}{eT}(\log N - \log(e^2/T\epsilon))$$

The decoding $\mathsf{Dec}(\rho, [\mathsf{aux}, E_{f(G)}, E_{f([N]-R)}, E_{f(R-G)}])$ is straight forward

- Invoke $A_{\mathsf{aux}}^f(.)$ on all $y \in \{f(G)\}$, answering all oracle queries (except the last one) using the mapping $[N] - R \to f([N] - R)$. The last query x of $A_{\mathsf{aux}}^f(y)$ can be recognized as it's the first query not in $[N] - R$, and we learn that $f(x) = y$.
- Output $f([N])$, which can be computed as the lists G, R and $f(G), f(R - G), f([N] - R)$ are all known.

References

[BBS06] Barkan, E., Biham, E., Shamir, A.: Rigorous bounds on cryptanalytic time/memory tradeoffs. In: Dwork, C. (ed.) CRYPTO 2006. LNCS, vol. 4117, pp. 1–21. Springer, Heidelberg (2006). https://doi.org/10.1007/11818175_1

[DFKP15] Dziembowski, S., Faust, S., Kolmogorov, V., Pietrzak, K.: Proofs of space. In: Gennaro, R., Robshaw, M. (eds.) CRYPTO 2015. LNCS, vol. 9216, pp. 585–605. Springer, Heidelberg (2015). https://doi.org/10.1007/978-3-662-48000-7_29

[DN93] Dwork, C., Naor, M.: Pricing via processing or combatting junk mail. In: Brickell, E.F. (ed.) CRYPTO 1992. LNCS, vol. 740, pp. 139–147. Springer, Heidelberg (1993). https://doi.org/10.1007/3-540-48071-4_10

[DTT10] De, A., Trevisan, L., Tulsiani, M.: Time space tradeoffs for attacks against one-way functions and prgs. In: Rabin, T. (ed.) CRYPTO 2010. LNCS, vol. 6223, pp. 649–665. Springer, Heidelberg (2010). https://doi.org/10.1007/978-3-642-14623-7_35

[FN91] Fiat, A., Naor, M.: Rigorous time/space tradeoffs for inverting functions, pp. 534–541 (1991)

[GT00] Gennaro, R., Trevisan, L.: Lower bounds on the efficiency of generic cryptographic constructions, pp. 305–313 (2000)

[Hel80] Hellman, M.E.: A cryptanalytic time-memory trade-off. IEEE Trans. Inf. Theory **26**(4), 401–406 (1980)

[PPK+15] Park, S., Pietrzak, K., Kwon, A., Alwen, J., Fuchsbauer, G., Gaži, P.: Spacemint: A cryptocurrency based on proofs of space. Cryptology ePrint Archive, Report 2015/528 (2015). http://eprint.iacr.org/2015/528

[RD16] Ren, L., Devadas, S.: Proof of space from stacked expanders. In: Hirt, M., Smith, A. (eds.) TCC 2016. LNCS, vol. 9985, pp. 262–285. Springer, Heidelberg (2016). https://doi.org/10.1007/978-3-662-53641-4_11

[Wee05] Wee, H.: On obfuscating point functions, pp. 523–532 (2005)

[Yao90] Yao, A.C.-C.: Coherent functions and program checkers (extended abstract), pp. 84–94 (1990)

The Sleepy Model of Consensus

Rafael Pass[(✉)] and Elaine Shi

Cornell Tech, Cornell University, New York, USA
`rafael@cs.cornell.edu`

Abstract. The literature on distributed computing (as well as the cryptography literature) typically considers two types of players—*honest* players and *corrupted* players. Resilience properties are then analyzed assuming a lower bound on the fraction of honest players. Honest players, however, are not only assumed to follow the prescribed the protocol, but also assumed to be *online* throughout the whole execution of the protocol. The advent of "large-scale" consensus protocols (e.g., the blockchain protocol) where we may have millions of players, makes this assumption unrealistic. In this work, we initiate a study of distributed protocols in a "sleepy" model of computation where players can be either *online* (awake) or *offline* (asleep), and their online status may change at any point during the protocol. The main question we address is:

> *Can we design consensus protocols that remain resilient under "sporadic participation", where at any given point, only a subset of the players are actually online?*

As far as we know, all standard consensus protocols break down under such sporadic participation, even if we assume that 99% of the online players are honest.

Our main result answers the above question in the affirmative. We present a construction of a consensus protocol in the sleepy model, which is resilient assuming only that a *majority of the online players are honest*. Our protocol relies on a Public-Key Infrastructure (PKI), a Common Random String (CRS) and is proven secure in the timing model of Dwork-Naor-Sahai (STOC'98) where all players are assumed to have weakly-synchronized clocks (all clocks are within Δ of the "real time") and all messages sent on the network are delivered within Δ time, and assuming the existence of sub-exponentially secure collision-resistant hash functions and enhanced trapdoor permutations. Perhaps surprisingly, our protocol significantly departs from the standard approaches to distributed consensus, and we instead rely on key ideas behind Nakamoto's blockchain protocol (while dispensing the need for "proofs-of-work"). We finally observe that sleepy consensus is impossible in the presence of a dishonest majority of online players.

Keywords: Blockchains · Distributed consensus · Protocol · Adaptive security

© International Association for Cryptologic Research 2017
T. Takagi and T. Peyrin (Eds.): ASIACRYPT 2017, Part II, LNCS 10625, pp. 380–409, 2017.
https://doi.org/10.1007/978-3-319-70697-9_14

1   Introduction

Consensus protocols are at the core of distributed computing and also provide a foundational building protocol for multi-party cryptographic protocols. In this paper, we consider consensus protocols for realizing a "linearly ordered log" abstraction—often referred to as *state machine replication* or *linearizability* in the distributed systems literature. Such protocols must respect two important resiliency properties, *consistency* and *liveness*. Consistency ensures that all honest nodes have the same view of the log, whereas liveness requires that transactions will be incorporated into the log quickly.

The literature on distributed computing as well as the cryptography literature typically consider two types of players—*honest* players and *corrupted/adversarial* players. The above-mentioned resiliency properties are then analyzed assuming a lower bound on the fraction of honest players (e.g., assuming that at least a majority of the players are honest). Honest players, however, are not only assumed to follow the prescribed the protocol, but also assumed to be *online* throughout the whole execution of the protocol. Whereas this is a perfectly reasonable assumption for the traditional environments in which consensus protocols typically were deployed (e.g., within a company, say "Facebuck", to support an application, say "Facebuck Credit", where the number of nodes/players is roughly a dozen), the advent of "large-scale" consensus protocols (such as e.g., the blockchain protocol)—where want to achieve consensus among *thousands* of players—makes this latter assumption unrealistic. For instance, in Bitcoin, only a small fraction of users having bitcoins are actually participating as miners. More generally, although it has not been explicitly articulated, an implicit desiderata for "permissioned blockchains" seems to be a notion of "availability-friendliness" where the blockchain should be able to confirm new transaction even if only a small fraction of participants are actually actively running the protocol.

1.1   The Sleepy Model of Consensus

Towards addressing this issue, and formalizing the notion of "availability-friendliness", we here initiate a study of distributed protocols in a "sleepy" model of computation. We here focus on the "standard" permission setting with a fixed number of player and the existence of a PKI. In the sleepy model, players can be either *online* ("awake/active") or *offline* ("asleep"), and their online status may change at any point during the protocol execution. The main question we address is:

> *Can we design consensus protocols that remain resilient under "sporadic participation"—where at any given point, only a subset of the players are actually online—assuming an appropriate fraction (e.g., majority) of the online players are honest?*

As far as we know, this question was first raised by Micali [19] in a recent manuscript[1]—he writes "*... a user missing to participate in even a single round is pessimistically judged malicious—although, in reality, he may have only experienced a network-connection problem, or simply taken a "break". [..] One possibility would be to revise the current Honest Majority of Users assumption so as it applies only to the "currently active" users rather than the "currently existing" users.*" In Micali's work, however, a different path is pursued.[2] In contrast, our goal here is to address this question. We believe that such a sleepy model of computation is the "right way" to formalize the availability-friendliness desiderata informally articulated for permissioned blockchains, and elucidates the resistence in the blockchain community to adopt classic consensus protocols for permissioned blockchains.

We note that it is easy to show that consensus in the sleepy model is impossible, even in a PKI model, unless we assume that at least a majority of the awake players are honest (if the set of awake players can arbitrarily change throughout the execution)—briefly, the reason for this is that a player that wakes up after being asleep for a long time cannot distinguish the real execution by the honest player and an emulated "fake" execution by the malicious players, and thus must choose the "fake" one with probability at least $\frac{1}{2}$. We formalize this in the online full version [25]. (Note that this simple result already demonstrates a sharp contrast with the classic (non-sleepy) model, where Dolev-Strong's [7] protocol can be used to realize a linearly ordered log of transactions assuming just the existence of a single honest player [25, 27].)

We then consider the following question:

*Can we design a consensus protocol that achieves consistency and liveness assuming only that a **majority of the online players** are honest?*

As far as we know, all standard consensus protocols break down in the sleepy model, even if we assume that 99% of the online players are honest! Briefly, the standard protocols can be divided into two types: (1) protocols that assume *synchronous communication*, where all messages sent by honest players are guaranteed to be received by all other honest nodes in the next round; or, (2) protocols handling *partially synchronous* or *asynchronous* communication, but in this case require knowledge of a tight bound on the number of actually participating honest players. In more detail:

- Traditional synchronous protocols (e.g., [7, 10, 12]) crucially rely on messages being delivered in the next round (or within a known, bounded delay Δ) to

---

[1] Although our paper is subsequent, at the original time of writing this paper, we were actually not aware of this; this discussion was present in the arXiv version from August 2016, but is no longer present in the most recent version of his manuscript.

[2] Briefly, rather than designing a protocol that remains resilient under this relaxed honesty assumption, he designs a protocol under an incomparable "honest-but-lazy" assumption, where honest players only are required to participate at infrequent but individually prescribed rounds (and if they miss participation in their prescribed round, they are deemed corrupted). Looking forward, the honest strategy in our protocols also satisfies such a laziness property.

reach agreement. By contrast, in the sleepy model, consider an honest player that falls asleep for a long time (greater than Δ) and then wakes up at some point in the future; it now receives all "old" messages with a significantly longer delay (breaking the synchrony assumption). In these protocols, such a player rejects all these old messages and would never reach agreement with the other players. It may be tempting to modify e.g., the protocol of [7] to have the players reach agreement on some transaction if some threshold (e.g., majority) of players have approved it—but the problem then becomes how to set the threshold, as the protocol is not aware of how many players are actually awake!

– The partially synchronous or asynchronous protocols (e.g., [3,5,8,18,21,28]) a-priori seem to handle the above-mentioned issue with the synchronous protocol: we can simply view a sleeping player as receiving messages with a long delay (which is allowed in the asynchronous model of communication). Here, the problem instead is the fact that the number of awake players may be significantly smaller than the total number of players, and this means that no transactions will ever be approved! A bit more concretely, these protocols roughly speaking approve transactions when a certain *number* of nodes have "acknowledged" them—for instance, in the classic BFT protocol of Castro and Liskov [5] (which is resilient in the standard model assuming a fraction $\frac{2}{3}$ of *all players* are honest), players only approve a transaction when they have heard $\frac{2N}{3}$ "confirmations" of some message where N is the total number of parties. The problem here is that if, say, only half of the N players are awake, the protocols stalls. (And again, as for the case of synchronous protocols, it is not clear how to modify this threshold without knowledge of the number of awake players.)

1.2   Main Result

Our main result answers the above question in the affirmative. We present constructions of consensus protocols in the sleepy model, which are resilient assuming only that a *majority of the awake players are honest*. Our protocols relies on the existence of a "bare" Public-Key Infrastructure (PKI)[3], the existence of a Common *Random* String (CRS)[4] and is proven secure in a simple version of the *timing model* of Dwork-Naor-Sahai [22] where all players are assumed to have weakly-synchronized clocks—all clocks are within Δ of the "real time", and all messages sent on the network are delivered within Δ time.

Our first protocol relies only on the existence of collision-resistant hash functions (and it is both practical and extremely simple to implement, compared to standard consensus protocols); it, however, only supports static corruptions and

---

[3] That is, players have some way of registering public keys; for honest players, this registration happens before the beginning of the protocol, whereas corrupted players may register their key at any point. We do not need players to e.g., prove knowledge of their secret-key.

[4] That is a commonly known truly random string "in the sky".

a static (fixed) schedule of which nodes are awake at what time step—we refer to this as a "static sleep schedule".

Theorem 1 (Informal). *Assume the existence of families of collision-resistant hash functions (CRH). Then, there exists a protocol for state-machine replication in the Bare PKI, CRS and in the timing model, which achieves consistency and liveness assuming a static sleep schedule and static corruptions, as long as at any point in the execution, a majority of the awake players are honest.*

Our next construction enhances the first one by achieving also resilience against an arbitrary adversarial selection of which nodes are online at what time; this protocol also handles adaptive corruptions of players. This new protocol, however, does so at the price of assuming subexponentially secure collision-resistant hash functions and enhanced trapdoor permutations (the latter are needed for the constructions of non-interactive zero-knowledge proofs); additionally, we here require using a large security parameter (greater than the bound on the total number of players), leading to a less efficient protocol.

Theorem 2 (Informal). *Assume the existence of families of sub-exponentially secure collision-resistant hash functions (CRH), and enhanced trapdoor permutations (TDP). Then, there exists a state-machine replication protocol in the Bare PKI, CRS and timing model, which achieves consistency and liveness under adaptive corruptions as long as at any point in the execution, a majority of the awake players are honest.*

We finally point out that if the CRS is selected after the public keys have been registered, and if we only need security w.r.t. static corruption but adaptive sleep schedules, we do not need to use a subexponential security or a large security parameter. (In fact, we can also get security w.r.t. adaptive corruptions with *erasure* if we additionally assuming the existence of a forward-secure signature, and consider a protocol in the random oracle model.)

Perhaps surprisingly, our protocol significantly departs from the standard approaches to distributed consensus, and we instead rely on key ideas behind Nakamoto's beautiful blockchain protocol [23], while dispensing the need for "proofs-of-work" [9]. As far as we know, our work demonstrates for the first time how the ideas behind Nakamoto's protocol are instrumental in solving "standard" problems in distributed computing; we view this as our main conceptual contribution (and hopefully one that will be useful also in other contexts).

Our proof will leverage and build on top of the formal analysis of the Nakamoto blockchain by Pass et al. [24], but since we no longer rely on proofs-of-work, several new obstacles arise. Our main technical contribution, and the bulk of our analysis, is a new combinatorial analysis for dealing with these issues.

We finally mention that ad-hoc solutions for achieving consensus using ideas behind the blockchain (but without proof-of-work) have been proposed [1,2,15], none of these come with an analysis, and it is not clear to what extent they improve upon standard state-machine replication protocols (and more seriously, whether they even achieve the standard notion of consensus).

1.3 Technical Overview

We start by providing an overview of our consensus protocol which only handles a static online schedule and static corruptions; we next show how to enhance this protocol to achieve adaptive security.

As mentioned, the design of our consensus protocols draws inspiration from Bitcoin's proof-of-work based blockchain [23]—the so-called "Nakamoto consensus" protocol. This protocol is designed to work in a so-called "permissionless setting" where anyone can join the protocol execution. In contrast, we here study consensus in the classic "permissioned" model of computation with a fixed set $[N]$ of participating players; additionally, we are assuming that the players can register public keys (whose authenticity can be verified). Our central idea is to eliminate the use of proofs of work in this protocol. Towards this goal, let us start by providing a brief overview of Nakamoto's beautiful blockchain protocol.

Nakamoto Consensus in a Nutshell. Roughly speaking, in Nakamoto's blockchain, players "confirm" transactions by "mining blocks" through solving some computational puzzle that is a function of the transactions and the history so far. More precisely, each participant maintains its own local "chain" of "blocks" of transactions—called the *blockchain*. Each block consists of a triple $(h_{-1}, \eta, \text{txs})$ where h_{-1} is a pointer to the previous block in chain, txs denotes the transactions confirmed, and η is a "proof-of-work"—a solution to a computational puzzle that is derived from the pair (h_{-1}, txs). The proof of work can be thought of as a "key-less digital signature" on the whole blockchain up until this point. At any point of time, nodes pick the *longest* valid chain they have seen so far and try to extend this longest chain.

Removing Proofs-of-Work. Removing the proof-of-work from the Nakamoto blockchain while maintaining provable guarantees turns out to be subtle and the proof non-trivial. To remove the proof-of-work from Nakamoto's protocol, we proceed as follows: instead of rate limiting through computational power, we impose limits on the type of puzzle solutions that are admissible for each player. More specifically, we redefine the puzzle solution to be of the form $(\mathcal{P}, t)$ where $\mathcal{P}$ is the player's identifier and t is referred to as the block-time. An honest player will always embed the current time step as the block-time. The pair $(\mathcal{P}, t)$ is a "valid puzzle solution" if $\mathsf{H}(\mathcal{P}, t) < D_p$ where H denotes a random oracle (for now, we provide a protocol in the random oracle model, but as we shall see shortly, the random oracle can be instantiated with a CRS and a pseudorandom function), and D_p is a parameter such that the hash outcome is only smaller than D_p with probability p. If $\mathsf{H}(\mathcal{P}, t) < D_p$, we say that $\mathcal{P}$ is *elected leader at time t*. Note that several nodes may be elected leaders at the same time steps.

Now, a node $\mathcal{P}$ that is elected leader at time step t can extend a chain with a block that includes the "solution" $(\mathcal{P}, t)$, as well as the previous block's hash h_{-1} and the transactions txs to be confirmed. To verify that the block indeed came from $\mathcal{P}$, we require that the entire contents of the block, (i.e., $(h_{-1}, \text{txs}, t, \mathcal{P})$), are signed under $\mathcal{P}$'s public key. Similarly to Nakamoto's protocol, nodes then choose the longest valid chain they have seen and extend this longest chain.

Whereas honest players will only attempt to mine solutions of the form $(\mathcal{P}, t)$ where t is the *current* time step, so far there is nothing that prevents the adversary from using incorrect block-times (e.g., time steps in past or the future). To prevent this from happening, we additionally impose the following restriction on the block-times in a valid chain:

1. A valid chain must have strictly increasing block-times;
2. A valid chain cannot contain any block-times in the "future" (where "future" is adjusted to account for nodes' clock offsets)

There are now two important technical issues to resolve. First, it is important to ensure that the block-time rules do not hamper liveness. In other words, there should not be any way for an adversary to leverage the block-time mechanism to cause alert nodes to get stuck (e.g., by injecting false block-times).

Second, although our block-time rules severely constrain the adversary, the adversary is still left with some wiggle room, and gets more advantages than honest nodes. Specifically, as mentioned earlier, the honest nodes only "mine" in the present (i.e., at the actual time-step), and moreover they never try to extend different chains of the same length. By contrast, the adversary can try to reuse past block-times in multiple chains. (In the proof of work setting, these types of attacks are not possible since there the hash function is applied also to the history of the chain, so "old" winning solutions cannot be reused over multiple chains; in contrast, in our protocol, the hash function is no longer applied to the history of the chain as this would give the attacker too many opportunities to become elected a leader by simply trying to add different transactions.)

Our main technical result shows that this extra wiggle room in some sense is insignificant, and the adversary cannot leverage the wiggle room to break the protocol's consistency guarantees. It turns out that dealing with this extra wiggle room becomes technically challenging, and none of the existing analysis for proof-of-work blockchains [11, 24] apply. More precisely, since we are using a blockchain-style protocol, a natural idea is to see whether we can directly borrow proof ideas from existing analyses of the Nakamoto blockchains [11, 24]. Existing works [11, 24] define three properties of blockchains—*chain growth* (roughly speaking that the chain grows at a certain speed), *chain quality* (that the adversary cannot control the content of the chain) and *consistency* (that honest players always agree on an appropriate prefix of the chain)—which, as shown in earlier works [24, 26] imply the consistency and liveness properties needed for state-machine replication. Thus, by these results, it will suffice to demonstrate that our protocol satisfies these properties.

The good news is that chain growth and chain quality properties can be proven in almost identically the same way as in earlier Nakamoto blockchain analysis [24]. The bad news is that the consistency proofs of prior works [11, 24] break down in our setting (as the attacker we consider is now more powerful as described above). The core of our proof is a new, and significantly more sophisticated analysis for dealing with this.

Removing the Random Oracle. The above-described protocol relies on a random oracle. We note that we can in fact instantiate the random oracle with a PRF whose seed is selected and made public in a common reference string (CRS). Roughly speaking, the reason for this is that in our proof we actually demonstrate the existence of some simple polynomial-time computable events—which only depend on the output of the random oracle/PRF—that determine whether *any* (even unbounded) attacks can succeed. Our proof shows that with overwhelming probability over the choice of the random oracle, these events do not happen. By the security of the PRF, these events thus also happen only with negligible probability over the choice of the seed of the PRF.

Dealing with Adaptive Sleepiness and Corruption. We remark that the above-described protocol only works if the choice of when nodes are awake is made before the PRF seed is selected. If not, honest players that are elected leaders could simply be put to sleep at the time step when they need to act. The problem is that it is *preditcable* when a node will become a leader. To overcome this problem, we take inspiration from a beautiful idea from Micali's work [19]—we let each player pick its own *secret seed* to a PRF and publish a commitment to the seed as part of its public key; the player can then evaluate its own private PRF and also prove in zero-knowledge that the PRF was correctly evaluated (so everyone else can verify the correctness of outputs of the PRF);[5]. Finally, each player now instantiates the random oracle with their own "private" PRF. Intuitively, this prevents the above-mentioned attack, since even if the adversary can adaptively select which honest nodes go to sleep, it has no idea which of them will become elected leaders before they broadcast their block.

Formalizing this, however, is quite tricky (and we will need to modify the protocol). The problem is that if users pick their own seed for the PRF, then they may be able to select a "bad seed" which makes them the leader for a long period of time (there is nothing in the definition of a PRF that prevents this). To overcome this issue, we instead perform a "coin-tossing into the well" for the evaluation of the random oracle: As before, the CRS specifies the seed k_0 of a PRF, and additionally, each user $\mathcal{P}$ commits to the seed $k[\mathcal{P}]$ of a PRF as part of their public key; node $\mathcal{P}$ can then use the following function to determine if it is elected in time t

$$\mathsf{PRF}_{k_0}(\mathcal{P}, t) \oplus \mathsf{PRF}_{k[\mathcal{P}]}(t) < D_p$$

where D_p is a difficulty parameter selected such that any single node is elected with probability p in a given time step. $\mathcal{P}$ additionally proves in zero-knowledge that it evaluated the above leader election function correctly in any block it produces.

But, have we actually gained anything? A malicious user may still pick its seed $k[\mathcal{P}]$ after seeing k_0 and this may potentially cancel out the effect of having $\mathsf{PRF}_{k_0}(\cdot)$ there in the first place! (For instance, the string $\mathsf{PRF}_{k_0}(\mathcal{P}, t) \oplus \mathsf{PRF}_{k[\mathcal{P}]}(t)$

---

[5] In essence, what we need is a VRF [20], just like Micali [19], but since we anyway have a CRS, we can rely on weaker primitives.

clearly is not random any more.) We note, however, that if the user seed $k[\mathcal{P}]$ is *significantly shorter* than the seed k_0, and the cryptographic primitives are subexponentially secure, we can rely on the same method that we used to replace the random oracle with a PRF to argue that even if $k[\mathcal{P}]$ is selected as a function of k_0, this only increases the adversary's success probability by a factor 2^L for each possibly corrupted user where $L := |k[\mathcal{P}]|$ is the bit-length of each user's seed (and thus at most 2^{NL} where N is the number of players) which still will not be enough to break security, if using a sufficiently big security parameter for the underlying protocol. We can finally use a similar style of a union bound to deal also with adaptive corruptions.

Better Efficiency and Assumption in Stronger Models. We note that the loss in efficiency due to the above-mentioned union bounds is non-trivial: the security parameter must now be greater than N; if we only require security with respect to static corruption, and allow the CRS to be selected *after* all public keys are registered—which would be reasonable in practice—then, we can deal with adaptive sleepiness without this union bound and thus without the loss in efficiency.

In fact, we can even deal with adaptive corruption of players in a model *with erasures* if we let players sign using a forward secure signature scheme (and at each step erase the old key). For technical reasons, we here, however, can only provide a proof of security in the random oracle model.[6] We defer the details of these approaches to the online full version [25].

1.4   Applications in Permissioned and Permissionless Settings

As mentioned, the variants of our protocols that deal with static corruption (and adaptive corruption with erasures) need not employ a large security parameter and can be implemented and adopted in real-world systems. We believe that our sleepy consensus protocol would be desirable in the following application scenarios.

Permissioned Setting: Consortium Blockchains. At the present, there is a major push where blockchain companies are helping banks across the world build "consortium blockchains". In a consortium blockchain, a consortium of banks each contribute some nodes and jointly run a consensus protocol, on top of which one can run distributed ledger and smart contract applications. Since enrollment is controlled, consortium blockchain falls in the classical "permissioned" model of consensus. Since the number of participating nodes may be large (e.g., typically involve hundreds of banks and possibly hundreds to thousands of nodes), many conjecture that classical protocols such as PBFT [5], Byzantine Paxos [16], and others where the total bandwidth scales quadratically w.r.t. the number of players might not be ideal in such settings. Our sleepy consensus protocol provides a

---

[6] Technically, the issue is that we need to rely on a PRF that is secure with respect to "selective-opening" and we can only construct this in the random oracle model.

compelling alternative in this setting—with sleepy consensus, tasks such as committee re-configuration can be achieved simply without special program paths like in classical protocols [17], and each bank can also administer their nodes without much coordination with other banks.

Permissionless Setting: Proof-of-Stake. The subsequent work Snow White by Daian et al. [6] adapted our protocol to a permissionless setting, and obtained one of the first provably secure proof-of-stake protocols. A proof-of-stake protocol is a permissionless consensus protocol to be run in an open, decentralized setting, where roughly speaking, each player has voting power proportional to their amount of "stake" in the cryptocurrency system (*c.f.* proof-of-work is where players have voting power proportional to their available computing power). Major cryptocurrencies such as Ethereum are eager to switch to a proof-of-stake model rather than proof-of-work to dispense with wasteful computation. To achieve proof-of-stake, the Snow White [6] extended the our sleepy consensus protocol by introducing a mechanism that relies on the distribution of stake in the system to periodically rotate the consensus committee.

Comparison with Independent Work. Although proof-of-stake is not a focus of our paper, we compare with a few independent works on proof-of-stake [14,19] due to the superficial resemblance of some elements of their protocol in comparison with ours. Specifically, the elegant work by Micali proposes to adapt classical style consensus protocols to realize a proof-of-stake protocol [19]; the concurrent and independent work by Kiayias et al. [14] proposes to use a combination of blockchain-style protocol and classical protocols such as coin toss to realize proof-of-stake. Both these works would fail in the sleepy model like any classical style protocol. We also point out that if we were to replace Kiayias et al.'s coin toss protocol with an ideal random beacon, their proof would still fail in the sleepy model. Other proof-of-stake protocols [1,2,15] may also bear superficial resemblance but they do not have formal security models or provable guarantees, and these protocols may also miss elements that turned out essential in our proofs. As far as we are aware, we are the first to formally show how to remove proof-of-work from Nakamoto's protocol in a provably secure way, while maintaining its desirable "availiability-friendliness" property.

1.5   Paper Organization

The remainder of the paper is organized as follows. In Sect. 2, we formally define the sleepy execution model. In Sect. 3, we define the abstraction realized by a blockchain protocol and a state machine replication protocol. Our goal in this paper is to realize state machine replication where a set of nodes agree on an growing log of transactions—but we achieve this through realizing a blockchain protocol—as previous works show [26], blockchains give a direct method of instantiating state machine replication.

Then, in Sect. 4, we describe the sleepy consensus protocol that implements the blockchain abstraction in the sleepy execution model, assuming a static sleep schedule and static corruptions. We present an overview of our proof for this statically secure protocol in Sect. 5, and defer the full proofs to the online full version [25].

Finally, in Sect. 6, we describe intuitively how to achieve adaptive security by leveraging additional cryptographic building blocks such as non-interactive zero-knowledge proofs and by relying on subexponential security and large security parameters (and the above-mentioned union bounds).

We defer the full presentation of the adaptively secure sleepy consensus protocol, as well as the more efficient variants mentioned above, to the online full version [25]. The full version also contain new lower bounds for the sleepy model.

2 Definitions

2.1 Protocol Execution Model

We assume a standard Interactive Turing Machine (ITM) model often adopted in the cryptography literature. A protocol specifies a set of instructions for the protocol participants to interact with each other. The protocol's execution is directed by an environment denoted $\mathcal{Z}$ who is in charge of activating a number of parties (also referred to as nodes) that will interact with each other.

Notations for Randomized Protocol Execution. We use the notation $\mathsf{view} \leftarrow_\$ \mathsf{EXEC}^\Pi (\mathcal{A}, \mathcal{Z}, \lambda)$ to denote a randomized execution of the protocol Π with security parameter λ and w.r.t. to an $(\mathcal{A}, \mathcal{Z})$ pair. Specifically, view is a random variable containing an ordered sequence of all inputs, outputs, and messages sent and received by all Turing Machines during the protocol's execution. We use the notation $|\mathsf{view}|$ to denote the number of time steps in the execution trace view.

Weak Clock Synchrony Assumptions. We assume the protocol execution proceeds in atomic time steps called rounds. Henceforth in the paper, we will use the terms "round" and "time" interchangably. We assume that all honest nodes are aware of the present round number (i.e., time). As is well-known, such a model can also capture the case of *weak clock synchrony* by treating the clock offset as part of the network delay. Specifically, in a setting where the maximum network delay and the maximum clock offset are both Δ (and if the clock offset is larger than the maximum network delay, we take the maximum of the two to be Δ), we can equivalently treat it as a setting with perfectly synchronized clocks but 3Δ maximum network delay. Our clock synchrony assumptions are similar to those described by Dwork et al. [8].

Public-Key Infrastructure. We assume the existence of a public-key infrastructure (PKI). Specifically, we adopt the same technical definition of a PKI as in the Universal Composition framework [4]. Specifically, we shall assume that the PKI is an ideal functionality $\mathcal{F}_{\mathrm{CA}}$ (available only to the present protocol instance) that does the following:

- On receive register(upk) from $\mathcal{P}$: remember (upk, $\mathcal{P}$) and ignore any future message from $\mathcal{P}$.
- On receive lookup($\mathcal{P}$): return the stored upk corresponding to $\mathcal{P}$ or $\perp$ if none is found.

In this paper, we will consider a Bare PKI model, nodes are allowed register their public keys with $\mathcal{F}_{CA}$ any time during the execution—although typically, the honest protocol may specify that honest nodes register their public keys upfront at the beginning of the protocol execution (nonetheless, corrupt nodes may still register late).

Corruption Model. At the beginning of any time step t, $\mathcal{Z}$ can issue instructions of the form

$$(\text{corrupt}, i) \text{ or } (\text{sleep}, i, t_0, t_1) \text{ where } t_1 \geq t_0 \geq t$$

$(\text{corrupt}, i)$ causes node i to become corrupt at the current time, whereas $(\text{sleep}, i, t_0, t_1)$ where $t_1 \geq t_0 \geq t$ will cause node i to sleep during $[t_0, t_1]$. Note that since corrupt or sleep instructions must be issued at the very beginning of a time step, $\mathcal{Z}$ cannot inspect an honest node's message to be sent in the present time step, and then retroactively make the node sleep in this time step and erase its message.

Following standard cryptographic modeling approaches, at any time, the environment $\mathcal{Z}$ can communicate with corrupt nodes in arbitrary manners. This also implies that the environment can see the internal state of corrupt nodes. Corrupt nodes can deviate from the prescribed protocol arbitrarily, i.e., exhibit byzantine faults. All corrupt nodes are controlled by a probabilistic polynomial-time adversary denoted $\mathcal{A}$, and the adversary can see the internal states of corrupt nodes. For honest nodes, the environment cannot observe their internal state, but can observe any information honest nodes output to the environment by the protocol definition.

To summarize, a node can be in one of the following states:

1. *Honest.* An honest node can either be *awake* or *asleep* (or sleeping/sleepy). Henceforth we say that a node is *alert* if it is honest and awake. When we say that a node is *asleep* (or sleeping/sleepy), it means that the node is honest and asleep.
2. *Corrupt.* Without loss of generality, we assume that all corrupt nodes are *awake*.

Henceforth, we say that corruption (or sleepiness resp.) is *static* if $\mathcal{Z}$ must issue all corrupt (or sleep resp.) instructions before the protocol execution starts. We say that corruption (or sleepiness resp.) is *adaptive* if $\mathcal{Z}$ can issue corrupt (or sleep resp.) instructions at any time during the protocol's execution.

Communication Model. The adversary is responsible for delivering messages between nodes. We assume that the adversary $\mathcal{A}$ can *delay* or *reorder* messages arbitrarily, as long as it respects the constraint that all messages sent from honest nodes must be received by all honest nodes in at most Δ time steps.

When a sleepy node wakes up, $(\mathcal{A}, \mathcal{Z})$ is required to deliver an unordered set of messages containing

- all the pending messages that node i would have received (but did not receive) had it not slept; and
- any polynomial number of adversarially inserted messages of $(\mathcal{A}, \mathcal{Z})$'s choice.

Henceforth in this paper, we assume that our protocol is aware of the maximum network delay Δ—in fact, we prove that no protocol without knowledge of Δ can reach consensus in the sleepy model (see our online version [25]).

2.2   Compliant Executions

Parameters of an Execution. Globally, we will use N to denote (an upper bound on) the total number of nodes, and N_{crupt} to denote (an upper bound on) the number of corrupt nodes, and Δ to denote the maximum delay of messages between alert nodes. More formally, we can define a $(N, N_{\mathrm{crupt}}, \Delta)$-respecting $(\mathcal{A}, \mathcal{Z})$ as follows.

Definition 1 $((N, N_{\mathrm{crupt}}, \Delta)$**-respecting** $(\mathcal{A}, \mathcal{Z}))$. *Henceforth, we say that $(\mathcal{A}, \mathcal{Z})$ is $(N, N_{\mathrm{crupt}}, \Delta)$-respecting w.r.t. protocol Π, iff the following holds: for any view in the support of $\mathsf{EXEC}^{\Pi}(\mathcal{A}, \mathcal{Z}, \lambda)$,*

- *$(\mathcal{A}, \mathcal{Z})$ spawns a total of N nodes in view among which N_{crupt} are corrupt and the remaining are honest.*
- *If an alert node i multicasts a message at time t in view, then any node j alert at time $t' \geq t + \Delta$ (including ones that wake up after t) will have received the message.*

Henceforth when the context is clear, we often say that $(\mathcal{A}, \mathcal{Z})$ is $(N, N_{\mathrm{crupt}}, \Delta)$-respecting omitting stating explicitly the protocol Π of interest.

Protocol-Specific Compliance Rules. A protocol Π may ensure certain security guarantees only in executions that respect certain compliance rules. Compliance rules can be regarded as constraints imposed on the $(\mathcal{A}, \mathcal{Z})$ pair. Henceforth, we assume that besides specifying the instructions of honest parties, a protocol Π will additionally specify a set of compliance rules. We will use the notation a

$$\Pi\text{-compliant } (\mathcal{A}, \mathcal{Z}) \text{ pair}$$

to denote an $(\mathcal{A}, \mathcal{Z})$ pair that respects the compliance rules of protocol Π—we also say that $(\mathcal{A}, \mathcal{Z})$ is compliant w.r.t. to the protocol Π.

2.3   Notational Conventions

Negligible Functions. A function $\mathsf{negl}(\cdot)$ is said to be *negligible* if for every polynomial $p(\cdot)$, there exists some λ_0 such that $\mathsf{negl}(\lambda) \leq \frac{1}{p(\lambda)}$ for every $\lambda \geq \lambda_0$.

Variable Conventions. In this paper, unless otherwise noted, all variables are by default functions of the security parameter λ. Whenever we say $\mathsf{var}_0 > \mathsf{var}_1$, this means that $\mathsf{var}_0(\lambda) > \mathsf{var}_1(\lambda)$ for every $\lambda \in \mathbb{N}$. Similarly, if we say that a variable var is positive or non-negative, it means positive or non-negative for every input λ. Variables may also be functions of each other. How various variables are related will become obvious when we define derived variables and when we state parameters' admissible rules for each protocol. Importantly, *whenever a parameter does not depend on λ, we shall explicitly state it by calling it a constant.*

Unless otherwise noted, we assume that all variables are non-negative (functions of λ). Further, unless otherwise noted, all variables are *polynomially bounded* (or *inverse polynomially bounded* if smaller than 1) functions of λ.

3   Problem Definitions

In this section, we first formally define a state machine replication protocol; roughly speaking, in state machine replication, nodes agree on a *linearly ordered log* over time, in a way that satisfies consistency and liveness. We next define a blockchain abstraction following Pass et al. [24] (which in turn relies on properties from Garay et al. [11]). For both of these definitions, we extend the definitions to the sleepy model of computation. Finally, as shown by Pass and Shi [26], any protocol satisfying the blockchain abstraction can be turned (by simply truncating the last few blocks) into a state machine replication protocol, and the same results applies also in the sleepy model. As a consequence, it will later suffice to simply prove that our protocol securely implements the blockchain abstraction.

3.1   State Machine Replication

We turn to formalizing the notion of state machine replication; we use the definition from [26] and extend it to the sleepy model by simply replacing honest nodes for alert nodes.

Inputs and Outputs. The environment $\mathcal{Z}$ may input a set of transactions txs to each alert node in every time step. In each time step, an alert node outputs to the environment $\mathcal{Z}$ a totally ordered LOG of transactions (possibly empty).

Security Definitions. Let T_{confirm} be a polynomial function in $\lambda, N, N_{\mathrm{crupt}}$, and Δ. We say that a state machine replication protocol Π is secure and has transaction conformation time T_{confirm} if for every Π-compliant $(\mathcal{A}, \mathcal{Z})$ that is $(N, N_{\mathrm{crupt}}, \Delta)$-respecting, there exists a negligible function negl such that for every sufficiently large $\lambda \in \mathbb{N}$, all but $\mathsf{negl}(\lambda)$ fraction of the views sampled from $\mathsf{EXEC}^{\Pi}(\mathcal{A}, \mathcal{Z}, \lambda)$ satisfy the following properties:

- *Consistency.* An execution trace view satisfies consistency if the following holds:
    - *Common prefix.* Suppose that in view, an alert node i outputs LOG to $\mathcal{Z}$ at time t, and an alert node j (same or different) outputs LOG$'$ to $\mathcal{Z}$ at time t', it holds that either LOG $\prec$ LOG$'$ or LOG$'$ $\prec$ LOG. Here the relation $\prec$ means "is a prefix of". By convention we assume that $\emptyset \prec x$ and $x \prec x$ for any x.
    - *Self-consistency.* Suppose that in view, a node i is alert at time t and $t' \geq t$, and outputs LOG and LOG$'$ at times t and t' respectively, it holds that LOG $\prec$ LOG$'$.
- *Liveness.* An execution trace view satisfies T_{confirm}-liveness if the following holds: suppose that in view, the environment $\mathcal{Z}$ inputs some set that contains tx to an alert node at time $t \leq |\mathsf{view}| - T_{\mathrm{confirm}}$, or that tx appears in some honest node's LOG at time t. Then, for any node i alert at any time $t' \geq t + T_{\mathrm{confirm}}$, let LOG be the output of node i at time t', it holds that any tx $\in$ LOG.

    Intuitively, liveness says that transactions input to an alert node get included in their LOGs within T_{confirm} time; further, if some transaction shows up in an honest node's LOG, it will show up in all other alert nodes' logs quickly.

3.2 Blockchain Formal Abstraction

In this section, we define the formal abstraction and security properties of a blockchain. Our definitions is identical to the abstraction of Pass et al. [24] (which in turn is based on earlier definitions from Garay et al. [11], and Kiayias and Panagiotakos [13]), and again simply replaces honest nodes with alert nodes.

Inputs and Outputs. We assume that in every time step, the environment $\mathcal{Z}$ provides a possibly empty input to every alert node. Further, in every time step, an alert node sends an output to the environment $\mathcal{Z}$. Given a specific execution trace view where $|\mathsf{view}| \geq t$, let i denote a node that is alert at time t in view, we use the following notation to denote the output of node i to the environment $\mathcal{Z}$ at time step t,

$$\text{output to } \mathcal{Z} \text{ by node } i \text{ at time } t \text{ in view}: \quad \mathsf{chain}_i^t(\mathsf{view})$$

where chain denotes an extracted ideal blockchain where each block contains an ordered list of transactions. Sleepy nodes stop outputting to the environment until they wake up again.

Chain Growth. The first desideratum is that the chain grows proportionally with the number of time steps. Let,

$$\mathsf{min\text{-}chain\text{-}increase}^{t,t'}(\mathsf{view}) = \min_{i,j} \left(|\mathsf{chain}_j^{t+t'}(\mathsf{view})| - |\mathsf{chain}_i^t(\mathsf{view})| \right)$$

$$\mathsf{max\text{-}chain\text{-}increase}^{t,t'}(\mathsf{view}) = \max_{i,j} \left(|\mathsf{chain}_j^{t+t'}(\mathsf{view})| - |\mathsf{chain}_i^t(\mathsf{view})| \right)$$

where we quantify over nodes i, j such that i is alert in time step t and j is alert in time $t + t'$ in view.

Let $\text{growth}^{t_0, t_1}(\text{view}, \Delta, T) = 1$ iff the following two properties hold:

- **(consistent length)** for all time steps $t \leq |\text{view}| - \Delta$, $t + \Delta \leq t' \leq |\text{view}|$, for every two players i, j such that in view i is alert at t and j is alert at t', we have that $|\text{chain}_j^{t'}(\text{view})| \geq |\text{chain}_i^{t}(\text{view})|$
- **(chain growth lower bound)** for every time step $t \leq |\text{view}| - t_0$, we have

$$\text{min-chain-increase}^{t, t_0}(\text{view}) \geq T.$$

- **(chain growth upper bound)** for every time step $t \leq |\text{view}| - t_1$, we have

$$\text{max-chain-increase}^{t, t_1}(\text{view}) \leq T.$$

In other words, growth^{t_0, t_1} is a predicate which tests that (a) alert parties have chains of roughly the same length, and (b) during any t_0 time steps in the execution, all alert parties' chains increase by at least T, and (c) during any t_1 time steps in the execution, alert parties' chains increase by at most T.

Definition 2 (Chain growth). *A blockchain protocol Π satisfies (T_0, g_0, g_1)-chain growth, if for all Π-compliant pair $(\mathcal{A}, \mathcal{Z})$, there exists a negligible function negl such that for every sufficiently large $\lambda \in \mathbb{N}$, $T \geq T_0$, $t_0 \geq \frac{T}{g_0}$ and $t_1 \leq \frac{T}{g_1}$ the following holds:*

$$\Pr\left[\text{view} \leftarrow EXEC^{\Pi}(\mathcal{A}, \mathcal{Z}, \lambda) : \text{growth}^{t_0, t_1}(\text{view}, \Delta, \lambda) = 1 \right] \geq 1 - \text{negl}(\lambda)$$

Additionally, we say that a blockchain protocol Π satisfies (T_0, g_0, g_1)-chain growth w.r.t. failure probability $\text{negl}(\cdot)$ if the above definition is satisfied when the negligible function is fixed to $\text{negl}(\cdot)$ for any Π-compliant $(\mathcal{A}, \mathcal{Z})$.

Chain Quality. The second desideratum is that the number of blocks contributed by the adversary is not too large.

Given a chain, we say that a block $B := \text{chain}[j]$ is honest w.r.t. view and prefix $\text{chain}[: j']$ where $j' < j$ if in view there exists some node i alert at some time $t \leq |\text{view}|$, such that (1) $\text{chain}[: j'] \prec \text{chain}_i^t(\text{view})$, and (2) $\mathcal{Z}$ input B to node i at time t. Informally, for an honest node's chain denoted chain, a block $B := \text{chain}[j]$ is honest w.r.t. a prefix $\text{chain}[: j']$ where $j' < j$, if earlier there is some alert node who received B as input when its local chain contains the prefix $\text{chain}[: j']$.

Let $\text{quality}^{T}(\text{view}, \mu) = 1$ iff for every time t and every player i such that i is alert at t in view, among any consecutive sequence of T blocks $\text{chain}[j+1..j+T] \subseteq \text{chain}_i^t(\text{view})$, the fraction of blocks that are honest w.r.t. view and $\text{chain}[: j]$ is at least μ.

Definition 3 (Chain quality). *A blockchain protocol Π has (T_0, μ)-chain quality, if for all Π-compliant pair $(\mathcal{A}, \mathcal{Z})$, there exists some negligible function* negl *such that for every sufficiently large* $\lambda \in \mathbb{N}$ *and every* $T \geq T_0$ *the following holds:*

$$\Pr \left[\textit{view} \leftarrow \textit{EXEC}^{\Pi}(\mathcal{A}, \mathcal{Z}, \lambda) : \textit{quality}^T(\textit{view}, \mu) = 1 \right] \geq 1 - \textsf{negl}(\lambda)$$

Additionally, we say that a blockchain protocol Π satisfies (T_0, μ)-chain quality w.r.t. failure probability $\textsf{negl}(\cdot)$ if the above definition is satisfied when the negligible function is fixed to $\textsf{negl}(\cdot)$ for any Π-compliant $(\mathcal{A}, \mathcal{Z})$.

Consistency. Roughly speaking, consistency stipulates common prefix and future self-consistency. Common prefix requires that all honest nodes' chains, except for roughly $O(\lambda)$ number of trailing blocks that have not stabilized, must all agree. Future self-consistency requires that an honest node's present chain, except for roughly $O(\lambda)$ number of trailing blocks that have not stabilized, should persist into its own future. These properties can be unified in the following formal definition (which additionally requires that at any time, two alert nodes' chains must be of similar length).

Let $\textsf{consistent}^T(\textsf{view}) = 1$ iff for all times $t \leq t'$, and all players i, j (potentially the same) such that i is alert at t and j is alert at t' in view, we have that the prefixes of $\textsf{chain}_i^t(\textsf{view})$ and $\textsf{chain}_j^{t'}(\textsf{view})$ consisting of the first $\ell = |\textsf{chain}_i^t(\textsf{view})| - T$ records are identical—this also implies that the following must be true: $\textsf{chain}_j^{t'}(\textsf{view}) > \ell$, i.e., $\textsf{chain}_j^{t'}(\textsf{view})$ cannot be too much shorter than $\textsf{chain}_i^t(\textsf{view})$ given that $t' \geq t$.

Definition 4 (Consistency). *A blockchain protocol Π satisfies T_0-consistency, if for all Π-compliant pair $(\mathcal{A}, \mathcal{Z})$, there exists some negligible function* negl *such that for every sufficiently large $\lambda \in \mathbb{N}$ and every $T \geq T_0$ the following holds:*

$$\Pr \left[\textit{view} \leftarrow \textit{EXEC}^{\Pi}(\mathcal{A}, \mathcal{Z}, \lambda) : \textit{consistent}^T(\textit{view}) = 1 \right] \geq 1 - \textsf{negl}(\lambda)$$

Additionally, we say that a blockchain protocol Π satisfies T_0-consistency w.r.t. failure probability $\textsf{negl}(\cdot)$ if the above definition is satisfied when the negligible function is fixed to $\textsf{negl}(\cdot)$ for any Π-compliant $(\mathcal{A}, \mathcal{Z})$.

Note that a direct consequence of consistency is that at any time, the chain *lengths* of any two alert players can differ by at most T (except with negligible probability).

3.3   Blockchain Implies State Machine Replication

Following [26], we note that a blockchain protocol implies state machine replication, if alert nodes simply output the stablized part of their respective chains (i.e., $\textsf{chain}[: -\lambda]$) as their LOG.

Lemma 1 (Blockchains imply state machine replication [26]). *If there exists a blockchain protocol that satisfies (T_G, g_0, g_1)-chain growth, (T_Q, μ)-chain quality, and T_C-consistency, then there exists a secure state machine replication protocol with confirmation time $T_{\text{confirm}} := O(\frac{T_G + T_Q + T_C}{g_0} + \Delta)$.*

Proof. This lemma was proved in the hybrid consensus paper [26] for a different execution model, but the same proof effectively holds in our sleepy execution model. Specifically, let $\Pi_{\text{blockchain}}$ be such a blockchain protocol. We can consider the following state machine replication protocol denoted Π': whenever an alert node is about to output chain to the environment $\mathcal{Z}$ in $\Pi_{\text{blockchain}}$, it instead outputs chain$[: -T_C]$. Further, suppose that Π''s compliance rules are the same as $\Pi_{\text{blockchain}}$'s. Using the same argument as the hybrid consensus paper [26], it is not hard to see that the resulting protocol is a secure state machine replication protocol with confirmation time $O(\frac{T_G+T_Q+T_C}{g_0} + \Delta)$.

As a consequence, henceforth, we will focus on realizing a blockchain protocol as this directly yields a state machine replication protocol.

4   Sleepy Consensus Under Static Corruptions

In this section, we will describe our basic Sleepy consensus protocol that is secure under static corruptions and static sleepiness. In other words, the adversary (and the environment) must declare upfront which nodes are corrupt as well as which nodes will go to sleep during which intervals. Furthermore, the adversary (and the environment) must respect the constraint that at any moment of time, roughly speaking the majority of *online* nodes are honest.

For simplicity, we will first describe our scheme pretending that there is a random oracle H; and then describe how to remove the random oracle assuming a common reference string.

4.1   Valid Blocks and Blockchains

Before we describe our protocol, we first define the format of valid blocks and valid blockchains.

We use the notation *chain* to denote a real-world blockchain. Our protocol relies on an extract function that extracts an ordered list of transactions from *chain* which alert nodes shall output to the environment $\mathcal{Z}$ at each time step. A blockchain is obviously a chain of blocks. We now define a valid block and a valid blockchain.

Valid Blocks. We say that a tuple $B := (h_{-1}, \text{txs}, \text{time}, \mathcal{P}, \sigma, h)$ is a valid block iff[7]

1. $\Sigma.\text{ver}_{\text{pk}}((h_{-1}, \text{txs}, \text{time}); \sigma) = 1$ where pk $:= \mathcal{F}_{\text{CA}}.\texttt{lookup}(\mathcal{P})$; an
2. $h = \text{d}(h_{-1}, \text{txs}, \text{time}, \mathcal{P}, \sigma)$, where d $: \{0,1\}^* \to \{0,1\}^\lambda$ is a collision-resistant hash function—technically collision resistant hash functions must be defined for a family, but here for simplicity we pretend that the sampling from the family has already been done before protocol start, and therefore d is a single function.

---

[7] Note that since corrupt nodes can register their public keys with $\mathcal{F}_{\text{CA}}$ late into the protocol, validity is actually defined w.r.t. a point of time during the execution.

Valid Blockchain. Let $\mathsf{eligible}^t(\mathcal{P})$ be a function that determines whether a party $\mathcal{P}$ is an eligible leader for time step t (see Fig. 1 for its definition). Let *chain* denote an ordered chain of real-world blocks, we say that *chain* is a valid blockchain w.r.t. $\mathsf{eligible}$ and time t iff

- $chain[0] = genesis = (\bot, \bot, \mathsf{time} = 0, \bot, \bot, h = \mathbf{0})$, commonly referred to as the genesis block;
- $chain[-1].\mathsf{time} \leq t$; and
- for all $i \in [1..\ell]$ where $\ell := |chain|$, the following holds:
  1. $chain[i]$ is a valid block;
  2. $chain[i].h_{-1} = chain[i-1].h$;
  3. $chain[i].\mathsf{time} > chain[i-1].\mathsf{time}$, i.e., block-times are strictly increasing; and
  4. let $t := chain[i].\mathsf{time}$, $\mathcal{P} := chain[i].\mathcal{P}$, it holds that $\mathsf{eligible}^t(\mathcal{P}) = 1$.

4.2   The Basic Sleepy Consensus Protocol

We present our basic Sleepy consensus protocol in Fig. 1. The protocol takes a parameter p as input, where p corresponds to the probability each node is elected leader in a single time step. All nodes that just spawned will invoke the init entry point. During initialization, a node generates a signature key pair and registers the public key with the public-key infrastructure $\mathcal{F}_{\mathrm{CA}}$.

Now, our basic Sleepy protocol proceeds very much like a proof-of-work blockchain, except that instead of solving computational puzzles, in our protocol a node can extend the chain at time t iff it is elected leader at time t. To extend the chain with a block, a leader of time t simply signs a tuple containing the previous block's hash, the node's own party identifier, the current time t, as well as a set of transactions to be confirmed. Leader election can be achieved through a public hash function H that is modeled as a random oracle.

Removing the Random Oracle. Although we described our scheme assuming a random oracle H, it is not hard to observe that we can replace the random oracle with a common reference string crs and a pseudo-random function PRF. Specifically, the common reference string $k_0 \leftarrow_\$ \{0,1\}^\lambda$ is randomly generated after $\mathcal{Z}$ spawns all corrupt nodes and commits to when each honest node shall sleep. Then, we can simply replace calls to $\mathsf{H}(\cdot)$ with $\mathsf{PRF}_{k_0}(\cdot)$.

Remark on How to Interpret the Protocol for Weakly Synchronized Clocks. As mentioned earlier, in practice, we would typically adopt the protocol assuming nodes have weakly synchronized clocks instead of perfect synchronized clocks. Section 2.1 described a general protocol transformation that allows us to treat weakly synchronized clocks as synchronized clocks in formal reasoning (but adopting a larger network delay). Specifically, when deployed in practice assuming weakly synchronized clocks with up to Δ clock offset, alert nodes would actually queue each received message for Δ time before locally delivering the message. This ensures that alert nodes will not reject other alert nodes' chains mistakenly thinking that the block-time is in the future (due to clock offsets).

Protocol $\Pi_{\text{sleepy}}(p)$

On input init() from $\mathcal{Z}$:
  let $(\mathsf{pk}, \mathsf{sk}) := \Sigma.\mathsf{gen}()$, register pk with $\mathcal{F}_{\text{CA}}$, let *chain* := *genesis*

On receive *chain'*:
  assert $|chain'| > |chain|$ and *chain'* is valid w.r.t. eligible and the current time t;
  chain := *chain'* and multicast *chain*

Every time step:
  - receive input transactions(txs) from $\mathcal{Z}$
  - let t be the current time, if $\mathsf{eligible}^t(\mathcal{P})$ where $\mathcal{P}$ is the current node's party identifier:
      let $\sigma := \Sigma.\mathsf{sign}(\mathsf{sk}, chain[-1].h, \mathsf{txs}, t)$, $h' := \mathsf{d}(chain[-1].h, \mathsf{txs}, t, \mathcal{P}, \sigma)$,
      let $B := (chain[-1].h, \mathsf{txs}, t, \mathcal{P}, \sigma, h')$, let *chain* := $chain\|B$ and multicast *chain*
  - output extract(*chain*) to $\mathcal{Z}$ where extract is the function outputs an ordered list containing the txs extracted from each block in *chain*

Subroutine $\mathsf{eligible}^t(\mathcal{P})$:
  return 1 if $\mathsf{H}(\mathcal{P}, t) < D_p$ and $\mathcal{P}$ is a valid protocol participant[a]; else return 0

---

[a] Without loss of generality, we may assume protocol participants are numbered 1 to N.

Fig. 1. The sleepy consensus protocol. The difficulty parameter D_p is defined such that the hash outcome is less than D_p with probability p. For simplicity, here we describe the scheme with a random oracle H—however as we explain in this section, H can be removed and replaced with a pseudorandom function and a common reference string.

Remark on Foreknowledge of Δ. Note that our protocol $\Pi_{\text{sleepy}}(p)$ is parametrized with a parameter p, that is, the probability that any node is elected leader in any time step. Looking ahead, due to our compliance rules explained later in Sect. 4.3, it is sufficient for the protocol to have foreknowledge of both N and Δ, then to attain a targeted resilience (i.e., the minimum ratio of alert nodes over corrupt ones in any time step), the protocol can choose an appropriate value for p based on the "resilience" compliance rules (see Sect. 4.3).

In our online version [?], we will justify why foreknowledge of Δ is necessary: we prove a lower bound showing that any protocol that does not have foreknowledge of Δ cannot achieve state machine replication even when all nodes are honest.

4.3   Compliant Executions

Our protocol can be proven secure as long as a set of constraints are expected, such as the number of alert vs. corrupt nodes. Below we formally define the complete set of rules that we expect $(\mathcal{A}, \mathcal{Z})$ to respect to prove security.

Compliant Executions. We say that $(\mathcal{A}, \mathcal{Z})$ is $\Pi_{\mathrm{sleepy}}(p)$-compliant if the following holds:

- *Static corruption and sleepiness.* $\mathcal{Z}$ must issue all `corrupt` and `sleep` instructions prior to the start of the protocol execution. We assume that $\mathcal{A}$ cannot query the random oracle H prior to protocol start.
- *Resilience.* There are parameters $(N, N_{\mathrm{crupt}}, \Delta)$ such that $(\mathcal{A}, \mathcal{Z})$ is $(N, N_{\mathrm{crupt}}, \Delta)$-respecting w.r.t. $\Pi_{\mathrm{sleepy}}(p)$, and moreover, the following conditions are respected:
  - There is a positive constant ϕ, such that for any view in the support of $\mathsf{EXEC}^{\Pi_{\mathrm{sleepy}}(p)}(\mathcal{A}, \mathcal{Z}, \lambda)$, for every $t \leq |\mathsf{view}|$,

$$\frac{\mathsf{alert}^t(\mathsf{view})}{N_{\mathrm{crupt}}} \geq \frac{1 + \phi}{1 - 2pN\Delta}$$

    where $\mathsf{alert}^t(\mathsf{view})$ denotes the number of nodes that are alert at time t in view.
  - Further, there is some constant $0 < c < 1$ such that $2pN\Delta < 1 - c$.

Informally, we require that at any point of time, there are more alert nodes than corrupt ones by a small constant margin.

Useful Notations. We define additional notations that will become useful later.

1. Let $N_{\mathrm{alert}} := N_{\mathrm{crupt}} \cdot \frac{1+\phi}{1-2pN\Delta}$ be a lower bound on the number of alert nodes in every time step;
2. Let $\alpha := pN_{\mathrm{alert}}$ be a lower bound on the expected number of alert nodes elected leader in any single time step;
3. Let $\beta := pN_{\mathrm{crupt}} \geq 1 - (1-p)^{N_{\mathrm{crupt}}}$ be the expected number of corrupt nodes elected leader in any single time step; notice that β is also an upper bound on the probability that some corrupt node is elected leader in one time step.

4.4   Theorem Statement

We now state our theorem for static corruption.

Theorem 3 (Security of Π_{sleepy} under static corruption). *Assume the existence of a common reference string (CRS), a bare public-key infrastructure (PKI), and that the signature scheme Σ is secure against any p.p.t. adversary. Then, for any constants $\epsilon, \epsilon_0 > 0$, any $0 < p < 1$, any $T_0 \geq \epsilon_0\lambda$, $\Pi_{sleepy}(p)$ satisfies (T_0, g_0, g_1)-chain growth, (T_0, μ)-chain quality, and T_0^2 consistency with $\exp(-\Omega(\lambda))$ failure probability for the following set of parameters:*

- *chain growth lower bound parameter* $g_0 = (1 - \epsilon)(1 - 2pN\Delta)\alpha;$
- *chain growth upper bound parameter* $g_1 = (1 + \epsilon)Np;$ *and*
- *chain quality parameter* $\mu = {}^{\prime}1 - \frac{1-\epsilon}{1+\phi}.$

where N, Δ, α *and* ϕ *are parameters that can be determined by* $(\mathcal{A}, \mathcal{Z})$ *as well as* p *as mentioned earlier.*

The proof of this theorem will be presented in Sect. 5.

Corollary 1 (Statically secure state machine replication in the sleepy model). *Assume the existence of a common reference string (CRS), a bare public-key infrastructure (PKI), and that the signature scheme Σ is secure against any p.p.t. adversary. For any constant $\epsilon > 0$, there exists a protocol that achieves state machine replication assuming static corruptions and static sleepiness, and that $\frac{1}{2} + \epsilon$ fraction of awake nodes are honest in any time step.*

Proof. Straightforward from Theorem 3 and Lemma 1.

5   Proofs for Static Security

In this section, we present the proofs for the basic sleepy consensus protocol presented in Sect. 4. We assume static corruption and static sleepiness and the random oracle model. Later in our paper, we will describe how to remove the random oracle, and further extend our protocol and proofs to adaptive sleepiness and adaptive corruptions.

We start by analyzing a very simple ideal protocol denoted Π_{ideal}, where nodes interact with an ideal functionality $\mathcal{F}_{\text{tree}}$ that keeps track of all valid chains at any moment of time. Later in our online version [25], we will show that the real-world protocol Π_{sleepy} securely emulates the ideal-world protocol.

5.1   Simplified Ideal Protocol Π_{ideal}

Ideal Protocol. Following [24], we first define a simplified protocol Π_{ideal} parametrized with an ideal functionality $\mathcal{F}_{\text{tree}}$—see Figs. 2 and 3. Looking forward, our ideal functionality, $\mathcal{F}_{\text{tree}}$, gives more power to the adversary than the ideal functionality used in [24] (i.e., our ideal functionality is *weaker* than the one in [24]); we provide a more detailed comparison below. As a consequence, our proof of security in this idealized model will be (significantly) more complicated than the one in [24]. We mention that the reason for using this weaker functionality is that we later aim to implement it *without using proofs-of-work* (and in particular, without assuming that a majority of the *computing power* is held by honest player, but rather under the assumption that a majority of the registered public keys are held by honest players).

Roughly speaking, $\mathcal{F}_{\text{tree}}$ flips random coins to decide whether a node is the elected leader for every time step, and an adversary $\mathcal{A}$ can query this information (i.e., whether any node is a leader in any time step) through the `leader` query

---

$$\mathcal{F}_{\text{tree}}(p)$$

On init: tree := genesis, time(genesis) := 0

On receive leader($\mathcal{P}, t$) from $\mathcal{A}$ or internally:

  if $\Gamma[\mathcal{P}, t]$ has not been set, let $\Gamma[\mathcal{P}, t] := \begin{cases} 1 & \text{with probability } p \\ 0 & \text{o.w.} \end{cases}$

  return $\Gamma[\mathcal{P}, t]$

On receive extend(chain, B) from $\mathcal{P}$: let t be the current time:

  assert chain $\in$ tree, chain$||$B $\notin$ tree, and leader($\mathcal{P}, t$) outputs 1
  append B to chain in tree, record time(chain$||$B) := t, and return "succ"

On receive extend(chain, B, t') from corrupt party $\mathcal{P}^*$: let t be the current time

  assert chain $\in$ tree, chain$||$B $\notin$ tree, leader($\mathcal{P}^*, t'$) outputs 1, and time(chain) < $t' \le t$
  append B to chain in tree, record time(chain$||$B) = t', and return "succ"

On receive verify(chain) from $\mathcal{P}$: return (chain $\in$ tree)

Fig. 2. Ideal functionality $\mathcal{F}_{\text{tree}}$.

interface. Finally, alert and corrupt nodes can call $\mathcal{F}_{\text{tree}}$.extend to extend known chains with new blocks—$\mathcal{F}_{\text{tree}}$ will then check if the caller is a leader for the time step to decide if the extend operation is allowed. $\mathcal{F}_{\text{tree}}$ keeps track of all valid chains, such that alert nodes will call $\mathcal{F}_{\text{tree}}$.verify to decide if any chain they receive is valid. Alert nodes always store the longest valid chains they have received, and try to extend it.

Observe that $\mathcal{F}_{\text{tree}}$ has two entry points named extend—one of them is the honest version and the other is the corrupt version. In this ideal protocol, alert nodes always mine in the present, i.e., they always call the honest version of extend that uses the current time t. In this case, if the honest node succeeds in mining a new chain denoted chain, $\mathcal{F}_{\text{tree}}$ records the current time t as chain's

---

Protocol Π_{ideal}

On init: chain := genesis

On receive chain': if |chain'| > |chain| and $\mathcal{F}_{\text{tree}}$.verify(chain') = 1: chain := chain', multicast chain

Every time step:

  – receive input B from $\mathcal{Z}$
  – if $\mathcal{F}_{\text{tree}}$.extend(chain, B) outputs "succ": chain := chain$||$B and multicast chain
  – output chain to $\mathcal{Z}$

Fig. 3. Ideal protocol Π_{ideal}

block-time by setting $\mathcal{F}_{\text{tree}}(\text{view}).\text{time}(\text{chain}) = t$. On the other hand, corrupt nodes are allowed to call a malicious version of extend and supply a past time step t'. When receiving an input from the adversarial version of extend, $\mathcal{F}_{\text{tree}}$ verifies that the new block's purported time t' respects the strictly increasing rule. If the corrupt node succeeds in mining a new block, then $\mathcal{F}_{\text{tree}}$ records the purported time t' as the chain's block-time.

Notations. Given some view sampled from $\text{EXEC}^{\Pi_{\text{ideal}}}(\mathcal{A}, \mathcal{Z}, \lambda)$, we say that a chain $\in \mathcal{F}_{\text{tree}}(\text{view}).\text{tree}$ has a block-time of t if $\mathcal{F}_{\text{tree}}(\text{view}).\text{time}(\text{chain}) = t$. We say that a node $\mathcal{P}$ (alert or corrupt) mines a $\text{chain}' = \text{chain}\|\mathsf{B}$ in time t if $\mathcal{P}$ called $\mathcal{F}_{\text{tree}}.\text{extend}(\text{chain}, \mathsf{B})$ or $\mathcal{F}_{\text{tree}}.\text{extend}(\text{chain}, \mathsf{B}, \_)$ at time t, and the call returned "succ". Note that if an alert node mines a chain at time t, then the chain's block-time must be t as well. By contrast, if a corrupt node mines a chain at time t, the chain's block-time may not be truthful—it may be smaller than t.

We say that $(\mathcal{A}, \mathcal{Z})$ is $\Pi_{\text{ideal}}(p)$-compliant iff the pair is $\Pi_{\text{sleepy}}(p)$-compliant. Since the protocols' compliance rules are the same, we sometimes just write compliant for short.

Theorem 4 (Security of Π_{ideal}). *For any constant $\epsilon_0, \epsilon > 0$, any $T_0 \geq \epsilon_0 \lambda$, Π_{sleepy} satisfies (T_0, g_0, g_1)-chain growth, (T_0, μ)-chain quality, and T_0^2-consistency against any Π_{ideal}-compliant, **computationally unbounded** pair $(\mathcal{A}, \mathcal{Z})$, with $\exp(-\Omega(\lambda))$ failure probability and the following parameters:*

- *chain growth lower bound parameter $g_0 = (1 - \epsilon)(1 - 2pN\Delta)\alpha$;*
- *chain growth upper bound parameter $g_1 = (1 + \epsilon)Np$; and*
- *chain quality parameter $\mu = 1 - \frac{1-\epsilon}{1+\phi}$.*

where N, Δ, α and ϕ are parameters that can be determined by $(\mathcal{A}, \mathcal{Z})$ as well as p as mentioned earlier.

In the remainder of this section, we will now prove the above Theorem 4.

Intuitions and Differences from the Ideal Protocol in [24]. The key difference between our ideal protocol and Nakamoto's ideal protocol as described by Pass et al. [24] is the following. In Nakamoto's ideal protocol, if the adversary succeeds in extending a chain with a block, he cannot reuse this block and concatenate it with other chains. Here in our ideal protocol, if a corrupt node is elected leader in some time slot, he can reuse the elected slot in many possible chains. He can also instruct $\mathcal{F}_{\text{tree}}$ to extend chains with times in the past, as long as the chain's block-times are strictly increasing.

Although our $\mathcal{F}_{\text{tree}}$ allows the adversary to claim potentially false block-times, we can rely on the following block-time invariants in our proofs: (1) honest blocks always have faithful block-times; and (2) any chain in $\mathcal{F}_{\text{tree}}$ must have strictly increasing block-times. Having observed these, we show that Pass et al.'s chain growth and chain quality proofs [24] can be adapted for our scenario.

Unfortunately, the main challenge is how to prove consistency. As mentioned earlier, our adversary is much more powerful than the adversary for the

Nakamoto blockchain and can launch a much wider range of attacks where he reuses the time slots during which he is elected. In our online version [25], we present new techniques for analyzing the induced stochastic process.

5.2   Overview of Ideal-World Proofs

As mentioned, chain growth and chain quality follow essentially in the same was as in [24], we thus here focus on giving an overview of the consistency proof.

Review: Consistency Proof for the Nakamoto Blockchain. We first review how Garay et al. [11] and Pass et al. [24] proved consistency for a proof-of-work blockchain, and explain why their proof fails in our setting. To prove consistency, [24] relies on a notion of a *convergence opportunity* (and [11] relies on an analog of this notion in the synchronous setting). Roughly speaking, a convergence opportunity is a period of time in which (1) there is a Δ-long period of silence in which no honest node mines a block; and (2) followed by a time step in which a *single* honest (or in our setting, alert) node mines a block; and (3) followed by yet another Δ-long period of silence in which no honest node mines a block.

Intuitively, convergence opportunities are a "good pattern" that helps with consistency. In particular, if the unique honest block mined during a convergence opportunity (henceforth denoted B^*) is at length ℓ, then the adversary must mine a block also at length ℓ, otherwise all honest nodes' chains must have a unique block, that is, B^* at length ℓ—and this forces convergence of the entire prefix of the chain up to block B^*.

Finally, to prove consistency, we need to show that in any sufficiently long window, there must be more convergence opportunities than there are adversarially mined blocks. [24] provides a lower bound on the number of convergence opportunities (this is the difficult part of the proof); in contrast, providing an upper bound on the number of adversarially mined blocks is easy and directly follows from a Chernoff bound due to the honest majority of computing power assumption.

Why the Proof Breaks Down in Our Setting. The above-mentioned bound on the number of adversarial blocks, however, relies on the fact that when an adversary successfully extends a chain with a block, he cannot simply transfer this block to some other chain: each mined block requires a separate "computational effort" (i.e., a successful mining), and thus the upperbound simply follows by upperbounding the number of "successful calls" to $\mathcal{F}_{\text{tree}}$.

In contrast, in our setting, if a corrupt node is elected leader in a certain time step t, he can now reuse this credit to extend *multiple* chains, *possibly even at different lengths*.

As such, the above argument (and the proofs of [11,24]) can no longer be applied: we cannot use an upperbound on the number of times the adversary is elected leader (the direct analogy of how many times an adversary mines a block in a proof-of-work blockchain) to get an upperbound on how many *chain lengths* the adversary can attack.

To overcome this, we devise a different proof strategy. The notion of a convergence opportunity will still be important to us (as well as the lower bound on the number of convergence opportunities), but our method for showing convergence will be more sophisticated.

Roadmap of Our Proof. We will define a good event called a (strong) *pivot* point. Roughly speaking, a (strong) pivot is a point of time t, such that if one draws any window of time $[t_0, t_1]$ that contains t, the number of adversarial time slots in that window, if non-zero, must be strictly smaller than the number of convergence opportunities in the same window. We will show the following:

- *A pivot forces convergence:* for any view where certain negligible-probability bad events do not happen: if there is such a pivot point t in view, then the adversary cannot have caused divergence prior to t.
- *Pivots happen frequently:* for all but negligible fraction of the views, pivot points happen frequently in time—particularly, in any sufficiently long time window there must exist such a pivot point. This then implies that if one removes sufficiently many trailing blocks from an alert node's chain (recall that by chain growth, block numbers and time roughly translate to each other), the remaining prefix must be consistent with any other alert node.

We defer the full proofs for chain quality, chain growth, and consistency for the ideal-world protocol Π_{ideal} to our online version [25].

5.3   Real-World Emulates Ideal-World

So far, we have argued security properties for the ideal world protocol. We next need to prove that the same properties, namely, chain growth, chain quality, and consistency translate to the real-world protocol as well. To show this, we rely on a standard simulation argument. For any real-world adversary $\mathcal{A}$, we construct an ideal-world adversary $\mathcal{S}$ (i.e., a simulator), such that no p.p.t. environment $\mathcal{Z}$ can distinguish whether it is in the real- or ideal-world. Recall that the security properties for our protocols are defined w.r.t. honest nodes' output to the environment $\mathcal{Z}$. Thus, such a simulation proof would immediately imply that all relevant security properties we have proven for the ideal world immediately hold in the real world as well. In essence, we prove the following lemma:

Lemma 2 (Real world emulates the ideal world). *For any p.p.t. adversary of the real-world protocol Π_{sleepy}, there exists a p.p.t. simulator $\mathcal{S}$ of the ideal-world protocol Π_{ideal}, such that for any p.p.t. environment $\mathcal{Z}$, for any $\lambda \in \mathbb{N}$, we have the following where $\overset{c}{\equiv}$ denotes computational indistinguishability.*

$$\{\text{view}_{\mathcal{Z}}(\text{EXEC}^{\Pi_{ideal}}(\mathcal{S}, \mathcal{Z}, \lambda))\}_\lambda \overset{c}{\equiv} \{\text{view}_{\mathcal{Z}}(\text{EXEC}^{\Pi_{sleepy}}(\mathcal{A}, \mathcal{Z}, \lambda))\}_\lambda$$

We defer the proof of this lemma to our online version [25].

5.4   Removing the Random Oracle

It is not difficult to modify our proof when the random oracle query $H(\mathcal{P}, t)$ is actually instantiated with $PRF_{k_0}(\mathcal{P}, t)$ where k_0 denotes a sufficiently long common reference string. Specifically, the formal proof introduces a hybrid world in which $\mathcal{F}_{tree}$'s internal random coins are replaced with outcomes from evaluating the PRF. Although the PRF key k_0 is observable by the adversary, we stress that our ideal-world protocol is secure against a computationally unbounded adversary. Moreover, our ideal-world protocol is secure as long as certain polynomial-time checkable properties, defined over the outcome of the random function or the PRF, are respected. Due to the pseudo-randomness of the PRF, if these polynomial-time checkable properties are respected for all but a negligible fraction of the random functions, then they are respected for all but a negligible fraction of the PRF family too. In our online version [25], we formalize this intuition and present a formal proof.

6   Achieving Adaptive Security

So far, we have assumed that the adversary issues both `corrupt` and `sleep` instructions statically upfront. In this section, we will show how to achieve adaptive security with complexity leveraging. It turns out even with complexity leveraging the task is non-trivial.

6.1   Intuition: Achieving Adaptive Sleepiness

To simplify the problem, let us first consider how to achieve adaptive sleepiness (but static corruption). In our statically secure protocol Π_{sleepy}, the adversary can see into the future for all honest and corrupt players. In particular, the adversary can see exactly in which time steps each honest node is elected leader. If `sleep` instructions could be adaptively issued, the adversary could simply put a node to sleep whenever he is elected leader, and wake up him when he is not leader. This way, the adversary can easily satisfy the constraint that at any time, the majority of the online nodes must be honest, while ensuring that no alert nodes are ever elected leader (with extremely high probability).

To defeat such an attack and achieve adaptive sleepiness (but static corruption), we borrow an idea that was (informally) suggested by Micali [19]. Basically, instead of computing a "leader ticket" η by hashing the party's (public) identifier and the time step t and by checking $\eta < D_p$ to determine if the node is elected leader, we will instead have an honest node compute a pseudorandom "leader ticket" itself using some secret known only to itself. In this way, the adversary is no longer able to observe honest nodes' future. The adversary is only able to learn that an honest node is elected leader in time step t when the node actually sends out a new chain in t—but by then, it will be too late for the adversary to (retroactively) put that node to sleep in t.

A Naïve Attempt. Therefore, a naïve attempt would be the following.

- Each node $\mathcal{P}$ picks its own PRF key $k[\mathcal{P}]$, and computes a commitment $c :=$ $\mathsf{comm}(k[\mathcal{P}]; r)$ and registers c as part of its public key with the public-key infrastructure $\mathcal{F}_{\mathrm{CA}}$. To determine whether it is elected leader in a time step t, the node computes $\mathsf{PRF}_{k[\mathcal{P}]}(t) < D_p$ where D_p is a difficulty parameter related to p, such that any node gets elected with probability p in a given time step.
- Now for $\mathcal{P}$ to prove to others that it is elected leader in a certain time step t, $\mathcal{P}$ can compute a non-interactive zero-knowledge proof that the above evaluation is done correctly (w.r.t. to the commitment c that is part of $\mathcal{P}$'s public key).

A Second Attempt. This indeed hides honest nodes' future from the adversary; however, the adversary may not generate $k[\mathcal{P}^*]$ at random for a corrupt player $\mathcal{P}^*$. In particular, the adversary can try to generate $k[\mathcal{P}^*]$ such that $\mathcal{P}^*$ can get elected in more time steps. To defeat such an attack, we include a relatively long randomly chosen string k_0 in the common reference string. For a node $\mathcal{P}$ to be elected leader in a time step t, the following must hold:

$$\mathsf{PRF}_{k_0}(\mathcal{P}, t) \oplus \mathsf{PRF}_{k[\mathcal{P}]}(t) < D_p$$

As before, a node can compute a non-interactive zero-knowledge proof (to be included in a block) to convince others that it computed the leader election function correctly.

Now the adversary can still adaptively choose $k[\mathcal{P}^*]$ after seeing the common reference string k_0 for a corrupt node $\mathcal{P}^*$ to be elected in more time steps; however, it can only manipulate the outcome to a limited extent: in particular, since k_0 is much longer than $k[\mathcal{P}^*]$, the adversary does not have enough bits in $k[\mathcal{P}^*]$ to manipulate to defeat all the entropy in k_0.

Parametrization and Analysis. Using the above scheme, we can argue for security against an adaptive sleepiness attack. However, as mentioned above, the adversary can still manipulate the outcome of the leader election to some extent. For example, one specific attack is the following: suppose that the adversary controls $O(N)$ corrupt nodes denoted $\mathcal{P}_0^*, \ldots, \mathcal{P}_{O(N)}^*$ respectively. With high probability, the adversary can aim for the corrupt nodes to be elected for $O(N)$ consecutive time slots during which period the adversary can sustain a consistency and a chain quality attack. To succeed in such an attack, say for time steps $[t : t + O(N)]$, the adversary can simply try random user PRF keys on behalf of $\mathcal{P}_0^*$ until it finds one that gets $\mathcal{P}_0^*$ to be elected in time t (in expectation only $O(\frac{1}{p})$ tries are needed); then the adversary tries the same for node $\mathcal{P}_1^*$ and time $t + 1$, and so on.

Therefore we cannot hope to obtain consistency and chain quality for $O(N)$-sized windows. Fortunately, as we argued earlier, since the adversary can only manipulate the leader election outcome to a limited extent given that the length of k_0 is much greater than the length of each user's PRF key, it cannot get corrupt nodes to be consecutively elected for too long. In our proof, we show that as long as we consider sufficiently long windows of N^c blocks in length (for an appropriate constant c and assuming for simplicity that $N = \omega(\log \lambda)$, then consistency and chain quality will hold except with negligible probability.

6.2 Intuition: Achieving Adaptive Corruption

Once we know how to achieve adaptive sleepiness and static corruption, we can rely on complexity leveraging to achieve adaptive corruption. This part of the argument is standard: suppose that given an adversary under static corruption that can break the security properties of the consensus protocol, there exists a reduction that breaks some underlying complexity assumption. We now modify the reduction to guess upfront which nodes will become corrupt during the course of execution, and it guesses correctly with probability $\frac{1}{2^N}$. This results in a 2^N loss in the security reduction, and therefore if we assume that our cryptographic primitives, including the PRF, the digital signature scheme, the non-interactive zero-knowledge proof, the commitment scheme, and the collision-resistant hash family have sub-exponential hardness, we can lift the static corruption to adaptive corruption.

6.3 Detailed Protocol and Proofs

We defer the detailed presentation of our adaptively secure protocol and proofs to our online version [25]—however, we state our main theorem for adaptive security below.

Theorem 5 (Adaptively secure state machine replication in the sleepy model). *Assume the existence of a Bare PKI, a CRS; the existence of sub-exponentially hard collision-resistant hash functions, and sub-exponentially hard enhanced trapdoor permutations. Then, for any constant $\epsilon > 0$, there exists a protocol that achieves state machine replication against adaptive corruptions and adaptive sleepiness, as long as $\frac{1}{2} + \epsilon$ fraction of awake nodes are honest in any time step.*

References

1. U. "BCNext": NXT (2014). http://wiki.nxtcrypto.org/wiki/Whitepaper:Nxt
2. Bentov, I., Gabizon, A., Mizrahi, A.: Cryptocurrencies without proof of work. In: Financial Cryptography Bitcoin Workshop (2016)
3. Cachin, C., Kursawe, K., Petzold, F., Shoup, V.: Secure and efficient asynchronous broadcast protocols. In: Kilian, J. (ed.) CRYPTO 2001. LNCS, vol. 2139, pp. 524–541. Springer, Heidelberg (2001). https://doi.org/10.1007/3-540-44647-8_31
4. Canetti, R.: Universally composable security: a new paradigm for cryptographic protocols. In: FOCS (2001)
5. Castro, M., Liskov, B.: Practical byzantine fault tolerance. In: OSDI (1999)
6. Daian, P., Pass, R., Shi, E.: Snow white: provably secure proofs of stake. https://eprint.iacr.org/2016/919.pdf

7. Dolev, D., Strong, H.R.: Authenticated algorithms for byzantine agreement. SIAM J. Comput. SIAMCOMP **12**(4), 656–666 (1983)
8. Dwork, C., Lynch, N., Stockmeyer, L.: Consensus in the presence of partial synchrony. J. ACM **35**, 288–323 (1988)
9. Dwork, C., Naor, M.: Pricing via processing or combatting junk mail. In: Brickell, E.F. (ed.) CRYPTO 1992. LNCS, vol. 740, pp. 139–147. Springer, Heidelberg (1993). https://doi.org/10.1007/3-540-48071-4_10
10. Feldman, P., Micali, S.: An optimal probabilistic protocol for synchronous byzantine agreement. SIAM J. Comput. **26**, 873–933 (1997)
11. Garay, J., Kiayias, A., Leonardos, N.: The bitcoin backbone protocol: analysis and applications. In: Oswald, E., Fischlin, M. (eds.) EUROCRYPT 2015. LNCS, vol. 9057, pp. 281–310. Springer, Heidelberg (2015). https://doi.org/10.1007/978-3-662-46803-6_10
12. Katz, J., Koo, C.-Y.: On expected constant-round protocols for byzantine agreement. J. Comput. Syst. Sci. **75**(2), 91–112 (2009)
13. Kiayias, A., Panagiotakos, G.: Speed-security tradeoffs in blockchain protocols. IACR Cryptology ePrint Archive 2015:1019 (2015)
14. Kiayias, A., Russell, A., David, B., Oliynykov, R.: Ouroboros: a provably secure proof-of-stake blockchain protocol. In: Katz, J., Shacham, H. (eds.) CRYPTO 2017. LNCS, vol. 10401, pp. 357–388. Springer, Cham (2017). https://doi.org/10.1007/978-3-319-63688-7_12
15. King, S., Nadal, S.: PPCoin: Peer-to-Peer Crypto-Currency with Proof-of-Stake, August 2012
16. Lamport, L.: Fast paxos. Distrib. Comput. **19**(2), 79–103 (2006)
17. Lamport, L., Malkhi, D., Zhou, L.: Vertical paxos and primary-backup replication. In: PODC, pp. 312–313 (2009)
18. Martin, J.-P., Alvisi, L.: Fast byzantine consensus. IEEE Trans. Dependable Secur. Comput. **3**(3), 202–215 (2006)
19. Micali, S.: Algorand: the efficient and democratic ledger (2016). https://arxiv.org/abs/1607.01341
20. Micali, S., Vadhan, S., Rabin, M.: Verifiable random functions. In: FOCS (1999)
21. Miller, A., Xia, Y., Croman, K., Shi, E., Song, D.: The honey badger of BFT protocols. In: ACM CCS (2016)
22. Mockapetris, P., Dunlap, K.: Development of the domain name system. In: SIGCOMM, Stanford, CA, pp. 123–133 (1988)
23. Nakamoto, S.: Bitcoin: a peer-to-peer electronic cash system (2008)
24. Pass, R., Seeman, L., Shelat, A.: Analysis of the blockchain protocol in asynchronous networks. https://eprint.iacr.org/2016/454
25. Pass, R., Shi, E.: The sleepy model of consensus (2016). https://eprint.iacr.org/2016/918
26. Pass, R., Shi, E.: Hybrid consensus: efficient consensus in the permissionless model. In: DISC (2017)
27. Pass, R., Shi, E.: Thunderella: blockchains with optimistic instant confirmation. Manuscript (2017)
28. Song, Y.J., van Renesse, R.: Bosco: one-step byzantine asynchronous consensus. In: Taubenfeld, G. (ed.) DISC 2008. LNCS, vol. 5218, pp. 438–450. Springer, Heidelberg (2008). https://doi.org/10.1007/978-3-540-87779-0_30

Instantaneous Decentralized Poker

Iddo Bentov[1]($\boxtimes$), Ranjit Kumaresan[2], and Andrew Miller[3]

[1] Cornell University, Ithaca, USA
iddobentov@cornell.edu
[2] Microsoft Research, Redmond, USA
vranjit@gmail.com
[3] University of Illinois at Urbana-Champaign, Champaign, USA
soc1024@illinois.edu

Abstract. We present efficient protocols for *amortized* secure multiparty computation with penalties and secure cash distribution, of which poker is a prime example. Our protocols have an initial phase where the parties interact with a cryptocurrency network, that then enables them to interact only among themselves over the course of playing many poker games in which money changes hands.

The high efficiency of our protocols is achieved by harnessing the power of stateful contracts. Compared to the limited expressive power of Bitcoin scripts, stateful contracts enable richer forms of interaction between standard secure computation and a cryptocurrency.

We formalize the stateful contract model and the security notions that our protocols accomplish, and provide proofs in the simulation paradigm. Moreover, we provide a reference implementation in Ethereum/Solidity for the stateful contracts that our protocols are based on.

We also adapt our off-chain cash distribution protocols to the special case of stateful duplex micropayment channels, which are of independent interest. In comparison to Bitcoin based payment channels, our duplex channel implementation is more efficient and has additional features.

1 Introduction

As demonstrated by Cleve [13], fair multiparty computation without an honest majority is impossible in the standard model of communication. Hence, there have been numerous attempts to circumvent this theoretical impossibility result, in particular by relying on techniques such as gradual release (cf. [35] for a survey) and optimistic fair exchange [4]. With the introduction of Bitcoin [33], the academic study of decentralized cryptocurrencies gave rise to a line of research that seeks to impose fairness in secure multiparty computation (MPC) by means of monetary penalties [8]. In this model, the participating parties make security deposits, and the deposits of parties who deviate from the protocol are used to compensate the honest parties.

Still, interacting with a Proof-of-Work based decentralized network entails long waiting times due to the need to be secure against reversal of the ledger history. A recent work by Kumaresan and Bentov [27] showed a Bitcoin based

© International Association for Cryptologic Research 2017
T. Takagi and T. Peyrin (Eds.): ASIACRYPT 2017, Part II, LNCS 10625, pp. 410–440, 2017.
https://doi.org/10.1007/978-3-319-70697-9_15

amortization scheme in which the parties run an initial setup phase requiring interaction with the cryptocurrency network, but thereafter they engage in many fair secure computation executions, communicating only among themselves for as long as all parties are honest.

1.1   Our Contributions

Asymptotic gains in amortized protocols. We present new protocols that rely on stateful contracts instead of Bitcoin transactions, and thereby improve upon the previous results in several ways. First, the setup phase of [27] requires the n parties to execute $O(n^2)$ PoW-based rounds of interaction with the cryptocurrency network, while our stateful protocols require $O(1)$ rounds. The protocols of [27] for secure MPC with penalties also require a security deposit of $O(n^2)$ coins per party, while our protocols require $O(n)$ coins per party. We use UC-style definitions [11] to formalize the security notions that are achieved by our amortized protocols, and provide proofs using the simulation paradigm.

Amortized SCD. Unlike the protocols in [27], our protocols support *secure cash distribution with penalties* (SCD), rather than only fair secure MPC with penalties. The distinction between SCD and fair MPC with penalties is that in SCD the inputs and outputs of the parties are comprised of both money and data, while fair MPC with penalties has only data for inputs and outputs (but uses money to compensate honest parties who did not learn the output).

Real poker. A canonical example of SCD is a mental poker game, where the outcome of the computation is not intrinsically useful, but rather determines how money should change hands. This means that following an on-chain setup phase, the parties can play any number of *instantaneous* poker games, for as long as no party has run out of money. Hence, while there is a large body of work on efficient mental poker schemes, to the best of our knowledge we are the first to provide a practical poker protocol with actual money transfers from the losers to the winners. Moreover, we accompany our poker protocol with an implementation for the Ethereum cryptocurrency.

Highly efficient payment channels. As a special case, our off-chain cash distribution protocols can also be used for stateful bi-directional payment channels. This use case does not require secure computation and yet it is particularly important. The reason for this is that micropayment channels can reduce the amount of transaction data that the decentralized cryptocurrency network maintains, and thus the long-term scalability pressures that a cryptocurrency faces can be relieved by well-functioning off-chain payment channels (see, e.g., [18] for further discussion). In comparison to Bitcoin based off-chain payment channels, our stateful approach yields better efficiency and extra features. Since micropayment channels are of independent interest, we provide a self-contained protocol and implementation of our stateful duplex off-chain channel.

1.2 Related Works

The first secure computation protocols that utilize Bitcoin to guarantee fairness are by Maxwell [30], Barber et al. [5], Andrychowicz et al. [2,3] and Bentov and Kumaresan [8]. Bitcoin based protocols for reactive cash distribution and poker were given by Kumaresan et al. [26]. The technique for amortized secure computation with penalties in the Bitcoin model was introduced by Kumaresan and Bentov [27]. Our protocols subsume and improve on these, providing both the amortization benefit of [27] with the cash distribution functionality of [26], and furthermore reduce the on-chain costs and the necessary amount of collateral. Several other works analyze fair protocols in rigorous models, in particular Kiayias et al. [24] and Kosba et al. [25] introduced formal cryptocurrency modeling and presented fair (non-amortized) protocols that improve upon the PoW-based round complexity and collateral requirements of the Bitcoin-based protocols in the prior works.

The cash distribution contract we present (see Sect. 2.3) is closely related to an ongoing proposal in the cryptocurrency community for "state channels" (cf. Coleman [14]), wherein a group of parties agrees on a sequence of "off-chain" state transitions, and resort to an on-chain reconciliation process only in the case that the off-chain communications break down. To our knowledge, no security definition has yet been provided for such applications. Furthermore, our application is much more expressive, since we can implement state transitions that depend on parties' private information, while still guaranteeing fairness.

The original mental poker protocol by Shamir et al. [36] relies on commutative encryption. However, their protocol was only for two parties and was found to have security vulnerabilities [15,29]. Following that, many different protocols for mental poker were proposed. For example, Crépeau presented secure poker protocols that are based on probabilistic encryption [16] and zero-knowledge proofs [17], but his constructions are rather inefficient. In 2003, a breakthrough by Barnett and Smart [6] gave a far more efficient poker protocol. Castellà-Roca et al. [12] utilized homomorphic encryption to construct a poker protocol that is similar to [6]. Bayer and Groth [7] later gave a secure and efficient shuffle procedure that can be integrated with [6].

The poker protocol that we integrate into our SCD implementation is by Wei and Wang [23], with a full version by Wei [22]. This protocol uses a proof of knowledge scheme that is slightly faster than [6,7,12], and provides a security proof using the simulation paradigm.

2 Overview

In Bitcoin, the full nodes maintain a data structure that is known as the "unspent transaction outputs set" (UTXO set), which represents the current state of the ledger. Each unspent output in the UTXO set incorporates a circuit (a.k.a. script or predicate), such that any party who can provide an input (a.k.a. witness) that satisfies the circuit can spend the coins amount of this output into a new unspent output. Therefore, if there is only one party who knows the witness for the circuit, then this party is in effect the holder of the coins.

Standard Bitcoin transactions use a signature as the witness. The signature is applied on data that also references the new unspent output, thereby binding the transaction to the specific receiver of the coins and thus prevents a man-in-the-middle attack by the nodes in the decentralized Bitcoin network.

However, Bitcoin allows the use of more complex circuits as well. Such circuits allow us to support quite elaborate protocols in which money changes hands, as opposed to using Bitcoin only for simple money transfers between parties.

Specifically, protocols for fair secure computation and fair lottery can be implemented with a blackbox use of an $\mathcal{F}_{CR}^{\star}$ functionality [8, 26, 28]. Essentially, $\mathcal{F}_{CR}^{\star}$ specifies that a "sender" P_1 locks her coins in accordance with some circuit ϕ, such that a "receiver" P_2 can gain possession of these coins if she supplies a witness w that satisfies $\phi(w) = 1$ before some predefined timeout, otherwise P_1 can reclaim her coins. As shown in [8, 27], the $\mathcal{F}_{CR}^{\star}$ functionality can be realized in Bitcoin, as long as the circuit ϕ can be expressed in the Bitcoin scripting language. In the aforementioned secure computation and lottery protocols [8, 26, 28], the particular circuit that is needed verifies a signature (just as in standard transactions) and a decommitment according to some arbitrary hardcoded value. Such a circuit can be realized by using a hash function for the commitment scheme (Bitcoin supports SHA1,SHA256,RIPEMD160). Since signature verification is an order of magnitude more complex than hash invocation, the complexity of an $\mathcal{F}_{CR}^{\star}$ transaction is only marginally higher than that of standard Bitcoin transactions.

Note that our underlying assumption is that an honest party can interact with the cryptocurrency network (within a bounded time limit) to ensure her monetary compensation. We also assume that the off-chain communication among the parties takes place in a separate point-to-point synchronous network. Given that the network is synchronous, our MPC protocols will be secure even if only one party is honest.

The $\mathcal{F}_{CR}^{\star}$ model can be regarded as a restricted version of the Bitcoin model, which is expressive enough for realizing multiparty functionalities that are impossible in the standard model. One may ask whether it is possible to design better protocols in a model that is more expressive than the Bitcoin model. In this work we will answer the question in the affirmative.

A possible extension to the Bitcoin transaction structure is covenants [32, 34], where each unspent output specifies not only the conditions on who can spend the coins (i.e., the circuit ϕ), but also conditions on who is allowed to receive the coins. Indeed, as shown in [32], covenants can be used to implement certain tasks that the current Bitcoin specifications do not support (e.g., vaults that protect against coin theft).

Generalizing further, each unspent output can maintain a *state*. That is, an unspent output will be comprised of a circuit ϕ and state variables, and parties can update the state variables by carrying out transactions that satisfy ϕ in accord with the current values of the state variables. Additionally, parties can deposit coins into the unspent output, and a party can withdraw some partial amount of the held coins by satisfying ϕ with respect to the state variables. This approach is used in Ethereum [10,38], where the notion of "outputs" is replaced with "user accounts" and automated "contract accounts".

With a slight abuse of terminology, the transaction format of the Bitcoin model can thus be described as "stateless". By this we mean that the coins of an unspent Bitcoin output are controlled by a hardcoded predicate that represents their current state, and anyone who can supply a witness that satisfies this predicate is able to spend these coins into an arbitrary new state.

Let us mention that the Bitcoin transaction format can still enable "smart contracts", in the sense of having coins that can be spent only if some other transaction took place (i.e., without relying on a third party). The technique for achieving this would generally involve multiple signed transactions that are prepared in advance and kept offline. Then, depending on the activity that occurs on the blockchain, some of the prepared transaction will become usable. However, in certain instances the amount of offline transactions may grow exponentially, as in the case of zero-collateral lotteries [31].

The protocols that we present in this work will be in a model that has stateful contracts. As described, this refers to unspent outputs that are controlled according to state variables. It should be emphasized that our protocols do not rely on a Turing-complete scripting language, as all the loops in the contracts that we design (in particular our poker contract) have a fixed number of iterations.

To justify our modeling choice, let us review the advantages of stateful contracts over stateless transactions. As a warmup, we begin by examining simple protocols for 2-party fair exchange.

2.1   Fair Exchange with Penalties Between Two Parties

Suppose that P_1, P_2 execute an unfair secure computation that generates secret shares of the output $x = x_1 \oplus x_2$ and commitments $T_1 = h(x_1), T_2 = h(x_2)$, and delivers (x_i, T_1, T_2) to party P_i. Consider the naive protocol for fair exchange of the shares x_1, x_2 via $\mathcal{F}_{CR}^\star$ transactions:

$$P_1 \xrightarrow[q,\tau]{T_2} P_2 \tag{1}$$

$$P_2 \xrightarrow[q,\tau]{T_1} P_1 \tag{2}$$

An arrow denotes an $\mathcal{F}_{CR}^\star$ transaction that lets P_i collect q coins from P_{3-i} before time τ, by revealing a decommitment y such that $h(y) = T_i$.

The above protocol is susceptible to an "abort" attack by a malicious P_2 that waits for P_1 to make the deposit transaction (i.e., Step 1), after which P_2 simply does not execute Step 2 to make a deposit transaction. Instead, P_2 claims the first transaction, and obtains q coins while P_1 obtains x_2. Now, P_2 simply aborts the protocol. In effect, this means that an honest P_1 paid q coins to learn P_2's share. Fairness with penalties requires that an honest party never loses any money, hence this naive approach does not work (cf. [8] for precise details).

The above vulnerability can be remedied via the following protocol:

$$P_1 \xrightarrow[q,\tau_2]{T_1 \wedge T_2} P_2 \qquad (3)$$

$$P_2 \xrightarrow[q,\tau_1]{T_1} P_1 \qquad (4)$$

For this improved protocol to be secure, two *sequential* PoW-based waiting periods are necessary. Otherwise, a corrupt P_2 may be able to reverse transaction (4) after P_1 claims it, so that P_1 would reveal her share and not be compensated.

By contrast, consider the 2-party fair exchange protocol that is based on a stateful contract, as illustrated in Fig. 1. Here, both parties should deposit q coins each, concurrently. If the $2q$ coins were not deposited into the contract before the timeout is reached, then an honest party who deposited into the contract can claim her coins back. In the case that the $2q$ coins were deposited, it triggers the contract to switch to a new state, where each party P_i can claim her deposit by revealing the hardcoded decommitment T_i. In contrast to the $\mathcal{F}_{\mathrm{CR}}^{\star}$ protocol, the stateful contract requires only one PoW-based waiting period before the honest parties may reveal their shares.

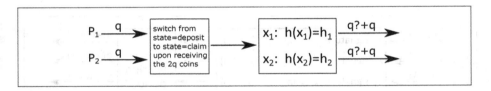

Fig. 1. Stateful contract for fair secure 2-party computation with penalties.

While the quantitative difference between stateful and stateless contracts in the above discussion may appear to be unimpressive, the distinction becomes more pronounced in the case of multiparty fair exchange (a.k.a. fair reconstruction [8]), and even more so in amortized protocols. Let us demonstrate the amortized multiparty case in the next section.

2.2 Amortized Multiparty Fair SFE with Penalties

We illustrate in Fig. 2 the stateful contract for n parties who wish to engage in amortized fair secure function evaluation (SFE) with penalties. The lifespan of this contract can be thought of as having three phases:

– *Deposit Phase:* All parties should deposit $q(n - 1)$ coins each. If $nq(n - 1)$ coins were deposited before the initial timeout is reached, then the contract switches into an "active" state. Otherwise, each honest party who deposited will claim her $q(n - 1)$ coins back.

- *Execution Phase:* While the state is "active", the n parties will not interact with the contract at all. Instead, they will engage in multiple executions of SFE. In the ith execution, the secure computation prepares secret shares $\{x_{i,j}\}_{j=1}^n$ of the output, as well as commitments $\{h_{i,j} = h(x_{i,j})\}_{j=1}^n$, and delivers $(x_{i,j}; h_{i,1}, h_{i,2}, \ldots, h_{i,n})$ to party P_j. Each party will then use her secret key (for which the corresponding public key is hardcoded in the contract) to create a signature $s_{i,j}$ for the tuple $(h_{i,1}, h_{i,2}, \ldots, h_{i,n})$, and send the signature $s_{i,j}$ to the other parties. Upon receiving all the signatures $\{s_{i,j}\}_{j=1}^n$, each honest P_j will send her secret share $x_{i,j}$ in the clear to the other parties.
- *Claim Phase:* In the case that a corrupt party P_c did not reveal her share $x_{i,c}$ during the execution phase, each honest party P_j will send $m_{i,j} = (x_{i,j}; s_{i,1}, s_{i,2}, \ldots, s_{i,n})$ to the contract, and thereby transition the contract into a "payout" state. The message $m_{i,j}$ also registers that P_j deserves to receive a compensation of q coins, in addition to her initial $q(n-1)$ coins deposit. Until a timeout, any party P_ℓ can avoid being penalized by sending $m_{i',\ell} = (x_{i',\ell}; s_{i',1}, s_{i',2}, \ldots, s_{i',n})$ with $i' \geq i$ to the contract. In case $i' > i$, this would invalidate the q coins compensation that was requested via $m_{i,j}$, and instead register that P_ℓ is owed q coins in compensation.

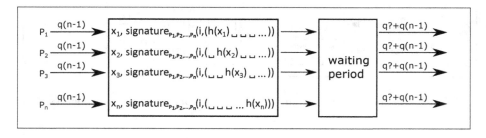

Fig. 2. Stateful contract for amortized multiparty fair SFE with penalties.

As can be observed, the n parties can engage in an unlimited amount of off-chain SFE executions (where the executions can compute different functions), and no interaction with the blockchain will take place as long as all parties are honest. When a corrupt party P_c deviates from this protocol, each honest party will receive q coins compensation, that is taken from P_c's initial security deposits of $q(n-1)$ coins. The actual protocol handles more technical issues, cf. Sect. 4.

By contrast, achieving the same guarantees in the $\mathcal{F}_{\text{CR}}^\star$ model is known to possible only via an intricate "see-saw" construction that requires $O(n^2)$ PoW-based rounds, and collateral of $O(qn^2)$ coins from each party [27]. Moreover, the stateless nature of Bitcoin transactions entail a global timeout after which the entire see-saw construction expires. Setting the global timeout to a high value enables many off-chain SFE executions, but also implies that a DoS attack by a corrupt party (who would abort before signing any secret shares of the output of the first execution) will cause each honest party to wait for a long time before being able to regain possession of her $O(qn^2)$ coins deposit. Due to the time value of money, this is obviously undesirable. The stateful approach does not require a global timeout that is measured in *absolute* terms. Instead, the contract remains operational for as long as all the parties wish to engage in the off-chain protocol, and transitioning the contract into the "payout" state will trigger an event whose expiration is *relative* to the time at which the transition occurred.

We stress that a corrupt party can always pretend to be inactive and force honest parties to interact with the cryptocurrency network. Thus, for example, this protocol can be combined with a reputation system. Further, our implementation uses a technique that shares the on-chain transaction fees among all the parties equally, so that corrupt parties always have to pay the fee (cf. Sect. 6). An Ethereum implementation of this contract is provided in [9, Fig. 18].

Can stateful contracts provide even more benefits? As the next section shows, the answer is "yes".

2.3   Stateful Off-Chain Cash Distribution Protocols

Suppose that the parties P_1, P_2 wish to play a multiple-round lottery game, such that either of them is allowed to quit after each round. Thus, P_1 enters the lottery with m coins, P_2 enters with w coins, and in the first round P_1 picks a random secret x_1 and commits to $\mathsf{com}(x_1)$, P_2 picks x_2 and commits to $\mathsf{com}(x_2)$, and then they decommit x_1, x_2 so that the least significant bit of $x_1 \oplus x_2$ decides whether P_1's balance is incremented to $m+1$ and P_2's balance is decremented to $w - 1$, or P_1's balance is decremented to $m - 1$ and P_2's balance is incremented to $w + 1$. If both of them wish to continue, then they will proceed to the next round and repeat this protocol.

Obviously, the parties must not be allowed to quit in the middle of a round without repercussions. That is, if P_1 reveals her decommitment x_1 and P_2 aborts, then P_1 should be compensated. Moreover, as in Sect. 2.2, it is better that the parties play each round without any on-chain interaction.

Therefore, in each round i, P_1 and P_2 will sign the current balance m_i, w_i together with the round index i and the commitments $\mathsf{com}(x_{i,1}), \mathsf{com}(x_{i,2})$. After the parties exchange these signed messages, they can safely send their decommitments $x_{i,1}, x_{i,2}$ in the clear. The logic of the stateful contract allows each party to send her decommitment along with the signed message, and thus finalize the game according to the current balances. If the other party does not reveal her decommitment during a waiting period, then the contract increments the balance of the honest party. If both parties reveal, then the contract computes $x_1 \oplus x_2$ to decide who won the last round, so that the balance of the winner is incremented and the balance of the loser is decremented. During the waiting period, an honest party can send a signed message with an index $i' > i$ and thereby invalidate the message that a corrupt party sent to the contract.

It should also be noted that an honest party should not continue to play after the balance of the other party reaches 0, since the contract cannot reward the winner more money than what was originally deposited.

We illustrate the contract in Fig. 3, and provide an Ethereum implementation in [9, Figs. 18–20]. The multiparty version of our lottery code is available at [1].

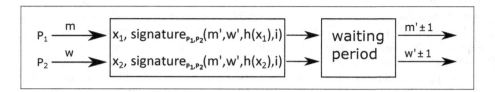

Fig. 3. Off-chain 2-party lottery.

Such a 2-party lottery is a very simple example of a *secure cash distribution with penalties* (SCD) functionality [26]. Another special case of SCD is a multiparty poker game in which money (i.e., coins of the cryptocurrency system that the parties hold) is transferred from losers to winners. In Sect. 3 and onwards we formulate the ideal functionalities and protocol for fair MPC and SCD, and in Sect. 6 we provide an efficient off-chain poker protocol with implementation.

As noted in Sect. 1, SCD can also be realized in the $\mathcal{F}_{CR}^{\star}$ model via a non-amortized (i.e., on-chain) protocol, though the construction requires a setup phase with $O(n^2)$ PoW-based rounds. To the best of our knowledge, there is no amortized SCD realization in the $\mathcal{F}_{CR}^{\star}$ model.

A slight variation can turn the above lottery contract into a bi-directional off-chain micropayment channel, see [9, Sect. 2.4].

3   Preliminaries

We say that a function $\mu(\cdot)$ is negligible in λ if for every polynomial $p(\cdot)$ and all sufficiently large λ's it holds that $\mu(\lambda) < 1/|p(\lambda)|$. A probability ensemble $X = \{X(t, \lambda)\}_{t \in \{0,1\}^*, n \in \mathbb{N}}$ is an infinite sequence of random variables indexed by a and $\lambda \in \mathbb{N}$. Two distribution ensembles $X = \{X(t, \lambda)\}_{\lambda \in \mathbb{N}}$ and $Y = \{Y(t, \lambda)\}_{\lambda \in \mathbb{N}}$ are said to be computationally indistinguishable, denoted $X \overset{c}{\equiv} Y$ if for every non-uniform polynomial-time algorithm D there exists a negligible function $\mu(\cdot)$ such that for every $t \in \{0, 1\}^*$,

$$|\Pr[D(X(t, \lambda)) = 1] - \Pr[D(Y(t, \lambda)) = 1]| \leq \mu(\lambda).$$

All parties are assumed to run in time polynomial in the security parameter λ. We prove security in the "secure computation with coins" (SCC) model proposed in [8]. Note that the main difference from standard definitions of secure computation [21] is that now the view of $\mathcal{Z}$ contains the distribution of coins. Let IDEAL$_{f, \mathcal{S}, \mathcal{Z}}(\lambda, z)$ denote the output of environment $\mathcal{Z}$ initialized with input z after interacting in the ideal process with ideal process adversary $\mathcal{S}$ and (standard or special) ideal functionality $\mathcal{G}_f$ on security parameter λ. Recall that our protocols will be run in a hybrid model where parties will have access to a (standard or special) ideal functionality $\mathcal{G}_g$. We denote the output of $\mathcal{Z}$ after interacting in an execution of π in such a model with $\mathcal{A}$ by $\text{HYBRID}^g_{\pi,\mathcal{A},\mathcal{Z}}(\lambda, z)$, where z denotes $\mathcal{Z}$'s input. We are now ready to define what it means for a protocol to SCC realize a functionality.

Definition 1. *Let $n \in \mathbb{N}$. Let π be a probabilistic polynomial-time n-party protocol and let $\mathcal{G}_f$ be a probabilistic polynomial-time n-party (standard or special) ideal functionality. We say that π SCC realizes $\mathcal{G}_f$ with abort in the $\mathcal{G}_g$-hybrid model (where $\mathcal{G}_g$ is a standard or a special ideal functionality) if for every non-uniform probabilistic polynomial-time adversary $\mathcal{A}$ attacking π there exists a non-uniform probabilistic polynomial-time adversary $\mathcal{S}$ for the ideal model such that for every non-uniform probabilistic polynomial-time adversary $\mathcal{Z}$,*

$$\{\text{IDEAL}_{f,\mathcal{S},\mathcal{Z}}(\lambda, z)\}_{\lambda \in \mathbb{N}, z \in \{0,1\}^*} \stackrel{c}{\equiv} \{\text{HYBRID}^g_{\pi,\mathcal{A},\mathcal{Z}}(\lambda, z)\}_{\lambda \in \mathbb{N}, z \in \{0,1\}^*}.$$

Definition 2. *Let π be a protocol and f be a multiparty functionality. We say that π securely computes f with penalties if π SCC-realizes the functionality $\mathcal{F}^\star_f$ according to Definition 1.*

Throughout this paper, we deal only with static adversaries and impose no restrictions on the number of parties that can be corrupted. Our schemes also make use of a digital signature scheme which we denote as (SigKeyGen, SigSign, SigVerify). Please see [8,26–28] for additional details on the model including the necessary modifications to UC. Please also see [24] which extensively treats these modifications, proposes alternative models, and uses protocol compilers different from GMW.

3.1 Ideal Functionalities

Secure computation with penalties—multiple executions. We now present the functionality $\mathcal{F}^*_{\text{MSFE}}$ which we wish to realize. Recall that secure computation with penalties guarantees the following.

- An honest party never has to pay any penalty.
- If a party aborts after learning the output and does not deliver output to honest parties, then *every* honest party is compensated.

See Fig. 4 for a formal description. Note that $\mathcal{F}^*_{\text{MSFE}}$ directly realizes multiple invocations of *non-reactive* secure computation with penalties. In the first phase referred to as the *deposit phase*, the functionality $\mathcal{F}^*_{\text{MSFE}}$ accepts safety deposits $\text{coins}(d)$ from each honest party and penalty deposit $\text{coins}(hq)$ from the adversary. Note that the penalty deposit suffices to compensate each honest party n the event of an abort. Once the deposits are made, parties enter the next phase referred to as the *execution phase* where parties can engage in unbounded number of secure function evaluations. In each execution, parties submit inputs and wait to receive outputs. As usual, the ideal adversary $\mathcal{S}$ gets to learn the output first

---

Notation: session identifier sid, parties $P_1, \ldots, P_n$, adversary $\mathcal{S}$ that corrupts $\{P_s\}_{s \in C}$, safety deposit d, penalty amount q. Set $H = [n] \setminus C$ and $h = |H|$.

DEPOSIT PHASE: Initialize flg $= \perp$. Wait to get message (setup, sid, $ssid$, j, coins(d)) from P_j for all $j \in H$. Wait to get message (setup, sid, $ssid$, coins(hq)) from $\mathcal{S}$.

EXECUTION PHASE: Set flg $= 0$. For id $= 1, 2, \cdots$, sequentially do:

- If a message (exit, sid) is received from P_j, then send (exit, sid, j) to all parties, and go to the claim phase.
- Wait to receive a message (input, sid, $ssid \| \mathrm{id}$, j, $y_j^{(\mathrm{id})}$, $g^{(\mathrm{id})}$) from P_j for all $j \in H$. Then send (function, sid, $ssid \| \mathrm{id}$, $g^{(\mathrm{id})}$) to all parties.
- Wait to receive a message (input, sid, $ssid \| \mathrm{id}$, $\{y_s^{(\mathrm{id})}\}_{s \in C}$, $g^{(\mathrm{id})}$) from $\mathcal{S}$. If no such message was received, then go to the claim phase.
- Compute $z^{(\mathrm{id})} \leftarrow g^{(\mathrm{id})}(y_1^{(\mathrm{id})}, \ldots, y_n^{(\mathrm{id})})$ and send (output, sid, $ssid \| \mathrm{id}$, $z^{(\mathrm{id})}$) to $\mathcal{S}$.
- If $\mathcal{S}$ returns (continue, sid, $ssid \| \mathrm{id}$), send (output, sid, $ssid \| \mathrm{id}$, $z^{(\mathrm{id})}$) to each P_j.
- Else if $\mathcal{S}$ returns (abort, sid, $ssid$), update flg $= 1$, and go to the claim phase.

CLAIM PHASE:

- If flg $= 0$ or $\perp$, send (return, sid, $ssid$, coins(d)) to P_j for $j \in H$. If flg $= 0$, send (return, sid, $ssid$, coins(hq)) to $\mathcal{S}$.
- Else if flg $= 1$, send (penalty, sid, $ssid$, coins($d + q + q_i$)) to P_i for all $i \in H$ where $q_i = 0$ unless $\mathcal{S}$ sent a message (extra, sid, $ssid$, $\{(i, q_i)\}_{i \in H}$, coins($\sum_{i \in H} q_i$)).

Fig. 4. Special ideal functionality $\mathcal{F}_{\mathrm{MSFE}}^*$ for multiple sequential SFE with penalties.

and then decide whether to deliver the output to all parties. If $\mathcal{S}$ decides to abort, then no further executions are carried out, parties enter the *claim phase*, and honest parties get coins($d + q$), i.e., their safety deposit plus the penalty amount. Now if $\mathcal{S}$ never aborts during a local execution, then the safety deposits are returned back to the honest parties, and $\mathcal{S}$ gets back its penalty deposit. Note that we explicitly allow an exit command that enables parties to exit the contract immediately. Prior works [26,27] required parties to wait for a pre-specified time out parameter before parties can reclaim their deposits.

Supporting cash distribution via $\mathcal{F}_{\mathrm{MSCD}}^*$. See Fig. 5 for a formal definition for $\mathcal{F}_{\mathrm{MSCD}}^*$. The definition for amortized secure cash distribution with penalties in the reactive setting $\mathcal{F}_{\mathrm{MSCD}}^*$ is identical to $\mathcal{F}_{\mathrm{MSFE}}^*$ except that we repeatedly evaluate reactive functions which is composed of multiple stage functions. Now, $\mathcal{S}$ can abort between different stages of a reactive function evaluation or within a single stage. In either case, the honest parties will be compensated via the penalty deposit coins(hq) submitted by $\mathcal{S}$ in the deposit phase. Furthermore, $\mathcal{F}_{\mathrm{MSCD}}^*$ also supports cash distribution which makes it useful for many applications. In particular, we allow parties to dynamically add deposits. The reactive functions that are evaluated take into account the current balance which is maintained in the variable $\boldsymbol{b}$, and the output of the evaluations update $\boldsymbol{b}$ to reflect how the cash is redistributed. That is, the amount of coins are specified as input to the reactive functions, and the output will influence how the coins are redistributed. Finally we note that unlike prior formulations [26, 27], we do not require an apriori upper bound on the *number of stages* that is common to the reactive functions supported by $\mathcal{F}^*_{\mathrm{MSCD}}$. Like most prior works (with the exception of [25]), we do not attempt to hide the (updated) balance vectors or the amount of coins.

3.2 Stateful Contract — Ideal Functionality

We now present the functionality $\mathcal{F}^*_{\mathrm{StCon}}$ which we use to abstract the smart contracts functionality provided by cryptocurrencies. At a high level, $\mathcal{F}^*_{\mathrm{StCon}}$ lets parties run a time-dependent stateful computation. In other words, the contract encodes a finite state computation where each transition could potentially be dependent on the *time of the transition*. Time-dependent transitions in our stateful contract functionality allow us to design protocols that can support early termination of contracts which could be a potentially critical feature in certain settings. Such features are not supported by claim-or-refund $\mathcal{F}^*_{\mathrm{CR}}$.

As is typical in the penalty model, we let parties make an initial deposit to the $\mathcal{F}^*_{\mathrm{StCon}}$ functionality. Following this, parties together specify a finite state computation denoted Prog along with an initial state st. The functionality then simply waits for a state transition to be triggered by any of the parties. Upon a valid trigger w (i.e., for which Prog produces non-$\perp$ output), the contract runs Prog on the tuple $(j, w, t; \mathsf{st})$ where j is the index of party P_j who supplies the trigger w at time t, and where st is the current state of Prog. Prog then outputs the current state which will be stored in the variable st and the amount of money e that P_j is supposed to obtain. Both st and e are revealed to all parties (i.e., $\mathcal{F}^*_{\mathrm{StCon}}$ has public state), and $\mathcal{F}^*_{\mathrm{StCon}}$ will distribute coins(e) to party P_j. The formal description of $\mathcal{F}^*_{\mathrm{StCon}}$ is given in Fig. 6. The functionality repeatedly accepts state transitions until it has distributed all the coins that were deposited to it in the initialization phase.

Analogous to the script complexity definition for $\mathcal{F}^*_{\mathrm{CR}}$-based protocols, we can define the *script complexity* (also referred to as "validation complexity") of a protocol in the $\mathcal{F}^*_{\mathrm{StCon}}$-hybrid model. See [9, Sect. 3.2] for more details.

Definition 3 ($\mathcal{F}^*_{\mathrm{StCon}}$ Script Complexity). *Let Π be a protocol among $P_1, \ldots, P_n$ in the $\mathcal{F}^*_{\mathrm{StCon}}$-hybrid model where $\mathcal{F}^*_{\mathrm{StCon}}$ is initialized with program Prog and initial state st. We say a trigger $T = (j, w, t)$ is valid iff $\mathsf{Prog}(j, w, t; \mathsf{st}) \neq \perp$ where st is the current state of $\mathcal{F}^*_{\mathrm{StCon}}$. We say a state st is valid iff there exists a valid sequence of triggers starting from some initial state st_0 that result in st becoming the current state of $\mathcal{F}^*_{\mathrm{StCon}}$. For a trigger T acting on input state st, we let $C(\mathsf{Prog}, T, \mathsf{st})$ denote the sum of the size of T, the size of the input states and the output states and the running time of Prog on input $(T; \mathsf{st})$. We define the $\mathcal{F}^*_{\mathrm{StCon}}$-validation complexity of Π, or in short transition validation complexity of Π as the maximum value of $C(\mathsf{Prog}, T, \mathsf{st})$ maximized over all possible choices of a valid trigger T and a valid state st.* ◇

Functionality $\mathcal{F}^*_{\mathrm{StCon}+}$. We also present another variant of $\mathcal{F}^*_{\mathrm{StCon}}$, which we call $\mathcal{F}^*_{\mathrm{StCon}+}$. The main difference is that $\mathcal{F}^*_{\mathrm{StCon}+}$ accepts coin deposits, via

Notation: session identifier sid, parties $P_1, \ldots, P_n$, adversary $\mathcal{S}$ that corrupts $\{P_s\}_{s \in C}$, *safety deposit* d, *penalty amount* q, set $H = [n] \setminus C$.

DEPOSIT PHASE: Initialize flg $= \bot$.

- Wait to receive a message (setup, $sid, ssid, j, \text{coins}(d)$) from P_j for all $j \in H$.
- Wait to receive a message (setup, $sid, ssid, \text{coins}(hq)$) from $\mathcal{S}$ where $h = |H|$.

EXECUTION PHASE: Initialize flg $= 0$ and $\boldsymbol{b} \leftarrow \boldsymbol{0}$. For id $= 1, 2, \ldots$, sequentially do:

- If a message (exit, $sid, ssid$) is received from P_j, then send (exit, $sid, ssid, j$) to all parties, and go to the claim phase.
- If a message (addmoney, $sid, ssid\|\text{id}, b_j, \text{coins}(b_j)$) is received from some P_j, and a message (addmoney, $sid, ssid\|\text{id}, b_j$) was received from every P_k with $k \neq j$, then send (addmoney, $sid, ssid\|\text{id}, b_j$) to all parties. Update $\boldsymbol{b} \leftarrow \boldsymbol{b} + (\cdots, 0, b_j, 0, \cdots)$.
- Initialize state $= \bot$. Wait to receive message (function, $sid, ssid\|\text{id}, g^{(\text{id})}$) from P_j for all $j \in H$. Then send (function, $sid, ssid\|\text{id}, g^{(\text{id})}$) to all parties.
- Parse $g^{(\text{id})} = \{g_k^{(\text{id})}\}_{k \in [\rho]}$. For $k = 1, \ldots, \rho$, sequentially do:
  - Wait to receive message (input, $sid, ssid\|\text{id}\|k, j, y_j'$) from P_j for all $j \in H$.
  - Wait to receive a message (input, $sid, ssid\|\text{id}\|k, \{y_s'\}_{s \in C}$) from $\mathcal{S}$. If no such message was received, update flg $= 1$ and go to the claim phase.
  - Compute $(z, \boldsymbol{b}', \text{state}) \leftarrow g_k^{(\text{id})}(y_1', \ldots, y_n'; \text{state}, \boldsymbol{b})$.
  - Send message (output, $sid, ssid\|\text{id}\|k, z, \boldsymbol{b}'$) to $\mathcal{S}$.
  - If $\mathcal{S}$ sends (continue, $sid, ssid\|\text{id}\|k$), send (out, $sid, ssid\|\text{id}\|k, z, \boldsymbol{b}'$) to all P_i.
  - If $\mathcal{S}$ returns (abort, $sid, ssid\|\text{id}\|k$), set flg $= 1$, and go to the claim phase.
  - Update $\boldsymbol{b} \leftarrow \boldsymbol{b}'$.

CLAIM PHASE:

- If flg $= 0$ or $\bot$, send (return, $sid, ssid, \text{coins}(d + \boldsymbol{b}_r)$) to all P_r for $r \in H$. If flg $= 0$, send (return, $sid, ssid, \text{coins}(hq + \sum_{s \in C} \boldsymbol{b}_s)$) to $\mathcal{S}$.
- Else if flg $= 1$, send (penalty, $sid, ssid, \text{coins}(d + q + \boldsymbol{b}_i + q_i)$) to P_i for all $i \in H$ where $q_i = 0$ unless $\mathcal{S}$ sent a message (extra, $sid, ssid, \{q_i\}_{i \in H}, \text{coins}(\sum_{i \in H} q_i)$). Send (remaining, $sid, ssid, \text{coins}(\sum_{s \in C} \boldsymbol{b}_s)$) to $\mathcal{S}$.

Fig. 5. Special ideal functionality $\mathcal{F}_{\text{MSCD}}^*$ for multiple sequential MPC with penalties.

an update command even during the execution phase. We note that dynamic updates to Prog is supported by our definition (although we do not rely on this feature in our protocols). We note that $\mathcal{F}_{\text{StCon+}}^*$ is also supported by Ethereum (Fig. 7).

Notation: session identifier sid, parties $P_1, \ldots, P_n$, *initial deposit vector* $(d_1, \ldots, d_n)$, program Prog and an initial state st. We assume that the initial deposit vector is specified in Prog.

INITIALIZATION PHASE: Wait to get message (init, sid, $ssid$, Prog, st, coins(d_j)) from P_j for all $j \in [n]$. Initialize $Q \leftarrow \sum_{j \in [n]} d_j$.

EXECUTION PHASE: Repeat until termination:

- Wait to receive a message (trigger, sid, $ssid$, w) from some party P_j at time t such that $\mathsf{Prog}(j, w, t; \mathsf{st}) \neq \bot$. Then, let $(\mathsf{st}, e) \leftarrow \mathsf{Prog}(j, w, t; \mathsf{st})$. If $e < Q$, then send (output, sid, $ssid$, j, w, t, st, e) to each P_k and send (pay, sid, $ssid$, coins(e)) to P_j. Update $Q \leftarrow Q - e$. If $Q = 0$, then terminate.

Fig. 6. Special ideal functionality $\mathcal{F}^*_{\mathrm{StCon}}$ for stateful contracts.

Notation: session identifier sid, set of parties $\{P_1, \ldots, P_n\}$, *initial deposit vector* $d_1 = \cdots d_n = (n-1)q$, where q is the penalty amount, a program Prog, an initial state st, and a validation function for updates Update. During the execution phase, parties will be able to add coins via Update.

INITIALIZATION PHASE: Wait to get (init, sid, $ssid$, Prog, Update, st, d_j, coins(d_j)) from each P_j. Initialize $Q \leftarrow \sum_{j \in [n]} d_j$.

EXECUTION PHASE: Repeat until termination:

- If a message (update, sid, $ssid$, u, b, coins(b)) is received from P_j at time t such that $\mathsf{Update}(j, w, t; \mathsf{st}) \neq \bot$, then set (Prog, Update, st) $\leftarrow \mathsf{Update}(j, u, t; \mathsf{Prog}, \mathsf{st})$, accept coins($b$), update $Q \leftarrow Q + b_j$, and send (update, sid, $ssid$, j, u, t, st, Prog, Update) to all parties.
- If a message (trigger, sid, $ssid$, w) is received from P_j at time t such that $\mathsf{Prog}(j, w, t; \mathsf{st}) \neq \bot$. Then, let $(\mathsf{st}, e) \leftarrow \mathsf{Prog}(j, w, t; \mathsf{st})$. If $e < Q$, then send (output, sid, $ssid$, j, w, t, st, e) to each P_k and send (pay, sid, $ssid$, coins(e)) to P_j. Update $Q \leftarrow Q - e$. If $Q = 0$, then terminate.

Fig. 7. Special ideal functionality $\mathcal{F}^*_{\mathrm{StCon+}}$ for stateful contracts.

4 Realizing $\mathcal{F}^*_{\mathrm{MSFE}}$ from $\mathcal{F}^*_{\mathrm{StCon}}$

In this section, we describe the protocols for amortized secure computation with penalties in the $\mathcal{F}^*_{\mathrm{StCon}}$-hybrid model. Due to lack of space we only give a brief overview. See [9, Appendix A] for full details.

The protocol for implementing $\mathcal{F}^*_{\mathrm{MSFE}}$ has three phases. In the first phase, parties interact with the on-chain stateful contract, i.e., the ideal functionality $\mathcal{F}^*_{\mathrm{StCon}}$. In particular, parties agree on setting contract parameters that fix the number of parties, the allowed state transitions of the contract, the time-out, and the compensation amounts to parties in case of an abort (cf. Fig. 9). Then in the second phase (cf. Fig. 8), parties perform the actual computation. This is done off-chain via local MPC executions. In addition to performing the computation, these MPC executions also provide hooks to the on-chain contract (to handle aborts). We describe the local executions first because they will introduce new variables which will serve as hooks to the contract via the contract parameters. In the next phase, we describe the process which honest parties use in case an off-chain local execution was aborted. In particular, in this phase, parties will go on to the on-chain contract to either continue the aborted local execution or claim their compensation. This phase occurs immediately following an abort in the local executions phase. Figure 10 describes how parties handle notifications received from $\mathcal{F}^*_{\mathrm{StCon}}$.

Local executions. The formal description of the local executions is in Fig. 8. In this section, we describe this phase in more detail. Suppose the parties are interested in computing a function $g^{(\mathrm{id})}$. At a high level, parties begin by running a standard secure computation protocol (that does not guarantee fairness) which computes $z = g^{(\mathrm{id})}(y_1, \ldots, y_n)$ where y_j represents the input of P_j. In addition this secure computation protocol also additively secret shares z into $z_1, \ldots, z_n$ and computes commitments h_j on each z_j using uniform randomness ω_j. Finally, the secure computation protocol outputs $x_j^{(\mathrm{id})} = (z_j; \omega_j)$ (i.e., the decommitment to h_j) and the value $\boldsymbol{h}^{(\mathrm{id})} = h_1 \| \cdots \| h_n$ to each party P_j. For the simulation to work, we need to use an honest-binding commitment scheme (cf. [9, Appendix D]).

Note that there could be aborts here and in every subsequent stage of the local execution. In each step, we assume that parties stop the protocol if they do not receive valid messages (i.e., including signatures) from any other party. Importantly, there is an additional (implicit) time-interval parameter δ which is used to detect aborts. In more detail, we say that a party aborted the protocol if (1) it is its turn to send a message, and (2) if the party does not send a valid protocol message within time-interval δ of the previous event. (We assume that all honest parties act immediately, i.e., within time-interval δ.) In the event of an abort (as defined above), parties go up to the on-chain contract for resolution (cf. Fig. 10). Once the local secure computation protocol ends, we ask each party P_j to compute a signature σ_j on the message $(\mathrm{id}, \boldsymbol{h}^{(\mathrm{id})})$ under its secret signing key sk_j and then broadcast it to all parties. In the next step, each party P_j broadcasts the decommitment $x_j^{(\mathrm{id})}$ to the value h_j contained in $\boldsymbol{h}^{(\mathrm{id})}$ (which in particular includes the secret share z_j of the output z). Once this step is completed, then all parties can recover the output of the id-th computation as $\bigoplus_{k \in [n]} z_k$. Please see [9, Appendix A] for more details.

Contract parameters. See Fig. 9 for a formal description. The state parameters are established in the following way. The variables $d_1, \ldots, d_n$ represent the amount $\mathsf{coins}((n-1)q)$ that the parties are expected to put in as the initial deposit to the contract.

State components and initialization. The variable st denotes the current state of the contract. The values $\boldsymbol{pk}$ and Δ are constant parameters to the contract and

1. (MPC step) Parties run a standard MPC protocol that
   - obtains inputs $y_1, \ldots, y_n$ from the parties; and
   - computes $z = g^{(\text{id})}(y_1, \ldots, y_n)$; and
   - secret shares z into $z_1, \ldots, z_n$; and
   - computes $h_j = \text{com}(x_j^{(\text{id})} = (z_j; \omega_j))$; where each ω_j is chosen uniformly at random; and
   - outputs $(x_j^{(\text{id})}, \boldsymbol{h}^{(\text{id})} = h_1\|\cdots\|h_n)$ to each P_j.
2. (Signature broadcast) Each P_j computes a signature σ_j on message $(\text{id}, \boldsymbol{h}^{(\text{id})})$ under the signing key sk_j and then broadcasts it. Let S_j denote the set of parties whose signature on message $(\text{id}, \boldsymbol{h}^{(\text{id})})$ was received by P_j. If $S_j = [n]$, then let $\boldsymbol{\sigma}^{(\text{id})} = \sigma_1\|\cdots\|\sigma_n$ and update $\text{best}_j \leftarrow (\text{id}, ((j, x_j^{(\text{id})}), \boldsymbol{h}^{(\text{id})}, \boldsymbol{\sigma}^{(\text{id})}))$. Else, parties abort the local execution and go to the on-chain contract for resolution (cf. Figure 10).
3. (Share broadcast) Each P_j broadcasts $x_j^{(\text{id})}$. Let $S_j^{(\text{id})}$ denote the set of parties whose share (authenticated against $\boldsymbol{h}^{(\text{id})}$) was received by P_j and let $X_j^{(\text{id})}$ denote the corresponding set of decommitments received by P_j. Each P_j updates $\text{best}_j \leftarrow (\text{id}, (X_j, \boldsymbol{h}, \boldsymbol{\sigma}))$. If $|S_j^{(\text{id})}| = [n]$, then P_j computes the output of the id-th local execution as $\bigoplus_{k=1}^{n} z_k$ where we parse $x_k^{(\text{id})} \in X_j^{(\text{id})}$ as $(z_k; \omega_k)$. Else, parties abort the local execution and go to the on-chain contract for resolution (cf. Figure 10).

Fig. 8. id-th off-chain local execution for implementing $\mathcal{F}^*_{\text{MSFE}}$.

are always maintained as part of the state. The parameter $\boldsymbol{pk}$ represents the set of public keys of all the parties. The parameter Δ represents the length of the time interval within which parties need to act in order to keep the contract from defaulting. In addition, each state variable has five components: (1) st.mode represents the current mode in which the contract is in, and is one of { "init", "exec", "exit", "payout", "abort", "inactive"}; (2) st.id represents the id of the execution that is being continued currently on the on-chain contract; (3) st.TT represents the current transcript of the execution that is being continued on the chain; (4) st.t represents the time when the on-chain contract was triggered and either (a) was moved to "exit" or (b) resulted in a change of the variable st.TT; (5) st.L is a boolean array that represents which parties have already withdrawn their deposits and compensations from $\mathcal{F}^*_{\text{StCon}}$.

We represent the state variable st as a five tuple (st.mode, st.id, st.TT, st.t, st.L). Also, st.TT is either $\perp$ (denoting the null transcript) or is a tuple of the form $(X, \boldsymbol{h}, \boldsymbol{\sigma})$. The initial state is st and its components are initialized in the following way: (1) st.mode = "init"; (2) st.id = -1; (3) st.TT = NULL; (4) st.$t = -1$; and (5) st.$L = (1, \ldots, 1)$. Recall that st also contains the list of all public keys of the participants $\boldsymbol{pk}$, and the global time-out parameter Δ.

Triggering state transitions. During course of the execution, the state of the contract would either (1) remain in the initial state with st.mode = "init"; or

(2) be in exit mode, i.e., with st.mode = "exit", where contract participants are trying to get their initial deposit out of the contract (and terminate the contract); or (3) be trying to continue an incomplete off-chain local execution by keeping track of the current state of the local execution computation, i.e., with st.mode = "exec"; or (4) be in payout mode with st.mode = "payout" where parties have successfully completed all executions so far and are waiting to get their initial deposits out of $\mathcal{F}_{\text{StCon}}^*$; or (5) be in "abort" mode where an execution was aborted and honest parties are waiting to get their initial deposits and compensation out of $\mathcal{F}_{\text{StCon}}^*$; or (6) be in "inactive" mode where $\mathcal{F}_{\text{StCon}}^*$ no longer accepts any further state transitions and in particular, has given out all the money that was deposited to it.

Transitions to different states will be triggered by a witness (j, w, t). Here j represents the party triggering the contract, i.e., party P_j. The value t represents the time at which the contract is triggered. Note that when st.mode = "payout"/"abort"/"inactive", the triggering witness w is simply the token value exit. As we will see later, the transitions from these states depends only on st.L and the triggering time t and st.t. The more interesting case is when st.mode = "init"/"exit"/"exec". In this case, the triggering witness w provides the most recent state of the current local execution. We will use a separate subroutine pred to determine the validity of a trigger (j, w, t) when the witness w represents a transcript of an execution.

Subroutine pred. The predicate pred essentially decides if a trigger to the contract is a valid continuation of the computation on the chain. Now, pred takes a trigger (j, w, t) and examines it in conjunction with the current state of the contract. First, pred parses the witness w as (id, TT = $(X, \boldsymbol{h}, \boldsymbol{\sigma})$) where id represents the (off-chain) execution that is being attempted to be continued on the chain by party P_j. The value TT = $(X, \boldsymbol{h}, \boldsymbol{\sigma})$ essentially provide (along with a proof) the most recent state of a computation (typically the last incomplete off-chain computation). In particular and in the context of non-reactive functionalities, the value X maintains the set of parties who have completed their step of the computation on the chain along with their broadcasted decommitments to the secret share of the final output. The values $\boldsymbol{h}$ and $\boldsymbol{\sigma}$ essentially authenticate to the contract that values in X are legitimate values corresponding to the id-th off-chain computation. In more detail, $\boldsymbol{h} = h_1 \| \cdots \| h_n$ is the set of commitments (that is public to all parties). The value $\boldsymbol{h}$ should be consistent with the broadcast values X in the sense that $\text{com}(X[k]) = h_k$ for all k such that $X[k] \neq \bot$. Likewise, the commitment values $\boldsymbol{h}$ need to be accompanied with $\boldsymbol{\sigma} = \sigma_1 \| \cdots \| \sigma_n$ where σ_i is the signature of party P_i attesting to the correctness of $\boldsymbol{h}$. Note that the signatures also tie the value of $\boldsymbol{h}$ to id.

Clearly, pred should output 1 if the witness is valid and if (j, w, t) happens to be the very first trigger to the contract. On the other hand, if (j, w, t) is not the first trigger to the contract, then we have to ensure that the trigger (j, w, t) provides a valid update to the contract state. Now the contract state could be in exit mode, i.e., st.mode = "exit", and in this case the trigger (j, w, t) with a valid witness w could be by an honest party to continue an incomplete

Notation. The variable st denotes the current state of the contract. We represent the state variable st as a five tuple (st.mode, st.id, st.TT, st.t, st.L). We omit $\boldsymbol{pk}$ and Δ from the state to keep the presentation simple. The initial deposits are $d_1 = \cdots = d_n = (n-1)q$. The initial state is ("init", -1, $\perp$, -1, $\mathbf{1}$).

Subroutine pred. Let $\mathsf{pred}(j, w, t; \mathsf{st}) = 1$ if
- w is parsed as (id, TT $= (X, \boldsymbol{h}, \boldsymbol{\sigma})$) with $\boldsymbol{h} = h_1 \| \cdots \| h_n$, $\boldsymbol{\sigma} = \sigma_1 \| \cdots \| \sigma_n$; and
- for each $k \in [n]$ it holds that $\mathsf{SigVerify}((\text{id}, \boldsymbol{h}), \sigma_k; pk_k) = 1$; and
- for each k such that $X[k] \neq \mathsf{NULL}$ it holds that $h_k = \mathsf{com}(X[k])$; and
- either (1) st.mode $=$ "init"; or (2) st.mode $=$ "exec"/"exit" and $t \leq$ st.$t + \Delta$ and either (2.1) id $>$ st.id or (2.2) id $=$ st.id and $X \not\subseteq$ st.TT.X.

State transitions. $\mathsf{Prog}(j, w, t; \mathsf{st})$: Initialize $e \leftarrow 0$.

- If $w = (\text{id}, \text{TT})$ and $\mathsf{pred}(j, w, t; \mathsf{st}) = 1$: If st.id $=$ id, then update st.TT.$X \leftarrow$ st.TT.$X \cup$ TT.X, else set st.TT $\leftarrow$ TT. Set st $\leftarrow$ ("exec", id, st.TT, t, st.L).
- Else if $w =$ exit:
  - If (1) st.mode $=$ "init" or (2) st.mode $=$ "exec" and $|\text{st.TT.}X| = n$: Set st.mode $\leftarrow$ "exit" and st.$t \leftarrow t$.
  - If st.mode $=$ "exec" or "abort", and $t >$ st.$t + \Delta$ and st.$L[j] = 1$ and $|\text{st.TT.}X| \neq n$: Then update st.$L[j] \leftarrow 0$ and st.mode $\leftarrow$ "abort" and st.$L[k] \leftarrow 0$ for all k such that st.TT.$X[k] = \perp$. Further, if st.TT.$X[j] \neq \perp$, then set $e \leftarrow n(n-1)q/|\text{st.TT.}X|$.
  - If st.mode $=$ "exit" or "payout", and $t >$ st.$t + \Delta$ and st.$L[j] = 1$: Set $e \leftarrow (n-1)q$ and update st.mode $\leftarrow$ "payout" and st.$L[j] \leftarrow 0$.

If st.$L[k] = 0$, for all $k \in [n]$ then we update st.mode $\leftarrow$ "inactive".

Fig. 9. $\mathcal{F}^*_{\text{StCon}}$ parameters for $\mathcal{F}^*_{\text{MSFE}}$.

off-chain execution. This is to ensure that a malicious party cannot subvert the continuation of the off-chain execution on the on-chain contract by trying to exit prematurely (i.e., when st.mode $=$ "init"). Likewise a malicious party might also submit an old completed execution (even while the current off-chain execution has not yet completed). Thus, we must have pred output 1 when the new id present in w is greater than st.id.

Now the contract could be in exec mode, i.e., st.mode $=$ "exec", in which case the contract is typically waiting for the on-chain execution to be completed. There are essentially two cases: (1) the current state does not correspond to or continue the most recent off-chain execution; in this case, the id in the new trigger must satisfy id $>$ st.id (i.e., the contract is essentially reset to the "correct last computation"), and (2) the new trigger continues the current state of the contract and for this id $=$ st.id must hold and also we need X to contain at least one value which is not in st.TT.X, i.e., there is some $k \in [n]$ such that $X[k] \neq \perp =$ st.TT.X. In either case, the new trigger must appear within the time interval Δ of the previous trigger (i.e., before time st.$t + \Delta$).

State transitions. The state transition function Prog takes as input the trigger (j, w, t) and the current state st. First, we check if the witness provided corresponds to an execution transcript. In this case, we invoke the predicate pred and if pred outputs 1, then we update st.TT depending on whether (1) st.id = id, in which case we update st.TT.X to include decommitments specified in X; or (2) st.id < id, in which case we update st.TT ← TT. If w does not correspond to an execution transcript, then we assume that it is a token value exit. There are effectively three cases to handle:

- If st.mode equals "init" or equals "exec" with a fully completed transcript, then we change st.mode to "exit" and store the triggering time t in st.t. This transition is provided to ensure that honest parties' deposits are not "locked in" and to enable them to withdraw their deposits from $\mathcal{F}^*_{\mathrm{StCon}}$.
- If st.mode = "exec"/"abort", then we check if $t >$ st.$t + \Delta$ and if $|$st.TT.$X| \neq n$ to confirm that the execution has indeed been aborted. In this case, if st.$L[j] = 1$, then we will allow P_j to take money out of $\mathcal{F}^*_{\mathrm{StCon}}$. We further need to check whether P_j was a malicious party that did not contribute to completing the execution. We do this by checking if st.TT.$X[j] \neq \bot$. If all checks pass, we let P_j to withdraw its initial deposit plus compensation, i.e., a total of $n(n-1)q/|$st.TT.$X|$ from $\mathcal{F}^*_{\mathrm{StCon}}$.
- If st.mode = "exit"/"payout", then we check if $t >$ st.$t + \Delta$. This is to prevent situations where a malicious party tries to subvert continuing the off-chain aborted execution on the chain. (That is, honest parties get an additional time Δ to get $\mathcal{F}^*_{\mathrm{StCon}}$ out of the exit mode.) If $t >$ st.$t + \Delta$ indeed holds, then we allow party P_j to take its initial deposit $(n-1)q$ out of the contract if it was not already paid before (i.e., st.$L[j] = 1$).

Main protocol. The formal description can be found in Fig. 10. In this section we will describe the main protocol that makes use of the local execution subprotocol of Fig. 8 and also how parties interact with $\mathcal{F}^*_{\mathrm{StCon}}$ according to the parameters described in Fig. 9. Parties basically start by initializing the $\mathcal{F}^*_{\mathrm{StCon}}$ parameters as in Fig. 9. Following this, they begin off-chain local executions. Recall that each party P_j maintains a variable best$_j$ which denotes the transcript corresponding to the latest *active* execution (both on-chain and off-chain) *according to the local view of party P_j* (see Fig. 8). This value will provide the necessary hook to the on-chain contract to handle off-chain aborts. In particular, the value best$_j$ will be submitted by party P_j in order to recover from aborted off-chain executions.

In our main protocol, we essentially deal with three different scenarios: (1) when parties want to exit the contract and get back their deposits and compensation, (2) when parties want to continue an aborted off-chain execution on the chain, and (3) when parties are notified of state changes in $\mathcal{F}^*_{\mathrm{StCon}}$.

Exiting the contract. First, we deal with the situation when parties would like to terminate the protocol and retrieve their initial deposits from the contract. To do so, we simply let parties submit a token value $w =$ exit to trigger and put

Parties initialize the parameters as in Figure 9. Then for $\mathsf{id} = 1, 2, \cdots$, parties run the local execution prescribed in Figure 8. Recall that each party P_j maintains a variable best_j during the local executions which is initialized as $\perp$.

1. If a party P_j wants to exit the contract and reclaim its initial deposit, then it sends $w = \mathsf{exit}$ to $\mathcal{F}^*_{\mathrm{StCon}}$.
2. If there is an abort during a local off-chain execution, then each party P_j triggers $\mathcal{F}^*_{\mathrm{StCon}}$ with the value best_j.
3. Each party P_j waits and responds to state changes in $\mathcal{F}^*_{\mathrm{StCon}}$ depending on the current state st:
   (a) If $\mathsf{st.mode} = $ "payout"/"abort" and $\mathsf{st}.L[j] = 1$, send $w = \mathsf{exit}$ to $\mathcal{F}^*_{\mathrm{StCon}}$.
   (b) If (1) $\mathsf{st.id} < \mathsf{best}_j.\mathsf{id}$, or (2) $\mathsf{st.id} = \mathsf{best}_j.\mathsf{id}$ and $\mathsf{st}.\mathrm{TT}.X[j] = \perp$, then submit best_j to $\mathcal{F}^*_{\mathrm{StCon}}$.
   (c) If $\mathsf{st.id} > \mathsf{best}_j.\mathsf{id}$, then submit $\mathsf{best}_j \leftarrow (\mathsf{st.id}, ((j, x_j^{(\mathsf{st.id})}), \mathsf{st}.\mathrm{TT}.\boldsymbol{h}, \mathsf{st}.\mathrm{TT}.\boldsymbol{\sigma}))$ to $\mathcal{F}^*_{\mathrm{StCon}}$.

Finally, parties also keep track of whether $\mathsf{st.mode} = $ "exit" or "exec" and the current time t is such that $t > \mathsf{st}.t + \Delta$. In this case, parties send $w = \mathsf{exit}$ in order to claim their payout or compensation.

Fig. 10. Main protocol for implementing $\mathcal{F}^*_{\mathrm{MSFE}}$.

the contract into exit mode. Note that malicious parties might revert $\mathcal{F}^*_{\mathrm{StCon}}$ to go into exec mode. In this case, $\mathcal{F}^*_{\mathrm{StCon}}$ will notify honest parties of the change. Honest parties will be able to recover from this and put the contract back into exit mode. This will be described when we discuss how parties react to notifications from $\mathcal{F}^*_{\mathrm{StCon}}$.

Continuing an aborted off-chain execution. This is where the value best_j comes in handy as it stores the most recent state of the off-chain executions. We instruct parties to trigger $\mathcal{F}^*_{\mathrm{StCon}}$ with the value best_j which includes P_j's decommitment $x_j^{(\mathsf{id})}$ which in turn ensures (by the logic in Fig. 9) that P_j will not be penalized.

Responding to notifications from $\mathcal{F}^*_{\mathrm{StCon}}$. First, if $\mathsf{st.mode} = $ "payout"/"abort", then parties send a token value $w = \mathsf{exit}$ to get their deposits out of $\mathcal{F}^*_{\mathrm{StCon}}$. In addition, if $\mathsf{st.mode} = $ "abort", then parties would also get compensation from $\mathcal{F}^*_{\mathrm{StCon}}$. Second, if the on-chain execution does not correspond to the most recent execution, then we ask parties to submit best_j to the contract. (This will also handle the case when honest parties try to exit the contract but a malicious party feeds an older execution to $\mathcal{F}^*_{\mathrm{StCon}}$.) Checking if the on-chain execution corresponds to the most recent execution is handled by checking first if $\mathsf{st.id} < \mathsf{best}_j.\mathsf{id}$ and then if $\mathsf{st.id} = \mathsf{best}_j.\mathsf{id}$ but $\mathsf{st}.\mathrm{TT}.X[j] = \perp$ (i.e., party P_j's decommitment is not yet part of the on-chain execution transcript). Finally, we also need to handle the corner case when $\mathsf{st.id} > \mathsf{best}_j.\mathsf{id}$. This scenario is actually possible when party P_j is honest but only when $\mathsf{st.id} = \mathsf{best}_j.\mathsf{id} + 1$. We now describe the sequence of events which lead to this case. Suppose in Step 2 of

Fig. 8 some malicious party did not broadcast its signature on $\boldsymbol{h}^{(\mathsf{id})}$, then party P_j will not update best_j. Thus $\mathsf{best}_j.\mathsf{id} = \mathsf{id} - 1$ where id is the execution id of the current execution. Note that each honest P_j would have submitted its signature on $\boldsymbol{h}^{(\mathsf{id})}$ in Step 2. Therefore, malicious parties would possess a valid $\boldsymbol{h}^{(\mathsf{id})}$ and $\boldsymbol{\sigma}^{(\mathsf{id})}$ for execution id. That is, a malicious party P_k is able to trigger $\mathcal{F}^*_{\mathsf{StCon}}$ with a witness $w = (\mathsf{id}, \mathsf{TT} = ((k, x_k^{(\mathsf{id})}), \boldsymbol{h}^{(\mathsf{id})}, \boldsymbol{\sigma}^{(\mathsf{id})})$ which will result in $\mathsf{pred}(k, w, t) = 1$. Thus, we need a mechanism to allow honest parties to continue the id-th execution (i.e., continue TT) and ensure that they don't get penalized. This is indeed possible since honest parties already obtain $x_j^{(\mathsf{st}.\mathsf{id})}$ from Step 1 of Fig. 8. That is, we let honest parties submit $w = (\mathsf{st}.\mathsf{id}, \mathsf{TT}_j = ((j, x_j^{(\mathsf{st}.\mathsf{id})}), \boldsymbol{h}^{(\mathsf{id})} = \mathsf{st}.\mathsf{TT}.\boldsymbol{h}, \boldsymbol{\sigma}^{(\mathsf{id})} = \mathsf{st}.\mathsf{TT}.\boldsymbol{\sigma}))$ to $\mathcal{F}^*_{\mathsf{StCon}}$.

Due to lack of space, we prove the following theorem in [9, Appendix A].

Theorem 1. *Let λ be a computational security parameter. Assume the existence of one-way functions. Then there exists a protocol that SCC-realizes (cf. Definition 1) $\mathcal{F}^*_{\mathsf{MSFE}}$ in the $(\mathcal{F}^*_{\mathsf{StCon}}, \mathcal{F}_{\mathsf{OT}})$-hybrid model whose script complexity (cf. Definition 3) is independent of the number of secure function evaluations and depends only on the length of outputs of the functions evaluated in $\mathcal{F}^*_{\mathsf{MSFE}}$ and is otherwise independent of them. Furthermore, in the optimistic case when all parties are honest, there are a total of $O(n)$ state transitions each having complexity $O(n\lambda)$.*

5   Realizing $\mathcal{F}^*_{\mathsf{MSCD}}$ from $\mathcal{F}^*_{\mathsf{StCon+}}$

We now discuss how to implement $\mathcal{F}^*_{\mathsf{MSCD}}$. Since we are now dealing with reactive functionalities, we make use of an MPC protocol that handles reactive functionalities (say GMW). Since we are dealing with cash distribution, we will let the MPC protocol take, in addition to regular inputs, values that represent the current balance of each party. Note that this balance could change at the end of each stage of the reactive function evaluation. We stress that the reactive functions take only strings as inputs/outputs (and do not handle coins), and the amount of coins and current balance are merely specified as strings. We assume that the updated balance vectors can be obtained directly from the transcript of the protocol implementing the reactive function.

The overall protocol structure closely mimics our protocol for implementing $\mathcal{F}^*_{\mathsf{MSFE}}$. We give a high level overview of the protocol and detail the main differences from the implementation of $\mathcal{F}^*_{\mathsf{MSFE}}$.

Local executions. See Fig. 11 for a formal description. The local executions begin by allowing parties to add coins to their deposit (which will be redistributed depending on the output of the stage functions of the reactive function) but only between different reactive function evaluations. In order to synchronize (in order to make the simulation go through) and ensure that coins are not added while a reactive function is being evaluated, we ask parties to obtain signatures from all parties. (Then we design $\mathcal{F}^*_{\mathsf{StCon+}}$ such that it accepts coins only when the submitting party has signatures from all participants.) Then, in the next step, we ask parties to agree on the transcript validation function for the reactive protocol π implementing $g^{(\mathrm{id})}$ that they are going to execute. That is, in the id-th local execution, parties agree on $\mathsf{tv}^{(\mathrm{id})}$ and each party signs this value under its signing key and broadcasts it to all parties. This is different from the previous case where while implementing $\mathcal{F}_{\mathrm{MSFE}}^*$, we needed parties to sign on $h^{(\mathrm{id})}$ in the id-th execution but *after* the secure computation protocol was run. Note that we also need parties to agree on the updated balance vector b before beginning the id-th local execution.

Once the signatures are done, parties begin evaluating each stage of the reactive computation in sequence. Parties then run a secure computation protocol for each stage sequentially until the entire reactive protocol completes. Note that a typical protocol for a reactive computation maintains state across different stages by secret sharing this value among the participants. That is, when parties are ready to begin the secure computation protocol for the next stage, they supply along with the inputs for the new stage also an authenticated secret share of the previous state. Note that the balance vectors are supplied as input to the reactive MPC (in order to calculate the updated balance). We abstract these details and assume that the authenticated secret shares of intermediate states corresponding to party P_j is part of its input y_j for this stage. The next message function nmf takes the current available transcript as input, along with the parties' input for this stage, the randomness, the current balance vector, and also the secret signing key of this party. Note that nmf and tv are such that for every partial transcript TT' such that $|\mathrm{TT}'| = (j-1) \bmod n$ and $\mathsf{tv}(\mathrm{TT}') = 1$, we have that $(m, \sigma) \leftarrow \mathsf{nmf}(\mathrm{TT}'; (y_j, \omega_j, b, sk_j))$ satisfies $\mathsf{tv}(\mathrm{TT}\|(m,\sigma)) = 1$. Observe that nmf produces signed messages that continue that transcript, and tv checks whether messages are signed appropriately, and if the newly extended transcript is valid according to the underlying reactive MPC. Such modifications to the underlying reactive MPC protocol (namely, adding signatures in nmf and verifying them in tv, and getting updated balance vectors) were also present in previous protocols that dealt with the reactive case [26,27]. Like in the implementation of $\mathcal{F}_{\mathrm{MSFE}}^*$, here too we ask each party P_j to maintain a value best_j which essentially maintains the transcript corresponding to the current execution. Note that best_j contains both $\mathsf{tv}^{(\mathrm{id})}$ as well signatures on it from all parties denoted by $\sigma^{(\mathrm{id})}$.

$\mathcal{F}_{\mathrm{StCon}+}^*$ **parameters.** See Fig. 12 for a formal description. The overall structure is similar to the $\mathcal{F}_{\mathrm{StCon}}^*$ parameters for $\mathcal{F}_{\mathrm{MSFE}}^*$, and in particular we interpret the state st has having multiple components which keep track of the current mode of the state, the current transcript, the current execution id, time of last exec mode transition, and which parties have already withdrawn money from $\mathcal{F}_{\mathrm{StCon}+}^*$. The main addition is now we also explicitly keep track of the transcript validation function of the current execution as part of the state. We denote this variable by st.tv. We also keep track of how much each party deposited in the variable st.B and the latest redistribution of cash (i.e., before st.id-th execution began) in the variable st.b. As with $\mathcal{F}_{\mathrm{StCon}}^*$ parameters for $\mathcal{F}_{\mathrm{MSFE}}^*$, here too we make use of a subroutine pred that effectively determines if the trigger witness w extends the

1. (Adding new coins) If some party P_j wants to add $\mathsf{coins}(b_j)$ to $\mathcal{F}^*_{\mathrm{StCon+}}$, then it sends $\boldsymbol{b}' \leftarrow \boldsymbol{b} + (\cdots, 0, b_j, 0, \cdots)$ to all parties. Each party P_k then generates a signature $\psi'_k \leftarrow \mathsf{Sign}(\mathsf{id}, \boldsymbol{b}')$ and broadcasts it. If P_j receives signatures from all parties, then P_j sends $u = (\mathsf{update}, \boldsymbol{b}', \boldsymbol{\psi}, \mathsf{coins}(b_j))$ to $\mathcal{F}^*_{\mathrm{StCon+}}$. Other parties wait to receive notification from $\mathcal{F}^*_{\mathrm{StCon+}}$ of the updated balance. If either (1) P_j did not obtain signatures from other parties; or (2) the remaining parties did not receive notification from $\mathcal{F}^*_{\mathrm{StCon+}}$, then all honest parties go to the on-chain contract with the intention of exiting the contract. See Figure 13. On the other hand, if the above step was successfully completed, then parties update $\boldsymbol{b} \leftarrow \boldsymbol{b}'$ and execute the next step.

2. (Parameter agreement) Initialize transcript $\mathrm{TT} = \bot$. Parties agree on the reactive function $g^{(\mathsf{id})} = \{g_k^{(\mathsf{id})}\}_k$ to be executed next. Parties also agree on a specific MPC protocol π using which they will securely compute $g^{(\mathsf{id})}$. Let the transcript verification predicate for this reactive MPC protocol be denoted by $\mathsf{tv}^{(\mathsf{id})}$. We use $|\mathsf{tv}^{(\mathsf{id})}|$ to denote the number of messages in a valid transcript that corresponds to a *completed* execution of $g^{(\mathsf{id})}$. Once they agree on $\mathsf{tv}^{(\mathsf{id})}$, each P_j computes $\sigma'_j \leftarrow \mathsf{Sign}((\mathsf{id}, \mathsf{tv}^{(\mathsf{id})}, \boldsymbol{b}); sk_j)$ and broadcasts this value to all parties. Each party sets $\boldsymbol{\sigma}^{(\mathsf{id})} = (\sigma'_1, \ldots, \sigma'_n)$. If not all signatures were obtained, then parties stop the local execution and go to the on-chain contract for resolution. Else each party P_j updates $\mathsf{best}_j \leftarrow (\mathsf{transcript}, \mathsf{id}, \bot, \mathsf{tv}^{(\mathsf{id})}, \boldsymbol{b}, \boldsymbol{\sigma}^{(\mathsf{id})})$

3. (Reactive MPC execution) Parse $g^{(\mathsf{id})} = \{g_k^{(\mathsf{id})}\}_{k \in [\rho]}$. For $k = 1, \ldots, \rho$:
   - Let $r_k^{(\mathsf{id})}$ denote the number of rounds in a standard (unfair) MPC protocol that implements $g_k^{(\mathsf{id})}$. We denote party P_j's inputs to $g_k^{(\mathsf{id})}$ by y_j and the corresponding randomness by ω_j. For $r = 1, \ldots, r_k^{(\mathsf{id})}$ sequentially:
     - Let $j = r \bmod n$. Party P_j computes its next message $(m_r, \sigma_r) \leftarrow \mathsf{nmf}(\mathrm{TT}; (y_j, \omega_j, \boldsymbol{b}, sk_j))$, and broadcasts (m_r, σ_r) to all parties.
     - If no message was received, then parties abort the local execution and go to the on-chain contract for resolution. Else, each P_j updates the transcript $\mathrm{TT} \leftarrow \mathrm{TT} \| (m_r, \sigma_r)$ and sets $\mathsf{best}_j \leftarrow (\mathsf{transcript}, \mathsf{id}, \mathrm{TT}, \mathsf{tv}^{(\mathsf{id})}, \boldsymbol{b}', \boldsymbol{\sigma})$.
   - Parties compute the output of $g_k^{(\mathsf{id})}$ using the completed transcript TT. Note that this output specifies the new balance vector $\boldsymbol{b}'$. Parties update $\boldsymbol{b} \leftarrow \boldsymbol{b}'$.

Fig. 11. The id-th off-chain local executions for implementing $\mathcal{F}^*_{\mathrm{MSCD}}$.

state of the current/latest off-chain execution. For the sake of presentation, we allow parties to submit a trigger witness w which extends $\mathsf{st.TT}$. Alternatively, and in cases where $\mathsf{st.id}$ does not correspond to the latest execution, we also let parties submit a trigger witness w which provides the entire transcript of an off-chain execution. In this case, pred outputs 1 if the trigger was submitted at time $t \leq \mathsf{st.}t + \Delta$ and if $\mathsf{id} > \mathsf{st.id}$ or $\mathsf{id} = \mathsf{st.id}$ but TT is a longer transcript than the (partial) transcript $\mathsf{st.TT}$.

Parameters. The variable st denotes the current state of the contract. The values pk and Δ are constant parameters to the contract and are always maintained as part of the state. We represent the state variable st as a seven tuple (st.mode, st.id, st.TT, st.t, st.L, st.tv, st.b, st.B). We omit pk and Δ and the total deposits so far from the state to keep the presentation simple. The initial deposits are $d_1 = \cdots = d_n = (n-1)q$. The initial state is ("init", -1, $\perp$, -1, $\mathbf{1}$, $\perp$, $\mathbf{0}$, $\mathbf{0}$). We also use a function cash (specified as part of st.tv) that takes a valid transcript TT and j as input, and outputs the amount of coins that need to be given to P_j.

Subroutine pred. Let $\mathsf{pred}(j, w, t; \mathsf{st}) = 1$ if

- w is parsed as (message, id, (m, σ)) with id = st.id, st.tv(st.TT$\|(m, \sigma)$) = 1; or
- w is parsed as (transcript, id, TT, tv, b, σ) with (1) tv(TT) = 1; and (2) $\sigma = (\sigma_1, \ldots, \sigma_n)$ and for each $k \in [n]$ it holds that $\mathsf{SigVerify}((\mathsf{id}, \mathsf{tv}, b), \sigma_k; pk_k) = 1$ and either (1) st.mode = "init"; or (2) st.mode = "exec"/"exit" and $t \leq$ st.$t + \Delta$ and either (2.1) id > st.id or (2.2) id = st.id and $|\mathsf{TT}| > |\mathsf{st.TT}|$.

Adding money. $\mathsf{Update}(j, u, t; \mathsf{Prog}, \mathsf{st})$ is defined as follows: If $u = (b', \psi, \mathsf{coins}(b_j))$, then parse ψ as $(\psi'_1, \ldots, \psi'_n)$ and verify whether $b_j \neq 0$ and $\mathsf{SigVerify}((j, b'), \psi_k; pk_k) = 1$ for all $k \in [n]$. Then check if $\sum_{k \in [n]} b'_k = b_j + \sum_{k \in [n]} \mathsf{st.}B[k]$. If all checks pass and if st.mode $\notin \{$"exit", "abort", "payout"$\}$, then update st.$B[j] \leftarrow$ st.$B[j] + b_j$. Else output $\perp$.

State transitions. $\mathsf{Prog}(j, w, t; \mathsf{st})$ is defined as follows: Initialize $e \leftarrow 0$.

- If $w = ($transcript, id, TT, tv, b, $\sigma)$ and $\mathsf{pred}(j, w, t; \mathsf{st}) = 1$: Set st $\leftarrow$ ("exec", id, TT, t, st.L, tv, b, st.B).
- Else if $w = ($message, id, $(m, \sigma))$ and $\mathsf{pred}(j, w, t; \mathsf{st}) = 1$: Update st.TT $\leftarrow$ st.TT$\|(m, \sigma)$.
- Else if $w = $ exit:
  • If (1) st.mode = "init", or (2) st.mode = "exec" and $|\mathsf{st.TT}| = |\mathsf{st.tv}|$: Set st.mode = "exit", st.$t \leftarrow t$.
  • If st.mode = "exec"/"abort", $t >$ st.$t + \Delta$, st.$L[j] = 1$, $|\mathsf{st.TT}| \neq |\mathsf{st.tv}|$ and $j \neq j_a = 1 + |\mathsf{st.TT}|$ mod n: Set $e \leftarrow nq +$ st.b_j, st.$L[j] \leftarrow 0$, st.$L[j_a] \leftarrow 0$ and st.mode $\leftarrow$ "abort".
  • If st.mode = "exit"/"payout", and $t >$ st.$t + \Delta$ and st.$L[j] = 1$: Set $e \leftarrow (n-1)q + \mathsf{cash}(j, \mathsf{st.TT})$, st.mode $\leftarrow$ "payout", and st.$L[j] \leftarrow 0$.

If at the end of a transition, it holds that st.$L[k] = 0$ for all $k \in [n]$, then we update st.mode $\leftarrow$ "inactive".

Fig. 12. $\mathcal{F}^*_{\mathrm{StCon+}}$ parameters for $\mathcal{F}^*_{\mathrm{MSCD}}$.

The state transition function Prog will make use of pred described above. If pred outputs 1 then, the contract moves into exec mode and records the trigger time and updates the transcript with the transcript contained in the trigger. If the trigger is a token value exit, then if the current mode is either "init" or "exec" and st.TT is a completed transcript (we check this by checking if $|\mathsf{st.TT}| = |\mathsf{st.tv}|$), then we put the contract into exit mode and record the time. The rest of the contract specification is quite similar to the $\mathcal{F}^*_{\text{MSFE}}$ case. In particular, when the trigger is a token value $w = \text{exit}$ and if $t > \text{st}.t + \Delta$ and the triggering party has not yet withdrawn money from $\mathcal{F}^*_{\text{StCon+}}$, we move the contract into "payout" or "abort" (depending on the current mode) and refund the deposit (plus current balance plus compensation if applicable) to the triggering party as long as it had not contributed to an off-chain/on-chain aborted execution. To detect whether the triggering party P_j aborted an execution, we simply check if $j = 1 + |\text{st}.\text{TT}| \bmod n$ and $|\text{st}.\text{TT}| \neq |\text{st}.\text{tv}|$ holds. In this case, we penalize the party by not giving its deposit back but instead distributing it among the remaining parties. This is slightly different from the $\mathcal{F}^*_{\text{MSFE}}$ case, where we could potentially penalize multiple corrupt parties depending on whether they contributed their output secret share to the execution. Here, on the other hand, we only penalize one party $j_a = 1 + |\text{st}.\text{TT}| \bmod n$. Note that we also redistribute the deposits made by the parties depending on the output of the latest complete execution. If the execution was completed on-chain, then we use the function cash_j (which we assume is a part of tv for simplicity) applied on $\text{st}.\text{TT}$ to determine how much money party P_j is supposed to obtain. On the other hand, if the on-chain execution was also aborted, then (in addition to paying compensation to the honest parties) we distribute the initial deposits depending on the latest balance prior to this execution (which is stored in variable $\text{st}.b_j$).

Finally, the Update function provides an interface for parties to add coins between different reactive MPCs. Upon receiving a witness $u = (\boldsymbol{b}', \boldsymbol{\psi}, \text{coins}(b_j))$, it checks if the provided coins b_j plus the coins already deposited with $\mathcal{F}^*_{\text{StCon+}}$ (i.e., sum of elements in $\text{st}.B$) matches the amount specified in the balance vector (i.e., sum of elements in $\boldsymbol{b}'$). Note that the signatures also include the party index j (to avoid situations where a different party abuses the broadcasted signatures). Also, note that the deposits $\text{st}.B$ only increase which ensures that there are no replay attacks. This is because once $\mathcal{F}^*_{\text{StCon+}}$ starts giving back coins (i.e., $\text{st}.\text{mode} \in \{\text{"payout"}, \text{"abort"}\}$), then it does not accept any more coins.

Main protocol: handling aborts and notifications from $\mathcal{F}^*_{\text{StCon+}}$. The formal description can be found in Fig. 13. As with implementing $\mathcal{F}^*_{\text{MSFE}}$, here too parties begin by initializing $\mathcal{F}^*_{\text{StCon+}}$ with the parameters as in Fig. 12, then continue executing off-chain as in Fig. 11. Each party P_j maintains a local variable best_j which represents the most recent transcript of the current off-chain execution. This value will be helpful while recovering from an aborted off-chain execution (cf. Step 2 of Fig. 13). While in the $\mathcal{F}^*_{\text{MSFE}}$ case, party P_j triggered $\mathcal{F}^*_{\text{StCon+}}$ with best_j when its decommitment did not appear in the transcript in $\mathcal{F}^*_{\text{StCon+}}$, here in the $\mathcal{F}^*_{\text{MSCD}}$ case, party P_j triggers $\mathcal{F}^*_{\text{StCon+}}$ with best_j when best_j contains a longer transcript than the one that is current on the contract. Like in the $\mathcal{F}^*_{\text{MSFE}}$ case, we need to handle the corner case when $\text{st}.\text{id} = \text{best}_j.\text{id} + 1$. This happens when honest parties have completed Step 1 of the $\text{st}.\text{id}$-th local execution phase but did not receive signatures on $\text{tv}^{(\text{st}.\text{id})}$ from all corrupt parties. In this case, each party P_j will choose new input and fresh randomness and continue the protocol from the transcript $\text{st}.\text{TT}$. Note that P_j responds only when $j = 1 + |\text{st}.\text{TT}| \bmod n$ (i.e., it is its turn) and when the execution has not already completed (i.e., $|\text{st.TT}| \neq |\text{st.tv}|$). Finally, there is one other case which is unique to $\mathcal{F}^*_{\text{MSCD}}$ implementation. Unlike the $\mathcal{F}^*_{\text{MSFE}}$ case, each party might have to send out multiple messages (corresponding to the reactive MPC protocol) within a single execution. In particular, once the aborted off-chain execution goes on-chain, it remains on-chain (i.e., parties have to respond within time Δ of the previous step in order to avoid paying a penalty) and needs to be completed by the parties.[1] This brings us to the final case where $\text{st.id} = \text{best}_j.\text{id}$ where in Step 3(d) of Fig. 13, honest party P_j uses the next message function in order to continue the transcript st.TT. We prove the following theorem in [9, Appendix C].

Theorem 2. *Let λ be a computational security parameter. Assume the existence of enhanced trapdoor permutations. Then there exists a protocol that* SCC*-realizes (cf. Definition 1) $\mathcal{F}^*_{\text{MSCD}}$ in the $(\mathcal{F}^*_{\text{StCon}+}, \mathcal{F}_{\text{OT}})$-hybrid model whose script complexity (cf. Definition 3) is independent of the number of secure computations. Furthermore, in the optimistic case (i.e., all parties are honest), there are a total of $O(n)$ state transitions (i.e., discounting updates) each of complexity $O(n\lambda)$.*

6   Efficient Poker Protocol

A tailor-made protocol for a poker game in which money changes hands was presented by Kumaresan et al. in [26, Sect. 6]. However, that protocol is not efficient enough in practice, due to two distinct reasons. The first reason is that [26, Sect. 6] was designed to work in the $\mathcal{F}^*_{\text{CR}}$ model, which incurs an expensive setup procedure and does not support off-chain amortization in its SCD variant (cf. Sect. 2). Furthermore, the $\mathcal{F}^*_{\text{CR}}$ verification circuits that [26, Sect. 6] uses are quite complex and not supported by the current Bitcoin scripting language. The second reason is that preprocessing shuffle protocol that [26, Sect. 6] employs is a generic secure MPC that delivers hash-based commitments to shares of the shuffled cards. It would be impractical to run a general-purpose secure MPC protocol (typically among 3 to 9 parties in a poker game) that performs the shuffle and the hash invocations, and indeed there are special-purpose poker protocols that perform much better.

See Sect. 1.2 for a survey of the various poker protocols. Our implementation uses the poker protocol of Wei and Wang [22,23], which improves an earlier work of Castellà-Roca et al. [12] by using a zero-knowledge proof of knowledge scheme instead of homomorphic encryption.

A potential disadvantage of special-purpose poker protocols is the on-chain verification cost: the generic secure MPC approach would allow us to define an

---

[1] Alternatively, when a contract goes on-chain, it is possible to make it come back off-chain right after getting the next message from the party that aborted the off-chain execution. Our protocol does not do this but can be easily modified to behave as described above. Note that this modification does not change our theorem statements.

Parties initialize the parameters as in Figure 12. Then for $\mathsf{id} = 1, 2, \cdots$: Parties run the local execution prescribed in Figure 11. Recall that each party P_j maintains a variable best_j which is initialized as $\bot$.

1. If a party P_j wants to exit the contract, then it sends $w = \mathsf{exit}$ to $\mathcal{F}^*_{\mathrm{StCon+}}$.
2. If there is an abort at any stage during a local off-chain execution, then parties do not continue any more local executions and instead trigger $\mathcal{F}^*_{\mathrm{StCon+}}$ with the value best_j if it is non-null.
3. Each party P_j waits and responds to state changes in $\mathcal{F}^*_{\mathrm{StCon+}}$ depending on the current state st:
   (a) If $\mathsf{st.mode} = $ "payout"/"abort" and $\mathsf{st}.L[j] = 1$, send $w = \mathsf{exit}$ to $\mathcal{F}^*_{\mathrm{StCon+}}$.
   (b) If (1) $\mathsf{st.id} < \mathsf{best}_j.\mathsf{id}$, or (2) $\mathsf{st.id} = \mathsf{best}_j.\mathsf{id}$ and $|\mathsf{best}_j.\mathrm{TT}| > |\mathsf{st.TT}|$, then submit best_j to $\mathcal{F}^*_{\mathrm{StCon+}}$.
   (c) If $\mathsf{st.id} = \mathsf{best}_j.\mathsf{id} + 1$ and $j = 1 + |\mathsf{st.TT}| \bmod n$, then choose input y_j and use fresh randomness ω_j and compute $(m, \sigma) \leftarrow \mathsf{nmf}(\mathsf{st.TT}; (y_j, \omega_j, \boldsymbol{b}, sk_j))$ and send $(\mathsf{message}, \mathsf{st.id}, (m, \sigma))$ to $\mathcal{F}^*_{\mathrm{StCon+}}$ and update $\mathsf{best}_j \leftarrow (\mathsf{transcript}, \mathsf{st.id}, \mathsf{st.TT}\|(m, \sigma), \mathsf{st.tv}, \mathsf{st}.\boldsymbol{b}, \mathsf{st}.\boldsymbol{\sigma})$.
   (d) If $\mathsf{st.id} = \mathsf{best}_j.\mathsf{id}$ and if $j = 1 + |\mathsf{st.TT}| \bmod n$ and if $|\mathsf{st.TT}| \neq |\mathsf{st.tv}|$, then compute $(m, \sigma) \leftarrow \mathsf{nmf}(\mathsf{st.TT}; (y_j, \omega_j, \boldsymbol{b}, sk_j))$ where y_j, ω_j are inputs to the current stage. Send $\mathsf{best}_j \leftarrow (\mathsf{message}, \mathsf{st.id}, (m, \sigma))$ to $\mathcal{F}^*_{\mathrm{StCon+}}$ and update $\mathsf{best}_j \leftarrow (\mathsf{transcript}, \mathsf{st.id}, \mathsf{st.TT}\|(m, \sigma), \mathsf{st.tv}, \mathsf{st}.\boldsymbol{b}, \mathsf{st}.\boldsymbol{\sigma})$.

Finally, parties also keep track of whether $\mathsf{st.mode} = $ "exit" or "exec" and the current time t is such that $t > \mathsf{st}.t + \Delta$. In this case, parties send $w = \mathsf{exit}$ in order to claim their payout or compensation and do not participate in any further local executions.

Fig. 13. Main protocol implementing $\mathcal{F}^*_{\mathrm{MSCD}}$.

on-chain predicate that verifies that a pre-image (corresponding to a share of a shuffled card) hashes to the commitment value, and penalize a corrupt party who would not supply the correct pre-image. By contrast, the efficient poker protocols rely on constructions that are algebraic in nature, which implies that the on-chain verification predicate would be significantly more complex.

In the case that all parties are honest, on-chain verification will never occur. In the case that corrupt parties deviate, they can force an honest party to send a transaction containing a witness that satisfies the complex on-chain predicate. The on-chain fallback procedure introduces an additional cost in the form of a transaction fee the party supplying the witness pays (to the miners). While the on-chain fallback also introduces a delay that all of the parties would bear, a malicious party may still cause an honest party to pay the fee. Fortunately, the cost of transaction fees is quite minor (cf. [9, Sect. 6]). Still, our implementation provides an improvement by employing a technique that shares the transaction fee across all parties. In Ethereum this is not straightforward, as the initiator of the transaction must provide all of the transaction fees up front; however, our technique compensates this party by paying funds collected from all of the parties in advance.

By using the efficient scheme of Wei and Wang, the main steps of our SCD poker protocol are as follows:

1. The parties will execute a deck preparation and shuffle protocols, that output group elements (cf. [23, Protocols 1, 3, 4]).
2. The parties will sign an off-chain message that commits to these group elements (cf. [9, Sect. 3.3]). This signed message could later be sent to the on-chain contract, in case that a corrupt party deviates from the protocol.
3. The parties will run the poker game according to the predefined rules, where in each round a specific party performs a valid action (e.g., raise/call/fold).
4. After each round, all the parties will sign an off-chain message that encodes the state of the poker table.
5. When a party draws a private card from the deck, the parties will execute the card drawing protocol of [23, Protocol 6].

Per the above discussion regarding the complexity of the on-chain predicate, it can be seen that the verification procedure for drawing a private card is dominated by a zero-knowledge proof of knowledge of equality of discrete logarithms (cf. [6,37]). To reduce the round complexity and avoid the HVZK concern, in Step 5 the parties will use a non-interactive proof of knowledge. While there are provably secure methods to obtain the NIZK (see [19]), our efficient implementation uses the Fiat-Shamir heuristic.

Our Ethereum implementation of the NIZK verifier is based on the Secp256k1 ellitpic curve group, the same used in Ethereum and in Bitcoin for digital signatures. The Ethereum language does not provide opcodes for working with elliptic curve points (though such native support is planned [20]). The current Ethereum scripting language features a dedicated opcode for secp256k1 signature verification, but this opcode is signature-specific and cannot (to our knowledge) be repurposed for the NIZK scheme. Thus we are forced to implement our NIZK the "hard way," making use of an Secp256k1 elliptic curve library (due to Andreas Olofsson) built from low-level Ethereum opcodes that implement the elliptic curve exponentiation (_mul) operation. Our Ethereum implementation, which is shown in [9, Fig. 22], bears an obvious resemblance to the high-level proof of knowledge of discrete logarithms protocol [6,37].

Using the pyethereum simulator framework, we found that the total gas cost of the NIZK verifier is 1.3 M, whereas an Ethereum block has a gas limit of 4.7 M. The cost of the verifier is dominated by the cost of four scalar multiplications. In contrast, the signature verification opcode costs only 3000 gas (a hundred times cheaper), despite performing a scalar multiplication anyway. Thus if Ethereum were modified to support more general elliptic curve arithmetic, we would anticipate a hundred-fold improvement with respect to the transaction fees.

Note that the off-chain signatures in Step 4 include only the current state and not the transcript history, because the proof of knowledge NIZKs do not branch. That is, at a specific round a party will need to provide a NIZK that depends

	NIZK verify	Scalar multiplication	Built-in instruction
Gas cost	1287858	303401	≤3000
Transactions per block	3	15	≥1500

only on public values: the intermediate result of the card drawing protocol, and the group elements that the parties committed in the first step.

The poker protocol of Wei and Wang that we deploy supports all the requirements that were suggested by Crépeau [16]. For example, complete confidentiality of strategy is supported, since the proof of knowledge verification would not reveal the cards at the end of the game. Thus, our implementation enables poker variants such as Texas hold 'em and five-card draw, where private cards are drawn after the game is already in progress. See Wei [22] for benchmarks that measure the running time of the initial shuffling (which is done off-chain).

Our poker implementation is available at [1].

References

1. https://github.com/amiller/instant-poker
2. Andrychowicz, M., Dziembowski, S., Malinowski, D., Mazurek, L.: Fair two-party computations via the bitcoin deposits. In: First Bitcoin Workshop, FC (2014)
3. Andrychowicz, M., Dziembowski, S., Malinowski, D., Mazurek, L.: Secure multiparty computations on bitcoin. In: IEEE Security and Privacy (2014)
4. Asokan, N., Shoup, V., Waidner, M.: Optimistic protocols for fair exchange. In: ACM CCS (1997)
5. Barber, S., Boyen, X., Shi, E., Uzun, E.: Bitter to better — how to make bitcoin a better currency. In: Keromytis, A.D. (ed.) FC 2012. LNCS, vol. 7397, pp. 399–414. Springer, Heidelberg (2012). https://doi.org/10.1007/978-3-642-32946-3_29
6. Barnett, A., Smart, N.P.: Mental poker revisited. In: Paterson, K.G. (ed.) Cryptography and Coding 2003. LNCS, vol. 2898, pp. 370–383. Springer, Heidelberg (2003). https://doi.org/10.1007/978-3-540-40974-8_29
7. Bayer, S., Groth, J.: Efficient zero-knowledge argument for correctness of a shuffle. In: Pointcheval, D., Johansson, T. (eds.) EUROCRYPT 2012. LNCS, vol. 7237, pp. 263–280. Springer, Heidelberg (2012). https://doi.org/10.1007/978-3-642-29011-4_17
8. Bentov, I., Kumaresan, R.: How to use bitcoin to design fair protocols. In: Garay, J.A., Gennaro, R. (eds.) CRYPTO 2014. LNCS, vol. 8617, pp. 421–439. Springer, Heidelberg (2014). https://doi.org/10.1007/978-3-662-44381-1_24
9. Bentov, I., Kumaresan, R., Miller, A.: Instantaneous decentralized poker. Technical report available at https://arxiv.org/abs/1701.06726 (2017)
10. Buterin, V.: (2013). https://github.com/ethereum/wiki/wiki/White-Paper
11. Canetti, R.: Universally composable security: a new paradigm for cryptographic protocols. In: FOCS (2001)
12. Castellà-Roca, J., Domingo-Ferrer, J., Riera, A., Borrell, J.: Practical mental poker without a TTP based on homomorphic encryption. In: Johansson, T., Maitra, S. (eds.) INDOCRYPT 2003. LNCS, vol. 2904, pp. 280–294. Springer, Heidelberg (2003). https://doi.org/10.1007/978-3-540-24582-7_21

13. Cleve, R.: Limits on the security of coin flips when half the processors are faulty (extended abstract). In: STOC, pp. 364–369 (1986)
14. Coleman, J.: State channels. http://www.jeffcoleman.ca/state-channels/
15. Coppersmith, D.: Cheating at mental poker. In: Williams, H.C. (ed.) CRYPTO 1985. LNCS, vol. 218, pp. 104–107. Springer, Heidelberg (1986). https://doi.org/10.1007/3-540-39799-X_10
16. Crépeau, C.: A secure poker protocol that minimizes the effect of player coalitions. In: Williams, H.C. (ed.) CRYPTO 1985. LNCS, vol. 218, pp. 73–86. Springer, Heidelberg (1986). https://doi.org/10.1007/3-540-39799-X_8
17. Crépeau, C.: A zero-knowledge Poker protocol that achieves confidentiality of the players' strategy *or* How to achieve an electronic Poker face. In: Odlyzko, A.M. (ed.) CRYPTO 1986. LNCS, vol. 263, pp. 239–247. Springer, Heidelberg (1987). https://doi.org/10.1007/3-540-47721-7_18
18. Croman, K., et al.: On scaling decentralized blockchains. In: Clark, J., Meiklejohn, S., Ryan, P.Y.A., Wallach, D., Brenner, M., Rohloff, K. (eds.) FC 2016. LNCS, vol. 9604, pp. 106–125. Springer, Heidelberg (2016). https://doi.org/10.1007/978-3-662-53357-4_8
19. Damgård, I., Fazio, N., Nicolosi, A.: Non-interactive zero-knowledge from homomorphic encryption. In: Halevi, S., Rabin, T. (eds.) TCC 2006. LNCS, vol. 3876, pp. 41–59. Springer, Heidelberg (2006). https://doi.org/10.1007/11681878_3
20. Ethereum EIP. https://github.com/ethereum/EIPs/pull/213
21. Goldreich, O.: Foundations of Cryptography, vol. 2. Cambridge University Press, Cambridge (2004)
22. Wei, T.J.: Secure and practical constant round mental poker. Inf. Sci. (2014)
23. Wei, T.J., Wang, L.-C.: A fast mental poker protocol. J. Math. Cryptology 6(1), 39–68 (2012)
24. Kiayias, A., Zhou, H.-S., Zikas, V.: Fair and robust multi-party computation using a global transaction ledger. In: Fischlin, M., Coron, J.-S. (eds.) EUROCRYPT 2016. LNCS, vol. 9666, pp. 705–734. Springer, Heidelberg (2016). https://doi.org/10.1007/978-3-662-49896-5_25
25. Kosba, A., Miller, A., Shi, E., Wen, Z., Papamanthou, C.: Hawk: the blockchain model of cryptography and privacy-preserving smart contracts. In: IEEE S&P (2016)
26. Kumaresan, R., Moran, T., Bentov, I.: How to use bitcoin to play decentralized poker. In: CCS (2015)
27. Kumaresan, R., Bentov, I.: Amortizing secure computation with penalties. In: CCS (2016)
28. Kumaresan, R., Vaikuntanathan, V., Vasudevan, P.N.: Improvements to secure computation with penalties. In: CCS (2016)
29. Lipton, R.: How to cheat at mental poker. In: AMS Short Course on Crypto (1981)
30. Maxwell, G.: (2011). https://en.bitcoin.it/wiki/Zero_Knowledge_Contingent_Payment
31. Miller, A., Bentov, I.: Zero-collateral lotteries in bitcoin and ethereum. In: IEEE S&B Workshop (2017). https://arxiv.org/abs/1612.05390
32. Möser, M., Eyal, I., Gün Sirer, E.: Bitcoin covenants. In: Clark, J., Meiklejohn, S., Ryan, P.Y.A., Wallach, D., Brenner, M., Rohloff, K. (eds.) FC 2016. LNCS, vol. 9604, pp. 126–141. Springer, Heidelberg (2016). https://doi.org/10.1007/978-3-662-53357-4_9
33. Nakamoto, S.: Bitcoin: A Peer-to-Peer Electronic Cash System (2008)
34. O'Connor, R., Piekarska, M.: Enhancing bitcoin transactions with covenants. In: Financial Cryptography Bitcoin Workshop (2017)

35. Pinkas, B.: Fair secure two-party computation. In: Biham, E. (ed.) EUROCRYPT 2003. LNCS, vol. 2656, pp. 87–105. Springer, Heidelberg (2003). https://doi.org/10.1007/3-540-39200-9_6
36. Shamir, A., Rivest, R., Adleman, L.: Mental poker. In: The Mathematical Gardener (1981)
37. Shoup, V., Alwen, J.: (2007). http://cs.nyu.edu/courses/spring07/G22.3220-001/lec3.pdf
38. Wood, G.: Ethereum: A secure decentralized transaction ledger (2014). http://gavwood.com/paper.pdf

Multi-party Protocols

More Efficient Universal Circuit Constructions

Daniel Günther, Ágnes Kiss[(✉)], and Thomas Schneider

TU Darmstadt, Darmstadt, Germany
guenther@rangar.de, {agnes.kiss,thomas.schneider}@crisp-da.de

Abstract. A universal circuit (UC) can be programmed to simulate any circuit up to a given size n by specifying its program bits. UCs have several applications, including private function evaluation (PFE). The asymptotical lower bound for the size of a UC is proven to be $\Omega(n \log n)$. In fact, Valiant (STOC'76) provided two theoretical UC constructions using so-called 2-way and 4-way constructions, with sizes $5n \log_2 n$ and $4.75n \log_2 n$, respectively. The 2-way UC has recently been brought into practice in concurrent and independent results by Kiss and Schneider (EUROCRYPT'16) and Lipmaa et al. (Eprint 2016/017). Moreover, the latter work generalized Valiant's construction to any k-way UC.

In this paper, we revisit Valiant's UC constructions and the recent results, and provide a modular and generic embedding algorithm for any k-way UC. Furthermore, we discuss the possibility for a more efficient UC based on a 3-way recursive strategy. We show with a counterexample that even though it is a promising approach, the 3-way UC does not yield an asymptotically better result than the 4-way UC. We propose a hybrid approach that combines the 2-way with the 4-way UC in order to minimize the size of the resulting UC. We elaborate on the concrete size of all discussed UC constructions and show that our hybrid UC yields on average 3.65% improvement in size over the 2-way UC. We implement the 4-way UC in a modular manner based on our proposed embedding algorithm, and show that our methods for programming the UC can be generalized for any k-way construction.

Keywords: Universal circuit · Private function evaluation · Function hiding

1 Introduction

Universal circuits (UCs) are Boolean circuits that can be programmed to simulate any Boolean function $f(x)$ up to a given size by specifying a set of program bits p_f. The UC then receives these program bits as input besides the input x to the functionality, and computes the result as $UC(x, p_f) = f(x)$. This means that the same UC can evaluate multiple Boolean circuits, only the different program bits are to be specified.

Valiant proposed an asymptotically size-optimal construction in [Val76] with size $\Theta(n \log n)$ and depth $\mathcal{O}(n)$, where n is the size of the simulated Boolean circuit description of $f(x)$. He provides two constructions, based on 2-way and

© International Association for Cryptologic Research 2017
T. Takagi and T. Peyrin (Eds.): ASIACRYPT 2017, Part II, LNCS 10625, pp. 443–470, 2017.
https://doi.org/10.1007/978-3-319-70697-9_16

4-way recursive structures. Recently, optimizations of Valiant's size-optimized construction appeared in concurrent and independent works of [KS16] and [LMS16]. Both works implement Valiant's 2-way recursive construction.

1.1 Applications of Universal Circuits

Size-optimized universal circuits have many applications. We review some of them here and refer to [KS16, LMS16] for further details.

Private Function Evaluation (PFE). Secure two-party computation or secure function evaluation (SFE) provides interactive protocols for evaluating a public function $f(x, y)$ on two parties' private inputs x and y. However, in some scenarios, the function f is a secret input of one of the parties. This setting is called private function evaluation (PFE). PFE of $f(x)$ can be achieved by running SFE of $UC(x, p_f)$, where the UC is a public function and the program bits p_f – and therefore f – are kept private due to the properties of SFE. Protocols designed especially for PFE such as [MS13, BBKL17] achieve the same asymptotic complexity $\mathcal{O}(n \log n)$ as PFE using UCs, where n is the size of the function f.[1] However, to the best of our knowledge, they have not yet been implemented, and they are not as generally applicable as PFE with UCs.

UC-based PFE can be easily integrated into any SFE framework and can directly benefit from recent optimizations. For instance, *outsourcing UC-based PFE* is directly possible with outsourced SFE [KR11]. The non-interactive secure computation protocol of [AMPR14] can also be generalized to obtain a *non-interactive PFE* protocol [LMS16].

One of the first applications for PFE was *privacy-preserving checking for credit worthiness* [FAZ05], where not only the loanee's data, but also the loaner's function needs to be kept private. PFE allows for running *proprietary software* on private data, such as privacy-preserving software diagnosis [BPSW07], medical programs [BFK+09], or privacy-preserving intrusion detection [NSMS14]. UCs can be applied to obliviously *filter remote streaming data* [OI05] and for hiding queries in *private database management systems* such as Blind Seer [PKV+14, FVK+15].

Applications Beyond PFE. Universal circuits can be applied for program obfuscation. Candidates for *indistinguishability obfuscation* are constructed using a UC as a building block in [GGH+13a, BOKP15], which can be improved using Valiant's UC implementation [KS16]. *Direct program obfuscation* was proposed in [Zim15], where the circuit is a secret key to a UC. [LMS16] mentions that UCs can be applied for secure two-party computation in the batch execution setting [HKK+14, LR15]. It can be applied for *verifiable computation* [FGP14], and

---

[1] There also exist PFE protocols with linear complexity $\mathcal{O}(n)$ which are based on public-key primitives [KM11, MS13, MSS14]. However, the concrete complexity of these protocols is worse than that of the protocols based on (mostly) symmetric-key primitives, i.e., the OT-based PFE protocols of [MS13, BBKL17] or PFE using UCs.

for multi-hop homomorphic encryption [GHV10]. Ciphertext-policy *Attribute-Based Encryption* was proposed in [Att14], where the policy circuit is hidden [GGH+13b].

1.2   Related Work on Universal Circuits

Valiant defined universal circuits in [Val76] and gave two size-optimized constructions. The constructions are based on so-called edge-universal graphs (EUGs) and utilize either a 2-way or a 4-way recursive structure, also called 2-way or 4-way UCs. Both achieve the asymptotically optimal size $\Theta(n \log n)$ [Val76, Weg87], where n is the size of the simulated circuit. The concrete complexity of the 4-way UC is $\sim 4.75n \log_2 n$ which is smaller than that of the 2-way UC of $\sim 5n \log_2 n$ [Val76].

The first modular UC construction was proposed by Kolesnikov and Schneider in [KS08b]. This construction achieves a non-optimal asymptotic complexity of $\mathcal{O}(n \log^2 n)$, and was the first implementation of UCs. A generalization of UCs for n-input gates was given in [SS08].

Recently, two independent works have optimized and implemented Valiant's 2-way UC [KS16, LMS16]. Kiss and Schneider in [KS16] mainly focus on the most prominent application of UCs, i.e., private function evaluation (PFE). Due to the free-XOR optimization of [KS08a] in the SFE setting, they optimize the size of the UC for the number of AND gates in the resulting UC implementation and provide a framework for PFE using UCs as public function. They also propose hybrid constructions for circuits with a large number of inputs and outputs, utilizing efficient building blocks from [KS08b]. Lipmaa et al. in [LMS16] also provide an (unpublished) implementation of the 2-way UC. While keeping the number of AND gates minimal, they additionally optimize for the total number of gates, i.e., include optimizations to also reduce the number of XOR gates. They adapt the construction to arithmetic circuits and generalize the design to a k-way construction in a modular manner, for $k \geq 2$.

Both papers utilize 2-coloring of the underlying graphs for defining the program bits p_f for any given functionality f. Generally, 2-coloring can be utilized for any 2^i-way construction. [LMS16] calculate the optimal value for k to be 3.147, and conclude that the two candidates for the most efficient 2^i-way constructions are the 2-way and 4-way UCs, of which the 4-way construction results in an asymptotically smaller size.

So far only Valiant's 2-way UC has been implemented and the not yet implemented 4-way UC was postulated to be the most efficient one.

1.3   Outline and Our Contributions

In summary, we provide the first implementation and detailed evaluation of Valiant's 4-way UC and propose an even more efficient hybrid UC. We elaborate on the size of the generalized k-way UCs for $k \neq 2$ and $k \neq 4$.

After revisiting Valiant's UC construction [Val76, KS16] and its k-way generalization [LMS16] in Sect. 2, we provide the following contributions:

Our modular programming algorithm (Sect. 3): We detail a modular algorithm that provides the description of the input function f as program bits p_f to the UC, both for Valiant's 4-way UC as well as for the k-way UC of Lipmaa et al. [LMS16].

New universal circuit constructions (Sect. 4): We start with a new 3-way UC. After providing modular building blocks for this UC, we show that it is asymptotically larger than Valiant's UCs. Then, we propose a *hybrid UC construction* that can efficiently combine k-way constructions for multiple values of k.[2] With this, we combine Valiant's 2-way and 4-way UCs to achieve the smallest UC known so far.

Size of UCs (Sect. 5): We compare the asymptotic and concrete sizes of Valiant's (2-way and 4-way) UCs and that of different k-way UCs. We show that of all k-way UCs, Valiant's 4-way UC provides the best results for large circuits. Moreover, our hybrid UC in most cases improves over the 2-way UC by up to around 4.5% in its size, and over the 4-way UC by up to 2% (for large input circuits). In Table 1 we compare the concrete communication of PFE using SFE and our new UC implementation to the previous works on special-purpose OT-based PFE protocols.

Implementation of Valiant's 4-way UC and experiments (Sect. 6): We implement Valiant's 4-way UC and describe how our implementation can directly be used in the PFE framework of [KS16]. We experimentally evaluate the performance of our UC generation and programming algorithm with a set of example circuits and compare it on the same platform with the 2-way UC compiler of [KS16].

Table 1. Comparison of overall communication between special-purpose PFE protocols and UC-based ones for simulated circuits of size n. The numbers are for 128 bit symmetric security. The underlying SFE protocol for UC-based PFE is Yao's protocol [Yao86] with the garbled row reduction optimization [NPS99] and X- and Y-switching blocks are instantiated using free XORs as described in [KS08a]. This yields one ciphertext per X- and Y-switching block, and three ciphertexts per universal gate.

n	Special-purpose PFE		UC-based PFE using Yao		
	[MS13]	[BBKL17]	2-way UC [KS16]	Our 4-way UC	Our hybrid UC
10^3	3.5 MB	2.0 MB	0.6 MB	0.6 MB	0.6 MB
10^4	44.8 MB	26.3 MB	8.4 MB	8.4 MB	8.2 MB
10^5	549.6 MB	324.0 MB	109.6 MB	107.8 MB	106.2 MB
10^6	6 509.9 MB	3 847.9 MB	1 360.3 MB	1 308.4 MB	1 308.4 MB
10^7	75 236.5 MB	44 562.1 MB	16 038.8 MB	15 677.7 MB	15 413.7 MB

---

[2] Our hybrid UC is orthogonal to that of [KS16], who combine Valiant's UC with building blocks from [KS08b] for the inputs and outputs.

2   Preliminaries

In this section, we summarize the existing UC constructions. We provide necessary background information in Sect. 2.1, explain Valiant's construction [Val76] in Sect. 2.2 and the improvements of [KS16, LMS16] on the 2-way, 4-way and k-way UCs in Sects. 2.3, 2.4 and 2.5, respectively.

2.1   Preliminaries to Valiant's UC Constructions

Let $G = (V, E)$ be a *directed graph* with set of *nodes* V and *edges* $E \subseteq V \times V$. The number of incoming [outgoing] edges of a node is called its *indegree* [*outdegree*]. A graph has *fanin* [*fanout*] d if the indegree [outdegree] of all its nodes is at most d. In the following, we denote by $\Gamma_d(n)$ the set of all acyclic graphs with fanin and fanout d having n nodes. Similarly, the fanin [fanout] of a circuit can be defined based on the maximal number of incoming [outgoing] wires of all its gates, inputs and outputs.

Let $G = (V, E) \in \Gamma_d(n)$. A mapping $\eta^G : V \to \{1, \ldots, n\}$ is called *topological order* if $(a_i, a_j) \in E \Rightarrow \eta^G(a_i) < \eta^G(a_j)$ and $\forall a_1, a_2 \in V : \eta^G(a_1) = \eta^G(a_2) \Rightarrow a_1 = a_2$. A topological order in $G \in \Gamma_d(n)$ can be found with computational complexity $\mathcal{O}(dn)$.

A circuit $C_{u,v}^{k^*}$ with u inputs, k^* gates and v outputs and fanin or fanout $d > 2$ can be reduced to a circuit with fanin and fanout 2. Shannon's expansion theorem [Sha49, Sch08] describes how gates with larger fanin can be reduced to gates with two inputs by adding additional gates. [Val76, KS16] describe adding copy gates in order to eliminate larger fanout and elaborate on the implied overhead ($k \leq 2k^* + v$). [KS08b, KS16] implement these methods and we thus assume that our input Boolean circuit $C_{u,v}^k$ has fanin and fanout 2 for all its u inputs, k gates and v outputs. We transform $C_{u,v}^k$ into a $\Gamma_2(n)$ graph G with $n = u + v + k$ by creating a node for each input, gate and output, and an edge for each wire in $C_{u,v}^k$.

Edge-embedding is a mapping from graph $G = (V, E)$ into $G' = (V', E')$ with $V \subseteq V'$ and E' containing a path for each $e \in E$, such that the paths are pairwise edge-disjoint. A graph $U_n(\Gamma_d) = (V_U, E_U)$ is an *Edge-Universal Graph* (EUG) for $\Gamma_d(n)$ if every graph $G \in \Gamma_d(n)$ can be edge-embedded into $U_n(\Gamma_d)$.[3] $U_n(\Gamma_d)$ has distinguished nodes called *poles* $\{p_1, \ldots, p_n\} \subseteq V_U$ where each node $a \in V$ is mapped to exactly one pole with a mapping φ, such that every node in G has a corresponding pole in $U_n(\Gamma_d)$. This mapping is defined by a concrete topological order η^G of the original graph G, i.e., $\varphi : V \to V_U$ with $\varphi(a) = p_{\eta^G(a)}$. Besides the poles, $U_n(\Gamma_d)$ might have additional nodes that enable the edge-embedding. For each edge $(a_i, a_j) \in E$ we then define a disjoint path between the corresponding poles $(\varphi(a_i), \ldots, \varphi(a_j)) = (p_{\eta^G(a_i)}, \ldots, p_{\eta^G(a_j)})$ in $U_n(\Gamma_d)$, i.e., without using any edge in $U_n(\Gamma_d)$ in more than one path.

Let $U_n(\Gamma_1)$ be an EUG for graphs in $\Gamma_1(n)$ with poles $P = \{p_1, \ldots, p_n\}$. The poles have fanin and fanout 1, while all other nodes have fanin and fanout 2. An

---

[3] For the sake of simplicity, we denote this graph with $U_n(\Gamma_d)$ instead of $U(\Gamma_d(n))$.

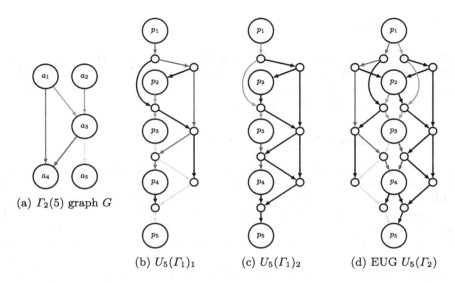

(a) $\Gamma_2(5)$ graph G (b) $U_5(\Gamma_1)_1$ (c) $U_5(\Gamma_1)_2$ (d) EUG $U_5(\Gamma_2)$

Fig. 1. (a) shows an example $\Gamma_2(5)$ graph G. (b)–(c) show the edge-embedding of G into two $U_5(\Gamma_1)$ instances with poles $(p_1, \ldots, p_5)$. (d) shows the edge-embedding of G into one $U_5(\Gamma_2)$ graph.

EUG $U_n(\Gamma_d)$ for $d \geq 2$ can be created by taking d instances of $U_n(\Gamma_1)$ EUGs, and merging each pole p_i with its multiple instances, allowing the poles to have fanin-fanout d. Let $U_n(\Gamma_d) = (V'_U, E'_U)$ be an EUG with fanin and fanout d, with $U_n(\Gamma_1)_1 = (V_1, E_1), \ldots, U_n(\Gamma_1)_d = (V_d, E_d)$. P contains the merged poles and $V'_U = P \cup_{i=1}^{d} V_i \backslash P_i$ and $E'_U = \cup_{i=1}^{d} E_i$.

We give an example for better understanding. Let $G = (V, E)$ be the graph with 5 nodes in Fig. 1a. Our aim is to edge-embed G into EUG $U_5(\Gamma_2)$. Therefore, we use two instances of $U_5(\Gamma_1)$: $U_5(\Gamma_1)_1$ in Fig. 1b and $U_5(\Gamma_1)_2$ in Fig. 1c. The edges $(a_1, a_4), (a_2, a_3)$ and (a_3, a_5) are embedded in $U_5(\Gamma_1)_1$, and the edges (a_1, a_3) and (a_3, a_4) in $U_5(\Gamma_1)_2$. Merging the poles of $U_5(\Gamma_1)_1$ and $U_5(\Gamma_1)_2$ produces $U_5(\Gamma_2)$ shown in Fig. 1d.

2.2   Valiant's UC Constructions

The size of a function f represented by a circuit $C_{u,v}^k$ with fanin and fanout 2 is $n = u + v + k$. In the following, we describe Valiant's UC construction [Val76, Weg87] that can be programmed to evaluate any function of size n. Circuit $C_{u,v}^k$ is represented as a graph $G \in \Gamma_2(n)$ (cf. Sect. 2.1).

Valiant's UC is based on an EUG $U_n(\Gamma_2) = (V_U, E_U)$ with fanin and fanout 2, which can be transformed to a Boolean circuit. $P \subseteq V_U$ contains the poles of $U_n(\Gamma_2)$ (cf. Sect. 2.1). Poles $\{1, \ldots, u\}$ correspond to the inputs, $\{(u + 1), \ldots, (u + k)\}$ to the gates, $\{(u + k + 1), \ldots, n\}$ to the outputs of $C_{u,v}^k$. The edges of the graph of the circuit $G = (V, E)$ have to be embedded into $U_n(\Gamma_2)$. After the transformations described in Sect. 2.1, every node in G has fanin and fanout 2, and we denote a topological order on V by η^G. We briefly describe the edge-embedding process in Sects. 2.3 and 3.

Translating a $U_n(\Gamma_2)$ into a Universal Circuit. Every node $w \in V_U$ fulfills a task when $U_n(\Gamma_2)$ is translated to a UC. Programming the UC means specifying its control bits along the paths defined by the edge-embedding and by the gates of circuit $C_{u,v}^k$. Depending on the number of incoming and outgoing edges and its type, a node is translated to:

G1. If w is a pole and corresponds to an input or output in G, then w is an *input or output* in $U_n(\Gamma_2)$ as well.

G2. If w is a pole and corresponds to a gate in G, w is programmed as a *universal gate* (UG). A 2-input UG supports any of the 16 possible gate types represented by the 4 control bits of the gate table (c_1, c_2, c_3, c_4). It implements function ug: $\{0,1\}^2 \times \{0,1\}^4 \to \{0,1\}$ that computes:

$$ug(x_1, x_2, c_1, c_2, c_3, c_4) = \overline{x_1 x_2} c_1 + \overline{x_1} x_2 c_2 + x_1 \overline{x_2} c_3 + x_1 x_2 c_4. \qquad (1)$$

Generally, a UG can be implemented with 3 AND and 6 XOR gates (resp. with a two-input gate when using Yao's protocol for SFE) [KS16].

G3. If w is no pole and has indegree and outdegree 2, w is programmed as an *X-switching block*, that computes $f_X : \{0,1\}^2 \times \{0,1\} \to \{0,1\}^2$ with $f_X((x_1, x_2), c) = (x_{1+c}, x_{2-c})$. This block can be implemented with 1 AND and 3 XORs (resp. a one-input gate with Yao) [KS08a].

G4. If w is no pole and has indegree 2 and outdegree 1, w is programmed as a *Y-switching block* that computes $f_Y : \{0,1\}^2 \times \{0,1\} \to \{0,1\}$ with $f_Y((x_1, x_2), c) = x_{1+c}$. This block can be implemented with 1 AND and 2 XORs (resp. a one-input gate with Yao) [KS08a].

G5. If w is no pole and has indegree 1 and outdegree 2, it has been placed to copy its input to its two outputs. Therefore, when translated to a UC, w is replaced by multiple outgoing wires in the parent node [KS16], since the UC itself does not have the fanout 2 restriction. In $U_n(\Gamma_2)$, w is added due to the fanout 2 restriction in the EUG.

G6. If w is no pole and has indegree and outdegree 1, w is removed and replaced by a wire between its parent and child nodes.

The nodes programmed as UG (G2), X-switching block (G3) or Y-switching block (G4) are so-called programmable blocks. This means that a programming bit or vector is necessary besides the two inputs to define their behavior as described above. These programming bits and vectors that build up the programming of the UC p_f are defined by the paths in the edge-embedding of G (the graph of circuit $C_{u,v}^k$ describing f) into $U_n(\Gamma_2)$.

Recursion Base. Valiant's construction is recursive, and the recursion base is reached when the number of poles is between 1 and 6. These recursion base graphs are shown in [Val76, KS16]. $U_1(\Gamma_1)$ is a single pole, $U_2(\Gamma_1)$ and $U_3(\Gamma_1)$ are two and three connected poles, respectively. $U_4(\Gamma_1)$, $U_5(\Gamma_1)$ and $U_6(\Gamma_1)$ are constructed with 3, 7 and 9 additional nodes, respectively.

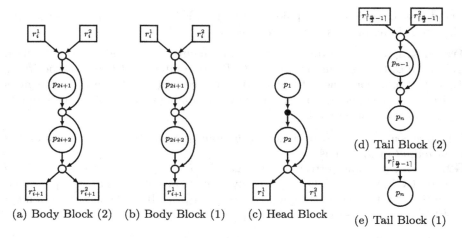

Fig. 2. (a) shows Valiant's 2-way EUG $U_n^{(2)}(\Gamma_1)$ [Val76]. (c) shows the corresponding head block, (b) and (d)–(e) show body and tail blocks, respectively, for different numbers of poles.

2.3 Valiant's 2-Way UC Construction

We described in Sect. 2.1 that a $U_n(\Gamma_d)$ EUG can be constructed of d instances of $U_n(\Gamma_1)$ EUGs. Therefore, Valiant provides an EUG for $\Gamma_1(n)$ graphs, two of which can build an EUG for $\Gamma_2(n)$ graphs. Let $P = \{p_1, \ldots, p_n\}$ be the set of poles in $U_n^{(2)}(\Gamma_1)$ that have indegree and outdegree 1. Valiant's 2-way EUG construction for $\Gamma_1(n)$ graphs of size $\sim 2.5n \log_2 n$ is shown in Fig. 2, where we emphasize the poles as large circles and the additional nodes as small circles or rectangles. The corresponding UC has twice the size $\sim 5n \log_2 n$, since it corresponds to the EUG for $\Gamma_2(n)$ graphs.

The rectangles are special nodes that build the set of poles in the next recursion step, i.e., $R^1_{\lceil \frac{n}{2}-1 \rceil} = \{r^1_1, \ldots, r^1_{\lceil \frac{n}{2}-1 \rceil}\}$, $R^2_{\lceil \frac{n}{2}-1 \rceil} = \{r^2_1, \ldots r^2_{\lceil \frac{n}{2}-1 \rceil}\}$. Another EUG is built with these poles which produces new subgraphs with size $\lceil \frac{\lceil \frac{n}{2}-1 \rceil}{2} - 1 \rceil$, s.t. we have four subgraphs at this level.

This construction is called the *2-way EUG or UC construction*. An open-source implementation of this construction optimized for PFE is provided in [KS16]. Independently, [LMS16] also implemented this 2-way UC, additionally optimizing for the total number of gates.

2.4 Valiant's 4-Way UC Construction

Valiant provides another, so-called *4-way EUG or UC construction* [Val76]. $U_n^{(4)}(\Gamma_1)$ has a 4-way recursive structure, i.e., nodes in special sets $R^1_{\lceil \frac{n}{4}-1 \rceil}$, $R^2_{\lceil \frac{n}{4}-1 \rceil}$, $R^3_{\lceil \frac{n}{4}-1 \rceil}$ and $R^4_{\lceil \frac{n}{4}-1 \rceil}$ are the poles in the next recursion step (cf. Fig. 4a). The recursion base is the same as in Sect. 2.2. This construction results in UCs of

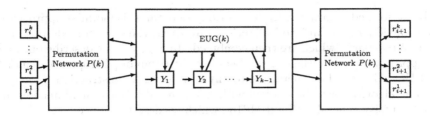

Fig. 3. k-way EUG construction $U_n^{(k)}(\Gamma_1)$ [LMS16].

smaller size $\sim 4.75n \log_2 n$ but has not been implemented before due to its more complicated structure.

2.5 Lipmaa et al.'s Generalized k-Way UC Construction

In [LMS16], Lipmaa et al. generalize Valiant's approach by providing a UC with any number of recursion points k, the so-called k-way EUG or UC construction. We note that their construction slightly differs from Valiant's EUG construction, since they do not consider the restriction on the fanout of the poles, i.e., the nodes in the EUG that correspond to universal gates or inputs (cf. Sect. 2.2). This optimization has also been included in [KS16] when translating an EUG to a UC, but including it in the block design leads to better sizes for the number of XOR gates.

The idea is to split $n = u + v + k$ in $m = \lceil \frac{n}{k} \rceil$ blocks as shown in Fig. 3. Every block i consists of k inputs $r_i^1, r_i^2, \ldots, r_i^k$ and k outputs $r_{i+1}^1, r_{i+1}^2, \ldots, r_{i+1}^k$ as well as k poles, except for the last block which has a number of poles depending on $n \bmod k$. For every $j \leq k$, the list of all r_i^j builds the poles of the jth subgraph of the next recursion step, i.e. we have k subgraphs. Additionally, every block begins and ends with a Waksman permutation network [Wak68] such that the inputs and outputs can be permuted to every pole. A Y-switching block is placed in front of every pole p_i which is connected to the ith output of the permutation network as well as the ith output of a block-intern EUG $U_k(\Gamma_1)$. Thus, [LMS16] reduce the problem of finding an efficient k-way EUG $U_n^{(k)}(\Gamma_2)$ to the problem of finding the smallest EUG $U_k(\Gamma_1)$. Their solution is to build the block-intern EUG with the UC construction of [KS08b], which was claimed to be more efficient for smaller circuits than [Val76]. However, they calculate the optimal k value to be around 3.147, which implies that the best solutions are found using small EUGs, for which Valiant provides hand-optimized solutions (i.e., for $k = 2, 3, 4, 5, 6$) [Val76].

3 Our Modular Edge-Embedding Algorithm

The detailed embedding algorithm and the open-source UC implementation of [KS16] was specifically built for the 2-way UC, dealing with the whole UC skeleton as one block. In contrast, based on the modular design of [LMS16], we modularize the edge-embedding task into multiple sub-tasks and describe how they can be performed separately. In this section, we detail this modular approach for edge-embedding a graph into Valiant's 4-way EUG: the edge-embedding can be split into two parts, which are then combined. In Sect. 3.1, we describe our modular approach based on the edge-embedding algorithm of [KS16] for Valiant's 2-way EUG. This can be generalized to any 2^i-way EUG construction. Moreover, the same algorithm can be applied with a few modifications for Lipmaa et al.'s k-way recursive generalization [LMS16], which we describe in Sect. 3.2.

3.1   Edge-Embedding in Valiant's 4-Way UC

Similar to the 2-way EUG construction (cf. Sect. 2.3), Valiant provides a 4-way EUG construction for $\Gamma_1(n)$ graphs which can be extended to an EUG for $\Gamma_2(n)$ graphs by utilizing two instances $U_n^{(4)}(\Gamma_1)_1$ and $U_n^{(4)}(\Gamma_1)_2$ as described in Sect. 2.1. The construction with our optimizations is visualized in Fig. 4. Valiant offers the main, so-called *Body Block* (Fig. 4a) consisting of 4 poles (large circles), 15 nodes (small circles) as well as 8 recursion points (squares). These body blocks are connected such that the 4 top [bottom] recursion points of one block are the 4 bottom [top] recursion points of the next block. Similarly to the 2-way EUG, 4 sets are created for n nodes, i.e., $R^1_{\lceil \frac{n}{4} \rceil - 1} = \{r_1^1, \ldots, r_{\lceil \frac{n}{4} \rceil - 1}^1\}$, $R^2_{\lceil \frac{n}{4} \rceil - 1} = \{r_1^2, \ldots, r_{\lceil \frac{n}{4} \rceil - 1}^2\}$, $R^3_{\lceil \frac{n}{4} \rceil - 1} = \{r_1^3, \ldots, r_{\lceil \frac{n}{4} \rceil - 1}^3\}$, and $R^4_{\lceil \frac{n}{4} \rceil - 1} = \{r_1^4, \ldots, r_{\lceil \frac{n}{4} \rceil - 1}^4\}$ which are the poles of 4 $U_{\lceil \frac{n}{2} \rceil - 1}(\Gamma_1)$ EUGs in the next recursion step. Then, these also create 4 subgraphs until the recursion base is reached, cf. Sect. 2.2.

We note that the top [bottom] block does not need the upper [lower] recursion points since its poles are the inputs [outputs] in the block. Therefore, we provide so-called Head and Tail Blocks. A *Head Block* occurs at the top of a chain of blocks (cf. Fig. 4e), it has 4 poles, no inputs, 4 output recursion points and 10 nodes, of which the first one (denoted by a filled circle) has one input and therefore translates to wires in the UC.

As a counterpart, *Tail Blocks* occur at the bottom of a chain of blocks, have at most 4 poles, 4 input recursion points, no outputs and at most 10 nodes depending on the number of poles. The 4 tail block constructions are depicted in Figs. 4f–i and are used, based on the remainder of n modulo 4, with the respective body or head blocks when $n \in \{5, 6, 7\}$, the lower parts of which are shown in Figs. 4a–d.

Block Edge-Embedding. In this first part of the edge-embedding process, we consider the 4 top [bottom] recursion points of the block as intermediate nodes where the inputs [outputs] of the block enter [leave]. The blocks are built s.t. any of these inputs can be forwarded to exactly one of the 4 poles of the block and the output of any pole can be forwarded to exactly one output or another pole having a higher topological order.

We formalize this behaviour as follows: In $U_n^{(4)}(\Gamma_1) = (V_U, E_U)$, let B be the block visualized in Fig. 4a with poles $p_{4i+1}, \ldots, p_{4i+4}$. Let mapping $\eta^U : V_U \to \mathbb{N}^+$ denote a topological order of V_U. Then, the nodes $r_i^1, \ldots, r_i^4$ and $r_{i+1}^1, \ldots, r_{i+1}^4$ denote the input and output recursion points of block B. Additionally, let $in =$

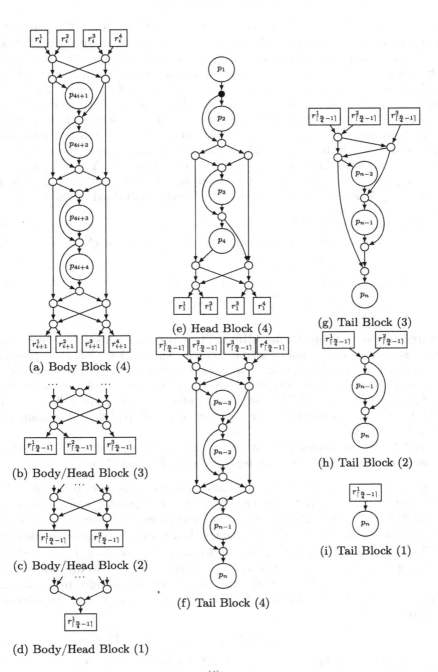

Fig. 4. (a) shows Valiant's 4-way EUG $U_n^{(4)}(\Gamma_1)$ [Val76]. (e) shows our head block construction, (a)–(d) and (f)–(i) show our body and tail block constructions, respectively, for different numbers of poles.

$(in_1, \ldots, in_4) \in \{0, \ldots, 4\}^4$ and $out = (out_1, \ldots, out_4) \in \{0, \ldots, 7\}^4$ denote the input and output vectors of B. The value 0 of the input and output vectors is a *dummy value* which is used if an input [a pole] is not forwarded to any pole [output] of B. The output vector has a larger value range, since a pole can be forwarded to another pole or an output recursion point. Therefore, we use values 1, 2 and 3 for poles p_2, p_3 and p_4 and values 4, 5, 6 and 7 for the output recursion points. Pole p_1 cannot be a destination for a path in B, since $\eta^U(p_1)$ is less than the topological order of any other pole in B. Additionally, the values of in and out need to be pairwise different or 0. Every combination of input and output vector covering the conditions formalized below in Eqs. 2–6 are valid for B. A pair $(r_i^l, p_j) \in \mathcal{P}$ or $(p_j, r_{i+1}^l) \in \mathcal{P}$ is a path from r_i^l to p_j or p_j to r_i^l in the set of all paths $\mathcal{P}$ in B. Then, $\mathcal{P}_B \subseteq \mathcal{P}$ denote the paths that are to be edge-embedded (cf. Sect. 6.1).

$$\forall l \in \{1, \ldots, 4\} : in_l \neq 0 \rightarrow (r_i^l, p_{in_l}) \in \mathcal{P}_B, \tag{2}$$

$$out_l \neq 0 \land out_l < 4 \rightarrow (p_j, p_{1+out_l}) \in \mathcal{P}_B \land \eta^U(p_j) < \eta^U(p_{1+out_l}) \tag{3}$$

$$out_l > 3 \rightarrow (p_j, r_{i+1}^{l-3}) \in \mathcal{P}_B. \tag{4}$$

$$\forall in_i, in_j \in in : i \neq j \rightarrow in_i = 0 \lor in_i \neq in_j. \tag{5}$$

$$\forall out_i, out_j \in out : i \neq j \rightarrow out_i = 0 \lor out_i \neq out_j. \tag{6}$$

Recursion Point Edge-Embedding. The block edge-embedding covers only the programming of the nodes within a block. Another task left is to program the recursion points. We use the supergraph construction of [KS16] which, in every step, splits a $\Gamma_2(n)$ graph in two $\Gamma_1(n)$ graphs, which are merged to two $\Gamma_2(\lceil \frac{n}{2} - 1 \rceil)$ graphs. [KS16] use this for defining the paths in Valiant's 2-way EUG. For Valiant's 4-way EUG, we use every second step of their algorithm with a minor modification.

Let $C_{u,v}^k$ be the Boolean circuit computing function f that our UC needs to compute, and $G \in \Gamma_2(n)$ its graph representation (cf. Sect. 2.2).

1. *Splitting $G \in \Gamma_2(n)$ in two $\Gamma_1(n)$ graphs G_1 and G_2*: As described in Sect. 2.1, Valiant's UC is derived from an EUG for $\Gamma_2(n)$ graphs, which consists of two EUGs for $\Gamma_1(n)$ graphs merged by their poles. Therefore, G is split into two $\Gamma_1(n)$ graphs G_1 and G_2. G_1 and G_2 then need to be edge-embedded into EUGs $(U_n^{(4)}(\Gamma_1))_1$ and $(U_n^{(4)}(\Gamma_1))_2$, respectively. $G = (V, E) \in \Gamma_2(n)$ is split by 2-coloring its edges as described in [Val76, KS16], which can always be done due to Kőnig's theorem [Kő31, LP09]. After 2-coloring, E is divided to sets E_1 and E_2, using which we build $G_1 = (V, E_1)$ and $G_2 = (V, E_2)$, with the following conditions:

$$\forall e \in E : (e \in E_1 \lor e \in E_2) \land \neg(e \in E_1 \land e \in E_2). \tag{7}$$

$$\forall e = (v_1, v_2) \in E_1 : \neg \exists e' = (v_3, v_4) \in E_1 : v_2 = v_4 \lor v_1 = v_3. \tag{8}$$

$$\forall e = (v_1, v_2) \in E_2 : \neg \exists e' = (v_3, v_4) \in E_2 : v_2 = v_4 \lor v_1 = v_3. \tag{9}$$

2. *Merging a $\Gamma_1(n)$ graph into a $\Gamma_2(\lceil \frac{n}{2} - 1 \rceil)$ graph*: In an EUG, the number of poles decreases in each recursion step and therefore, merging a $\Gamma_1(n)$

graph into a $\Gamma_2(\lceil \frac{n}{2} - 1 \rceil)$ graph provides information about the paths to be taken. Let $G_1 = (V, E) \in \Gamma_1(n)$ be a topologically ordered graph and $G_m = (V', E') \in \Gamma_2(\lceil \frac{n}{2} \rceil)$ be a graph with nodes $v'_1, \ldots, v'_{\lceil \frac{n}{2} \rceil}$. We define two labellings η_{in} and η_{out} on G_m with $\eta_{\text{in}}(v_i) = i$ and $\eta_{\text{out}}(v_i) = \eta_{\text{in}}(v_i) - 1 = i - 1$. Additionally, we define a mapping θ_V that maps a node $v_i \in V$ to a node $v_j \in V'$ with $\theta_V(v_i) = v'_{\lceil \frac{i}{2} \rceil}$. That means two nodes in G_1 are mapped to one node in G_m. At last, we define a mapping θ_E that maps an edge $e_i = (v_i, v_j) \in E$ to an edge $e_j \in E'$ with $\theta_E((v_i, v_j)) = (v_{\eta_{\text{in}}(\theta_V(v_i))}, v_{\eta_{\text{out}}(\theta_V(v_j))})$. That means every edge in G_1 is mapped to an edge in G_m as follows: $e = (v_i, v_j) \in E$ is mapped to $e' = (v'_k, v'_l) \in E'$, s.t. $v'_k = \theta_V(v_i)$, but v'_l is not the new node of v_j in G_m but v'_{l+1}. G_m is built as follows: $V' = \{v'_1, \ldots, v'_{\lceil \frac{n}{2} \rceil}\}$ and $E' = \bigcup_{e \in E} \theta_E(e)$. Then for all $e = (v'_i, v'_j) \in E'$ and $j < i$, e is removed from E', along with the last node $v_{\lceil \frac{n}{2} \rceil}$ (due to the definition of θ_E, it does not have any incoming edges). The resulting G_m is a topologically ordered graph in $\Gamma_2(\lceil \frac{n}{2} - 1 \rceil)$.

3. *The supergraph for Valiant's 4-way EUG construction:* In the first step, G is split to two $\Gamma_1(n)$ graphs G_1 and G_2. G_1 and G_2 contain all the edges that should be embedded as paths between poles in the first and second EUGs for $\Gamma_1(n)$, respectively. We now explain how to edge-embed the $\Gamma_1(n)$ graph G_1 into an EUG $U_n^{(4)}(\Gamma_1)$ (for G_2 it is the same).

For embedding in a 2-way UC, G_1 is firstly merged to a $\Gamma_2(\lceil \frac{n}{2} \rceil)$ graph G_m. G_m is then 2-colored and split into two $\Gamma_1(\lceil \frac{n}{2} \rceil)$ graphs G_1^1 and G_1^2 [KS16]. These get merged to two $\Gamma_2(\lceil \frac{\lceil \frac{n}{2} - 1 \rceil}{2} - 1 \rceil)$ graphs G_m^1 and G_m^2. G_1^1 is the first and G_1^2 is the second subgraph of G_1. Then $G_1^{\psi \circ 1}$ and $G_1^{\psi \circ 2}$ denote the first and second subgraph of G_1^{ψ}, respectively. These steps are repeated until the Γ_1 subgraphs have at most 4 nodes.

In Valiant's *4-way EUG construction* [Val76], a supergraph that creates 4 subgraphs in each step is necessary. We require a merging method where a $\Gamma_1(n)$ graph is merged to a $\Gamma_4(\lceil \frac{n}{4} - 1 \rceil)$ graph where 4 nodes build a new node, and 4-color this graph to retrieve 4 subgraphs. However, this can directly be solved by using the method described above from [KS16]: after repeating the 2-coloring and the merging twice, we gain 4 subgraphs (G_1^{11}, G_1^{12}, G_1^{21} and G_1^{22}). These can be used as if they were the result of 4-coloring the graph obtained by merging every 4 nodes into one.

However, there is a modification in this case: the first 2-coloring is a preprocessing step, which does not map to an EUG recursion step. Therefore, we have to define another labelling $\eta_{\text{out}_P}(v) = \eta_{\text{in}}(v)$, since in this preprocessing step we need to keep node $v_{\lceil \frac{n}{2} \rceil}$. Then the creation of the supergraph for the 4-way EUG construction works as follows: We merge G_1 to a $\Gamma_2(\lceil \frac{n}{2} \rceil)$ graph with labelling η_{in} and η_{out_P} and get G_m. After that, we split G_m into two $\Gamma_1(\lceil \frac{n}{2} \rceil)$ graphs G_1^1 and G_1^2. These get merged to $\Gamma_2(\lceil \frac{n}{4} \rceil - 1)$ graphs G_m^1 and G_m^2 using the η_{in} and η_{out} labellings. Finally, these two graphs get splitted into 4 $\Gamma_1(\lceil \frac{n}{4} - 1 \rceil)$ graphs G_1^{11}, G_1^{12}, G_1^{21} and G_1^{22}. These are the relevant graphs for the first recursion

Listing 1. Edge-embedding algorithm for Valiant's 4-way EUG

```
1   procedure edge-embedding (U, G₁ = (V,E))
2     Let S be the set of the 4 Γ₁ subgraphs of G₁ in the supergraph
3     Let R be the 4 recursion step graphs
4     Let B be the set of blocks in U
5     for all e = (vᵢ,vⱼ) ∈ E do
6       Let i' and j' denote the positions of vᵢ and vⱼ in their blocks
7       bᵢ ← ⌈i/4⌉, bⱼ ← ⌈j/4⌉ // number of block in which vᵢ and vⱼ are
8       Let out [r₁] denote the output vector [recursion points] of B_{bᵢ}
9       Let in [r₀] denote the the input vector [recursion points] of B_{bⱼ}
10      if bᵢ = bⱼ do // vᵢ and vⱼ are in the same block
11        if vᵢ ≠ vⱼ do
12          out_{i'} ← j' - 1
13        end if
14      else // vᵢ and vⱼ are in different blocks
15        Let s = (V',E') ∈ S denote the Γ₁ graph with e' = (p_{bᵢ}, p_{b_{j-1}}) ∈ E' and
               ↪ e' is not marked
16        Mark e'
17        Let x denote the number with s = Sₓ
18        Set the control bit of r₀ˣ to 1
19        if bⱼ = bᵢ + 1 do // bⱼ and bᵢ are neighbours
20          y ← 0
21        else
22          y ← 1
23        end if
24        Set the control bit of r₁ˣ to y
25        out_{i'} ← x + 4, inₓ ← j'
26      end if
27    end for
28    Edge-embed all blocks in U // edge-embed all sub-blocks
29    for i = 1 to 4 do
30      if Sᵢ exists do
31        call edge-embedding(Rᵢ, Sᵢ)
32      end if
33    end for
34  end procedure
```

step in Valiant's 4-way EUG construction. Now we continue for all 4 subgraphs until we reach the recursion base with 4 or less nodes.

4-Way Edge-Embedding Algorithm. In Listing 1, we combine block edge-embedding and recursion point edge-embedding:

Let U denote the part of $U_n^{(4)}(\Gamma_1)$ without recursion steps (the main chain of blocks) and $G_1 = (V, E)$ be the $\Gamma_1(n)$ graph which is to be edge-embedded in $U_n^{(4)}(\Gamma_1)$. S denotes the set of the 4 subgraphs of G_1 in the supergraph, i.e. $S = \{G_1^{11}, G_1^{12}, G_1^{21}, G_1^{22}\}$. A *recursion step graph* of U is one of the graphs having one of the 4 sets of recursion points as poles (e.g. $r_1^1, \ldots, r_{\lceil \frac{n}{4} - 1 \rceil}^1$) without the recursion steps. R denotes the set of all 4 recursion step graphs of U, and B denotes the set of all blocks in U.

We give a brief explanation of Listing 1 that describes the edge-embedding process. For any edge $e = (v_i, v_j) \in E$ in G_1, b_i and b_j denote the block numbers in which v_i and v_j are. There are 2 cases:

1. v_i **and** v_j **are in the same block:** $b_i = b_j$. The edge-embedding can be solved within the block and no recursion points have to be programmed for this path. Therefore, vector out of block B_{b_i} is set accordingly.
2. v_i **and** v_j **are in different blocks:** $b_i \neq b_j$. There exists an edge $e' = (b_i, b_{j-1})$ in one of the four $\Gamma_1(\lceil \frac{n}{4} - 1 \rceil)$ subgraphs of G_1 that is not yet used for an edge-embedding. This determines that the path in the next recursion step has to be between poles p_{b_i} and $p_{b_{j-1}}$. We denote with $s \in S$ the subgraph of G_1 which contains e', and x denotes its number in S, i.e. $S_x = s$. This implies in which of the 4 recursion step graphs we need to edge-embed the path from p_{b_i} to $p_{b_{j-1}}$, and so which recursion points we need to program. We first set the programming bit of the x-th input [output] recursion points to 1 since the path between the poles with labelling i and j enters [leaves] the next recursion step over this recursion point. A special case to be considered here is when blocks B_{b_i} and B_{b_j} are neighbours (i.e. $b_j = b_i + 1$). Then, the path enters and leaves the next recursion step graph at the same node, whose programming bit thus has to be 0. The output vector of block B_{b_i} is the i'^{th} value to the xth recursion point and the input vector of block B_{b_j} is the xth value to the j'^{th} pole in this block.

We repeat these steps for all edges $e \in E$. Since all in- and output vectors of all blocks in B are set, they can be embedded with the block edge-embedding. For all 4 subgraphs of G_1 in the supergraph and in the EUG, we call the same procedure with $S_i \in S$, $R_i \in R$, $1 \leq i \leq 4$.

3.2   Edge-Embedding in Lipmaa et al.'s k-Way UC

In this section, we extend the recent work of [LMS16] by providing a detailed and modular embedding mechanism for any k-way EUG construction described in Sect. 2.5. We provide the main differences to the edge-embedding of the 4-way EUG construction detailed in Sect. 3.1.

k-**Way Block Edge-Embedding.** In this setting, our main block is a programmable block B of size x with k poles $p_1, \ldots, p_k$, and k input [output] recursion points $r_0^1, \ldots, r_0^k$ $[r_1^1, \ldots, r_1^k]$. B is topologically ordered with mapping η^U as defined in Sect. 2.1. Vectors $in = (in_1, \ldots, in_k) \in \{0, \ldots, k\}^k$, and $out = (out_1, \ldots, out_k) \in \{0, \ldots, 2k - 1\}^k$ denote the input and output vectors of B, respectively. Values $k, \ldots, 2k - 1$ in out denote the recursion point targets $r_1^1, \ldots, r_1^k$ (cf. Sect. 3.1). We formalize the setting of in and out in Eqs. 10–14. We denote with $\mathcal{P}$ the set of all paths in B, and the $\mathcal{P}_B \subseteq P$ the paths that get edge-embedded in B.

$$\forall i \in \{1, \ldots, k\} : in_i \neq 0 \rightarrow (r_0^i, p_{in_i}) \in \mathcal{P}_B, \tag{10}$$

$$out_i \neq 0 \wedge out_i < k \rightarrow (p_i, p_{1+out_i}) \in \mathcal{P}_B \wedge \eta^U(p_i) < \eta^U(p_{1+out_i}) \tag{11}$$

$$out_i > k - 1 \rightarrow (p_i, r_1^{i-k+1}) \in \mathcal{P}_B. \tag{12}$$

$$\forall in_i, in_j \in in : i \neq j \rightarrow in_i = 0 \vee in_i \neq in_j. \tag{13}$$

$$\forall out_i, out_j \in out : i \neq j \rightarrow out_i = 0 \vee out_i \neq out_j. \tag{14}$$

k-Way Recursion Point Edge-Embedding. $G \in \Gamma_2(n)$ denotes the transformed graph of a Boolean circuit $C_{u,v}^k$, where $n = u + k + v$.

1. *Splitting $G \in \Gamma_2(n)$ in two $\Gamma_1(n)$ graphs G_1 and G_2:* Similarly as in Sect. 3.1, we first split G into two $\Gamma_1(n)$ graphs G_1 and G_2 with 2-coloring.
2. *Merging a $\Gamma_1(n)$ graph into a $\Gamma_k(\lceil \frac{n}{k} - 1 \rceil)$ graph:* $G_1 = (V, E) \in \Gamma_1(n)$ is merged into a $\Gamma_k(\lceil \frac{n}{k} - 1 \rceil)$ graph $G_m = (V', E')$ (same for G_2). Therefore, we redefine mapping θ_V (cf. Sect. 3.1) that maps node $v_i \in V$ to node $v_j \in V'$. In this scenario, k nodes in V build one node in V', so $\theta_V(v_i) = v_{\lceil \frac{i}{k} \rceil}$. The mapping of the edges θ_E is the same as in the 4-way EUG construction, and $(v_i', v_j') \in E'$ where $j < i$ edges are removed along with $v_{\lceil \frac{n}{k} \rceil}$ in the end. G_m is then a topologically ordered graph in $\Gamma_1(\lceil \frac{n}{k} - 1 \rceil)$.
3. *The supergraph for Lipmaa et al.'s k-way EUG construction:* The next step is to split $G_m \in \Gamma_1(\lceil \frac{n}{k} - 1 \rceil)$ into k $\Gamma_1(\lceil \frac{n}{k} - 1 \rceil)$ graphs. This is done with k-coloring: a directed graph $K = (V, E)$ can be k-colored, if k sets $E_1, \ldots, E_k \subseteq E$ cover the following conditions:

$$\forall i, j \in \{1, \ldots, k\} : i \neq j \rightarrow E_i \cap E_j = \emptyset. \tag{15}$$

$$\forall e \in E : \exists i \in \{1, \ldots, k\} : e \in E_i. \tag{16}$$

$$\forall i \in \{1, \ldots, k\} : \forall e = (v_1, v_2) \in E_i :$$
$$\neg \exists e' = (v_3, v_4) \in E_i : v_2 = v_4 \vee v_1 = v_3. \tag{17}$$

According to Kőnig's theorem [Kő31,LP09], $\Gamma_k(n)$ graphs can always be k-colored efficiently (cf. full version [GKS17, Appendix A] for details). The rest of the supergraph construction and the way it is used for edge-embedding is the same as for the 4-way EUG construction as described in Sect. 3.1.

k-Way Edge Embedding Algorithm. The edge-embedding algorithm is the same as shown in Listing 1, after replacing every 4 with k.

4   New Universal Circuit Constructions

Here, we describe our ideas for novel, potentially more efficient, UC constructions. Firstly, in Sect. 4.1, we describe modular building blocks for a *3-way UC*. We show that Valiant's optimized $U_3(\Gamma_1)$ cannot directly be applied as a building block in the construction due to the fact that it must have an additional node to be a generic EUG. We prove that the EUG without this node is not a valid EUG by showing a counterexample. Therefore, it actually results in a worse asymptotic size than Valiant's 2-way and 4-way UC constructions. Secondly, in Sect. 4.2, we propose a *hybrid UC construction*, utilizing both Valiant's 2-way and 4-way UC constructions so that the overall size of the resulting hybrid UC is minimized, and is at least as efficient as the better construction for the given size.

4.1   3-Way Universal Circuit Construction

The optimal k value for minimizing the size of the k-way UC was calculated to be 3.147 in [LMS16]. We describe our idea of a 3-way UC construction. Intuitively, based on an optimization by Valiant [Val76], this UC should result in the best asymptotic size. The asymptotic size of any k-way UC depends on the size of its modular body block B_k (e.g., Fig. 4a for the 4-way UC). Once it is determined, the size of the UC is $\text{size}(U_n^{(k)}(\Gamma_2)) = 2 \cdot \text{size}(U_n^{(k)}(\Gamma_1)) \approx 2 \cdot \frac{\text{size}(B_k)}{k} n \log_k n = 2 \cdot \frac{\text{size}(B_k)}{k \log_2(k)} n \log_2 n$. The modular block consists of two permutation networks $P(k)$, an EUG $U_k(\Gamma_1)$, and $(k-1)$ Y-switching blocks (cf. Sect. 2.5, [LMS16]).

Size of Body Block B_3 with Valiant's Optimized $U_3(\Gamma_1)$. According to Valiant [Val76], an EUG $U_3(\Gamma_1)$ with 3 poles contains only 3 connected poles (used as recursion base in Sect. 2.2). An optimal permutation network $P(3)$ that achieves the lower bound has 3 nodes as well. This implies that $\text{size}(B_3) = 2 \cdot P(3) + \text{size}(U_3(\Gamma_1)) + (3-1) = 11$. Then, the size of the UC becomes $\approx 2 \cdot \frac{11}{3 \log_2 3} n \log_2 n \approx 4.627 n \log_2 n$, which means an asymptotically by around 2.5% smaller size than that of the 4-way UC.

However, there is a flaw in this initial design. Valiant's $U_3(\Gamma_1)$ only works as an EUG for 3 nodes under special conditions, e.g., when it is a subgraph within a larger EUG construction. There are 3 possible edges in a topologically ordered graph $G = (V, E)$ in $\Gamma_1(3)$: $(1,2), (2,3)$ and $(1,3)$. $(1,2)$ and $(2,3)$ can be directly embedded in $U_3(\Gamma_1)$ using (p_1, p_2) and (p_2, p_3), respectively. $(1,3)$, however, has to be embedded as a path *through* node 2, i.e., as a path $((p_1, p_2), (p_2, p_3))$. When $U_3(\Gamma_1)$ is a subgraph of a bigger EUG, this is possible by programming p_2 accordingly. However, when we use this $U_3(\Gamma_1)$ as a building block in our EUG construction, it cannot directly be applied. A generic $U_3(\Gamma_1)$ that can embed $(1,3)$ without going through p_2 as before has an additional Y-switching block.

We depict in Fig. 5a the 3-way body block that uses Valiant's optimized $U_3(\Gamma_1)$ in the k-way block design of [LMS16]. Assume that the output of pole p_{3i+1} has to be directed to pole p_{3i+3}. Then, it needs to go through pole p_{3i+2}, which means that the edge going in to p_{3i+2} is used by this path. However, there might be another edge coming from the permutation network as an input to p_{3i+2}, e.g., from p_{3i} from the preceding block. This cannot be directed to p_{3i+2} anymore as shown in Fig. 5a. Therefore, in the 3-way body block construction, it does not suffice to use Valiant's optimized $U_3(\Gamma_1)$ [Val76].

Size of Body Block B_3 with Our Generic $U_3(\Gamma_1)$. In Fig. 5b, we show the 3-way body block with the generic $U_3(\Gamma_1)$ that allows the output from p_{3i+1} to be directed to p_{3i+3} without having to go *through* p_{3i+2}. This results in $\text{size}(B_3) = 2 \cdot P(3) + \text{size}(U_3(\Gamma_1)) + (3-1) = 12$, which implies that the asymptotic size of the UC is $\approx 2 \cdot \frac{12}{3 \log_2 3} n \log_2 n \approx 5.047 n \log_2 n$. Unfortunately, this is worse than the asymptotic size of the 2-way construction, and we therefore conclude that the asymptotically most efficient known UC construction is Valiant's 4-way UC construction.

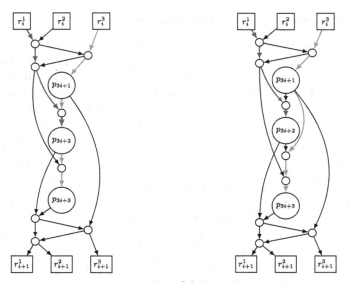

(a) Body Block with Valiant's $U_3(\Gamma_1)$ (b) Body Block with our generic $U_3(\Gamma_1)$

Fig. 5. Body block construction for our 3-way EUG $U_n^{(3)}(\Gamma_1)$.

4.2   Hybrid Universal Circuit Construction

In this section, we detail our hybrid UC that minimizes its size based on Valiant's 2-way and 4-way UCs, which are asymptotically the smallest UCs to date. Given the size of the input circuit $C_{u,v}^k$, i.e., $n = u + k + v$, we can calculate at each recursion step if it is better to create 2 subgraphs of size $\lceil \frac{n}{2} - 1 \rceil$ and utilize the 2-way recursive skeleton, or it is more beneficial to create a 4-way recursive skeleton with 4 subgraphs of size $\lceil \frac{n}{4} - 1 \rceil$.

We assume that for every n, we have an algorithm that computes the size $(\text{size}(U_n^{\text{hybrid}}(\Gamma_1)))$ of the hybrid construction for sizes smaller than n. We give details on how it is computed in Sect. 5. Then, Listing 2 describes the algorithm for constructing a hybrid UC, at each step based on which strategy is more efficient. We note that our hybrid construction is generic, and given multiple k-way UC constructions as parameter K ($K = \{2, 4\}$ in our example), it minimizes the concrete size of the resulting UC.

5   Size of UC Constructions

Lipmaa et al.'s k-way UC construction is depicted in a modular manner in [LMS16, Fig. 12] and discussed briefly in Sect. 2.5 and Fig. 3. They show that a k-way body block consists of two permutation networks $P(k)$, an EUG for k nodes, i.e., $U_k(\Gamma_1)$, and additionally, $(k - 1)$ Y-switching blocks. In this section, we recapitulate the sizes (Table 2) of the k-way EUG and give an estimate for the

Listing 2. Hybrid construction algorithm

```
1   procedure hybrid (p₁,...,pₙ , K = {2,4})
2     Let size(U_{n'}^{hybrid}(Γ₁)) be the function calculating the size of the
          ↪ smaller hybrid constructions with size n' ≤ n
3     for all k ∈ K do // Number of poles in the last block for all k
4       if n | k do
5         m_k ← k
6       else
7         m_k ← n mod k
8       end if
9       s_k ← size(Head_k(k)) + (⌈n/k⌉ − 3) · size(Body_k(k)) + size(Body_k(r_k)) + size(Tail_k(m_k))+
          ↪   m₂ · size ( size(U_{⌈n/2⌉−1}^{hybrid}(Γ₁)) ) + ((k − m_k) · size ( size(U_{⌊n/k⌋−1}^{hybrid}(Γ₁)) )
10    end for
11    s_i ← min(s_k : k ∈ K) // Choose the better construction
12    Create skeleton for i−way construction with n poles
13    call hybrid (r₁¹,...,r_{⌈n/i⌉−1}¹, K) , ..., hybrid (r₁^{m_i},...,r_{⌈n/i⌉−1}^{m_i}, K)
14    if (i − m_i) > 0 do
15      call hybrid (r₁^{m_i},...,r_{⌊n/i⌋−1}^{m_i}, K) , ..., hybrid (r₁^i,...,r_{⌊n/i⌋−1}^i, K)
16    end if
17  end procedure
```

leading constant for Lipmaa et al.'s EUG construction with size $\mathcal{O}(n \log_2 n)$, for $k \in \{2,\ldots,8\}$. For a detailed discussion on the depth of the UCs, the reader is referred to the full version of this paper [GKS17, Sect. 5]. We conclude that the best asymptotic size is achieved by Valiant's 4-way UC. This result does not exclude the possibility for a more efficient UC in general, but it shows that the most efficient UC using Lipmaa et al.'s k-way UC from [LMS16] is the 4-way UC. Two k-way EUGs for $\Gamma_1(n)$ graphs build up an EUG for $\Gamma_2(n)$ graphs as described in Sect. 2.1. Therefore, the leading constant for the size of the UC is twice that of the EUG $U_n^{(k)}(\Gamma_1)$, which is summarized in the same table.

5.1   Asymptotic Size of k-Way UC Constructions

We review the sizes of the building blocks of a k-way body block, i.e., the size of an EUG $U_k(\Gamma_1)$ for k, and the size of a permutation network $P(k)$ with k inputs and outputs, as well as the size of the resulting UCs.

Edge-Universal Graph with k Poles. Valiant optimized EUGs up to size 6 by hand in [Val76]: for $k = 2$, $U_2(\Gamma_1)$ has two poles, for $k = 3$ we discussed in Sect. 4.1 that an additional node is necessary. For $k \in \{4,5,6\}$ the sizes are $\{6,10,13\}$, as shown in [KS16, Fig. 1] (note that the nodes noted as empty circles disappear in the UC). For $k = 7$ and $k = 8$, we observe that Valiant's 2-way UC results in a better size than that of the 4-way UC due to the smaller permutation network and less recursion nodes. Therefore, we use these constructions to compute the size of $U_7(\Gamma_1)$ and $U_8(\Gamma_1)$. As mentioned in [LMS16], another possibility is to use the UC of [KS08b] instead of these EUGs since they have better sizes for small circuits. These UCs $U^{KS08}(k)$ are built from two smaller $U^{KS08}(\frac{k}{2})$, a $P(\frac{k}{2})$ and $\frac{k}{2}$ Y switches. It results in a smaller size of 21 for $k = 8$.

Table 2. The leading factors of the asymptotic $\mathcal{O}(n \log_2 n)$ **size** for k-way edge-universal graphs $(U_n^{(k)}(\Gamma_1))$ and universal circuits (UC) for $k \in \{2, \ldots, 8\}$. n denotes the size of the input $\Gamma_2(n)$ circuit, $U_k(\Gamma_1)$ the size of Valiant's edge-universal graph with k poles, $U^{\text{KS08}}(k)$ the size of the UC of [KS08b], $P^{\text{l}}(k)$ the lower bound for the size of a permutation network for k nodes, and $P^{\text{W}}(k)$ the size of Waksman's permutation network [Wak68]. B_k^W is the size of the body block.

k	$U_k(\Gamma_1)$	$U^{\text{KS08}}(k)$	$P^{\text{l}}(k)$	$P^{\text{W}}(k)$	B_k^W	$U_n^{(k)}(\Gamma_1)$ $(\cdot n \log_2 n)$	UC $(\cdot n \log_2 n)$
2	2	2	1	1	5	2.5	5
3	4	6	3	3	12	≈ 2.524	≈ 5.047
4	6	7	5	5	19	2.375	4.75
5	10	11	7	8	30	≈ 2.584	≈ 5.168
6	13	14	10	11	40	≈ 2.579	≈ 5.158
7	19	19	13	14	53	≈ 2.697	≈ 5.394
8	23	21	16	17	62	≈ 2.583	≈ 5.167

Permutation Networks. Waksman in [Wak68] showed that the lower bound for the size of a permutation network is $\lceil \log_2(k!) \rceil$ for k elements. We present this lower bound in Table 2 as $P^{\text{l}}(k)$. The permutation network with the smallest size is Waksman's permutation network $P^{\text{W}}(k)$ [Wak68,BD02]. For $k \in \{2, 3, 4\}$ its size reaches the lower bound, but for larger k values, his permutation network utilizes additional nodes. Since these are the smallest existing permutation networks, we use these when calculating the size of the UC. Even with the lower bound $P^{\text{l}}(k)$, for $k \in \{5, 6, 7, 8\}$ we would have the respective leading terms $\{4.824, 4.900, 5.190, 5\}$, which are larger than 4.75 for $k = 4$.

Body Blocks. A body block B_k^W is built of $(k - 1)$ Y-switching blocks, an EUG for k nodes, and two permutation networks [LMS16] (cf. Fig. 3). The size of B_k^W is the sum of the sizes of its building blocks, i.e., $\text{size}(B_k^W) = \min\left(\text{size}(U_k(\Gamma_1)), \text{size}(U^{\text{KS08}}(k))\right) + 2 \cdot \text{size}(P^W(k)) + k - 1$.

Edge-Universal Graphs and Universal Circuits with n Poles. The asymptotic size of EUG $U_n^{(k)}(\Gamma_1)$ is determined as $\text{size}(U_n^{(k)}(\Gamma_1)) = \frac{\text{size}(B_k^W)}{k \log_2 k} n \log_2 n$ and the leading factor for a UC is twice this number.

5.2   Concrete Size of UC Constructions

The size of Lipmaa et al.'s k-way universal circuits depends on the size of their building blocks [LMS16]. More concretely, finding either better edge-universal graphs for small number of nodes or optimal permutation networks could improve the sizes of these UCs. Lipmaa et al. calculated the optimal k value for minimizing the size of a k-way UC to be 3.147 [LMS16].

Table 2 shows that the smallest sizes are achieved by the 4-way, followed by the 2-way UCs. The 3-way UC, as mentioned in Sect. 4.1, is less efficient due to the additional node in $U_3(\Gamma_1)$. We observe that the sizes grow with increasing k values due to the permutation networks and EUGs.

Concrete Sizes of 4-Way and 2-Way UCs. Based on the parity (2-way UC) and the remainder modulo 4 (4-way UC), not only the size of the outest skeleton, but also that of the smaller subgraphs can be optimized. It was considered in [KS16] for the 2-way UC, and we now generalize the approach for k-way UCs. We provide a recursive formula for the concrete size of the optimized k-way EUG as follows. Let m_k be defined as

$$m_k := \begin{cases} n \mod k & \text{if } k \nmid n, \\ k & \text{if } k \mid n. \end{cases} \tag{18}$$

Then, given the designed Head, Body and Tail blocks with sizes shown in Table 3, we can compute the size by calculating the size of all the components of the outest skeleton, and the sizes of the smaller subgraphs with the recursive formula shown in Eq. 19.[4]

$$\begin{aligned} \text{size}(U_n^{(k)}(\Gamma_1)) = {} & \text{size(Head}(k)) + \left(\left\lceil \frac{n}{k} \right\rceil - 3\right) \cdot \text{size(Body}(k)) \\ & + \text{size(Body}(m_k)) + \text{size(Tail}(m_k)) \\ & + m_k \cdot \text{size}\left(U_{\lceil \frac{n}{k} \rceil - 1}^{(k)}(\Gamma_1)\right) + (k - m_k) \cdot \text{size}\left(U_{\lfloor \frac{n}{k} \rfloor - 1}^{(k)}(\Gamma_1)\right). \end{aligned} \tag{19}$$

Concrete Size of Our Hybrid UC. We provide a hybrid UC in Sect. 4.2 for minimizing the size of the resulting UC. This construction chooses at each step the skeleton that results in the smallest size and therefore, we provide the recursive algorithm for determining its size in Eq. 20. size(Head$_k(i)$), size(Tail$_k(i)$) and

Table 3. The sizes of building blocks of the 2-way and 4-way UCs (cf. Figs. 2 and 4).

Block	Head				Body				Tail			
k\Poles	4	3	2	1	4	3	2	1	4	3	2	1
Fig.	-	-	2c	-	-	-	2a	2b	-	-	2d	2e
2-way	-	-	4	-	-	-	5	5	-	-	4	1
Fig.	4e	4g	4h	4i	4a	4b	4c	4d	4f	4g	4h	4i
4-way	14	14	13	12	19	19	18	17	14	9	4	1

---

[4] We note that for $k \geq 3$, there exist Head$(k-1), \ldots,$ Head(1) blocks. These are used for one n, e.g., Head(1) when $n = k+1$, and Head$(k-1)$ when $n = 2k$. For simplicity, we consider these as special recursion base numbers in our calculations.

size(Body$_k(i)$) are the values from Table 3 for $k = 2$ and $k = 4$. The size of the hybrid UC is minimized as

$$
\text{size}(U_n^{\text{hybrid}}(\Gamma_1)) = \min \Big(\text{size}(\text{Head}_k(k)) + \Big(\Big\lceil \frac{n}{k} \Big\rceil - 3\Big) \cdot \text{size}(\text{Body}_k(k))
$$
$$
+ \text{size}(\text{Body}_k(m_k)) + \text{size}(\text{Tail}_k(m_k)) + m_k \cdot \text{size}\Big(U_{\lceil \frac{n}{k}-1\rceil}^{\text{hybrid}}(\Gamma_1)\Big)
$$
$$
+ (k - m_k) \cdot \text{size}\Big(U_{\lfloor \frac{n}{k}-1\rfloor}^{\text{hybrid}}(\Gamma_1)\Big); \quad k \in \{2, 4\}\Big), \tag{20}
$$

which can be computed using a dynamic programming algorithm.

Improvement of 4-Way Construction. The bottom (blue) line in Fig. 6 shows the concrete improvement in percentage of the 4-way UC construction over the 2-way UC construction up to ten million nodes in the simulated input circuit. From the asymptotic leading factors in Table 2, we expect an improvement of up to $1 - \frac{4.75}{5} = 5\%$. For the smallest n values ($n \leq 15$), the 2-way UC is up to 33.3% better than the 4-way UC. However, from $n = 212$ on, the 4-way UC construction is better, except for some short intervals as shown in Fig. 6 (the difference in these intervals, however, is at most 3.45%). From here on, the 4-way UC is on average 3.12% better in our experiments, where the biggest improvement is 4.48%. Moreover, from $n = 10885$ on, the 4-way UC always outperforms the 2-way UC.

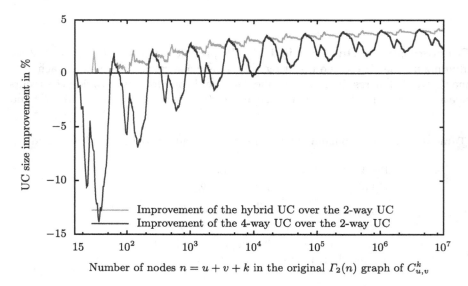

Fig. 6. Improvement of our hybrid and Valiant's 4-way UC over Valiant's 2-way UC for $15 \leq n \leq 10^7$ with logarithmic x axis. (Color figure online)

Improvement of Hybrid Construction. The improvement achieved by our hybrid construction (cf. Sect. 4.2) is depicted in the same Fig. 6, as the top (green) line. For some n values the hybrid UC achieves the same size as the 2- or 4-way UCs, but due to its nature, it is never worse. This means that the improvement of our hybrid UC is always nonnegative, and greater than or equal to the improvement achieved by the 4-way UC. Moreover, in most cases the hybrid UC results in better sizes than any of the other two constructions: this means that some subgraphs are created for an n for which the 2-way UC is smaller, and therefore the 2-way recursive structure is utilized. The overall improvement for all n values is on average 3.65% and at most 4.48% over the 2-way UC construction.

6 Implementation and Evaluation

The first implementation of Valiant's 2-way UC, along with a toolchain for PFE (cf. Sect. 1.1) was given in [KS16]. The 4-way UC has smaller asymptotic size $\sim 4.75n \log_2 n$, but has not been implemented before due to its more complicated structure and embedding algorithm.

In this work, we improve the implementation of the open-source framework of [KS16] by using the 4-way UC construction that can directly be applied in the PFE framework. Our improved implementation is available at http://encrypto.de/code/UC. Firstly, the functionality is translated to a Boolean circuit using the Fairplay compiler [MNPS04,BNP08]. This is then transformed into a circuit in $\Gamma_2(n)$, i.e., with at most two incoming and outgoing wires for each gate, input and output. This is done in a preprocessing step of the framework in [KS16]. The input circuit description of our UC implementation is the same as that of the UC compiler of [KS16], and we also adapt our output UC format to that of [KS16] that includes the gate types described in Sect. 2.2. This format is compatible with the ABY framework [DSZ15] for secure function evaluation.

We discuss our implementation of Valiant's 4-way UC in Sect. 6.1 and give experimental results in Sect. 6.2. For a description on how the hybrid UC can be implemented, the reader is referred to the full version [GKS17, Sect. 6.3].

6.1 Our 4-Way Universal Circuit Implementation

The architecture of our UC implementation is the same as that of [KS16], and therefore, we describe our UC design based on the steps described in [KS16, Fig. 6]. Our implementation gets as input a circuit with u inputs, v outputs and k gates, and outputs a 4-way UC with size $n = u + k + v$, as well as the programming p_f corresponding to the input circuit (cf. Sect. 1).

Transforming Circuit $C_{u,v}^k$ into $\Gamma_2(u + k + v)$ Graph G. As a first step, we transform the circuit $C_{u,v}^k$ into a $\Gamma_2(n)$ graph $G = (V, E)$ with $n = u + k + v$ (cf. Sect. 2.1). Then, we define a topological order η^G on the nodes of G s.t. every input node v_i has a topological order of $1 \leq \eta^G(v_i) \leq u$ and every output node v_j is labelled with $u + k + 1 \leq \eta^G(v_j) \leq u + k + v$.

Table 4. Comparison of the sizes of the UCs (2-way, 4-way, and hybrid) for sample circuits from [TS15]. Bold numbers denote if the 2-way or the 4-way UC is smaller; the smallest size is always achieved by our hybrid UC. The UC generation time is given for both implemented UCs.

Circuit	n	Circuit size (#AND gates)			UC generation (ms)	
	$u + v + k$	2-way UC [KS16]	Our 4-way UC	Our Hybrid UC	2-way UC [KS16]	Our 4-way UC
AES-non-exp	46 847	$2.96 \cdot 10^6$	$\mathbf{2.93 \cdot 10^6}$	$2.86 \cdot 10^6$	9 008.9	10 325.8
AES-exp	38 518	$2.38 \cdot 10^6$	$\mathbf{2.38 \cdot 10^6}$	$2.31 \cdot 10^6$	6 961.7	8 361.3
DES-non-exp	31 946	$1.96 \cdot 10^6$	$\mathbf{1.92 \cdot 10^6}$	$1.89 \cdot 10^6$	5 563.8	6 599.5
DES-exp	32 207	$19.8 \cdot 10^6$	$\mathbf{19.4 \cdot 10^6}$	$1.90 \cdot 10^6$	5 654.0	6 765.0
md5	66 497	$4.42 \cdot 10^6$	$\mathbf{4.26 \cdot 10^6}$	$4.26 \cdot 10^6$	14 805.5	14 897.8
sha-256	201 206	$1.49 \cdot 10^7$	$\mathbf{1.46 \cdot 10^7}$	$1.44 \cdot 10^7$	81 889.1	57 439.0
add_32	342	$9.58 \cdot 10^3$	$\mathbf{9.55 \cdot 10^3}$	$9.44 \cdot 10^3$	29.6	35.3
add_64	674	$\mathbf{2.21 \cdot 10^4}$	$2.27 \cdot 10^4$	$2.17 \cdot 10^4$	53.9	89.6
comp_32	216	$\mathbf{5.53 \cdot 10^3}$	$5.54 \cdot 10^3$	$5.49 \cdot 10^3$	17.7	21.2
mult_32x32	12 202	$6.54 \cdot 10^5$	$\mathbf{6.50 \cdot 10^5}$	$6.35 \cdot 10^5$	1 639.2	2 177.1
Branching_18	200	$\mathbf{4.92 \cdot 10^3}$	$5.07 \cdot 10^3$	$4.88 \cdot 10^3$	21.0	24.2
CreditChecking	82	$\mathbf{1.50 \cdot 10^3}$	$1.51 \cdot 10^3$	$1.49 \cdot 10^3$	3.1	12.7
MobileCode	160	$\mathbf{3.65 \cdot 10^3}$	$3.88 \cdot 10^3$	$3.61 \cdot 10^3$	10.6	29.0

Creating an EUG $U_n^{(4)}(\Gamma_2)$ for $\Gamma_2(n)$ Graphs. An EUG $U_n^{(4)}(\Gamma_2)$ is constructed by creating two instances of $U_n^{(4)}(\Gamma_1)$ as shown in Sect. 2.2. The two instances get merged to $U_n^{(4)}(\Gamma_2)$ so that one builds the left inputs and outputs and the other builds the right inputs and outputs of the gates (based on the two-coloring of G). We create the EUGs with Valiant's 4-way EUG [Val76] with our optimized blocks from Sect. 3.1 (cf. Fig. 4).

Programming $U_n^{(4)}(\Gamma_2)$ to Compute $C_{u,v}^k$. We edge-embed graph G into $U_n^{(4)}(\Gamma_2)$ as described in Sect. 3.1. [KS16] use their supergraph construction to define the paths between the poles uniquely for Valiant's 2-way EUG. We modify this supergraph as described in Sect. 3.1 for Valiant's 4-way EUG and perform the edge-embedding as described in Listing 1. The programming bits of the nodes are set during the edge-embedding process along the paths between the poles. The block edge-embedding is done by analyzing the possible input values and defining the valid paths as described in Sect. 3.1.

Outputting a Universal Circuit with Its Programming. As a final step, EUG $U_n^{(4)}(\Gamma_2)$ is topologically ordered and output in the UC format of [KS16]. The programming bits p_f defined by the embedding are also output in a separate file based on the topological order.

6.2 Our Experimental Results

In order to show the improvement of our method, we ran experiments on a Desktop PC, equipped with an Intel Haswell i7-4770K CPU with 3.5 GHz and 16 GB RAM, and provide our results in Table 4. To compare with the runtime of the UC compiler of [KS16], we ran the same experiments on the same platform using their 2-way UC implementation.

As [KS16], we use a set of real-life circuits from [TS15] for our benchmarks, and compare the sizes of the resulting circuits and the generation and embedding runtimes. We can see that from the 2-way and 4-way UC constructions, the 4-way UC, as expected, is always smaller for large circuits than the 2-way UC. However, it is sometimes better even for small circuits, e.g., for 32-bit addition with $n = 342$. The hybrid construction always provides the smallest UC for our example circuits.

In the last two columns, we report the runtime of the UC compiler of [KS16] and our 4-way UC implementation for generating and programming the universal circuit corresponding to the example circuits. Table 4 shows that the differences in runtime are not significant, and due to its more complicated structure, the 4-way UC takes more time to generate and program in general. However, we can see from the largest example with more than 200000 nodes that asymptotically, the 4-way UC results in a runtime improvement as well, as less nodes need to be programmed.

Acknowledgements. This work has been co-funded by the German Federal Ministry of Education and Research (BMBF) and the Hessen State Ministry for Higher Education, Research and the Arts (HMWK) within CRISP and by the DFG as part of project E3 within CROSSING. We thank the reviewers of ASIACRYPT'17 for their helpful comments.

References

[AMPR14] Afshar, A., Mohassel, P., Pinkas, B., Riva, B.: Non-interactive secure computation based on cut-and-choose. In: Nguyen, P.Q., Oswald, E. (eds.) EUROCRYPT 2014. LNCS, vol. 8441, pp. 387–404. Springer, Heidelberg (2014). https://doi.org/10.1007/978-3-642-55220-5_22

[Att14] Attrapadung, N.: Fully secure and succinct attribute based encryption for circuits from multi-linear maps. Cryptology ePrint Archive, Report 2014/772 (2014). http://ia.cr/2014/772

[BBKL17] Bicer, O., Bingol, M.A., Kiraz, M.S., Levi, A.: Towards practical PFE: an efficient 2-party private function evaluation protocol based on half gates. Cryptology ePrint Archive, Report 2017/415 (2017). http://ia.cr/2017/415

[BD02] Beauquier, B., Darrot, É.: On arbitrary size Waksman networks and their vulnerability. Parallel Proces. Lett. **12**(3–4), 287–296 (2002)

468   D. Günther et al.

[BFK+09] Barni, M., Failla, P., Kolesnikov, V., Lazzeretti, R., Sadeghi, A.-R., Schneider, T.: Secure evaluation of private linear branching programs with medical applications. In: Backes, M., Ning, P. (eds.) ESORICS 2009. LNCS, vol. 5789, pp. 424–439. Springer, Heidelberg (2009). https://doi.org/10.1007/978-3-642-04444-1_26

[BNP08] Ben-David, A., Nisan, N., Pinkas, B.: FairplayMP: a system for secure multi-party computation. In: CCS 2008, pp. 257–266. ACM (2008)

[BOKP15] Banescu, S., Ochoa, M., Kunze, N., Pretschner, A.: Idea: benchmarking indistinguishability obfuscation – a candidate implementation. In: Piessens, F., Caballero, J., Bielova, N. (eds.) ESSoS 2015. LNCS, vol. 8978, pp. 149–156. Springer, Cham (2015). https://doi.org/10.1007/978-3-319-15618-7_12

[BPSW07] Brickell, J., Porter, D.E., Shmatikov, V., Witchel, E.: Privacy-preserving remote diagnostics. In: CCS 2007, pp. 498–507. ACM (2007)

[DSZ15] Demmler, D., Schneider, T., Zohner, M.: ABY - a framework for efficient mixed-protocol secure two-party computation. In: NDSS 2015. The Internet Society (2015). Code: http://encrypto.de/code/ABY

[FAZ05] Frikken, K.B., Atallah, M.J., Zhang, C.: Privacy-preserving credit checking. In: Electronic Commerce (EC 2005), pp. 147–154. ACM (2005)

[FGP14] Fiore, D., Gennaro, R., Pastro, V.: Efficiently verifiable computation on encrypted data. In: CCS 2015, pp. 844–855. ACM (2014)

[FVK+15] Fisch, B., Vo, B., Krell, F., Kumarasubramanian, A., Kolesnikov, V., Malkin, T., Bellovin, S.M.: Malicious-client security in blind seer: a scalable private DBMS. In: IEEE S&P 2015, pp. 395–410. IEEE (2015)

[GGH+13a] Garg, S., Gentry, C., Halevi, S., Raykova, M., Sahai, A., Waters, B.: Candidate indistinguishability obfuscation and functional encryption for all circuits. In: FOCS 2013, pp. 40–49. IEEE (2013)

[GGH+13b] Garg, S., Gentry, C., Halevi, S., Sahai, A., Waters, B.: Attribute-based encryption for circuits from multilinear maps. In: Canetti, R., Garay, J.A. (eds.) CRYPTO 2013. LNCS, vol. 8043, pp. 479–499. Springer, Heidelberg (2013). https://doi.org/10.1007/978-3-642-40084-1_27

[GHV10] Gentry, C., Halevi, S., Vaikuntanathan, V.: i-hop homomorphic encryption and rerandomizable Yao circuits. In: Rabin, T. (ed.) CRYPTO 2010. LNCS, vol. 6223, pp. 155–172. Springer, Heidelberg (2010). https://doi.org/10.1007/978-3-642-14623-7_9

[GKS17] Günther, D., Kiss, Á., Schneider, T.: More efficient universal circuit constructions. Cryptology ePrint Archive, Report 2017/798 (2017). http://ia.cr/2017/798

[HKK+14] Huang, Y., Katz, J., Kolesnikov, V., Kumaresan, R., Malozemoff, A.J.: Amortizing garbled circuits. In: Garay, J.A., Gennaro, R. (eds.) CRYPTO 2014. LNCS, vol. 8617, pp. 458–475. Springer, Heidelberg (2014). https://doi.org/10.1007/978-3-662-44381-1_26

[Kő31] Kőnig, D.: Gráfok és mátrixok. Matematikai és Fizikai Lapok **38**, 116–119 (1931)

[KM11] Katz, J., Malka, L.: Constant-round private function evaluation with linear complexity. In: Lee, D.H., Wang, X. (eds.) ASIACRYPT 2011. LNCS, vol. 7073, pp. 556–571. Springer, Heidelberg (2011). https://doi.org/10.1007/978-3-642-25385-0_30

[KR11] Kamara, S., Raykova, M.: Secure outsourced computation in a multi-tenant cloud. In: IBM Workshop on Cryptography and Security in Clouds (2011)

[KS08a] Kolesnikov, V., Schneider, T.: Improved garbled circuit: free XOR gates and applications. In: Aceto, L., Damgård, I., Goldberg, L.A., Halldórsson, M.M., Ingólfsdóttir, A., Walukiewicz, I. (eds.) ICALP 2008. LNCS, vol. 5126, pp. 486–498. Springer, Heidelberg (2008). https://doi.org/10.1007/978-3-540-70583-3_40

[KS08b] Kolesnikov, V., Schneider, T.: A practical universal circuit construction and secure evaluation of private functions. In: Tsudik, G. (ed.) FC 2008. LNCS, vol. 5143, pp. 83–97. Springer, Heidelberg (2008). https://doi.org/10.1007/978-3-540-85230-8_7

[KS16] Kiss, Á., Schneider, T.: Valiant's universal circuit is practical. In: Fischlin, M., Coron, J.-S. (eds.) EUROCRYPT 2016. LNCS, vol. 9665, pp. 699–728. Springer, Heidelberg (2016). https://doi.org/10.1007/978-3-662-49890-3_27

[LMS16] Lipmaa, H., Mohassel, P., Sadeghian, S.S.: Valiant's universal circuit: improvements, implementation, and applications. Cryptology ePrint Archive, Report 2016/017 (2016). http://ia.cr/2016/017

[LP09] Lovász, L., Plummer, M.D.: Matching Theory. AMS Chelsea Publishing Series. American Mathematical Society, Providence (2009)

[LR15] Lindell, Y., Riva, B.: Blazing fast 2PC in the offline/online setting with security for malicious adversaries. In: CCS 2015, pp. 579–590. ACM (2015)

[MNPS04] Malkhi, D., Nisan, N., Pinkas, B., Sella, Y.: Fairplay - secure two-party computation system. In: USENIX Security 2004, pp. 287–302. USENIX (2004)

[MS13] Mohassel, P., Sadeghian, S.: How to hide circuits in MPC an efficient framework for private function evaluation. In: Johansson, T., Nguyen, P.Q. (eds.) EUROCRYPT 2013. LNCS, vol. 7881, pp. 557–574. Springer, Heidelberg (2013). https://doi.org/10.1007/978-3-642-38348-9_33

[MSS14] Mohassel, P., Sadeghian, S., Smart, N.P.: Actively secure private function evaluation. In: Sarkar, P., Iwata, T. (eds.) ASIACRYPT 2014. LNCS, vol. 8874, pp. 486–505. Springer, Heidelberg (2014). https://doi.org/10.1007/978-3-662-45608-8_26

[NPS99] Naor, M., Pinkas, B., Sumner, R.: Privacy preserving auctions and mechanism design. In: ACM Conference on Electronic Commerce (EC 1999), pp. 129–139. ACM (1999)

[NSMS14] Niksefat, S., Sadeghiyan, B., Mohassel, P., Sadeghian, S.S.: ZIDS: a privacy-preserving intrusion detection system using secure two-party computation protocols. Comput. J. 57(4), 494–509 (2014)

[OI05] Ostrovsky, R., Skeith, W.E.: Private searching on streaming data. In: Shoup, V. (ed.) CRYPTO 2005. LNCS, vol. 3621, pp. 223–240. Springer, Heidelberg (2005). https://doi.org/10.1007/11535218_14

[PKV+14] Pappas, V., Krell, F., Vo, B., Kolesnikov, V., Malkin, T., Geol Choi, S., George, W., Keromytis, A.D., Bellovin, S.: Blind seer: a scalable private DBMS. In: IEEE S&P 2014, pp. 359–374. IEEE (2014)

[Sch08] Schneider, S.: Practical secure function evaluation. Master's thesis, University Erlangen-Nürnberg, Germany, February 2008

[Sha49] Shannon, C.: The synthesis of two-terminal switching circuits. Bell Labs Tech. J. 28(1), 59–98 (1949)

[SS08] Sadeghi, A.-R., Schneider, T.: Generalized universal circuits for secure evaluation of private functions with application to data classification. In: Lee, P.J., Cheon, J.H. (eds.) ICISC 2008. LNCS, vol. 5461, pp. 336–353. Springer, Heidelberg (2009). https://doi.org/10.1007/978-3-642-00730-9_21

[TS15]   Tillich, S., Smart, N.: Circuits of basic functions suitable for MPC and FHE (2015). https://www.cs.bris.ac.uk/Research/CryptographySecurity/MPC/

[Val76] Valiant, L.G.: Universal circuits (preliminary report). In: STOC 1976, pp. 196–203. ACM (1976)

[Wak68] Waksman, A.: A permutation network. J. ACM **15**(1), 159–163 (1968)

[Weg87] Wegener, I.: The complexity of Boolean functions. Wiley-Teubner (1987)

[Yao86] Yao, A.C.-C.: How to generate and exchange secrets (extended abstract). In: FOCS 1986, pp. 162–167. IEEE (1986)

[Zim15] Zimmerman, J.: How to obfuscate programs directly. In: Oswald, E., Fischlin, M. (eds.) EUROCRYPT 2015. LNCS, vol. 9057, pp. 439–467. Springer, Heidelberg (2015). https://doi.org/10.1007/978-3-662-46803-6_15

Efficient Scalable Constant-Round MPC via Garbled Circuits

Aner Ben-Efraim[1]([⊠]), Yehuda Lindell[2], and Eran Omri[3]

[1] Ben-Gurion University, Be'er Sheva, Israel
anermosh@post.bgu.ac.il
[2] Bar-Ilan University, Ramat Gan, Israel
lindell@biu.ac.il
[3] Ariel University, Ariel, Israel
omrier@ariel.ac.il

Abstract. In the setting of secure multiparty computation, a set of mutually distrustful parties carry out a joint computation of their inputs, without revealing anything but the output. Over recent years, there has been tremendous progress towards making secure computation practical, with great success in the two-party case. In contrast, in the multiparty case, progress has been much slower, even for the case of semi-honest adversaries.

In this paper, we consider the case of constant-round multiparty computation, via the garbled circuit approach of BMR (Beaver et al., STOC 1990). In recent work, it was shown that this protocol can be efficiently instantiated for semi-honest adversaries (Ben-Efraim et al., ACM CCS 2016). However, it scales very poorly with the number of parties, since the cost of garbled circuit evaluation is *quadratic* in the number of parties, per gate. Thus, for a large number of parties, it becomes expensive. We present a new way of constructing a BMR-type garbled circuit that can be evaluated with only a *constant* number of operations per gate. Our constructions use key-homomorphic pseudorandom functions (one based on DDH and the other on Ring-LWE) and are concretely efficient. In particular, for a large number of parties (e.g., 100), our new circuit can be evaluated faster than the standard BMR garbled circuit that uses only AES computations. Thus, our protocol is an important step towards achieving concretely efficient large-scale multiparty computation for Internet-like settings (where constant-round protocols are needed due to high latency).

Keywords: Garbled circuits · Constant round MPC · Key-homomorphic PRFs · Concrete efficiency

The first author was supported by ISF grants 544/13 and 152/17, by a grant from the BGU Cyber Security Research Center, and by the Frankel Center for Computer Science. The second author is supported by the European Research Council under the ERC consolidators grant agreement n. 615172 (HIPS) and by the BIU Center for Research in Applied Cryptography and Cyber Security in conjunction with the Israel National Cyber Bureau in the Prime Minister's Office. The third author is supported by a grant from the Israeli Science and Technology ministry and by a Israel Science Foundation grant 544/13.

T. Takagi and T. Peyrin (Eds.): ASIACRYPT 2017, Part II, LNCS 10625, pp. 471–498, 2017.
https://doi.org/10.1007/978-3-319-70697-9_17

1   Introduction

1.1   Background

Protocols for secure multiparty computation enable a set of parties to carry out a joint computation on private inputs, without revealing anything but the output. In the 1980s, powerful feasibility results were presented, showing that any polynomial-time function can be securely computed [9,18,37]. These feasibility hold both for *semi-honest* adversaries (who follow the protocol specification, but try to learn more than allowed by inspecting the transcript), and for *malicious* adversaries who can run any arbitrary adversarial strategy. Furthermore, protocols for constant-round secure computation were demonstrated both for the two-party case [37] and for the multiparty case [5]. These constant-round protocols work by constructing a garbled circuit, which is essentially an encrypted version of the circuit, that can be evaluated obliviously.

Over the past decade, there has been a major research effort to improve the efficiency of secure computation, with great success. For the two-party case, there are highly efficient protocols for both the semi-honest and malicious cases, and following both the garbled-circuit and secret sharing paradigms (see [2,6, 20,21,21,23,30,34,38] for just a handful of examples). As a result, it is possible to run secure two-party computation protocols in practice, for many real-world problems. We remark that a significant portion of the research effort to achieve efficient secure two-party computation focused on the simpler case of semi-honest adversaries. The results for this case proved to be crucial for obtaining efficient protocols for malicious adversaries as well. Thus, the study of efficiency for semi-honest adversaries has proved itself important in the goal of achieving stronger security as well.

In contrast to the aforementioned success for the specific case of two parties, in the setting of *multiparty* secure computation, with strictly more than two parties, progress has been much slower. In particular, for the case of constant-round protocols for many parties, no honest majority, and semi-honest adversaries, the only work has been in [7,8].[1] The recent work of [8] shows that constant-round secure multiparty computation can be achieved with good performance for the case of semi-honest adversaries. However, the technique of BMR [5] for obtaining constant-round protocols via a multiparty garbled circuit has an inherent scalability problem.

---

[1] There has been work for the malicious setting, with no honest majority [14,25,27], but these protocols are of course much more expensive. Very recently, the work of [19] showed how to extend the results of [8] to the malicious setting with very little overhead, and [36] also presented similar results for the malicious setting using a different protocol. These results suffer from the same scalability problem of [8] that we describe. We argue that it is important to go back to the semi-honest setting and improve efficiency, in order to enable future improvements for the malicious setting.

In addition, there has been work—e.g., in [13]—for the semi-honest setting that follows the GMW paradigm [18] and so has a number of rounds equal to the depth of the circuit being computed. Such protocols can perform very well in very fast networks, but not in Internet-like networks with high latency.

In order to understand this, we first remark that the BMR protocol can be divided into two phases: In the first phase, the parties run a secure protocol to construct a multiparty garbled circuit. This phase can be run even before the inputs are provided, and involves relatively heavy computation in order to securely build the garbled circuit. Then, in the second online phase, after the parties receive their inputs, the parties merely send keys on the input wires associated with their inputs, and then each party can locally evaluate the garbled circuit and obtain output. This paradigm is very attractive since the online phase requires almost no communication, and efficient local computation only. Note that the evaluation of the garbled circuit requires symmetric decryption operations only which are extremely fast in practice using AES-NI instructions. Despite the above, [8] discovered that even the local evaluation computation can become very expensive when the number of parties is large. The reason for this is that each gate requires $O(n^2)$ AES operations, when the number of parties is n. Thus, for a large number of parties—say $n = 100$—the number of AES operations per gate is 10,000. Therefore, the cost of evaluating a BMR garbled circuit for 100 parties is about *10,000 times higher* than the cost of evaluating a two-party (Yao) garbled circuit.

1.2   Our Results

Motivated by the results in [8] and the inherent scalability problem with BMR garbled circuits, our aim in this paper is to construct a variant of the BMR garbled circuit that scales well as the number of parties grow. We remark that if one only focuses on the problem of the cost of the online phase, then the scalability problem can be easily solved. Specifically, one can use any generic multiparty protocol like that of [18] to securely compute a standard Yao two-party garbled circuit, with no party knowing any of the actual keys on the wires (and the parties receiving secret shares of the keys on the input wires). Then, in the online phase, they merely need to exchange shares on the input wires, and can compute the output by evaluating a standard two-party garbled circuit. From a theoretical perspective, this method has many attractive properties; amongst other things, the online time is independent of the number of parties. However, if we are interested in constructing *concretely efficient* protocols that can be implemented and run in practice, then this approach completely fails. The reason for this is that constructing a Yao garbled circuit via multiparty computation is completely unrealistic in practice. This is because the encryption function itself must be computed inside the secure computation, multiple times for every gate.

The above leads us to the following important research goal:

Design a new BMR-type garbled circuit that can be constructed securely with concrete efficiency in the offline phase, and can be efficiently evaluated in the online phase at a cost that is independent of the number of parties.

As discussed above, our goal is *concrete efficiency*, and thus we are interested in obtaining constructions that can be implemented and run in practice, and are faster than previous approaches. Thus, our goal is to obtain a method that is strictly faster than the optimized version of [8] for the case of a large number of parties.

Our method for achieving the above goal utilizes *key-homomorphic pseudo-random functions* (KHPRF), introduced by Naor et al. [31] in the random oracle model, and by Boneh et al. [10] in the standard model. Informally speaking, a pseudorandom function is key-homomorphic if there exist appropriate operations $\widetilde{+}$, $\widetilde{\cdot}$ such that for every pair of keys k_1, k_2 and every input x, it holds that $F_{k_1 \widetilde{+} k_2}(x) = F_{k_1}(x) \widetilde{\cdot} F_{k_2}(x)$. Intuitively, this means that n parties with independent keys $k_1, \ldots, k_n$ can compute $F_K(x)$ for $K = k_1 \widetilde{+} \cdot \widetilde{+} k_n$, by each locally computing $F_{k_i}(x)$, and then using secure computation to compute $F_K(x) = F_{k_1}(x) \widetilde{\cdot} \cdots \widetilde{\cdot} F_{k_n}(x)$. We now informally explain how this can be used to construct a scalable BMR-type circuit.

In a BMR garbled circuit, for every wire w, each party P_i chooses two keys $k_{w,i}^0, k_{w,i}^1$. Then, a garbled gate with input wires u, v and output wire w is constructed by masking all of the keys on the output wire with the appropriate keys on the input wires. For example, in an AND gate, the values $(k_{w,1}^0, \ldots, k_{w,n}^0)$ need to be masked with the combinations of $\left((k_{u,1}^0, \ldots, k_{u,n}^0), (k_{v,1}^0, \ldots, k_{v,n}^0) \right)$, $\left((k_{u,1}^0, \ldots, k_{u,n}^0), (k_{v,1}^1, \ldots, k_{v,n}^1) \right)$, and $\left((k_{u,1}^1, \ldots, k_{u,n}^1), (k_{v,1}^0, \ldots, k_{v,n}^0) \right)$, whereas the values $(k_{w,1}^1, \ldots, k_{w,n}^1)$ need to be masked with $\left((k_{u,1}^1, \ldots, k_{u,n}^1), (k_{v,1}^1, \ldots, k_{v,n}^1) \right)$. This ensures that if the parties have the appropriate keys on the input wires of the gate, then they will obtain the appropriate keys on the output wire of the gate. Now, in order to ensure security, it must be that every *single* key on the input wires suffices to mask the output. Thus, for example, each of $k_{u,i}^0$ and $k_{v,i}^0$ must mask all of $(k_{w,1}^0, \ldots, k_{w,n}^0)$. This is achieved by setting the ciphertext $C_{0,0}$, associated with inputs $(0,0)$, to be

$$
C_{0,0} = \left(\bigoplus_{i=1}^{n} F_{k_{u,i}^0}(g\|1)\| \ldots \|F_{k_{u,i}^0}(g\|n) \right) \oplus \left(\bigoplus_{i=1}^{n} F_{k_{v,i}^0}(g\|1)\| \ldots \|F_{k_{v,i}^0}(g\|n) \right)
$$
$$
\oplus \left(k_{w,1}^0\| \ldots \|k_{w,n}^0 \right),
$$

where $\|$ denotes concatenation, g is the gate identity, and F is a pseudorandom function (in practice, AES). Each gate is then constructed as four ciphertexts, for all of the four combinations of input values. Observe that using this method, if party P_i alone is honest, then its single key suffices for masking the output (because the pseudorandom function is used to obtain a long pseudorandom string which masks the keys on the output wire).

Given the above, it is now clear that in order to evaluate a garbled gate, the parties need to invoke the pseudorandom function $2n^2$ times. Specifically, given keys on the inputs wires $\left(k_{u,1}^0\| \ldots \|k_{u,n}^0 \right)$ and $\left(k_{v,1}^0\| \ldots \|k_{v,n}^0 \right)$, the pseudorandom function is invoked n times for each of the $2n$ keys. Concretely, for $n = 100$, this means that 20,000 pseudorandom computations are made *for every gate*. In the two-party case, only two invocations are needed (or one, using the fixed-key method of [6]).

Now, consider the possibility of constructing ciphertexts as above, but using a key-homomorphic pseudorandom function instead. Concretely, we now define the ciphertext $C_{0,0}$ as follows:

$$C_{0,0} = \left(\widetilde{\varPi}_{i=1}^{n}(F_{k_{u,i}^0}(g))\right) \,\widetilde{\cdot}\, \left(\widetilde{\varPi}_{i=1}^{n}(F_{k_{v,i}^0}(g))\right) \,\widetilde{\cdot}\, \left(k_{w,1}^0 \,\widetilde{+}\, \ldots \,\widetilde{+}\, k_{w,n}^0\right),$$

where $\widetilde{\cdot}$, $\widetilde{+}$ are the key-homomorphic operations informally defined above and $\widetilde{\varPi}_{i=1}^{n}(y_i) \overset{\text{def}}{=} y_1 \,\widetilde{\cdot}\, \ldots \,\widetilde{\cdot}\, y_n$. Intuitively, such a ciphertext could be securely computed in the offline phase at a cost that is comparable to the original BMR ciphertext, by replacing $\oplus$ with the $\widetilde{\cdot}$ operation. Of course, in order to fulfill our goal, the offline phase must also be concretely efficient and thus we do indeed show that this equation can be efficiently computed securely. Now, the important observation is that the ciphertext $C_{0,0}$ above is actually *equal* to

$$C_{0,0} = \left(F_{K_u^0 \,\widetilde{+}\, K_v^0}(g)\right) \,\widetilde{\cdot}\, K_w^0$$

where $K_u^0 = k_{u,1}^0 \,\widetilde{+}\, \ldots \,\widetilde{+}\, k_{u,n}^0$, $K_v^0 = k_{v,1}^0 \,\widetilde{+}\, \ldots \,\widetilde{+}\, k_{v,n}^0$, and $K_w^0 = k_{w,1}^0 \,\widetilde{+}\, \ldots \,\widetilde{+}\, k_{w,n}^0$. Thus, the result is a garbled circuit that can be evaluated using only one invocation of the pseudorandom function, *irrespective of the number of parties*. It is important to note that key-homomorphic pseudorandom functions are much more expensive to compute than a plain pseudorandom function. Nevertheless, by implementing and running a comparison with the code of [8], we show that for a large number of parties—say $n = 100$—it is faster to compute a key-homomorphic pseudorandom function than $2n^2$ AES computations (even using AES-NI with a fixed key, as first proposed in [6]).

We mention that [1] used key-homomorphic properties of the LWE-based fully-homomorphic encryption schemes of [11,12] to obtain a secure multiparty protocol (in the CRS model) with only three rounds of interaction and communication complexity that is independent of the underlying function. However, their construction utilizes fully homomorphic encryption, and thus requires parties to locally preform very heavy computation. Furthermore, the intensive part of the computation requires the encryption of the inputs, and therefore done in the online phase. Thus, it is less relevant for the goal of concrete efficiency in the online phase that we consider in this work.

Instantiations and implementation. We present two concrete instantiations of our protocol, using two different key-homomorphic pseudorandom functions; the first is secure under the DDH assumption, and the second is secure assuming Ring-LWE. For each instantiation, we describe a concretely efficient protocol for securely generating the appropriate multiparty garbled circuit in the offline phase. We implemented the online version of our protocols, which is dominated by the local evaluation of the garbled circuit. In Sect. 6, we describe our implementation and results. Figure 1 contains a graph of the online circuit evaluation time for different schemes: BMR refers to the original BMR circuit and is clearly quadratic; all of the other lines are different versions of our protocol, and the

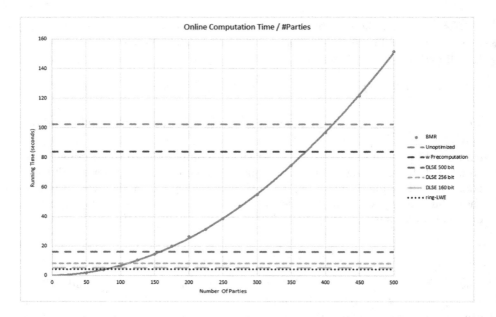

Fig. 1. Online computation time

circuits have evaluation time that is independent of the number of parties. As we mentioned above, this clearly demonstrates that even though the original BMR circuit uses only very fast primitives (AES), it runs slower than all other schemes when enough parties participate.

We remark that the "unoptimized" and "with precomputation" lines refer to a construction based on the standard DDH assumption with a 1024-bit safe prime, whereas the DLSE lines refer to an instantiation based on the assumption that DDH is still hard with *short exponents*; larger primes were also tested and the results appear in Sect. 6. In [24], it was proven that the hardness of DDH with short exponents is implied by the standard DDH assumption and the assumption that the discrete log problem is hard with short exponents. We prove variants of this that are needed for our optimized key-homomorphic pseudorandom function. The "Ring-LWE" refers to our instantiation based on Ring-LWE.

Paper organization. In Sect. 2, we recall the basic definitions required in this work, including those of secure multiparty computation, pseudo-random functions, and the Ring-LWE, DDH and DLSE problems. In Sect. 3, we describe our general paradigm construction, prove its correctness and state our main security theorem. The proof of security will appear in the full version. In Sect. 4, we show how to instantiate our general paradigm based on the DDH problem: In Sect. 4.1, we review the DDH and DLSE problems and prove a few statements. In Sect. 4.2, we describe an instantiation in the random oracle model. In Sect. 4.3, we show an instantiation without a random oracle, and a significantly optimized instantiation based on DDH and DLSE. In Sect. 4.4, we describe two possible offline protocols, one based on the BGW protocol [9] assuming an honest majority, and another that is secure up to $n - 1$ corrupt parties in the OT-hybrid model. In Sect. 5, we explain an instantiation based on Ring-LWE: in Sect. 5.1 we explain the online phase. In Sect. 5.2 we explain a "tailored" offline protocol that runs in time quasilinear in the number of parties. In Sect. 6, we provide experimental results of the online computation time of our protocols and a comparison with the online computation time of BMR.

2   Preliminaries

A function $\mu : \mathbb{N} \to \mathbb{N}$ is **negligible** if for every positive polynomial $p(\cdot)$ and all sufficiently large κ it holds that $\mu(\kappa) < \frac{1}{p(\kappa)}$. We use the abbreviation PPT to denote probabilistic polynomial-time. A *probability ensemble* $X = \{X_\kappa\}_{\kappa \in \mathbb{N}}$ is an infinite sequence of random variables indexed by κ. Two distribution ensembles $X = \{X_\kappa\}_{\kappa \in \mathbb{N}}$ and $Y = \{Y_\kappa\}_{\kappa \in \mathbb{N}}$ are **computationally indistinguishable**, denoted $X \stackrel{c}{\equiv} Y$, if for every PPT D, there exists a negligible function $\mu(\cdot)$ such that for every $\kappa \in \mathbb{N}$, $\left| \Pr\left[D(X_\kappa, 1^\kappa)) = 1\right] - \Pr\left[D(Y_\kappa, 1^\kappa) = 1\right] \right| < \mu(\kappa)$.

For a distribution D over a finite set A, we let $x \leftarrow D$ denote the selection of an element $x \in A$ according to distribution D. If D is the uniform distribution over A, we simply write $x \leftarrow A$. For $i \in \mathbb{N}$, we denote by U_i the random variable defined by $x \leftarrow \{0,1\}^i$. For an integer ℓ, we denote $[\ell] = \{1, \dots, \ell\}$.

2.1   The DLSE and the DDH Problems over Short Exponents

The DLSE (discrete logarithm over short exponents) problem was first introduced in [32]. Following the presentation of this problem in [24], we provide a parameterized version of it. Let $\kappa \in \mathbb{N}$ be the security parameter, let q be a κ-bit prime, and let $c \in \mathbb{N}$ is such that $0 \le c < \kappa$.

Notation 1. *For* $0 \le c \le \kappa$, *let* $R_{\kappa-c} = \{2^{\kappa-c}u \mid 0 \le 2^{\kappa-c}u < q\}$.

Specifically, $R_\kappa = \mathbb{Z}_q$. As we will see below, the set $R_{\kappa-c}$ will denote the domain from which exponents are chosen in the short discrete log and DDH problems.

Let $\mathcal{G}$ be a generation algorithm that on input 1^κ returns a triplet $(\mathbb{G}, q, h)$, where $\mathbb{G}$ is a cyclic group of order q (with q of length κ), and h is a generator of $\mathbb{G}$. The discrete-log over short exponents problem, denoted DLSE_c, is defined as follows:

Definition 1 (discrete logarithm over short exponents). *Let* $c \in \mathbb{N}$ *be a constant. The* DLSE_c *problem is hard with* $\mathcal{G}$ *if for all* PPT *algorithms* $\mathcal{A}$, *there exists a negligible function* $\mu(\cdot)$, *such that*

$$\Pr_{(\mathbb{G},q,h) \leftarrow \mathcal{G}(1^\kappa), v \leftarrow R_{\kappa-c}} [\mathcal{A}(\mathbb{G}, q, h, h^v) = v] \le \mu(\kappa).$$

The standard DL problem is DLSE_c with $c = \kappa$, and the hardness of DLSE_c clearly depends on c. Using similar notation as above, we define the DDH (decisional Diffie-Hellman) problem over short exponents, as first considered by Koshiba and Kurosawa [24].

Definition 2 (decisional Diffie-Hellman over short exponents). *Let* $c_1, c_2 \in \mathbb{N}$ *be constants. Then, the* DDH_{c_1, c_2} *problem is hard with* $\mathcal{G}$, *if*

$$\{(\mathbb{G}, q, h, h^x, h^y, h^{xy})\}_{\kappa \in \mathbb{N}} \overset{c}{\equiv} \{(\mathbb{G}, q, h, h^x, h^y, h^z)\}_{\kappa \in \mathbb{N}},$$

where the distributions are generated by choosing $(\mathbb{G}, q, h) \leftarrow \mathcal{G}(1^\kappa)$, *and then choosing* $x, y, z \leftarrow R_{\kappa - c_1} \times R_{\kappa - c_2} \times \mathbb{Z}_q$.

The DDH_{c_1, c_2} is also referred to as the (short-short)-DDH if $c_1 = c_2$ are much smaller than κ, and as the (short-full)-DDH if c_1 is much smaller than κ and $c_2 = \kappa$ (i.e., y is uniform in $\mathbb{Z}_q$). The standard DDH problem is $\text{DDH}_{\kappa, \kappa}$.

In [24], it was shown that if both DDH and DLSE are hard then so are the the (short-full) -DDH and (short-short)-DDH, and further that the converse also holds. That is, amongst other things, they proved the following:

Theorem 2 *[24, Theorem 2]. Let* $\mathcal{G}$ *be a generation algorithm and let* $c \in \mathbb{N}$. *If both* DDH *and* DLSE_c *are hard with* $\mathcal{G}$, *then* $\text{DDH}_{c, \kappa}$ *is hard with* $\mathcal{G}$.

2.2   The Ring-LWE Problem

We briefly state a simple variation of the Ring-LWE hardness assumption. A more complete definition can be found in [28]. Let $p = 2N + 1$ be a prime, where N, called the dimension or security parameter, is a power of two. We fix the polynomial ring $\mathcal{R}_p = \mathbb{Z}_p[X]/(X^N + 1)$, i.e., the polynomials over $\mathbb{Z}_p$ modulo $X^N + 1$.

Definition 3. *The* decisional Ring-LWE *hardness assumption states that it is hard to distinguish between the following two sets of pairs:*

1. $\{(a_i, b_i)\}_{i \in I}$
2. $\{(a_i, a_i \cdot k + e_i)\}_{i \in I}$

where $\{a_i\}_{i \in I}, \{b_i\}_{i \in I}$, *and* k *are chosen uniformly at random from the ring, and* $\{e_i\}_{i \in I}$ *are chosen from a spherical Gaussian distribution. Furthermore, by transforming to the "Hermite normal form", the assumption holds also if* k *is chosen from a spherical Gaussian distribution as well.*

In general, it is necessary to bound $|I|$, i.e., the number of samples, usually by some $\ell = O(1)$ or $\ell = O(\log N)$. More information, generalizations, and improvements can be found in [28,29].

2.3 Key Homomorphic Pseudorandom Functions

Recall that a function is pseudorandom if no PPT adversary given oracle access to either the function with a randomly chosen key, or to a truly random function, can distinguish between the cases. A weak pseudorandom function is as above, except that the adversary only receives the function output on randomly chosen inputs.

We next recall the definition of key-homomorphic PRFs. Key-homomorphic PRFs were introduced by Naor et al. [31]. The formal definition and first key-homomorphic PRF without a random oracle, based on LWE, were introduced by Boneh et al. [10].

Definition 4. *A family of functions* $\{F_k \colon \mathcal{X} \to \mathcal{G}\}_{k \in \mathcal{D}}$ *is a family of* key-homomorphic functions *if the key domain,* $\mathcal{D}$*, and the image,* $\mathcal{G}$*, are equipped with group operations,* $(\mathcal{D}, \widetilde{+})$ *and* $(\mathcal{G}, \widetilde{\cdot})$*, respectively, such that for every* $k_1, k_2 \in \mathcal{D}$ *and* $x \in \mathcal{X}$ *it holds that* $F_{k_1 \widetilde{+} k_2}(x) = F_{k_1}(x) \widetilde{\cdot} F_{k_2}(x)$*. A* key-homomorphic PRF (KHPRF) *(resp.,* key-homomorphic weak PRF(KHWPRF)*) is a pseudorandom (resp., weak pseudorandom) function that is key-homomorphic.*

Let $\{F_k \colon \mathcal{X} \to \mathcal{G}\}_{k \in \mathcal{D}}$ be a family of key-homomorphic functions. Then, the inverse of an element $h \in \mathcal{G}$ is denoted $(h)^{-1}$. We denote $\widetilde{\Sigma}_{i=1}^m (k_i) \overset{\text{def}}{=} k_1 \widetilde{+} \cdots \widetilde{+} k_m$ and $\widetilde{\Pi}_{i=1}^m (h_i) \overset{\text{def}}{=} h_1 \widetilde{\cdot} \cdots \widetilde{\cdot} h_m$.

In general, it may be the case that $\mathcal{D} \neq \mathcal{G}$. As we shall see, this can pose some difficulty because keys from $\mathcal{D}$ are encrypted in $\mathcal{G}$. Furthermore, we also need to encrypt the bit necessary for point-and-permute. Therefore, we will assume throughout the existence of an efficiently computable function $f \colon \mathcal{D} \times \{0, 1\} \to \mathcal{G}$, with an efficiently computable inverse f^{-1}. We note that for all known KHPRFs, such a function exists.

2.4 Secure Multiparty Computation

We follow the standard definition of secure multiparty computation for semi-honest adversaries, as it appears in [17]. In brief, an n-party protocol π is defined by n interactive probabilistic polynomial-time Turing machines $P_1, \ldots, P_n$, called parties. The parties hold the security parameter 1^κ as their joint input and each party P_i holds a private input x_i. The computation proceeds in rounds. In each round j of the protocol, each party sends a message to each of the other parties (and receives messages from all other parties). The number of rounds in the protocol is expressed as some function $r(\kappa)$ in the security parameter.

The view of a party in an execution of the protocol contains its private input, its random string, and the messages it received throughout this execution. The random variable $\text{view}_{P_i}^\pi (\boldsymbol{x}, 1^\kappa)$ describes the view of P_i when executing π on inputs $\boldsymbol{x} = (x_1, \ldots, x_n)$ (with security parameter κ). Here, x_i denotes the input of party P_i. The output an execution of π on $\boldsymbol{x}$ (with security parameter κ) is described by the random variable $\text{Output}^\pi (\boldsymbol{x}, 1^\kappa) = \left(\text{Output}_{P_1}^\pi (\boldsymbol{x}, 1^\kappa), \ldots, \text{Output}_{P_n}^\pi (\boldsymbol{x}, 1^\kappa) \right)$, where $\text{Output}_P^\pi (\boldsymbol{x}, 1^\kappa)$ is the output of party P in this execution, and is implicit in the view of P.

Similarly, for a set of parties with indices $I \subseteq [n]$, we denote by $\boldsymbol{x}_I$ the set of their inputs, by $\mathrm{view}_I^\pi(\boldsymbol{x}, 1^\kappa)$ their joint view, and by $\mathrm{Output}_I^\pi(\boldsymbol{x}, 1^\kappa)$ their joint output. In the setting of this work, it suffices to consider deterministic functionalities. We therefore provide the definition of security only for deterministic functionalities; see [17] for a motivating discussion regarding the definition.

Definition 5 (security for deterministic functionalities). *A protocol π t-securely computes a deterministic functionality $f\colon (\{0,1\}^*)^n \mapsto (\{0,1\}^*)^n$ in the presence of semi-honest adversaries if the following hold:*

Correctness: *For every $\kappa \in \mathbb{N}$ and every n-tuple of inputs $\boldsymbol{x} = x_1, \ldots, x_n$, it holds that*

$$\Pr\left[\mathrm{Output}^\pi(\boldsymbol{x}, 1^\kappa) = f(\boldsymbol{x})\right] = 1, \tag{1}$$

where the probability is taken over the random coins of the parties.

Privacy: *There exists a probabilistic polynomial-time (in the security parameter) algorithm $\mathcal{S}$ (called "simulator"), such that for every subset $I \subseteq [n]$ of size at most t:*

$$\left\{\mathcal{S}_A\left(\boldsymbol{x}_I, f_I(\boldsymbol{x}), 1^\kappa\right)\right\}_{\boldsymbol{x}\in(\{0,1\}^*)^n;\kappa\in\mathbb{N}} \overset{C}{\equiv} \left\{\mathrm{view}_I^\pi\left(\boldsymbol{x}, 1^\kappa\right)\right\}_{\boldsymbol{x}\in(\{0,1\}^*)^n;\kappa\in\mathbb{N}}. \tag{2}$$

3   Multiparty Garbled Circuits via Key-Homomorphic PRFs

In this section, we describe a general paradigm for constructing garbled circuits for many parties via key-homomorphic PRFs. As explained above, the goal is to allow parties to efficiently construct a multiparty garbled circuit in which the number of decryptions per gate and the size of the decryption keys are independent of the number of parties.

In Sect. 3.1 we define our main offline functionality, $\mathcal{F}_{GC}$, that constructs a garbled circuit from a key-homomorphic PRF. In Sect. 3.2 we describe an online protocol for computing outputs using $\mathcal{F}_{GC}$. In Sect. 3.3 we show the correctness of the protocol and we state conditions when $\mathcal{F}_{GC}$ is secure, i.e., when a secure implementation of F_{GC}, along with the online protocol, can be used as a secure multiparty protocol for computing any boolean circuit. The proof of security will appear in the full version.

In the following, let C denote the Boolean circuit and let $|C|$ denote the number of gates in C. For every gate g in C, we let $g(\alpha, \beta)$ denote the gate operation on $\alpha, \beta \in \{0, 1\}$. We further abuse notation by also using g to denote the index of the gate in a fixed topological ordering of C.

The parties are denoted $P_1, \ldots, P_n$, where n denotes the number of parties. The set of all wires in the circuit is denoted by W. For each wire $\omega \in W$ each party P_i will hold two keys $k_{\omega,0}^i, k_{\omega,1}^i \in \mathcal{D}$, which we will call the *individual keys*. The "summations" of the individual keys, $k_{w,0} \overset{\text{def}}{=} \widetilde{\Sigma}_{i=1}^n(k_{w,0}^i)$ and $k_{w,1} \overset{\text{def}}{=} \widetilde{\Sigma}_{i=1}^n(k_{w,1}^i)$, will be termed the *joint keys*.

We associate a hidden bit λ_ω to each wire $\omega \in W$, generally unknown to all parties. During the online phase, the parties reveal an external value bit e_ω for each wire ω. It will later become clear that $e_\omega = \lambda_\omega \oplus t_\omega$, where t_ω is the real bit of the wire ω in an ungarbled computation of C for the inputs provided by the parties.

3.1 The Offline Phase

In this section, we describe the main functionality of the offline phase of our protocol, Functionality $\mathcal{F}_{GC}$. This functionality constructs a garbled circuit from a family of key-homomorphic PRFs.

Functionality $\mathcal{F}_{GC}$ receives as public input a circuit C with wires W, a family of key-homomorphic functions $\{F_k\}$, and a set $x_0, \ldots, x_{4|C|-1} \in \mathcal{X}$. The functionality also receives from each party P_i two individual keys, $k_{\omega,0}^i$ and $k_{\omega,1}^i$, for each wire $\omega \in W$. The functionality associates to each wire $\omega \in W$ a hidden bit λ_ω, that is generally not revealed to any of the parties. The functionality then computes and outputs to the parties a version of Yao's garbled circuit, in which the keys for the encryption are the joint keys. More precisely, for each gate $g \in C$ with input wires u, v and output wire w and for every $\alpha, \beta \in \{0, 1\}$, the functionality secretly computes $e_{w,\alpha,\beta} \overset{\text{def}}{=} g((\lambda_u \oplus \alpha), (\lambda_v \oplus \beta)) \oplus \lambda_w$ and

$$KEY_{w,\alpha,\beta} \overset{\text{def}}{=} \begin{cases} k_{w,0} & e_{w,\alpha,\beta} = 0 \\ k_{w,1} & e_{w,\alpha,\beta} = 1. \end{cases}$$ The value $e_{w,\alpha,\beta}$ is equal to the external value

of the wire that would be revealed to the parties during evaluation, if the external values of u and v are α and β respectively. $KEY_{w,\alpha,\beta}$ is the joint key that corresponds to this external value.

The functionality outputs to all parties the garbled gates

$$\tilde{g}_{\alpha,\beta} = \left(F_{k_{u,\alpha} \mp k_{v,\beta}} (x_{4g+2\alpha+\beta}) \right)^{\widetilde{-1}} \cdot f(KEY_{w,\alpha,\beta}, e_{w,\alpha,\beta}).$$

The functionality also outputs to each party the λ's associated with its input and output wires. The full details of Functionality $\mathcal{F}_{GC}$ appear in Fig. 2.

3.2 The Online Phase

In this section, we describe our online phase protocol. In this phase, the parties exchange garbled inputs and compute the output of the garbled circuit. The general flow of the online phase is the same as in the BMR protocol. Specifically, it contains two short rounds of communication, in which the parties learn the keys for input wires. From then on, all computations are done locally.

In the first step of the protocol, each party computes and broadcasts the external values $e_\omega = \lambda_\omega \oplus b_\omega$ for each of its input wires ω. This is possible, as each party knows the real value b_ω of each of its input wires, and the party knows λ_ω of its input wires from the output of functionality $\mathcal{F}_{GC}$.

In the next step, the parties broadcast their individual keys corresponding to the external values for each of the input wires and compute the joint keys

---

Functionality $\mathcal{F}_{GC}$ for Constructing the Garbled Circuit

Public Inputs: a circuit C with wires W, a family of key-homomorphic functions $\{F_k\}$ and values $x_0, \ldots, x_{4|C|-1} \in \mathcal{X}$.
Private Inputs: Each party P_i gives as input two keys $k^i_{w,0}, k^i_{w,1} \in \mathcal{D}$ for every wire $w \in W$.
Denote $k_{w,0} \stackrel{def}{=} \widetilde{\Sigma}^n_{i=1}(k^i_{w,0})$ and $k_{w,1} \stackrel{def}{=} \widetilde{\Sigma}^n_{i=1}(k^i_{w,1})$.
Random Input: To each wire $\omega \in W$ of the circuit, a hidden random bit $\lambda_\omega \in_R \{0, 1\}$ is associated.
Output: The functionality outputs to all parties the garbled circuit GC – for every gate $g \in C$, with input wires u, v, and output wire w, and for every $\alpha, \beta \in \{0, 1\}$, it outputs:

$$\tilde{g}_{\alpha,\beta} = \left(F_{k_{u,\alpha} \,\widetilde{+}\, k_{v,\beta}}(x_{4g+2\alpha+\beta}) \right)^{\widetilde{-}1} \,\widetilde{\cdot}\, f(KEY_{w,\alpha,\beta}, e_{w,\alpha,\beta}) \tag{3}$$

Where

$$e_{w,\alpha,\beta} \stackrel{def}{=} g((\lambda_u \oplus \alpha), (\lambda_v \oplus \beta)) \oplus \lambda_w \tag{4}$$

and

$$KEY_{w,\alpha,\beta} \stackrel{def}{=} k_{w,e_{w,\alpha,\beta}}. \tag{5}$$

Notice that

$$F_{k_{u,\alpha} \,\widetilde{+}\, k_{v,\beta}}(x_{4g+2\alpha+\beta}) = F_{\widetilde{\Sigma}^n_{i=1}(k^i_{u,\alpha} \,\widetilde{+}\, k^i_{v,\beta})}(x_{4g+2\alpha+\beta})$$
$$= \widetilde{\Pi}^n_{i=1}(F_{k^i_{u,\alpha} \,\widetilde{+}\, k^i_{u,\beta}}(x_{4g+2\alpha+\beta})). \tag{6}$$

In addition, each party P_i receives the λ's corresponding to its input and output wires.

Fig. 2. Functionality $\mathcal{F}_{GC}$

for these wires. Then, each party locally decrypts the correct row at each gate in topological order, and recovers the key and external value for the output wire of that gate. In this way, at the end of this step the parties recover the external values of the output wires. Finally, the parties then recover the real outputs of the function by XORing the output external values with the corresponding λ's, which they know from the output of functionality $\mathcal{F}_{GC}$.

3.3 Correctness and Security

In this section we show the correctness of using the online protocol in the $\mathcal{F}_{GC}$ hybrid model, and state condition under which this results in a secure multiparty protocol for any Boolean circuit. The proof of security is deferred to the full version.

Online Protocol

Public Inputs: a circuit C with wires W, a family of key-homomorphic functions $\{F_k\}$ and values $x_0, \ldots, x_{4|C|-1} \in \mathcal{X}$.

Private Inputs: All parties hold their input to the functionality F_{GC} and the output they received from it.

In addidition, each party holds its input bit b_ω for each of its input wires ω.

Computation:

1. Each party broadcasts $e_\omega = \lambda_\omega \oplus b_\omega$ for the each of its input wires ω.
2. Each party P_i broadcasts k^i_{ω, e_ω} for each input wire of the circuit.
3. Each party does the following computation locally:
   (a) Computes $k_{\omega, e_\omega} = \widetilde{\Sigma}^n_{i=1}(k^i_{\omega, e_\omega})$ for every input wire ω of the circuit.
   (b) In topological order, for each gate $g \in C$, with input wires u, v and output wire w computes

   $$(k_{w, e_w}, e_w) = f^{-1}(F_{k_{u, e_u} \; \widetilde{+} \; k_{v, e_v}}(x_{4g + 2e_u + e_v}) \; \widetilde{\cdot} \; \tilde{g}_{e_u, e_v}), \qquad (7)$$

   and recovers the external value e_w and corresponding key k_{w, e_w} for the output wire w of the gate.
   (c) For each of its output wires ω computes the true value of the output $e_\omega \oplus \lambda_\omega$.

Output: Each party recovers the true value of the output, $e_\omega \oplus \lambda_\omega$, for each of its output wires ω.

Fig. 3. Online protocol

Correctness. We now show that the outputs received by the parties from the online phase corresponds to the correct outputs. By Step 3c of the online protocol, it follows from the following claim:

Claim 3. *For each $\omega \in W$, the external value e_ω revealed by the parties in the online protocol is equal to $\lambda_\omega \oplus t_\omega$, where t_ω is the true value of the wire ω in an ungarbled computation with the same inputs.*

The proof is by induction on the topological ordering of C. We give a sketch of the proof, omitting the proof of correctness of the decryption in Step 3b.

Proof Sketch. For the input wires, $e_\omega = \lambda_\omega \oplus t_\omega$ follows from Step 1 of the online protocol. For any other wire $\omega \in W$ that is the output wire of gate g, with input wires u, v, we have from Step 3b of the online protocol, Eqs. (3) and (4) in functionality $\mathcal{F}_{GC}$, and the induction assumption, that

$$e_\omega = e_{\omega, e_u, e_v} = g((\lambda_u \oplus e_u), (\lambda_v \oplus e_v)) \oplus \lambda_\omega = g(t_u, t_v) \oplus \lambda_\omega = t_\omega \oplus \lambda_\omega. \qquad \square$$

Security. We state conditions when a secure implementation of $\mathcal{F}_{GC}$, along with the online protocol, can be used as a secure multiparty protocol to compute any Boolean circuit C. We first give two definitions. Let $\{F_k: \mathcal{X} \to \mathcal{G}\}_{k \in \mathcal{D}}$ be a family of functions, and let $R: \mathcal{X} \to \mathcal{G}$ be random functions.

Definition 6. *For $x_1, \ldots, x_n \in \mathcal{X}$, we say that Property $I(x_1, \ldots, x_n)$ holds for $\{F_k\}$ if for a randomly chosen $k \in \mathcal{D}$, it holds that*

$$\{(x_i, F_k(x_i))\}_{i=1}^n \stackrel{c}{\equiv} \{(x_i, R(x_i))\}_{i=1}^n. \tag{8}$$

Note that if $\{F_k\}$ is a PRF then Property $I(x_1, \ldots, x_n)$ holds for *any* choice of $x_1, \ldots, x_n$. If $x_1, \ldots, x_n$ are *random*, then Property $I(x_1, \ldots, x_n)$ holds also if $\{F_k\}$ is a weak PRF.

Let C be a Boolean circuit with set of wires W.

Definition 7. *For $x_0, \ldots, x_{4|C|-1} \in \mathcal{X}$ (possibly with $x_i = x_j$ for $i \neq j$), we say that Property $J(x_0, \ldots, x_{4|C|-1})$ holds for C if for every gate $g \in C$ it holds that $\{x_{4g}, x_{4g+1}, x_{4g+2}, x_{4g+3}\}$ are all distinct and for every wire $\omega \in W$ that is an input wire to two different gates $g_1, g_2 \in C$, it holds that*

$$\{x_i\}_{i=4g_1}^{4g_1+3} \cap \{x_i\}_{i=4g_2}^{4g_2+3} = \emptyset. \tag{9}$$

The idea of Definition 7 is that if two gates share a common input wire, they do not share the same x's. This is to ensure that the same PRF is not queried twice with the same input. We are now ready to state our main security theorem.

Theorem 4. *If Property $I(x_0, \ldots, x_{4|C|-1})$ holds for $\{F_k\}$ and Property $J(x_0, \ldots, x_{4|C|-1})$ holds for C, then the online protocol in Fig. 3 securely computes C in the semi-honest $\mathcal{F}_{GC}$-hybrid model with up to $n-1$ corrupt parties.*

Theorem 4 can be proved in a standard way, similar to the proofs of [26,27]. We will give a slightly different proof in the full version.

Remark 1. An important observation from the proof is that the security of the protocol relies on the individual keys, and not the joint keys. Intuitively, this is because the adversary knows many of the individual keys, and could thus, in certain circumstances, learn partial information on the joint key. For example, this observation is important in our DDH instantiation, where we restrict the individual keys. It is also important in our LWE instantiation, as we shall see in Sect. 5.

4 Explicit Instantiation Based on DDH

In this section we describe an explicit instantiation of our key-homomorphic PRF based MPC protocol that relies on the DDH assumption. The resulting encryption scheme used in the garbling can be seen as a variant of the ElGamal scheme. However, using the ElGamal scheme naïvely can be insecure. We show a simple example demonstrating this in the full version.

In Sect. 4.1, we prove some properties of the DDH problem and the DLSE (Discrete Log over Short Exponents) problem. In Sect. 4.2, we explain our instantiation relying on DDH and a Random Oracle. In Sect. 4.3, we explain how to remove the Random Oracle and also describe some optimizations. Some of these optimizations explicitly rely on the hardness of DLSE. In Sect. 4.4, we describe two possible offline protocols for our DDH instantiation: in Sect. 4.4.1 we describe a protocol based on BGW that is secure when assuming an honest majority. In Sect. 4.4.2 we describe an OT based protocol that is secure against up to $n-1$ corrupt parties.

4.1   Some Properties of DDH and DLSE

In this section we state a couple of properties of DDH and DLSE that we use in our instantiation. The proofs use theorems from [24,31], and will be supplied in the full version.

Notation 5. *For $0 \le c \le \kappa$, let $\widehat{R}_{\kappa-c} = \{u \mid 0 \le 2^{\kappa-c}u < q\}$.*

Notice that this notation differs from Notation 1 because here the most significant bits are zeros. However, it turns out that in prime order groups, the hardness of DDH with short exponenents is equivalent using Notation 5 or Notation 1. This enables us to prove the following variations of a theorem from [31].

Theorem 6. *Assuming the DDH problem is hard,*

1. *If $c = O(\log \kappa)$ then $\left\{F_k(x) = x^k\right\}_{k \in \widehat{R}_{\kappa-c}}$ is a key-homomorphic weak PRF.*
2. *If the DLSE_c problem is hard then $\left\{F_k(x) = x^k\right\}_{k \in \widehat{R}_{\kappa-c}}$ is a key-homomorphic weak PRF.*

Furthermore, if H is a random oracle whose images are generators of the group, then $\left\{F_k(x) = H(x)^k\right\}_{k \in \widehat{R}_{\kappa-c}}$ is a key-homomorphic PRF in the above two cases.

4.2   Concrete Instantiation

In this section we describe a concrete instantiation of our protocol, based on the hardness of DDH. We first assume also that using SHA256 as described below is a random oracle. In Sect. 4.3, we explain how to avoid the random oracle assumption.

Our concrete implementation of our protocol is as follows. Let $p = 2q + 1$ be a safe prime with κ bits, and denote by $\mathbb{G}_q$ the subgroup of order q of the multiplicative group $\mathbb{Z}_p^*$, i.e., the group of quadratic residues. We let our PRF family $\{F_k\}$ be $F_k(x) = (H(x))^k \mod p$, where H is a random oracle modeled by $H(x) \stackrel{\text{def}}{=} (\mathrm{SHA}(r\|x\ell)\| \cdots \|\mathrm{SHA}(r\|(x\ell + (\ell - 1))))^2 \mod p$ with $\ell = \left\lceil \frac{\kappa}{256} \right\rceil$, and r an agreed random nonce. Notice that $H(x)$ is a generator of $\mathbb{G}_q$.[2] The domain of the individual keys is initially $\mathbb{Z}_q$, but we slightly modify this below. We define $x_i = i$, $a \tilde{\ } b = a \cdot b \mod p$, $a + b = a + b \mod q$, and $f(a) = a^2 \mod p$, which is in $\mathbb{G}_q$ for any non-zero element $a \in \mathbb{Z}_p$.

We note some difficulties for using the instantiation above.

---

[2] The only possible exceptions are the values 0 and 1, but this happens with negligible probability.

1. The input of the function f should contain both the key and the external value.
2. The squaring function has 2 inverses for every entry. Therefore, as f needs to be invertible, there must be some way for the parties to decide which of the 2 inverses is correct.
3. The joint key is not well defined. Initially, the individual keys are drawn from $\mathbb{Z}_q$. The summation of the keys in the message to be encrypted will be modulo p, while the summation of the keys in the exponent will be modulo q (i.e., the size of the subgroup of the generator).

    Since the keys are chosen randomly, the summation of the individual keys of the parties in the message, which is the joint key revealed during the online phase, and the joint key in the exponent, which is the real key used for encryption/decryption, will, in general, not be equal. This will cause errors in the computation.

We solve the above problems by letting the individual keys be drawn from $\left\{0, \ldots, \frac{q-1}{4n}\right\} \subset \mathbb{Z}_q$. First, notice that the summation of the keys is now well defined as $k_{w,0} + k_{w,1} = \Sigma_{i=1}^{n} k_{w,0}^i + \Sigma_{i=1}^{n} k_{w,1}^i < q < p$ so it is equal modulo q and modulo p. Therefore, the key used for the decryption is equal to the sum of the keys that are revealed from the decryption of the input wires. By Theorem 6, Item 1, this choice of keys still renders our protocol secure.

To insert the external value bit, we multiply the key by two and add the external value bit, i.e., $f(KEY_{w,\alpha,\beta}, e_{w,\alpha,\beta}) = (2KEY_{w,\alpha,\beta} + e_{w,\alpha,\beta})^2$. Notice that $2KEY_{w,\alpha,\beta} + e_{w,\alpha,\beta} < \frac{q-1}{2}$. Restricted to this domain, the squaring function has a unique inverse in $\mathbb{Z}_p$, which can be computed in polynomial time.

The instantiated functionality $\mathcal{F}_{GC}$ and the online protocol are simply applying the above concrete instantiation, i.e., the key-homomorphic PRF based on DDH, to the respective figures in Sect. 3.

4.3 Removing the Random Oracle and Optimizations

In this section we first show how to remove the Random Oracle assumption. Then, we describe optimizations, some of which need to further assume DLSE_c for sufficiently small c.

Removing the random oracle. Recall that the use of the random oracle was necessary to show that $\{F_k\}$ is a PRF family. However, Theorem 4 does not explicitly require that $\{F_k\}$ be a PRF. If $x_i = h_i$ are agreed *random distinct* generators of $\mathbb{G}_q$, then defining $F_k(a) = a^k$ is sufficient: $\{F_k\}$ is a weak PRF, and thus Property $I(x_1, \ldots, x_{4|C|-1})$ holds because the generators are random. Property $J(x_1, \ldots, x_{4|C|-1})$ also holds because all the generators are distinct.

Furthermore, it is possible to require less than $4|C|$ distinct random generators – we can reuse the same random generators (i.e., $x_i = h_{r_i}$ where $r_i = r_j$ may happen for $i \neq j$), as long as Property $J(x_1, \ldots, x_{4|C|-1})$ also holds. I.e., as long as for every random generator h and individual key k, the pseudorandom value $F_k(h)$ is used only once for encryption (Eq. (3)) in all gates.

We note that the same individual key is always used more than once, both in two rows of each gate, and also in any other gate with the same input wire. However, we can bound the number of necessary random generators we need by 8 times the maximal fan-out of the circuit. The proof will be given in the full version. The proof gives a simple deterministic (with respect to the circuit and topological ordering) algorithm for deciding which generators to use at each gate. Furthermore, these agreed generators can be chosen once and used repeatedly for many different circuits.

We now move on to describing some optimizations to the above protocol. There are two main bottlenecks for the online computation time – computing the exponentiations h^k, and computing square root modulo p to recover the key and external value. We next show how to optimize these steps.

Precomputation. We point out that the set of generators used is already known in the offline phase. Thus, to speed up the exponentiations, the parties can precompute $h^2, h^4, , ..., h^{2^{\lfloor \log q \rfloor}}$ for each generator h. If the number of generators is small, (e.g., in the version without the random oracle described above), this will also significantly speed up the offline phase. If the same generators are used for multiple circuits then this precomputation can be done only once for all the circuits.

Optimizations based on hardness of DLSE. We now show how, by further assuming the $DLSE_c$ assumption for sufficiently small c, we can significantly improve the computation time. Clearly, assuming the keys are short significantly improves the time of exponentiations – if the domain of the keys is R_{n-c} then the exponentiations are approx. $\frac{\kappa}{c}$ times faster.

A second and less obvious optimization is that assuming $DLSE_c$ with $c < \frac{\kappa}{2} - \log n - 2$ significantly shortens the time of modular square root. This follows from the following observation.

Observation 7. *If $m \in \mathbb{N}$ is d bits long with $d < \frac{\kappa}{2}$, then $m^2 \mod p = m^2$.*

Following Observation 7, if the keys are short enough, we can replace taking square root modulo p by taking regular square root. The time of computing square root is several orders of magnitude faster than computing square root modulo p. Thus, this practically removes the time it takes to compute modular square root.

Remark 2. If we assume both DLSE and that the set of generators is small, then for computing $(h^k)^{-1}$ for a generator h and key k (such as needed in our offline protocols in Sect. 4.4), it is more efficient to compute exponentiation by a short key (k). This can be done via the equality $(h^k)^{-1} = (h^{-1})^k$, i.e., precomputing the tables also for the inverse of every generator h (of course, this makes sense only if the number of generators is small, as in the version without the random oracle).

4.4   Offline Phase Protocols for DDH Based Implementation

We now describe two different secure protocols for computing the offline phase of the instantiation based on DDH, one based on BGW [9] that requires an honest majority and another based on oblivious transfer that is secure up to $n-1$ corrupt parties.

The offline protocol for computing the version without the random oracle is identical except that in all locations $H(x_i)$ is replaced by h_{r_i} as explained in Sect. 4.3. If one assumes $DLSE_c$, then the individual keys should be drawn from $\{0, \ldots, \frac{q}{2^{\kappa-c}}\}$.

The current bottleneck in both suggested protocols is computing unbounded fan-in multiplication, i.e., computing Shamir or additive shares of $\Pi_{i=1}^{n} m_i$, where n is the number of parties and m_i is known only to party i. Thus, any improvement to protocols computing unbounded fan-in multiplication immediately implies an improvement to our offline protocols. Currently, the best constant round protocol for computing unbounded fan-in multiplication is the protocol given by Bar-Ilan and Beaver [4].

4.4.1   BGW Based Offline Protocol for the DDH Based F_{GC}

In this section we describe an offline protocol for our DDH instantiation that is based on the BGW protocol [9], which requires an honest majority. The running time of the described protocol is comparable to the running time of BGW based protocols for the offline phase of the BMR circuit, e.g., [7,8], when the number of parties is *large*.

Our BGW offline protocol is achieved by secret-sharing both the individual keys k^i and the exponentiations g^{-k_i} in Shamir secret-sharing, and then using the BGW protocol to compute the garbled gates

$$\tilde{g}_{\alpha,\beta} = \left(\left((H(4g + 2\alpha + \beta))^{k_{u,\alpha}+k_{v,\beta}} \right)^{-1} \cdot (2KEY_{w,\alpha,\beta} + e_{w,\alpha,\beta})^2 \right. \tag{10}$$

The main protocol is given in Fig. 4. The subprotocols are standard using the BGW protocol. Note that $\Pi_{i=1}^{n} m_i$ is computed in constant rounds using [4]. The protocol in Fig. 4 is based entirely on BGW, and therefore securely computes the functionality F_{GC} in the semi-honest model, assuming an honest majority.

4.4.2   OT Based Offline Protocol for the DDH Based F_{GC}

In this Section we describe a protocol for computing the DDH based functionality F_{GC} that is secure up to $n-1$ corrupt parties in the OT-hybrid model.

The basic observation for the OT protocol is that if two (not necessarily disjoint) sets of parties $\mathcal{P}_1$ and $\mathcal{P}_2$ hold additive shares (in $\mathbb{Z}_p$) of secrets $s_1, s_2 \in \mathbb{Z}_p$ respectively, then using one OT round, the parties can compute an additive sharing of $s_1 \cdot s_2$ amongst the set of parties $\mathcal{P}_1 \cup \mathcal{P}_2$.

Offline Protocol for DDH Instantiation Based on BGW

Inputs: All parties hold the circuit C, the number of parties n, and the prime field $\mathbb{Z}_p$.

Computation: The parties perform the following computations, where all shares and computations, including those in the sub-protocols, are done over $\mathbb{Z}_p$.

1. For every $\omega \in W$
   (a) Each party P_i randomly selects its individual keys $k_{w,0}^i, k_{w,1}^i \in \{0, \ldots, \frac{q-1}{4n}\} \subset \mathbb{Z}_q \subset \mathbb{Z}_p$. We denote the joint keys by $k_{w,0}, k_{w,1} \in \mathbb{Z}_q$.
   (b) Using two rounds of interaction, the parties run a coin-tossing protocol and receive Shamir shares of the hidden permutation bit $\lambda_w \in \{0, 1\}$[7].
2. For every gate $g \in C$ with input wires u, v and output wire w, and for every $\alpha, \beta \in \{0, 1\}$, do:
   (a) Locally compute $m_i = (H(4g+2\alpha+\beta))^{-\left(k_{u,\alpha}^i + k_{v,\beta}^i\right)}$, and secret share m_i in a t-out-of-n Shamir secret-sharing scheme..
   (b) Secret share $k_{w,0}^i, k_{w,1}^i$ in a t-out-of-n Shamir secret-sharing scheme. By summing the received shares, each party recovers a share of $k_{w,0}$ and $k_{w,1}$.
   (c) Using standard sub-protocols, compute Shamir shares of $\Pi_{i=1}^n m_i = \left((H(4g + 2\alpha + \beta))^{k_{u,\alpha}+k_{v,\beta}}\right)^{-1}$ and of $(2KEY_{w,\alpha,\beta} + e_{w,\alpha,\beta})^2$.
   (d) Using a single BGW round, compute shares of

   $$\tilde{g}_{\alpha,\beta} = \left((H(4g + 2\alpha + \beta))^{k_{u,\alpha}+k_{v,\beta}}\right)^{-1} \cdot (2KEY_{w,\alpha,\beta} + e_{w,\alpha,\beta})^2.$$

   (e) Reconstruct $\tilde{g}_{\alpha,\beta}$, i.e., broadcast shares of $\tilde{g}_{\alpha,\beta}$ and interpolate.
3. Reconstruct λ_ω for every output wire ω of the circuit.

Outputs: $\tilde{g}_{\alpha,\beta}$ for each gate $g \in C$ and every $\alpha, \beta \in \{0, 1\}$ and λ_ω for every output wire ω of the circuit.

Fig. 4. BGW protocol

To show this, we denote $s_1 = s_1^{i_1} + \cdots + s_1^{i_{|\mathcal{P}_1|}}$ and $s_2 = s_2^{j_1} + \cdots + s_2^{j_{|\mathcal{P}_2|}}$. The observation then follows by noticing that

$$
\begin{aligned}
s_1 \cdot s_2 &= (s_1^{i_1} + \cdots + s_1^{i_{|\mathcal{P}_1|}})(s_2^{j_1} + \cdots + s_2^{j_{|\mathcal{P}_2|}}) \\
&= s_1^{i_1} \cdot s_2^{j_1} + \cdots + s_1^{i_1} \cdot s_2^{j_{|\mathcal{P}_2|}} + \cdots + s_1^{i_{|\mathcal{P}_1|}} \cdot s_2^{j_{|\mathcal{P}_2|}}.
\end{aligned}
$$

Any multiplication of the form $s_1^i \cdot s_2^j$ can be performed using $\log p$ string OTs between parties P_i and P_j, as explained in [22].

The OT protocol for the offline phase is similar to the BGW based protocol, except that shares are additive, and the multiplications are computed using the above observation. As in the BGW based protocol, the main bottleneck of the protocol is to compute unbounded fan-in multiplication.

5   Instantiation Based on Ring-LWE

Boneh et al. [10] and Banerjee and Peikert [3] constructed almost key-homomorphic PRFs from LWE and ring-LWE respectively. It seems quite possible that one can use these almost key-homomorphic PRFs in our construction.

We go in a slightly different route – we build a protocol based directly on decisional ring-LWE hardness assumption. The function we use is not a real PRF, as it is not deterministic. However, by the decision ring-LWE assumption it is indistinguishable from random, which we show is sufficient for our construction.

Let $p = 2N + 1$ be a prime where N is a power of two, and denote $\mathcal{R}_p = \mathbb{Z}_p[X]/(X^N + 1)$. We define $\mathcal{F} = \{f_k \colon \mathcal{R}_p \to \mathcal{R}_p | f_k(a) = a \cdot k + e\}$, where a, k, and e are polynomials in the ring and the coefficients of e come from a gaussian distribuition $\mathcal{D}$. Assuming decision ring-LWE, for a *bounded constant* number of *distinct random* inputs, the output of f_k is indistinguishable from random. Furthermore, the above holds also if the keys themselves come from Gaussian distribution. Notice that the output of f_k is not deterministic due to the error.

Since in this protocol the key domain is a subset of the image of f_k, it might seem at first that the function f used to map the keys into the image of f_k can be the identity. However, this would be problematic due to the error. To avoid the error, the function f multiplies the coefficients of the key by $\lceil \sqrt{p} \rceil$, see Sect. 5.1.

The proof of security for the LWE instantiation is similar to the proof of the DDH instantiation without the random oracle. Notice that the proof did not require the PRF to be deterministic, only that the function is indistiguishable from random, and that the decrypted messages can be recovered correctly. For the latter point, by using correct parameters, this happens with overwhelming probability. It is important to note that for the encryption, the function is never queried twice on the same input.

In order to encrypt also the external values, we set the last coordinate of the key to be 0. Thus, we lose one dimension of the key, which slightly reduces security. We give a more detailed explanation on the security in the full version.

In the following protocols, using the ideas explained in Sect. 4.3 for removing the random-oracle, we let $a_1, \ldots, a_{8 \cdot f_{out}}$ be public random elements of the ring (which is also public), where f_{out} is the maximal fan-out of the circuit. We denote by $A(g, \alpha, \beta)$ the random element associated with row (α, β) of gate g, such that any two gates that share an input wire do not share any of the random elements (cf. Definition 7). The full description of functionality F_{GC}, instantiated as described above, appears in Fig. 5. Notice that for security, it must hold that $8 \cdot f_{out}$ is less than the bound on the number of samples.

5.1   Online Phase for Ring-LWE Based Instantiation

The online phase of our LWE based instantiation follows the general online phase, except that after each decryption, the error needs to be eliminated before the hidden key and external value can be recovered. The main idea for eliminating the error is that both the error and the key come from a Gaussian distribution. Thus, they will be far from the mean with only negligible probability.

Instantiated Functionality $\mathcal{F}_{GC}$ Based on LWE

Private Inputs: Each party P_i gives as input two keys $k_{w,0}^i, k_{w,1}^i \in \mathcal{D}^{N-1} \times \{0\}$ and four noise vectors $E_{g,\alpha,\beta}^i \leftarrow \mathcal{D}^N$ for every wire $w \in W$, where $\alpha, \beta \in \{0,1\}$.
Denote $k_{w,0} \overset{\text{def}}{=} \Sigma_{i=1}^n k_{w,0}^i$ and $k_{w,1} \overset{\text{def}}{=} \Sigma_{i=1}^n k_{w,1}^i$.
Random Input: To each wire $\omega \in W$ of the circuit, a hidden random bit $\lambda_\omega \in_R \{0,1\}$ is associated.
Output: The functionality outputs to all parties the garbled circuit GC – for every gate $g \in C$ with input wires u, v, and output wire w, and for every $\alpha, \beta \in \{0,1\}$, it outputs:

$$\tilde{g}_{\alpha,\beta} = A(g,\alpha,\beta) \cdot (k_{u,\alpha} + k_{v,\beta}) + E_{g,\alpha,\beta} + (\lceil \sqrt{p} \rceil \cdot (KEY_{w,\alpha,\beta} \| e_{w,\alpha,\beta})), \quad (11)$$

where

$$e_{w,\alpha,\beta} \overset{\text{def}}{=} g((\lambda_u \oplus \alpha),(\lambda_v \oplus \beta)) \oplus \lambda_w \quad (12)$$

$$KEY_{w,\alpha,\beta} \overset{\text{def}}{=} k_{w,0} + ((k_{w,1} - k_{w,0}) \cdot e_{w,\alpha,\beta}). \quad (13)$$

and $E_{g,\alpha,\beta}$ is the cumulative error, i.e.,

$$E_{g,\alpha,\beta} = \Sigma_{i=1}^n E_{g,\alpha,\beta}^i. \quad (14)$$

In addition, each party P_i receives the λ's corresponding to its input and output wires.

Fig. 5. Instantiated functionality $\mathcal{F}_{GC}$

If the mean is $\frac{\sqrt{p}}{2}$ and the standard deviation is sufficiently small, then with overwhelming probability the error will be in the range $[0, \sqrt{p}]$. Therefore, if the message is multiplied by $\sqrt{p}$ before encrypting, then dividing by $\sqrt{p}$ gets rid of the error and recovers the message.

Using this method, the probability that the protocol will output correctly is the probability that both the cumulative error and encrypted message are in the range $[0, \sqrt{p}]$ in all coordinates of all decrypted rows. If the parameters are correctly chosen, this happens with overwhelming probability. The online phase protocol for our ring-LWE based instantiation is given in Fig. 6.

5.2 Tailored Offline Phase Protocol for Ring-LWE Based Implementation

We now describe a "tailored" secure protocol for computing the offline phase for our ring-LWE based instantiation. The protocol is based on oblivious transfer and is secure up to $n-1$ corrupt parties. Other protocols for computing this functionality are also possible, e.g., using BGW if one assumes an honest majority.

The protocol we describe here is asymptotically better – the amount of work done by each party grows only quasilinearly in the number of parties. Furthermore,

---

Online Protocol for LWE Based Instantiation

Private Inputs: All parties hold their input to the functionality F_{GC} and the output they received from it.
In addidition, each party holds its input bit b_ω for each of its input wires ω.
Computation:

1. Each party broadcasts $e_\omega = \lambda_\omega \oplus b_\omega$ for the each of its input wires ω.
2. Each party P_i broadcasts k^i_{ω,e_ω} for each input wire of the circuit.
3. Each party does the following computation locally:
   (a) Computes $k_{\omega,e_\omega} = \Sigma^n_{i=1} k^i_{\omega,e_\omega}$ for every input wire ω of the circuit.
   (b) In topological order, for each gate $g \in C$, input wires u, v and output wire w computes

$$- A(g, \alpha, \beta) \cdot (k_{u,e_u} + k_{v,e_v}) + \tilde{g}_{\alpha,\beta} = E_{g,\alpha,\beta} + (\lceil \sqrt{p} \rceil \cdot (k_{w,e_w} \| e_w)).$$

   Dividing by $\sqrt{p}$ to get rid of the error, the party recovers the external value e_w and corresponding key k_{w,e_w} for the output wire w of the gate.
   (c) For each of its output wires ω computes the true value of the output $e_\omega \oplus \lambda_\omega$.

Output: Each party recovers the true value of the output, $e_\omega \oplus \lambda_\omega$, for each of its output wires ω.

Fig. 6. Online protocol for LWE instantiation

the PRFs are only computed *locally* and not in MPC, and therefore the circuit for computing the garbled circuit is independent of the complexity of the PRF.

The main idea of the protocol is that each party will encrypt its share, by adding the PRF, and broadcast. The sum of the received encryptions will be the garbled gate. This works due to the fact that both the shares and the (randomized) PRFs are additively homomorphic. Thus, it is reasonable to assume that a similar protocol exists for all key-homomorphic PRFs in which the operation on the keys is the same as the operation on the image of the PRFs.

To give more detail, the parties compute additive shares of the keys to encrypt using a sub-protocol. Then, each party encrypts its share and broadcasts. The parties sum the received broadcasts and recover the garbled gates. Intuitively, the security follows from Remark 1 and the following:

– For rows that are not decrypted during the online phase, at least one of the individual keys is unknown to the adversary. Thus, the encryption hides the share of the party.
– For rows that are decrypted during the online phase, the shares that are revealed (if the adversary is able to recover them, e.g., if the adversary controls $n-1$ parties) can be simulated by what the adversary can already learn from his shares and the decrypted key.

Offline Protocol for LWE Instantiation

Inputs: All parties hold the circuit C, the number of parties n, and the prime field $\mathbb{Z}_p$.

Computation: The parties perform the following computations, where all shares and computations are done in $\mathcal{R}_p{}^a$.

1. For every $\omega \in W$
   (a) Each party P_i selects its individual keys $k_{w,0}^i, k_{w,1}^i \leftarrow \mathcal{D}^N$, and sets the last coordinate of each key to 0. We denote the joint keys by $k_{w,0}, k_{w,1}$. Note that the last coordinate of the joint keys is also zero.
   (b) The parties run a coin-tossing protocol and receive additive shares of the hidden permutation bit $\lambda_w \in \{0, 1\}$ for every wire w.
2. For every gate $g \in C$ with input wires u, v and output wire w, and for every $\alpha, \beta \in \{0, 1\}$, the parties do the following:
   (a) Using a standard sub-protocol, compute additive shares of $k_{w, e_{w,\alpha,\beta}} \| e_{w,\alpha,\beta}$. Denote the share of party P_i by $(k_{w, e_{w,\alpha,\beta}} \| e_{w,\alpha,\beta})^i$.
   (b) Choose a Gaussian noise vector $E_{g,\alpha,\beta}^i \leftarrow \mathcal{D}^N$ and locally compute the encryption

$$A(g, \alpha, \beta) \cdot (k_{u,\alpha}^i + k_{v,\beta}^i) + E_{g,\alpha,\beta}^i + \left(\lceil \sqrt{p} \rceil \cdot (k_{w, e_{w,\alpha,\beta}} \| e_{w,\alpha,\beta})^i \right). \quad (15)$$

   (c) Broadcast the encryption and sum, thus recovering

$$\tilde{g}_{\alpha,\beta} = A(g, \alpha, \beta) \cdot (k_{u,\alpha} + k_{v,\beta}) + E_{g,\alpha,\beta} + \left(\lceil \sqrt{p} \rceil \cdot (k_{w, e_{w,\alpha,\beta}} \| e_{w,\alpha,\beta}) \right). \quad (16)$$

3. Reconstruct λ_ω for every output wire ω of the circuit.

Outputs: $\tilde{g}_{\alpha,\beta}$ for each gate $g \in C$ and every $\alpha, \beta \in \{0, 1\}$ and λ_ω for every output wire ω of the circuit.

---

^{a}As vector spaces, $\mathcal{R}_p \cong \mathbb{Z}_p^N$, but the multiplication is different

Fig. 7. LWE offline protocol

The offline protocol is described in detail in Fig. 7. The sub-protocol for computing additive shares of $KEY_{w,\alpha,\beta} \| e_{w,\alpha,\beta}$ is straightforward using the observation in Sect. 4.4.2.

6 Implementation Details and Experimental Results

In this section we give details on our implementations and report our experimental results for online computation time with and without our various optimizations above.

We wrote our code for the online computation in C++ using the NTL library [35] for modular arithmetic and OpenSSL [33] for SHA. The code for our Ring-LWE based instantiation used the building blocks of Gaussian sampling, NTT transform, and arithmetic operations from [15]. Our code is publically available at https://github.com/cryptobiu/Protocols.

We tested our code on a circuit with 100,000 AND gates and 0 XOR gates (see Remark 3 below). Times are the average of 5 runs and reported in seconds. The experiments were run on Ubuntu 14.04.4 operating system using a single core of Intel(R) Core(TM) i7-5600U CPU @ 2.60 GHz processor. Clearly, times of all versions can be improved using parallelization, but we leave this to further research.

DDH Instantiation. The unoptimized version refers to the initial instantiation described in Sect. 4.2. The precomputation version refers to version without the random oracle that utilizes the precomputation of the exponent tables, described in Sect. 4.3. The $DLSE_{500}$, $DLSE_{256}$ and $DLSE_{160}$ refer to the optimizations relying on DLSE described in Sect. 4.3, with the individual keys having 500, 256, and 160 bits respectively. Since the security relies on the number of bits of the individual keys, and the size of the joint keys has $\log n$ more bits, we fixed the number of these bits to be 10, corresponding to up to 1023 parties. I.e., in $DLSE_{500}$ the joint keys were 510 bits in $DLSE_{256}$ 266 bits and in $DLSE_{160}$ 170 bits[3]. All versions were used with a 1024 bit safe prime. Thus, $c \approx 500$ is the maximal $DLSE_c$ that can still use our sqrt. optimization. We provide a table to show the efficiency of the improvements, and they are also depicated in Fig. 1. The real security of $DLSE_c$ is unclear as far as we know. But there are known attacks that need only $\sim O(2^{\frac{c}{2}})$ exponentiations. Thus, if one aims for 80 bits of security, then one should use at least the $DLSE_{160}$ version.

Version	Unoptimized	Precomp.	$DLSE_{500}$	$DLSE_{256}$	$DLSE_{160}$
Online Computation Time (sec.)	102.3	83.9	16.15	8.4	5.4

We also ran tests using larger safe primes with 1536, 2048, and 3072 bits under the $DLSE_{256}$ assumption. The results are given in the following table and in Fig. 8.

Number of bits in prime ($DLSE_{256}$)	$\kappa = 1024$	$\kappa = 1536$	$\kappa = 2048$	$\kappa = 3072$
Online Computation Time (sec.)	8.4	14.9	21.65	43.2

LWE Instantiation. Our code for the ring-LWE instantiation used the following parameters, suggested in [16]: $p = 1051649, N = 512, \sigma = \frac{8}{\sqrt{2}}$. We chose these parameter to allow enough room for error, thus allowing a larger number of parties to participate. Note that for this choice of parameters, the total probability

---

[3] For the smaller number of parties the joint keys will have a few less bits, e.g., for <128 parties and $DLSE_{160}$ the joint key should have only 167 instead of 170 bits. Thus, the time could possibly be very slightly better for the smaller number of parties.

the entire protocol errs on our chosen circuit is $<2^{-40}$ for up to 300 parties. For 500 parties, the total probability for error is $\approx 2^{-15.5}$, so changing the parameters should be considered for this number of parties.

The online computation time of our ring-LWE instantiation beat even the $DLSE_{160}$ version using a 1024 bit prime, with an average online computation time of approx. 4.45 s. For comparison, the result is depicated in Figs. 1 and 8.

Comparison with BMR. For comparison, we also measured the online computation time of a state of the art BMR implementation of [8] on the same circuit and hardware, with a varying number of parties. The code of [8] is highly optimized and uses AES-NI with pipelining for the encryption/decryption. The results are depicated in Figs. 1 and 8 for comparison.

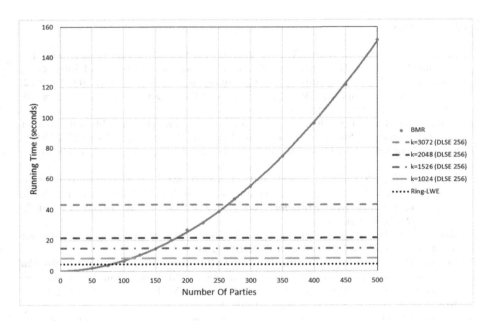

Fig. 8. Online Computation Time with $DLSE_{256}$ and Varying Primes

As expected, for a large enough number of parties, our protocols' online computation time is faster than the respective time of the BMR protocol. With a 1024 bit prime: for the unoptimized version, the cutoff point is between 400 and 450 parties. For the precomputation version the cutoff is between 350 and 400 parties. For the $DLSE_{500}$ and $DLSE_{256}$ versions the cutoff points are between 150 and 175 and between 100 and 125 respectively. The cutoff points for the $DLSE_{160}$ and the ring-LWE version are near 75 parties. Cutoff points for larger primes are at a slightly higher number of parties.

We expect that on different hardware the times and cutoff points will differ, but the overall conclusion should remain the same. We also note that we did not use any dedicated hardware optimizations, while the BMR code of [8] uses pipelined fixed-key AES-NI.

Remark 3. The BMR protocol of [8] contains a free-XOR optimization, while our code computes XOR gates similarly to AND gates. Thus, for a true comparison on a specific circuit C with X XOR & XNOR gates and A non-XOR[4] gates, the time t of the BMR protocol should be adjusted to $t \cdot \frac{A}{A+X}$. However, for any fixed circuit, this will only change the exact location of the cutoff points, and not the overall conclusion.

Acknowledgements. We would like to thank Shalev Keren, Moria Farbstein and Lior Koskas for helping with the code, and to thank Shai Halevi and Vadim Lyubashevsky for helpful discussions on LWE.

References

1. Asharov, G., Jain, A., López-Alt, A., Tromer, E., Vaikuntanathan, V., Wichs, D.: Multiparty computation with low communication, computation and interaction via threshold FHE. In: Pointcheval, D., Johansson, T. (eds.) EUROCRYPT 2012. LNCS, vol. 7237, pp. 483–501. Springer, Heidelberg (2012). https://doi.org/10.1007/978-3-642-29011-4_29
2. Asharov, G., Lindell, Y., Schneider, T., Zohner, M.: More efficient oblivious transfer and extensions for faster secure computation. In: Proceedings of the 2013 ACM SIGSAC conference on Computer & Communications Security, pp. 535–548. ACM (2013)
3. Banerjee, A., Peikert, C.: New and improved key-homomorphic pseudorandom functions. In: Garay, J.A., Gennaro, R. (eds.) CRYPTO 2014. LNCS, vol. 8616, pp. 353–370. Springer, Heidelberg (2014). https://doi.org/10.1007/978-3-662-44371-2_20
4. Bar-Ilan, J., Beaver, D.: Non-cryptographic fault-tolerant computing in constant number of rounds of interaction. In: PODC (1989)
5. Beaver, D., Micali, S., Rogaway, P.: The round complexity of secure protocols. In: Proceedings of the Twenty-Second Annual ACM Symposium on Theory of Computing, STOC 1990, pp. 503–513. ACM, New York (1990). https://doi.org/10.1145/100216.100287. ISBN 0-89791-361-2
6. Bellare, M., Hoang, V.T., Keelveedhi, S., Rogaway, P.: Efficient garbling from a fixed-key blockcipher. In: 2013 IEEE Symposium on Security and Privacy, SP 2013, Berkeley, CA, USA, 19–22 May 2013, pp. 478–492 (2013)
7. Ben-David, A., Nisan, N., Pinkas, B.: Fairplaymp: a system for secure multi-party computation. In: Proceedings of the 15th ACM Conference on Computer and Communications Security, pp. 257–266. ACM (2008)
8. Ben-Efraim, A., Lindell, Y., Omri, E.: Optimizing semi-honest secure multiparty computation for the internet. In: 23rd ACM Conference on Computer and Communications Security (ACM CCS) 2016 (2016). To appear
9. Ben-Or, M., Goldwasser, S., Wigderson, A.: Completeness theorems for noncryptographic fault-tolerant distributed computations. In: Proceedings of the 20th ACM Symposium on the Theory of Computing, pp. 1–10 (1988)

---

[4] NOT gates can be eliminated even without free-XOR optimization, by modifying the circuit.

10. Boneh, D., Halevi, S., Hamburg, M., Ostrovsky, R.: Circular-secure encryption from decision Diffie-Hellman. In: Wagner, D. (ed.) CRYPTO 2008. LNCS, vol. 5157, pp. 108–125. Springer, Heidelberg (2008). https://doi.org/10.1007/978-3-540-85174-5_7

11. Brakerski, Z., Gentry, C., Vaikuntanathan, V.: (leveled) fully homomorphic encryption without bootstrapping. In: Innovations in Theoretical Computer Science 2012, Cambridge, MA, USA, 8–10 January 2012, pp. 309–325 (2012)

12. Brakerski, Z., Vaikuntanathan, V.: Fully homomorphic encryption from ring-LWE and security for key dependent messages. In: Rogaway, P. (ed.) CRYPTO 2011. LNCS, vol. 6841, pp. 505–524. Springer, Heidelberg (2011). https://doi.org/10.1007/978-3-642-22792-9_29

13. Choi, S.G., Hwang, K.-W., Katz, J., Malkin, T., Rubenstein, D.: Secure multiparty computation of boolean circuits with applications to privacy in on-line marketplaces. In: Dunkelman, O. (ed.) CT-RSA 2012. LNCS, vol. 7178, pp. 416–432. Springer, Heidelberg (2012). https://doi.org/10.1007/978-3-642-27954-6_26

14. Damgård, I., Keller, M., Larraia, E., Pastro, V., Scholl, P., Smart, N.P.: Practical covertly secure MPC for dishonest majority – or: breaking the SPDZ limits. In: Crampton, J., Jajodia, S., Mayes, K. (eds.) ESORICS 2013. LNCS, vol. 8134, pp. 1–18. Springer, Heidelberg (2013). https://doi.org/10.1007/978-3-642-40203-6_1

15. de Clercq, R., Roy, S.S., Vercauteren, F., Verbauwhede, I.: Efficient software implementation of ring-LWE encryption. In: 2015 Design, Automation Test in Europe Conference Exhibition (DATE), pp. 339–344, March 2015. https://doi.org/10.7873/DATE.2015.0378

16. Ghosh, S., Kate, A.: Post-quantum forward-secure onion routing. In: Malkin, T., Kolesnikov, V., Lewko, A.B., Polychronakis, M. (eds.) ACNS 2015. LNCS, vol. 9092, pp. 263–286. Springer, Cham (2015). https://doi.org/10.1007/978-3-319-28166-7_13

17. Goldreich, O.: Foundations of Cryptography. Basic Applications, vol. II. Cambridge University Press, Cambridge (2004)

18. Goldreich, O., Micali, S., Wigderson, A.: How to play any mental game. In: Proceedings of the 19th ACM Symposium on the Theory of Computing, pp. 218–229 (1987)

19. Hazay, C., Scholl, P., Soria-Vazquez, E.: Low cost constant round MPC combining BMR and oblivious transfer. Cryptology ePrint Archive, Report 2017/214 (2017). http://eprint.iacr.org/2017/214

20. Huang, Y., Evans, D., Katz, J., Malka, L.: Faster secure two-party computation using garbled circuits. In: Proceedings of the 20th USENIX Security Symposium, San Francisco, CA, USA, 8–12 August 2011

21. Ishai, Y., Kilian, J., Nissim, K., Petrank, E.: Extending oblivious transfers efficiently. In: Boneh, D. (ed.) CRYPTO 2003. LNCS, vol. 2729, pp. 145–161. Springer, Heidelberg (2003). https://doi.org/10.1007/978-3-540-45146-4_9

22. Kiltz, E., Leander, G., Malone-Lee, J.: Secure computation of the mean and related statistics. In: Kilian, J. (ed.) TCC 2005. LNCS, vol. 3378, pp. 283–302. Springer, Heidelberg (2005). https://doi.org/10.1007/978-3-540-30576-7_16

23. Kolesnikov, V., Schneider, T.: Improved garbled circuit: free XOR gates and applications. In: Aceto, L., Damgård, I., Goldberg, L.A., Halldórsson, M.M., Ingólfsdóttir, A., Walukiewicz, I. (eds.) ICALP 2008. LNCS, vol. 5126, pp. 486–498. Springer, Heidelberg (2008). https://doi.org/10.1007/978-3-540-70583-3_40

24. Koshiba, T., Kurosawa, K.: Short exponent Diffie-Hellman problems. In: Bao, F., Deng, R., Zhou, J. (eds.) PKC 2004. LNCS, vol. 2947, pp. 173–186. Springer, Heidelberg (2004). https://doi.org/10.1007/978-3-540-24632-9_13

25. Larraia, E., Orsini, E., Smart, N.P.: Dishonest majority multi-party computation for binary circuits. In: Garay, J.A., Gennaro, R. (eds.) CRYPTO 2014. LNCS, vol. 8617, pp. 495–512. Springer, Heidelberg (2014). https://doi.org/10.1007/978-3-662-44381-1_28

26. Lindell, Y., Pinkas, B.: A proof of security of yao's protocol for two-party computation. J. Cryptology **22**(2), 161–188 (2009)

27. Lindell, Y., Pinkas, B., Smart, N.P., Yanai, A.: Efficient constant round multiparty computation combining BMR and SPDZ. In: Gennaro, R., Robshaw, M. (eds.) CRYPTO 2015. LNCS, vol. 9216, pp. 319–338. Springer, Heidelberg (2015). https://doi.org/10.1007/978-3-662-48000-7_16

28. Lyubashevsky, V., Peikert, C., Regev, O.: On ideal lattices and learning with errors over rings. In: Gilbert, H. (ed.) EUROCRYPT 2010. LNCS, vol. 6110, pp. 1–23. Springer, Heidelberg (2010). https://doi.org/10.1007/978-3-642-13190-5_1

29. Lyubashevsky, V., Peikert, C., Regev, O.: A toolkit for ring-LWE cryptography. In: Johansson, T., Nguyen, P.Q. (eds.) EUROCRYPT 2013. LNCS, vol. 7881, pp. 35–54. Springer, Heidelberg (2013). https://doi.org/10.1007/978-3-642-38348-9_3

30. Malkhi, D., Nisan, N., Pinkas, B., Sella, Y., et al.: Fairplay-secure two-party computation system. In: USENIX Security Symposium, San Diego, CA, USA, vol. 4. (2004)

31. Naor, M., Pinkas, B., Reingold, O.: Distributed pseudo-random functions and KDCs. In: Stern, J. (ed.) EUROCRYPT 1999. LNCS, vol. 1592, pp. 327–346. Springer, Heidelberg (1999). https://doi.org/10.1007/3-540-48910-X_23

32. van Oorschot, P.C., Wiener, M.J.: On Diffie-Hellman key agreement with short exponents. In: Maurer, U. (ed.) EUROCRYPT 1996. LNCS, vol. 1070, pp. 332–343. Springer, Heidelberg (1996). https://doi.org/10.1007/3-540-68339-9_29

33. OpenSSL Project. OpenSSL project (2006). http://www.openssl.org/

34. Schneider, T., Zohner, M.: GMW vs. Yao? efficient secure two-party computation with low depth circuits. In: Sadeghi, A.-R. (ed.) FC 2013. LNCS, vol. 7859, pp. 275–292. Springer, Heidelberg (2013). https://doi.org/10.1007/978-3-642-39884-1_23

35. Shoup, V.: NTL: A library for doing number theory (2003). http://www.shoup.net/ntl

36. Wang, X., Ranellucci, S., Katz, J.: Global-scale secure multiparty computation. In: 24th ACM Conference on Computer and Communications Security (ACM CCS) 2017 (2017, to appear)

37. Yao, A.C.: How to generate and exchange secrets. In: Proceedings of the 27th IEEE Symposium on Foundations of Computer Science, pp. 162–167 (1986)

38. Zahur, S., Rosulek, M., Evans, D.: Two halves make a whole. In: Oswald, E., Fischlin, M. (eds.) EUROCRYPT 2015. LNCS, vol. 9057, pp. 220–250. Springer, Heidelberg (2015). https://doi.org/10.1007/978-3-662-46803-6_8

Overlaying Conditional Circuit Clauses for Secure Computation

W. Sean Kennedy, Vladimir Kolesnikov[✉], and Gordon Wilfong

Bell Labs, Murray Hill, NJ, USA
{william.kennedy,vladimir.kolesnikov,gordon.wilfong}@nokia-bell-labs.com

Abstract. We improve secure function evaluation (SFE) by optimizing circuit representation of the function and designing new SFE protocols.

(1) We propose a heuristic for constructing a circuit C_0, universal for a given set of Boolean circuits $S = \{C_1, ..., C_k\}$. Namely, given each C_i, we view it as a directed acyclic graph (DAG) D_i by ignoring the Boolean gate functions of C_i. We embed $D_1, ..., D_k$ in a new DAG D_0, such that each C_i can be obtained by a corresponding programming of D_0 (i.e. by assignment of Boolean gates to the nodes of D_0). DAG D_0, viewed as a Boolean circuit with unprogrammed gates, is the S-universal circuit C_0.

(2) Our heuristic often produces C_0 significantly smaller than Valiant's universal circuit or a circuit incorporating all $C_1, ..., C_k$. Exploiting this, we construct new Garbled Circuit (GC) and GMW-based SFE protocols, which are particularly efficient for circuits with if/switch clauses.

Our GMW protocol evaluates 8-input Boolean gates at the same cost as the usual 2-input gates. This advances general GMW-based SFE, and is particularly useful for circuits with if/switch conditional clauses.

Experimentally, for a switch containing 32 simple circuits, our construction resulted in $\approx 6.1\times$ smaller circuit C_0. This directly translates into $\approx 6.1\times$ improvement in GMW SFE computing this switch. Recent state-of-the-art generic circuit optimizations from hardware design adapted to SFE report $10 - 20\%$ circuit (garble table) reduction.

Our SFE is in the semi-honest model, and is compatible with Free-XOR. We further show that optimal embedding is NP-hard.

Keywords: Set-universal circuit · secure computation · garbled circuit · GMW

1 Introduction

Eliminating costs imposed by the circuit representation of Garbled Circuits (GC) and Goldreich, Micali and Wigderson (GMW) techniques has been an important open problem since the introduction of GC/GMW, with moderate success to date. There are two natural redundancies: GC/GMW must unroll the

© International Association for Cryptologic Research 2017
T. Takagi and T. Peyrin (Eds.): ASIACRYPT 2017, Part II, LNCS 10625, pp. 499–528, 2017.
https://doi.org/10.1007/978-3-319-70697-9_18

loops and evaluate *all* conditional (if/switch) clauses so as to hide which was evaluated. (There is a third redundancy, protecting memory access patterns at the expense of processing entire input/array/data structure, which is applicable to both circuit and random-access representation. It is addressed by the influential work on Oblivious RAM (ORAM), started by [13] with then-"impractical" $\log^4 n$ factor overhead.)

Our work aims to solve the second kind of circuit redundancy. Constructing a small circuit C_0, universal for k given circuits, will allow to garble, transfer and evaluate just C_0, when computing switch on the k circuits. We believe this could be a useful tool in Secure Function Evaluation (SFE) compiler design.

On the cost of SFE and OT rounds. Our GC protocol *may* add a round of communication for each switch statement. We argue that the associated latency cost is negligible in many practical scenarios. This is because often the latency-related idling will be productively used for computation and communication in the same or another SFE instance. This is the case, e.g., in larger-scale SFE deployments, where many instances will be run in parallel, and where SFE *throughput* is a far more important parameter than *latency*. Note, in our GMW protocols there is no increase in the number of rounds due to switch.

1.1 Motivating Applications

Functions with switch statements. In Blind Seer [7,25], a GC-based private database (DB) system, private DB search is achieved by two players jointly securely evaluating the query match function on the search tree of the data. Blind Seer does not fully protect query privacy: it leaks the query circuit topology as the full universal circuit is not practical, as admitted by the authors. Applying our solution to that work would hide this important information, cheaply. Indeed, say, by policy the DB client is allowed to execute one of several (say, 2–50) types of queries. The privately executed SQL query can then be a switch of the number of clauses, each corresponding to an allowed query type. In Blind Seer the clause is selected by client's input, omitting some of the machinery. As a result, our Blind Seer application is particularly effective bringing improvements with as little as two clauses. Most of the cost of this DB system is in running SFE of the query match function at a large scale, so improvement to the query circuit will directly translate to overall improvement. We note that the core of the Blind Seer system is in the semi-honest model, but a malicious client is considered in [7].

(Part of) our work can be viewed as a heuristic aiming to construct a circuit universal for a set of functions $S = \{C_1, ..., C_k\}$ (S-universal circuit) at the cost less than that of full universal circuit. Thus (cf. next motivating example), our work may improve applications where we want to evaluate and hide which function/query was chosen by a player (say, which one of several functions allowed by policy or known because of auxiliary information).

SFE of semi-private functions (SPF-SFE) (see additional discussion in Sect 1.3) is a notion introduced in [26], bridging the gap between expensive private function SFE (PF-SFE) based on Universal Circuit [15,19,22,30], and regular SFE (via GC) that does not hide the evaluated function. SPF-SFE partially hides the evaluated function; namely, given a set of functions, the evaluator will not learn which specific function was evaluated. Indeed, often only specific subroutines are sensitive, and it is they that might be sufficiently protected by S-universal circuit for an appropriate set of circuits S. [26] presents a convincing example of privacy-preserving credit checking, where the check function itself needs to be protected, and shows that using S-universal circuits as building blocks is an effective way of approaching this. Further, [26] builds a compiler which assembles GC from the S-universal building blocks (which they call PPB, Privately Programmable Blocks). While [26] provides only a few very simple hand-designed blocks (see our discussion in Sect. 1.3), our work can be viewed as an efficient general way of constructing such blocks. We stress that in the SPF-SFE application, GC generator knows the computed function (clause selection), and our constructions are particularly efficient.

CPU/ALU emulation. Extending the idea of SPF-SFE, one can imagine a general approach where the players privately emulate the CPU evaluating a sequence of complex instructions from a fixed instruction set (instruction choice implemented as a GC `switch`). Additionally, if desired, instructions' inputs can be protected by employing the selection blocks of [19]. Such an approach can be built within a suitable framework (e.g., that of [26]) from S-universal circuits provided by this work. We note that circuit design and optimization is tedious, and not likely to be performed by hand except for very simple instances, such as those considered in [26]. Instead, a tool, such as the one we are proposing, will be required.

In a recent work [32], a secure and practically efficient MIPS ALU is proposed, where the ALU is implemented as a `switch` over 37 currently supported ALU instructions evaluated on ORAM-stored data. TinyGarble [29] also design and realize a garbled processor, using the MIPS I instruction set, for private function evaluation. Our constructions would work with [32] in a drop-in replacement manner.

1.2 Technical Contribution

We improve secure function evaluation (SFE) by optimizing circuit representation of the function and designing new SFE protocols.

Our contribution consists of several complementary technical advances. Our most technically involved contribution is a novel algorithm to embed any k circuits in a new circuit/graph C_0. The embedding is such that a GC/GMW evaluation of any of $C_1, ..., C_k$ could be implemented by a corresponding evaluation of C_0. The size of C_0 could be smaller (6.1× smaller in our experiment) than the sum of the sizes of $C_1, ..., C_k$.

In the SPF-SFE case, when the evaluated clause C_i is circuit generator's private input, generator G simply sends the garbling implementing C_i.

For the general GC case, where clause is selected by an internal variable, we construct a new GC protocol with total communication cost $3kn_0 + 22n_0s$, where s is the computational security parameter and $n_0 = |C_0|$. For efficient embeddings, this compares favorably to state-of-the-art GC. The half-gates GC [34] of the above k clauses will cost $2ns$, where $n = \sum_j |C_j|$. We show how GC branches can be nested, and we can apply our construction on each nesting level.

GMW SFE is more interesting. We make a novel observation that the cost of evaluation of the GMW gates is similar to that for a moderate number of boolean inputs, as that of a two-input gate. We exploit this to obtain an efficient GMW protocol for circuits with clauses whose cost per gate is similar to that of standard GMW.

Our approach is heuristic. We show that solving the graph embedding problem exactly is NP-hard.

Experimental validation and performance. For our experiment we considered 32 simple circuits implementing variants of several basic functions and generated an embedding 6.1× smaller than the standard circuit implementing the clauses. SPF-SFE and GMW of the corresponding `switch` is also improved by 6.1×.

For the GC case when clause is selected by internal variable, our 6.1× smaller embedding results in communication cost similar to that of classical Yao [20,33], due to per-gate overheads. We expect our protocol to overtake optimal GC [34] for embeddings of slightly greater number of circuits or with further heuristic improvements.

1.3   Background and Related Work

Garbled Circuit, GMW, OT and Universal Circuit. Significant part of SFE research focuses on minimizing the size of the basic GC of Yao [20,33], such as garbled row reduction techniques Free-XOR [18] and its enhancements FleXOR [17] and half-gates [34]. In contrast, in this work, we effectively eliminate the need for evaluation of entire subcircuits.

The GMW protocol [11,12] had received less attention in the 2-party SFE literature than GC. In GMW, the two parties interact to compute the circuit gate-by-gate as follows. Players start with 2-out-of-2 additively secret-shared input wire values of the gate and obtain corresponding secret shares of the output wire of the evaluated gate. Addition (XOR) gates are done locally, simply by adding the shares. Multiplication (AND) gates are done using 1-out of-4 OT. For binary circuits, there are four possible combinations of each of the player's shares. Thus an OT is executed, where one player (OT receiver) selects one of the four combinations, and the other player (OT sender) provides/OT-sends the corresponding secret shares of the output wire. In the semi-honest model, the secret shares can be as short as a single bit. As in the GC approach, our work greatly reduces the size of the evaluated circuit.

Asymptotically, the best way to embed a large number of sub-circuit graphs into one circuit graph would be using the universal circuits [19,30]. Respectively, for sub-circuits of size n, the size of the universal circuit generated by [19,30] is $\approx 19n \log n$, and $\approx 1.5n \log^2 n + 2.5n \log n$. Very recent works [15,22] polish and implement Valiant's construction. They report a more precise estimate of the cost (in universal gates) of Valiant's UC of between $\approx 5n \log n$ and $10n \log n$. Programming of universal gates may each cost 3 AND and 6 XOR gates (but will not be needed in PFE and applications we discuss in this work). In sum, universal circuit approach becomes competitive for a number of clauses far larger than a typical `switch`. In a universal circuit embedding [15,19,22,30], gates are embedded in gates and wires are embedded in pairwise disjoint chains of wires (with possible intermediate gates). Our embedding is more general allowing the chains of wires to overlap in a controlled way, leading to smaller container circuits.

Another technique for Private Function Evaluation (PFE) was proposed by Mohassel and Sadeghian [23]. They propose an alternative (to the universal circuit) framework of SFE of a function whose definition is private to one of the players. Their approach is to map each gate outputs to next gate outputs by considering a mapping from all circuit inputs to all outputs, and evaluate it obliviously. For GC, they achieve a factor 2 improvement as compared to Valiant [30] and a factor $3 - 6$ improvement as compared to Kolesnikov and Schneider [19]. Similarly to [19,30], [23] will not be cost-effective for a small number of clauses.

Thus, (part of) our work can be viewed as heuristically constructing a circuit universal for a set of functions $\mathcal{S} = \{C_1, ..., C_k\}$ at the cost less than that of full universal circuit.

One of our contributions is an improved 1-out of-k OT algorithm to deliver the garbled `switch` clause to the evaluator. This is a special case of PIR (private information retrieval). We note existing sublinear in k work on computational PIR (CPIR) of 1 out of k ℓ-bit strings, e.g., [2,21,24]. Note, a symmetric CPIR (CSPIR) is needed for our application. CSPIR of [21] achieves costs $\Theta(s \log^2 k + \ell \log k)$, where s is a possibly non-constant security parameter. However, the break-even points where the OT sublinearity brings benefit are too high. For example, [2] costs more in communication than the naive linear-in-k OT for $k \leq 2^{40}$. Further, known CPIR protocols heavily (at least linearly in k) rely on expensive public-key operations, such as, in case of [21], length-flexible additive-homomorphic encryption (LFAH) of Damgård and Jurik [3,4].

We also mention, but do not discuss in detail, that hardware design considers circuit minimization problems as well. However, their typical goal is to minimize chip area while allowing multiple executions of the same (sub)circuit. Current state-of-the-art in applying to MPC the powerful tool chains from hardware design is producing $10 - 20\%$ circuit (garble table) reduction [5,6,29], while our targeted circuit optimizations may result in better performance ($\approx 6.1\times$ circuit reduction achieved in our experiment.)

Semi-private function SFE (SPF-SFE) [26]. As discussed in the Introduction, SPF-SFE is a convincing trade-off between efficiency and the privacy of the evaluated function. Our work on construction of container circuits corresponds to that of privately programmable blocks (PPB) of [26], which were hand-optimized in that work. In our view, the main contribution of [26] is in identifying and motivating the problem of SPF-SFE and building a framework capable of integrating PPBs into a complete solutions. They provide a number of very simple (but nevertheless useful) PPBs. In our notation, they consider the following sets for $\mathcal{S}$-universal circuit: $\mathcal{S}_{COMP} = \{<, >, \leq, \geq, \neq\}, \mathcal{S}_{ADD,SUB} = \{+, -\}, \mathcal{S}_{MULT} = \{\text{input} * \text{constant}\}, \mathcal{S}_{BOOLGATE} = \{\vee, \wedge, \oplus, NAND, NOR, XNOR\}, \mathcal{S}_{UC} = \{\text{all circuits}\}$, as well as the following sets recast from [19]: $\mathcal{S}_{SEL} = \{\text{input select circuits}\}, \mathcal{S}_{IN\_PERM} = \{\text{input permute circuits}\}, \mathcal{S}_{SEL} = \{\text{Y bit selector}\}, \mathcal{S}_{SEL} = \{\text{X bit selector}\}$. Each of these sets only consists of functions with already identical or near-identical topology; this is what enabled hand-optimization and optimal sizes of the containers. Other than the universal circuit PPB, no attempt was made to investigate construction PPBs of circuits of *a priori* differing topology.

In contrast, we can work with *any set* $\mathcal{S}$ of circuits for $\mathcal{S}$-universal circuit and heuristically improve on the full universal circuit, and on the standard option of evaluating of all $\mathcal{S}$ circuits and selecting the output.

GMW for multi-input gates. In independent and concurrent work, Dessouky et al. [6] discovered the same method of obtaining cheap GMW gates with multi-valued inputs by using the OT extension of [16] (multiple boolean inputs and multi-valued inputs are easily interchangeable due to [16]). In their work, Dessouky et al. make several performance optimizations to the usage of [16]. They also show in detail that for some functions, (e.g., AES), multi-input GMW gates are advantageous. In their notation, this approach is called lookup-table (LUT)-based secure computation. Our work focuses on different application of LUT-based computation, circuit clause overlay, and may achieve, in its domain, higher performance improvement.

1.4   Notation

Let f be the function we want to evaluate and C a boolean circuit representing f. We consider a `switch` statement inside f, evaluating one of k clauses depending on the internal variable or input of f. Let $C_1, ..., C_k$ be the subcircuits of C corresponding to the k clauses of f. We will often use the terms "clause" and "subcircuit" interchangeably, and their meaning will be clear from the context. For simplicity we will often discuss clauses of the same size n, although in the evaluation section we consider concrete examples with different clause sizes.

We define directed acyclic graphs (DAGs) $D_1, ..., D_k$ from circuits $C_1, ..., C_k$ where, with the exception of auxiliary nodes representing circuit inputs and outputs, the graph's nodes represent circuit gates and the graph's directed edges represent circuit wires. These graphs represent the *topology* or the *wiring* of the corresponding circuits. When the meaning is obvious from context we may interchangeably refer to these graphs/circuits as D_i or C_i. From DAGs $D_1, ..., D_k$ we will build a *container DAG* D_0, with the property that any of $C_1, ..., C_k$ can be implemented from D_0 by assigning corresponding gate functions to the nodes of D_0. We will usually call this *programming* of D_0. We note that for efficiency we may produce partially programmed D_0, i.e. one where some of the gates are already fixed. We will interchangeably refer to this container graph/circuit as D_0 and C_0. Circuits C_0 are, of course, generated for circuit-based secure computation. We specifically discuss GC and GMW protocols. We will often unify our references to the use of GC and GMW. For example, when clear from the context, by "garbling C_0" we will mean using C_0 in either GC or GMW.

Other standard variables we will use are s, which is the computational security parameter, and n_0, which is the size of D_0. Circuits C_0 will then be evaluated In the GC protocols there are two players, GC constructor, which we will denote P1, and GC evaluator, or P2.

2   Technical Solution Overview

In this section, our goal is to describe the complete intuition behind our approach. Having this big-picture view should help put in perspective the formalizations and details that follow in the next sections.

Consider the SFE of a circuit C, and inside it a `switch` statement with k clauses/subcircuits $C_1, ..., C_k$, only one of which is evaluated based on a player's input or an internal variable. In this overview we focus on the more complex and more general second scenario (internal variable), while pointing out the very efficient solution to the first scenario as well.

Our starting point is the widely known observation that in some GC variants (e.g. in classical Yao [20,33]), the evaluator will not learn the logic of any gate, but only the structure of the wiring of the circuit. We start by supposing that all our subcircuits already have the same wiring, i.e. the underlying DAGs are the same. We provide intuition on how to unify the wiring in the following Sect. 2.3.

2.1   Improved GC for `switch` of Identically-Wired Clauses

If all k clauses/subcircuits had the same topology/wiring, all that is needed is for the circuit generator to generate and deliver to the evaluator the garbling of the right subcircuit.

SPF-SFE. In the important special case where `switch` clause is selected by a player's private input, this is trivial and has no extra overhead: this player will be the GC generator and he simply sends the set of garbled tables programming the clause which corresponds to his input.

General case. Consider the case where `switch` is selected by an internal variable. One natural way to deliver the garbling would be to execute a 1-out-of-k OT on the clauses. Unfortunately, this, under the hood, would require sending garblings of each of $C_1, ..., C_k$ to the evaluator[1], which would not improve over the standard GC.

---

[1] See related work in Sect. 1.3 for discussion on the high costs of sublinear PIR for smaller-size DBs.

We can do better. To sketch the main idea, we let each C_i be a $\{\vee, \wedge, \oplus\}$-circuit[2]. (As all C_i are identically wired, their DAG representations D_i are the same, and the container DAG D_0 is equal to D_i. Recall, in our notation, $|D_0| = n_0$.) For now do not consider Free-XOR; it will be clear later that our approach works with Free-XOR. Now, enumerate the gates in each C_i and let d_i be a string of length n_0 defining the sequence of gates in C_i (in our construction, each symbol in d_i will denote one of a five possible gates – $\{\vee, \wedge, \oplus\}$, as well as an auxiliary left and right input wire pass-through gates L and R). Perform 1-out-of-k OT on the strings d_i to deliver to the evaluator the right circuit definition string. Then for each gate, the players will run 1-out-of-5 OT, where the generator's input will be the five possible gate garblings, and the evaluator will use the previously obtained d_i to determine its OT choices.

Notice that each string d_i reveals to the evaluator precisely which circuit has been transferred. This is easy to hide: for each gate g_j, the GC constructor selects a random permutation π_j on the five types of gates and applies π_j to the j-th symbol of d_i during d_i construction. He also applies π_j to permute his OT input of five garbled tables. Finally, sending to the evaluator d_i based on the internal state is easy. For a switch with two clauses, the generator simply sends d_1 encrypted with the 0-key of the selection wire, and d_2 encrypted with the corresponding 1-key. For a switch with k clauses, each string d_i will be encrypted with the key derived from the wire labels corresponding to the choice of the i-th clause.

For the reader familiar with the details of standard GC, it should be clear that the above switch-evaluation algorithm can be readily plugged into the standard GC protocol. Let s be the computational security parameter. Following cost calculations presented after Theorem 1 and in Sect. 8, the communication cost of evaluating the switch on the k clauses will be approximately $3kn_0 + 22n_0 s$. In contrast, standard GC would require sending all k garblings at the cost of $4ns$ ($2ns$ using recent half-gate garbling [34]), where $n = \sum_i |C_i|$. The $4ns$ term is the most expensive term; reducing it to $22n_0 s$ and making it independent of n is the contribution of our GC protocol. We again stress that if clause is selected by GC generator, we can use all GC optimizations, and our GC cost is $2n_0 s$. Finally, we note that in above calculations we did not account for the cost of circuitry selecting the output of the right clause and ignoring outputs of other clauses. This circuit is linear in ko, where o is the number of outputs in each clause. This circuit needs to be evaluated in the state-of-the-art GC, but not in our solution.

We further note that switch clauses can be nested. We discuss this in Sect. 4.

---

[2] We could reduce the set of considered circuit gates, e.g., by eliminating the OR gate and implementing it with AND and XOR gates. This would present us with the trade off between more efficient gate processing and potentially larger circuits. Alternatively, we could consider a larger gate basis and have costlier gate processing. We defer exploration of these trade-offs as future work.

2.2   Improved GMW for switch of Identically-Wired Clauses

An approach similar to the one described above in Sect. 2.1 can be particularly efficiently applied in the GMW setting. We will take advantage of our novel observation on the cost of multi-input GMW gates under the OT extension of Kolesnikov and Kumaresan [16].

As in our GC protocol above, we consider the circuit definition strings d_i. As in the GC protocol, for each gate g_j, one player selects a random permutation (or mask) π_j on the five types of gates and applies π_j to the j-th symbol of d_i during d_i construction. This masked definition string is transferred to the other player via OT.

In contrast with GC, we will not do the expensive 1-out of-5 OT on garbled gates. In GMW, we will evaluate gates on three input wires: two circuit wires and one 5-valued wire selecting the gate function ($\{\vee, \wedge, \oplus, L, R\}$). The players thus will run 1-out of-20 OT (the 20 possibilities are the five gate functions, each with four wire input possibilities) to obtain the secret share of the output.

Our simple but critical observation is that with using [16] OT, and because the GMW secret shares are a single bit each, the evaluation of multi-input gate, for moderate number of inputs, *costs approximately the same* as that of the two-input gate. Indeed, the main cost of the OT is the [16] rows transfer. Sending the encryptions of the actual secrets, while exponential in the number of inputs, is dominated by the OT matrix row transfer for gates with up to about 8 binary inputs. In our case, sending of 20 secrets requires only 20 bits (one bit per secret) in addition to the OT matrix transfer. Thus, additional communication as compared to standard 1-out of-4 GMW OT extension (also implemented via [16]) is only $20 - 4 = 16$ bits!

As a result, the circuit reduction achieved by embedding several clauses into one container is directly translated into the overall improvement for semi-honest GMW protocol.

2.3   Efficient Circuit Embedding to Obtain Identically-Wired Clauses

We now describe the intuition behind our graph/circuit embedding algorithm, as well as summarize its performance in terms of the size of the embedding graph. In Sect. 3, we describe a circuit embedding algorithm, which takes as input the set of k circuits $C_1, ..., C_k$ and returns an (unprogrammed) container circuit C_0 capable of embedding each of these circuits, as well as the programming strings needed to generate the garblings of C_0 which implement/garble each C_i.

Our approach is graph theoretic. Assume for simplicity that we have exactly two input circuits. As a first step, we translate each circuit C_i to a directed acyclic graph (DAG) D_i (see Fig. 1 for example and Sect. 3 for a formal definition). The problem of finding a "small" container circuit embedding both C_1 and C_2 is now reduced to finding a "small" DAG which "contains" D_1 and D_2. Informally, a DAG D 'contains' another DAG D' if through a series of node deletions, edge

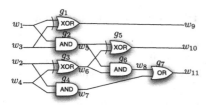
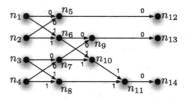

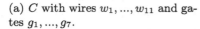

(a) C with wires $w_1, ..., w_{11}$ and gates $g_1, ..., g_7$.

(b) D with nodes $n_1, ...n_{14}$ corresponding to C's gates, inputs and outputs.

Fig. 1. A 2-bit adder circuit C and corresponding circuit DAG D. Edges of D are assigned weights to facilitate Free XOR optimizations (cf. Sect. 3).

deletions and replacing each 3-node path uvw where v has in-degree and out-degree 1 with a 2-node path, i.e., an edge, uw in D, one can recover a graph isomorphic to D'.

We start by showing that if the input DAGs are restricted to have out-degree at most one, then there exists a polynomial time algorithm (Algorithm 1) to find a DAG D_0, also of out-degree at most one, of minimum size. We remark that our approach is closely related to the classical polynomial time algorithm for testing whether or not two trees are isomorphic [31], though it is more difficult than this. Indeed, the operation which replaces 3-node paths with 2-node paths is closely related to edge contraction of graph minors [28].

Restricting DAGs to having out-degree at most one corresponds to restricting circuits to having fan out at most one and is, of course, unrealistic. To develop a general algorithm (see Fig. 2 for a toy example), we observe that the nodes of every DAG D with r sinks can be covered by a set of r DAGs each with out-degree at most one, i.e. *subtrees*. For each pair of such subtrees (one from D_1 and one from D_2) we first apply Algorithm 1 to determine the minimum cost (roughly, the minimum $|D_0|$) of co-embedding the pair. We use these costs to weight an auxiliary complete bipartite graph: roughly, one part is labeled by the subtrees of D_1, one part is labeled by the subtrees of D_2, and the weight of the edge is the minimum cost of co-embedding the subtrees corresponding to the edge's endpoints. The minimum weight perfect matching in this graph corresponds to a valid container circuit that can be easily constructed. In generality, only considering subtrees covering the nodes of D_1, D_2 may leave out some edges, which we then appropriately reinsert into D_0 to guarantee that D_0 will be universal for both D_1, D_2.

We now turn to the performance of our algorithm. Clearly any circuit embedding of circuits C_1 and C_2 has size at least $\max\{|C_1|, |C_2|\}$ and needs to have size at most $|C_1| + |C_2|^3$. Our experimental validation (see Sect. 8) embeds two circuits into a circuit whose size is on average 15.1 percent of the way between

---

³ For some pairs of circuits our heuristic may produce a container circuit of size greater than $|C_1| + |C_2|$. In this case, we will simply take C_0 to include both C_1 and C_2.

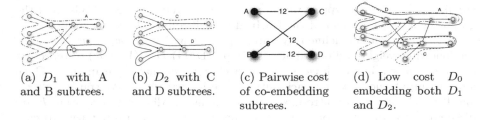

(a) D_1 with A and B subtrees.

(b) D_2 with C and D subtrees.

(c) Pairwise cost of co-embedding subtrees.

(d) Low cost D_0 embedding both D_1 and D_2.

Fig. 2. Determining a low cost (circuit) DAG embedding two input (circuit) DAGs.

these trivial lower and upper bounds. *Assuming this embedding performance,* by divide-and-conquer repeated embedding we would obtain an embedding of k circuits of size n into a circuit of size $1.151^{\log k}n = k^{0.203}n$ (Lemma 1).

2.4   NP-hardness of Graph Embedding

In Sect. 7 we show that the problem of finding a minimum-cost circuit C_0 into which two given circuits C_1 and C_2 can be embedded is NP-complete. The proof uses a reduction from the well-known NP-complete problem 3-SAT [10]. In fact, the reduction shows the somewhat stronger result that says that the problem remains NP-complete even when one of C_1 or C_2 is a tree (and the other a DAG) and both have bounded in-degree and out-degree.

Intuitively, the idea of the reduction is that the DAG, say C_1, represents all possible truth assignments of the variables and all possible ways to satisfy each clause in the 3-SAT instance while the tree C_2 represents the requirement that each variable must be set to true or false and the requirement that each clause has (at least) one resulting literal that is true. Then we show that an embedding of C_2 into C_1 is possible if and only if there is a satisfying assignment for the 3-SAT instance. Clearly such an embedding has minimum cost. When such an embedding exists, it can easily be interpreted as a particular truth assignment to the variables and an "assignment" of one resulting true literal to each clause.

3   Definition of Circuit Embedding

In this section, we bridge circuit-based SFE and graph theory. In particular, we describe the CIRCUIT EMBEDDING ALGORITHM, which takes as input the set of k circuits $C_1, ..., C_k$ and returns the desired n_0-gate *container circuit* C_0 together with *definition (programming) strings* $d_1, ..., d_k$. Specifically, a container circuit is an unprogrammed circuit, that is, a collection of gates each of whose function is unspecified, though the wire connections between these gates are fixed. The function of these gates is then specified by choosing a programming stream, which is a mapping from the gates to the functions $\{\vee, \wedge, \oplus, L, R\}$, where L (resp. R) is the left (resp. right) wire pass through gate. To describe the CIRCUIT EMBEDDING ALGORITHM, we start by describing a mapping between circuits and a specific type of weighted directed acyclic graphs (DAGs) and the graph theoretic equivalent of the Garbled Circuit approach we are proposing.

Let C be a circuit defined by gates $g_1, ..., g_n$ and wires $w_1, ..., w_m$. We use the following weighted directed acyclic graph (DAG) $D = (V, A, w)$ to represent it. The node set V has three parts: for each wire w_i that is an input to C we add an "input" node n_i, for each output wire w_i, we add an "output" node n_i, and for each gate g_i, we introduce a "gate" node n_i. All directed edges in E are directed in the direction of evaluation. Specifically, for each input wire to gate g_i there is an edge from its corresponding "input" node to the "gate" node n_i. For each output wire from gate g_i there is an edge from the "gate" node n_i to its corresponding "output" node. For each wire from gate g_i to gate g_j, there is an edge from n_i to n_j. Finally, for simplicity in dealing with free-XORs and the cost of circuit, we give each edge a weight. For a gate node g_i corresponding to an XOR-gate, we give all in-edges e of g_i weight $w_e = 0$; for output nodes n_i, we give all in-edges e of n_i weight $w_e = 0$; for all other edges e receive weight $w_e = 1$. See Fig. 1 for an example. We call such a DAG, the circuit DAG. We remark that given a circuit DAG we can always determine an unprogrammed circuit corresponding to it.

The cost of a circuit is the total size of the truth tables needed to represent it, i.e., $\sum_{\text{non-XOR } g_i} 2^{\{\text{fan in of gate } g_i\}}$, where XOR-gates add zero cost [18]. This translates to the corresponding circuit DAG as $\text{cost}(D) := \sum_{u \in D} 2^{\sum_{v \in N_D^-(u)} w_{vu}}$, where $N_D^-(u)$ is the set of in-neighbors of node $d \in D$.

We are interested in the minimum cost container circuit C_0 that can be used to embed circuits $C_1, ..., C_k$. Necessarily, this requires that for each C_j there is a 1-1 mapping f from the gates of C_i to C_0, such that, for each wire of C_j between gate g_i and $g_{i'}$ there is a set of wires linking $f(g_i)$ and $f(g_{i'})$. Moreover and as we now describe, the flow of information of C_j must be preserved in C_0.

An *out-arborescence* is a directed acyclic graph that is weakly connected[4] and every node has in-degree at most one. We define the *source* of an out-arborescence T, denoted source(T), as the unique vertex with in-degree zero. Let $D' = (V', A', w')$ and $D = (V, A, w)$ be DAGs.

Definition 1. An *embedding* of D' into D is a mapping f from **nodes** of V' to **out-arborescences of** D and from **(weighted) directed-edges of** A' to **(weighted) directed-edges of** A satisfying

1. for all $u' \neq v' \in V'$, $f(u') \cap f(v') = \emptyset$,
2. for $u'v' = e' \in A'$, $\exists x \in f(u')$ such that $f(e')$ starts at x and ends at the source of $f(v')$, and
3. for $u'v' = e' \in A'$, $w'_{e'} \leq w_{f(e')}$.

It follows immediately from the definition that there is a 1-1 mapping between nodes of D' and the sources of the out-arborescences in D specified by f. Moreover, for every node n' of D' and source of $f(n') = n$, f is a mapping such that

---

[4] A directed graph is weakly connected if replacing all edges with undirected edges yields a connected graph, that is, every pair of nodes in the graph is connected by some path.

for each in-edge e' of n' and there is a unique in-edge $e = f(e')$ of n such that $w'_{e'} \le w_e$. From this it follows that the sum of the weights on the in-edges of n is at least as large as the sum of the weights on the in-edges of n'. Hence, we have the following observation.

Observation 1. $\text{cost}(D) \ge \text{cost}(D')$.

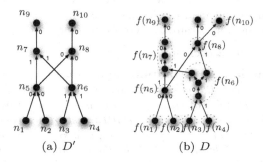

(a) D' (b) D

Fig. 3. An example embedding f of D' in D.

We now are in a position to describe the CIRCUIT EMBEDDING ALGORITHM. Let $C_1, ..., C_k$ be the set of k-input circuits. First, we find the corresponding circuits DAGs $D_1, ..., D_k$. Second, given this set of circuits DAGs, we determine a *low cost* circuit DAG D_0 that embeds each of $D_1, ..., D_k$ with functions $f_1, ..., f_k$. The heuristic we describe in Sect. 6 is one approach to solve this second step. Third, we determine the container circuit C_0 as the circuit corresponding to D_0. Finally, we determine the programming string d_i for each i. To do so, we need only specify the function of each gate node in D_0. For a specific embedding f_i, each gate node v of D_0 is either A) a source or B) a non-source node of some out-arborescence. In the former case, $d_i(v)$ is equal to either AND, OR or XOR depending on the function of the pre-image of the out-arborescence rooted at v. In the latter case, $d_i(v)$ is equal to L as the left input wire pass-through.

4   GC Protocol for Overlaying Subcircuits

In this section we will formalize the intuition of Sect. 2.1. Namely, we will present a full GC protocol with processing of k *identically wired* switch clauses at approximately the cost of *one* such clause, and prove its security. Of course, identically wired clauses are not typical in circuits. In Sect. 6 we show how to embed a number of arbitrary circuits into a single container circuit, so that each of the circuits could be implemented by a corresponding programming of the gates of the container circuit. Our approach is heuristic, but we of course can always fall back on running standard GC/GMW protocols if the performance of the heuristic does not outperform.

Our approach can be instantiated using a number of GC garbling techniques. For simplicity of presentation, and because it is a standard GC trick, in the following presentation we omit assigning and processing the wire key pointers which will tell the evaluator which garbled table row to decrypt. Also, we do not include Free-XOR in this algorithm. We will argue later that our construction allows to take full advantage of Free-XOR. Finally, for ease of presentation and w.l.o.g., our construction is for functions with a single switch.

Consider an $\{\vee, \wedge, \oplus\}$ circuit C with a switch $(C_1, ..., C_k)$ statement, which evaluates one of subcircuit clauses $C_1, ..., C_k$ based on an internal variable. Let Enc, Dec be a semantically secure encryption scheme.

Protocol 1. *(GC with switch statements)*

1. **Once-per-function Precomputation.** *Parse C, identify* switch *$(C_1, ..., C_k)$, and call the graph embedding algorithm on $C_1, ..., C_k$. Obtain the container circuit C_0 of size n_0 as well as k circuit programming strings $d_1, ..., d_k$, each of size n_0. Each d_i will consist of symbols $\{\vee, \wedge, \oplus, L, R\}$, where L (resp. R) is the left (resp. right) input wire pass-through gate. Denote the j-th symbol of d_i by $d_{i,j}$. Let C' be the C with the switch $(C_1, ..., C_k)$ replaced with C_0. C' is assumed known to both players before the computation.*

2. *For each wire W_i of C', GC generator randomly generates two wire keys w_i^0, w_i^1.*

3. *For each gate g_i of $C' \setminus C_0$ in topological order, GC generator garbles g_i to obtain the garbled gate table. For each of 2^2 possible combinations of g_i's input values $v_a, v_b \in \{0, 1\}$, set*

$$e_{v_a, v_b} = H(k_a^{v_a} || k_b^{v_b} || i) \oplus w_c^{g_i(v_a, v_b)}$$

Garbled table of g_i is a randomly permuted set $\{e_{v_a, v_b}\}, v_a, v_b \in \{0, 1\}$.

4. *GC generator sends all generated garbled tables to GC evaluator. Garblings of inputs of C' are sent to GC evaluator directly and via OT, as is standard in GC.*

5. *GC generator generates n_0 random permutations π_i over $\{\vee, \wedge, \oplus, L, R\}$.*

6. *GC generator computes the following. Let $W_{j_1}, ... W_{j_t}$ be the wires defining the switch choice, $t = \lceil \log k \rceil$. For $i = 1$ to k, set $\tilde{d}_i = \pi_1(d_{i,1}), ..., \pi_{n_0}(d_{i,n_0})$. Now, each $\tilde{d}_i$ looks random as an independent random permutation π_j was applied to each symbol $d_{i,j}$. Let $ED = Enc_{key_1}(\tilde{d}_1), ..., Enc_{key_k}(\tilde{d}_k)$. Here key key_i is derived from wire keys of $W_{j_1}, ... W_{j_t}$, corresponding to switch selection i, by setting $key_i = H(\text{"switchkey"}, w_{j_1}^{i_t}, ... w_{j_t}^{i_t})$.*

7. *GC generator sends encrypted circuit definition strings ED to GC evaluator, in a random order.*

8. *GC evaluator evaluates in topological order all gates that are possible[5]; in particular, the wire keys defining the values of the switch statement will be known to GC evaluator.*

---

[5] Recall, for simplicity we did not explicitly include the standard permute-and-point table row pointers in our protocol. We assume the evaluator knows decryption of which row to use, e.g. via using the standard permute-and-point technique.

9. GC evaluator derives the decryption key for $Enc_{key_i}(\tilde{d}_i)$ and decrypts to obtain $\tilde{d}_i$, the (permuted) definition string for the clause to be evaluated. Evaluator will know which string to decrypt by including an additional pointer bit in the wire labels of $W_{j_1}, ... W_{j_t}$ (point-and-permute).

10. For each gate $g_i \in C_0$, in topological order

   (a) GC generator prepares five garbled tables, $\{T_\vee, T_\wedge, T_\oplus, T_L, T_R\}$ implementing one each of gate functions $\{\vee, \wedge, \oplus, L, R\}$, i.e. OR, AND, XOR, Left wire pass-through, Right wire pass-through. Note that all five garbled tables are constructed with respect to the same input/output wire labels of gate g_i.

   (b) The two players execute in parallel n_0 semi-honest 1-out-of-5 OT protocols, for j from 1 to n_0. Here GC generator's input is $\pi_j(\{T_\vee, T_\wedge, T_\oplus, T_L, T_R\})$, and GC evaluator's input is the symbol of the programming string obtained in Step 9, i.e. $\pi_j(\{\vee, \wedge, \oplus, L, R\})$. As a result, GC evaluator receives garbled gate tables of the remaining gates.

11. GC evaluator evaluates in topological order all remaining gates of C' and sends output wire keys to generator for decryption.

Observation 2. *For simplicity of presentation and to focus on the novel contribution, we omitted explicitly writing out some standard GC techniques, such as permute-and-point.*

Observation 3 (Free-XOR compatibility). *We presented the protocol without regard to free-XOR. However, it is easy to see that our construction is compatible with it. Indeed, as is also argued in discussion on the circuit embedding heuristic in Sect. 6, the generated container circuit will have many gates fixed to be XOR gates, rather than placeholders for one of $\{\vee, \wedge, \oplus, L, R\}$. It is easy to see that since any of k clauses could be implemented in the container circuit, and "permanently" fixing some of its gates to be XOR is done in circuit pre-processing, this will not affect security.*

In our circuit embedding heuristics, we aimed to maximize the number of such gates so as to take the full advantage of free-XOR.

We note that it is not immediately clear how to use the 3-row garbled-row reduction (GRR3) of [27] in our approach. This is because the GRR3 idea is to define one of the garbled rows as a function of garbled values of the two input wires which result in this output value (and omit that row from the table). However, in our setting, we don't know which gate function will be used. Hence, the 0–1 semantics of the implicitly defined gate output label may be different for different gate functions which may program this specific gate of C_0. This will cause problems for garbling subsequent gates. We thus do not use GRR3, and use standard 4-row tables. We discuss optional inclusion of NOT gates in our circuits in the full version.

Theorem 1. *Let OT protocol be secure in the semi-honest model. Let Enc be semantically-secure encryption. Let H be a hash function modeled by a random oracle. Then Protocol 1 is secure two-party computation protocol in the semi-honest model.*

The proof (with respect to the standard security definition of secure computation) is deferred to the full version, due to the lack of space.

Cost calculation. As compared to plain GC of C, our protocol uses additional OT instances. This comes cheap due to the Ishai et al.'s OT extension [14] and follow-up optimizations, such as [1,16]. Further, an extension of [14] for 1-out of-k OT of Kolesnikov and Kumaresan [16] can be effectively used for our 1-out of-5 OTs.

In detail, let s be the computational security parameter, and take the size of each garbled table as $4s$. Then the communication cost of evaluating the `switch` on the k clauses embedded in container C_0 of size n_0 will be approximately $3kn_0 + 22n_0s$.

Indeed, 1-out of-k OT of circuit programming strings will take approximately $3kn_0$ bits (k encryptions of $3n_0$-bit long strings, plus a 1-out of-k OT on short decryption keys of size s, whose cost is small and is ignored.) Running 1-out of-5 OT on gate tables of size $4s$ is done via [16]. (Recall, [16] shows how to do 1-out of-5 OT for only double the cost of 1-out of-2 OT.) The cost consists of sending 5 encryptions each of length $4s$, and running 1-out of-4 OT on random secrets of size s, which costs approximately $2s$, i.e. one OT extension matrix row of [16]. Summing up, we get our cost approximately $3kn_0 + 22n_0s$.

Ignoring lower order term $3kn_0$, *we can view our communication cost per gate as approximately factor* 5.5 *of that of the standard Yao-gate* , and factor 11 of that of the optimal garbling of Zahur et al. [34]. We note that in cases where clause is selected by the input of a player (GC generator), our cost of each gate is the same as that of [34]. We finally note that we, in contrast with all prior GC protocols, *do not* need to include the circuitry selecting the output of the right clause and ignoring outputs of other clauses.

We discuss experimental results, which depend on the quality of embedding, in Sect. 8.

Nesting `switch` statements

We observe that a natural implementation of `switch` nesting will be secure and cheap. Intuitively, this is because the vast majority of the cost – OTs of the gates – will remain unaffected by sub-switches, and *only the programming strings management will need to be adjusted*.

Due to the lack of space, we present the detailed nesting construction in the full version.

5 GMW Protocol for Overlaying Subcircuits

Our GMW protocol is a natural recasting of our GC protocol into the GMW approach, with the exception of us making and exploiting the novel observation that multi-input gates in GMW are cheap. In GMW, we will program a gate simply by viewing it as having an additional 5-ary function definition input. The circuit programming string will be secret-shared among the players with a $(2,2)$

secret sharing, just like regular GMW wire values. Thus our programmable gate evaluation is just a slight generalization of the GMW evaluation.

Protocol 2 *(GMW with* `switch` *statements, sketch)*

1. **Once-per-function Precomputation.** *Parse* C, *identify* `switch` $(C_1, ..., C_k)$, *and call the graph embedding algorithm on* $C_1, ..., C_k$. *Obtain the container circuit* C_0 *of size* n_0 *as well as* k *circuit programming strings* $d_1, ..., d_k$, *each of size* n_0. *Each* d_i *will consist of symbols* $\{\vee, \wedge, \oplus, L, R\}$, *where* L *(resp.* R*) is the left (resp. right) input wire pass-through gate. Denote the* j-th *symbol of* d_i *by* $d_{i,j}$. *Let* C' *be the* C *with the* `switch` $(C_1, ..., C_k)$ *replaced with* C_0. C' *is assumed known to both players before the computation.*
2. *Beginning with the secret sharing of the inputs, for each gate* g_i *of* $C' \setminus C_0$ *in topological order, players evaluate the gates according to the GMW protocol. In particular, wires defining the the the* `switch` *choice will be processed.*
3. *Let* $W_{j_1}, ... W_{j_t}$ *be the wires defining the* `switch` *choice,* $t = \lceil \log k \rceil$. *Players use OT to generate a* $(2, 2)$ *secret sharing of the selected programming string as follows.*
   (a) *GC generator generates* n_0 *random permutations* $\pi = \{\pi_j\}$ *over* $\{\vee, \wedge, \oplus, L, R\}$.
   (b) *GC generator computes the following. For* $i = 1$ *to* k, *set* $\tilde{d}_i = \pi_1(d_{i,1}), ..., \pi_{n_0}(d_{i,n_0})$. *Now, each* $\tilde{d}_i$ *looks random as an independent random permutation* π_j *was applied to each symbol* $d_{i,j}$.
   Player P_2 *uses his shares of* $W_{j_1}, ... W_{j_t}$ *to obtain (via OT with* P_1*) the permuted programming string* $\tilde{d}_i$.
4. *Players proceed to evaluate all remaining gates of* C'. *The gates in* C_0 *have an additional input specifying the gate function. This input is taken from the circuit programming string* d. *Note that* d *is already secret-shared among the two players:* P_1 *has* π, P_2 *has* $\tilde{d}_i$. *Each (two-input Boolean) gate of* C_0 *is evaluated by a slight generalization of GMW, where 1-out of-4 · 5 OT is run by the players.*
   Specifically, for each of the five possible gate functions $\{\vee, \wedge, \oplus, L, R\}$ *for gate* $g_j \in C_0$, P_1 *prepares four corresponding GMW OT secrets. Then* P_1 *permutes the five groups of four GMW OT secrets according to* π_j. *Then* P_1 *and* P_2 *run 1-out of-4 · 5 OT, where* P_2*'s input is* $\tilde{d}_{i,j}$ *and the GMW shares of the wire values.*
5. *Players combine their shares on the output wires of* C' *and reconstruct the output.*

Theorem 2. *Let* OT *protocol be secure in the semi-honest model. Then Protocol 2 is a secure two-party computation protocol in the semi-honest model.*

Proof. The proof of security of this protocol in the semi-honest model is trivial. Indeed, assuming the security of OT, it is easy to check that players ever receive only secret shares of the wire values. We omit this simple proof.

Cost calculation. As compared to standard GMW protocol, our protocol requires OT of the programming strings. Additionally, it evaluates multi-input gates, resulting in 1-out of-20 OT. As discussed above, in particular in GC cost calculation, the OT of programming strings is a low-order cost term and can be ignored. Further, as explained in detail in Sect. 2.2, the cost of 1-out of-20 OT using [16] is only 16 bits greater than that of the 1-out of-4 OT, and hence this difference can also be swept under the rug. We conclude that the above Protocol 2 (GMW with switch statements) implements the oblivious circuit programming at the cost similar to that of the standard GMW protocol.

Nesting switch statements. The discussion and results of GC-based nesting (Sect. 4) directly applies to the GMW setting.

6   Embedding Circuits of Bounded Fan In

In this section, we sketch a heuristic algorithm which given a set of k circuit DAGs, $D_1, ..., D_k$, returns a circuit DAG D_0 such that for each D_i there exists an embedding f_i into D_0, and D_0 has as small of cost as possible. We proceed in two main steps.

First, in Sect. 6.1, we restrict our attention to circuits that have fan out one and fan in bounded by 2, though a straightforward generalization leads to fan in bounded by any constant. These are commonly referred to as *in-arborescences of bounded in-degree* 2, but for ease of exposition we will call them *tree circuits*. We describe a polynomial time **exact** algorithm that given two circuit trees T_1 and T_2 finds a circuit tree T of minimum cost embedding both T_1 and T_2. Specifically, we prove the following.

Definition 2. The *cost* of embedding a set of circuit DAGs $D_1, ..., D_k$, denoted $cost(D_1, ..., D_k)$, is the cost of a circuit DAG D_0 of minimum cost such that there is an embedding of D_i into D_0 for all $i = 1..k$.

Theorem 3. *Let T_1 and T_2 be tree circuits. There exists an $O(|T_1||T_2|)$ algorithm to determine an optimal, i.e. minimum cost, tree circuit T embedding both T_1 and T_2.*

Second, in Sect. 6.2, we remove the bound on the fan out, only requiring that the input circuits have fan in bounded by 2. We describe an algorithm which relies on the algorithm of Sect. 6.1 as a subroutine. Letting D_1 and D_2 be circuit DAGs of fan out 2, we describe a polynomial time heuristic algorithm to a determine circuit DAG D_0 embedding both D_1 and D_2.

A straightforward approach, using this heuristic as a subroutine, then allows for k-circuit inputs. The following lemma describes an algorithm which returns a circuit whose size grows sublinearly in the number of input circuits (assuming certain performance of the heuristic). Because we condition it on the heuristic, we stress that the Lemma cannot be universally applied. Rather, we use the Lemma to formalize the embedding algorithm for larger number of clauses, as well as to discuss its projected performance.

Lemma 1. *Let $\tau > 1$. Assume there exists an algorithm which takes as input circuit DAGs D' and D'', each of size exactly n, and returns a circuit DAG*

D_0 *embedding both* D' *and* D'' *whose size is at most* τn. *Then there exists an algorithm which takes as input circuit DAGs* $D_1, ..., D_k$, *each of size exactly* n, *and returns* D_0 *of size at most* $k^{\log_2 \tau} n$.

Proof. Let $D_1, .., D_k$ be a set of circuit DAGs and assume $k = 2^\ell$. We first apply the heuristic of Sect. 6 to determine a circuit DAGs D_i^2 embedding both D_{2i-1} and D_{2i} for each $i = 1, ..., k/2$. Iterating this for $j \geq 2$, for each $i = 1, ..., k/2^j$ we determine D_i^j from D_{2i-1}^{j+1} and D_{2i}^{j+1}. We then return $D = D_1^\ell$.

We prove the size bound by induction, where the base case is assumed. By induction, assume that the algorithm returns D_i^{j-1}, $i = 1, ..., k/2^j$, each of whose size is at most $\tau^{j-1} n$. Hence, the size of $D_{i'}^j$ is at most $(\tau) * \tau^{j-1} n = \tau^j n$. Setting $j = \ell = \log k$ yields the desired result.   □

In Sect. 8, we evaluate the performance of the heuristic, which, in our experiments on average achieves a τ value of 1.151. Assuming this τ value, Lemma 1 would imply the k-circuit size is at most $k^{0.203} \times \max\{|D_1|, |D_2|, ..., |D_k|\}$.

6.1 Tree Circuits

In order to prove Theorem 3, we use dynamic programming and *match* pairs of vertices of T_1 and T_2 as follows. For simplicity, we omit dealing with Free-XOR for now. Let $\delta^-(v)$ be the in-degree of a node.

Definition 3. For circuit DAG D and $t \in D$, let $D[t]$ be the circuit DAG induced on vertices v such that there exists a directed path from from v to t in D.

Definition 4. Define the *matchcost* of $a \in T_1$ and $b \in T_2$ as the minimum cost of a tree T such that there exists a mapping f_1 that embeds $T_1[a]$ into T and a mapping f_2 that embeds $T_2[b]$ into T where $f_1(a) = f_2(b)$. Denote this minimum cost by match(a, b).

Consider computing cost(T_1, T_2) where a is the root of T_1 and b is the root of T_2. Clearly, there is no advantage, with respect to cost, to mapping a and b to disjoint subtrees of T and so either (i) $f_1(a) \in T[f_2(b)]$, or (ii) $f_2(b) \in T[f_1(a)]$. From this it follows that we can compute cost(T_1, T_2) by considering $O(|T_1| + |T_2|)$ matchcosts.

Definition 5. Let T_1 and T_2 be tree circuits with roots a and b, respectively. Define:

(i) $\text{cost}_2(T_1, T_2) := \min_{t \in T_2} (\text{cost}(T_2) - \text{cost}(T_2[t]) + \text{match}(a, t))$.
(ii) $\text{cost}_1(T_1, T_2) := \min_{t \in T_1} (\text{cost}(T_1) - \text{cost}(T_1[t]) + \text{match}(t, b))$.

Lemma 2. *Let* T_1 *and* T_2 *be tree circuits with roots* a *and* b, *respectively. Let* T *be a minimum cost tree circuit with* f_1 *embedding* T_1 *and* f_2 *embedding* T_2.

(i) If $f_1(a) \in T[f_2(b)]$, *then* $\text{cost}(T_1, T_2) = \text{cost}_2(T_1, T_2)$,
(ii) If $f_2(b) \in T[f_1(a)]$, *then* $\text{cost}(T_1, T_2) = \text{cost}_1(T_1, T_2)$.

Proof. Without loss of generality, assume that $f_1(a) = t' \in T[f_2(b)]$ and consider the minimum cost and minimum edge tree circuit T. The root r of T is equal to $f_2(b)$ (by minimality) and there exists $t \in T_2$ such that $f_2(t) = t'$. We have that $\mathrm{cost}(T_1, T_2)$ is equal to the cost of embedding the tree $T_2 - T_2[t]$ plus the minimum cost of a tree T' that embeds both $T_2[t]$ and T_1 given that a and t are mapped to the root of T'. Hence, $\mathrm{cost}(T_1, T_2) = \mathrm{cost}(T_2 - T_2[t]) + \mathrm{match}(a, t) = \mathrm{cost}(T_2) - \mathrm{cost}(T_2[t]) + \mathrm{match}(a, t) = \mathrm{cost}_2(T_1, T_2)$. The lemma follows. $\square$

Corollary 1. $\mathrm{cost}(T_1, T_2)$ *is equal to the minimum of* $\mathrm{cost}_1(T_1, T_2)$, *and* $\mathrm{cost}_2(T_1, T_2)$.

In order to achieve the runtime of Theorem 3, we observe that we can determine these costs using the children of a and b together with a single match.

Lemma 3. *Let T_1 and T_2 be tree circuits with roots a and b, respectively. Then,*

$$\mathrm{cost}_1(T_1, T_2) = \min\Big\{ \mathrm{match}(a, b),$$

$$\min_{a' \in N_{T_1}^-(a)} (\mathrm{cost}(T_1) - \mathrm{cost}(T_1[a']) + \mathrm{cost}_1(T_1[a'])(T_2)) \Big\},$$

$$\mathrm{cost}_2(T_1, T_2) = \min\Big\{ \mathrm{match}(a, b),$$

$$\min_{b' \in N_{T_2}^-(b)} (\mathrm{cost}(T_2) - \mathrm{cost}(T_2[b']) + \mathrm{cost}_2(T_1)(T_2[b'])) \Big\}.$$

Proof. We have that

$$\min_{t \in T_1[a']} (\mathrm{cost}(T_1) - \mathrm{cost}(T_1[t]) + \mathrm{match}(t, b))$$

$$= \mathrm{cost}(T_1) - \mathrm{cost}(T_1[a']) + \min_{t \in T_1[a']} (\mathrm{cost}(T_1[a'])\mathrm{cost}(T_1[t]) + \mathrm{match}(t, b))$$

$$= \mathrm{cost}(T_1) - \mathrm{cost}(T_1[a']) + \mathrm{cost}_1[T_1[a']][T_2].$$

Hence, $\mathrm{cost}_1(T_1, T_2)$

$$= \min_{t \in T_1} (\mathrm{cost}(T_1) - \mathrm{cost}(T_1[t]) + \mathrm{match}(t, b))$$

$$= \min\Big\{ \mathrm{match}(a, t), \min_{a' \in N_{T_1}^-(a)} \min_{t \in T_1[a']} (\mathrm{cost}(T_1) - \mathrm{cost}(T_1[t]) + \mathrm{match}(t, b)) \Big\}$$

$$= \min\Big\{ \mathrm{match}(a, b), \min_{a' \in N_{T_1}^-(a)} (\mathrm{cost}(T_1) - \mathrm{cost}(T_1[a']) + \mathrm{cost}_1[T_1[a']][T_2]) \Big\},$$

completing the proof of the lemma. $\square$

From Lemma 2, in order to determine $\mathrm{cost}(T_1, T_2)$, it remains to show how to determine $\mathrm{match}(a, b)$. Since the mapping of a and b are fixed, matchcosts are easier to compute. Indeed, we can assume $f_1(a) = f_2(b)$ is the root of T. Moreover, if either $T_1[a]$ or $T_2[b]$ is a singleton then $\mathrm{match}(a, b)$ can be determined in a straightforward way.

Observation 4. *If $T_1[a]$ is a singleton, then for all $b \in T_2$, match$(a, b) =$ cost$(T_1[a], T_2[b]) =$ cost$(T_2[b])$. If $T_2[b]$ is a singleton, then for all $a \in T_1$, match$(a, b) =$ cost$(T_1[a], T_2[b]) =$ cost$(T_1[a])$.*

From Observation 4 it is trivial to determine match(a, b) whenever either a is a leaf of T_1 or b is a leaf of T_2. Specifically, in the case that b is a leaf, we have match$(a, b) = \sum_{t \in T_1[a]2} \Sigma_{v \in N^-_{T_1}(t)} w_{vt}$ and when a is a leaf, match$(a, b) = \sum_{t \in T_2[a]2} \Sigma_{v \in N^-_{T_2}(t)} w_{vt}$.

We therefore can assume that $T_1[a]$ and $T_2[b]$ each have at least three vertices. To determine match(a, b) we simply consider all possible pairings of the children.

Lemma 4. *For $a \in T_1$ with in-neighbors a_0, a_1 and $b \in T_2$ with in-neighbors b_0, b_1 we have*

$$\text{match(a, b)} = 2^2 + \min_{i \in \{0,1\}} \min_{j \in \{0,1\}} \left(\text{cost}(T_1[a_i], T_2[b_j]) + \text{cost}(T_1[a_{1-i}], T_2[b_{1-j}]) \right).$$

Proof. Since $\delta(a) = \delta^-(b) = 2$, the minimum cost of a tree circuit T embedding both a and b is 2^2 plus the minimum cost of embedding the subtrees $T_1[a_0], T_1[a_1], T_2[b_0]$, and $T_2[b_1]$. We only need to check which of the four possible feasible combinations achieves the minimum. □

We now can finish the proof of Theorem 3 whose pseudo code is given as Algorithm 1.

Proof (Proof of Theorem 3). Consider Algorithm 1. We note that by proceeding in a reverse BFS-ordering of both $V(T_1)$ and $V(T_2)$ we ensure that we can compute cost$_1$, cost$_2$ and match in Lines 7, 8 and 9. Hence, the correctness of this algorithms follows from Lemmas 3 and 4 and Corollary 1. Clearly the run time is equal to $O(|T_1||T_2|)$ times the runtime of determining $M[a_i, b_j]$ and $C[a_i, b_j]$. We consider these two parts separately. First, by Observations 4 and Lemma 4, determining $M[a_i, b_j]$ takes constant time. Hence, the total time taking determining the $|T_1| \times |T_2|$ array is $O(|T_1||T_2|)$. By Lemma 3, determining $C_1[a_i, b_j]$ takes $O(\delta^-(a) + 1)$ time. Hence, the total time determining C_1 is $\sum_{a_i} \sum_{b_j} O(\delta^-(a_i) + 1) = |T_2| \sum_{a_i} O(\delta^-(a_i) + 1) = O(|T_2||T_1|)$. Similarly, the total time determining C_2 is $O(|T_1||T_2|)$. The runtime now follow. Finally, determining an optimal tree is now trivial given the choices made by Algorithm 1. □

We finish this section by noting that to deal with XOR-gates, which are *free*, when two XOR gates are mapped to the same node in T, we ensure zero addition cost is added. With these modifications, it follows that Algorithm 1 can also be used to compute cost(T_1, T_2) in this more general case.

```
1  Input: Binary tree circuits T₁, T₂
2  Output: cost(T₁, T₂)
3  let a₁, ..., aₙ₁ be a BFS-ordering of V(T₁)
4  let b₁, ..., bₙ₂ be a BFS-ordering of V(T₂)
5  for i = n₁ down to 1:
6  ... for j = n₂ down to 1:
7  ...... determine M[aᵢ, bⱼ] = match(aᵢ, bⱼ).
8  ...... determine C₁[aᵢ, bⱼ] = cost₁(T₁[aᵢ], T₂[bⱼ]).
9  ...... determine C₂[aᵢ, bⱼ] = cost₂(T₁[aᵢ], T₂[bⱼ]).
10 ...... set C[aᵢ, bⱼ] = min(C₁[aᵢ, bⱼ], C₂[aᵢ, bⱼ]).
11 return C(T₁[a₁], T₂[b₁])
```

$$\textbf{Algorithm 1}: \text{Determining cost}(T_1, T_2)$$

6.2   General Circuits

Heuristic Algorithm. We develop a polynomial time heuristic algorithm using the machinery of Sect. 6.1. We then finish by sketching the proof of correctness. In Sect. 8, we present the results of our experimental validation for this algorithm. For simplicity, assume every non-leaf node of T_1 and T_2 has weighted in-degree exactly two and we omit dealing with Free-XOR for now. We remark that again the ideas are easily extended to the general case.

We start by considering a related question. Let $D_1 = (V_1, E_1)$ and $D_2 = (V_2, E_2)$ be input circuit DAGs, each with exactly one output wire node. Let T_1 be a spanning in-arborescence subgraph of D_1 and let T_2 be a spanning in-arborescence subgraph of D_2. We determine a minimum cost circuit DAG D_0 embedding both D_1 and D_2 subject to the restriction that there must be a spanning in-arborescence subgraph T of D_0 such that (A) both T_1 and T_2 embed in T, and (B) leaves of T_1, resp. T_2, map to leaves of T. Denote the minimum cost of such a DAG by $\text{cost}(D_1|_{T_1}, D_2|_{T_2})$ We remark that there always exists an appropriate choice of T_1 and T_2 such that D_0 will be a optimal embedding of D_1 and D_2. Further we remark, that we can essentially ignore Condition (B), since given any embedding of T_i, it is always possible to extend the out-arborescence of any leaf node of T_i down to a leaf node of T.

Analogous to Lemma 2, we can determine $\text{cost}(D_1[a]|_{T_1[a]}, D_2[b]|_{T_2[b]})$ for $a \in T_1$ and $b \in T_2$ by considering $O(|T_1| + |T_2|)$ matchs.

Definition 6. Define the match* of $a \in D_1$ and $b \in D_2$ as the minimum cost of a circuit DAG D_0 such that there exists a mapping f_1 that embeds $D_1[a]$ into D_0 and a mapping f_2 that embeds $D_2[b]$ into D_0 such that $f_1(a) = f_2(b)$ and there exists a spanning in-arborescence subgraph T of D_0 such that (A) and (B) hold.

Definition 7. Let r be the root of circuit DAG D with gate nodes G. Further assume T is an in-arborescence subgraph of D containing r. Define the *cost of D on vertices of T* as $\text{cost}_{T(D)} := \sum_{v \in V(T) \cap G} 2^{\text{ffi}^-(v)} \cdot$

Definition 8. Let T_1 and T_2 be circuit DAGs with roots a and b, respectively. Define:

(i) $\mathrm{cost}_2(D_1|_{T_1}, D_2|_{T_2}) := \min_{t \in T_2}(\mathrm{cost}_{T_2}(D_2)$
$-\mathrm{cost}_{T_2[t]}(D_2[t]) + \mathrm{match}^*(a, t))$.

(ii) $\mathrm{cost}_1(D_1|_{T_1}, D_2|_{T_2}) := \min_{t \in T_1}(\mathrm{cost}_{T_1}(D_1)$
$-\mathrm{cost}_{T_1[t]}(D_1[t]) + \mathrm{match}^*(t, b))$.

Lemma 5. *Let D_1 and D_2 be circuit DAGs. For $a \in D_1$ let T_1 be an in-arborescence subgraph of $D_1[a]$ containing a and for $b \in D_2$ let T_2 be a in-arborescence of $D_2[b]$ containing b. Then,*

$$\mathrm{cost}(D_1[a]|_{T_1[a]}, D_2[b]|_{T_2[b]})$$
$$= \min\{\mathrm{cost}_1(D_1|_{T_1}, D_2|_{T_2}), \mathrm{cost}_2(D_1|_{T_1}, D_2|_{T_2})\}.$$

From Lemma 5, in order to determine $\mathrm{cost}(D_1[a]|_{T_1[a]}, D_2[b]|_{T_2[b]})$, it remains to show how to determine $\mathrm{match}^*(a, b)$. As before, if either $T_1[a]$ or $T_2[b]$ is a singleton then $\mathrm{match}^*(a, b)$ is as follows.

Observation 5. *If $T_1[a]$ is a singleton, then for all $b \in T_2$, $\mathrm{match}^*(a, b) = \mathrm{cost}_{T_2[b]}(D_2[b])$. If $T_2[b]$ is a singleton, then for all $a \in T_1$, $\mathrm{match}^*(a, b) = \mathrm{cost}_{T_1[a]}(D_1[a])$.*

When neither $T_1[a]$ nor $T_2[b]$ is a singleton then whenever $\delta^-(a) = \delta^-(b) = 2$ we determine $\mathrm{match}^*(a, b)$ as follows. For $a \in T_1$ with in-neighbors a_0, a_1 and $b \in T_2$ with in-neighbors b_0, b_1 we have $\mathrm{match}^*(a, b)$ is equal to:

$$2^2 + \min_{i \in \{0,1\}} \min_{j \in \{0,1\}} (\mathrm{cost}(D_1[a_i]|_{T_1[a_i]}, D_2[b_j]|_{T_2[b_j]})$$
$$+ \mathrm{cost}(D_1[a_{1-i}]|_{T_1[a_{1-i}]}, D_2[b_{1-j}]|_{T_2[b_{1-j}]})).$$

The case when the degrees do not match up is more complicated. Indeed, either the node a is incident to an edge which goes between two subtrees of $\mathcal{F}_1$ or the node b is incident to an edge which goes between two subtrees of $\mathcal{F}_2$. In this case the match^* is undefined. To get beyond this, we consider two cases separately. First, if a is a leaf node in T_1 and b is a leaf node in T_2 then we need to create a dummy gate node which takes as input $f_1(a)$ and $f_2(b)$. Such a construction has $\mathrm{match}^* = 12$ since we suffer cost 4 for each of $f_1(a)$, $f_2(b)$ and the dummy gate. Second, assume a is not a leaf node in T_1, the case when b is not a leaf in T_2 is symmetric. Our heuristic then sets match^* equal to the minimum cost of a tree such that $f_1(a)$ to be the in-neighbor of $f_2(b)$.

We now can determine D_0 using the following variant of Algorithm 1.

CIRCUIT DAG EMBEDDING ALGORITHM

1. Chose a spanning in-arborescence forest $\mathcal{F}_1$ of D_1 such that one in-arborescence of $\mathcal{F}_1$ contains each output node of D_1. Similarly, choose $\mathcal{F}_2$ of D_2. In our implementation, we will focus on choosing such forests uniformly at random. Such forests can be found by choosing a single edge from each of the out-edges for each node of D_1 and D_2; we omit further details.

1 Input: D_1, D_2, T_1 and T_2
2 Output: $\mathrm{cost}(D_1|_{T_1}, D_2|_{T_2})$
3 let $a_1, ..., a_{n_1}$ be a BFS-ordering of $V(T_1)$
4 let $b_1, ..., b_{n_2}$ be a BFS-ordering of $V(T_2)$
5 for $i = n_1$ **down to** 1:
6 ... **for** $j = n_2$ **down to** 1:
7 ...... **determine** $M[a_i, b_j] = \mathrm{match}^*(a_i, b_j)$.
8 ...... **determine** $C[a_i, b_j] = \mathrm{cost}(D_1[a_i]|_{T_1[a_i]}, D_2[b_j]|_{T_2[b_j]})$.
9 return $C(T_1[a_1], T_2[b_1])$

Algorithm 2: Determining $\mathrm{cost}(D_1|_{T_1}, D_2|_{T_2})$

2. For each $T_1 \in \mathcal{F}_1$ and $T_2 \in \mathcal{F}_2$ compute $\mathrm{cost}(D_1'|_{T_1}, D_2'|_{T_2})$, where D_1', respectively D_2', is DAG found by taking the union of all edges of D_1, respectively D_2, with at least one end-point in T_1, respectively T_2. Here we can apply Algorithm 2 directly.

3. Using the costs computed in Step 2, we compute an optimal pair of in-arborescences as follows. Let G be the weighted bipartite graph with bipartition (A, B) defined as follows. Let $m := \max\{|\mathcal{F}_1|, |\mathcal{F}_2|\}$. The set A contains a node labeled by each in-arborescences of $\mathcal{F}_1$ plus $m - |\mathcal{F}_1|$ 'dummy' nodes. Similarly, the set B contains a node labeled by each in-arborescences of $\mathcal{F}_2$ plus $m - |\mathcal{F}_2|$ 'dummy' nodes. G is a complete bipartite graph, where an edge between a node of A labeled by T_1 and node B labeled by T_2 has weight $\mathrm{cost}(D_1'|_{T_1}, D_2'|_{T_2})$, and an edge between a dummy node and node of A labeled by T_1 has weight $\mathrm{cost}_{T_1}(D_1')$, respectively node of B labeled by T_2 has weight $\mathrm{cost}_{T_2}(D_2')$.

   Since G is complete, it has a perfect matching.

   Moreover, any perfect matching corresponds to a pairing of output nodes in D_1 with output nodes in D_2, where nodes matched to 'dummy' nodes have no partner and are embedded as a copy of themselves. Hence, it follows that a minimum cost matching in B corresponds to a minimum cost pairing of output nodes. We remark, that computing such a minimum cost perfect matching in time polynomial in $|A| + |B|$ is a classical result (see for e.g. [9]).

4. We now determine the final circuit DAG. For each $T_1^i - T_2^j$ pairing from the minimum cost perfect matching, we construct a tree circuit T^{i-j} and embeddings $f_1^{i-j} : T_1^i \to T^{i-j}$ and $f_2^{i-j} : T_2^j \to T^{i-j}$, where 'dummy' pairings are the identity embedding.

   Let $T = \bigcup_{i,j} T^{i-j}$. An embedding f_1 of $\mathcal{F}_1$ into T is found by taking the union of the f_1^{i-j} over all $T_1^i - T_2^j$ pairings (including 'dummy' nodes). An embedding f_2 of $\mathcal{F}_2$ into T is found in a similar way. Let D_0 be the DAG found by taking a copy of T. First, we add an edge from the source of $f_1(x)$ to the source of $f_1(y)$ of weight w'_{xy} for each edge $xy \in E_1 - E(\mathcal{F}_1)$. For each edge $xy \in E_2 - E(\mathcal{F}_2)$ we do the same though adding these edges might cause cycles. Before adding xy, we test if there exists a directed path from y to x in D_0. If such a path P exists, then there must exists an edge of P only used by the circuit D_1. By splitting the path up to this edge, we can insure that

D_0 plus xy is acyclic. We then update f_1 and f_2 to include these additional edge mappings.

We complete the proof by showing that D_0 is a feasible solution.

Theorem 4. *The* CIRCUIT DAG EMBEDDING ALGORITHM *finds a feasible circuit DAG.*

Proof. Without loss of generality, it is enough to show that f_1 is a valid embedding of D_1 into D_0. By construction f_1 is a mapping from nodes of D_1 to out-arborescences of D_0 and from edges of D_1 to edges of D_0. We need only verify that Conditions 1, 2 and 3 of Definition 1 hold. Since Condition 1 holds for f_1^i and the perfect matching ensures that every vertex of D_1 is in exactly one paired embedding, Condition 1 holds for f_1. Conditions 2 and 3 hold, since either an edge is mapped by some f_1^i, satisfying Conditions 2 and 3, or the edge goes between trees of $\mathcal{F}_1$, where Step 4 adds these edges between sources of out-arborescences of weight satisfying Condition 3. It now follows that D_0 is a feasible solution. □

7 Optimally Embedding Graphs Is NP-complete

Here we consider the complexity of the problem of finding a minimum sized container digraph D_0 embedding two digraphs D_1 and D_2. The problem is seen to be NP-complete via a reduction from 3-SAT. The full details of the proof are deferred to the full version.

Theorem 5. *The problem of finding a digraph D_0 such that embedding two digraphs D_1 and D_2 into D_0 has cost at most k is NP-complete even when the in-degree and out-degree of each node in D_1 and D_2 is bounded by 2 and at most one of D_1 or D_2 is a tree.*

8 Experimental Evaluation and Validation

Here we report on our experimental evaluation of the heuristic given in Sect. 6, as well as on the resulting efficiency of Protocol 1 in comparison with standard GC and Protocol 2 in comparison with standard GMW.

Evaluation methodology. The main metric we use to compare our approach to GC is the *total* bandwidth required, consumed by *all* OT instances and garbled gates transfers. We do not penalize ourselves for the potential increase in latency due to additional round per `switch`, associated with our approach. As discussed in the Introduction, this is because in large circuit/batch execution round trip delays may overlap with data transmission, and, if so, latency will not impact performance. Of course, in some scenarios (e.g. large network latency, small circuit/single execution) latency may dominate. We leave full implementation and parameters tuning as important future work to address these settings.

We stress that for the SPF-SFE and GMW case, where we report significant concrete improvement in our experiments, we do not require additional rounds as compared to standard GC.

We validate our approach with the experiments on a set of circuits which we built using circuit compiler CBMC-GC [8], summarized in Table 1. We constructed these 32 circuits by exploiting variations and combinations of a number of available arithmetic and bit-operation circuits. Because of this, there are commonalities among the input circuits/DAGs (which is typical in practice), and which may be affecting performance. We stress that our algorithms are not aware of the commonalities in the circuits and apply generically. We further note that these circuits are not hand-optimized for the functions they compute. Indeed, our goal is not to find the best circuit for a specific function, but to validate our heuristic and to understand its behavior. We do this by running it on a set of simple circuits of varying sizes and similarity for our experiments. In many applications (e.g., private DB policies) the clauses would be more similar, and we expect even better performance.

Results. Firstly, we stress that our heuristics are still *highly unoptimized*. Even with this, we are able to determine container circuit C_0 containing all 32 input circuits $C_1, ..., C_{32}$ whose size is 0.1637 times the size of all circuits taken together. To explain further, we note that the size of C_0 is trivially at least $\max_{i=1..32}\{|C_i|\}$ and at most $\sum_{i=1}^{32}|C_i|$. Here $|C_i|$ denotes the cost of a circuit including free-XOR[6]. As we will explain, the size of C_0 compared to these bounds yields an important metric for the performance of the algorithm. Formally, we define the *expansion metric, or EM* as $m = \frac{|C_0| - \max_{i=1..32}\{|C_i|\}}{\sum_{i=1}^{32}|C_i|}$[7]. Clearly, $m \in [0, 1]$ where values closer to 0 indicate better performance of the algorithm.

Starting with the 32 input circuits, we first heuristically determine over 100 random trials the smallest circuit containing each of the $\binom{32}{2}$ pairs. For a particular pair of circuits C_i, C_j, we define the *round EM* to be the minimum over all random trials of $\frac{|C_0| - \max\{|C_i|, |C_j|\}}{|C_i| + |C_j|}$. Given all these container circuits, we choose the pairing of circuits of minimum total size. We use these 16 resultant circuits as the input circuits for the next round and repeat the process.

Table 2 compares the total number of non-free gates for a S-Universal Circuit, $S = \{C_1, \ldots, C_{32}\}$, using existing approaches and our work. In Fig. 4, we report the total size of circuits in each of the five rounds, resulting in total size reduction of $6.1\times$.

---

[6] We remark that in all our experiments we use circuits that have fan in at most 2. A standard reduction allows us to eliminate gates of fan in exactly 1. In our experiments we use cost and size interchangeably, since they are closely related.

[7] It would be more general to include the size of Valiant's universal circuit in the expansion metric definition. For s defined as the size of a universal circuit for all circuits of size up to $\max\{|C_i|\}$, set EM $m = \frac{|C_0| - \max\{|C_1|, ..., |C_{32}|\}}{\min\{s, |C_1| + .. + |C_{32}|\}}$. However, in our experiments and clause numbers, s is much larger than $|C_1| + .. + |C_{32}|$, so we omit this complication in this writeup.

Table 1. Circuits used for heuristic evaluation.

Circuit Test #	Function	Total # of Gates	# XOR Gates
C_1	$A + B$ (32 bit)	154	123
C_{17}	$A + B$ (16 bit)	74	59
C_{16}	$A - B$ (16 bit)	103	74
C_{32}	$A - B$ (32 bit)	215	154
C_{18}	$A < B$ (32 bit)	191	127
C_{19}	$A < B$ (16 bit)	74	63
C_2	$A \leq B$ (32 bit)	191	127
C_3	$A \leq B$ (16 bit)	95	63
C_4	Hamming (32 bit)	1610	1223
C_{20}	Hamming (16 bit)	775	587
C_5	Integer Division (32 bit)	3283	1925
C_{21}	Integer Division (16 bit)	3225	1830
C_7	$A * B$ (32 bit)	3283	1925
C_{22}	counting loop (16 bit)	1490	494
C_{25}	$A + B < 230$ and $A - B > 20$ (32 bit)	487	333
C_9	$A + B < 230$ and $A - B > 20$ (16 bit)	231	157
C_{27}	$A * B > 200$ (32 bit)	3368	1982
C_{11}	$A * B > 200$ (16 bit)	904	478
C_{23}	$A * B$ (16 bit)	867	453
C_{24}	$A + B < 100$ (32 bit)	183	122
C_8	$A + B < 100$ (16 bit)	87	58
C_{26}	$B > 1020$ and $A * B > 10$ (32 bit)	3458	2037
C_{10}	$B > 1020$ and $A * B > 10$ (16 bit)	946	501
C_{28}	$A * B > B + 10 * A$ (32 bit)	3881	2311
C_{12}	$A * B > B + 10 * A$ (16 bit)	1145	631
C_{29}	$B * A + 555$ (32 bit)	3343	1956
C_{13}	$B * A + 555$ (16 bit)	895	468
C_{30}	$B^2 + A^2 > 1$ (32 bit)	5613	3881
C_{14}	$B^2 + A^2 > 1$ (16 bit)	1373	905
C_{31}	$B^2 + A * B + A^2$ (32 bit)	8660	5809
C_{15}	$B^2 + A * B + A^2$ (16 bit)	2132	1361
C_6	leading bit (16 bit)	221	74

Table 2. Total non-free gate counts for a $\mathcal{S}$-Universal Circuit, $\mathcal{S} = \{C_1, \ldots, C_{32}\}$.

Combined Circuit	Valiant Universal Circuit	Our construction
20,543	562,900	3,363

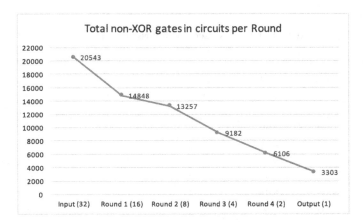

Fig. 4. Determining a container circuit C_0 containing all 32 input circuits $C_1, \ldots, C_{32}$. Reported is the total non-XOR gates in the intermediate container circuits. Figure 5 reports the actual pairings of circuits in each round.

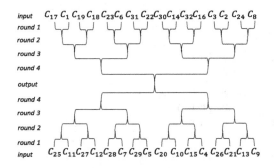

Fig. 5. The circuit pairing found.

In the full version, we include additional discussion on the experiment results and anticipated behavior.

Acknowledgment. The second author was supported by the Office of Naval Research (ONR) contract number N00014-14-C-0113.

References

1. Asharov, G., Lindell, Y., Schneider, T., Zohner, M.: More efficient oblivious transfer and extensions for faster secure computation. In: Sadeghi, A-R., Gligor, V.D., Yung, M. (eds.) ACM CCS 2013, pp. 535–548. ACM Press, November 2013
2. Cachin, C., Micali, S., Stadler, M.: Computationally private information retrieval with polylogarithmic communication. In: Stern, J. (ed.) EUROCRYPT 1999. LNCS, vol. 1592, pp. 402–414. Springer, Heidelberg (1999). https://doi.org/10.1007/3-540-48910-X_28

3. Damgård, I., Jurik, M.: A length-flexible threshold cryptosystem with applications. In: Safavi-Naini, R., Seberry, J. (eds.) ACISP 2003. LNCS, vol. 2727, pp. 350–364. Springer, Heidelberg (2003). https://doi.org/10.1007/3-540-45067-X_30
4. Damgård, I., Jurik, M.: A generalisation, a simplification and some applications of paillier's probabilistic public-key system. In: Kim, K. (ed.) PKC 2001. LNCS, vol. 1992, pp. 119–136. Springer, Heidelberg (2001). https://doi.org/10.1007/3-540-44586-2_9
5. Demmler, D., Dessouky, G., Koushanfar, F., Sadeghi, A-R., Schneider, T., Zeitouni, S.: Automated synthesis of optimized circuits for secure computation. In: Ray, I., Li, N., Kruegel, C. (eds.) ACM CCS 2015, pp. 1504–1517. ACM Press, October 2015
6. Dessouky, G., Koushanfar, F., Sadeghi, A.-R., Schneider, T., Zeitouni, S., Zohner, M.: Pushing the communication barrier in secure computation using lookup tables. In: 24 Annual Network and Distributed System Security Symposium (NDSS 2017). The Internet Society, 26 February–1 March 2017. (to appear)
7. Fisch, B.A., Vo, B., Krell, F., Kumarasubramanian, A., Kolesnikov, V., Malkin, T., Bellovin, S.M.: Malicious-client security in blind seer: a scalable private DBMS. In: 2015 IEEE Symposium on Security and Privacy, pp. 395–410. IEEE Computer Society Press, May 2015
8. Franz, M., Holzer, A., Katzenbeisser, S., Schallhart, C., Veith, H.: CBMC-GC: an ANSI C compiler for secure two-party computations. In: Cohen, A. (ed.) CC 2014. LNCS, vol. 8409, pp. 244–249. Springer, Heidelberg (2014). https://doi.org/10.1007/978-3-642-54807-9_15
9. Fredman, M.L., Tarjan, R.E.: Fibonacci heaps and their uses in improved network optimization algorithms. J. ACM **34**(3), 596–615 (1987)
10. Garey, M.R., Johnson, D.S.: Computers and Intractability: A Guide to the Theory of NP-Completeness. W.H. Freeman and Co., San Francisco (1979)
11. Goldreich, O.: Foundations of Cryptography: Volume 2, Basic Applications, 2nd edn. Cambridge University Press, Cambridge (2009)
12. Goldreich, O., Micali, S., Wigderson, A.: How to play ANY mental game or A completeness theorem for protocols with honest majority. In: Aho, A. (ed.) 19th ACM STOC, pp. 218–229. ACM Press, May 1987
13. Goldreich, O., Ostrovsky, R.: Software protection and simulation on oblivious RAMs. J. ACM **43**(3), 431–473 (1996)
14. Ishai, Y., Kilian, J., Nissim, K., Petrank, E.: Extending oblivious transfers efficiently. In: Boneh, D. (ed.) CRYPTO 2003. LNCS, vol. 2729, pp. 145–161. Springer, Heidelberg (2003). https://doi.org/10.1007/978-3-540-45146-4_9
15. Kiss, Á., Schneider, T.: Valiant's universal circuit is practical. In: Fischlin, M., Coron, J.-S. (eds.) EUROCRYPT 2016. LNCS, vol. 9665, pp. 699–728. Springer, Heidelberg (2016). https://doi.org/10.1007/978-3-662-49890-3_27
16. Kolesnikov, V., Kumaresan, R.: Improved OT extension for transferring short secrets. In: Canetti, R., Garay, J.A. (eds.) CRYPTO 2013. LNCS, vol. 8043, pp. 54–70. Springer, Heidelberg (2013). https://doi.org/10.1007/978-3-642-40084-1_4
17. Kolesnikov, V., Mohassel, P., Rosulek, M.: FleXOR: flexible garbling for XOR gates that beats free-XOR. In: Garay, J.A., Gennaro, R. (eds.) CRYPTO 2014. LNCS, vol. 8617, pp. 440–457. Springer, Heidelberg (2014). https://doi.org/10.1007/978-3-662-44381-1_25
18. Kolesnikov, V., Schneider, T.: Improved garbled circuit: free XOR gates and applications. In: Aceto, L., Damgård, I., Goldberg, L.A., Halldórsson, M.M., Ingólfsdóttir, A., Walukiewicz, I. (eds.) ICALP 2008. LNCS, vol. 5126, pp. 486–498. Springer, Heidelberg (2008). https://doi.org/10.1007/978-3-540-70583-3_40

19. Kolesnikov, V., Schneider, T.: A practical universal circuit construction and secure evaluation of private functions. In: Tsudik, G. (ed.) FC 2008. LNCS, vol. 5143, pp. 83–97. Springer, Heidelberg (2008). https://doi.org/10.1007/978-3-540-85230-8_7

20. Lindell, Y., Pinkas, B.: A proof of security of Yao's protocol for two-party computation. J. Cryptol. 22(2), 161–188 (2009)

21. Lipmaa, H.: An oblivious transfer protocol with log-squared communication. In: Zhou, J., Lopez, J., Deng, R.H., Bao, F. (eds.) ISC 2005. LNCS, vol. 3650, pp. 314–328. Springer, Heidelberg (2005). https://doi.org/10.1007/11556992_23

22. Lipmaa, H., Mohassel, P., Sadeghian, S.: Valiant's universal circuit: Improvements, implementation, and applications. Cryptology ePrint Archive, Report 2016/017 (2016). http://eprint.iacr.org/2016/017

23. Mohassel, P., Sadeghian, S.: How to hide circuits in MPC an efficient framework for private function evaluation. In: Johansson, T., Nguyen, P.Q. (eds.) EUROCRYPT 2013. LNCS, vol. 7881, pp. 557–574. Springer, Heidelberg (2013). https://doi.org/10.1007/978-3-642-38348-9_33

24. Ostrovsky, R., Skeith, W.E.: A survey of single-database private information retrieval: techniques and applications. In: Okamoto, T., Wang, X. (eds.) PKC 2007. LNCS, vol. 4450, pp. 393–411. Springer, Heidelberg (2007). https://doi.org/10.1007/978-3-540-71677-8_26

25. Pappas, V., Krell, F., Vo, B., Kolesnikov, V., Malkin, T., Choi, S.G., George, W., Keromytis, A.D., Bellovin, S.: Blind seer: a scalable private DBMS. In: 2014 IEEE Symposium on Security and Privacy, pp. 359–374. IEEE Computer Society Press, May 2014

26. Paus, A., Sadeghi, A.-R., Schneider, T.: Practical secure evaluation of semi-private functions. In: Abdalla, M., Pointcheval, D., Fouque, P.-A., Vergnaud, D. (eds.) ACNS 2009. LNCS, vol. 5536, pp. 89–106. Springer, Heidelberg (2009). https://doi.org/10.1007/978-3-642-01957-9_6

27. Pinkas, B., Schneider, T., Smart, N.P., Williams, S.C.: Secure two-party computation is practical. In: Matsui, M. (ed.) ASIACRYPT 2009. LNCS, vol. 5912, pp. 250–267. Springer, Heidelberg (2009). https://doi.org/10.1007/978-3-642-10366-7_15

28. Robertson, N., Seymour, P.D.: Graph minors. I. excluding a forest. J. Comb. Theor. Ser. B 35(1), 39–61 (1983)

29. Songhori, E.M., Hussain, S.U., Sadeghi, A-R., Schneider, T., Koushanfar, F.: TinyGarble: Highly compressed and scalable sequential garbled circuits. In: 2015 IEEE Symposium on Security and Privacy, pp. 411–428. IEEE Computer Society Press, May 2015

30. Valiant, L.G.: Universal circuits (preliminary report). In: STOC, pp. 196–203. ACM Press, New York (1976)

31. Verma, R.M., Reyner, S.W.: An analysis of a good algorithm for the subtree problem, correlated. SIAM J. Comput. 18(5), 906–908 (1989)

32. Wang, X., Gordon, S.D., McIntosh, A., Katz, J.: Secure computation of MIPS machine code. In: Askoxylakis, I., Ioannidis, S., Katsikas, S., Meadows, C. (eds.) ESORICS 2016. LNCS, vol. 9879, pp. 99–117. Springer, Cham (2016). https://doi.org/10.1007/978-3-319-45741-3_6

33. Yao, A.C-C.: How to generate and exchange secrets (extended abstract). In: 27th FOCS, pp. 162–167. IEEE Computer Society Press, October 1986

34. Zahur, S., Rosulek, M., Evans, D.: Two halves make a whole. In: Oswald, E., Fischlin, M. (eds.) EUROCRYPT 2015. LNCS, vol. 9057, pp. 220–250. Springer, Heidelberg (2015). https://doi.org/10.1007/978-3-662-46803-6_8

JIMU: Faster LEGO-Based Secure Computation Using Additive Homomorphic Hashes

Ruiyu Zhu[✉] and Yan Huang

Indiana University, Bloomington, USA
zhu52@indiana.edu

Abstract. LEGO-style cut-and-choose is known for its asymptotic efficiency in realizing actively-secure computations. The dominant cost of LEGO protocols is due to *wire-soldering*—the key technique enabling to put independently generated garbled gates together in a bucket to realize a logical gate. Existing wire-soldering constructions rely on homomorphic commitments and their security requires the majority of the garbled gates in every bucket to be correct.

In this paper, we propose an efficient construction of LEGO protocols that does not use homomorphic commitments but is able to guarantee security as long as at least one of the garbled gate in each bucket is correct. Additionally, the faulty gate detection rate in our protocol doubles that of the state-of-the-art LEGO constructions. With moderate additional cost, our approach can even detect faulty gates with probability 1, which enables us to run cut- and-choose on larger circuit gadgets rather than individual AND gates. We have implemented our protocol and our experiments on several benchmark applications show that the performance of our approach is highly competitive in comparison with existing implementations.

1 Introduction

Since 1980s, significant effort has been devoted to making secure computation protocols practical. This include novel garbling schemes [4,5,16,47], programming tools [17,19,30–32], and their applications [18,21,34,38,44,45]. While these works are restricted in the *passive* (honest-but-curious) threat model, which is fairly weak to model real-world adversaries, security against active adversaries is often more desirable.

The most practical approach for building actively-secure two-party computation protocols by far is the *cut-and-choose* paradigm. With cut-and-choose, roughly speaking, one party generates κ garbled circuits where κ depends on the statistical security parameter s; some fraction of those are "checked" by the other party—who aborts if any misbehavior is detected—and the remaining fraction are evaluated with the results being used to derive the final output. A rigorous analysis of the cut-and-choose paradigm was first given by Lindell and Pinkas [27], which required setting κ to roughly $3s$, and was later optimized

© International Association for Cryptologic Research 2017
T. Takagi and T. Peyrin (Eds.): ASIACRYPT 2017, Part II, LNCS 10625, pp. 529–572, 2017.
https://doi.org/10.1007/978-3-319-70697-9_19

to $\kappa = s$ [7,26] since it suffices to have only one honest circuit used for evaluation (hence we call them *SingleCut* protocols).

For better asymptotic efficiency, Nielsen and Orlandi proposed LEGO [36], which exploited the circuit evaluator's randomness to group individual NAND gates (as opposed to *circuits* in the batched-execution setting) to thwart active attacks. This idea evolved to MiniLEGO [11], which is compatible with the free-XOR technique. Their recent independent work [37] provides an implementation of their protocol [12], demonstrating the practical efficiency of LEGO approach. However, they use homomorphic commitments and the security of their protocol depends on the *majority* of the gadgets in every bucket being correct. In contrast, our work shows a different construction of LEGO protocols with highly competitive performance.

Researchers have also exploited the idea of batched cut-and-choose (hence we call BatchedCut) to efficiently execute a batch of N computational instances of the same function f between the same two parties using possibly different inputs [20,28,29]. It was believed that BatchedCut allows to reduce κ to $O(s/\log N)$. However, we will show in Sect. 6.4 that this should really be $2 + O(s/\log N)$ and 2 is actually a *tight* bound on the complexity of any BatchedCut protocols.

1.1 Contribution

New Techniques. We propose two new optimizations for constructing efficient LEGO protocols:

1. The main bottleneck of LEGO protocols is *wire-soldering*, which converts, in a privacy-preserving way, a wire-label of a logical-gate bucket to a wire-label on a garbled gate to enable combining multiple independently garbled gates to realize a logical gate. To achieve high performance wire-soldering, we introduce a new cryptographic primitive called *XOR-homomorphic interactive hash* (IHash) to replace the XOR-homomorphic commitments used in prior works. We propose a simple construction of IHash by integrating Reed-Solomon codes, pseudorandom generators (PRG), and a single invocation of a w-out-of-n oblivious transfer protocol (Sect. 4.2). We proved the security of our interactive hash construction (Sect. 4.3). IHash can be a primitive of independent interest, e.g., it may also be used to efficiently solder circuits in other BatchedCut protocols.
2. Using IHash, we are able to improve existing LEGO-based cut-and-choose mechanism in two more aspects:
   (a) Our protocols guarantee security assuming a *single* correctly garbled gate exists in every bucket. In contrast, existing LEGO-based protocols [11, 12,36,37] require *majority correctness* in every bucket. This enhancement allows us to roughly reduce the number of gadgets in every bucket by $1/2$ when offering 40-bit statistical security.

(b) We can increase the faulty gate detection rate from $1/4$ with previous works [11,12,37] to $1/2$. At moderate additional cost, we can even detect faulty gates with probability 1. This technique allows us to run cut-and-choose on larger circuit components rather than individual ANDs.

The above optimizations combined not only simplify the construction of LEGO-protocols but also the analysis for deriving the cut-and-choose parameters. Due to these benefits, our approach is adopted to work with pools for building highly scalable reactive secure computation services against active attacks [49]. Independent of this work, Duplo [24] is the first to apply the fully-check technique with homomorphic commitments, while our work shows how the idea works with homomorphic hashes, enabling simpler parameter analysis.

Implementation and Evaluation. We have implemented our protocol and experimentally evaluated its performance with several representative computations. In particular, our protocol exhibits very attractive performance in handling the target function's input and output wires: $0.57\,\mu s$ per garbler's input-wire and $8.24\,\mu s$ per evaluator's input-wire, and $0.02\,\mu s$ per output-wire, which are roughly 24x, 2.4x, and 600x faster than WMK [42]'s highly optimized designs (Fig. 9). Without exploiting parallelism, our protocol is able to execute 105.3M logical XOR gates per second and (when bucket size is 5) 45.5K logical AND per second on commodity hardware (two Amazon EC2 `c4.2xlarge` instances over LAN). We show, for the first time, that by cut-and-choosing SubBytes, even small applications such as a single AES could run 2x faster and consume 2x less bandwidth than cut-and-choosing ANDs.

Finally, we prove an asymptotic tight bound on the duplication factor κ of BatchedCut protocols (Sect. 6.4). This bound turns out to be overlooked in prior works [11,12,28,36].

2 Technical Overview

Notations. We assume P_1 and P_2, holding x and y respectively, want to securely compute a function $f(x, y)$. We use the standard definition of actively-secure two-party computation [15]. Throughout this paper, we assume P_1 is the circuit generator (who is also the IHash sender) and P_2 is the circuit evaluator (who is also the IHash receiver). For simplicity, we assume that only P_2 will receive the final result $f(x, y)$. We assume f can be represented as a circuit C containing N AND gates while the rest are all XORs. All vectors in this paper are by default column vectors. We summarize the list of variables in Fig. 1.

2.1 LEGO Protocols

LEGO protocols belong to the BatchedCut category of cut-and-choose-based secure computation protocols [50]. For a Boolean circuit C of N logical gates, the high-level steps of a LEGO protocol to compute C are,

s	The statistical security parameter.	ℓ_w	The symbol-length of wire-labels
k	The computational security parameter.	n_w	The symbol-length of wire-label encoding.
N	The number of logical AND gates (i.e., buckets) in the circuit.	w_w	The number of watched symbols in a wire-label encoding.
T	Total number of garbled AND gates generated by P_1.	σ_w	The bit-length of symbols used in wire-label encoding.
B	Bucket size, i.e., the number of garbled AND gates in a bucket.	ℓ_p	The symbol-length of permutation messages.
Δ	The global secret delta between a 0-label and its corresponding 1-label.	n_p	The symbol-length of wire-label encoding.
$\langle m \rangle$	The i-hash of a message m.	w_p	The number of watched symbols in a wire-label encoding.
ρ	The permutation message whose parity bit, p, is the permutation bit. Each wire has a freshly sampled ρ.	σ_p	The bit-length of symbols used in permutation message encoding.

Fig. 1. Variables and their meanings.

1. **Generate.** P_1 generates a total of T garbled gates.
2. **Evaluate.** P_2 randomly picks $B \cdot N$ gates and groups them into N buckets. Each bucket will realize a gate in C. P_2 evaluates every bucket by first translating wire-labels on the bucket's input-wires to wire-labels on individual garbled gate's input-wires, evaluating every garbled gate in the bucket, and then translating the obtained wire-labels on the garbled gates' output-wires back to a wire-label on the bucket's output-wire. (The wire-label translation, also called *wire-soldering*, is explained in more detail below.)
3. **Check.** P_2 checks each of the rest $T - BN$ garbled gates for correctness. If any of these gates was found faulty, P_2 aborts. Though, due to the randomized nature of the checks, P_2 will not always be able to detect it when checking a faulty gate.
4. **Output.** P_1 reveals the secret mapping on the circuit's final output-wires so that P_2 is able to map the final output-wire labels into their logical bit values.

The first construction [36] was based on NANDs and require public key operations for wire-soldering. Fredericksen et al. [11] later proposed a LEGO scheme that is compatible with the notable free-XOR optimization [25] using XOR-Homomorphic commitments as a black box. Under this paradigm, it suffices to assume all the garbled gates are ANDs since all XORs can be securely computed locally and no extra treatment is needed to ensure correct behavior on processing XORs. However, due to the use of the global secret Δ for free-XOR, a garbled AND can't be fully opened for check purpose. Instead, a random one of the four possible pairs of inputs to a binary gate is picked to check correctness.

Wire-Soldering. As depicted in Fig. 5, each bucket realizes a logical gate, thus has input and output wires like the logical gate it realizes. In order to evaluate an independently generated garbled gate assigned to a bucket, an input-wire of the bucket (with wire-labels w^0_{bucket} and $w^1_{\mathsf{bucket}} = w^0_{\mathsf{bucket}} \oplus \Delta$ denoting 0 and 1) needs to be connected to the corresponding input-wire (with wire-labels w^0_{gate} and $w^1_{\mathsf{gate}} = w^0_{\mathsf{gate}} \oplus \Delta$) of the garbled gate to evaluate. This is done by requiring P_1 to send $d = w^0_{\mathsf{bucket}} \oplus w^0_{\mathsf{gate}}$ and P_2 to xor d with the wire-label on the bucket (either w^0_{bucket} or w^1_{bucket}) he obtained from evaluating the previous bucket. To prevent a malicious P_1 from sending a forged d, existing protocols used XOR-Homomorphic commitments to let P_1 commit Δ, w^0_{bucket}, and w^0_{gate} (which allows P_2 to derive the commitment of d homomorphically), so that P_2 can verify the validity of d from its decommitment without learning any extra information about w^0_{bucket} and w^0_{gate}.

2.2   Our Optimizations

Below we sketch the intuition behind our optimization ideas of LEGO protocols.

XOR-Homomorphic Interactive Hash. XOR-homomorphic Interactive Hash (IHash) is a cryptographic protocol involving two participants, which we call the *sender* and the *receiver*, respectively. The design of IHash is directly motivated by the security goals of wire-soldering:

1. **Binding.** Every i-hash of a secret message uniquely identifies the message with all but a negligible probability, so that the message holder cannot modify a secret message once its i-hash is sent.
2. **Hiding.** The i-hash receiver does not learn any extra information about the secret message other than the i-hash itself. For a uniform-randomly sampled message, it is guaranteed that certain entropy remains after its i-hash is sent because by definition an i-hash needs to be shorter than the original message.
3. **XOR-Homomorphism.** Given the i-hashes of two messages m_1 and m_2, the receiver can locally compute the i-hash of $m_1 \oplus m_2$. This enables the receiver (circuit evaluator) to solder wire-labels from independently garbled gates using a verifiable label-difference supplied by the circuit generator.

Unlike a commitment scheme which requires the committer's cooperation to match messages with commitments, IHash allows the receiver *alone* to verify if any message matches with an i-hash (like with traditional hashes). In addition, a commitment hides every bit of its message whereas i-hashes allow leaking arbitrary information about its message through the i-hash itself, up to the length of the i-hash. Nevertheless, we find that this somewhat weaker primitive suffices to solder wires in LEGO protocols.

Figure 2 illustrates our construction of the XOR-homomorphic IHash scheme. The high-level idea is to let the IHash sender encode his/her secret message m using a $[n, \ell, n - \ell + 1]_{2^\sigma}$ Reed-Solomon code and let the receiver secretly watch (soon we will detail how to watch secretly) w of the n symbols in $\mathsf{Encode}(m)$. Recall that a $[n, \ell, n - \ell + 1]_{2^\sigma}$-code is one that takes in an ℓ-symbol message and

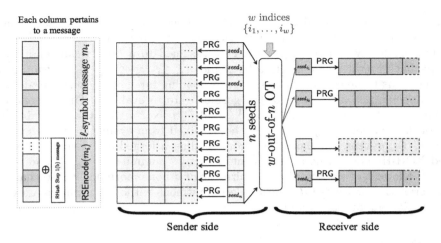

Fig. 2. Interactive Hash based on OT and Reed-Solomon code

outputs an n-symbol encoding (where symbols are of σ-bit) so that the minimal distance between the codewords is $n - \ell + 1$ symbols. The w watched symbols are the i-hash of m. The receiver can verify a particular message matches with its i-hash by encoding the message and making sure all values at the w watched positions on the encoding coincide with the i-hash it holds. As a result, with respect to binding, if the sender forges a message m', at least $n - \ell + 1$ symbols in the codewords have to be different, hence, the i-hash receiver can detect the forgery with all but $\binom{\ell-1}{w} / \binom{n}{w}$ probability. With respect to hiding, although the i-hash reveals $w\sigma$ bits entropy in m to the receiver, if the original message has $\ell\sigma$ bits entropy, then the rest $(\ell - w)\sigma$ bits entropy remains perfectly hidden to the receiver. Therefore, to guarantee hiding, we can set ℓ sufficiently large based on the security parameter. Finally, the additive (i.e., XOR) homomorphic property of i-hashes is inherent in the linearity of Reed-Solomon codes.

The "secret watch" above can be realized by a w-out-of-n oblivious transfer protocol. Moreover, we only need to invoke this oblivious transfer *once*. The key idea is to let the sender pick n random seeds and obliviously transfer w seeds of the receiver's choice. Later, the sender sends correction messages $\mathsf{Encode}(m)_i \oplus \mathsf{PRG}(seed_i)$ to the receiver where $\mathsf{Encode}(m)_i$ is the ith symbol of m's encoding and $i \in \{1, \ldots, n\}$. Thus, learning $seed_i$ allows the receiver to see the corresponding symbols in m's codeword. We also notice that the input-wire labels of all garbled gates are uniformly random. Therefore, setting the ith symbol of the jth wire-label to be $\mathsf{PRG}(seed_i, j)$ where $i \in \{1, \ldots, \ell\}$ while using a systematic code will reduce the work to only send the corrections on the last $n - \ell$ symbols, i.e., $m_i \oplus \mathsf{PRG}(seed_i, j)$ for $i \in \{\ell + 1, \ldots, n\}$.

Fast Wire-Soldering. Wire-soldering is one of the most challenging efficiency barriers in LEGO protocols. Recall that P_1 garbles all the AND gates independently. Thus, in the circuit evaluation phase where B random AND gates are grouped into a bucket to evaluate a logical AND gate, P_2 needs to "translate" an input wire-label of a bucket to its corresponding input wire-label on a garbled gate in the bucket. To this end, we require, at the garbling stage, that, for every garbled gate, P_1 i-hash one wire-label on every wire of the garbled gate to P_2; and, at the gate evaluation stage, that P_1 send the xor-differences between every pair of the source and target wire-labels. The validity of the xor-differences can be verified against their i-hashes. Note that even if an i-hash leaks entropy, we can increase the length of the wire-labels to ensure enough entropy remain in the labels to guarantee the needed computational security.

Moreover, for benefits that will be clear soon, we require P_1 also to i-hash the global Δ (required by the free-XOR technique) to P_2. Recall that all the wire-labels at the bucket level also need to be i-hashed to P_2. To prevent P_2 from learning logical values of the intermediate wire-labels, P_1 will i-hash either the 0-label or the 1-label of each wire with equal probability. Without extra treatment, however, this will allow a malicious P_1 to surreptitiously flip a wire-label's logical value. We fix this issue by adding a random permutation message ρ to each wire and use ρ's parity bit to bind the plaintext bit the i-hashed wire-label represents. For integrity of ρ, we require P_1 to i-hash ρ to P_2 so that P_2 can verify the ρ values of each wire at the garbled gate checking stage. We stress that the value of ρ for all intermediate wires will never be revealed.

To achieve fast wire-soldering for practically efficient LEGO protocols, we found the following two optimizations indispensable.

1. The Reed-Solomon encoding process can be viewed as multiplying the public encoding matrix $A_{n \times \ell} = [a_{i,j}]_{1 \leq i \leq n, 1 \leq j \leq \ell}$ with the message vector $\boldsymbol{m}_{\ell \times 1} = [m_1, \ldots, m_\ell]$ where m_i's are σ-bit symbols. To ensure security of LEGO protocols, the (n, ℓ, σ) values would be $(n_w, \ell_w, \sigma_w) = (86, 32, 8)$ for wire-labels and $(n_p, \ell_p, \sigma_p) = (44, 20, 6)$ for permutation messages. A naïve implementation of the encoding process will require more than 2700 Galois Field (GF) multiplications per wire-label and 900 GF multiplications per permutation message, which amounts to more than 10K multiplications per garbled gate. Even if field multiplications are realized as table-lookups, 10K memory accesses per gate is already 40× slower than AESNI-based garbling itself, making LEGO approach noncompetitive in practice.

   Our key idea to speedup encoding process is to pack many symbols into operands (e.g., _m128, _m256, _m512) of vector instructions and leverage Intel Intrinsic instructions [1] to enable efficient message encoding. Below we illustrate the idea with an example where $n = 96, \ell = 32, \sigma = 8$, i.e., encoding a 32-symbol message $\boldsymbol{m}$ into a 96-symbol codeword where symbols are of 8-bit. First, we use a systematic code so that it suffices to compute the last $n - \ell = 64$ symbols of the codeword since the first $\ell = 32$ symbols are identical to the original message. Thus, we can restrict our attention to the last 64 rows of the encoding matrix A, call it A'. Let $\boldsymbol{a}_{.,1} = [a_{33,1}, \ldots, a_{96,1}]^T$ be the first column of the matrix A' and we can store the 64 symbols of $\boldsymbol{a}_{.,1}$ in a single _m512 register. Let m_i be the ith symbol of $\boldsymbol{m}$. The last 64 symbols in the encoding of $\boldsymbol{m}$ can be computed as $\boldsymbol{c} = \sum_{i=1}^{\ell} m_i \boldsymbol{a}_{.,i} = \sum_{i=1}^{\ell} [m_i a_{33,i}, \ldots, m_i a_{96,i}]^T$. Since $m_i \in \mathrm{GF}(2^8)$

and all column vectors $\boldsymbol{a}_{.,i}$ are publicly fixed, each $m_i \boldsymbol{a}_{.,i}$ can thus be efficiently derived with a single lookup into a table of 256 entries of `__m512` values and the sum can be computed with $\ell - 1 = 31$ `__mm512_xor` instructions. This optimization would reduce $32 \times 96 = 3072$ field multiplications per encoding down to just 32 memory reads, about a hundredth cost of the implementation based on naïve table-lookups.

2. We observe that, although the hiding of i-hashes for wire-labels is *computational* (since an adversary could use the garbled truth table to search the "right" label offline), the hiding on the permutation message ρ is *perfect* because no additional constraints are provided to allow offline search for the permutation bit (i.e., the parity of ρ). Thus, it suffices to require only 1-bit of entropy remain in ρ after i-hashing each ρ for perfectly hiding which wire-labels on a wire was i-hashed. This observation allows us to select much more efficient parameters to i-hash ρ, i.e., $(n_p, \ell_p, \sigma_p) = (44, 20, 6)$ as opposed to $(86, 32, 8)$ for i-hashing wire-labels.

Increased Faulty Gate Detection Rate. In existing protocols [11,12,37], a faulty gate being checked will only be detected with probability $1/4$. This is because the garbled gates are produced with respect to the global secret Δ (the xor-difference between a 0-label and its 1-label) required by the free-XOR technique, hence only one out of the four garbled rows of a binary gate can be opened for checking. In contrast, our protocol allows a faulty gate to be detected with probability $1/2$. We achieve this by integrating the Half-Gate garbling technique [47] which requires only two garbled rows per gate Since each check opens one of the two garbled rows, this allows to detect a faulty gate with probability 50% without revealing Δ, which is formally proved in Lemma 5.3.

To detect faulty gates with 100% probability, the idea is to allow fully open a garbled gate at gate-checking time. This requires garbling each gate with respect to a freshly sampled Δ and sending the i-hash of this Δ with the gate. To solder two wires garbled with different Δ, additional verifiable XOR messages are also needed. We detail this special soldering procedure at the end of Sect. 5.1.

Dealing with Faulty Gates Used for Evaluation. Our protocol is able to guarantee security as long as a single correctly garbled gate exists in every bucket. This improvement is due to a combination of the IHash and the free-XOR techniques. Denote the i-hash of a message $m \in \{0,1\}^*$ by $\langle m \rangle$. We let the circuit evaluator to learn $\langle \Delta \rangle$ where Δ is the global secret. On each wire, the evaluator also learns an i-hash $\langle w \rangle$ where w defines either the 0-label or the 1-label on that wire (but the evaluator doesn't know which). Recall that the evaluator can locally verify the validity of a wire-label using the IHash's Verify algorithm. Therefore, if at least one gate in a bucket is good, evaluating all the garbled gates in the bucket will give one or more valid output wire-labels. When translating these wire-labels to the bucket output wire-label, one of the following two cases has to happen:

1. They all match with the same i-hash, either $\langle w \rangle$ or $\langle w \rangle \oplus \langle \Delta \rangle$. Since all *valid* wire-labels are consistent on the plaintext bit they represent and one of them

is known to be correct, the evaluator can directly proceed with this valid wire-label to evaluate the subsequent buckets.

2. Some of them translate to $\langle w \rangle$ whereas others translate to $\langle w \rangle \oplus \langle \Delta \rangle$. In this case, the evaluator can simply xor the two valid labels to recover Δ. Once Δ is known, the evaluator can use it to recover the circuit generator's private input x and locally computes and outputs $f(x, y)$.

Hence, in either case our protocol can be proved secure.

GenAND(i, Δ, w_l^0, w_r^0)

Require: $i \in \mathbb{N}$. $\Delta, w_l^0, w_r^0 \in \{0, 1\}^{256}$.

$q_l := \mathsf{lsb}\,(w_l^0)$

$q_r := \mathsf{lsb}\,(w_r^0)$

$j := 2i$

$j' := 2i + 1$

$T_G := H(w_l^0, j) \oplus H(w_l^0 \oplus \Delta, j) \oplus q_r \Delta$

$w_g^0 := H(w_l^0, j) \oplus q_l T_G$

$T_E := H(w_r^0, j') \oplus H(w_r^0 \oplus \Delta, j') \oplus w_l^0$

$w_e^0 := H(w_r^0, j') \oplus q_r(T_E \oplus w_l^0)$

$w_o^0 := w_g^0 \oplus w_e^0$

return (w_o^0, T_G, T_E)

EvlAND(i, w_l, w_r, T_G, T_E)

Require: $i \in \mathbb{N}$. $w_l, w_r \in \{0, 1\}^{256}$.

$s_l := \mathsf{lsb}\,(w_l)$; $s_r := \mathsf{lsb}\,(w_r)$

$j := 2i$; $j' := 2i + 1$

$w_g := H(w_l, j) \oplus s_l T_G$

$w_e := H(w_r, j') \oplus s_r(T_E \oplus w_l)$

$w_o := w_g \oplus w_e$

return w_o

H(m, j)

$w := \mathsf{Compress}(m)$

$k := 2j$; $k' := 2j + 1$

$w := m \oplus k$; $w' := m \oplus k'$

return $\mathsf{AES}(w) \oplus w \parallel \mathsf{AES}(w') \oplus w'$

Fig. 3. Garbling with 256-bit wire-labels. AES ($\cdot$) denotes calling AES with a fixed, publicly-known key. Compress(m) essentially computes $A'm$ where A' is a rank-16, 16×32 matrix over $\mathrm{GF}(2^8)$. A' is randomly picked by the circuit generator after the evaluator chose its watch symbols.

Entropy Extraction for Efficient Garbling. With fixed-key AESNI instructions, the state-of-the-art garbling technique is able to produce 20 million garbled rows per second, which is about $10\times$ faster than a SHA256-based garbling scheme [4,33,47]. However, the wire-labels in our protocol need to be longer than 128 bits to ensure enough entropy (e.g., more than 80 bits) remains even part of a wire-label is leaked to the evaluator through its i-hash. Although it is straightforward to use SHA256 to implement garbling to accommodate longer wire-labels in the random oracle model, *a priori*, it is unclear how this can be efficiently realized only assuming fixed-key AES is an ideal cipher.

Our intuition is to compress a longer wire-label down to a 128-bit label while preserving as much entropy as possible, and then run existing fixed-key AES-based garbling with the compressed labels. As a concrete example, wire-labels in our protocol are 256-bit (i.e. 32 8-bit symbols). During i-hashing, a 32-symbol wire-label will be encoded into an 86-symbol codeword; and the evaluator randomly picks 21 of the 86 symbols in the codewords to watch. Since the "watch" reveals $8 \times 21 = 168$ bits entropy, 88 bits of entropy remains in each wire-label. To compress a 256-bit wire-label m to a 128-bit m' while carrying over the entropy, our strategy is to randomly sample a 16×32, rank-16 matrix $A' = [a'_{i,j}]$ where $a'_{i,j} \in \mathrm{GF}(2^8)$ and compute $m' = A'm$ (in other words, A' represents a set of 16 linearly independent row-vectors in the vector space $\mathrm{GF}(2^8)^{32}$). Note that A' is sampled only *after* the evaluator has chosen its watched symbols. Intuitively, this compression preserves entropy because the chances are extremely low that any one of the 16 row-vectors happens to be a linear combination of a set of 21 row-vectors of $\mathrm{GF}(2^8)^{32}$ that are picked randomly and independently by the evaluator. We present a formal analysis of the entropy loss in Sect. 6.2.

Given this Compress algorithm (i.e., essentially a matrix multiplication as described above), we can formalize our garbling scheme based on that of Half-Gates [47]. The main difference lies in the function H. Our H, specified in Fig. 3, maps $\{0,1\}^{256} \times \mathbb{N}$ to $\{0,1\}^{256}$ and involves two calls to a fixed-key AES cipher to produce a 256-bit pseudorandom mask to encrypt a longer output wire-label.

2.3   Related Work

TinyLEGO. As an independent and concurrent work, TinyLEGO [12,37] explored ways to improve LEGO protocols in the single-execution setting. To save on wire-soldering, Duplo [24] also explored checking multi-gate gadgets. However, our approach is different in several aspects:

1. To solder the wires, previous work used additive homomorphic commitment, whereas we propose the notion of IHash and give a highly efficient construction of IHash using PRG, Reed-Solomon code and Intel Intrinsics [1]. We show that, despite IHash being leakier than homomorphic commitments, it suffices the purpose of constructing efficient LEGO protocols. Our construction of IHash shares some similarity with that of XOR-Homomorphic commitments in [13]. However, the two schemes differ in the way OTs are used, the selection of error correcting codes, and the way to pick critical protocol parameters.
2. TinyLEGO involves cut-and-choosing two types of garbled gadgets (i.e., ANDs and wire-authenticators) and requires correct *majority* in the total number of garbled gadgets in each bucket, whereas our protocol only uses garbled ANDs and the security holds as long as a single correctly garbled AND exists in each bucket. In addition, the faulty gate detection rate in our protocol is twice of that in TinyLEGO.
3. Because of our optimizations, our protocol can run more than 2x faster in a LAN and be highly competitive over a WAN. However, their protocols are about 20–50 more efficient in bandwidth thus would be advantageous in some bandwidth-stringent network environments. See Sect. 7 for detailed performance comparisons.

NNOB [35] and SPDZ [10]. Both NNOB and SPDZ require a linear number of rounds and expensive pre-processing, our protocol has only a small constant rounds and lightweight setup. Thus, ours performs better when the network latency cannot be ignored, or when resources are limited in the preparation phase.

Lindell-Riva [29] and Rindal-Rosulek [41]. In the offline/online setting, Lindell-Riva and Rindal-Rosulek provided very efficient prototypes of secure computation protocols. Although the high-level idea of cut-and-choose resembles that of LEGO, their results are not applicable if only one (or just very few) executions is needed. For example, with [29], one AES can be computed in about 74 ms amortized time assuming a 75,000 ms offline delay is acceptable for preparing 1024 executions. Since wire-soldering is much less of an issue, their technique would be far less efficient when carried out to cut-and-choose individual-gates.

Wang-Malozemoff-Katz. Wang et al. [42] recently designed and implemented by far the most efficient SingleCut secure computation protocol. Thanks to a (mostly) symmetric-key cryptography based garbler's input consistency enforcement mechanism and their careful use of SSE instructions for preventing selective failure attacks, a single AES instance can be computed in 65 ms (with input/output wires processed at roughly $20\,\mu s$ per wire). However, even compared with their optimized protocol, processing input/output wires in our protocol can still be much faster, hence will be competitive in computing shallow circuits with many input/output wires (see Sect. 7 for detailed performance comparisons). Moreover, due to the advantages of LEGO protocols in supporting actively secure RAM-based secure computations, it would be interesting future work to develop better homomorphic hash constructions and plug it into our framework to obtain improved LEGO protocols.

Wang-Ranellucci-Katz. Wang et al. [43] proposed a secure two-party computation scheme based on collaborative garbling over authenticated multiplicative triples. Their approach would be advantageous in speed when compared to state-of-the-art LEGO protocols [37]. However, ideas of this work can be extended to allow fully open a garbled gate at the verification stage and applied at a circuit-level to produce protocols that are more efficient in bandwidth. Hence, our framework is still interesting in designing efficient protocols in low-bandwidth settings.

Parallelism. Noting the *embarrassingly parallelizable* nature of the protocols in this domain (including ours), we follow the convention of many existing works [42, 47] to restrict our attention to the single-threaded model and treat computation as an energy-consuming scarce resource.

3   Preliminaries

3.1   Oblivious Transfer

We use 1-out-of-2 oblivious transfers to send wire labels corresponding to the evaluator's input, and two k-out-of-n oblivious transfers for wire soldering. A k-out-of-n oblivious transfer protocol takes n messages $m_1, \ldots, m_n$ from the sender and a set of k indices $i_1, \ldots, i_k \in [1, n]$ from the receiver, and outputs nothing but the k messages $m_{i_1}, \ldots, m_{i_k}$ to the receiver. Composably secure 1-out-of-2 OT can be efficiently instantiated from dual-mode cryptosystems [39] and efficiently extended [2,23] from inexpensive symmetric operations plus a constant number of base OTs. Camenisch et al. [8] proposed an efficient and simulatable k-out-of-n OT in the Random Oracle Model.

3.2   Garbled Circuits

First proposed by Yao [46], garbled circuits were later formalized as a cryptographic primitive of its own interest [5]. Bellare et al. have carved out three security notions for garbling: *privacy, obliviousness,* and *authenticity.* We refer readers to their paper for the formal definitions. In the past few years, many optimizations have been proposed to improve various aspects of garbled circuits, such as bandwidth [40,47], evaluator's computation [40], memory consumption [19], and using dedicated hardware [5]. Our protocol leverages Half-Gates garbling recently proposed by Zahur et al. [47] which offers the simulation-based definition of privacy, obliviousness, and authenticity under a *circular correlation robustness* assumption of the hash function H. We summarize their garbling algorithms GenAND and EvlAND in Fig. 3.

More formally, a *garbling scheme* $\mathcal{G}$ is a 5-tuple (Gb, En, Ev, De, f) of algorithms, where Gb is an efficient randomized *garbler* that, on input $(1^k, f)$, outputs (F, e, d); En is an *encoder* that, on input (e, x), outputs X; Ev is an *evaluator* that, on input (F, X), outputs Y; De is a *decoder* that, on input (d, Y), outputs y. The *correctness* of $\mathcal{G}$ requires that for every $(F, e, d) \leftarrow$ Gb$(1^k, f)$ and every x,

$$\mathsf{De}(d, \mathsf{Ev}(F, \mathsf{En}(e, x))) = f(x).$$

Let $\varPhi$ be a prefixed function modeling the acceptable information leak and "$\approx$" symbolizes *computational indistinguishability. Privacy* of $\mathcal{G}$ implies that there exists an efficient simulator $\mathcal{S}$ such that for every x,

$$\left\{ (F, X, d) : \begin{array}{l} (F, e, d) \leftarrow \mathsf{Gb}(1^k, f), \\ X \leftarrow \mathsf{En}(e, x). \end{array} \right\} \approx \{\mathcal{S}(1^k, f, \varPhi(f))\}.$$

Obliviousness of $\mathcal{G}$ implies that there exists an efficient simulator $\mathcal{S}$ such that for every x,

$$\{(F, e, d) \leftarrow \mathsf{Gb}(1^k, f), X \leftarrow \mathsf{En}(e, x) : (F, X)\} \approx \{\mathcal{S}(1^k, f)\}.$$

4 Homomorphic Interactive Hash

In this section, we describe XOR-homomorphic interactive hash, a new primitive that enables multiple enhancements in our LEGO protocol.

4.1 Definition

It involves two parties, known as the *sender* (P_1) and the *receiver* (P_2), to *compute* an interactive hash (i-hash) while the receiver can locally *verify* a message against an i-hash that it holds. The ideal functionality of $\mathcal{F}_{\text{XorIHash}}$ is described in Fig. 4, where Hash is an efficient two-party probabilistic algorithm that takes a message m from P_1 and outputs an i-hash of m (denoted as $\langle m \rangle$) to P_2 without revealing any additional information to either party; and Verify is an efficient algorithm (locally computable by P_2) that takes an i-hash $\langle m \rangle$ and a message m' and outputs a bit b indicating whether $m = m'$. Like conventional hashes, we require $|\langle m \rangle| < |m|$ and that for any two distinct messages m_1 and m_2, $\langle m_1 \rangle \neq \langle m_2 \rangle$ except for a negligible probability. Finally, we require the hashes to be XOR-homomorphic, i.e., $\langle m_1 \rangle \oplus \langle m_2 \rangle = \langle m_1 \oplus m_2 \rangle$.

- **Hash.** Upon receiving (Hash, $m_1, \ldots, m_\nu$) ($\nu \geq 1$) from P_1: for every i, if there is a recorded value (cid_i, m_i), generate a delayed output (Receipt, cid_i, $\langle m_i \rangle$) to P_2 where $\langle m_i \rangle$ denotes the hash of m_i and $|\langle m_i \rangle| < |m_i|$; otherwise, pick a fresh number cid_i, record (cid_i, m_i) and generate a delayed output (Receipt, cid_i, $\langle m_i \rangle$) to P_2.

- **Verify.** Upon receiving (Verify, $cid_1, \ldots, cid_\nu$, d) from P_2: if there are recorded values (cid_1, m_1), $\ldots$, (cid_ν, m_ν) (otherwise do nothing), set $z = 1$ if $m_1 \oplus \cdots \oplus m_\nu = d$, and $z = 0$, otherwise; generate a delayed output (VerifyResult, z) to R.

Fig. 4. Ideal XOR-homomorphic interactive hashes. (P_1 is the hash sender and P_2 the hash receiver. "Send a delayed output x to party P" reflects a standard treatment of fairness, i.e., "send (x, P) to the adversary; when receiving ok from the adversary, output x to P.")

Note that Fig. 4 actually describes a family of ideal functionalities for $\mathcal{F}_{\text{XorIHash}}$, as it leaves the exact definition of $\langle m_i \rangle$ unspecified (other than requiring $|\langle m_i \rangle| < |m_i|$). Along with a specific definition of $\langle m_i \rangle$, the ideal functionality defined in Fig. 4 will yield a concrete XOR-homomorphic interactive hash scheme. For example, our construction given in Sect. 4.2 realize a concrete version of IHash in which $\langle m_i \rangle$ is defined as $m_i' * v$ where m_i' is the Reed-Solomon encoding of m_i, v is a binary vector supplied by P_2 containing exact w 1-bits and '$*$' denotes pair-wise multiplication of two equal-length vectors but leaving out all entries (in the product vector) corresponding to the 0-entries in v.

Interactive hash offers certain *"hiding"* and *"binding"* properties. That is, with all but negligible probability, the receiver of $\langle m \rangle$ learns nothing about m except for what can be efficiently computed from $\langle m \rangle$; and the sender of $\langle m \rangle$ can't claim a different message m' to be the preimage of $\langle m \rangle$. However, unlike cryptographic commitments, with IHash, (1) some entropy in m can be leaked to the receiver yet the rest remains; (2) the message owner can't compute the hash on its own; and (3) the hash receiver can verify *on its own* whether a message matches with a hash.

4.2   Construction

Figure 2 illustrates the high-level idea behind our construction. Let OT_n^w be an ideal functionality for a w-out-of-n oblivious transfer. $\mathsf{Encode}_{\ell,n,d}(\cdot)$ denotes the encoding algorithm of $[n, \ell, n-\ell+1]_{2^\sigma}$ Reed-Solomon *systematic* code, i.e., (over σ-bit symbols) ℓ-symbol messages are encoded into n-symbol *codewords* with minimal distance of $n - \ell + 1$ symbols. Let PRG be a pseudorandom generator and s, k are the statistical and computational security parameters.

To allow the receiver to obliviously watch the set of w positions on every message's n-symbol codeword without invoking an OT instance per message, we let the sender generate n secret seeds and call *only once* a w-out-of-n OT to allow the receiver learn w of these seeds (Step 1 of IHash.Setup). These seeds are then used as keys to a PRG to create n *rows* of pseudorandom symbols, of which the receiver is able to recover w rows. When a message m is ready to be i-hashed, the sender simply encodes m and sends the xor-difference between m's codeword and the next *column* of n pseudorandom symbols generated from the seeds (Step 1b of IHash.Hash) so that the receiver can record the symbols for which it watched the corresponding keys.

To obtain active-security, our protocol actually generates $n + \xi$ i-hashes when i-hashing n messages, then uses the extra ξ i-hashes to verify that the sender followed the protocol honestly (Step 2 of IHash.Hash). In our protocol, we set $\xi = \left\lceil -\frac{1}{\sigma} \log \left(2^{-s} - \binom{\ell-1}{w} / \binom{n}{w} \right) \right\rceil$ to bound the failure probability of our simulator $\mathcal{S}$ in the security proof of Theorem 4.1 by 2^{-s}.

The detailed construction steps are as follows.

- $\mathsf{IHash}_{\ell,\sigma}.\mathsf{Setup}(\{seed_1, \ldots, seed_n\}; \{i_1, \ldots, i_w\})$
  Note $\{seed_1, \ldots, seed_n\}$ is P_1's secret input and $\{i_1, \ldots, i_w\}$ is P_2's secret inputs.
  1. P_1 and P_2 run OT_n^w where P_1 is the sender with inputs $seed_1, \ldots, seed_n$, and P_2 is the receiver with inputs $i_1, \ldots, i_w$. At the end of this step, P_2 learns $seed_{i_1}, \ldots, seed_{i_w}$.
- $\mathsf{IHash}_{\ell,\sigma}.\mathsf{Hash}(\boldsymbol{m}_1, \ldots, \boldsymbol{m}_\nu)$
  1. For $t = 1, \ldots, \nu + \xi$,
     (a) For $1 \leq i \leq \ell$, P_1 computes $x_i = \mathsf{PRG}(seed_i, t)$, where $x_i \in \{0,1\}^\sigma$, and sets $\boldsymbol{m}_t' := x_1 \| \ldots \| x_\ell$.

(b) For $\ell < i \leq n$, P_1 sends to P_2

$$x_i' := \mathsf{PRG}(seed_i, t) \oplus \mathsf{Encode}(\boldsymbol{m}_t')[i]$$

where $\mathsf{Encode}(\boldsymbol{m}_t')[i]$ denotes the ith symbol of $\boldsymbol{m}_t'$'s *systematic* codeword.

(c) $\forall i \in \{i_1, \ldots, i_w\}$, P_2 computes

$$x_i := \begin{cases} \mathsf{PRG}(seed_i, t), & \text{if } 1 \leq i \leq \ell \\ \mathsf{PRG}(seed_i, t) \oplus x_i', & \text{if } \ell < i \leq n \end{cases}.$$

Then P_2 sets $\langle \boldsymbol{m}_t' \rangle = (x_{i_1}, \ldots, x_{i_w})$.

2. For $t = 1, \ldots, \xi$,
   (a) P_2 randomly picks $y \leftarrow \{0,1\}^{\sigma \cdot \nu}$ and sends it to P_1.
   (b) P_1 sends $\hat{\boldsymbol{m}}'_t := \sum_{i=1}^{\nu} y_i \boldsymbol{m}_i' + \boldsymbol{m}_{\nu+t}'$ where $y_i \in \{0,1\}^\sigma$ to P_2.
   (c) P_2 runs $\mathsf{IHash}_{\ell,\sigma}.\mathsf{Verify}\left(\sum_{i=1}^{\nu} y_i \langle \boldsymbol{m}_i' \rangle + \langle \boldsymbol{m}_{\nu+t}' \rangle, \hat{\boldsymbol{m}}'_t \right)$ where $y_i \in \{0,1\}^\sigma$ and aborts if it fails.

3. For $i = 1, \ldots, \nu$, P_1 sends $\boldsymbol{x}_i := \boldsymbol{m}_i \oplus \boldsymbol{m}_i'$ to P_2, who then computes

$$\langle \boldsymbol{m}_t \rangle := \Big\|_{i=i_1}^{i_w} \left(\langle \boldsymbol{m}_t' \rangle[i] \oplus \mathsf{Encode}(\boldsymbol{x}_t)[i] \right)$$

where $\|_{i=1}^{n} a_i$ means $a_1 \| \cdots \| a_n$ and $\mathsf{Encode}(\boldsymbol{x}_t)[i]$ denotes the ith symbol of $\boldsymbol{x}_t$'s codeword. P_1 outputs nothing and P_2 outputs $\langle \boldsymbol{m}_1 \rangle, \ldots, \langle \boldsymbol{m}_\nu \rangle$.

– $\mathsf{IHash}_{\ell,\sigma}.\mathsf{Verify}\left(\langle \boldsymbol{m}_1 \rangle, \ldots, \langle \boldsymbol{m}_\nu \rangle, \bigoplus_{i=1}^{\nu} \boldsymbol{m}_i' \right)$
   1. P_2 computes $\langle \boldsymbol{m} \rangle := \bigoplus_{i=1}^{t} \langle \boldsymbol{m}_i \rangle$ and let $\boldsymbol{m}' = \bigoplus_{i=1}^{\nu} \boldsymbol{m}_i'$.
   2. P_2 parses $\langle \boldsymbol{m} \rangle$ into $(x_{i_1}, \ldots, x_{i_w}) \in \{0,1\}^{\sigma \cdot w}$ and returns 1 if for all $i \in \{i_1, \ldots, i_w\}$, $\mathsf{Encode}(\boldsymbol{m}')[i] = x_i$; and 0, otherwise.

   (Setting $t = 1$ allows P_2 to verify any single messages.)

Optimization. If the goal is only to i-hash random messages as it is used in our main protocol, it suffices to treat the $\boldsymbol{m}_i'$s (generated by calling PRG in Step 1a) as the random messages to i-hash, hence no need to send the first ℓ symbols of $\boldsymbol{x}_i$ in Step 3 (where $\boldsymbol{x}_i$ is the xor-differences between an input message $\boldsymbol{m}_i$ and a random message $\boldsymbol{m}_i'$), saving $\ell\sigma$ bits per i-hashed message.

4.3   Proof of Security

Theorem 4.1. *Assuming there exists a secure OT, the protocol described in Sect. 4.2 securely realizes an XOR-homomorphic interactive hash.*

Due to page limit, we move the proof to Appendix A.1 of the full paper [48].

Our proof uses two lemmas that we state below but prove in Appendices A.2 and A.3 of the full paper [48].

Lemma 4.2. *If $\exists i \in \{1, \ldots, k\}$ such that $m_i' \neq m_i$, then Step 2 of IHash.Hash has to abort except with 2^{-s} probability.*

Lemma 4.3. *Let $\mathbf{H}_{min}$ be the min-entropy function. For $\mathbf{H}_{min}(m) = \sigma \cdot \ell$ where $m \in \{0,1\}^{\sigma \cdot \ell}$ and $\langle m \rangle \in \{0,1\}^{\sigma \cdot w}$,*

1. $\mathbf{H}_{min}(m|\langle m \rangle) = (\ell - w)\sigma$. *I.e., $(\ell - w)\sigma$ entropy remains even if P_2 learns $\langle m \rangle$.*
2. *For every m_1 and m_2 where $m_1 \neq m_2$,*

$$\Pr\left[\langle m_1 \rangle = \langle m_2 \rangle\right] \leq \binom{\ell - 1}{w} \bigg/ \binom{n}{w}.$$

5   The Main Protocol

5.1   Protocol Description

Assume P_1 (the generator) and P_2 (the evaluator) wish to compute f over secret inputs x, y, where f is realized as a boolean circuit C that has only AND and XOR gates. The protocol proceeds as follows.

0. **Setup.** The parties decide the public parameters $\ell_w, \sigma_w, \ell_p, \sigma_p$ from the security parameters s, k (see Sects. 6.1 and 6.2 for the detailed discussion).
   (a) P_1 randomly picks $seed_1, \ldots, seed_{n_w}$; P_2 randomly picks $i_1, \ldots, i_{n_w}$. P_1 (as the sender using $seed_1, \ldots, seed_{n_w}$) and P_2 (as the receiver using $i_1, \ldots, i_{n_w}$) run $\mathsf{IHash}_{\ell_w, \sigma_w}$.Setup to initialize the IHash scheme for i-hashing wire-labels.
   (b) P_1 randomly picks $\Delta \in \{0,1\}^{\lambda_w}$ where $\lambda_w = \ell_w \sigma_w$ and calls $\mathsf{IHash}_{\ell_w, \sigma_w}$.Hash to send $\langle \Delta \rangle$ to P_2.
   (c) P_1 (using seeds $H(\Delta, 1), \ldots, H(\Delta, n_p)$) and P_2 (using freshly sampled indices $i_1', \ldots, i_{n_p}'$) run $\mathsf{IHash}_{\ell_p, \sigma_p}$.Setup to initialize the IHash scheme for wire permutation strings.
   (d) P_1 sends $H(H(\Delta, 1)), \ldots, H(H(\Delta, n_p))$ to P_2.

   Then, P_1 randomly select 16 linearly-independent vectors $a_1, \ldots, a_{16}$ from $GF(8)^{\ell_w}$, which will be row vectors of the matrix to be left multiplied with a wire-label to realize the Compress function (compressing a 256-bit wire-label into 128-bit, see Fig. 3). P_1 sends $a_1, \ldots, a_{16}$ to P_2.

1. **Circuit Initialization.** Let n_w be the total number of wires in C. P_1 picks $m_1, \ldots, m_{n_w} \in \{0,1\}^{\lambda_w}$ where $\lambda_w = \ell_w \sigma_w$; then run $\mathsf{IHash}_{\ell_w, \sigma_w}$.Hash with P_2 to send $\langle m_1 \rangle, \ldots, \langle m_{n_w} \rangle$ to P_2. Then, P_1 samples $\rho^1, \ldots, \rho^{n_w} \in \{0,1\}^{\lambda_p}$ where $\lambda_p = \ell_p \sigma_p$, then run $\mathsf{IHash}_{\ell_p, \sigma_p}$.Hash with P_2 to send $\langle \rho^{i_1} \rangle, \ldots, \langle \rho^{i_w} \rangle$ to P_2. For all $1 \leq i \leq n_w$, P_1 sets $p_i = \rho_1^i \oplus \cdots \oplus \rho_{\lambda_p}^i$ (where ρ_j^i denotes the j-th bit of ρ^i) and $w_i^0 := m_i \oplus p_i \Delta$. Let $w_i^1 = m_i \oplus \bar{p}_i \Delta$, hence $m_i = w_i^{p_i}$. Then, P_1 and P_2 process the initial input-wires as follows.

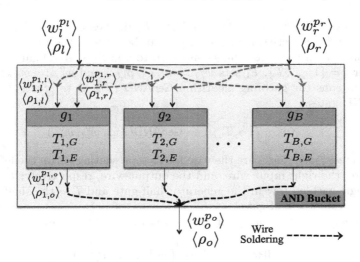

Fig. 5. A bucket of B garbled gates. (Wire labels and hashes exist at both the bucket-level (e.g. $w_l^{p_l}$) and gate-level (e.g. $w_l^{p_{i,l}}$).)

(a) For $1 \leq i \leq n_I^{P_1}$, let (w_i^0, w_i^1) be the pair of wire labels on the wire associated with x_i, P_1 sends $w_i^{x_i}$ to P_2.

(b) For every input-wire W_i associated with P_2's private input y_i:

   i. W_i is $\oplus$-split into s wires $W_{i,1}, \ldots, W_{i,s}$.

   ii. P_1 picks $m_1, \ldots, m_s$ and run $\mathsf{IHash}_{\ell_w, \sigma_w}$.Hash with P_2 to send $\langle m_1 \rangle, \ldots, \langle m_s \rangle$ to P_2. For $1 \leq j \leq s$, P_1 sets $w_{i,j}^0 = m_j$ and $w_{i,j}^1 = w_{i,j}^0 \oplus \Delta$.

   iii. P_2 samples $y_{i,1} \leftarrow \{0,1\}, \ldots, y_{i,s} \leftarrow \{0,1\}$ such that $y_{i,1} \oplus \cdots \oplus y_{i,s} = y_i$.

   iv. For $1 \leq j \leq s$, P_2 retrieves $w_{i,j}^{y_{i,j}}$ from P_1 through oblivious transfer, and verifies $w_{i,j}^{y_{i,j}}$ against $\langle w_{i,j}^{y_{i,j}} \rangle$ (note P_2 can compute $\langle w_{i,j}^{y_{i,j}} \rangle :=$ $\langle w_{i,j}^0 \rangle \oplus y_{i,j} \langle \Delta \rangle$). Any verification failure will result in P_2's delayed abort at Step 5.

   v. P_2 sets $w_i^{y_i} := w_{i,1}^{y_{i,1}} \oplus \cdots \oplus w_{i,s}^{y_{i,s}}$, $\langle w_i^{y_i} \rangle := \langle w_{i,1}^{y_{i,1}} \rangle \oplus \cdots \oplus \langle w_{i,s}^{y_{i,s}} \rangle$, and $p_i = 0$.

2. **Generate.** P_2 randomly picks $\mathcal{J} \in \{0,1\}^k$ and commits it to P_1. $\mathcal{J}$ will be used as the randomness for cut-and-choose later.

(a) P_1 picks $2T$ random λ_w-bit messages ($\lambda_w = \ell_w \sigma_w$) and run $\mathsf{IHash}_{\ell_w, \sigma_w}$.Hash with P_2 to send the i-hashes of the $2T$ random messages. Denote the $2T$ messages by $\{m_{i,l}, m_{i,r}\}_{i=1}^T$, and their i-hashes by $\{\langle m_{i,l} \rangle, \langle m_{i,r} \rangle\}_{i=1}^T$.

(b) P_1 picks $3T$ random λ_p-bit ($\lambda_p = \ell_p \sigma_p$) messages and run $\mathsf{IHash}_{\ell_p, \sigma_p}$.Hash with P_2 to send the i-hashes of these $3T$ random messages. Denote these messages and i-hashes by $\{\rho^{i,l}, \rho^{i,r}, \rho^{i,o}\}_{i=1}^T$ and $\{\langle \rho^{i,l} \rangle, \langle \rho^{i,r} \rangle, \langle \rho^{i,o} \rangle\}_{i=1}^T$. P_1 computes $p_{i,l} = \rho_1^{i,l} \oplus \cdots \oplus \rho_{\lambda_p}^{i,l}$, where $\rho_j^{i,l}$

denotes the jth bit of $\rho^{i,l}$. Similarly, P_1 derives $p_{i,r}$ and $p_{i,o}$ from $\rho^{i,r}$ and $\rho^{i,o}$, respectively. ($p_{i,l}, p_{i,r}, p_{i,o}$ will be used as the i-hash permutation bits on the three wires connected to a garbled AND gate.)

(c) For $i = \{1, \ldots, T\}$, P_1 sets $w_{i,l}^0 := m_{i,l} \oplus p_{i,l}\Delta$ and $w_{i,r}^0 := m_{i,r} \oplus p_{i,r}\Delta$, then runs the garbling algorithm GenAND (Fig. 3) to create T garbled AND gates:

$$(w_{i,o}^0, T_{i,G}, T_{i,E}) \leftarrow \mathsf{GenAND}(i, \Delta, w_{i,l}^0, w_{i,r}^0)$$

where $w_{i,l}^0, w_{i,r}^0, w_{i,o}^0$ are the wire labels representing 0's on the left input-wire, the right input wire, and the output-wire, respectively; $T_{i,G}$ is the single garbled row in the generator half-gate and $T_{i,E}$ the single row in the evaluator half-gate.

(d) Let $w_{i,o}^1 := w_{i,o}^0 \oplus p_{i,o}\Delta$ for all $1 \leq i \leq T$. P_1 and P_2 run

$$\mathsf{IHash}_{\ell_w, \sigma_w}.\mathsf{Hash}\left(w_{1,o}^{p_{1,o}}, \ldots, w_{T,o}^{p_{T,o}}\right)$$

so that P_2 learns $\left\langle w_{1,o}^{p_{1,o}}\right\rangle, \ldots, \left\langle w_{T,o}^{p_{T,o}}\right\rangle$.

3. **Evaluate.** P_2 opens to P_1 the cut-and-choose randomness $\mathcal{J}$, which is used to select and group $B \cdot N$ garbled ANDs into N buckets.

Recall that for every logical gate in C, P_2 has obtained from step 1 two wire-labels w_l^a, w_r^b, which correspond to secret values a, b on the input-wires and i-hashes $\langle \rho^l \rangle, \langle w_l^{p_l} \rangle, \langle \rho^r \rangle, \langle w_r^{p_r} \rangle, \langle \rho^o \rangle, \langle w_o^{p_o} \rangle$. P_1 and P_2 follow an identical topological order to process the logical gates as follows: For every XOR, P_2 sets $w_o := w_l^a \oplus w_r^b$; For every logical AND (Fig. 5), we denote the B garbled AND gates by $g_1, \ldots, g_B$,

(a) P_2 sets $\mathcal{O}$ to an empty set and executes the following for $1 \leq i \leq B$ (note that P_2 always continues execution until Step 5 even if any check failed),

  i. Let $p_{i,l}$ be the i-hash permutation bit of the left input-wire of g_i, i.e., $p_{i,l} = g_i.p_l$. Let $p_{i,r}, p_{i,o}, \rho^{i,l}, \rho^{i,r}, \rho^{i,o}$ be similarly defined. P_1 sends $\rho^l \oplus \rho^{i,l}$, $\rho^r \oplus \rho^{i,r}$ and $\rho^o \oplus \rho^{i,o}$ to P_2, who verifies them against their i-hashes and computes

$$p_l \oplus p_{i,l} := \bigoplus_{1 \leq j \leq \lambda_p} (\rho_j^l \oplus \rho_j^{i,l})$$

$$p_r \oplus p_{i,r} := \bigoplus_{1 \leq j \leq \lambda_p} (\rho_j^r \oplus \rho_j^{i,r})$$

$$p_o \oplus p_{i,o} := \bigoplus_{1 \leq j \leq \lambda_p} (\rho_j^o \oplus \rho_j^{i,o}).$$

  ii. For $b \in \{0, 1\}$, define $w_{i,l}^b$ be the wire-label representing signal b on gate g_i's left input-wire, and let $w_{i,r}^b$, $w_{i,o}^b$ be similarly defined with

g_i's right input-wire and output-wire. P_1 sends

$$\delta_l := w_l^{p_l} \oplus w_{i,l}^{p_{i,l}} \oplus (p_l \oplus p_{i,l})\Delta$$

$$\delta_r := w_r^{p_r} \oplus w_{i,r}^{p_{i,r}} \oplus (p_r \oplus p_{i,r})\Delta$$

$$\delta_o := w_o^{p_o} \oplus w_{i,o}^{p_{i,o}} \oplus (p_o \oplus p_{i,o})\Delta$$

to P_2, who verifies them against their hashes and computes $w_{i,l}^a :=$ $w_l^a \oplus \delta_l$ and $w_{i,r}^b := w_r^b \oplus \delta_r$.

   iii. Recall that $T_{i,G} = g_i.T_G$, $T_{i,E} = g_i.T_E$. P_2 runs $w_{i,o} :=$ EvlAND$(w_l^a, w_r^b, T_{i,G}, T_{i,E})$, and sets $w_o := w_{i,o} \oplus \delta_o$.

   iv. P_2 verifies w_o against $\langle w_o^{p_o} \rangle$ and $\langle w_o^{p_o} \rangle \oplus \langle \Delta \rangle$. If either verification succeeds, P_2 adds w_o to $\mathcal{O}$.

(b) If $\mathcal{O}$ contains two different labels, say w and w'. P_2 computes $\Delta^* := w \oplus w'$, and uses Δ^* to recover P_1's private inputs x and computes $f(x, y)$. Otherwise, $\mathcal{O} = \{w\}$ so P_2 sets $w_o = w$.

4. **Check.** P_2 verifies the correctness of the rest $T - BN$ garbled AND gates. For every check-gate parsed into

$$\left(\langle \rho^l \rangle, \langle w_l^{p_l} \rangle, \langle \rho^r \rangle, \langle w_r^{p_r} \rangle, \langle \rho^o \rangle, \langle w_o^{p_o} \rangle, T_G, T_E\right),$$

(a) P_2 samples $a \leftarrow \{0, 1\}, b \leftarrow \{0, 1\}$, sends them to P_1.

(b) P_1 sends ρ^l, ρ^r, ρ^o to P_2. P_2 verifies them with $\langle \rho^l \rangle, \langle \rho^r \rangle$, and $\langle \rho^o \rangle$. Let $p_l = \rho_1^l \oplus \cdots \oplus \rho_{\lambda_p}^l$, $p_r = \rho_1^r \oplus \cdots \oplus \rho_{\lambda_p}^r$, $p_o = \rho_1^o \oplus \cdots \oplus \rho_{\lambda_p}^o$,

   i. P_1 sends $w_l^a = w_l^{p_l} \oplus (a \oplus p_l)\Delta$ to P_2, who verifies it against $\langle w_l^{p_l} \rangle \oplus (a \oplus p_l)\langle \Delta \rangle$.

   ii. P_1 sends $w_r^b = w_r^{p_r} \oplus (b \oplus p_r)\Delta$ to P_2, who verifies it against $\langle w_r^{p_r} \rangle \oplus (b \oplus p_r)\langle \Delta \rangle$.

   iii. Let $z = a \wedge b$. P_1 sends $w_o^z = w_o^{p_o} \oplus (z \oplus p_o)\Delta$ to P_2, who verifies it against $\langle w_o^{p_o} \rangle \oplus (z \oplus p_o)\langle \Delta \rangle$.

(c) P_2 checks $w_o^z = $ EvlAND(w_l^a, w_r^b, T_G, T_E).

5. **Output determination.**

(a) If any check failed in steps 3 and 4, P_2 aborts.

(b) P_1 proves in zero knowledge that it executes the Step 0b and Step 0c honestly. Namely, the double-hashes received in Step 0d, the i-hash $\langle \Delta \rangle$ hold by P_2, and the watched seeds of IHash$_{\ell_p, \sigma_p}$.Setup are all respect to the same Δ. P_2 aborts if the ZK proof fails.

(c) Otherwise, P_2 outputs $f(x, y)$, either from a recovered Δ or interpreted final output labels.

Remarks. In practice, Step 5b can be done efficiently with ZK proof techniques [14, 47], costing $2n_p$ semi-honest garbled circuit executions of SHA256.

Solder Gates Garbled with Different Δs. The protocol given above assumes all the gates are garbled with respect to the same Δ and allows detecting faulty gates with probability 50%. To further increase the faulty gate detection rate to 100%, we can let P_1 to garble each gate with a freshly sampled secret Δ, so that when a garbled gate is chosen to be checked, P_1 can fully open the gate without leaking the Δ used in other garbled gates.

The procedure to solder a wire associated with $(\langle \Delta_1 \rangle, \langle \rho_1 \rangle, \langle w_1^{p_1} \rangle)$ (where p_1 is the xor-sum of all bits of ρ_1) to another wire associated with $(\langle \Delta_2 \rangle, \langle \rho_2 \rangle, \langle w_2^{p_2} \rangle)$ (where p_2 is the xor-sum of all bits of ρ_2) is also a bit different from that described in the main protocol. First, P_1 reveals $\Delta' := \Delta_1 \oplus \Delta_2$ and $\rho' := \rho_1 \oplus \rho_2$ to P_2, who validates Δ' and ρ' using the corresponding i-hashes. Let p' be the xor-sum of all bits of ρ'. If $p' = 0$, which implies $\langle w_1^{p_1} \rangle$ and $\langle w_2^{p_2} \rangle$ are hashes of wire-labels denoting the same plaintext signal, then P_1 reveals $w' := w_1^{p_1} \oplus w_2^{p_2}$ to P_2; otherwise, if $p' = 1$, then P_1 reveals $w' := w_1^{p_1} \oplus w_2^{p_2} \oplus \Delta_2$ to P_2. With a wire-label w to be translated, P_2 will validate w', then output $w \oplus w'$ as the translated wire-label if $\langle w \rangle = \langle w_1^{p_1} \rangle$; and output $w \oplus w' \oplus \Delta'$ if $\langle w \rangle = \langle w_1^{p_1} \rangle \oplus \langle \Delta_1 \rangle$.

Figure 6 shows wire-label conversion for every possible combination of p_1, p_2, w, and verifies P_2 can always output a label that denotes the same plaintext signal as w. To see the security of this soldering scheme, we note that to a malicious P_2, (Δ', ρ', w') is always indistinguishable from a tuple of random messages.

p_1	p_2	Input (w)	p'	w'	$w \overset{?}{=} w_1^{p_1}$	Output
0	0	w_1^0	0	$w_1^0 \oplus w_2^0$	Yes	w_2^0
0	0	w_1^1	0	$w_1^0 \oplus w_2^0$	No	w_2^1
0	1	w_1^0	1	$w_1^0 \oplus w_2^1 \oplus \Delta_2$	Yes	w_2^0
0	1	w_1^1	1	$w_1^0 \oplus w_2^1 \oplus \Delta_2$	No	w_2^1
1	0	w_1^0	1	$w_1^1 \oplus w_2^0 \oplus \Delta_2$	No	w_2^0
1	0	w_1^1	1	$w_1^1 \oplus w_2^0 \oplus \Delta_2$	Yes	w_2^1
1	1	w_1^0	0	$w_1^1 \oplus w_2^1$	No	w_2^0
1	1	w_1^1	0	$w_1^1 \oplus w_2^1$	Yes	w_2^1

On rows where $w = w_1^{p_1}$, Output $:= w \oplus w'$. On rows where $w \neq w_1^{p_1}$, Output $:= w \oplus w' \oplus \Delta'$ where $\Delta' = \Delta_1 \oplus \Delta_2$.

Fig. 6. Soldering wires associated with different Δs.

5.2   Proof of Security

First, we show that for any N and security parameter s, k, it is possible to set parameters T, B, n, ℓ, w such that our protocol securely computes $f(x, y)$ (except with probability 2^{-s}). For concrete values of s, k, N, we detail how to optimize T, B, n, ℓ, w for performance in Sect. 6.3.

Lemma 5.1. *For any N, s, k, there exist T, B such that if P_2 does not abort at Step 5, then with all but 2^{-s} probability every bucket has at least one correctly garbled gate.*

The validity of Lemma 5.1 is implied by our cut-and-choose parameter selection strategy described in Sect. 6.3.

Lemma 5.2. *If every bucket has at least one correctly garbled gate, P_2 will output $f(x, y)$ at Step 5 except with negligible probability.*

Lemma 5.3. *Given $\langle \Delta \rangle$, a garbled AND gate (T_G, T_E), and the i-hashes of its wire-labels and permutation messages $(\langle \rho^l \rangle, \langle w_l^{p_l} \rangle, \langle \rho^r \rangle, \langle w_r^{p_r} \rangle, \langle \rho^o \rangle, \langle w_o^{p_o} \rangle)$, where $p_l = \rho_1^l \oplus \ldots \oplus \rho_{\lambda_p}^l$, $p_r = \rho_1^r \oplus \ldots \oplus \rho_{\lambda_p}^r$, $p_o = \rho_1^o \oplus \ldots \oplus \rho_{\lambda_p}^o$. If any of the following is not satisfied (see Fig. 3 for EvlAND),*

$$\mathsf{EvlAND}(w_l^0, w_r^0, T_G, T_E) = w_o^0; \quad \mathsf{EvlAND}(w_l^0, w_r^1, T_G, T_E) = w_o^0;$$
$$\mathsf{EvlAND}(w_l^1, w_r^0, T_G, T_E) = w_o^0; \quad \mathsf{EvlAND}(w_l^1, w_r^1, T_G, T_E) = w_o^1,$$

then P_2 will detect this with probability at least $1/2$ at Step 4.

Due to page limit, we moved the proof of Lemmas 5.2 and 5.3 to Appendices A.4 and A.5 of the full paper [48].

Theorem 5.4. *Under the assumptions outlined in Sect. 3, the protocol in Sect. 5.1 securely computes f in the presence of malicious adversaries.*

Due to page limit, we moved the proof to Appendix A.6 of the full paper [48].

6   Parameters and Bounds

6.1   IHash-ing Permutation Messages

Here the goal is to decide the best parameters $n_p, \ell_p, w_p, \sigma_p$ that are used to i-hash the wire permutation messages, i.e., the ρ's used in the main protocol, to achieve the necessary binding and hiding properties. This can be framed into a constrained optimization problem:

$$(n_p, \ell_p, w_p, \sigma_p) = \arg \min \, \mathsf{cost}(n, \ell, w, \sigma)$$

subject to:

$$\binom{\ell - 1}{w} \Big/ \binom{n}{w} \leq 2^{-s} \tag{1}$$

$$\sigma(\ell - w) \geq k \tag{2}$$
$$2^\sigma \geq n$$
$$n, \ell, \sigma, w \in \mathbb{Z}^+$$

where inequality (1) ensures s-bit statistical binding, inequality (2) ensures k-bit hiding, and the target cost function can depend on a number of deployment-specific tradeoffs between bandwidth and computation.

We stress that hiding for the permutation messages is *perfect* because there is no additional information revealed to allow a malicious evaluator to verify its guess on the hidden bits (comparing to the fact that garbled truth table can be used to verify guesses about the wire-labels). Once the cost function is fixed, an efficient solver through aggressive pruning can be constructed. In our experiment, we set $n_p = 44, \ell_p = 20, w_p = 19, \sigma_p = 6$, which provides 40-bit statistical binding (verify this by plugging them into (1)) and 6-bit perfect hiding since $(\ell_p - w_p) \cdot \sigma_p = 6$ (although only 1-bit perfect hiding is needed).

6.2   lHash-ing and Compress-ing Wire-Labels

Here our goal is to determine the best parameters $n_w, \ell_w, w_w, \sigma_w$ to process the wire-labels so that s-bit statistical security and k-bit computational security can be guaranteed for the main protocol. Note the entropy hiding on the wire-labels downgrades to *computational* because a malicious evaluator could run *offline* tests on its guesses of the wire-labels using the garbled rows.

To ensure 40-bit binding and at least 80-bit hiding, $\ell_w \cdot \sigma_w$ (the wire-labels' length) has to be more than 128 bits. This poses a challenge to efficient garbling using fixed-key AES assembly instructions since AES only works on 128-bit blocks. We solve this challenge by making $(\ell_w - w_w)\sigma_w$ slightly larger (i.e., by a factor of $1 + \varepsilon$) than k, followed by a linear Compress function to derive 128-bit compressed labels that each carries more than 80-bit entropy from the original, watched wire-labels. Namely, for wire-labels, we replace constraint (2) by

$$\sigma_w(\ell_w - w_w) \geq (1 + \varepsilon)k$$

where $\varepsilon > 0$ compensates the entropy loss during the compression. We choose $n_w = 86, \ell_w = 32, w_w = 21$ and $\sigma_w = 8$.

To compress wire-labels, the generator samples $128/\sigma_w = 16$ linear-independent vectors (over $GF(2^8)^{32}$) once and left-multiply them to 256-bit wire-labels to obtain 128-bit compressed labels.

Recall that these 16 linear-independent vectors are declared only after the evaluator chose its 21 watch symbols. The entropy analysis of this Compress function can be done by considering the following experiment:

1. P_1 randomly samples 32 symbols, $m_1, \ldots, m_{32} \in GF(2^8)$.
2. P_2 randomly chooses 21 linear-independent vectors $\boldsymbol{W} = (W_1, \ldots, W_{21})$ from $GF(2^8)^{32}$ and thus learns $(m_1, \ldots, m_{32}) \cdot W_i$ for all $1 \leq i \leq 21$.
3. P_1 randomly chooses and sends 16 linear-independent vectors $\boldsymbol{T} = (T_1, \ldots, T_{16}) \in GF(2^8)^{32}$, then outputs $v_1, \ldots, v_{16}$, where $v_i = (m_1 \ldots m_{32}) \cdot T_i$ for all $1 \leq i \leq 16$.

The question is: how much entropy in the output $v_1, \ldots, v_{16}$ remains hidden to P_2? In other words, for every $\mathcal{A}$, every rank-21 matrix $\boldsymbol{W} \in GF(2^8)^{32 \times 21}$ and every rank-16 matrix $T \in \mathrm{GF}(2^8)^{32 \times 16}$, define

$$Q = \Pr\left(m \leftarrow \mathrm{GF}(2^8)^{32} \; : \; \mathcal{A}(W, T, m \cdot W) = m \cdot T \right).$$

We want to know the *min-entropy* of $m \cdot T$, which is essentially $\log(1/Q)$. We answer this question with Lemma 6.1 and elaborate the analysis in its proof.

Lemma 6.1. *Setting* $n_w = 86, \ell_w = 32, w_w = 21$ *and* $\sigma_w = 8$ *ensures 40-bit statistical binding and more than 87.999 bits hiding in the compressed wire-labels; while setting* $n_w = 88, \ell_w = 48, w_w = 32$ *and* $\sigma_w = 8$ *ensures 40-bit statistical binding and more than 127 bits hiding in the compressed wire-labels.*

Proof. Following Lemma 4.3, 40-bit statistical binding for both settings can be verified by computing $-\log\left[\binom{\ell_w - 1}{w_w} / \binom{n_w}{w_w}\right] = -\log\left[\binom{31}{21} / \binom{86}{21}\right] > 40$ and $-\log_2\left[\binom{\ell_w - 1}{w_w} / \binom{n_w}{w_w}\right] = -\log_2\left[\binom{47}{32} / \binom{88}{32}\right] > 40$.

Next, we examine the hiding aspect. Let $T = (T_1, T_2, \cdots, T_{16})$ be an 32×16 matrix over $\mathrm{GF}(2^8)$ of rank 16, and $W = (W_1, W_2, \cdots, W_{21})$ be an 32×21 matrix over $\mathrm{GF}(2^8)$ of rank 21. We want to show that for every adversary $\mathcal{A}$,

$$-\log \Pr\left(m \leftarrow \mathrm{GF}(2^8)^{32} \; : \; \mathcal{A}(W, T, m \cdot W) = m \cdot T \right) > 87.999.$$

Define $D = \dim(T \oplus W) - \dim(W)$ where "$\oplus$" denotes *direct sum* and "$\dim(\cdot)$" denotes the *dimension* of a given vector space. For every rank-t matrix T and every rank-w matrix W, we note that

$$\Pr\left(m \leftarrow \mathrm{GF}(2^8)^{32} \; : \; \mathcal{A}(W, T, m \cdot W) = m \cdot T \right) = 2^{-8D}.$$

Thus, our goal is to show that $-\log \mathbb{E}(2^{-8D}) > 87.999$. The concept of *expectation* is introduced because 2^{-8D} itself is a random variable over the random choices of picking T and W.

Define $d_{i,j} = \dim(\mathrm{Span}(T_{i+1}, T_{i+2}, \cdots, T_t, W)) - \dim(W)$ under the condition that $\dim(\mathrm{Span}(T_1, T_2, \cdots, T_i) \cap W) = j$. Thus, $D = d_{0,0}$ and we can derive $d_{i,j}$ from $d_{i+1,j+1}$ and $d_{i+1,j}$ using recursion. Vector T_{i+1} has to fall into one of the two cases:

1. $T_{i+1} \in W$: Thus $\mathrm{Span}(T_{i+1}, T_{i+2}, \cdots, T_t, W) = \mathrm{Span}(T_{i+2}, T_{i+3}, \cdots, T_t, W)$. In addition, because $T_{i+1} \notin \mathrm{Span}(T_1, T_2, \cdots, T_i)$ and $\mathrm{Span}(T_1, T_2, \cdots, T_{i+1}) \cap W = (\mathrm{Span}(T_1, T_2, \cdots, T_i) \cap W) \oplus Span(T_{i+1})$, we know,

$$\dim(\mathrm{Span}(T_1, T_2, \cdots, T_{i+1}) \cap W) = \dim(\mathrm{Span}(T_1, T_2, \cdots, T_i) \cap W) + 1$$
$$= j + 1$$

Note the probability of $T_{i+1} \in W$, conditioned on $T_{i+1} \notin \mathrm{Span}(T_1, \ldots, T_i)$, is $(2^{8 \cdot (21-j)} - 1)/(2^{8 \cdot (32-i)} - 1)$. This is because there are $2^{8 \cdot (32-i)} - 1$ possible non-zero T_{i+1} that satisfy $T_{i+1} \notin \mathrm{Span}(T_1, \ldots, T_i)$; and $2^{8 \cdot (21-j)} - 1$ non-zero choices of T_{i+1} such that $T_{i+1} \in W$ but $T_{i+1} \notin \mathrm{Span}(T_1, \ldots, T_i) \cap W$, since $\dim(\mathrm{Span}(T_1, \ldots, T_i) \cap W) = j$.

2. $T_{i+1} \notin \boldsymbol{W}$: Thus $\mathrm{Span}(T_{i+1}, T_{i+2}, \cdots, T_t, \boldsymbol{W}) = \mathrm{Span}(T_{i+2}, T_{i+3}, \cdots, T_t, W) \oplus \mathrm{Span}(T_{i+1})$, and

$$\dim(\mathrm{Span}(T_1, T_2, \cdots, T_{i+1}) \cap \boldsymbol{W}) = \dim(\mathrm{Span}(T_1, T_2, \cdots, T_i) \cap \boldsymbol{W}) = j.$$

This happens with probability $1 - (2^{8 \cdot (21-j)} - 1)/(2^{8 \cdot (32-i)} - 1)$.

Therefore,

$$d_{i,j} = \frac{2^{8 \cdot (21-j)} - 1}{2^{8 \cdot (32-i)} - 1} d_{i+1,j+1} + \left(1 - \frac{2^{8 \cdot (21-j)} - 1}{2^{8 \cdot (32-i)} - 1}\right)(d_{i+1,j} + 1)$$

Moreover,

$$\mathbb{E}(2^{-8 \cdot d_{i,j}}) = \frac{2^{8 \cdot (21-j)} - 1}{2^{8 \cdot (32-i)} - 1} \mathbb{E}(2^{-8 \cdot d_{i+1,j+1}}) + \left(1 - \frac{2^{8 \cdot (21-j)} - 1}{2^{8 \cdot (32-i)} - 1}\right) \mathbb{E}\left(2^{-8 \cdot (d_{i+1,j}+1)}\right)$$

$$= \frac{2^{8 \cdot (21-j)} - 1}{2^{8 \cdot (32-i)} - 1} \mathbb{E}(2^{8 \cdot d_{i+1,j+1}}) + \left(1 - \frac{2^{8 \cdot (21-j)} - 1}{2^{8 \cdot (32-i)} - 1}\right) \frac{1}{2^8} \mathbb{E}(2^{-8 \cdot d_{i+1,j}})$$

Finally, the base cases for bootstrapping the recursive calculation are: (1) $d_{16,j} = 0$ for all j; and (2) $d_{i,21} = i$ for all i. It is easy to calculate $d_{0,0}$ in full precision with a computer program. Thus,

$$-\log_2 \mathbb{E}(2^{-8 \cdot D}) = \log_2 \mathbb{E}(2^{8 \cdot d_{0,0}})$$
$$= \log_2 \frac{340282366920938463463374607431768211457}{1099511627777}$$
$$> 87.9999999999986 > 88 - 10^{-11}.$$

When setting $n_w = 88, \ell_w = 48, w_w = 32$ and $\sigma_w = 8$, we can derive a similar recurrence as above and solve it for a different $d_{0,0}$ and verify that $\log_2 \mathbb{E}(2^{8 \cdot d_{0,0}}) > 127$. $\qquad\square$

Remark. As an alternative, the generic Leftover Hash Lemma [3,22] could be used to solve the wire-label entropy extraction problem. However, the potentially large entropy loss $(2\log(1/2^{-80}) + O(1) = 160 + O(1)$ bits) makes it unsuitable in our case. In contrast, less than 10^{-11} bit of entropy (out of the $8 \cdot (32 - 21) = 88$ bits remaining entropy before Compress-ing) will actually be lost due to our Compress function (see Lemma 6.1's proof)!

6.3   LEGO Cut-and-Choose Parameters

While existing analysis of LEGO cut-and-choose rely on empirical point trials in a likely area of the parameter space, we show that the search can be fully guided to efficiently identify the best cut-and-choose parameters in practical scenarios.

Recall the goal is to determine the best T, B to ensure s-bit statistical security for computing a circuit of N gates. Assume out of the total T garbled gates, b of them are faulty gates. Let P_c be the probability of selecting a particular set of

$T - BN$ gates in which t gates are faulty. Then $P_c = \binom{b}{t}\binom{T-b}{T-NB-t} \big/ \binom{T}{T-NB}$. Let P_e be the probability that at least one bucket is filled entirely by faulty gates, then $P_e \leq N\binom{b}{B} \big/ \binom{NB}{B}$ where the equality is approached from below when $N \gg B$. Since the overall failure probability $P_{\text{overall}} \leq \max_b \sum_{t=1}^{b} 2^{-t} P_c P_e$, it suffices to find the smallest T such that $\max_b \sum_{t=1}^{b} 2^{-t} P_c P_e \leq 2^{-s}$. Note that we only need to consider b values up to a upper bound slightly larger than s because $P_c P_e < 1$ and $\sum 2^{-t}$ converges as $t \to \infty$. In addition, we observe that the smallest T for any fixed B (we call T_B) can be quickly determined since P_{overall} strictly decreases when T grows. Thus, T, B can be efficiently determined through pruning when examining $B = 2, 3, \dots \lceil T_B/N \rceil$.

6.4 A Fallacy and a Tight Bound

Define $\kappa = T/N$. Prior works claimed $O(\kappa) = O(T/N) = O((skN/\log N)/N) = O(sk/\log N)$, implying $\kappa \to 0$ when $N \to +\infty$ [11,12,36]. However, we found that this is not the case. More precisely, T should be $O(\kappa \cdot N + skN/\log N)$ and *2 is a tight bound of κ*. That is, $\kappa \leq 2$ cannot be achieved without compromising security while any $\kappa > 2$ is securely achievable (using our protocol) if N is large enough. However, the formal proof of this seemingly intuitive result is nontrivial.

Theorem 6.2. *Let $\kappa = T/N$ and $\Pr_{overall}$ be the overall success rate of P_1 attacking LEGO-style cut-and-choose.*

1. *If $\kappa \leq 2$, there exists a constant $c > 0$ and an integer N_0 such that $\Pr_{overall} > c$ for all $N > N_0$.*
2. *For every statistical security parameter s and computational security parameter k, there exists an integer N_0 such that the protocol of Sect. 5.1 with a $\kappa > 2$ securely computes all circuits of size $N > N_0$.*

Due to page limit, the proof of Theorem 6.2 is moved to Appendix B of the full paper [48].

7 Evaluation

Measurement Methodology. We ran experiments on Amazon EC2 (instance type: c4.2xlarge) running Ubuntu Linux in both the LAN (2.5 Gbps, < 1 ms latency) and WAN (200 Mbps, 20 ms latency) network settings. Our IHash parameters are listed in Fig. 7. As our comparison baseline, we chose WMK [42] and TinyLEGO [12,37], two implementations of single-execution setting protocols representing the state-of-the-art. All comparisons are aligned on the same hardware and network environment, based on single-threaded executions. We include results for both 88- and 127-bit computational security for our protocols, but anticipate the performance numbers for 127-bit computational security would drop significantly if processors with AVX512 instructions become available.

Microbenchmarks. We first measured the performance of our protocol over seven basic tasks (see Fig. 8):

	n_w	ℓ_w	σ_w	w_w	n_p	ℓ_p	σ_p	w_p
$s = 40, k = 88$	86	32	8	21	44	20	6	19
$s = 40, k = 127$	88	48	8	32	44	20	6	19

Fig. 7. IHash parameters.

	Garble		Check		Evaluate		Solder	
	CPU	BW	CPU	BW	CPU	BW	CPU	BW
$k = 88$	1.93	327	1.16	127	0.75	106	0.37	159
$k = 127$	2.37	333	1.40	159	0.79	138	0.48	207

(a) Cost per garbled gate.

	Per P_1's Input		Per P_2's Input		Per Output	
	CPU	BW	CPU	BW	CPU	BW
$k = 88$	0.29	86	4.11	5360	0.028	32
$k = 127$	0.34	88	4.18	6400	0.031	48

(b) Cost per input-/output-wire.

Fig. 8. Microbenchmarks. (The unites are either microsecond or byte. CPU timings do not include network time. Timings are averaged over 10^6 executions.)

1. **Garble:** This is to generate three wires (including the wire permutation messages and corresponding i-hashes) and a garbled truth table for AND.
2. **Check:** This includes evaluating a garbled gate and verifying the results with the three revealed wire permutations bits.
3. **Evaluate:** This includes evaluating a garbled truth table.
4. **Solder:** This includes solder three wires of a garbled AND gate.
5. **P_1's Input:** This includes generating a fresh wire and sending the wire-label associated to an input bit of P_1.
6. **P_2's Input:** This includes generating 40 fresh wires without the wire permutation messages and running 40 extended OTs.
7. **Output:** This includes revealing the wire permutation messages of an output wire if P_2 should learn the output (or sending the output wire-label obtained from evaluation if P_1 should learn the output).

We have also compared the performance of the basic procedures between our approach and WMK's [42] (Fig. 9). While our speed of processing logical AND gates is about 2.5–5x slower, we can outperform WMK's highly optimized circuit input/output handling mechanism by 2–200x in LAN setting and 3–75

	P_1's Input		P_2's Input		Output		Logical AND	
	WMK	**Ours**	WMK	**Ours**	WMK	**Ours**	WMK	**Ours**
LAN	13.8	**0.57**	19.7	**8.24**	12.3	**0.02**	4.13	**21**
WAN	158.3	**3.8**	111.3	**36.4**	105.1	**1.4**	51.6	**135**

Fig. 9. Microbenchmark comparisons with WMK [42]. The units are either microsecond/wire or microsecond/gate. Timings are wall-clock time. Security parameters are aligned at $s = 40, k = 127$, and assuming $B = 5$ in our protocol.

in the WAN setting, which demonstrates a clear advantage of LEGO approach over traditional cut-and-choose protocols. Note that as the communication cost becomes the bottleneck in the WAN setting, the performance gap of a task will approach the bandwidth requirement ratio of the task between the two protocols.

Applications. Figure 10 shows how our protocol compares to the baselines over several end-to-end oblivious applications running with 2^{-40} statistical security. These include

1. **AES.** It encrypts one block using AES. The circuit takes a 128-bit input from each party and computes $N = 6800$ AND gates. We set $B = 5$, $T = 39535$.
2. **DES.** It encrypts one block using DES. The circuit takes a 768-bit key from P_1 and a 64-bit message from P_2. Since $N = 18175$, we set $B = 5$, $T = 97593$.
3. **Comparison.** It compares two 10K-bit integers so it takes 10K bits from P_1 and 10K bits from P_2 and outputs 1 bit to indicate which number is larger. It involves $N = 10K$ AND gates, so we set $B = 5$ and $T = 55973$.
4. **Hamming Distance.** This computes the Hamming distance between two 2048-bit strings. The circuit takes a 2048-bit input from each party and computes $N = 4087$ AND gates, so we set $B = 5$, $T = 25432$.

Our approach is more than 2x faster than TinyLEGO [37] in the LAN setting. Our measurements of their AES and DES protocols are in line with the numbers reported in their paper, though we noticed that theirs run much slower on applications with longer inputs and outputs such as Compare and Hamming (whose performance numbers were not included in their paper). We suspect (and get confirmed by one of the TinyLEGO implementers) that this is probably due to some implementation issues and the use of *input authenticators* mechanism required in TinyLEGO. On the flip side, we note that TinyLEGO uses 20–50% less bandwidth.

In comparison with WMK, our protocol is about 4–5x slower when running AES and DES over the LAN. However, for input/output intensive applications like Compare and Hamming, the overall performance of our protocol is very close, and can even be 30–80% faster than WMK, especially in the WAN setting.

Cut-and-Choose Larger Circuit Components. We used AES as an example application to study the potential benefit of JIMU's support of cut-and-choosing

	AES		DES		Compare		Hamming	
	LAN	WAN	LAN	WAN	LAN	WAN	LAN	WAN
WMK [42]	39	580	66	1200	266	3479	70	946
TinyLEGO [37]	241	1055	561	1883	1345	3927	353	1459
Ours ($k = 88$)	**97**	**827**	**254**	**2182**	**170**	**1705**	**61.5**	**733**
Ours ($k = 127$)	**119**	**1060**	**300**	**2589**	**202**	**1921**	**75**	**858**

(a) Time. (Numbers are in millisecond. Timings are end-to-end wall-clock time excluding that of *one-time* setup work such as the base OTs.)

	AES	DES	Compare	Hamming
WMK [42]	10.2	27.2	78.9	18.8
TinyLEGO [37]	15.3	30.5	65.6	21.4
Ours ($k = 88$)	**22.8**	**55.6**	**84.7**	**25.2**
Ours ($k = 127$)	**26.5**	**64.6**	**89.8**	**27.6**

(b) Bandwidth. (Numbers are in MB.)

Fig. 10. Applications performance. (Measurements are averaged over 10 executions. The numbers don't include the setup cost, i.e., for the base OTs and ZK proof of Step 5b.)

larger circuit components. We adopted the strategy of Huang et al. [19] with which the only components using non-XOR gates are SubBytes (each has 34 ANDs if Boyar and Peralta's SubByte circuit is used [6]). We compared the approach where SubBytes are the basic unit of cut-and-choose with that of cut-and-choosing individual ANDs. Figure 11 shows the detailed comparison on how the protocol parameters and performance are affected by increasing the size of basic cut-and-choose units. For running the same application (a column of Fig. 11), cut-and-choosing SubBytes may require using larger B than cut-and-choosing ANDs because N will be 34x smaller. However, much overhead for enabling wire-soldering can be saved for all internal wire connections. We observe 45%–60% savings in time and 50%–60% savings in bandwidth when SubBytes are treated as the basic cut-and-choose units while the savings increase as the number of AES circuits involved in the application increases.

C&C Unit		1 AES		32 AES		64 AES		128 AES	
		Time	BW	Time	BW	Time	BW	Time	BW
AND	Params	$N = 5440$ $B = 5$ $T = 33643$		$N = 174K$ $B = 4$ $T = 763K$		$N = 348K$ $B = 4$ $T = 1470K$		$N = 696K$ $B = 4$ $T = 2871K$	
	LAN	0.12	21.9	3.89	495.88	7.65	956.1	15.28	1869.8
	WAN	0.86		20.51		40.27		79.91	
SubByte	Params	$N = 160$ $B = 7$ $T = 1401$		$N = 5120$ $B = 5$ $T = 28222$		$N = 10240$ $B = 5$ $T = 54200$		$N = 20480$ $B = 4$ $T = 100410$	
	LAN	**0.06**	**11.1**	**1.66**	**230.5**	**3.24**	**447.1**	**6.07**	**802.8**
	WAN	**0.32**		**9.82**		**19.07**		**34.69**	

Unit of timings is second. Unit of bandwidths is MB.

Fig. 11. Benefits of cut-and-choosing larger components.

Acknowledgement. We thank Phillip Rogaway for hints on compressing wire-labels and Feng-Hao Liu for helpful discussions. We are also grateful to the numerous comments from Xiao Wang and Darion Cassel on the implementation. This work is supported by NSF award #1464113 and NIH award 1U01EB023685-01.

A   Proofs

A.1   Proof of Theorem 4.1

Proof. We examine every interface of IHash in an OT-hybrid world.

1. For a corrupted sender P_1, observing that P_2 doesn't have secret inputs to the IHash functionality, we define
   Experiment 0: This is the real world execution between P_1 and P_2, running IHash.Hash followed by IHash.Verify.
   Experiment 1: The simulator S interacts with P_1 like P_2, except that it extracts the seeds $seed_i$ that P_1 picked (through the ideal OT functionality). In particular, S proceeds with the steps below to recover $m_1, \ldots, m_\nu$:
   (a) S runs the checks of Step 2 as if it is P_2 in the real world protocol and aborts if P_2 would.
   (b) If S did not abort in the step above but does detect P_1 cheating on certain symbol positions (that happen to be not watched by P_2) in the check step, S marks all these overlooked symbol positions. Let n_{mark} be the total number of the marked positions.

(c) After the checks of Step 2, $\mathcal{S}$ can solve a system of homogeneous linear equations to recover $\boldsymbol{m}_i$ for all $1 \leq i \leq \nu$.

Finally, $\mathcal{S}$ sends these recovered messages to the ideal IHash functionality and outputs whatever P_1 outputs.

To see the two experiments are indistinguishable, we note that n_{mark} will be less than $n - \ell$ except for $\binom{n-n_{mark}}{w}/\binom{n}{w}$, and $\binom{n-n_{mark}}{w}/\binom{n}{w} < \binom{\ell-1}{w}/\binom{n}{w}$ and the latter is guaranteed to be smaller than 2^{-s} by the parameter selection strategy (Sect. 6.1). Further, note that P_1 can't cheat on any of the unmarked symbol positions without being caught in the check step except for 2^{-s} probability. Combining the two observations, we know that there has to be more than ℓ valid symbols left unmarked to allow $\mathcal{S}$ to recover a unique $\boldsymbol{m}_i$ for every i except for a negligible probability. That said, applying Lemma 4.2 by naming the messages i-hashed in Experiment 0 as $\boldsymbol{m}_i$ and those in Experiment 1 as $\boldsymbol{m}'_i$, we know if neither experiment aborts, it must be the same messages that are i-hashed in both experiments, except for a negligible probability.

2. For a corrupted P_2, we note that P_1 in the ideal world outputs nothing. Consider

   Experiment 0: This is the real world execution between P_1 and P_2, running IHash.Hash followed by IHash.Verify.

   Experiment 1: The simulator calls the ideal IHash on the outset to learn ihashes $\langle \boldsymbol{m}_i \rangle$, then $\mathcal{S}$ interacts with P_2 like P_1 in experiment 0 but using $\boldsymbol{m}'_i$ as its input messages where $\langle \boldsymbol{m}'_i \rangle = \langle \boldsymbol{m}_i \rangle$ (such an $\boldsymbol{m}'_i$ is always easy to find through solving a system of homogeneous linear equations). $\mathcal{S}$ outputs whatever P_2 outputs.

   It is easy to see the above two experiments are indistinguishable based on the security of the PRG.

This completes the proof.

A.2   Proof of Lemma 4.2

Assume there exists a nonempty set I such that

$$\boldsymbol{m}'_i \neq \mathsf{PRG}(seed_1, i)\| \ldots \|\mathsf{PRG}(seed_\ell, i), \text{ if } i \in I.$$

Hence, with y randomly picked from $\{0,1\}^\nu$ and a fixed t, the probability that $\sum_{i \in I} y_i \boldsymbol{m}'_i = \hat{\boldsymbol{m}}'_t - \sum_{i \in [\nu]/I} y_i \boldsymbol{m}'_i - \boldsymbol{m}'_t$ is $2^{-\sigma}$. That is,

$$\sum_{i \in I} y_i \boldsymbol{m}'_i + \sum_{i \in [\nu]/I} y_i \boldsymbol{m}'_i + \boldsymbol{m}'_t = \sum_{i \in [\nu]} y_i \boldsymbol{m}'_i + \boldsymbol{m}'_t = \hat{\boldsymbol{m}}'_t$$

can hold with probability at most $2^{-\sigma}$. By Part 2 of Lemma 4.3,

$$\sum_{i \in [\nu]} y_i \langle \boldsymbol{m}'_i \rangle + \langle \boldsymbol{m}'_t \rangle = \left\langle \hat{\boldsymbol{m}}'_t \right\rangle \iff \sum_{i \in [\nu]} y_i \boldsymbol{m}'_i + \boldsymbol{m}'_t = \hat{\boldsymbol{m}}'_t$$

except for $\binom{\ell-1}{w}/\binom{n}{w}$ probability. Because Step 2c checks $\sum_{i\in[\nu]} y_i\langle m'_i\rangle + \langle m'_t\rangle = \langle \hat{m}'_t\rangle$, it can pass with probability at most $2^{-\sigma}+(1-2^{-\sigma})\cdot\binom{\ell-1}{w}/\binom{n}{w}$. Repeating step 2c t times renders the probability passing all checks below $2^{-t\sigma}+(1-2^{-t\sigma})\cdot\binom{\ell-1}{w}/\binom{n}{w}$. Since Step 2c repeats $\left\lceil -\frac{1}{\sigma}\log\left(2^{-s}-\binom{\ell-1}{w}/\binom{n}{w}\right)\right\rceil$ times, a bad m' can be i-hashed without triggering a failure with probability at most 2^{-s}. $\qquad\square$

A.3 Proof of Lemma 4.3

1. Recall $\mathbf{H}_{min}(m) = \ell\cdot\sigma$ and $\langle m\rangle$ is the product of m and w linearly independent vectors. Thus, in P_2's view, for any fixed m_0, $\Pr(m=m_0|\langle m\rangle=\langle m_0\rangle)=2^{-(\ell-w)\sigma}$. Therefore,

$$\mathbf{H}_{min}(m|\langle m\rangle)$$
$$= -\sum_{m_0}\Pr(m=m_0|\langle m\rangle=\langle m_0\rangle)\cdot\log\Pr(m=m_0|\langle m\rangle=\langle m_0\rangle)$$
$$= -\sum_{m_0}2^{-(\ell-w)\sigma}\cdot[-(\ell-w)\sigma]$$
$$= 2^{-(\ell-w)\sigma}\cdot[(\ell-w)\sigma]\cdot 2^{(\ell-w)\sigma}=(\ell-w)\sigma.$$

2. If $m_1\neq m_2$, their $[n,\ell,n-\ell+1]_{2^\sigma}$ Reed-Solomon codewords can have at most $\ell-1$ identical symbols. Since a random w (out of n) symbols of the encoding are watched, with probability at least $1-\binom{\ell-1}{w}/\binom{n}{w}$, the ihashes of the two messages will be different. $\qquad\square$

A.4 Proof of Lemma 5.2

Proof. Evaluating an incorrectly garbled gate will yield an output wire-label that either matches $\langle w_o^{p_o}\rangle$, or $\langle w_o^{p_o}\rangle\oplus\langle\Delta\rangle=\langle w_o^{p_o}\oplus\Delta\rangle$, or none of them.

1. If the output label matches with neither i-hashes, the evaluation result will be ignored (Step 3(a)iv);
2. If the output label matches with an i-hash and represents the same plain-text value as the output label obtained from evaluating the correctly garbled gate in the same bucket, the corrupted garbled gate does not affect the evaluation.
3. If the output label matches with an i-hash but represents the opposite plain-text value as the output label obtained from evaluating the correctly garbled gate, P_2 learns Δ at step 3b. Then P_2 will be able to learn P_1's input x by examining the buckets that take bits of x as immediate inputs. Note that P_2 can derive all the wire permutation string from Δ, hence, along with the wire-labels representing P_1's input (whose validity is guaranteed by the corresponding i-hashes), P_2 is able to precisely infer every bits of x.

In all cases, P_2 can correctly output $f(x,y)$ with all but negligible probability. $\qquad\square$

A.5   Proof of Lemma 5.3

Proof. By the definition of EvlAND, the four equations above are essentially

$$H(w_l^0) \oplus \mathsf{lsb}\,(w_l^0)T_G \oplus H(w_r^0) \oplus \mathsf{lsb}\,(w_r^0)(T_E \oplus w_l^0) = w_o^{p_o}$$
$$H(w_l^1) \oplus \mathsf{lsb}\,(w_l^1)T_G \oplus H(w_r^0) \oplus \mathsf{lsb}\,(w_r^0)(T_E \oplus w_l^1) = w_o^{p_o}$$
$$H(w_l^0) \oplus \mathsf{lsb}\,(w_l^0)T_G \oplus H(w_r^1) \oplus \mathsf{lsb}\,(w_r^1)(T_E \oplus w_l^0) = w_o^{p_o}$$
$$H(w_l^1) \oplus \mathsf{lsb}\,(w_l^1)T_G \oplus H(w_r^1) \oplus \mathsf{lsb}\,(w_r^1)(T_E \oplus w_l^1) = w_o^{p_o} \oplus \Delta.$$

That is,

$$\mathsf{lsb}\,(w_l^0)T_G \oplus w_o^{p_o} \oplus \mathsf{lsb}\,(w_r^0)T_E = H(w_l^0) \oplus H(w_r^0) \oplus \mathsf{lsb}\,(w_r^0)w_l^0$$
$$\mathsf{lsb}\,(w_l^1)T_G \oplus w_o^{p_o} \oplus \mathsf{lsb}\,(w_r^0)T_E = H(w_l^1) \oplus H(w_r^0) \oplus \mathsf{lsb}\,(w_r^0)w_l^1$$
$$\mathsf{lsb}\,(w_l^0)T_G \oplus w_o^{p_o} \oplus \mathsf{lsb}\,(w_r^1)T_E = H(w_l^0) \oplus H(w_r^1) \oplus \mathsf{lsb}\,(w_r^1)w_l^0$$
$$\mathsf{lsb}\,(w_l^1)T_G \oplus w_o^{p_o} \oplus \mathsf{lsb}\,(w_r^1)T_E = H(w_l^1) \oplus H(w_r^1) \oplus \mathsf{lsb}\,(w_r^1)w_l^1 \oplus \Delta,$$

which can be viewed as a linear system of four equations over three variables T_G, T_E, and $w_o^{p_o}$. Note that all coefficients on the left-hand side and all constants on the right-hand side of the equations are fixed by the seven i-hashes known to P_2. Also note that any three out of the four equations are linearly independent except with negligible probability, because H is modeled as a random oracle and $\mathsf{lsb}\,(w_l^0) \oplus \mathsf{lsb}\,(w_l^1) = 1$, $\mathsf{lsb}\,(w_r^0) \oplus \mathsf{lsb}\,(w_r^1) = 1$, $w_l^0 \oplus w_l^1 = w_r^0 \oplus w_r^1 = \Delta$. Thus, if any three of the four equations hold, the fourth one will be automatically satisfied as it is simply a linear combination of the other three. In addition, we know there must be one solution to the system of four equations if P_1 follows the specification of GenAND. Therefore, if $T_G, T_E, w_o^{p_o}$ take some corrupted values such that any one equation does not hold, there has to be at least one other equation that does not hold (otherwise, $T_G, T_E, w_o^{p_o}$ have to satisfy all four equations). Therefore, if the gate is corrupted, at least two of the equations fail to hold, allowing P_2 to detect it with probability at least $1/2$ when it randomly checks one equations at step 4c.   □

A.6   Proof of Theorem 5.4

Proof. We analyze the protocol in a hybrid world where the parties have access to ideal functionalities for 1-out-of-2 oblivious transfer, commitment, and interactive hash. The standard composition theorem [9] implies security when the sub-routines are instantiated with secure implementations of these functionalities.

If P_1 is corrupted. We construct a polynomial-time simulator $\mathcal{S}$ that interacts with the corrupted P_1 as P_2 with input $y = 0$ in the protocol of Sect. 5.1, except for the following changes:

1. All invocations of interactive hash protocol is replaced with calls to the ideal interactive hash functionality simulated by $\mathcal{S}$.

2. At Step 1a, on receiving $w_i^{x_i}$, S learns P_1's input x_i for all $1 \leq i \leq n_I^{P_1}$, using its knowledge of $\rho^i, p_i, w_i^{p_i}$ extracted from the ideal interactive hash functionality.

3. At Step 5 of **Output determination**, if S does not abort, (instead of outputs $f(x,0)$), S sends x to the trusted party and receives in return $z = f(x,y)$. S rewrites P_1's output $f(x,0)$ with z.

We can infer $\text{REAL}^{P_1,P_2}(x,y) \approx \text{IDEAL}^{T,S,P_2}(x,y)$ for every x,y, where REAL and IDEAL are defined canonically as in [15, Definition 7.2.6], from two basic observations:

1. If P_2 aborts in the real execution, S will also abort in the ideal execution as the changes in S from P_2 does not affect their abort behavior.

2. If P_2 does not abort in the real execution, by Lemmas 5.1 and 5.2, it will output $f(x,y)$ in the ideal execution except for negligible probability. Over exactly identical inputs and random tapes, S will not abort either and will send the extracted x to the trusted party in the ideal world to obtain the same outcomes as the real execution.

If P_2 is corrupted. An efficient simulator S can be constructed that interacts with the corrupted P_2 as P_1 with input $x = 0$ using the Sect. 5.1 protocol, except for the following changes:

1. At Step 1b, an ideal 1-out-of-2 OT functionality is used for P_2 to obtain wire labels on the split input-wires. This allows S to extract $y_{i,1}, \ldots, y_{i,s}$, thus learn P_2's effective input $y_i := y_{i,1} \oplus \cdots \oplus y_{i,s}$ for all $1 \leq i \leq n_I^{P_2}$.

2. At Step 2, an ideal commitment functionality (simulated by S) is used to commit $\mathcal{J}$. This allows S to extract the cut-and-choose string $\mathcal{J}$. S sends the y to the trusted party and receive in return $z = f(x,y)$. Then, S generates the T garbled AND gates such that all gates to be checked are correct; and all gates filling the bucket of the ith final output-wire evaluate to constant z_i.

We can infer $\text{REAL}^{P_1,P_2}(x,y) \approx \text{IDEAL}^{T,P_1,S}(x,y)$ for every x,y from the following observations:

1. If P_1 aborts in the real execution, S will also abort in the ideal execution. The real model P_2 learns (at the best) a transcript of garbled circuits, which, by the privacy and obliviousness properties of garbling scheme, is computationally indistinguishable from that generated by S.

2. If P_1 does not abort in the real execution, P_1 will output $f(x,y)$ in the ideal execution as well (except with negligible probability). Over exactly identical inputs and random tapes, S will not abort either and will extract y to obtain $f(x,y)$ from the trusted party.

B   Proof of Theorem 6.2

Here we generalize the bucket size B to decimals, implying buckets can be of several different integer sizes.

Proof of Theorem 6.2 **Part** 1: We first show the theorem for the case $\kappa = 2$. The validity of Theorem 6.2 for $\kappa = 2$ can be established based on Lemmas B.1 and B.2:

1. According to Lemma B.1, the protocol cannot be secure if the buckets are of sizes 1 or 2;
2. According to Lemma B.2, fixing $\kappa = 2$ and T, N, using buckets of size greater than 2 will only make it even more vulnerable to attacks.

For the case of $\kappa < 2$, we know that, compared to the case when $\kappa = 2$, T has to be reduced (for any fixed N). This implies one of the following three has to be true:

1. less garbled gates can be used for verification;
2. less garbled gates be used for evaluation;
3. both the above two happen.

Since any of above three facts will make it easier for the cheating P_1 to succeed, Part 1 of Theorem 6.2 holds for $\kappa < 2$ too.

Proof of Theorem 6.2 **Part** 2: The proof is the same as that of Theorem 5.4 except that the use of Lemmas 5.1 and 5.2 are replaced by Lemma B.6.   □

Remark. The conclusion that "$\kappa \leq 2$ can't be secure" is meaningful only in the *asymptotic* sense. A $\kappa \leq 2$ could still offer certain *concrete* security, e.g., a $\kappa \leq 2$ could provide 3 bits of statistical security. However, our proof shows that a $\kappa \leq 2$ cannot provide statistical security of more than 5 bits for any realistic application that needs more than three AND buckets. To see this, simply set $N_0 = 3$ and plugging in $c_2 = 0.48, c_3 = 0.13, c_1 = 0.85$ (so $c > 0.05$, that is, $\log 0.05^{-1} \approx 4.32$ bits of security at best) in the proof of Claim B.5.

Lemma B.1. *Assume $B \leq 2$. Let* $\mathrm{Pr}_{overall}(N, x, b)$ *be the probability that P_1, with b bad gates, succeeds in attacking a protocol computing a circuit of N buckets, of which x fraction is size 2 and $(1 - x)$ fraction is size 1. If $\kappa = 2$, there exists a constant $c > 0$ and an integer N_0 such that* $\mathrm{Pr}_{overall}(N, x, b) > c$ *for all $N > N_0$.*

Proof. Assume for simplicity that in the gate verification stage, if a faulty gate is indeed selected for verification, P_2 is able to detect it with probability 1, i.e., $\tau = 1$. (If $\tau < 1$, it is easier for P_1 to succeed.)

Let $\mathrm{Pr}_c(N, x, b)$ be the probability that P_1 survives the gate verification stage; and $\mathrm{Pr}_e(N, x, b)$ be the probability that P_1 succeeds in the evaluation stage given that it passes the verification stage. Because $\tau = 1$, in a successful attack, no bad gates are "consumed" in the verification stage. Therefore, there exists a positive constant c and a constant N_0 such that for all $N > N_0$,

$$\mathrm{Pr}_{overall}(N, x, b) = \mathrm{Pr}_c(N, x, b) \cdot \mathrm{Pr}_e(N, x, b)$$

$$\geq \left(\frac{1+x}{2}\right)^b \left[1 - \frac{(1-x)b^2}{(1+x)(2N-b)}\right] \cdot \mathrm{Pr}_e(N, x, b) \qquad \text{[Claim B.3]}$$

$$\geq \left(\frac{1+x}{2}\right)^b \left[1 - \frac{(1-x)b^2}{(1+x)(2N-b)}\right] \left(1 - \left(\frac{2x}{1+x}\right)^b\right) \qquad \text{[Claim B.4]}$$

$$> c \qquad \text{[Claim B.5]}$$

This completes the proof. $\qquad\qquad\qquad\qquad\qquad\qquad\qquad\qquad\qquad\qquad\qquad\Box$

Lemma B.2. *Fixing T, N and $\kappa = 2$, setting $B \leq 2$ leads to lower successful attacks rates than setting $B > 2$.*

Proof. First, we show that, if $B > 2$, a scheme S_1 that used a size-1 bucket and a size-B bucket performs no better (in terms of preventing attacks to cut-and-choose) than a scheme S_2 that replaces the two buckets by a size-2 bucket and a size-$(B-1)$ bucket (note this change preserves the total numbers of garbled gates and the only difference between S_1 and S_2 is this pair of buckets). We can derive this fact from counting the number of arrangements of b' bad gates over this pair of buckets that *foil* attacks to cut-and-choose:

1. If $b' < B - 1$, S_1 has $\binom{B}{b'}$ foiling arrangements (as long as no bad gates go to the size-1 bucket) while S_2 has $\binom{B-1}{b'} + \binom{2}{1}\binom{B-1}{b'-1}$ (the bad gates either all go to the size-$(B-1)$ bucket, or only $b'-1$ of them go to the size-$(B-1)$ bucket and the rest goes to the size-2 bucket). Because $\binom{B-1}{b'} + \binom{2}{1}\binom{B-1}{b'-1} = \binom{B}{b'} + \binom{B-1}{b'-1} > \binom{B}{b'}$, S_2 works better.
2. If $b' = B - 1$, the count is B for S_1 and $2(B-1)$ for S_2. Because $2(B-1) > B$, S_2 works better than S_1.
3. If $b' > B - 1$, the counts are 0 for both S_1 and S_2. Hence, S_2 works no worse than S_1.

Through recursively applying the argument above, we always end up with a scheme that either uses (1) only buckets of size 2 or above, which implies $\kappa > 2$ because at least 1 gate needs to be used for verification, hence contracting the assumption $\kappa = 2$; Or (2) only buckets of size 1 and 2. $\qquad\qquad\Box$

Claim B.3. *Fix $\kappa = 2$ and N. Let x fraction of the buckets are of size 2 and the rest $(1 - x)$ are of size 1. Let $\mathrm{Pr}_c(N, x, b)$ be the probability that P_1 survives the gate verification stage with b bad gates. Then*

$$\mathrm{Pr}_c(N, x, b) \geq \left(\frac{1+x}{2}\right)^b \left[1 - \frac{(1-x)b^2}{(1+x)(2N-b)}\right].$$

Proof. Since a total of $(1 + x)N$ garbled gates are used in evaluation while the rest $T - (1 + x)N$ gates are used for checking, we have

$$\mathrm{Pr}_c(N, x, b) = \binom{T-b}{T-(1+x)N} \Big/ \binom{T}{T-(1+x)N}$$

$$= \binom{2N-b}{2N-(1+x)N} \Big/ \binom{2N}{2N-(1+x)N} \qquad\qquad (3)$$

$$= \frac{[(1+x)N]}{2N} \cdot \dots \cdot \frac{[(1+x)N - b + 1]}{[2N - b + 1]} \geq \left[\frac{(1+x)N - b}{2N - b} \right]^b \quad (4)$$

$$= \left(\frac{1+x}{2} \right)^b \left(1 - \frac{(1-x)b}{(1+x)(2N-b)} \right)^b \geq \left(\frac{1+x}{2} \right)^b \left(1 - \frac{(1-x)b^2}{(1+x)(2N-b)} \right) \quad (5)$$

where equality (3) holds because $\kappa = T/N = 2$; the inequality (4) holds because every of the b fractions is larger than or equal to $[(1+x)N - b]/(2N - b)$; and (5) can be derived from the binomial inequality (i.e., $\forall x \in \mathbb{R}, x > -1$, and $\forall n \in \mathbb{N}, (1+x)^n \geq 1 + nx$) and the fact that $\frac{(1-x)b}{(1+x)(2N-b)} < 1$ when $0 \leq b \leq \kappa N = 2N$, $\frac{1}{N} \leq x \leq 1 - \frac{1}{N}$. $\qquad \square$

Claim B.4. *Fix N. Let x represents the fraction of the buckets are of size 2, so the rest $(1-x)$ are of size 1. Let $\mathrm{Pr}_e(N, x, b)$ be the probability of P_1 successfully cheats in the evaluation stage, with b bad evaluation gates. Then $\mathrm{Pr}_e(N, x, b) \geq 1 - \left(\frac{2x}{1+x} \right)^b$.*

Proof. P_1's attack fails if every bucket has at least one good gate (achievable using our proposed protocol in Sect. 5). Throwing $(1+x)N$ gates (b of which is bad) into N buckets, with probability $2^b \binom{xN}{b} / \binom{(1+x)N}{b}$ every bucket will contain at least one good gate as $2^b \binom{xN}{b}$ is the number of ways to place b bad gates into the xN size-2 buckets subject to at most one bad gate per bucket, and $\binom{(1+x)N}{b}$ is the total number of ways to group all gates without any restriction. Therefore,

$$\mathrm{Pr}_e(N, x, b) = 1 - \frac{2^b \binom{xN}{b}}{\binom{(1+x)N}{b}} = 1 - \frac{2^b [xN - (b-1)][xN - (b-2)] \cdots [xN]}{[(1+x)N - (b-1)][(1+x)N - (b-2)] \cdots [(1+x)N]} \geq 1 - \left(\frac{2x}{1+x} \right)^b$$

where the inequality holds because every of the b fractions is greater than or equal to $(2x)/(1+x)$. $\qquad \square$

Claim B.5. *There exists a constant $c > 0$ and a constant N_0 such that for all integer $N > N_0, \forall x \in [\frac{1}{N}, 1 - \frac{1}{N}]$, there exists a positive integer b such that*

$$\left(\frac{1+x}{2} \right)^b \left(1 - \frac{(1-x)b^2}{(1+x)(2N-b)} \right) \left(1 - \frac{2x}{1+x} \right)^b > c.$$

Proof. It suffices to show that there exists $c_1, c_2, c_3 > 0$ and $N_0 > 0$ such that $\forall N > N_0, \forall x \in [1/N, 1 - 1/N]$, there exists a positive integer b that satisfies all of the three inequality below,

$$[(1+x)/2]^b > c_1 \quad (6)$$

$$1 - (1-x)b^2/[(1+x)(2N-b)] > c_2 \quad (7)$$

$$1 - [2x/(1+x)]^b > c_3 \quad (8)$$

Because (6) holds as long as $b < \log c_1 / \log \frac{1+x}{2}$, (7) holds if

$$b < \sqrt{2N\frac{1+x}{1-x}(1-c_2) + \frac{1}{4}\left(\frac{1+x}{1-x}\right)^2(1-c_2)^2} - \frac{1}{2}\cdot\frac{1+x}{1-x}\cdot(1-c_2),$$

(8) holds as long as $b > \log(1 - c_3)/\log\frac{2x}{1+x}$, and b needs to be a positive integer, it suffices to show that there exist positive $c_1, c_2, c_3,$ and N_0 such that the following two inequalities hold for all $N > N_0$,

$$\frac{\log(1 - c_3)}{\log\frac{2x}{1+x}} + 1 < \frac{\log c_1}{\log\frac{1+x}{2}} \tag{9}$$

$$\frac{\log(1-c_3)}{\log\frac{2x}{1+x}} + 1 < -\frac{1}{2}\cdot\frac{1+x}{1-x}\cdot(1-c_2) + \sqrt{2N\frac{1+x}{1-x}(1-c_2) + \frac{1}{4}\cdot\left(\frac{1+x}{1-x}\right)^2(1-c_2)^2}$$

$$\tag{10}$$

We note that (9) is equivalent to

$$\frac{1}{\log c_1}\log\frac{1+x}{2x} > \frac{1}{\log(1-c_3)}\log\frac{2}{1+x} + \log\frac{1+x}{2x}\log\frac{2}{1+x},$$

which will always hold as long as

$$\frac{1}{\log c_1} - \frac{1}{\log(1-c_3)} - \log 2 > 0 \tag{11}$$

because $\frac{1+x}{2x} > \frac{2}{1+x}$ and $\log\frac{2}{1+x} < \log 2$ hold for all $x \in [1/N, 1 - 1/N]$. Since (11) doesn't involve x, it is easy to find a c_1 based on the value of c_3 such that (11) is satisfied.

Next, we note that (10) is equivalent to

$$\frac{2N\frac{1+x}{1-x}(1-c_2)}{\sqrt{2N\frac{1+x}{1-x}(1-c_2) + \frac{1}{4}(\frac{1+x}{1-x})^2(1-c_2)^2} + \frac{1}{2}\cdot\frac{1+x}{1-x}\cdot(1-c_2)} > 1 + \frac{\log\frac{1}{1-c_3}}{\log\frac{1+x}{2x}},$$

which can be simplified, by defining $c_2' = 1 - c_2$ and $y = (1 + x)/(1 - x)$, to

$\left(\frac{2N}{\sqrt{\frac{2N}{yc_2'} + \frac{1}{4}} + \frac{1}{2}} - 1\right)\log\left(1 + \frac{1}{y-1}\right) > \log\frac{1}{1-c_3}$. Now we analyze this inequality

in two cases:

Case I ($y < 3$): If $y < 3$,

$$\left(\frac{2N}{\sqrt{\frac{2N}{yc_2'} + \frac{1}{4}} + \frac{1}{2}} - 1\right)\log\left(1 + \frac{1}{y-1}\right) \geq \left(\frac{2N}{\sqrt{\frac{2N}{c_2'} + \frac{1}{4}} + \frac{1}{2}} - 1\right)\log\left(1 + \frac{1}{y-1}\right)$$

$$\geq \left(\frac{2N}{\sqrt{\frac{2N}{c_2'} + 1}} - 1\right)\log\left(1 + \frac{1}{y-1}\right)$$

$$\geq \left(\frac{2N}{\sqrt{\frac{2N}{c_2'} + 1}} - 1\right)\log\frac{3}{2}.$$

Because there exists an integer N_0 such that for all $N > N_0$,

$$\left(\frac{2N}{\sqrt{\frac{2N}{c_2'}+1}} - 1\right) \log \frac{3}{2} > \log \frac{1}{1-c_3} \tag{12}$$

(10) can also be satisfied when $N > N_0$, regardless of the values of c_2' and c_3.

Case II ($y \geq 3$): If $y \geq 3$, then $0 < 1/(y-1) < 1/2$, and (because $y = \log(1+x)$ is concave function when $x \in [0, 1/2]$) we have $\log[1 + 1/(y-1)] \geq [2\log(3/2)]/(y-1) > [2\log(3/2)]/y$. In addition, for all $N > N_0$ (where N_0 is defined as above), we have $\frac{2N}{\sqrt{\frac{2N}{yc_2'}+\frac{1}{4}+\frac{1}{2}}} - 1 > \frac{2N}{\sqrt{\frac{2N}{c_2'}+1}} - 1 > 0$. Thus, for all $N > N_0$, we know

$$\left(\frac{2N}{\sqrt{\frac{2N}{yc_2'}+\frac{1}{4}+\frac{1}{2}}} - 1\right) \log\left(1 + \frac{1}{y-1}\right) \geq 2\log(3/2)\left(\frac{2N}{\sqrt{\frac{2Ny}{c_2'}+\frac{1}{4}y^2+\frac{1}{2}y}} - \frac{1}{y}\right)$$

$$\geq 2\log(3/2)\left(\frac{2N}{\sqrt{\frac{4N^2}{c_2'}+N^2+N}} - \frac{1}{3}\right) = 2\log(3/2)\left(\frac{2}{1+\sqrt{\frac{4}{c_2'}+1}} - \frac{1}{3}\right),$$

where the second inequality holds because $3 \leq y \leq 2N$. Therefore, we can find c_3 based on (the value of) c_2' to satisfy

$$2\log(3/2)\left(\frac{2}{1+\sqrt{\frac{4}{c_2'}+1}} - \frac{1}{3}\right) > \log \frac{1}{1-c_3}, \tag{13}$$

which will guarantee (10) holds.

To sum up, we have shown that if we pick an arbitrary positive number c_2 then find c_3 based on (13) and c_2, find c_1 based on (11) and c_3, and finally find N_0 based on (12) and c_2, c_3, then for all $N > N_0$, all three inequalities, (6), (7), (8) should hold. This completes the proof.   □

Lemma B.6. *Fix $B = 2$. Let $\mathrm{Pr}_{overall}(N, b)$ be the probability that a malicious P_1 succeeds when generating b bad gates. For any $\kappa > 2$ and any $\varepsilon > 0$, there exists N_0 such that*

$$\mathrm{Pr}_{overall}(N, b) < \varepsilon, \quad (\forall N > N_0)$$

Proof. Let $0 < \tau \leq 1$ be the probability that P_2 detects that g is faulty (through checking) conditioned that g is indeed faulty. We have

$$\mathrm{Pr}_{overall}(N, b) = \sum_{i=0}^{b}(1-\tau)^i \frac{\binom{b}{i}\binom{T-b}{T-2N-i}}{\binom{T}{T-2N}} \mathrm{Pr}_e(N, b-i)$$

where $(1-\tau)^i \binom{b}{i} \binom{T-b}{T-2N-i} / \binom{T}{T-2N}$ is the probability that P_1, with b bad gates initially, surviving the checking stage losing i bad gates (due to verification, while P_2 detecting none of them). Because $\exists i_0$ such that $(1-\tau)^{i_0} < \varepsilon/2$,

$$\mathrm{Pr}_{overall}(N, b)$$

$$= \sum_{i=0}^{b} (1-\tau)^i \frac{\binom{b}{i}\binom{T-b}{T-2N-i}}{\binom{T}{T-2N}} \mathrm{Pr}_e(N, b-i)$$

$$= \sum_{i=0}^{i_0} (1-\tau)^i \frac{\binom{b}{i}\binom{T-b}{T-2N-i}}{\binom{T}{T-2N}} \mathrm{Pr}_e(N, b-i) + \sum_{i=i_0+1}^{b} (1-\tau)^i \frac{\binom{b}{i}\binom{T-b}{T-2N-i}}{\binom{T}{T-2N}} \mathrm{Pr}_e(N, b-i)$$

$$\leq \sum_{i=0}^{i_0} (1-\tau)^i \frac{\binom{b}{i}\binom{T-b}{T-2N-i}}{\binom{T}{T-2N}} \mathrm{Pr}_e(N, b-i) + (1-\tau)^{i_0} \sum_{i=i_0+1}^{b} \frac{\binom{b}{i}\binom{T-b}{T-2N-i}}{\binom{T}{T-2N}} \mathrm{Pr}_e(N, b-i)$$

$$\leq \sum_{i=0}^{i_0} (1-\tau)^i \frac{\binom{b}{i}\binom{T-b}{T-2N-i}}{\binom{T}{T-2N}} \mathrm{Pr}_e(N, b-i) + \frac{\varepsilon}{2} \sum_{i=i_0+1}^{b} \frac{\binom{b}{i}\binom{T-b}{T-2N-i}}{\binom{T}{T-2N}}$$

$$\leq \sum_{i=0}^{i_0} (1-\tau)^i \frac{\binom{b}{i}\binom{T-b}{T-2N-i}}{\binom{T}{T-2N}} \mathrm{Pr}_e(N, b-i) + \frac{\varepsilon}{2} \sum_{i=1}^{b} \frac{\binom{b}{i}\binom{T-b}{T-2N-i}}{\binom{T}{T-2N}}$$

$$\leq \sum_{i=0}^{i_0} (1-\tau)^i \frac{\binom{b}{i}\binom{T-b}{T-2N-i}}{\binom{T}{T-2N}} \mathrm{Pr}_e(N, b-i) + \frac{\varepsilon}{2} \cdot 1$$

$$\leq \sum_{i=0}^{i_0} (1-\tau)^i \binom{b}{i} \left(\frac{T-2N}{T}\right)^i \left(\frac{2N}{T-i}\right)^{b-i} \mathrm{Pr}_e(N, b-i) + \frac{\varepsilon}{2} \qquad \text{[Claim B.8]}$$

$$\leq \sum_{i=0}^{i_0} (1-\tau)^i \binom{b}{i} \left(\frac{T-2N}{T}\right)^i \left(\frac{2N}{T-i}\right)^{b-i} \mathrm{Pr}_e(N, b) + \frac{\varepsilon}{2}$$

$$\leq \sum_{i=0}^{i_0} (1-\tau)^i \binom{b}{i} \left(\frac{T-2N}{T}\right)^i \left(\frac{2N}{T-i_0}\right)^{b-i} \mathrm{Pr}_e(N, b) + \frac{\varepsilon}{2}$$

$$\leq \sum_{i=0}^{b} (1-\tau)^i \binom{b}{i} \left(\frac{T-2N}{T}\right)^i \left(\frac{2N}{T-i_0}\right)^{b-i} \mathrm{Pr}_e(N, b) + \frac{\varepsilon}{2}$$

$$= \left((1-\tau)\frac{T-2N}{T} + \frac{2N}{T-i_0}\right)^b \mathrm{Pr}_e(N, b) + \frac{\varepsilon}{2}.$$

holds for any $N > N_1$ (where N_1 is determined according to the proof of Claim B.7).

Since $T = \kappa N$, we have

$$\lim_{N \to \infty} \frac{(1-\tau)(T-2N)}{T} + \frac{2N}{T-i_0}$$

$$= \lim_{N \to \infty} \frac{(1-\tau)(\kappa-2)}{\kappa} + \frac{2}{\kappa - i_0/N} = 1 - \frac{\tau(\kappa-2)}{\kappa} < 1.$$

Therefore, there exists N_1 such that for all $N > N_1$, $(1-\tau)(T-2N)/T + 2N/(T-i_0) < 1$. Hence, for every $\varepsilon > 0$, we can find a b_0 such that,

1. for all $b > b_0$,

$$\mathrm{Pr}_{overall}(N, b)$$
$$\leq \left[(1 - \tau)(T - 2N)/T + 2N/(T - i_0)\right]^b \mathrm{Pr}_e(N, b) + \varepsilon/2$$
$$< \varepsilon \mathrm{Pr}_e(N, b)/2 + \varepsilon/2 < \varepsilon/2 + \varepsilon/2 = \varepsilon.$$

2. for all $b \leq b_0$, Claim B.7 shows how to further find an integer N_2 such that for all $N > N_2$,

$$\mathrm{Pr}_{overall}(N, b)$$
$$\leq \left[(1 - \tau)(T - 2N)/T + 2N/(T - i_0)\right]^b \mathrm{Pr}_e(N, b) + \varepsilon/2$$
$$< \left[1 - \tau(\kappa - 2)/\kappa\right]^b \varepsilon/2 + \varepsilon/2 < \varepsilon/2 + \varepsilon/2 = \varepsilon.$$

Thus, setting $N_0 = \max(N_1, N_2)$ completes the proof. $\square$

Lemma B.7. *Let* $\mathrm{Pr}_e(N, b)$ *be the probability that a malicious* P_1 *who survives the gate checking stage succeeds with* b *bad gates selected for evaluation. Then*

1. *For any fixed* N, $\mathrm{Pr}_e(N, b)$ *strictly increases with* b.
2. *For any* b *and* $\varepsilon > 0$, $\exists N_0$ *such that* $\mathrm{Pr}_e(N, b) < \varepsilon$.

Proof. Since all buckets are of size 2, the following can be derived similarly to the proof of Claim B.4 (by setting $x = 1$),

$$\mathrm{Pr}_e(N, b) = 1 - \frac{2^b \binom{N}{b}}{\binom{2N}{b}}$$
$$= 1 - \frac{2N - 2(b-1)}{2N - (b-1)} \cdot \ldots \cdot \frac{2N}{2N}$$
$$\leq 1 - \left(\frac{2N - 2b + 2}{2N - b + 1}\right)^b = 1 - \left(\frac{2 - 2(b-1)/N}{2 - (b-1)/N}\right)^b \quad (14)$$

First, note that (14) implies that $\mathrm{Pr}_e(N, b)$ increases strictly with b, because the greater b is, the more multiplicative fractions (all smaller than 1) are in the product.

Second, note that for any b, ε, $\left(\frac{2N - 2b + 2}{2N - b + 1}\right)^b$ goes arbitrarily close to 1 when N is sufficiently large. Therefore, $\forall b, \varepsilon, \exists N_1$ such that for all $N > N_1$, $\mathrm{Pr}_e(N, b) < \varepsilon$. $\square$

Lemma B.8. *If* T, N, b, i *are non-negative integers such that* $T > 2N$, $T \geq b$, *and* $i \leq b$, *then* $\frac{\binom{T-b}{T-2N-i}}{\binom{T}{T-2N}} \leq \left(\frac{T-2N}{T}\right)^i \left(\frac{2N}{T-i}\right)^{b-i}$.

Proof.

$$
\frac{\binom{T-b}{T-2N-i}}{\binom{T}{T-2N}} = \frac{(T-b)!(T-2N)!(2N)!}{T!(T-2N-i)!(2N-b+i)!}
$$

$$
= \frac{\big[(T-2N-i+1)\cdots(T-2N)\big]\big[(2N-b+i+1)\cdots 2N)\big]}{(T-b+1)(T-b+2)\cdots T}
$$

$$
= \frac{(T-2N-i+1)\cdots(T-2N)}{(T-i+1)\cdots T} \cdot \frac{(2N-b+i+1)\cdots 2N}{(T-b+1)\cdots(T-i)}
$$

$$
\leq \left(\frac{T-2N}{T}\right)^{i}\left(\frac{2N}{T-i}\right)^{b-i}
$$

□

References

1. Intel: Intel Intrinsics Guide. Latest version: 3.0.1. Released on 23 July 2013
2. Asharov, G., Lindell, Y., Schneider, T., Zohner, M.: More efficient oblivious transfer extensions with security for malicious adversaries. In: Oswald, E., Fischlin, M. (eds.) EUROCRYPT 2015. LNCS, vol. 9056, pp. 673–701. Springer, Heidelberg (2015). https://doi.org/10.1007/978-3-662-46800-5_26
3. Barak, B., Dodis, Y., Krawczyk, H., Pereira, O., Pietrzak, K., Standaert, F.-X., Yu, Y.: Leftover hash lemma, revisited. In: Rogaway, P. (ed.) CRYPTO 2011. LNCS, vol. 6841, pp. 1–20. Springer, Heidelberg (2011). https://doi.org/10.1007/978-3-642-22792-9_1
4. Bellare, M., Hoang, V.T., Keelveedhi, S., Rogaway, P.: Efficient garbling from a fixed-key blockcipher. In: IEEE Symposium on Security and Privacy (2013)
5. Bellare, M., Hoang, V.T., Rogaway, P.: Foundations of garbled circuits. In: CCS (2012)
6. Boyar, J., Peralta, R.: A small depth-16 circuit for the AES S-box. In: Gritzalis, D., Furnell, S., Theoharidou, M. (eds.) SEC 2012. IAICT, vol. 376, pp. 287–298. Springer, Heidelberg (2012). https://doi.org/10.1007/978-3-642-30436-1_24
7. Brandão, L.T.A.N.: Secure two-party computation with reusable bit-commitments, via a cut-and-choose with forge-and-lose technique. In: Sako, K., Sarkar, P. (eds.) ASIACRYPT 2013. LNCS, vol. 8270, pp. 441–463. Springer, Heidelberg (2013). https://doi.org/10.1007/978-3-642-42045-0_23
8. Camenisch, J., Neven, G., Shelat, A.: Simulatable adaptive oblivious transfer. In: Naor, M. (ed.) EUROCRYPT 2007. LNCS, vol. 4515, pp. 573–590. Springer, Heidelberg (2007). https://doi.org/10.1007/978-3-540-72540-4_33
9. Canetti, R.: Security and composition of multiparty cryptographic protocols. J. Cryptol. **13**, 143–202 (2000)
10. Damgård, I., Pastro, V., Smart, N., Zakarias, S.: Multiparty computation from somewhat homomorphic encryption. In: Safavi-Naini, R., Canetti, R. (eds.) CRYPTO 2012. LNCS, vol. 7417, pp. 643–662. Springer, Heidelberg (2012). https://doi.org/10.1007/978-3-642-32009-5_38
11. Frederiksen, T.K., Jakobsen, T.P., Nielsen, J.B., Nordholt, P.S., Orlandi, C.: MiniLEGO: efficient secure two-party computation from general assumptions. In: Johansson, T., Nguyen, P.Q. (eds.) EUROCRYPT 2013. LNCS, vol. 7881, pp. 537–556. Springer, Heidelberg (2013). https://doi.org/10.1007/978-3-642-38348-9_32

12. Frederiksen, T., Jakobsen, T., Nielsen, J., Trifiletti, R.: TinyLEGO: an interactive garbling scheme for maliciously secure two-party computation. ePrint/2015/309 (2015)
13. Frederiksen, T.K., Jakobsen, T.P., Nielsen, J.B., Trifiletti, R.: On the complexity of additively homomorphic UC commitments. In: Kushilevitz, E., Malkin, T. (eds.) TCC 2016. LNCS, vol. 9562, pp. 542–565. Springer, Heidelberg (2016). https://doi.org/10.1007/978-3-662-49096-9_23
14. Frederiksen, T.K., Nielsen, J.B., Orlandi, C.: Privacy-free garbled circuits with applications to efficient zero-knowledge. In: Oswald, E., Fischlin, M. (eds.) EURO-CRYPT 2015. LNCS, vol. 9057, pp. 191–219. Springer, Heidelberg (2015). https://doi.org/10.1007/978-3-662-46803-6_7
15. Goldreich, O.: Foundations of Cryptography: Basic Applications, vol. 2. Cambridge University Press, Cambridge (2004)
16. Gueron, S., Lindell, Y., Nof, A., Pinkas, B.: Fast garbling of circuits under standard assumptions. In CCS (2015)
17. Holzer, A., Franz, M., Katzenbeisser, S., Veith, H.: Secure two-party computations in ANSI C. In: CCS (2012)
18. Huang, Y., Evans, D., Katz, J.: Private set intersection: are garbled circuits better than custom protocols? In: NDSS (2012)
19. Huang, Y., Evans, D., Katz, J., Malka, L.: Faster secure two-party computation using garbled circuits. In: USENIX Security Symposium (2011)
20. Huang, Y., Katz, J., Kolesnikov, V., Kumaresan, R., Malozemoff, A.J.: Amortizing garbled circuits. In: Garay, J.A., Gennaro, R. (eds.) CRYPTO 2014. LNCS, vol. 8617, pp. 458–475. Springer, Heidelberg (2014). https://doi.org/10.1007/978-3-662-44381-1_26
21. Huang, Y., Malka, L., Evans, D., Katz, J.: Efficient privacy-preserving biometric identification. In: NDSS (2011)
22. Impagliazzo, R., Levin, L.A., Luby, M.: Pseudo-random generation from one-way functions (extended abstracts). In: STOC (1989)
23. Keller, M., Orsini, E., Scholl, P.: Actively secure OT extension with optimal overhead. In: Gennaro, R., Robshaw, M. (eds.) CRYPTO 2015. LNCS, vol. 9215, pp. 724–741. Springer, Heidelberg (2015). https://doi.org/10.1007/978-3-662-47989-6_35
24. Kolesnikov V., Nielsen, J.B., Rosulek, M., Trieu, N., Trifiletti, R.: DUPLO: unifying cut-and-choose for garbled circuits. Technical report, ePrint/2017/344 (2017)
25. Kolesnikov, V., Schneider, T.: Improved garbled circuit: free XOR gates and applications. In: Aceto, L., Damgård, I., Goldberg, L.A., Halldórsson, M.M., Ingólfsdóttir, A., Walukiewicz, I. (eds.) ICALP 2008. LNCS, vol. 5126, pp. 486–498. Springer, Heidelberg (2008). https://doi.org/10.1007/978-3-540-70583-3_40
26. Lindell, Y.: Fast cut-and-choose based protocols for malicious and covert adversaries. In: Canetti, R., Garay, J.A. (eds.) CRYPTO 2013. LNCS, vol. 8043, pp. 1–17. Springer, Heidelberg (2013). https://doi.org/10.1007/978-3-642-40084-1_1
27. Lindell, Y., Pinkas, B.: An efficient protocol for secure two-party computation in the presence of malicious adversaries. In: Naor, M. (ed.) EUROCRYPT 2007. LNCS, vol. 4515, pp. 52–78. Springer, Heidelberg (2007). https://doi.org/10.1007/978-3-540-72540-4_4
28. Lindell, Y., Riva, B.: Cut-and-choose yao-based secure computation in the online/offline and batch settings. In: Garay, J.A., Gennaro, R. (eds.) CRYPTO 2014. LNCS, vol. 8617, pp. 476–494. Springer, Heidelberg (2014). https://doi.org/10.1007/978-3-662-44381-1_27

29. Lindell, Y., Riva, B.: Blazing fast 2PC in the offline/online setting with security for malicious adversaries. In: CCS (2015)
30. Liu, C., Huang, Y., Shi, E., Katz, J., Hicks, M.W.: Automating efficient RAM-model secure computation. In: IEEE Symposium on S&P (2014)
31. Liu, C., Wang, X., Nayak, K., Huang, Y., Shi, E.: ObliVM: a programming framework for secure computation. In: IEEE Symposium on S&P (2015)
32. Malkhi, D., Nisan, N., Pinkas, B., Sella, Y.: Fairplay – a secure two-party computation system. In: USENIX Security Symposium (2004)
33. Wang, X., Malozemoff, A., Katz, J.: EMP-toolkit: efficient multiparty computation toolkit. https://github.com/emp-toolkit
34. Nayak, K., Wang, X., Ioannidis, S., Weinsberg, U., Taft, N., Shi, E.: GraphSC: parallel secure computation made easy. In: IEEE Symposium on S&P (2015)
35. Nielsen, J.B., Nordholt, P.S., Orlandi, C., Burra, S.S.: A new approach to practical active-secure two-party computation. In: Safavi-Naini, R., Canetti, R. (eds.) CRYPTO 2012. LNCS, vol. 7417, pp. 681–700. Springer, Heidelberg (2012). https://doi.org/10.1007/978-3-642-32009-5_40
36. Nielsen, J.B., Orlandi, C.: LEGO for two-party secure computation. In: Reingold, O. (ed.) TCC 2009. LNCS, vol. 5444, pp. 368–386. Springer, Heidelberg (2009). https://doi.org/10.1007/978-3-642-00457-5_22
37. Nielsen, J.B., Schneider, T., Trifiletti, R.: Constant round maliciously secure 2PC with function-independent preprocessing using LEGO. ePrint/2016/1069 (2017)
38. Nikolaenko, V., Weinsberg, U., Ioannidis, S., Joye, M., Boneh, D., Taft, N.: Privacy-preserving ridge regression on hundreds of millions of records. In: IEEE Symposium on S&P (2013)
39. Peikert, C., Vaikuntanathan, V., Waters, B.: A framework for efficient and composable oblivious transfer. In: Wagner, D. (ed.) CRYPTO 2008. LNCS, vol. 5157, pp. 554–571. Springer, Heidelberg (2008). https://doi.org/10.1007/978-3-540-85174-5_31
40. Pinkas, B., Schneider, T., Smart, N.P., Williams, S.C.: Secure two-party computation is practical. In: Matsui, M. (ed.) ASIACRYPT 2009. LNCS, vol. 5912, pp. 250–267. Springer, Heidelberg (2009). https://doi.org/10.1007/978-3-642-10366-7_15
41. Rindal, P., Rosulek, M.: Faster malicious 2-party secure computation with online/offline dual execution. In: USENIX Security Symposium (2016)
42. Wang, X., Malozemoff, A.J., Katz, J.: Faster two-party computation secure against malicious adversaries in the single-execution setting. In: EUROCRYPT (2017)
43. Wang, X., Ranellucci, S., Katz, J.: Authenticated garbling and efficient maliciously secure two-party computation. In: CCS (2017)
44. Wang, X., Huang, Y., Chan, T.-H., Shelat, A., Shi, E.: SCORAM: oblivious RAM for secure computation. In: CCS (2014)
45. Wang, X., Huang, Y., Zhao, Y., Tang, H., Wang, X., Bu, D.: Efficient genome-wide, privacy-preserving similar patient query based on private edit distance. In: CCS (2015)
46. Yao, A.: How to generate and exchange secrets (extended abstract). In: FOCS (1986)
47. Zahur, S., Rosulek, M., Evans, D.: Two halves make a whole—reducing data transfer in garbled circuits using half gates. In: Oswald, E., Fischlin, M. (eds.) EUROCRYPT 2015. LNCS, vol. 9057, pp. 220–250. Springer, Heidelberg (2015). https://doi.org/10.1007/978-3-662-46803-6_8

48. Zhu, R., Huang, Y.: JIMU: faster LEGO-based secure computation using additive homomorphic hashes. In: Takagi, T., Peyrin, T. (eds.) ASIACRYPT 2017, Part II. LNCS, vol. 10625, pp. 529–572. Springer, Cham (2017). ePrint/2017/226
49. Zhu, R., Huang, Y., Cassel, D.:. Pool: scalable on-demand secure computation service against malicious adversaries. In: CCS (2017)
50. Zhu, R., Huang, Y., Shelat, A., Katz, J.: The cut-and-choose game and its application to cryptographic protocols. In: USENIX Security Symposium (2016)

Operating Modes Security Proofs

Analyzing Multi-key Security Degradation

Atul Luykx[1,2]([✉]), Bart Mennink[3,4], and Kenneth G. Paterson[5]

[1] imec-COSIC, KU Leuven, Leuven, Belgium
atul.luykx@esat.kuleuven.be
[2] Department of Computer Science, University of California, Davis,
One Shields Ave, Davis, CA 95616, USA
[3] Digital Security Group, Radboud University, Nijmegen, The Netherlands
b.mennink@cs.ru.nl
[4] CWI, Amsterdam, The Netherlands
[5] Information Security Group, Royal Holloway, University of London,
Egham, Surrey TW20 0EX, UK
kenny.paterson@rhul.ac.uk

Abstract. The multi-key, or multi-user, setting challenges cryptographic algorithms to maintain high levels of security when used with many different keys, by many different users. Its significance lies in the fact that in the real world, cryptography is rarely used with a single key in isolation. A folklore result, proved by Bellare, Boldyreva, and Micali for public-key encryption in EUROCRYPT 2000, states that the success probability in attacking any one of many independently keyed algorithms can be bounded by the success probability of attacking a single instance of the algorithm, multiplied by the number of keys present. Although sufficient for settings in which not many keys are used, once cryptographic algorithms are used on an internet-wide scale, as is the case with TLS, the effect of multiplying by the number of keys can drastically erode security claims. We establish a sufficient condition on cryptographic schemes and security games under which multi-key degradation is avoided. As illustrative examples, we discuss how AES and GCM behave in the multi-key setting, and prove that GCM, as a mode, does not have multi-key degradation. Our analysis allows limits on the amount of data that can be processed per key by GCM to be significantly increased. This leads directly to improved security for GCM as deployed in TLS on the Internet today.

Keywords: Multi-key · Multi-user · Multi-oracle · AES · GCM · TLS · Weak keys

1 Introduction

A crucial aspect to analyzing cryptographic algorithms is modeling real-world settings. These models should not only accurately reflect the limits imposed by the environments and the security properties desired, but they should also

© International Association for Cryptologic Research 2017
T. Takagi and T. Peyrin (Eds.): ASIACRYPT 2017, Part II, LNCS 10625, pp. 575–605, 2017.
https://doi.org/10.1007/978-3-319-70697-9_20

produce meaningful ways to estimate how security deteriorates with use. In particular, in practice, algorithms are fixed, and hence so are key sizes, block sizes, groups, and various other parameters. Therefore it is important to be able to compute adversarial success probabilities relative to their resources as precisely as possible.

For example, block ciphers have traditionally been analyzed in a setting where adversaries are given access to the encryption and decryption oracles keyed with a value chosen uniformly at random, unknown to the adversary. For many purposes using a block cipher which is secure in this model is sufficient, barring easy access to side channel information. Estimates for adversarial success are obtained by analyzing the best known attacks against the block cipher, relative to both *computational* complexity, or the cost of running the attack as measured according to, say, time and memory, and *data* complexity, or the amount of data the adversary receives from the oracles, measured in, for example, bits. Taking a concrete example, AES [22], one can map the cost needed to recover a key, as is done in Fig. 1a. For the 128-bit key version of full round AES, the best known attacks have computational complexity improving over brute-force search by a factor of 2 to 4, and arbitrarily increasing data complexity does not allow one to reduce computational complexity much.

The analysis of block ciphers contrasts sharply with that of *modes of operation* for block ciphers, which are algorithms that repeatedly use block cipher calls to achieve security properties beyond what a block cipher can provide on its own. As an important (but by no means the only) example, the Authenticated Encryption with Associated Data (AEAD) [46] mode of operation GCM [38] uses a block cipher to achieve data confidentiality and authenticity simultaneously, formalized in a setting where adversaries are given access to keyed encryption and decryption oracles. The security of GCM is proved by showing that any AEAD adversary against the mode can be converted into an adversary against the pseudo-randomness of the underlying block cipher [31,43]. Thus, if GCM were to be used with AES, then AES-GCM is secure in the AEAD sense under the assumption that AES is secure as a pseudo-random permutation (PRP).

However, the quality of the reduction from AES to AES-GCM deteriorates with use. Following the concrete security paradigm [7], this degradation has been quantified to be roughly $\sigma^2/2^{128}$ [31,43], where σ is the number of blocks of ciphertext seen by the adversary, or its data complexity. This is depicted in Fig. 1b. Therefore, quantifying AES-GCM's security relies not only on understanding AES's security, but also on how GCM as a mode degrades security.

Note that in the case of AES-GCM, an understanding of how adversarial computational and data complexity affect security can be built by looking at AES and GCM separately. AES's security degrades as computational resources increase, but increased data complexity does not seem to introduce better attacks. GCM's security as a mode degrades as data complexity increases, but computational complexity does not play a role.

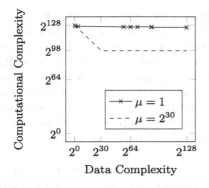
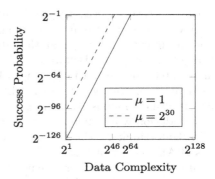

(a) Key recovery attacks against full AES-128. All attacks have success probability one. Data on single-key attacks from [17, 18, 48].

(b) Upper bound on attack success probability against the mode GCM, based on the equation $\mu\sigma^2/2^{128}$, where σ is the data complexity.

Fig. 1. Comparison of how data complexity affects attacks against AES-128 and GCM in the single ($\mu = 1$) and multi-key ($\mu = 2^{30}$) settings. Note that the AES graph depicts attacks, whereas the GCM graph depicts upper bounds on attack success probability.

1.1   From Single-Key to Multi-key

The security models described earlier for block ciphers and modes gave the adversaries access to encryption and decryption oracles operating under a single key. However, in practice cryptographic algorithms are used by many different users, each potentially with many different keys. For example, AES-GCM is now widely used in TLS to protect web traffic via HTTPS,[1] and is currently used by millions, or perhaps billions, of users daily. Hence it is important to understand what happens to security in the so-called *multi-key* setting, where adversaries are successful if they compromise the security of one out of many users, meaning their winning condition is a disjunction of single key winning conditions.

For block ciphers the picture changes both quantitatively and qualitatively. Whereas in the single-key setting, the best attacks against AES do not improve with increased data complexity, in the multi-key setting they do, as depicted in Fig. 1a. As observed first by Biham [14], and later refined as a time-memory-data trade-off by Biryukov, Mukhopadhyay, and Sarkar [15], one can take advantage of the fact that recovering a key out of a large group of keys is much easier than targeting one key. The same observation can be applied to any deterministic symmetric-key algorithm, as is done for MACs by Chatterjee, Menezes, and Sarkar [21].

More generally, a folklore result guarantees that the attack success probability increases by at most a factor μ when moving from the single-key to the multi-key setting with μ keys. In the case of key recovery against AES, the fact

---

[1] The latest figures from the ICSI Certificate Notary (https://notary.icsi.berkeley.edu/) suggest that more than 70% of all TLS connections use AES-GCM.

that this increase is necessary can be illustrated with an actual attack. For the mode GCM, a security bound involving a factor μ is easily established using a hybrid argument, meaning that the adversarial success probability is bounded by roughly $\mu \cdot \sigma^2/2^{128}$. Bellare and Tackmann [11] were the first to formalize authenticated encryption in the multi-key setting and to analyze countermeasures against multi-key attacks in the context of TLS 1.3. Their work similarly establishes bounds containing a μ-factor. This leads to a significant security degradation when there are many GCM instances present, as illustrated in Fig. 1b. Unfortunately, this is exactly the situation faced in large-scale deployments of AES-GCM such as TLS.

Unlike block ciphers, there are no known attacks which establish the tightness of the $\mu \cdot \sigma^2/2^{128}$ security bound for the GCM mode. Assuming there were such an attack, then the bound would say that, using the *same* amount of resources σ as a single-key adversary, a multi-key adversary would be able to *increase* its success probability by a factor of μ. Therefore a successful multi-key adversary against the GCM mode would be able to use its resources much more efficiently than a single-key attacker would.

Quantifying this difference, in order for a single-key adversary to be able to achieve the same bound $\mu\sigma^2/2^{128}$ using σ_1 resources, $\sigma_1^2/2^{128}$ must equal $\mu\sigma^2/2^{128}$, or in other words, $\sigma_1 = \mu^{1/2}\sigma$. In particular, $\sigma/\mu = \mu^{-3/2}\sigma_1$, and so a multi-key adversary's per-key cost would decrease proportional to $\mu^{-3/2}$ relative to a single-key adversary's per-key cost, while achieving the same success probability. So, if there were a multi-key adversary interacting with, say, ten thousand GCM instances, and matching the generic bound, then in order for a single-key adversary to match the multi-key adversary's success probability, it must spend a factor of one million more than a multi-key adversary has to spend per key. Note that even in the case of AES, the best known multi-key attack does not make better use of its data resources: it achieves the same success probability as the single-key attacks with roughly the same per-key data cost, namely, one plaintext-ciphertext pair per key.

1.2 Overview and Contributions

We set out to understand why there are seemingly no attacks matching the multi-key bounds established by the folklore result, and by formal proofs in certain cases, against modes such as GCM. To do so, we systematically analyze the transition from games in which adversaries are given access to oracles representing a single, keyed algorithm, to games where adversaries are given access to multiple oracles representing different, independently keyed instances of an algorithm.

The fact that the folklore result is the best generic reduction possible has already been established by Bellare, Boldyreva, and Micali [5], where they construct a public-key encryption scheme which necessarily has the μ-degradation. However, we take the informal guidance provided by such special cases a step further in Sect. 2, and point out that the multi-key setting is the natural one in which to consider *weak keys*, by illustrating how they can allow multi-key adversaries to make better use of resources in comparison with single-key adversaries.

In Sect. 3 we continue by distilling a sufficient condition under which adversaries gain no advantage in the multi-key setting over the single-key setting. Informally, the condition states that it should always be better to attack an instance of an algorithm for which the adversary is given more information, as measured by the number of queries made to the instance. Note that this condition is not satisfied for algorithms with weak keys: if the adversary knows that an instance uses a strong key, then it might be better for it to take its chances with an instance for which it has little information, but where it might get to attack a weak key.

Although intuitively appealing, the condition that we extract can be difficult to use as a criterion in security analyses. Therefore in Sect. 4 we compare various methods for proving the absence of multi-key degradation, such as for the PRP-PRF switch and for Wegman-Carter MACs [50]. Finally, we prove that GCM has security bounds that are independent of μ using our sufficient condition.

1.3 Interpretation

Our claim that GCM enjoys a multi-key security bound that does not depend on μ might seem counter-intuitive. After all, GCM uses a block cipher, and, as illustrated above with an attack, all block ciphers necessarily have security that degrades with μ. It seems natural that one can apply a similar attack to GCM thereby establishing μ-degradation.

The result concerning GCM is a statement made once the underlying block cipher is replaced by a uniformly distributed random permutation, which is a standard technique used to reduce the block cipher's insecurity to GCM's insecurity when used with that block cipher. Stated as an imprecise formula, for a single key, we have that

$$\mathsf{Insecurity}(GCM, E) \leq \mathsf{Insecurity}(GCM, \pi) + \mathsf{Insecurity}(E), \qquad (1)$$

where E is the keyed block cipher, and π is the random permutation. Passing to the multi-key setting means that one now considers the insecurity of GCM with multiple independently keyed block ciphers $E_{K_1}, E_{K_2}, \ldots, E_{K_\mu}$, which are then replaced with independent uniformly distributed random permutations $\pi_1, \pi_2, \ldots, \pi_\mu$. Saying that GCM as a mode does not degrade with μ is a statement about how the insecurity of GCM with $\pi_1, \pi_2, \ldots, \pi_\mu$ does not degrade as the number of independent permutations increases, and as a result, the reduction from the insecurity of the underlying block ciphers $E_{K_1}, \ldots, E_{K_\mu}$ to the insecurity of $(GCM, E_{K_1}, \ldots, E_{K_\mu})$ does not deteriorate according to μ. However, any multi-key attack against E still holds, and is taken into account when considering the term corresponding to the insecurity of E in the multi-key version of (1).

In other words, what we are able to show is that security does not degrade "doubly", once for GCM and once again for the block cipher, when the number of keys increases. More importantly, one can conclude that in order to understand the multi-key security of AES-GCM, one can focus on the multi-key security of AES.

1.4   Practical Implications

This insight has an immediate and important practical consequence. Recently the TLS Working Group of the IETF has been considering data limits for the AEAD schemes to be used in TLS 1.3, the new version of TLS under development. Amongst these schemes is AES-GCM. Luykx and Paterson provided an analysis of the safe data limits for AES-GCM.[2] They did this by first analyzing the known bounds for AES-GCM in the single-key setting and then applying a factor μ in order to obtain bounds for the multi-key setting. The safe data limits for AES-GCM turned out to be surprisingly small, especially in the multi-key case: the current draft of TLS 1.3 states that, in the single-key setting, only $2^{24.5}$ full-size records may be encrypted on a given connection while keeping a safety margin of approximately 2^{-57}. Following the analysis of Luykx and Paterson, one would infer that the safety margin decreases proportionately with μ in the multi-key case. This analysis prompted the TLS Working Group to mandate a key updating mechanism for TLS 1.3. Our multi-key analysis for AES-GCM shows that this additional feature, which adds complexity to an already complex protocol, may be unnecessary.

1.5   Other Work Reducing Multi-key Degradation

The approach outlined above is that of the *standard model*. Bellare and Tackmann [11] use the *ideal cipher model* in order to understand how different modifications to GCM improve resistance against key recovery in the multi-key setting. Their goal is not to establish μ-independence, but to rather extend the effective key length of GCM over that of the underlying block cipher in order to make key recovery more difficult. However, for GCM, they end up with a factor of μ in their security bounds as a consequence of their method of analysis, whereas our results show that this is not inevitable.

In special cases the dependence on μ disappears. Bellare, Bernstein, and Tessaro show that this is the case with AMAC [4]. Hoang and Tessaro (HT) [28] establish a similar result for key-alternating ciphers, and even show more generally that if a construction has transcripts satisfying some special properties, then μ no longer appears when considering bounds on indistinguishability. The HT-condition is a useful sufficient condition because it only places a requirement on how an upper bound on the difference between the probabilities of two transcripts behaves. However, its applicability is limited, as we will illustrate in Sect. 4.4, because it does not provide a meaningful bound when considering integrity. In concurrent work, Hoang and Tessaro [29] generalize their previous approach, and apply it to double encryption. Their transcript-driven approach

---

[2] https://mailarchive.ietf.org/arch/msg/tls/M-fcRtoeCtMxDNtMsPrUsBV5rgk.

provides different insight into how to prove the lack of multi-key security degradation, and can be applied equally well to GCM to arrive at the same conclusion as we do.

1.6 Further Related Work

A significant amount of work has gone into understanding what happens when security properties are analyzed in the multi-key setting in a variety of different contexts. These include public key encryption [5], key establishment protocols [9,16], signatures [39], message authentication codes [4,21], tweakable block ciphers [27,52], and hybrid encryption [20,51]. Bader et al. recently established impossibility results showing that a loss of a factor μ is inevitable when moving to the multi-key setting for a range of public-key primitives [2]. Most recently, Shrimpton and Terashima [47] introduced a new model in order to bridge gaps between standard and ideal model bounds to analyze settings where the standard model bounds provide little assurance of security, like the multi-key setting. Other research on security of block ciphers in the multi-key setting includes the works by Andreeva et al. [1], Mouha and Luykx [40], Tessaro [49], and Fouque et al. [24]. However, there is no systematic treatment of the problem like that provided in our work.

2 Weak Key Attacks

Bellare, Boldyreva, and Micali (BBM) [5] give an example of a public-key encryption scheme which illustrates that the factor μ is necessary in any generic bound. The example creates a new public-key encryption scheme from an existing one by introducing a "bad" event into the construction which occurs with some fixed probability and allows adversaries to win easily. When interacting with a single instance, the bad event occurs with low probability. However, by working with multiple instances, one can increase the chances of triggering the bad event.

The BBM example illustrates a type of attack one can perform against algorithms in the multi-key setting that is different from the time-memory-data trade-off applied to AES [15]. The multi-key attack against AES precomputes the encryption of a plaintext under a large set of keys, and hopes for a collision between the precomputed values and the oracles in order to immediately recover keys. This attack can be applied to any block cipher, no matter how secure it is.

An analogue of the BBM example in the block cipher setting is a block cipher with *weak keys*, these being keys under which one can attack the block cipher much more efficiently than expected. For example, the recently introduced block cipher Midori64 [3] has a class of 2^{32} weak keys [26] out of 2^{128}, which when identified (which can be done with a single query), can lead to key recovery within computational complexity 2^{16} and data complexity 2. When analyzed in the single-key setting, attackers either get a strong key, in which case key recovery presumably still takes roughly $2^{128} - 2^{32}$ time, or a weak key, leading to a speed-up. When analyzed in the multi-key setting, the chances of finding a weak key are much higher, and adversaries can allocate their resources more efficiently.

A good strategy for a multi-key adversary attacking an algorithm with weak keys would be to first spend some resources across its μ oracles to detect if one of them is using a weak key, and then to allocate as many resources as necessary to attack the weak key. If P is the probability that a key is weak, C_W the cost to break the algorithm with probability one given that it is using a weak key, and C_D the cost to detect a weak key, then in cost at most $C_W + \mu C_D$, the success probability of breaking the algorithm can be improved by a factor

$$\frac{1 - (1 - P)^\mu}{P} = 1 + (1 - P) + (1 - P)^2 + \cdots + (1 - P)^{\mu-1}, \qquad (2)$$

which is the probability of finding at least one weak key out of μ over the probability of a single key being weak. If P is small, then this means the success probability increases by a factor almost linear in μ. Plugging in the numbers for Midori64, we have that a multi-key adversary interacting with $\mu = 2^{16}$ keys, with computational complexity 2^{17} and data complexity $\mu + 2$ has success probability a factor of approximately 2^{16} better than the single-key attack, which has computational complexity 2^{16} and data complexity 2.

When formally analyzing modes of operation, time-memory-data key recovery attacks are usually taken out of consideration because the block cipher is replaced with a uniformly random permutation. Instead, attacks that might improve in the multi-key setting are those that take advantage of bad events in security proofs. Consider the following pathological example, which does not use keyed functions, thereby making key-recovery attacks inapplicable. Let O be an oracle from X to X which with probability ϵ equals a constant function, and with probability $1 - \epsilon$ equals the identity function, and let W be the set of transcripts containing collisions. Then single-oracle adversaries attempting to find collisions in O win with probability at most ϵ, whereas multi-oracle adversaries can increase their advantage significantly.

3   When Multiple Oracles Do Not Benefit Adversaries

In this section we introduce and prove the sufficient condition characterizing when adversaries have no advantage with multiple oracles over a single oracle. We start by introducing basic notation and definitions used throughout the section, and then review the generic folklore bound. We end the section by showing how the condition is sufficient.

3.1   Notation

Given a set X, $\mathsf{X}^{\leq q}$ denotes the set of non-empty sequences of X of length less than or equal to q, and X^+ denotes the set of non-empty arbitrary length sequences of elements of X. Given $\boldsymbol{x} \in \mathsf{X}^+$, $|\boldsymbol{x}|$ denotes its length, and $\lfloor \boldsymbol{x} \rfloor_q$ denotes the first q elements of $\boldsymbol{x}$, that is, $(x_1, \ldots, x_q)$, and all of $\boldsymbol{x}$ if $q \geq |\boldsymbol{x}|$. If $\mathsf{W} \subset \mathsf{X}^+$, then $\lfloor \mathsf{W} \rfloor_q$

consists of $\lfloor x \rfloor_q$ for $x \in \mathsf{W}$. The concatenation of two sequences $x, x' \in \mathsf{X}^+$ is denoted $x \| x'$.

A prefix of a sequence x is a sequence x' where $x' = \lfloor x \rfloor_i$ for some $i \leq |x|$. An extension of a sequence x is a sequence x' such that x is a prefix of x'.

3.2 Games and Adversaries

We use elements of Maurer's random systems formalization [35, 36] with slightly different notation.

An (X, Y)-system [36] takes inputs from X and generates outputs in Y which can depend probabilistically on all previous outputs. A game G from X to Y is a tuple $(\mathbf{O}, \mathsf{W})$ consisting of an (X, Y)-system $\mathbf{O}$ and a random variable $\mathsf{W} \subset (\mathsf{X} \times \mathsf{Y})^+$ which may depend on $\mathbf{O}$, representing the "winning" transcripts.

For example, key recovery of some keyed function $F_k : \mathsf{X}_F \rightarrow \mathsf{Y}_F$ could be modelled as follows. Let $G : K \rightarrow \{0\}$ be an output oracle for the adversary, meaning when the adversary has a guess for the key, it queries G on that guess. Then $\mathbf{O}$ is defined as $\mathbf{O}(0, x) = F_k(x)$ and $\mathbf{O}(1, x) = G(x)$, with k chosen uniformly at random from K. Then $\mathbf{O}$ is a $(\{0\} \times \mathsf{X}_F \cup \{1\} \times K, \mathsf{Y}_F \cup \{0\})$-system, and W is defined to be all transcripts containing a query of the form $\mathbf{O}(1, k) = 0$. Note that this example illustrates how W need not be independent of $\mathbf{O}$.

We write $\mathbf{O}\langle t \rangle$ to mean the event that

$$(\mathbf{O}(x_1), \mathbf{O}(x_2), \ldots, \mathbf{O}(x_\ell)) = (y_1, y_2, \ldots, y_\ell), \tag{3}$$

where $t = ((x_1, y_1), \ldots, (x_\ell, y_\ell))$. Note that the order of the queries in the transcript is important since $\mathbf{O}$ could be stateful.

An adversary $\mathbf{A}$ interacting with $G = (\mathbf{O}, \mathsf{W})$ is a (Y, X)-system, which produces a sequence of inputs $(x_1, x_2, \ldots) \in \mathsf{X}^+$, where x_i is generated using $y_1, y_2, \ldots, y_{i-1}$ with $y_j = \mathbf{O}(x_j)$ for $j = 1, \ldots, i-1$; note that x_1 is generated without any $\mathbf{O}$-output. We let $\mathbf{A}^{\mathbf{O}}$ denote the sequence $((x_1, y_1), (x_2, y_2), \ldots) \in (\mathsf{X} \times \mathsf{Y})^+$, which is a random variable. We say that a transcript $\mathbf{A}^{\mathbf{O}}$ wins if $\mathbf{A}^{\mathbf{O}} \in \mathsf{W}$, and write $\mathbf{A}\langle t \rangle$ for $t = ((x_1, y_1), (x_2, y_2), \ldots, (x_q, y_q)) \in (\mathsf{X} \times \mathsf{Y})^+$ to denote the event that $\mathbf{A}$ produces x_i as the ith oracle input when given $(y_1, y_2, \ldots, y_{i-1})$ as oracle outputs, for $i = 1, \ldots, q$.

Let q be a non-negative integer, then the advantage of an adversary $\mathbf{A}$ winning game G within q queries is

$$\operatorname*{adv}_{G, q} \mathbf{A} := \mathbb{P}\left[\mathbf{A}^{\mathbf{O}} \in \lfloor \mathsf{W} \rfloor_q\right], \tag{4}$$

which is the probability that $\mathbf{A}^{\mathbf{O}}$'s interaction is contained in the set of winning transcripts W of length at most q. For example, Eq. 4 says that the advantage of an adversary in recovering the key to F_k from above is the probability that $\mathbf{A}$ produces a transcript when interacting with $\mathbf{O}$ which is of length at most q, and contains at least one query of the form $G(k) = 0$.

Ultimately, the quantity we are interested in is

$$\sup_{\mathbf{A}} \operatorname{adv}_{G,q} \mathbf{A}. \tag{5}$$

Without loss of generality, we may focus on deterministic adversaries, since for all $\mathbf{A}$,

$$\operatorname{adv}_{G,q} \mathbf{A} = \mathbb{P}\left[\mathbf{A}^{\mathbf{O}} \in \lfloor \mathsf{W} \rfloor_q\right] \tag{6}$$

$$= \sum_{A \in \mathbb{D}} \mathbb{P}\left[\mathbf{A}^{\mathbf{O}} \in \lfloor \mathsf{W} \rfloor_q \mid \mathbf{A} = A\right] \cdot \mathbb{P}\left[\mathbf{A} = A\right] \tag{7}$$

$$\leq \sup_{A \in \mathbb{D}} \mathbb{P}\left[A^{\mathbf{O}} \in \lfloor \mathsf{W} \rfloor_q\right], \tag{8}$$

where $\mathbb{D}$ represents all deterministic adversaries. Furthermore, we generally assume that the input and output spaces of our oracles are finite. This means there are finitely many optimal choices for adversaries to make, hence the above supremum is attained, and can be described as a maximum. For this reason we can speak of *optimal* adversaries, that is, any adversary that attains the maximum advantage given a particular oracle, game, and query bound.

Unless specified otherwise, we only consider games that are *monotone*, that is, $t \in \mathsf{W}$ implies that all extensions t' of t are in W. In monotone games it is also useful to consider the first query which triggers the winning event: before this query is made the adversary has not yet won, and this is the first query for which one can say that the adversary has won.

3.3   Multi-oracle Games and an Existing Bound

Consider an adversary $\mathbf{A}$ interacting with multiple independent games

$$\{G_i = (\mathbf{O}_i, \mathsf{W}_i)\}_{i \in I}, \tag{9}$$

with as goal to win the disjunction of the G_i. Letting X_i denote the domain of oracle $\mathbf{O}_i$ and X the set of elements (i, x) such that $x \in \mathsf{X}_i$, the game $G = (\mathbf{O}, \mathsf{W})$ that $\mathbf{A}$ plays can be defined with the single oracle $\mathbf{O}(i, x) = \mathbf{O}_i(x)$, and by W where $t \in \mathsf{W}$ if the projection $\Pi_i t$ of t onto the $\mathbf{O}_i$-queries is in W_i for some i.

If we know the security bounds for each G_i, then there is a simple way of bounding $\mathbf{A}$'s advantage without computing it from scratch: for each $i \in I$ construct an adversary $\mathbf{A}_i$ which runs $\mathbf{A}$, plays game G_i, and simulates all the other games independently. The adversary $\mathbf{A}_i$ perfectly simulates $\mathbf{A}$'s game precisely because game G_i is independent of all other games. Moreover, $\mathbf{A}_i$ wins if $\mathbf{A}$ does in game G_i. Then, taking a union bound over i, $\mathbf{A}$'s advantage within q queries can be bounded by the sum of the advantages of the $\mathbf{A}_i$ for $i \in I$, or

$$\operatorname{adv}_{G,q} \mathbf{A} \leq \sum_{i \in I} \operatorname{adv}_{G_i,q} \mathbf{A}_i. \tag{10}$$

The setting we focus on is when the G_i are independent instances of the same game G_1. Given a game $G = (\mathbf{O}, \mathsf{W})$ from X to Y, define $\overline{G} = (\overline{\mathbf{O}}, \overline{\mathsf{W}})$ to be the game giving access to the family $\{\mathbf{O}_i\}_\mathbb{N}$, which is a family of independently distributed copies of $\mathbf{O}$ indexed by $\mathbb{N}$, and where $t \in \overline{W}$ if $\Pi_i t \in W$ for some $i \in \mathbb{N}$. In this case the generic multi-key bound simplifies to

$$\operatorname*{adv}_{G,q} \mathbf{A} \leq \mu \cdot \operatorname*{adv}_{G_1,q} \mathbf{A}_1, \tag{11}$$

where μ is the size of I, or a bound on the number of different oracles that $\mathbf{A}$ queries. This bound can be applied to any game, and has been in the case of public-key encryption [5] and PRFs [4,6].

Definition 1. *The oracle $\mathbf{O}$ does not exhibit multi-key security degradation with respect to $G = (\mathbf{O}, W)$ and $Q > 0$, if for all $0 < q \leq Q$*

$$\sup_{\mathbf{A}} \operatorname*{adv}_{\overline{G},q} \mathbf{A} \leq \sup_{\mathbf{A}} \operatorname*{adv}_{G,q} \mathbf{A}. \tag{12}$$

3.4   Sufficient Condition

Since the goal of multi-oracle adversaries is to win any of the single-oracle games it is given, finding the optimal strategy is a question of targeting those single-oracle games for which it has the highest chance of winning, relative to its query allotment. The information that the adversary can work with is the transcripts produced from each single-oracle game and how many queries it has left. So, for example, a good strategy for an adversary might be to query each oracle once, and to estimate based on all of the transcripts which oracle is the weakest, and then to focus on the weakest one.

Conversely, if all of the oracles are equally strong, then, intuitively, one might think that it does not make a difference that the adversary can work with more than one oracle, since there is little difference between the various oracles, and the adversary's best strategy would seem to be to focus its effort on just one of them. However, to formally establish this we require an additional condition: it must be the case that when an optimal single-oracle adversary is given additional knowledge about the oracle, then its chance of winning the game does not *decrease* relative to an optimal single-oracle adversary given less knowledge. Now, if an adversary is interacting with multiple oracles, and it has more information about one oracle over the others, then its best strategy is to stick to that oracle instead of switching to another one.

This condition breaks down, for instance, if a construction has weak keys: if an adversary has the knowledge that its oracle is using a weak key, then it might have better advantage in winning the game versus an oracle where there is still a chance of interacting with a strong key.

Below, we formalize the idea of giving adversaries additional knowledge via *games with advice*, which is equivalent to the concept of projected systems and their advantage by Gaži and Maurer [25]. Gaži and Maurer's projected systems explicitly define new conditional probability distributions which explain the behavior of the system from a given starting transcript. For our purposes we do not need to use the definition of a projected system directly, only the associated advantage definition.

Definition 2. *Let $G = (\mathbf{O}, \mathsf{W})$ be a game and $t \in (\mathsf{X} \times \mathsf{Y})^+$ be a transcript. Then G with advice t, denoted G^t, is defined as $(\mathbf{O}, \mathsf{W}^t)$, where $s \in \mathsf{W}^t$ if and only if t is a prefix of s and $s \in \mathsf{W}$. The advantage of adversary $\mathbf{A}$ in winning game $G^t = (\mathbf{O}, \mathsf{W}^t)$ within q queries is*

$$\operatorname*{adv}_{G^t, q} \mathbf{A} := \mathbb{P}\left[\mathbf{A}^{\mathbf{O}} \in \lfloor \mathsf{W}^t \rfloor_q \mid \mathbf{O}\langle t \rangle,\, t \notin \mathsf{W}\right]. \tag{13}$$

Note that any analysis of an adversary's advantage in winning a game with advice is only meaningful if $\mathbb{P}\left[\mathbf{O}\langle t \rangle,\, t \notin \mathsf{W}\right] > 0$.

The definition below contains the additional condition we need in order to show in Theorem 1 that multi-oracle adversaries do not gain any advantage relative to single-oracle adversaries. Note that it only looks at single-oracle adversaries, meaning if a game satisfies the condition, then one can conclude something about multi-oracle adversaries just by looking at single-oracle adversaries.

Informally, the condition states the following. Take a game G, a transcript t, and *any* shorter transcript t' — it does not have to be a prefix of t. Then two settings are compared: one in which adversaries are given t as starting information, and one in which adversaries are given t' as starting information. In both settings adversaries are allotted the same number of queries left to make, computed as $q - |t|$ in the condition. Then the condition states that optimal adversaries starting with t should have advantage greater than or equal to optimal adversaries starting with t', and this should hold for all t which can result from the interaction between an optimal adversary and the game, and all t' shorter than t. Even though the condition might seem strong, the proof of Theorem 1 is non-trivial. In Lemma 3 we show that GCM's underlying polynomial hash satisfies it.

The other details in the condition are there to remove pathological situations, for example removing transcripts t which could never occur, or to remove situations that do not need to be taken into account in the condition in order for the proof to hold, for example removing transcripts t and t' for which adversaries are guaranteed to win. For this purpose, define transcript t to be $(\mathbf{A}, G)$-*meaningful* if

$$\mathbb{P}\left[\lfloor \mathbf{A}^{\mathbf{O}} \rfloor_{|t|} = t, t \notin \mathsf{W}\right] > 0. \tag{14}$$

Definition 3 (Progressive Games). *Let $G = (\mathbf{O}, \mathsf{W})$ be a monotone game from X to Y and Q be any non-negative integer. Suppose that for all $q \leq Q$, all optimal adversaries $\mathbf{A}$, all $(\mathbf{A}, G)$-meaningful t such that $q' := q - |t| \geq 0$, we have that, for all transcripts t' with $|t'| < |t|$ that are meaningful with respect to some adversary,*

$$\sup_{\mathbf{C}} \operatorname*{adv}_{G^t, q} \mathbf{C} \geq \sup_{\mathbf{B}} \operatorname*{adv}_{G^{t'}, q' + |t'|} \mathbf{B}. \tag{15}$$

Then G is said to be progressive.

Theorem 1. *Let $\mathbf{O}$ be an oracle and $G = (\mathbf{O}, \mathsf{W})$ be a progressive game. Then $\mathbf{O}$ does not exhibit multi-key security degradation.*

3.5   Proof of Theorem 1

Notation. Let $[0,1]$ be the unit interval, and let $\cdot$ denote the dot product of two equal-length elements of $[0,1]^+$, i.e.

$$\boldsymbol{x} \cdot \boldsymbol{y} = \sum_i x_i y_i. \tag{16}$$

Let $\boldsymbol{x} \in [0,1]^+$, then $\mathbf{1} \cdot \boldsymbol{x}$ denotes the dot product of $\boldsymbol{x}$ with a vector consisting of $|\boldsymbol{x}|$ ones (or put simply, $\mathbf{1} \cdot \boldsymbol{x}$ is the sum of the components in $\boldsymbol{x}$).

Decision Trees. The interaction between a game G and a deterministic adversary $\mathbf{A}$ can be viewed as a *decision tree* as follows. The adversary produces a first input $x_1 \in \mathsf{X}$ to the oracle $\mathbf{O}$, which represents the root of the tree. The oracle produces an output $y_1 \in \mathsf{Y}$, and depending upon the output, $\mathbf{A}$ decides its next oracle input. Each of the possible oracle outputs $y_1 \in \mathsf{Y}$ results in an edge extending from the root to a child node, which contains $\mathbf{A}$'s second oracle query, assuming (x_1, y_1) has occurred. Then, starting from a child node, we extend the tree further by adding edges according to the second oracle output, connecting them to the third oracle inputs. Without loss of generality, we may restrict ourselves to decision trees where each edge has a non-zero chance of occurring: if the output y_1 is not possible with input x_1, then we do not include that edge in the tree.

Consider for example some adversary $\mathbf{A}_H$ playing a game $H = (\mathbf{R}, \mathsf{V})$ where the oracle $\mathbf{R}$'s output domain is $\{\alpha, \beta\}$, and V is some arbitrary set of winning transcripts. Then the root of $\mathbf{A}_H$'s decision tree will contain some value x representing an input to $\mathbf{R}$, and is connected by two edges, labeled by α and β respectively, to two child nodes. The child node connected to x via α represents the adversary's second oracle input assuming the first oracle output was α, and similarly for the other child node. Figure 2a illustrates what the tree looks like for this example with deterministic adversaries making three queries. Throughout this section we use the notation $\mathbf{A}_H$ and $H = (\mathbf{R}, \mathsf{V})$ to refer to this running example, and the notation $\mathbf{A}$ and $G = (\mathbf{O}, \mathsf{W})$ to refer to a generic adversary and game.

The *level* of the root node equals one, and a child of a node with level ℓ has level $\ell + 1$. Each node in the tree is connected by a unique path to the root. Let x_i be a node with path $x_1 \xrightarrow{y_1} x_2 \xrightarrow{y_2} \cdots x_{i-1} \xrightarrow{y_{i-1}} x_i$ connecting it to the root. Then the transcript associated to the node x_i is $((x_1, y_1), (x_2, y_2), \ldots, (x_{i-1}, y_{i-1}))$.

Probability Labeling. Starting from a decision tree T for adversary $\mathbf{A}$ and game G, we construct a labeling P_T consisting of probabilities from which one can compute the adversary's advantage. The root node in T is labeled with the probability that the adversary wins on the first query. If y denotes the label of an edge emanating from the root node in T, then the corresponding label in P_T is the probability that the first query does not win, and the output of the first

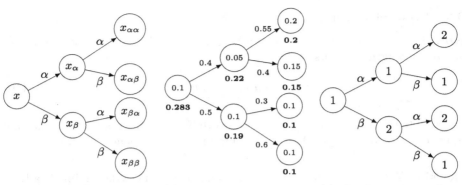

(a) Decision tree of $\mathbf{A}_H$. Each edge is labeled by an oracle output, and each node is labeled by an oracle input.

(b) Probability label P_T for tree depicted left. Written in bold underneath each node is its value ν.

(c) Decision tree of a multi-oracle adversary $A_{\overline{H}}$ making queries to two different oracles.

Fig. 2. An example of how a decision tree is constructed (left) along with a possible probability labeling (center) from the game $H = (\mathbf{R}, \mathsf{V})$ with adversary $\mathbf{A}_H$, as well the decisions made by a possible multi-oracle adversary $\mathbf{A}_{\overline{H}}$.

query is y. The node at the end of this edge is then labeled by the probability that the second query wins, given that the first query does not win and the output of the first query is y. Note that the sum of the label of the root node and all its edges must equal one, since either the first query wins, or the first query does not win, and the edges split up the event that the first query does not win according to the output of the first query.

The labeling P_T is then extended to the entire tree T using a similar process. Given a node x_i and its associated transcript t, the node x_i is labeled by the probability that the ith query x_i wins given that the preceding transcript t does not win, i.e. $t \notin \mathsf{W}$, and t has occurred, i.e. $\lfloor \mathbf{A}^{\mathbf{O}} \rfloor_{|t|} = t$, or in other words, letting $P_T(x_i)$ denote the labeling of node x_i,

$$P_T(x_i) := \mathbb{P}\left[\lfloor \mathbf{A}^{\mathbf{O}} \rfloor_i \in \mathsf{W} \mid \lfloor \mathbf{A}^{\mathbf{O}} \rfloor_{i-1} = t, t \notin \mathsf{W}\right]. \qquad (17)$$

In the same way, an edge $x_i \xrightarrow{y_i} x_{i+1}$ is labeled in P_T by

$$P_T(x_i \xrightarrow{y_i} x_{i+1}) := \mathbb{P}\left[\lfloor \mathbf{A}^{\mathbf{O}} \rfloor_i = t', t' \notin \mathsf{W} \mid \lfloor \mathbf{A}^{\mathbf{O}} \rfloor_{i-1} = t, t \notin \mathsf{W}\right], \qquad (18)$$

where $t' = t \| ((x_i, y_i))$. The resulting labeling P_T maintains the property that the sum of the labels on any non-leaf node and all edges emanating from it equals one.

In Fig. 2b we illustrate a probability labeling associated to $\mathbf{A}_H$ and H. In this case the probability that $\mathbf{A}_H$ wins on its first query is 0.1. The probability that $\mathbf{A}_H$ does not win on its first query and $\mathbf{R}(x_\alpha) = \alpha$, is 0.4, etc.

Given a probability labeling P_T for T, we can assign a value ν to each node in T. If the node v is a leaf node, then its value is the labeling of the node,

$P_T(v)$. Otherwise, let $c_1, c_2, \ldots, c_k$ denote v's children, where the label of the edge connecting v to c_i is e_i. Letting $\boldsymbol{c} := (\nu(c_1), \nu(c_2), \ldots, \nu(c_k))$ and $\boldsymbol{e} = (e_1, e_2, \ldots, e_k)$, the value of the node v is defined as

$$\nu(v) := P_T(v) + \boldsymbol{e} \cdot \boldsymbol{c}. \tag{19}$$

The value of a tree T for an adversary $\mathbf{A}$ is defined as the value of the root node. It is easy to see by an inductive argument across the levels of T that the value of T equals the advantage of $\mathbf{A}$. Figure 2b displays the values of the nodes associated to the labeling of $\mathbf{A}_H$ and H.

Probability Labeling of Multi-oracle Trees. The nodes in a decision tree T corresponding to a deterministic multi-oracle adversary $\mathbf{A}$ playing game $\overline{G}$ fix the oracles that $\mathbf{A}$ queries at each step. This fact can be used to simplify the labeling P_T for multi-oracle adversaries. Given a node x_i in T, we know that $P_T(x_i)$ equals

$$\mathbb{P}\left[\lfloor \mathbf{A}^{\overline{\mathbf{O}}}\rfloor_i \in \overline{W} \mid \lfloor \mathbf{A}^{\overline{\mathbf{O}}}\rfloor_{i-1} = t, t \notin \overline{W}\right], \tag{20}$$

where t is the transcript of length $i-1$ associated to x_i. Say that x_i is a query to oracle $\mathbf{O}_j$. Then we can interpret $\mathbf{A}$ interacting with $\overline{\mathbf{O}}$ during this query as being equivalent to a single-query adversary $\mathbf{B}$ interacting with only $\mathbf{O}_j$, such that

$$\mathbb{P}\left[\lfloor \mathbf{A}^{\overline{\mathbf{O}}}\rfloor_i \in \overline{W} \mid \lfloor \mathbf{A}^{\overline{\mathbf{O}}}\rfloor_{i-1} = t, t \notin \overline{W}\right] = \mathbb{P}\left[t \| \lfloor \mathbf{B}^{\mathbf{O}_j}\rfloor_1 \in \overline{W} \mid \overline{\mathbf{O}}\langle t\rangle, t \notin \overline{W}\right], \tag{21}$$

where we have replaced the event $\lfloor \mathbf{A}^{\overline{\mathbf{O}}}\rfloor_{i-1} = t$ by $\overline{\mathbf{O}}\langle t\rangle$ since $\mathbf{A}$ is deterministic. Simplifying further, note that $t \| \lfloor \mathbf{B}^{\mathbf{O}_j}\rfloor_1 \in \overline{W}$ if and only if $\Pi_j\left(t \| \lfloor \mathbf{B}^{\mathbf{O}_j}\rfloor_1\right) \in W_j$ conditioned on the fact that $t \notin \overline{W}$, which means we can focus on

$$\mathbb{P}\left[\Pi_j\left(t \| \lfloor \mathbf{B}^{\mathbf{O}_j}\rfloor_1\right) \in W_j \mid \overline{\mathbf{O}}\langle t\rangle, t \notin \overline{W}\right]. \tag{22}$$

The event on the left hand side above is independent of all games except for G_j, and so the above probability equals

$$\mathbb{P}\left[\Pi_j\left(t \| \lfloor \mathbf{B}^{\mathbf{O}_j}\rfloor_1\right) \in W_j \mid \mathbf{O}_j\langle \Pi_j t\rangle, \Pi_j t \notin W_j\right]. \tag{23}$$

This means that the label of a node x_i only depends on the particular oracle that to which x_i is queried. We call $\Pi_j t$ the *effective transcript* associated to x_i, since those are the only queries from the transcript which affect $P_T(x_i)$.

From Multi-oracle to Single-Oracle Trees. Consider Fig. 2c, which depicts the decision tree of an optimal multi-oracle adversary $\mathbf{A}_{\overline{H}}$ playing $\overline{H}$, the multi-oracle version of H. Instead of placing the oracle-input values in each node, we now write the index of the oracle that the adversary queries, so a node containing 2 is a query to $\mathbf{R}_2$. Since all oracles share the same output domain $\{\alpha, \beta\}$, the edges remain the same as in Fig. 2a. In particular, we will continue to name the nodes by their labels in Fig. 2a.

Consider query $x_{\alpha\alpha}$ in Fig. 2c. Since $\mathbf{A}_{\overline{H}}$ has decided to make it a query to $\mathbf{R}_2$, but this is the first query to $\mathbf{R}_2$ on the path containing $x_{\alpha\alpha}$, the effective transcript of that node is empty. In contrast, if $x_{\alpha\alpha}$ would have been an $\mathbf{R}_1$-query, then its effective transcript would have had length two, since $x_{\alpha\alpha}$'s associated transcript contains only $\mathbf{R}_1$-queries. Assuming H is progressive, then there is an optimal adversary $\mathbf{C}_H$ making a single query to $\mathbf{R}_1$ with advantage greater than or equal to the value of node $x_{\alpha\alpha}$. Therefore, we can construct an adversary where $x_{\alpha\alpha}$ is replaced by a query to $\mathbf{R}_1$ without decreasing $\nu(x_{\alpha\alpha})$.

The same reasoning does not hold for the query $x_{\beta\alpha}$, since the effective transcript of that node has length one regardless of whether $\mathbf{R}_1$ or $\mathbf{R}_2$ is queried. However, we do know that if x_β had been an $\mathbf{R}_1$-query, then an optimal choice for $x_{\beta\alpha}$ would have been to query $\mathbf{R}_1$ again since the effective transcript of $\mathbf{R}_1$ would have been longer than the effective transcript of $\mathbf{R}_2$. In particular, consider the decision tree U_H constructed as follows: stick to oracle $\mathbf{R}_1$, and behave as $\mathbf{A}_{\overline{H}}$ does until $\mathbf{A}_{\overline{H}}$ no longer queries $\mathbf{R}_1$, then for each node, compute an optimal choice of oracle input based on the associated transcript up to that point. Assuming H is progressive, we have that

1. $P_{T_H}(x) = P_{U_H}(x)$, $P_{T_H}(x_\alpha) = P_{U_H}(x_\alpha)$, and $P_{T_H}(x_{\alpha\beta}) = P_{U_H}(x_{\alpha\beta})$, since U_H is the same as T_H for those queries, and
2. $P_{T_H}(x_{\alpha\alpha}) \leq P_{U_H}(x_{\alpha\alpha})$ and

$$P_{T_H}(x_\beta) \leq P_{U_H}(x_\beta), P_{T_H}(x_\beta) \leq P_{U_H}(x_\alpha), P_{T_H}(x_{\beta\alpha}) \leq P_{U_H}(x_{\beta\alpha}),$$
$$P_{T_H}(x_{\beta\beta}) \leq P_{U_H}(x_{\beta\beta}), P_{T_H}(x_{\beta\beta}) \leq P_{U_H}(x_{\beta\alpha}), \text{ and } P_{T_H}(x_{\beta\alpha}) \leq P_{U_H}(x_{\beta\beta}), \quad (24)$$

since the effective transcripts of these nodes in U_H are always longer than their effective transcripts in T_H.

In short, for each subtree S of T_H starting with an $\mathbf{R}_2$-query, the value of each node in a given level ℓ of the corresponding subtree V in U_H is greater than or equal to the value of each node in the same level ℓ of S. Using Lemma 2 below, we can conclude that the value of x_β in T_H is less than or equal to the value of x_β in U_H, and finally that T_H's value is never greater than U_H's value.

The above reasoning can be extended to arbitrary decision trees for a progressive game G. Consider a multi-oracle decision tree T and a single-oracle decision tree U which is the same as T but departs from T the moment T does not make an $\mathbf{O}_1$-query; from that point on U optimizes its next queries only based on effective transcripts. Without loss of generality assume that T contains an $\mathbf{O}_1$-query for its root node. Let S be a subtree of T such that its root node is not an $\mathbf{O}_1$-query, and is the only non-$\mathbf{O}_1$ query on its path connecting it to the root of T. Let V be the corresponding subtree in U. Then, by virtue of G being progressive, given a node s in S and v in V at level ℓ, we know that the effective transcript of s is longer than that of v, and applying Eq. (15), we know that $P_U(s) \geq P_T(v)$. Therefore the probability label of each node in V in that level is greater than or equal to all probability labels in S at the same level. Applying Lemma 2 below, we get our desired result.

To establish Lemma 2, we first need the following result.

Lemma 1. *Let* $a \in [0, 1]$ *and* $\boldsymbol{a}^1 \in [0, 1]^+$ *be such that* $a + 1 \cdot \boldsymbol{a}^1 = 1$; *define* b *and* $\boldsymbol{b}^1$ *similarly. Say that* $a \geq b$. *Let* $\boldsymbol{a}^2, \boldsymbol{b}^2 \in [0, 1]^+$ *with* $\min_i a_i^2 \geq \max_i b_i^2$, *then*

$$a + \boldsymbol{a}^1 \cdot \boldsymbol{a}^2 \geq b + \boldsymbol{b}^1 \cdot \boldsymbol{b}^2. \tag{25}$$

Proof. Let $a^* = \min_i a_i^2$ and $b^* = \max_i b_i^2$, then $\boldsymbol{b}^1 \cdot \boldsymbol{b}^2 \leq b^* 1 \cdot \boldsymbol{b}^1$, and $a^* 1 \cdot \boldsymbol{a}^1 \leq \boldsymbol{a}^1 \cdot \boldsymbol{a}^2$. Therefore,

$$b + \boldsymbol{b}^1 \cdot \boldsymbol{b}^2 \leq b + b^* 1 \cdot \boldsymbol{b}^1 = b + b^*(1 - b) = b^* + (1 - b^*)b \tag{26}$$

$$\leq b^* + (1 - b^*)a = a + b^* 1 \cdot \boldsymbol{a}^1 \leq a + a^* 1 \cdot \boldsymbol{a}^1 \leq a + \boldsymbol{a}^1 \cdot \boldsymbol{a}^2. \tag{27}$$

$\square$

Lemma 2. *Let* S *and* V *be decision trees with the same number of levels. Let* $v_1^\ell, v_2^\ell, \ldots$ *and* $s_1^\ell, s_2^\ell, \ldots$ *denote the nodes of* V *and* S *in level* ℓ, *respectively. Say that for all levels* ℓ, *we have that* $\min_i P_V(v_i^\ell) \geq \max_j P_S(s_j^\ell)$. *Then* $\nu(V) \geq \nu(S)$.

Proof. We induct by level of the tree. Our inductive hypothesis is that $\min_i \nu(v_i^\ell) \geq \max_j \nu(s_j^\ell)$, and we want to show that it holds for level $\ell - 1$. However, applying Lemma 1, we get the desired result. $\square$

4 Proving the Absence of Multi-key Degradation

4.1 Notation and Definitions

We continue to use the notation and definitions from Sect. 3, along with the following.

We use the prefix "multi" to refer to the multi-key setting of the algorithms in question. So, for example, the PRP-PRF switch becomes the multi-PRP-PRF switch, and GCM becomes multi-GCM.

An adversary is *non-adaptive* if the oracle inputs it generates are independent of all oracle outputs. We identify such adversaries with sequences $\boldsymbol{x} \in \mathsf{X}^+$ and write $adv_G \, \boldsymbol{x}$ to mean the advantage of the non-adaptive adversary which queries $\boldsymbol{x}$ to win game G.

A *distinguisher* $\mathbf{D}$ is an adversary $\mathbf{A}$ together with a random variable $\mathsf{W} \subset (\mathsf{X} \times \mathsf{Y})^+$, where $\mathbf{A}$ interacts with oracles from X to Y. The advantage of $\mathbf{D}$ in distinguishing oracles $\mathbf{O}_1$ and $\mathbf{O}_2$ is given by

$$\underset{\mathbf{D}}{\Delta}(\mathbf{O}_1 \,;\, \mathbf{O}_2) := \left| \mathbb{P}\left[\mathbf{A}^{\mathbf{O}_1} \in \mathsf{W}\right] - \mathbb{P}\left[\mathbf{A}^{\mathbf{O}_2} \in \mathsf{W}\right] \right|. \tag{28}$$

Note that this definition is equivalent to the usual definition, where the distinguisher's output bit has been changed to the set W, which is some random variable that may depend on $\mathbf{A}$ but is independent of the oracle: $\mathbf{A}^y \in \mathsf{W}$ if and only if $\mathbf{A}^y$ outputs one, for all possible sequences of oracle outputs y. Concretely, if $\mathbf{D}_1 = (\mathbf{A}, \mathsf{W})$ is a distinguisher, then the corresponding conventional distinguish $\mathbf{D}_2$ runs $\mathbf{A}^{\mathsf{O}}$, outputs 1 if the transcript is in W and 0 otherwise; conversely, if $\mathbf{D}_2$ is a conventional distinguisher, then $\mathbf{A}$ is constructed by running $\mathbf{D}_2$ with $\mathbf{A}$'s oracle, and outputting $\mathbf{D}_2$'s oracle inputs, and W is defined as the transcripts for which $\mathbf{D}_2$ outputs 1.

We extend single-oracle distinguishing games to their multi-oracle counterparts as follows: instead of being given access to only one instance, either $\mathbf{O}_1$ or $\mathbf{O}_2$, distinguishers are given the task of distinguishing any number of instances of $\mathbf{O}_1$, $\mathbf{O}_1^1, \mathbf{O}_1^2, \ldots$, with any number of instances of $\mathbf{O}_2$, $\mathbf{O}_2^1, \mathbf{O}_2^2, \ldots$.

A uniformly distributed random function (URF) with domain X and range Y is a random variable that is uniformly distributed over the set of all functions from X to Y. A uniformly distributed random permutation (URP) with domain X is a random variable that is uniformly distributed over the set of all permutations on X.

4.2   Non-adaptivity and the Multi-PRP-PRF Switch

The PRP-PRF switching lemma bounds the distinguishing advantage between a URP π with domain X and a URF ϕ with domain and range X. The lemma states that for all distinguishers $\mathbf{D}$ making no more than q queries,

$$\underset{\mathbf{D}}{\Delta}\left(\pi\,;\,\phi\right) \leq \frac{q^2}{2\,|\mathsf{X}|}. \tag{29}$$

Various papers have proofs of this statement, such as [10, 19, 36]. The corresponding multi-oracle indistinguishability game is

$$\underset{\mathbf{D}}{\Delta}\left(\{\pi_i\}_{i \in I}\,;\,\{\phi_i\}_{i \in I}\right). \tag{30}$$

Using the generic bound from Sect. 3.3 we get

$$\underset{\mathbf{D}}{\Delta}\left(\{\pi_i\}_{i \in I}\,;\,\{\phi_i\}_{i \in I}\right) \leq |I|\,\frac{q^2}{2\,|\mathsf{X}|}, \tag{31}$$

which deteriorates according to the number of oracles present, $|I|$.

Adaptivity does not help adversaries in distinguishing a URP from a URF, as shown for example by Maurer [36]. However, this does not help to prove that there is no degradation in the multi-oracle setting, since non-adaptivity being optimal in the single-oracle setting does not imply that non-adaptivity is still optimal in the multi-oracle setting; Demay et al. [23] construct an example to illustrate this fact, and it can also be seen by considering the weak key example from Sect. 2, where the best strategy in the single-oracle setting is non-adaptive.

Demay et al. [23] also prove that if the oracles in the indistinguishability game satisfy some condition (*conditional equivalence*), which URPs and URFs do, then optimality of non-adaptivity in the multi-oracle setting can be established. However, even if non-adaptive adversaries are optimal in the multi-oracle setting, they can still gain advantage over single-oracle adversaries. Consider for example some game G where adversaries win with probability $1/2^{i+100}$ on the ith query, regardless of what the queries are, and independently of the other queries. In the single-oracle setting adaptivity does not help, and the advantage of any adversary is roughly $2^{-100}(2^q - 1)/2^q$. Similarly, in the multi-oracle setting adaptivity does not help, but an adversary with access to μ oracles can achieve an advantage of roughly $\mu 2^{-100}(2^{q/\mu} - 1)/2^{q/\mu}$ if they make q/μ queries to each oracle, which approaches $\mu 2^{-100}$ if q/μ is relatively large.

Nevertheless, assuming non-adaptivity in the multi-oracle setting allows us to identify a simpler requirement on games than being progressive. The following result establishes exactly when multi-oracle adversaries have no advantage over single-oracle adversaries when adaptivity does not help.

Proposition 1. *Suppose that $G = (\mathbf{O}, \mathsf{W})$ is a game with optimal non-adaptive adversaries in the multi-oracle setting. Suppose also that for all q and $q' \leq q$,*

$$\sup_{\mathbf{A}} \operatorname*{adv}_{G,q'} \mathbf{A} + \sup_{\mathbf{A}} \operatorname*{adv}_{G,q-q'} \mathbf{A} \leq \sup_{\mathbf{A}} \operatorname*{adv}_{G,q} \mathbf{A}. \tag{32}$$

Then adversaries gain no advantage in interacting with multiple independent instances of G.

Proof. Let $\mathbf{A}$ be a non-adaptive multi-oracle adversary. Let $\mathbf{A}_i := \Pi_i(\mathbf{A})$ and say that $q_i = |\mathbf{A}_i|$, where q_i is not necessarily positive. Then we can bound $\mathbf{A}$'s advantage with

$$\operatorname*{adv}_{\overline{G},q} \mathbf{A} \leq \sum_{i=1}^{Q} \operatorname*{adv}_{G,q_i} \mathbf{A}_i, \tag{33}$$

since $\mathbf{A}$ queries at most Q different oracles. By assumption we know that there is a single-oracle adversary $\mathbf{B}_{1,2}$ making $q_1 + q_2$ queries such that

$$\operatorname*{adv}_{G,q_1} \mathbf{A}_1 + \operatorname*{adv}_{G,q_2} \mathbf{A}_2 \leq \operatorname*{adv}_{G,q_1+q_2} \mathbf{B}_{1,2}. \tag{34}$$

The same can be done with $\mathbf{B}_{1,2}$ and $\mathbf{A}_3$ to create adversary $\mathbf{B}_{1,2,3}$, and so on, resulting in a single-oracle adversary which has advantage greater than or equal to $\mathbf{A}$. $\square$

Maurer [36] proved conditional equivalence of URPs and URFs. Therefore adaptivity does not help distinguishers in the single-oracle PRP-PRF switch. Demay et al. [23] proved that conditional equivalence in the single-oracle setting translates to conditional equivalence in the multi-oracle setting. Therefore multi-oracle URPs and URFs are conditionally equivalent, and hence adaptivity does not help in distinguishing multiple URPs from multiple URFs. In particular, distinguishing URPs from URFs is equivalent to finding collisions in URFs [36], which translates indistinguishability into a collision finding game G. Since the advantage in finding a collision in a URF equals the probability that there is some collision among q independent, uniformly distributed elements, the condition in

Eq. (32) is satisfied, hence there is no multi-oracle degradation for the PRP-PRF switch.

Note that the transition from an indistinguishability-based game to one which is not, is in general non-trivial. It is straightforward to show equivalence between the PRP-PRF switch and finding collisions in a PRF, however the general case requires a more involved transformation. See for example Lemma 5 of Maurer, Pietrzak, and Renner [37].

4.3   Hoang and Tessaro's Technique and an Improvement

Instead of using Proposition 1, one can prove a similar result about the multi-PRP-PRF switch by using the technique of Hoang and Tessaro (HT) [28]. The HT-condition requires understanding the difference in transcript probabilities between a URP and a URF. Let t be a transcript of length q, and say there exists a function $\epsilon(q)$ such that

$$\mathbb{P}\left[\pi\langle t\rangle\right] \geq \mathbb{P}\left[\phi\langle t\rangle\right] \cdot (1 - \epsilon(q)). \tag{35}$$

Hoang and Tessaro call this ϵ-*point-wise proximity* of π and ϕ, and we say that ϕ is ϵ-point-wise close to π. If $\epsilon(q') + \epsilon(q - q') \leq \epsilon(q)$ and $\epsilon(q) \leq 0.5$, then their Lem. 2 establishes that the analogous difference in multi-oracle transcripts is at most $2 \cdot \epsilon(q)$. Following either Hoang and Tessaro's [28] or Chang and Nandi's [19] proof for the PRP-PRF switch, the HT-condition establishes that multi-oracle adversaries have at most a factor of two gain over single-oracle adversaries.

In fact, with only the requirement that $\epsilon(q') + \epsilon(q - q') \leq \epsilon(q)$, one can prove that adversaries gain *no* — not even a factor 2 — advantage in the multi-oracle setting relative to ϵ.

Proposition 2. *Suppose that* $\mathbf{R}$ *and* $\mathbf{S}$ *are* ϵ-*point-wise close and that for all* q *and* $q' \leq q$, $\epsilon(q') + \epsilon(q - q') \leq \epsilon(q)$. *Then* $\overline{\mathbf{R}}$ *and* $\overline{\mathbf{S}}$, *which are oracles giving adversaries access to arbitrarily many independent instances of* $\mathbf{R}$ *and* $\mathbf{S}$, *are* ϵ-*point-wise close as well.*

Proof. It suffices to prove that for all t such that $|t| = q$ and $\mathbb{P}\left[\overline{\mathbf{S}}\langle t\rangle\right] > 0$,

$$\mathbb{P}\left[\overline{\mathbf{R}}\langle t\rangle\right] / \mathbb{P}\left[\overline{\mathbf{S}}\langle t\rangle\right] \geq 1 - \epsilon(q). \tag{36}$$

Let I be the set of instances queried in t, and say that $q_i = |\Pi_i t|$, then

$$\frac{\mathbb{P}\left[\overline{\mathbf{R}}\langle t\rangle\right]}{\mathbb{P}\left[\overline{\mathbf{S}}\langle t\rangle\right]} = \prod_{i \in I} \frac{\mathbb{P}\left[\mathbf{R}_i\langle\Pi_i t\rangle\right]}{\mathbb{P}\left[\mathbf{S}_i\langle\Pi_i t\rangle\right]} \geq \prod_{i \in I}(1 - \epsilon(q_i)) \geq 1 - \sum_{i \in I}\epsilon(q_i) \geq 1 - \epsilon(q). \tag{37}$$

$\square$

An important difference between our setting and Hoang and Tessaro's is that our oracles are independent of each other, whereas Hoang and Tessaro also consider oracles which are built using some shared underlying ideal primitive, which is why Proposition 2 cannot be applied to their setting.

The condition that $\epsilon(q') + \epsilon(q - q') \leq \epsilon(q)$ looks very similar to the condition of Proposition 1 required in order to achieve no multi-oracle degradation when adaptivity does not help, since ϵ is an upper bound on the success probability of single-oracle adversaries. However, Proposition 2 is a statement about the computed bounds, and it might be the case that multi-oracle adversaries have some advantage gain over single-oracle adversaries, but that this difference is not visible with a particular upper bound ϵ; after all, setting $\epsilon(q) = q$ is true for all pairs of oracles, but then Proposition 2 becomes meaningless. In contrast, satisfying the hypotheses of Proposition 1, and, more generally, a game being progressive, establishes something inherent about the oracle in question sufficient to prove that multi-oracle adversaries gain nothing over single-oracle adversaries.

4.4   Integrity and the Inapplicability of Point-Wise Proximity

Finding meaningful ϵ which establishes point-wise-proximity is impossible in some cases, as we illustrate for MAC (Message Authentication Code) schemes and integrity. Our focus is on stateful MAC schemes, although the same observations can be applied to deterministic MAC schemes.

Definition 4. *A nonce-based MAC scheme from* $N \times M$ *to* T *is a pair of algorithms* (F, V), *where* F, *the tagging algorithm, maps a tuple of a nonce from* N *and message from* M *to tags in* T, *and* V, *the verification algorithm, maps inputs from* $N \times M \times T$ *to either* $\top$ *or* $\bot$, *indicating validity or invalidity of an input.*

A secure MAC scheme is one in which it is difficult to construct a new input to V such that V outputs $\top$. We translate Bellare and Namprempre's strong unforgeability [8] into our formalization.

Definition 5. *Let* (F, V) *be a nonce-based MAC scheme. The integrity game* G *with respect to* (F, V) *is defined as* $(\mathbf{O}, \mathsf{W})$, *with* $\mathbf{O}$ *an oracle giving adversaries access to* F *and* V, *and* W *defined as the set of transcripts consisting of* F*-queries where each nonce-input is unique, and containing at least one* V*-query* (n, m, t) *where* $V(n, m, t) = \top$, *and* $F(n, m) = t$ *is not in the preceding transcript.*

Recall that adversarial advantage is defined as in Eq. (4).

In order to use pointwise proximity in an integrity game G, it needs to be written as an indistinguishability game, which is done as follows:

$$\Delta\left(F, V\,;\,F, \bot\right), \tag{38}$$

with $\bot$ an algorithm always outputting $\bot$. Establishing ϵ-point-wise proximity between (F, V) and $(F, \bot)$ means finding an ϵ such that for all transcripts t of length q,

$$\mathbb{P}\left[(F, V)\langle t\rangle\right] \geq (1 - \epsilon) \cdot \mathbb{P}\left[(F, \bot)\langle t\rangle\right], \tag{39}$$

where we write (F, V) and $(F, \bot)$ as shorthands for oracles. Letting O denote either V or $\bot$, the transcript consisting of $O(n, m, t) = \bot$ followed by $F(n, m) = t$ has zero probability with (F, V) and non-zero probability with $(F, \bot)$, meaning ϵ

must equal one. Swapping (F, V) and $(F, \perp)$ in Eq. (39) causes the same problem with any transcript containing an $O(n, m, t) = \top$ query. Therefore, ϵ-point-wise proximity can only hold for $\epsilon = 1$, making the bounds obtained with ϵ-point-wise proximity vacuous.

4.5   Bernstein's Theorem in the Multi-oracle Setting

Rather than considering indistinguishability, ϵ-pointwise proximity can be directly applied to games themselves, as is done by Bernstein [12,13], where ϵ-pointwise proximity is called interpolation probability. Bernstein shows that the probability that an adversary outputs 1 when interacting with an oracle which is ϵ-pointwise close to a URF, is at most $(1 - \epsilon)^{-1}$ times the probability the adversary outputs one when interacting with a URF. Bernstein replaces the use of the PRP-PRF switch with his result when computing integrity bounds for MACs, thereby significantly improving them. Iwata, Ohashi, and Minematsu apply this technique to GCM as well [32, Sect. 7.5 and Appendix C].

Although Bernstein only considers the special case in which one of the oracles is a URF, it can be easily generalized to any oracle. We state the result in terms of distinguishers, which is equivalent to considering adversaries with binary output. Note that this means the result is only applicable to games where W is independent of the oracle O.

Theorem 2. *Let* $\mathbf{D} = (\mathbf{A}, \mathsf{W})$ *be any distinguisher and* q *a positive integer, then if* $\mathbf{O}_1$ *is* ϵ-*pointwise close to* $\mathbf{O}_2$,

$$\mathbb{P}\left[\mathbf{A}^{\mathbf{O}_1} \in \lfloor \mathsf{W} \rfloor_q\right] \leq (1 - \epsilon(q))^{-1} \cdot \mathbb{P}\left[\mathbf{A}^{\mathbf{O}_2} \in \lfloor \mathsf{W} \rfloor_q\right]. \tag{40}$$

Proof. Without loss of generality, assume that $\mathbf{A}$ makes exactly q queries, as one can always consider a distinguisher $\mathbf{D}'$ instead which runs $\mathbf{A}$, makes exactly q queries, and ignores the additional query-outputs.

$$\mathbb{P}\left[\mathbf{A}^{\mathbf{O}_1} \in \lfloor \mathsf{W} \rfloor_q\right] = \sum_{|t|=q} \mathbb{P}\left[\mathbf{A}\langle t \rangle, t \in \mathsf{W}\right] \cdot \mathbb{P}\left[\mathbf{O}_1\langle t \rangle\right] \tag{41}$$

$$\leq (1 - \epsilon(q))^{-1} \cdot \sum_{|t|=q} \mathbb{P}\left[\mathbf{A}\langle t \rangle, t \in \mathsf{W}\right] \cdot \mathbb{P}\left[\mathbf{O}_2\langle t \rangle\right] \tag{42}$$

$$= (1 - \epsilon(q))^{-1} \cdot \mathbb{P}\left[\mathbf{A}^{\mathbf{O}_2} \in \lfloor \mathsf{W} \rfloor_q\right]. \tag{43}$$

$\square$

Bernstein's theorem can be applied to the multi-oracle setting using Proposition 2: if $\mathbf{O}_1$ is ϵ-pointwise close to $\mathbf{O}_2$, and ϵ satisfies the hypothesis of Proposition 2, then the above result can be applied to $\overline{\mathbf{O}}_1$ and $\overline{\mathbf{O}}_2$. For example, this holds in the case of URPs and URFs, hence Bernstein's theorem can be applied to multi-URPs and multi-URFs as well.

Corollary 1. *Let* $\mathbf{D} = (\mathbf{A}, \mathsf{W})$ *be any distinguisher and* q *a positive integer. Let* π *denote a URP and* ϕ *a URF, with* $\overline{\pi}$ *and* $\overline{\phi}$ *their multi-oracle counterparts, then*

$$\mathbb{P}\left[\mathbf{A}^{\overline{\pi}} \in \lfloor \mathsf{W} \rfloor_q\right] \leq (1 - \epsilon(q))^{-1} \cdot \mathbb{P}\left[\mathbf{A}^{\overline{\phi}} \in \lfloor \mathsf{W} \rfloor_q\right], \qquad (44)$$

where ϵ *is the proximity function of* π *and* ϕ.

4.6 Multi-Wegman-Carter Security

Wegman-Carter authenticators [50] are nonce-based MAC schemes mapping messages in M to tags in T. The tagging algorithm takes a nonce $n \in \mathsf{N}$ and a message $m \in \mathsf{M}$, and maps (n, m) to $\phi(n) + h(m)$, where T is a group, ϕ is a URF, and $h : \mathsf{M} \to \mathsf{T}$ is a random function for which it is difficult to find collisions. The verification algorithm takes a nonce $n \in \mathsf{N}$, a message $m \in \mathsf{M}$, and a tag $t \in \mathsf{T}$, and checks whether (n, m) maps to t; it outputs $\top$ if this is the case, and $\bot$ otherwise.

Usually the security of Wegman-Carter authenticators is proved [33, 45, 50] relative to

$$\sup_{m_1 \neq m_2, t} \mathbb{P}\left[h(m_1) - h(m_2) = t\right], \qquad (45)$$

however we will need to describe h's collision resistance differently in order to characterize when Wegman-Carter authenticators exhibit no multi-oracle degradation.

Definition 6. *Let* $h : \mathsf{M} \to \mathsf{T}$ *be a random function with* T *a group. Define the collision game* $G = (\mathbf{O}, \mathsf{W})$ *where* $\mathbf{O} : \mathsf{M}^2 \times \mathsf{T} \to \{\top, \bot\}$ *outputs* $\top$ *on input* (m_1, m_2, t) *if* $h(m_1) - h(m_2) = t$, *and* $\bot$ *otherwise, and* W *consists of all transcripts containing an* $\mathbf{O}$-*query* (m_1, m_2, t) *with* $m_1 \neq m_2$ *and* $\mathbf{O}(m_1, m_2, t) = \top$.

Proposition 3. *Consider adversaries which make no more than* $|\mathsf{N}|$ *queries. Then Wegman-Carter authenticators exhibit no multi-oracle degradation with respect to the integrity game from Definition 5 if the underlying random function* h *exhibits no multi-oracle degradation with respect to the collision game in Definition 6.*

Proof. Let (F, V) denote the Wegman-Carter authenticator and let G be its associated integrity game. Let $\mathbf{A}$ be a multi-oracle adversary playing $\overline{G}$.

First we establish that adversaries gain no advantage by choosing their nonces adaptively. Let $\boldsymbol{n}^i = (n_1^i, n_2^i, \ldots)$ be an enumeration of N, one for each possible oracle $i \in \mathbb{N}$. Then we construct adversary $\mathbf{A}_n$ from $\mathbf{A}$ as follows. $\mathbf{A}_n$ runs $\mathbf{A}$, and maintains a mapping $\iota : \mathbb{N} \times \mathsf{N} \to \mathbb{N}$ which keeps track of the order in which a particular nonce $n \in \mathsf{N}$ was queried for oracle $i \in \mathbb{N}$; for example if $(3, X)$ is the fifth nonce queried to the third oracle, then $\iota(3, X) = 5$. Each time $\mathbf{A}$ makes an F-query (n, m) to oracle i, $\mathbf{A}_n$ makes the F-query $(n_{\iota(i,n)}^i, m)$ to oracle i and returns the response to $\mathbf{A}$. Similarly, each time $\mathbf{A}$ makes a V-query (n, m, t) to oracle i, $\mathbf{A}_n$ makes the V-query $(n_{\iota(i,n)}^i, m, t)$ to oracle i and returns the response to $\mathbf{A}$. Since the mapping $n \mapsto n_{\iota(i,n)}^i$ is bijective for each i, $\mathbf{A}_n$'s advantage is at least that of $\mathbf{A}$ since the URF ϕ underlying oracle i is indistinguishable from the URF $n \mapsto \phi(n^i_{\iota(i,n)})$. Therefore we restrict our attention to adversaries which choose their nonces non-adaptively.

Consider an adversary interacting in the multi-oracle integrity game. Since adaptivity does not help when picking nonces, and the total number of queries is not greater than $|\mathsf{N}|$, we can force the adversary to pick distinct nonces to query. This allows us to replace all the URFs from each Wegman-Carter authenticator by a single URF, since the inputs to the URF will always be distinct. Therefore, we may restrict our attention to adversaries interacting with multiple Wegman-Carter authenticators using the same URF.

For each nonce n, we let m_n denote the associated message input, and t_n F's output under n, so that $F(n, m_n) = t_n$. To each nonce n we can associate two sets of pairs $R_n, S_n \subset \mathsf{M} \times \mathsf{T}$ where $(m, t) \in R_n$ if there is a V-query $V(n, m, t)$ before the F-query using n as nonce is made, and S_n is all pairs (m, t) from V-queries after the F-query using n as nonce is made; S_n is empty if there is no such F-query. Without loss of generality we can assume that for all queried nonces, $R_n \cup S_n \neq \emptyset$, since otherwise $F(n, m_n)$ is independent of the adversary winning.

A nonce wins if one of its associated verification queries results in $\top$, meaning there exists $(m, t) \in R_n \cup S_n$ such that $V(n, m, t) = \top$. Note that $\phi(n) = t - h(m)$ for $(m, t) \in S_n$ if and only if $h(m_n) - h(m) = t_n - t$, and similarly $h(m_n) - h(m) = t_n - t$ for $(m, t) \in R_n$ if and only if $\phi(n) = t - h(m)$. Therefore, a nonce n wins only if

$$\phi(n) \in \{t - h(m) \mid (m, t) \in R_n\} \tag{46}$$

or there exists $(m, t) \in S_n$ such that

$$h(m_n) - h(m) = t_n - t. \tag{47}$$

We call a verification query $V(n, m, t)$ a guess if it occurs before the corresponding F-query with nonce n, and a collision attempt if it occurs after the F-query. A guess succeeds only if Eq. (46) is satisfied, and a collision attempt succeeds only if Eq. (47) is satisfied.

Let $\mathbf{A}$ be an adversary interacting with multiple Wegman-Carter authenticators using the same URF (always querying distinct nonces to the authenticators), and different random functions h_i for $i \in \mathbb{N}$. The adversary $\mathbf{A}$ either wins with a guess, or a collision attempt.

Say that it is given that $\mathbf{A}$ does not win with a guess. This means that for all n,

$$\phi(n) \notin \{t - h_i(m) \mid (m, t) \in R_n\}, \tag{48}$$

and $\mathbf{A}$ wins only if there is a nonce n for which Eq. (47) is satisfied, meaning $\mathbf{A}$ has found a collision for h. We construct an adversary $\mathbf{B}$ playing the multi-oracle collision game with h_i. The adversary $\mathbf{B}$ runs $\mathbf{A}$, responds to $\mathbf{A}$'s guesses with $\bot$, it responds to $\mathbf{A}$'s F-queries by uniformly sampling an element from

$$\mathsf{T} \setminus \{t \mid (m, t) \in R_n\}, \tag{49}$$

and $\mathbf{B}$ responds to $\mathbf{A}$'s collision attempts $V(n, m, t)$ by querying $(m_n, m, t_n - t)$ to the appropriate oracle ($\mathbf{O}_i$ if h_i was queried), where $F(n, m_n) = t_n$. Then, given that $\mathbf{A}$ does not win with a guess, $\mathbf{B}$ perfectly simulates $\mathbf{A}$'s game since all of $\mathbf{A}$'s guesses fail, F is distributed correctly given that all of $\mathbf{A}$'s guesses fail, and $\mathbf{A}$'s collision attempts are passed directly to the collision oracles.

By hypothesis, we know that for every $i > 0$ there is an adversary $\mathbf{C}_i$ playing the collision game with one random function h such that $\mathrm{adv}_i \, \mathbf{C}_i \geq \mathrm{adv}_i \, \mathbf{B}$, and in particular $\mathrm{adv}_i \, \mathbf{C}_i$ is greater than or equal to the probability that $\mathbf{B}$ wins and makes i queries.

Using $\mathbf{C}_i$ and $\mathbf{A}$, we construct a single-oracle adversary $\mathbf{A}_1$ playing the Wegman-Carter integrity game. First $\mathbf{A}_1$ runs $\mathbf{A}$ and responds to $\mathbf{A}$'s queries using its own independently simulated Wegman-Carter authenticators. Once $\mathbf{A}$ is finished, $\mathbf{A}_1$ takes all of $\mathbf{A}$'s guesses and forwards them to its own oracle. Then, if $\mathbf{A}_1$ does not win with a guess, it computes how many queries i it has remaining, and then runs $\mathbf{C}_i$. The probability that $\mathbf{A}$ wins with a guess equals the probability that $\mathbf{A}_1$ wins with a guess, since it is the probability that ϕ gets mapped into the sets defined in Eq. (46).

The probability that $\mathbf{A}$ makes i non-guess queries and wins, given that $\mathbf{A}$'s guesses fail, is bounded by the probability that $\mathbf{B}$ wins and makes i queries, which in turn is bounded by $\mathrm{adv}_i \, \mathbf{C}_i$. Therefore the probability that $\mathbf{B}$ wins is bounded by the sum of $p_i \cdot \mathrm{adv}_i \, \mathbf{C}_i$, where p_i is the probability that $\mathbf{A}_1$ has i queries remaining after its guesses. Since the sum of the p_i is 1, we know that the probability that $\mathbf{A}_1$ wins given that its guesses fail is greater than or equal to the probability that $\mathbf{B}$ wins. Therefore we have shown that the single-oracle adversary $\mathbf{A}_1$ has no less advantage than the multi-oracle adversary $\mathbf{A}$. $\square$

4.7   Multi-GCM Security

Given the results in the previous sections, it is straightforward to prove that GCM does not have bounds which increase as μ increases. We give a brief description of GCM with 96 bit nonces, which is the one used by TLS; a complete description of GCM can be found in the original document [38] or the analysis by Iwata, Ohashi, and Minematsu [31].

Our definitions of confidentiality and integrity extend those of Iwata et al. to the multi-oracle setting. Concretely, we consider the multi-oracle counterpart of the IND-CPA distinguishing game as explained in Sect. 4.1. Translating INT-CTXT into our notation means we arrive at an integrity definition which is nearly identical to Definition 5, with the only modification being the restriction that former outputs by the encryption oracle cannot be used for decryption.

GCM uses a block cipher $E : \mathsf{K} \times \mathsf{X} \rightarrow \mathsf{X}$, where $\mathsf{X} = \{0, 1\}^{128}$, however, using standard arguments, we can focus on GCM using a URP π over X instead. $\mathrm{GCM}[\pi]$ consists of an encryption enc and a decryption algorithm dec where

$$\mathsf{enc} : \mathsf{N} \times \mathsf{H} \times \mathsf{M} \rightarrow \mathsf{C}, \tag{50}$$

$$\mathsf{dec} : \mathsf{N} \times \mathsf{H} \times \mathsf{C} \rightarrow \mathsf{M} \cup \{\bot\}, \tag{51}$$

with N the nonce space, H the associated data, M the plaintexts, C the ciphertexts, and $\perp$ an error symbol.

On input of (n, a, m), enc generates unique inputs to π, $n^0, n^1, \ldots, n^\ell$. The values n^i for $i > 0$ are used to run CTR mode [42] in order to encrypt the plaintext m. The resulting ciphertext c is then used together with the associated data a, and run through a polynomial hash function $h : \mathbf{A} \times \mathbf{C} \to \mathbf{X}$, also called GHASH. GHASH's output is then XORed together with the output of π under n^0 to create a Wegman-Carter-style authenticator. The polynomial hash h uses $L := \pi(0^{128})$ as a key. GCM with 96 bit nonces ensures that every time π is called by the encryption oracle, π receives a different input.

By applying a PRP-PRF switch to GCM, π is replaced with a URF ϕ, and so the confidentiality of GCM can be bounded by the PRP-PRF switch, as illustrated by Iwata et al. [31]. In the multi-oracle setting a multi-PRP-PRF switch can be performed, thereby establishing that multi-GCM's confidentiality is bounded above by the multi-PRP-PRF switch. As shown previously, the multi-PRP-PRF switch is independent of the number of keys, hence multi-GCM's confidentiality bound is independent of the number of keys.

Rather than applying a PRP-PRF switch for integrity, we can apply Bernstein's theorem, as Iwata, Ohashi, and Minematsu did [32, Sect. 7.5 and Appendix C]. As a result, one can show that GCM's integrity can be bounded by the integrity of the Wegman-Carter authenticator using GHASH. This is because π is replaced by a URF ϕ, and the inputs to ϕ used in the underlying CTR mode are always distinct from the inputs to ϕ used in the underlying Wegman-Carter authenticator. Therefore, the underlying Wegman-Carter authenticator becomes independent of the underlying CTR mode, and GCM with ϕ is just an Encrypt-then-MAC [8,41] style authenticated encryption algorithm, meaning its integrity bound is bounded above by the integrity of the underlying MAC.

In the same way, by applying Corollary 1, the integrity of multi-GCM can bounded by that of a multi-Wegman-Carter authenticator. Therefore, establishing that GCM's integrity bound does not degrade in the multi-oracle setting can be done by proving that GHASH with respect to the collision game of Definition 6 is progressive. In the lemma below we do exactly this, although we drop out the padding and input formatting from GHASH since it does not significantly affect the analysis below.

Lemma 3. *Let* X *be a finite field and let* $h : \mathsf{X} \times \mathsf{X}^{\leq \ell} \to \mathsf{X}$ *be the function defined by*

$$h(k, \boldsymbol{x}) = \sum_{i=1}^{q} k^i x_{\ell-i}, \tag{52}$$

where $|\boldsymbol{x}| = q \leq \ell$, *then if* k *is a uniformly distributed random key over* X, $h(k, \cdot)$ *with respect to the collision game* G *of Definition 6 is progressive.*

Proof. Let $\mathbf{A}$ be an adversary playing G, and say that it makes queries

$$(m_1, m_1', t_1), \ldots, (m_q, m_q', t_q), \tag{53}$$

then **A**'s advantage is given by the probability that for some i,

$$h(k, m_i) - h(k, m_i') = t_i . \tag{54}$$

The value $h(k, m_i) - h(k, m_i')$ is a polynomial in k of degree $\max\{|m_i|, |m_i'|\}$, hence Eq. (54) defines a set of keys K_i for which the equation holds. In particular, Eq. (54) holds if and only if $k \in K_i$, therefore **A**'s advantage is the probability that $k \in K_1 \cup \cdots \cup K_q$. A non-winning transcript is a set of inputs for which Eq. (54) does not hold, therefore conditioning on a non-winning transcript of length q' is the same as saying that $k \notin K_1' \cup \cdots \cup K_{q'}'$.

In particular, we can remove some adaptivity from optimal single-oracle adversaries as follows. For each query (m, m', t) which does not result in a collision, the adversary eliminates a set of potential keys, and increases the set B of non-keys, that is, $k \notin B$. Therefore the optimal adversary selects (m, m', t) such that ℓ keys are eliminated for each query (where ℓ is the maximum degree possible of the polynomial). In order to do so, the adversary can just pick elements $r_1, r_2, \ldots, r_{q\ell}$ outside of $\mathsf{X} \setminus B$, reconstruct the polynomials $(\mathbf{k} - r_{(i-1)\ell+1})(\mathbf{k} - r_{(i-1)\ell+2}) \cdots (\mathbf{k} - r_{i\ell})$ for $i = 1, \ldots, q$, where $\mathbf{k}$ is a formal symbol, and from these polynomials reconstruct the corresponding h-queries. In particular, any transcript of length i which is meaningful will eliminate exactly $i \cdot \ell$ keys.

Furthermore, the game is progressive because the longer the transcript given to an optimal adversary, the larger the set of keys which are eliminated, and the greater the chance that a collision occurs.   □

This allows us to conclude that GCM's integrity bound does not exhibit multi-oracle degradation, and as a result, we have the following proposition.

Proposition 4. *The confidentiality and integrity bounds for GCM with 96 bit nonces in the multi-key setting are the same as those in the single-key setting as established by Iwata et al. in [31, Corollary 3] and [31, Sect. 7.5 and Appendix C], respectively.*

5   Future Work

Although we have been able to establish that GCM does not exhibit multi-key degradation, there are still many other widely deployed algorithms for which there are as yet no results. Our approach has been to extract an abstract condition which could be applied to any algorithm and which is sufficient for proving the absence of multi-key security degradation. However the condition seems to be quite strong, and there might be other conditions which exactly capture when an algorithm suffers from multi-key degradation and when it does not, possibly applying to restricted classes of schemes. For example, our condition makes no restriction on whether the algorithm is stateful or stateless, while a condition for stateless algorithms might be simpler, or more powerful. How useful such conditions are remains to be seen, but they would at least fundamentally advance our understanding of the analysis of algorithms, and at best allow us to categorize algorithms according to their multi-key degradation.

Acknowledgments. The authors would like to thank the anonymous reviewers. Furthermore, the authors would also like to thank Tomer Ashur, and Phil Rogaway for their comments, and in particular John Steinberger for pointing out a flaw in a previous version of this manuscript. This work was supported in part by the Research Council KU Leuven: GOA TENSE (GOA/11/007). Atul Luykx is supported by a Fellowship from the Institute for the Promotion of Innovation through Science and Technology in Flanders (IWT-Vlaanderen) and by NSF grants 1314885 and 1717542. Bart Mennink is supported by a postdoctoral fellowship from the Netherlands Organisation for Scientific Research (NWO) under Veni grant 016.Veni.173.017. Kenneth Paterson is supported in part by a research programme funded by Huawei Technologies and delivered through the Institute for Cyber Security Innovation at Royal Holloway, University of London, and in part by EPSRC grant EP/M013472/1.

References

1. Andreeva, E., Daemen, J., Mennink, B., Assche, G.V.: Security of keyed sponge constructions using a modular proof approach. In: Leander [35], pp. 364–384
2. Bader, C., Jager, T., Li, Y., Schäge, S.: On the impossibility of tight cryptographic reductions. In: Fischlin, M., Coron, J.-S. (eds.) EUROCRYPT 2016. LNCS, vol. 9666, pp. 273–304. Springer, Heidelberg (2016). https://doi.org/10.1007/978-3-662-49896-5_10
3. Banik, S., Bogdanov, A., Isobe, T., Shibutani, K., Hiwatari, H., Akishita, T., Regazzoni, F.: Midori: a block cipher for low energy. In: Iwata and Cheon [30], pp. 411–436
4. Bellare, M., Bernstein, D.J., Tessaro, S.: Hash-function based PRFs: AMAC and its multi-user security. In: Fischlin, M., Coron, J.-S. (eds.) EUROCRYPT 2016. LNCS, vol. 9665, pp. 566–595. Springer, Heidelberg (2016). https://doi.org/10.1007/978-3-662-49890-3_22
5. Bellare, M., Boldyreva, A., Micali, S.: Public-key encryption in a multi-user setting: security proofs and improvements. In: Preneel, B. (ed.) EUROCRYPT 2000. LNCS, vol. 1807, pp. 259–274. Springer, Heidelberg (2000). https://doi.org/10.1007/3-540-45539-6_18
6. Bellare, M., Canetti, R., Krawczyk, H.: Pseudorandom functions revisited: the cascade construction and its concrete security. In: FOCS 1996, pp. 514–523. IEEE Computer Society (1996)
7. Bellare, M., Desai, A., Jokipii, E., Rogaway, P.: A concrete security treatment of symmetric encryption. In: FOCS 1997, pp. 394–403. IEEE Computer Society (1997)
8. Bellare, M., Namprempre, C.: Authenticated encryption: relations among notions and analysis of the generic composition paradigm. In: Okamoto, T. (ed.) ASIACRYPT 2000. LNCS, vol. 1976, pp. 531–545. Springer, Heidelberg (2000). https://doi.org/10.1007/3-540-44448-3_41
9. Bellare, M., Rogaway, P.: Entity authentication and key distribution. In: Stinson, D.R. (ed.) CRYPTO 1993. LNCS, vol. 773, pp. 232–249. Springer, Heidelberg (1994). https://doi.org/10.1007/3-540-48329-2_21
10. Bellare, M., Rogaway, P.: Code-based game-playing proofs and the security of triple encryption. Cryptology ePrint Archive, Report 2004/331 (2004)
11. Bellare, M., Tackmann, B.: The multi-user security of authenticated encryption: AES-GCM in TLS 1.3. In: Robshaw and Katz [44], pp. 247–276

12. Bernstein, D.J.: Stronger security bounds for permutations (2005). https://cr.yp.to/antiforgery/permutations-20050323.pdf. Accessed 31 Oct 2017

13. Bernstein, D.J.: Stronger security bounds for Wegman-Carter-Shoup authenticators. In: Cramer, R. (ed.) EUROCRYPT 2005. LNCS, vol. 3494, pp. 164–180. Springer, Heidelberg (2005). https://doi.org/10.1007/11426639_10

14. Biham, E.: How to decrypt or even substitute DES-encrypted messages in 2^{28} steps. Inf. Process. Lett. **84**(3), 117–124 (2002)

15. Biryukov, A., Mukhopadhyay, S., Sarkar, P.: Improved time-memory trade-offs with multiple data. In: Preneel, B., Tavares, S. (eds.) SAC 2005. LNCS, vol. 3897, pp. 110–127. Springer, Heidelberg (2006). https://doi.org/10.1007/11693383_8

16. Blake-Wilson, S., Johnson, D., Menezes, A.: Key agreement protocols and their security analysis. In: Darnell, M. (ed.) Cryptography and Coding 1997. LNCS, vol. 1355, pp. 30–45. Springer, Heidelberg (1997). https://doi.org/10.1007/BFb0024447

17. Bogdanov, A., Chang, D., Ghosh, M., Sanadhya, S.K.: Bicliques with minimal data and time complexity for AES. In: Lee, J., Kim, J. (eds.) ICISC 2014. LNCS, vol. 8949, pp. 160–174. Springer, Cham (2015). https://doi.org/10.1007/978-3-319-15943-0_10

18. Bogdanov, A., Khovratovich, D., Rechberger, C.: Biclique cryptanalysis of the full AES. In: Lee, D.H., Wang, X. (eds.) ASIACRYPT 2011. LNCS, vol. 7073, pp. 344–371. Springer, Heidelberg (2011). https://doi.org/10.1007/978-3-642-25385-0_19

19. Chang, D., Nandi, M.: A short proof of the PRP/PRF switching lemma. Cryptology ePrint Archive, Report 2008/078 (2008)

20. Chatterjee, S., Koblitz, N., Menezes, A., Sarkar, P.: Another look at tightness II: practical issues in cryptography. Cryptology ePrint Archive, Report 2016/360 (2016)

21. Chatterjee, S., Menezes, A., Sarkar, P.: Another look at tightness. In: Miri, A., Vaudenay, S. (eds.) SAC 2011. LNCS, vol. 7118, pp. 293–319. Springer, Heidelberg (2012). https://doi.org/10.1007/978-3-642-28496-0_18

22. Daemen, J., Rijmen, V.: The Design of Rijndael: AES - The Advanced Encryption Standard. Information Security and Cryptography. Springer, Heidelberg (2002). https://doi.org/10.1007/978-3-662-04722-4

23. Demay, G., Gaži, P., Maurer, U., Tackmann, B.: Optimality of non-adaptive strategies: The case of parallel games. In: IEEE International Symposium on Information Theory, pp. 1707–1711. IEEE (2014)

24. Fouque, P.-A., Joux, A., Mavromati, C.: Multi-user collisions: applications to discrete logarithm, Even-Mansour and PRINCE. In: Sarkar, P., Iwata, T. (eds.) ASIACRYPT 2014. LNCS, vol. 8873, pp. 420–438. Springer, Heidelberg (2014). https://doi.org/10.1007/978-3-662-45611-8_22

25. Gaži, P., Maurer, U.: Free-start distinguishing: combining two types of indistinguishability amplification. In: Kurosawa, K. (ed.) ICITS 2009. LNCS, vol. 5973, pp. 28–44. Springer, Heidelberg (2010). https://doi.org/10.1007/978-3-642-14496-7_4

26. Guo, J., Jean, J., Nikolić, I., Qiao, K., Sasaki, Y., Sim, S.M.: Invariant subspace attack against Midori64 and the resistance criteria for S-box designs. IACR Trans. Symmetric Cryptol. **1**(1), 19 (2017)

27. Guo, Z., Wu, W., Liu, R., Zhang, L.: Multi-key analysis of tweakable even-mansour with applications to minalpher and OPP. IACR Trans. Symmetric Cryptol. **1**, 19 (2017)

28. Hoang, V.T., Tessaro, S.: Key-alternating ciphers and key-length extension: exact bounds and multi-user security. In: Robshaw, M., Katz, J. [45], pp. 3–32

29. Hoang, V.T., Tessaro, S.: The multi-user security of double encryption. In: Coron, J.-S., Nielsen, J.B. (eds.) EUROCRYPT 2017. LNCS, vol. 10211, pp. 381–411. Springer, Cham (2017). https://doi.org/10.1007/978-3-319-56614-6_13

30. Iwata, T., Cheon, J.H. (eds.): ASIACRYPT 2015. LNCS, vol. 9453. Springer, Heidelberg (2015). https://doi.org/10.1007/978-3-662-48800-3

31. Iwata, T., Ohashi, K., Minematsu, K.: Breaking and repairing GCM security proofs. In: Safavi-Naini, R., Canetti, R. (eds.) CRYPTO 2012. LNCS, vol. 7417, pp. 31–49. Springer, Heidelberg (2012). https://doi.org/10.1007/978-3-642-32009-5_3

32. Iwata, T., Ohashi, K., Minematsu, K.: Breaking and repairing GCM security proofs. IACR Cryptology ePrint Archive 2012/438 (2012)

33. Krawczyk, H.: LFSR-based hashing and authentication. In: Desmedt, Y.G. (ed.) CRYPTO 1994. LNCS, vol. 839, pp. 129–139. Springer, Heidelberg (1994). https://doi.org/10.1007/3-540-48658-5_15

34. Leander, G. (ed.): FSE 2015. LNCS, vol. 9054. Springer, Heidelberg (2015). https://doi.org/10.1007/978-3-662-48116-5

35. Maurer, U.: Indistinguishability of random systems. In: Knudsen, L.R. (ed.) EUROCRYPT 2002. LNCS, vol. 2332, pp. 110–132. Springer, Heidelberg (2002). https://doi.org/10.1007/3-540-46035-7_8

36. Maurer, U.: Conditional equivalence of random systems and indistinguishability proofs. In: IEEE International Symposium on Information Theory, pp. 3150–3154. IEEE (2013)

37. Maurer, U., Pietrzak, K., Renner, R.: Indistinguishability amplification. In: Menezes, A. (ed.) CRYPTO 2007. LNCS, vol. 4622, pp. 130–149. Springer, Heidelberg (2007). https://doi.org/10.1007/978-3-540-74143-5_8

38. McGrew, D.A., Viega, J.: The security and performance of the Galois/Counter Mode (GCM) of operation. In: Canteaut, A., Viswanathan, K. (eds.) INDOCRYPT 2004. LNCS, vol. 3348, pp. 343–355. Springer, Heidelberg (2004). https://doi.org/10.1007/978-3-540-30556-9_27

39. Menezes, A., Smart, N.P.: Security of signature schemes in a multi-user setting. Des. Codes Cryptography **33**(3), 261–274 (2004)

40. Mouha, N., Luykx, A.: Multi-key security: the even-mansour construction revisited. In: Gennaro, R., Robshaw, M. (eds.) CRYPTO 2015. LNCS, vol. 9215, pp. 209–223. Springer, Heidelberg (2015). https://doi.org/10.1007/978-3-662-47989-6_10

41. Namprempre, C., Rogaway, P., Shrimpton, T.: Reconsidering generic composition. In: Nguyen, P.Q., Oswald, E. (eds.) EUROCRYPT 2014. LNCS, vol. 8441, pp. 257–274. Springer, Heidelberg (2014). https://doi.org/10.1007/978-3-642-55220-5_15

42. National Institute of Standards and Technology: DES Modes of Operation. FIPS 1981, December 1980

43. Niwa, Y., Ohashi, K., Minematsu, K., Iwata, T.: GCM security bounds reconsidered. In: Leander [35], pp. 385–407

44. Robshaw, M., Katz, J. (eds.): CRYPTO 2016. LNCS, vol. 9814. Springer, Heidelberg (2016). https://doi.org/10.1007/978-3-662-53018-4

45. Rogaway, P.: Bucket hashing and its application to fast message authentication. J. Cryptol. **12**(2), 91–115 (1999)

46. Rogaway, P.: Authenticated-encryption with associated-data. In: Atluri, V. (ed.) ACM CCS 2002, pp. 98–107. ACM (2002)

47. Shrimpton, T., Terashima, R.S.: Salvaging weak security bounds for blockcipher-based constructions. In: Cheon, J.H., Takagi, T. (eds.) ASIACRYPT 2016. LNCS, vol. 10031, pp. 429–454. Springer, Heidelberg (2016). https://doi.org/10.1007/978-3-662-53887-6_16

48. Tao, B., Wu, H.: Improving the biclique cryptanalysis of AES. In: Foo, E., Stebila, D. (eds.) ACISP 2015. LNCS, vol. 9144, pp. 39–56. Springer, Cham (2015). https://doi.org/10.1007/978-3-319-19962-7_3

49. Tessaro, S.: Optimally secure block ciphers from ideal primitives. In: Iwata, T., Cheon, J.H. (eds.) [31], pp. 437–462

50. Wegman, M.N., Carter, L.: New hash functions and their use in authentication and set equality. J. Comput. Syst. Sci. **22**(3), 265–279 (1981)

51. Zaverucha, G.: Hybrid encryption in the multi-user setting. Cryptology ePrint Archive, Report 2012/159 (2012)

52. Zhang, P., Hu, H.: On the provable security of the tweakable even-mansour cipher against multi-key and related-key attacks. Cryptology ePrint Archive, Report 2016/1172 (2016)

Full-State Keyed Duplex
with Built-In Multi-user Support

Joan Daemen[1,2(✉)], Bart Mennink[1,3], and Gilles Van Assche[2]

[1] Digital Security Group, Radboud University, Nijmegen, The Netherlands
{joan,b.mennink}@cs.ru.nl
[2] STMicroelectronics, Diegem, Belgium
gilles.vanassche@st.com
[3] CWI, Amsterdam, The Netherlands

Abstract. The keyed duplex construction was introduced by Bertoni et al. (SAC 2011) and recently generalized to full-state absorption by Mennink et al. (ASIACRYPT 2015). We present a generalization of the full-state keyed duplex that natively supports multiple instances by design, and perform a security analysis that improves over that of Mennink et al. in terms of a more modular security analysis and a stronger and more adaptive security bound. Via the introduction of an additional parameter to the analysis, our bound demonstrates a significant security improvement in case of nonce-respecting adversaries. Furthermore, by supporting multiple instances by design, instead of adapting the security model to it, we manage to derive a security bound that is largely independent of the number of instances.

Keywords: Duplex construction · Full-state · Distinguishing bounds · Authenticated encryption

1 Introduction

Bertoni et al. [8] introduced the sponge construction as an approach to design hash functions with variable output length (later called extendable output functions (XOF)). The construction faced rapid traction in light of NIST's SHA-3 competition, with multiple candidates inspired by the sponge methodology. Keccak, the eventual winner of the competition and now standardized as SHA-3 [27], internally uses the sponge construction. The sponge construction found quick adoption in the area of lightweight hashing [19,32]. Also beyond the area of hash functions various applications of the sponge construction appeared such as keystream generation and MAC computation [12], reseedable pseudorandom sequence generation [10,30], and authenticated encryption [11,14]. In particular, the ongoing CAESAR competition for the development of a portfolio of authenticated encryption schemes has received about a dozen sponge-based submissions.

At a high level, the sponge construction operates on a state of b bits. This is split into an inner part of size c bits and an outer part of size r bits,

© International Association for Cryptologic Research 2017
T. Takagi and T. Peyrin (Eds.): ASIACRYPT 2017, Part II, LNCS 10625, pp. 606–637, 2017.
https://doi.org/10.1007/978-3-319-70697-9_21

where $b = c + r$. Data absorption and squeezing is done via the outer part, r bits at a time, interleaved with evaluations of a b-bit permutation f. Bertoni et al. [9] proved a bound on the security of the sponge construction in the indifferentiability framework of Maurer et al. [37]. While it was clear from the start that this birthday-type bound in the capacity is tight for the unkeyed use cases, i.e., hashing, for the keyed use cases of the sponge it appeared that a higher level of security could be achieved. This has resulted in an impressive series of papers on the generic security of keyed versions of the sponge, with bounds improving and the construction becoming more efficient.

1.1   Keyed Sponge and Keyed Duplex

Keyed Sponge. Bertoni et al.'s original keyed sponge [13] was simply the sponge with input $(K\|M)$ where K is the key. Chang et al. [21] suggested an alternative where the initial state of the sponge simply contains the key in its inner part. Andreeva et al. [2] generalized and improved the analyses of both the outer- and inner-keyed sponge, and also considered security of these functions in the multi-target setting. In a recent analysis their bounds were improved by Naito and Yasuda in [42]. All of these results, however, stayed relatively close to the (keyless) sponge design that absorbs input in blocks of r bits in the outer part of the state. It turned out that, thanks to the secrecy of part of the state after key injection, one can absorb data over the full state, and therewith achieve maximal compression. Full-state absorbing was first explicitly proposed in a variant of sponge for computing MACs: donkeySponge [14]. It also found application in various recent sponge-inspired designs, such as Chaskey [41].

Nearly tight bounds for the full-state absorbing keyed sponge were given by Gaži et al. [29] but their analysis was limited to the case of fixed output length. Mennink et al. [38] generalized and formalized the idea of the full-state keyed sponge and presented an improved security bound for the general case where the output length is variable.

Keyed Duplex. Whereas the keyed sponge serves message authentication and stream encryption, authenticated encryption is mostly done via the keyed duplex construction [11]. This is a stateful construction that consists of an initialization interface and a duplexing interface. Initialization resets the state and a duplexing call absorbs a data block of at most $r - 1$ bits, applies the underlying permutation f and squeezes at most r bits. Bertoni et al. [11] proved that the output of duplexing calls can be simulated by calls to a sponge, a fortiori making duplex as strong as sponge.

Mennink et al. [38] introduced the full-state keyed duplex and derived a security bound on this construction with dominating terms:

$$\frac{\mu N}{2^k} + \frac{M^2}{2^c}. \tag{1}$$

Here M is the data complexity (total number of initialization and duplexing calls), N the computational complexity (total number of offline calls to f),

$\mu \leq 2M$ is a term called the "multiplicity," and k the size of the key. This security bound was derived by describing the full-state keyed duplex in terms of the full-state keyed sponge. A naive bounding of μ (to cover the strongest possible adversary) yields a dominating term of the form $2MN/2^k$, implying a security strength erosion of $\log_2 M$ with respect to exhaustive key search.

The duplex construction finds multiple uses in the CAESAR competition [20] in the embodiment of the authenticated encryption mode SpongeWrap [11] or close variants of it. The recent line of research on improving bounds of sponge-inspired authenticated encryption schemes, set by Jovanovic et al. [35], Sasaki and Yasuda [46], and Reyhanitabar et al. [44], can be seen as an analysis of a specific use case of the keyed duplex. The Full-State SpongeWrap [38], an authenticated encryption scheme designed from the full-state keyed duplex, improves over these results. Particularly, the idea already found application in the Motorist mode of the CAESAR submission Keyak [16].

Trading Sponge for Duplex. As said, the duplex can be simulated by the sponge, but not the other way around. This is the case because duplex pads each input block and cannot simulate sponge inputs with, e.g., long sequences of zeroes. It is therefore natural that Mennink et al. [38] derived a security bound on the full-state keyed duplex *by viewing it as an extension to the full-state keyed sponge*. However, we observe that the introduction of full-state absorption changes that situation: the full-state keyed duplex can simulate the full-state keyed sponge. All keyed usages of the sponge can be described quite naturally as application of the keyed duplex and it turns out that proving security of keyed duplex is easier than that of keyed sponge. Therefore, in keyed use cases, the duplex is preferred as basic building block over the sponge.

1.2   Multi-target Security

The problem of multi-target security of cryptographic designs has been acknowledged and analyzed for years. Biham [17] considered the security of blockciphers in the multi-target setting and shows that the security strength can erode to half the key size if data processed by sufficiently many keys is available. Various extensions have subsequently appeared [7,18,34]. It has been demonstrated (see, e.g., [5] for public key encryption and [22] for message authentication codes) that the security of a scheme in the multi-target setting can be reduced to the security in the single-target setting, at a security loss proportional to the number of keys used.

However, in certain cases, a dedicated analysis in the multi-target setting could render improved bounds. Andreeva et al. [2] considered the security of the outer- and inner-keyed sponge in the multi-target setting, a proof which internally featured a security analysis of the Even-Mansour blockcipher in the multi-target setting. The direction of multi-target security got subsequently popularized by Mouha and Luykx [40], leading to various multi-target security results [4,33] with security bounds (almost) independent of the number of targets involved.

1.3   Our Contribution

We present a generalization of the full-state keyed duplex that facilitates multiple instances by design (Sect. 2.2). This generalization is realized via the formalization of a state initialization function that has access to a key array $\mathbf{K}$ consisting of u keys of size k, generated following a certain distribution. Given as input a key index δ and an initialization vector iv, it initializes the state using iv and the δth key taken from $\mathbf{K}$. We capture its functional behavior under the name of an *extendable input function* (XIF) and explicitly define its idealized instance.

Unlike the approach of Mennink et al. [38], who viewed the full-state keyed duplex as an extension to the full-state keyed sponge, our analysis is a dedicated analysis of the full-state keyed duplex. To accommodate bounds for different use cases, we have applied a re-phasing to the definition of the keyed duplex. In former definitions of the (keyed) duplex, a duplexing call consisted of input injection, applying the permutation f, and then output extraction. In our new definition, the processing is as follows: first the permutation f, then output extraction, and finally input injection. This adjustment reflects a property present in practically all modes based on the keyed duplex, namely that the user (or adversary) must provide the input before knowing the previous output. The re-phasing allows us to prove a bound on keyed duplex that is tight even for those use cases. The fact that, in previous definitions, an adversary could see the output before providing the input allowed it to force the outer part of the state to a value of its choice, and as such gave rise to a term in the bound at worst $MN/2^c$ and at best $\mu N/2^c$, where μ is a term that reflects a property of the transcript that needs to be bound by out-of-band reasonings.

Alongside the re-phasing, we have eliminated the term μ and express the bound purely as a function of the adversary's capabilities. Next to the total offline complexity N, i.e., the number queries the adversary can make to f and the total online complexity M, i.e., the total number of construction queries (to keyed duplex or ideal XIF), we introduce two metrics: L and Ω, both reflecting the ability of the adversary to force the outer part of the state to a value of its choice. The metric L counts the number of construction queries with repeated path (intuitively, a concatenation of all data blocks up to a certain permutation call), which may typically occur in MAC functions and authenticated encryption schemes that do not impose nonce uniqueness. The metric Ω counts the number of construction queries where the adversary can overwrite the outer part of the state. Such a possibility may occur in authenticated encryption schemes that release unverified decrypted ciphertext (cf., [1]). A comparison of the scheme analyzed in this work with those in earlier works is given in Table 1.

We prove in Sect. 4 a bound on the advantage of distinguishing a full-state keyed duplex from an ideal XIF in a multi-target setting. We here give the bound for several settings, all of which having multiple keys sampled uniformly at random without replacement. For adversaries with the ability to send queries with repeated paths and queries that overwrite the outer part of the state, the

Table 1. Comparison of the schemes analyzed in earlier works and this work. By "pure bound" we mean that the derived security bound is expressed purely as a function of the adversary's capabilities. Differences in bounds are not reflected by the table.

	Full state absorption	Extendable output	Multi-target	Pure bound
Bertoni et al. [13]	—	✓	—	✓
Bertoni et al. [11]	—	✓	—	✓
Chang et al. [21]	—	✓	—	✓
Andreeva et al. [2]	—	✓	✓	—
Gaži et al. [29]	✓	—	—	✓
Mennink et al. [38]	✓	✓	—	—
Naito and Yasuda [42]	—	✓	—	✓
This work	✓	✓	✓	✓

dominating terms in our bound are:

$$\frac{q_{\mathrm{iv}}N}{2^k} + \frac{(L+\Omega)N}{2^c}. \tag{2}$$

The metric q_{iv} denotes the maximum number of sessions started with the same iv but different keys. For adversaries that cannot send queries with repeated paths or send queries that overwrite the outer part of the state, one of the dominating terms depends on the occurrence of multicollisions via a coefficient $\nu_{r,c}^M$ that is fully determined by the data complexity M and parameters r and c (see Sect. 6.5, and particularly Fig. 4). For wide permutations we can have large rates (i.e., $r > 2\log_2(M) + c$) and the dominating terms in our bound are:

$$\frac{q_{\mathrm{iv}}N}{2^k} + \frac{N}{2^{c-1}}. \tag{3}$$

For relatively small rates the data complexity can be such that $M > 2^{r-1}$ and for that range the dominating terms are upper bounded by (assuming $\nu_{r,c}^{2M} \le \frac{bM}{2^{r+1}}$):

$$\frac{q_{\mathrm{iv}}N}{2^k} + \frac{bMN}{2^b} + \frac{M^2}{2^b}. \tag{4}$$

For the case in-between where M is in the range $2^{(r-c)/2} < M \le 2^{r-1}$, the bound becomes (assuming $\nu_{r,c}^{2M} \le \min(b/\log\frac{2^r}{2M}, b/4)$):

$$\frac{q_{\mathrm{iv}}N}{2^k} + \frac{bN}{\max(4, r-1-\log_2 M)2^{c-1}}. \tag{5}$$

This bound is valid for permutation widths of 200 and above. These bounds are significantly better than that of [38].

Concretely, in implementations of duplex-based authenticated encryption schemes that respect the nonce requirement and do not release unverified plaintext, we have $L = \Omega = 0$. Assuming keys are randomly sampled without replacement, the generic security is governed by (3), (4), or (5). Depending on the

Table 2. Application of our analysis to Ketje, Ascon, NORX, and Keyak. For the nonce misuse case, we consider $L + \Omega = M/2$. A "Strength" equal to s means that it requires a computational effort 2^s to distinguish. Here, $a = \log_2(Mr)$.

Scheme		Parameters			Respecting (Eqs. (3)–(5))		Misuse (Eq. (2))
		b	c	r	Strength	Equation	Strength
Ketje [15]	Jr	200	184	16	$\min\{196 - a, 177\}$	(4), (5)	$189 - a$
	Sr	400	368	32	$\min\{396 - a, 360\}$	(4), (5)	$374 - a$
Ascon [24]	128	320	256	64	$\min\{317 - a, 248\}$	(4), (5)	$263 - a$
	128a	320	192	128	$\min\{318 - a, 184\}$	(4), (5)	$200 - a$
NORX [3]	32	512	128	384	127	(3)	$137 - a$
	64	1024	256	768	255	(3)	$266 - a$
Keyak [16]	River	800	256	544	255	(3)	$266 - a$
	Lake	1600	256	1344	255	(3)	$267 - a$

parameters, a scheme is either in case (3), or case (4)–(5), where a transition happens for $M = 2^{r-1}$. Table 2 summarizes achievable security strengths for the duplex-based CAESAR contenders.

Our general security bound, covering among others a broader spectrum of key sampling distributions, is given in Theorem 1. It is noteworthy that, via the built-in support of multiple targets, we manage to obtain a security bound that is largely independent of the number of users u: the only appearance is in the key guessing part, $q_{iv}N/2^k$, which shows an expected decrease in the security strength of exhaustive key search by a term $\log_2 q_{iv}$. Note that security erosion can be avoided altogether by requiring iv to be a global nonce, different for each initialization call (irrespective of the used key).

Our analysis improves over the one of [38] in multiple aspects. First, our security bound shows less security erosion for increasing data complexities. Whereas in (1) security strength erodes to $k - \log_2 M$, in (2) this is $c - \log_2(L + \Omega)$ with $L + \Omega < M$. By taking $c > k + \log_2 M_{\max}$ with $M_{\max}$ some upper bound on the amount of data an adversary can get its hands on, one can guarantee that this term does not allow attacks faster than exhaustive key search.

Second, via the use of parameters L and Ω our bound allows for a more flexible interpretation and a wide range of use cases. For example, in stream ciphers, $L = \Omega = 0$ by design. This also holds for most duplex-based authenticated encryption schemes in the case of nonce-respecting adversaries that cannot obtain unverified decrypted ciphertexts.

Third, even in the general case (with key size taken equal to c bits and no nonce restriction on iv), our bound still improves over the one of [38] by replacing the multiplicity metric, that can only be evaluated a posteriori, by the metrics L and Ω, that reflect what the adversary can do.

Fourth, in our approach we address the multi-key aspect natively. This allows us to determine the required set of properties on the joint distribution of all keys under attack. Theorem 1 works for arbitrary key sampling techniques with individual keys of sufficient min-entropy and the probability that two keys in the array collide is small enough, and demonstrates that the full-state keyed duplex remains secure even if the keys are not independently and uniformly randomly distributed.

Finally, we perform an analysis on the contribution of outer-state multi-collisions to the bound that is of independent interest. This analysis strongly contributes to the tightness of our bounds, as we illustrate in the Stairway to Heaven graph in Fig. 4.

1.4   Notation

For an integer $n \in \mathbb{N}$, we denote $\mathbb{Z}_n = \{0, \ldots, n-1\}$ and by $\mathbb{Z}_2^n$ the set of bit strings of length n. $\mathbb{Z}_2^*$ denotes the set of bit strings of arbitrary length. For two bit strings $s, t \in \mathbb{Z}_2^*$, their bitwise addition is denoted $s + t$. The expression $\lfloor s \rfloor_\ell$ denotes the bitstring s truncated to its first ℓ bits. A random oracle [6] $\mathcal{RO}$: $\mathbb{Z}_2^* \to \mathbb{Z}_2^n$ is a function that maps bit strings of arbitrary length to bit strings of some length n. In this paper, the value of n is determined by the context. We denote by $(x)_{(y)}$ the falling factorial power $(x)_{(y)} = x(x-1)\cdots(x-y+1)$.

Throughout this work, b denotes the width of the permutation f. The parameters c and r denote the capacity and rate, where $b = c + r$. For a state value $s \in \mathbb{Z}_2^b$, we follow the general convention to define its outer part by $\overline{s} \in \mathbb{Z}_2^r$ and its inner part by $\widehat{s} \in \mathbb{Z}_2^c$, in such a way that $s = \overline{s} \| \widehat{s}$. The key size is conventionally denoted by k, and the number of users by u. Throughout, we assume that $u \leq 2^k$, and regularly use an encoding function $\mathsf{Encode} : \mathbb{Z}_u \to \mathbb{Z}_2^k$, mapping integers from $\mathbb{Z}_u$ to k-bit strings in some injective way.

2   Constructions

In Sect. 2.1, we will discuss the key sampling technique used in this work. The keyed duplex construction is introduced in Sect. 2.2, and we present its "ideal equivalent," the ideal extendable input function, in Sect. 2.3. To suit the security analysis, we will also need an in-between hybrid, the randomized duplex, discussed in Sect. 2.4.

2.1   Key Sampling

Our keyed duplex construction has built-in multi-user support, and we start with a formalization of the key sampling that we consider. At a high level, our formalization is not specific for the keyed duplex, and may be of independent interest for modeling multi-target attacks.

In our formalization the adversary can invoke a keyed object (block cipher, stream cipher, PRF, keyed sponge, ...) with a key selected from a key array $\mathbf{K}$ containing u keys, each of length k bits:

$$\mathbf{K} = (\mathbf{K}[0], \ldots, \mathbf{K}[u-1]) \in \left(\mathbb{Z}_2^k\right)^u.$$

These keys are sampled from the space of k-bit keys according to some distribution $\mathcal{D}_K$. This distribution can, in theory, be anything. In particular, the distribution of the key with index δ may depend on the values of the δ keys sampled before.

Two plausible examples of the key distribution are *random sampling with replacement* and *random sampling without replacement*. In the former case, all keys are generated uniformly at random and pairwise independent, but it may cause problems in case of accidental collisions in the key array. The latter distribution resolves this by generating all keys uniformly at random from the space of values excluding the ones already sampled. A third, more extreme, example of $\mathcal{D}_K$ generates $\mathbf{K}[0]$ uniformly at random and defines all subsequent keys as $\mathbf{K}[\delta] = \mathbf{K}[0] + \mathsf{Encode}(\delta)$.

Different distributions naturally entail different levels of security, and we define two characteristics of a distribution that are relevant for our analysis. Note that the characteristics take u as implicit parameter. The first characteristic is the min-entropy of the individual keys, defined as

$$H_{\min}(\mathcal{D}_K) = -\log_2 \max_{\delta \in \mathbb{Z}_u, a \in \mathbb{Z}_2^k} \Pr(\mathbf{K}[\delta] = a), \qquad (6)$$

or in words, minus the binary logarithm of the probability of the key value to be selected with the highest probability. The three example samplings outlined above have min-entropy k, regardless of the value u.

The second characteristic is related to the maximum collision probability between two keys in the array:

$$H_{\mathrm{coll}}(\mathcal{D}_K) = -\log_2 \max_{\substack{\delta, \delta' \in \mathbb{Z}_u \\ \delta \neq \delta'}} \Pr(\mathbf{K}[\delta] = \mathbf{K}[\delta']). \qquad (7)$$

Uniform sampling with replacement has maximum collision probability equal to 2^{-k} and so $H_{\mathrm{coll}}(\mathcal{D}_K) = k$. Sampling without replacement and our third example clearly have collision probability zero, giving $H_{\mathrm{coll}}(\mathcal{D}_K) = \infty$.

2.2 Keyed Duplex Construction

The full-state keyed duplex (KD) construction is defined in Algorithm 1, and it is illustrated in Fig. 1.

It calls a b-bit permutation f and is given access to an array $\mathbf{K}$ consisting of u keys of size k bits. A user can make two calls: initialization and duplexing calls.

In an initialization call it takes as input a key index δ and a string iv $\in \mathbb{Z}_2^{b-k}$ and initializes the state as $f(\mathbf{K}[\delta]\|\mathrm{iv})$. In the same call, the user receives an r-bit output string Z and injects a b-bit string σ. A duplexing call just performs the latter part: it updates the state by applying f to it, returns to the user an r-bit output string Z and injects a user-provided b-bit string σ.

Both in initialization and duplexing calls, the output string Z is taken from the state prior to the addition of σ to it, but the user has to provide σ before

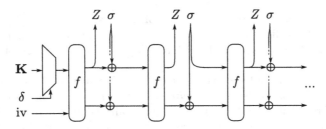

Fig. 1. Full-state keyed duplex construction KD_K^f. In this figure, the sequence of calls is Z = KD.Init(δ, iv, σ, false), Z = KD.Duplexing(σ, true), and Z = KD.Duplexing(σ, false).

receiving Z. This is in fact a re-phasing compared to the original definition of the duplex [11] or of the full-state keyed duplex [38], and it aims at better reflecting typical use cases. We illustrate this with the SPONGEWRAP authenticated encryption scheme [11] and its more recent variants [38]. In this scheme, each plaintext block is typically encrypted by (i) applying f, (ii) fetching a block of key stream, (iii) adding the key stream and plaintext blocks to get a ciphertext block, and (iv) adding the plaintext block to the outer part of the state. By inspecting Algorithm 3 in [11], there is systematically a delay between the production of key stream and its use, requiring to buffer a key stream block between the (original) duplexing calls. In contrast, our re-phased calls better match the sequence of operations.

The flag in the initialization and duplexing calls is required to implement decryption in SPONGEWRAP and variants. In that case, the sequence of operations is the same as above, except that step (iii) consists of adding the key stream and ciphertext blocks to get a plaintext block. However, a user would need to see the keystream block before being able to add the plaintext block in step (iv). One can see, however, that step (iv) is equivalent to overwriting the outer part of the state with the ciphertext block. Switching between adding the plaintext block (for encryption) and overwriting with the ciphertext block (for decryption) is the purpose of the flag. The usage of the flag, alongside the re-phasing is depicted in Fig. 1.

Note that in Algorithm 1 in the case that the flag is true, the outer part of the state is overwritten with $\overline{\sigma}$. For consistency with the algorithms of constructions we will introduce shortly, this is formalized as bitwise adding Z to $\overline{\sigma}$ before its addition to the state if flag is true. Alternatively, one could define an authenticated encryption mode that does not allow overwriting the state with the ciphertext block C. For example, encryption would be $C = P + \mathbf{M} \times Z$, with P the plaintext block and $\mathbf{M}$ a simple invertible matrix. Upon decryption, the outer part of the state then becomes $C + (\mathbf{M} + \mathbf{I}) \times Z$. If $\mathbf{M}$ is chosen such that $\mathbf{M} + \mathbf{I}$ is invertible, the adversary has no control over the outer part of the state after the duplexing call. This would require changing "$\overline{\sigma} \leftarrow \overline{\sigma} + Z$" into "$\overline{\sigma} \leftarrow \overline{\sigma} + \mathbf{M} \times Z$" in Algorithm 1.

Algorithm 1. Full-state keyed duplex construction $\mathrm{KD}_{\mathbf{K}}^{f}$

Require: $r < b,\ k \leq b$

Instantiation: $\mathrm{KD} \leftarrow \mathrm{KD}_{\mathbf{K}}^{f}$ with $\mathbf{K}$ an array of u keys of size k

    Key array: $\mathrm{KD}.\mathbf{K} \xleftarrow{\mathcal{D}_{\mathrm{K}}} \mathbf{K}$

Interface: $Z = \mathrm{KD}.\mathsf{Init}(\delta, \mathrm{iv}, \sigma, \mathsf{flag})$ with $\delta \in \mathbb{Z}_u$, $\mathrm{iv} \in \mathbb{Z}_2^{b-k}$, $\sigma \in \mathbb{Z}_2^b$, $\mathsf{flag} \in$
    $\{\mathrm{true, false}\}$, and $Z \in \mathbb{Z}_2^r$
    $s \leftarrow f(\mathbf{K}[\delta] \| \mathrm{iv})$
    $Z \leftarrow \lfloor s \rfloor_r$
    if $\mathsf{flag} = \mathrm{true}$ **then** $\overline{\sigma} \leftarrow \overline{\sigma} + Z$
    $s \leftarrow s + \sigma$
    return Z

Interface: $Z = \mathrm{KD}.\mathsf{Duplexing}(\sigma, \mathsf{flag})$ with $\sigma \in \mathbb{Z}_2^b$, $\mathsf{flag} \in \{\mathrm{true, false}\}$, and $Z \in \mathbb{Z}_2^r$
    $s \leftarrow f(s)$
    $Z \leftarrow \lfloor s \rfloor_r$
    if $\mathsf{flag} = \mathrm{true}$ **then** $\overline{\sigma} \leftarrow \overline{\sigma} + Z$
    $s \leftarrow s + \sigma$
    return Z

2.3 Ideal Extendable Input Function

We define an ideal extendable input function (IXIF) in Algorithm 2. It has the same interface as KD, but instead it uses a random oracle $\mathcal{RO} : \mathbb{Z}_2^* \rightarrow \mathbb{Z}_2^r$ to generate its responses. In particular, every initialization call initializes a Path as $\mathsf{Encode}(\delta) \| \mathrm{iv}$. In both initialization and duplexing calls, an r-bit output is generated by evaluating $\mathcal{RO}(\mathsf{Path})$ and the b-bit input string σ is absorbed by appending it to the Path. Figure 2 has an illustration of IXIF (at the right).

Note that IXIF properly captures the random equivalent of the full-state keyed duplex: it simply returns random values from $\mathbb{Z}_2^r$ for every new path, and repeated paths result in identical responses. IXIF is in fact almost equivalent to the duplex as presented by Mennink et al. [38]: as a matter of fact, when (i) not considering multiple keys for our construction and (ii) avoiding overlap of the iv with the key (as possible in the construction of [38]), the ideal functionalities are the same. In our analysis, we do not consider overlap of the iv with the key as (i) it unnecessarily complicates the analysis and (ii) we discourage it as it may be a security risk if the keys in the key array $\mathbf{K}$ are not independently and uniformly randomly distributed.

2.4 Randomized Duplex Construction

To simplify our security analysis, we introduce a hybrid algorithm lying in-between KD and IXIF: the full-state randomized duplex (RD) construction. It is defined in Algorithm 3. It again has the same interface as KD, but the calls to the permutation f and the access to a key array $\mathbf{K}$ have been replaced by

---

Algorithm 2. Ideal extendable input function $\text{IXIF}^{\mathcal{RO}}$

Instantiation: $\text{IXIF} \leftarrow \text{IXIF}^{\mathcal{RO}}$
  Path: IXIF.Path $\leftarrow$ empty string

Interface: $Z = \text{IXIF.Init}(\delta, \text{iv}, \sigma, \text{flag})$ with $\delta \in \mathbb{Z}_u$, $\text{iv} \in \mathbb{Z}_2^{b-k}$, $\sigma \in \mathbb{Z}_2^b$, flag $\in$
  $\{\text{true}, \text{false}\}$, and $Z \in \mathbb{Z}_2^r$
  Path $\leftarrow$ Encode$(\delta)\|\text{iv}$
  $Z \leftarrow \mathcal{RO}(\text{Path})$
  if flag = true **then** $\overline{\sigma} \leftarrow \overline{\sigma} + Z$
  Path $\leftarrow$ Path$\|\sigma$
  return Z

Interface: $Z = \text{IXIF.Duplexing}(\sigma, \text{flag})$ with $\sigma \in \mathbb{Z}_2^b$ flag $\in \{\text{true}, \text{false}\}$, and $Z \in \mathbb{Z}_2^r$
  $Z \leftarrow \mathcal{RO}(\text{Path})$
  if flag = true **then** $\overline{\sigma} \leftarrow \overline{\sigma} + Z$
  Path $\leftarrow$ Path$\|\sigma$
  return Z

---

two primitives: a uniformly random injective mapping $\phi : \mathbb{Z}_u \times \mathbb{Z}_2^{b-k} \to \mathbb{Z}_2^b$, and a uniformly random b-bit permutation π. The injective mapping ϕ replaces the keyed state initialization by directly mapping an input (δ, iv) to a b-bit state value. The permutation π replaces the evaluations of f in the duplexing calls. In our use of RD, ϕ and π will be secret primitives. Figure 2 has an illustration of RD (at the left).

3 Security Setup

The security analysis in this work is performed in the *distinguishability framework* where one bounds the advantage of an adversary $\mathcal{A}$ in distinguishing a real system from an ideal system.

Definition 1. *Let $\mathcal{O}, \mathcal{P}$ be two collections of oracles with the same interface. The advantage of an adversary $\mathcal{A}$ in distinguishing $\mathcal{O}$ from $\mathcal{P}$ is defined as*

$$\Delta_{\mathcal{A}}(\mathcal{O} \; ; \; \mathcal{P}) = \left| \Pr\left(\mathcal{A}^{\mathcal{O}} \to 1\right) - \Pr\left(\mathcal{A}^{\mathcal{P}} \to 1\right) \right|.$$

Our proofs in part use the H-coefficient technique from Patarin [43]. We will follow the adaptation of Chen and Steinberger [23]. Consider any information-theoretic deterministic adversary $\mathcal{A}$ whose goal is to distinguish $\mathcal{O}$ from $\mathcal{P}$, with its advantage denoted

$$\Delta_{\mathcal{A}}(\mathcal{O} \; ; \; \mathcal{P}) .$$

The interaction of $\mathcal{A}$ with its oracle, either $\mathcal{O}$ or $\mathcal{P}$, will be stored in a *transcript* τ. Denote by $D_{\mathcal{O}}$ (resp. $D_{\mathcal{P}}$) the probability distribution of transcripts that can be obtained from interaction with $\mathcal{O}$ (resp. $\mathcal{P}$). Call a transcript τ *attainable* if

---

Algorithm 3. Full-state randomized duplex construction $\mathrm{RD}^{\phi,\pi}$

Require: $r < b$

Instantiation: $\mathrm{RD} \leftarrow \mathrm{RD}^{\phi,\pi}$
  State: $\mathrm{RD}.s \leftarrow 0^b$

Interface: $Z = \mathrm{RD}.\mathrm{Init}(\delta, \mathrm{iv}, \sigma, \mathrm{flag})$ with $\delta \in \mathbb{Z}_u$, $\mathrm{iv} \in \mathbb{Z}_2^{b-k}$, $\sigma \in \mathbb{Z}_2^b$, flag $\in$ {true, false}, and $Z \in \mathbb{Z}_2^r$
  $s \leftarrow \phi(\delta, \mathrm{iv})$
  $Z \leftarrow \lfloor s \rfloor_r$
  if flag = true **then** $\overline{\sigma} \leftarrow \overline{\sigma} + Z$
  $s \leftarrow s + \sigma$
  return Z

Interface: $Z = \mathrm{RD}.\mathrm{Duplexing}(\sigma, \mathrm{flag})$ with $\sigma \in \mathbb{Z}_2^b$ flag $\in$ {true, false}, and $Z \in \mathbb{Z}_2^r$
  $s \leftarrow \pi(s)$
  $Z \leftarrow \lfloor s \rfloor_r$
  if flag = true **then** $\overline{\sigma} \leftarrow \overline{\sigma} + Z$
  $s \leftarrow s + \sigma$
  return Z

---

it can be obtained from interacting with $\mathcal{P}$, hence if $\Pr(D_\mathcal{P} = \tau) > 0$. Denote by $\mathcal{T}$ the set of attainable transcripts, and consider any partition $\mathcal{T} = \mathcal{T}_{\mathrm{good}} \cup \mathcal{T}_{\mathrm{bad}}$ of the set of attainable transcripts into "good" and "bad" transcripts. The H-coefficient technique states the following [23].

Lemma 1 (H-coefficient Technique). *Consider a fixed information-theoretic deterministic adversary $\mathcal{A}$ whose goal is to distinguish $\mathcal{O}$ from $\mathcal{P}$. Let ε be such that for all $\tau \in \mathcal{T}_{\mathrm{good}}$:*

$$\frac{\Pr(D_\mathcal{O} = \tau)}{\Pr(D_\mathcal{P} = \tau)} \geq 1 - \varepsilon. \tag{8}$$

Then, $\Delta_\mathcal{A}(\mathcal{O} \,;\, \mathcal{P}) \leq \varepsilon + \Pr(D_\mathcal{P} \in \mathcal{T}_{\mathrm{bad}})$.

The H-coefficient technique can thus be used to neatly bound a distinguishing advantage in the terminology of Definition 1, and a proof typically goes in four steps: (i) investigate what transcripts look like, which gives a definition for $\mathcal{T}$, (ii) define the partition of $\mathcal{T}$ into $\mathcal{T}_{\mathrm{good}}$ and $\mathcal{T}_{\mathrm{bad}}$, (iii) investigate the fraction of (8) for good transcripts and (iv) analyze the probability that $D_\mathcal{P}$ generates a bad transcript.

4 Security of Keyed Duplex Construction

We prove that the full-state keyed duplex construction (KD) is sound. We do so by proving an upper bound for the advantage of distinguishing the KD calling a random permutation f from an ideal extendable input function (IXIF). Both in the real and ideal world the adversary gets additional query access to f and f^{-1}, simply denoted as f.

The main result is stated in Sect. 4.2, but before doing so, we specify the resources of the adversary in Sect. 4.1.

4.1   Quantification of Adversarial Resources

We will consider information-theoretic adversaries that have two oracle interfaces: a construction oracle, $\mathrm{KD}_\mathbf{K}^f$ or $\mathrm{IXIF}^{\mathcal{RO}}$, and a primitive oracle f. For the construction queries, it can make initialization queries or duplexing queries. Note that, when querying $\mathrm{IXIF}^{\mathcal{RO}}$, every query has a *path* Path associated to it. To unify notation, we also associate a Path to each query (initialization or duplexing) to $\mathrm{KD}_\mathbf{K}^f$. This Path is defined the straightforward way: it simply consists of the concatenation of $\mathsf{Encode}(\delta)$, iv of the most recent initialization call and all σ-values that have been queried after the last initialization but before the current query. Using this formalization, every initialization *or* duplexing call that the adversary makes to $\mathrm{KD}_\mathbf{K}^f$ or $\mathrm{IXIF}^{\mathcal{RO}}$ can be properly captured by a tuple

$$(\mathsf{Path}, Z, \sigma),$$

where, intuitively, Path is all data that is used to generate response $Z \in \mathbb{Z}_2^r$, and $\sigma \in \mathbb{Z}_2^b$ is the input string (slightly abusing notation; $\sigma = \sigma$ if flag = false and $\sigma = \sigma + (Z||0^c)$ if flag = true).

Following Andreeva et al. [2], we specify adversarial resources that impose limits on the transcripts that any adversary can obtain. The basic resource metrics are quantitative: they specify the number of queries an adversary is allowed to make for each type.

- **N:** the number of primitive queries. It corresponds to computations requiring no access to the (keyed) construction. It is usually called the **time or offline complexity**. In practical use cases, N is only limited by the computing power and time available to the adversary.
- **M:** the number of construction queries. It corresponds to the amount of data processed by the (keyed) construction. It is usually called the **data or online complexity**. In many practical use cases, M is limited.

We remark that identical calls are counted only once. In other words, N only counts the number of primitive queries, and M only counts the number of unique tuples (Path, σ).

It is possible to perform an analysis solely based on these metrics, but in order to more accurately cover practical settings that were not covered before (such as the multi-key setting or the nonce-respecting setting), and to eliminate the multiplicity (a metric used in all earlier results in this direction), we define a number of additional metrics.

- **q:** the total number of different initialization tuples $(\mathsf{Encode}(\delta), \mathsf{iv})$. Parameter q corresponds to the number of times an adversary can start a *fresh* initialization of KD or IXIF.
- **q_{iv}:** iv multiplicity, the maximum number of different initialization tuples $(\mathsf{Encode}(\delta), \mathsf{iv})$ with same iv, maximized over all iv values.
- **Ω:** the number of queries with flag $=$ true.
- **L:** equals the number of queries M minus the number of distinct paths. It corresponds to the number of construction queries that have the same Path as some prior query.

In many practical use cases, q is limited, but as it turns out re-initialization queries give the adversary more power. The metric q_{iv} is relevant in multi-target attacks, where clearly $q_{\mathsf{iv}} \leq u$. The relevance of Ω and L is the following. In every query with flag equal to true, the adversary can force the outer part of the input to f in a later query to a chosen value α by taking $\overline{\sigma} = \alpha$. Note that, as discussed in Sect. 2.2, by adopting authenticated encryption schemes with a slightly non-conventional encryption method, Ω can be forced to zero. Similarly, construction queries with the same path return the same value Z, and hence allow an adversary to force the outer part of the input to f in a later query to a chosen value α by taking σ such that $\overline{\sigma} = Z + \alpha$. An adversary can use this technique to increase the probability of collisions in $f(s) + \sigma$ and to speed up inner state recovery. By definition, $L \leq M - 1$ but in many cases L is much smaller. In particular, if one considers KD in the nonce-respecting setting, where no $(\mathsf{Encode}(\delta), \mathsf{iv})$ occurs twice, the adversary never generates a repeating path, and $L = 0$.

4.2 Main Result

Our bound uses a function that is defined in terms of a simple balls-into-bins problem.

Definition 2. *The* multicollision limit function *$\nu_{r,c}^{M}$, with M a natural number, returns a natural number and is defined as follows. Assume we uniformly randomly distribute M balls in 2^r bins. If we call the number of balls in the bin with the highest number of balls μ, then $\nu_{r,c}^{M}$ is defined as the smallest natural number x that satisfies:*

$$\Pr\left(\mu > x\right) \leq \frac{x}{2^c} \, .$$

In words, when uniformly randomly sampling M elements from a set of 2^r elements, the probability that there is an element that is sampled more than x times is smaller than $x2^{-c}$.

Theorem 1. *Let f be a random permutation and $\mathcal{RO}$ be a random oracle. Let $\mathbf{K}$ be a key array generated using a distribution $\mathcal{D}_{\mathbf{K}}$. Let $\mathrm{KD}_{\mathbf{K}}^{f}$ be the construction of Algorithm 1 and $\mathrm{IXIF}^{\mathcal{RO}}$ be the construction of Algorithm 2 and let $\nu_{r,c}^{M}$*

be defined according to Definition 2. For any adversary $\mathcal{A}$ with resources as discussed in Sect. 4.1, and with $N + M \leq 0.1 \cdot 2^c$,

$$\Delta_{\mathcal{A}}(\mathrm{KD}_{\mathbf{K}}^f, f \; ; \; \mathrm{IXIF}^{\mathcal{RO}}, f) \leq \frac{(L+\Omega)N}{2^c} + \frac{2\nu_{r,c}^{2(M-L)}(N+1)}{2^c} + \frac{\binom{L+\Omega+1}{2}}{2^c}$$
$$+ \frac{(M-q-L)q}{2^b - q} + \frac{M(M-L-1)}{2^b}$$
$$+ \frac{(M-q-L)q}{2^{H_{\min}(\mathcal{D}_{\mathrm{K}})+\min\{c,b-k\}}} + \frac{q_{\mathrm{iv}}N}{2^{H_{\min}(\mathcal{D}_{\mathrm{K}})}} + \frac{\binom{u}{2}}{2^{H_{coll}(\mathcal{D}_{\mathrm{K}})}}.$$

The proof is given in Sect. 4.3.

For the case where $k + c \leq b - 1$, and where $\mathcal{D}_{\mathrm{K}}$ corresponds to uniform sampling without replacement, the bound simplifies to

$$\Delta_{\mathcal{A}}(\mathrm{KD}_{\mathbf{K}}^f, f \; ; \; \mathrm{IXIF}^{\mathcal{RO}}, f) \leq \frac{(L+\Omega)N}{2^c} + \frac{2\nu_{r,c}^{2(M-L)}(N+1)}{2^c} + \frac{\binom{L+\Omega+1}{2}}{2^c}$$
$$+ \frac{q_{\mathrm{iv}}N}{2^k} + \frac{(M-q-L)q}{2^{k+c-1}} + \frac{M(M-L-1)}{2^b}.$$

The behavior of the function $\nu_{r,c}^M$ is discussed in Sect. 6.5 and illustrated in the Fig. 4, which we refer to as the *Stairway to Heaven* graph.

4.3   Proof of Theorem 1

Let $\mathcal{A}$ be any information-theoretic adversary that has access to either, in the real world $(\mathrm{KD}_{\mathbf{K}}^f, f)$, or in the ideal world $(\mathrm{IXIF}^{\mathcal{RO}}, f)$. Note that, as $\mathcal{A}$ is information-theoretic, we can without loss of generality assume that it is deterministic, and we can apply the technique of Sect. 3. By the triangle inequality,

$$\Delta_{\mathcal{A}}(\mathrm{KD}_{\mathbf{K}}^f, f \; ; \; \mathrm{IXIF}^{\mathcal{RO}}, f)$$
$$\leq \Delta_{\mathcal{B}}(\mathrm{KD}_{\mathbf{K}}^f, f \; ; \; \mathrm{RD}^{\phi,\pi}, f) + \Delta_{\mathcal{C}}(\mathrm{RD}^{\phi,\pi}, f \; ; \; \mathrm{IXIF}^{\mathcal{RO}}, f), \qquad (9)$$

where $\mathrm{RD}^{\phi,\pi}$ for random injection function ϕ and random permutation π is the construction of Algorithm 3, and where $\mathcal{B}$ and $\mathcal{C}$ have the same resources $(N, M, q, q_{\mathrm{iv}}, L, \Omega)$ as $\mathcal{A}$.

In the last term of (9), RD calls an ideal injective function ϕ and a random permutation π, both independent of f, and IXIF calls a random oracle $\mathcal{RO}$, also independent of f. The oracle access to f therefore does not "help" the adversary in distinguishing the two, or more formally,

$$\Delta_{\mathcal{C}}(\mathrm{RD}^{\phi,\pi}, f \; ; \; \mathrm{IXIF}^{\mathcal{RO}}, f) \leq \Delta_{\mathcal{D}}(\mathrm{RD}^{\phi,\pi} \; ; \; \mathrm{IXIF}^{\mathcal{RO}}). \qquad (10)$$

where $\mathcal{D}$ is an adversary with the same construction query parameters as $\mathcal{A}$, but with no access to f.

The two remaining distances, i.e., the first term of (9) and the term of (10), will be analyzed in the next lemmas. The proof of Theorem 1 directly follows.

Lemma 2. *For any adversary $\mathcal{D}$ with resources as discussed in Sect. 4.1 but with no access to f,*

$$\Delta_{\mathcal{D}}(\mathrm{RD}^{\phi,\pi} ; \mathrm{IXIF}^{\mathcal{RO}}) \leq \frac{\binom{L+\Omega+1}{2}}{2^c} + \frac{M(M-L-1)}{2^b}. \tag{11}$$

Lemma 3. *For any adversary $\mathcal{B}$ with resources as discussed in Sect. 4.1,*

$$\Delta_{\mathcal{B}}(\mathrm{KD}_{\mathbf{K}}^f, f ; \mathrm{RD}^{\phi,\pi}, f) \leq \frac{(L+\Omega)N}{2^c} + \frac{2\nu_{r,c}^{2(M-L)}(N+1)}{2^c} + \frac{(M-q-L)q}{2^b-q}$$
$$+ \frac{(M-q-L)q}{2^{H_{\min}(\mathcal{D}_{\mathbf{K}})+\min\{c,b-k\}}} + \frac{q_{\mathrm{iv}}N}{2^{H_{\min}(\mathcal{D}_{\mathbf{K}})}} + \frac{\binom{u}{2}}{2^{H_{coll}(\mathcal{D}_{\mathbf{K}})}}. \tag{12}$$

The proof of Lemma 2 is given in Sect. 5, and the proof of Lemma 3 is given in Sect. 6.

5 Distance Between RD and IXIF

In this section we bound the advantage of distinguishing the randomized duplex from an ideal extendable input function, (11) of Lemma 2. The distinguishing setup is illustrated in Fig. 2. The derivation is performed using the H-coefficient technique.

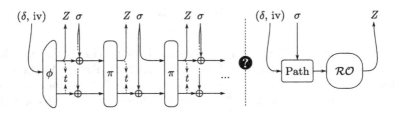

Fig. 2. Distinguishing experiment of RD and IXIF.

Description of Transcripts. The adversary has only a single interface, $\mathrm{RD}^{\phi,\pi}$ or $\mathrm{IXIF}^{\mathcal{RO}}$, but can make both initialization and duplexing queries. Following the discussion of Sect. 4.1, we can unify the two different types of queries, and summarize the conversation of $\mathcal{D}$ with its oracle in a transcript of the form

$$\tau_{\mathrm{C}} = \{(\mathsf{Path}_j, Z_j, \sigma_j)\}_{j=1}^M.$$

The values Z_j correspond to the outer part of the state just before σ_j gets injected. To make the analysis easier, we give at the end of the experiment for each query the inner value of the state at the moment Z_j is extracted (in the real world). We denote this as $t_j = \bar{t}_j \| \hat{t}_j$ with $\bar{t}_j = Z_j$. In the IXIF, $\hat{t}_j$ is a value that is randomly generated for each path Path and can be expressed as $\mathcal{RO}'(\mathsf{Path})$ for some random oracle $\mathcal{RO}'$ with c-bit output. We integrate those values in the transcript, yielding:

$$\tau = \{(\mathsf{Path}_j, t_j, \sigma_j)\}_{j=1}^M.$$

Definition of Good and Bad Transcripts. We define a transcript τ as *bad* if it contains a t-collision or an s-collision, where $s = t + \sigma$. A t-collision is defined as equal t values despite different Path values:

$$\exists(\mathsf{Path}, t, \sigma), (\mathsf{Path}', t', \sigma') \in \tau \text{ with } (\mathsf{Path} \neq \mathsf{Path}') \text{ AND } (t = t'). \tag{13}$$

An s-collision is defined as equal s values despite different (Path, σ') values:

$$\exists(\mathsf{Path}, t, \sigma), (\mathsf{Path}', t', \sigma') \in \tau \text{ with}$$
$$((\mathsf{Path}, \sigma) \neq (\mathsf{Path}', \sigma')) \text{ AND } (t + \sigma = t' + \sigma'). \tag{14}$$

In case the oracle is $\mathrm{RD}^{\phi,\pi}$, a t-collision is equivalent to two different inputs to π with identical outputs; an s-collision corresponds to the case of two identical inputs to f where the outputs are expected to be distinct. By considering these transcripts as bad, all queries properly define input-output tuples for ϕ and π.

Bounding the H-coefficient Ratio for Good Transcripts. Denote $\mathcal{O} = \mathrm{RD}^{\phi,\pi}$ and $\mathcal{P} = \mathrm{IXIF}^{\mathcal{RO}}$ for brevity. Consider a good transcript $\tau \in \mathcal{T}_{\mathrm{good}}$. For the real world $\mathcal{O}$, the transcript defines exactly q input-output pairs for ϕ and exactly $M - q - L$ input-output pairs for π. It follows that $\Pr(D_{\mathcal{O}} = \tau) = 1/((2^b)_{(q)}(2^b)_{(M-q-L)})$. For the ideal world $\mathcal{P}$, every different Path defines exactly one evaluation of $\mathcal{RO}(\mathsf{Path})\|\mathcal{RO}'(\mathsf{Path})$, so $\Pr(D_{\mathcal{P}} = \tau) = 2^{-(M-L)b}$. We consequently obtain that $\dfrac{\Pr(D_{\mathcal{O}} = \tau)}{\Pr(D_{\mathcal{P}} = \tau)} \geq 1$.

Bounding the Probability of Bad Transcripts in the Ideal World. In the ideal world, every t is generated as $\mathcal{RO}(\mathsf{Path})\|\mathcal{RO}'(\mathsf{Path})$. As the number of distinct Path's in τ is $M - L$, there are $\binom{M-L}{2}$ possibilities for a t-collision, each occurring with probability 2^{-b}. The probability of such a collision is hence $\dfrac{\binom{M-L}{2}}{2^b}$.

There are $\binom{M}{2}$ occasions for an s-collision. Denote by S the size of the subset of these occasions for which the adversary can (in the worst case) force the outer part of $s = t + \sigma$ to be a value of its choice. Note that $S \leq \binom{L+\Omega+1}{2}$. In the worst case, in these S occasions the outer part of s always has the same value and s-collision probability is 2^{-c}. For the $\binom{M}{2} - S$ other occasions the s-collision probability is 2^{-b}. Thus, the probability of an s-collision is upper bound by (using our bound on S):

$$\frac{\binom{M}{2} - S}{2^b} + \frac{S}{2^c} \leq \frac{\binom{M}{2} - \binom{L+\Omega+1}{2}}{2^b} + \frac{\binom{L+\Omega+1}{2}}{2^c} \leq \frac{\binom{M}{2} - \binom{L+1}{2}}{2^b} + \frac{\binom{L+\Omega+1}{2}}{2^c}.$$

The total probability of having a bad transcript is hence upper bound by:

$$\frac{\binom{M-L}{2}}{2^b} + \frac{\binom{M}{2} - \binom{L+1}{2}}{2^b} + \frac{\binom{L+\Omega+1}{2}}{2^c} = \frac{M(M-L-1)}{2^b} + \frac{\binom{L+\Omega+1}{2}}{2^c}.$$

As the H-coefficient ratio is larger than 1, this is the bound on the distinguishing advantage and we have proven Lemma 2.

6   Distance Between KD and RD

In this section we bound the advantage of distinguishing the keyed duplex from a randomized duplex, (12) of Lemma 3. The analysis consists of four steps. In Sect. 6.1, we revisit the KD-vs-RD setup, and exclude the case where the queries made by the adversary result in a forward multiplicity that exceeds a certain threshold T_{fw}. Next, in Sect. 6.2 we convert our distinguishing setup to a simpler one, called the *permutation setup* and illustrated in Fig. 3. In this setup, the adversary has direct query access to the primitives ϕ and π of the randomized duplex, and at the keyed duplex side, we define two constructions on top of f that turn out to be hard to distinguish from ϕ and π. We carefully translate the resources of the adversary $\mathcal{B}$ in the KD-vs-RD setup to those of the adversary $\mathcal{C}$ in the permutation setup. In Sect. 6.3 we subsequently prove a bound in this setup. This analysis in part depends on a threshold on backward multiplicities T_{bw}. In Sect. 6.4 where we return to the KD-vs-RD setup and blend all results. Finally, in Sects. 6.5 and 6.6 we analyze the function $\nu_{r,c}^M$ that plays an important role in our analysis.

We remark that forward and backward multiplicity appeared before in Bertoni et al. [10] and Andreeva et al. [2], but we resolve them internally in the proof. There is a specific reason for resolving forward multiplicity *before* the conversion to the permutation setup and backward multiplicity *after* this conversion. Namely, in the permutation setup, an adversary could form its queries so that the forward multiplicity equals $M - q$, leading to a non-competitive bound, while the backward multiplicity cannot be controlled by the adversary as it cannot make inverse queries to the constructions. It turns out that, as discussed in Sect. 6.4, we can bound the thresholds as functions of M, L, and Ω.

6.1   The KD-vs-RD Setup

As in Sect. 4.1, we express the conversation that $\mathcal{B}$ has with $\text{KD}_{\mathbf{K}}^f$ or $\text{RD}^{\phi,\pi}$ in a transcript of the form:

$$\tau_C = \{(\text{Path}_j, Z_j, \sigma_j)\}_{j=1}^M.$$

We denote by μ_{fw} the maximum number of occurrences in this transcript of a value $Z_j + \overline{\sigma}_j$ over all possible values:

$$\mu_{\text{fw}} = \max_\alpha \#\{(\text{Path}_j, Z_j, \sigma_j) \in \tau_C \mid Z_j + \overline{\sigma}_j = \alpha\}. \tag{15}$$

We now distinguish between two cases: μ_{fw} above some threshold T_{fw} and below it. Denoting $\mathcal{O} = (\text{KD}_{\mathbf{K}}^f, f)$ and $\mathcal{P} = (\text{RD}^{\phi,\pi}, f)$, we find using a hybrid argument,

$$\Delta_{\mathcal{B}}(\mathcal{O}\ ;\ \mathcal{P}) = \left|\Pr\left(\mathcal{B}^{\mathcal{O}} \to 1\right) - \Pr\left(\mathcal{B}^{\mathcal{P}} \to 1\right)\right|$$
$$\leq \left|\Pr\left(\mathcal{B}^{\mathcal{O}} \to 1 \wedge \mu_{\mathrm{fw}} \leq T_{\mathrm{fw}}\right) - \Pr\left(\mathcal{B}^{\mathcal{P}} \to 1 \wedge \mu_{\mathrm{fw}} \leq T_{\mathrm{fw}}\right)\right|$$
$$+ \left|\Pr\left(\mathcal{B}^{\mathcal{O}} \to 1 \wedge \mu_{\mathrm{fw}} > T_{\mathrm{fw}}\right) - \Pr\left(\mathcal{B}^{\mathcal{P}} \to 1 \wedge \mu_{\mathrm{fw}} > T_{\mathrm{fw}}\right)\right|$$
$$\leq \left|\Pr\left(\mathcal{B}^{\mathcal{O}} \to 1 \wedge \mu_{\mathrm{fw}} \leq T_{\mathrm{fw}}\right) - \Pr\left(\mathcal{B}^{\mathcal{P}} \to 1 \wedge \mu_{\mathrm{fw}} \leq T_{\mathrm{fw}}\right)\right|$$
$$+ \max\left\{\Pr\left(\mu_{\mathrm{fw}} > T_{\mathrm{fw}} \text{ for } \mathcal{O}\right), \Pr\left(\mu_{\mathrm{fw}} > T_{\mathrm{fw}} \text{ for } \mathcal{P}\right)\right\}. \quad (16)$$

As we will find out (and explicitly mention) in Sect. 6.4, the bound we will derive on $\Pr\left(\mu_{\mathrm{fw}} > T_{\mathrm{fw}}\right)$ in fact applies to both $\mathcal{O}$ and $\mathcal{P}$, and for brevity denote the maximum of the two probabilities by $\Pr_{\mathcal{O},\mathcal{P}}\left(\mu_{\mathrm{fw}} > T_{\mathrm{fw}}\right)$.

6.2 Entering the Permutation Setup

To come to our simplified setup we define two constructions: the Even-Mansour construction and a "state initialization construction." The original Even-Mansour construction builds a b-bit block cipher from a b-bit permutation f and takes two b-bit keys K_1 and K_2 [25,26], and is defined as $f(x + K_1) + K_2$. We consider a variant, where $K_1 = K_2 = 0^r||\kappa$ with κ a c-bit key, and define

$$E_{\kappa}^{f}(x) = f(x + (0^r||\kappa)) + (0^r||\kappa). \quad (17)$$

The state initialization construction is a dedicated construction of an injective function that maps an iv and a key selected from a key array $\mathbf{K}$ to a b-bit state and that takes a c-bit key κ.

$$I_{\kappa,\mathbf{K}}^{f}(\delta, \mathrm{iv}) = f(\mathbf{K}[\delta]||\mathrm{iv}) + (0^r||\kappa). \quad (18)$$

Now, let $\kappa \xleftarrow{\$} \mathbb{Z}_2^c$ be any c-bit key. We call κ the *inner masking key*. Using the idea of bitwise adding the inner masking key twice in-between every two primitive evaluations [2,21,38], we obtain that: $\mathrm{KD}_{\mathbf{K}}^{f} = \mathrm{RD}^{I_{\kappa,\mathbf{K}}^{f}, E_{\kappa}^{f}}$. We thus obtain for (16), leaving the condition $\mu_{\mathrm{fw}} \leq T_{\mathrm{fw}}$ implicit:

$$\Delta_{\mathcal{B}}(\mathrm{KD}_{\mathbf{K}}^{f}, f\ ;\ \mathrm{RD}^{\phi,\pi}, f) = \Delta_{\mathcal{B}}(\mathrm{RD}^{I_{\kappa,\mathbf{K}}^{f}, E_{\kappa}^{f}}, f\ ;\ \mathrm{RD}^{\phi,\pi}, f)$$
$$\leq \Delta_{\mathcal{C}}(I_{\kappa,\mathbf{K}}^{f}, E_{\kappa}^{f}, f\ ;\ \phi, \pi, f). \quad (19)$$

Clearly an adversary $\mathcal{B}$ can be simulated by an adversary $\mathcal{C}$ as any construction query can be simulated by queries to the initialization function $\mathcal{O}_{\mathrm{i}}$ ($I_{\kappa,\mathbf{K}}^{f}$ in the real world and ϕ in the ideal world) and the duplexing function $\mathcal{O}_{\mathrm{d}}$ (E_{κ}^{f} in the real world and π in the ideal world). Hence, we can quantify the resources of adversary $\mathcal{C}$ in terms of the resources of adversary $\mathcal{B}$, making use of the threshold T_{fw} on the multiplicity (cf., (16)). This conversion will be formally performed in Sect. 6.4.

6.3 Distinguishing Bound for the Permutation Setup

We now bound $\Delta_{\mathcal{C}}(I_{\kappa,\mathbf{K}}^{f}, E_{\kappa}^{f}, f\ ;\ \phi, \pi, f)$. The permutation setup is illustrated in Fig. 3. The derivation is performed using the H-coefficient technique.

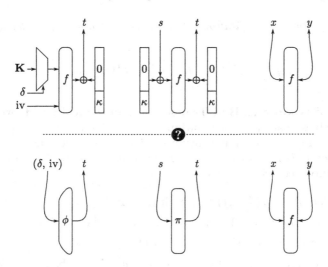

Fig. 3. Permutation setup.

Description of Transcripts. The adversary has access to either $(I^f_{\kappa,\mathbf{K}}, E^f_\kappa, f)$ or (ϕ, π, f). The queries of the adversary and their responses are assembled in three transcripts $\tau_\mathrm{f}, \tau_\mathrm{d}$, and τ_i.

$\tau_\mathrm{f} = \{(x_j, y_j)\}^N_{j=1}$ The queries to f and f^{-1}. The transcript does not code whether the query was $y = f(x)$ or $x = f^{-1}(y)$.

$\tau_\mathrm{i} = \{(\delta_i, \mathrm{iv}_i, t_i)\}^{q'}_{i=1}$ The queries to the initialization function $\mathcal{O}_\mathrm{i}$, $I^f_{\kappa,\mathbf{K}}$ in the real world and ϕ in the ideal world.

$\tau_\mathrm{d} = \{(s_i, t_i)\}^{M'}_{i=1}$ The queries to the duplexing function $\mathcal{O}_\mathrm{d}$, E^f_κ in the real world and π in the ideal world.

The resources of $\mathcal{C}$ are defined by the number of queries in each transcript: N, M', and q', as well as $q_\mathrm{iv} = \max_\alpha \#\{(\delta, \mathrm{iv}, t) \in \tau_\mathrm{i} \mid \mathrm{iv} = \alpha\}$. In addition, the resources of $\mathcal{C}$ are limited on τ_d, for which the *forward multiplicity* must be below the threshold T_fw:

$$\max_\alpha \#\{(s_i, t_i) \in \tau_\mathrm{d} \mid \bar{s}_i = \alpha\} \le T_\mathrm{fw}.$$

To ease the analysis, we will disclose the full key array $\mathbf{K}$ and the inner masking key κ at the end of the experiment (in the ideal world, κ and the elements of $\mathbf{K}$ will simply be dummy keys). The transcripts are thus of the form $\tau = (\mathbf{K}, \kappa, \tau_\mathrm{f}, \tau_\mathrm{i}, \tau_\mathrm{d})$. Note that it is fair to assume that none of the transcripts contains duplicate elements (i.e., the adversary never queries f twice on the same value). Additionally, as we consider attainable transcripts only and ϕ, π, f are injective mappings, τ does not contain collisions.

We define the backward multiplicity as characteristic of the transcript τ:

Definition 3. *In the permutation setup, the backward multiplicity μ_{bw} is defined as:*

$$\mu_{bw} = \max_{\alpha} \left(\#\{(s_i, t_i) \in \tau_d \mid \bar{t}_i = \alpha\} + \#\{(\delta, iv, t_i) \in \tau_i \mid \bar{t}_i = \alpha\} \right).$$

Definition of Good and Bad Transcripts. In the real world, every tuple in (τ_f, τ_i, τ_d) defines exactly one evaluation of f. We define a transcript τ as *bad* if it contains an input or output collision of f or if the backward multiplicity is above some limit T_{bw}. In other words, τ is bad if one of the following conditions is satisfied. Input collisions between:

$$\tau_f \text{ and } \tau_i : \exists (x, y) \in \tau_f, (\delta, iv, t) \in \tau_i \text{ with } \left(x = \mathbf{K}[\delta]\|iv\right); \tag{20}$$

$$\tau_f \text{ and } \tau_d : \exists (x, y) \in \tau_f, (s, t) \in \tau_d \text{ with } \left(x = s + 0^r\|\kappa\right); \tag{21}$$

$$\tau_i \text{ and } \tau_d : \exists (\delta, iv, t) \in \tau_i, (s', t') \in \tau_d \text{ with } \left(\mathbf{K}[\delta]\|iv = s' + 0^r\|\kappa\right); \tag{22}$$

$$\text{within } \tau_i : \exists (\delta, iv, t), (\delta', iv', t') \in \tau_i \text{ with } \left(\delta \neq \delta'\right) \text{ AND } \left(\mathbf{K}[\delta]\|iv = \mathbf{K}[\delta']\|iv'\right). \tag{23}$$

Output collisions between:

$$\tau_f \text{ and } \tau_i : \exists (x, y) \in \tau_f, (\delta, iv, t) \in \tau_i \text{ with } \left(y = t + 0^r\|\kappa\right); \tag{24}$$

$$\tau_f \text{ and } \tau_d : \exists (x, y) \in \tau_f, (s, t) \in \tau_d \text{ with } \left(y = t + 0^r\|\kappa\right); \tag{25}$$

$$\tau_i \text{ and } \tau_d : \exists (\delta, iv, t) \in \tau_i, (s', t') \in \tau_d \text{ with } \left(t + 0^r\|\kappa = t' + 0^r\|\kappa\right). \tag{26}$$

Finally, τ is bad if the backward multiplicity μ_{bw} is above the threshold T_{bw}:

$$\mu_{bw} > T_{bw}. \tag{27}$$

Note that output collisions within τ_i are excluded by attainability of transcripts. Similarly, collisions (input or output) within τ_f as well as collisions within τ_d are excluded by attainability of transcripts.

Bounding the H-coefficient Ratio for Good Transcripts. Denote $\mathcal{O} = (I^f_{\kappa, \mathcal{D}_K}, E^f_\kappa, f)$ and $\mathcal{P} = (\phi, \pi, f)$ for brevity. Consider a good transcript $\tau \in \mathcal{T}_{good}$.

In the real world $\mathcal{O}$, the transcript defines exactly $q' + M' + N$ input-output pairs of f, so $\Pr(D_\mathcal{O} = \tau) = 1/(2^b)_{(q'+M'+N)}$. In the ideal world $\mathcal{P}$, every tuple in τ_f defines exactly N input-output pairs for f, every tuple in τ_i defines exactly q' input-output pairs for ϕ, and every tuple in τ_d defines exactly M' input-output pairs for π. It follows that $\Pr(D_\mathcal{P} = \tau) = 1/((2^b)_{(N)}(2^b)_{(q')}(2^b)_{(M')})$. We consequently obtain that $\dfrac{\Pr(D_\mathcal{O} = \tau)}{\Pr(D_\mathcal{P} = \tau)} \geq 1$.

Bounding the Probability of Bad Transcripts in the Ideal World. In the ideal world, κ is generated uniformly at random. The key array $\mathbf{K}$ is generated according to distribution $\mathcal{D}_K$, cf., Sect. 2.1. We will use the min-entropy and maximum collision probability definitions of (6) and (7).

For (20), fix any $(x, y) \in \tau_f$. There are at most q_{iv} tuples in τ_i with iv equal to the last $b - k$ bits of x. For any of those tuples, the probability that the first k bits of x are equal to $\mathbf{K}[\delta]$ is at most $2^{-H_{\min}(\mathcal{D}_K)}$, cf., (6). The collision probability is hence at most $q_{iv} N / 2^{H_{\min}(\mathcal{D}_K)}$.

For (21), fix any $(x, y) \in \tau_f$. There are at most T_{fw} tuples in τ_d with $\overline{x} = \overline{s}$. For any of those tuples, the probability that $\widehat{x} = \widehat{s} + \kappa$ is 2^{-c}. The collision probability is hence at most $T_{fw} N / 2^c$.

For (24) or (25), we will assume $\neg(27)$. Fix any $(x, y) \in \tau_f$. There are at most T_{bw} tuples in $\tau_i \cup \tau_d$ with $\overline{y} = \overline{t}$. For any of those tuples, the probability that $\widehat{y} = \widehat{t} + \kappa$ is 2^{-c}. The collision probability is hence at most $T_{bw} N / 2^c$.

For (22), fix any $(\delta, iv, t) \in \tau_i$ and any $(s', t') \in \tau_d$. Any such combination sets (22) if $0^k \| iv + s' = \mathbf{K}[\delta] \| 0^{b-k} + 0^r \| \kappa$. Note that the randomness of $\mathbf{K}[\delta]$ may overlap the one of κ. If $k + c \leq b$, the two queries satisfy the condition with probability at most $2^{-(H_{\min}(\mathcal{D}_K)+c)}$, cf., (6). On the other hand, if $k > b - c$, the first $b - c$ bits of $\mathbf{K}[\delta]$ has a min-entropy of at least $H_{\min}(\mathcal{D}_K) - (k - (b - c))$. In this case, the two queries satisfy the condition with probability at most

$$2^{-(H_{\min}(\mathcal{D}_K)-(k-(b-c))+c)} = 2^{-(H_{\min}(\mathcal{D}_K)+b-k)}.$$

The collision probability is hence at most $\frac{M' q'}{2^{H_{\min}(\mathcal{D}_K)+\min\{c, b-k\}}}$, using that τ_i contains q' elements and τ_d contains M' elements.

For (26), fix any $(\delta, iv, t) \in \tau_i$ and any $(s', t') \in \tau_d$. As ϕ and π are only evaluated in forward direction, and ϕ is queried at most q' times, the probability that $t = t'$ for these two tuples is at most $1/(2^b - q')$. The collision probability is hence at most $M' q' / (2^b - q')$.

For (23), a collision of this form implies the existence of two distinct δ, δ' such that $K[\delta] = K[\delta']$. This happens with probability at most $\binom{u}{2}/2^{H_{coll}(\mathcal{D}_K)}$, cf., (7).

The total probability of having a bad transcript is at most:

$$\frac{(T_{fw} + T_{bw}) N}{2^c} + \Pr_{\mathcal{P}} (\mu_{bw} > T_{bw}) + \frac{M' q'}{2^b - q'}$$
$$+ \frac{M' q'}{2^{H_{\min}(\mathcal{D}_K)+\min\{c, b-k\}}} + \frac{q_{iv} N}{2^{H_{\min}(\mathcal{D}_K)}} + \frac{\binom{u}{2}}{2^{H_{coll}(\mathcal{D}_K)}}. \tag{28}$$

As the H-coefficient ratio is larger than 1, Eq. (28) is the bound on the distinguishing advantage.

6.4   Returning to the KD-vs-RD Setup

The resources of $\mathcal{C}$ can be computed from those of $\mathcal{B}$ (see Sect. 4.1) in the following way:

- $q' \leq q$: for every query to $\mathcal{O}_i$ there must be at least one initialization query.

- $M' \leq M - q - L$: The minus L is there because queries with repeated paths just give duplicate queries to $\mathcal{O}_i$ and the q initialization queries do not give queries to $\mathcal{O}_d$.

The remaining resources have the same meaning for $\mathcal{B}$ and $\mathcal{C}$. Filling in these values in Eq. (28) and combining with Eq. (16) yields:

$$\Delta_{\mathcal{B}}(\mathcal{O} \; ; \; \mathcal{P}) \leq \left(\frac{T_{\mathrm{fw}} N}{2^c} + \mathrm{Pr}_{\mathcal{O}, \mathcal{P}} \left(\mu_{\mathrm{fw}} > T_{\mathrm{fw}} \right) \right) \tag{29a}$$

$$+ \left(\frac{T_{\mathrm{bw}} N}{2^c} + \mathrm{Pr}_{\mathcal{P}} \left(\mu_{\mathrm{bw}} > T_{\mathrm{bw}} \right) \right) \tag{29b}$$

$$+ \frac{(M - q - L)q}{2^b - q} + \frac{(M - q - L)q}{2^{H_{\min}(\mathcal{D}_{\mathrm{K}}) + \min\{c, b - k\}}} \tag{29c}$$

$$+ \frac{q_{\mathrm{iv}} N}{2^{H_{\min}(\mathcal{D}_{\mathrm{K}})}} + \frac{\binom{u}{2}}{2^{H_{\mathrm{coll}}(\mathcal{D}_{\mathrm{K}})}} . \tag{29d}$$

Clearly $\mu_{\mathrm{fw}} \leq M - q - L$ and $\mu_{\mathrm{bw}} \leq M - L$. So by taking $T_{\mathrm{fw}} = T_{\mathrm{bw}} = M - L$, lines (29a)–(29b) reduce to $2(M - L)N/2^c$. However, much better bounds can be obtained by carefully tuning T_{fw} and T_{bw}.

Although the probabilities on the μ_{fw} and μ_{bw} are defined differently (the former in the KD-vs-RD setup, the latter in the permutation setup), in essence they are highly related and we can rely on multicollision limit function of Definition 2 for their analysis. There is one caveat. Definition 2 considers balls thrown uniformly at random into the 2^r bins, hence a bin is hit with probability $1/2^r$. In Lemma 6 in upcoming Sect. 6.6, we will prove that for non-uniform bin allocation where the probability that a ball hits any particular bin is upper bounded by $y2^{-r}$, the multicollision limit function is at most $\nu_{r,c}^{yM}$. In our case the states are generated from a set of size at least $2^b - M - N$ (for both $\mathcal{O}$ and $\mathcal{P}$) and thus its outer part is thrown in a bin with probability at most $2^c/(2^b - M - N)$, where we use that $M + N \leq 2^{b-1}$. Using the fact that $\nu_{r,c}^M$ is a monotonic function in M and that $2^b/(2^b - M - N) < 2$ for any reasonable value of $M + N$, we upper bound the multicollision limit function by $\nu_{r,c}^{2(M-L)}$

We first look at (29b) and treat μ_{bw}. As it is a metric of the responses of queries to π and ϕ, it is a stochastic variable. It corresponds to the multicollision limit function of Definition 2, where $M - L$ balls are distributed over 2^r bins, and each bin is hit with probability at most $2/2^r$. Using above observation, we take $T_{\mathrm{bw}} = \nu_{r,c}^{2(M-L)}$, and (29b) becomes

$$\frac{\nu_{r,c}^{2(M-L)} N}{2^c} + \frac{\nu_{r,c}^{2(M-L)}}{2^c} = \frac{\nu_{r,c}^{2(M-L)}(N + 1)}{2^c} .$$

The case of μ_{fw} in (29c) is slightly more complex. As discussed in Sect. 4.1, the adversary can enforce the outer part $Z_j + \overline{\sigma}_j$ to match a value α in case Path_j is a repeating path. Moreover, for queries with flag = true, it can also enforce the outer part to any chosen value. These total to $L + \Omega$ queries. For the remaining queries, for simplicity upper bound by $M - L$ here, the adversary

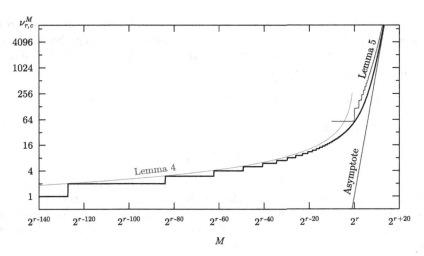

Fig. 4. Stairway to Heaven graph: $\nu_{r,c}^M$ computed with (33) for $r + c = 256$, with upper bounds and asymptote for $M \to \infty$.

has no control over the outer part. Therefore, if take $T_{\mathrm{fw}} = L + \Omega + \nu_{r,c}^{2(M-L)}$, we have $\Pr\left(\mu_{\mathrm{fw}} > T_{\mathrm{fw}}\right) = \frac{\nu_{r,c}^{2(M-L)}}{2^c}$. Namely, this is the probability that in the (at most) $M - L$ queries where the adversary has no control over the outer part, the multiplicity is above $\nu_{r,c}^{2(M-L)}$ assuming that the $L + \Omega$ queries are manipulated to hit the same outer value as those $\nu_{r,c}^{2(M-L)}$ queries. Eq. (29a) now becomes:

$$\frac{(L + \Omega + \nu_{r,c}^{2(M-L)})N}{2^c} + \frac{\nu_{r,c}^{2(M-L)}}{2^c} = \frac{(L + \Omega)N}{2^c} + \frac{2\nu_{r,c}^{2(M-L)}(N+1)}{2^c}.$$

Plugging these two bounds into (29a)–(29b) yields the bound of Lemma 3.

6.5   Bounds on $\nu_{r,c}^M$

We will upper bound $\nu_{r,c}^M$ by approximating the term $\Pr(\mu > x)$ in Definition 2 by simpler expressions that are strictly larger.

In Definition 2, μ is the maximum of the number of balls over all 2^r bins. If we model the number of balls in a particular bin as a stochastic variable X_i with some distribution function $D_i(x) = \Pr(X_i \le x)$, clearly, the distribution function of the maximum over all bins is the product of the distribution functions: $D_{\max}(x) = \prod_i D_i(x)$. Assuming all variables have the same distribution and that they are independent, we hence obtain:

$$\Pr(\mu > x) = 1 - \Pr(\mu \le x) = 1 - (\Pr(X \le x))^{2^r}. \tag{30}$$

The distributions that are of interest here are the number of balls in a bin, and they are not independent as they must sum to M. This means that if we know one distribution is high, the others are somewhat lower than if they would be independent. This makes that the value obtained by taking the product of factors $\Pr(X \leq x)$ slightly underestimates the probability $\Pr(\mu \leq x)$. Using the inequality $(1 - \epsilon)^y \geq 1 - \epsilon y$, we obtain

$$(\Pr(X \leq x))^{2^r} = (1 - \Pr(X > x))^{2^r} \geq 1 - 2^r \Pr(X > x),$$

and we obtain for (30):

$$\Pr(\mu > x) < 2^r \Pr(X > x). \tag{31}$$

We will now upper bound $\Pr(X > x)$. The number of balls x in any particular bin has a binomial distribution. If the number of bins 2^r and the total number of balls M are large enough, for $x > \lambda$ this is (tightly) upper bounded by a Poisson distribution with $\lambda = M2^{-r}$. The probability that a Poisson-distributed variable X is larger than x is given by:

$$\Pr(X > x) = \sum_{i \geq x} \frac{e^{-\lambda}\lambda^i}{i!} = \frac{e^{-\lambda}\lambda^x}{x!} \sum_{i \geq 0} \frac{\lambda^i}{(i+x)_{(i)}} < \frac{e^{-\lambda}\lambda^x}{x!} \sum_{i \geq 0} \frac{\lambda^i}{x^i} = \frac{xe^{-\lambda}\lambda^x}{(x-\lambda)x!}.$$

This yields for (31):

$$\Pr(\mu > x) < 2^r \frac{xe^{-\lambda}\lambda^x}{(x-\lambda)x!}.$$

From Definition 2, we obtain that $\nu_{r,c}^M$ is upper bounded by the smallest value x that satisfies

$$\frac{2^b e^{-\lambda}\lambda^x}{(x-\lambda)x!} \leq 1, \tag{32}$$

with $\lambda = M2^{-r}$. Remarkably, the dependence of $\nu_{r,c}^M$ on r, c and M is only via $b = r + c$ and $\lambda = M2^{-r}$. Hence, it is a function in two variables b and λ rather than three. Taking the logarithm of (32), applying Stirling's approximation $(\ln(x!) \geq \frac{1}{2}\ln(2\pi x) + x(\ln(x) - 1))$ and rearranging the terms gives:

$$x\left(\ln(x) - \ln(\lambda) - 1\right) + \ln(x - \lambda) + \frac{1}{2}\ln(2\pi x) + \lambda \geq \ln(2)b. \tag{33}$$

We will now derive expressions from (32) and (33) that give insight in the behavior of this function for the full range of λ.

Case $\lambda < 1$. If we consider Eq. (33) with value of x given and we look for the maximum value of x such that it holds. This gives the value of λ where $\nu_{r,c}^M$ transitions from $x - 1$ to x. We can now prove the following lemma.

Lemma 4. *The value of λ where $\nu_{r,c}^M$ transitions from $x - 1$ to x is lower bounded by $2^{-b/x}$.*

Proof. We need to prove that for $\lambda = 2^{-b/x}$, inequality (33) holds:

$$x(\ln(x) - 1) + \ln(x - 2^{-b/x}) + \frac{1}{2}\ln(2\pi x) + 2^{-b/x} \geq 0.$$

For $x > e$ all terms in the left hand side of this equation are positive and hence the equation is satisfied. The only other relevant value is $x = 2$ and it can be verified by hand that this is satisfied for all b.   □

If we substitute λ by $M2^{-r}$, this gives bounds on M for which $\nu_{r,c}^M$ achieves a certain value. If we denote by M_x the value where $\nu_{r,c}^M$ transitions from $x - 1$ to x, we have $M_x \geq 2^{r-b/x} = 2^{((x-1)r-c)/x}$. In particular $M_2 \geq 2^{(r-c)/2}$. It follows that $\nu_{r,c}^M$ is 1 for $M \leq 2^{(r-c)/2}$. Clearly, M must be an integer value, so the value of $\nu_{r,c}^M$ for $M = 1$ will be above 1 if $r < c + 2$.

Case $\lambda = 1$. Equation (33) for $\lambda = 1$ reads

$$x\left(\ln(x) - 1\right) + \ln(x - 1) + \frac{1}{2}\ln(2\pi x) + 1 \geq \ln(2)b,$$

and $\nu_{r,c}^M$ is upper bounded by the smallest x such that this inequality holds, or equivalently, such that

$$x \geq \frac{\ln(2)b - 1 - \ln(x - 1) - \frac{1}{2}\ln(2\pi x)}{\ln(x) - 1}.$$

The right hand side of this equation is upper bounded by $\frac{\ln(2)b}{\ln(x)-1}$. Therefore, $\nu_{r,c}^M$ is certainly upper bounded by the smallest x such that

$$x \geq \frac{\ln(2)b}{\ln(x) - 1}.$$

This expression can be efficiently evaluated for all values of b, and it turns out that the value of $\nu_{r,c}^{2^r}$ increases from $b/4$ for values of b close to 200 to values $b/6$ for values of b close to 2000.

Case $\lambda > 1$. For large λ, Eq. (33) becomes numerically unstable. We derive a formula for integer values of λ, or equivalently values of M that are a multiple of 2^r (w.l.o.g.). By a change of variable from x to $x = \lambda + y$ we obtain for the left hand side of (32):

$$\frac{2^b e^{-\lambda}\lambda^x}{(x - \lambda)x!} = \frac{2^b e^{-\lambda}\lambda^{\lambda+y}}{y(\lambda + y)!} = \frac{2^b \lambda^y}{y(\lambda + y)_y}\frac{(\lambda/e)^\lambda}{\lambda!} \leq \frac{2^b \lambda^y}{y\sqrt{2\pi\lambda}(\lambda + y)_y}$$

using Stirling's approximation. Now (32) holds provided that

$$\frac{2^b \lambda^y}{y\sqrt{2\pi\lambda}(\lambda + y)_y} = \frac{2^b}{y\sqrt{2\pi\lambda}\prod_{i=1}^y(1 + \frac{i}{\lambda})} \leq 1.$$

Taking the logarithm:

$$\sum_{i=1}^{y} \ln\left(1 + \frac{i}{\lambda}\right) + \ln(y) + \frac{1}{2}\ln(2\pi\lambda) \geq \ln(2)b. \tag{34}$$

This equation allows efficiently computing $\nu_{r,c}^M$ for $M > 2^r$ and also to prove a simple upper bound for the range $\lambda > 1$.

Lemma 5. *For M a nonzero integer multiple of 2^r, we have*

$$\nu_{r,c}^M \leq \frac{M}{2^r} + \nu_{r,c}^{2^r}\left\lceil\sqrt{\frac{M}{2^r}}\right\rceil.$$

Proof. First of all, note that for $\lambda = 1$, (34) is satisfied for $y = \nu_{r,c}^{2^r} - 1$. Therefore, we have

$$\Xi := \sum_{i=1}^{\nu_{r,c}^{2^r}-1} \ln(1+i) + \ln(\nu_{r,c}^{2^r} - 1) + \frac{1}{2}\ln(2\pi) - \ln(2)b \geq 0.$$

Our goal is to prove that (34) holds for $y = \nu_{r,c}^{2^r}\lceil\sqrt{\lambda}\rceil$. Since $\Xi \geq 0$, we will in fact prove that

$$\sum_{i=1}^{\nu_{r,c}^{2^r}\lceil\sqrt{\lambda}\rceil} \ln\left(1 + \frac{i}{\lambda}\right) + \ln(\nu_{r,c}^{2^r}\lceil\sqrt{\lambda}\rceil) + \frac{1}{2}\ln(2\pi\lambda) - \ln(2)b \geq \Xi.$$

Note that

$$\sum_{i=1}^{\nu_{r,c}^{2^r}\lceil\sqrt{\lambda}\rceil} \ln\left(1 + \frac{i}{\lambda}\right) + \ln(\nu_{r,c}^{2^r}\lceil\sqrt{\lambda}\rceil) + \frac{1}{2}\ln(2\pi\lambda) - \ln(2)b - \Xi$$

$$\geq \sum_{i=1}^{\nu_{r,c}^{2^r}\lceil\sqrt{\lambda}\rceil} \ln\left(1 + \frac{i}{\lambda}\right) - \sum_{i=1}^{\nu_{r,c}^{2^r}-1} \ln(1+i).$$

This can be rewritten as

$$\sum_{i=0}^{\nu_{r,c}^{2^r}-1}\left(\sum_{j=1}^{\lceil\sqrt{\lambda}\rceil} \ln\left(1 + \frac{i\lceil\sqrt{\lambda}\rceil + j}{\lambda}\right) - \ln(1+i)\right),$$

and our claim holds if we can prove that the summand is at least 0 for all $i = 0, \ldots, \nu_{r,c}^{2^r} - 1$. This is easily verified as

$$\sum_{j=1}^{\lceil\sqrt{\lambda}\rceil} \ln\left(1 + \frac{i\lceil\sqrt{\lambda}\rceil + j}{\lambda}\right) \geq \sum_{j=1}^{\lceil\sqrt{\lambda}\rceil} \ln\left(1 + \frac{i\lceil\sqrt{\lambda}\rceil}{\lambda}\right) = \ln\left(\left(1 + \frac{i\lceil\sqrt{\lambda}\rceil}{\lambda}\right)^{\lceil\sqrt{\lambda}\rceil}\right),$$

which is at least

$$\ln\left(1 + \frac{i\lceil\sqrt{\lambda}\rceil^2}{\lambda}\right) \geq \ln(1 + i),$$

as in general $(1 + x)^y \geq 1 + xy$.   □

Clearly, for large M, $\nu_{r,c}^M$ asymptotically converges to $M/2^r$.

6.6   Dealing with Non-uniform Sampling

In this section we address the non-uniform balls-and-bins problem. We consider the balls-and-bins problems for some values r and c where the probability that a ball hits a particular bin (of the 2^r bins) is not 2^{-r}. In other words, the distribution is not uniform. In general the probability distribution for the n-th ball depends on how the previous $n - 1$ balls were distributed. We denote this distribution by D and define $D(i \mid s)$ as the probability that a ball falls in bin i given the sequence s of bins in which the previous $n - 1$ balls fell. We denote by $\nu_{r,c}^{D,M}$ the variant of the function with the same name with the given distribution.

Definition 4. *The* multicollision limit function *for some distribution D, $\nu_{r,c}^{D,M}$, with M a natural number, returns a natural number and is defined as follows. Assume we independently distribute M balls in 2^r bins according to a distribution D. If we call the number of balls in the bin with the highest number of balls μ, then $\nu_{r,c}^{D,M}$ is defined as the smallest natural number x that satisfies:*

$$\Pr\left(\mu > x\right) < \frac{x}{2^c}.$$

We can now prove the following lemma.

Lemma 6. *If for every bin, according to the distribution D the probability for a ball to end up in that bin satisfies $|D(i \mid s) - 2^{-r}| \leq y2^{-r}$ for some $y \leq 0.1$ and any i and s, then $\nu_{r,c}^{D,M} \leq \nu_{r,c}^{2M}$, provided $M \leq y2^c$ and $r \geq 5$.*

Before proving Lemma 6, note that in our application of the lemma in Sect. 6.4, the ith ball hits a certain bin with probability

$$\frac{2^c - (i - 1)}{2^b - (i - 1)} \leq p \leq \frac{2^c}{2^b - (i - 1)}.$$

Assuming that $i - 1 \leq y2^c$ and $y \leq 0.1$, we obtain that $(1 - y) \cdot 2^{-r} \leq p \leq (1 + y) \cdot 2^{-r}$, and that the condition imposed by Lemma 6 is satisfied. As in our setup there are in total $M + N$ queries to f, this is satisfied if $M + N \leq 0.1 \times 2^c$.

Proof. Consider the following two experiments:

- Experiment 1: we drop $2M$ balls into 2^r bins and the distribution is uniform.
- Experiment 2: we drop M balls into 2^r bins and the probability for a ball to land in any particular bin is between $(1 - y) \cdot 2^{-r}$ and $(1 + y) \cdot 2^{-r}$.

We need to prove that $\nu_{r,c}^{2M}$ of the first experiment is at least $\nu_{r,c}^{D,M}$ of the second experiment. The general strategy is as follows. First, we prove that $\nu_{r,c}^{2M}$ is lower bounded by some threshold t. Then, if for all $x \geq t$, we have $\Pr\left(\mu^{\exp 1} > x\right) \geq \Pr\left(\mu^{\exp 2} > x\right)$, then $\nu_{r,c}^{D,M} \leq \nu_{r,c}^{2M}$ because $x = \nu_{r,c}^{2M}$ satisfies the equation $\Pr\left(\mu^{\exp 2} > x\right) < \frac{x}{2^c}$. Clearly, the condition above is satisfied if for all $x \geq t$ and for all bins i, we have $\Pr\left(X_i^{\exp 1} > x\right) \geq \Pr\left(X_i^{\exp 2} > x\right)$, where X_i is the number of balls in bin i. And in turn, it is satisfied if for all $x \geq t$ and for all bins i, we have $\Pr\left(X_i^{\exp 1} = x\right) \geq \Pr\left(X_i^{\exp 2} = x\right)$.

First, by the pigeonhole principle, in experiment 1 there is always a bin with $\max\{2M/2^r, 1\}$ balls, and $\nu_{r,c}^{2M}$ is at least this value: $\nu_{r,c}^{2M} \geq \max\{2M/2^r, 1\}$. Then, consider any bin and the probability that it contains exactly x balls. In experiment 1, the bin contains exactly x balls if in the sequence of $2M$ balls, x balls fall into the particular bin and $2M - x$ fall in another bin, and thus:

$$\Pr\left(X_i^{\exp 1} = x\right) = \binom{2M}{x} \cdot (2^{-r})^x \cdot (1 - 2^{-r})^{2M-x}.$$

For experiment 2 we likewise obtain, using the fact that the ith ball ends in the bin with probability $(1 - y) \cdot 2^{-r} \leq p \leq (1 + y) \cdot 2^{-r}$ for any i:

$$\Pr\left(X_i^{\exp 1} = x\right) \leq \binom{M}{x} \cdot ((1 + y) \cdot 2^{-r})^x \cdot (1 - (1 - y) \cdot 2^{-r})^{M-x}.$$

Using that $\binom{2M}{x}/\binom{M}{x} \geq 2^x$ and $(1 - 2^{-r})^2 \geq 1 - 2 \cdot 2^{-r}$, the condition certainly holds if

$$\left(\frac{2}{1+y} \frac{1 - (1 - y) \cdot 2^{-r}}{1 - 2^{-r}}\right)^x \geq \left(\frac{1 - (1 - y) \cdot 2^{-r}}{1 - 2 \cdot 2^{-r}}\right)^M,$$

which in turn certainly holds if

$$\left(\frac{2}{1+y}\right)^x \geq \left(1 + \frac{1+y}{2^r - 2}\right)^M,$$

which in turn certainly holds if

$$\left(\frac{2}{1+y}\right)^x \geq \left(1 + \frac{1+y}{2^r - 2}\right)^{\max\{M, 2^{r-1}\}}. \tag{35}$$

We have to prove that this condition holds for all $x \geq \max\{2M/2^r, 1\}$. The left hand side is increasing in x, whereas the right hand side is constant in x, and we therefore only have to prove it for $x = \max\{2M/2^r, 1\}$ (w.l.o.g., assuming that x is integral). Therefore, our goal now is to prove that

$$\frac{2}{1+y} \geq \left(1 + \frac{1+y}{2^r - 2}\right)^{2^{r-1}}.$$

Using that $1 + a \le e^a$ and $r \ge 5$, the condition is satisfied if

$$\frac{2}{1+y} \ge e^{(1+y)\frac{16}{30}},$$

which in turn holds for $y \le 0.1$. $\square$

Acknowledgement. Bart Mennink is supported by a postdoctoral fellowship from the Netherlands Organisation for Scientific Research (NWO) under Veni grant 016.Veni.173.017.

References

1. Andreeva, E., Bogdanov, A., Luykx, A., Mennink, B., Mouha, N., Yasuda, K.: How to securely release unverified plaintext in authenticated encryption. In: Sarkar and Iwata [45], pp. 105–125
2. Andreeva, E., Daemen, J., Mennink, B., Van Assche, G.: Security of keyed sponge constructions using a modular proof approach. In: Leander [36], pp. 364–384
3. Aumasson, J., Jovanovic, P., Neves, S.: NORX v3.0 (2016). Submission to CAESAR Competition. https://competitions.cr.yp.to/round3/norxv30.pdf
4. Bellare, M., Bernstein, D.J., Tessaro, S.: Hash-function based PRFs: AMAC and its multi-user security. In: Fischlin and Coron [28], pp. 566–595
5. Bellare, M., Boldyreva, A., Micali, S.: Public-key encryption in a multi-user setting: security proofs and improvements. In: Preneel, B. (ed.) EUROCRYPT 2000. LNCS, vol. 1807, pp. 259–274. Springer, Heidelberg (2000). https://doi.org/10.1007/3-540-45539-6_18
6. Bellare, M., Rogaway, P.: Random oracles are practical: a paradigm for designing efficient protocols. In: Denning, D.E., Pyle, R., Ganesan, R., Sandhu, R.S., Ashby, V. (eds.) ACM CCS 1993, pp. 62–73. ACM (1993)
7. Bernstein, D.J.: The Poly1305-AES message-authentication code. In: Gilbert, H., Handschuh, H. (eds.) FSE 2005. LNCS, vol. 3557, pp. 32–49. Springer, Heidelberg (2005). https://doi.org/10.1007/11502760_3
8. Bertoni, G., Daemen, J., Peeters, M., Van Assche, G.: Sponge functions. In: ECRYPT Hash Workshop 2007, May 2007
9. Bertoni, G., Daemen, J., Peeters, M., Van Assche, G.: On the indifferentiability of the sponge construction. In: Smart, N. (ed.) EUROCRYPT 2008. LNCS, vol. 4965, pp. 181–197. Springer, Heidelberg (2008). https://doi.org/10.1007/978-3-540-78967-3_11
10. Bertoni, G., Daemen, J., Peeters, M., Van Assche, G.: Sponge-based pseudorandom number generators. In: Mangard, S., Standaert, F.-X. (eds.) CHES 2010. LNCS, vol. 6225, pp. 33–47. Springer, Heidelberg (2010). https://doi.org/10.1007/978-3-642-15031-9_3
11. Bertoni, G., Daemen, J., Peeters, M., Van Assche, G.: Duplexing the sponge: single-pass authenticated encryption and other applications. In: Miri and Vaudenay [39], pp. 320–337
12. Bertoni, G., Daemen, J., Peeters, M., Van Assche, G.: The Keccak reference, January 2011
13. Bertoni, G., Daemen, J., Peeters, M., Van Assche, G.: On the security of the keyed sponge construction. In: Symmetric Key Encryption Workshop, February 2011

636   J. Daemen et al.

14. Bertoni, G., Daemen, J., Peeters, M., Van Assche, G.: Permutation-based encryption, authentication and authenticated encryption. In: Directions in Authenticated Ciphers, July 2012
15. Bertoni, G., Daemen, J., Peeters, M., Van Assche, G., Van Keer, R.: CAESAR submission: KETJE V2, September 2016
16. Bertoni, G., Daemen, J., Peeters, M., Van Assche, G., Van Keer, R.: CAESAR submission: KETJE V2, document version 2.2, September 2016
17. Biham, E.: How to decrypt or even substitute DES-encrypted messages in 2^{28} steps. Inf. Process. Lett. **84**(3), 117–124 (2002)
18. Biryukov, A., Mukhopadhyay, S., Sarkar, P.: Improved time-memory trade-offs with multiple data. In: Preneel, B., Tavares, S. (eds.) SAC 2005. LNCS, vol. 3897, pp. 110–127. Springer, Heidelberg (2006). https://doi.org/10.1007/11693383_8
19. Bogdanov, A., Knežević, M., Leander, G., Toz, D., Varıcı, K., Verbauwhede, I.: SPONGENT: a lightweight hash function. In: Preneel, B., Takagi, T. (eds.) CHES 2011. LNCS, vol. 6917, pp. 312–325. Springer, Heidelberg (2011). https://doi.org/10.1007/978-3-642-23951-9_21
20. CAESAR: Competition for authenticated encryption: security, applicability, and robustness, November 2014. http://competitions.cr.yp.to/caesar.html
21. Chang, D., Dworkin, M., Hong, S., Kelsey, J., Nandi, M.: A keyed sponge construction with pseudorandomness in the standard model. In: NIST SHA-3 Workshop, March 2012
22. Chatterjee, S., Menezes, A., Sarkar, P.: Another look at tightness. In: Miri and Vaudenay [39], pp. 293–319
23. Chen, S., Steinberger, J.: Tight security bounds for key-alternating ciphers. In: Nguyen, P.Q., Oswald, E. (eds.) EUROCRYPT 2014. LNCS, vol. 8441, pp. 327–350. Springer, Heidelberg (2014). https://doi.org/10.1007/978-3-642-55220-5_19
24. Dobraunig, C., Eichlseder, M., Mendel, F., Schläffer, M.: Ascon v1.2 (2016). Submission to CAESAR Competition. http://ascon.iaik.tugraz.at
25. Even, S., Mansour, Y.: A construction of a cipher from a single pseudorandom permutation. In: Imai, H., Rivest, R.L., Matsumoto, T. (eds.) ASIACRYPT 1991. LNCS, vol. 739, pp. 210–224. Springer, Heidelberg (1993). https://doi.org/10.1007/3-540-57332-1_17
26. Even, S., Mansour, Y.: A construction of a cipher from a single pseudorandom permutation. J. Cryptol. **10**(3), 151–162 (1997)
27. FIPS 202: SHA-3 Standard: Permutation-Based Hash and Extendable-Output Functions (2015)
28. Fischlin, M., Coron, J.-S. (eds.): EUROCRYPT 2016, Part I. LNCS, vol. 9665. Springer, Heidelberg (2016). https://doi.org/10.1007/978-3-662-49890-3
29. Gaži, P., Pietrzak, K., Tessaro, S.: The exact PRF security of truncation: tight bounds for keyed sponges and truncated CBC. In: Gennaro and Robshaw [31], pp. 368–387
30. Gaži, P., Tessaro, S.: Provably robust sponge-based PRNGs and KDFs. In: Fischlin and Coron [28], pp. 87–116 (2016)
31. Gennaro, R., Robshaw, M. (eds.): CRYPTO 2015, Part I. LNCS, vol. 9215. Springer, Heidelberg (2015). https://doi.org/10.1007/978-3-662-47989-6
32. Guo, J., Peyrin, T., Poschmann, A.: The PHOTON family of lightweight hash functions. In: Rogaway, P. (ed.) CRYPTO 2011. LNCS, vol. 6841, pp. 222–239. Springer, Heidelberg (2011). https://doi.org/10.1007/978-3-642-22792-9_13

33. Hoang, V.T., Tessaro, S.: Key-alternating ciphers and key-length extension: exact bounds and multi-user security. In: Robshaw, M., Katz, J. (eds.) CRYPTO 2016, Part I. LNCS, vol. 9814, pp. 3–32. Springer, Heidelberg (2016). https://doi.org/10.1007/978-3-662-53018-4_1

34. Hong, J., Sarkar, P.: New applications of time memory data tradeoffs. In: Roy, B. (ed.) ASIACRYPT 2005. LNCS, vol. 3788, pp. 353–372. Springer, Heidelberg (2005). https://doi.org/10.1007/11593447_19

35. Jovanovic, P., Luykx, A., Mennink, B.: Beyond $2^{c/2}$ security in sponge-based authenticated encryption modes. In: Sarkar and Iwata [45], pp. 85–104

36. Leander, G. (ed.): FSE 2015. LNCS, vol. 9054. Springer, Heidelberg (2015). https://doi.org/10.1007/978-3-662-48116-5

37. Maurer, U., Renner, R., Holenstein, C.: Indifferentiability, impossibility results on reductions, and applications to the random oracle methodology. In: Naor, M. (ed.) TCC 2004. LNCS, vol. 2951, pp. 21–39. Springer, Heidelberg (2004). https://doi.org/10.1007/978-3-540-24638-1_2

38. Mennink, B., Reyhanitabar, R., Vizár, D.: Security of full-state keyed sponge and duplex: applications to authenticated encryption. In: Iwata, T., Cheon, J.H. (eds.) ASIACRYPT 2015, Part II. LNCS, vol. 9453, pp. 465–489. Springer, Heidelberg (2015). https://doi.org/10.1007/978-3-662-48800-3_19

39. Miri, A., Vaudenay, S. (eds.): SAC 2011. LNCS, vol. 7118. Springer, Heidelberg (2012). https://doi.org/10.1007/978-3-642-28496-0

40. Mouha, N., Luykx, A.: Multi-key security: the Even-Mansour construction revisited. In: Gennaro and Robshaw [31], pp. 209–223

41. Mouha, N., Mennink, B., Van Herrewege, A., Watanabe, D., Preneel, B., Verbauwhede, I.: Chaskey: an efficient MAC algorithm for 32-bit microcontrollers. In: Joux, A., Youssef, A. (eds.) SAC 2014. LNCS, vol. 8781, pp. 306–323. Springer, Cham (2014). https://doi.org/10.1007/978-3-319-13051-4_19

42. Naito, Y., Yasuda, K.: New bounds for keyed sponges with extendable output: independence between capacity and message length. In: Peyrin, T. (ed.) FSE 2016. LNCS, vol. 9783, pp. 3–22. Springer, Heidelberg (2016). https://doi.org/10.1007/978-3-662-52993-5_1

43. Patarin, J.: The "Coefficients H" technique. In: Avanzi, R.M., Keliher, L., Sica, F. (eds.) SAC 2008. LNCS, vol. 5381, pp. 328–345. Springer, Heidelberg (2009). https://doi.org/10.1007/978-3-642-04159-4_21

44. Reyhanitabar, R., Vaudenay, S., Vizár, D.: Boosting OMD for almost free authentication of associated data. In: Leander [36], pp. 411–427

45. Sarkar, P., Iwata, T. (eds.): ASIACRYPT 2014, Part I. LNCS, vol. 8873. Springer, Heidelberg (2014). https://doi.org/10.1007/978-3-662-45611-8

46. Sasaki, Y., Yasuda, K.: How to incorporate associated data in sponge-based authenticated encryption. In: Nyberg, K. (ed.) CT-RSA 2015. LNCS, vol. 9048, pp. 353–370. Springer, Cham (2015). https://doi.org/10.1007/978-3-319-16715-2_19

Improved Security for OCB3

Ritam Bhaumik[(⊠)] and Mridul Nandi

Indian Statistical Institute, Kolkata, India
bhaumik.ritam@gmail.com, mridul.nandi@gmail.com

Abstract. OCB3 is the current version of the OCB authenticated
encryption mode which is selected for the third round in CAESAR. So far
the integrity analysis has been limited to an adversary making a single
forging attempt. A simple extension for the best known bound establishes
integrity security as long as the total number of query blocks (including
encryptions and forging attempts) does not exceed the birthday-bound.
In this paper we show an improved bound for integrity of OCB3 in terms
of the number of blocks in the forging attempt. In particular we show that
when the number of encryption query blocks is not more than birthday-
bound (an assumption without which the privacy guarantee of OCB3
disappears), even an adversary making forging attempts with the num-
ber of blocks in the order of $2^n/\ell_{\mathrm{MAX}}$ (n being the block-size and ℓ_{MAX}
being the length of the longest block) may fail to break the integrity of
OCB3.

Keywords: OCB · OCB3 · Authenticated encryption · Integrity · Multiple verification query

1 Introduction

Authenticated encryption schemes [Rog02], which target both data confidentiality and integrity simultaneously, have received considerable attention in recent
years. The increased interest is in part due to the ongoing CAESAR competition [cae], which aims to deliver a portfolio of state-of-the-art authenticated
encryption schemes covering a spectrum of security and efficiency trade-offs.
While other possibilities exist, it is natural to build AE schemes from blockciphers, employing some mode of operation. Some of the known blockcipher
based authenticated encryptions are OCB [RBB03, Rog04], GCM [MV04, MV05],
COPA [ABL+13], ELmD [DN14] and AEZ [HKR15]. Due to the CAESAR competition, many designs have appeared in literature. Moreover, some designs have
been refined for better performance and improved security. OCB3 (submitted to
CAESAR and now a third round candidate) is one such example, an enhancement of the well known construction OCB.

OCB and OCB3. OCB is a blockcipher-based mode of operation that achieves
authenticated encryption in almost the same amount of time as the fastest conventional mode, CTR mode [WHF02], which achieves privacy alone in that time.

© International Association for Cryptologic Research 2017
T. Takagi and T. Peyrin (Eds.): ASIACRYPT 2017, Part II, LNCS 10625, pp. 638–666, 2017.
https://doi.org/10.1007/978-3-319-70697-9_22

Despite this, OCB is simple and clean, and easy to implement in either hardware or software. For every message, it uses two blockcipher calls to process a constant block 0^n and a nonce, one blockcipher call for each message block and one additional blockcipher call for the checksum.

The refined OCB or OCB3 [KR11] aims to shave off one AES encipherment per message encrypted about 98% of the time. The nonce here is used as a counter, i.e., in a given session, its top segment (of 122 bits) stays fixed, while, with each successive message, the bottom segment (of 6 bits) gets bumped up by one. This is the approach recommended in RFC 5116 [McG08, Sect. 3.2].

Known Security Results of OCB3. Though the original OCB has already been proved to be secure, the security bound provided by [KR11], in particular the authenticity bound, does not show the standard birthday-bound security when the adversary is allowed to make multiple verification queries. More formally, the original bound is $O(\sigma_T^2/N) + O(1/2^\tau)$ where σ_T denotes the total number of input blocks in q encryption queries and τ denotes the tag length, if the number of verification queries is one. This bound generally implies $O(q'\sigma_T^2/N) + O(q'/2^\tau)$ when the number of verification queries is $q' \geq 1$, hence the provable security is degraded.

Another adaptation of the proof can be applied to obtain a bound of the form $O((\sigma_T + \sigma_T')^2/N + q'/2^\tau)$, where σ_T' is the blocks in the decryption queries. It still remains birthday-bound in σ_T'. All known attacks exploit the collision in the input blocks for all encryption queries and hence σ_T^2/N is tight. But no matching attack with advantage $\sigma_T'^2/N$ is known. A recent collision attack on PHash [GPR17] (used to process the associated data) can be applied to obtain an integrity attack with advantage $O(\sigma_T'/N)$. (See [BGM04].)

Our Contribution. We show that the existing attack is the best possible by improving the integrity advantage. We follow the combined AE security distinguishing game [RS06] to bound the integrity security of OCB3. We use Patarin's coefficients H technique [Pat08] to bound the AE distinguishing game.

Theorem 1 (Main Result). *Let $\mathcal{A}$ be an adversary that makes q encryption queries consisting of σ message blocks in all with at most ℓ_{MAX} blocks per query, and α associated data blocks in all, and q' decryption queries in a nonce-respecting authenticated encryption security game with associated data against a real oracle $\mathcal{O}_1$ representing OCB3 and an ideal oracle $\mathcal{O}_0$ representing an ideal nonce-based authenticated encryption function. Then*

$$\mathbf{Adv}^{naead}_{\mathcal{O}_1, \mathcal{O}_0}(\mathcal{A}) \leq \frac{5\sigma_T^2}{N} + \frac{2\sigma^4}{N^2} + \frac{64q'\ell_{MAX} + 15q'}{N},$$

where $\sigma_T = \sigma + \alpha + q$ is the total number of blocks queried in the encryption queries (including messages, associated data and nonces).

2   Preliminaries

$\mathbb{N}$ denotes the set of non-negative integers. For $n \in \mathbb{N}$, $[n]$ denotes the set $\{1, \ldots, n\}$ ($[0]$ is thus the empty set). For $m, n \in \mathbb{N}$, $[m..n]$ denotes the set $\{m, \ldots, n\}$ (which is the empty set when $m > n$). For a binary string $x \in \{0, 1\}^*$, $|x|$ will denote the number of bits in x. We fix an arbitrary block-length $n \in \mathbb{N} \setminus \{0\}$. If $|x| = n$, we call x a complete block; if $|x| < n$, we call x an incomplete block; if $|x| = 0$ (the null string), we call x the empty block. (By convention, the empty block is also an incomplete block.) $\oplus$ and $\cdot$ denote the field addition (XOR) and field multiplication respectively over the finite field $\{0, 1\}^n$. During calculations, for two block x and y, we will simply write $x + y$ to denote $x \oplus y$.

For $i \in [|x|]$, $x[i]$ denotes the i-th bit of x (we begin all indexing from 1, so $x[1]$ is the first bit of x). For $i \geq 1$ and $j \leq |x|$, $x[i..j]$ denotes the $(j - i + 1)$-bit contiguous substring of x starting at the i-th bit when $i \leq j$, and the empty string otherwise. For two strings x and y, $x||y$ denotes the concatenation of x and y. For a bit b, b^m denotes an m-bit string with each bit equal to b.

Any $x \in \{0, 1\}^*$ can be mapped uniquely to a sequence $(x_1, \ldots, x_\ell, x_*)$, where $\ell \in \mathbb{N}$, $x_1, \ldots, x_\ell$ are complete blocks, and x_* is an incomplete (possibly empty) block, such that

$$x = x_1 || \cdots || x_\ell || x_*.$$

For this mapping we take $\ell = \lfloor |x|/n \rfloor$, $x_i = x[n(i - 1) + 1..ni]$ for $i \in [\ell]$, and $x_* = x[n\ell + 1..|x|]$. For an incomplete block x, $\mathsf{pad}(x)$ denotes the complete block

$$x||10^* = x||1||0^{n-|x|-1}.$$

For a complete block x, $\mathsf{chop}_k(x)$ denotes the incomplete block $x[1..k]$. For some $m \in \mathbb{N}$, for $x \in \{0, 1\}^m, k \in [m]$, $x \gg k$ denotes x rotated b bits to the right, i.e., $0^b||x[1..m-b]$, while $x \ll b$ denotes x rotated b bits to the left, i.e., $x[b+1..m]||0^b$.

We say a function $f : \{0, 1\}^n \longrightarrow \{0, 1\}^n$ is *partially determined* if we know the values of f on a strict subset of $\{0, 1\}^n$. This subset is called $\mathsf{Dom}(f)$. A partially determined state of f can be viewed as a restriction of f to $\mathsf{Dom}(f)$. The range of this restricted function is called $\mathsf{Ran}(f)$. We will treat a partially determined function as *updatable*: for some $x \in \{0, 1\}^n \setminus \mathsf{Dom}(f)$ and some $y \in \{0, 1\}^n$, (x, y) may be added to f, so that $\mathsf{Dom}(f)$ expands to $\mathsf{Dom}(f) \cup \{x\}$, and $\mathsf{Ran}(f)$ becomes $\mathsf{Ran}(f) \cup \{y\}$. We say f is *permutation-compatible* if $|\mathsf{Dom}(f)| = |\mathsf{Ran}(f)|$.

For a set $\mathcal{S}$, we write $x \xleftarrow{\$} \mathcal{S}$ to denote that x is sampled from S uniformly. For a given domain $\mathcal{D}$ and a given co-domain $\mathcal{R}$, $\mathsf{Func}[\mathcal{D}, \mathcal{R}]$ will denote the set of all functions from $\mathcal{D}$ into $\mathcal{R}$. We say f^* is an ideal random function from $\mathcal{D}$ to $\mathcal{R}$ to indicate that $f^* \xleftarrow{\$} \mathsf{Func}[\mathcal{D}, \mathcal{R}]$. If f^* is an ideal random function, it can be viewed as a *with-replacement* sampler from $\mathcal{R}$: for distinct inputs $x_1, \ldots, x_m \in \mathcal{D}$, $f^*(x_i) \xleftarrow{\$} \mathcal{R}$ for $i \in [m]$, and $f^*(x_1), \ldots, f^*(x_m)$ are all independent. Similarly $\mathsf{Perm}[\mathcal{D}]$ will denote the set of all permutations on $\mathcal{D}$. We say π^* is an ideal random permutation on $\mathcal{D}$ to indicate that $\pi^* \xleftarrow{\$} \mathsf{Perm}[\mathcal{D}]$. If f^* is an ideal random permutation, it can be viewed as a *without-replacement* sampler from $\mathcal{D}$: for distinct inputs $x_1, \ldots, x_m \in \mathcal{D}$, $(\pi^*(x_1), \ldots, \pi^*(x_m)) \xleftarrow{\$} \mathcal{D}^{\underline{s}}$ where $\mathcal{D}^{\underline{s}}$ denotes the set of all s-tuples of distinct elements from $\mathcal{D}$.

2.1   Some Basic Results

We briefly state some results which would be used in our security analysis.

Property-1. Suppose $X_1, \ldots, X_s$ is a random without-replacement sample from $\mathcal{D}$. Then for any $1 \leq i_1 < \cdots < i_r \leq s$, $X_{i_1}, \ldots, X_{i_r}$ is also a random without-replacement sample from $\mathcal{D}$. In other words, the joint distribution of a without-replacement sample is independent of the ordering of the sample.

Property-2. Suppose A is a binary full row rank matrix of dimension $nd \times ns$ for some positive integers n, d and s. Let $X_1, \ldots, X_s$ be a without-replacement sample from $\{0, 1\}^n$, and X be the column vector $(X_1, \ldots, X_n)$. Then

$$\Pr\left[AX = c\right] \leq \frac{1}{(2^n - s + r) \cdots (2^n - s + 1)}$$

for any d dimensional binary vector c. Moreover if c is not in the column space of A then this probability is zero.

2.2   Distinguishing Advantage

For two oracles $\mathcal{O}_0$ and $\mathcal{O}_1$, an algorithm $\mathcal{A}$ trying to distinguish between $\mathcal{O}_0$ and $\mathcal{O}_1$ is called a distinguishing adversary. $\mathcal{A}$ plays an interactive game with $\mathcal{O}_b$ for some bit b unknown to $\mathcal{A}$, and then outputs a bit $b_{\mathcal{A}}$. The winning event is $[b_{\mathcal{A}} = b]$. The distinguishing advantage of $\mathcal{A}$ is defined as

$$\mathbf{Adv}_{\mathcal{O}_1, \mathcal{O}_0}(\mathcal{A}) := \left| \Pr\left[b_{\mathcal{A}} = 1 \mid b = 1\right] - \Pr\left[b_{\mathcal{A}} = 1 \mid b = 0\right] \right|.$$

Let $\mathbf{A}[q, t]$ be the class of all distinguishing adversaries limited to q oracle queries and t computations. We define

$$\mathbf{Adv}_{\mathcal{O}_1, \mathcal{O}_0}[q, t] := \max_{\mathcal{A} \in \mathbf{A}[q, t]} \mathbf{Adv}_{\mathcal{O}_1, \mathcal{O}_0}(\mathcal{A}).$$

When the adversaries in $\mathbf{A}[q, t]$ are allowed to make both encryption queries and decryption queries to the oracle, this is written as $\mathbf{Adv}_{\pm \mathcal{O}_0, \pm \mathcal{O}_1}[q, q', t]$, where q is the maximum number of encryption queries allowed and q' is the maximum number of decryption queries allowed. Enc_b and Dec_b denote respectively the encryption and decryption function associated with $\mathcal{O}_b$.

$\mathcal{O}_0$ conventionally represents an ideal primitive, while $\mathcal{O}_1$ represents either an actual construction or a mode of operation built of some other ideal primitives. Typically the goal of the function represented by $\mathcal{O}_1$ is to emulate the ideal primitive represented by $\mathcal{O}_0$. We use the standard terms *real oracle* and *ideal oracle* for $\mathcal{O}_1$ and $\mathcal{O}_0$ respectively. A security game is a distinguishing game with an optional set of additional restrictions, chosen to reflect the desired security goal. When we talk of distinguishing advantage with a specific security game G in mind, we include G in the superscript, e.g., $\mathbf{Adv}_{\mathcal{O}_1, \mathcal{O}_0}^{G}(\mathcal{A})$.

2.3   The Authenticated Encryption Security Game

A *nonce-based authenticated encryption scheme with associated data* consists of a key space $\mathcal{K}$, a message space $\mathcal{M}$, a tag space $\mathfrak{T}$, a nonce space $\mathcal{N}$ and an associated data space $\mathfrak{A}$, along with two functions $\mathsf{Enc} : \mathcal{K} \times \mathcal{N} \times \mathfrak{A} \times \mathcal{M} \longrightarrow \mathcal{M} \times \mathfrak{T}$ and $\mathsf{Dec} : \mathcal{K} \times \mathcal{N} \times \mathfrak{A} \times \mathcal{M} \times \mathfrak{T} \longrightarrow \mathcal{M} \cup \{\bot\}$, with the correctness condition that for any $K \in \mathcal{K}, \mathsf{N} \in \mathcal{N}, \mathsf{A} \in \mathfrak{A}, \mathsf{M} \in \mathcal{M}$, we have

$$\mathsf{Dec}(K, \mathsf{N}, \mathsf{A}, \mathsf{Enc}(K, \mathsf{N}, \mathsf{A}, \mathsf{M})) = \mathsf{M}.$$

In addition, in most popular authenticated encryption schemes (including OCB3), the map $p_{\mathcal{M}} \circ \mathsf{Enc}(K, \mathsf{N}, \mathsf{A}, \cdot)$ for fixed $K, \mathsf{N}, \mathsf{A}$ is a length-preserving permutation, where $p_{\mathcal{M}} : \mathcal{M} \times \mathfrak{T} \longrightarrow \mathcal{M}$ is the projection on $\mathcal{M}$.

In the nonce-respecting authenticated encryption security game with associated data naead, Enc_1 and Dec_1 of the real oracle are the encryption function $\mathsf{Enc}(K, \cdot, \cdot, \cdot)$ and decryption function $\mathsf{Dec}(K, \cdot, \cdot, \cdot, \cdot)$ respectively of the authenticated encryption scheme under consideration for a key K randomly chosen from $\mathcal{K}$; in the ideal oracle, $\mathsf{Enc}_0 : \mathcal{N} \times \mathfrak{A} \times \mathcal{M} \longrightarrow \mathcal{M} \times \mathfrak{T}$ is an ideal random function from $\mathcal{N} \times \mathfrak{A} \times \mathcal{M}$ to $\mathcal{M} \times \mathfrak{T}$, and $\mathsf{Dec}_0 : \mathcal{N} \times \mathfrak{A} \times \mathcal{M} \times \mathfrak{T} \longrightarrow \mathcal{M} \cup \{\bot\}$ is the constant function that returns $\bot$ irrespective of the input. We henceforth refer to $(\mathsf{Enc}_0, \mathsf{Dec}_0)$ as the *ideal nonce-based authenticated encryption scheme*. The distinguishing adversary operates under the following restrictions:

– no two encryption queries can have the same nonce;
– if an encryption query $(\mathsf{N}, \mathsf{A}, \mathsf{M})$ yields (C, T), a decryption query $(\mathsf{N}, \mathsf{A}, \mathsf{C}, \mathsf{T})$ is not allowed.

The distinguishing advantage of the adversary in the nonce-respecting authenticated encryption security game with associated data will be denoted $\mathbf{Adv}_{\mathcal{O}_1, \mathcal{O}_0}^{\mathsf{naead}}(\mathcal{A})$. Note that security under this formulation covers the two standard security goals of authenticated encryption:

– *(Privacy)* Security against an adversary who tries to distinguish the construction from an ideal prf $f^* : \mathcal{M} \longrightarrow \mathcal{M} \times \mathfrak{T}$, and
– *(Integrity)* Security against an adversary who tries to make a successful forging attempt on the construction.

2.4   Coefficients H Technique

Consider a security game G where the adversary can make both encryption queries and decryption queries. The part of the computation visible to the adversary at the time of choosing its final response is known as a view. This includes the queries and the responses, and may also include any additional information the oracle chooses to reveal to the adversary at the end of the query-response phase of the game. The probability of the security game with an oracle $\mathcal{O}$ resulting in a given view V is known as the interpolation probability of V, denoted $\mathsf{ip}_{\mathcal{O}}[V]$.

Note that for a view to be realised, two things need to happen:

- The adversary needs to make the queries listed in the view;
- The oracle needs to make the corresponding responses.

Of these, the former is deterministic; the latter, probabilistic. Thus when we talk of interpolation probability, we are only concerned with the oracle responses, with the assumption that the adversary's queries are consistent with the view. For any other adversary, the interpolation probability is trivially 0. Thus $\mathsf{ip}_{\mathcal{O}}[V]$ depends only on the oracle $\mathcal{O}$ and the view V and not on the adversary; hence the notation.

We extend the notation of interpolation probability to a set $\mathcal{V}$ of views: $\mathsf{ip}_{\mathcal{O}}[\mathcal{V}]$ denotes the probability that the security game with $\mathcal{O}$ results in a view $V \in \mathcal{V}$. Now we state a theorem, due to Jacques Patarin, known as the Coefficient H Technique.

Theorem 2 (Coefficient H Technique). *[Pat08] Suppose there is a set $\mathcal{V}_{bad}$ of views satisfying the following:*

- $\mathsf{ip}_{\mathcal{O}_0}[\mathcal{V}_{bad}] \leq \epsilon_1$;
- *For any $V \notin \mathcal{V}_{bad}$,*

$$\frac{\mathsf{ip}_{\mathcal{O}_1}[V]}{\mathsf{ip}_{\mathcal{O}_0}[V]} \geq 1 - \epsilon_2.$$

Then for an adversary $\mathcal{A}$ trying to distinguish between $\mathcal{O}_1$ and $\mathcal{O}_0$, we have the following bound on its distinguishing advantage:

$$\mathbf{Adv}^G_{\pm\mathcal{O}_1,\pm\mathcal{O}_0}(\mathcal{A}) \leq \epsilon_1 + \epsilon_2.$$

3   OCB3 Construction

The OCB3 encryption and decryption algorithms are described in Algorithm 1. We take block length $n = 128$. E_K denotes a call to blockcipher, and $\mathcal{H}_K$ denotes a call to a hash function based on the stretch-then-shift xor-universal hash H_κ, as described below in Subsect. 3.1. Note that they share the same key K. Hashing of associated data A is done through parallel masked calls to E_K, which are added to get the authentication key auth. The message space $\mathcal{M}$ consists of all messages with at least one full block, i.e., all strings of 128 bits or more, and the nonce space N consists of all 128-bit strings whose first 122 bits are not all 0. The message M is encrypted in ECB mode, with masking that incorporates the nonce N. If there is an incomplete block at the end, it is added after a 10* padding to an encrypted masking key. Finally, a checksum of the message blocks is masked and encrypted through E_K, and auth is added to it to produce the authentication tag T. We ignore here the last step of OCB3, where a tag of a desired length τ is obtained by chopping T as required. A schematic view of the encryption is illustrated in Fig. 1, which treats each masked blockcipher call as a call to a tweakable blockcipher. The masking scheme corresponding to the various tweakable blockcipher calls is given in Table 1. The important thing to note here is that *all the masking coefficients are distinct*.

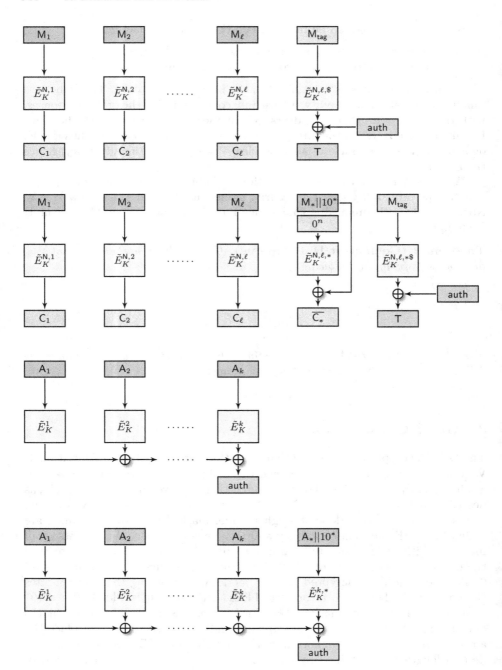

Fig. 1. A schematic view of the OCB3 construction. **Top to Bottom:** encrypting M when $|M_*| = 0$; encrypting M when $|M_*| > 0$; hashing A when $|A_*| = 0$; hashing A when $|A_*| > 0$.

Table 1. Masking scheme corresponding to the various blockcipher calls in the schematic view of OCB3. $L = E_K(0)$; $Q = \mathcal{H}_K(N)$, $\mathcal{H}_K$ being the hash function described in Subsect. 3.1, and N the nonce; and the tweak space is $\mathcal{N} \times ((\mathbb{N} \setminus \{0\}) \cup (\mathbb{N} \times \{*, \$, *\$\})) \cup (\mathbb{N} \setminus \{0\} \cup (\mathbb{N} \times \{*\}))$. For $i \in \mathbb{N}$, $a(i) := \bigoplus_{j \leq i}(1 \ll \mathsf{ntz}(j))$ denotes the Gray-code representation of i, $\mathsf{ntz}(j)$ being the number of trailing 0's in the binary representation of j.

Call		Definition		Coeff.	Def.
$\tilde{E}_K^{N,i}$	(x)	$E_K(Q \oplus \lambda_i \cdot L \oplus x) \oplus Q \oplus \lambda_i \cdot L$, $i \in \mathbb{N} \setminus \{0\}$		λ_i	$4a(i)$
$\tilde{E}_K^{N,i,*}$	(x)	$E_K(Q \oplus \lambda_i^* \cdot L \oplus x)$,	$i \in \mathbb{N}$	λ_i^*	$4a(i) + 1$
$\tilde{E}_K^{N,i,\$}$	(x)	$E_K(Q \oplus \lambda_i^\$ \cdot L \oplus x)$,	$i \in \mathbb{N}$	$\lambda_i^\$$	$4a(i) + 2$
$\tilde{E}_K^{N,i,*\$}$	(x)	$E_K(Q \oplus \lambda_i^{*\$} \cdot L \oplus x)$,	$i \in \mathbb{N}$	$\lambda_i^{*\$}$	$4a(i) + 3$
$\tilde{E}_K^i$	(x)	$E_K(\lambda_i \cdot L \oplus x)$,	$i \in \mathbb{N} \setminus \{0\}$		
$\tilde{E}_K^{i,*}$	(x)	$E_K(\lambda_i^* \cdot L \oplus x)$,	$i \in \mathbb{N}$		

3.1   Stretch-then-Shift Hash

In this subsection we describe the hash function $\mathcal{H}_K$ used in OCB3 to process the nonce N. It is based on an xor-universal hash function H_κ with a 128-bit key and a 6-bit input, defined as

$$H_\kappa(x) := ((\kappa \| (\kappa \oplus (\kappa \ll 8))) \ll x) \gg 128.$$

This is a linear function of κ, and thus can be described as left multiplication with a matrix $\mathbf{H}[x]$ as

$$H_\kappa(x) := \mathbf{H}[x] \cdot \kappa.$$

It is easy to show that when $\kappa \xleftarrow{\$} \{0,1\}^{128}$, for any $x \in \{0,1\}^6$, we have $H_\kappa(x) \xleftarrow{\$} \{0,1\}^{128}$. The authors show with a computer-aided exhaustive search that when κ is uniform over $\{0,1\}^{128}$, for any $x, x' \in \{0,1\}^6, x \neq x'$ and any $\delta \in \{0,1\}^{128}$, we have

$$\Pr\left[H_\kappa(x) \oplus H_\kappa(x') = \delta\right] = \frac{1}{2^{128}}.$$

We describe here a generalised hash $\mathcal{H}[\pi]$, based on an arbitrary permutation π. We begin by splitting N into two parts:

$$\mathsf{TN} = \tau(\mathsf{N}) := (\mathsf{N} \gg 6) \ll 6,$$
$$\mathsf{BN} = \beta(\mathsf{N}) := \mathsf{N} \oplus \tau(\mathsf{N}),$$

so as BN denotes the last 6 bits of N, and TN denotes the first 122 bits, with 6 0's appended at the end (Note that as long as $\mathsf{N} \in \mathcal{N}$, TN cannot be 0). Next we define

$$\mathsf{KN} = \kappa(\mathsf{N}) := \pi(\tau(\mathsf{N})).$$

Finally, we define

$$Q = \mathcal{H}[\pi](N) := H_{\kappa(N)}(\beta(N)) = \mathbf{H}[\beta(N)] \cdot \kappa(N).$$

The $\mathcal{H}_K$ used in OCB3 is an instantiation $\mathcal{H}[\pi]$ with $\pi = E_K$, i.e.,

$$\mathcal{H}_K(N) := \mathcal{H}[E_K](N).$$

4   Security Result

We present the main security result of the paper, along with an overview of our proof approach. Consider a nonce-based authenticated encryption security game with associated data involving $\mathsf{OCB3}[\pi]$, an ideal version of OCB3 where E_K is replaced by a random permutation π. Recall Theorem 1 from Sect. 1.

Theorem 1. *Let $\mathcal{A}$ be an adversary that makes q encryption queries consisting of σ message blocks in all with at most ℓ_{MAX} blocks per query, and α associated data blocks in all, and q' decryption queries in a nonce-respecting authenticated encryption security game with associated data against the oracles $\mathcal{O}_1$ and $\mathcal{O}_0$, where $\mathcal{O}_1$ simulates $\mathsf{OCB3}[\pi]$, and $\mathcal{O}_0$ simulates an ideal nonce-based authenticated encryption scheme with associated data. Then*

$$\mathbf{Adv}_{\mathcal{O}_1,\mathcal{O}_0}^{naead}(\mathcal{A}) \leq \frac{5\sigma_T^2}{N} + \frac{2\sigma^4}{N^2} + \frac{64q'\ell_{MAX} + 15q'}{N},$$

where $\sigma_T = \sigma + \alpha + q$ is the total number of blocks queried in the encryption queries (including messages, associated data and nonces).

4.1   Proof Approach

Before delving into the details of the proof, we give an overview of it. There are two parts to this security bound: the privacy bound, represented by the term $5\sigma_T^2/N + 2\sigma^4/N^2$, and the integrity bound, represented by the term $(64q'\ell_{MAX}+15q')/N$. The privacy bound is birthday in the number of encryption-query blocks, and and relies on the simple requirement that every blockcipher output is distinct. The integrity bound, being beyond-birthday in the number of decryption-query blocks (as long as ℓ_{MAX} is within a reasonable bound) is trickier to obtain, and is the main contribution of the paper.

   We consider a slightly modified game where we let the real oracle $\mathcal{O}_1$ reveal the inputs and outputs of all internal blockcipher calls in the encryption queries at the end of the query phase. Thus, it becomes necessary for the ideal oracle $\mathcal{O}_0$ to sample these values. In Subsect. 4.3, we describe the sampling order for $\mathcal{O}_0$, which proceeds in four steps. Step 1 takes place during the encryption-query phase itself, when $\mathcal{O}_0$ behaves as in a standard naead game, sampling the ciphertext and tag blocks on the fly. (In the decryption query phase, $\mathcal{O}_0$ always outputs $\perp$.) After the query phase, in Step 2 and Step 3, the inputs and

Encryption

input : N, A, M
output: C, T

begin

    $L \leftarrow E_K(0)$
    $Q \leftarrow \mathcal{H}_K(N)$
    $\text{auth} \leftarrow 0$
    for $j \leftarrow 1$ **to** k **do**
        $U_j \leftarrow \lambda_j \cdot L \oplus A_j$
        $V_j \leftarrow E_K(U_j)$
        $\text{auth} \leftarrow \text{auth} \oplus V_j$
    end for
    if $|A_*| > 0$ **then**
        $\overline{A_*} \leftarrow \text{pad}(A_*)$
        $U_* \leftarrow \lambda_k^* \cdot L \oplus \overline{A_*}$
        $V_* \leftarrow E_K(U_*)$
        $\text{auth} \leftarrow \text{auth} \oplus V_*$
    end if
    $M_{\text{tag}} \leftarrow 0$
    for $j \leftarrow 1$ **to** ℓ **do**
        $X_j \leftarrow Q \oplus \lambda_j \cdot L \oplus M_j$
        $Y_j \leftarrow E_K(X_j)$
        $C_j \leftarrow Q \oplus \lambda_j \cdot L \oplus Y_j$
        $M_{\text{tag}} \leftarrow M_{\text{tag}} \oplus M_j$
    end for
    $\lambda_{\text{tag}}^i \leftarrow \lambda_\ell^\$$
    if $|M_*| > 0$ **then**
        $\overline{M_*} \leftarrow \text{pad}(M_*)$
        $X_* \leftarrow Q \oplus \lambda_\ell^* \cdot L$
        $Y_* \leftarrow E_K(X_*)$
        $\overline{C_*} \leftarrow Y_* \oplus \overline{M_*}$
        $C_* \leftarrow \overline{C_*} \gg (n - |M_*|)$
        $M_{\text{tag}} \leftarrow M_{\text{tag}} \oplus \overline{M_*}$
        $\lambda_{\text{tag}}^i \leftarrow \lambda_\ell^{*\$}$
    end if
    $X_{\text{tag}} \leftarrow Q \oplus \lambda_{\text{tag}}^i \cdot L \oplus M_{\text{tag}}$
    $Y_{\text{tag}} \leftarrow E_K(X_{\text{tag}})$
    $T \leftarrow \text{auth} \oplus Y_{\text{tag}}$
    return C
    return T
end

Decryption

input : N', A', C', T'
output: M' or $\perp$

begin

    $L' \leftarrow E_K(0)$
    $Q' \leftarrow \mathcal{H}_K(N')$
    $\text{auth}' \leftarrow 0$
    for $j \leftarrow 1$ **to** k' **do**
        $U'_j \leftarrow \lambda_j \cdot L' \oplus A'_j$
        $V'_j \leftarrow E_K(U'_j)$
        $\text{auth}' \leftarrow \text{auth}' \oplus V'_j$
    end for
    if $|A'_*| > 0$ **then**
        $\overline{A'_*} \leftarrow \text{pad}(A'_*)$
        $U'_* \leftarrow \lambda_{k'}^* \cdot L' \oplus \overline{A'_*}$
        $V'_* \leftarrow E_K(U'_*)$
        $\text{auth}' \leftarrow \text{auth}' \oplus V'_*$
    end if
    $M'_{\text{tag}} \leftarrow 0$
    for $j \leftarrow 1$ **to** ℓ' **do**
        $Y'_j \leftarrow Q' \oplus \lambda_j \cdot L' \oplus C'_j$
        $X'_j \leftarrow E_K^{-1}(Y'_j)$
        $M'_j \leftarrow Q' \oplus \lambda_j \cdot L' \oplus X'_j$
        $M'_{\text{tag}} \leftarrow M'_{\text{tag}} \oplus M'_j$
    end for
    $\lambda'^i_{\text{tag}} \leftarrow \lambda_{\ell'}^\$$
    if $|C'_*| > 0$ **then**
        $\overline{C'_*} \leftarrow \text{pad}(C'_*)$
        $X'_* \leftarrow Q' \oplus \lambda_{\ell'}^* \cdot L'$
        $Y'_* \leftarrow E_K(X'_*)$
        $\overline{M'_*} \leftarrow Y'_* \oplus \overline{C'_*}$
        $M'_* \leftarrow \overline{M'_*} \gg (n - |C'_*|)$
        $M'_{\text{tag}} \leftarrow M'_{\text{tag}} \oplus \text{pad}(M'_*)$
        $\lambda'^i_{\text{tag}} \leftarrow \lambda_{\ell'}^{*\$}$
    end if
    $Y'_{\text{tag}} \leftarrow \text{auth}' \oplus T'$
    $X'_{\text{tag}} \leftarrow E_K^{-1}(Y'_{\text{tag}})$
    $M''_{\text{tag}} \leftarrow Q' \oplus \lambda^i_{\text{tag}} \cdot L' \oplus X'_{\text{tag}}$
    if $M'_{\text{tag}} = M''_{\text{tag}}$ **then**
        return M'
    else
        return $\perp$
    end if
end

Algorithm 1. The OCB3 algorithm. A_*, M_*, A'_*, C'_* are incomplete (possibly empty) blocks at the end of A, M, A', C' respectively. The hash function $\mathcal{H}_K$ is described in Subsect. 3.1

outputs of the internal blockcipher calls in all the encryption queries are sampled. Finally, in Step 4, the inputs and outputs of the internal blockcipher calls in the decryption queries which are not yet determined are sampled, completing the sampling process.

During this sampling process, we keep checking the sampled values for various bad events. badA occurs at the end of Step 1 if there are certain undesirable collisions or multicollisions in the sampled ciphertext and tag blocks. badB or badC occurs at the end of Step 2 or Step 3 respectively if there are certain collisions in the inputs or sampled outputs of the internal blockcipher calls. Finally, badD[i] occurs at the end of Step 4 if after sampling the inputs and outputs of the internal blockcipher calls in the i-th decryption query it turns out that the correct output of $\mathcal{O}_0$ should not have been $\perp$. badD[i] corresponds to the violation of integrity security, and the bounding of the probability of badD[i], which is done by carefully selecting the specific collisions we need to ban, forms the heart of this paper.

In Subsect. 4.5, we calculate the probabilities of the various bad cases. The calculations for badA, badB and badC are straightforward. For badD[i], we look at several cases, and establish a bound for badD[i] based on some lemmas the proof of which we defer to Sect. 5. By Property-1 in Subsect. 2.1, we can reorder the sampling phase in Step 4 to first sample the blockcipher outputs required for badD[i]. Finally, we bound the probability of $\cup_{i=1}^{q'}$ badD[i] by the union-bound. In Sect. 5, we prove the lemmas through an exhaustive case-analysis.

4.2 Notation for Adversary Interactions

First we set up the notation for the adversary interactions in the game described in Theorem 1. The i-th encryption query consists of

- a message M^i, consisting of $\ell^i \geq 1$ complete blocks and an incomplete (possibly empty) block M_*^i at the end;
- associated data A^i, consisting of $k^i \geq 1$ blocks and an incomplete (possibly empty) block A_*^i at the end;
- a nonce block N^i, with the first 122 bits not all zero, such that for any $i' \in [i-1]$, $\mathsf{N}^i \neq \mathsf{N}^{i'}$.

Following the notation for $\mathcal{H}[\pi]$ described in Subsect. 3.1, we define $\mathsf{TN}^i := \tau(\mathsf{N}^i), \mathsf{BN}^i := \beta(\mathsf{N}^i), \mathbf{H}^i := \mathbf{H}[\mathsf{BN}^i]$. The corresponding output consists of

- a ciphertext C^i, consisting of ℓ^i complete blocks and an incomplete block C_*^i at the end, with $|\mathsf{C}_*^i| = |\mathsf{M}_*^i|$;
- a tag block T^i.

The i-th decryption query consists of

- a ciphertext C'^i, consisting of $\ell'^i \geq 1$ complete blocks and an incomplete (possibly empty) block C'^i_* at the end;
- a tag block T'^i;

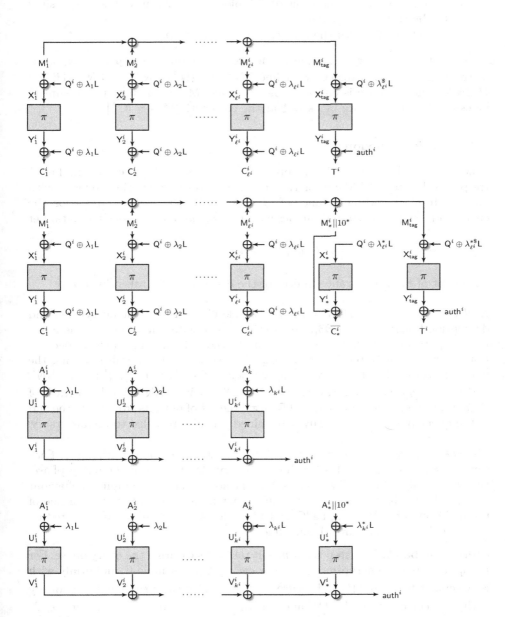

Fig. 2. The OCB3[π] construction: notation for the i-th encryption query. L denotes $\pi(0)$ and Q^i denotes $\mathcal{H}[\pi](N^i)$. **Top to Bottom:** encrypting M^i when $|M_*^i| = 0$; encrypting M^i when $|M_*^i| > 0$; hashing A^i when $|A_*^i| = 0$; hashing A^i when $|A_*^i| > 0$.

- associated date A'^i, consisting of k'^i blocks and an incomplete (possibly empty) block A'^i_* at the end;
- a nonce block N'^i, with the first 122 bits not all zero.

(Note that in the decryption queries, nonces are allowed to repeat.) Again, as in the i-th encryption query, we define $TN'^i := \tau(N'^i), BN'^i := \beta(N'^i), \mathbf{H}'^i := \mathbf{H}[BN'^i]$. The response is either $\perp$, or a message M'^i consisting of ℓ'^i complete blocks and an incomplete block M'^i_* at the end, with $|M'^i_*| = |C'^i_*|$.

4.3   Oracle Behaviour

Now we describe the oracles involved in the game in greater detail. Let $\mathcal{I}$ (resp. $\mathcal{I}'$) denote the indices for the encryption (resp. decryption) queries with incomplete-block messages, and let $\mathcal{J}$ (resp. $\mathcal{J}'$) denote the indices for the encryption (resp. decryption) queries with incomplete-block associated data. Let

$$\mathcal{F} := \left\{ i \in [q] \mid (\nexists i' < i)(TN^{i'} = TN^i) \right\}$$

be the set of first-appearance indices of the distinct values taken by TN^i.

Real Oracle. Enc_1 and Dec_1 of the real oracle $\mathcal{O}_1$ represent the encryption and decryption functions of $OCB3[\pi]$ respectively. The notation we use for the internal computations of $\mathcal{O}_1$ while responding to the i-th encryption (resp. decryption) query is illustrated in Fig. 2 (resp. Fig. 3). In addition, still following the notation from Subsect. 3.1, for $i \in [q]$ we define $KN^i := \pi(TN^i), Q^i := \mathbf{H}^i \cdot KN^i$, and for $i \in [q']$ we define $KN'^i := \pi(TN'^i), Q'^i := \mathbf{H}'^i \cdot KN'^i$. We keep track of $Dom(\pi)$, the set of inputs to π, and $Ran(\pi)$, the set of outputs from π. At the end of the query phase, the partially determined π is also revealed to the adversary.

Ideal Oracle. Dec_0 of the ideal oracle is the constant function returning $\perp$. Enc_0 samples and returns (C^i, T^i) for the i-th query. At the end of the query phase, the $\mathcal{O}_0$ partially samples π and gives it to the adversary. The sampling behaviour followed by $\mathcal{O}_0$ is described in the subsequent paragraphs. (Note that if one of the bad events badA, badB, badC, or badD[i] for some $i \in [q']$ is encountered by $\mathcal{O}_0$, its behaviour thereafter is undefined.)

Step 1 and badA. This step is *online*—it takes place during the query phase. For $i \in [q]$, on the i-th encryption query, for each $j \in [\ell^i]$, sample C^i_j uniformly with replacement from $\{0,1\}^n$ and return C^i_j to the adversary; sample T^i uniformly with replacement from $\{0,1\}^n$ and return T^i to the adversary; and if $i \in \mathcal{I}$, sample $\overline{C^i_*}$ uniformly with replacement from $\{0,1\}^n$, set $C^i_* = \text{chop}_{|M^i_*|}(\overline{C^i_*})$; and return C^i_* to the adversary.

badA occurs when we have

$$C^{i_1}_{j_1} + C^{i_1}_{j'_1} = C^{i_2}_{j_2} + C^{i_2}_{j'_2} = C^{i_3}_{j_2} + C^{i_3}_{j'_2}$$

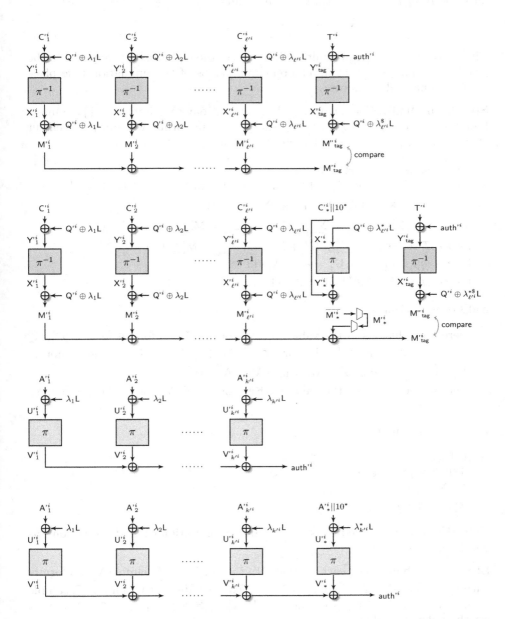

Fig. 3. The OCB3[π] construction: notation for the i-th decryption query. L denotes $\pi(0)$ and Q'^i denotes $\mathcal{H}[\pi](N'^i)$. **Top to Bottom:** decrypting C'^i when $|C'^i_*| = 0$; decrypting C'^i when $|C'^i_*| > 0$; hashing A'^i when $|A'^i_*| = 0$; hashing A'^i when $|A'^i_*| > 0$.

for some $i_1, i_2, i_3 \in [q]$ and three distinct pairs $(j_1, j_1'), (j_2, j_2'), (j_3, j_3')$ satisfying

$$\lambda_{j_1} + \lambda_{j_1'} = \lambda_{j_2} + \lambda_{j_2'} = \lambda_{j_3} + \lambda_{j_3'}.$$

This restriction on certain multi-collisions over the ciphertexts is required in the proof of Lemma 5 in Sect. 5. The remaining steps of the simulation take place after the query phase is over.

Step 2 and badB. Begin with $\pi = \{\}$ (so that $\mathsf{Dom}(\pi) = \mathsf{Ran}(\pi) = \{\}$). Sample L uniformly from $\{0,1\}^n$. For $i \in \mathcal{F}$, sample KN^i uniformly without replacement from $\{0,1\}^n \setminus \{\mathsf{L}\}$. Next set the following values:

- for $i \in [q]$, set $\mathsf{Q}^i = \mathbf{H}^i \cdot \mathsf{KN}^i$;
- for $i \in [q], j \in [\ell^i]$ set $\mathsf{X}_j^i = \mathsf{M}_j^i + \mathsf{Q}^i + \lambda_j \cdot \mathsf{L}$ and $\mathsf{Y}_j^i = \mathsf{C}_j^i + \mathsf{Q}^i + \lambda_j \cdot \mathsf{L}$;
- for $i \in \mathcal{I}$ set $\mathsf{X}_*^i = \mathsf{Q}^i + \lambda_{\ell^i}^* \cdot \mathsf{L}$ and $\mathsf{Y}_*^i = \mathsf{M}_*^i \| 10^* + \overline{\mathsf{C}_*^i}$;
- for $i \in [q] \setminus \mathcal{I}$ set $\mathsf{M}_{\mathsf{tag}}^i = \sum_{i=1}^{\ell^i} \mathsf{M}_j^i$ and $\mathsf{X}_{\mathsf{tag}}^i = \mathsf{M}_{\mathsf{tag}}^i + \mathsf{Q}^i + \lambda_{\ell^i}^\$ \cdot \mathsf{L}$;
- for $i \in \mathcal{I}$ set $\mathsf{M}_{\mathsf{tag}}^i = \sum_{i=1}^{\ell^i} \mathsf{M}_j^i + \mathsf{M}_*^i$ and $\mathsf{X}_{\mathsf{tag}}^i = \mathsf{M}_{\mathsf{tag}}^i + \mathsf{Q}^i + \lambda_{\ell^i}^{*\$} \cdot \mathsf{L}$;
- for $i \in [q], j \in [k^i]$ set $\mathsf{U}_j^i = \mathsf{A}_j^i + \lambda_j \cdot \mathsf{L}$;
- for $i \in \mathcal{J}$ set $\mathsf{U}_*^i = \mathsf{A}_*^i \| 10^* + \lambda_{k^i}^* \cdot \mathsf{L}$.

badB occurs when:

- there are collisions in the values 0, TN^i for $i \in \mathcal{F}$, X_j^i for $i \in [q], j \in [\ell^i]$, X_*^i for $i \in \mathcal{I}$, $\mathsf{X}_{\mathsf{tag}}^i$ for $i \in [q]$, U_j^i for $i \in [q], j \in [k^i]$, U_*^i for $i \in \mathcal{J}$, not counting the trivial collisions $\mathsf{U}_j^i = \mathsf{U}_j^{i'}$ when $\mathsf{A}_j^i = \mathsf{A}_j^{i'}$; or
- there are collisions in the values L, KN^i for $i \in \mathcal{F}$, Y_j^i for $i \in [q], j \in [\ell^i]$, Y_*^i for $i \in \mathcal{I}$.

Add the following to π:

- $(0, \mathsf{L})$;
- $(\mathsf{TN}^i, \mathsf{KN}^i)$ for $i \in \mathcal{F}$;
- $(\mathsf{X}_j^i, \mathsf{Y}_j^i)$ for $i \in [q], j \in [\ell^i]$;
- $(\mathsf{X}_*^i, \mathsf{Y}_*^i)$ for $i \in \mathcal{I}$;

Note that the π sampled thus far remains permutation-compatible as long as badB does not occur.

Step 3 and badC. For each distinct $\mathsf{U}_j^i, i \in [q], j \in [k^i]$, sample V_j^i uniformly without replacement from $\{0,1\}^n \setminus \mathsf{Ran}(\pi)$. For each distinct U_*^i for $i \in \mathcal{J}$, sample V_*^i uniformly without replacement from $\{0,1\}^n \setminus (\mathsf{Ran}(\pi) \cup \{\mathsf{V}_j^i \mid j \in [k^i]\})$. Next set the following values:

- for $i \in [q] \setminus \mathcal{J}$ set $\mathsf{auth}^i = \sum_{j=1}^{k^i} \mathsf{V}_j^i$;
- for $i \in \mathcal{J}$ set $\mathsf{auth}^i = \sum_{j=1}^{k^i} \mathsf{V}_j^i + \mathsf{V}_*^i$;
- for $i \in [q]$ set $\mathsf{Y}_{\mathsf{tag}}^i = \mathsf{T}^i + \mathsf{auth}^i$.

badC occurs when:

- $Y^i_{tag} \in \text{Ran}(\pi)$ for some $i \in [q]$;
- $Y^i_{tag} = V^i_j$ for some $i \in [q], j \in [k^i]$;
- $Y^i_{tag} = V^i_*$ for some $i \in \mathcal{I}$; or
- $Y^i_{tag} = Y^{i'}_{tag}$ for some $i, i' \in [q]$.

Add the following to π:

- (X^i_{tag}, Y^i_{tag}) for $i \in [q]$;
- (U^i_j, V^i_j) for $i \in [q], j \in [k^i]$;
- (U^i_*, V^i_*) for $i \in \mathcal{J}$;

Note that the π sampled thus far remains permutation-compatible as long as neither of badB and badC occurs.

Step 4 and badD$[i]$. In this step we keep updating π (and hence $\text{Dom}(\pi)$ and $\text{Ran}(\pi)$) on the fly. For each $i \in [q']$, set M'^i_{tag} and M''^i_{tag} as follows:

- If $TN'^i \in \text{Dom}(\pi)$, set $KN'^i = \pi(TN'^i)$, otherwise sample KN'^i uniformly without replacement from $\{0,1\}^n \setminus \text{Ran}(\pi)$, and add (TN'^i, KN'^i) to π;
- For $j \in [\ell'^i]$, set $Y'^i_j = C'^i_j + Q'^i + \lambda_j \cdot L$; if $Y'^i_j \in \text{Ran}(\pi)$, set $X'^i_j = \pi^{-1}(Y'^i_j)$, otherwise sample X'^i_j uniformly without replacement from $\{0,1\}^n \setminus \text{Dom}(\pi)$, and add (X'^i_j, Y'^i_j) to π; finally, set $M'^i_j = X'^i_j + Q'^i + \lambda_j \cdot L$;
- If $i \in \mathcal{I}'$, set $X^i_* = Q'^i + \lambda^*_{\ell'i} \cdot L$; if $X^i_* \in \text{Dom}(\pi)$, set $Y^i_* = \pi(X^i_*)$, otherwise sample Y^i_* uniformly without replacement from $\{0,1\}^n \setminus \text{Ran}(\pi)$, and add (X^i_*, Y^i_*) to π; finally, set $\overline{M'^i_*} = C'^i_* \| 10^* + Y^i_*$;
- If $i \notin \mathcal{I}'$, set $M'^i_{tag} = \sum_{j=1}^{\ell'i} M'^i_j$;
- If $i \in \mathcal{I}'$, set $M'^i_{tag} = \sum_{j=1}^{\ell'i} M'^i_j + M'^i_*$;
- For $j \in [k'^i]$, set $U'^i_j = A'^i_j + \lambda_j \cdot L$; if $U'^i_j \in \text{Dom}(\pi)$, set $V'^i_j = \pi(U'^i_j)$, otherwise sample V'^i_j uniformly without replacement from $\{0,1\}^n \setminus \text{Ran}(\pi)$, and add (U'^i_j, V'^i_j) to π;
- If $i \in \mathcal{J}'$, set $U^i_* = A^i_* \| 10^* + \lambda^*_{\ell'i} \cdot L$; if $U^i_* \in \text{Dom}(\pi)$, set $V^i_* = \pi(U^i_*)$, otherwise sample V^i_* uniformly without replacement from $\{0,1\}^n \setminus \text{Ran}(\pi)$, and add (U^i_*, V^i_*) to π;
- If $i \notin \mathcal{J}'$, set $auth'^i = \sum_{j=1}^{k'i} V'^i_j$;
- If $i \in \mathcal{J}'$, set $auth'^i = \sum_{j=1}^{k'i} V'^i_j + V'^i_*$;
- Set $Y'^i_{tag} = T'^i + auth'^i$; if $Y'^i_{tag} \in \text{Ran}(\pi)$, set $X'^i_{tag} = \pi^{-1}(Y'^i_{tag})$, otherwise sample X'^i_{tag} uniformly without replacement from $\{0,1\}^n \setminus \text{Dom}(\pi)$, and add (Y'^i_{tag}, X'^i_{tag}) to π; finally, set $M''^i_{tag} = X'^i_{tag} + Q'^i + \lambda^{*\$}_{\ell'i} \cdot L$;
- Return π to the adversary.

badD$[i]$ occurs when $M'^i_{tag} = M''^i_{tag}$.

4.4   Notation for the Proof

Before we begin the proof, we introduce some more notation. Let $\mathcal{P}'^i$ denote the set of positions for the i-th decryption query, defined as

$$\mathcal{P}'^i := \begin{cases} [\ell'^i] \cup \{\mathsf{tag}\}, & i \notin \mathcal{I}', \\ [\ell'^i] \cup \{*, \mathsf{tag}\}, & i \in \mathcal{I}'. \end{cases}$$

Further, let $\mathcal{P}'^i(-) := \mathcal{P}'^i \setminus \{\mathsf{tag}\}$. For $p \in \mathcal{P}'^i$, let $\Delta_p'^i$ denote the masking key for position p in i-th decryption query, defined as

$$\Delta_p'^i := \begin{cases} \lambda_p \cdot \mathsf{L} + \mathsf{Q}'^i, & p \in [\ell'^i], \\ \lambda_{\mathsf{tag}}'^i \cdot \mathsf{L} + \mathsf{Q}'^i, & p = \mathsf{tag}, \\ \lambda_{\ell'^i}^* \cdot \mathsf{L} + \mathsf{Q}'^i, & p = *, i \in \mathcal{I}', \end{cases}$$

where

$$\lambda_{\mathsf{tag}}'^i := \begin{cases} \lambda_{\ell'^i}^\$, & i \notin \mathcal{I}', \\ \lambda_{\ell'^i}^{*\$}, & i \in \mathcal{I}'. \end{cases}$$

For convenience, we will abuse the notation of set membership and extend it to sequences. Thus, for a sequence S and a block Z, $\mathsf{Z} \in \mathsf{S}$ will imply that Z occurs somewhere in S.

4.5   Proof of Theorem

Let $\mathcal{V}_{\mathsf{bad}}$ consist of all those transcripts where one of badA, badB, badC or $\mathsf{badD}[i]$ for some $i \in [q']$ has been encountered. Then

$$\mathsf{ip}_{\mathcal{O}_0}[\mathcal{V}_{\mathsf{bad}}] \leq \Pr_{\mathcal{O}_0}[\mathsf{badA}] + \Pr_{\mathcal{O}_0}[\mathsf{badB}] + \Pr_{\mathcal{O}_0}[\mathsf{badC}] + \sum_{i=1}^{q'} \Pr_{\mathcal{O}_0}[\mathsf{badD}[i]].$$

We make the following claim:

Claim. We have the following bounds on the bad events under $\mathcal{O}_0$:

$$\Pr_{\mathcal{O}_0}[\mathsf{badA}] \leq \frac{\sigma^4}{N^2}, \tag{1}$$

$$\Pr_{\mathcal{O}_0}[\mathsf{badB}] \leq \frac{3\sigma_T^2}{N}, \tag{2}$$

$$\Pr_{\mathcal{O}_0}[\mathsf{badC}] \leq \frac{2q\sigma_T}{N}, \tag{3}$$

$$\Pr_{\mathcal{O}_0}[\mathsf{badD}[i]] \leq \frac{64\ell_{\mathrm{MAX}} + 15}{N} + \frac{32\sigma_T^2}{N^2}. \tag{4}$$

From the claim we have

$$\mathsf{ip}_{\mathcal{O}_0}[\mathcal{V}_{\mathsf{bad}}] \leq \frac{\sigma^4}{N^2} + \frac{3\sigma_T^2}{N} + \frac{2q\sigma_T}{N} + \sum_{i=1}^{q'} \left(\frac{64\ell_{\mathrm{MAX}} + 15}{N} + \frac{32\sigma_T^2}{N^2} \right)$$

$$\leq \frac{5\sigma_T^2}{N} + \frac{2\sigma^4}{N^2} + \frac{64q'\ell_{\mathrm{MAX}} + 15q'}{N}.$$

Consider a view $V \notin \mathcal{V}_{\mathsf{bad}}$. In the real oracle, to obtain V, exactly $\sigma_T + |\mathcal{F}| + 1$ calls are made to π: one for each message block, one for each position-wise distinct associated data block, one for each distinct TN^i, one for 0, and one for each tag. We know that these are all distinct because neither of badB and badC has been encountered. Hence

$$\mathsf{ip}_{\mathcal{O}_1}[V] = \frac{1}{N^{\sigma_T + |\mathcal{F}| + 1}}.$$

In the ideal oracle, in Step 1, the $\sigma + q$ online outputs are sampled uniformly with replacement. In Steps 2 and 3, $|\mathcal{F}| + 1 + \alpha$ outputs are sampled uniformly without replacement. Finally, since $\mathsf{badD}[i]$ was not encountered for any $i \in [q']$, all decryption queries in V must have returned $\perp$, which $\mathcal{O}_0$ always returns. Hence

$$\mathsf{ip}_{\mathcal{O}_0}[V] = \frac{1}{N^{\sigma + q}} \cdot \frac{1}{N^{|\mathcal{F}| + 1 + \alpha}} \leq \frac{1}{N^{\sigma_T + |\mathcal{F}| + 1}} = \mathsf{ip}_{\mathcal{O}_1}[V].$$

Theorem 2 then gives us the required result.   □

Proof of Claim. Suppose badA is encountered. Then we have

$$\mathsf{C}_{j_1}^{i_1} + \mathsf{C}_{j_1'}^{i_1} = \mathsf{C}_{j_2}^{i_2} + \mathsf{C}_{j_2'}^{i_2} = \mathsf{C}_{j_2}^{i_3} + \mathsf{C}_{j_2'}^{i_3},$$

$$\lambda_{j_1} + \lambda_{j_1'} = \lambda_{j_2} + \lambda_{j_2'} = \lambda_{j_3} + \lambda_{j_3'},$$

for some $i_1, i_2, i_3 \in [q]$ and three distinct pairs $(j_1, j_1'), (j_2, j_2'), (j_3, j_3')$. Now for fixed $i_1, i_2, i_3, j_1, j_1', j_2, j_2', j_3, j_3'$, this probability is at most $1/N^2$. Notice that if for any choice of (j_1, j_2, j_3, j_1'), there is at most one choice of j_2' and at most one choice of j_3'. For any choice of i_1, there are at most σ^2 choices for (i_2, j_2) and (i_3, j_3), and at most $(\ell^i)^2$ choices for (j_1, j_1'). Summing over i gives us

$$\Pr_{\mathcal{O}_0}[\mathsf{badA}] \leq \frac{\sigma^2}{N^2} \sum_{i=1}^{q} (\ell^i)^2 \leq \frac{\sigma^4}{N^2},$$

establishing (1). For badB, since we are now sampling without replacement, each collision event has probability at most $1/(N-1)$. There are at most $(q+1)$ values among 0, TN^i for $i \in \mathcal{F}$, and they are all distinct by sampling; there are at most α distinct values among U_j^i for $i \in [q], j \in [k^i]$, U_*^i for $i \in \mathcal{J}$; and there are $\sigma + q$ values among X_j^i for $i \in [q], j \in [\ell^i]$, X_*^i for $i \in \mathcal{I}$ and $\mathsf{X}_{\mathrm{tag}}^i$ for $i \in [q]$. These give us at most $(q+1)(\sigma + \alpha + q) + (\sigma + \alpha + q)^2/2$ possible collision pairs. Similarly, among L, KN^i for $i \in \mathcal{F}$, Y_j^i for $i \in [q], j \in [\ell^i]$, and Y_*^i for $i \in \mathcal{I}$, there are at most $(q+1)\sigma + \sigma^2/2$ possible collision pairs. Thus we have

$$\Pr_{\mathcal{O}_0}[\mathsf{badB}] \leq \frac{(q+1)(\sigma+\alpha+q)}{N-1} + \frac{(\sigma+\alpha+q)^2}{2(N-1)} + \frac{(q+1)\sigma}{N-1} + \frac{\sigma^2}{2(N-1)}$$

$$\leq \frac{3(\sigma+\alpha+q)^2}{N} = \frac{3\sigma_T^2}{N},$$

establishing (2). For badC, since at this point $|\mathrm{Ran}(\pi)| = \rho := \sigma + |\mathcal{F}| + 1$, the probability of each collision event is at most $1/(N-\rho-1)$. Since there are at most $q\rho + q\alpha + q^2/2$ possible collision pairs, we have

$$\Pr_{\mathcal{O}_0}[\mathsf{badC}] \leq \frac{q\rho}{N-\rho-1} + \frac{q\alpha}{N-\rho-1} + \frac{q^2}{2(N-\rho-1)} \leq \frac{2q\sigma_T}{N},$$

establishing (3). To prove (4) we need to bound the probability of $\mathsf{badD}[i]$, and for that we consider several cases. For now we assume that $i \in \mathcal{I}'$. Then $\mathcal{P}'^i(-)$ is simply $[\ell'^i]$. For some $p \in \mathcal{P}'^i$, we say X'_p^i is *trivially determined* if the adversary can deduce the value of X'_p^i from the transcript of the encryption queries. This can happen in two ways:

- When $p \in [\ell'^i]$, X'_p^i is trivially determined if for some i' we have $\mathsf{N}'^i = \mathsf{N}^{i'}$ and $\mathsf{C}'_p^i = \mathsf{C}_p^{i'}$, which forces X'_p^i to equal $\mathsf{X}_p^{i'}$ — then we say X'_p^i is i'-trivial;
- $\mathsf{X}'_{\mathsf{tag}}^i$ is trivially determined if for some j we have $\mathsf{A}'^i = \mathsf{A}^j$ and $\mathsf{T}'^i = \mathsf{T}^j$, which forces $\mathsf{X}'_{\mathsf{tag}}^i$ to equal $\mathsf{X}_{\mathsf{tag}}^j$ — then we say $\mathsf{X}'_{\mathsf{tag}}^i$ is j-trivial.

We look at five cases:

- **Case 1.** X'_p^i is trivially determined for all $p \in \mathcal{P}'^i$;
- **Cases 2 and 3.** For some $p_0 \in \mathcal{P}'^i$, $\mathsf{X}'_{p_0}^i$ is not trivially determined, and X'_p^i is trivially determined for all $p \in \mathcal{P}'^i \setminus \{p_0\}$:
  - **Case 2.** $p_0 \in [\ell'^i]$;
  - **Case 3.** $p_0 = \mathsf{tag}$;
- **Case 4.** For some $p_0 \in [\ell'^i]$, $\mathsf{X}'_{p_0}^i$ and $\mathsf{X}'_{\mathsf{tag}}^i$ are not trivially determined, and X'_p^i is trivially determined for all $p \in [\ell'^i] \setminus \{p_0\}$;
- **Case 5.** For some distinct $p_0, p_1 \in [\ell'^i]$, $\mathsf{X}'_{p_0}^i$ and $\mathsf{X}'_{p_1}^i$ are not trivially determined.

Note that the decryption query satisfies one of the five cases above, and moreover that this case can be chosen in advance by the adversary, by appropriately setting the query parameters. Accordingly, we can divide $[q']$ into five disjoint subsets $\mathcal{S}[1], \dots, \mathcal{S}[5]$, such that for $k \in [5]$, $\mathcal{S}[\langle k \rangle]$ denotes the set of decryption queries which fall under **Case** $\langle k \rangle$ above.

We now state five lemmas for the five separate cases, the proofs of which we defer to Sect. 5.

Lemma 1. *For $i \in \mathcal{S}[1]$, $\Pr_{\mathcal{O}_0}[\mathsf{badD}[i]] \leq \dfrac{2}{N}$.*

Lemma 2. *For $i \in \mathcal{S}[2]$, $\mathrm{Pr}_{\mathcal{O}_0}[badD[i]] \leq \dfrac{64\ell_{MAX} + 15}{N}$, as long as $2\sigma_T \leq N$.*

Lemma 3. *For $i \in \mathcal{S}[3]$, $\mathrm{Pr}_{\mathcal{O}_0}[badD[i]] \leq \dfrac{10}{N}$, as long as $2\sigma_T \leq N$.*

Lemma 4. *For $i \in \mathcal{S}[4]$, $\mathrm{Pr}_{\mathcal{O}_0}[badD[i]] \leq \dfrac{2}{N} + \dfrac{32\sigma_T^2}{N^2}$.*

Lemma 5. *For $i \in \mathcal{S}[5]$, $\mathrm{Pr}_{\mathcal{O}_0}[badD[i]] \leq \dfrac{64\ell_{MAX} + 4}{N} + \dfrac{32\sigma_T^2}{N^2}$.*

Taking the maximum over these bounds gives (4), and completes the proof of the claim. □

5   Proof of Lemmas

We recall that for the i-th decryption query, we first sample/set the inputs and outputs of π, and then define M'^i_p as

$$
\mathsf{M}'^i_p := \begin{cases} \mathsf{X}'^i_p + \varDelta'^i_p, & p \in [\ell'^i], \\ \mathrm{chop}_{|\mathsf{C}'^i_p|}(\mathsf{Y}'^i_p + \overline{\mathsf{C}'^i_p})\|10*, & p = *, i \in \mathcal{I}'. \end{cases}
$$

Finally, we set

$$
\mathsf{M}'^i_{\mathrm{tag}} := \sum_{p \in \mathcal{P}'^i(-)} \mathsf{M}'^i_p,
$$

$$
\mathsf{M}''^i_{\mathrm{tag}} := \mathsf{X}'^i_{\mathrm{tag}} + \varDelta'^i_{\mathrm{tag}},
$$

and $badD[i]$ is triggered when $\mathsf{M}'^i_{\mathrm{tag}} = \mathsf{M}''^i_{\mathrm{tag}}$.

The Subcase Tree. In Subsect. 4.5, we divide the set $[q']$ of decryption queries into five subsets $\mathcal{S}[1], \ldots, \mathcal{S}[5]$, depending on which of five cases a particular decryption query satisfies. We divide each of the cases, except **Case 1**, into various sub-cases. Whenever a X'^i_p is trivially determined for $p \in [\ell'^i]$, we let i' be such that X'^i_p is i'-trivial, and whenever $\mathsf{X}'^i_{\mathrm{tag}}$ is trivially determined, we let j be such that $\mathsf{X}'^i_{\mathrm{tag}}$ is j-trivial. (Note that there can be exactly one choice for each of i' and j.)

- **Case 2.** Here $\mathsf{X}'^i_{p_0} = \pi^{-1}(\mathsf{C}'^i_{p_0} + \varDelta^{i'}_{p_0})$, so we branch based on $\mathsf{C}'^i_{p_0} + \varDelta^{i'}_{p_0}$:
  - *Subcase 2(a).* $\mathsf{C}'^i_{p_0} + \varDelta^{i'}_{p_0} \notin \mathrm{Ran}(\pi)$;
  - *Subcase 2(b).* $\mathsf{C}'^i_{p_0} + \varDelta^{i'}_{p_0} = \mathsf{KN}^{j'}$ for some $j' \in [q]$;
  - *Subcase 2(c).* $\mathsf{C}'^i_{p_0} + \varDelta^{i'}_{p_0} = \mathsf{V}^{j'}_{s_0}$ for some $j' \in [q], s_0 \in [k^{j'}]$;
  - *Subcase 2(d).* $\mathsf{C}'^i_{p_0} + \varDelta^{i'}_{p_0} = \mathsf{C}^{j'}_{p_1} + \varDelta^{j'}_{p_1}$ for some $j' \in [q], p_1 \in [\ell'^{j'}]$;
  - *Subcase 2(e).* $\mathsf{C}'^i_{p_0} + \varDelta^{i'}_{p_0} = \mathrm{auth}^{j'} + \mathsf{T}^{j'}$ for some $j' \in [q]$.

- **Case 3.** Here $X'^i_{p_0} = X'^i_{\text{tag}} = \pi^{-1}(\text{auth}'^i + T'^i)$, so we branch based on $\text{auth}'^i + T'^i$:
  - *Subcase 3(a).* $\text{auth}'^i + T'^i \notin \text{Ran}(\pi)$;
  - *Subcase 3(b).* $\text{auth}'^i + T'^i = \text{KN}^{j'}$ for some $j' \in [q]$;
  - *Subcase 3(c).* $\text{auth}'^i + T'^i = V^{j'}_{s_0}$ for some $j' \in [q], s_0 \in [k^{j'}]$;
  - *Subcase 3(d).* $\text{auth}'^i + T'^i = C^{j'}_{p_1} + \Delta^{j'}_{p_1}$ for some $j' \in [q], p_1 \in [\ell^{j'}]$;
  - *Subcase 3(e).* $\text{auth}'^i + T'^i = \text{auth}^{j'} + T^{j'}$ for some $j' \in [q]$.
- **Case 4.** Here $X'^i_{p_0} = \pi^{-1}(C'^i_{p_0} + \Delta^{i'}_{p_0})$ and $X'^i_{\text{tag}} = \pi^{-1}(\text{auth}'^i + T'^i)$, so we can branch based on $C'^i_{p_0} + \Delta^{i'}_{p_0}$ and $\text{auth}'^i + T'^i$ and get seventeen cases here: one covering either of them being randomly sampled, and the other sixteen a Cartesian product between *Subcases 2(b)–2(e)* and *Subcases 3(b)–3(e)*. However, most of this cases can be settled using near-identical arguments, so we make the case division to reflect the interesting cases:
  - *Subcase 4(a).* $C'^i_{p_0} + \Delta^{i'}_{p_0} \notin \text{Ran}(\pi)$ or $\text{auth}'^i + T'^i \notin \text{Ran}(\pi)$;
  - *Subcase 4(b).* $C'^i_{p_0} + \Delta^{i'}_{p_0} \in \text{Ran}(\pi)$, $\text{auth}'^i + T'^i = V^{j'}_{s_0}$ for some $j' \in [q], s_0 \in [k^{j'}]$;
  - *Subcase 4(c).* $C'^i_{p_0} + \Delta^{i'}_{p_0} \in \text{Ran}(\pi)$, $\text{auth}'^i + T'^i \in \text{Ran}(\pi)$, $\text{auth}'^i + T'^i \neq V^{j'}_{s_0}$ for all $j' \in [q], s_0 \in [k^{j'}]$.
- **Case 5.** Here $X'^i_{p_0} = \pi^{-1}(C'^i_{p_0} + \Delta^{i'}_{p_0})$ and $X'^i_{p_1} = \pi^{-1}(C'^i_{p_1} + \Delta^{i'}_{p_1})$, so we can branch based on $C'^i_{p_0} + \Delta^{i'}_{p_0}$ and $C'^i_{p_1} + \Delta^{i'}_{p_1}$:
  - *Subcase 5(a).* $C'^i_{p_0} + \Delta^{i'}_{p_0} \notin \text{Ran}(\pi)$ or $C'^i_{p_1} + \Delta^{i'}_{p_1} \notin \text{Ran}(\pi)$;
  - *Subcase 5(b).* $C'^i_{p_0} + \Delta^{i'}_{p_0} = C^{j'}_{p_2} + \Delta^{j'}_{p_2}, C'^i_{p_1} + \Delta^{i'}_{p_1} = C^{j''}_{p_3} + \Delta^{j''}_{p_3}$ for some $j', j'' \in [q], j' \neq j'', p_2 \in [\ell^{j'}], p_3 \in [\ell^{j''}]$;
  - *Subcase 5(c).* $C'^i_{p_0} + \Delta^{i'}_{p_0} = C^{j'}_{p_2} + \Delta^{j'}_{p_2}, C'^i_{p_1} + \Delta^{i'}_{p_1} = C^{j'}_{p_3} + \Delta^{j'}_{p_3}$ for some $j' \in [q], p_2, p_3 \in [\ell^{j'}]$;
  - *Subcase 5(d).* $C'^i_{p_0} + \Delta^{i'}_{p_0} \in \text{Ran}(\pi)$ and $C'^i_{p_1} + \Delta^{i'}_{p_1} \in \text{Ran}(\pi)$, either $C'^i_{p_1} + \Delta^{i'}_{p_1} \neq C^{j'}_{p_2} + \Delta^{j'}_{p_2}$ for any $j' \in [q], p_2 \in [\ell^{j'}]$ or $C'^i_{p_1} + \Delta^{i'}_{p_1} \neq C^{j'}_{p_2} + \Delta^{j'}_{p_2}$ for any $j' \in [q], p_2 \in [\ell^{j'}]$.

Now we turn to the proof of the lemmas. We make the following observations:

- When X'^i_p is i'-trivial for some $p \in [\ell'^i]$, $M'^i_p = M^{i'}_p$;
- When X'^i_{tag} is j-trivial, $M'^i_{\text{tag}} + \Delta'^i_{\text{tag}} = M^j_{\text{tag}} + \Delta^j_{\text{tag}}$.

For brevity, when we write the collision equation(s) that need to be satisfied for $\text{badD}[i]$ to occur, the random variables that contribute to the subsequent probability calculation are indicated thus.

5.1   Proof of Lemma 1

Lemma 1. *For $i \in \mathcal{S}[1]$, $\Pr_{\mathcal{O}_0}[\text{badD}[i]] \leq \dfrac{2}{N}$.*

Proof. When $i \in \mathcal{S}[1]$, X'^i_p is trivially determined for all $p \in \mathcal{P}'^i$. From observations above, for $\mathsf{badD}[i]$ to occur, we must have

$$\sum_{p \in [\ell i']} \mathsf{M}^{i'}_p + \varDelta'^i_{\mathsf{tag}} = \mathsf{M}^j_{\mathsf{tag}} + \varDelta^j_{\mathsf{tag}},$$

or $$\mathsf{Q}^{i'} + \mathsf{Q}^j + (\lambda^\$_{\ell'i} + \lambda^\$_{\ell j}) \cdot \mathsf{L} = \sum_{p \in [\ell i']} \mathsf{M}^{i'}_p + \mathsf{M}^j_{\mathsf{tag}}.$$

If $\ell^j \neq \ell'^i$, the equation is

$$\mathbf{H}^{i'} \cdot \mathsf{KN}^{i'} + \mathbf{H}^j \cdot \mathsf{KN}^j + (\lambda^\$_{\ell'i} + \lambda^\$_{\ell j}) \cdot \underbrace{\mathsf{L}} = \sum_{p \in [\ell i']} \mathsf{M}^{i'}_p + \mathsf{M}^j_{\mathsf{tag}},$$

where the coefficient of L is non-zero. The probability of this $\leq 1/(N-2) \leq 1/2N$. If $\ell^j = \ell'^i$, so $j \neq i'$ (the decryption query has to be non-trivial), but $\mathsf{TN}^j = \mathsf{TN}^{i'}$, then $\mathbf{H}^j + \mathbf{H}^{i'}$ is full-rank, so

$$(\mathbf{H}^{i'} + \mathbf{H}^j) \cdot \underbrace{\mathsf{KN}^j} = \sum_{p \in [\ell i']} \mathsf{M}^{i'}_p + \mathsf{M}^j_{\mathsf{tag}}.$$

The probability of this $\leq 1/N$. And finally when $\ell^j = \ell'^i$ and $\mathsf{TN}^j \neq \mathsf{TN}^{i'}$, we have

$$\mathbf{H}^{i'} \cdot \underbrace{\mathsf{KN}^{i'}} + \mathbf{H}^j \cdot \mathsf{KN}^j = \sum_{p \in [\ell i']} \mathsf{M}^{i'}_p + \mathsf{M}^j_{\mathsf{tag}}.$$

The probability of this $\leq 1/(N-1) \leq 2/N$. This completes the proof. $\square$

5.2   Proof of Lemma 2

Lemma 2. *For $i \in \mathcal{S}[2]$, $\mathrm{Pr}_{\mathcal{O}_0}\left[\mathit{badD}[i]\right] \leq \dfrac{64\ell_{MAX} + 15}{N}$, as long as $2\sigma_T \leq N$.*

Proof. When $i \in \mathcal{S}[2]$, for some $p_0 \in [\ell'^i]$, X'^i_p is trivially determined for all $p \in \mathcal{P}'^i \setminus \{p_0\}$, but $\mathsf{X}'^i_{p_0}$ is not trivially determined . The equation for $\mathsf{badD}[i]$ becomes

$$\sum_{p \in [\ell i'] \setminus \{p_0\}} \mathsf{M}^{i'}_p + \pi^{-1}(\mathsf{C}'^i_{p_0} + \varDelta^{i'}_{p_0}) + \varDelta^{i'}_{p_0} + \varDelta^{i'}_{\mathsf{tag}} = \mathsf{M}^j_{\mathsf{tag}} + \varDelta^j_{\mathsf{tag}}$$

or $$\pi^{-1}(\mathsf{C}'^i_{p_0} + \varDelta^{i'}_{p_0}) + \mathsf{Q}^j + (\lambda_{p_0} + \lambda^\$_{\ell i'} + \lambda^\$_{\ell j}) \cdot \mathsf{L} = \mathsf{MM},$$

where $$\mathsf{MM} := \sum_{p \in [\ell i'] \setminus \{p_0\}} \mathsf{M}^{i'}_p + \mathsf{M}^j_{\mathsf{tag}}.$$

Based on the value of $\mathsf{C}'^i_{p_0} + \varDelta^{i'}_{p_0}$, we look at the subcases listed in the tree at the beginning of this section. Note that

$$\varDelta^{i'}_{p_0} + \varDelta^{i'}_{\mathsf{tag}} + \varDelta^j_{\mathsf{tag}} = \mathsf{Q}^j + (\lambda_{p_0} + \lambda^\$_{\ell i'} + \lambda^\$_{\ell j}) \cdot \mathsf{L}.$$

- *Subcase 2(a).* $C'^i_{p_0} + \Delta'^i_{p_0} \notin \mathrm{Ran}(\pi)$, so $X'^i_{p_0}$ is sampled. The equation for badD[i] is

$$X'^i_{p_0} + \mathbf{H}^j \cdot KN^j + (\lambda_{p_0} + \lambda^\$_{\ell i'} + \lambda^\$_{\ell j}) \cdot L = MM.$$

The probability of this $\leq 1/(N-2) \leq 2/N$.

- *Subcase 2(b).* $C'^i_{p_0} + \Delta'^i_{p_0} = KN^{j'}$ for some $j' \in [q]$, so $X'^i_{p_0} = TN^{j'}$, and a second equation comes from the condition for badD[i]. When $\ell^{i'} = \ell^j$ and $TN^{i'} = TN^{j'}$, these two equations become

$$(\mathbf{I} + \mathbf{H}^{i'}) \cdot \underbrace{KN^{j'}} + \lambda_{p_0} \cdot \underbrace{L} = C'^i_{p_0},$$

$$\mathbf{H}^j \cdot KN^j + \lambda_{p_0} \cdot \underbrace{L} = TN^{j'} + MM.$$

Since there are at most 2^6 choices for j', this probability $\leq 64/(N-1)(N-2) \leq 256/N^2 \leq 1/N$. When $\ell^{i'} = \ell^j$ and $TN^{i'} \neq TN^{j'}$, the two equations become

$$\underbrace{KN^{j'}} + \mathbf{H}^{i'} \cdot KN^{i'} + \lambda_{p_0} \cdot L = C'^i_{p_0},$$

$$\mathbf{H}^j \cdot KN^j + \lambda_{p_0} \cdot L = TN^{j'} + MM.$$

Here there are q choices for j', and this probability $\leq q(N-2)(N-3) \leq 4q/N^2$. When $\ell^{i'} \neq \ell^j$, the two equations become

$$\underbrace{KN^{j'}} + \mathbf{H}^{i'} \cdot KN^{i'} + \lambda_{p_0} \cdot \underbrace{L} = C'^i_{p_0},$$

$$\mathbf{H}^j \cdot KN^j + (\lambda_{p_0} + \lambda^\$_{\ell i'} + \lambda^\$_{\ell j}) \cdot \underbrace{L} = TN^{j'} + MM.$$

Here too there are q choices for j', and this probability $\leq q/(N-2)(N-3) \leq 4q/N^2$. Thus, the probability of badB and *Subcase 2(b)* simultaneously happening is at most $2/N$, as long as $2q \leq N$.

- *Subcase 2(c).* $C'^i_{p_0} + \Delta'^i_{p_0} = V^{j'}_{s_0}$ for some $j' \in [q], s_0 \in [k^{j'}]$, so $X'^i_{p_0} = U^{j'}_{s_0}$, and a second equation comes from the condition for badD[i]. Here the two equations are

$$\underbrace{V^{j'}_{s_0}} + \mathbf{H}^{i'} \cdot KN^{i'} + \lambda_{p_0} \cdot L = C'^i_{p_0},$$

$$\mathbf{H}^j \cdot \underbrace{KN^j} + (\lambda_{s_0} + \lambda_{p_0} + \lambda^\$_{\ell i'} + \lambda^\$_{\ell j}) \cdot L = A^{j'}_{s_0} + MM.$$

There are α choices for (j', s_0), so the probability of this $\leq \alpha/(N-2)(N-3) \leq 4\alpha/N^2 \leq 2/N$, as long as $2\alpha \leq N$.

– *Subcase 2(d).* $\mathsf{C'}^i_{p_0} + \Delta^{i'}_{p_0} = \mathsf{C}^{j'}_{p_1} + \Delta^{j'}_{p_1}$ for some $j' \in [q], p_1 \in [\ell^{j'}]$, so $\mathsf{X'}^i_{p_0} = \mathsf{X}^{j'}_{p_1}$, and a second equation comes from the condition for badD[i]. The two equations are

$$\mathsf{Q}^{i'} + \mathsf{Q}^{j'} + (\lambda_{p_1} + \lambda_{p_0}) \cdot \mathsf{L} = \mathsf{C}^{j'}_{p_1} + \mathsf{C'}^i_{p_0},$$
$$\mathsf{Q}^{j} + \mathsf{Q}^{j'} + (\lambda_{p_1} + \lambda_{p_0} + \lambda^{\$}_{\ell^{i'}} + \lambda^{\$}_{\ell^j}) \cdot \mathsf{L} = \mathsf{M}^{j'}_{p_1} + \mathsf{MM}.$$

When $i' = j$, the two equations become

$$\mathbf{H}^{i'} \cdot \mathsf{KN}^{i'} + \mathbf{H}^{j'} \cdot \mathsf{KN}^{j'} + (\lambda_{p_1} + \lambda_{p_0}) \cdot \underbrace{\mathsf{L}} = \mathsf{C}^{j'}_{p_1} + \mathsf{C'}^i_{p_0},$$
$$\mathbf{H}^{i'} \cdot \mathsf{KN}^{i'} + \mathbf{H}^{j'} \cdot \mathsf{KN}^{j'} + (\lambda_{p_1} + \lambda_{p_0}) \cdot \underbrace{\mathsf{L}} = \mathsf{M}^{j'}_{p_1} + \mathsf{MM},$$

which is actually a single equation with the constraint that $\mathsf{C}^{j'}_{p_1} + \mathsf{C'}^i_{p_0} = \mathsf{M}^{j'}_{p_1} + \mathsf{MM}$, which implies that j' must satisfy $\mathsf{M}^{j'}_{p_1} + \mathsf{C}^{j'}_{p_1} = \mathsf{C'}^i_{p_0} + \mathsf{MM}$. Since there are no collisions on $\mathsf{M}^{j'}_{p_1} + \mathsf{C}^{j'}_{p_1}$ over all pairs (j', p_1), there is at most one choice of (j', p_1) satisfying this. Thus, the probability of this $\leq 1/(N-2) \leq 2/N$. When $i \neq j'$ but $\mathsf{KN}^{i'} = \mathsf{KN}^j$ and $\ell^{i'} = \ell^j$, the two equations become

$$\mathbf{H}^{i'} \cdot \underbrace{\mathsf{KN}^{i'}} + \mathbf{H}^{j'} \cdot \mathsf{KN}^{j'} + (\lambda_{p_1} + \lambda_{p_0}) \cdot \underbrace{\mathsf{L}} = \mathsf{C}^{j'}_{p_1} + \mathsf{C'}^i_{p_0},$$
$$\mathbf{H}^{j} \cdot \underbrace{\mathsf{KN}^{i'}} + \mathbf{H}^{j'} \cdot \mathsf{KN}^{j'} + (\lambda_{p_1} + \lambda_{p_0}) \cdot \underbrace{\mathsf{L}} = \mathsf{M}^{j'}_{p_1} + \mathsf{MM}.$$

There are at most σ choices for (j', p_1), so the probability $\leq \sigma/(N-1)(N-2) \leq 4\sigma/N^2$. When $i \neq j'$, $\mathsf{KN}^{i'} = \mathsf{KN}^j \neq \mathsf{KN}^{j'}$, $\ell^{i'} \neq \ell^j$, the two equations become

$$\mathbf{H}^{i'} \cdot \underbrace{\mathsf{KN}^{i'}} + \mathbf{H}^{j'} \cdot \mathsf{KN}^{j'} + (\lambda_{p_1} + \lambda_{p_0}) \cdot \underbrace{\mathsf{L}} = \mathsf{C}^{j'}_{p_1} + \mathsf{C'}^i_{p_0},$$
$$\mathbf{H}^{j} \cdot \underbrace{\mathsf{KN}^{i'}} + \mathbf{H}^{j'} \cdot \mathsf{KN}^{j'} + (\lambda_{p_1} + \lambda_{p_0} + \lambda^{\$}_{\ell^{i'}} + \lambda^{\$}_{\ell^j}) \cdot \underbrace{\mathsf{L}} = \mathsf{M}^{j'}_{p_1} + \mathsf{MM}.$$

There are at most σ choices for (j', p_1), so the probability of this $\leq \sigma/(N-1)(N-2) \leq 4\sigma/N^2$. When $i \neq j'$, $\mathsf{KN}^{i'} = \mathsf{KN}^j = \mathsf{KN}^{j'}$, $\ell^{i'} \neq \ell^j$, the two equations become

$$(\mathbf{H}^{i'} + \mathbf{H}^{j'}) \cdot \underbrace{\mathsf{KN}^{i'}} + (\lambda_{p_1} + \lambda_{p_0}) \cdot \underbrace{\mathsf{L}} = \mathsf{C}^{j'}_{p_1} + \mathsf{C'}^i_{p_0},$$
$$(\mathbf{H}^{j} + \mathbf{H}^{j'}) \cdot \underbrace{\mathsf{KN}^{i'}} + (\lambda_{p_1} + \lambda_{p_0} + \lambda^{\$}_{\ell^{i'}} + \lambda^{\$}_{\ell^j}) \cdot \underbrace{\mathsf{L}} = \mathsf{M}^{j'}_{p_1} + \mathsf{MM}.$$

They may be multiples of the same equation, in which case we have at most 64 choices for j' and at most ℓ_{MAX} choices for p_1. Then this probability $\leq 64\ell_{\mathrm{MAX}}/N$. When they are different equations, this probability $\leq \sigma/N(N-1) \leq 2\sigma/N^2$. So the probability of badD[i] and *Subcase 2(d)* simultaneously happening is at most $(64\ell_{\mathrm{MAX}} + 7)/N$ as long as $2\sigma \leq N$.

– *Subcase 2(e).* $C'^i_{p_0} + \Delta'^i_{p_0} = \text{auth}^{j'} + T^{j'}$ for some $j' \in [q]$, so $X'^i_{p_0} = X^{j'}_{\text{tag}}$, and a second equation comes from the condition for $\text{badD}[i]$. When $j = j'$, the two equations become

$$\underbrace{V^j_1} + \sum_{s \in [2..k^j]} V^j_s + \mathbf{H}^{i'} \cdot KN^{i'} + \lambda_{p_0} \cdot \underbrace{L} = T^{j'} + C'^i_{p_0},$$

$$(\lambda_{p_0} + \lambda^\$_{\ell i'}) \cdot \underbrace{L} = M^{j'}_{\text{tag}} + MM,$$

and the probability of this $\leq 1/(N - k^j - 1)(N - k^j - 2) \leq 4/N^2$. When $j \neq j'$, the two equations become

$$\underbrace{V^{j'}_1} + \sum_{s \in [2..k^{j'}]} V^{j'}_s + \mathbf{H}^{i'} \cdot KN^{i'} + \lambda_{p_0} \cdot L = T^{j'} + C'^i_{p_0},$$

$$\mathbf{H}^j \cdot KN^j + \mathbf{H}^{j'} \cdot \underbrace{KN^{j'}} + (\lambda^\$_{\ell j'} + \lambda_{p_0} + \lambda^\$_{\ell i'} + \lambda^\$_{\ell j}) \cdot L = M^{j'}_{\text{tag}} + MM.$$

and the probability of this $\leq q/(N - k^{j'} - 3)(N - k^{j'} - 4) \leq 4q/N^2$. Thus, the probability of badB and *Subcase 2(e)* simultaneously happening is at most $2/N$, as long as $2q \leq N$.

Summing over the subcases completes the proof.   □

5.3   Proof of Lemma 3

Lemma 3. *For $i \in S[3]$, $\text{Pr}_{\mathcal{O}_0}[badD[i]] \leq \dfrac{10}{N}$, as long as $2\sigma_T \leq N$.*

Proof. When $i \in S[3]$, X'^i_p is trivially determined for all $p \in [\ell'^i]$, but X'^i_{tag} is not trivially determined. The equation for $\text{badD}[i]$ becomes

$$\pi^{-1}(\text{auth}'^i + T'^i) + \Delta'^i_{\text{tag}} = \sum_{p \in [\ell^{i'}]} M^{i'}_p.$$

Based on the value of $\text{auth}'^i + T'^i$, we look at the subcases listed in the tree at the beginning of this section.

– *Subcase 3(a).* $\text{auth}'^i + T'^i \notin \text{Ran}(\pi)$, so X'^i_{tag} is sampled. The equation for $\text{badD}[i]$ is

$$\underbrace{X'^i_{\text{tag}}} + \mathbf{H}^{i'} \cdot KN^{i'} + \lambda^\$_{\ell i'} \cdot L = \sum_{p \in [\ell^{i'}]} M^{i'}_p.$$

The probability of this $\leq 1/(N - 2) \leq 2/N$.

- *Subcase 3(b).* $\mathsf{auth'}^i + \mathsf{T'}^i = \mathsf{KN}^{j'}$ for some $j' \in [q]$, so $\mathsf{X'}^i_{\mathsf{tag}} = \mathsf{TN}^{j'}$, and a second equation comes from the condition for $\mathsf{badD}[i]$. The two equations are

$$\underbrace{\mathsf{KN}^{j'}} + \sum_{s \in [k'^i]} \mathsf{V'}^i_s = \mathsf{T'}^i,$$

$$\mathbf{H}^{i'} \cdot \mathsf{KN}^{i'} + \lambda^{\$}_{\ell^{i'}} \cdot \underbrace{\mathsf{L}} = \mathsf{TN}^{j'} + \sum_{p \in [\ell^{i'}]} \mathsf{M}^{i'}_p.$$

The probability of this $\leq 4q/N^2 \leq 2/N$, as long as $2q \leq N$.

- *Subcase 3(c).* $\mathsf{auth'}^i + \mathsf{T'}^i = \mathsf{V}^{j'}_{s_0}$ for some $j' \in [q]$, $s_0 \in [k^{j'}]$, so $\mathsf{X'}^i_{p_0} = \mathsf{U}^{j'}_{s_0}$, and a second equation comes from the condition for $\mathsf{badD}[i]$. The two equations are

$$\underbrace{\mathsf{V}^{j'}_{s_0}} + \sum_{s \in [k'^i]} \mathsf{V'}^i_s = \mathsf{T'}^i,$$

$$\mathbf{H}^{i'} \cdot \mathsf{KN}^{i'} + (\lambda_{s_0} + \lambda^{\$}_{\ell^{i'}}) \cdot \underbrace{\mathsf{L}} = \mathsf{A}^{j'}_{s_0} + \sum_{p \in [\ell^{i'}]} \mathsf{M}^{i'}_p.$$

When the top equation vanishes, the probability of this $\leq 1/(N-1) \leq 2/N$. When both equations are there, the probability of this $\leq 4\alpha/N^2$. So the probability of $\mathsf{badD}[i]$ and *Subcase 3(c)* simultaneously happening is at most $2/N$, as long as $2\alpha \leq N$.

- *Subcase 3(d).* $\mathsf{auth'}^i + \mathsf{T'}^i = \mathsf{C}^{j'}_{p_1} + \Delta^{j'}_{p_1}$ for some $j' \in [q], p_1 \in [\ell^{j'}]$, so $\mathsf{X'}^i_{\mathsf{tag}} = \mathsf{X}^{j'}_{p_1}$, and a second equation comes from the condition for $\mathsf{badD}[i]$. The two equations are

$$\underbrace{\mathsf{V'}^i_1} + \sum_{s \in [2..k'^i]} \mathsf{V'}^i_s + \mathbf{H}^{j'} \cdot \mathsf{KN}^{j'} + \lambda_{p_1} \cdot \mathsf{L} = \mathsf{C}^{j'}_{p_1} + \mathsf{T'}^i,$$

$$\mathbf{H}^{j'} \cdot \mathsf{KN}^{j'} + \mathbf{H}^{i'} \cdot \mathsf{KN}^{i'} + (\lambda_{p_1} + \lambda^{\$}_{\ell^{i'}}) \cdot \underbrace{\mathsf{L}} = \mathsf{M}^{j'}_{p_1} + \sum_{p \in [\ell^{i'}]} \mathsf{M}^{i'}_p.$$

The probability of this $\leq 4\sigma/N^2 \leq 2/N$, as long as $2\sigma \leq N$.

- *Subcase 3(e).* $\mathsf{auth'}^i + \mathsf{T'}^i = \mathsf{auth}^{j'} + \mathsf{T}^{j'}$ for some $j' \in [q]$, so $\mathsf{X'}^i_{\mathsf{tag}} = \mathsf{X}^{j'}_{\mathsf{tag}}$, and a second equation comes from the condition for $\mathsf{badD}[i]$. If $\mathsf{A}^{j'} = \mathsf{A'}^i$, we have $\mathsf{T}^{j'} \neq \mathsf{T'}^i$ (by definition of this case), or $\mathsf{auth}^{j'} \neq \mathsf{auth'}^i$, a contradiction. So $\mathsf{A}^{j'} \neq \mathsf{A'}^i$. If we can find $s_0 \leq k^{j'}$ such that either $s_0 > k'^i$, or $\mathsf{A}^{j'}_{s_0} \neq \mathsf{A'}^i_{s_0}$, then the equations are

$$\underbrace{\mathsf{V}^{j'}_{s_0}} + \sum_{s \in [k'^i]} \mathsf{V'}^i_s + \sum_{s \in [k^{j'}] \backslash \{s_0\}} \mathsf{V}^{j'}_s = \mathsf{T'}^i + \mathsf{T}^{j'},$$

$$\mathbf{H}^{j'} \cdot \mathsf{KN}^{j'} + \mathbf{H}^{i'} \cdot \mathsf{KN}^{i'} + (\lambda^{\$}_{\ell^{j'}} + \lambda^{\$}_{\ell^{i'}}) \cdot \underbrace{\mathsf{L}} = \mathsf{M}^{j'}_{\mathsf{tag}} + \sum_{p \in [\ell^{i'}]} \mathsf{M}^{i'}_p.$$

Otherwise we can find $s_0 \leq k'^i$ such that either $s_0 > k^{j'}$, or $\mathsf{A'}^i_{s_0} \neq \mathsf{A}^{j'}_{s_0}$, then the equations are

$$\underbrace{\mathsf{V'}^i_{s_0}}_{} + \sum_{s \in [k'^i] \setminus \{s_0\}} \mathsf{V'}^i_s + \sum_{s \in [k^{j'}]} \mathsf{V}^{j'}_s = \mathsf{T'}^i + \mathsf{T}^{j'},$$

$$\mathbf{H}^{j'} \cdot \mathsf{KN}^{j'} + \mathbf{H}^{i'} \cdot \mathsf{KN}^{i'} + (\lambda^\$_{\ell^{j'}} + \lambda^\$_{\ell^{i'}}) \cdot \underbrace{\mathsf{L}}_{} = \mathsf{M}^{j'}_{\text{tag}} + \sum_{p \in [\ell^{i'}]} \mathsf{M}^{i'}_p.$$

The probability of either of these does not exceed $2/N$ as long as $2\sigma \leq N$.

Summing over the subcases completes the proof.   □

5.4   Proof of Lemma 4

Lemma 4. *For* $i \in \mathcal{S}[4]$, $\Pr_{\mathcal{O}_0} [badD[i]] \leq \dfrac{2}{N} + \dfrac{32\sigma_T^2}{N^2}$.

Proof. The bounds for **Case 4** are simple to derive:

- *Subcase 4(a).* We get a bound of $2/N$ on the probability as in *Subcase 2(a)* or *Subcase 3(a)*.

- *Subcase 4(b).* We treat this separately because the equation

$$\mathsf{auth}'^i + \mathsf{T}'^i = \mathsf{V}^{j'}_{s_0}$$

can vanish. When it does not vanish, we can find two independent collision equations that need to be satisfied, which can occur together with a probability of at most $2/N^2$. When it does vanish, we proceed as in *Subcase 3(c)* and use instead the equation where $\mathsf{Q}^{i'} + (\lambda_{s_0} + \lambda^\$_{\ell^{i'}}) \cdot \mathsf{L}$ is equated to a constant. There are four possibilities here based on *Subcases 2(b)-2(e)*. In each possibility, there are at most σ_T^2 choices for these collision indices. Thus, each possibility with $badD[i]$ has a probability of at most $2\sigma_T^2/N^2$, and *Subcase 4(b)* has a probability of at most $8\sigma_T^2/N^2$.

- *Subcase 4(c).* There are twelve possibilites here, based on various combinations of *Subcases 2(b)-2(e)* and *Subcases 3(b),3(d),3(e)*. In each of these possibilities, we can find two independent collision equations that need to be satisfied, which can occur together with a probability of at most $2/N^2$. There are at most σ_T^2 choices for these collision indices. Thus, each possibility with $badD[i]$ has a probability of at most $2\sigma_T^2/N^2$, and *Subcase 4(c)* has a probability of at most $24\sigma_T^2/N^2$.

Summing over the three subcases completes the proof.   □

5.5   Proof of Lemma 5

Lemma 5. *For $i \in \mathcal{S}[5]$, $\mathrm{Pr}_{\mathcal{O}_0}[badD[i]] \leq \dfrac{64\ell_{MAX} + 4}{N} + \dfrac{32\sigma_T^2}{N^2}.$*

Proof. We look one by one at the subcases listed in the tree at the beginning of this section.

– *Subcase 5(a).* Here we get a bound of $2/N$ on the probability as in *Subcase 2(a)*.
– *Subcase 5(b).* This is the case when

$$\mathsf{C'}_{p_0}^{i} + \mathsf{Q}^{i'} + \lambda_{p_0} \cdot \mathsf{L} = \mathsf{C}_{p_2}^{j'} + \mathsf{Q}^{j'} + \lambda_{p_2} \cdot \mathsf{L},$$
$$\mathsf{C'}_{p_1}^{i} + \mathsf{Q}^{i'} + \lambda_{p_1} \cdot \mathsf{L} = \mathsf{C}_{p_3}^{j''} + \mathsf{Q}^{j''} + \lambda_{p_3} \cdot \mathsf{L}$$

for some $j', j'' \in [q], p_2 \in [\ell^{j'}], p_3 \in [\ell^{j''}]$ with $j' \neq j''$. If $\mathsf{TN}^{j'} = \mathsf{TN}^{i'}$, the equations may become dependent on each other. But here there are at most 64 choices for j', since the nonce is distinct in every encryption query, and only 64 distinct values of $\mathsf{N}^{j'}$ can yield the same $\mathsf{TN}^{j'}$. Thus the probability of this does not exceed $64\ell_{\mathrm{MAX}}/N$. Otherwise, we always get two independent equations, and the bound of $2\sigma_T^2/N^2$ holds. Thus, the probability of *Subcase 5(b)* with $badD[i]$ does not exceed $64\ell_{\mathrm{MAX}}/N + 2\sigma_T^2/N^2$.
– *Subcase 5(c).* This is trickier. Here too these two equations may become the same equation. Since the equations can be rewritten as

$$\mathsf{Q}^{i'} + \mathsf{Q}^{j'} + (\lambda_{p_0} + \lambda_{p_2}) \cdot \mathsf{L} = \mathsf{C'}_{p_0}^{i} + \mathsf{C}_{p_2}^{j'},$$
$$\mathsf{Q}^{i'} + \mathsf{Q}^{j'} + (\lambda_{p_1} + \lambda_{p_3}) \cdot \mathsf{L} = \mathsf{C'}_{p_1}^{i} + \mathsf{C}_{p_3}^{j'},$$

they become the same equation when $\lambda_{p_0} + \lambda_{p_2} = \lambda_{p_1} + \lambda_{p_3}$ and $\mathsf{C'}_{p_0}^{i} + \mathsf{C'}_{p_2}^{i} = \mathsf{C'}_{p_1}^{i} + \mathsf{C'}_{p_3}^{i}$. Thus any valid choice of (p_2, p_3) must satisfy

$$\lambda_{p_2} + \lambda_{p_3} = \lambda_{p_0} + \lambda_{p_1},$$
$$\mathsf{C'}_{p_2}^{i} + \mathsf{C'}_{p_3}^{i} = \mathsf{C'}_{p_0}^{i} + \mathsf{C'}_{p_1}^{i},$$

i.e., for each such choice of (p_2, p_3), $\lambda_{p_2} + \lambda_{p_3}$ takes the same fixed value, and $\mathsf{C'}_{p_2}^{i} + \mathsf{C'}_{p_3}^{i}$ take the same fixed value. Since $badA$ has not occurred, we know there are at most 2 such choices of (p_2, p_3). Thus the probability of this does not exceed $2/N$.
– *Subcase 5(d).* There can be fifteen possibilities, depending on various combinations of the subcases of **Case 2**. In each of these, we can find two independent collision equations that need to be satisfied, which can occur together with a probability of at most $2/N^2$. There are at most σ_T^2 choices for these collision indices. Thus, each of those possibilities with $badD[i]$ has a probability of at most $2\sigma_T^2/N^2$, and *Subcase 5(d)* has a probability of at most $30\sigma_T^2/N^2$.

Summing over the four subcases completes the proof.   □

More detailed proofs of Lemmas 4 and 5 can be found in the full version of the paper at the IACR eprint archive, at the url https://eprint.iacr.org/2017/845.pdf.

References

[ABL+13] Andreeva, E., Bogdanov, A., Luykx, A., Mennink, B., Tischhauser, E., Yasuda, K.: Parallelizable and authenticated online ciphers. In: Sako, K., Sarkar, P. (eds.) ASIACRYPT 2013. LNCS, vol. 8269, pp. 424–443. Springer, Heidelberg (2013). https://doi.org/10.1007/978-3-642-42033-7_22

[BGM04] Bellare, M., Goldreich, O., Mityagin, A.: The power of verification queries in message authentication and authenticated encryption. Cryptology ePrint Archive, Report 2004/309 (2004). http://eprint.iacr.org/2004/309

[cae] Caesar competition (2013). https://competitions.cr.yp.to/caesar.html

[DN14] Datta, N., Nandi, M.: ELmE: A misuse resistant parallel authenticated encryption. In: Susilo, W., Mu, Y. (eds.) ACISP 2014. LNCS, vol. 8544, pp. 306–321. Springer, Cham (2014). https://doi.org/10.1007/978-3-319-08344-5_20

[GPR17] Gaži, P., Pietrzak, K., Rybár, M.: The exact security of PMAC. IACR Trans. Symmetric Cryptol. 2016(2), 145–161 (2017)

[HKR15] Hoang, V.T., Krovetz, T., Rogaway, P.: Robust authenticated-encryption AEZ and the problem that it solves. In: Oswald, E., Fischlin, M. (eds.) EUROCRYPT 2015. LNCS, vol. 9056, pp. 15–44. Springer, Heidelberg (2015). https://doi.org/10.1007/978-3-662-46800-5_2

[KR11] Krovetz, T., Rogaway, P.: The software performance of authenticated-encryption modes. In: Joux, A. (ed.) FSE 2011. LNCS, vol. 6733, pp. 306–327. Springer, Heidelberg (2011). https://doi.org/10.1007/978-3-642-21702-9_18

[McG08] McGrew, D.: An interface and algorithms for authenticated encryption (2008). https://buildbot.tools.ietf.org/html/rfc5116

[MV04] McGrew, D., Viega, J.: The security and performance of the galois/counter mode (GCM) of operation. In: Canteaut, A., Viswanathan, K. (eds.) INDOCRYPT 2004. LNCS, vol. 3348, pp. 343–355. Springer, Heidelberg (2004). https://doi.org/10.1007/978-3-540-30556-9_27

[MV05] McGrew, D., Viega, J.: The galois/counter mode of operation (gcm). In: NIST Modes Operation Symmetric Key Block Ciphers (2005)

[Pat08] Patarin, J.: The "Coefficients H" technique. In: Avanzi, R.M., Keliher, L., Sica, F. (eds.) SAC 2008. LNCS, vol. 5381, pp. 328–345. Springer, Heidelberg (2009). https://doi.org/10.1007/978-3-642-04159-4_21

[RBB03] Rogaway, P., Bellare, M., Black, J.: OCB: A block-cipher mode of operation for efficient authenticated encryption. ACM Trans. Inf. Syst. Secur. 6(3), 365–403 (2003)

[Rog02] Rogaway, P.: Authenticated-encryption with associated-data. In: Proceedings of the 9th ACM Conference on Computer and Communications Security, CCS 2002, NY, USA, pp. 98–107. ACM, New York (2002)

[Rog04] Rogaway, P.: Efficient instantiations of tweakable blockciphers and refinements to modes OCB and PMAC. In: Lee, P.J. (ed.) ASIACRYPT 2004. LNCS, vol. 3329, pp. 16–31. Springer, Heidelberg (2004). https://doi.org/10.1007/978-3-540-30539-2_2

[RS06] Rogaway, P., Shrimpton, T.: A provable-security treatment of the key-wrap problem. In: Vaudenay, S. (ed.) EUROCRYPT 2006. LNCS, vol. 4004, pp. 373–390. Springer, Heidelberg (2006). https://doi.org/10.1007/11761679_23

[WHF02] Whiting, D., Housley, R., Ferguson, N.: AES encryption & authentication using CTR mode & CBC-MAC. IEEE P802, 11 (2002)

The Iterated Random Function Problem

Ritam Bhaumik[1], Nilanjan Datta[2], Avijit Dutta[1], Nicky Mouha[3,4](✉),
and Mridul Nandi[1]

[1] Indian Statistical Institute, Kolkata, India
bhaumik.ritam@gmail.com, avirocks.dutta13@gmail.com,
mridul.nandi@gmail.com
[2] Indian Institute of Technology, Kharagpur, India
nilanjan_isi_jrf@yahoo.com
[3] National Institute of Standards and Technology, Gaithersburg, MD, USA
nicky@mouha.be
[4] Project-team SECRET, Inria, Paris, France

Abstract. At CRYPTO 2015, Minaud and Seurin introduced and studied the *iterated random permutation* problem, which is to distinguish the r-th iterate of a random permutation from a random permutation. In this paper, we study the closely related *iterated random function* problem, and prove the first almost-tight bound in the adaptive setting. More specifically, we prove that the advantage to distinguish the r-th iterate of a random function from a random function using q queries is bounded by $O(q^2 r (\log r)^3 / N)$, where N is the size of the domain. In previous work, the best known bound was $O(q^2 r^2 / N)$, obtained as a direct result of interpreting the iterated random function problem as a special case of CBC-MAC based on a random function. For the iterated random function problem, the best known attack has an advantage of $\Omega(q^2 r / N)$, showing that our security bound is tight up to a factor of $(\log r)^3$.

Keywords: Iterated random function · Random function · Pseudorandom function · Password hashing · Patarin · H-coefficient technique · Provable security

1 Introduction

Take any n-bit hash function h. Assuming that this hash function can be modelled as a random function, the probability that the outputs of h collide given $q \ll 2^{n/2}$ distinct inputs is about $q^2/2^n$: the well-known birthday attack.

Now let us consider another hash function g, defined as the r-th iterate of h, i.e. $g(m) = h(h(\ldots h(m)))$, where h is applied r times. For the same number of

---

Certain algorithms and commercial products are identified in this paper to foster understanding. Such identification does not imply recommendation or endorsement by NIST, nor does it imply that the algorithms or products identified are necessarily the best available for the purpose.

© International Association for Cryptologic Research 2017
T. Takagi and T. Peyrin (Eds.): ASIACRYPT 2017, Part II, LNCS 10625, pp. 667–697, 2017.
https://doi.org/10.1007/978-3-319-70697-9_23

queries $q \ll 2^{n/2}$, the birthday attack has about an r times higher probability to succeed for g than for h (see e.g. Preneel and van Oorschot [18, Lemma 2]).

Iteration is of fundamental importance in many cryptographic constructions. For example, a "possibly weak" function may be iterated to improve its resistance against various cryptanalysis attacks, or a password hashing function may be iterated to slow down dictionary attacks. But quite surprisingly, the security of iterating a random function is not yet a well-understood problem.

In the aforementioned (non-adaptive) birthday attack, the distinguishing advantage between a random function and an iterated random function increases by about a factor r. But what happens if we consider adaptive collision-finding attacks as well? Or in general, what if we want to consider any adaptive attack, not necessarily a collision-finding attack? Could there be more efficient attacks that have not yet been discovered?

Recently at CRYPTO 2015, Minaud and Seurin [15] put this possibility to rest for the iterated random permutation problem. They proved that the advantage to distinguish an iterated random permutation from a random permutation using q queries is bounded by $O(qr/N)$, where N is the size of the domain, and showed that their bound is almost tight by providing a matching attack.

In this paper, we will do the same for the iterated random function problem. Whereas the best bound in previous work is $O(q^2 r^2/N)$, we will prove a bound of $O(q^2 r (\log r)^3/N)$, where log is the logarithm to the base e. Our bound is tight up to a factor of about $(\log r)^3$, and thereby rules out the possibility of better attacks.

NOTE. We will focus on asymptotic bounds for large r, as this is parameter range where large improvements over the currently best-known bounds can be achieved. Although our bounds hold for any $r \geq 2$, we will apply generous relaxations to derive an easy-to-see bound that only improves the currently-known bounds for larger, but nevertheless practically-relevant values of r. Also, we will only consider the iteration of a uniformly random function in an information-theoretic setting. A simple hybrid argument can be used to extend this result to the pseudorandom function (prf) advantage in a computational setting, as shown by Minaud and Seurin [15, Theorem 1] for the iterated random permutation problem.

APPLICATIONS. In spite of the frequent use of iterated random functions in practice, this paper is the first to study this problem without relying on the trivial CBC-MAC bound. The most obvious application of iterated random functions is in password hashing, where a hash function is iterated in order to slow down brute force attacks. This idea is used in PKCS #5's PBKDF1 and PBKDF2. In typical password-based key derivation functions, the iteration count is often quite high, ranging from several hundreds of thousands [9], to even ten million [19], as suggested by NIST for critical keys. To analyse the effect of iteration in these constructions, it is common to model the secret low-entropy password as a random-but-known key [11], or even an adversarially-chosen input [20]. But also small values of r, such as $r = 2$, appear in practical applications. In the book "Practical cryptography" [13], Ferguson and Schneier suggest to use

SHA-256(SHA-256(m)) to avoid length-extension attacks. They use this construction in their RSA encryption implementation, as well as in their Fortuna random number generator. Interestingly, about 2^{64} evaluations of SHA-256(SHA-256(m)) are performed *every second* as part of bitcoin mining [21].

RELATED WORK. The security of an iterated random function was first analysed by Yao and Yin [22,23], when they analysed the security of the password-based key derivation functions PBKDF1 and PBKDF2. Their work is parallel to that of Wagner and Goldberg [20], who analysed the security of an iterated random permutation in the context of the Unix password hashing algorithm. Bellare et al. [4] extended these results, and also pointed out some problems in the proofs of Yao and Yin.

As Wagner and Goldberg explain in [20], it is possible to interpret the iterated random permutation problem as a special case of CBC-MAC where the iteration count r equals the number of message blocks, and all message blocks except for the first one are all-zero. The same holds for the iterated random function problem, except that a random function instead of a random permutation is used inside the CBC-MAC construction.

A first proof of the security of CBC-MAC was given by Bellare et al. in [1,2]. For CBC-MAC with a random function, they prove that the advantage of an information-theoretic adversary that makes at most q queries is upper bounded by $1.5r^2q^2/N$. Using the well-known prp-prf switching lemma [5], they derive from this an upper bound of $2r^2q^2/N$ for CBC-MAC with a random permutation. The simplicity of CBC-MAC makes it a good test case for various proof techniques. Of particular interest is the short proof of CBC-MAC by Bernstein [7]. For a more detailed proof using the same technique, we refer to Nandi [16].

In [3], Bellare et al. proved a security bound that is linear in r, instead of quadratic in r as in previous proofs. They point out that their analysis only applies to CBC-MAC with a random permutation, and not with a random function: such a bound is ruled out by an attack by Berke [6]. However, Berke's attack cannot be translated to the iterated random function problem, as the number of message blocks for each of the queries in the attack is not constant.

The iterated random function problem is similar to the nested iterated (NI) construction that Gaži et al. [14] analysed at CRYPTO 2014. However, the analysis of the NI construction critically relies on the use of two *different* random functions, or more precisely on the use of a pseudo-random function (prf) with two different keys. Our analysis applies to the case where only *one* random function is iterated. As we will show, the iterated random function problem will require a more complicated analysis of collision probabilities, in order to avoid ending up with a bound that is quadratic in r.

MAIN RESULTS. The main results of this paper are the proofs of two theorems. Theorem 1 bounds the success probability of a common class of collision adversaries, and Theorem 2 bounds the advantage of distinguishing an iterated random function from a random function. In these theorems, the function $\phi(q, r)$ is defined as

$$\phi(q,r) := 2\left(\frac{q^2\sqrt{r}}{N}\right) + 2\sqrt{\frac{q^2 r \log r}{N}} + 16\left(\frac{q^2 r \log r}{N}\right)^2 + 49(\log r)^2\left(\frac{q^2 r \log r}{N}\right).$$

Theorem 1. *Let f be a random function, and let $\mathcal{A}$ be a collision-finding adversary that makes q queries to f^r as follows: every query is either chosen from a set (of size $m \le q$) of predetermined points, or is the response of a previous query. Under the assumption that $N \log r > 90$, the following bound holds for the success probability $cp^r[q]$ of $\mathcal{A}$:*

$$cp^r[q](\mathcal{A}) \le \phi(q,r).$$

Theorem 2. *Let f be a random function, and let $\mathcal{A}$ be an adversary trying to distinguish f^r from f through q queries. Then, under the assumption that $N \log r > 90$, we have*

$$\mathbf{Adv}_{f,f^r}(q) \le \frac{q^2 r}{N} + \frac{2q^2}{N} + \phi(q,r).$$

A NOTE ON THE SETTING. We should point out that our results are in an indistinguishability setting. Our goal is to distinguish, in a black-box way, between an iterated random function and a random function. In the indifferentiability setting, the adversary also has access to the underlying random function, or to a simulator that tries to mimic its behaviour. Dodis et al. [12] proved that indifferentiability for an iterated random function holds only with poor concrete security bounds, as they provide a lower bound on the complexity of any successful simulator.

OUTLINE. Notation and preliminaries are introduced in Sect. 2. We study the probabilities to find various types of collisions in a random function in Sect. 3. These results are used in Sect. 4 to bound the probabilities of single-trail attacks and two-trail collision attacks, and eventually to also bound a more general collision attack on an iterated random function. The advantage of distinguishing an iterated random function from a random function is bounded in Sect. 5. For readability, we defer the technical proof of Lemma 7 of Sect. 4 to Sect. 6. We conclude the paper in Sect. 7.

2   Notation and Preliminaries

In this section, we will state some simple lemmas without proof. The proofs of these lemmas can be found in the full version of this paper [8].

FUNCTIONS. Let $f : \mathcal{D} \to \mathcal{D}$ be a function over a domain $\mathcal{D}$ of size N. A collision for a function f is defined as a pair $(x, x') \in \mathcal{D}$ with $x \ne x'$ such that $f(x) = f(x')$. A three-way collision is a triple (x, x', x'') such that $f(x) = f(x') = f(x'')$ for distinct x, x' and x''. For a positive integer r, the r-th iterate f^r of a function f is defined inductively as follows:

$$f^1 = f,$$
$$f^r = f \circ f^{r-1}, r > 1.$$

By convention, let f^0 be the identity function. In the remainder of this paper, we will assume that $r \geq 2$. Let a random function denote a function that is drawn uniformly at random from the set of all functions of the same domain and range.

FALLING FACTORIAL POWERS AND THE β FUNCTION. We use the falling factorial powers notation, where for a non-negative integer $i \leq N$, $N^{\underline{i}}$ is defined as

$$N^{\underline{i}} := \frac{N!}{(N-i)!} = N(N-1)\cdots(N-i+1). \tag{1}$$

Note that $N^{\underline{i}}$ denotes the number of permutations of N items taken i at a time, or the number of ways to choose a sample of size i without replacement from a population of size N. When $i > N$, we define $N^{\underline{i}} := 0$. We also define a function $\beta(i)$ that we will frequently encounter:

$$\beta(i) := \frac{N^{\underline{i}}}{N^i}. \tag{2}$$

Again, we define $\beta(i) := 0$ for $i > N$. We derive below a simple bound on $\beta(i)$.

Lemma 1. *Let $\alpha > 0$ be a real number. Then, for $i \geq \sqrt{2\alpha N} + 1$, we have*

$$\beta(i) \leq e^{-\alpha}.$$

PARTIAL SUMS OF THE HARMONIC SERIES. The divergent infinite series

$$\sum_{i=1}^{\infty} \frac{1}{i} = 1 + \frac{1}{2} + \frac{1}{3} + \frac{1}{4} + \cdots$$

is known as the harmonic series. We will be interested in partial sums of the series of the form

$$\sum_{i=a+1}^{b} \frac{1}{i} = \frac{1}{a+1} + \frac{1}{a+2} + \cdots + \frac{1}{b-1} + \frac{1}{b}.$$

We will use the following simple bound for this sum. Throughout this paper, let log denote the natural logarithm, that is the logarithm to the base e.

Lemma 2. *For any two positive integers a and b with $b \geq a$,*

$$\sum_{i=a+1}^{b} \frac{1}{i} \leq \log\left(\frac{b}{a}\right).$$

COUNTING DIVISORS. For a positive integer a and an integer b we use the notation $a|b$ to denote a *divides* b, i.e., $ak = b$ for some integer k. We write $a \nmid b$ when a does not divide b. The number of divisors of b is denoted $\mathsf{d}(b)$. We will use the following simple bound on $\mathsf{d}(b)$.

Lemma 3. *For any positive integer b,*

$$d(b) < 2\sqrt{b}.$$

THE σ FUNCTION. The function $\sigma(b)$ defined as

$$\sigma(b) := \sum_{a|b} a$$

denotes the sum of the divisors of b. We will use the following simple lemma about $\sigma(b)$.

Lemma 4. *For any positive integer b,*

$$\sum_{a|b} \frac{b}{a} = \sigma(b).$$

A simple bound on $\sigma(b)$ can be obtained as follows.

Lemma 5. *For any positive integer $b \geq 2$,*

$$\sigma(b) < 3b \log b.$$

3   Random Function Collisions

In this section, we look at different approaches to find collisions on a random function f. We will bound their success probabilities, and use them in Sect. 4 to get bounds on the success probabilities of collision attacks on an iterated random function f^r.

3.1   Single-Trail Attack

SINGLE-TRAIL ATTACK. Let $[q]$ denote the set $\{1, \ldots, q\}$. The single-trail attack works by starting with an arbitrary initial point x and producing a *trail* of points, hoping to find a collision. A trail is uniquely defined by q queries $f^{i-1}(x)$ for $i \in [q]$, where the i-th query $f^{i-1}(x)$ has response $f^i(x)$. We assume that the attack does not stop when a collision is found, but makes q queries and then checks for collisions. If a collision is found, it will appear as a rho-shaped trail, as illustrated in Fig. 1. Therefore, a collision obtained through a single-trail attack will be called a ρ-collision.

TERMINOLOGY. Suppose the q-query single-trail attack finds a collision. For some t, c, suppose it takes $t + c$ queries to find this collision, so that

$$f^{t+c}(x) = f^t(x),$$

i.e., the output of the $(t+c)$-th query is identical to the output of the t-th query. Then, t is called the tail length of the ρ-collision, and c is called the cycle length.

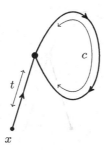

Fig. 1. Single-trail attack starting from x, resulting in a ρ collision with tail length t and cycle length c. We call the probability of this collision $\mathsf{cp}_\rho(t, c)$.

For fixed t, c, we want to bound the probability that a q-query single-trail attack gives a ρ-collision on f with tail length t and cycle length c. Call this probability $\mathsf{cp}_\rho[q](t, c)$.

BOUNDING $\mathsf{cp}_\rho[q](t, c)$. To get a ρ-collision on f with tail length t and cycle length c, we need to call f at $t + c$ distinct values. Thus, if $q < t + c$, $\mathsf{cp}_\rho[q](t, c) = 0$. So suppose $q \geq t + c$. Out of these $t + c$ calls to f, the first $t + c - 1$ give distinct outputs, and the last coincides with the t-th output. Thus, the number of different ways this can happen is $N^{\underline{t+c-1}}$, out of the total N^{t+c} possible outcomes for the $t + c$ calls to f. Thus,

$$\mathsf{cp}_\rho[q](t, c) = \frac{N^{\underline{t+c-1}}}{N^{t+c}} = \frac{\beta(t + c - 1)}{N}.$$

This is just a function of t and c (since the queries made after the collision is found are of no consequence), so we will use the simpler notation $\mathsf{cp}_\rho(t, c)$, with the implicit assumption that $q \geq t + c$. For a fixed real $\alpha > 0$, when $t + c \geq \sqrt{2\alpha N} + 2$, Lemma 1 gives us the bound

$$\mathsf{cp}_\rho(t, c) \leq \frac{e^{-\alpha}}{N}. \tag{3}$$

When $t + c < \sqrt{2\alpha N} + 2$, we will simply use the bound

$$\mathsf{cp}_\rho(t, c) \leq \frac{1}{N}. \tag{4}$$

3.2   Two-Trail Attack

TWO-TRAIL ATTACK. In the two-trail attack, we start with two different points x_1 and x_2, and produce two trails: the trail $f^{i-1}(x_1)$ for $i \in [q_1]$, and the trail $f^{i-1}(x_2)$ for $i \in [q_2]$, hoping to find a collision. In total $q_1 + q_2$ queries are made, where the i-th query for $i \in [q_1]$ is $f^{i-1}(x_1)$, with response $f^i(x_1)$, and the $(q_1 + i)$-th query for $i \in [q_2]$ is $f^{i-1}(x_2)$, with response $f^i(x_2)$. If a collision is

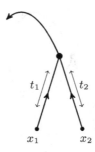

Fig. 2. Two-trail attack starting from x_1 and x_2, resulting in a λ-collision with foot lengths t_1 and t_2, respectively. We call the probability of this collision $\mathsf{cp}_\lambda(t_1, t_2)$.

found, the two trails will form a lambda shape, as illustrated in Fig. 2. Therefore, a collision obtained through a two-trail attack will be called a λ-collision.

TERMINOLOGY. Suppose the (q_1, q_2)-query two-trail attack finds a λ-collision, regardless of whether a ρ-collisions has occurred on either trail. Suppose that a λ-collision is found after making t_1 queries along the first trail and t_2 queries along the second, i.e.,

$$f^{t_1}(x_1) = f^{t_2}(x_2).$$

t_1 and t_2 are called the foot lengths of the λ-collision. For fixed t_1, t_2, we want to bound the probability that a (q_1, q_2)-query two-trail attack finds a λ-collision with foot lengths t_1 and t_2. Denote this probability as $\mathsf{cp}_\lambda[q_1, q_2](t_1, t_2)$.

BOUNDING $\mathsf{cp}_\lambda[q_1, q_2](t_1, t_2)$. To get a λ-collision on f with foot lengths t_1 and t_2, we need to call f at t_1 distinct values on the first trail and t_2 distinct values on the second trail. Thus, if $q_1 < t_1$ or $q_2 < t_2$, $\mathsf{cp}_\lambda[q_1, q_2](t_1, t_2) = 0$. So we assume $q_1 \geq t_1$ and $q_2 \geq t_2$. Out of these $t_1 + t_2$ queries, the first $t_1 - 1$ in one trail and the first $t_2 - 1$ in the other trail give distinct outputs, and the last calls on the two trails coincide on a value distinct from all the earlier ones, i.e., the $t_1 + t_2$ calls lead to $t_1 + t_2 - 1$ distinct outputs, and one collision. Thus, the number of different ways this can happen is $N^{t_1+t_2-1}$, out of the total $N^{t_1+t_2}$ possible outcomes for the $t_1 + t_2$ calls to f. Thus,

$$\mathsf{cp}_\lambda[q_1, q_2](t_1, t_2) = \frac{N^{t_1+t_2-1}}{N^{t_1+t_2}} = \frac{\beta(t_1 + t_2 - 1)}{N}.$$

Again, this is only a function of t_1 and t_2 (since the queries made after the collision is found are of no consequence), so we will use the simpler notation $\mathsf{cp}_\lambda(t_1, t_2)$, with the implicit assumption that $q_1 \geq t_1$ and $q_2 \geq t_2$. For our purposes it will be enough to use the bound

$$\mathsf{cp}_\lambda(t_1, t_2) \leq \frac{1}{N}. \tag{5}$$

3.3 A $\lambda\rho$-Double-Collision on a Two-Trail Attack

When a two-trail attack leads to two collisions, a double-collision is said to occur. In Sect. 4, in addition to the above bounds, we also need a bound on the probability of two closely related double-collisions. We deal with a $\lambda\rho$-double-collision in this section, and a ρ'-double-collision in the next. A $\lambda\rho$-double-collision takes place when a two-trail attack leads to a λ-collision, and then the combined trail becomes the tail of a ρ-collision, as shown in Fig. 3.[1]

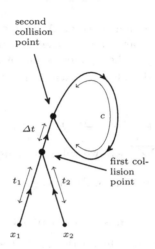

Fig. 3. Two-trail attack starting from x_1 and x_2, resulting in a $\lambda\rho$-collision. First, there is a λ-collision with foot lengths t_1 and t_2, respectively. Then, the combined trail continues for Δt queries, and completes a cycle of length c, after which a ρ-collision occurs. We call the probability of this double-collision $\mathsf{cp}_{\lambda\rho}(t_1, t_2, \Delta t, c)$.

TERMINOLOGY. We assign four parameters to this collision: the foot lengths t_1 and t_2 of the λ, the intervening length Δt between the two collisions, and the cycle length c of the ρ. Note that Δt can be seen as the tail length of the ρ-collision if we imagine it to have resulted from a single-trail attack beginning at the point of the λ-collision. For fixed $t_1, t_2, \Delta t, c$ we want to find the probability that a (q_1, q_2)-query two-trail attack finds a $\lambda\rho$-double-collision with foot lengths t_1 and t_2, intervening length Δt and cycle length c. Call this probability $\mathsf{cp}_{\lambda\rho}[q_1, q_2](t_1, t_2, \Delta t, c)$.

BOUNDING $\mathsf{cp}_{\lambda\rho}[q_1, q_2](t_1, t_2, \Delta t, c)$. To get a λ-collision on f with foot lengths t_1 and t_2, we need to call f at t_1 distinct values on the first trail, and t_2 distinct values on the second trail; and to get a ρ-collision on f with tail length Δt and

---

[1] Note that we only call it a double-collision if both trails continue up to the point of second collision.

cycle length c, we need to call f at Δt common values on each trail, and a further c points on the first trail; this adds up to $t_1 + t_2 + \Delta t + c$ distinct values in all. Thus, when $q_1 < t_1 + \Delta t + c$ or $q_2 < t_2 + \Delta t$, $\mathsf{cp}_{\lambda\rho}[q_1, q_2](t_1, t_2, \Delta t, c) = 0$. So we assume $q_1 \geq t_1 + \Delta t + c$ and $q_2 \geq t_2 + \Delta t$. These $t_1 + t_2 + \Delta t + c$ calls lead to $t_1 + t_2 + \Delta t + c - 2$ distinct outputs, and two collisions. Thus, the number of different ways this can happen is $N^{t_1+t_2+\Delta t+c-2}$, out of the total $N^{t_1+t_2+\Delta t+c}$ possible outcomes for the $t_1 + t_2 + \Delta t + c$ calls to f. Thus,

$$\mathsf{cp}_{\lambda\rho}[q_1, q_2](t_1, t_2, \Delta t, c) = \frac{N^{t_1+t_2+\Delta t+c-2}}{N^{t_1+t_2+\Delta t+c}} = \frac{\beta(t_1 + t_2 + \Delta t + c - 2)}{N^2}.$$

As before, this is only a function of $t_1, t_2, \Delta t$ and c (since the queries made after the ρ collision is found are of no consequence), so we use the simpler notation $\mathsf{cp}_{\lambda\rho}(t_1, t_2, \Delta t, c)$, with the implicit assumption that $q_1 \geq t_1 + \Delta t + c$ and $q_2 \geq t_2 + \Delta t$. For a fixed real $\alpha > 0$, when $t_1 + t_2 + \Delta t + c \geq \sqrt{2\alpha N} + 3$, Lemma 1 gives us the bound

$$\mathsf{cp}_{\lambda\rho}(t_1, t_2, \Delta t, c) \leq \frac{e^{-\alpha}}{N^2}. \tag{6}$$

When $t_1 + t_2 + \Delta t + c < \sqrt{2\alpha N} + 3$, we will simply use the bound

$$\mathsf{cp}_{\lambda\rho}(t_1, t_2, \Delta t, c) \leq \frac{1}{N^2}. \tag{7}$$

3.4   A ρ'-Double-Collision on a Two-Trail Attack

A ρ'-double-collision takes place when a two-trail attack leads to a ρ with two tails. This is shown in Fig. 4. We will allow $\Delta t = 0$, in which case a three-way collision occurs.

TERMINOLOGY. As before, we assign four parameters to this collision: the tail lengths t_1 and t_2 of the ρ, the intervening length Δt between the two collisions, and the cycle length c of the ρ. For fixed $t_1, t_2, \Delta t, c$ we want to find the probability that a two-trail attack with sufficiently many queries finds a ρ'-double-collision with tail lengths t_1 and t_2, intervening length Δt, and cycle length c. Call this probability $\mathsf{cp}_{\rho'}[q_1, q_2](t_1, t_2, \Delta t, c)$.

BOUNDING $\mathsf{cp}_{\rho'}[q_1, q_2](t_1, t_2, \Delta t, c)$. The bounding of $\mathsf{cp}_{\rho'}[q_1, q_2](t_1, t_2, \Delta t, c)$ is almost identical to that of $\mathsf{cp}_{\lambda\rho}[q_1, q_2](t_1, t_2, \Delta t, c)$. To get a ρ'-double-collision with tail lengths t_1 and t_2, intervening length Δt, and cycle length c, we need to call f at $t_1 + c - \Delta t$ distinct values on the first trail, t_2 distinct values on the second trail, and Δt common values on each trail, resulting in calls at $t_1 + t_2 + c$ distinct values in all. Thus, when $q_1 < t_1 + c$ or $q_2 < t_2 + \Delta t$, $\mathsf{cp}_{\rho'}[q_1, q_2](t_1, t_2, \Delta t, c) = 0$. So we assume $q_1 \geq t_1 + c$ and $q_2 \geq t_2 + \Delta t$. These $t_1 + t_2 + c$ calls lead to $t_1 + t_2 + c - 2$ distinct outputs. Thus, the number of different ways this can happen is $N^{t_1+t_2+c-2}$, out of the total $N^{t_1+t_2+c}$ possible outcomes for the $t_1 + t_2 + c$ calls to f. Thus,

$$\mathsf{cp}_{\rho'}[q_1, q_2](t_1, t_2, \Delta t, c) = \frac{N^{t_1+t_2+c-2}}{N^{t_1+t_2+c}} = \frac{\beta(t_1 + t_2 + c - 2)}{N^2}.$$

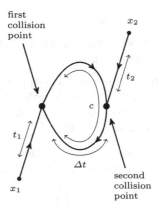

Fig. 4. Two-trail attack starting from x_1 and x_2, resulting in a ρ'-collision with tail lengths t_1 and t_2, intervening length Δt, and cycle length c. We will allow $\Delta t = 0$, in which case a three-way collision occurs. We call the probability of this double-collision $\mathsf{cp}_{\rho'}(t_1, t_2, \Delta t, c)$.

As before, this is only a function of $t_1, t_2, \Delta t$ and c (since the queries made after the ρ collision is found are of no consequence), so we use the simpler notation $\mathsf{cp}_{\lambda\rho}(t_1, t_2, \Delta t, c)$, with the implicit assumption that $q_1 \geq t_1 + \Delta t + c$ and $q_2 \geq t_1 + \Delta t$. Recalling that

$$\mathsf{cp}_{\lambda\rho}(t_1, t_2, 0, c) = \frac{\beta(t_1 + t_2 + c - 2)}{N^2},$$

we conclude that

$$\mathsf{cp}_{\rho'}(t_1, t_2, \Delta t, c) = \mathsf{cp}_{\lambda\rho}(t_1, t_2, 0, c). \tag{8}$$

4 Iterated Random Function Collisions

In this section we revisit the two types of collision attacks described in Sect. 3, and analyse their success probabilities when applied to f^r. The main proof in this paper relies heavily on the results obtained in this section.

A CAUTIONARY NOTE. At first glance, this section may appear to be similarly organised as Sect. 3. It is important to keep in mind that we are now interested in something entirely different. In Sect. 3, we looked at the probabilities of specific ρ- and λ-collisions with fixed parameters. In this section, instead, we focus on the probabilities that single-trail attacks and two-trail attacks of some specified number of queries succeed in finding collisions on f^r. By reducing these collisions to collisions on f, we can use the union bound on the bounds obtained in Sect. 3 to get the desired bounds. To distinguish from the collision probabilities on f, which we denoted $\mathsf{cp}[\cdot]$, we now use the notation $\mathsf{cp}^r[\cdot]$ for the collision probabilities on f^r.

4.1   Single-Trail Attack

We want to bound the probability that a q-query single-trail attack finds a collision on f^r. Call this probability $\mathsf{cp}_\rho^r[q]$.

REDUCING TO COLLISION ON f. Suppose the q-query single-trail attack finds a ρ-collision on f^r with tail length t' and cycle length c'. Observe that this collision necessarily arises out of a ρ-collision on f, with tail length t and cycle length c for some t, c. This can happen in two ways:

– DIRECT COLLISION. This happens when r divides c. Then, define k such that rk is the first multiple of r that is not less than t, i.e.,

$$k := \left\lceil \frac{t}{r} \right\rceil,$$

then $rk + c$ is also a multiple of r, and since $f^{t+c}(x) = f^t(x)$, and $rk \geq t$, we also have

$$f^{rk+c}(x) = f^{rk}(x).$$

Writing

$$k' = \frac{c}{r},$$

we have

$$(f^r)^{k+k'}(x) = (f^r)^k(x),$$

our ρ-collision on f^r. Note that according to this notation,

$$t' = k = \left\lceil \frac{t}{r} \right\rceil, c' = k' = \frac{c}{r}.$$

Loosely speaking, in a direct collision, the first collision on f arrives *in phase* with r, i.e.,

$$t = t + c \bmod r,$$

so that this first collision on f leads immediately to a collision on f^r at the next multiple of r.

– DELAYED COLLISION. A *delayed collision* occurs when r does not divide c, i.e., the first collision arrives *out of phase*. Then we need to keep cycling about the ρ of f till the phase is adjusted, and only then we arrive at the next multiple of r and find a collision on f^r. Suppose it cycles around η times. For the phase to be adjusted, $c\eta$ should be a multiple of r. The smallest value of η that satisfies this is

$$\eta = \frac{r}{d}$$

where $d = \gcd(c, r)$ is the greatest common divisor of c and r. Let $k = \left\lceil \frac{t}{r} \right\rceil$ as before, and let

$$k' = \frac{c}{d}.$$

As before, since we have $f^{t+cn}(x) = f^t(x)$, and $rk \geq t$, we have

$$f^{rk+cn}(x) = f^{rk}(x),$$

which gives us the ρ-collision

$$(f^r)^{k+k'}(x) = (f^r)^k(x),$$

as before. Again, according to this notation,

$$t' = k = \left\lceil \frac{t}{r} \right\rceil, c' = k' = \frac{c}{d}.$$

REQUIRED CONDITIONS. Observing that a direct collision can be seen as a special case of delayed collision, where $d = \gcd(c,r) = r$, we can summarise the above as follows: a ρ-collision on f with tail length t and cycle length c *eventually* leads to a ρ-collision on f^r with tail length t' and cycle length c' where

$$t' = k = \left\lceil \frac{t}{r} \right\rceil, c' = k' = \frac{c}{d},$$

with $d = \gcd(c,r)$ as before. Thus, for a ρ-collision on f to result in a ρ-collision on f^r, the only required condition is that q is sufficiently large, i.e.,

$$t' + c' \leq q.$$

In terms of t and c, this becomes

$$\left\lceil \frac{t}{r} \right\rceil + \frac{c}{d} \leq q.$$

Recall that we are trying to bound the probability $\mathsf{cp}_\rho^r[q]$ of finding a ρ-collision on f^r in q queries. This is equivalent to the probability of finding a ρ-collision on f with the parameters t and c satisfying the above condition. Recall that in Sect. 3, we bounded this probability for a fixed (t,c), which we called $\mathsf{cp}_\rho(t,c)$. We can now use the union bound to get a bound on $\mathsf{cp}_\rho^r[q]$.

USING THE UNION BOUND ON $\mathsf{cp}_\rho^r[q]$. Let $\mathcal{S}$ be the set of (t,c) values that satisfy the requirement

$$\left\lceil \frac{t}{r} \right\rceil + \frac{c}{\gcd(c,r)} \leq q.$$

For a fixed $\alpha > 0$, we can split $\mathcal{S}$ into two parts:

$$\mathcal{S}^+[\alpha] := \left\{ (t,c) \in \mathcal{S} \mid t + c \geq \sqrt{2\alpha N} + 2 \right\},$$

$$\mathcal{S}^-[\alpha] := \left\{ (t,c) \in \mathcal{S} \mid t + c < \sqrt{2\alpha N} + 2 \right\}.$$

Applying the union bound with bounds (3) and (4) obtained for $\mathsf{cp}_\rho(t,c)$ gives

$$
\begin{aligned}
\mathsf{cp}_\rho^r[q] &\leq \sum_{\mathcal{S}} \mathsf{cp}_\rho(t,c) \\
&= \sum_{\mathcal{S}^+[\alpha]} \mathsf{cp}_\rho(t,c) + \sum_{\mathcal{S}^-[\alpha]} \mathsf{cp}_\rho(t,c) \\
&\leq \sum_{\mathcal{S}^+[\alpha]} \frac{e^{-\alpha}}{N} + \sum_{\mathcal{S}^-[\alpha]} \frac{1}{N} \\
&= \#\mathcal{S}^+[\alpha] \cdot \frac{e^{-\alpha}}{N} + \#\mathcal{S}^-[\alpha] \cdot \frac{1}{N}
\end{aligned}
\tag{9}
$$

BOUNDING $\#\mathcal{S}^-[\alpha]$. We observe that whenever $(t,c) \in \mathcal{S}^-[\alpha]$,

$$
t < \sqrt{2\alpha N} + 2,
$$

and

$$
c < q \cdot \gcd(c,r).
$$

If we count the number of (t,c) satisfying these conditions, it will give us an upper bound on $\#\mathcal{S}^-[\alpha]$. There are at most $\sqrt{2\alpha N} + 2$ values of t satisfying $t < \sqrt{2\alpha N} + 2$. For a fixed $d = \gcd(c,r)$, c has to be a multiple of d not exceeding qd. The number of such values of c is q. Since d must be a factor of r, we get the total number of values of c satisfying $c < q \cdot \gcd(c,r)$ to be at most $q \cdot \mathsf{d}(r)$. Putting it all together we get

$$
\#\mathcal{S}^-[\alpha] \leq (\sqrt{2\alpha N} + 2) \cdot q \cdot \mathsf{d}(r).
\tag{10}
$$

BOUNDING $\#\mathcal{S}^+[\alpha]$. For $(t,c) \in \mathcal{S}^+[\alpha]$, it will be enough for our purposes to consider the bounds

$$
t \leq qr,
$$

and

$$
c < q \cdot \gcd(c,r).
$$

Using the same reasoning as before, the number of values of c that satisfy $c < q \cdot \gcd(c,r)$ is at most $q \cdot \mathsf{d}(r)$. For t there are now at most qr values. Thus, we obtain the bound

$$
\#\mathcal{S}^+[\alpha] \leq q^2 r \cdot \mathsf{d}(r).
\tag{11}
$$

FINAL BOUND FOR $\mathsf{cp}_\rho^r[q]$. We can now plug (10) and (11) into (9):

$$
\begin{aligned}
\mathsf{cp}_\rho^r[q] &\leq \#\mathcal{S}^+[\alpha] \cdot \frac{e^{-\alpha}}{N} + \#\mathcal{S}^-[\alpha] \cdot \frac{1}{N} \\
&\leq q^2 r \cdot \mathsf{d}(r) \cdot \frac{e^{-\alpha}}{N} + (\sqrt{2\alpha N} + 2) \cdot q \cdot \mathsf{d}(r) \cdot \frac{1}{N}
\end{aligned}
$$

for any real $\alpha > 0$. We will simplify it by plugging in a suitable value of α.

SIMPLIFYING THE BOUND. We know from Lemma 3 that

$$\mathsf{d}(r) < 2\sqrt{r}.$$

We put $\alpha = \log r$. Then we have

$$\sqrt{2\alpha N} = \sqrt{2N \log r},$$

and

$$e^{-\alpha} = \frac{1}{r}.$$

When $N \log r \geq 16$, we have

$$
\begin{aligned}
\sqrt{2\alpha N} + 2 &= \sqrt{2N \log r} + 2 \\
&= 2\sqrt{N \log r} - \left[(2 - \sqrt{2}) \cdot \sqrt{N \log r} - 2 \right] \\
&\leq 2\sqrt{N \log r} - \left[(2 - \sqrt{2}) \cdot 4 - 2 \right] \\
&= 2\sqrt{N \log r} - \left[6 - \sqrt{2} \right] \\
&< 2\sqrt{N \log r}.
\end{aligned}
$$

Thus,

$$\mathsf{cp}_\rho^r[q] \leq 2 \cdot \left(\frac{q^2 \sqrt{r}}{N} \right) + 2 \cdot \sqrt{\frac{q^2 r \log r}{N}}.$$

This gives us a bound for the success probability of a q-query single-trail attack on f^r. We state the result as a lemma.

Lemma 6. *Under the assumption that $N \log r \geq 16$, we have*

$$\mathsf{cp}_\rho^r[q] \leq 2 \cdot \left(\frac{q^2 \sqrt{r}}{N} \right) + 2 \cdot \sqrt{\frac{q^2 r \log r}{N}}.$$

4.2   Two-Trail Attack

We want to bound the probability that a (q_1, q_2)-query two-trail attack finds a λ-collision on f^r. Call this probability $\mathsf{cp}_\lambda^r[q_1, q_2]$.

REDUCING TO COLLISION ON f. Suppose the (q_1, q_2)-query two-trail attack finds a λ-collision on f^r with foot lengths t_1' and t_2'. As in the case of the ρ-collision on f^r, this can only arise from a λ-collision on f, say with foot lengths t_1 and t_2, which can again happen in two ways:

- DIRECT COLLISION. A direct collision takes place when the two f-trails collide in phase, i.e.,

$$t_1 = t_2 \bmod r.$$

When this happens, the two trails continue till the next multiple of r, where they give a λ-collision on f^r. This collision takes place at

$$t_1' = \left\lceil \frac{t_1}{r} \right\rceil, t_2' = \left\lceil \frac{t_2}{r} \right\rceil.$$

– DELAYED COLLISION. A delayed collision takes place when the two f-trails collide out of phase, i.e.,

$$t_1 \neq t_2 \bmod r.$$

If one of the trails results in a ρ-collision on f^r, this implies that a successful single-trail attack has been carried out on f^r. Here, we will only focus on the scenario where a λ-collision on f^r can still happen. But then one of the two f-trails must have entered into a cycle, otherwise both f-trails will remain out of phase. This can only happen in one of two ways:

- After the λ-collision on f, the combined trail forms the tail of a ρ collision on f, that is, they form a $\lambda\rho$-collision on f as in Fig. 3. One of the trails, say the one from x_1, cycles around the ρ enough number of times to adjust the phase, and then the two f-trails continue to the next multiple of r, giving a λ-collision on f^r;[2]
- After the λ-collision on f, one of the two f-trails, say the one from x_1, continues and collides with the trail from x_2, that is, they form a ρ'-collision on f as in Fig. 4. When $\Delta t = 0$, a three-way collision on f occurs. The trail from x_1 cycles around the ρ enough number of times to adjust the phase, giving a λ-collision on f^r.

In our calculations, we assume that it is the trail from x_1 that cycles multiple times, while the one from x_2 waits for the collision on f^r to happen. We obtain a bound which is symmetric over q_1 and q_2, and thus also holds for the case when the two trails reverse roles. Let τ_1 and τ_2 be the respective lengths of the two trails till *the point of waiting*, i.e., the point of ρ-collision of the trail from x_1. Calling Δt the distance between the two collision points, we simply have

$$\tau_1 = t_1 + \Delta t, \tau_2 = t_2 + \Delta t$$

for the $\lambda\rho$-collision, and

$$\tau_1 = t_1, \tau_2 = t_2 + \Delta t$$

for the ρ'-collision. Let the cycle length of this ρ be c (note that its tail length is τ_1 with respect to this trail). Suppose this trail cycles η times about the ρ in order to adjust the phase difference. Then η is the smallest number that satisfies

$$\tau_1 + c\eta = \tau_2 \bmod r.$$

Suppose k is such that

$$\tau_1 + c\eta = \tau_2 + rk.$$

---

[2] This is indeed a (delayed) λ-collision on f^r: from the point of view of f^r, neither of the two trails could be seen to enter into a cycle.

Also, let

$$k_2 = \left\lceil \frac{\tau_2}{r} \right\rceil.$$

From our definition of τ_1 and τ_2, we have that

$$f^{\tau_1}(x_1) = f^{\tau_2}(x_2),$$

and from the ρ-collision $f^{\tau_1+c}(x_1) = f^{\tau_1}(x_1)$, it follows that

$$f^{\tau_1+c\eta}(x_1) = f^{\tau_1}(x_1).$$

From these two we get

$$f^{\tau_1+c\eta}(x_1) = f^{\tau_2}(x_2).$$

From the definition of k we have

$$f^{\tau_2+rk}(x_1) = f^{\tau_2}(x_2).$$

Continuing on to rk_2, we get a λ-collision on f^r as

$$(f^r)^{k+k_2}(x_1) = (f^r)^{k_2}(x_2).$$

According to this notation we have a λ-collision on f^r with foot lengths t_1' and t_2', such that

$$t_1' = k + k_2 = \left\lceil \frac{\tau_1 + c\eta}{r} \right\rceil, t_2' = k_2 = \left\lceil \frac{\tau_2}{r} \right\rceil.$$

When this comes from a $\lambda\rho$-collision, we have

$$t_1' = \left\lceil \frac{t_1 + \Delta t + c\eta}{r} \right\rceil, t_2' = \left\lceil \frac{t_2 + \Delta t}{r} \right\rceil.$$

When this comes from a ρ'-collision, we have

$$t_1' = \left\lceil \frac{t_1 + c\eta}{r} \right\rceil, t_2' = \left\lceil \frac{t_2 + \Delta t}{r} \right\rceil.$$

We will treat these two cases separately, even though they are closely related.

REQUIRED CONDITIONS. Again, we observe that the direct collision is a special case of the delayed collision with $\Delta t = 0$ and $\eta = 0$. *However, there is an important difference.* For the delayed λ-collision, we require *two* collisions on f, unlike all other collisions we have seen so far, which need only one. This case corresponds to the $\lambda\rho$-double-collision and the ρ'-double-collision from Sect. 3, and requires some special treatment, as we will see in the course of our calculations. The condition needed here is that both trails continue long enough for the collision to happen, i.e.,

$$t_1' \le q_1, t_2' \le q_2.$$

In terms of $t_1, t_2, \Delta t, c, \eta$, this translates to

$$\left\lceil \frac{t_1 + \Delta t + c\eta}{r} \right\rceil \le q_1, \left\lceil \frac{t_2 + \Delta t}{r} \right\rceil \le q_2$$

for the $\lambda\rho$-double-collision and

$$\left\lceil \frac{t_1 + c\eta}{r} \right\rceil \le q_1, \left\lceil \frac{t_2 + \Delta t}{r} \right\rceil \le q_2$$

for the ρ'-double-collision. Recall that we are trying to calculate $\mathsf{cp}_\lambda^r[q_1, q_2]$, the probability of getting a λ-collision on f^r with a (q_1, q_2)-query two-trail attack starting from x_1 and x_2. Based on our observations above, this can happen in two ways:

- A DIRECT λ-COLLISION ON f. This is the direct collision scenario, where the collision is in phase. The foot lengths t_1 and t_2 have the constraints

$$\left\lceil \frac{t_1}{r} \right\rceil \le q_1, \left\lceil \frac{t_2}{r} \right\rceil \le q_2, t_1 = t_2 \bmod r.$$

  For fixed t_1, t_2, we recall that the probability of this collision is $\mathsf{cp}_\lambda(t_1, t_2)$.
- A $\lambda\rho$-DOUBLE-COLLISION ON f. This is the first case of the delayed collision scenario, where the collision is out of phase. Here, t_1 and t_2 are the foot lengths of the λ, Δt is the distance between the two collision points, c is the cycle length of the ρ, and η is the number of cycles necessary around the ρ. Recall that one of the trails circles around the ρ, while the other waits for the λ-collision on f^r to happen. We continue with our assumption that the one from x_1 does the cycling and the one from x_2 waits, since we will eventually count over all pairs of trails. Now $t_1, t_2, \Delta t, c, \eta$ have the constraints

$$\left\lceil \frac{t_1 + \Delta t + c\eta}{r} \right\rceil \le q_1, \left\lceil \frac{t_2 + \Delta t}{r} \right\rceil \le q_2, t_1 + c\eta = t_2 \bmod r.$$

  For fixed $t_1, t_2, \Delta t, c, \eta$, we recall that the probability of this $\lambda\rho$-double-collision is $\mathsf{cp}_{\lambda\rho}(t_1, t_2, \Delta t, c)$.
- A ρ'-DOUBLE-COLLISION ON f. This is the second case of the delayed collision scenario. Here, t_1 and t_2 are the lengths of the two tails of the ρ, Δt is the distance between the two collision points, c is the cycle length of the ρ, and η is the number of cycles necessary around the ρ. Again, the trail from x_1 circles around the ρ, while the trail from x_2 waits for the λ-collision on f^r to happen. Thus, $t_1, t_2, \Delta t, c, \eta$ have the constraints

$$\left\lceil \frac{t_1 + c\eta}{r} \right\rceil \le q_1, \left\lceil \frac{t_2 + \Delta t}{r} \right\rceil \le q_2, t_1 + c\eta = t_2 + \Delta t \bmod r.$$

Our strategy for bounding $\mathsf{cp}_\lambda^r[q_1, q_2]$ will be similar to the one we used for bounding $\mathsf{cp}_\rho^r[q]$: to take the bounds on $\mathsf{cp}_\lambda(t_1, t_2)$ for fixed t_1, t_2,

$\mathsf{cp}_{\lambda\rho}(t_1, t_2, \Delta t, c)$ for fixed $t_1, t_2, \Delta t, c$ and $\mathsf{cp}_{\rho'}(t_1, t_2, \Delta t, c)$ for fixed $t_1, t_2, \Delta t, c$ obtained in Sect. 3, and then use the union bound over all possible values these parameters can take.

APPLYING THE UNION BOUND TO $\mathsf{cp}_{\lambda}^r[q_1, q_2]$. Let $\mathcal{S}_1$ be the set of (t_1, t_2) values that satisfy the constraints

$$\left\lceil \frac{t_1}{r} \right\rceil \leq q_1, \quad \left\lceil \frac{t_2}{r} \right\rceil \leq q_2, t_1 = t_2 \bmod r,$$

and let

$$\mathsf{p}_1 := \sum_{\mathcal{S}_1} \mathsf{cp}_{\lambda}(t_1, t_2).$$

Let $\mathcal{S}_2$ be the set of $(t_1, t_2, \Delta t, c, \eta)$ values that satisfy the constraints

$$\left\lceil \frac{t_1 + \Delta t + c\eta}{r} \right\rceil \leq q_1, \quad \left\lceil \frac{t_2 + \Delta t}{r} \right\rceil \leq q_2, t_1 + c\eta = t_2 \bmod r,$$

and let

$$\mathsf{p}_2 := \sum_{\mathcal{S}_2} \mathsf{cp}_{\lambda\rho}(t_1, t_2, \Delta t, c).$$

Let $\mathcal{S}_3$ be the set of $(t_1, t_2, \Delta t, c, \eta)$ values that satisfy the constraints

$$\left\lceil \frac{t_1 + c\eta}{r} \right\rceil \leq q_1, \quad \left\lceil \frac{t_2 + \Delta t}{r} \right\rceil \leq q_2, t_1 + c\eta = t_2 + \Delta t \bmod r,$$

and let

$$\mathsf{p}_3 := \sum_{\mathcal{S}_3} \mathsf{cp}_{\rho'}(t_1, t_2, \Delta t, c).$$

In addition, for the case where the trails reverse roles, we define $\mathcal{S}_4$ as the set of $(t_1, t_2, \Delta t, c, \eta)$ values that satisfy the constraints

$$\left\lceil \frac{t_1 + \Delta t}{r} \right\rceil \leq q_1, \quad \left\lceil \frac{t_2 + \Delta t + c\eta}{r} \right\rceil \leq q_2, t_1 = t_2 + c\eta \bmod r,$$

and

$$\mathsf{p}_4 := \sum_{\mathcal{S}_4} \mathsf{cp}_{\lambda\rho}(t_1, t_2, \Delta t, c).$$

Similarly, we define $\mathcal{S}_5$ as the set of $(t_1, t_2, \Delta t, c, \eta)$ values that satisfy the constraints

$$\left\lceil \frac{t_1 + \Delta t}{r} \right\rceil \leq q_1, \quad \left\lceil \frac{t_2 + c\eta}{r} \right\rceil \leq q_2, t_1 + \Delta t = t_2 + c\eta \bmod r,$$

and

$$\mathsf{p}_5 := \sum_{\mathcal{S}_5} \mathsf{cp}_{\rho'}(t_1, t_2, \Delta t, c).$$

We state here the following bounds on $\mathsf{p}_1, \mathsf{p}_2, \mathsf{p}_3$, the proof of which we defer to Sect. 6:

Lemma 7. *Under the assumption that* $N \log r > 90$,

$$p_1 \leq \frac{q_1 q_2 r}{N},$$

$$p_2 \leq 8 \cdot (\log r)^2 \cdot \left(\frac{q_1 q_2 r}{N}\right)^2 + 24 \cdot (\log r)^3 \cdot \left(\frac{q_1 q_2 r}{N}\right),$$

$$p_3 \leq 8 \cdot (\log r)^2 \cdot \left(\frac{q_1 q_2 r}{N}\right)^2 + 24 \cdot (\log r)^3 \cdot \left(\frac{q_1 q_2 r}{N}\right).$$

FINAL BOUND FOR $\mathsf{cp}_\lambda^r[q_1, q_2]$. We observe that the bounds for p_2 and p_3 in Lemma 7 are symmetric over q_1 and q_2. Thus, we have

$$p_4 \leq 8 \cdot (\log r)^2 \cdot \left(\frac{q_1 q_2 r}{N}\right)^2 + 24 \cdot (\log r)^3 \cdot \left(\frac{q_1 q_2 r}{N}\right),$$

$$p_5 \leq 8 \cdot (\log r)^2 \cdot \left(\frac{q_1 q_2 r}{N}\right)^2 + 24 \cdot (\log r)^3 \cdot \left(\frac{q_1 q_2 r}{N}\right).$$

Using the union bound, we get

$$\mathsf{cp}_\lambda^r[q_1, q_2] \leq p_1 + p_2 + p_3 + p_4 + p_5.$$

This gives us the required bound, which we state next in the form of a lemma.

Lemma 8. *When* $N \log r > 90$,

$$\mathsf{cp}_\lambda^r[q_1, q_2] \leq 32 \cdot \left(\frac{q_1 q_2 r \log r}{N}\right)^2 + 97 \cdot (\log r)^2 \cdot \left(\frac{q_1 q_2 r \log r}{N}\right).$$

Proof. As $r \geq 2$, we can relax the bound of p_1 as

$$p_1 \leq \frac{q_1 q_2 r}{N} \leq \frac{q_1 q_2 r}{N} \cdot (\log r)^3.$$

The rest follows from Lemma 7.   □

4.3   A More General Collision Attack

Previously, we looked at two main approaches for a collision attack: the single-trail attack and the two-trail attack, and we bounded their success probabilities. Now, we will bound the success probability of a more general collision attack. More specifically, we consider collision attack subject to the restriction that is given in the statement of Theorem 1 in Sect. 1: every query is either chosen from a set of size m (with $m \leq q$) of predetermined starting points, or is the response of a previous query. First, let us introduce the notion of a transcript.

TRANSCRIPT. Let us consider any adversary $\mathcal{A}$ that interacts with an oracle $\mathcal{O}$. This interaction can be represented as a transcript, that is, as a list of queries made and answers returned. Let the transcript tr be defined as the q-tuple of input-output pairs $\mathsf{tr} = ((x_1, y_1), (x_2, y_2), \ldots, (x_q, y_q))$. Without loss of generality, we do not consider adversaries here that repeat the same query, i.e., all q queries are distinct.

SOURCES AND TRAILS. For $j, j' \in [q], j \neq j'$, we say that $x_{j'}$ is a *predecessor* of x_j if

$$f(x_{j'}) = x_j.$$

We call x_j a *source* if it does not have a predecessor. If there exists a non-empty subset of the queries for which every query has a predecessor that is in the same subset, and no query has a predecessor outside the set, we call this subset a *permutation cycle*. Note that a permutation cycle forms a rho-shape with a tail of length zero. For a permutation cycle, we define the query x_j of the permutation cycle with the smallest index j to be a *source*.

Suppose that there are m sources along the q queries, which we call $z_1, \ldots, z_m$. Then we can see the attack as an m-trail attack, with the m trails starting from $z_1, \ldots, z_m$ and of lengths $q_1, \ldots, q_m$ respectively. Thus, each point that is not a source must be on one of these m trails.

If the collision attack is successful, then for some $i, i' \in [q]$ with $i \neq i'$, we have

$$f(x_i) = f(x_{i'}).$$

In that case, one of the following must hold:

- x_i and $x_{i'}$ are on the same trail, say the one from z_p – in this case, a successful q_p-query single-trail attack starting from z_p has occurred;
- x_i and $x_{i'}$ are on different trails, say the ones from z_p and $z_{p'}$ respectively – in this case, a successful $(q_p, q_{p'})$-query two-trail attack starting from $(z_p, z_{p'})$ has occurred.

A WORD ON THE CHOICE OF $q_1, \ldots, q_m$. We note here that since we are allowing the trails to collide and merge with each other, the trail lengths $q_1, \ldots, q_m$ are not necessarily unique, since the queries on the merged trail can be counted on either trail, or both. We can get around this by choosing to count each merged trail as part of any one of the pre-merging trails, while the other is thought to stop at the point of collision. This way, we ensure that $\sum_{j=1}^m q_j = q$.

To bound the success probability of this more general collision attack, we can use the previously obtained bounds on the success probabilities of single-trail attacks and two-trail attacks along with the union bound. With notation as above we recall the following bounds:

- SINGLE-TRAIL ATTACK. For a q-query single-trail attack, Lemma 6 gives us the bound

$$\mathsf{cp}_\rho^r[q] \leq 2 \cdot \left(\frac{q^2 \sqrt{r}}{N} \right) + 2 \cdot \sqrt{\frac{q^2 r \log r}{N}}.$$

- TWO-TRAIL ATTACK. For a (q_1, q_2)-query two-trail attack, Lemma 8 gives us the bound

$$\mathsf{cp}_\lambda^r[q_1, q_2] \leq 32 \cdot \left(\frac{q_1 q_2 r \log r}{N} \right)^2 + 97 \cdot (\log r)^2 \cdot \left(\frac{q_1 q_2 r \log r}{N} \right).$$

Let $\mathsf{cp}^r[q](\mathcal{A})$ denote the probability that the collision adversary $\mathcal{A}$ making q queries finds a collision on f^r. For $q_1, \ldots, q_m$, with

$$\sum_{i=1}^m q_i = q,$$

and let $\mathsf{cp}^r[q](q_1, \ldots, q_m)$ denote the probability that a collision attack with m trails of lengths $q_1, \ldots, q_m$ finds a collision on f^r. Thus,

$$\mathsf{cp}^r[q](\mathcal{A}) \le \max_{\sum q_i = q} \mathsf{cp}^r[q](q_1, \ldots, q_m).$$

By the union bound, we have

$$\mathsf{cp}^r[q](q_1, \ldots, q_m) \le \sum_{i=1}^m \mathsf{cp}^r_\rho[q_i] + \sum_{i=1}^{m-1} \sum_{j=i+1}^m \mathsf{cp}^r_\lambda[q_i, q_j].$$

We bound the two terms separately.

$$\sum_{i=1}^m \mathsf{cp}^r_\rho[q_i] = \sum_{i=1}^m \left[2 \cdot \left(\frac{q_i^2 \sqrt{r}}{N} \right) + 2 \cdot \sqrt{\frac{q_i^2 r \log r}{N}} \right]$$

$$= 2 \cdot \left(\frac{\sqrt{r}}{N} \right) \cdot \sum_{i=1}^m q_i^2 + 2 \cdot \sqrt{\frac{r \log r}{N}} \cdot \sum_{i=1}^m q_i$$

$$\le 2 \cdot \left(\frac{\sqrt{r}}{N} \right) \cdot q^2 + 2 \cdot \sqrt{\frac{r \log r}{N}} \cdot q$$

$$= 2 \cdot \left(\frac{q^2 \sqrt{r}}{N} \right) + 2 \cdot \sqrt{\frac{q^2 r \log r}{N}};$$

$$\sum_{i=1}^{m-1} \sum_{j=i+1}^m \mathsf{cp}^r_\lambda[q_i, q_j] = \sum_{i=1}^{m-1} \sum_{j=i+1}^m \left[32 \cdot \left(\frac{q_i q_j r \log r}{N} \right)^2 \right.$$

$$\left. + 97 \cdot (\log r)^2 \cdot \left(\frac{q_i q_j r \log r}{N} \right) \right]$$

$$= 32 \cdot \left(\frac{r \log r}{N} \right)^2 \cdot \sum_{i=1}^{m-1} \sum_{j=i+1}^m q_i^2 q_j^2$$

$$+ 97 \cdot (\log r)^2 \cdot \left(\frac{r \log r}{N} \right) \cdot \sum_{i=1}^{m-1} \sum_{j=i+1}^m q_i q_j$$

$$\le 16 \cdot \left(\frac{r \log r}{N} \right)^2 \cdot q^4 + 49 \cdot (\log r)^2 \cdot \left(\frac{r \log r}{N} \right) \cdot q^2$$

$$= 16 \cdot \left(\frac{q^2 r \log r}{N} \right)^2 + 49 \cdot (\log r)^2 \cdot \left(\frac{q^2 r \log r}{N} \right).$$

Since these bounds are free of $q_1, \ldots, q_m$, this proves Theorem 1 of the paper.

5   Bounding the Advantage of Distinguishing f and f^r

5.1   Security Game

THE SETUP. An oracle $\mathcal{O}$ imitating a function g takes q queries $\{x_i \mid i \in [q]\}$ and returns

$$\{y_i = g(x_i) \mid i \in [q]\}.$$

The q-tuple of input-output pairs of the oracle is called the transcript, denoted as

$$\mathsf{tr} = ((x_1, y_1), (x_2, y_2), \ldots, (x_q, y_q)).$$

Both the real oracle $\mathcal{O}_{\mathrm{REAL}}$ and the ideal oracle $\mathcal{O}_{\mathrm{IDEAL}}$ will initially select a uniformly random function f. Then, $\mathcal{O}_{\mathrm{REAL}}$ goes on to imitate f^r, while $\mathcal{O}_{\mathrm{IDEAL}}$ imitates f itself. For any adversary $\mathcal{A}$, we want to bound its advantage, defined as

$$\mathbf{Adv}_{f,f^r}(q) = \left| \Pr\left[\mathcal{A}^{\mathcal{O}_{\mathrm{IDEAL}}}(q) \to 1\right] - \Pr\left[\mathcal{A}^{\mathcal{O}_{\mathrm{REAL}}}(q) \to 1\right] \right|.$$

As in the collision attack of Sect. 4.3, we can view the transcript tr as m trails of lengths $q_1, \ldots, q_m$ with sources $z_1, \ldots, z_m$, possibly with collisions, such that no query is counted in more than one trail, and hence

$$\sum_{j=1}^{m} q_j = q.$$

For $i \in [m]$, we shall use the notation

$$z_{i,1} := \mathcal{O}(z_i),$$
$$z_{i,j} := \mathcal{O}(z_{i,j-1}), 2 \leq j \leq q_i.$$

GOOD AND BAD TRANSCRIPTS. We partition the set of attainable transcripts into a set $\mathcal{T}_{\mathsf{good}}$ of good transcripts, and a set $\mathcal{T}_{\mathsf{bad}}$ of bad transcripts. We say $\mathsf{tr} \in \mathcal{T}_{\mathsf{bad}}$ if either of the following holds:

– For some $i \in [m]$,

$$z_{i,q_i} = z_i,$$

that is, the i-th trail forms a permutation cycle. Note that, by our construction of the trails, $z_{i_1,j}$ cannot equal z_{i_2} unless $i_1 = i_2$.
– For some $i_1, i_2 \in [m], j_1 \in [q_{i_1}], j_2 \in [q_{i_2}]$ with $(i_1, j_1) \neq (i_2, j_2)$, we have

$$z_{i_1,j_1} = z_{i_2,j_2},$$

that is, there is a ρ-collision on one of the trails ($i_1 = i_2$), or there is a λ-collision on two of the trails ($i_1 \neq i_2$).

5.2 Applying the H-Coefficient Technique

Let us denote the probability distribution of the transcripts in the real world by $\mathrm{Pr}_{\mathcal{O}_{\mathrm{REAL}}}$, and in the ideal world by $\mathrm{Pr}_{\mathcal{O}_{\mathrm{IDEAL}}}$. Our proof will use Patarin's H-coefficient technique [17].

Lemma 9 (H-Coefficient Technique). *Let $\mathcal{A}$ be an adversary, and let $\mathcal{T} = \mathcal{T}_{good} \cup \mathcal{T}_{bad}$ be a partition of the set of attainable transcripts. Let ε_1 be such that for all $tr \in \mathcal{T}_{good}$:*

$$\frac{\mathrm{Pr}_{\mathcal{O}_{\mathrm{REAL}}}\,[tr]}{\mathrm{Pr}_{\mathcal{O}_{\mathrm{IDEAL}}}\,[tr]} \geq 1 - \varepsilon_1.$$

Furthermore, let $\varepsilon_2 = \mathrm{Pr}_{\mathcal{O}_{\mathrm{IDEAL}}}\,[tr \in \mathcal{T}_{bad}]$. Then $\mathbf{Adv}_{f,f^r}(q) \leq \varepsilon_1 + \varepsilon_2$.

Proof. For a proof and a detailed explanation of this technique, see Chen and Steinberger [10]. □

PROBABILITY OF BAD TRANSCRIPTS IN IDEAL MODEL. We can easily bound the probability that a transcript tr from the ideal oracle $\mathcal{O}_{\mathsf{IDEAL}}$ is in $\mathcal{T}_{\mathsf{bad}}$. Suppose all of the q responses lie outside $\{z_i \mid i \in [m]\}$, and there is no collision between any of the responses. When this happens, tr cannot be in $\mathcal{T}_{\mathsf{bad}}$. The probability of this is at least $1 - \dfrac{2q^2}{N}$: two responses collide with probability at most $\dfrac{q^2}{N}$; and a response collides with a z_i with probability at most $\dfrac{q^2}{N}$, since there are m different values of z_i, and $m \leq q$. Thus,

$$\varepsilon_2 := \mathrm{Pr}_{\mathcal{O}_{\mathrm{IDEAL}}}\,[tr \in \mathcal{T}_{\mathsf{bad}}] \leq \frac{2q^2}{N}.$$

PROBABILITY OF GOOD TRANSCRIPTS. We now focus only on transcripts in $\mathcal{T}_{\mathsf{good}}$. Let us consider a good and attainable transcript $tr \in \mathcal{T}_{\mathsf{good}}$. For the ideal oracle, as the number of distinct inputs is q, we have

$$\mathrm{Pr}_{\mathcal{O}_{\mathrm{IDEAL}}}\,[tr] = \frac{1}{N^q}.$$

Now we bound $\mathrm{Pr}_{\mathcal{O}_{\mathrm{REAL}}}\,[tr]$ for $tr \in \mathcal{T}_{\mathsf{good}}$. Consider a $(q_1, \ldots, q_m)$-query m-trail collision attack on f^r, with sources $z_1, \ldots, z_m$ respectively. Theorem 1 tells us that this attack fails with probability at least $1 - \phi(q, r)$, where

$$\phi(q, r) := 2\left(\frac{q^2\sqrt{r}}{N}\right) + 2\sqrt{\frac{q^2 r \log r}{N}} + 16\left(\frac{q^2 r \log r}{N}\right)^2 + 49(\log r)^2\left(\frac{q^2 r \log r}{N}\right).$$

We now observe that when this attack fails, the attack transcript is either isomorphic as a graph to tr, or contains a permutation cycle.[3] A permutation cycle

---

[3] Note that the graph isomorphism follows from a simple relabeling of inputs and outputs, starting with the sources of every trail. This is possible because excluding collisions and permutation cycles means that no two inputs will have the same output, and outputs never correspond to a source.

occurs when queries of f^r collide with a source z_i, which has probability at most $\dfrac{q^2 r}{N}$, since there are m different values of z_i and $m \leq q$. Thus, the attack transcript is isomorphic to tr with probability at least

$$1 - \phi(q,r) - \frac{q^2 r}{N}.$$

Now the graph of this attack transcript has $q+m$ nodes, all distinct. Of these, the m sources are already fixed. The rest can take values in $N^{\underline{q}}$ ways. Now all of these $N^{\underline{q}}$ graphs are equally likely to occur in the scenario described above, i.e., when the m-trail attack fails and does not contain a permutation cycle. One of the equally likely $N^{\underline{q}}$ graphs is the graph of tr. Thus,

$$\Pr\nolimits_{\mathcal{O}_{\text{REAL}}}[\text{tr}] \geq \left(1 - \phi(q,r) - \frac{q^2 r}{N}\right) \cdot \frac{1}{N^{\underline{q}}}.$$

APPLYING THE H-COEFFICIENT TECHNIQUE. Let $R(\text{tr})$ be the ratio of the probabilities of $\text{tr} \in \mathcal{T}_{\text{good}}$ under $\mathcal{O}_{\text{REAL}}$ and $\mathcal{O}_{\text{IDEAL}}$ respectively. Then we have shown above that

$$R(\text{tr}) \geq \left(1 - \phi(q,r) - \frac{q^2 r}{N}\right) \cdot \frac{1}{\beta(q)}.$$

From Lemma 1, we have

$$\beta(q) \leq 1.$$

Thus,

$$R(\text{tr}) \geq 1 - \varepsilon_1$$

where

$$\varepsilon_1 := \phi(q,r) + \frac{q^2 r}{N}.$$

Hence, by the H-coefficient technique of Lemma 9, we have

$$\mathbf{Adv}_{f,f^r}(q) \leq \varepsilon_1 + \varepsilon_2.$$

This proves Theorem 2 of the paper.

6   Proof of Lemma 7

RECALLING THE SETUP. In Sect. 4 we defined three sets $\mathcal{S}_1$, $\mathcal{S}_2$, and $\mathcal{S}_3$. $\mathcal{S}_1$ is the set of (t_1, t_2) values that satisfy the constraints

$$\left\lceil \frac{t_1}{r} \right\rceil \leq q_1, \left\lceil \frac{t_2}{r} \right\rceil \leq q_2, t_1 = t_2 \bmod r;$$

$\mathcal{S}_2$ is the set of $(t_1, t_2, \Delta t, c, \eta)$ values that satisfy the constraints

$$\left\lceil \frac{t_1 + \Delta t + c\eta}{r} \right\rceil \leq q_1, \left\lceil \frac{t_2 + \Delta t}{r} \right\rceil \leq q_2, t_1 + c\eta = t_2 \bmod r,$$

$\mathcal{S}_3$ is the set of $(t_1, t_2, \Delta t, c, \eta)$ values that satisfy the constraints

$$\left\lceil \frac{t_1 + c\eta}{r} \right\rceil \leq q_1, \quad \left\lceil \frac{t_2 + \Delta t}{r} \right\rceil \leq q_2, t_1 + c\eta = t_2 + \Delta t \text{ mod } r.$$

We further defined the following:

$$\mathsf{p}_1 = \sum_{\mathcal{S}_1} \mathsf{cp}_\lambda(t_1, t_2);$$

$$\mathsf{p}_2 = \sum_{\mathcal{S}_2} \mathsf{cp}_{\lambda\rho}(t_1, t_2, \Delta t, c);$$

$$\mathsf{p}_3 = \sum_{\mathcal{S}_3} \mathsf{cp}_{\rho'}(t_1, t_2, \Delta t, c).$$

Lemma 7 claimed the following bounds for p_1, p_2 and p_3 (as long as $N \log r > 90$):

$$\mathsf{p}_1 \leq \frac{q_1 q_2 r}{N},$$

$$\mathsf{p}_2 \leq 6 \cdot (\log r)^2 \cdot \left(\frac{q_1 q_2 r}{N} \right)^2 + 18 \cdot (\log r)^3 \cdot \left(\frac{q_1 q_2 r}{N} \right),$$

$$\mathsf{p}_3 \leq 6 \cdot (\log r)^2 \cdot \left(\frac{q_1 q_2 r}{N} \right)^2 + 18 \cdot (\log r)^3 \cdot \left(\frac{q_1 q_2 r}{N} \right).$$

In this section, we establish these bounds.

BOUNDING p_1. For this we need to bound $\#\mathcal{S}_1$. This case is very simple. We observe the $t_1 \leq q_1 r$, so there are at most $q_1 r$ choices for t_1. Once t_1 is fixed, given the constraints $t_1 = t_2 \text{ mod } r$ and $t_2 \leq q_2 r$, there are at most q_2 choices for t_2. Thus, we have

$$\#\mathcal{S}_1 \leq q_1 q_2 r,$$

which, using (5), gives the bound

$$\mathsf{p}_1 = \sum_{\mathcal{S}_1} \mathsf{cp}_\lambda(t_1, t_2) \leq \#\mathcal{S}_1 \cdot \frac{1}{N} \leq \frac{q_1 q_2 r}{N}.$$

TOWARDS BOUNDING p_2: COUNTING OVER t_1, t_2 AND Δt. This is the most involved part of the calculations. For simplicity of notation we define the function

$$\zeta(\alpha) := (\sqrt{2\alpha N} + 3)^2 = 2\alpha N + 6\sqrt{2\alpha N} + 9.$$

Recall that $\mathcal{S}_2$ is the set of all $(t_1, t_2, \Delta t, c, \eta)$ satisfying

$$\left\lceil \frac{t_1 + \Delta t + c\eta}{r} \right\rceil \leq q_1, \quad \left\lceil \frac{t_2 + \Delta t}{r} \right\rceil \leq q_2, t_1 + c\eta = t_2 \text{ mod } r.$$

We begin by fixing a choice of c and η. We want to bound the number of choices for $(t_1, t_2, \Delta t)$. For this we relax the constraints a little. Let $\mathcal{S}_2' = \mathcal{S}_2'(c, \eta)$ be the set of values for $(t_1, t_2, \Delta t)$ satisfying

$$t_1 \leq q_1 r, \Delta t \leq q_2 r, t_2 \leq q_2 r, t_1 + c\eta = t_2 \text{ mod } r.$$

Now we fix a real number $\alpha > 0$, and split $\mathcal{S}'_2$ into two disjoint sets:

$$\mathcal{S}'^+_2[\alpha] := \left\{ (t_1, t_2, \Delta t) \in \mathcal{S}'_2 \mid \max(t_1, \Delta t) \geq \sqrt{2\alpha N} + 3 \right\},$$

$$\mathcal{S}'^-_2[\alpha] := \left\{ (t_1, t_2, \Delta t) \in \mathcal{S}'_2 \mid \max(t_1, \Delta t) < \sqrt{2\alpha N} + 3 \right\}.$$

For $\mathcal{S}'^+_2[\alpha]$, there are at most $q_1 r$ choices for t_1 and at most $q_2 r$ choices for Δt, and for each of these choices, we have at most q_2 choices for t_2. Thus,

$$\#\mathcal{S}'^+_2[\alpha] \leq q_1 q_2^2 r^2.$$

For $\mathcal{S}'^-_2[\alpha]$, there are at most $\sqrt{2\alpha N} + 3$ choices for t_1 and at most $\sqrt{2\alpha N} + 3$ choices for Δt, and for each of these choices, since choosing t_1 also fixes $t_2 \bmod r$, we have at most q_2 choices for t_2. Thus,

$$\#\mathcal{S}'^-_2[\alpha] \leq (\sqrt{2\alpha N} + 3)^2 \cdot q_2 = \zeta(\alpha) \cdot q_2.$$

When $(t_1, t_2, \Delta t) \in \mathcal{S}'^+_2[\alpha]$,

$$t_1 + t_2 + \Delta t + c\eta \geq \sqrt{2\alpha N} + 3,$$

so that according to (6):

$$\mathsf{cp}_{\lambda\rho}(t_1, t_2, \Delta t, c) \leq \frac{e^{-\alpha}}{N^2}.$$

When $(t_1, t_2, \Delta t) \in \mathcal{S}'^-_2[\alpha]$, (7) gives us

$$\mathsf{cp}_{\lambda\rho}(t_1, t_2, \Delta t, c) \leq \frac{1}{N^2}.$$

Let

$$\mathsf{p}_2(c, \eta) := \sum_{\mathcal{S}'_2} \mathsf{cp}_{\lambda\rho}(t_1, t_2, \Delta t, c)$$

$$= \sum_{\mathcal{S}'^+_2[\alpha]} \mathsf{cp}_{\lambda\rho}(t_1, t_2, \Delta t, c) + \sum_{\mathcal{S}'^-_2[\alpha]} \mathsf{cp}_{\lambda\rho}(t_1, t_2, \Delta t, c)$$

$$\leq q_1 q_2^2 r^2 \cdot \frac{e^{-\alpha}}{N^2} + \zeta(\alpha) \cdot q_2 \cdot \frac{1}{N^2}$$

$$= \frac{q_2}{N^2} \cdot \left[q_1 q_2 r^2 \cdot e^{-\alpha} + \zeta(\alpha) \right].$$

TOWARDS BOUNDING p_2: COUNTING OVER c AND η. We next bound the number of choices for (c, η) that satisfy the constraints. Again, we relax the constraints a little. Let $\mathcal{T}$ be the set of (c, η) values such that

$$c\eta \leq q_1 r.$$

Next we fix $d = \gcd(c, r)$. Let $\mathcal{T}[d]$ denote the set

$$\{(c, \eta) \in \mathcal{T} \mid \gcd(c, r) = d\}.$$

c now takes values over multiples of d. We split the counting into two parts:

- When $c \leq q_1 d$, we recall that η is defined as the smallest solution to $t_1 + c\eta = t_2 \bmod r$. From elementary number theory, we have $\eta \leq \dfrac{r}{d}$. Thus, there are q_1 choices of c and for each there are $\dfrac{r}{d}$ choices for η, so in all there are $\dfrac{q_1 r}{d}$ such choices for η and c.

- When $c > q_1 d$, we use the bounds $c \leq q_1 r$ and $\eta \leq \dfrac{q_1 r}{c}$. Let $z = \dfrac{c}{d}$. Thus, as c runs over all multiples of d from $(q_1 + 1) \cdot d$ to $q_1 r$, z takes all integer values from $q_1 + 1$ to $\dfrac{q_1 r}{d}$. Thus, the number of choices for η and c with $c > q_1 d$ is

$$\sum_{z=q_1+1}^{\frac{q_1 r}{d}} \frac{q_1 r}{zd} = \frac{q_1 r}{d} \cdot \sum_{z=q_1+1}^{\frac{q_1 r}{d}} \frac{1}{z} \leq \frac{q_1 r}{d} \cdot \log\left(\frac{r}{d}\right),$$

the last step following from Lemma 2.

Putting these two together, we get

$$\#\mathcal{T}[d] \leq \frac{q_1 r}{d} \cdot \left(1 + \log\left(\frac{r}{d}\right)\right).$$

Now, d can take values over all factors of r, so we have

$$\#\mathcal{T} = \sum_{d|r} \#\mathcal{T}[d] \leq \sum_{d|r} \frac{q_1 r}{d} \cdot \left(1 + \log\left(\frac{r}{d}\right)\right)$$

$$\leq \sum_{d|r} \frac{q_1 r}{d} \cdot (1 + \log r) \leq q_1 \cdot (1 + \log r) \sum_{d|r} \frac{r}{d}$$

$$\leq q_1 \cdot (1 + \log r) \cdot \sigma(r),$$

the last step coming from Lemma 4.

Finally, we observe that whenever $(t_1, t_2, \Delta t, c, \eta) \in \mathcal{S}_2$, we have $(t_1, t_2, \Delta t) \in \mathcal{S}_2'(c, \eta)$, and $(c, \eta) \in \mathcal{T}$. Hence,

$$\mathsf{p}_2 = \sum_{\mathcal{S}_2} \mathsf{cp}_{\lambda\rho}(t_1, t_2, \Delta t, c) \leq \sum_{\mathcal{T}} \sum_{\mathcal{S}_2'} \mathsf{cp}_{\lambda\rho}(t_1, t_2, \Delta t, c) = \sum_{\mathcal{T}} \mathsf{p}_2(c, \eta).$$

This gives us the bound

$$\mathsf{p}_2 \leq \frac{q_1 q_2}{N^2} \cdot (1 + \log r) \cdot \sigma(r) \cdot \left[q_1 q_2 r^2 \cdot e^{-\alpha} + \zeta(\alpha)\right]. \tag{12}$$

BOUNDING P3. Recall that $\mathcal{S}_3$ is the set of all $(t_1, t_2, \Delta t, c, \eta)$ satisfying

$$\left\lceil \frac{t_1 + c\eta}{r} \right\rceil \leq q_1, \quad \left\lceil \frac{t_2 + \Delta t}{r} \right\rceil \leq q_2, t_1 + c\eta = t_2 + \Delta t \bmod r.$$

The set $\mathcal{S}_3$ is almost identical to the set $\mathcal{S}_2$. However, the counting arguments are identical to those for p_2, as the relaxation of the constraints is valid for p_2 as well as p_3. Combined with (8), we have

$$\mathsf{p}_3 = \sum_{\mathcal{S}_3} \mathsf{cp}_{\rho'}(t_1, t_2, \Delta t, c) = \sum_{\mathcal{S}_3} \mathsf{cp}_{\lambda\rho}(t_1, t_2, 0, c) \leq \sum_{\mathcal{T}} \sum_{\mathcal{S}_2'} \mathsf{cp}_{\lambda\rho}(t_1, t_2, 0, c).$$

Thus, we have

$$\mathsf{p}_3 \leq \frac{q_1 q_2}{N^2} \cdot (1 + \log r) \cdot \sigma(r) \cdot \left[q_1 q_2 r^2 \cdot e^{-\alpha} + \zeta(\alpha) \right].$$

SIMPLIFYING THE BOUNDS. Now we make a series of generous relaxations to get a simple easy-to-see bound for p_2 and p_3. Under the assumption that $\sqrt{2\alpha N} + 3 \leq \sqrt{3\alpha N}$, we have $\zeta(\alpha) \leq 3\alpha N$. The assumption can be written as

$$(\sqrt{3} - \sqrt{2}).\sqrt{\alpha N} \geq 3.$$

In other words,

$$\alpha N \geq 9(\sqrt{3} + \sqrt{2})^2 = 9(5 + 2\sqrt{6}).$$

Now, $2\sqrt{6} < 5$, so a sufficient condition to ensure this is $\alpha N \geq 90$. We now put $\alpha = \log r$, and observe in passing that the ensuing assumption that $N \log r \geq 90$ is quite reasonable. For this choice of α, we have

$$\zeta(\alpha) \leq 3N \log r, \tag{13}$$

and

$$e^{-\alpha} = \frac{1}{r}. \tag{14}$$

Since $(5/3) \cdot \log r \geq 1$ for $r \geq 2$, we have

$$1 + \log r < \frac{5}{3} \log r + \log r = \frac{8}{3} \log r. \tag{15}$$

Finally, to bound $\sigma(r)$, we use Lemma 5, which gives us

$$\sigma(r) < 3r \log r. \tag{16}$$

Plugging (13)–(16) into (12), we have

$$\mathsf{p}_2 \leq \frac{q_1 q_2}{N^2} \cdot 3r \log r \cdot \frac{8}{3} \log r \cdot \left(q_1 q_2 r^2 \cdot \frac{1}{r} + 3N \log r \right)$$
$$= 8 \cdot (\log r)^2 \cdot \left(\frac{q_1 q_2 r}{N} \right)^2 + 24 \cdot (\log r)^3 \cdot \left(\frac{q_1 q_2 r}{N} \right).$$

Similarly,

$$\mathsf{p}_3 \leq 8 \cdot (\log r)^2 \cdot \left(\frac{q_1 q_2 r}{N} \right)^2 + 24 \cdot (\log r)^3 \cdot \left(\frac{q_1 q_2 r}{N} \right).$$

This completes the proof of Lemma 7.

7   Conclusion and Future Work

We studied the iterated random function problem, and proved the first bound in this setting that is tight up to a factor of $(\log r)^3$. In previous work, the iterated random function problem was seen as a special case of CBC-MAC based on a random function f. We obtained our bound by analysing the probability of a common class of collision attacks, and applying Patarin's H-coefficient technique to bound the advantage of distinguishing f^r from f. Trying to improve the $(\log r)^3$ factor in the security bound is an interesting topic for future work.

References

1. Bellare, M., Kilian, J., Rogaway, P.: The security of cipher block chaining. In: Desmedt, Y.G. (ed.) CRYPTO 1994. LNCS, vol. 839, pp. 341–358. Springer, Heidelberg (1994). https://doi.org/10.1007/3-540-48658-5_32

2. Bellare, M., Kilian, J., Rogaway, P.: The security of the Cipher Block Chaining message authentication code. J. Comp. Syst. Sci. **61**(3), 362–399 (2000)

3. Bellare, M., Pietrzak, K., Rogaway, P.: Improved security analyses for Cipher Block Chaining Message Authentication Codes. In: Shoup, V. (ed.) CRYPTO 2005. LNCS, vol. 3621, pp. 527–545. Springer, Heidelberg (2005). https://doi.org/10.1007/11535218_32

4. Bellare, M., Ristenpart, T., Tessaro, S.: Multi-instance security and its application to password-based cryptography. In: Safavi-Naini, R., Canetti, R. (eds.) CRYPTO 2012. LNCS, vol. 7417, pp. 312–329. Springer, Heidelberg (2012). https://doi.org/10.1007/978-3-642-32009-5_19

5. Bellare, M., Rogaway, P.: The security of triple encryption and a framework for code-based game-playing proofs. In: Vaudenay, S. (ed.) EUROCRYPT 2006. LNCS, vol. 4004, pp. 409–426. Springer, Heidelberg (2006). https://doi.org/10.1007/11761679_25

6. Berke, R.: On the security of iterated MACs. Ph.D. thesis, ETH Zürich (2003)

7. Bernstein, D.J.: A short proof of the unpredictability of cipher block chaining, January 2005. http://cr.yp.to/antiforgery/easycbc-20050109.pdf

8. Bhaumik, R., Datta, N., Dutta, A., Mouha, N., Nandi, M.: The Iterated Random Function Problem. ePrint Report 2017/892 (2017). full version of this paper

9. Bossi, S., Visconti, A.: What users should know about Full Disk Encryption based on LUKS. In: Reiter, M., Naccache, D. (eds.) CANS 2015. LNCS, vol. 9476, pp. 225–237. Springer, Cham (2015). https://doi.org/10.1007/978-3-319-26823-1_16

10. Chen, S., Steinberger, J.: Tight Security Bounds for Key-Alternating Ciphers. In: Nguyen, P.Q., Oswald, E. (eds.) EUROCRYPT 2014. LNCS, vol. 8441, pp. 327–350. Springer, Heidelberg (2014). https://doi.org/10.1007/978-3-642-55220-5_19

11. Dodis, Y., Gennaro, R., Håstad, J., Krawczyk, H., Rabin, T.: Randomness extraction and key derivation using the CBC, Cascade and HMAC modes. In: Franklin, M. (ed.) CRYPTO 2004. LNCS, vol. 3152, pp. 494–510. Springer, Heidelberg (2004). https://doi.org/10.1007/978-3-540-28628-8_30

12. Dodis, Y., Ristenpart, T., Steinberger, J., Tessaro, S.: To hash or not to hash again? (In) differentiability results for H^2 and HMAC. In: Safavi-Naini, R., Canetti, R. (eds.) CRYPTO 2012. LNCS, vol. 7417, pp. 348–366. Springer, Heidelberg (2012). https://doi.org/10.1007/978-3-642-32009-5_21

13. Ferguson, N., Schneier, B.: Practical Cryptography. Wiley, New York (2003)
14. Gaži, P., Pietrzak, K., Rybár, M.: The exact PRF-Security of NMAC and HMAC. In: Garay, J.A., Gennaro, R. (eds.) CRYPTO 2014. LNCS, vol. 8616, pp. 113–130. Springer, Heidelberg (2014). https://doi.org/10.1007/978-3-662-44371-2_7
15. Minaud, B., Seurin, Y.: The iterated random permutation problem with applications to cascade encryption. In: Gennaro, R., Robshaw, M. (eds.) CRYPTO 2015. LNCS, vol. 9215, pp. 351–367. Springer, Heidelberg (2015). https://doi.org/10.1007/978-3-662-47989-6_17
16. Nandi, M.: A simple and unified method of proving indistinguishability. In: Barua, R., Lange, T. (eds.) INDOCRYPT 2006. LNCS, vol. 4329, pp. 317–334. Springer, Heidelberg (2006). https://doi.org/10.1007/11941378_23
17. Patarin, J.: The "Coefficients H" technique. In: Avanzi, R.M., Keliher, L., Sica, F. (eds.) SAC 2008. LNCS, vol. 5381, pp. 328–345. Springer, Heidelberg (2009). https://doi.org/10.1007/978-3-642-04159-4_21
18. Preneel, B., van Oorschot, P.C.: MDx-MAC and building fast MACs from hash functions. In: Coppersmith, D. (ed.) CRYPTO 1995. LNCS, vol. 963, pp. 1–14. Springer, Heidelberg (1995). https://doi.org/10.1007/3-540-44750-4_1
19. Turan, M.S., Barker, E., Burr, W., Chen, L.: Recommendation for key derivation using pseudorandom functions (Revised). NIST Special Publication 800–132, National Institute of Standards and Technology (NIST), December 2010
20. Wagner, D., Goldberg, I.: Proofs of security for the Unix password hashing algorithm. In: Okamoto, T. (ed.) ASIACRYPT 2000. LNCS, vol. 1976, pp. 560–572. Springer, Heidelberg (2000). https://doi.org/10.1007/3-540-44448-3_43
21. Wuille, P.: Bitcoin network graphs (2017). http://bitcoin.sipa.be/
22. Yao, F.F., Yin, Y.L.: Design and Analysis of Password-Based Key Derivation Functions. In: Menezes, A. (ed.) CT-RSA 2005. LNCS, vol. 3376, pp. 245–261. Springer, Heidelberg (2005). https://doi.org/10.1007/978-3-540-30574-3_17
23. Yao, F.F., Yin, Y.L.: Design and analysis of password-based key derivation functions. IEEE Trans. Inf. Theor. **51**(9), 3292–3297 (2005)

Author Index

Printed in the United States
By Bookmasters